AN
ENCYCLOPÆDIA OF
METALLURGY AND
MATERIALS

AN
ENCYCLOPÆDIA OF
METALLURGY AND
MATERIALS

C. R. Tottle

M.Met., M.Sc., C.Eng., F.I.M., F. Inst.P.
Professor Emeritus, University of Bath

The Metals Society

Macdonald and Evans

Macdonald & Evans Ltd
Estover, Plymouth PL6 7PZ

Originally published in volume form as
A Dictionary of Metallurgy 1958

First published jointly with The Metals Society as
An Encyclopaedia of Metallurgy and Materials 1984

British Library Cataloguing in Publication Data

Tottle, C. R.
    An encyclopaedia of metallurgy and materials.
    1. Metallurgy—Dictionaries
    I. Title
    669'.0321 TN665

ISBN 0-7121-0571-9

Printed and bound in Great Britain by
The Pitman Press, Bath

Preface

The original *Dictionary of Metallurgy* was first compiled by Dr. A.D. Merriman, for publication in monthly instalments in the journal *Metal Treatment and Drop Forging.* In 1958, as a result of voluminous correspondence from readers of the journal, Macdonald & Evans Ltd, undertook to publish the Dictionary as a single volume, following suitable revision and enlargement by the author. In the mid-1960s, a *Concise Encyclopaedia of Metallurgy* was created from the Dictionary, in appropriate form, to act as a reference work for students and technologists.

A decade later, the importance of materials other than metals in engineering work suggested a further revision, so that a reference work suitable for all students and practitioners of materials science and technology, in addition to engineers, became the aim of the present volume. Recognition that metals and alloys still provide the great bulk of engineering materials, led to the retention of "Metallurgy" in the title, but no attempt at emphasis has been made on any one class of materials. The original text formed the basis of revision, but the more scientific terms, especially in materials science, have been treated at much greater length for the benefit of students and hence "Encyclopaedia" is preferred to "Dictionary".

Terms from metallurgical practice which have fallen out of use have been cut, or eliminated, and replaced by additional material, which is partly amplification of metallurgical science, but mostly consists of the inclusion of details on ceramics, glass, polymers and composites, with relevant treatment of topics now mostly referred to as "materials science".

Definitions have accordingly been tailored to the varying needs of potential readers, so that for the non-technical reader, the most likely terms of interest are tackled as simply as possible without causing confusion, but for the student, the more scientific aspects are given wider treatment, assuming prior understanding of physics, chemistry or engineering. Mineralogical information has been pruned, but the names and descriptions of minerals involved in preparation of metal or materials, or used in materials technology have been retained.

Within this context, the new edition aims to provide a ready reference book for a wide range of readers, particularly those in developing countries, where industrialisation must proceed by practical application as rapidly as possible. Businessmen, politicians and similar non-technical people, frequently need to refer to reasonably comprehensive definitions of technical terms, but many students also welcome a source of quick reference, especially when time is pressing and a full textbook treatment may not be required. The book does not set out to replace existing textbooks, but rather to provide a compendium of several of them.

The tables have been considerably amplified, to bring together useful information in a more readily accessible form and cross-reference between alternative possible materials may also be made easier by this presentation. SI units have been used but when these may assist understanding, or SI units are inappropriate, Imperial units have been retained.

The author is grateful for an alphabet of only twenty-six letters, and the fact that some are less commonly used than others!

1984   C.R. TOTTLE

Acknowledgments

The author wishes to thank the publishers for their agreement in the use of material from his book *The Science of Engineering Materials,* Heinemann Educational Books, 1966.

He is indebted to the publishers mentioned below for permission to incorporate the following items in some of the entries dealing with materials science.

The Institution of Metallurgists and the Metals Society for use of portions of articles on Non-Destructive Testing in their journals *The Metallurgist,* July 1976 and *Metals and Materials,* March 1976, respectively.

Butterworths & Co. Ltd., for extracts from the book *Fibres, Films, Plastics and Rubbers,* 1971, by W.J. Roff and J.R. Scott, included in the entry on polymerisation, and trade names and general characteristics of polymer groups incorporated in Table 10.

John Wiley and Sons Ltd., for extracts on the relationship between fluids and solids in terms of viscosity; viscosity of glass; sols and gels, and the behaviour of clay, pastes and sands; stress relaxation and fluidity; and Fig. 7.13, taken from *The Mechanical Properties of Matter,* by A.H. Cottrell, 1964, and included in the entry on viscosity.

Contents

Preface   i

Acknowledgments   iii

List of Tables   vii

How to Use this Encyclopædia   ix

The Encyclopædia   1

List of Tables

1	SI and Imperial units	xi
2	Physical properties of the elements	xiv
3	Physical properties of typical alloys, ceramics and polymers	xvi
4	Trade names, composition and applications of aluminium-based alloys	xix
5	Trade names, composition and applications of copper-based alloys	xxvi
6	Trade names, composition and applications of ferrous alloys	xlvi
7	Trade names, composition and applications of nickel-based alloys	lvii
8	Trade names, composition and applications of low-melting point alloys	lxiv
9	Trade names, composition and applications of other non-ferrous alloys	lxxii
10	Trade names, composition and applications of polymeric materials	lxxxvi
11	Etchants used in metallography	xcii
12	Arrest points of carbon steel	10
13	Electronic structure of the elements	12
14	Brinell hardness numbers and conversion table	29
15	Dimensions of standard wire gauges	31
16	Classification of coals using Seyler system	45
17	Relative humidity (per cent) (revised for SI units)	71
18	Electrode potentials at 25 °C, or electromotive series (volts)	76
19	Approximate fusing currents for common fuse wire	85
20	Details of electro-analysis of metals	85
21	Electrode potential at 25 °C of common metals	86
22	Solutions for electrolytic polishing of metals	87
23	Properties of annealed and cold-worked ferritic stainless steels	102
24	Chief constituents of the more common ferro-alloys	102
25	Flame test colours	108
26	Typical composition of some gaseous fuels	113
27	Radioactive isotopes—half life	128
28	Ignition temperatures	139
29	Details of various types of impact tests	140
30	Comparison of grey and white irons	148
31	Arrest points associated with iron-carbon alloys	149
32	Main divisions of Jurassic rocks	154
33	Speed of gases at 0 °C	158
34	Properties of liquid fuels	167
35	Colours corresponding to various temperatures	169
36	Absorption coefficients of some common materials	178
37	Moh's hardness scale	186
38	Commonly used mould paints	189
39	The periodic classification of the elements (after Bohr)	221
40	Composition of plating baths	231
41	Plating baths for the deposition of various metals	232
42	Maximum working temperature and e.m.f. produced by various thermocouple elements	250
43	Energy quanta corresponding to different wavelengths	251
44	Relative cooling rates of quenching media	252
45	Wavelengths and frequencies of various radiations	254
46	Some useful radioactive isotopes	257
47	Refractive indices of laboratory materials	264
48	Composition of common refractory materials	264
49	Temperature coefficients of electrical resistance	266
50	Rock forming mineral groups	272
51	Types of Rockwell hardness	272
52	Melting points of salt baths for heat treatment	279
53	Formulae for calculating the section modulus (Z) of different shapes	286
54	Softening temperature of Seger cones	287
55	Common soldering fluxes	301
56	Method of stripping one metal from another	315
57	Percentage of carbon in steel for various types of temper	322
58	Temper colours	323
59	Tempering temperatures of some alloy steels	323
60	Circular section test pieces B.S.S. 18.	325
61	Non-proportional rectangular test pieces B.S.S. 18	325
62	Universal electro-polishing bath	344
63	Viscosities of some common liquids and gases at 20 °C	354
64	Latent heat of vaporisation of common substances	356
65	Melting points of some common waxes	365
66	Welding methods	366

How to Use this Encyclopædia

Items are arranged in conventional alphabetical order with accompanying figures to supplement the text where appropriate. Tables of numerical data are also arranged within the text when they refer to a specific item, e.g. the electronic structure of the elements in Table 13 accompanies the textual description of the atom, to emphasise the relationship between the atoms in terms of the basic "building bricks". The periodic table as such is situated in the text under "periodic classification" since this is what it is.

The details in Tables 2 to 11 refer mainly to properties, grouped for comparison, and separated from the text since they can be more easily consulted for purely factual information, and by the same token, more easily contrasted. The physical properties of the elements for example, can be seen in one table (Table 2) although a general description of each element by name occurs in the text, with cross references to Table 2. Similarly, metallic alloys, ceramics and polymers are referred to in the text in the chemical sense, but physical properties and uses are found grouped in Tables 3-10. The microscopic examination of metals is dealt with in the text under headings such as "Microscopy", "Polishing" and "Etching", but details of the many etchant media are given in Table 11, at the beginning. It is hoped, by this arrangement, that readers requiring regular reference to physical properties or facts will find it simpler to use Tables 2-11, without having to seek the appropriate point in the text. Equally, the text carries appropriate information specific to that item in tables within the text, or cross-referenced to the main tables 2-11 at the front of the book. A list of all tables is given in the contents, to assist those cases where only factual information is required.

Information on the SI unit system is placed in Table 1 before all other tables, for obvious reasons, since it is relevant to everything which follows.

TABLE 1. Symbols and Units of the S.I. System: a Summary

QUANTITY	UNIT NAME	UNIT SYMBOL
Basic units		
length	metre	m
mass	kilogram	kg
time	second	s
electric current	ampere	A
thermodynamic temperature	kelvin	K (supersedes $^{\circ}$K and deg. K)
amount of substance	mole	mol
luminous intensity	candela	cd

NOTE: When using Celsius scale for temperature, write $^{\circ}$C to distinguish from C, the symbol for coulomb. Mole, symbol mol, is defined as the amount of substance containing the same number of elementary units as atoms in 0.012 kg of nuclide carbon-12. The elementary unit must be specified with mol, e.g. atom, molecule, ion, electron, photon, etc.

QUANTITY	UNIT NAME	UNIT SYMBOL
Dimensionless (supplementary) units		
plane angle	radian	rad
solid angle	steradian	sr
Derived units		
force	newton	$N = kg\ m\ s^{-2}$
pressure, stress	pascal	$Pa = N\ m^{-2} = m^{-1}\ kg\ s^{-2}$
work, energy, heat quantity	joule	$J = N\ m = m^2\ kg\ s^{-2}$
power	watt	$W = J\ s^{-1} = m^2\ kg\ s^{-3}$
quantity of electricity, electric charge	coulomb	$C = A\ s$
electric potential, electromotive force, potential difference	volt	$V = W\ A^{-1} = m^2\ kg\ s^{-3}\ A^{-1}$
capacitance	farad	$F = A\ s\ V^{-1} = m^{-2}\ kg^{-1}\ s^4\ A^2$
electrical resistance	ohm	$\Omega = V\ A^{-1} = m^2\ kg\ s^{-3}\ A^{-2}$
electrical conductance	siemens	$S = A\ V^{-1} = m^{-2}\ kg^{-1}\ s^3\ A^2$
frequency	hertz	$Hz = s^{-1}$
magnetic flux	weber	$Wb = V\ s = A\ H = m^2\ kg\ s^{-2}\ A^{-1}$
magnetic flux density	tesla	$T = Wb\ m^{-2} = kg\ s^{-2}\ A^{-1}$
inductance	henry	$H = Wb\ A^{-1} = V\ s\ A^{-1} = m^2\ kg\ s^{-2}\ A^{-2}$
luminous flux	lumen	$lm = cd\ sr$
illuminance	lux	$lx = lm\ m^{-2} = m^{-2}\ cd\ sr$

QUANTITY	UNIT NAME	UNIT SYMBOL
Other derived units		
area		m^2
volume		m^3
speed, velocity		$m\ s^{-1}$
acceleration		$m\ s^{-2}$
wave number		m^{-1}
density, mass density		$kg\ m^{-3}$
current density		$A\ m^{-2}$
magnetic field strength		$A\ m^{-1}$
concentration		$mol\ m^{-3}$
radioactive activity		s^{-1}
specific volume		$m^3\ kg^{-1}$
luminance		$cd\ m^{-2}$
angular velocity		$rad\ s^{-1}$
angular acceleration		$rad\ s^{-2}$
viscosity — dynamic		$Pa\ s = m^{-1}\ kg\ s^{-1}$
viscosity — kinematic		$m^2\ s^{-1}$
moment of force (torque)		$N\ m = m^2\ kg\ s^{-2}$
surface tension		$N\ m^{-1} = kg\ s^{-2}$
surface energy		$J\ m^{-2} = N\ m^{-1} = kg\ s^{-2}$
heat flux density		$W\ m^{-2} = kg\ s^{-3}$
heat capacity, entropy		$J\ K^{-1} = m^2\ kg\ s^{-2}\ K^{-1}$

TABLE 1.

QUANTITY	UNIT NAME	UNIT SYMBOL
specific heat capacity, specific entropy		$J\ kg^{-1}\ K^{-1} = m^2\ s^{-2}\ K^{-1}$
specific energy, latent heat		$J\ kg^{-1} = m^2\ s^{-2}$
thermal conductivity		$W\ m^{-1}\ K^{-1} = m\ kg\ s^{-3}\ K^{-1}$
energy density		$J\ m^{-3} = m^{-1}\ kg\ s^{-2}$
electric field strength		$V\ m^{-1} = m\ kg\ s^{-3}\ A^{-1}$
electric charge density		$C\ m^{-3} = m^{-3}\ s\ A$
electric flux density		$C\ m^{-2} = m^{-2}\ s\ A$
permittivity		$F\ m^{-1} = m^{-3}\ kg^{-1}\ s^4\ A^2$
permeability		$H\ m^{-1} = m\ kg\ s^{-2}\ A^{-2}$
molar energy, enthalpy, free energy, activation energy		$J\ mol^{-1} = m^2\ kg\ s^{-2}\ mol^{-1}$
molar entropy, molar heat capacity		$J\ mol^{-1}\ K^{-1} = m^2\ kg\ s^{-2}\ K^{-1}\ mol^{-1}$
radiant intensity		$W\ sr^{-1} = m^2\ kg\ s^{-3}\ sr^{-1}$
radiance		$W\ m^{-2}\ sr^{-1} = kg\ s^{-3}\ sr^{-1}$
momentum		$N\ s = kg\ m\ s^{-1}$
angular momentum		$N\ s\ m = kg\ m^2\ s^{-1}$
thermal expansion coefficient		K^{-1}

Other units in use with SI units

minute	min	$1\ min = 60\ s$
hour	h	$1\ h = 60\ min = 3600\ s$
day	d	$1\ d = 24\ h = 86\ 400\ s$
degree	°	$1° = (\pi/180)\ rad$
minute	′	$1′ = (1/60)° = (\pi/10\ 800)\ rad$
second	″	$1″ = (1/60)′ = (\pi/648\ 000)\ rad$
litre	l	$1\ l = 1\ dm^3 = 10^{-3}\ m^3$
tonne	t	$1\ t = 10^3\ kg$

Values obtained experimentally

electronvolt	eV	$1\ eV \approx 1.602\ 19 \times 10^{-19}\ J$
unified atomic mass unit	u	$1\ u \approx 1.660\ 53 \times 10^{-27}\ kg$
astronomical unit	AU*	$1AU \approx 149\ 600 \times 10^6\ m$
parsec	pc	$1\ pc \approx 206\ 265\ AU = 30\ 857 \times 10^{12}\ m$

* English abbreviation only, can be confused with old ångström unit.

Conversion of non SI units to SI equivalent

length	thou	$10^{-3}\ in = 25.4\ \mu m$
	inch	$1\ in = 25.4\ mm$
	foot	$1\ ft = 0.3048\ m$
	ångström unit	$1\ A.U.$ or $1Å = 10^{-10}\ m$
area	square inch	$1\ in^2 = 645.16\ mm^2$
	square foot	$1\ ft^2 = 0.092\ 903\ m^2$
	hectare	$1\ ha = 10^4\ m^2$
	decare	$1\ deca = 10^3\ m^2$
	are	$1\ a = 100\ m^2$
volume	cubic inch	$1\ in^3 = 16.387\ 1 \times 10^{-6}\ m^3$
	cubic foot	$1\ ft^3 = 28.316\ 8 \times 10^{-3}\ m^3$
	UK gallon	$1\ gal = 4.546\ 092 \times 10^{-3}\ m^3$
mass	pound	$1\ lb = 0.453\ 592\ kg$
	ton	$1\ ton = 1\ 016.05\ kg$
velocity	miles per hour	$1\ m.p.h. = 1.609\ 34\ km\ h^{-1}$
	feet per second	$1\ ft.s^{-1} = 0.304\ 8\ m\ s^{-1}$
density	pounds per cubic foot	$1\ lb.ft^{-3} = 16.0185\ kg\ m^{-3}$

TABLE 1.

QUANTITY	UNIT NAME	UNIT SYMBOL
force	pounds force	1 lb. f = 4.448 22 N
	ton force	1 ton f = 9.964 02 kN
pressure, stress	pounds per square inch	1 p.s.i. = 6 894.76 N m^{-2}
	tons per square inch	1 ton in^{-2} = 15.444 3 MN m^{-2} = 1.544 kN cm^{-2} = 1.544 h bar
	bar	1 bar = 10^5 N m^{-2}
	atmosphere	1 atm = 101.325 kN m^{-2}
	millimetres Hg	1 mm Hg = 133.322 N m^{-2}
	torr	1 torr = 133.322 N m^{-2}
viscosity — dynamic kinematic	centipoise centistoke	1 CP = 0.001 N s m^{-2} 1 CST = 10^{-6} m^2 s^{-1}
energy	British thermal unit	1 B.Th.U. or B.t.u. = 1.05506 kJ
	therm	1 Th = 105.506 MJ
	thermie	1 th = 4.185 5 MJ
	calorie (I.T.)	1 cal$_{I.T.}$ = 4.186 8 J
	calorie (15° cal)	1 cal^{15} = 4.185 5 J
	erg	1 erg = 10^{-7} J
power	horsepower	1 h.p. = 0.745 7 kW
	metric horsepower	1 ch. or PS = 735.499 W
heat flow	British thermal unit per hour	1 B.t.u per hr. = 0.293 071 W
	calories per second	1 cal.s^{-1} = 4.1868 J s^{-1} or W
calorific value	British thermal unit per cubic foot	1 B.t.u. ft^{-3} = 37.258 9 k J m^{-3}
thermal capacity per unit mass		1 cal.g^{-1} °C^{-1} = 4.186 8 J g^{-1} °C^{-1}
thermal conductivity		1 cal.cm.cm.$^{-2}$ s^{-1}.°C^{-1} = 418.68 J m m^{-2} s^{-1} °C^{-1} or W m m^{-2} °C^{-1}
magnetic field strength		1 oersted = 10^3/4πA m^{-1}
magnetic flux density		1 gauss = 10^{-4} T

Physical constants

	molar gas constant	R = 8.314 3 J mol^{-1} K^{-1}
	Boltzmann constant	k = 1.380 54 × 10^{-23} J K^{-1}
	Avogadro constant	N_A = 6.02252 × 10^{23} mol^{-1}
	Planck constant	h = 6.625 6 × 10^{-34} J s
	Faraday constant	$F = N_A e$ = 9.648 70 × 10^4 C mol^{-1}
	mass of hydrogen atom	m_H = 1.673 43 × 10^{-27} kg
	mass of proton	m_p = 1.672 52 × 10^{-27} kg
	mass of neutron	m_n = 1.674 82 × 10^{-27} kg
	mass of electron	m_e = 9.109 1 × 10^{-31} kg
	speed of light	c = 2.997 925 × 10^8 m s^{-1}
	Stefan-Boltzmann constant	σ = 5.669 7 × 10^{-8} W m^{-2} K^{-4}
	molar volume of ideal gas at S.T.P. (101 325 N m^{-2} and 273.15 K)	V_m = 2.241 36 × 10^{-2} m^3 mol^{-1}

TABLE 2. Physical Properties of the Elements 1 — 94 (at 25 °C unless otherwise stated)

Element		Atomic Weight	Density, 10^3 kg m^{-3}	Melting Point, °C	Boiling Point, °C	Specific Heat, J g^{-1} °C^{-1}	Thermal Conductivity, W cm^{-1} °C^{-1}	Electrical Resistivity, $\mu\Omega$ cm	Modulus of Elasticity, GN m^{-2}	Linear Co-efficient of Thermal Expansion, °C^{-1} × 10^6	Crystal structure
Actinium	Ac	227	10.07	1050	3200	–	–	–	–	–	cub.
Aluminium	Al	26.98	2.702	660.37	2467	0.900	2.37	2.655	69	23.7	f.c.c.
Antimony	Sb	121.7	6.684	630.74	1750	0.205	0.185	39.0	78	10.8*	(8–N)
Argon	Ar	39.94	1.784 g l^{-1}	−189.2	−185:7	0.519	0.18 × 10^{-3}	–	–	–	–
Arsenic	As	74.92	5.727	817	613 s	0.329	–	33.3	–	–	(8–N)
Astatine	At	210	–	302	337	–	–	–	–	–	
Barium	Ba	137.3	3.51	725	1640	0.293	–	0.5	–	–	b.c.c.
Beryllium	Be	9.01	1.85	1278	2970	1.825	2.18	4.0	302	12.6*	hex.
Bismuth	Bi	208.98	9.80	271.3	1560	0.124	0.084	106.8	31.6	13.3*	(8–N)
Boron	B	10.81	2.34	2300	2550	1.026	–	1.8 × 10^{12}	440	8.3	comp.
Bromine	Br	79.90	3.12	−7.2	58.78	0.473	0.04 × 10^{-3}	7.8 × 10^{18}	–	–	–
Cadmium	Cd	112.40	8.642	320.9	765	0.232	0.93	6.83	30	30*	comp.
Calcium	Ca	40.08	1.54	839	1484	0.653	1.21	3.91	–	22.0	f.c.c.
Carbon-Diamond	C	12.011	3.51	–	–	0.519	–	5 × 10^{20}	–	1.2	d.
Carbon-Graphite	C	12.011	2.25	3550	4827	0.712	0.238	1375	4.84*	7.9*	hex.
Cerium	Ce	140.12	6.78	798	3257	0.205	0.19	75.0	30.0	8.5	f.c.c.
Cesium	Cs	132.91	1.879	28.4	678.4	0.239	0.14	20.29	–	97.0	b.c.c.
Chlorine	Cl	35.45	3.214 g l^{-1}	−100.98	−34.6	0.477	0.09 × 10^{-3}	10 (−70 °C)	–	–	–
Chromium	Cr	51.996	7.20	1857	2672	0.448	0.91	12.9	290	6.2	b.c.c.
Cobalt	Co	58.93	8.90	1495	2870	0.456	0.69	6.24	207	12.3*	hex.
Copper	Cu	63.55	8.96	1083.4	2567	0.385	3.98	1.673	117.5	16.6	f.c.c.
Dysprosium	Dy	162.50	8.556	1409	2335	0.173	0.11	92.6	63.2	8.6	hex.
Erbium	Er	167.26	9.045	1522	2510	0.168	0.096	86.0	73.2	9.2	hex.
Europium	Eu	151.96	5.253	822	1597	0.176	–	91.0	14.7	32.0	b.c.c.
Fluorine	F	18.998	1.69 g l^{-1}	−219.6	−188.1	0.825	0.27 × 10^{-3}	–	–	–	–
Francium	Fr	223	–	27	677	–	–	–	–	–	
Gadolinium	Gd	157.2	7.898	1312	3233	0.230	0.088	131.0	56.5	6.4	hex.
Gallium	Ga	69.72	5.904	29.78	1983	0.41	0.33	17.4	–	18.0	comp.
Germanium	Ge	72.5	5.35	937.4	2830	0.322	–	46 × 10^6	–	6.0	d.
Gold	Au	196.966	19.3	1063	2807	0.129	3.15	2.35	2.44	14.2	f.c.c.
Hafnium	Hf	178.4	13.31	2227	4602	0.147	0.223	35.1	138	1.1	hex.
Helium	He	4.0026	0.179 g l^{-1}	−272.2	−268.6	5.19	1.51 × 10^{-5}	–	–	–	–
Holmium	Ho	164.93	8.781	1470	2720	0.165	–	81.4	67.1	9.5(400° C)	hex.
Hydrogen	H	1.008	0.0899 g l^{-1}	−259.14	−252.87	14.28	1.87 × 10^{-3}	–	–	–	–
Indium	In	114.82	7.30	156.61	2080	0.234	0.25	8.37	10.8	24.8*	comp.
Iodine	I	126.90	4.93	113.5	184.35	0.427	–	1.3 × 10^{15}	–	–	comp.
Iridium	Ir	192.2	22.42	2410	4130	0.133	1.49	5.3	513	6.5	f.c.c.
Iron	Fe	55.84	7.86	1535	2750	0.444	0.803	9.71	212	11.7	b.c.c.

TABLE 2.

Element		Atomic Weight	Density, 10^3 kg m^{-3}	Melting Point, °C	Boiling Point, °C	Specific Heat, J g^{-1} °C^{-1}	Thermal Conductivity, W cm^{-1} °C^{-1}	Electrical Resistivity, $\mu\Omega$ cm	Modulus of Elasticity, GN m^{-2}	Linear Co-efficient of Thermal Expansion, °C^{-1} × 10^6	Crystal structure
Krypton	K	83.80	3.736 g l^{-1}	−156.6	−152.30	0.247	0.099 × 10^{-3}	−	−	−	−
Lanthanum	La	138.91	6.166	920	3454	0.197	0.138	57	38.4	4.9	hex.
Lead	Pb	207.2	11.34	327.5	1740	0.159	0.346	20.65	16.4	29.1	f.c.c.
Lithium	Li	6.94	0.534	180	1347	3.56	0.712	8.55	−	55.8	hex.
Lutetium	Lu	174.97	9.84	1656	3315	0.155	−	58.2	84.5	12.5(400° C)	hex.
Magnesium	Mg	24.30	1.74	649	1090	1.02	1.59	4.45	44.7	25.8*	hex.
Manganese	Mn	54.938	7.20	1244	1962	0.477	1.59	185.0	158	22.8	comp.
Mercury	Hg	200.5	13.55	−38.87	356.58	0.139	0.084	98.4	−	−	comp.
Molybdenum	Mo	95.94	10.2	2617	4612	0.250	1.4	5.2	335	5.44	b.c.c.
Neodymium	Nd	144.24	6.80	1010	3067	0.205	0.13	64.3	38.7	6.7	comp.
Neon	Ne	20.18	0.9 g l^{-1}	−248.67	−245.92	1.03	0.48 × 10^{-3}	−	−	−	−
Neptunium	Np	237.05	20.25	640	3902	−	−	−	−	−	comp.
Nickel	Ni	58.7	8.90	1453	2732	0.444	0.90	6.84	202	13.3	f.c.c.
Niobium	Nb	92.91	8.57	2468	4742	0.268	0.52	12.5	105	7.1	b.c.c.
Nitrogen	N	14.01	1.25 g l^{-1}	−209.86	−195.8	1.04	0.26 × 10^{-3}	−	−	−	−
Osmium	Os	190.2	22.57	3045	5027	0.131	0.61	9.5	620	4.6	hex.
Oxygen	O	15.999	1.429 g l^{-1}	−218.4	−183	0.917	0.27 × 10^{-3}	−	−	−	−
Palladium	Pa	106.4	12.02	1552	3140	0.245	0.728	10.8	120	11.7	f.c.c.
Phosphorus	P	30.97	2.34	590	ignites	0.758	−	10^{17}	−	−	(8−N)
Platinum	Pt	195.0	21.45	1772	3827	0.133	0.73	10.6	152	8.9	f.c.c.
Plutonium	Pu	244	19.84	641	3232	0.134	0.08	141.4	99	54	comp.
Polonium	Po	210	9.4	254	962	0.126	−	−	−	−	cub.
Potassium	K	39.10	0.86	63.65	774	0.754	0.99	6.15	−	83	b.c.c.
Praseo-dymium	Pr	140.91	6.782	931	3122	0.193	0.117	68.0	32.4	4.8	hex.
Promethium	Pm	145	−	1080	2460	0.185	−		42.2	−	
Proactinium	Pa	231.04	15.37	1230	−	0.121	−	−	−	−	d.
Radium	Ra	226.03	5	700	1140	0.121	−	−	−	−	−
Radon	Rn	222	9.73 g l^{-1}	−71	−61.8	0.094	−	−	−	−	−
Rhenium	Re	186.2	21.02	3180	5627	0.138	−	19.3	462	−	hex.
Rhodium	Rh	102.91	12.4	1966	3900	0.244	1.50	4.51	320	8.5	f.c.c.
Rubidium	Rb	85.467	1.53	38.89	688	0.360	0.29	12.5	−	90	b.c.c.
Ruthenium	Ru	101.07	12.3	2310	3900	0.239	−	7.6	465	−	hex.
Samarium	Sm	150.4	7.54	1072	1778	0.180	−	105.0	34.5	−	comp.
Scandium	Sc	44.96	2.99	1539	2832	0.557	−	50.9	79.4	12	hex.
Selenium	Se	78.96	4.79	217	685	0.321	0.005	12.0	58.0	36.8	(8−N)
Silicon	Si	28.09	2.33	1410	2355	0.703	0.835	10.0	107	4.68	d.
Silver	Ag	107.87	10.5	961.9	2212	0.234	4.27	1.59	76.2	18.8	f.c.c.
Sodium	Na	22.99	0.97	97.8	883	1.23	1.34	4.2	−	70	b.c.c.
Strontium	Sr	87.62	2.54	769	1384	0.73	−	23.0	−	−	f.c.c.

TABLE 2.

Element		Atomic Weight	Density, 10^3 kg m^{-3}	Melting Point, °C	Boiling Point, °C	Specific Heat, J g^{-1} °C^{-1}	Thermal Conductivity, W cm^{-1} °C^{-1}	Electrical Resistivity, $\mu\Omega$ cm	Modulus of Elasticity, GN m^{-2}	Linear Co-efficient of Thermal Expansion, °C^{-1} × 10^6	Crystal structure
Sulphur	S	32.06	2.07	112.8	444.6	0.733	–	2×10^{23}	–	–	comp.
Tantalum	Ta	180.95	16.6	2996	5425	0.140	0.54	12.45	187	6.5	b.c.c.
Technetium	Tc	98.91	11.50	2172	4877	0.243	–	–	–	–	–
Tellurium	Te	127.6	6.25	449.5	1990	0.201	0.059	4.36×10^5	–	16.75	(8–N)
Terbium	Tb	158.93	8.23	1360	3041	0.183	–	114.5	57.6	7.0	hex.
Thallium	Tl	204.4	11.85	303.5	1457	0.129	0.373	18.0	–	30.2	hex.
Thorium	Th	232.04	11.7	1750	4790	0.113	0.37	13.0	71	12	f.c.c.
Thulium	Tm	168.93	9.31	1545	1950	0.160	–	67.6	75.4	11.6(400° C)	hex.
Tin	Sn	118.69	7.28	231.9	2270	0.222	0.64	11.0	40.7	20*	d.
Titanium	Ti	47.9	4.5	1660	3287	0.523	0.2	42.0	107	8.5*	hex.
Tungsten	W	183.85	19.35	3410	5660	0.133	1.78	5.65	406	4.5	b.c.c.
Uranium	U	238.03	19.05	1132	3818	0.116	0.25	30.0	186	13.4*	comp.
Vanadium	V	50.94	6.11	1890	3380	0.486	0.60	25.6	131	8.95	b.c.c.
Xenon	Xe	131.30	5.89 g l^{-1}	–111.9	–107.1	0.158	–	–	–	–	–
Ytterbium	Yb	173.0	6.98	824	1193	0.145	–	25.1	17.8	25.0	f.c.c.
Yttrium	Y	88.91	4.46	1523	3337	0.285	0.12	59.6	65.1	10.8(400° C)	hex.
Zinc	Zn	63.57	7.13	419.6	907	0.389	1.15	5.92	82.7	29.2*	hex.
Zirconium	Zr	91.22	6.51	1852	4377	0.281	0.209	40.0	94.5	5.9*	hex.

s = sublimes; cub. = cubic; b.c.c. = body-centred cubic; f.c.c. = face-centred cubic; hex. = hexagonal; (8–N) = obeys (8–N) rule (text); comp. = complex; * = mean figure for polycrystal of anisotropic material; d. = diamond.

TABLE 3. Physical Properties of some Typical Alloys, Ceramics and Polymers (at 25 °C unless otherwise stated)
(All figures are approximate, and dependent on purity and thermal and mechanical history)

Metallic Alloys

Material	Density, 10^3 kg m^{-3}	Modulus of Elasticity, GN m^{-2}	Tensile strength, MN m^{-2}	Elongation, %	Linear co-efficient of thermal expansion, °C^{-1} × 10^6	Thermal conductivity, W cm^{-1} °C^{-1}	Electrical resistivity, $\mu\Omega$ cm	Impact strength, J	Hardness
Aluminium Bronze 90% Cu	7.5	82-103	415-520 (C)	15-25 (C)			4.78	27-40	90-100(B)
Brass 70 Cu 30 Zn	8.53	100	320 (A)	70 (A)	18.8	1.26	6.2		65(D.P.)
Brass 60 Cu 40 Zn	8.40	104	380 (A)	45 (A)	19.2	1.20	6.0		75(D.P.)
Bronze 90 Cu 10 Sn	8.80	104	450 (A)	65 (A)	18.0	0.505	9.5		55(R.B.)
Duralumin 40% Cu	2.8	72.5	555 (H.T.)	8 (H.T.)	21.6	1.680	3.8		180(D.P.)
Inconel 600 75% Ni	8.42	218	620 (A)	47 (A)	13.0	0.134	103	165	65-85 (R.B.)
Iron Cast Grey 3.25% C	7.3	134	150-250	0.5	10.5	0.45		12	180-240(B)
Iron Cast S.G. 3.5% C	7.35	165	600-750	3	10.8	0.45		50	240-290(B)
Meehanite GA	7.33	140	350-400	0.8	9.7	0.45		9.5	235(B)

TABLE 3.

Material	Density, 10^3 kg m^{-3}	Modulus of Elasticity, GN m^{-2}	Tensile strength, MN m^{-2}	Elongation, %	Linear coefficient of thermal expansion, °C$^{-1} \times 10^6$	Thermal conductivity, W cm^{-1} °C^{-1}	Electrical resistivity, $\mu\Omega$ cm	Impact strength, J	Hardness
Monel 67 Ni 30 Cu	8.83	179	480-620 (A)	60-35 (A)	14.4	0.218	51.0	120-160	60-80 R.B.
Nimonic 80 A	8.22	222(H.T.)	1240(H.T.)	24(H.T.)	11.1	0.13			290-370 (D.P.)
Steel 0.2% C	7.8	205	450 (N)	35 (N)	11.7	0.503	17.0	60	140(D.P.)
Steel 0.4% C	7.8	205	600 (N)	25 (N)	11.35	0.482		15	200(D.P.)
Steel 0.8% C	7.8	205	850 (N)	15 (N)	10.8	0.461		5	250(D.P.)
Steel 1.0% C	7.8	205	950 (N)	10 (N)				3	300(D.P.)
Steel 3% Ni	7.85	204	690(H.T.)	22(H.T.)	11.35	0.344	25.6	55	210(B)
Steel Austenitic 18/8	7.93	200	590-650 (A)	55-60	19.0	0.147	70.0	157-164	145-160(B)
Stellite (No. 23)	8.54		775	9	16.0		160		400(B)

A — Annealed;   N — Normalised;   H.T. — Heat-treated;   C — Cast;   R.B. — Rockwell B scale;   D.P. — Diamond Pyramid;   B — Brinell

Polymers

Type	Density, 10^3 kg m^{-3}	Modulus of Elasticity, GN m^{-2}	Tensile strength, MN m^{-2}	Elongation (in tension), %	Linear coefficient of thermal expansion, °C$^{-1} \times 10^6$	Thermal conductivity, mW cm^{-1} °C^{-1}	Electrical resistivity*, $\mu\Omega$ cm	Impact strength, J m^{-1} (notched)	Dielectric constant, MHz
Acrylic	1.17-1.20	3.1	47-75	3-10	50-90	1.68-2.52	$> 10^{20}$	0.41-0.68	2.2-3.2
Cellulose Acetate	1.23-1.43	0.45-2.7	13-59	6-7	80-160	1.68-3.36	10^{16}-10^{19}	0.54-7.1	3.2-7.0
Cellulose Acetate Butyrate	1.15-1.22	0.345-1.38	18-48	40-90	110-170	1.68-3.36	10^{16}-10^{18}	1.08-8.5	3.2-6.2
Epoxide — Rigid	1.0-3.2	1.4-4.15	34-82	5-10	50-90	1.68-2.10	$> 10^{21}$	0.41-1.22	3-4
Epoxide — Flexible	1.2		7-28	10-100	50-90	1.68-2.10	$> 10^{21}$	0.67-2.04	3-4
Polytetrafluoroethylene	2.1-2.2	0.345-0.62	17-41	250-600	90-220	2.52	10^{25}	3.38-5.4	2.1
Melamine-Formaldehyde (cellulose filled)	1.47-1.52	8.9	48-89	0.6-0.9	40	2.94-4.2	10^{18}-10^{20}	0.32-0.48	7.2-8.2
Nylon 6	1.13	1.03-2.48	70-82	90-320	80-130	2.18-2.42	10^{18}-10^{21}	1.34-4.9	3-7
Phenol-Formaldehyde (unfilled)	1.25-1.3	5.2-6.9	48-55	1.0-1.5	25-60	1.26-2.52	10^{17}-10^{18}	0.27-0.49	4.5-5.0
Phenol-Formaldehyde (cellulose filled)	1.32-1.45	5.5-8.2	45-59	0.4-0.8	30-45	1.68-2.94	10^{15}-10^{19}	0.32-0.81	4-7
Polyacetal	1.425	2.84	69	15-75	80	2.30	6×10^{20}	1.9-3.1	3.7
Polycarbonate	1.2	2.22	59-66	60-100	70	1.93	2.1×10^{20}	16.4-21.6	2.6
Polyethylene — low density	0.91-0.93	0.117-0.24	6.9-16	90-650	160-180	3.35	$> 10^{22}$	no break	2.25-2.35
Polyethylene — high density	0.94-0.965	0.55-1.03	21-38	50-800	110-130	4.6-5.2	$> 10^{22}$	2.0-16.3	2.25-2.35
Polypropylene	0.90-0.91	0.89-1.38	29-38	50-600	110	1.38	$> 10^{22}$	0.54-9.4	2.25-2.30
Polystyrene	1.04-1.11	2.76-4.2	34-83	1.0-2.5	60-80	1.01-1.38	$> 10^{19}$	0.27-0.68	2.4-3.1
Polystyrene — toughened	0.98-1.10	1.73-3.1	17-48	7-60	34-210	0.42-1.26	$> 10^{19}$	4.1-16.3	2.4-3.8
Styrene-Acrylonitrile	1.075-1.10	2.76-3.9	65-83	1.5-3.5	68	1.22	$> 10^{19}$	0.47-0.68	2.75-3.10
Acrylonitrile-Butadiene Styrene	0.98-1.10	0.69-2.82	17-62	10-140	60-130	0.62-3.6	2.7×10^{22}	1.36-4.1	2.70-4.75
Urea-Formaldehyde (cellulose filled)	1.47-1.52	10.3	41-89	0.5-1.0	27	2.92-4.2	10^{18}-10^{19}	0.36-0.46	6.4-6.9

TABLE 3.

Type	Density, 10^3 kg m^{-3}	Modulus of Elasticity, GN m^{-2}	Tensile strength, MN m^{-2}	Elongation, (in tension) %	Linear co-efficient of thermal expansion, °C^{-1} × 10^6	Thermal conductivity, mW cm^{-1} °C^{-1}	Electrical resistivity*, $\mu\,\Omega$ cm	Impact strength, J m^{-1} (notched)	Dielectric constant, MHz
Polyvinyl chloride rigid	1.38-1.4	2.42	59	2-40	50	1.47	10^{22}	1.36-4.05	3.0
Polyvinyl chloride flexible	1.35	4.15	41	—	50	1.89	10^{19}	22 unaged	—
Vinylidene chloride	1.65-1.72	0.34-0.55	20.34	up to 250	190	1.26	10^{20}-10^{22}	0.41-1.36	3-4

*at 50% Relative Humidity.

Ceramics

Type	Density, 10^3 kg m^{-3}	Modulus of Elasticity, GN m^{-2}	Tensile strength, MN m^{-2}	Compression strength, MN m^{-2}	Linear co-efficient of thermal expansion, °C^{-1} × 10^6	Thermal Conductivity, W cm^{-1} °C^{-1}	Electrical resistivity, $\mu\,\Omega$ cm	Moh Hardness Number (new)	Melting point, °C
Alumina Al_2O_3	3.99	345	700	3000	13.5	0.29		12.0	2 050
Beryllia BeO	3.01	375	175	1500	7.5	2.1		11.2	2 570
Calcined Magnesite MgO	3.65	83			10.8	0.22	10^{11} (1100 °C)	5.5	2 800
Chromite $FeO.Cr_2O_3$	4.49				8.0	0.16		5.5	2 180
Concrete	2.2-2.40	13-30	2.5	20-40	10-15	0.02			d
Firebrick $Al_2O_3.SiO_2$	2.10	15-25	5-8	25	4.5-5.0	0.01	1.4×10^{14}	5.5	> 1 400
Fused Silica SiO_2	2.20	70	100		0.5	0.014	10^{26}	7.0	1 600
Graphite Brick	2.25	6.9	25 m	35-75 m	3.5 m	0.20 m	10^3	1.5	3 800
Magnesia Spinel $MgO.Al_2O_3$	3.60				8.5	0.019		8.0	2 130
Mullite $3Al_2O_3 . 2SiO_2$	3.03	140	100		5.5	0.063			1 800
Porcelain K_2O, SiO_2, Al_2O_3	2.30	65	45	300	3.5	0.01			> 1 400
Quartz (brick) SiO_2	2.65	50			12.5	0.08	1.9×10^{14}	8.0	1 400
Silicon carbide SiC	3.20	400	250	600	4.5	1.2	2.5×10^6 (1100 °C)	13.0	2 250
Silicon nitride Si_3N_4	3.18	200	100	400	3.0	0.25			1 910 S
Sillimanite $Al_2O_3 SiO_2$	3.24					0.01		7.5	1 800
Steatite MgO, Al_2O_3, SiO_2	2.65	100	75	550	8.1	0.03			> 1 400
Thoria ThO_2	9.85					0.01		6.5	3 050
Tungsten carbide	15.63	710	350	550	5.0	0.03	2×10^4	10.4	2 700
Zircona ZrO_2	5.55				11.1	0.03		11.0	2 500

d = disintegrates;   S = sublimes;   m = mean for anisotropic material.

Miscellaneous

Type	Density, 10^3 kg m^{-3}	Modulus of Elasticity, GN m^{-2}	Tensile strength, MN m^{-2}	Compression strength, MN m^{-2}	Linear co-efficient of thermal expansion, °C^{-1} × 10^6	Thermal Conductivity, W cm^{-1} °C^{-1}	Electrical resistivity, $\mu\,\Omega$ cm	Moh Hardness Number (new)	Melting point, °C
Glass Pyrex	2.23	34	up to 150		3.0	0.03	10^{22}		1 500*
Glass Soda	2.40	70	up to 175		8.5	0.01	5×10^{17}		1 500*
Hardwood (Oak)	0.72	9.5		15-50	4.5*	0.0014			
Softwood (Pine)	0.50	8.3		11.4 w.g. 2.1 a.g.	4.5*	0.014			

w.g. = parallel to grain;   a.g. = perpendicular to grain;   * = approximate.

TABLE 4. Trade Names, Composition and Applications of Aluminium-based Alloys

TRADE NAME or other designation		COMPOSITION, % by weight					SPECIAL PROPERTIES	USES
		Al	Cu	Si	Other	Other		
AEROLITE		Balance	1.2	0.5	1 Fe	0.4 Mg	Good strength/weight ratio	Sandwich metal in clad metal (see cladding). Automobile parts
ALAR	00.5	Balance	0.1	4.5-6.0	0.6 Fe 0.5 Mn	0.1 Mg 0.1 Ni 0.1 Zn	Resistance to atmospheric corrosion. Ductility. Good casting properties.	Architectural castings. Chemical plant. Domestic utensils
	21	do.	2-4	4-6	0.8 Fe 0.4 Mn 0.2 Ti	0.3 Zn	Tensile strength. Good casting properties. Age hardening	Covers, housings, crankcases, gear cases.
	00.12	do.	—	10-13	0.7 Fe 0.5 Mn 0.2 Ti	0.15 Mg 0.1 Ni 0.2 Zn	Corrosion resistance. Harder, less ductile. Good casting properties	General purpose. Machined castings
ALCAN (Generic name for alloys by Aluminium Company of Canada)	16S	Balance	2.5	0.7 max	0.7 Fe max	0.3 Mg	Heat treatable	Aircraft rivets
	17S	do.	3.8-4.5	—	0.5 Fe max 0.5-0.7 Mn	0.5-0.8 Mg	do.	Stressed parts in aircraft and structures
	26S	do.	3.9-5	0.5-1	0.7 Fe max 0.4-1.2 Mn	0.2-0.8 Mg 0.2 Zn max 0.1 Cr max	do.	do.
	50S	do.	0.1 max	0.3-0.7	0.4 Fe max	0.4-0.9 Mg 0.2 Zn max	do.	Architectural members (window frames), road transport
	B54S	do.	—	—	0.3 Mn	4.4 Mg	Formability. Corrosion resistance	Ship-building. Structural members
	75S	do.	1.5	—	0.3 Cr 0.2 Ti max	2.5 Mg 5.6 Zn	Heat treatable	Aircraft structures
	GB117	do.	3	5	0.5 Mn	—	do.	Gear boxes, clutch housings, pump covers
	D135	do.	0.1 max	6.5-7.5	0.5 Fe max 0.3 Mn max	0.2-0.45 Mg 0.1 Ni max	Castability. Heat treatable	Engine cylinder blocks. Chemical and food plant.
	160	do.	0.1	10-13	0.6 Fe max 0.5 Mn max	0.1 Mg max 0.1 Ni 0.1 Zn 0.2 Ti	Castability. Not heat treatable	Large castings, marine and electrical. Valve bodies, radiators, grills
	GB162	do.	1 Cu	11.5	2 Ni	1 Mg	Low thermal expansion. Good creep strength	Pistons, engine parts
	340	do.	—	—	—	8 Mg	Corrosion resistant	Die castings
ALCOA (Generic name for alloys by Aluminium Corporation of America) Similar range to ALCAN	A13	Balance	—	11-13	—	—	Castability	Low stressed castings
	122	do.	10	0.5	1.2 Fe	0.2 Mg	Wear resistance. Age hardening	Pistons, valve guides, camshaft bearings
	2025	do.	4.5	0.8	—	—	Age hardening	Forged propellor blades
ALMASILIUM		Balance.	—	2	—	1 Mg	Corrosion resistance	General purpose alloy

TABLE 4.

TRADE NAME or other designation		Al	Cu	Si	Other	Other	SPECIAL PROPERTIES	USES
ALMINAL	2	Balance	0.7-2.5	9.0-11.5	0.2 Ti 1 Fe 0.5 Mn	0.3 Mg 1 Ni 1.2 Zn	Strength	Die casting
Cast Alloys	6	do.	0.1	10-13	0.5 Mn	0.1 Mg 0.1 Ni 0.1 Zn	Corrosion resistance	Die casting
	7	do.	1.0-2.5	1.5-3.5	0.5-1.7 Ni 0.3-1.4 Fe	0.2 Mg 0.1 Mn	Pressure tight when cast	Castings
	8	do.	0.1	4.5-6.0	0.3-0.7 Mn 0.1 Ni 0.6 Fe	0.4 Mg 0.1 Zn	—	General purpose
	9	do.	0.1	10-13	0.3-0.7 Mn 0.6 Fe 0.1 Zn 0.1 Ni	0.2-0.6 Mg	Strength. Corrosion resistance	General purpose
	10	do.	—	0.35	—	9.5-11 Mg	Strength	Die casting
	11	do.	4.0-5.0	0.25	0.25 Fe	0.2 Ti	Strength	General purpose
	20	do.	3.0	9.0	1 Fe 0.5 Zn	0.2 Mg	Strength	Die casting
Wrought Alloys	4	do.	0.15	0.6	0.7 Fe 0.5 Mn 0.2 Ti	1.75 Mg	Corrosion resistance	Structural members
	5	do.	0.15	0.6	0.7 Fe 1 Mn	3 Mg 0.2 Ti	do.	do. Shipbuilding
	6	do.	0.15	0.6	0.7 Fe 1 Mn	4.5 Mg	do.	do.
	9	do.	0.15	0.3	0.6 Fe 0.2 Ti	0.4 Mg	Age hardening	do.
	10	do.	0.15	0.3	0.6 Fe 1 Mn	0.4-1.5 Mg 0.2 Ti	Corrosion resistance, strength	do.
	11	do.	1.0	0.75-1.25	0.75 Fe 1 Mn	0.5-1.25 Mg 0.3 Ti	—	General purpose. Shipbuilding
	15	do.	3.5-4.8	1.5	1 Fe 1.2 Mn	0.6 Mg 0.3 Ti	Strength	do.
ALPAX (Silumin)		Balance	—	10-13.5	—	Modified	Corrosion resistance. Casting properties	Castings, good strength/weight ratio
ALUMAL		do.	—	—	—	1.25 Mn	High strength when strain hardened	Sheet, bar for components
ANTICORODAL (Cast) 5		do.	—	4-6	0.5-1 Mn	0.4-1 Mg	Corrosion resistance	Architecture. Shipbuilding. General purpose
(Wrought) 15		do.	—	0.5-1.5	0.2-1 Mn	0.5-1 Mg	Strength when heat treated	Instrument housings. Crankcases
ARDAL		do.	2	—	1.7 Fe	0.6 Ni	Strength. Not heat treatable	Bearings, pistons, wire, tubes
AVIONAL		do.	4	0.3-0.7	—	0.5-0.7 Mn 0.5-1.0 Mg	Strength when heat treated	Sheet or bar
BIRMABRIGHT		do.	—	—	—	3-6 Mg 0.25-0.75 Mn	Corrosion resistance in sea water	Marine components, boats, fittings
BIRMASIL (Special)		do.	0.1 max	10-13	2.5-3.5 Ni 0.2 Ti	0.6 Fe max 0.5 Mn max 0.6 Mg	Strength. Corrosion resistance	Aircraft. General castings for automobile engines
BIRMASTIC		do.	Small	12	2.5-3.5 Ni	Small Fe Small Mn	Strength	Castings
BIRMETAL AM503		Balance	—	—	1.5 Mn	—	Weldability	Fuel and oil tanks. Trailer bodies. Welded structures
B.B.016		do.	0.25	0.6	—	1 Mg	Age hardening	Structural components
BMB551		do.	4.5	1	0.5 Mn	2.75 Mg 5.5 Zn	Corrosion resistance. Heat treatable	do.
BMB2308		do.	1.75	—	0.2 Mn 0.15 Cr	2 Mg 6.5 Zn	Heat treatable	Aircraft structures

TABLE 4.

TRADE NAME or other designation		COMPOSITION, % by weight					SPECIAL PROPERTIES	USES
		Al	Cu	Si	Other	Other		
BIRMID	D8	do.	2-4	4-6	0.3-0.7 Mn 0.35 Ni	0.3 Zn 0.2 Ti	Age hardening. Castability	Heavy duty light castings
	300	do.	—	0.25	0.1 Mn	9.5-11 Mg 0.2 Ti	do.	High strength castings
	428	do.	6-8	2-4	0.6 Mn max 1 Fe max	2-4 Zn	Castability	Housings, gear cases
CALLOY		do.	—	—	—	10-25 Ca	Deoxidant	Steelmaking
CERALUMIN (similar to R.R.50)	B	Balance	1-2.5	1.5-3.5	0.05-0.2 Mg 0.05-0.3 Ce	0.3-1.4 Fe 0.5-1.7 Ni	Strength in cast state	High strength/weight ratio. Aircraft parts
(similar to R.R.53B)	C	do.	2-3	1-1.4	0.5-1 Mg 0.05-0.2 Ce	0.8-1.4 Fe 1-2 Ni 0.01 max Mn 0.05-0.3 Ti and/or Nb	At elevated temperatures	Pistons, cylinder heads, gear boxes
CHROMET		90	—	10	—	—	Easily cast	Bearings for hard shafts
CINDAL		Balance	—	—	0.1-0.5 Cr 0.1-0.3 Mg	0.5-1 Zn	Heat treatable. Strength when modified by chlorine	—
DURALUMIN		do.	3.5-4.5	0 to 1	0.4-1.2 Fe	0-0.8 Mn 0.6 Mg	Heat treatable to vary strength and ductility. High strength. Higher temperature strength. Age hardening	Aircraft and all structural applications for high strength to weight ratio
	LC	do.	0.6	—	0.5 Mn 0.05 Cr	2.6 Mg 6 Zn		
	W	do.	4	0.2	0.3 Fe	1.5 Mg 1 Ni		
DURICILIUM	W	Balance	0.15	0.3-0.7	0.7 Mg 0.2 Ti	0.6 Fe	Heat treatable to vary strength	Structural members. Sheet and forgings for aircraft
		do.	4	0.5	0.7 Mg	0.5 Mn		
HIDUMINIUM ALLOYS (Generic name for alloys by High Duty Alloys Ltd.)	10	Balance	0.1 max	10-13	0.2 Ti max	0.6 Fe max	Fluidity	Castings of intricate shape
	40	do.	—	4.5-6	0.25-0.8 Mg	0.25-0.8 0.2 Ti max	High strength heat treated. Fluidity	Castings, sand and die for marine use
	44	do.	—	0.75-1.3	0.5-1.2 Mg 0.6 Fe max	0.5-1 Mn	Corrosion resistance	Forgings for marine use
	66	do.	3.5-4.8	0.5-1.5	0.5-1 Mg 1 Fe	0.5-1 Mn	High strength. Heat treatable	Forgings for aircraft
	80	do.	4-5	0.3 max	0.2 Fe max	0.1-0.25 Ti	do. Hot short	Castings sand and die. Brackets, levers, housings
	90	do.	0.15 max	0.25 max	9.5-10.5 Mg	—	Corrosion resistance. Age hardening	do. Gasoline flow meters
	R.R.50	do.	0.8-2.0	1.5-2.8	0.05-2 Mg 0.8-1.75 Ni	0.8-1.4 Fe 0.2 Ti	High strength	Cylinder blocks
	R.R.53B	do.	1.5	0.7-0.9	0.7-1 Mg 1-2 Ni	0.8-1.5 Fe	do.	Pistons, textile and food machinery
	R.R.56	do.	1.8-2.5	0.55-1.25	0.65-1.2 Mg 0.6-1.4 Ni	0.6-1.2 Fe 0.05-0.15 Ti	do.	Crankcases. Connecting rods. Aircraft parts. Supercharger rotors
	R.R.58	Balance	1.5-3	0-0.3	1.2-1.8 Mg 0.5-1.5 Ni	1.0-1.5 Fe 0.2 Ti	High strength	Large pressings and forgings
	R.R.59	do.	2.2	0.7-1.3	1.25 Mg 1.35 Ni	1.35 Fe 0.08 Ti	do.	Pistons. Structural components used at elevated temperatures

TABLE 4.

TRADE NAME or other designation	COMPOSITION, % by weight					SPECIAL PROPERTIES	USES
HIDUMINIUM ALLOYS (contd.)	Al	Cu	Si	Other	Other		
R.R.77	do.	1.5	0.6	2-4 Mg 1 Ni max	0.6 Fe 0.3 Ti 4-6 Zn	do.	Structural components requiring strength greater than mild steel
R.R.88	do.	3 max	0.6	2–4 Mg 1 Cr max	0.6 Fe 4–8 Zn	do.	do.
HYDRONALIUM HY25	Balance	—	0.2-1	3-12 Mg 0.2-0.5 Mn	—	Corrosion resistance, particularly to sea water	Sheet metal for marine applications, e.g. boats
KUPFERSILUMIN	do.	0.8	13	0.3 Mn	—	Castability	Castings for wheels, rolls, engine blocks
LAUTAL 1	do.	4.5-5.5	0.2-0.5	—	—	Strength with ductility	Engineering structures of high strength/weight ratio
2	do.	4.5	0.75	0.75 Mn	—		
LO-EX ALLOYS	Balance	0.9	14.0	2 Ni	1 Mg	Low coefficient of thermal expansion $(19.5 \times 10^{-6}$ $^{\circ}C^{-1})$	Pistons in internal combustion engines
	do.	0.9	12.0	2 Ni	1 Mg		
	do.	0.75-1.25	13-15	1.5-2.5 Ni	0.75-1.25 Mg		
LUBRAL alloys G-AE6CN	do.	1.1	0.5 max	5.5-7 Sn 0.7-1.3 Ni 0.1 Mg max	0.4 Fe max 0.15 Te max	Low friction. High fatigue strength. Developed in Italy	Automobile engine bearings of long life with high r.p.m. engines
G-AE6CNS	do.	0.7-1.3	1-2	5.5-7 Sn 0.3-0.7 Ni 0.1 Mg max	0.7 Fe max 0.15 Te max	do.	
MACH'S alloy	Balance	—	—	5 Mg	—	Medium strength	Aircraft components
MAGNALITE	do.	2.5	—	1.5 Mg	2 Ni 0.5 Zn	Strength at elevated temperature	Pistons of internal combustion engines
MAGNALIUM	do.	1.75	—	1-5.5 Mg	0-1.2 Ni	Strength	Castings for aircraft
MAHLE alloy	83-85	1	11-13	1.2 Mg 1 Ni	0.5 Fe 0.08 Mn	Strength at elevated temperature	Pistons of internal combustion engines
MIRALITE	95	—	0.3	4.1 Ni	0.4 Fe 0.04 Pb 0.04 Na	Corrosion resistance	Trim for railway coaches, automobiles, boats, aircraft. Pistons
MOCK SILVER	84.2	Balance	—	10.2 Sn	0.1 P	Colour. Tarnish resistance	Substitute for silver in cheap jewellery
MONTEGAL	Balance	—	0.8	1 Mg	0.2 Ca	Non age hardening. Improved electrical conductivity	Electrical machines
NEOMAGNAL A	do.	—	—	5 Mg	5 Zn	Low friction	Bearings
NICKELOY	94	4.5	—	1.5 Ni	—	Heat treatable	Engineering components at elevated temperature
NICRAL A	Balance	0.25-1	—	0.1 Ni	0.25-0.5 Cr 0.25-0.5 Mg	Heat treatable to vary strength and ductility	General engineering
B	do.	0.25	—	0.5 Ni	0.25 Cr 0.25 Mg		
NONCORALIUM	do.	—	—	16 Sb	0.7-2.5 Zn 1.5-3.5 Mg	Corrosion resistance	Cast and wrought components
NORAL alloys (wrought) 3S	Balance	0.15	0.6	1-1.5 Mn	—	Work hardening	Building, packaging, holloware
A56S	do.	0.15	0.6	5 Mg 1 Mn	0.75 Fe 0.5 Cr 0.2 Ti	Toughness. High strength. High corrosion resistance	Marine superstructures

TABLE 4.

TRADE NAME or other designation		COMPOSITION, % by weight					SPECIAL PROPERTIES	USES
		Al	Cu	Si	Other	Other		
NORAL alloys (contd.)	26S	do.	4.25	0.75	0.75 Mn	0.5 Mg	Heat treatable for high strength	Aircraft and highly stressed structures
	51S	do.	0.15	1	0.65 Mg 1 Mn	0.6 Fe	Heat treatable for strength. Good corrosion resistance	Structures exposed to corrosive environment, architectural and building fittings
	C77S	do.	1.75	—	2 Mg	7 Zn 0.13 Cr	Heat treatable for high strength/weight ratio	Aircraft structures
NORAL alloys (cast)	125	do.	1.25	5	0.5 Mg	0.2 Ti max	Heat treatable. Easily cast	Pressure tight castings
	226	do.	4.5	0.9 max	0.25 Ti max	0.7 Fe max	Heat treatable, easily machined	Strong castings
	350	do.	—	0.35	10 Mg 0.25 Ti	0.35 Fe	Exceptional corrosion resistance, shock resistance, machineability	Marine castings
NOVALITE		85	12.5	0.5	1.4 Ni 0.3 Mg	0.5 Fe	Strength at elevated temperatures	Cast pistons for internal combustion engines
NUVAL	Sheet	92	—	0.6	7 Mg	0.4 Mn	Heat treatable. Good corrosion resistance	Marine structures
	(43) Casting	93.8	—	5	1.2 Fe	—	Castability	Light castings
OSMAGAL		Balance	—	—	—	1.8 Mn	Work hardening	Packaging, holloware. Heat exchangers
PACK alloy		do.	4	1.5	4 Ni	—	Heat treatable	Die castings
PALIUM Z		do.	4.5	—	0.6 Mg 4 Pb	0.3 Mn 2.6 Sn 0.3 Zn	Low friction	Heavy duty bearings
PANSERI alloy		82-83	Balance	11.5	0.4 Mg	4.5 Ni	Castability. Strength at elevated temperatures	Pistons for internal combustion engines
PANTAL	Wrought	Balance	—	0.5-1	0.8-2 Mg	0.5-1.4 Mn 0.3 Ti	Heat treatable. High strength and good corrosion resistance	Food handling machinery. Rubber and plastics processing
		do.	—	0.8	0.8 Mg	0.7 Mn 0.3 Fe		
	Cast	93.8	—	5	0.6 Mg	0.6 Mn		
		93.7	—	5	0.6 Mg	0.6 Mn 0.1 Ti		
PERALUMIN	(1) Wrought	Balance	—	—	0.5-1.5 Mg	0.5-1 Mn	Heat treatable for varying strength according to alloy content. Good corrosion resistance	Structures and machinery exposed to corrosive environment (marine)
	(3)	do.	—	—	2-3 Mg	0-0.4 Mn		
	(5)	do.	—	—	4-6 Mg	0-0.4 Mn		
	(3) Cast	do.	—	—	2-4 Mg	0.3 Mn		
	(5)	do.	—	—	4-6 Mg	0.3 Mn		
QUARZAL	5	Balance	5	—	0-1 Ni 0-0.5 Ti	0-1 Fe	Strength at elevated temperatures	Bearings
	15	do.	15	—	6 Mn 0-1 Ni	0-1 Fe		Pistons
RANEY'S alloy		Balance	—	—	30 Ni	—	Hardness	Bearings
RECIDAL		Balance	4	0.7	1.5 Fe < 0.25 Ni 0.2 Ti < 0.1 Mn 0.6-0.14 Pb + Cd	0.6 Mg < 0.2 Zn < 0.1 Sn	Heat treatable to high strength. Free cutting when machined	Automobile and aircraft engines

TABLE 4.

TRADE NAME or other designation		COMPOSITION, % by weight					SPECIAL PROPERTIES	USES
		Al	Cu	Si	Other	Other		
REFLECTAL		Balance	–	–	0.3-1 Mg	–	High purity. High specular reflectivity, corrosion resistance	Mirrors, optical instruments, imitation jewellery, trophies
RENYX		91.5	4	0.5	4 Ni	–	Castability. Corrosion resistance	Die castings
REYNOLDS	alloy 13	Balance	≯ 1	12	≯ 1 Ni + Fe	–	Strength. Castability	Die castings
	380	do.	3.5	8.5	≯ 1.5 Ni + Fe	–	Machineability	Die castings
	R301	do.	4.5	1	1 Fe trace Cr	0.8 Mn trace Mg	Heat treatable for strength. Corrosion resistance	Fabricated sections. Extensions. Aircraft wings
ROMANIUM		Balance	0.25	–	1.7 Ni 0.17 W	0.25 Sb 0.15 Sn	Stronger than pure aluminium	Extrusions
RUSELITE		Balance	4	–	2 Cr 2 Mo	0.2 Ti	Corrosion resistance	Die castings. Hardware
S.A.M. alloy		Balance	1.5	2	11 Mischmetal or Ce	1.25 Ni 1 Mn 0.3 Cr 0.02 Ti	High temperature strength and oxidation resistance	Internal combustion engine components, aircraft and vehicles
SIBLEY alloy		67	–	–	33 Zn	–	Strength. Ductility	Sheet and thin sections
SICAL		22-29	–	50-55	2-4 Ti 0.2 C 0.2 Mn	1 Ca Balance Fe	Low melting point	Master alloy for introduction of silicon into aluminium alloys. Deoxidiser
SILAFONT	2	Balance	–	12-13	0.3-0.6 Mn	0.2-0.4 Mg	Castability strength. Heat treatable	Castings for chemical industry, ships and automobiles
SILAL-V		do.	–	0.5	0.8 Mn	1.25 Mg 0.3 Ti	Strength when heat treated	Automobile and aircraft parts
SILMELEC		do.	–	1	0.6 Mn	0.6 Mg	High electrical conductivity	Electrical conductors and wire
SILUMIN	A	do.	–	9-10	0.4 Co	0.25 Mg	Heat treatable for high strength. Good corrosion resistance. Excellent casting properties. Some suitable for all types of casting process	Automobile and aircraft castings, in internal combustion engines, suspensions, landing gear, etc.
	B	do.	–	7-10	0.5 Co	0.25 Mg		
	C	do.	0.06	11	0.3 Mn	0.45 Mg 0.3 Fe		
	D	do.	–	12	0.4 Mn	0.5 Mg		
	E	do.	–	12.5	0.5 Mn	0.3 Mg		
	F	do.	–	13	0.5 Mn	0.2 Mg		
	G	do.	0.8	8	0.8 Mn	8 Zn	(modified by Na) do.	General castings
	β/γ	do.	–	9.5	0.3-0.4 Mn	0.5 Mg 0.3-0.4 Fe	do.	Automobile castings
	γ	do.	–	11.7	0.3-0.4 Mn	0.3-0.4 Mg	(modified by Na) do.	do.
SIMGAL		Balance	–	0.5	0.5 Mg	0.6 Fe	Ductility	Extruded sections. Architectural trim
STALANIUM		do.	–	–	7 Mg	0.5 Sb	Work hardening. Exceptional corrosion resistance	Marine environments. Boat building
STAY alloy		89	11	–	0.1 Ti	–	Strength. Fluidity when molten	Die castings
STEWART alloy	No. 16	Balance	0.25 max	1 max	4 Ni	–	Heat resistance	Pistons
	No. 10	91	4	5	–	–	–	Cylinder heads
	No. 1	87.5	–	12.5	–	–	Castability	Light castings
	No. 35	95	–	5	–	–	–	–

TABLE 4.

TRADE NAME or other designation		COMPOSITION, % by weight				SPECIAL PROPERTIES	USES
	Al	Cu	Si	Other	Other		
STUDAL	97.7	–	–	1.3 Mn	1 Mg	Strength	Sheets. Forgings
SUPER DURALUMIN	Balance	4.3	0.8	0.5 Mn	0.5 Mg	Heat treatable	Rolling sections. Forgings
SUPERMAGALUMA	94.35	–	–	0.15 Mn	5.5 Mg	High strength. Corrosion resistance	Sheet forms
SUSINI	Balance	1.5-4.5	–	1-8 Mn	0.5-1.5 Zn	High strength. Non heat treatable	Sheet, rod sections
SYLCUM	do.	7.5	9	0.5 Mn	1.4 Ni 0.5 Fe	High strength at elevated temperature. Corrosion resistance	Internal combustion engine pistons
TELECTAL alloy	do.	–	1.5	–	0.1 Li	Non heat treatable. High electrical conductivity	Light alloy parts
TENUAL No. 226	Balance	9.2-10.8	–	1-1.5 Fe	0.15-0.35 Mg	Heat resistance	Cast pistons. Valve tappets. Camshaft bearings
TENZALOY	Balance	0.8	–	8 Zn	0.4 Mg	Machineability. Good strength	Castings
THALASSAL	Balance	–	–	2.5 Mn	2.25 Mg 0.2 Sb	Strength. Sea water corrosion resistance	Aircraft and marine parts
TIERS ARGENT	66.7	–	–	33.3 Ag	–	Colour. Corrosion resistance	Decorative ware
TITANAL	82	12.2	0.5	0.8 Mg	0.7 Fe	Heat treatable	Pistons
TITAN alloy	94.5	–	–	1.5 Mn	4 Mg	Corrosion resistance	Marine fittings
ULMAL	Balance	–	1	0.5 Mn	10 Mg	Strength and ductility	Structural components
ULTRALUMIN	do.	4	0.3-1.5	0-1.5 Mn	0.5-2 Mg	Corrosion resistance. Heat treatable for mechanical strength	do.
VANADIUM-Aluminium Master Alloy	50	–	–	50 V	–	Lower melting point than pure V. Imparts toughness	Additions of vanadium to aluminium alloys
VANALIUM	80	5	–	14 Zn	0.25 V 0.75 Fe	Corrosion resistance	Castings. Aircraft and automobile engines
VERILITE No. 1	96	2.5	0.4	0.3 Ni	0.8 Fe	Corrosion resistance. Heat treatable	Structural applications Aircraft cylinder heads
No. 2	95.5	1	–	1.5 Cr 1.5 Ni	0.5 Mn		
	95.5	1	–	1.5 Cr 1.5 Ni	0.5 Mg		
VITAL	Balance	1	0.6	1.2 Zn	–	Heat treatable	General engineering
VIVAL	98.6	–	0.8	0.6 Mg	–	Strength. Ductility	Forgings
	98	–	0.5	1 Mg	0.5 Mn		Rolled sections
WEBBITE	Balance	–	–	5-7 Ti	–	Deoxidant. Grain refiner	Master alloy for additions of titanium to aluminium
WILMIL	do.	0.1	13.5	trace Na	0.5 Mn 0.1 Mg	Corrosion resistance. Castability (modified)	Castings. Marine
WOLFRAMINIUM	97.6	0.36	–	1.4 Sb 1 Sn	0.2 Fe 0.05 W	Corrosion resistance	General engineering

TABLE 4.

TRADE NAME or other designation	COMPOSITION, % by weight					SPECIAL PROPERTIES	USES
	Al	Cu	Si	Other	Other		
X-alloy	Balance	3.5	0.6	1.25 Fe	0.6 Mg 0.6 Ni	Strength. Heat treatable	General engineering. Pistons
	do.	3.5	—	6 Fe	0.6 Mg	Low friction	Bearings
Y-alloy	do.	3.5-4.5	0.2-0.6	1.2-1.7 Mg	1.8-2.3 Ni 0-0.6 Fe	Heat resistance. Low coefficient of thermal expansion. Tensile strength 185 MN m^{-2}	Cast pistons, cylinder heads. Wrought sections, bar, sheet or strip
Z-alloy	93	—	—	6.5 Ni	0.5 Ti	Low friction. Hardness	Medium heavy bearings
ZICRAL	Balance	1-2	0.7	up to 0.4 Cr 7-9 Zn	0.1-0.7 Mn 1.5-3 Mg	High strength	Wrought sections for structural purposes
ZIMALIUM	70-93	—	—	3-20 Zn	4-11 Mg	Strength with ductility	do. Aircraft parts
ZIRKONAL	Balance	15	0.5	8 Mn	—	Machineability. Strength	Structural members
ZISIUM	do.	1-3	—	0-1 Sn	15 Zn	do.	Forgings

TABLE 5. Copper-based Alloys

TRADE NAME or other designation	Composition, % by weight					SPECIAL PROPERTIES	USES
	Cu	Zn	Sn	Ni	Other		
ACID BRONZE	Balance	-	8-10	0-1.5	2-17 Pb	Corrosion resistance	Pumping equipment
ADMIRALTY BRASS	70	29	1	-	-	Corrosion resistance	Condenser tubes
ADMIRALTY GUNMETAL	88	2	10	-	-	Strength and resistance to marine corrosion	Steam and pressure tight castings, bearings, valves, pump pistons
ADNIC	69-70	-	1	29	0.1-2.5 Mn	Corrosion and heat resistance	Laundry and paper making machinery, plumbing
ADVANCE METAL (Similar to CONSTANTAN)	56	-	-	42.5	1.5 Mn	Negligible temperature coefficient of resistance	Electrical precision instruments, low temperature pyrometers
AITCH'S METAL	60	38.2	-	-	1.8 Fe	Good casting properties	Pumps, valves, pipe flanges
AJAX METAL (Leaded bronzes for bearing applications)	77	-	12	-	11 Pb	Standard alloy	Bearings
	64	-	5	1	30 Pb	Bearings	Bearings
	79.4	-	10	-	10 Pb 0.6 P	Higher strength	Bearings
	90	5	-	-	5 Si	Good casting properties	Bearings

TABLE 5.

TRADE NAME or other designation	COMPOSITION, % by weight					SPECIAL PROPERTIES	USES
	Cu	Zn	Sn	Ni	Other		
ALBATRA METAL	Balance	20	-	20	Up to 1.25 Pb	Strength and ductility, corrosion resistance	Hardware, domestic utensils, auto trim
ALFENIDE METAL	60	30	-	10	1 Fe	Corrosion. Resistance to dilute acids	Chemical plant, ornamental uses
ALLAN RED BRONZE	55	-	5	-	40 Pb Small S	Low friction	Bearings
ALPHA BRASS (Generic) (Gilding metal)	Balance	Up to 38	-	-	-	Homogeneous solid solution	Easily cold worked
	do.	10-15	-	-	-	High ductility, deep colour	
ALPHA BRONZE (Generic)	Balance	-	Up to 14	-	Deoxidised	Homogeneous solid solution	Easily cold worked, condenser tubes
	do.	-	4-5	-	with	High ductility	Condenser tubes
	do.	-	8-10	-	Phosphorus	Higher strength	Springs
ALUMINIUM BRASS	76	22	-	-	2 Al	Corrosion resistance	Condenser tubes, heat exchangers
ALUMINIUM BRONZE (Generic) (Alpha)	Balance	Possible	Possible	Possible	Up to 16 Al Possibly Mn	Corrosion and oxidation resistance, high strength, non-magnetic, non-spark, golden colour and ductile	Architectural and ornamental
	do.	-	-	-	5 Al		
	do.	-	-	-	7 Al	do.	Tubes, marine fittings
(Duplex)	do.	-	-	Some	9 + Al Some Fe	Heat treatable	Castings
	80	-	-	5	10 + Al 5 Fe	High strength	Valve seats, marine propellers, gears and tools
AMBRAC	75	5	-	20	-	Corrosion resistance	Plumbing hardware
	65	5	-	30	-	do.	Condenser tubes
ANTIMONY BRONZE	Balance	-	-	1.5-2.5	7-8 Sb	Strength	Gears
ATLAS ALLOY (90)	do.	-	-	-	9 Al 1 Fe	Heat resistance, strength when heat treated	Aero engines, die pressing
ATLAS BRONZE	do.	-	-	-	9 Al 9 Pb	Machineable with ease	
BARRONIA	83	12.5	4	-	0.5 Pb	Corrosion resistance, oxidation resistance	Severe service steam conditions at high temperature
BASIS BRASS	Balance	27	-	-	-	Ductility with strength	Sheet and strip for pressing
BASIS QUALITY BRASS	do.	23.5	-	-	-	do.	do.
BATH METAL No. 1	83	17	-	-	-	Yellow brass, cheap	Tableware, bath fixtures
No. 2	55	45	-	-	-		
BAZAR METAL	Balance	30	-	8-10	-	Ductile	Domestic hardware, ornamental
BEARING BRASS no. 600	60	36.5	-	-	1.5 Al 1 Fe 1 Mn	Bearing metal useless above 300 °C	Cheap bearings, slow moving machinery
BEARING BRONZES	Balance	-	5-20	-	0-20 Pb Small P	Anti-friction, strength	Worm wheels, gears, journal bearings

TABLE 5.

TRADE NAME or other designation	COMPOSITION, % by weight					SPECIAL PROPERTIES	USES
	Cu	Zn	Sn	Ni	Other		
BELL METAL	do.	<1	20-25	-	-	Strength, heat treatment varies ductility	Musical instruments, heavy duty bearings
BENEDICT METAL (Plate)	57	20	2	12	9 Pb	Corrosion resistance	Domestic ware
BERYLLIUM BRONZE	95.5	-	2	-	2.5 Be	Anti-friction	Springs, contacts
BETA BRASS	Balance	36-45	-	-	-	Brittle	Forged and rolled to bar sections
BINDING BRASS	63-64	35	-	-	1-2 Pb	Machineable	Automatic machining processes, screw cutting
BOBIERRE METAL	Balance	34-42	-	-	-	Strength	Extruded sections, nuts and bolts
BRAZING SOLDERS, Common	50	50	-	-	-	Melting point 870-900 °C	Brazing joints
Silver	16-50	4-38	-	-	10-80 Ag	Melting point 675-820 °C	Joining silver or copper alloys
BROWN METAL	85	15	-	-	-	Ductility	Sheet, bar, wire products
BUSH METAL	81	5	14	-	-	Anti-friction	Railway rolling stock bearings
CAP COPPER	96.5	3.5	-	-	-	Ductility	Percussion caps
CAROBRONZE Wrought	91.2	-	8.5	-	0.3 P	Corrosion resistance and strength	Bushes, chemical plant
	Balance	-	7.5-9.0	-	0.1-0.4 P		
CARTRIDGE BRASS	70	30	-	-	-	Best combination of ductility and strength	Cartridge cases, deep drawn products
CARTRIDGE GILDING METAL	93	7	-	-	-	Ductility, easily formed	Strip, sheet, ornamental
CHAMET BRONZE A	60	39	1	-	-	Corrosion, ductility	Sheet, tube products
CHINESE ART METAL	Balance	10	1	-	15-20 Pb	Ductile, machinable	Ornamental
CHINESE BRONZE	78	-	22	-	-	Casting properties	Musical instruments
CHROMAX BRONZE	66.6	12	-	15.2	3 Cr	Greater strength than equivalent brass alloy, greater corrosion resistance	Structural components in harsh atmospheres, bearings
					3 Al		
CLOCK BRASS	61	37	-	-	2 Pb	Machinable, tough	Small gears, plates, as in clocks
COINAGE BRONZE	95	1	4	-	-	Malleable	Coinage
CONSTANTAN	55	-	-	45	-	Low temperature co-efficient of resistance, high thermal e.m.f.	Resistance coils, thermocouples
CORRONIUM	80	15	5	-	-	Tough and strength corrosion resistance	Structural components
CROTORITE	88-90	-	-	7	3 Al 0.3 Mn	Heat resistance and strength	Plate, strip
	88-90	-	-	-	9-9.75 Al 0.2-0.6 Mn 0.2-2 Fe	Corrosion resistance	Forgings
CUNICO I	50	-	-	21	29 Co	No iron, non spark	Permanent magnets
II	35	-	-	41	24 Co		

TABLE 5.

TRADE NAME or other designation	Cu	Zn	Sn	Ni	Other	SPECIAL PROPERTIES	USES
	COMPOSITION, % by weight						
CUNIFE	60	-	-	20	20 Fe	Ductile, magnetic	Permanent magnets
CUPALOY	99.4	-	-	-	0.5 Cr 0.1 Ag	Strength, high electrical conductivity	Bar, strip, slip rings, terminal studs
CUPRALITH	Balance	-	-	-	1-10 Li	Deoxidant, grain refinement	Degassing copper alloys, little change in conductivity
CUSILOY	do.	-	1-2	-	1-3 Si 0.7-1 Fe	Ductile, corrosion resistance	Wire drawing
DAIRY BRONZE	64	8	4	20	4 Pb	Silver white colour, easily sterilised	Milk containers and plant
DAMAXINE	Balance	-	9.2-11.2	-	0.3-1.3 P up to 7 Pb	Strength, toughness	Bearings
DAVIS BRONZE (METAL)	65	-	-	30	4 Fe 1 Mn	Oxidation resistance	Turbine blades, high temperature valves
DAWSON'S BRONZE	84	-	15.9	-	0.1 Pb 0.05 As	Good casting properties	Heavy duty journal bearings
DELTA BRASS	58.3	39	0.06	1.4	1.4 Fe Pb, Mn	Ductility with strength	Press work
DELTA BRONZE	54-56	Balance	-	-	1-3 Mn 1-2 Pb	Machineability	Machined fittings (marine)
DELTA METAL	55	41	-	-	4 Fe	High tensile strength, corrosion resistance do.	Sheet, forgings, marine bearings,
	60	36	2	-	1 Fe 1 Pb	U.T.S. > 31 kN cm^{-2} elongation 20%	Castings, bearings, ship propellers
DEVARDA'S ALLOY	50	5	-	-	45 Al	Reduces alkaline, sodium nitrate or nitrite to give pure ammonia	Production of pure ammonia in special applications, e.g. heat treatment atmospheres
DURANA METAL	65	30	2	-	1.5 Al 1.5 Fe	High tensile strength, corrosion resistance	General engineering, chemical plant
DUREX	Balance	-	10	-	4-5 Graphite	Sintered, porous	Bearings for oil impregnation
DURONZE I	97	-	0-2	-	1-3 Si		Cold heading processes, screw headed parts
707	91	-	-	-	2 Si 7 Al		
DUTCH GOLD	Balance	20	-	-	-	Ductility, colour	Foil for "gilding"
DUTCH METAL	do.	20-24	-	-	-	do.	Bronzing, foil, gold leaf
ENGRAVER'S BRASS	62.5	37-75	-	-	1.75 Pb	Hard, machineable	Hardware, clock parts
EUREKA (see Constantan)	55-60	-	-	45-40	-	Negligible temperature coefficient of resistance	Resistance coils, instruments
EVERBRITE	60-65	-	-	30	3-8 Fe	Corrosion resistance	Turbine blades, valve discs, seats, nozzles
EVERDUR 6552	95	-	-	-	4 Si 1 Mn	Corrosion resistance combined with strength of mild steel	Pickling tanks, pumps for acids, pipe fittings used in chemical plant
661	95.75	-	-	-	3 Si 1 Mn 0.25 Pb		
655	95.8	-	-	-	3.1 Si 1.1 Mn		

TABLE 5.

TRADE NAME or other designation	COMPOSITION, % by weight					SPECIAL PROPERTIES	USES
	Cu	Zn	Sn	Ni	Other		
651	98.5	-	-	-	1.5 Si 0.25 Mn		
FERRY (see Constantan)	55-60	-	-	40-45	-	Negligible temperature coefficient of resistance	Low resistance coils in electrical instruments
GENELITE	70	Balance	13-14	-	9 Pb 5-6 Graphite	Low friction, strength	Sintered bearings, self lubricating for aero engines
GERMAN SILVER (also Nickel silver, white copper, argentan, mungoose metal)	52-80	10-35	-	5-35	-	Ductile and malleable, cheaper than silver	Domestic utensils, especially when silver plated, substitute for silver
GILDING METAL	Balance	5-10	-	-	-	Malleable, easily rolled to foil	Coatings, foil
GOLD BRONZE	do.	-	-	-	3-5 Al	Colour, strength	Decorative articles, architectural components
GRAPHITE BRONZE	50	-	-	-	50 Graphite	Low friction	Bearings
	79	10	-	-	11 Graphite		
GUILLAUME'S METAL	Balance	-	-	35.7	-	Corrosion resistance	Tape, chemical plant
GUN METALS No.1 (Admiralty)	88	2	10	-	-	Tensile strength, corrosion resistance	Bearings, steam pipe fittings, marine castings
General	88	4	8	-	-		
No. 2	85	5	5	-	5 Pb		
No. 3	86	5	7	-	2 Pb		
HECNUM	55-60	-	-	40-45	-	Negligible temperature coefficient of resistivity	Resistance coils in electrical instruments
HERCULES BRONZE A	85.5	2	10	-	2.5 Al	Corrosion resistance	Chemical plant
HERCULES METAL	61	37.5	-	-	1.5 Al	do.	Marine applications
HERCULOY	96-98	-	0-0.6	-	1.75-3 Si 0.25-1 Mn	Corrosion resistance with good mechanical properties	Good for welded structures, welding rods for Cu alloys
HEUSLER ALLOYS No. 1	50-72	-	-	-	18-26 Mn 10-25 Al	All are ferromagnetic	Magnetic applications where ferrous alloys disadvantageous, e.g. where sparks must be avoided, as in mines, or where rusting must be avoided
No. 3	67	-	-	-	20 Mn 13 Al	Quenched and annealed B_{max} approx. 0.8 T	
No. 4	61	-	-	-	26 Mn 13 Al	Cu_2MnAl approx. B_{max} 0.55 T , similar to nickel, but lower permeability	
HIDURAX ALLOYS No. 4	Balance	-	-	0.5-5	1.5-6 Fe 0-6 Mn 8.5-10.5 Al	Corrosion resistance, wear resistance, strength	Pump parts, gear wheels
No. 13A	do.	-	-	12-16	2-4 Al 1-3 Fe	do.	Valve spindles
HIGH STRENGTH BRASS (Manganese bronze)	55-60	Balance		-	0.4-2 Fe 0.5-1.5 Al 1.5 Mn	High strength with good corrosion resistance	Marine applications, ship's propellers

TABLE 5.

TRADE NAME or other designation	COMPOSITION, % by weight					SPECIAL PROPERTIES	USES
	Cu	Zn	Sn	Ni	Other		
HIGH STRENGTH BRONZE (strictly a brass)	60-68	do.	-	-	2-4 Fe 3-7.5 Al 2.5-5 Mn	do.	do.
HOLFOS BRONZE	Balance	-	11-12	-	0.1-0.2 P 0.25 Pb	Fine grain size	Castings
HYDRAULIC BRONZE	82-83.75	5-8	3.25-4.25	-	5-7 Pb	Machineability	Components requiring much machining or mass production
HYTENSYL BRONZE	Balance	23	-	-	3 Mn 3 Fe 4 Al	High tensile strength	Trunnion bearings for bridges, large worn wheels
IDEAL	55-60	-	-	45-40	-	Negligible temperature coefficient of resistance	Standard resistances, instrument coils
IMMADIUM Wrought I	55	42	1	-	1.6 Fe 1.4 Al	High tensile strength and good corrosion resistance	Pump rods, valve spindles, stay bolts, propeller shafts
Cast IV	56	40	1	-	1.5 Fe 0.3 Al 0.06 As		
Forged II	66	28.2	1	-	1.5 Fe 3 Al 0.3 Mn		
V	67.5	25.6	-	-	1.6 Fe 5 Al 0.3 Mn 1 Pb		
VI	70	25	-	-	2 Fe 3 Mn		
ISABELLIN	84	-	-	-	13 Mn 3 Al	Negligible temperature coefficient of resistance	Electrical resistors, instruments
JACKSON ALLOY	63-64	30-35	-	-	2-5 Sb	Malleable	Stampings, buttons
JACOB S ALLOY	94.9	-	-	-	4 Si 1.1 Mn	High strength, toughness, corrosion resistance, low electrical conductivity	Castings
JEWELLER'S BRONZE	89	9	2	-	-	Malleable, ductile, corrosion resistant	Beaten metal work, ornamental stampings
JOHNSON'S BRONZE No. 40	90	9.5	0.5	-		Low friction, toughness	Bearings, sheet metal
No. 44	86	3-5	3-4.5	-	Balance Pb	do.	do.
JOURNAL BRASS	65	-	5-10	-	5-30 Pb	Low friction, toughness	Railway coach bearings
JOURNAL BRONZE Locomotives	86	-	14	-	-	Low friction, other properties to suit application	Bearings
Wagon	84-88	2-4	10-12	-	-		
General	80	-	10	-	10 Pb		
Automobiles (U.S.A.)	65-75	3	4.5-6.5	-	24 Pb		
KARAKANE ALLOYS 2	71.4	-	14.2	-	14.3 Pb	Low damping capacity, easily cast and machined	Bells, musical instruments
3	70.0	3.0	19.0	-	8.0 Pb		
4	70.0	3.5	17.3	-	10.4 Pb		
5	64.0	9.0	24.0	-	3 Fe		
6	61.0	6.0	18.0	-	12.0 Pb 3 Fe		

TABLE 5.

TRADE NAME or other designation	COMPOSITION, % by weight					SPECIAL PROPERTIES	USES
	Cu	Zn	Sn	Ni	Other		
KEEN'S ALLOY	75	2.3	2.8	16	2 Co 0.5 Al 1.5 Fe	Toughness, corrosion resistance	Chemical plant components
KELMET BRONZE	67.7-70.5	—	6.5	0.04	22.5-25.5 Pb 0.03 S	Machineability	Screws, linkages, threaded parts, bearings
KERMET	Balance	-	-	2	33-37 Pb	Low friction	Bearings, aircraft engines
KINGHORN METAL	58.5	39.3	0.95	-	1.15 Fe	Corrosion resistance	Sheet products and tubes, screws and nuts
KUMIUM	99.5	-	-	-	0.5 Cr	Good electrical conductivity	Electrical machines, contacts
KUNIAL	71.7-72.9	20	-	5.8-6.3	1-1.8 Al	High modulus of elasticity, heat treatable	Springs, valves, primers, fuse bodies
	72.5	20	-	6.0	1.5 Al		
	64.0	20	-	13.5	2.5 Al		
KUNIFER 5	Balance	-	-	5	1 Fe 1 Mn	Corrosion resistance, strength and ductility combined	Marine applications
10	do.	-	-	10	1 Fe 1 Mn		do.
30A	do.	-	-	30	2 Fe max 2 Mn max		Condenser tubes
KUTERN	Balance	-	-	-	0.3 to 0.7 Fe	Free machining, high electrical and thermal conductivity	Electrical machines
LANCASHIRE BRASS	73	27	-	-	-	Malleable and ductile	Strip formed products
LAVELSSIERE BRONZE	61	38	1	-	-	Slightly harder than 60/40 brass, corrosion resistance	General engineering, condenser tubes, marine parts
LEAD BRONZE BEARING ALLOYS	78	-	up to 2	up to 2	20 Pb	Low friction, high fatigue strength on thermal cycling, easily bonded to steel shells	High speed engine bearings, particularly aircraft
	74	-	up to 2	-	24 Pb		
	69	-	-	-	30 Pb 0.6 Ag		
	70-79	-	6-10	-	15-20 Pb		
	59	-	-	-	40 Pb up to 1 Ag		
LECHESNE ALLOY	90-60	-	-	10-40	0.2 Al	Corrosion resistance, strength	Sheet and tube products, chemical plant
LEDRITE	61	35.6	-	-	3.4 Pb	Machineability, strength	Cast, stamped and extruded products with machined threads, bores, etc.
LEMARGUAND	39	37	9	7	8 Co	Good corrosion resistance	Chemical plant, cheap jewellery
LOW BRASS	80	20	-	-	-	Ductile and malleable, excellent cold working properties	Ornamental metalwork, flexible hose, clock dials, medallions, buttons
MCGILL METAL	Balance	-	-	-	9 Al 2 Fe	Mechanical strength, ductility, machineability, corrosion resistance	Castings and forgings, pump liners, gears

TABLE 5.

TRADE NAME or other designation	Cu	Zn	Sn	Ni	Other	SPECIAL PROPERTIES	USES
	COMPOSITION, % by weight						
MACHT'S METAL	60	38-38.5	-	-	1.5-1.8 Fe	Strength	Forgings
MAILLECHORT	65-67	13-14	-	16-20	1-3.5 Fe	Corrosion resistance	Sheet and bar products, hardware
MALLET ALLOY	74.6	25.4	-	-	-	Ductility	Sheet, foil, wire
MALLORY METAL	63-68	Balance	-	-	3 Fe up to 4 Mn up to 4 Al	High strength	Wrought products, worm gears, propeller blades
100	100	Balance	-	-	2.6 Co 0.4 Be	High strength, high electrical conductivity	Electrical components, springs
MANGANESE BRONZE Wrought alloys	56-59	Balance	0.75-1.25	-	1.25 Fe 2 Mn 0.2 Al 0.5 Pb	Tensile strength 54 kN cm^{-2} (35 tons in^{-2})	Bars, sheet, strip, forgings for marine and general industrial use
	do.	do.	0.5	-	0.7-1.2 Fe 0.5-1.2 Mn 0.2-1.2 Al 1 Pb max	Tensile strength 46 kN cm^{-2} (30 tons in^{-2}) Elongation 20%	
	do.	do.	do.	-	0.7-1.2 Fe 0.5-2 Mn 2-3 Al 1 Pb max	Tensile strength 54 kN cm^{-2} (35 tons in^{-2}) Elongation 15%	
	64-66	Balance	0.75-1.25	-	0.8-1.25 Fe 0.2-0.5 Mn 2.5-5 Al 0.25 Pb max	Tensile strength 54 kN cm^{-2} (35 tons in^{-2}) Elongation 15%	
Castings	56-62	Balance	0.5	-	0.5-1 Fe 0.5-1.5 Mn 0.25 Al 2 Pb max	Tensile strength 31 kN cm^{-2} (20 tons in^{-2}) Elongation 12%	Marine propellers, pumps, gear wheels, worms
	57-63	do.	0.5	-	0.5-1 Fe 2 Mn 0.5-2 Al 1 Pb max	Tensile strength 46 kN cm^{-2} (30 tons in^{-2}) Elongation 15%	
	50-70	Balance	0.25	-	1-2 Fe 0.5-2.5 Mn 2-5 Al 0.25 Pb max	Tensile strength 68 kN cm^{-2} (44 tons in^{-2}) Elongation 12%	Marine propellers, pumps, gear wheels, worms
	82	5	8	-	2 Mn 3 Pb max	High tensile strength	Hydraulic components, structures or general components
	62	26.5	-	-	3 Fe 3.5 Mn 5 Al		
	Balance	19-22	-	-	2-3 Fe 3.5-4.5 Mn 4-6 Al	Strength, low friction	Bearings
	60	20.5	-	16.5	3 Mn	Corrosion resistance	Chemical plant
MANGANIN (also called MINALPHA)	84	-	-	4	12 Mn	Low temperature co-efficient of resistivity, high resistivity	Resistance wire, electrical instruments, springs in strip form
MANILA GOLD	85	12	-	-	2 Pb	Colour, ductility	Cheap jewellery
MANNHEIM GOLD	83	10	7	-	-	do.	do.
MARKET BRASS (Common brass)	65	35	-	-	-	Cold working ductility and strength, corrosion resistance	Strip or sheet for pressing or stamping, condenser tubes

TABLE 5.

TRADE NAME or other designation		COMPOSITION, % by weight					SPECIAL PROPERTIES	USES
		Cu	Zn	Sn	Ni	Other		
McKECHNIE'S BRONZE		57	41	1	-	1 Fe 0.5-1 Pb	Strength, machinability	Castings for pumps, hydraulic equipment, nuts and bolts
MERCO BRONZE		88	-	10	-	2 Pb	Strength, machinability	General engineering
MERCOLOY		60	10	1	25	2 Fe 2 Pb	Corrosion resistance, strength	General engineering, valves
MINARGENT		56.8	-	-	39.8	2.8 W Balance Al	Colour, corrosion resistance	Substitute for silver, silver solder
MIRA		74-75	0-0.6	1-8	0.25-1	16 Pb 0-6.8 Sb 0.04 Fe	Machineability, corrosion resistance	Valves, pipes for acids
MIRROR ALLOYS (Speculum metals)	Birr	70.3	-	29.7	-	-	High specular reflectivity, corrosion resistance	Mirrors for astronomical telescopes, scientific instruments and optical use generally
	Chinese	80.8	-	10.7	-	8.5 Sb		
	Cooper	57-58	3.5	27.5	-	9-10 Pt 1-3 As		
	Edwards	69.8	2.6	25.0	-	2.5 As		
	Mudge	68.8	-	31.2	-	-		
	Richardson	65.3	0.7	30.0	-	2 As 2 Si		
	Ross	68.2	-	31.8	-	-		
	(Sallit) Dollit	64.6	-	31.3	4.1	-		
	General	66-88	-	34.12	-	-		
MOCK GOLD No. 1		71	4	-	-	25 Pt	Colour, corrosion resistance	Substitute for gold in jewellery
MONTANA GOLD		89	10.5	-	-	0.5 Al	Colour, ductility	do.
MORRISON'S BRONZE		91	-	9	-	-	Ductility	Fabricated rod, strip, sheet
MOSAIC GOLD		65	35	-	-	-	Colour, ductility	Cheap jewellery
MUNTZ METAL		58-61	38.5-42	-	-	0.35-0.9 Pb	Ductility when not worked, machineability	Condenser tubes and plates, tubing for threading
		do.	do.	-	-	1.1 Pb		
NADA NT		91.75	-	3.75	3.75	0.75 Pb	Tarnish resistance, castability	Small castings
NARITE		79-81	-	-	1	13-15 Al 5 Fe	Toughness, castability	Deep drawing dies, forming tools
NAVAL BRASS		59-62	Balance	0.5-1	-	up to 0.6 Pb	Corrosion resistance, toughness	Condenser plates, marine fittings, propeller shafts, pistons, valve stems, welding rods
NAVY BRONZE		86-90	Balance	5.5-6.5	-	1-2 Pb	Corrosion resistance	Stem pipe fittings, valve bodies, gears, bushes, bearings.
NEEDLE BRONZE		84.5	5.5	8	-	2 Pb	Machineability, corrosion resistance	Complex machined parts subject to corrosion
NEOGEN		58	27	2	12	0.5 Al 0.5 Bi	Corrosion resistance, strength	Highly stressed parts subject to corrosion
NERGANDIN		70	28	-	-	2 Pb	Ductility, seawater corrosion resistance	Condenser tubes

TABLE 5.

TRADE NAME or other designation		COMPOSITION, % by weight					SPECIAL PROPERTIES	USES
		Cu	Zn	Sn	Ni	Other		
NEWLOY		64	-	1	35	-	Corrosion resistance	Tableware, resistance wire
NIAG		46.7	40.7	-	9.1	2.8 Pb 0.3 Mn	Corrosion resistance, formability	Architectural trim, stampings, rods
NICKEL BRASS		50.5-57	38-45	-	1-2.8	0.5-2 Fe 0-1.4 Al 0-0.3 Mn	Corrosion resistance	Turbine blades
		44-46	41.5-44	-	10	2-2.5 Pb	Ductility	Stampings, rods
		46	44	-	8	2 Mn	Strength	General engineering
		44	43	-	12	0.6 Fe		Steam turbine parts
NICKEL BRONZE		83.9	-	8	2	6 Pb 0.1 P	Corrosion resistance to acids	Chemical plant
		71-82.5	2	7	8-20	-	Strength, age hardening	Engineering components
		76-81	1	6-12	1	5-16 Pb 0.2 P	Low friction	Heavy duty bearings
		85-86	-	11-12	2-3	0.1-0.2 P	Toughness	Roller bearings
		Balance	0-2	5-11	1.5-5	0.03-0.15 P	Castability	Gears and wheels
		do.	1-8	4-6	14-40	1 Fe 1 Mn 0.2 Si	Corrosion resistance	Steam valves
		63	0.5	5	1.5	30 Pb 0.01 P	Low friction	Light duty bearings
NICKELIN	II	67	-	-	30-31	2-3 Mn	High electrical resistivity	Resistors
	I	54	20	-	26	-	Corrosion resistance	Sections for trim, tableware
NICKELOID		60-55	-	-	40-45	-	High electrical resistivity, corrosion resistance	Resistors, food equipment
NICKEL OREIDE		63-66.5	30.5-32.75	-	2-6	-	Ductility, strength, corrosion resistance	Wrought sections, hardware
NICKEL SILVER ALLOYS		60	20	-	20	-	Corrosion resistance	Marine fittings
		60	22	-	18	-	do.	Cutlery spoons and forks
		42-50	20-41	1-2	14-35	1-4 Pb	do.	Die castings
		70	10	-	20	-	do.	Steam plant
		50-55	25	-	20-25	-	do.	Tableware
		60-65	Balance	-	10-20	0.04 Pb 0.3-0.5 Mn	do.	Made in 10, 12, 15, 18, 20% nickel for food equipment, architectural use
		70	8	4	14	-	do.	Cast valve seats
		Balance	5	4	20	-	do.	Builders' hardware
NICLA		40-46	39-41	-	12-15	1.75-2.5 Pb Balance Al	Ductility, corrosion resistance	Castings and wrought products
NICO METAL		90	-	-	10	-	Corrosion resistance	Tubes, sheet, rod
NIDA		91-92	-	8-9	-	Trace P	Ductility	Drawn tubes

TABLE 5.

TRADE NAME or other designation		COMPOSITION, % by weight					SPECIAL PROPERTIES	USES
		Cu	Zn	Sn	Ni	Other		
NI-VEE	A	88	2	5	5	-	Toughness, strength	Pumps, valves, castings, where pressure tightness required, statuary, bearings for high stress
	B	87	2	5	5	1 Pb		
	(C) L5Z5	80	5	5	5	5 Pb	Low friction, high fatigue strength, free machining	
	L2	87	0.5	5	5	2.5 Pb		
	(D) L10	80	-	5	5	10 Pb		
NOIL		80	-	20	-	-	Toughness	Cast piston rings, non-lubricated conditions
NON-GRAN		87	2	11	-	-	Strength	Castings, pulleys
NOVOCONSTANT		82.5	-	-	-	12 Mn 4 Al 1.5 Fe	High electrical resistivity	Electrical resistors
NU-BRONZE		95.8	-	-	3.25	0.25 Mn 0.73 Si	Good electrical conductivity, wear resistance	Electrical machinery castings
NU-GOLD		87.8	12.2	-	-	-	Colour, ductility	Substitute for gold in cheap jewellery
NUREMBERG GOLD		90	-	-	-	7.5 Al 2.5 Au	Colour, tarnish resistance, ductility	Jewellery
ODA METAL		45-65	-	-	27-45	1-10 Mn 0.5-3 Fe	Corrosion resistance	Castings and wrought sections for steam and chemical plant
OHMAL		87.5	-	-	3.5	9 Mn	Electrical resistivity	Electrical resistance wire standard coils
OILITE		88-90	-	10-12	-	0.1-0.2 P Graphite powder 1 Fe (max)	Porosity when compacted and sintered, absorbs 35% of volume as oil	Self lubricating bearings in instruments, high speed machinery, clocks
OKER-CAST		72	24.5	-	-	2.32 Fe 1.1 Pb	Machineability, ductility	Mass production fittings, threaded stock
OLDSMOLOY		45	39	2	14	-	Corrosion resistance, strength	Castings for steam and chemical plant
OLYMPIC BRONZE	A	96	1	-	-	3 Si	High strength ductile alloys, resistant to season cracking, low thermal and electrical conductivity	Chemical plant construction, welding rods, hardware
	B	97.5	1	-	-	1.5 Si		
	C	94.75	1	-	-	4.25 Si		
ORANIUM BRONZE	H	90	-	-	-	10 Al	Corrosion resistance	Hardware, tools
	HH	88.5	-	-	-	11.5 Al	Strength	Gears, bearings
	HX	89	-	-	-	11 Al		
	M	95	-	-	-	5 Al	Toughness	Propellers
	MH	91.5	-	-	-	8.5 Al		
	S	97	-	-	-	3 Al		
OREIDE		87.25	11.5	1.25	-	-	Ductility, malleability	Hardware, pressings and stampings
	B	85.5	14.5	-	-	-		
		80.5	14.5	4.85	-	0.1 Pb		

TABLE 5.

TRADE NAME or other designation		COMPOSITION, % by weight					SPECIAL PROPERTIES	USES
OREIDE (contd.)		Cu	Zn	Sn	Ni	Other		
		80.5	19.5	-	-	-		
		90	10	-	-	-		
	A	68.2	31.5	0.3	-	-		
ORMULU	Gold	90.5	3	6.5	-	-	Rich golden colour, corrosion resistance	Ornamental work in statuary, metal applique to stone and wood
	do.	94.2	-	5.8	-	-		
	Yellow	58	25.5	16.5	-	-		
OUNCE METAL (leaded gunmetal)		84-86	4-6	4-6	1 max	4-6 Pb	Castability	Pumps, valves, general castings, statuary
PACKFONG (also Pai-t'ung, Paktong, Paaktong, Packtong, Chinese nickel silver, Tombac)	Chinese	26-40	41-32	-	16-37	0-2.5 Fe	Ductility, malleability, corrosion resistance	Sheet metal products, stampings, pressings, tubes, cutlery, holloware
	German	54	28	-	18	-		
	Cookson	40.9	45	-	11.1	2.5 Fe 0.23 Pb 0.16 Co		
	Sheffield	57	25	-	13	3 Fe		
	Birmingham	62-63	17-26	-	10-19	-		
PARIS METAL		80	5	2	6-16	1 Co 1-5 Fe	Strength, corrosion resistance	Statuary
PARKER'S ALLOY (Parker's Chrome)		60	20	-	10	10 Cr	Corrosion resistance	Marine castings, pen nibs
PARSON'S BRONZE Wrought		57-60	Balance	1-1.5	-	0.2-1 Mn 0.6-1.2 Fe 0-1 Al	High tensile strength, corrosion resistance	Castings replacing malleable iron for improved corrosion resistance and non-magnetic properties, (marine), propellers, valves, engine frames
Cast		56-58	do.	0.7-1.1	-	0.2-0.4 Mn 0.8-1.1 Fe 0.2-0.4 Al		
PATTERN METAL		83	10	4	-	3 Pb	Toughness, machine-ability, good surface	Metal patterns for moulding
PELLUX		Balance	-	-	-	1-10 Cr 0.5 Fe	Deoxidants	Additions of alloying elements as master alloys, used in granular form
	do.	-	-	-	10-50 Mn 1 Fe		
	do.	-	-	10-20	1 Fe		
PEN METAL		85	13	2	-	-	Moderate hardness with ductility	Pen nibs, strip for trim
PERKING BRASS		76-80	-	20-24	-	-	Hardness, brittleness	Castings for musical instruments (bells, gongs), bearings, reflectors
PERMET		49	-	-	21	30 Co	High value of BH max	Permanent magnets
PHILSIM		86.25	6.35	7.4	-	-	Strength, corrosion resistance	Castings
PHONO-BRONZE		98.6 97.6	-	1.4 1.4	-	trace Cd 1 Si trace Cd	Corrosion and abrasion resistant	Electrical work as wire
PHOSPHOR BRONZE		96	-	3.75	-	0.1 P	High modulus of elasticity, corrosion and corrosion fatigue resistance	Springs, instrument suspensions and components, contacts
(wrought)		94.5	-	5.25	-	do.		

TABLE 5.

TRADE NAME or other designation	COMPOSITION, % by weight					SPECIAL PROPERTIES	USES
	Cu	Zn	Sn	Ni	Other		
PHOSPHOR BRONZE (contd.)	93	-	6.75	-	do.		
	94	-	5.5	-	do.		
	91.5	-	8.25	-	0.25 P	do., hardness	Bearings
(cast)	Balance	0.5 max	9.5	-	0.1-0.4 P 0.25 Pb max	do.	General purpose castings
	do.	0.5 max	10 min	-	0.5 P min 0.25-0.75 Pb	do.	Bearings
	do.	0.3 max	12	0.5 max	0.15 P min 0.5 Pb max	do.	Gears and bearings
	do.	2 max	7.5	1 max	0.3 P min 3.5 Pb	do.	Light bearings
	do.	-	14	-	trace P	do., high hardness low ductility	Distortion free conditions
	do.	-	-	18	1 P max	do.	Heavy duty bearings, bridges and turntables
PHOS-COPPER	Balance	-	-	-	7-10 P	Low melting point (720 °C), high electrical conductivity	Brazing copper conductors to preserve electrical conductivity
PHOSPHOR-COPPER	Balance	-	-	-	10-15 P	Deoxidises	Master alloy for phosphorus additions, deoxidant
PINCHBECK (Guinea gold)	85-88	12-15	-	-	-	Colour, ductility	Cheap jewellery
PINKUS METAL Brass	88.1	6.9	2.5	0.3	0.33 Sb 1.8 Pb	Ductility	Rolled sections for machining
Bronze	72.5	1.5	14.7	-	2.5 Sb 8.8 Pb	Hardness, castability	Castings
PIN METAL	62	38	-	-	-	Toughness, hardness	Common pins, in drawn condition
PLASTIC BRONZES	64.1	-	5.05	0.75	30.1 Pb	Lower hardness and strength than normal bronze, but better than white metal	Heavy bearings with large diameter axles, e.g. railway coaches, crane jibs
	63.6	0.9	5.6	1	28.9 Pb		
	65	1	4	-	30 Pb		
	64	-	5	1	30 Pb		
	67.4	1	5	-	26.6 Pb		
	67.7	0.5	5	-	26.8 Pb		
PLATER'S BRASS	80-90	Balance	up to 2	-	-	High purity and uniform texture	Anodes for plating brass on to other articles
PLATINOID (A) M	60	24	-	14	2 W	High electrical resistivity	Electrical resistance wires, M type also as jewellery. Coinage, thermocouples
B	62	22	-	15	1 W		
	54	20.5	-	24.8	0.2 Mn 0.5 Fe		
PLATINOID SOLDER	47	42	-	11	-	Lower melting point than Platinoid alloy	Brazing connections to Platinoid
PLUMBER'S WHITE	58	25	-	15	1 Fe 1 Pb 0.3 Mn	Ductility, toughness	Plumbing fittings, pipework
PLUMRITE	85	15	-	-	-	Ductility, malleability, corrosion resistance	Tube and pipework, drawing and stamping

TABLE 5.

TRADE NAME or other designation		COMPOSITION, % by weight				SPECIAL PROPERTIES	USES
	Cu	Zn	Sn	Ni	Other		
P.M.G.	Balance	2	-	-	2 Fe 3-4 Si	Corrosion resistance, heat treatable, age hardening	Forgings
PRINCE'S METAL	61-83	17-39	-	-	-	Range of ductility and strength	As for brasses of similar composition
PRONIETHIUM	67	30	-	-	3 Al	Corrosion resistance	Condenser tubes, steam fittings
RAKEL'S ALLOY	Balance	1	-	10	1 Mn	do.	do.
RAKEL'S METAL	do.	1	-	-	10 Al 1 Mn	do.	do.
RED BRASS Casting	82-85	7-10	3-5	-	5-6 Pb	Good resistance to atmospheric corrosion, good machineability with Pb present	Engineering components as described by name in first column
Valves	72-85	5-10	5	-	5-10 Pb		
Wrought	82-85	15-18	-	-	-		
Condenser tubes	84	Balance	-	-	0.075 Pb		
Rods	83-86	do.	-	-	1.5-2.3 Pb		
	87-90	do.	-	1.25	1.25-2.3 Pb		
REDFORD'S ALLOY	85.7	1.8	10	-	2.5 Pb	Corrosion resistance	Steam fittings, stuffing boxes
REED BRASS	69	30	1	-	-	Low damping capacity	Reeds for musical instruments, pipes, condenser tubes
REICH'S BRONZE	85.2	-	-	-	7 Al 7.5 Fe 0.2 Pb 0.6 Mn	Toughness, corrosion resistance	Rods, sections, for chemical plant, marine applications
REITH'S ALLOY	75	-	11	-	5 Sb 9 Pb	Low friction	Bearings
RESISCO	90.5-91	-	-	2	7-7.5 Al 0-0.1 Mn	Ductility, corrosion resistance	Condenser tubes
RESISTAL	88-90	-	-	-	9-10 Al 1-2 Fe	Corrosion resistance	Pump castings, chemical plant
RESISTEN 1	86.5	-	-	-	11.7 Mn 1.8 Fe	High electrical resistivity with low temperature co-efficient	Electrical standards and instruments
2	85	-	-	-	12 Mn 2 Fe		
3	84.3	-	-	-	13.7 Mn 2 Fe		
4	84.3	-	-	-	13.5 Mn 2 Fe 0.2 Si		
RETZ ALLOY	75	-	10	-	10 Pb 5 Sb	Low friction	Bearings
REVALON	76	22	-	-	2 Al	Corrosion resistance	Condenser tubes
RHEOTAN I	84	4	-	-	12 Mn	Electrical resistivity	Resistors
II	52	18	-	25	5 Fe		Coinage
RICHARD'S BRONZE	55	42	-	-	2 Al 1 Fe	Corrosion resistance	Castings
ROCON COPPER	Balance	-	-	-	0.5 As	Harder than pure copper, corrosion resistance	Sheets, roofing sections
ROMAN BRASS	60	39	1	-	-	Toughness, corrosion resistance	Plumbing and water fittings
ROMAN BRONZE	58-60	-	Balance	-	1.1 Al 1 Fe 0.2 Mn	Strength, corrosion resistance	Castings, statuary

TABLE 5.

TRADE NAME or other designation	COMPOSITION, % by weight					SPECIAL PROPERTIES	USES
	Cu	Zn	Sn	Ni	Other		
ROSS'S ALLOY	68	-	32	-	-	Strength, corrosion resistance	Castings
RUBEL BRONZE	80	-	-	4.5	4.5 Fe 10 Al 1 Mn	High tensile strength, corrosion resistance	Castings and wrought products for marine and exterior applications
	56.5	40	-	2	1.2 Fe 0.3 Al		
	57	39	-	2	1.5 Fe 0.5 Al		
	53	40	-	-	5 Mn 2 Fe		
	52	40	-	2	3.3 Mn 2.3 Fe 0.3 Al		
	51	40	-	2	4 Mn 2.5 Fe 0.5 Al		
RULE BRASS	62.5	35	-	-	2.5 Pb	Free machining	Mass produced products, screws, stampings
SAMBRAS	83	10	-	-	5 Si 1 Mn 1 Al	High strength	Castings
SCEPTRE BRASS 1	61.7	35.8	-	-	1 Al 1.5 Fe 0.7 Pb	High tensile strength	Extrusions
2	64.5	33	0.45	-	1 Al 1 Fe		
SCREW BRASS	58	40	-	-	2 Pb	Free machining	Screws and mass produced threads
SCREW BRONZE	93	5	1	-	1 Pb	do.	do.
SEAWATER BRONZE	45	5.5	16	32.5	1 Bi	Corrosion resistance	Marine components
SECRETAN	90-50	-	-	-	5-9 Al 1.5 Mg 0.5 P	Corrosion resistance	Chemical plant
SEMI PLASTIC BRONZE	76.5-79.5	0-4	5-7	-	14.5-17.5 Pb 0-0.4 Fe 0-0.4 Sb	Low friction	Bearings
SEN ALLOY	40	-	-	20	40 Fe	Strength	Unknown
SEYMOURITE	64	18	-	18	-	Corrosion resistance	Cutlery, domestic ware, architectural trim
SHAKUDO	Balance	-	-	-	1-4.2 Au 0-2 Ag 0-0.1 Pb	Colour, tarnish resistance	Decorative and artistic work
SHEATHING BRONZE	45	6	16	32	1 Bi	Toughness, corrosion resistance	Cladding on cables, junction boxes and components for weather protection
SHIBU-ICHI	51-86	-	-	-	Balance Ag 0.08-0.12 Au	Colour, tarnish resistance	Decorative and artistic work
SIGNAL BRONZE	98.5	-	1.5	-	-	Reasonable electrical resistivity, stronger than copper	Wire products
SILCURDUR	Balance	-	-	-	2.2 Si 0.5-0.7 Mn	Corrosion resistance	Chemical plant
SIFBRONZE	59	37-5-38.5	0.5-2.5	0-1.75	0-1 Pb Up to 0.8 Fe	Low melting point	Oxyacetylene welding of copper alloys
SILFOS	80	-	-	-	15 Ag 5 P	Low melting point (645 °C)	Silver soldering

TABLE 5.

TRADE NAME or other designation	COMPOSITION, % by weight					SPECIAL PROPERTIES	USES
	Cu	Zn	Sn	Ni	Other		
SILICON BRASS	81.5	14	-	-	4.5 Si	Low damping capacity	Bell metal
	79-81	Balance	-	-	3-3.8 Si 0-0.6 Fe	Corrosion resistance	Chemical plant
	81	15	-	-	4 Si	Castability	Castings
	85-95	4-13	-	-	1-2 Si	Ductility, reasonable electrical conductivity	Resistance welding
	80	16.5	-	-	3.5 Si	Ductility	Wire products
	70	27	-	-	1 Si 1 Al 1 Fe	High modulus of elasticity	Springs
	67	32	-	-	1 Si	Low friction	Bearings
	83	10	-	-	5 Si 1 Mn 1 Al	Low melting point	Die castings
SILICON BRONZE	Balance	8	-	-	3-4 Si 1-1.5 Fe 2-3 Pb 3 Mn	Toughness, low friction	Bearings
	do.	5	-	-	2.5-3 Si 1 Fe 2.5 Mn 1.5-2 Pb	Toughness	Gears
	90	2.5	-	-	3-4.5 Si 3 Fe 0.3-1.5 Mn	do.	Castings
	Balance	-	-	-	3 Si 0.2-0.5 Te	Free machining	Mass produced components
	99	-	0.8	-	0.03 Si	Ductility	Wire products
	Balance	-	5	-	3 Si 0.75 Fe 1.5 Cd 0.76 Mn 3.5 Al	Good electrical conductivity, ductility	Hard drawn wire products for electrical conductors
SILICON COPPER	Balance	-	-	-	10-50 Si	Low melting point	Master alloy
	99	-	-	-	1 Si	Ductility	Wire
	97	-	-	-	3 Si	Corrosion resistance	Sheet products
	95	-	-	-	5 Si	Low melting point, fluidity	Castings
SILLIMAN BRONZE	86.5	-	-	-	9.5 Al 4 Fe	Strength, corrosion and oxidation resistance	Rods, sections and forged products
SILMET (see Nickel Silver)	Balance	29	-	8	-	Corrosion resistance, colour	Cutlery, decorative trim, domestic table-ware and culinary utensils
	63	27	-	10	-		
	63	22	-	15	-		
	62	20	-	18	-		
	62	18	-	20	-		
	57	18	-	25	-		
	57	13	-	30	-		
SILVEL	68-73	12-16	-	0-6.5	7-12 Mn 0.5 Pb 2 Fe 0.2 Al	Strength, corrosion resistance	Unknown

TABLE 5.

TRADE NAME or other designation	COMPOSITION, % by weight					SPECIAL PROPERTIES	USES
	Cu	Zn	Sn	Ni	Other		
SILVER BRONZE	67.5	Balance	-	-	18 Mn 1.25 Al 0.25 Si	Strength, corrosion resistance	Applications where moderately elevated temperatures apply, e.g. steam plant
SILVERINE	72-80	1-8	1-3	16-17	1-2 Co 1-1.5 Fe	Corrosion resistance, strength	Chemical plant
SILVEROID	54	-	-	45	1 Mn	Corrosion resistance, strength, ductility	Cutlery, culinary equipment, electrical devices
	48	25	-	26	1 Pb		Ornamental trim
SILVER SOLDER ALLOYS (copper base) (see precise precious alloy Table 9 for silver-based types)	52	38	-	-	10 Ag	Melting point 790 °C	Soldering steel
	45	35	-	-	20 Ag	do. 780 °C	do. Copper alloys
	38	32	-	-	30 Ag	do. 740 °C	do. do.
	50	17	-	-	33 Ag		do. Cast iron
	45	30	-	-	20 Ag 5 Cd		Yellow solder
SILVORE	62	19.2	-	18.5	0.3 Pb	Corrosion resistance	Decorative trim, sections, extrusions
SILZIN	75-85	10-20	-	-	4.5-5.5 Si	do.	Chemical plant
SIMILAR ALLOYS	80	20	-	-	-	Ductility, stronger when tin present	General engineering sections, sheet, rods, etc.
	83.5	9.5	7	-	-		
	89.5	10	0.5	-	-		
SINCHU	66.5	33.4	-	-	0.1 Fe	Strength and ductility	Extrusions, sections, rod
SPELTER BRONZE	Balance	45	3-5	-	-	Melting point 860 °C	Brazing of copper alloys and steels
SPELTER SOLDER	50-53	Balance	-	-	0.5 Pb max	Melting range 870-882 °C	do.
SPRING BRASS	70-72	Balance	-	-	-	Used after heavy cold work to raise elastic limit	Springs
STATUARY BRONZE	75-94	1-19	3-10	Up to 0.7	1-6 Pb up to 0.3 P	Corrosion resistance, castability, colour, ability to form a surface "patina"	Castings for statues, ornaments and plaques
STEAM BRONZE	88	4.5	6	-	1.5 Pb	Corrosion resistance, strength, pressure tightness	Steam valve body castings, gears, bearings
STEEL BRONZE	52-59	36-42	-	-	2-3 Mn 1 Fe 1 Al	High strength, corrosion resistance particularly to sea water	Ship propellers
STERLIN	68.5	12.8	-	17.9	0.8 Pb	Corrosion resistance, malleable	Tableware, usually silver plated (E.P.N.S.)
STERLING BRASS	66	33	-	-	1 Pb	Machineability	Machined components, unions, tap fittings
STERLITE	54.5	16.5	0.5 approx.	27	1.75 Fe + Pb + Sn 0.2 Mn	Corrosion resistance, colour	Electrical fittings
STIRLING METAL	66.2	33.2	-	-	0.6 Fe 0.02 Pb	High strength	Castings, extrusions, hot pressings

TABLE 5.

TRADE NAME or other designation	COMPOSITION, % by weight					SPECIAL PROPERTIES	USES
	Cu	Zn	Sn	Ni	Other		
STONE'S BRONZE	Balance	0-1.5	11-12	-	0.1-0.3 P	Toughness, strength	Centrifugal castings for gear wheels, worm wheels, also sand and chill castings
SUMET BRONZE	Balance	-	0-5	-	17-30 Pb	Low friction	Bearings
SUPERSTON	do.	-	-	4-6	8.5-10.5 Al 4-6 Fe	Corrosion resistance, high tensile strength (650 MN m^{-2})	Chemical and steam plant, conditions of structural strength with corrosive environment
TALMI GOLD	86-90	9-12	0-1	-	up to 0.3 Au replacing Sn	Colour, corrosion resistance	Decorative articles, trim
TAMTAM	78	-	22	-	-	Castability	Castings
TANTALUM BRONZE	Balance	-	-	-	10 Al 1.2 Mo 0.2 Ta	High strength, hardness	Castings, steam valves
TAURUS BRONZE	86-89.5	-	10-14	-	0-0.3 P	Castability, machinability	Sand and centrifugal castings, gears, bearings
	84-88	4-6	2-8	-	0-4 Pb		
	83	6	6	-	-		
	77-82	-	10-12	1	5-12 Pb 0.3 P		
TELCUMAN	85	-	-	3	12 Mn	Low temperature coefficient of resistivity	Precision electrical instruments
TELLURIUM BRONZE	Balance	-	1.5	-	1 Te	Free machining, high strength	Machined components
TEMPALOY	89-96	-	-	3-5	0.8-1 Si 0-4.7 Al	Corrosion resistance, heat treatable for variations in hardness and ductility	Chemical plant components, structural parts
TENSILITE	64	29.5	-	-	3 Al 2.5 Mn 1 Fe	Corrosion resistance, high tensile strength	Castings
TETMEJER	Copper	-	-	-	5-10 Al 2.75 Si some Fe	Low friction	Bearings
THERLO	85	-	-	-	2-5.5 Al 9.5-13 Mn	Low temperature coefficient of resistivity	Coils for electrical resistance standards
THURSTON'S BRASS	55	44.5	0.5	-	-	Low melting point	
TINCOSIL ALLOYS	42	41	-	16	1 Pb	Corrosion resistance	Wrought products, die castings
	42	38-43	-	15-20	-		
	46.75	41.75	-	10	1.5 Pb		
TISSIER'S METAL	97	2	Balance	-	or Balance As	Malleability	Domestic hardware, bearings
TITAN ALLOYS	59	37	-	-	2 Al 0.5 Si 0.5 Fe	High strength	Wrought products
	59	38.8	0.7	-	0.4 Mn 1.1 Fe	Strength at slightly elevated temperature	Bolts, stays
	53.5	46	-	-	0.2 Al 0.3 Fe	Castability	Castings
	59.3	40.1	0.6	-	-	Low melting point	Welding rods

TABLE 5.

TRADE NAME or other designation		COMPOSITION, % by weight					SPECIAL PROPERTIES	USES
		Cu	Zn	Sn	Ni	Other		
TOBIN BRONZE (Naval brass)		Balance	38-40	0.5-2.5	-	up to 0.15 Fe 0.35 Pb	Strength with ductility, corrosion resistance	Marine applications
TOMBAC	Chinese	85	15	-	-	-	Malleable and ductile, easily cold worked	Buttons, small presswork, spinning, hydraulic forming, e.g. bellows
	German	85	6	8	-	0.3 Sb 0.3 As 0.1 Bi		
	French	80	20	-	-	-		
	do.	80	17	3	-	-		
	Canadian	82	17.5	0.5	-	-		
	Red	97.5	2.5	-	-	-		
	Golden	82	17.5	0.5	-	-		
	Button alloy	99	1	-	-	-		
	do.	84	16	-	-	-		
	Sheet	88-92	8-12	-	-	-		
	Common	71.5	28.5	-	-	-		
TOMBASIL		67-75	21-31	-	-	1.75-5 Si	Toughness, abrasion resistance	Marine equipment, handling of powdered materials where ferrous alloys unacceptable
TOUCAS METAL		35.6	Balance	7	28.6	7 Fe 7 Pb 7 Sb	Ductility, colour	Ornamental ware
TOURNAYS BRASS		82.5	17.5	-	-	-	Malleability	Buttons and ornaments
TRODALOY		97	-	-	-	0.4 Be Balance Co	Strength, good electrical conductivity, heat treatable	Contact springs in electrical machines and instruments
		99.5	-	-	-	0.1 Be Balance Cr		
TRUMPET METAL		81	17.8	1.2	-	-	Low damping capacity, ductility, malleability	Tubing for musical instruments
TUNGSTEN BRASS		60	34	0.1	0.1	2 W 2.8 Al trace Mn	Toughness, stronger than other alloy brasses	Presswork, stampings, castings
		60	22	-	14	4 W		
TUNGSTEN BRONZE		90-95	-	0-3	-	2-10 W	Stronger than pure tin bronzes	Castings
TUNGUM		81-86	Balance	-	0.8-1.4	0.7-1.2 Al 0.8-1.3 Si	Corrosion resistance, mechanical strength	Marine applications, chemical plant
TURBADIUM		50	44	0.5	2	1 Fe 1.75 Mn	Castability	Marine castings, ship propellers
TURBISTON		56-61	33-40	0-1.5	-	0.2-2.5 Al 0.2-2 Mn 0.5-2 Fe	High strength, corrosion resistance	do.
UCHATIUS BRONZE		92	-	8	-	-	Low friction	Bearings
ULCONY METAL		65	-	-	-	35 Pb	do.	do.
VALVE BRONZE (Steam bronze)	UK	Balance	3-9	2-10	-	3-6 Pb	Toughness, castability	Valve bodies, oil pumps, gears, bearings and bearing shells
	USA	85	5	5	-	5 Pb		

TABLE 5.

TRADE NAME or other designation	COMPOSITION, % by weight					SPECIAL PROPERTIES	USES
	Cu	Zn	Sn	Ni	Other		
VALVE METAL	81	9	3	-	7 Pb	Toughness	Low pressure valves, plumber's fittings and fixtures
VANADIUM BRASS	70	29.5	-	-	0.5 V	Stronger than plain brass	Marine applications
	58.5	38.5	-	-	0.03 V 0.5 Mn 1 Fe 1.5 Al	do.	
VANADIUM BRONZE	61	38.5	-	-	0.5 V	High strength, corrosion resistance	Aircraft propeller bushings
VANADIUM COPPER	Balance	-	-	-	10-15 V	Corrosion resistance	Master alloy for additions of vanadium. Submarine and ship's parts, especially propeller bushings
VIALBRA	76	22	-	-	2 Al	Corrosion resistance	Condenser tubes
VICTOR BRONZE	58.5	39	-	-	1.5 Al 1 Fe 0.05 V	Corrosion resistance, strength	Pipes and fittings
VICTOR METAL	50	35	-	15	up to 0.1 Al up to 0.3 Fe	Colour, castability, machinability	Cast fittings
VOLVIT	91	-	9	-	-	Low friction	Bearings
VULCAN BRONZE	59	38.5	-	0.5	1 Si 1 Fe	do.	do.
VULCAN METAL	81	-	0.4	1.5	11 Al 0.7 Cr 4.4 Fe 1 Si	Corrosion resistance	Chemical plant
WESSEL'S ALLOY	Balance	12-17	-	19-32	0-2 Ag	do.	do.
WHEEL BRASS	68	30	-	-	2 Pb	Machineability	Machined parts, screws
WHITE BENEDICT METAL	60	18	1	16.5	4.5 Pb	Corrosion resistance	Condenser tubes, chemical plant
	20	5	-	75	-	do. Hardness	Chemical plant for severe corrosive environment
WHITE BRASS	70	10	-	-	20 Mn	Colour, ductility	Rolled sections
Bristol	58	37	5	-	-		Plaques
Forbes	54	46	-	-	-		do.
WHITE BRONZE	54	42	-	4	-	Hardness, colour	Ornamental work
WIEGOLD	66	34	-	-	-	Corrosion resistance	Dental prosthetics
WINN'S BRONZE	62-68	28-35	-	2-2.3	0.5-1 Pb 0-0.5 Fe	Corrosion resistance	Superheater tubes in steam plant
WOLFRAM BRASS	60	22	-	14	4 W	Strength	Castings
WOLFRAM BRONZE	90	-	-	-	10 W	do.	do.
WYMDALOY	60	-	-	20	20 Mn	Corrosion, wear and abrasion resistance, heat treatable to high strength	Forgings
XANTAL	81-90	0-1	-	0-4	8-11 Al 0-4 Fe	Strength and ductility	Castings, wrought sections
YALE BRONZE	89-91	7.5-8	0.5-1.5	-	up to 1 Pb	Strength, machinability	Nuts and bolts

TABLE 5.

TRADE NAME or other designation	COMPOSITION, % by weight					SPECIAL PROPERTIES	USES
	Cu	Zn	Sn	Ni	Other		
YORCALBRO	76	22	-	-	2 Al	Corrosion resistance	Condenser tubes
YOREALINIC	91	-	-	2	7 Al	do.	do.
ZODIAC	64	16	-	20	-	Electrical resistivity	Electrical resistors

TABLE 6. Ferrous Alloys

TRADE NAME or other designation	COMPOSITION, % by weight					SPECIAL PROPERTIES	USES
	Fe	Ni	Cr	Other	Other		
ALCOMAX (in USA ALNICO V)	Balance	15	-	25 Co	10 Al Small Cu, Ti, Nb	Strongly ferromagnetic, high remanence	Permanent magnets
ALCRESS	70-75	-	20	-	5-10 Al	High specific resistivity, oxidation resistance	Electrical resistors
ALDECOR (Steel)	Balance	0.2	0.5-1.25	(0.05 S max)	0.15 C max 0.25-1.3 Cu 0.08-0.28 Mo	High strength	Engineering components
ALFENOL	do.	-	-	-	16 Al	High magnetic permeability, low hysteresis loss	Electromagnets, core laminations
ALFER	do.	-	-	-	11-13 Al	High magnetostriction	Transducers, electro-mechanical
ALNI	51	32	-	-	13 Al 4 Cu Small Ti, Nb	Strongly ferromagnetic, high remanence	Permanent magnets
ALNICO	Balance	Up to 20	-	Up to 20 Co	Up to 15 Al Some Ti	Strongly ferromagnetic, high remanence	Permanent magnets
ALSIFER	40	-	-	-	40 Si 20 Al	Deoxidant	Master alloy for Al alloys
ALSIMIN	40	-	-	-	45 Si 15 Al	Deoxidant	Steelmaking, master alloy
ARMCO IRON (American Rolling Mill Company)	99.9	-	-	-	<0.1 C	Very pure iron, soft and malleable	Bar, rod, tube, hammer welding
AUDIOLLOY	51.5	48	-	-	-	High magnetic, permeability	Electromagnetic instruments
AUER METAL	35	-	-	-	65 Misch-metall	Pyrophoric	Lighter flints
BADIN METAL	Balance	-	-	-	18-20 Si 8-10 Al 4-6 Ti	Deoxidant	Steelmaking

TABLE 6.

TRADE NAME or other designation	COMPOSITION, % by weight					SPECIAL PROPERTIES	USES
	Fe	Ni	Cr	Other	Other		
BOHLER (Silver steel)	do.	-	-	0.25-0.45 Mn	1-1.25 C 0.35 S max 0.35 P max	Hardness, capable of fine surface finish	Precision tool making, instrument pivots and shafts
BREARLEY STEEL (Cutlery stainless)	do.	-	12-14	-	0.38 C	Corrosion resistance, reasonable strength	Cutlery
CALITE	50	40	5.5	-	4.5 Al	Heat resistance, fails in sulphurous atmospheres	Furnace components, up to 1000 °C service
CALMET	Balance	12	25	-	5 Al	Oxidation resistant including moderate sulphurous atmosphere	Furnace components, up to 1050 °C service
CAMLOY	50	25-35	10-20	-	-	Corrosion and heat resistance	-
C.C.S.	Balance	-	5.25	-	4.25 W 1.15 Si 0.4 C	Hardness at elevated temperature	Tools, dies, press liners
CHISEL STEEL	Balance	-	-	-	0.85-0.95 C	Hardness, toughness	Cold chisels, punches
	do.	-	1-2	-	0.3-0.4 C		
	do.	3	-	-	0.4 C		
CHROMAX	50	35	15	-	-	Heat resistance	Furnace components
CHRO-MOW	do.	-	5	-	1.25 W 1 Si 1.35 Mo 0.3 C	Hardness at elevated temperature	Die holders, forging tools, extrusion punches
COMOL	do.	-	-	12 Co	17 Mo 0.06 C max	Hot formed, solution treated 1250 °C, readily machineable	Permanent magnets
COMPENSATOR ALLOY	Balance	34	-	-	-	Magnetic, permeability varies with ambient temperature	Instruments
CONPERNIK	do.	50	-	-	-	Constant permeability over wide range of flux density	Electromagnetic devices
COR-TEN	do.	0.65 max or absent	0.75	0.75 Si 0.15 P	0.4 Cu 0.25 Mn 0.1 C	High strength	Engineering machinery
CROLOY	do.	-	1.2-9.5	0.5-1.5 Mo Al, Ti often	0.08-0.13 C 0.2-0.6 Si may be 1.5 Si	High temperature, strength	Boiler tubes, steam plant
CROMANSIL	do.	-	0.5	1.2 Mn	0.2 C 0.7 Si	High temperature resistance	Engine valves and parts
CRONITE ALLOYS	do. do. do. do.	55 38 20 12	18 20 25 23	0.75- 3 W optional	0.2-0.75 C	Heat and corrosion resistance	High temperature applications, furnace and engine parts
DEAD SOFT STEEL	do.	-	-	Up to 0.15 C	-	Easily fabricated when fully annealed	General engineering, construction or manufacture
DELTAMAX	50	50	-	-	-	Rectangular hysteresis loop, low coercive force	Magnets, choke cores
DISCALOY	55	25	13	3 Mo 0.7 Si	0.7 Mn 0.5 Al 0.05 C	Creep strength	Gas turbine discs

TABLE 6.

TRADE NAME or other designation		COMPOSITION, % by weight					SPECIAL PROPERTIES	USES
		Fe	Ni	Cr	Other	Other		
DURIMET		Balance	29 (15-35)	20	2.5 Mo 0.07 C max	3.5 Cu 1 Si	Corrosion and erosion resistant. Austenitic	Pumps, valves, particularly for caustic liquor
DURIRON		do.	-	-	14.5 Si 0.8 C	0.35 Mn	Corrosion resistance to strong acids	Sulphuric acid plant, chemical plant
ECONOMET		do.	29-31	8-10	-	-	Heat and corrosion resistance	Castings
ELINVAR		53-61	33-35	4-5	1-3 W 0.5-2 Si	0.5-2 Mn 0.5-2 C	Low coefficient of thermal expansion, modulus of elasticity unaffected by temp.	Hairsprings for watches, balance wheels, tuning forks, instruments
EMMEL IRON		Balance	-	-	2.3-2.8 C 0.1-0.3 P	2.1-2.7 Si 0.8-1.4 Mn 0.09-1.15 S	Tensile strength, hardness. Pearlitic	Castings
ENDURO ALLOYS	4-6	Balance	-	4-6	0.1-0.25 C	-	Corrosion and heat resistance	Chemical plant, turbines, etc.
	S	do.	-	11.5-13	0.12 C max	—		
	FC	do.	-	12-15	do.	0.5-0.65 Mo		
	AA	do.	-	16-18	do.	-	do.	
	HC	do.	-	23-30	0.35 C max	-		
	16-6	do.	6 max	16 max	0.15-0.2 C	4 Mn		
	18-8	do.	7-9	17-19	0.08-0.2 C	-		
	18-8-B	do.	do.	do.	do.	2-3 Si		
	18-8-Mo	do.	7-11	do.	0.11 C max	2-4 Mo		
	19-9	do.	8-10	18-20	0.11 C max	-		
	HCN	do.	11-13	24-26	0.2 C max	-		
	NC-3	do.	19-21	24-26	0.25 C max	-		
ENDURON		Balance	-	16.5	1.5 Mn	1.5 Si 2.2 C	Heat resistance	Furnace parts
ERA STEELS	HR1	Balance	7.6	20	4 W	1.5 Si 0.35 C	Heat and corrosion resistance	Turbines, furnace parts
	HR2	do.	7.5	21	-	1.5 Si 0.13 C		
FEBRITE		48	35	17	-	-	do.	do.
FERNICHROME		37	30	8	25 Co	-	Low thermal expansion	Glass to metal seals in electrical components
FERNICO		54	28	-	18 Co	-	Low thermal expansion	Glass to metal seals in electrical components
FLASHKUT		Balance	-	4	1 C	-	Hardness	Clipping tools used by drop forgers
GLOBELOY		do.	-	4	2 total C	6 Si 0.5 Mn	Heat resistance	Castings
GRAINAL		do.	-	-	13-25 V	15-20 Ti 10-20 Al	Deoxidises liquid steel, adds alloying elements	Addition to steel for hardenability and toughness
GRAPHIDOX		do.	-	-	48-52 Si	9-11 Ti 5-7 Ca	Graphitises cast iron, deoxidises liquid steel, adds alloying elements	Iron and steelmaking

TABLE 6.

TRADE NAME or other designation	COMPOSITION, % by weight					SPECIAL PROPERTIES	USES
	Fe	Ni	Cr	Other	Other		
GRAPHITIC NITRALLOY	do.	-	0.2-0.4	0.25 Mo 0.5 Mn	1.2-1.3 C 1.35-1.5 Al	Graphitises on oil quench from 900 °C, temper 745-760 °C, 0.3-0.5 carbon remains combined	Nitrided steel, heat treatable
GUNITE K	do.	-	-	2.3 C 0.7 Mn	1 Si 0.15 P 0.08 S	Malleable cast iron	Castings requiring some ductility
HADFIELD'S STEEL	do.	-	-	11.5-13 Mn	1-1.3 C	Austenitic, cold work causes transformation with increased hardness, high resistance to abrasion	Ore crushers, dredger buckets, railway switch points and crossings
HALCOMB 218	Balance	-	5	0.4 C 0.35 V	1 Si 1.35 Mo	Hardness, heat resistance, abrasion resistance	Forging and extrusion dies
236	do.	-	12	0.3 C 0.35 Mn 1 V	0.5 Si 12 W		
HALVAN	do.	-	1	0.4 C 0.2 V	0.7 Mn	Hardness	Cutting tools
HIGH SPEED STEEL	do.	-	4	18 W	1 V	Hardness, retained to high temperature (up to about 700 °C)	Cutting tools, operating at very high speed
	do.	-	do.	20 W	1 V		
	do.	-	do.	14 W	2 V		
	do.	-	do.	2-4 W	1 V 6-8 Mo		
	do.	-	do.	18 W	2 V 0.4-0.9 Mo		
	do.	-	do.	20 W	2 V 0.75 Mo 10-13 Co		
HIPERCO	Balance	-	-	30 Co	-	High magnetic saturation value (B_{max} approx. 2.4 Tesla)	Electromagnetic circuits
HIPERNIK	do.	45-50	-	-	-	High magnetic permeability, soft magnetic material	do.
HIPERSIL	do.	-	-	3-3.5 Si	0.03 C max 0.02 S max 0.02 P max 0.1 Mn max	Soft magnetic alloy of low hysteresis loss	Transformer cores
HYBNICKEL	do.	18	8	up to 3 Al	-	Heat resistance, machineability, good oxidation resistance to scaling	High temperature applications
HYDRA METALS E	do.	-	3.5	9-10 W	0.3 C	Toughness, hardness, abrasion resistance	Extrusion dies, pressure pads
Z	do.	2.5	3.0	9-10 W	0.26 C 0.5 Mo		
HYNICAL	do.	32	-	12 Al	-	High BH max	Permanent magnets
HYNICO	do.	20	-	10 Al	6 Cu 13 Co	do.	do.
HY-TUF STEEL	Balance	1.8	-	0.25 C 0.4 Mo	1.3 Mn 1.5 Si	High tensile strength and high impact strength (Tensile strength 1.6 GN m^{-2})	High strength parts capable of heat treatment
INOXARGENT	do.	13	13	0.07 C	-	Corrosion resistance	Chemical plant

TABLE 6.

TRADE NAME or other designation	COMPOSITION, % by weight					SPECIAL PROPERTIES	USES
	Fe	Ni	Cr	Other	Other		
INVAR (Nilvar similar)	do.	36	-	1 C+Mn+Si	-	Negligible coefficient of thermal expansion (-0.3 to $+2.5 \times 10^{-6}$)	Measuring instruments and standards, chronometers, clocks, watches, compensating mechanisms, thermostatic strip
ISOELASTIC	do.	36	8	0.5 Mo	-	Modulus of elasticity	Springs
IZETT STEEL	do.	-	-	0.05 Al	-	Non-age hardening	Stable structures
JALTEN	do.	-	-	0.25 C	1.5 Mn 0.25 Si 0.4 Cu	Weldability, higher strength than mild steel	Structures
KAHLBAUM IRON	do.	0.004	-	-	0.001 Mn 0.015 Si 0.0014 Cu trace S	Commercially pure iron	Hammer welded and shaped products, chain and slings
KANTHAL (also called Megapyr)	Balance	-	25	5 Al	3 Co	Corrosion resistance oxidation resistance at high temperature, low creep strength above 1000 °C	Electrical resistor furnace heating elements, ribbon or wire, up to 1300 °C intermittently, 1150 °C continuously
KONIK	do.	0.35	0.12	0.1 C 0.25 Cu	0.08 Si 0.35 Mn	Strength, low cost	Where higher strength than mild steel required
KOVAR	do.	23-30	-	17-30 Co	0.6-0.8 Mn	Low coefficient of thermal expansion, similar to hard glass	Glass to metal wire seals
KRUPP AUSTENITIC (K.A. steel)	do.	7	20	-	-	Corrosion resistance	Chemical plant
KRUPP TRIPLE STEEL	do.	-	3.5-4	0.9 C 0.4 Mn	3.5-4 W 2.5 Mo 2.5 V	Hardness at high temperature	Twist drills
LANZ IRON	do.	-	-	2.8-3.2 C 0.3 P 0.13 S max	0.8-1.2 Si 0.6-0.8 Mn	Pearlitic structure when produced in hot mould, good strength, low elongation	High duty iron castings
LEDLOY	do.	-	-	-	0.1-0.25 C 0.25 Pb	Machineability	Capstan lathe and similar mass production machined products
LOW HYSTERESIS STEEL	do.	-	-	-	2.5-4 Si	High magnetic permeability and high electrical resistivity, low hysteresis loss in a.c. circuits	Transformer laminations, generator parts, electrical machines generally
MANGANAL	do.	3	-	12 Mn	0.6-0.9 C	High strength, abrasion resistance	Hot rolled plates for crushers, stamping mills
MANGANESE STEELS	Balance	-	-	0.3-0.9 Mn	Low carbon	Deoxidant and to fix sulphur as harmless inclusions	Conventional steel, to improve rolling properties
	do.	-	-	0.5-0.9 Mn	0.72-0.89 C	Strength	Hard steel rails
	do.	-	-	0.7-1 Mn	0.13-0.17 C	Weldability, good grain refinement	Boiler plate, structural steel
	do.	-	-	0.6-0.9 Mn	0.45-0.6 C	High modulus and strength	Leaf springs

1

TABLE 6.

TRADE NAME or other designation	COMPOSITION, % by weight					SPECIAL PROPERTIES	USES
MANGANESE STEELS (contd.)	Fe	Ni	Cr	Other	Other		
	do.	-	-	1.5-1.75 Mn	0.8-0.95 C	Dimensional stability after hardening	Tool gauges
	Balance	-	-	1.1-1.4 Mn	0.2-0.3 C	High strength	Castings
	do.	-	-	1.5 Mn	<0.5 C	Toughness	Rolled products
	do.	-	-	1.75 Mn	0.3-0.55 C	do.	Axles, gas cylinders, steam valves
Hadfields	do.	-	-	10 Mn	1-1.3 C	Abrasion resistance, work hardening	Stone crushers
	do.	-	-	10-15 Mn	1.1-1.5 C	do.	Railway points and cross-overs
MARTINEL STEEL	do.	-	-	0.75 Mn	0.24 C 0.1 Si	Heat treatable to vary strength and ductility	Ship building
MAXITE	do.	-	4	18 W	1 V 4 Co	Hardness at high temperature	Cutting tools
MAXTACK	do.	-	2	10 W 1 Si	2.5 Mn 2.25 C	Self hardening	Dies for nail and tack production
MEGAPYR	Balance	-	20	10 Al	-	High resistivity	Electrical resistors
MINOVAR (Ni-Resist Type 5)	Balance	34-36	-	1.5-2.75 Si 0.08 P max 0.1 Cr max	2.4 C max 0.5 Mn max	Low thermal expansion, corrosion resistance	Machine tool castings, forming dies, expansion joints
N 155	do.	20-21	21	3 Mo 1.5 Mn	0-20.5 Co 2.2 W	Heat resistance	Gas turbine parts
NEEDLED STEEL	do.	-	-	1.5 Mn	0.4-0.5 C 0.0002 B	Hardenability through steel section	Parts requiring heat-treated structure to be uniform
NICALOI (Nicaloy)	53-51	47-49	-	-	-	High magnetic permeability	Permanent magnets
NICKELOY	50	50	-	-	-	Corrosion and heat resistant, magnetic	Chemical plant and electrical equipment
NICROSILAL	Balance	17.5-18.5	2-2.5	4.5-6 Si	0.6-0.7 Mn 1.8 C Low S and P	Heat and corrosion resistance	Castings for high temperature service
NI-HARD	do.	3-4.5	0.75-1.5	0.7 Si 0.8 Mn	3 C Low S and P	Abrasion resistance, (martensitic)	Pump bodies, impellers, grinding balls and grinding plates
NILO-ALLOYS 36	do.	36	-	-	-	Low coefficient of thermal expansion	Pendulum rods, thermostats
40	do.	40	-	-	-	do.	Thermostats, heating elements
42	do.	42	-	-	-	do.	Core of copper clad wires in metal/glass seals
475	do.	47	5	-	-	do.	Metal seals in soft lead and soda glass
48	do.	48	-	-	-	do.	Thermostats, metal seals into electronic valves and tubes
(also known as Radio metal) 50	do.	50	-	-	-	do.	do.
K	do.	29	-	-	17 Co	do.	Metal/glass seals in X-ray tubes

TABLE 6.

TRADE NAME or other designation	COMPOSITION, % by weight					SPECIAL PROPERTIES	USES
	Fe	Ni	Cr	Other	Other		
NI-RESIST Type I Cast Iron (AUS 101 A) (Flake Graphite)	Balance	13.5-17.5	1.75-2.5	5.5-7.5 Cu	1-1.5 Mn 1-2.8 Si 3 C max	Corrosion, wear, heat resistance	Chemical plant handling mineral acids, piston ring inserts in aluminium alloy pistons
Type IB (AUS 101 B)	do.	do.	2.75-3.5	do.	do.	Superior corrosion resistance to Type I	Similar to Type I
Type 2 (AUS 102 A)	do.	18-22	1.75-2.5	0.5 Cu max	0.8-1.5 Mn 1-2.8 Si 3 C max	Heat and oxidation resistance, corrosion resistance	Steam plant to 700 °C. Caustic solution plant, food handling, rayon ammoniacal solutions
Type 2 B (AUS 102 B)	do.	18-22	3-6	0.5 Cu max	0.8-1.5 Mn 1-2.8 Si 3 C max	Heat and abrasion resistance	Firebox castings, exhaust manifolds, turbo charger castings, to 800 °C
Type 3 (AUS 105)	do.	28-32	2.5-3.5	do.	0.4-0.8 Mn 1-2 Si 2.6 C max	Low thermal expansion, corrosion resistance	Thermal shock to 230 °C, exhaust manifolds, turbo-superchargers to 800 °C, chemical slurries, wet steam
Type 4	do.	29-32	4.5-5.5	do.	0.4-0.8 Mn 5-6 Si 4.5-5.5 C	Corrosion resistance, good oxidation, erosion and wear resistance	Food handling plant, chemical plant handling salts and acids
Type 5	do.	34-36	0.1 max	do.	0.4-0.8 Mn 1-2 Si 2.4 C max	Minimum thermal expansion, corrosion resistance	Machine tool parts, dies, expansion joints, scientific instruments
(Spheroidal Graphite) Type D-2 (AUS 202 A)	do.	18-22	1.75-2.5	0.08 P max	0.7-1 Mn 1.75-3 Si 3 C max	Corrosion, erosion, frictional wear and heat resistance	Dishwasher components, centrifugal pumps, impellers, filter bodies, compressors, brake drums, service up to 760 °C
Type D 2 B (AUS 202 B)	do.	do.	2.75-4	do.	do.	Superior corrosion resistance to Type D2, erosion and oxidation resistance better	Chemical plant handling neutral or reducing salts, seawater filters
Type D 2 C (AUS 203)	do.	21-24	0.5 max	do.	1.8-2.4 Mn 2-3 Si 2.9 C max	Toughness at sub zero temperature, high ductility, lower corrosion resistance than Types D2 and D2B	Bearing supports, jet engines, similar to above where lower temperatures involved
Type D 2 M	do.	21.5-24	0.2 max	do.	3.75-4.5 Mn 1.5-2.6 Si 2.7 C max	Toughness at cryogenic temperatures (−196 °C)	All plant handling low temperature gases, transport and storage
Type D 3 (AUS 205)	do.	28-32	2.5-3.5	do.	0.5 Mn max 1.5-2.8 Si 2.6 C max	Resistance to thermal shock, corrosion and erosion by wet slurries, steam	Steam plant at low pressure end, chemical plant handling corrosive slurries
Type D 3 A	do.	do.	1-1.5	do.	do.	Increased resistance to wear over Type D3	Metal to metal contact, galling and wear problems
Type D 4	do.	29-32	4.5-5.5	do.	0.5 Mn max 5-6 Si 2.6 C max	Superior corrosion, erosion and oxidation resistance	Plant handling sulphurous gases up to 540 °C, otherwise up to 800 °C in the absence of sulphur

TABLE 6.

TRADE NAME or other designation	COMPOSITION, % by weight					SPECIAL PROPERTIES	USES
	Fe	Ni	Cr	Other	Other		
NI-RESIST (Spheroidal Graphite) *(contd.)* Type D 5	do.	34-36	0.1 max	do.	0.5 Mn max 1.5-2.75 Si 2.4 C max	Low thermal expansion	High temperatures involving thermal stress (cycling)
Type D 5 B	do.	do.	2-3	do.	do.	As D5 but with improved oxidation and creep resistance	Compressor stators in gas turbines, moulds, machine parts exposed to heat or oxidation
NI-SPAN	Balance	42	5-6	2-3 Ti	-	Temperature coefficient of modulus of elasticity constant	Springs for clocks, watches, balances and weighing machines, frequency standards, diaphragm gauges
NITRALLOY	do	-	0.9-1.8	0.15-1 Mo 0.85-1.2 Al	0.4-0.7 Mn 0.2-0.4 Si 0.2-0.45 C	High resistance to wear when nitrided	Shafts in rotating machinery subject to wear and high surface stress
EZ	do.	-	do.	do. 0.15-0.25 Se	0.5-1.1 Mn 0.2-0.4 Si 0.2-0.45 C		
N	do.	3.25-3.75	1-1.3	0.2-0.3 Mo 1.1-1.4 Al	0.4-0.7 Mn 0.2-0.4 Si 0.2-0.27 C	Tougher core than other alloys when nitrided, precipitation hardened	Gears
NO-MAG	Balance	11	-	7 Mn	1.5 Si 3 C	High electrical resistivity, non-magnetic, good mechanical properties	Electrical machinery castings, switch covers, resistance grids
OMEGA	do.	-	-	0.7 Mn 0.45 Mo	1.85 Si 0.2 V 0.6 C	High impact resistance when quenched and tempered	Chisels, shear blades, swaging tools
PARAGON STEEL	do.	-	0.75	1.6 Mn	0.25 V	Non shrinking, retains shape on heat treatment	Dies, gauges, precision tools
PLATINITE	do.	49	-	-	-	Low coefficient of thermal expansion, matching common glass	Lead in wires sealed into evacuated or gas filled bulbs
RADIO METAL — *see* NILO							
RED FOX R.F. 31	do.	16	25	0.6 Mn	1.5 Si 0.2 C	Heat resistance	Gas turbines and related components at high temperatures
R.F. 33	do.	31	20	0.7 Mn	0.8 Si 0.1 C		
REFRACTALOY B	35.5	30	24.4	2 Mn	8 Mo	Heat resistance	High temperature service
REMALLOY	Balance	-	-	12 Co	17 Mo	High *BH* max	Permanent magnets
RESISTA	do.	-	-	0.2 Cu	0.2 P	Better corrosion resistance than iron or mild steel	Domestic boilers, pipes and ducts
REZISTAL STEELS Type 303	Balance	7-9	17-19	0.08-0.2 C	-	Corrosion resistance	Chemical plant
Type 307	do.	9-12	19-22	do.	-	Ductility	Domestic ware
Type 309B	do.	10-13	21-26	0.2 C max	2-3 Si	Heat resistance	Furnace plant, gas turbine components
Type 311	do.	24-26	19-21	0.25 C max	do.	do.	Food processing
Type 325	do.	21-23	7-10	do.	1-1.5 Cu	do.	

TABLE 6.

TRADE NAME or other designation	COMPOSITION, % by weight					SPECIAL PROPERTIES	USES
	Fe	Ni	Cr	Other	Other		
RHOMETAL	do.	36-45	2-5	2-3 Si	-	High magnetic permeability, low electrical losses at high frequencies	Magnetic cores in electromagnetic circuits
SICROMA	do.	-	-	0.5 Mn 0.45-0.65 Mo	1.15-1.65 Si 0.15 C	Toughness, strength	General engineering
SICROMAL	do.	-	up to 24	up to 3.5 Al	1 Si small Mo	Corrosion resistance, particularly acids	Chemical plant
SILCAZ	do.	-	-	35-40 Si 7 Al	10 Ca 10 Ti 4 Zn 0.5 B	Low melting point, strong deoxidant	Master alloy, deoxidant
SILCHROME	do.	-	8	3.9 Si	0.4 C	Oxidation and wear resistance but attacked by lead in hydrocarbon fuels	Exhaust valves for cool running engines and on leaded fuels
SILICOMANGANESE	do. do.	- -	- -	70 Mn 20 Si 65 Mn 18 Si	0.5-1 C do.	Deoxidant	Finishing addition in steelmaking
SILICOSPIEGEL	do.	-	-	25 Mn	5-6 Si up to 4 C	Deoxidant	do. First used to reduce blowholes in casting
SILMO	do.	-	-	0.5-2 Si 0.5 Mn	0.45-0.65 Mo 0.15 C max	Oxidation resistance, strength	Heat resisting components, valves, dampers
SILVAX	do.	-	-	35-40 Si 6 Zr 0.5 B	10 Ti 10 V	Low melting point, deoxidant	Master alloy for alloying additions
SILVER STEEL	do.	-	0.5 max	0.25-0.45 Mn	0.3 Si max 0.95-1.25 C 0.045 S max 0.045 P max	Hardness, supplied ground or polished, to specified limits of accuracy	General purpose steel for machining to reasonable precision for shafts, pivots, mechanisms
SIMANAL	Balance	-	-	20 Mn	20 Al 20 Si	Deoxidant, grain refinement	Finishing addition in steelmaking
SIRIUS	do.	16	17	3 W 12 Co	2 Ti 0.25 C	Heat and corrosion resistance	High temperature applications, turbine discs and blades
SKHL STEEL (USSR)	do.	0.3-0.7	0.4-0.8	0.3-0.5 Cu	0.2 C	Strength	Structural steelwork
SMITH'S ALLOY	55	-	37.5	7.5 Al	-	Heat resistance, high electrical resistivity	Electrical resistors, heating elements and furnace parts
SMZ ALLOY	Balance	-	-	60-65 Si	5-7 Mn 5-7 Zr	Deoxidant, grain refinement, innoculant	Treatment of cast iron for graphitisation (nodular)
SPIEGEL	do.	-	-	15-32 Mn	4.5-5.5 C	Low in sulphur and phosphorus, low melting point, deoxidant	Addition in steelmaking, to vary manganese content and deoxidise
SPRING STEEL Plain carbon	Balance	-	-	0.5-0.8 Mn	0.1-0.4 Si 0.5-0.85 C	Work hardening and heat treatable steels with high modulus of elasticity and elastic limit	Springs
Silico-manganese	do.	-	-	0.6-1 Mn	1.5-2 Si 0.45-0.65 C		
Chromium	do.	-	1	0.5 Mn	0.5 Si 0.6 S		
Chromium Vanadium	do.	-	1	do.	do. 1 V		

TABLE 6.

TRADE NAME or other designation	COMPOSITION, % by weight					SPECIAL PROPERTIES	USES
	Fe	Ni	Cr	Other	Other		
STAINLESS IRON	Balance	-	11-14	1 Mn	1 Si <0.15 C	Malleability, corrosion resistance, weldability	Chemical plant, steam plant and structural plant subject to severe atmospheric corrosion
STAINLESS STEEL Ferritic	Balance	-	16-30	-	<0.1 C	Corrosion resistance, not susceptible to heat treatment	Steam plant, chemical plant, domestic appliances, car trim, cutlery, culinary utensils, type needs to be carefully chosen for complete range of conditions
do.	do.	-	11-14	-	0.1 C	Corrosion resistance, malleability	
Martensitic	do.	-	11-14	-	0.1-0.45 C	Corrosion resistance, capable of heat treatment for edge tools, when 0.25-0.3 C used	
do.	do.	2	17	-	0.25 C	do.	
Austenitic	do.	6-20	16-26	-	0.1 to 0.2 C	Highest corrosion resistance, some compositions susceptible to welding problems and instability at temperatures 600-900 °C	
STENTOR STEEL	Balance	-	-	1.6 Mn	0.25 Si 0.09 C	Dimensional stability in heat treatment, hardens from 780 °C	Gauges, dies
SUPERLOY	do.	-	30	8 Co 8 Mo	5 W 0.05 B 0.2 C	Hardness when deposited by welding	Hard facings, (granules of alloy packed into mild steel tube for use as welding rod)
SUPERNILVAR	do.	31	-	4-6 Co	-	Zero coefficient of thermal expansion when hot rolled	Instruments
SWEETALLOY	do.	36	18	0.5 Mn	0.3 C	Corrosion resistance, heat resistance	Chemical plant
TAM ALLOY	do.	-	-	15-21 Ti	3.5-8 C	High melting point, deoxidant, grain refinement	Steelmaking, particularly grain refinement
TELCOSEAL	54	29	-	17 Co	-	Low coefficient of thermal expansion	Glass to metal seals, in thermionic valves, X-ray tubes, cathode ray tubes
T.E. ALLOY	38.7	30	20	4 Mo 0.7 Mn	4 W 1.9 Ta 0.1 C 0.5 Si	Heat resistance	Gas turbine components
TICONAL	Balance	12-20	-	15-30 Co 2-7 Cu	0-10 Ti 5-10 Al	Strongly ferromagnetic, high remanence	Permanent magnets
TIMANG	Balance	-	-	15 Mn	-	Work hardening	Wire for rock screens
TIMIDUR	do.	30	15	1.7 Ti 0.8 Mn	0.5 Si 0.2 Al 0.15 C	Heat resistance	Gas turbine blades
TISCO STEEL	do.	35-40	-	15 Mn	-	Work hardening, abrasion resistance	Steel rails for crossovers and points, on rail and tramways
TOOL STEELS	Balance	-	-	-	1.3-1.4 C	Hardness, toughness	Razors
do.	do.	-	-	-	1.2-1.3 C		Metal saws, lathe tools

TABLE 6.

TRADE NAME or other designation	COMPOSITION, % by weight					SPECIAL PROPERTIES	USES
TOOL STEELS (*contd.*)	Fe	Ni	Cr	Other	Other		
	do.	-	-	-	1-1.2 C		Woodworking tools
	do.	-	1	4 W 0.3 V	1.2 C		Fine cutting finishing tools
	do.	-	-	-	0.9-1.1 C		Axes, large chisels, rock drills
	do.	-	0.6	0.6 W 0.1 V	0.32 Si 0.9 C		Gauges
	do.	-	-	-	0.9-0.95 C		Cutlery, taps and dies, needles
	do.	-	-	-	0.8-0.85 C		Punches, heavy tools, hammers
	do.	-	-	0.2 V	0.85 V		Heavy shears
	do.	-	4	5.75 W 1.5 V	5 Co 4.75 Mo 0.8 C		High speed cutting tools
	do.	-	4	1.5-18 W 1 V	0-5 Co 0.4-8.5 Mo 0.75 C		do.
	do.	-	0.9	0.2 V	0.7 C		Cold work shears
	do.	-	1.8	2 W	0.4 C		Chisels and hot shears
	do.	-	3	10 W 0.3 V	0.3 Mn 0.3 C		Hot die steel
UNILOY	Balance	0.5	12	-	0.1 C Small Mo	Heat treatable for high strength, hardness	General engineering, bar stock
UTALOY	do.	35	12	-	0.2 C max	Heat resistance	Furnace equipment for heat treatment, salt baths
UTILOY	do.	29	20	3 Cu 1.75 Mo	1 Si 0.07 C max	Corrosion resistance particularly mineral acids	Chemical plant
V ALLOYS 1	31-40	-	38-42	8-11 Mn	14-16 Si	Lower melting point than base metals	Master alloys to add chromium to iron and steel
2	31-43	-	28-32	14-16 Mn	15-21 Si		
3	34	-	40	6 Mn	18 Si		
VANADIUM – Iron Master alloy (Grainal)	Balance	-	-	25 V 10 Al	15 Ti 0.2 B	Low melting point, deoxidant, grain refiner	Additions of vanadium to steel
VANADIUM STEEL (Permendur)	49	-	-	2 V	49 Co	High magnetic permeability	Electromagnetic circuits
VASCO STEELS	Balance	-	0.8	0.2 V	0.5-1 C	Hardness	Tools
	do.	-	3.5	9.25 W 0.45 V	1 C	do. Toughness	Hot dies
	do.	2	4	15.5 W 3 Mo	-		Extrusion dies
	do.	1.5	4.75	5 V 0.25 Mn	12.5 W		High speed tools
VIRGO	do.	8	18	4.4 W	0.24 C	Heat and corrosion resistance	Chemical plant
	do.	14	13	2.2 W 0.7 Mo	0.48 C		
WORTLE PLATE	do.	-	-	-	2.5 C	Hardness, wear resistance	Drawing dies for wire

TABLE 6.

TRADE NAME or other designation	COMPOSITION, % by weight					SPECIAL PROPERTIES	USES
	Fe	Ni	Cr	Other	Other		
Z-METAL (Pearlitic malleable cast iron)	Balance	-	-	0.75-1.25 Mn 1 Si up to 1 Cu	2-2.6 C total 0.3-0.8 C combined	Strength 470 MN m^{-2} after heat treatment	Pumps and chemical plant
ZORITE	Balance	35	15	1.75 Mn	0.5 C	Heat resistance	Furnace parts, chemical plant

TABLE 7. Nickel-based Alloys

TRADE NAME or other designation	COMPOSITION, % by weight					SPECIAL PROPERTIES	USES
	Ni	Cr	Cu	Fe	Other		
ACCOLOY	38-68	12-18	-	Balance	Small Mo Small Ti	Heat resistance	Castings
ALUMEL	94.5	-	-	-	2 Al 1 Si 2.5 Mn	Oxidation resistance	Thermocouples
BRIGHTRAY	80	20	-	-	-	Oxidation resistance	Heating elements (as wire), operates up to 1200 °C
B.T.G. ALLOY	60	12	-	Balance	1-4 W 2 Mn 0.5 C	Heat resistance	Heating elements
CALMALLOY	69	-	29	2	-	High magnetic temperature coefficient 0-100 °C	Temperature compensation in electrical instruments
CALOMIC	65	15	-	20	-	Oxidation resistance	Heating elements up to 1000 °C
CALORITE	65	12	-	15	8 Mn	Heat resistance	-
CHLORIMET 3	60	18	-	3 (max)	18 Mo 0.07 C (max)	Corrosion resistance	Chemical plant
CHROMEL A	80	20	-	-	-	Heat resistance Oxidation resistance	Heating elements to 1090 °C
CHROMEL B	85	15	-	-	-		
CHROMEL C	64	11	-	25	-		Heating elements to 950 °C
CHROMEL P	90	10	-	-	-		Thermocouples
CHRONITE	63.5	13.5	-	Balance	1 Al 1 Mn 0.4 Si	Heat and corrosion resistance	Burners, valves, annealing boxes
COMPENSATOR ALLOY	55-60	-	Balance	-	-	Magnetic permeability varies with ambient temperature	Instruments
COPEL (CONSTANTAN)	55	-	45	-	-	Low temperature coefficient of resistance	Resistance coils in electrical instruments
CORRONEL	66	-	-	6	28 Mo	Corrosion resistance to mineral acids (not HNO$_3$)	Chemical plant

TABLE 7.

TRADE NAME or other designation		COMPOSITION, % by weight					SPECIAL PROPERTIES	USES
		Ni	Cr	Cu	Fe	Other		
CORRONIL		70	-	26	-	4 Mn	Corrosion resistance, similar to Monel	do.
D NICKEL		95-96	-	-	-	4-5 Mn	Harder than pure nickel	Chemical plant components
DURANICKEL (formerly Z Nickel)		93.7	-	0.05	0.35	4.4 Al 0.5 Si 0.4 Ti 0.3 Mn	Age hardening, corrosion resistance	Chemical plant
EVANOHM		75	20	2.5	-	2.5 Al	Low temperature coefficient of resistance, easily soldered	Electrical instruments
FAHRITE	N5	60	12	-	28	-	Heat resistance	Castings
	N6	65	15	-	20	-		
	N6 Mo	65	15	-	15	5 Mo		
FERNITE	No. 1	65	15	-	20	-	Heat and corrosion resistance	Furnace parts subjected to corrosive gases
	No. 2	35	15	-	50	-		
	No. 4	22	28	-	50	-		
GALLIMORE METAL		45	-	28	2	25 Zn 2 Si 2 Mn	Corrosion resistance	Stampings, aircraft and wrought products
GENALLOY		31-40	12-21	-	Balance	0.5 C	Heat resistance	Furnace components
GLAIVORY		65	15	-	20	-	Heat resistance	Electric heating elements
HARDITE		35-93	10-18	-	Balance	1-2 Mn 2-7 Si	do.	Furnace parts
HASTELLOY	A	55.5-59.5	-	-	18-22	18-22 Mo 0.04-0.15 C	Corrosion resistance, good mechanical properties at high temperature, particularly B	Castings for furnace applications, gas turbine blades and components, carburising boxes, thermocouple fittings
	B	62.5-66.5	-	-	4-7	26-30 Mo 0.4-0.15 C		
	C	54.5-59.5	13-16	-	4-7	15-19 Mo 0.4-0.15 C 3.5-5.5 W		
	D	85	-	3	-	10 Si 2 Al		
HYDREX		65	15	-	Balance	0.75-1.25 Mn 0.2 C 0.5 Si	Heat resistance	do.
ILIUM	G	56	24	8	-	4 Mo 1 Si 1.5 Mn 2 W	Corrosion resistance particularly to HNO$_3$, H$_2$SO$_4$, up to 15% HCl	Chemical plant, vessels and pipework
	R	62	21	3	8	5 Mo 0.4 Si 0.5 Mn	Organic acids and organic compounds	do.
		56	22	6	-	6 Mo 1 W	do.	do.
		62.5	21	6.5	1	5 Mo 1 Al 1 Mn 2 W	Heat resistance	Thermocouple and pyrometer tubes
INCOLOY	DS	36-39	17-19	0.25 max	Balance	0.15 C max 0.9-1.3 Mn 2-2.5 Si	Heat resistance	Furnace retorts, thermocouple tubes, fanshafts, oil burners furnace supports and boxes
	800	30-35	19-23	0.75 max	do.	0.1 C max 1.5 Mn max 1 Si max 0.15-0.6 Al 0.15-0.6 Ti	High strength, oxidation and carburisation resistance	Hydrocarbon cracking furnaces, heating element sheaths, furnace parts

TABLE 7.

TRADE NAME or other designation		COMPOSITION, % by weight					SPECIAL PROPERTIES	USES
		Ni	Cr	Cu	Fe	Other		
INCOLOY (contd.)	825	38-46	19.5-23.5	1.5-3	do.	0.05 C max 1 Mn max 0.5 Si max 0.6-1.2 Ti 0.2 Al max 2.5-3.5 Mo	Resistance to oxidising corrosive conditions and stress corrosion cracking	Heat exchangers, boiler feed water heaters, pickling tanks, acid and alkali tanker drums, filtration plant
INCONEL	600	Balance	14-17	0.5 max	6-10	0.15 C max 1 Mn max 0.5 Si max	Excellent resistance to oxidation, nitriding, carburisation, good strength at high temperatures	Furnace parts in low sulphur atmospheres, carburising and nitriding boxes, trays ammonia crackers, muffles, heating element sheaths, steam generation equipment
	625	60	21.5	-	5 max	0.1 C max 9 Mo 0.5 Mn max 0.5 Si max 3.65 Nb	Corrosion resistance in addition to oxidation	Condensers in chemical plant, reactor vessels, paper pulp digesters
	X750	73	15	0.05 max	6.75	0.04 C max 0.7 Mn 0.3 Si 2.5 Ti 0.8 Al 0.85 Nb	do. especially involving oxygen and fluorine compounds	Gas turbine parts, springs at high temperature
JAE METAL		70	-	30	-	-		Magnetic shunts
JOFO		60	20	-	Balance	-	Heat resistance	Castings for service up to 1250 °C
K42B		42-46	18-19	-	7.15-13	22-25 Co 2.5 Ti 0-0.06 C 0-0.03 Si 0-0.7 Mn	Heat resistance, creep resistance	High temperature machine parts
KARMA ALLOY		Balance	20	-	3	3 Al 0.3 Si 0.15 Mn 0.06 C	High electrical resistance, heat resistance, high strength	Heating elements, furnace parts
KONAL		70-73	-	-	7.5	17-19 Co 2.5-2.8 Ti 0.5 Si	Heat resistance	High temperature components
KONEL		Balance	19	-	7.5	25 Co 2.5 Ti	Heat resistance	Valve stems in internal combustion engines, radio filament cores as substitute for platinum
KROMAX		80	20	-	-	-	Heat resistance	Electrical heating elements
KROMORE		85	15	-	-	-	do.	do.
LANGALOY	4R	63	-	-	5	30 Mo 0.75 Mn 0.75 Si	Heat resistance and corrosion resistance	Castings for chemical plant or general engineering involving corrosion
	5R	56	15	-	5	17 Mo 0.75 Si 0.75 Mn 5 W	do.	
	6R	85	-	3	-	10 Si	do.	
MAGNO		Balance	-	-	-	5 Mn	Low temperature coefficient of resistivity	Resistance wire

TABLE 7.

TRADE NAME or other designation		Ni	Cr	Cu	Fe	Other	SPECIAL PROPERTIES	USES
MANGONIC		97	-	-	-	3 Mn	High resistivity	Thermocouple elements, sparking plug electrodes, support wires in electric lamps and thermionic valves
MEGAPERM		65	-	-	25	10 Mn	High magnetic permeability	Electromagnetic circuits
MISCO ALLOYS	B	12	24	-	Balance	Some C	Heat and corrosion resistance	Furnace parts, thermocouple sheaths
	C	10	30	-	do.	do.		
	HN-1	65	15	-	do.	do.		
	HN-2	60	12	-	do.	do.		
MOCK GOLD		64	-	12	-	12 Pt 12 Ag	Colour, corrosion resistance	Substitute for gold in jewellery, etc.
MONEL	400 Wrought	63-67	-	Balance	2.5 max	2 Mn max 0.5 Si max 0.3 C max 0.024 S max	Corrosion resistance, moderate heat resistance (400 °C)	Condenser tubes, turbine blades, pump bodies, laundry and chemical plant, catering fittings, marine uses
	Cast H	63-68	-	do.	3 max	0.5-1.5 Mn 2.5-3 Si 0.12 C max	do. but greater	
	Cast S	63-68	-	do.	3 max	0.5-1.5 Mn 3.5-4 Si 0.12 C max	Strength and hardness	
	K	63-70	-	do.	2 max	2-4 Al 1.5 Mn max 1 Si max 0.25 C max 0.01 S max	do. Non-magnetic, high resistance to vibrational and impact stress	
MO-PERMALLOY		78.5	-	-	17.7	3.8 Mo	Magnetic permeability	Electromagnetic circuits
MO-PERMINVAR		45	-	-	22.5	7.5 Mo 25 Co	Magnetic permeability	do.
MT 17 ALLOY		30	21	-	19	21 Co 2.2 W 3 Mo 1.6 Ti 0.5 Si 0.6 C 1.5 Mn	Heat resistance, creep strength	Furnace parts
MUMETAL		77.4	-	5	13.8	3.8 Mo	High magnetic permeability, low hysteresis	Magnetic shielding in instruments
NICHROME ALLOYS		65	15	-	20	0.05 C	Heat resistance to 900 °C	Heating elements, resistors
Pyromic, Kromax, Brightray		80	20	-	-	-	do. to 1150 °C	Thermocouple sheaths
Rayo, Kromore		85	15	-	-	-	do. do.	Furnace parts
NICHROSI		Balance	15-30	-	-	16-18 Si	Heat and oxidation resistance	Furnace parts
NICROBRAZ		65-70	13-20	-	-	trace B	Low melting point	Brazing metal for austenitic and chromium alloy steels

TABLE 7.

TRADE NAME or other designation	COMPOSITION, % by weight					SPECIAL PROPERTIES	USES
	Ni	Cr	Cu	Fe	Other		
NIMONIC ALLOYS 75	Balance	18-21	0.5 max	5 max	1 Mn max 1 Si max 0.2-0.6 Ti 0.08-0.15 C	Heat resistance, creep resistance	Gas and steam turbine blades and fittings, furnace parts, all high temperature highly stressed components especially in prime movers
80A	do.	18-21	0.2 max	3 max	1 Mn max 1 Si max 1.8-2.7 Ti 1-1.8 Al 2 Co max 0.008 B max 0.01 C max 0.015 S max	do.	
90	do.	18-21	-	3 max	1 Mn max 1.5 Si max 1.8-3 Ti 0.8-2 Al 15-21 Co 0.005 Pb max 0.13 C max	do.	
105	do.	13.5-15.75	0.5 max	2 max	1 Mn max 1 Si max 0.9-1.5 Ti 4.5-4.9 Al 18-22 Co 4.5-5.5 Mo 0.005 Pb max 0.2 C max	do.	
115	do.	14-16	0.2 max	1 max	1 Mn max 1 Si max 3.5-4.5 Ti 4.5-5.5 Al 13.6-16.5 Co 3-5 Mo 0.005 Pb max 0.2 C max	do.	
PE11	38 (Ni + Co = 39)	17-19	-	Balance	0.2 Mn max 0.3 Si max 2.1-2.5 Ti 0.6-1 Al 1 Co max 4.75-5.75 Mo 0.03 B max 0.25 Zr max 0.1 C max 0.015 S max	do.	
PE16	40-43 (Ni + Co = 42-45)	15-18	-	Balance	0.2 Mn max 0.3 Si max 0.9-1.5 Ti 0.9-1.5 Al 2 Co max 2.5-4 Mo 0.005 B max 0.05 Zr max 0.1 C max 0.015 S max	do.	
PK33	Balance	16-20	0.2 max	1 max	0.5 Mn max 0.5 Si max 1.5-3 Ti 1.7-2.5 Al 12-16 Co 5-9 Mo 0.005 B max 0.06 Zr max 0.07 C max 0.015 S max	do.	

TABLE 7.

TRADE NAME or other designation		COMPOSITION, % by weight					SPECIAL PROPERTIES	USES
		Ni	Cr	Cu	Fe	Other		
PARR METAL (Illium)	A	80	15	5	-	-	Corrosion resistance, high density, particularly resistant to mineral acids	Chemical plant, pumps
	B	66	18	8.5	-	1 Mn 3.5 W 2 Al 0.2 Ti 0.2 B		
	C	60.6	21	6.5	-	1 Mn 2.1 W 1 Al 1 Si 4.6 Mo		
PEERLESS		78.5	16.5	-	3	2 Mn	High electrical resistivity, oxidation resistance	Electrical resistors
PERMAFY		80	-	-	20	-	High magnetic permeability, high electrical resistance	Electrical machinery
PERMALLOY	C	77	-	5	14	4 Mo	Greater magnetic permeability than iron, readily magnetised in weak fields	Electromagnetic circuits involving low currents, cladding of submarine cables, transformers, shields for cathode ray tubes
	B	45	-	-	55	-		
	D	36	-	-	64	-		
	F	65	-	-	35	-	Low hysteresis loss	
		78	-	-	22	-	High magnetic permeability	Magnetic cores
		78.2	-	-	21.4	0.37 Co		
		78.5	3.8	-	17.7	-		
		70	-	-	30	-		
		78.5	-	-	17.7	3.8 Mo	do.	a.c. circuits
		78.5	-	-	21.5	-	do.	d.c. circuits
		81	-	-	19	-	do.	Loading coils
		81	-	-	17	2 Mo	do.	do.
PERMINVAR	A	20-75	-	-	Balance	5-40 Co	Very low magnetic hysteresis, constant permeability at low flux density, high initial magnetic permeability, especially when Mo present	Electromagnetic circuits, especially magnetic cores
	B	45	-	-	23	25 Co 7 Mo		
	C	45	-	-	27	25 Co 2.5 Mo 0.35 Mn		
	D	45	-	-	30	25 Co		
	E	70	-	-	22.5	7.5 Co		
PHOSPHOR-NICKEL		Balance	-	-	-	20 P	Deoxidises	Master alloy for adding phosphorus, deoxidant
PIONEER METAL		38	20	-	35	3 Mo 4 Si	Corrosion resistance	Severe corrosion conditions
		35	25	-	35	5 Mo		
PLACET		60	15	-	20	5 Mn	Heat resistance	Electrical resistors, heating elements
PYROS		82	7	-	3	5 W 3 Mn	Linear temperature coefficient of thermal expansion from 0 to 1000 °C	Dilatometers, instruments involving rise in temperature
Q-ALLOY	A	66-68	15-19	-	Balance	-	Heat resistance, oxidation resistance	Furnace and other high temperature service
	B	60	12	-	do.	-		
REDRAY		85	15	-	-	-	Heat resistance	Electrical resistance heating elements

TABLE 7.

TRADE NAME or other designation		COMPOSITION, % by weight					SPECIAL PROPERTIES	USES
		Ni	Cr	Cu	Fe	Other		
REFRACTALOY	26	37	17.9	-	18.4	20 Co 3 Mo 3 Ti 0.7 Mn	Heat resistance	High temperature service, gas turbine blades
RESISTO		69	10	-	19	1 Si 0.4 Co 0.5 Mn	High electrical resistivity, oxidation resistance	Heating elements, resistors
ROSEIN		40	-	-	-	30 Al 20 Sn 10 Ag	Non-magnetic	Unknown
SOFT-WELD		65.3	-	32.2	1.3	1.2 Mn	Low melting point, toughness	Welding cast iron
SUHLER COPPER		31.5	-	4.4	-	25.5 Zn 2.6 Pb	Corrosion resistance, colour	Trim
SUPERMALLOY		79	-	-	15	5 Mo 0.5 Mn 0.5 C + Si + S max	High magnetic permeability at low field strengths	Electromagnetic circuits
THERMALLOY	A	65	20	-	15	-	Heat and corrosion resistance	Chemical and furnace plant in absence of sulphur
	B	40	18	-	42	-	do.	Retorts, muffles, lead pots
	40	12	26	-	Balance	0.2-0.5 C	do. up to 800 °C	Dampers, valves, chain links
	50	33-37	14-16	-	do.	1 Mn 2 Si 0.4-0.55 C	do.	Hearth plates, rollers, grids, chain belts
	72	58-63	12-14	-	do.	1 Mn 2 Si 0.4-0.6 C	do. up to 1100 °C	Heat treatment boxes, retorts, muffles
	EA	66.5	-	30	2	Balance C + Si	Temperature sensitive magnetic properties, permeability decreases with rise in temperature	Magnetic shunts in watt-meters and similar instruments
TOPHET	A	80	20	-	-	-	Heat resistance Corrosion resistance	Resistance wires up to 1150 °C
	B	65-75	16-23	-	Balance	0.15 C		Chemical plant
	C	60	24	-	16	-		Resistance wires up to 1000 °C
	D	34	20	-	46	-		do. to 850 °C
VAC-MELT	AA	77.5	20	-	0.5	2 Mn	Heat resistance	Resistance wire up to 1200 °C
	A	74.5	20	-	1.5	1 Mn 1 Mo	do.	do. up to 1090 °C
	B	60	15	-	16	2 Mn 7 Mo	do.	do. up to 1060 °C
	C	61	18.5	-	16.5	4 Mn	do.	do. up to 1040 °C
VIKRO		64	15-20	-	Balance	1 Mn 1 C 0.5-1 Si	Heat resistance	Furnace parts
WESTEECO		55-60	15-18	-	Balance	-	do.	do.
WORTHITE		24	20	1.75	do.	3 Mo 3.25 Si	Corrosion resistance	Chemical plant, steam plant
X ITE		37-40	17-21	-	do.	-	Heat resistance	Furnace parts
Z-NICKEL		Balance	-	0.02	0.2	0.2 Si 0.2 Mn	Strength 620 MN m^{-2}, age hardened	Hot rolled sections, cold drawn sections

TABLE 8. Low Melting Point Alloys

TRADE NAME or other designation	COMPOSITION, % by weight					SPECIAL PROPERTIES	USES
	Sn	Pb	Bi	Sb	Other		
ACCUMULATOR METAL	-	Balance	-	-	0.1 Ca	High corrosion strength, improved strength	Battery plates
	-	Balance	-	Up to 10			
ALGER METAL (Algiers Metal)	Balance	-	-	10	0-0.3 Cu	Good casting properties, low melting point	Intricate castings in decorative work
ALUMINIUM SOLDER	50-85	-	-	-	15-50 Zn	Low melting point	Soft solders for aluminium and its alloys
	73-87	-	-	-	8-15 Zn 5-12 Al		
	65	-	-	-	30 Zn 5 Bi		
	89	-	-	-	11 Cd		
	55	-	-	-	11 Al 1 Cu 33 Zn		
	56	-	-	-	3 Al 3 Cu 38 Zn		
	36	-	-	-	64 Bi		
	67.5	-	-	-	16.5 Zn 15 Al 1 Cu		
	36	44	-	-	20 Zn		
ANATOMICAL ALLOY	19	17	-	-	53.5 Bi 10.5 Hg	Low melting point (60 °C)	Anatomical models
ANTIMONIAL LEAD	-	Balance	-	Up to 30	-	Corrosion resistance	Battery plates, chemical plant linings
ARGENTINE METAL	85	-	-	15	-	Fluid, good casting properties	Toys (castings)
ASHBURY METAL	80	-	-	14	3 Ni 2 Cu 1 Zn	Ductility, moderate strength	Sheet
BABBITT METAL Original Loco	84	-	-	10	6 Cu	Low friction	Bearings, where thin layer of Babbitt metal is coated on to a harder shell
Engine	90	-	-	7	3 Cu		
BAHN METAL	-	Balance	-	-	0.7 Ca trace Ni 0.6 Na	Soft anti-friction	Railway bearings
BRITANNIA METAL ENGLISH	90-95	-	-	4.5-9	1 Cu	Silvery white, ductile	Domestic tableware, ornamental plate
CABLE SHEATHING ALLOY	1-3	Balance	-	-	-	Ductile	Covering wire cables, lead foil, condensers, fuses
CAPSULE METAL	8	92	-	-	-	Ductile	Impact extrusion of tubular containers
CAZIN	-	-	-	-	17.4 Zn 82.6 Cd	Eutectic, melts 263 °C	Solder for steel cable
CERRO-ALLOYS CERROBASE	-	44.5	55.5	-	-	Eutectic, melts 124 °C	Fusible alloys for foundry patterns, tube bending and fixture use
CERROBEND	13.3	26.7	50.0	-	10 Cd	do. melts 70 °C	
CERROLAW 147	12.77	25.63	48.0	-	4.0 In 9.6 Cd	Melt range 61-65 °C	

TABLE 8.

DE NAME or designation		COMPOSITION, % by weight					SPECIAL PROPERTIES	USES
		Sn	Pb	Bi	Sb	Other		
ALLOYS OLAW	136B	15.0	18.0	49.0	-	18.0 In	do. 57-69 °C	
	136	12.0	18.0	49.0	-	21.0 In	Eutectic, melts 57 °C	
	140	12.6	25.4	47.5	-	5.0 In 9.5 Cd	Melt range 56-65 °C	
	117	8.3	22.6	44.7	-	19.1 In 5.3 Cd	Eutectic, melts 47 °C	
	117B	11.3	22.6	44.7	-	16.1 In 5.3 Cd	Melt range 47-52 °C	
	105	7.97	21.7	42.91	-	18.33 In 4.0 Hg 5.09 Cd	do. 38-43 °C	
CERROMATRIX		14.5	28.5	48.0	9.0	-	Melt range 103-227 °C	
		34.5	-	44.5	-	21 Cd	do. 103-120 °C	
		25.0	-	50.0	-	25 Cd	do. 103-113 °C	
CERROSAFE		11.3	37.7	42.5	-	8.5 Cd	Melt range 70-90 °C	
		24.5	17.9	45.3	-	12.3 Cd	do. 70-88 °C	
		13.0	37.0	40.0	-	10 Cd	do. 70-85 °C	
		13.0	35.0	42.0	-	10 Cd	do. 70-80 °C	
CERROSEAL 35		50.0	-	-	-	50 In	Melt range 117-127 °C	
		1.0	44.0	55.0	-	-	do. 117-120 °C	
CERROTRIC		42.0	-	58.0	-	-	Melts 138.5 °C	Fusible alloys for foundry patterns, tube bending and fixture use
CERROTRU		43.0	-	57.0	-	-	do. 138 °C	
		41.6	1.0	57.4	-	-	Melt range 134-135 °C	
COMSOL		0.25	97	-	-	2.5 Ag 0.25 Cu	Melts 296 °C, strong to 150 °C	Soldering electrical components
CONDENSER FOIL		82	15	-	2	-	Malleable to thin sheet	Condensers for electronics
DANDELION METAL		10	72	-	18	-	Low friction	Bearings
D'ARGET'S ALLOY		25	25	50	-	-	Melting point 93 °C	Fusible alloy, sprinklers
ELECTRICIAN'S SOLDER		94.5	5.5	-	-	-	Low melting point	Solder for instrument and circuit use
ELECTROTYPE METAL		2.5-4	Balance	-	2.5-3	-	Ductility	Backing for printer's electrotypes
EUTECTIC SOLDER		62	38	-	-	-	Melting point 183 °C	High speed machine soldering, fine soldering for electronics
EVEREST METAL (Thermit metal)		5-7	Balance	-	14-16	0.8-1.2 Cu 0.7-1.5 Ni 0.3-0.8 As 0.7-1.5 Cd	Low friction but good strength	Bearings
FAHRIG'S METAL (FAHRY'S; Reverse Bronze)		90	-	-	-	10 Cu	Low friction	do.
FINE SOLDER		67	33	-	-	-	Low melting point	Soldered joints
FOUNDRY TYPE METAL		10-20	54-70	-	20-28	Small Cu	Low melting point, hardness	Printing type metal
FRARY METAL		-	Balance	-	-	2 Ba 1 Ca	do.	Bearings

TABLE 8.

TRADE NAME or other designation	COMPOSITION, % by weight					SPECIAL PROPERTIES	USES
	Sn	Pb	Bi	Sb	Other		
FUSIBLE ALLOYS	99.25	-	-	-	0.75 Cu	Melts 227 °C	Foundry patterns, tube bending, seals, fixtures
	96.5	-	-	-	3.5 Ag	do. 221 °C	
Binary Eutectics	-	-	-	-	17.0 Cd 83.0 Tl	do. 203 °C	
	92.0	-	-	-	8.0 Zn	do. 199 °C	
	-	-	47.5	-	52.5 Tl	do. 188 °C	
	62.0	38.0	-	-	-	do. 183 °C	
	67.0	-	-	-	33.0 Cd	do. 176 °C	
	56.5	-	-	-	43.5 Tl	do. 170 °C	
Ternary Eutectic	51.2	30.6	-	-	18.2 Cd	do. 145 °C	
Ternary Alloy	40.0	42.0	-	-	18.0 Cd	Melt range 145-160 °C	
Binary Eutectic	-	-	60.0	-	40.0 Cd	Melts 144 °C	
Ternary Alloy	48.8	41.0	10.2	-	-	Melt range 142-166 °C	
Ternary Eutectic	40.0	-	56.0	-	4.0 Zn	Melts 130 °C	
do.	46.0	-	-	-	17.0 Cd 37.0 Tl	do. 128 °C	
do.	25.9	-	53.9	-	20.2 Cd	do. 103 °C	
Ternary Alloy	33.0	33.0	34.0	-	-	Melt range 96-143 °C	
Ternary Eutectic	15.5	32.0	52.5	-	-	Melts 96 °C	
do.	-	40.2	51.7	-	8.1 Cd	do. 92 °C	
Quaternary Alloy	15.4	30.8	38.4	-	15.4 Cd	Melt range 70-97 °C	
Quaternary Eutectic	13.1	27.3	49.5	-	10.1 Cd	Melts 70 °C	
Quinternary Alloy	13.2	26.3	49.3	-	9.8 Cd 1.4 Ga	Melt range 65-66 °C	
Binary Eutectic	8.0	-	-	-	92.0 Ga	Melts 20 °C	
GEMMA'S AXLE METAL	Balance	-	-	11	5.5 Cu	Low friction, easily cast	Bearings
GLIEVER BEARING METAL	8	76.5	-	14	1.5 Fe	Low friction	do.
GLYCO	Up to 8	Balance	-	22	-	Low melting, strength	do.
GRAPHITE METAL	15	68	-	17	-	Low friction	do.
HANOVER METAL	87	-	-	8	5 Cu	Low friction, hardness	do.
HARD LEAD	90	-	-	8	2 Cu	do.	do.
HARD TIN	99.6	-	-	-	0.4 Cu	Ductility, stronger than pure tin	Collapsible tubes, foils
HAUSER'S ALLOY	-	50	33.3	-	16.7 Cd	Low melting	See Fusible alloys
HOYLE'S ALLOY	46	42	-	12	-	Low friction	Bearings
HOYT'S METAL	91	-	-	6.8	2.2 Cu	Low friction, tough, ductile	Connecting rod bearings in aero engines and diesel engined air compressors
HUGHES' METAL	14	76	-	10	-	Low friction	Bearings
HUSMAN'S METAL	Balance	10	-	11	4.5 Cu 0.4 Zn	Tough, ductile, low friction	do.
JACANA METAL	10	71	1	18	-	Low friction	do.

TABLE 8.

TRADE NAME or other designation	COMPOSITION, % by weight					SPECIAL PROPERTIES	USES
	Sn	Pb	Bi	Sb	Other		
JACOBI'S ALLOYS — Tin base	85	-	-	10	5 Cu	Ductility with reasonable strength	Tableware, ornamental work, machine parts
Lead base	5	85	-	10	-	Low friction	Bearings
	27	62	-	10	-	do.	do.
KAISERZINN	93	-	-	5.5	1.5 Cu	Malleable	Sheet and formed products
KARMARSCH'S ALLOY	85	-	1.5	5	3.5 Cu 1.5 Zn	Malleable and ductile	Sheet and formed products
	Balance	1.2	-	-	12.5 Cu	Low friction	Bearings
	do.	-	-	7.5	3.7 Cu		do.
KATZENSTEIN ALLOY	7.5	Balance	-	16.5	0.5 Cu	Low friction	Bearings
KOELLER'S ALLOY	Balance	-	1.8	10.5	1 Cu	Malleable and ductile	Sheet forming, ornamental parts
KUPPER'S SOLDER	7.15	Balance	-	9.5-7	-	Low melting, saving in tin	Solder wire
LEAD BATTERY METAL	0.1-0.5	Balance	-	7-12	-	Corrosion resistance in H_2SO_4, strength, castability	Lead acid accumulators and similar batteries
LEY (LAY)	75-80	25-20	-	-	-	Low melting	Solder
LICHTENBERG'S ALLOY (Onion's alloy)	20	30	50	-	-	Melts 92 °C	Fine castings
LINOTYPE ALLOY	3-5	84-86	-	11-12	-	Rapid solidification, almost eutectic alloy, liquidus 240-250 °C, solidus 240 °C	Die casting machines for printing, casting a slug containing one line of type; e.g. Linotype, Intertype, Ludlow casters
LION METAL	70-89	-	-	8-3	8-3 Cu	Melts 205 °C, strong, low friction	Main bearings for locomotives, marine engines, electrical machinery
LIPOWITZ'S ALLOYS	13	27	50	-	10 Cd	Fusible alloy, melts 72 °C	Sealant plugs in fire sprinklers, fine castings, plug material
	11	22	41	-	8 Cd 18 In	do., melts 46.5 °C	
LOTUS METAL	10	75	-	15	-	Low friction	Bearings
LUDENSCHEIDT PLATE	72	-	-	24	4 Cu	Malleable, low friction	Sheet metal work, bearings
MAGNOLIA METAL	6	Balance	-	15	-	do.	Bearings
MALOTT METAL	34	20	46	-	-	Melts 95 °C	Fine castings, fusible seals
MATHESIUS METAL	-	Balance	-	-	Up to 1 Ca or Sr	Forms intermetallic compounds increasing strength of lead (170 MN m^{-2} 11 tons in.$^{-2}$)	Bearing metal
MATRIX ALLOY	14.5	28.5	48	9	-	Expands on solidification, low melting point, low strength	Duplicate master patterns in foundries, proof casting of forging dies, type casting, location in jigs, plugging holes in castings
MELOTTE'S ALLOY	31	19	50	-	-	Melts 99 °C	Fusible seals

TABLE 8.

TRADE NAME or other designation	COMPOSITION, % by weight					SPECIAL PROPERTIES	USES
	Sn	Pb	Bi	Sb	Other		
METAL ARGENTINE	Balance	-	up to 1	2-5	0.5 Cu	Malleable	Tableware
MINOFAR	do.	-	-	17-20	9-10 Zn 3-4 Cu	Malleable	Tableware, ornaments
MONOTYPE METAL	5	80	-	15	-	Rapid solidification, strength	Single letter casting machines
MOTA METAL	85-87	-	-	8.5-9.5	4-6 Cu	Strength, low friction	High stressed bearings in C.I. and I.C. engines
NATKE'S ALLOY	73	-	-	18	9 Cu	Low friction	Bearings
NEWTON'S ALLOY	18.8	31.2	50	-	-	Melts 95 °C	Fusible plugs and seals
NICKELINE	85.4	0.43	-	8.8 trace Fe, As	4.75 Cu 0.28 Zn	Ductility	Packing rings in steam equipment, stuffing boxes
NICO I	4-5	71-69	-	23	2-3 Ni	Low melting point	Small castings
II	10	79	-	10	1 Ni	do.	do.
OCPAN	80-90	<0.25	-	10-15	2.5 Cu	By-product of tin smelting in the USA	Master alloy for bearing metal casting
ONION'S ALLOY (*see* LICHTENBERG'S)							
PALID No. 5	-	90	-	5	5 As	Ductility, malleability	Sheet, strip, foil, extruded tube, chemical plant
No. 6	-	86	-	8	6 As		
No. 7	-	82	-	11	7 As		
PARSON'S BRASS (White Star Brass)	74-76	14-15	-	7-8	3-4.5 Cu	Low friction	Bearings
PARSON'S MOTA METAL	86-92	-	-	4.5-9	3-5 Cu	do.	do.
PATTERN METAL	40	30	-	-	30 Zn	Machineability, good surface finish	Metal patterns for moulding
	15	42	-	-	3 Cu 40 Zn		
PEWTER Roman (100 AD)	70	30	-	-	-	High malleability with reasonable strength after working to shape, toxicity higher with antimony and lead present, although harder	Holloware, sheet and foil, complex beaten metal plaques and decorative work, domestic articles of aesthetic beauty when highly polished
Tudor	90	10	-	-	-		
(16th cent.)	91	-	-	9	-		
Common	Balance	5-15	-	-	-		
	81.2	11.5	-	-	5.7 Cu 1.6 Zn		
	84.9	-	6	1.9	6.8 Cu		
Best	86	-	-	14	-		
Common	89.3	1.6	-	7.5	1.6 Cu		
	80	-	-	20	-		
English (medieval)	81.2	11.5	-	5.7	1.6 Cu		
(modern)	90-92	-	-	2-5	3.8 Cu		
Ornamental	80	20	-	-	-		
Tankard	88.5	-	-	7.2	3.5 Cu 0.8 Zn		
Brittania metal (common)	Balance	-	-	5-10	1-3 Cu		

TABLE 8.

TRADE NAME or other designation	COMPOSITION, % by weight					SPECIAL PROPERTIES	USES
PEWTER (*contd.*)	Sn	Pb	Bi	Sb	Other		
(tableware)	95	2.5	-	-	2.5 Cu		
Dutch (white metal)	81	-	-	9	10 Cu		
Hanover (white metal)	87	-	-	8	5 Cu		
Ludenshield plate	72	-	-	24	4 Cu		
Minofor A	20.3	-	-	64	6 Zn		
do. B	66	-	-	20	4 Cu 9 Zn 1 Fe		
do. C	68.5	-	-	18.2	3.3 Cu 10 Zn		
do. D	72	-	-	24	4 Cu		
do. E	84	-	-	9	2 Cu 5 Zn		
Queen's metal A	50.5	16.5	-	16.5	16.5 Zn		
do. B	73.4	8.9	-	8.9	8.8 Zn		
do. C	87	-	-	8.5	3.5 Cu 1 Zn		
do. D	88.5	-	-	7.1	3.5 Cu 0.9 Zn		
do. E	88.5	-	1	7	3.5 Cu		
Titania metal	91.4	7.6	-	-	0.75 Cu 0.25 Zn		
do. cast	92.4	0.32	-	4.6	2.5 Cu 1.13 Fe		
do. domestic	80	-	-	16	2.7 Cu 1.3 Zn		
do. plate	90	6	-	-	2.7 Cu 1.3 Zn		
PHOSPHOR-TIN	95	-	-	-	5 P	Deoxidises	Master alloy for addition of phosphorus to tin base alloys, bronzes and gun-metals
	90	-	-	-	10 P		
	95.5	0.32	-	0.07	0.33 Al 3.8 P		
	95	0.65	-	0.24	0.2 Cu 3.6 P		
	96.5	-	-	-	3.5 P		
PLASTIC METAL	85	-	-	10	5 Cu	Reasonable hardness and ductility	Bearings
	78	-	-	15	7 Cu		
	83	-	-	9	8 Cu		
	78	-	-	11	11 Cu		
PLUMBER'S SOLDER	50	50	-	-	-	Melts 182 °C	Wiping joints in lead pipe
PRINCE'S METAL	84.8	-	-	15.2	-	Ductility, malleability	Holloware
REAMUR ALLOY	-	-	-	70	30 Fe	Unknown	Unknown
REGAL METAL	83.3	-	-	11	5.7 Cu	Low friction	Bearings

TABLE 8.

TRADE NAME or other designation		COMPOSITION, % by weight					SPECIAL PROPERTIES	USES
		Sn	Pb	Bi	Sb	Other		
REGULUS METAL (Hard lead)		-	Balance	-	6-12	-	Hardness much higher than pure lead, cast-ability	Storage battery plates, linings for chemical tanks, cable coverings, type metal, bullets
RELY ALLOYS	1	84	-	-	10	6 Cu	Low friction, varying strength with amount of lead and copper	Bearings
	2	72	12	-	10	6 Cu		
	3	40	42	-	15	3 Cu		
	4	31	50	-	16	3 Cu		
	5	15	70	-	14	1 Cu		
	6	6	80	-	14	-		
RICHARD'S BEARING ALLOY		82	-	-	10	8 Cu	Low friction	Bearings
ROSE'S ALLOY	1	22	28	50	-	-	Low melting point, range 96-110 °C	Safety devices in fire alarms, fusible plugs
	2	30	35	35	-	-		
SATCO METAL	1	1	98	-	-	0.5 Ca 0.25 Hg 0.05 Al 0.04 K 0.075 Mg 0.04 Li	Low friction	Axle box bearings
	2	1	98.5	-	-	0.05 Ca 0.25 Hg 0.05 Al 0.04 K 0.075 Mg 0.04 Li		
	3	2.4	97.4	-	-	0.15 Ca 0.07 K	do.	Big-end bearings
	4	2.56	95.5	-	-	0.62 Ca 0.15 Mg 0.54 Na	do.	Diesel engine bearings
	5	1	98	-	-	0.5 Ca 0.07 Al 0.08 Mg	do.	
SEIFERT'S SOLDER		73	5	-	-	21 Zn	Low melting point	Soldering aluminium
SODIUM LEAD		-	Balance	-	-	2 Na	Deoxidant	Treatment for low melting alloys for deoxidation
SOFT SOLDER Common		39-40	Balance	-	2-2.5	-	Low melting point, high diffusion rate of tin into other metals and alloys	Soldering of a wide range of copper alloys; also used for ferrous alloys when high strength in joint unnecessary
Plumber's		29-30	do.	-	1-1.5	-		
do.		33.33	66.66	-	-	-		
do.		38	60	-	2	-		
do.		39	60	-	1	-		
Tinsmith's		44-45	Balance	-	2.5	-		
do.		60	40	-	-	-		
		20	40	40	-	-	Melting point 113 °C	
		16	54	30	-	-	do. 170 °C	
		39.5	39.5	-	-	19 Cd 2 Zn	Melting range 136-165 °C	
		20	64	16	-	-	Melting point 212 °C	

TABLE 8.

TRADE NAME or other designation	COMPOSITION, % by weight					SPECIAL PROPERTIES	USES
SOFT SOLDER (*contd.*)	Sn	Pb	Bi	Sb	Other		
	9	65	-	-	26 Cd	Melting range 150-225 °C	
	14	70	16	-	-	Melting point 238 °C	
	14-15	Balance	-	-	1.5-2 Ag 0.5 In	Melting range 170-270 °C	
	15	77.5	5	1	1.5 Ag	Melting point 258 °C	
	20	78.25	0.5	-	1.25 Ag	do. 270 °C	
	10	87.75	-	-	2.25 Ag	do. 290 °C	
	-	96	-	-	3 Ag 1 In	do. 310 °C	
SPERRY'S METAL	35	Balance	-	15	-	Low friction	Bearings
STANNIOL	Balance	0.7-2.4	-	-	0.33-1 Cu 0.1 Fe or Ni optional	Corrosion resistance, harder than pure tin	Applications where pure tin is too soft, e.g. in foils for capacitors and coatings
STAR ALLOY	9-10.5	Balance	-	17-19	1 Cu	Low friction	Bearings
STEREOTYPEMETAL	4	83.75	-	11.75	0.5 Cu	Low melting point, good castability	Production of stereo-type plates in printing
	4	Balance	-	13-18	-	do.	
	10	do.	-	15	-	do.	
STERLING'S aluminium solder	62.3	8	-	1.2	15 Zn 11 Al 2.5 Cu	Low melting point	Soldering aluminium
TANDEM METAL	6	Balance	-	17	-	Low friction	Bearings
TEA LEAD	2	98	-	-	-	Slightly harder than pure lead	Lining tea chests, foil
TEGO	1-3	78-83	-	15-18	1-2 Cu 0.5 As	Low friction	Bearings
TELLEDIUM	-	Balance	-	-	0.1 Te	Slightly harder than pure metal	Cold rolled pipes and tubes, cable sheathing (extruded)
TERMITE	5.75	73.5-74	-	14.5	2.5-3 Cu 2 Cd 1 As	Low friction	Bearings
TERNE METAL (*see* Terne plate in text)	18	80-80.5	-	1.5-2	-	Low friction, corrosion resistance	Bearings, coating for steel in fuel tanks for petroleum
THERMIT	6.3	75	0.01	15.6	1 Ni	Low friction	Bearings
	5-7	72-78	-	14-16	0.8-1.2 Cu 0.8-1.5 Ni 0.3-0.8 As		
	4	72	-	20	1 Cu 3 Ni		
	4	71	-	22	0.5 Cu 0.5 As 0.05 P		
THURSTON'S METAL	Balance	-	-	19	10 Cu	Low friction	Bearings
TINFOIL	Balance	-	-	-	8.5 Zn 0.15 Ni	Malleability, non-toxic	Thin sheets for food wrap, electrical capacitors, (largely superseded by Al), sheets 0.5-0.2 mm thick
	do.	0-8	-	0.5-3.2	0-4 Cu		
TINMAN'S SOLDER (Tertiarium)	33.3	66.7	-	-	-	Low melting point	Soldering tinplate, copper alloys

TABLE 8.

TRADE NAME or other designation	COMPOSITION, % by weight					SPECIAL PROPERTIES	USES
	Sn	Pb	Bi	Sb	Other		
TINSEL	60	40	-	-	-	Colour, malleability	Thin strip for decoration, (largely replaced by metallised plastic film)
TINSMITH'S SOLDER (Slicker solder)	66	34	-	-	-	Low melting point	Soldering kitchen utensils
TOURUN METAL	90	-	-	-	10 Cu	Low friction	Bearings
TRABUK	87.5	-	2	5	Balance Ni	Ductility, corrosion resistance	Replacement alloy for nickel silver in ornamental articles
TUTANIA ALLOYS	92.5	0.32	-	4.6	2.6 Cu	Low friction	Bearings
	91.5	7.65	-	-	0.7 Cu 0.25 Zn	do.	do.
	80	-	-	16	2.7 Cu 1.3 Zn	Colour, malleability	Domestic ware
	90	6	-	-	2.7 Cu 1.3 Zn	do.	Plate ware
ULCO METAL	-	Balance	-	-	1-2 Ba 0.5-1 Ca	Low friction	Bearings
UNION METAL	-	do.	-	-	0.2 Cu 1.5 Mg	do.	do.
VARRENTROP'S ALLOY	66	-	-	-	33 Cd dissolved in Hg	Low melting point	Dental fillings
WAGNER'S ALLOY	Balance	-	0.8	10	3 Zn 1 Cu	Ductility, malleability	Hollow ware, tableware, ornaments
WARNE'S METAL	37	-	26	-	26 Ni 11 Co	do.	do. Jewellery
WELCH'S ALLOY	52	-	-	-	48 Ag	Low melting point	Dental prosthetics (fillings)
WHITE BRASS	65	-	-	-	29 Zn 6 Cu	Low friction	Bearings, automobile
	66	-	-	31	3 Cu	do.	do., general
WOOD'S METAL (alloy)	12.5	25	50	-	12.5 Cd	Melting range 70-72 °C	Sprinkler heads
	13.3	26.7	50	-	10 Cd	Melts 70 °C	Master patterns
YAMATO METAL	Balance	1	-	3-6	3-6 Cu 1 Ni 1 Cd max	Low friction	Bearings for internal combustion engines
ZINN (Stanniol)	99	0.7	-	-	0.3 Cu	do.	Hard service bearings

TABLE 9. Other Non-ferrous Alloys

TRADE NAME or other designation	COMPOSITION, % by weight					SPECIAL PROPERTIES	USES
Chromium-based alloys	Cr	Ni	Fe	Mo	Other		
C.M. alloy 469	60	-	14	25	0.03 C	Heat resistance	Castings
Cr-Fe	60	-	40	-	-	do.	do.
Cr-Ta	97	-	-	-	2 Ta 0.1 C	do.	do.

TABLE 9.

TRADE NAME or other designation		COMPOSITION, % by weight					SPECIAL PROPERTIES	USES
Cobalt-based alloys		Co	Cr	Ni	W	Other	Non-magnetic	Mainsprings of watches, instrument suspensions
ELGILOY		40	20	15	-	2 Mn 7 Mo 0.04 Be 0.15 C Balance Fe		
HAYNES STELLITE (Similar compositions are products of different manufacturers)	No. 3 (Cast)	Balance	30	3 max	12	2.5 C 3 Fe	Wear resistance, hardness at high temperature	Cutting tools
	4 (Cast)	do.	30	3 max	14	0.8 C 8 Fe	Corrosion and abrasion resistance	
	6 (Cast)	do.	30	3 max	4	1 C 3 Fe	Abrasion resistance, corrosion resistance	Hard facing coatings, valves, liners when wrought
(Molybdenum Mo and Tungsten W often interchangeable)	6K (Wrought)	do.	31	3	4	1.6 C 3 Fe	do.	Knives, scraper blades
	12 (Cast)	do.	30	3	8	1.3 C 3 Fe	do.	Hard facings
	19 (Cast)	do.	31	-	10	1.7 C 3 Fe	do. Wear resistance	Cutting tools, nozzle discs and bearings in gas turbines
	25 (Wrought)	do.	20	10	15	0.1 C 3 Fe	Heat treatable for strength and heat resistance	Turbine blades
	36 (Cast)	do.	18.5	10	15	0.4 C 0.03 B	do.	do.
	98M2 (Cast)	do.	30	3.5	18	2 C 2 Fe	Abrasion and wear resistance	Cutting tools
	150 (Wrought)	do.	28	-	-	20 Fe 0.7 Si 0.1 C	Heat resistance	Furnace parts
	151 (Cast)	do.	20	-	12.5	0.05B 0.5 C	Heat treatable	Nozzle discs
	152 (Cast)	do.	20	1	11	2 Ta 2 Nb 0.4 C	do.	Turbine blades
	302 (Cast)	do.	21.5	-	10	9 Ta 0.2 Zr 0.85 C	do.	do.
	Star J	do.	32	2.5	17	2.5 C 2 Fe	Abrasion resistance, high temperature hardness	Cutting tools
	HS 21 (Cast)	do.	27	3	-	1 Fe 0.25 C 5 Mo	Heat resistance, creep strength	Jet engine impellers and parts, turbine blades
	HS 23	do.	24	2	6	0.4 C	do.	do.
(Superalloy)- (HA 25) (L-605)	HS 25	do.	20	10	15	0.1 C	do.	do.
	HS 27	do.	25	3.2	6	0.4 C	do.	do.
	HS 30	do.	26	15	-	1 Fe 0.45 C 6 Mo	do.	do.
(Superalloy)-	HS 31 (Cast)	do.	25	10	7.5 (or Mo)	1.5 Fe 0.5 C	do.	do.
	HA 36	do.	19	10	15	1 Fe 0.03 B 0.4 C	do.	do.
J alloy		60	23	-	-	2 C 6 Mo 2 Ta 1 Mn	Heat resistance, creep resistance	Gas turbines, furnace parts

TABLE 9.

TRADE NAME or other designation		COMPOSITION, % by weight					SPECIAL PROPERTIES	USES
Cobalt-based alloys		Co	Cr	Ni	W	Other		
METALINE		35	-	-	-	30 Cu 25 Al Balance Fe	Corrosion resistance	Master alloy for preparation of fine powder magnets
NIRANIUM		64.2	28.8	4.3	2	0.7 Al 0.2 C 0.1 Si	Corrosion resistance	Dental bridges, caps and plates
OCTANIUM		40	20	15.5	-	15 Fe 2 Mn 0.15 C 0.03 Be 7 Mo	Non-magnetic, low thermal expansion, corrosion and fatigue resistant	Springs, surgical instruments, fountain pen nibs
REFRACTALOY	70	30.2	20.3	20.1	3.8	15.3 Fe 8.3 Mo 2 Mn	Heat resistance	Gas turbine blades and components
	80	30	20	20	5	14 Fe 0.6 Mn 10 Mo		
REFRACTORY		30	20	21	4	14 Fe 8 Mo	do.	do.
STELLITE (see also Haynes Stellite; Haynes was the original inventor. Stellite alone is a trade name used by other companies in recent years, and the alloys may be different)	1	Balance	33	-	13	2.5 C	Abrasion and wear resistance	Hard facing
	3	do.	30	-	13	2.4 C	do.	Rotary seal rings, pump sleeves, wear pads, bearing sleeves
	4	do.	31	-	13	1 C	Heat and wear resistance	do.
	6	do.	26	-	5	1 C	do. Corrosion resistance	Steam and chemical valves, shear blades, tong bits
	7	do.	26	-	6	0.4 C	Heat resistance	Gas turbine blades, extrusion dies
	8	do.	30	-	-	0.2 C 6 Mo	do.	do.
	12	do.	29	-	9	1.8 C	High creep strength	Hard facing
	20	do.	3	-	18	2.5 C	Abrasion and corrosion resistance	do.
	21	do.	27	2.5	-	2 Fe 0.007 B 0.25 C 5.5 Mo	Heat treatable	Turbine blades
	31	do.	25.5	10.5	5	0.5 C	Heat resistance	do.
	100	do.	34	-	19	2 C	Hardness at high temperature, abrasion and wear resistance	Tool bits, milling cutters, blades
(Superalloy)	X40 (Cast)	do.	25	10	7	0.3 C	Heat resistance, resistance to thermal shock	Jet engine parts
(Superalloy)	SA40	65	27	-	4.5	1.2 C Balance Fe	Hardness	Hard facing
	V36	Balance	25	20	2	4 Mo 2 Nb 3 Fe 1 Mn 0.4 Si 0.25 C	do.	do.
SUN BRONZE		50-60	-	-	-	30-40 Cu 10 Al	Strength, toughness	Castings
SUPERALLOYS (see also under Haynes Stellite, for HS31, HS21, HA25, and under Stellite for X40, SA40)	S816	Balance	20	20	4	4 Mo 4 Nb 4 Fe 1.2 Mn 0.4 Si 0.4 C	Heat treatable for creep strength, heat and corrosion resistance	Jet engine components, turbine blades
	HE 1049 (Cast)	do.	26	10	15	0.4 B 3 Fe max 0.4 C	do.	do.

TABLE 9.

TRADE NAME or other designation	COMPOSITION, % by weight					SPECIAL PROPERTIES	USES
Cobalt-based alloys	Co	Cr	Ni	W	Other		
SUPERALLOYS (contd.) UMCo-50	do.	28	-	-	21 Fe 0.1 C	Heat resistance	Furnace parts
ML 1700 (Cast)	do.	25	-	15	0.4 B 0.2 C	Creep strength, heat and corrosion resistance	Turbine blades
J 1570	do.	20	28	7	4 Ti 2 Fe 0.2 C	do.	do.
J 1650	do.	19	27	12	2 Ta 3.8 Ti 0.02 B 0.2 C	Heat treatable for creep strength, do.	do.
WI-52 (Cast)	do.	21	1 max	11	2 Nb 2 Fe 0.45 C	do.	do.
HA 151 (Cast)	Balance	20	-	12.7	0.05 B 0.5 C	Heat treatable for creep strength, heat and corrosion resistance	Gas turbine components, turbine blades
MAR-M-302 (Cast)	do.	21.5	-	10	9 Ta 0.005 B 0.2 Zr 0.85 C	do.	do.
MAR-M-322 (Cast)	do.	21.5	-	9	4.5 Ta 0.75 Ti 2.25 Zr 1 C	do.	do.
X-45 (Cast)	do.	25.5	10.5	7.5	0.01 B 2 Fe 0.25 C	do.	do.
NASA-CoWRe	do.	3	-	25	1 Ti 1 Zr 2 Re	do.	Rocket components do.
Ai-Resist 13 (Cast)	do.	21	1 max	11	2 Nb 2.5 Fe max 3.5 Al 0.1 Y 0.45 C	do.	do.
MAR-M-509 (Cast)	do.	24	10	7	3.5 Ta 0.2 Ti 0.5 Zr 0.6 C	do.	do.
UMCo-51	do.	28	-	-	2.1 Nb 19 Fe 0.3 C	Heat resistance	Furnace components
Ai-Resist 213	do.	19	-	4.7	6.5 Ta 0.15 Zr 3.5 Al 0.1 Y 0.18 C	Heat treatable for creep, heat and corrosion resistance	Jet engines
MAR-M-918	do.	20	20	-	7.5 Ta 0.1 Zr 0.05 C	do.	do.
HA 188	do.	22	22	14	1.5 Fe 0.08 La 0.1 C	do.	do.
Ai-Resist 215 (Cast)	do.	19	-	4.5	7.5 Ta 0.13 Zr 4.3 Al 0.17 Y 0.35 C	do.	do.
CM-7	do.	20	15	15	1.3 Ti 0.5 Al 0.1 C	do.	do.
FSX-414 (Cast)	do.	29.5	10.5	7	0.01 B 2 Fe 0.35 C	do.	do.
TICONIUM	32.5	27.5	31.4	-	5.2 Mo 1.6 Fe 0.35 Si 0.7 Mn 0.18 C	Corrosion resistance, strength, abrasion resistance	Dental prostheses
TIMKEN X	30.7	16.8	28.6	-	11 Fe 10.5 Mo 1.4 Mn 0.75 Si 0.13 C	Heat resistance	Gas turbine components
VICALLOY I	54	-	-	-	32 Fe 14 V	Ductility, high permeability	Magnetic circuits, magnetic recorder tape
II	52	-	-	-	35 Fe 13 V		

TABLE 9.

TRADE NAME or other designation		COMPOSITION, % by weight					SPECIAL PROPERTIES	USES
Cobalt-based alloys		Co	Cr	Ni	W	Other		
VICALLOY (contd.)	III	52	-	-	-	38.5 Fe 9.5 V		
	IV	62	-	-	-	30 Fe 8 V		
VITALLIUM		62.5	31.2	-	-	5.1 Mo 0.5 Mn 0.3 Si 0.4 C	Heat resistance	Castings for high temperature service
		64.5	28.7	-	-	5.6 Mo 1 Fe 0.24 C		
		63.6	27	2	-	5 Mo 0.6 Mn 1 Fe 0.6 Si 0.4 C		Turbo supercharger blades
		62	31	-	-	5.1 Mo 0.5 Mn 0.7 Fe 0.3 Si 0.4 C		Dental prostheses
		65	30	-	-	5 Mo		Medical prostheses
Hard alloys								
AKRIT		16 W	37.5 Co	30 Cr	4 Mo	10 Ni 2.5 C	Hardness, strength at high temperature	Cutting tips for lathe tools
COLMONOY		68-85 Ni	9 Fe, Si, C. Balance chromium boride				Hardness	Surfacing for wear resistance
DIAMONDITE		95.6 W	3.9 C				Hardness at high temperature, contains W_2C and WC	Sintered tool tips
ELKONITE		74.25 W		Balance Cu	-	-	Hardness, reasonable electrical conductivity	Electrodes
ELMARIT		92 W	3 Co	5 C			Hardness	Sintered tool tips
HEAVY ALLOY		79-90 W	1-16 Ni	3-20 Cu	-	-	High density	Made by powder compacting, balance weights on engine crankshafts
PROLITE		Balance WC	3-15 Co	3-15 TiC	-	-	Hardness, wear	Coal boring, percussion drill bits, dies
RAMER		-	-	Balance TaC	-	8 Ni	do.	Lathe tool bits
RENIK'S METAL		94 W	6 Ni	-	-	-	Hardness at elevated temperatures	Cutting tools
TIZIT		40-85 W	4-15 Ti	3-5 Cr	1-5 Ce	2-4 C 3-40 Fe	Hardness at elevated temperatures	do.
TURBIDE		Balance TiC	10 CrC	5 Ni 5 Co	-	-	Heat resistance, creep strength, resistance to thermal shock	Sintered parts for gas turbines and furnace parts
VOLOMIT		93.5 W	2 Fe	4.5 C	-	-	Hardness at elevated temperatures	Cutting tools
WIDIA		84-87 W	6-13 Co	3-5.7 C	-	-	do.	Sintered 1500 °C in inert atmosphere for cutting tools
WIMET		Balance WC	0-15 TiC	up to 11 Co	-	-	do.	Tips for cutting tools, drawing and extrusion dies, wear resistance parts

TABLE 9.

TRADE NAME or other designation		COMPOSITION, % by weight					SPECIAL PROPERTIES	USES
Magnesium based alloys		Mg	Al	Mn	Zn	Other		
DOW METAL		Balance	10 9 8 6	0.13 0.13 0.15 0.2	- 0.6 - -	0.5 Si 0.2 Si up to 0.3 Si up to 0.3 Si	Castability, light weight	Die castings
ECLIPS ALLOY		Balance	1.25	1	-	-	Castability	Die castings
ELEKTRON ALLOYS								
Cast-	A8	Balance	8	0.3	0.5	-	Heat treatable	General purpose castings
	AZ91	do.	9.5	0.3	0.5	-	do.	do. Pressure die castings
	AZ91X	do.	9.4	0.3	0.4	0.0015 Be	do.	Die castings
	C	do.	7.5-9.5	0.15 (min.)	0.3-1.5	-	do.	Cheap general purpose alloy
	MSR-A	do.	1.7 fractionated Rare Earth			0.7 Zr 2.5 Ag	High yield when heat treated	Pressure tight components
	MSR-B	do.	2.5 fractionated Rare Earth			0.7 Zr 2.5 Ag		do.
	MTZ	do.	-	-	-	0.7 Zr 3 Th	Creep resistant to 350 °C	do.
	RZ5	do.	1.2 Rare Earth		4	0.7 Zr	Strength, weldable	do.
	TZ6	do.	-	-	5.5	0.7 Zr 1.8 Th	do.	do.
	Z5Z	do.	-	-	4.5	0.7 Zr	Strength to 150 °C	General structures
	ZRE1	do.	2.7 Rare Earth		2.2	0.6 Zr	Creep resistant to 250 °C	Pressure tight components
	ZT1	do.	-	-	2.2	0.7 Zr 3 Th	Creep resistant to 350 °C	do.
Wrought-	A2855	Balance	8	0.3	0.4	-	Strength	Press forgings
	AM503	do.	-	1.5	-	-	Corrosion resistance	Extrusions, tubes, sheets
	AZ31	do.	3.0	0.3	1	-	Formability	do.
	AZM	do.	6.0	0.3	1	-	Strength, weldability	do. forgings
	ZTY	do.	-	-	0.5	0.6 Zr 0.7 Th	Creep resistant to 350 °C	Structures at temperature
	ZW1	do.	-	-	1.3	0.6 Zr	No stress corrosion	Sheet, extrusions
	ZW3	do.	-	-	3.0	0.6 Zr	do., weldability	do., forgings
	ZW6	do.	-	-	5.5	0.6 Zr	Strength, not weldable	do., do.
EUREKA RZ5 (see Elektron RZ5)		Balance	1.25 Rare Earth		4.5	0.6 Zr	Strength up to 100 °C	Structural components
MAGNUMINIUM ALLOYS	133	Balance	0.05 max	1-2	0.03 max	0.05 Si 0.02 Cu 0.03 Fe	Weldable	Castings and wrought products
	155	do.	3.5-5	0.15-0.4	1 max	0.05 Si 0.15 Cu 0.05 Fe	Strength/weight ratio	Sheet and tube
	166	do.	4-9	0.15-0.4	1.5 max	do.	do.	Forging
	181	do.	7.5-9	0.15-0.4	1 max	0.3 Si 0.15 Cu 0.05 Fe 0.01 Ni	do.	Sand casting
	220	do.	9-10.5	0.15-0.4	1 max	do.	do.	Die castings
	226	do.	5.5-8.5	0.2-0.4	1.5 max	0.05 Si 0.03 Cu 0.03 Fe	do.	Extruded products

TABLE 9.

TRADE NAME or other designation	COMPOSITION, % by weight					SPECIAL PROPERTIES	USES
Magnesium based alloys	Mg	Al	Mn	Zn	Other		
MAGNUMINIUM 288A ALLOYS (contd.)	do.	7.5-9.5	0.15-0.4	1 max	0.05 Si 0.03 Fe	do.	Wrought products
299	do.	2.5-4.5	0.1-0.4	-	4.5-7.5 Sn 0.3 Si 0.1 Cu 0.05 Fe 2 Ag	do.	Sand and die castings
MAZLO ALLOYS	Balance	6	0.15	3	-	Corrosion resistance to salt water, heat treated for higher strength	Marine castings
SELEKTRON	Balance	-	-	2-3	1-4 Cd 0-2 Ca	Strength	Rolled sections, sheet, rod
Precious metal alloys	Au	Ag	Pt	Other	Other		
AMERICAN GOLD	90	-	-	-	10 Cu	Corrosion resistance	US coinage
COIN SILVER	-	90	-	-	10 Cu	Corrosion resistance	US coinage
EASY-FLO	-	50	-	16.5 Zn	15.5 Cu 18 Cd	Melting point 630 °C	Silver soldering
FINE SILVER (see Sterling silver)	-	99.9	-	0.1 probably Cu		Standard of quality taken as "pure" silver	
JAPANESE SILVER	-	50	-	50 Al	-	Low cost	Jewellery
LEVAL'S ALLOY	-	71.5	-	-	28.5 Cu	Eutectic alloy, higher creep resistance than pure copper	Electrical instruments and standards, as solder
PALAU	80	-	-	20 Pd	-	Colour and corrosion resistance	Substitute for platinum
PALLADIUM - COPPER	-	Balance	-	70 Pd	25 Cu up to 1 Ni	Non-magnetic, good corrosion resistance	Precision clock and watch wheels
PLATINO	89	-	11	-	-	Colour and corrosion resistance	Jewellery and ornaments
PLATINUM ALLOYS	-	-	96	4 Ru	-	Colour, corrosion resistance, high melting point, good electrical resistivity characteristics	Jewellery and ornaments
	10	3.5	83.5	3 Pd	-		
	20	4	50	26 Pd	-		
	25	-	42	33 Pd	-		
	-	-	70	25 Pd	-	do.	Fuse wires
	-	-	95	5 Ir	-	do.	Jewellery
	-	-	77	-	23 Co	do. Ferromagnetic	Magnetic circuits
	-	-	90	10 Ir	-	Colour, corrosion resistance, high melting point, good electrical resistivity characteristics	Thermocouples
	-	-	80	20 Rh	-		Furnace windings
	60	-	40	-	-		White alloy for soldering
	-	37.5	50	-	12.5 Cu		Pen metal
	-	73	27	-	-		Soldering Pt
	-	30-50	Balance	-	-		Watch cases
	-	-	95	-	5 W		
READY FLO	-	56	-	22 Cu	17 Zn 5 Sn	Low melting point	Brazing of silver or copper alloys
SILANCA	-	92-94.5	-	4-4.5 Sb	1-3 Cd up to 2.5 Zn	Tougher than pure silver	Soldering silver products

TABLE 9.

TRADE NAME or other designation	COMPOSITION, % by weight					SPECIAL PROPERTIES	USES
Precious metal alloys	Au	Ag	Pt	Other	Other		
SILMANAL	-	87	-	8.5 Mn	4.5 Al	Ferromagnetic	Electromagnetic circuits
SILVER SOLDER ALLOYS	-	72	-	28 Cu	-	Melting point 780 °C	Soldering copper
	-	40	-	36 Cu	24 Zn	do. 720 °C	do.
	-	45	-	30 Cu	25 Zn	do. 680 °C	do.
Readyflo	-	56	-	22 Cu	17 Zn 5 Sn	do. 630 °C	do. and brass
Bureau of Standards	-	40	-	14 Cu	6 Zn 40 Sn		do.
	-	80	-	13.5 Cu	6.5 Zn	Hardness	Hard joints in copper alloys and silver articles
	-	60	-	20 Cu	15 Zn 5 Ni	High strength	do.
	-	50	-	33 Cu	17 Zn		Soldering brass
	-	45	-	17 Cu	18 Zn 20 Cd	Melting point 625 °C	do. Cast iron
	-	70	-	20 Cu	10 Zn	do. 725 °C	do. Steel
	-	85	-	-	15 Mn		High temperature use
	-	65	-	20 Cu	15 Zn		Soldering nickel
STANDARD GOLD UK	91.66	-	-	Balance Cu	-	Corrosion resistance	Legal gold coinage
USA	90	-	-	Balance Cu	-		
STANDARD SILVER (*see also* Coinage UK Silver)	-	92.5	-	7.5 Cu	-	Corrosion and wear resistance	Legal coinage standards
USA	-	90	-	10 Cu	-		
STERLING SILVER UK	-	92.5	-	7.5 Cu	-	do.	Legal standard for tableware and decorative metalwork, assayed and hallmarked
WHITE GOLD	90	-	10	-	-	Colour resembling platinum, corrosion resistance	Jewellery
	90	-	-	10 Pd	-		
High grade-	75	-	5	20 Pd	-		
	50	35	-	15 Pd	-		
Electrum	40-50	50-60	-	-	-		
18 carat	75	18.5	-	5.5 Zn	1 Cu		
14 carat	58.5	-	-	7.5 Zn	17 Ni 17 Cu		
12 carat	50	-	-	10 Zn	20 Ni 20 Cu		
10 carat	42	-	-	8 Zn	25 Ni 25 Cu		
9 carat	37.5	-	-	17.5 Zn	17.5 Ni 17.5 Cu		
High strength	80	-	-	5 Zn	15 Ni		
	78	-	-	5 Zn	15 Ni 2 Mn		
	70	-	-	-	15 Ni 15 Cr		
WHITE GOLD SOLDER	30-80	small	-	up to 15 Zn	up to 15 Ni small Cu	Low melting point 690-850 °C	Soldering jewellery

TABLE 9.

TRADE NAME or other designation	COMPOSITION, % by weight					SPECIAL PROPERTIES	USES
Rare earth alloys	Ce	Other	Other	Other	Other		
LAN-CER-AMP	45-50	30 La	Balance (Pr Y)	-	-	Powerful deoxidant and desulphuriser; combines readily with nitrogen at high temperatures	Ladle addition in steelmaking
PYROPHORIC ALLOYS Auer	35	24 La	35 Fe	4 Yt	2 Er	Rapid oxidation when finely divided	Lighter flints, gas lighter devices, pyrotechnics
Huber	85	15 Mg	-	-	-		
Mischmetall	40-50	20-40 La	-	Balance Yt + Er			
Welsbach	70	-	30 Fe	-	-		
Welsbach	60	-	30 Fe	Balance La + Yt + Er			
No. 1A	57	3 Mg	-	40 Cd	-		
No. 2	67	3 Mg	-	30 Sn	-		
No. 3	67	3 Mg	-	30 Zn	-		
No. 4	67	3 Mg	-	30 Pb	-		
	87	-	1.5 Fe	Balance Zn	-		
	Balance	-	15 Fe	-	-		
	60	-	40 Fe	-			
	50	-	-	Balance Mg	-		Explosive
Sensitive	75	-	-	Balance Pt			
WELSBACH'S ALLOY	70	-	30 Fe	-	-	Pyrophoric	Gas lighter flints
Titanium based alloys	Ti	Cr	Fe	Al	Other	High strength to weight ratio, with reasonable ductility, tensile strengths 830-1080 MN m^{-2}, elongation 10-15%, reduction of area 30-35%	Aircraft and structures requiring strength and low weight penalty
	Balance	5	-	3	-		
	do.	2	2	5	2 Mo		
	do.	2-7	1-4	-	-		
	do.	-	-	-	8 Mn		
	do.	-	-	4	4 Mn		
	do.	-	-	5	2.5 Sn		
	do.	4	2	-	-		
	do.	2	2	-	2 Mo		
	do.	-	-	-	7 Mn		
	do.	-	-	-	2.5 Cu		
	do.	-	-	8	8 Zr 1 Nb		
	do.	-	-	2.3	11 Sn 5 Zr 1 Mo		
	do.	-	-	6	4 V		
	do.	11	-	3	13 V		

TABLE 9.

TRADE NAME or other designation	COMPOSITION, % by weight					SPECIAL PROPERTIES	USES
Titanium based alloys	Ti	Cr	Fe	Al	Other		
TITANIUM MASTER ALLOYS Ti-Cu	90	-	-	-	10 Cu	Lower melting point than base metal, deoxidants, grain refiners	Additions to alloys of other metals to add Ti, to deoxidise and perform grain refinement
	80	-	-	-	20 Cu		
Ti-Al	50	-	-	50	-		
	25	-	-	75	-		
	15	-	-	85	-		
	5	-	-	95	-		
Ti-Cu-Al	10	-	-	65	25 Cu		
	15	-	-	60	25 Cu		
	20	-	-	30	50 Cu		
Ti-Mn-Al	20	-	-	30	50 Mn		
	25	-	-	25	50 Mn		
Ti-Ni-Al	20	-	-	40	40 Ni		
	25	-	-	35	40 Ni		
Tungsten based alloys	W	Co	Cu	Ni	Other		
TUNGSTEN-COBALT	75-95	25-5	-	-	-	Wear resistance, hardness	Dies for wire drawing
TUNGSTEN-COPPER Master alloys	10-15	-	Balance	-	-	Lower melting point than base metal	Additions to alloys to introduce W into bronze and brass
	40-50	-	do.	-	-		
TUNGSTEN-NICKEL	90	-	-	10	-	Wear and abrasion resistance	Contacts
	90-75	-	-	10-25	-	Heat resistance	Furnace components
Zinc based alloys	Zn	Sn	Pb	Cu	Other		
BIDERY METAL	88.5	-	5.6	5.6	-	Ductile	Domestic hardware (India)
BINDING METAL	93.5	2.8	-	-	3.7 Pb	Good casting properties	Binding wire rope ends for slings
BIRMINGHAM PLATINA (or PLATINUM)	75	-	-	25	-	Brittle, silvery white	Buttons, costume jewellery
DI-METAL	96	-	-	-	4 Al	Castability	Die castings
DOLER	93	-	-	3	4 Al	do.	do.
DURAK	Balance	0.001	0.003	0.75-1.25	0.03 Mn 0.03 Mg 3.9-4.3 Al 0.075 Fe	Castability	Die castings
ERHARD'S METAL	Balance	0.2	1	11	-	Castability	Bearings
ERAYDO	98	-	-	2	0.1 Ag	Ductility	Radio shields
FENTON BEARING METAL	80	14.5	-	5.5	-	Low friction, with strength	Bearings
	80	14.5	-	-	5.5 Sb		

TABLE 9.

TRADE NAME or other designation	COMPOSITION, % by weight					SPECIAL PROPERTIES	USES
Zinc-based alloys	Zn	Sn	Pb	Cu	Other		
FOUNTAINE MOREAU	92	-	1	6	1 Fe	Colour	Ornamental work
	93.5	3	-	-	3.5 Sb		Binding metal
GERMANIA BEARING METAL	Balance	10	5	4.5	0.8 Fe	do.	Bearings
GLIEVER BEARING METAL	73.3	7	5	4.2	9 Sb 1.5 Cd	Low friction	Bearings
GLYKO METAL	85.3	5	-	2.5	4.7 Sb 2 Al	Low melting point	Castings
HAMITON METAL	Balance	-	3	3	1.5 Sb	Easily cast	Castings
IRIDIUM BEARING METAL A	83	15.75	-	1.25	-	Low friction	Bearings
B	77	22.0	-	1.0	-		
KEMLER METAL	76	-	-	9	15 Al	Castability, toughness	Die castings
KIRSITE	Balance	-	-	4	3.5-5 Al 1 Mg	Toughness, low cost	Dies for blanking and forming metal parts
KRUPP BEARING ALLOYS	90	1.5	-	7	1.5 Sb	Low friction	Bearings
	88	2.0	-	8	2.0 Sb		
	77	-	15	-	8.0 Sb		
	46	-	45	-	9.0 Sb		
LEDDEL ALLOY	Balance	-	-	5-6.5	5-6.5 Al	Low friction	Bearings
LEDEBUR'S BEARING METAL	Balance	17.5	-	5.5	-	do.	do.
LUMEN	86	-	-	10	4 Al	Low friction	Bearings
MAZAK ALLOYS 3	Balance	0.001	0.003	0.03	3.9-4.3 Al 0.03-0.06 Mg 0.075 Fe max 0.003 Cd max 0.001 Sn max	Reasonable strength, cheap Castability	Die castings
5	do.	do.	do.	0.75-1.25	3.9-4.3 Al 0.03-0.06 Mg 0.075 Fe max 0.003 Cd max 0.001 Sn max	do.	
MOCK PLATINUM	55	-	-	45	-	Colour, corrosion resistance	Substitute for platinum in cheap jewellery
MOURAY'S SOLDER	91-80	-	-	3-8	6-12 Al	Low melting point	Soldering aluminium
PATTERN METAL	40	40	20	-	-	Machinability, good surface finish	Metal patterns for moulding
PHOSPHOR-ZINC	Balance	-	-	-	10 P	Deoxidises	Master alloy for addition of phosphorus to brasses and gun metals
PIEROTT'S METAL	Balance	7.6	3	2.3	3.8 Sb	Low friction	Bearings
PLATINA	75	-	-	25	-	White colour	Cheap cast buttons

TABLE 9.

TRADE NAME or other designation	COMPOSITION, % by weight					SPECIAL PROPERTIES	USES
Zinc-based alloys	Zn	Sn	Pb	Cu	Other		
PLATINE	57	-	-	43	-	White colour, brittleness	Cheap cast articles
RENYX AZN	92	-	-	3	4 Al <1 Mn	Castability	Die castings
REVERSED BRASS (Fontaine moreau alloy)	90-92	-	-	7-8	⊁3 Pb + Fe	Castability	Die castings
SALGE METAL	Balance	10	1	4	-	Low friction	Bearings
SAXONIA METAL	do.	5	3	6	0.2 Al	do.	do.
SCHROMBERG ALLOY	Balance	10	-	3	-	Low friction	Bearings
	do.	40	0.2	0.4	0.15 Fe		
SCHULZ ALLOY	91	-	-	6	3 Al	Low friction	Bearings
SIEMEN'S ALLOY	48	-	-	-	47 Cd 5 Sb	do.	do.
SODIUM ZINC	Balance	-	-	-	2 Na	Deoxidation	Treatment of molten zinc alloys to deoxidise
SOREL	80	-	-	10	10 Fe	Colour	Ornamental
SULZER'S ALLOY	Balance	10	1-2	4	-	Low friction	Bearings
TEXAS METAL	do.	4.5	1.2	2.5	-		Die castings
THURSTON'S ALLOY	80	14	-	6	-	Castability	Castings
UNBREAKABLE METAL	Balance	0.005 max	0.007 max	0.1 max	5.25-5.75 Al 0.005 Cd max 0.1 Fe max	Castability, cheapness	Slush cast lighting fixtures
VAUCHER'S ALLOY	75	18	4.5	-	2.5 Sb	Low friction	Bearings
WHITE BRASS	60	-	-	40	-	Castability	Ornamental castings
	55	-	-	45	-	do.	do.
	93	-	-	3	4 Al	Ductility	Jewellery
	80	10	-	10	-	do.	Soft alloy
	60	-	-	20	20 Ni	do.	Building, plaques
	72	-	-	28	-	do.	Plating alloy
WUEST'S SOLDER	75	-	-	Balance	20 Al	Low melting point	Soldering aluminium
ZAMAK (US name for Mazak alloys)	Balance	0.005 max	-	2.5-3.5	3.5-4.5 Al 0-0.1 Fe 0-0.5 Mg 0.005 Cd max	Castability, resistance to atmospheric corrosion	Die castings
	do.	-	-	Up to 1	10 Al	Ductility	Wrought products
	do.	-	-	0.4	0.8 Al 0.2 Mg	do.	Wire
ZELCO	83	-	-	Balance	15 Al	Low melting point	Aluminium solder
ZILLOY	Balance	-	-	1	0-0.025 Mn up to 0.8 Cd 0.1 Mg trace Pb	Ductility	Corrugated sheets for roofing
ZIMAL	Balance	-	-	2.5-3.3 Cu	up to 0.15 Mn 0.01 Cd <0.2 Pb	Strength	Gravity and pressure die casting
ZINC TINSEL	60	-	40	-	-	Ductility, malleability	Strip for decoration

TABLE 9.

TRADE NAME or other designation	COMPOSITION, % by weight					SPECIAL PROPERTIES	USES
Zinc-based alloys	Zn	Sn	Pb	Cu	Other		
ZISCON	67-75	-	-	-	25-33 Al		Castings
Zirconium-based alloys	Zr	Sn	Fe	Cr	Other		
ZIRCALLOY 2	Balance	1.5	0.12	0.1	0.05 Ni	Corrosion resistance, low neutron absorption	Fuel element cladding in water cooled nuclear reactors
3	do.	0.2-0.3	0.2-0.3	-	-	Creep strength to 400 °C	Nuclear reactor applications
4	do.	1.5	0.2	0.1	-	Corrosion resistance	Structural components in pressurised water reactors
ZIRCONIUM AJR	Balance	-	-	-	1 Cu 1.5 Mo	do. to CO_2 gas	Nuclear reactors with gas cooling by CO_2
ZIRCONIUM GR11	Balance	-	0.1	0.1	0.01 N	Corrosion resistance	Chemical plant

TABLE 10. Polymeric Materials

GENERIC TYPE: ACRYLIC POLYMERS

REPRESENTATIVE UNIT: PMMA (Polymethylmethacrylate) PMA (Polymethylacrylate). (Polymers of propenoic acid methyl ester and propenoic acid 2 methyl methyl ester).

Trade names: Acco, Acrylite, Asterite, Bavik, Crystalex, Diakon, Lucite, Oroglas, Perlac, Perspex, Plexiglas, Plexigum, Polypenco.

Process: Monomer cast into sheets and blocks with some polymer added and initiator (benzoyl peroxide). Heating slowly causes polymerisation. Solution polymerisation produces surface coating lacquers. Emulsion polymerisation for paint bases. Suspension polymerisation (or granular) for powders.

Properties: Transparent. Optical clarity 92% transmission. Easily formed, stable to outdoor exposure and dimensionally. Molecular weight 5×10^5 to 10^6 typical. Amorphous structure, little crystallinity. Reasonable electrical insulator. Slow burning rate. Soluble in ketones, esters, chlorinated hydrocarbons. Resistant to alcohols at low temperature, relatively unaffected by aliphatic hydrocarbons and alkalis. Attacked by oxidising mineral acids and H_2SO_4.

Applications: Control knobs, dials, handles. Brush backs, mirror backs. Hospital equipment. Display material, signs, light fittings, skylights, aircraft canopies, telephone fittings, sanitary ware. Lenses. TV implosion guards.

REPRESENTATIVE UNIT: Polyacrylic acids and their salts.

Trade names: Methacrol, Plexileim, Syncol, Texigel.

Process: Aqueous dispersions or solutions for coatings.

Properties: As for acrylic plastics, but capacity to form thin films.

Applications: Thickening paint, spreading on textiles, soil aggregation, drilling mud.

REPRESENTATIVE UNIT: Polyacrylate rubber.

Trade names: Cyanacryl, Hycar, Krynac, Lactoprene, Poly FBA, Paracril, Thiacril.

Process: Cross-linked with amines or peroxides with fluorine containing monomer yields vulcanisable elastomer. Copolymer of acrylic ester with monomer containing −Cl, −CN, or −COOH.

Properties: Resists cracking in flexure, stable outdoors and to sulphur bearing hydrocarbons and lubricants. Low permeability to gases. Service range −40 °C to +200 °C. Low resilience. Low brittleness temperature unless plasticised by esters. Poor resistance to steam and water, alcohols and alkalis. Reasonable electrical insulator.

Applications: Gaskets, seals. Coatings for glass and asbestos textiles. Printing rolls, air bags, wire and cable insulation. As latex, on non-woven textiles, pigment binder in textile and paper printing. Adhesives.

TABLE 10.

REPRESENTATIVE UNIT: Cyanoacrylate adhesives.

Trade names: Eastman 910 Superglue.

Process: Plasticised, thickened and stabilised to inhibit polymerisation.

Properties: Polar liquids, polymerise and set rapidly on pressing into thin film between surfaces having mildly alkaline base or adsorbed water films. Polymerisation retarded by acids. Resists many solvents, but slowly attacked by dilute acids, alkalis, hot water, steam. Range of adhesion −15 °C to +80 °C. READILY BONDS HUMAN AND ANIMAL SKIN!!

Applications: Joining metals, ceramics, rubber, and plastics which have basic reactions or adsorb water. VERY RAPID SET.

REPRESENTATIVE UNIT: Anaerobic adhesives.

Trade names: Loctite.

Process: Free radical catalyst and accelerator added to acrylic esters of aliphatic diols.

Properties: Remain fully fluid, even in air, until confined to close fitting surfaces (ANAEROBIC). Polymerise at room temperature when forced into crevices. Cross-linking produces very strong bonds.

Applications: Locking nuts on to bolts and studs in place of locknuts. General adhesion of close fitting shafts, tubes, pulleys, gears.

REPRESENTATIVE UNIT: PAN (Polyacrylonitrile or Polyvinylcyanide). (Polypropenoic acid, nitrile).

Trade names: Acrilan, Courtelle, Orlon, Acribel, Creslan, Crylor, Dolan, Dralon, Leacril, Nitron, Nymcrylon, Redon, Tacryl, Wolcrylon, Zefran.

Process: Polymerisation of monomer in aqueous solution, suspension or emulsion, or in organic solvents, initiated by peroxides or azo compounds. Commonly copolymer to improve dye fastness. Fibre spinning dry or wet. Hot stretching or annealing follows spinning. Two component fibres produce crimping fibres when heated. Moulding of PAN requires high pressure and 200 °C-300 °C. Plasticised by castor oil.

Properties: Resistant to sunlight and ageing. Discolour when heated (>100 °C). Unaffected by living organisms. Soft resilient fibres, good crease resistance and retention (pleats). Molecular weight 5×10^4 to 10^5, amorphous but partially crystalline when stretched and oriented as fibres. Abrasion resistance better than cellulose but inferior to nylon. Burning rate slow, decomposes above 250 °C. Fibres ironed up to 150 °C but discolour with time. Unaffected by dry cleaning fluids, oils, greases, dilute alkalis and acids. Decomposed by concentrated alkali, nitric and sulphuric acids. Fibres wetted by water but only on surface, hence quick drying. Difficult to dye through fibre, dyeing affects crystallisation in spinning copolymer.

Applications: Substitute or diluent for wool, yielding warm light fabrics. Ladies garments, protective clothing, sails for yachts. Hydrolysed, forms soil conditioner (Krilium). Used as internal plasticiser for vinyl polymers, oil resistant rubber, and as copolymer with styrene and butadiene in ABS.

REPRESENTATIVE UNIT: PAN copolymerised with PVC.

Trade names: Dynel, Vinyon N.

Process: 40% acrylonitrile copolymerised with vinyl chloride (MODACRYLIC).

REPRESENTATIVE UNIT: PAN copolymerised with PVDC.

Trade names: Saniv, Teklan, Venel.

Process: Copolymerisation with vinylidene chloride.

Process: Both groups can be polymerised in aqueous emulsion or organic solvent then spun. Stretching after spinning.

Properties: Resistant to sunlight and water, cheaper than straight PAN. Flame proof. Swollen or dissolved by organic solvents more than straight PAN. Soften at lower temperatures.

Applications: Children's and invalid's nightwear. Curtains and fabrics exposed to more than average fire risk.

REPRESENTATIVE UNIT: Carbon fibres derived from PAN.

Trade names: Celion, Carboform, Fortafil, Grafil, Hyfil, Kureha, Magnamite, Panex, RK 30, RK 35, Thornel, Torayca.

Process: Acrylonitrile fibres decomposed in air, vacuum, or inert gas by heating, in presence of organic acids or bases. Five minutes at 250 °C causes blackening with some oxidation. 1000 °C in inert atmosphere and final graphitising treatment at 1500-3000 °C leads to crystalline end product.

Properties: Exceptionally high elastic modulus 340 GN m^{-2} and tensile strength 2 GN m^{-2}.

Applications: Reinforcement in composite materials bonded with polyesters or epoxy resins. High stiffness/weight ratio.

TABLE 10.

GENERIC TYPE: ALKENE POLYMERS

REPRESENTATIVE UNIT: Polyethylene (polyethene).

Trade names: Agilene, Alathon, Alkathene, Ampacet, Brea, Carlona, Courlene, Dur-X, Dylan, Epolene, Ethafoam, Gerpak, Hi-fax, Hilex, Hostalen, Lypolen, Marlex, Orizon, Petrothene, Poly-Eth, Polythene, Riblene, Rigidex, Rotene, SeilonETH, Staflen, Super Dylan, Supreme, Vestolen.

Process: Ethylene obtained by cracking natural gas or petroleum light fractions. Low density polymer by free radical catalysis, at 200-350 GN m^{-2} pressure and 200 °C, exothermic reaction. Copolymers with acrylates or vinyl acetate by same method. High density polymer by low pressure processes, *(a)* Ziegler catalyst in inert hydrocarbon solvent at 50-75 °C, atmospheric pressure, *(b)* Phillips catalyst at 2.75-3.5 GN m^{-2} and 100-175 °C, suspended in cyclohexane.

Properties: Tough, flexible, chemically inert, good electrical insulator. Low density branched chains (2-3% branches) high density more linear. Molecular weight up to 50 000 low density, up to 200 000 high (Ziegler). Degree of crystallinity about 60% low density, 95% high. Cross-linked by irradiation or chemically, more amorphous structure. Soluble in hydrocarbon and chlorinated hydrocarbon solvents only at high temperatures (above 50-60 °C) relatively unaffected by polar solvents. Decomposed by strong oxidising agents. Subject to environmental stress cracking in polar liquids or vapours (greatest with high density). Mechanical properties dependent on density and crystallinity.

Applications: Packaging, domestic hardware. Cable dielectric. Flexible bottles, paper coatings, toys, tubing. In solution or emulsions for polishes and lubricants.

REPRESENTATIVE UNIT: Chlorosulphonated polyethylene.

Trade name: Hypalon.

Process: Gaseous chlorine and sulphur dioxide are bubbled through a solution of polyethylene in a hot hydrocarbon solvent, aided by ultra-violet radiation or initiators. The product can then be cross-linked to produce rubber.

Properties: Outstanding resistance to ozone, weathering, heat ageing and chemical attack. Decomposed by hot oxidising acids and in aromatic or chlorinated solvents, and esters. Good mechanical properties without fillers.

Applications: Additions for flooring, chemical plant liners, footwear, fabric proofing, cables, tyres. Paints for marine use, and lacquers for rubber articles.

REPRESENTATIVE UNIT: Polypropylene (polypropene).

Trade names: Agile, Alpha, Avisun, Boltaron, Cadco, Carlona P, Catalin, El Rex, Escon, Gering, Grex, Maplen, Marlex, Moplefan, Napryl, Novolen, Oleform, Petrothene, Pro-fax, Propathene, Propylex, Seilon PRO, Tenite, Thermoseal, Vitaline. Fibres: Herculon, Meraklon, Ulstron.

Process: Monomer obtained by cracking natural gas or C$_3$ fractions of petroleum distillation. Liquid monomer at low pressure fed into naphtha or similar inert solvent with stereospecific catalyst (Ziegler-Natta) at 30-100 °C. Atactic polymer (soluble) centrifuged off, leaving isotactic polymer.

Properties: Translucent, high tensile strength, higher softening temperature than polyethylene, free from environmental stress cracking, but brittle at only moderate low temperatures. More readily oxidised than polyethylene, otherwise chemically similar. Injection moulding including narrow web allows integral hinge, flexing introducing molecular orientation normal to the hinge axis. Biaxially orientated films of high strength and clarity obtained by bubble blowing extruded tubing. Fibres can be melt spun, linear oriented films can be fibrillated. Molecular weight over 80 000. Crystallinity 65-70%.

Applications: Automobile parts, sanitary ware, domestic appliances, soil pipe systems, footwear, hospital equipment. Ducts, expansion tanks, large tanks for chemical plant. Water pipes, packaging, cordage, netting, blankets, carpets and brushes.

REPRESENTATIVE UNIT: Ethylene-Propylene rubber (Ethene-Propene rubber)(EPM/EPR/EPDM/EPT).

Trade names: C23, Dutral, Intalon, Keltan, Nordel, Royalene, Vistalon.

Process: Copolymerisation with stereospecific catalysts with or without addition of a diene. Preferential polymerisation of ethylene must be avoided. Without diene addition, a saturated copolymer results which can be cross-linked by peroxide. With dicyclopentadiene, (2%) a terpolymer results, the unsaturated bonds then allow cross-linking with sulphur.

Properties: Virtually without odour, when vulcanised resist heat, polar liquids, ozone. Good flexibility at low temperature, good abrasion resistance. Molecular weight 100 to 200 000. Excellent electrical properties. Reinforcement by fillers necessary for tensile strength.

Applications: Tyres. Insulation and sheathing for electric cables. Footwear, fabric proofing, sealing. Medical prostheses.

GENERIC TYPE: CELLULOSIC POLYMERS (POLYCELLOBIOSE)

REPRESENTATIVE UNIT: Polyanhydroglucose. Poly (1,4 anhydro β -D-glucopyranose).

Natural forms: Pure cotton, Flax, Ramie, Wood pulp, Kapok, Coir, Hemp, Jute, Sisal.

TABLE 10.

Regenerated forms:	Viscose	- Fibro, Triple A, Evlan, Sarille, Tenasco, Corval, Topel, Rayolanda.
(Rayon fibre)	Cuprammonium	- Bemberg, Cuprama, Cupresa.
	Polynosic	- Avril, Polyflox, Toramomen, Vincel, Zantrel.
	High strength	- Avron, Durafil, Tenasco Super, Tyrex, Fortisan.
	Films	- Cellophane, Cuprophane.

Process: Cellulose is the chief structural material of plants. The purest forms are already fibrous, as in cotton, ramie, flax and hemp, but of short staple. Regeneration of cellulose allows spinning of continuous fibre, but reduces molecular weight by so doing. In the cuprammonium process, pure cellulose (usually cotton) is dissolved in cuprammonium hydroxide, then regenerated by spinning into warm water followed by an acid bath to remove hydroxyl ions. In the viscose process, cellulose is dissolved in sodium hydroxide, treated with carbon disulphide to form cellulose xanthate then spun into an acid bath. High tenacity rayon is obtained by stretching in an acid bath, and polynosic by varying the xanthate bath and stretching in a low concentration acid bath.

Properties: Relatively inert to most organic liquids. Swollen by water and concentrated alkali. Decomposes on heating in air without melting. Affected by living organisms especially when damp. Molecular weight changes on processing from 300 000 to 10^6 in natural fibres, to 50-150 000 in regenerated cellulose. The material is highly crystalline, with fibrils some 5-10 nm wide aggregated into bundles about 25 nm wide, forming crystallites 30-60 nm long.

Applications: Textiles, paper, fillers for reinforcing thermosetting plastics and rubber. Clear films for packaging. Cords and thread.

REPRESENTATIVE UNIT: Cellulose Acetate.

Trade names: Acetylcellulose, Bexoid, Celastoid, Cellastine, Cellomold, Celluloid (safety), Clarifoil, Dexel, Eanplast, Fibestos, Kodapak, Nixon C/A, Plastacele, Rhodoid, Sicalit, Tenite, Vuepak. Fibres — Arnel, Tricel (triacetate), Celafibre, Fibroceta, Dicel, Lansil, Cotopa, Alon.

Process: Cellulose fibres pretreated with acetic acid, then esterified with acetic anhydride in acetic acid and catalyst (H_2SO_4 or $HClO_4$). Exothermic reaction requires cooling to prevent degradation. Hydrolysed to form triacetate or secondary acetate. Fibres can be esterified in acetic anhydride vapour (Alon, Cotopa). Copolymers of cellulose acetate with propionate, butyrate, are processed similarly. Tougher and more weather resistant than acetate.

Properties: White fibrous material obtainable as film or fibre. Soluble in ketones, alcohols, acetic acid, some esters, to degree dependent on acetyl content. Triacetate has restricted range of solvents. Decomposed by acids and alkalis in moderate concentration. Soften at moderate temperatures with secondary acetate, much higher for triacetate (220 °C). Triacetate can be heat set for pleating, dimensional stability and resistance to water absorption. Reasonably resistant to sunlight, ageing and weathering when stabilised. Unaffected by living organisms.

Applications: Telephones, lampshades, spectacle frames, cartons, buttons, machinery guards. Packing, photographic film base, glazing, colour filters. Solutions used as lacquers for yarn and fabric, leather. Fibres have good handling quality but relatively low strength.

REPRESENTATIVE UNIT: Cellulose nitrate.

Trade names: Amerith, Celluloid, Collodion (solution), Guncotton, Herculoid, Nitrocellulose, Pyroxylin, Xylonite.

Process: Cellulose fibre treated in concentrated nitric and sulphuric acid mixture, washed to remove acid. Degree of substitution depends on anhydrous conditions and low temperature to prevent degradation.

Properties: Soluble in alcohol and ether mixtures, ketones, esters. Highly flammable, but phosphate plasticisers reduce danger. Easily decomposed by heat, affected by sunlight. Hard and tough in thin sheets, but flexible.

Applications: Explosives use only highly nitrated forms as guncotton and cordite. Plastics used for knife handles, piano keys. Surface coatings for leather, bookbinding cloth, adhesives and lacquer.

GENERIC TYPE: EPOXY RESINS

REPRESENTATIVE UNIT: Ethoxyline, epoxide.

Trade names: Araldite, Bondstrand, Cadco, Chemtite, Corvel, Devran, Durcon, Epikote, Epiphen, Epi-Rez, Epi-Tex, Epolite, Epon, Epoxical, Epoxylite, Fibrecast, Gaco, Gen-Epoxy, Hystran, Maraset, Nuklad, Pfaudlon 201, Rezklad, RX, Synthane, Unox.

Process: Epoxide type compound, commonly epichlorohydrin is reacted with diphenylolpropane (Bisphenol) or a similar compound to form a syrup or brittle resin of relatively low molecular weight. Curing is effected by polyamines at room temperature, or carboxylic acid anhydrides at higher temperatures. Modifications to the cure can be carried out by various additions.

Properties: The cured resins are insoluble, infusible, tough and inert solids. Relatively unaffected by water, hydrocarbons, alcohols, and medium concentrations of inorganic acids. Decomposed by strong acids. Usable up to 150 °C, good electrical insulation.

Applications: Encapsulation of electronic components, impregnation of windings. Castings for electrical equipment, jig tools and patterns. Adhesives for metal bonding (aircraft and automobile) glass, ceramics and polar materials. Bonding of reinforcing laminates, surface coatings for metals in corrosive environments.

TABLE 10.

GENERIC TYPE: FLUORINATED HYDROCARBON

REPRESENTATIVE UNIT: Polytetrafluoroethylene (polytetrafluorethene)/polytrifluorochlorethylene/fluorinated ethylene propylene.

Trade names: Cadco TF/Chempro, Ebolene, Fluon, Fluoray, Genetran, Halon, Hostaflan TF, Teflon, Temp-R-Tape, Tetran.

Process (PTFE): Chloroform is reacted with hydrogen fluoride to yield an intermediate monochlorodifluoromethane, then cracked thermally at $600°$-$800°C$ in platinum tubes to yield CF_2 monomer. Polymerisation takes place slowly with time, but is accelerated by a radical initiator such as aqueous persulphate.

Properties: Very high molecular weight, 500 000 to 5×10^6. Highly crystalline, amorphous over $327°C$. Excellent electrical properties and chemically inert. High thermal resistance, range of use $-100°C$ to $+350°C$. Very low friction and non-adhesive properties. Difficult to fabricate due to high M.W. Unaffected by sunlight, micro-organisms. Suffers from creep under stress.

Applications: Unique properties extend uses. High temperature electrical insulator, dielectric in electronic components, film base for printed circuits. Chemical plant gaskets, diaphragms, seals, tubing. Expansion bearings and low sliding friction applications. Non-stick surfaces in cookery.

GENERIC TYPE: PHENOL AND AMINO-FORMALDEHYDE RESINS (HYDROXYBENZENE AND AMINO-METHANAL)

REPRESENTATIVE UNIT: Phenol-formaldehyde.

Trade names: Alberit, Alresin, Avis, Azolone, Bakelite, Cegeite, Chierolo, Dendrodene, Durez, Durophen, Elo, EpokR, Fabrolite, Featalak, Fiberite, Fluosite, Fudow, G.E., Gederite, Kerit, Lerite, Metholon, Moldesite, Mouldrite, Nestorite, Phenall, Progilite, Resart, Resinol, Rockite, Rogers, Sarvis, Setalict, Sirfen, Sternite, Synmold, Tessilite, Tessilplast, Trolitan. Modified resins: Arochem, Beckacite, Caladene. Cast resins: Catalin. Laminates: Aroborite, Celisol, Celotex, Paxolin, Tufnol. Ion exhange resin: Amberlite. Surface coating resin: Acrophen, Amberlac, Amberol, Becophen, Compregnite.

Process: Direct addition of phenol to formaldehyde with excess formaldehyde catalysed by strong bases leads to compounds of relatively low molecular weight (A stage or resol). Further heating causes condensation and lightly cross-linked structures (B stage or resitol). Continued heating continues condensation to rigid non softening molecular networks (C stage or resite). Fillers and pigments are added at the end of stage A.

An excess of phenol and an acid catalyst leads to direct condensation and loss of water, a compound of low molecular weight mainly linear in structure. When water is removed and M.W. increased, the product is called a novolac, and this can be blended with fillers, etc., and cross-linked in a second stage, to form a strongly cross-linked network.

Properties: Novolacs are brittle, fusible and soluble in most organic liquids. Cast resins are infusible and insoluble after thermal hardening, but machinable, strong and tough. Moulded resins are similar. Laminates with paper or cloth fabric are tough but machinable. None are affected by most aqueous liquids, but all decompose in hot concentrated acids and alkalis. Very good electrical insulators and abrasion resistant.

Applications: Novolacs used for lacquer and varnish, coating paper, metal cans and electrical equipment. Cast resins for handles, knobs and ornamental objects. Moulded resins for electrical fittings and containers, similar to cast. Laminates for decorative surfaces, bearings, gear wheels for quietness, electrical panels, wear resistant components. Resins also used for bonding wood (plywood), brake linings, grinding wheels, sand cores in foundries, travel goods, ion-exchange.

REPRESENTATIVE UNIT: Urea-formaldehyde/melamine-formaldehyde.

Trade names: Aerolite, Avisco, Becamine, Beetle, Cascamite, Cibanoid, Cymel, EpokU, Formica, Gabrite, Iporca, Kaurit, Melantine, Melmac, Melmex, Melolam, Melopas, Micarta, Mouldrite, Nestorite, Paralac, Permelite, Plaskon, Resimene, Scarab, Setamine, Sylplast, Uformite.

Process: Urea dissolves in aqueous formaldehyde to form a series of compounds which finally cross-link, either with mildly acid conditions or with heat. Fillers and pigments can be added at intermediate stages and stabilisation before cure allows storage. Melamine similarly reacts to form condensed cross-linked products in the presence of acid or by heating. These compounds are more complex than those of urea-formaldehyde.

Properties: Variety of products possible with process conditions. Viscous syrups can be hardened by catalysts to form adhesives, particularly for wood. Pre-condensates can be impregnated then cured in situ to coat paper and textiles. Solvent based material can be evaporated and thermosetting resins used for mouldings and laminates. Both U-F and M-F are light fast, U-F attacked by alkalis and boiling water. M-F resistant to heat, water, acids, bacteria, fungi and insects. Good mechanical strength, electrical insulation. Maximum service temperature U-F $75°C$ M-F $100°C$.

Applications: Crockery, electrical fittings. Laminated board surfaces, particularly M-F. Adhesives for wood, fabric bonding. Flame retardant finishes on textiles, crease resistance and wet strength. Stoving enamels. U-F foam for thermal insulation in cavity walls.

GENERIC TYPE: POLYACETAL (POLYFORMALDEHYDE, POLYMETHANAL)

REPRESENTATIVE UNIT: Acetal.

Trade names: Delrin, Kematal. Copolymers: Alkon, Celecon, Hostaform.

Processes: In aqueous solution, formaldehyde consists largely of methyleneglycol $CH_2(OH)_2$. When cooled, polymerisation occurs spontaneously. Commercially, gaseous formaldehyde is led into an inert hydrocarbon solvent containing an ionic catalyst. Trioxan $(CH_2O)_3$ can also be polymerised in the melt or in solution with a catalyst. Stabilisation by acetylation or methylation of the hydroxyl end groups is necessary. Radical inhibitors such as amines or formaldehyde absorbing compounds are also added.

Properties: Hard, tough, resilient thermoplastic with good dimensional stability and low moisture absorption. Resembles polyethylene but higher melting point and elasticity. Dissolves only in hot solvents, particularly bases, otherwise relatively unaffected by organic liquids. Decomposed by mineral acids. Good electrical insulator, relatively high modulus of elasticity. Good abrasion resistance, but less than nylon 6.6.

TABLE 10.

Applications: Snap-fit applications, gear wheels, valve seats, impellers, often replacing metal die-castings.

GENERIC TYPE: POLYAMIDE (NYLON)

REPRESENTATIVE UNIT: Polycaprolactum or polycaproamide.

Trade names: Nylon 6, Akulon, Caprolan, Celon, Durethan BK, Enkalon, Grilon, Perlon L.

REPRESENTATIVE UNIT: Polyundecanoamide.

Trade names: Nylon 11, Rilsan B.

REPRESENTATIVE UNIT: Polyhexamethylene adipamide.

Trade names: Nylon 6.6, Bri-Nylon, Chemstrand-Nylon, Maranyl, Zytel.

REPRESENTATIVE UNIT: Polyhexamethylene sebacamide.

Trade names: Nylon 6.10, Brulon, Perlon N.

REPRESENTATIVE UNIT: Polymerised fatty acid polyamides.

Trade names: Versalon, Versamid.

Process: Self condensation of amino acids or polymerisation of their lactams or condensation between diamines and dicarboxylic acids. The first type are known by a single number, e.g. Nylon 6, and the second by two, e.g. Nylon 6.6, the numbers representing the number of carbon atoms in the monomer molecule(s) involved. There are several possible processing methods with each type.

Properties: Plastics material tough and horn like. Fibres strong and abrasion resistant, transparent. Organic alcohols and bases attack them and concentrated mineral acids. Relatively inert to hydrocarbons, esters, ethers, oils and concentrated alkalis. Molecular weight 20 000 with fibres, up to 30 000 with moulding and 100 000 by extrusion. Stretch spun fibres highly crystalline with folded chains. Unstretched, either amorphous or random polycrystalline. Oxidise by discolouration above 150 °C rapidly above 200 °C. Absorb moisture, prolonged immersion leads to 10% with Nylon 6 and 6.6, 2-3% Nylon 6.10 or 11. Good mechanical properties, but strength falls with temperature rise or humidity increase. Good electrical properties, but affected by moisture content.

Applications: Mechanical applications, gears, oil resistant bearings, Nylon 11 and 6.10 often filled with chopped glass fibre or graphite. Fibres used for marine hawsers, climbing ropes, parachute and tyre cords. Nylon 6 and 6.6 in tarpaulins, carpets, apparel fabrics. Nylon stockings can be heat set to shape and have good abrasion resistance. Nylon 11 and 6.10, lowest moisture absorption, used as monofilaments in surgical sutures, fishing lines, bristles. Soluble copolymers used in adhesives and varnish. Surface coatings for paper and printing bases, inks, heat sealants. Reduce shrinkage of woollen fabrics, and modify thixotropic paint.

GENERIC TYPE: POLYESTERS

REPRESENTATIVE UNIT: Alkyd resins.

Trade names: Glyptal, Alkydal, Beckosol, Beetle (BA), Cretalkyd, Durecol, Epok, Mitchalac, Paralac, Plastokyd, Plusol, Scoplas, Scopolux, Soalkyd, Synolac, Synresate, Vilkyd, Wresinol.

Process: Alkyd is derived from ALCohol and acID, and the resins are linear polyesters modified by branching or cross-linking. Condensation between a polyhydric alcohol (e.g. glycerol) and a poly basic acid (e.g. phthalic acid or anhydride) is carried out at 200-280 °C, with an unsaturated fatty acid or glyceride of the acid, known as a drying oil. The latter performs the branching and cross-linking. Alternatively, the drying oil and polyol are reacted at 250 °C with a catalyst (litharge) and then esterified with a dibasic acid. Water has to be continually removed, either directly or with a refluxing liquid. Various combinations of base and acid can be employed to modify properties. The drying oil increases solubility and flexibility of the polyester, and the latter increases the rate of drying of the oil and improves gloss.

Properties: Varying colour and hardness, initially readily soluble in hydrocarbons, ketones, esters and alcohols. When spread as thin films, and solvent evaporated, the resin becomes insoluble, due to radical polymerisation of the unsaturated portion by oxidation or rise in temperature. Thin films are flexible, adhere well, and are not unduly affected by moisture or ultra violet radiation.

Applications: Paints and coatings. Water dispersed alkyds form emulsion paints. Additions are made to create special properties, e.g. chlorinated rubber for fire resistance, polyamides for thixotropy. Also used in printing ink, foundry core binders, floor coverings, adhesives.

REPRESENTATIVE UNIT: Polyethylene terephthalate.

Trade names: Terylene, Dacron, Diolen, Lavsan, Tergal, Terlenka, Teron. Fibres: Trevira. Films: Melinex, Hostaphan, Mylar, Terphane, Videne. Modified: A-Tell, Grilene, Kodel, Vestan, Vycron. Moulding: Arnite.

TABLE 10.

Process: Terephthalic acid and methanol are esterified to produce dimethyl terephthalate. This is reacted with ethylene glycol at 150 °C and the product condensed under vacuum at 275 °C, with numerous additives claimed to accelerate. The methanol is removed and recycled. Various other additions are included in the final stage to improve dyeing characteristics.

Properties: Strong fibres and tough biaxially oriented films. High softening temperature, can be heat set to shape, little affected by moisture and light. Relatively unaffected by organic liquids except phenols and related compounds. Decomposed by hot alkali. Molecular weight up to 30 000, amorphous when rapidly cooled from melt, crystalline on heating above 80 °C. High modulus of elasticity and tensile strength. Very good electrical insulator in stretched films, good dielectric.

Applications: Quick drying light weight fabrics, stretch type and pleat retaining. Filter cloth, sieves, paper making felts. Belting for conveyors, Vee belts, fish nets, tarpaulins, sailcloth. Films for packaging, including boiling bags, electrical capacitors, diaphragms, magnetic tape, movie film base, typewriter ribbon, hose linings. Metallised for mirrors, Lurex thread, electrical purposes.

REPRESENTATIVE UNIT: Polycarbonate.

Trade names: Lexan, Lexel (fibre), Makrolon, Merlon, Panlite.

Process: Diphenylolalkanes and an organic carbonate undergo ester interchange under reduced pressure at 200 °C. The vacuum is increased and temperature raised to 300 °C to remove phenol. Alternatively, the phenyl compound is dissolved in aqueous alkali and treated at room temperature with phosgene, with organic solvent for the polymer and a catalyst.

Properties: Soluble in chlorinated hydrocarbon, swollen by ketones, benzene, but relatively unaffected by alcohols, dilute acids or oxidising agents. Decomposed by hot alkali. Good dimensional stability, high mechanical strength and good electrical properties. Particularly good impact strength.

Applications: Gears and bearings. Capacitor dielectrics up to 130 °C. Photographic film base. Automobile trim.

REPRESENTATIVE UNIT: Unsaturated polyesters. Moulding and laminating resins.

Trade names: Artrite, Cellobond, Crystic, Filabond, Hetron, Laminac, Marco, Orkast, Palatal, Paraplex, Polylite, Synres, Vibrin.

Process: Dihydric alcohol and dibasic acid is esterified with elimination of water. (The unsaturated bonds are commonly in the acid.) The ester is then blended with a reactive monomer, commonly styrene, with an inhibitor such as hydroquinone. Water is removed by reflux solvent in esterification, and some time is required to produce a reasonable yield. Common starting materials are a glycol and maleic or fumaric acids, with some saturated acid for flexibility (phthalic). The resin is supplied as a syrup.

Properties: Uncured resin is usually in solution in styrene. As such, soluble in ketones, and many esters. Cured resins are insoluble. Relatively unaffected by aliphatic hydrocarbons, organic acids, salt solutions, alcohol, but not so resistant as epoxy resins. Mechanical and physical properties very dependent on fabrication, particularly in fibre reinforced polyester. Electrical properties are good, both as encapsulating resin or filled. Low temperature properties are good, heat distortion varies from 80-130 °C for unfilled resin, to 100-250 °C for laminates. Cross-linking increases with initial rise in temperature.

Applications: Free radical initiator such as benzoyl or methyl ethyl ketone peroxide causes polymerisation and cross-linking to a rigid network. Accelerators such as cobalt naphthenate and dimethylaniline speed up the reaction at room temperature, but as the reaction is exothermic, temperature soon rises. Laminates usually employ glass fibre mat, cloth, or filaments wound on to a former, but asbestos and sisal are also used. Mineral fillers such as metal powder, slate, chalk, are employed for casting resins. Pressures up to 3.5 MN m^{-2} and temperatures up to 130 °C may be used to improve density and uniformity. Uses are so widespread, cost of mould making and hand lay up is determining factor. Marine uses from dinghies to large superstructures on naval vessels. Chemical plant uses from tubular reactors (filament wound) to ducts. Mechanical uses in automotive bodywork, mechanical handling plant, building industry, aircraft, missiles. Electrical laminates and equipment for encapsulation.

===

GENERIC TYPE: POLYSTYRENE AND COPOLYMERS. (POLYETHENYL BENZENE)

REPRESENTATIVE UNIT: Polyvinylbenzene or polyphenylethylene.

Trade names: Afcolene, Ampacet, Amphenol, Bextrene, Carinex, Distrene, Dylene, El Rex, Erinoid, Fostalite, Fastarene, Grace, Lacqrene, Lorkalene, Luran, Lustrex, Nypene, Piccolastic, Pliolite, Restirolo, Starex, Styron. Expanded: Dylite, Pelaspan, Santofome, Styrofoam.

Process: Ethylene (from natural gas or petroleum) is reacted with benzene (from coal or petroleum) in a Friedel-Crafts reaction to yield ethylbenzene, which is then dehydrogenated, to styrene. Styrene may also be synthesised directly from acetylene or acetylene and benzene. Styrene polymerises on heating or exposure to u.v. light. Oxygen inhibits this, but during storage an inhibitor such as sulphur or quinol is used. Mass polymerisation leads to variable molecular weight, solution or suspension methods give better control but affect the clarity of the product. Emulsion polymerisation is used with copolymers. Low temperatures give tough polymers of high M.W. but are difficult to process. High temperatures cause low M.W. and brittleness.

Properties: Glass-like material, easily fabricated, and relatively cheap, but brittle unless modified. Excellent moulding properties, giving detailed replication of complex shapes. Soluble in aromatic and chlorinated hydrocarbons, ketones, and some esters. Relatively unaffected by water, alcohols, alkalis and non oxidising acids. Burns fairly readily with sooty flame. Good electrical insulator. Rigidity of the network induces residual stress during fabrication, which affects the properties. Molecular weight up to 10^6. Softens above 80 °C (second order transition) very soft above 100 °C. Affected by sunlight, reducing mechanical properties.

Applications: Packaging, particular disposable cups and containers, easily formed to shapes, e.g. eggs. Toys, electrical equipment (particularly high frequency). Domestic appliance casings and liners (refrigerators, radio receivers), building wall tiles and panels. The expanded variety is used in thermal insulation and refrigerated vehicles, acoustic tiles and shock absorbing packaging.

TABLE 10.

GENERIC TYPE: HIGH IMPACT BLENDS AND COPOLYMERS

REPRESENTATIVE UNIT: Styrene/acrylonitrile (SA, SAN).

Trade names: Restil, Terluran, Tyril, Zerlon.

REPRESENTATIVE UNIT: Styrene/acrylonitrile/butadiene (ABS).

Trade names: Abson, Bexan, Blendex, Boltaron, Cadco, Cycolac, Kralastic, Lustran, Novodur, Seilon S3.

Process: Natural or butadiene rubber improves the brittleness of polystyrene but graft copolymerisation is essential for best results. Small particles of rubber blended into product assist in preventing rapid crack propagation. Copolymers of styrene and acrylonitrile give advantages of both, being harder, less prone to crazing, more resistant to heat and organic liquids. Impact is improved, but is still low. Butadiene graft copolymerised with styrene/acrylonitrile gives an excellent combination (ABS). Poly blends of the three constituents are also good.

Properties: ABS is tough and resilent, chemically resistant and easily processed. Impact strength is acceptable.

Applications: Injection mouldings, extruded pipe and sections, thermoplastic sheet for vacuum forming. Used for same purposes in engineering as polystyrene, automobile bodywork, heavier domestic appliances, building panels, instrument casings, fan covers.

GENERIC TYPE: RUBBER

REPRESENTATIVE UNIT: cis-Polyisoprene (natural rubber) poly (2-methylbuta-1, 3-diene).

Trade names: Latex: DeLaval, Englatex, Hectolex, Hiltex, Lacentex, Laconvertex, Lacretex, Lanortex, Positex, Prevul, Qualitex, Revertex, Revultex, Vuljex. Raw or finished: Dynat, Heveacrumb, Kualakep, Natcom.

Process: Natural rubber obtained from the bark of Hevea brasiliensis. The latex is used as such or coagulated with acetic acid and dries as sheets (crepe) or compressed from crumbs into blocks (Heveacrumb). Then usually mixed with fillers, pigments, and cross-linked (vulcanised) with sulphur, sulphur monochloride or organic peroxides. Accelerators may be incorporated to speed vulcanisation.

Properties: Raw rubber dissolved by hydrocarbons, chlorinated hydrocarbons, many ketones and esters. Vulcanised rubber insoluble, but swells in almost all organic liquids, including oils. Relatively unaffected by water, dilute acids and alkalis. Mechanically the high elastic recovery from deformation is unique. Electrical properties are good, but microorganisms attack the protein in natural rubber, and oxidation eventually destroys the elasticity (ageing) due to cross-linking. Oxidation is accelerated by elevated temperature. When fully vulcanised, a hard tough plastic results, ebonite, of very good electrical properties. Molecular weight 10^5 to 10^6.

Application: Unvulcanised for shoe soles, pressure sensitive adhesives and electrical insulation. Vulcanised — pneumatic tyres account for 65% usage; conveyor and transmission belts, electrical insulation, footwear, hose and tubing, flooring, mats, road surfacing, surgical goods, sports accessories, clothing, inflatable articles, shock absorbers, seals, packings and valves, for the rest.

REPRESENTATIVE TYPE: Butadiene rubber.

Trade names: Ameripol CB, Astyr, Budene, Bunas 85, 115 and CB, Cariflex BR, Cis-4, Cisdene, Diene, Duragen, Europrene-cis, Intene, JSR-BRO1, Plioflex 5000 and 5000s, SKB, SKBM, SKD, SKLD, SKV, Synpol E-BR, Synpol 8407, Taktene.

Process: Butadiene is derived from C_4 fractions of petroleum distillation. Liquid butadiene polymerises in the presence of alkali metals, an antioxidant being added. In solution in hydrocarbons, a stereospecific catalyst is used. In water containing an emulsifier (fatty acid soap) and a catalyst, polymerisation can be cold (5 °C) or hot (50 °C).

Properties: Vulcanisates resemble natural rubber but superior in heat resistance, less ageing, better abrasion resistance and better at low temperatures. Mechanical properties are most inferior to natural rubber, but better at low temperatures.

Applications: Partial replacement for natural rubber in tyres to improve wear and grip on ice and snow. Less rolling resistance and groove cracking. Also to replace natural rubber in belts, footwear, hose, flooring. Addition to styrene copolymers.

REPRESENTATIVE UNITS: Styrene-butadiene rubber/nitrile rubber/chloroprene rubber/polysulphide rubber. Details of these synthetic materials require careful study before consideration of application, and specialist literature should be consulted.

GENERIC TYPE: SILICONES

REPRESENTATIVE UNIT: Organo-silicon oxide polymers (Polysiloxane).

Trade names: Rubber: Adrub, LS-53, LS-63, NSR, Polysil, Silastic, Silastomer, Silcoset, Silicol, Sil-O-Flex, SKT (and variants). Resin: Dri-film, Sylgard (and variants).

Process: Organo-silicon chlorides or organochlorosilanes are hydrolysed and polymerised by condensation with loss of water. When a difunctional compound is hydrolysed with dilute HCl, liquid cyclic polysiloxanes result which, purified by vacuum distillation and polymerised at 150-200 °C with an alkaline catalyst, yield silicone rubber. In an organic solvent hydrolysis yields a product which polymerised with a catalyst at 120-200 °C produces a resin. Vulcanisates for silicone rubber are organic peroxides, but some rubber will cross-link with sulphur, metal salts, acids or alkalis.

TABLE 10.

Properties: Exceptional resistance to heat, oxidation and ozone as vulcanised rubber. Chemical resistance high, electrical properties excellent. Mechanical properties are inferior to organic rubber at normal temperatures, but change little with rise in temperature and so become superior between +70 and +120 °C after which silicone rubber also deteriorates. Cured resins are stable to prolonged heating in air with good electrical properties and moisture repellant. Strength is inferior to many organic resins.

Applications: Rubber — High cost detracts from some uses. Seals, gaskets, for low or high temperature use, casting and encapsulation, coatings, de-icing equipment on aircraft, electrical insulation at high risk points, justify the expense. Contact with foodstuffs, surgical prostheses also justified. Resins used in impregnation of paper and fabric, electrical equipment. Lamination of glass fibre, bonding mica and asbestos, release agents.

GENERIC TYPE: VINYL CHLORIDE AND RELATED COMPOUNDS (CHLOROETHENE; 1, 1, DI-CHLOROETHENE)

REPRESENTATIVE UNIT: Polychloroethene (polyvinyl chloride) (PVC).

Trade names: Agilide, Breon, Carina, Corvic, Darvic, Deckor, Ekavyl, Exon, Geon, Gobanyl, Halvic, Hefa, Hostalit, Insular, Kanevinyl, Lonza, Lutofan, Marvinol, Novon, Opalon, PCU, Pechiney, Pevikon, Pliovic, Policloro, Rhodopas, Sicron, Solvay, Solvic, Ultron, Vestolit, Vinatex, Vinnol, Vipla, Vibak, Vygen, Welvic. Fibres: Clevyl, Fibravyl, Isovyl, Khlorun, Movyl, Rhovyl, Thermovyl.

Process: Bulk polymerisation yields the purest product, initiated by benzoyl peroxide, 17 hours at 55 °C to give 80% conversion of monomer. Suspension polymerisation uses suspension agents such as polyvinyl alcohol and benzoyl peroxide, heat affecting polymerisation. Emulsion processes are very similar, using emulsifying in place of suspension agents. High molecular weight results from the emulsion method, but is more expensive.

Properties: Rigid PVC is a hard horn-like material, easily fabricated, with fillers and pigments added to give a wide range of properties. Electrical properties are good, and chemical resistance, but ketones and solvent mixtures dissolve it. Unaffected by water, concentrated alkali, aliphatic hydrocarbons. Slowly attacked by concentrated oxidising acids. Heated above 100 °C in air evolves hydrogen chloride, so fires may do more damage by attack of HCl on metalwork than fire itself. Plasticised PVC (phthalates, sebacates, fatty acid derivatives) is less chemically resistant since plasticiser may leach out. Flexible down to −50 °C and serviceable to +75 °C. Better abrasion resistance than vulcanised rubber.

Applications: Rigid PVC— pipes, rods, tubes, sheet. Corrugated roofing, guttering, ducts, wall cladding, lining for chemical tanks. Plasticised PVC — flexible hose and tubing, curtains, protective clothing, cable insulation, footwear, wallcovering (fabric backed), flooring. Imitation leather for furniture and automobile seats, handbags, gloves. Dip coating for protection.

REPRESENTATIVE UNIT: Polyvinylidene chloride. (Poly 1, 1, dichloroethene) (PVDC).

Trade names: Lumite, Polidene, Saran, Tygan, Velon, Viclan, Zetek.

Process: Dichloroethylene monomer polymerises with radical type initiators or in u.v. light. Commonly polymerised in aqueous suspension or emulsion, with oxygen excluded. To facilitate processing, copolymerisation with vinylchloride or acrylonitrile is common.

Properties: High resistance to many chemicals, unaffected by living organisms. Abrasion resistant, but affected by sunlight and u.v., unless copolymerised and stabilised. Decomposed by prolonged contact with ammonia. Produced as monofilaments or as films, which are highly crystalline and oriented. In fabrication, contact with common metals must be avoided, as decomposition occurs of hot polymer, solutions or dispersions.

Applications: Mouldings for water or solvent handling. Monofilament and yarn for seat covering, blinds, filters, screens, fibre for non-flammable imitation hair (dolls). Highly impermeable films to gas and vapour for packaging. Heat shrinkable. Good coating as latex, in flame retardant paint, and high gloss paint.

TABLE 11. Etching Reagents for Metals and Alloys

NAME	COMPOSITION AND CONDITIONS OF USE	RELEVANT APPLICATION	SPECIAL EFFECTS AND RESULTS
Abel's Reagent	10% aqueous chromic acid, ambient temperature	Corrosion resisting steels e.g. Cr-Ni type	Attacks carbides
Aqua Regia	25% HNO_3 (conc.) 75% HCl (conc.)	Noble metals, gold, platinum, etc.	Noble metals, dissolve as chlorides
Alkaline Sodium Picrate	25 g NaOH in 100 cm³ water + 2 g picric acid, boil 30 minutes, decant, use hot for 5-15 minutes	Ferrous metals	Blackens cementite (Fe_3C)
Baucke's Reagent	5% nitric acid in alcohol, 1 hour ambient	Lead	Etches lead without pitting

TABLE 11.

NAME	COMPOSITION AND CONDITIONS OF USE	RELEVANT APPLICATION	SPECIAL EFFECTS AND RESULTS

Belaieff's Etch	1. 3% HNO_3 in alcohol
	2. 1 g $CuCl_2$
	   4 g $MgCl_2$
	   2 cm³ HCl
	   18 cm³ water
	   100 cm³ absolute alcohol,
	   (Le Chatelier No. 1 reagent),
	   dissolve copper from surface,
	   with strong ammonia
		High manganese steel (Hadfield's)	Reveals structure of austenitic steel containing 12% Mn

Belynski's Reagent	1% $CuSO_4$ (aqueous)	Macro-etching iron and steel	Dendritic structures in ingots, castings
Benedick's Reagent	5% solution of 3, nitrobenzene sulphonic acid	Steel	Colours martensite more than austenite
Bingham's Etch	20 drops HCl in 50 cm³ of ethanol, electrolyse	Zinc base alloys with Cu and Al	Etches grain structure
Bolton's Reagent	78 cm³ saturated picric acid in alcohol, 20 cm³ water, 2 cm³ HNO_3	Grey cast iron	Microstructure revealed
Boylston's Reagent	5% HNO_3 in ethanol or methanol	Carbon steels	General structure
Canfield's Reagent	1.5 g $CuCl_2$ 5 g hydrated $Ni(NO_3)_2$ 6 g $Fe\,Cl_3$ 12 cm³ water Few cm³ nitric acid speeds up 150 cm³ ethanol slows down 90 seconds normal etch	Iron and steel	Reveals segregation of phosphorus, as white areas on brown to yellow background; photographic paper soaked in 5% aqueous potassium iron III cyanide, then pressed on to etched surface, gives phosphorus "print"
Carapella's Reagent	5 g $FeCl_3$ 96 cm³ ethanol plus 2 cm³ HCl	Non-ferrous alloys, manganese steels	General etching
Carbide Etching	5% NaOH or KOH solution, boil 5-10 minutes	Ferrous metals	Reveals iron carbide, does not work cold
Curran's Reagent	10 g $FeCl_3$, 30 cm³ HCl (conc.) 120 cm³ water, swab on with cotton wool, 30 seconds	Corrosion resistant steels (Ni-Cr)	Grain boundaries, twins outlined
Czochralski's Reagents 1.	10-20% aqueous ammonium persulphate, swab 1-2 minutes soak	Macro-etching iron and steel	Dendritic structure, porosity, inclusions
2.	10-20% aqueous HF, immerse until matt, clear in conc. HCl	Macro-etching aluminium and alloys	Grain structure, porosity inclusions
3.	5-20% aqueous chromic acid, immerse	Magnesium-aluminium alloys	Reveals Mg_2Al_3 compound
4.	5% aqueous acetic acid	Lead and alloys	Grain contrast etch
5.	Perhydrol added to 4	Sodium-lead alloys	Grain contrast
Daeve's Reagent	20 g potassium ferricyanide 10 g potassium hydroxide 100 cm³ water Cold — 20 seconds for most carbides Boiling — 5 minutes for Fe_3C	Chromium bearing steels, high speed steel, tool steels	Carbides and tungsten compounds attacked. Cementite, Fe_3C only when boiling
Flick's Solution	10 cm³ HF 15 cm³ HCl (conc.) 90 cm³ water	Macroscopic examination of aluminium, microscopic etching of Al alloys	Grain boundary etched, dark film forms on surface, removed by 1 second immersion in strong HNO_3 or chromic acid
Fry's Reagent No. 1	1.5 g $CuCl_2$ 30 cm³ HCl 95 cm³ water 30 cm³ alcohol	Macroscopic examination of steel	Good grain contrast, shows phosphorus segregation
No. 3	5 g $CuCl_2$ 40 cm³ HCl 30 cm³ water 25 cm³ alcohol	do. Soft iron and mild steel in particular	Good grain contrast, strain lines and precipitation hardening

TABLE 11.

NAME	COMPOSITION AND CONDITIONS OF USE	RELEVANT APPLICATION	SPECIAL EFFECTS AND RESULTS
General Etching Reagents for Specific Metals or Alloys	0.5% aqueous solution HF, heat in warm water, swab with cotton wool soaked in etchant	Aluminium	General structure
	Light etching with aqueous HF 0.01 to 10%, accentuates colour of constituents	do.	Identification of phases in alloys
	10 cm^3 HF 10 cm^3 HNO$_3$ 30 cm^3 glycerine (propanetriol)	do.	Grain boundary etch, slow at normal temperatures, gentle heating in water before etching assists
	5-20% aqueous solution of NaOH Remove dark film by 1 second immersion in strong HNO$_3$ or chromic acid	Aluminium and aluminium alloys	Grain boundaries and general structure revealed CuAl$_2$, NiAl$_2$ are coloured brown, FeAl$_3$ unaffected
	1-25% HNO$_3$ in water at 70 °C, then quench in cold water	do.	Similar to above, distinguishes alpha Al-Fe-Si from FeAl$_3$
	1-10% HNO$_3$ in alcohol	Al-Mg alloys	Mg$_2$Al$_3$ brown in 1% solution
	4% picric acid (2, 4, 6 trinitrophenol)	Al alloys	Leaves smoother surface CuAl$_2$ darkened more rapidly than FeAl$_3$ or other phases
	H$_2$SO$_4$ at 70 °C	do.	CuAl$_2$; NiAl$_2$ not affected, FeAl$_3$ brown, Al-Cu-Fe-Mn blackened, Al-Fe-Mn and Al-Cu-Fe not affected
	See also Flick's solution, Keller's etch, Tucker and Vilella	do.	
	Strong NH$_4$OH, accelerated by addition of 3% H$_2$O$_2$, swab with cotton wool	Copper and its alloys	Grain boundary attacked
	25-75 cm^3 NH$_4$OH(conc.) 25 cm^3 H$_2$O$_2$ (3%)	Alpha solid solutions of copper alloys, and beta brass	Colours grain, equal proportions of constituents etches grain boundary, increasing peroxide increases grain contrast
	As above, followed by immersion in Grard's etch	do.	Alpha phase slightly darker, eutectoid remains bluish white
	NH$_4$OH Ammonium persulphate	Cu and its alloys, cupro-nickel in particular	General etchant
	NH$_4$OH $\atop$ KMnO$_4$ equal parts diluted to 0.5%	Cu and its alloys	General etchant
	Trial and error needed with the above solutions, ammonium or potassium oxalate (ethanedioate) may also be added, as weak oxidising agents		
	5 g CuCl$_2$. 2NH$_4$Cl 120 cm^3 water	do.	Grain contrast
	2 g K$_2$Cr$_2$O$_7$ 4 cm^3 NaCl sat. solution 8 cm^3 H$_2$SO$_4$ (conc.) 100 cm^3 water saturated chromic acid solution works similarly	do.	Grain boundaries and oxide inclusions shown
	50 cm^3 CrO$_3$ (15% soln.) 2 drops HCl (conc.)	do.	Grain boundaries attacked
	1 cm^3 H$_2$SO$_4$ (conc.) 20 cm^3 H$_2$O$_2$ (3% soln.)	Cast copper	Attacks oxide inclusions
	10 cm^3 K$_2$Cr$_2$O$_7$ (sat. soln.) 1 cm^3 H$_2$SO$_4$	do.	Attacks oxide inclusions

TABLE 11.

NAME	COMPOSITION AND CONDITIONS OF USE	RELEVANT APPLICATION	SPECIAL EFFECTS AND RESULTS
General Etching Reagents for Specific Metals or Alloys *(contd.)*	1 g ammonium persulphate 90 cm³ water Add few drops ammonia just before etching	Cu and alloys	Relief etching very pronounced
	10% solution HNO_3	do.	Grain boundaries emphasised
	30 g HCl (conc.) 10 g $FeCl_3$ 120 cm³ alcohol or water	Cu alloys, particularly brass and bronze	Grain contrast, but roughens surface
	With above solutions; those containing strong oxidising agents	Cu and alloys including Al bronze	Beta phase attacked in alpha-beta brass and bronze
	Mild oxidising agents	do.	Beta grains coloured
	Etchants containing ammonia	do.	Beta crystals appear white on dark background
	10 cm³ NH_4OH 1 cm³ H_2O_2 (3%)	Brasses with more than 40% zinc	Good grain contrast
	10 cm³ NH_4OH (conc.) 30 cm³ Amm. oxalate (ethanedioate) (sat.)	do.	Slight grain contrast, good grain boundary etching
	5 cm³ HNO_3 20 g CrO_3 75 cm³ water or 20 cm³ HCl 10 g $FeCl_3$ 120 cm³ water	Al bronze (wrought)	Grain boundary etching
	50% HCl or 10% HNO_3	Lead and alloys Lead-antimony alloys	General etchant do.
	5% acetic (ethanoic) acid in alcohol or 5% picric acid (2, 4, 6 trinitrophenol) in alcohol or acetone (2 propanone) 3% H_2O_2 addition accelerates attack	Lead and alloys	Good grain contrast
	10 cm³ acetic (ethanoic) acid glacial 10 cm³ HNO_3 40 cm³ glycerol (propanetriol)	Lead-antimony alloys Lead-calcium alloys Lead-tin alloys of low tin content	General etching
	10 g $FeCl_3$ 30 cm³ HCl 120 cm³ water	Lead antimony alloys Babbitt bearing metals	General grain contrast
	10% acetic (ethanoic) acid 10% tartaric (L, 2, 3, dihydroxybutane dioic) acid or 2% nitric acid	Magnesium	Macro-etching for structure, segregation and porosity
	75 cm³ ethylene glycol (1, 2 ethanediol) 1 cm³ HNO_3 (conc.) 24 cm³ water	Cast magnesium alloys	Macro-etching
	60 cm³ ethylene glycol (1, 2 ethanediol) 20 cm³ acetic (ethanoic) acid glacial 1 cm³ HNO_3 (conc.) 19 cm³ water	Heat treated magnesium alloy castings or wrought magnesium alloys	do.
	0.7 cm³ H_3PO_4 (S.G. 1.7) 4 cm³ picric acid (2, 4, 6 trinitrophenol) 100 cm³ ethanol	Magnesium alloys	Darkens magnesium solid solution areas

TABLE 11.

NAME	COMPOSITION AND CONDITIONS OF USE	RELEVANT APPLICATION	SPECIAL EFFECTS AND RESULTS
General Etching Reagents for Specific Metals or Alloys *(contd.)*	8 g $FeCl_3$ 25 cm³ HCl 100 cm³ water Add 5 cm³ of above stock solution to 100 cm³ alcohol for use	do.	Grain contrast
	75 cm³ Diethylglycol (1, 2 ethanediol) 1 cm³ HNO_3 (conc.) 24 cm³ water	do.	do.
	5 g citric acid 95 cm³ water	do.	do.
	Strong HNO_3	Ni and alloys	Structure, flow lines and porosity in macro-etching
	25 cm³ HNO_3 (conc.) 75 cm³ HCl (*Aqua regia*)	do.	General macro-etching
	See also Marble's reagent and Carapella's etch	Ni and Ni-Cu alloys	
	50 cm³ NaCN (10% soln.) 50 cm³ $(NH_4)_2S_2O_8$ (10% soln.)	do.	Does not remove graphite and sulphide inclusions
	45 cm³ HCl (conc.) 15 cm³ HNO_3 (conc.) 15 cm³ glycerine (propanetriol) or 3 cm³ HF 10 cm³ glycerine (propanetriol) 85 cm³ water	Ni-Cr alloys	General etchant
	10% ammonium persulphate solution in water	Cast Ni	do.
	3 g CrO_3 10 cm³ HNO_3 (conc.) 3 g NH_4Cl 90 cm³ water	Monel (Ni-Cu alloy)	Grain contrast
	50 cm³ HNO_3 (conc.) 25 cm³ acetic (ethanoic) acid glacial 25 cm³ water	Ni and Ni alloys	General etchant
	10 HNO_3 (conc.) 5 cm³ acetic (ethanoic) acid glacial 85 cm³ water Electrolyse for 20-60 seconds with 1.5 volts P.D.	do.	Good contrast
	10 g oxalic (ethanedioic) acid 100 cm³ water Electrolyse 20-60 seconds with 1.5 volts P.D.	Ni and Ni alloys	Grain contrast
	75 cm³ HCl (conc.) 25 cm³ HNO_3 (conc.) 100 cm³ glycerine (propanetriol)	Ni-Fe alloys	Grain contrast
	25 cm³ HNO_3 (conc.) 125 cm³ HCl (conc.) 150 cm³ water	Precious metals Au, Ag, Pt, etc.	General etchant
	20% KCN in water Electrolyse with 5-10 volts P.D.; a.c.	Platinum alloys, osmium alloys, iridium alloys	Grain contrast
	10% ammonium persulphate 10% KCN solution	Palladium alloys, silver	do.
	Concentrated solution of sodium hypochlorite	Ruthenium	do.

TABLE 11.

NAME	COMPOSITION AND CONDITIONS OF USE	RELEVANT APPLICATION	SPECIAL EFFECTS AND RESULTS
General Etching Reagents for Specific Metals or Alloys (contd.)	Glacial acetic (ethanoic) acid in alcohol	Ag	Grain contrast
	75 cm³ NH₄OH 25 cm³ H₂O₂ (3%)	do.	do.
	Dilute K₂Cr₂O₇ few drops H₂SO₄	do.	do.
	Electrolytic etching especially recommended using a.c. at low voltage. Gold is difficult, except by a.c. electrolysis, *aqua regia*, or iodine solution		
	3% picric acid (2, 4, 6 trinitrophenol) in alcohol 1% picric acid (2, 4, 6 trinitrophenol) used hot	Steels (macro-etching)	Segregations show up, areas high in P stain first
	10-15% HNO₃ 2% HNO₃ initially, then 10% HNO₃	do. Large ingots of steel	Segregations show weak contrast. Improves contrast
	10-20% H₂SO₄ used hot	Steels	Preferentially attacks sulphide inclusions, but less effective than HNO₃
	50% HCl used hot 50% cold, for 5 seconds	do.	Heterogeneity revealed, but etching very rapid, blackens soft spots on hardened steel
	2 parts H₂SO₄ added to 1 part HCl, used hot	do.	General defects, also suitable for deep etching
	10-20% ammonium persulphate	Iron and steels	Segregations, and heterogeneity. Differences in heat treatment show
	8% CuCl₂ . 2NH₄Cl neutral solution, 4% CuCl₂ . 2NH₄Cl with short time immersions and repeated washings, wash off Cu layer with 0.5% ammonium persulphate	Iron and mild steel, not suitable for graphitic cast iron, highly alloyed, or hardened steels	Porous spongy layer of copper deposited, washed off, showing dark areas where C and P high. Purer areas generally unattacked
	1% CuSO₄	Iron and steel	Reveals dendritic structure with carbon content, 0.1-0.3%, seldom used now
	120 g CuCl₂ .2NH₄Cl 50 cm³ HCl (conc.) 1 litre water	do.	Deep etch
	For additional material on macro-etching of iron and steel, *see* Rosenhain, Stead, Le Chatelier, Oberhoffer, Fry, Canfield, Heyn, Baumann, in this table, or in main text.		
(NITAL)	1 to 5 cm³ HNO₃ (S.G.1.42) 100 cm³ ethanol	Steel (micro-etching)	Reveals grain boundaries in ferrite, darkens pearlite. Reveals case depth in case hardened or nitrided steel. Good contrast between ferrite and cementite in pearlite, and between ferrite or austenite and harder phases such as martensite or troosite. At high magnification, laminations of ferrite and cementite in pearlite show up well, ferrite darkens.
	1 to 5 cm³ HNO₃ (S.G.1.42) 100 cm³ amyl alcohol (pentanol)	do.	Good grain boundary etch at low temperature, grain contrast at high
(PICRAL)	2-5 g picric acid (2, 4, 6 trinitrophenol) 95 cm³ ethanol	Plain carbon and low alloy steels	Good grain contrast, above 4% picral develops pearlite structure, reveals grain size
	do. Add 3% HCl	Hardened and tempered steels	Reveals grain size
	do. Wash, dip into dilute ammonium persulphate	Plain carbon, low alloy steels	Colours ferrite grains
	Saturated solution of picric acid (2, 4, 6 trinitrophenol) in alcohol, plus 1% HNO₃	Steels	Good grain boundary etchant

TABLE 11.

NAME	COMPOSITION AND CONDITIONS OF USE	RELEVANT APPLICATION	SPECIAL EFFECTS AND RESULTS
General Etching Reagents for Specific Metals or Alloys (contd.)	30 cm³ HCl (conc.) 10 cm³ HNO₃ (conc.)	"Stainless" steels	Rapid etching on occasions, reveals general structure
	10 cm³ HNO₃ (conc.) 20 cm³ HCl (conc.) 30 cm³ glycerine (propanetriol)	Fe-Cr alloys containing one or all of Al, W, V, Mo. Austenitic steels	Reveals general structure
	5 g FeCl₃ 50 cm³ HCl (conc.) 100 cm³ water	Austenitic "stainless" steel	General structure
	FeCl₃ saturated solution in HCl, few drops HNO₃	do.	do.
	50 cm³ HCl 1-100 cm³ CrO₃ 10% solution in water	Heat-treated "stainless" steels	General structure, proportion of CrO₃ governs vigour of reaction
	10 cm³ HNO₃ 10-30 cm³ HCl 20-30 cm³ glycerol (propanetriol) or 10 cm³ HNO₃ 20 cm³ HCl 20 cm³ glycerol (propanetriol) 10 cm³ H₂O₂ (3%) or 10 cm³ HNO₃ 20 cm³ HF 30 cm³ glycerol (propanetriol) or 60 cm³ HNO₃ 40 cm³ acetic acid or 1 g picric acid (2, 4, 6 trinitrophenol) 0.5 cm³ HCl 100 cm³ alcohol	Austenitic manganese steels Fe-Cr base alloys High speed steels Also for high Si steels and alloys	General structure revealed in very corrosion resistant alloys
	10 g CrO₃ 100 cm³ water	"Stainless" steels	Reveals structure and attacks carbides
	10 cm³ 4% HNO₃ in alcohol 30 cm³ saturated solution, orthonitrophenol (2 nitro-hydroxybenzene) in alcohol	Hardened steels	Differentiates between constituents
	50% solution NaOH added to cold saturated solution picric acid (2, 4, 6 trinitrophenol)	Case-hardened steels	Reveals cementite
	do. Boil 5-10 minutes or electrolyse, specimen as anode 6 volts, ½-1 minute	Alloy steels	Differentiates between carbides Cementite attacked Cr carbides unaffected, carbides of W-Fe and iron -tungstide stained more rapidly than Fe₃C. WC, unaffected.
	20 cm³ NaOH (10% aqueous) 10 cm³ H₂O₂ (3%)	Fe-W alloys with carbon absent do. carbon present	Tungstides darkened Tungsten carbide darkened
	1-5 g K₃Fe(CN)₆ 10 g KOH 100 cm³ water freshly prepared, boil 15 minutes	Alloy steels	Differentiates between carbides and nitrides
	10% oxalic (ethanedioic) acid solution electrolyse 1.5-6 volts, 10-60 seconds, specimen anode	"Stainless" steels	Carbides and general structure, lower voltage preferred
	See also Marble, Fry, Stead, Pilling, Kourbatoff, Murakami, Daeves, in this table for micro-etching of steels.		
	2 g FeCl₃ 5 cm³ HCl (conc.) 30 cm³ water 60 cm³ absolute alcohol	Sn and alloys, Sn-coated metals	General structure, but not in those containing lead

TABLE 11.

NAME	COMPOSITION AND CONDITIONS OF USE	RELEVANT APPLICATION	SPECIAL EFFECTS AND RESULTS
	5 g ammonium persulphate 100 cm³ water	Sn and alloys	Reveals grain boundaries
(Nital)	1-5% HNO_3 in alcohol	Sn alloys, bearing metals, pewter	Reveals Sn-Sb; and Sn-Cu compounds
	2% HCl in alcohol	Pure Sn	Reveals grain structure
(Picral)	2-5% picric acid (2, 4, 6 trinitrophenol) in alcohol	Sn coated steel or iron (tinplate), steel backed bearings with Sn base alloys	Reveals Sn-Fe compounds at interface
	10 cm³ HNO_3 30 cm³ acetic (ethanoic) acid 50 cm³ glycerine (propanetriol)	Sn coated metals, Sn-Pb alloys	General structure
	50 cm³ NH_4OH 20-50 cm³ H_2O_2 (3%)	Sn-coated Cu and Cu alloys	Reveals Sn-Cu compounds at interface
	20 g Na_2S 100 cm³ water 1-2 cm³ HCl	Sn-Sb-Cu alloys	H_2S generated, differentiates between Sn-Cu and Sn-Sb compounds
	50 cm³ acetic (ethanoic) acid 50 cm³ water 1 drop H_2O_2	Soldered joints (high Sn solder)	Reveals alloy structure
	5 g $SnCl_2$ 10 cm³ HCl	Sn and Sn-rich alloys	General etching
	5 g $KClO_3$ 10 cm³ HCl	Sn and Sn rich alloys	Good grain contrast
	See also Romig and Rowland		
	See Kroll's etch for Ti alloys		
	20 g CrO_3 1.5 g Na_2SO_4 100 cm³ water, rinse immediately in 20% solution of CrO_3 and thoroughly wash in running water	Zn and alloys	Alternate polish and etch required to reveal general structure, staining results unless wash carried out after etching
	0.5-10% HCl in water	do.	General etchant
	1-10% HNO_3 in water	do.	do.
	10 g I_2 30 g KI 100 cm³ water	do.	do.
	See also Palmerton, Timofeef in this table		
Grard's Reagent	0.5 g $FeCl_3$ 50 cm³ HCl 100 cm³ water, use following swab with ammonia-hydrogen peroxide	Bronzes, monel	Distinguishes eutectoid areas in dendrites of bronzes
Heyn Sulphur Printing	10 g $HgCl_2$ 20 cm³ HCl (conc.) 100 cm³ water Lay silk sheet over specimen, paint with above solution; alternative to silk is gelatined paper	Steels	Sulphur inclusions evolve H_2S which blackens mercuric salt, washed silk gives mirror image of sulphur concentrations, useful for local distribution of sulphur, phosphorus produces yellow stain on silk
Humphrey's Reagent	120 g $CuCl_2.2NH_4Cl$ 50 cm³ HCl (conc.) 1 litre water	Steel	Produces strong relief, neutral solution used to remove scratches and marks, then acid added. When complete, copper washed off, light emery polish leaves high relief, printer's ink can be used to obtain record

TABLE 11.

NAME	COMPOSITION AND CONDITIONS OF USE	RELEVANT APPLICATION	SPECIAL EFFECTS AND RESULTS
Keller	$2.5 \, cm^3$ HNO_3 (conc.) $1.5 \, cm^3$ HCl (conc.) $1 \, cm^3$ HF $95 \, cm^3$ water immerse 10-20 seconds, wash in warm water, dip in HNO_3 (conc.)	Al and alloys particularly duralumin	General structure
Klemm	Saturated $KMnO_4$ plus 10% KOH etch 1 minute at $70 \, ^{\circ}C$	Steels	Faint dark dots develop with temper brittleness, best viewed in dark field illumination
Kourbatoff Etch	$20 \, cm^3$ methanol $20 \, cm^3$ ethanol $20 \, cm^3$ ISO amyl alcohol (diethylcarbinol) $10 \, cm^3$ butyl alcohol (1-butanol) use $70 \, cm^3$ above mixture plus $30 \, cm^3$ 4% HNO_3 in acetic (ethanoic) anhydride, very quick immersion	Hardened steels	Colours sorbite and troosite
Kroll	1-2% HF do. plus 2% HNO_3	Ti and alloys	Structural detail with little relief, develops grain boundaries and cracks
Le Chatelier No. 1 Reagent	$1 \, g$ $CuCl_2$ $4 \, g$ $MgCl_2$ $2 \, cm^3$ HCl (conc.) $18 \, cm^3$ water $100 \, cm^3$ absolute alcohol	Steel macro-etching	Differentiates areas containing phosphorus in low P steels, contrast increased by electrolysis
Marble	$20 \, g$ $CuSO_4$ $100 \, cm^3$ HCl (conc.) $100 \, cm^3$ water	Ni and Ni-Cu alloys, "stainless" steels, nitrided steels	Macro-etching to reveal general structure, reveals case depth of nitrided steels
Matwieff	0.2% oxalic (ethanedioic) acid (aqueous) 20-30 seconds	Steels	Darkens sulphide inclusions
McCance	Silver nitrate solution acidified with H_2SO_4, in gelatine	Steel	Detects sulphur bearing constituents
Merica	$20-40 \, cm^3$ HNO_3 $30-40 \, cm^3$ acetic (ethanoic) acid $30-40 \, cm^3$ acetone (2-propanone)	Ni and Ni-Cu alloys	General structure
Meyer and Eichholz	$6 \, g$ $CuCl_2$ $6 \, g$ $FeCl_3$ $10 \, cm^3$ HCl (conc.) $100 \, cm^3$ alcohol	Steel	Reveals general macrostructure and strain lines in most ferrous metals
Murakami	$10 \, g$ $K_3Fe(CN)_6$ $10 \, g$ KOH (or $9 \, g$ NaOH) $100 \, cm^3$ water Used cold. Used boiling for 5 minutes	Steels	Carbides in Cr steels and ternary carbides Fe_3WC and Fe_4W_2C in high speed tool steels etch in about 20 seconds Fe_3C coloured
Oberhoffer	$1 \, g$ $CuCl_2$ $0.5 \, g$ $SnCl_2$ $30 \, g$ $FeCl_3$ $30 \, cm^3$ HCl (conc.) $500 \, cm^3$ water $500 \, cm^3$ ethanol, after polishing, clean and dry with alcohol, leave to stand for 30 minutes then etch	Steels, macro-etching	Slow action reveals dendrites and grain boundaries, very uniform in effect, deposition of copper on pure ferrite does not occur
Palmerton	$10 \, g$ CrO_3 $1.5 \, g$ Na_2SO_4 $100 \, cm^3$ water	Zn and alloys	Alternate polishing and etching reveals structure

c

TABLE 11.

NAME		COMPOSITION AND CONDITIONS OF USE	RELEVANT APPLICATION	SPECIAL EFFECTS AND RESULTS
Pilling		4% picric acid (2, 4, 6 trinitrophenol) in alcohol followed by 5% HNO_3 in amyl alcohol (1-pentanol)	Hardened steels	Troosto-sorbite coloured brown, martensite blue, austenite orange
Romig and Rowland		1 drop HNO_3 2 drops HF 24 cm^3 glycerine (propanetriol)	Sn coatings on iron and steel	Reveals Sn-Fe compounds at interface
Rosenhain		30 g $FeCl_3$ 100 cm^3 HCl (conc.) 1 g $CuCl_2$ 0.5 g $SnCl_2$ 1 litre of water, highly polished surface needed, solution used only once	Steel macro-etching	Pearlitic steel with uniform P distribution deposits Cu on ferrite which appears dark, uneven P distribution shows clear areas of P segregation
Sauveur		20 cm^3 H_2SO_4 (conc.) 10 cm^3 HCl (conc.) 30 cm^3 water 30 minutes near boiling point	Iron and steel	Develops macrostructure
Stead	No. 1	10 g $CuCl_2$ 40 g $MgCl_2$ 20 cm^3 HCl (conc.) 1 litre alcohol	Steel	Low P steel shows up distribution of P Primary dendrites
	No. 2	5 g $CuCl_2$ 4 g $MgCl_2$ 1 cm^3 HCl (conc.) 20 cm^3 water 1 litre alcohol	do.	High P steel, shows up distribution of P
Timofeef		94 cm^3 HNO_3 (conc.) 6 g CrO_3 A few drops added to 100 cm^3 water before use	Zn and alloys	General structure
Tucker		15 cm^3 HF (48%) 45 cm^3 HCl (conc.) 15 cm^3 HNO_3 (conc.) 25 cm^3 water	Al alloys macro-etching	General structure
Vilella		33 cm^3 HF (conc.) 17 cm^3 HNO_3 (conc.) 50 cm^3 glycerine (propanetriol)	Al alloys macro-etching	General structure, assists in identification of many structural constituents which remain unattacked and retain original colours

Abnormal steels. These are case-hardening or alloy steels in which the degree of hardness which could be expected from their composition is not developed by carburising or heat-treatment.

Abrasive. A scouring and polishing material used for grinding hard substances such as metals, glass, etc. The commonest natural abrasives are quartz sand and sandstone. Diamond, the hardest of the natural abrasives, is used in the form of powder for cutting and drilling rocks and stones, for engraving and for machining the hardest metals. For these purposes, industrial diamonds known as carbonado (black diamonds) are chiefly employed, the fine particles, which are exceedingly sharp, being usually used in conjunction with copper or steel into whose surfaces the diamond dust readily embeds itself. Somewhat softer than the diamond, but still very hard, are the minerals emery, corundum and garnet. Emery is an impure form of corundum and is extensively used in the form of powder, cloth and emery wheels. In the latter, the emery is held by a suitable bonding material.

The most important artificial abrasive is carborundum, a compound of silicon and carbon (silicon carbide) produced by fusing coke and sand in an electric furnace. It is somewhat harder than emery but has similar applications. Other high-grade abrasives include manufactured alumina products such as Aloxite, Alundum, Lionite, Diamantine, etc.

Quartz sand is chiefly used in sand-blasting machines, which have a wide range of applications. The so-called sandpaper is made by coating paper with powdered glass, and is more correctly known as glass-paper. Sandstone is used for grindstones, and other low-grade abrasives include such materials as pumice powder, honestone or whetstone (a fine-textured even-grained sedimentary rock), diatomaceous earth and tripoli.

Polishing of metals is usually effected by using hard substances such as emery and carborundum, in the finest possible powdered form for the initial grinding, and following up with alumina, magnesia, rouge (iron oxide), chromium (II) oxide (chromic oxide) (sometimes called green rouge) or putty powder (tin oxide).

Absorption cross-section (nuclear). A quantitative measure of the probability of a neutron being absorbed into nuclei of a material through which that neutron is travelling, without causing instability of the nucleus. An alternative description is "capture" cross-section, and the original unit of measurement, the barn (q.v.), equivalent to an actual cross-sectional area of 10^{-28} m^2. A nuclear cross-section of a few barns implies a nucleus of such dimensions that the neutron will travel through 10^8 atoms on average before striking a nucleus, but the collision may simply result in deflection of both neutron and nucleus. Absorption, or capture following collision, requires particular conditions of nuclear structure, and is a function of neutron energy and velocity. Light elements tend to have low absorption, heavy elements a high absorption, but isotopes of the same element can have widely different cross-sections, e.g. boron 10 isotope has high absorption, and is used as a control material in nuclear reactors, even though it is an element of low atomic number (light).

Accelerator. A term used to describe machines that are, in fact, devices for accelerating particles such as protons, deuterons or alpha particles by means of electrical fields with the aim of attaining very high energies. The higher the energy of a particle the greater is the probability of hitting and disintegrating the nucleus of an atom. The machines are thus sometimes referred to as atom-smashing machines. They are:

(a) electrostatic generators, in which alpha particles or protons are accelerated to high energies by machines developing very high d.c. voltage connected to a large discharge tube;

(b) betatron. In this machine the accelerating force is provided by an electric field induced by a rapidly varying magnetic field. The electrons move in a circular orbit perpendicular to the plane of the magnetic field;

(c) cyclotron. An apparatus for accelerating ions or charged atomic particles by causing them to traverse a spiral path between two flat, hollow, semi-circular electrodes (or dees). An oscillating voltage is applied between the dees so that at each half-revolution of the particle, the latter receives an increase in energy and so moves in a spiral path. The magnetic field is perpendicular to the plane of the dees. The process is also referred to as multiple acceleration;

(d) synchrotron. An apparatus for accelerating electrons to very high energies. It combines the essential features of the betatron and the cyclotron.

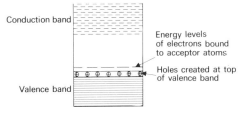

Fig. 1 *Energy diagram of a p-type semiconductor.*

Acceptor (semi-conductor). An atom added to a semiconductor material, with fewer electrons than the host element, such that it can accept an electron from a nearby host atom, to complete the normal covalent bond characteristic of the host element. The transfer of an electron leaves behind an excess of positive charge, known as a "positive hole", and the acceptance of the extra electron by the addition or "impurity" atom leaves that atom with an excess negative charge. The passage of electrons is now affected by the discrete localised energy levels around the impurity atom and the positive hole. At low temperatures, these are close together, but as temperature rises and atoms are excited, then there is mobility of the charges, and conductivity increases. Boron and gallium are trivalent elements, and when added to silicon or germanium with four electrons, per atom, can act as acceptors, or p-type semiconductors. (*See also* **Extrinsic semiconductor** and Fig. 1).

Accumulator. A device for storing energy for future use. In the hydraulic accumulator a vertical ram works in a cylinder; the load upon the ram regulates the pressure of water that the device supplies constantly to the machinery it is designed to operate. In a steam accumulator the device is modified so that steam pressure is substituted for the load on the ram. The pneumatic accumulator is similar to the steam type, except that air pressure is substituted in place of steam. An electrical accumulator or storage battery is a device for storing electrical energy. The plates are made in the form of grids and are filled with lead oxide. The electrolyte is sulphuric acid. On passing a current through the cell, lead peroxide forms on the anode, while at the cathode spongy lead is formed. On joining the plates externally a current flows in the wire and the chemical changes are reversed. The nickel-iron accumulator has plates in the form of tubes, made of steel and nickel plated. One set is filled with nickel hydroxide and the other with iron oxide mixed with graphite. The electrolyte is caustic soda. The e.m.f. of the lead accumulator is 2.1 volts, and of the nickel-iron battery 1.33 volts. The capacity of a battery depends on the area of the plates.

Acetone (propanone) $CH_3.CO.CH_3$. One of the family of organic materials called "ketones", which always incorporate a grouping of $-CO-$. A powerful solvent, particularly for cellulose derivatives, and used extensively in polymer chemistry.

Acetylene (ethyne) C_2H_2. The initial member of a group of organic materials formerly called "acetylenes" but now more commonly known as "alkynes". All contain unsaturated carbon bonds, i.e. bonds between carbon atoms themselves rather than between carbon and another atom.

Acheson furnace. An electric furnace primarily designed for the production of carborundum. The constituents (coke and sand) are packed round a central core of coke through which the current is passed.

Acicular. Having an elongated needle-like crystalline structure.

Acid dip. A process in which brass is dipped in a solution of mixed acids (nitric and sulphuric) to produce a bright surface. (*See* **Brass finishing.**)

Acid lining. Refractory materials, used in the hearth and linings of a melting furnace, which combine readily with basic oxides during the melting process. The materials include silica bricks, sand, flint, ganister and nearly all fireclays.

Acid rock. An igneous rock containing over 60% silica SiO_2. Characteristic minerals are quartz and orthoclase. The typical rock is granite.

Acid steel. Steel produced in a furnace with an acid (i.e. siliceous) hearth and lining and under a slag which is predominantly acid. Pig-iron suitable for treatment by the acid open-hearth process (which is able to remove only carbon, silicon and manganese from the melt) must be low in sulphur and phosphorus. Since this limitation of the process necessitates careful selection of raw materials, acid open-hearth steel is sometimes considered superior to basic steel.

Acid value. This is denoted by the volume (in m^3) of a 0.1 Molar caustic soda or potash solution required to neutralise 10^3 kg of the substance in question, e.g. oil or resin.

Acoustic emission. *See* **Non-destructive testing** and **Emission (acoustic).**

Acrylic resins. A generic term for polymers derived from acrylic (propenoic) and/or methacrylic (2 methyl-propenoic) acid, principally esters of lower alcohols. Also includes acrylonitrile (vinyl cyanide) units, where the -CN group replaces -COOR. (*See* **Polyacrylics.**)

Actinides. The name given to a group of elements following the metal actinium in the periodic table. They are characterised by the energy levels of their outer electrons, where the normal progression from one 6d electron in actinium is interrupted, by the next electrons entering the intermediate energy level 5f at thorium and progressing through protoactinium and uranium, to the man-made "trans-uranic elements", to lawrencium, element 103, with 14 electrons in the 5f level. They resemble, in this electronic structure, the make-up of the lanthanides or rare earths, where the 4f levels are involved, before the 5d level has been completed.

Due to the outer "valency" electrons, involved in chemical reactions, being 6d in each case, the actinides (like the lanthanides also) are chemically very similar, and difficult to separate one from another. As the outer electrons are all close together in energy level, it is extremely likely that interchange between electrons in 7s, 6d and 5f energy levels takes place readily. This gives rise to multiple valency chemical combinations, and often highly coloured compounds. The actinides as elements have unstable nuclei, and include many fissile materials involved in nuclear reactions. This constitutes their main importance, but really involves only thorium, uranium and plutonium at the present time. (*See* "Electronic structure of the elements" under **Atom.**)

Actinium. A radioactive metal accompanying thorium and extracted from pitchblende. Actinium emanation renders substances with which it comes into contact temporarily radioactive. (*See* Table 2.)

Activated alumina. The name of the product obtained by partly dehydrating aluminium hydroxide by heat. In the activated condition it can be used for absorbing moisture from gases and for dehydrating organic liquids. At moderately elevated temperatures it acts as a catalyst in a number of chemical processes.

Activated carbon. A form of carbon, produced by carbonising vegetable matter in the absence of air. It is capable of absorbing large quantities of gases. It can be used for clarifying liquids and for the purification of electroplating solutions.

Activation. In ore-dressing the term refers to the effect produced when certain chemical compounds are introduced into froth flotation (q.v.) to modify the surface response (e.g. wettability) of minerals, and so aid the process of collection into the froth. When, for example, lead-zinc sulphide ores are finely ground and treated in water suspension by bubbling air through the liquid, the addition of sodium sulphite reduces the wettability of the zinc sulphide and enables the lead sulphide to be separated in the froth and floated off. At a later stage copper sulphate is added because this increases the response of the zinc sulphide to separation by the air bubbles. The zinc sulphide is thus separated from the gangue materials. In this example, sodium sulphite (Na_2SO_3) is a depressor and copper sulphate is an activator for zinc sulphide. The term is also used in powder metallurgy in reference to the acceleration of the rate of sintering.

Activity. If a pure substance, such as zinc, is dissolved in another, such as copper, its properties are different from those which it exhibits in its pure form. In this case the vapour pressure of the zinc in the copper is less than that of pure zinc; the ease with which it can be converted to oxide, chloride or other compounds of zinc is correspondingly decreased. This change in the behaviour of a dissolved substance is commonly represented by the thermodynamic function called "activity". This is defined precisely by the equation:

$$RT \log_e a_A = \overline{G}_{A\,(A_xB_yC_z\ etc.)}$$

where R is the gas constant, T is the absolute temperature and a_A is the activity of component A.

$$\overline{G}_{A\,(A_xB_yC_z\ etc.)}$$

is the maximum work or free energy change which is associated with the reversible transfer, at constant pressure, of 1 mole of component A from the solution (at a fixed composition, $A_xB_yC_z$ etc.) to some other standard condition of A, such as the pure state. a_A is the activity of A in the solution *relative to this same standard condition*. Although developed originally for the thermodynamic treatment of solutions, the activity function can be used equally well, with precisely the same definition, in the treatment of solid, liquid and gaseous systems, whether homogeneous or heterogeneous.

Adamite. An abrasive material consisting of fine aluminium oxide.

Addition agents. Substances added to electroplating baths for the purpose of modifying the characteristics of the deposits without appreciably altering the electrical properties of the electrolyte.

Addition polymerisation. The process of binding together groups of molecules by simple addition linkage at the ends, to form a "polymer" from a number of simple units, the "monomer" without the elimination of any by-product molecules. When each "mer" is a single identical unit, a chain-like compound is formed. When the units are different, variations can occur forming a "copolymer" bearing some resemblance in its properties to a solid solution of metals or inorganic compounds. Addition polymerisation may require heat, pressure, and the presence of a catalyst. Initiators are used to stabilise the ends of long chains also, and variation in the length of chain molecules, and hence molecular weight and other related properties becomes possible (*see* Fig. 2).

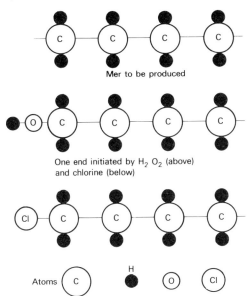

Fig. 2 Initiation of mers for polymerisation.

Adsorptiometric size determination. The title given to a method of determining particle size in a liquid suspension. The adsorptiometer measures the ratio of the incident light falling on a cell containing the suspension, to the transmitted light. If C is the concentration of a sample in kgm^{-3}, and S is the specific surface of the powder in $m^2 kg^{-1}$, then for a cell in which the light path is L m.

$$C = \frac{4}{SL} \log_e I_1/I_2$$

I_1 and I_2 being the intensity of light through the clear liquid and through the suspension respectively.

Adsorption. The condensation, in the form of a thin film of molecules, of a gas, or of a dissolved or suspended substance, upon the surface of a solid. The phenomenon is due to unbalanced surface forces resulting in a certain amount of latent surface energy. In the absence of chemical action the only force holding the adsorbed atom is the unsatisfied crystal force, and thus thermal agitation removes the adsorbed layer which is normally of unimolecular thickness. This is physical adsorption. Chemical adsorption (chemisorption) occurs when gas molecules are held on the surface by chemical bonds. In such cases (e.g. oxygen on tungsten) the heat of adsorption is high and the reversal of the process is achieved only with difficulty.

Thicker adsorbed layers may be built up by gases at temperatures near their critical points. This is utilised in the production of high vacua.

Adsorbed substances frequently show enhanced activity, and on this is based the theory that catalysis can be regarded as an adsorption phenomenon.

The use of certain dye-stuffs in volumetric analysis and the techniques of chromatography are based on adsorption phenomena.

Aerosiderite. A mass of meteoric iron.

Aeruginous. Partaking of, pertaining to, or resembling verdigris, the green corrosion product of copper.

Afterblow. The further period of blow, after the flame drops, in the Bessemer process. The object of the afterblow is to remove the phosphorus after the carbon has been eliminated. The period must be carefully controlled to avoid oxidation of the iron.

After-contraction (or expansion). The change in dimensions of a refractory material which has been reheated to a given temperature for a given time. It is usually expressed as a percentage of the original length.

Agaric mineral. A white earthy variety of calcite $CaCO_3$, softer than chalk, which is deposited from solution. Often found in caverns.

Agate. A more or less pure form of silica deposited from a natural aqueous solution. It is a variegated chalcedony, or incompletely crystalline form of silica, and usually shows a banded or ribbon-like structure due to pauses in growth. It is hard enough to be used in the bearings of delicate balances and for mortars.

Age-hardening. This spontaneous hardening occurs in a wide range of alloys, both ferrous and non-ferrous, during storage at atmospheric temperatures. It was first observed in the aluminium alloy, duralumin. In the case of a very mild steel quenched from about $700\,°C$, there results a marked increase in both the hardness and the strength. With certain alloys, including copper-iron, molybdenum-iron and nickel-base high-temperature alloys, heat-treatment at moderately high temperatures is necessary to produce age-hardening. (*See also* **Ageing**.)

Ageing. A precipitation process, often submicroscopic, which occurs when a supersaturated solid solution is allowed to rest at atmospheric temperature after quenching. The effect, which results in a spontaneous modification of the properties of the metal, is known as age-hardening. Ageing can be induced by exposure to a favourable temperature after (*a*) rapid cooling from an elevated temperature (quench ageing) or (*b*) by a limited degree of cold working (strain ageing). If reheating is necessary to induce precipitation, the treatment is known as "artificial ageing", but is preferably referred to as "tempering", the effect produced being "temper hardening".

Ageing (of thermoplastic polymers). Oxygen from the atmosphere can diffuse into some polymeric materials, to cause cross-linking, and consequent loss of elasticity and increased hardening. This is particularly important in elastomers such as rubber, and gives rise to the common expression "perishing". Ultra-violet light accelerates both the oxygen diffusion and the cross-linking. Elements such as sulphur are used intentionally to cross link rubber (*see* **Vulcanisation**) but sulphurous gases will diffuse into materials in service, as with oxygen, and cause further hardening. Being a time-dependent process, the word "ageing" is relevant. Further ageing can result in growth or coarsening of particles, usually leading to a degree of softening.

Agglomerate. A mixture of materials, with a distribution of particle size, shape and constitution, bound together to form artefacts larger than the individual units. The binder can be natural or synthetic.

Aggregate. A mass consisting of rock from mineral fragments. The inert material such as sand, gravel and crushed stone that is to be mixed with cement. In reference to metals and alloys, the term is applied to mechanical mixtures of two or more phases. Quenched steel, for example, is an aggregate of three phases: solid solution of carbon in gamma-iron, alpha-iron and iron carbide.

Composite materials are referred to as aggregates when the addition (filler) material is bulky in nature, rather than fibrous, and the matrix binder may be polymeric, glassy, metallic or ceramic in nature. Concrete is a composite material in which generally natural minerals (stone, sand, gravel) are bound with a synthetic "cement", such as calcium aluminium silicate. Tarmacadam is similar, but the aggregate is bound with natural bitumen pitch, or tar from coal or oil distillation.

Airbond. A resin-type core binder possessing air-drying properties.

Air-drying. A method of treating the air-blast of a blast furnace prior to heating. Since a furnace works better with dry air, several methods of extracting the moisture have been proposed. The old method of refrigeration is too costly, and modern apparatus now uses silica gel as the drying agent.

Air-hardening steel. An alloy steel that can be hardened by cooling in air from above its critical temperature range, and which does not require quenching from above that temperature to produce hardening. A typical example is a nickel-chromium steel containing 0.3% C, 4% Ni, 1.5% Cr.

Airy points. The points of support of a horizontal bar which ensure that the bending shall be a minimum. If l be the length of the bar and n the number of supports, then the distance between supports is given by $l/\sqrt{n^2 - 1}$.

Ajax-Northrup furnace. (EFCO-Northrup furnace). This is a coreless high-frequency induction furnace in which standard crucibles may be used. The metal to be melted is contained in a crucible, around which is located a water-cooled coil carrying the current. The high frequency in the coil induces eddy currents in the metal, generating sufficient heat to melt it.

Alabaster. A very fine-grained and compact variety of gypsum $CaSO_4 . 2H_2O$.

Alaskite. A pegmatic granite containing a high proportion of quartz and used in the production of "glass spar" for the making of enamels for use on metals.

Albion metal. A tin-clad lead foil produced by rolling a sheet of lead sandwiched between sheets of tin under sufficient pressure to cause the sheets to weld together and produce a composite clad sheet. This metal was used in making certain types of metal buttons, toys, cheap jewellery, wrapping foil for tobacco and non-food products, and collapsible tubes where non-toxic properties are not essential.

Albite. A soda felspar, consisting mainly of sodium aluminium silicate $Na_2O.Al_2O_3.6SiO_2$, found chiefly in acid and intermediate igneous rocks (see also **Anorthite**).

Alclad. A composite sheet formed by coating an aluminium alloy of the duralumin type with pure aluminium in order to combine the strength of the alloy with the corrosion resistance of the pure metal coating.

Aldip process. A process for coating cast iron and steel with aluminium. After cleaning, pickling, rinsing and furnace drying the part is heated in a bath of pre-heating salt and then transferred to a bath of molten aluminium. After a short immersion, the part is returned to a heated salt bath from which it is slowly raised. Air-blasting is used to remove the excess of aluminium.

Alexandrite. A dark green variety of chrysoberyl (beryllium aluminate, $BeO.Al_2O_3$). It is found in the Ural Mountains and prized as a gemstone. It appears to change colour in artificial light — the light transmitted in certain directions appearing reddish in colour.

Al-fin process. The name of a process developed for coating cast iron and steel parts with aluminium. After degreasing, pickling, rinsing and drying, the parts are immersed in a molten flux and then quickly transferred to a bath of molten aluminium. The excess of aluminium is shaken off and the parts are quenched in oil or water.

The process is not applicable to finished parts. The principal applications are for fixing aluminium cooling fins on aero-engine cylinders, making steel-backed aluminium bearings, cylinder liners and heads, brake drums and heat exchangers. On immersion in the molten aluminium or aluminium alloy a bi-metallic compound $FeAl_3$ is formed, and this can provide the intermediate layer that will ensure good bonding between the iron and the aluminium that is subsequently cast about it. After being coated with the aluminium the part is quickly placed in a prepared mould and the aluminium casting alloy poured round it so that the cast metal fuses with the still molten alloy coating on the iron. The casting may subsequently be machined as desired.

Alkali. The name derived from the Arabic for the ashes of saltwort. The name is now applied to soluble bases, the commonest of which are soda, potash and ammonia. They neutralise acids to form salts. The carbonates of sodium and potassium are known as mild alkalis, while the hydroxides of these metals are the caustic alkalis.

Alkali metals. The alkali metals are the most basic and active of all the metals. They readily tarnish in air and react with water to form bases (alkalis). The metals may be prepared by the electrolysis of the fused hydroxides or chlorides, or by the interaction of the carbonate (except lithium) with carbon at high temperatures. In ascending order of atomic weight this family of elements contains lithium (6.94), sodium (22.99), potassium (39.1), rubidium (85.5) and caesium (132.9).

Alkaline-earth bearing alloys. These consist of lead hardened by calcium and barium. They are of the precipitation hardening type and are prepared by electrolysing fused calcium chloride and/or barium chloride with lead as the cathode material. They include Bahn metal, Fraiy metal and Union metal.

Alkaline-earth metals. This family contains the elements calcium (at.wt.40.1), strontium (87.6), barium (137.4). Radium is also a member of this family, but, owing to its unusual properties and its relation to uranium, it is usually considered apart. The oxides of these metals, lime, strontia, baryta, were originally called the alkaline earths on account of their resemblance to the alkalis and to the oxides of the earths (e.g. those of aluminium and iron). All the metals are silvery white. They readily tarnish in air and react with water. They may be prepared by the electrolysis of the fused chloride of the metal. (See also **Calcium, Barium, Strontium**.)

Alligation. The term means "binding together" and is applied in metallurgy to ore deposits in which two metallic ores are closely associated, e.g. cassiterite (tin oxide) and wolfram (tungsten oxide) in acid igneous rocks.

Alligator skin. See **Orange-peel effect.**

All-mine pig-irons. Pig-irons produced from charges consisting only of ore, flux and fuel.

Allomeric. Used of substances which have the same crystalline form but different chemical composition.

Allomorphous. Used of substances having the same chemical composition but different crystalline forms.

Allophane. A hydrated silicate of alumina $Al_2O_3.SiO_2.5H_2O$.

Allotriomorphic. A term applied to crystals which have taken their shape from their surroundings and have not been allowed to develop the contours prescribed by their crystalline form.

Allotropy. A term used to describe the property possessed by certain chemical elements (e.g. carbon, sulphur, iron, etc.) existing in two or more distinct forms which are identical chemically but have different physical properties (e.g. density, crystalline form, solubility, etc.). Under certain conditions (e.g. temperature and pressure) one allotropic modification may be converted into another form.

Alloy. A substance possessing metallic properties and composed of two or more elements of which at least one must be a metal. The term is usually reserved for those cases where there is an intentional addition to a metal for the purpose of improving certain properties. Though pure metals may possess certain useful properties, such as high thermal and electrical conductivity, they seldom possess the strength required for industrial applications. Copper is practically the only metal used in bulk in the commercially pure state. In the case of most other metals, alloying elements are added to increase the hardness, strength and toughness of the basic metal and to obtain properties that are not found in the pure metals.

The influence of an alloying element depends on the manner in which it is present in the resultant alloy, i.e. on the nature and constitution of the alloy. This can best be understood by reference to the simplest case of a binary alloy system. Most metals are completely miscible in the molten state, and such an intimate blending of metals may be regarded as a solution of the constituent metals in one another. Metals which are immiscible in the molten state rarely provide useful alloys. In the molten state some metals can enter into chemical combination, and when this occurs the melt may be regarded as a mutual solution of the intermetallic compound and the excess component metal. Apart from pure intermetallic compounds and eutectic alloys, which, like pure metals, have a definite solidification temperature, all alloys solidify over a range of temperature. The behaviour of the mutual solution of the molten metals during solidification influences the properties of the alloy. In extreme cases the state of mutual solution may remain undisturbed and only one constituent be present in the solidified metal. The structure of an alloy of this type resembles that of a pure metal and consists of polyhedral grains, each of which has the same *average* composition as that of the molten metal. Such crystals are known as solid solutions. Only a few pairs of metals, such as copper and nickel, can form solid solutions throughout the whole range of possible compositions.

At the other extreme, the two component metals may separate completely during solidification, and in this case the solid metal will consist of a mixture of the crystals of the two pure metals. Intermediate between these two types are those alloys in which the state of mutual solution is partially maintained in the solid state. Most metals can contain a certain limited amount of another metal in solution, and alloys containing less than this limiting amount of the second constituent solidify as solid solutions without separation of the components. Alloys containing more than the limiting proportion of the second constituent undergo a partial separation during solidification. In these alloys the crystals are not pure metals but consist of saturated solid solutions of each of the metals in the other.

Other transformations may take place in the solid state and lead to the formation of new phases. The most comprehensive and satisfactory means of describing the nature and constitution of alloys in any given system is provided by the "equilibrium diagram", or "constitutional diagram", which is based primarily upon a study of thermal data.

Intermetallic compounds are usually hard and brittle. They have a hardening effect upon the alloy and are a very desirable feature of many alloys, particularly bearing metals, but when present in large amounts have an embrittling effect upon the metal. As a general rule, solid solutions containing relatively small amounts of alloying element possess the characteristic properties of the predominant metal, but hardness and strength are increased. (*See also* Tables 3-9.)

Alloy cast iron. A term applied to cast irons containing alloying elements such as nickel, chromium, molybdenum, etc., which have been added to modify the properties of the metal. These properties depend very largely on the manner in which the carbon is present in the iron — whether it exists as free carbon (e.g. flake graphite, nodular graphite, etc.) or in combination as carbides. For many purposes, a pearlitic structure which combines resistance to wear with good machinability is desirable, but under normal conditions this demands the use of a low-silicon iron. Nickel and aluminium, like silicon, tend to increase the formation of graphite. Chromium, molybdenum and vanadium increase the tendency to form carbides. Alloying elements also have an effect on the structure and properties of the pearlite and when present in sufficient quantities produce changes similar to those observed in alloy steels.

Alloy, non-ferrous. An alloy based on metals other than iron. The usual basis metals in non-ferrous alloys are copper, lead, zinc, aluminium, tin, magnesium and nickel.

Alloy steel. All steels contain carbon and small amounts of silicon, sulphur, manganese and phosphorus. Steels which contain intentional additions of elements other than these or in which silicon and manganese are present in large amounts are termed alloy steels. The alloying elements are deliberately added to produce certain properties in the product. In government statistics, alloy steel figures relate to steel, other than high-speed steel, containing any of the elements named below and in the quantities indicated: Cr or Ni 0.4% or more, Mo, W or V 0.1% or more, Mn 10% or more.

Alloy system. A term used to describe all alloys which can be produced by alloying the component metals, i.e. 0-100% of each metal. When two metals are mixed, the resulting system is termed a binary alloy system. The terms ternary, quaternary and quinary alloy systems refer to mixtures of three, four and five metals respectively.

Alluvial deposits. These are deposits formed by the action of running water. The deposits are found where the velocity (and hence the burden) of the flowing water has been decreased. They consist of sandy or gravelly material which becomes finer as the velocity of the water is progressively reduced. In alluvial deposits, minerals from the primary source are deposited in fractions according to their specific gravity. Heavy minerals like gold, wolfram and cassiterite are usually deposited in the higher reaches of the river or stream, while the lighter materials are carried farther downstream.

Almandine. A deep red variety of garnet, sometimes referred to as "precious garnet" $(Fe,Mg)_3.Al_2(SiO_4)_3$. It often shows a violet tinge and resembles rubies in most properties, but is less hard and has a higher specific gravity. It exhibits single refraction and no dichroism. Found in Sri Lanka, Brazil, Tyrol, the USA and Greenland.

Alpha iron. An allotropic modification of iron which crystallises in the body-centred cubic system and is stable below 910 °C. The crystallographic change is not instantaneous, and consequently there is some difference between the transition temperatures on heating and cooling. In cast iron and steels the transition temperature varies with the carbon content. The phase is commonly known as ferrite.

Alpha-particles. These are positively charged particles that are emitted by certain radioactive substances. They are helium atoms that have lost two electrons. When an electric discharge is passed through helium at low pressure, doubly-ionised helium is a product that is identical with α particles.

Alplate process. A patented process for applying coatings of aluminium, magnesium and beryllium to steel, nichrome alloys, etc. The surfaces of the steel are saturated with hydrogen at 1,000 °C and the steel is then dipped in a bath of molten aluminium at a lower temperature. The hydrogen evolved at the moment of contact with the cooler molten aluminium reduces surface films of oxide, prevents the formation of aluminium oxide and enables a good coating to be obtained.

Alrak process. A chemical treatment for protecting the surface of aluminium and aluminium-base alloys. It involves immersion in a boiling solution of 5% sodium carbonate and 1% sodium chromate.

Alstonite (Bromlite). A double carbonate of calcium and barium which occurs as gangue material in lead veins, associated with galena and barytes. It is used in sugar-beet refining and in pottery.

Alumilite process. An anodic oxidation process used for light metals. The electrolyte is sulphuric acid solution.

Alumina. An earthy mineral which is essentially oxide of aluminium Al_2O_3. It is a constituent of all slates, clays and shales, and in the pure state is found naturally as ruby, sapphire and amethyst and as corundum. Electrically fused alumina is an important abrasive, and ranks close to diamond in hardness.

Pure alumina is a suitable polishing medium for general use in metallographic laboratories. It is sold under the name of diamantine, but is easily prepared in the following manner. Pure ammonium alum is heated in a porcelain dish to drive off the water of crystallisation. It is then heated for 4-5 hours in a muffle furnace at 1,000 °C, the crust being broken down at frequent intervals to facilitate the escape of the acid fumes. On cooling, the product is crushed and bottled for use. Artificial gems are now produced from molten alumina (m.p. 1,900 °C).

Aluminised screens. A term used to refer to screens of cathode-ray tubes in which a thin smooth layer of aluminium is laid down on the phosphor layers. The layer must be thin enough not to absorb the electron beam to any serious extent, but thick enough to stop gas ions which would damage the screen by producing ion spots. When smooth, the aluminium layer acts as a mirror surface and increases the light output of the tube, and at the same time allows the screen to conduct away the electronic current without using secondary-emission effects.

Aluminising. The name applied to a process for impregnating the surface of carbon or alloy steel with aluminium in order to obtain protection from oxidation and corrosion — usually from flue gases at high temperature. The iron-aluminium alloy layer is normally 0.64-0.76 mm deep.

Aluminium. This metal ranks third in abundance among all elements and, in the form of silicates, comprises nearly one-tenth of the weight of the earth's crust. It is never found native, but all clays contain aluminium silicate, usually in association with the silicates of iron, calcium and magnesium. The extraction of aluminium from clay, however, is not at present an economic proposition and the most important source of aluminium is bauxite — a hydrated aluminium oxide $Al_2O_3.2H_2O$, which remains as a residue when aluminium silicates have decayed and lost their silica by prolonged tropical weathering. Sulphurous acid has been used to extract aluminium from clays in Germany and Russia. The metal has also been extracted from Labradorite (a felspar), corundum Al_2O_3, leucite $K_2O.2Al_2O_3.4SiO_2$ and from nepheline $K_2O.3Na_2O.4Al_2O_3.9SiO_2$.

The metal was first isolated in 1827, but until the closing years of the last century production was on a very small scale, and the metal was regarded as an expensive curiosity of a very light "precious" metal.

As the oxide is very refractory and cannot be reduced to the metal by any of the usual smelting processes, the metal did not become of industrial importance until electrochemical methods of extraction were developed from the Heroult (France) and Hall (USA) processes about 1886. Even the highest grades of bauxite contain appreciable impurities, chiefly oxides of iron and titanium, and must be purified before the metal is extracted. The ore is mixed with caustic soda solution and heated in pressure chambers so that the alumina may form the soluble compound sodium aluminate. This is separated from the insoluble residues by filtration. Aluminium hydroxide is then precipitated from the solution and heated at a temperature of 1,500 °C to obtain practically pure alumina. This latter is then dissolved in molten cryolite $3NaF.AlF_3$ and reduced to aluminium by electrolysis.

An ample supply of water power, for generating the necessary electrical energy, is a great advantage and consequently the largest producers of aluminium are Canada and the USA, where hydroelectric power is available.

Aluminium has a bluish silvery-white appearance and is capable of taking a high polish. Its most valuable property is lightness, the specific gravity (2.7) being about one-third that of steel. It is a good conductor of electricity, the conductivity of the pure metal being about 62% that of pure silver and, though this is modified by the presence of impurities and by the condition of the metal, the weight of aluminium required for a conductor of similar performance is about 50% that of copper.

Aluminium has a high thermal conductivity, a high specific heat and a latent heat of fusion of about 377 kJ kg^{-1}. It melts at 660 °C and its boiling point is estimated at about 2 500 °C. It is a comparatively soft metal, with a Brinell hardness number of 24, and possesses good ductility. It flows readily under pressure and can be rolled into foil, drawn into wire or extruded into almost any desired section. Though it is a very reactive metal it has a high resistance to atmospheric corrosion, because its surface is always protected by a very thin but extremely tenacious film of oxide. This film can be thickened (up to about 0.025 mm) by an anodising process in which the metal becomes the anode of an electrolytic cell containing an electrolyte of chromic or sulphuric acid. This film, which is able to react with and absorb dyes, can be tinted with many attractive colours.

Commercial aluminium is marketed in four grades.

(a) A1. This is the super quality and contains not less than 99.99% Al. It has excellent resistance to corrosion, especially under marine conditions. It is much more expensive than the other grades and is used more exclusively for search-light and other reflectors.

(b) A2. This contains 99.8% Al and is used in chemical plant. Because of its high ductility it is used in collapsible tubes, in foils for bottle caps, wrapping, etc., and for components made by impact extrusion.

(c) A3. This contains 99.6% Al. It possesses high resistance to corrosion and is used in the form of bar, sheet and wire for components in which a high degree of purity is not essential. This grade is used for electrical conductors, e.g. hard-drawn aluminium wire, and steel-cored aluminium conductor-cable for overhead power transmission purposes.

(d) A4. This contains 99% Al and is most extensively used in the wrought form as hollow ware, mouldings, panelling sheets, food containers and similar uses where ductility rather than strength is required.

The strength of pure aluminium is relatively low but may be increased by cold work or by alloying. The strength of certain cast aluminium alloys may be still further improved by heat-treatment, and in wrought alloys the mechanical properties may be enhanced by cold work, by heat-treatment or by suitable combination of these.

The most important alloying elements used for increasing the strength of aluminium are copper, magnesium, manganese, nickel, silicon and zinc. The alloys may be considered in two main groups: cast alloys which are cast directly into the required form by sand casting, die casting or pressure casting; and wrought alloys which are cast into ingots or billets and subsequently fabricated in the desired form by extrusion, rolling, drawing or forging. All the alloys show generally good resistance to atmospheric corrosion and to many of the corrosive media usually encountered, but the alloys containing copper are inferior in this respect to those containing silicon. Alloys containing magnesium, or magnesium and silicon, are frequently coated with pure aluminium to obtain the corrosion-resistant properties of the pure cladding and the strength of the alloy core, e.g. Alclad.

A large proportion of aluminium alloy castings are used in the "as cast" condition, but certain of them, e.g. AC7 (0.8-2% Cu, 0.05-0.2% Mg, 0.75-2.8% Si, 0.8-1.75% Ni, 0.25-1.4% Fe), are given a low-temperature precipitation or artificial ageing treatment. Others, such as AC10 (10% Mg), are subjected to a high-temperature "solution" heat-treatment, while alloys such as AC14 (4% Cu, 1.5% Mg, 2% Ni) are given a combination of the two forms of treatment. One of the best known groups of the aluminium alloys known as duralumin, which was developed from an alloy containing 3.5% Cu and 0.5% Mg, has an interesting historical significance as the first heat-treatable light alloy. In 1906 Wilm discovered that the hardness and strength of this alloy increased gradually during the first four or five days of storage after quenching (ageing). Duralumin was largely used during the 1914-18 war as sheet, strip, girders, rivets, etc., for airships and other types of aircraft. The commonest duralumin alloys contain 4% Cu, 0.5% Mg, 0.7% Mn, 0.7% Si, but the composition varies within fairly narrow limits.

Most of the heat-treatable aluminium alloys are susceptibl. to ageing or precipitation hardening after quenching from about 500 °C, and in certain cases the benefits of increased strength can be enhanced by heating for some hours at about 165 °C (artificial ageing q.v.).

The "Y" alloys, containing about 4% Cu, 2% Ni, 1.5% Mg and 0.5% Si, are particularly suitable for forging and are often used for pistons. Their properties are similar to those of duralumin at atmospheric temperature, but are superior at elevated temperatures.

Aluminium-silicon alloys containing 10-14% silicon with secondary additions of copper, magnesium and nickel are known under the names of "Silumin" or "Alpax" and are mainly used as castings. They are "modified" by fluxing with molten sodium in order to refine the grain structure and improve strength, ductility and soundness.

More complex alloys have been developed to meet the exacting needs of aircraft. Among these are the RR series of high-duty alloys, one of which, RR 77, contains 2.5-3% Cu, 2-4% Mg, 4-6% Zn, together with small amounts of silicon, iron, manganese and titanium, and has a tensile strength of about 620 MN m^{-2}.

Aluminium-tin alloys (5-10% Sn) have found applications in bearings. Aluminium paint is favoured for its good covering powers and its high reflectivity. The metal is also used as a deoxidiser for steels and as an alloying element with certain other metals, principally with copper in aluminium bronze and with steel in heat-resistant aluminium steels. Aluminium powder is used in the Goldschmidt (thermit) process for reducing metallic oxides to metals (see Tables 2 and 4).

Aluminium bronze. This is the name conferred on a series of copper-aluminium alloys which may or may not also contain additions of iron, manganese, nickel, tin and zinc. As tin is seldom present in appreciable amounts, the alloys are not strictly true bronzes. The aluminium content may be up to about 16%, but is more generally between 5% and 12%. The most common alloys contain about 10%. Up to about 9.4% Al can be retained in solid solution in copper, but for normal cold-working practice the limiting content is about 8%, and alpha aluminium bronzes containing from 5% to 7.5% Al are wrought into the form of sheet, strip, bar, rod, wire and tube. The alloys possess good powers of resistance to corrosion and are used for condenser tubes, marine fittings and as a substitute for steel where non-magnetic properties are required.

Aluminium bronze containing about 5% Al, which has a rich golden-yellow colour and does not tarnish readily on exposure to air, is used for ornaments and for decorative architectural fittings. It has been used for making imitation gold cigarette cases and trinkets.

Aluminium bronzes containing more than 9% Al have a duplex structure and are used in the sand-cast, die-cast and wrought forms. These alloys almost invariably contain iron, and nickel is frequently added. Up to 5% of each element, together with 10% Al and 80% Cu gives particularly good mechanical properties.

A high shrinkage and a tendency to form dross cause certain difficulties in casting, but when correctly cast the alloys possess good resistance to corrosion and high strength, which may be further improved by heat-treatment. Aluminium bronze castings and forgings are used for gears, valve seat-inserts, propellers, engineering and ornamental castings. For die-casting the most usual composition consists of approximately 9% Al with about 2% Fe. (See also Table 4.)

Aluminium silicate $xAl_2O_3.ySiO_2$. A compound of aluminium oxide and silica, in which the relative proportions x and y may vary within limits. When water molecules are present, to form a hydrated aluminium silicate, the material is clay, and in the purest form of clay mineral, kaolinite, $x = 1$ and $y = 2$, with 2 molecules of water. When clay is fired, at high temperature, the water is lost and cannot easily be recombined, hence the compound is an important material of construction. Aluminium silicates are extremely widely distributed in nature, the three elements involved being the most abundant in the earth's crust.

Aluminothermy. When aluminium powder is mixed with oxide of iron (smithy scales) and ignited (in a crucible), the great evolution of heat resulting from the oxidation of the aluminium (and the simultaneous reduction of the iron oxide) causes the iron to melt. The process is used for welding certain articles *in situ*, e.g. rails, ships' propellers. The molten iron is tapped from below. Chromium and manganese oxides can be reduced to the metallic state in the same manner.

Alums. Sulphate compounds, of which the earliest recorded was "alumstone" (q.v.), a double sulphate of potassium and aluminium, with water of crystallisation. The crystals form well-shaped examples of regular octahedra, often to very large sizes. The crystals of some alums form local electric charge dipoles, which can be influenced by the application of an electric field (*see* **Ferroelectricity**) and are therefore important in the design of circuits, as special "switches", for computer use, or in capacitors.

Alumstone (Alunite). A hydrated sulphate of aluminium and potassium $K_2O.3Al_2O_3.4SO_4.6H_2O$.

Alundum. A basic refractory substance made by fusing natural aluminium oxide (bauxite) in an electric furnace. The impurities are removed by settling and the product is crushed with clay and felspar, and fired at 1,500 °C.

Alunite. *See* **Alumstone.**

Alzak process. An electrolytic brightening process used for the production of aluminium reflectors. The process involves treatment in an acid electrolyte, usually fluoboric acid, followed by immersion in a hot alkaline solution. The brightening surface is subsequently given an anodic oxide film in sulphuric acid.

Amalgam. Mercury attacks and dissolves many metals forming alloys called amalgams, which are liquid when the mercury is in excess. If the excess mercury is removed by squeezing through chamois leather, a pasty, semi-solid amalgam is usually obtained. Gold, silver, tin, zinc, lead, antimony, bismuth, copper and the alkali metals may be amalgamated by solution in mercury, but are not easily attacked unless the surface of the metal is clean. The presence of free acid which removes the oxide films facilitates the formation of the amalgam. Copper amalgam is usually produced by decomposing a mercury salt with copper. Iron amalgam may be obtained by electrolysis of iron II chloride (ferrous chloride) solution using mercury as the cathode. Amalgams of iron, nickel, cobalt, manganese and platinum can be obtained by the action of sodium amalgam on the salts of these metals.

Amalgams find many varied uses in industry. The zinc plates of electric batteries are amalgamated to minimise attack on the metal and to prevent wastage of zinc by local action. Tin amalgam is used for "silvering" mirrors and gold amalgam is sometimes used for gilding. Amalgams of gold, silver and alloys containing varying proportions of silver, tin and copper are used for dental fillings. The fact that gold and silver dissolve readily in mercury is the basis of the amalgamation process for the extraction of these metals.

Amalgamation. The term applied to a process of extracting gold and silver from their ores. After crushing and pulverising in stamp mills, the pulverised ore runs over mercury or over amalgamated copper plates. The ore particles amalgamate with the mercury and at intervals the plates are scraped. Mercury is removed from the scrapings by distillation, leaving the gold or silver in the retorts. The tailings of auriferous ores are usually further treated by the cyanide process to improve the extraction of gold.

Amblygonite. An aluminium-lithium phosphate, used as a lithium ore.

Amethyst. A purple or violet transparent form of quartz SiO_2 which owes its colour probably to the presence of traces of manganese. Used as a gemstone.

Amianthus. The name given to the best and more silky kinds of asbestos. It consists of long flexible fibres and, because of some resemblance to flax, it is sometimes called mountain flax. It is used in making incombustible curtains and board for fireproof purposes. It is also used for lampwicks and as a filling for gas fires. The ancients called asbestos "amianthus", which means undefiled, in allusion to the ease with which fabric woven from the mineral was cleaned by fire.

Ammonia carburising. The simultaneous nitriding and carburising of steel in a gaseous atmosphere containing both ammonia and carburising gases.

Ammonium dihydrogen phosphate. A material which is piezo-electric, i.e. produces an electric field when strained, and vice versa. Used in mechanical-electrical transducers.

Amorphous carbide. In the production of carborundum in the electric furnace the silicon carbide is usually surrounded by a layer of siloxicon, which is thought to be a solid solution of silica in silicon carbide. This is referred to as amorphous carbide and is used as a refractory. It is said not to be affected by iron or iron oxide at a white heat.

Amorphous material. Material which does not display the usual characteristics of a crystalline substance, i.e. the regularity of atomic arrangement is of short range order only. Some materials can be prepared in either form, depending on applied conditions. The X-ray diffraction pattern of an amorphous material resembles that of a liquid, and a liquid shows only short range order. Not all materials lacking geometrical arrangements of atoms in the solid state are amorphous however (*see* Fig. 3).

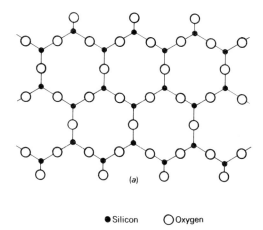

● Silicon   ○ Oxygen

Fig. 3(a) Crystalline silica.

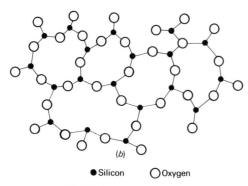

● Silicon ○ Oxygen

Fig. 3(b) Vitreous (amorphous) silica.

Amosite. A form of asbestos (q.v.).

Amphiboles. A group of minerals containing silicates of iron, magnesium, calcium and sometimes sodium and potassium, with or without silicate of aluminium.

Amphigene. A white or ash-grey silicate of aluminium and potassium $K_2O.Al_2O_3.4SiO_2$ sometimes known as white garnet (leucite).

Amphoteric. A term applied to oxides and hydroxides which can act basic towards strong acids and acidic towards strong alkalis. Substances which can dissociate electrolytically to produce hydrogen or hydroxyl ions according to conditions.

Anatase. A variety of titanium oxide TiO_2 crystallising in slender, nearly transparent tetragonal prisms.

Andrade creep. *See* **Creep.**

Andradite. Common garnet, calcium-iron silicate $3CaO.Fe_2O_3.3SiO_2$.

Angle of loss. In an alternating system, such as vibration of a material, the driving force is out of phase (ahead) of the displacement occurring in the material. If the force and displacement are plotted graphically as simple harmonic motion (sine waves) then the angle between two comparable points and the zero, is the phase difference or angle of loss, usually given the symbol δ. In vibration, δ is a measure of the capacity of the material to damp out the vibration. If the applied stress is σ and the displacement ϵ then $\sigma = \sigma_m \sin \omega t$ where σ_m is the maximum stress, ω the frequency, t the time and

$$\epsilon = \frac{\sigma_m}{E} \sin (\omega t - \delta)$$

Where E = modulus of elasticity of the material.

Angle of nip. A term used in connection with rolling mills. It is the angle between the radius of the work rolls at the point of initial contact and the normal to the sheet surface at this point. Sometimes termed "angle of bite" (*see* Fig. 4).

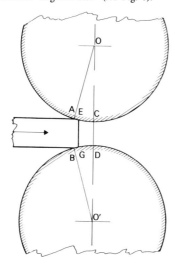

Fig. 4 *Diagram showing the angle of nip (AOC and BO'D) of a pair of rolls.*

Anglesite. A sulphate of lead $PbSO_4$ crystallising in the orthorhombic system. It is usually associated with galena and results from the decomposition of that mineral in the upper parts of lead veins. When found in quantity it is a valuable lead ore.

Angstrom unit. A unit named in honour of the Swedish physicist, A.J. Angstrom. One angstrom unit is equal to one ten-millionth of a millimetre (= 10^{-10} metre). The usual abbreviations are Å or A.U. (not an SI unit).

Angular momentum. With reference to electrons moving in the vicinity of atomic nuclei, the product of the linear momentum and the radius of the path of motion, assumed circular, i.e. the "moment" of momentum. Moving electrons behave in a manner which can be associated with a wave motion, as well as resembling a particle. To express the angular momentum associated with their motion, arbitrary quantum numbers are used. The magnitude of the angular momentum of an electron can be expressed as $(h/2\pi) [\ell (\ell + 1)]^{1/2}$ where h = Planck's constant, ℓ is the secondary quantum number associated with angular momentum, having values from 0 to $(n - 1)$ where n is the principal quantum number, a measure of electron energy in elementary terms. Angular momentum of electrons produces several effects in atoms, principally a magnetic field, associated with the moving electric charge.

Angus Smith process. A rust-preventing process applied to sanitary ironwork. Immediately after casting, the metal is heated to about 315 °C and then plunged into a hot mixture consisting of 4 parts coal tar, 3 parts prepared oil and 1 part paranaphthalene.

Anhydrite. Anhydrous calcium sulphate $CaSO_4$ which occurs as a saline residue associated with gypsum and rock salt. It is used in the manufacture of ammonium sulphate fertiliser.

Anion. A term invented by Faraday to designate the particles which move in an electrolyte, under the guidance of the current, towards the anode.

Anisotropy. The reverse of isotropy; a reduction in symmetry of the atomic arrangement of crystals, usually variation in the separation distance of neighbouring atoms in particular directions. When stressed, anisotropic materials deform to different extents in particular directions and show differences in coefficients of linear thermal expansion in those directions. Cubic crystals, with atoms at the corners of a cube, are almost isotropic in some properties. Tetragonal crystals, with atoms at the corners of a rectangular prism, are anisotropic, because one side of the prism is longer than the other two perpendicular to it. Several unusual phenomena in materials can be attributed to anisotropy, e.g. thermal fatigue, "growth" in graphite and similar materials (*see* Fig. 5).

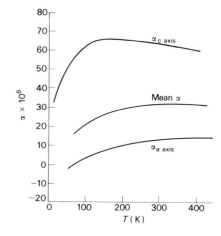

Fig. 5 *Anisotropic expansion of a hexagonal metal—zinc.*

Ankerite. A variety of dolomite containing some iron (II) carbonate (ferrous) besides calcium and magnesium carbonates.

Annabergite. A hydrated arsenate of nickel $Ni_3(AsO_4)_2.8H_2O$

resulting from the decomposition of kupfernickel, which occurs as a coating of apple-green crystals and is known as "nickel bloom".

Annealing. On account of the slowness with which changes occur in solid metals, many alloys are in a condition of imperfect equilibrium both chemically and structurally even when very slowly cooled from the liquid state. Castings and cast ingots have a pronounced dendritic structure which may gradually be eliminated to produce a more homogeneous equiaxed structure, with improved mechanical properties, by suitable heat-treatment. A similar type of treatment may be used to remove the effects of hot or cold deformation. Annealing is a heat-treatment process which usually involves a relatively slow cooling after holding the material for some time at the annealing temperature. The annealing temperature and the rate of cooling depend on the material being treated and the purpose for which the treatment is carried out. These purposes may include the following: (a) to induce softness in the material; (b) to remove internal stresses resulting from casting or from previous heat-treatments; (c) to refine the crystal structure; (d) to modify physical properties such as ductility, toughness and electrical or magnetic properties; (e) to eliminate dissolved gases; and (f) to produce a definite micro-structure. In the heat-treatment of iron and steel, the term annealing covers the following processes: full annealing, process annealing, normalising, spheroidising, tempering, malleabilising, graphitising. The process of annealing ingots is called soaking.

Annealing carbon. A finely divided, probably amorphous, form of carbon which separates from white cast iron and certain steels during prolonged annealing (as in malleable cast iron).

Anode effect. A term used in electrolytic processes to denote the sudden drop in current due to the formation of a film of gas on the surface of the anode.

Anode mud. The insoluble residue which is left on or falls from the anode in electrolytic refining processes.

Anodic. In a general sense means above hydrogen in the electro-chemical series, but more specifically may signify a relatively more positive potential.

Anodic oxidation. Oxidation caused by nascent oxygen liberated at the anodes of an electrolytic cell. A process for producing a hard corrosion-resistant film of oxide on the surface of aluminium or light alloys, by making the metal the anode in an electrolytic cell containing chromic or sulphuric acid. Chromic acid gives a grey opaque coating, while sulphuric acid produces a clear, translucent finish. The process is also known as anodic treatment or anodising.

Anodising. *See* **Anodic oxidation.**

Anolyte. That portion of the electrolyte immediately surrounding the anode, which is changed in chemical composition and concentration by the reactions at the anode.

Anorthite. A lime felspar containing mainly calcium aluminium silicate $Ca.Al_2.Si_2O_8$ with sodium aluminium silicate. It occurs as a primary constituent of basic igneous rocks and crystallises in the triclinic system. With albite, it forms an isomorphous series. As the proportion of albite decreases the refractive index of the mineral increases (100% albite, $\mu = 1.53$; 100% anorthite, $\mu = 1.58$).

Anorthoclase. A sodium potassium felspar consisting of aluminium silicate and varying amounts of sodium and potassium silicates.

Anthophyllite. A silicate of iron and magnesium $(Mg.Fe)O.SiO_2$. It is a variety of asbestos.

Anthracite. A black or brownish-black variety of coal. It is sometimes iridescent; its lustre is usually brilliant and it breaks with a conchoidal or uneven fracture. Its streak is black and it does not soil the fingers. Anthracite contains up to 95% C, the elements hydrogen, oxygen and nitrogen being present in small quantities. It is less easily kindled than most fuels and burns slowly and with little flame or smoke, but emitting considerable heat during combustion. It occurs locally in many coalfields and is mined chiefly in South Wales, Scotland, Pennsylvania and Canada. (*See also* **Coal.**)

Anthraconite. A dark-coloured limestone containing bituminous matter which emits a fetid odour when struck (also known as

Stinkstone).

Anticathode. This is the "target" inserted in an X-ray tube on to which the cathode rays or high-speed electrons are directed. The anticathode is connected electrically, and outside the tube, with the anode. In a Coolidge tube the anticathode is also the anode. The impact of the electrons on the anticathode causes the latter to emit X-radiation characteristic of the substance of which it is made.

Antiferroelectric. The electrical analogue of antiferromagnetism, in which ions in the structure align their charges in antiparallel fashion, i.e. they are polarised, but the alignment in opposite directions of an equal number of equal and opposite charges produces no net electric field.

Antiferromagnetic. When the magnetic moments of individual atoms in a material are distributed with as many pointing in one direction as in the opposite direction, so that the effects cancel out exactly, the material is said to be antiferromagnetic. This occurs in some metals of the transition group, e.g. chromium and manganese (q.v.), in many intermetallic compounds and in some oxides.

Anti-fouling compound. A substance applied as a paint to ships' bottoms and to steel structures which are subject to the action of sea-water in order to discourage the formation of marine growths. These marine paints usually contain metallic soaps, including copper, arsenic and mercury.

Antifriction metal. A name applied to Babbitt's and some other alloys used in bearings. They are usually alloys of copper, lead, tin, antimony and zinc and are characterised by the low friction power losses when used for bearings.

Antimonial nickel. A mineral consisting mainly of antimony and nickel, frequently also containing considerable amounts of lead sulphide. It is found in the Harz district of Germany (Andreasberg). It may be regarded as a nickel antimonide, NiSb, and is also known as Breithauptite.

Antimonite. Antimony trisulphide, Sb_2S_3. It is also known as stibnite, antimony glance and grey antimony. It is the chief source of antimony.

Antimony. A lustrous metal with a bluish silvery-white appearance and a crystalline (rhombohedral) structure, which may be coarsely laminated or granular, according to the rate of cooling. It is very brittle and can easily be pulverised to a fine powder (antimony black). The metal melts at 630 °C and boils at about 1,635 °C but volatilisation occurs at much lower temperatures. During solidification antimony expands very slightly and this property is exploited in certain alloys. The metal does not tarnish readily on exposure to air and thus has been used as a decorative coating. It is a poor conductor of heat and electricity.

Native antimony, sometimes with traces of silver, iron and arsenic, occurs in small deposits in Borneo, but the chief source of the metal is stibnite Sb_2S_3. The metal may also be produced from the oxide ores cervantite $Sb_2O_3.Sb_2O_5$ and valentinite Sb_2O_3. Stibnite has for centuries been used as a medicine and by women of the East as a cosmetic for darkening the eyebrows. China is normally the chief producer of the metal from ore deposits in Minesin and other areas. At one time Bolivia produced about one-third of the world's supply.

Antimony is usually extracted from sulphide ores found in France, Sweden, Borneo, Bolivia, Mexico, China and Japan. The pulverised stibnite concentrates are heated with iron scrap in plumbago crucibles under a flux of salt or nitre. Poor ores may be concentrated by a preliminary "liquation treatment", in which the ore is placed in perforated pots and heated until the sulphide melts and runs out into a receiver.

Antimony ores may also be treated by roasting, to convert the sulphide into oxide, and subsequently reducing the oxide by carbon in small reverberatory furnaces in the presence of an alkaline flux such as soda ash or salt cake. The metal produced by these processes is subsequently refined to eliminate iron, sulphur and arsenic. A considerable amount of volatilisation takes place during smelting and refining, and this volatilised metal is collected in chambers for refining.

The metal exhibits allotropy, the normal rhombohedral variety, or β antimony, being the most stable. Yellow or α antimony is formed when antimony hydrite SbH_3 is in contact with air at −90 °C. This passes into the black antimony on exposure to light and on heating.

Finely powdered antimony (or black antimony) is used to impart a metallic lustre to plaster casts and for certain types of bric-a-brac, but the pure metal is too brittle for most commercial applications. Its main uses are in alloys, many of which inherit the property of expanding slightly on solidification and are therefore able to produce fine impressions. The most important alloys are the Babbitt bearing metals, which consist of tin, antimony and copper, with or without additions of lead; Britannia metal and pewter, which consist mainly of tin, antimony and copper; and type metals, which are lead-rich lead-tin-antimony alloys. Certain types of solder, which are unsuitable for use on zinc or zinc-containing alloys or in automatic soldering machines, may contain antimony up to 6% of the tin content. Antimony is used as a hardening agent in tin-base die-casting alloys for syphon tops, etc. Copper-base antimony-nickel alloys have been developed in Canada for worm wheels and gears. Alloys of antimony and lead are used in the making of bullets, battery plates and toys and in chemical plants. Antimony oxides and sulphides are used as pigments in the manufacture of paints and enamels. (See also Tables 2 and 8.)

Antioxidant. Material added to polymers, particularly elastomers and some thermoplastics, to prevent cross-linking by the ingress of oxygen or sulphur (see **Ageing**). Carbon black is used as a filler in rubber to exclude light, since ultra-violet radiation enhances the cross-linking and therefore hardening.

Apatite. The double phosphate and fluoride of calcium

$$3Ca_3(PO_4)_2 . CaF_2$$

or the double phosphate and chloride of calcium. In the latter case the mineral is called chlorapatite. Usually the chloride and fluoride are present together. It occurs as a primary constituent of igneous rocks in veins of pneumatolytic origin in Canada, Norway, many Pacific islands, Florida and Utah. It is a source of phosphorus and of the fertiliser superphosphate.

Aqua fortis. Strong nitric acid.

Aquamarine. A pale blue variety of beryl $3BeO.Al_2O_3.6SiO_2$ used as a gemstone. The crystals are hexagonal: sp. gr. 2.65-2.9, hardness 7.5 on the Mohs scale (see **Beryl**).

Aragonite. Calcium carbonate is dimorphous, the two forms being calcite (q.v.) and aragonite. The latter was first found in Aragon and hence its name. Aragonite is the orthorhombic variety of calcium carbonate which occurs in beds of gypsum with iron and sometimes small amounts of strontium carbonate.

Arborescent. A term signifying "tree-like" and applied to crystallites having the appearance of fir trees, which sometimes occurs in cavities in metals (cf. dendritic).

Arc blow. A phenomenon occurring during arc-welding that is associated with the tendency of the arc to wander. The deflection of the arc from the shortest path is attributed to the production of self-induced magnetic fields, and the effect is likely to be greater when direct current is used with ferromagnetic materials.

Arctic bronze. The name of a series of leaded bearing bronzes that have a fine grain structure produced by rapid cooling in (chill cast) metal moulds.

Argentiferous galena. A sulphide of lead PbS in which silver is present as sulphide in sufficient quantities to justify economic extraction.

Argentite. A sulphide of silver Ag_2S containing 87% Ag and one of the most important sources of this metal. It crystallises in the cubic system, has a specific gravity of 7.2-7.36 and a hardness of 2-2.5 (Mohs scale). It is found in Montana, Nevada, Ontario, New South Wales, Mexico, Peru, Bolivia, Chile, Saxony, Bohemia and Norway. The mineral is also called silver glance.

Argon. An inert gas, present to the extent of about 1% in atmospheric air. It is prepared from air by liquefaction of the latter. The more volatile nitrogen is allowed to distil off first and the oxygen is separated in a rectifying column, since the boiling point of argon at −186 °C lies between those of oxygen and nitrogen, which are respectively −182.7 °C and −196 °C. The gas does not combine with any known element. It is used in "gas-

filled" incandescent lamps and in Argonarc welding. It has been shown that some of the "rare" gases can be brought into combination with fluorine either by reaction with fluorine alone or with platinum hexafluoride, and that the compounds XF_2, XF_4 and XF_6 can be obtained. Analogous compounds with krypton have been prepared but they are unstable above about −180 °C. Combination of fluorine and argon will not easily be effected.

Armour plate. An alloy steel in plate form for the protection of warships and tanks and for the making of armaments. Typical composition: C 0.2-0.4%, Cr 1-3.5%, Ni 1.5-3.5%, Mo 0-0.5%.

Arrest points. A discontinuity or break in the continuity of the heating or cooling curves of a metal, due to an absorption of heat on heating or to an evolution of heat on cooling. These discontinuities occur at particular temperatures and indicate that the metal is undergoing a change. The letter A, taken from the French arrêt, is used to indicate an arrest point, and Ac is the abbreviation for the arrest point on a heating (chauffage) curve and Ar for the arrest point on a cooling (refroidissement) curve. The various arrest points in the heating and cooling curves of metals are distinguished by numbers written after the letters, being numbered as they occur with increasing temperature. The nomenclature is illustrated in the case of plain carbon steel in Table 12 (see Fig. 6).

TABLE 12. *Arrest Points of Carbon Steel*

Arrest	Temperature	Significance
A_0	200 °C	A magnetic change in cementite.
A_1	723 °C	The eutectoid transformation austenite ⇌ ferrite and cementite.
A_2	768 °C	A change in the magnetic and certain other properties of α iron (not a phase change).
A_3	723-910 °C	The beginning of separation of ferrite from austenite on cooling or the completion of solution of ferrite in austenite on heating.
A_3	910 °C	The temperature at which pure iron undergoes a crystallographic change from a body-centred cubic (α-iron) to a face-centred cubic lattice (γ-iron).
A_4	1,400 °C	The transformation γ ⇌ δ iron.

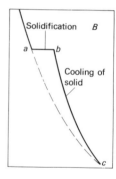

Fig. 6 *Typical cooling curves showing an arrest point. A shows cooling without change of state. B shows arrest point ab. The horizontal part ab represents the process of solidification and the curve bc the cooling of the solid metal.*

Arrhenius plot. Arrhenius explained the kinetics of reactions in terms of "active" molecules which could undergo reaction, and whose number was dependent on temperature. He evolved the expression

$$\frac{d \log_e k}{dt} = \frac{E}{RT^2}$$

where k is the velocity coefficient of reaction, t time, T temperature, R the gas constant, and E another constant, later given the name "activation energy". Integrating the equation, $\log_e k = I − E/RT$, where I is an integration constant. By plotting values of $\log_e k$ against $1/T$, a straight line is obtained, the value of the slope being E, the activation energy.

Arsenic. Ordinary arsenic is a steel-grey element with a brilliant metallic lustre, which crystallises in the rhombohedral system. It occurs native in veins associated with lead and silver ores in certain parts of Saxony, Bohemia, etc. The fresh fracture of arsenic is tin white, but the surface quickly tarnishes to grey.

In combination with other elements it is widely distributed and arsenic occurs in most pyritic minerals chiefly as oxide As_2O_3, sulphides (orpiment As_2S_3, realgar As_2S_2 and nickel glance NiAsS), sulpharsenides (mispickel FeAsS and cobaltite CoAsS) and arsenides (kupfernickel NiAs and smaltite $CoAs_2$).

Arsenic exhibits allotropy. Alpha (α) arsenic, or yellow arsenic, is produced by the rapid condensation of arsenic vapour, but passes quickly into the grey (γ) variety on exposure to light. Beta (β) or black arsenic is formed by the slow condensation of arsenic vapour.

Grey (γ) arsenic is brittle. If heated under pressure it melts at $817\,°C$, but under ordinary conditions it sublimes without melting at temperatures in excess of $100\,°C$. Commercial arsenic is obtained by the sublimation of native arsenic; as a by-product of other metal-smelting processes; by heating mispickel in earthenware vessels in absence of air to sublime the arsenic; by reducing white arsenic As_2O_3 with charcoal.

Arsenic is used chiefly in the manufacture of arsenic compounds for weed-killers, insecticides, anti-fouling paints, etc., and finds limited applications in alloys. Small amounts of arsenic are used to prevent segregation in certain lead-base bearing alloys, and for increasing the hardness of lead alloys, as in "chilled shot".

Copper can retain up to about 7.25% As in solid solution. Most commercial copper contains traces of arsenic, the most pronounced effect of arsenic being to reduce the electrical conductivity. Arsenical (tough-pitch) copper, in which 0.35-0.55% As is present as an alloying element to increase the tenacity of the metal, withstands flame heat and scaling better than the high-conductivity copper and is used for locomotive fire-box plates and stays and for calico-printing rollers (*see* Table 2).

Arsenolite. Arsenic trioxide formed as a decomposition product of arsenical ores. It usually occurs as a white incrustation.

Artificial ageing. When an alloy is rapidly cooled from a solid solution, segregation is inhibited, and at room temperature the alloy is initially in a metastable condition, and segregation can proceed only at a very slow rate. Segregation proceeding in this way is known as ageing, and artificial ageing is induced by raising the material to an elevated temperature. (*See also* **Ageing.**)

Asbestos. The term commercial asbestos is applied to certain minerals which occur in the form of long fibres, and possess good heat-resisting properties. Under this generic term are included three mineral groups: *(a)* anthophyllite-orthorhombic amphibole (Mg.Fe)O.SiO$_2$; *(b)* hornblende-tremolite CaO. $3MgO.4SiO_2$; *(c)* serpentine-chrysotile, hydrated silicate of magnesia $3MgO.2SiO_2.2H_2O$. The amphibole variety is usually found in veins in association with serpentine and other rocks rich in magnesia. The chrysotile variety is the most important commercially, since it can be woven into rope or cloth. These are used as fireproof materials for filtering acids and corrosive substances, for covering machinery and hot water and steam piping. The amphibole variety is used chiefly in making filter-pads (being more resistant than other varieties) and in cements and plasters and insulating materials. All varieties have very high melting point and low electrical and heat conductivity. The main supplies come from Canada, South and Central Africa, the USSR and the USA. (*See also* **Amianthus.**)

Asbolan. Sometimes referred to as earthy cobalt. It is a decomposition product of manganese minerals, is of variable composition, but may contain up to 32% cobalt oxide.

Assaying. The term given to the chemical analysis of mineral ores and metals.

Asterism. As its name implies, this is an effect, roughly star-shaped, that may be seen in X-ray spectrograms. It is usually taken to indicate the presence of internal stresses in the material that is being examined.

Aston process. A process for producing wrought iron that may be considered an alternative to the puddling process. A slag made from iron oxide and silica is melted in an open-hearth furnace and transferred to a ladle, where Bessemerised pig iron is added slowly. When the gas evolution ceases, the contents of the ladle are allowed to settle. The slag rises to the top and is poured off. The remaining metal is treated by pressing and rolling to give a product resembling puddled iron.

Atacamite. A hydrated oxychloride of copper, crystallising in the orthorhombic system, which occurs in the zone of weathering of copper lodes. Found in South America, Spain, Australia and Cornwall (UK).

Atacticity. The arrangement of the substituent groups of polymers at random above and below the backbone chain. The opposite of stereospecificity.

Atom. A term originally devised to denote the smallest indivisible portion of an element that could take part in a chemical reaction. The first part of this description concerning the indivisibility of an atom is no longer tenable.

Atoms consist of a number of elementary particles, chiefly electrons, protons and neutrons, arranged in a pattern which characterises the physical and chemical properties of the individual element. The protons and neutrons are concentrated in the centre space of the atom, occupying a very small volume, but constituting most of the mass. The electrons are grouped around the nucleus, forming a "cloud" of negative charge, of very low mass. The behaviour of these electrons is responsible for physical phenomena and chemical reaction, particularly those electrons of the highest energy level, furthest in space from the nucleus. This behaviour can sometimes be represented best by the assumption of a roughly spherical shaped particle, but in other cases the electrons are assumed to behave as a "wave motion", which leads to the concept of an electron "cloud" around each atomic nucleus.

The energy levels of the electrons were originally conceived to explain atomic spectra, produced by excitation of atoms (by heat, electric discharges) which caused the electrons to acquire higher energy than their ground state, following which, release of that energy as emitted radiation led to a return to the original energy states. Since the quantum theory required energy take-up or release to follow a systematic pattern (the quanta of energy) it followed that translations of energy between levels must obey a quantum relationship. Bohr, in the first truly acceptable atomic model, conceived the electrons moving in orbits, similar to planets round the sun. If the energy of a ground state orbit is E_1, and that of a higher state E_2, then return from E_2 to E_1 levels causes the emission of a spectral line according to the equation: $E_2 - E_1 = Ah\nu$ where h is Planck's constant, ν is the frequency of radiation emitted, and A is a constant for the particular change involved. From this, Bohr showed that the energy levels of electron orbits would be related inversely to the squares of natural numbers.

Modern quantum theory refined the Bohr model, to describe each electron energy level in terms of four quantum numbers.

(a) n is the principal quantum number, a measure of electron energy in elementary terms, varying from 1 upwards.

(b) ℓ is the secondary quantum number, a measure of the angular momentum of the electron, i.e. characterising the forces which control the motion in space and of value 0 to $(n-1)$.

(c) m_ℓ is a third quantum number, to measure a component of angular momentum in the direction of a superimposed weak magnetic field. The values are $+\ell$ to $-\ell$ through 0.

(d) m_s is the fourth quantum number, to indicate a component of magnetic moment due to the electron spinning on its own axis. The value is $\pm\,½$, to indicate the direction of spin.

The values of ℓ, the secondary quantum number, are frequently denoted by the shorthand $\ell = 0 = s; \ell = 1 = p; \ell = 2 = d; \ell = 3 = f$. As the values of n build up, the energy levels indicated by n and ℓ therefore increase according to the principle that the lowest state of each value of ℓ is 1s; 2p; 3d; 4f.

The energy levels of the electrons can then be built up from these lowest states, so that in a heavy atom, with a large number of electrons, several states of energy will be occupied by electrons. The Pauli exclusion principle states that each energy state governed by a particular combination of values of all four quantum numbers can be occupied by only one electron. This means that values of n and ℓ must have variations in m_ℓ and m_s, i.e. the magnetic field influence, or the spin, must vary. Because of the restrictions on the values of ℓ, 0 to $(n-1)$, there are restrictions of m_ℓ, but there can always be two electrons for a set of values n, ℓ, m_ℓ, since they can be of opposite spin ($m_s = \pm\,½$). A set of rules can then be stated, in general terms, with reference to these variations.

(a) As the principal quantum number n varies 1, 2, 3, 4, etc. the number of electrons that can be contained is 2, 8, 18, 32, respectively, or $2n^2$.

(b) For a given value of n, the maximum number of electrons when $\ell = 0$, the s state, is 2; for $\ell = 1$, the p state, 6; for $\ell = 2$, the d state, 10; and for $\ell = 4$, the f state, 14 electrons.

The build up of the periodic table of elements, in order of atomic number (or total number of extranuclear electrons) can then be seen in terms of the four quantum numbers. Table 13 shows the values of n and ℓ only, for simplicity, and the old-fashioned "chemist's" indication of electron "shells", labelled K, L, M, etc., referring to values of principal quantum number n. The fine structure, involving m_ℓ and m_s, is not too difficult to work out if the rules are used as described above.

Atomic arc welding. An arc welding process in which a jet of hydrogen is played on an arc drawn between two tungsten electrodes.

Atomic co-ordination. The number of nearest neighbouring atoms in a crystal lattice, which is an indication of the closeness of packing. The greatest number of atoms which can be packed into a given volume results in a co-ordination number of 12, close packed hexagonal or face-centred cubic, the lowest number found is 4, in the tetrahedral diamond structure.

Atomic diffusion. The relative movement which occurs between atoms, particularly in the solid state. Atoms can migrate through crystal lattices over considerable distances, but the mechanism varies according to the materials involved and the ambient conditions. (*See also* **Diffusion**.)

Atomic heat. The quantity of heat required to raise the temperature of 1 mole of an element through 1 K, i.e. atomic heat = atomic weight $\times$ specific heat of an element. Atomic weights and specific heats of elements are given in Table 2.

Atomic hydrogen arc. *See* **Atomic arc welding**.

Atomic magnetic moment. Moving electric charges generate magnetic fields, and the electrons in atoms are no exception.

TABLE 13. *Electronic Structure of the Elements*

Shell		K	L		M			N				O				P			
	At. No.	$n=1$	2		3			4				5				6			7
Element		$\ell=0$ 1s	0 2s	1 2p	0 3s	1 3p	2 3d	0 4s	1 4p	2 4d	3 4f	0 5s	1 5p	2 5d	3 5f	0 6s	1 6p	2 6d	0 7s
H	1	1																	
He	2	2																	
Li	3	2	1																
Be	4	2	2																
B	5	2	2	1															
C	6	2	2	2															
N	7	2	2	3															
O	8	2	2	4															
F	9	2	2	5															
Ne	10	2	2	6															
Na	11	2	2	6	1														
Mg	12	2	2	6	2														
Al	13	2	2	6	2	1													
Si	14	2	2	6	2	2													
P	15	2	2	6	2	3													
S	16	2	2	6	2	4													
Cl	17	2	2	6	2	5													
A	18	2	2	6	2	6													
K	19	2	2	6	2	6		1											
Ca	20	2	2	6	2	6		2											
Sc	21	2	2	6	2	6	1	2											
Ti	22	2	2	6	2	6	2	2											
V	23	2	2	6	2	6	3	2											
Cr	24	2	2	6	2	6	4	2											
Mn	25	2	2	6	2	6	5	2											
Fe	26	2	2	6	2	6	6	2											
Co	27	2	2	6	2	6	7	2											
Ni	28	2	2	6	2	6	8	2											
Cu	29	2	2	6	2	6	9	2											
Zn	30	2	2	6	2	6	10	2											
Ga	31	2	2	6	2	6	10	2	1										
Ge	32	2	2	6	2	6	10	2	2										
As	33	2	2	6	2	6	10	2	3										
Se	34	2	2	6	2	6	10	2	4										
Br	35	2	2	6	2	6	10	2	5										
Kr	36	2	2	6	2	6	10	2	6										
Rb	37	2	2	6	2	6	10	2	6			1							
Sr	38	2	2	6	2	6	10	2	6			2							
Y	39	2	2	6	2	6	10	2	6	1		2							
Zr	40	2	2	6	2	6	10	2	6	2		2							
Nb	41	2	2	6	2	6	10	2	6	3		2							
Mo	42	2	2	6	2	6	10	2	6	4		2							
Tc	43	2	2	6	2	6	10	2	6	5		2							
Ru	44	2	2	6	2	6	10	2	6	6		2							
Rh	45	2	2	6	2	6	10	2	6	7		2							
Pd	46	2	2	6	2	6	10	2	6	8		2							
Ag	47	2	2	6	2	6	10	2	6	9		2							
Cd	48	2	2	6	2	6	10	2	6	10		2							
In	49	2	2	6	2	6	10	2	6	10		2	1						
Sn	50	2	2	6	2	6	10	2	6	10		2	2						
Sb	51	2	2	6	2	6	10	2	6	10		2	3						
Te	52	2	2	6	2	6	10	2	6	10		2	4						
I	53	2	2	6	2	6	10	2	6	10		2	5						
Xe	54	2	2	6	2	6	10	2	6	10		2	6						

The angular momentum of the electron creates a field, and so does the electron spin.

If ℓ is the secondary quantum number, a measure of the angular momentum of the electron, ($\ell = 0$ to $n - 1$ where n is the principal quantum number) then the magnetic moment due to angular momentum is represented by

$$\frac{eb}{4\pi mc}\,[\ell\,(\ell + 1)]^{1/2}$$

where e is the charge on the electron, m its mass, b Planck's constant and c the velocity of light. The unit

$$\frac{eb}{4\pi mc}$$

is significant, since ℓ is the variable in respect of each and every electron in the atom. The unit is called the Bohr magneton, a convenient expression for the unit of electron magnetic moment.

The spinning electron generates a magnetic moment given by

$$\frac{2e}{2mc}\cdot\frac{b}{2\pi}\,[s\,(s + 1)]^{1/2},$$

where $s = \frac{1}{2}$. The fourth quantum number m_s indicates the value of the spin moment, equal to 1 Bohr magneton, and $m_s = \pm\frac{1}{2}$ to indicate the direction of spin will be opposed where two electrons have the same value of m_s, m_ℓ, ℓ and n, the four quantum numbers.

As the magnetic moments due to angular momentum will depend on the number of electrons present in the atom, reference is made to the "dipole moment", since each electron causes behaviour similar to a block of material carrying circulating electric currents. The atomic magnetic moment will be the sum of all individual dipole moments, and if these are exactly equal and opposite, as in the rare gases with complete electron shell configurations, no resultant magnetic moment can remain. In the majority of atoms however, the electrons are not equal and opposite, so that some net magnetic moment results. It is the magnitude of these moments, and their dependence on angular momentum or spin of the electrons, which creates the variety of magnetic behaviour. This behaviour is carried over into compounds of the elements, where unusual combinations can result, giving rise to many valuable technological materials. The subject is further discussed in the case of solid elements and compounds, under headings ferro, dia and paramagnetism, ferri, antiferro and antiferrimagnetism.

Atomic number. Usually given the symbol Z, where Z equals the number of positive charges (protons) on the nucleus, and the number of extranuclear electrons in the neutral atom. When

TABLE 13. (contd.) *Electronic Structure of the Elements*

Shell		K	L		M			N				O				P			Q
		$n=1$	2		3			4				5				6			7
	At. No.	$\ell=0$	0	1	0	1	2	0	1	2	3	0	1	2	3	0	1	2	0
Element		1s	2s	2p	3s	3p	3d	4s	4p	4d	4f	5s	5p	5d	5f	6s	6p	6d	7s
Cs	55	2	2	6	2	6	10	2	6	10		2	6			1			
Ba	56	2	2	6	2	6	10	2	6	10		2	6			2			
La	57	2	2	6	2	6	10	2	6	10		2	6	1		2			
Ce	58	2	2	6	2	6	10	2	6	10	1	2	6	1		2			
Pr	59	2	2	6	2	6	10	2	6	10	2	2	6	1		2			
Nd	60	2	2	6	2	6	10	2	6	10	3	2	6	1		2			
Il	61	2	2	6	2	6	10	2	6	10	4	2	6	1		2			
Sm	62	2	2	6	2	6	10	2	6	10	5	2	6	1		2			
Eu	63	2	2	6	2	6	10	2	6	10	6	2	6	1		2			
Gd	64	2	2	6	2	6	10	2	6	10	7	2	6	1		2			
Tb	65	2	2	6	2	6	10	2	6	10	8	2	6	1		2			
Ds	66	2	2	6	2	6	10	2	6	10	9	2	6	1		2			
Ho	67	2	2	6	2	6	10	2	6	10	10	2	6	1		2			
Er	68	2	2	6	2	6	10	2	6	10	11	2	6	1		2			
Tm	69	2	2	6	2	6	10	2	6	10	12	2	6	1		2			
Yb	70	2	2	6	2	6	10	2	6	10	13	2	6	1		2			
Lu	71	2	2	6	2	6	10	2	6	10	14	2	6	1		2			
Hf	72	2	2	6	2	6	10	2	6	10	14	2	6	2		2			
Ta	73	2	2	6	2	6	10	2	6	10	14	2	6	3		2			
W	74	2	2	6	2	6	10	2	6	10	14	2	6	4		2			
Re	75	2	2	6	2	6	10	2	6	10	14	2	6	5		2			
Os	76	2	2	6	2	6	10	2	6	10	14	2	6	6		2			
Ir	77	2	2	6	2	6	10	2	6	10	14	2	6	7		2			
Pt	78	2	2	6	2	6	10	2	6	10	14	2	6	8		2			
Au	79	2	2	6	2	6	10	2	6	10	14	2	6	9		2			
Hg	80	2	2	6	2	6	10	2	6	10	14	2	6	10		2			
Tl	81	2	2	6	2	6	10	2	6	10	14	2	6	10		2	1		
Pb	82	2	2	6	2	6	10	2	6	10	14	2	6	10		2	2		
Bi	83	2	2	6	2	6	10	2	6	10	14	2	6	10		2	3		
Po	84	2	2	6	2	6	10	2	6	10	14	2	6	10		2	4		
At	85	2	2	6	2	6	10	2	6	10	14	2	6	10		2	5		
Rn	86	2	2	6	2	6	10	2	6	10	14	2	6	10		2	6		
Fr	87	2	2	6	2	6	10	2	6	10	14	2	6	10		2	6		1
Ra	88	2	2	6	2	6	10	2	6	10	14	2	6	10		2	6		2
Ac	89	2	2	6	2	6	10	2	6	10	14	2	6	10		2	6	1	2
Th	90	2	2	6	2	6	10	2	6	10	14	2	6	10	1	2	6	1	2
Pa	91	2	2	6	2	6	10	2	6	10	14	2	6	10	2	2	6	1	2
U	92	2	2	6	2	6	10	2	6	10	14	2	6	10	3	2	6	1	2
Np	93	2	2	6	2	6	10	2	6	10	14	2	6	10	4	2	6	1	2
Pu	94	2	2	6	2	6	10	2	6	10	14	2	6	10	5	2	6	1	2
Am	95	2	2	6	2	6	10	2	6	10	14	2	6	10	6	2	6	1	2
Cm	96	2	2	6	2	6	10	2	6	10	14	2	6	10	7	2	6	1	2
Bk	97	2	2	6	2	6	10	2	6	10	14	2	6	10	8	2	6	1	2
Cf	98	2	2	6	2	6	10	2	6	10	14	2	6	10	9	2	6	1	2
Es	99	2	2	6	2	6	10	2	6	10	14	2	6	10	10	2	6	1	2
Fm	100	2	2	6	2	6	10	2	6	10	14	2	6	10	11	2	6	1	2
Md	101	2	2	6	2	6	10	2	6	10	14	2	6	10	12	2	6	1	2
No	102	2	2	6	2	6	10	2	6	10	14	2	6	10	13	2	6	1	2
Lr	103	2	2	6	2	6	10	2	6	10	14	2	6	10	14	2	6	1	2

arranged in order of Z, the elements fall into a pattern, known as the periodic table, because of the grouping which results (see Atom and Table 13). The atomic weight of the elements does not follow this pattern exactly, since the number of neutrons (high mass) on the nucleus may vary with the isotopes involved.

Atomic planes. The sheets of atoms arranged in a crystal lattice, with particular spacings between atoms in the sheet (plane) and between planes, which are repeated indefinitely in a perfect crystal when related to three mutually perpendicular axes. These planes constitute a particular type of bond between the atoms, both within the planes and between planes, and account for much of the mechanical and thermal behaviour of the material.

Atomic radii. Atoms are not "solid" objects, but mainly empty space, in which are assembled the nuclei (with a complex structure of their own) and surrounding electrons. The electrons are not "solid" either, behaving sometimes as particles and sometimes as a wave motion. Since the electrons have angular momentum and carry a negative charge, they are arranged in equilibrium with each nucleus to create a steady state, in each "free" atom. When the atoms are combined, in the solid state, the energy levels which are fixed in the free atom, can vary, for the outer electrons of highest energy level to accommodate a new equilibrium, between attraction and repulsion forces of individual atoms. The atomic radius is the representation in space of the outermost energy level of an electron associated with that atom measured from the centre of the nucleus, as if the atom were a sphere, and the highest energy level electron orbited the nucleus as if on the surface of that sphere.

Atomic radii are quoted normally for the solid state of crystals, assuming 12 nearest neighbours in the lattice (co-ordination number 12), the closest packing attainable, or as half the internuclear distance between identical atoms in a covalent bond. The covalent radius is often used to calculate the bond length between dissimilar atoms. When ionised, the atom has lost or gained one or more electrons and this results in a decrease in radius for cations (lost electrons) and an increase for anions (gained electrons).

Atomic reactor. See **Nuclear reactor.**

Atomic vibration. The atoms in any material, whether free or combined, possess free kinetic energy above the point of absolute zero, and this is manifest as continual movement or vibration. The vibration is proportional to some function of temperature, depending on the state of the atom, in combination or free, and the type of combination. The capacity to absorb or lose kinetic energy of vibration is the specific heat, the increased or decreased mobility is measured as thermal expansion or contraction and the transmission of energy through the vibrating mass is the thermal conductivity.

Atomic volume. The volume occupied by 1 mole of an element. In this case one mole refers to one gram atom.

Atomic weight. The weight of an atom of any element relative to the weight of a standard element. The standard was originally hydrogen atomic weight 1, but then changed to oxygen. The chemical scale of atomic masses is a standard 16, for the naturally occurring mixture of oxygen isotopes. The physical scale gives 16 as the atomic mass of the most commonly occurring oxygen isotope, and atomic weights on this scale are 1.000275 times those on the chemical scale. The atomic mass unit (a.m.u.) is one sixteenth the mass of the oxygen 16 atom, and thus 1 a.m.u. = 1.66×10^{-24} g. If the atomic weight of an element is measured in grammes, then this represents 1 mole of atoms of the element. (See **Avogadro's number.** The internationally recognised unit is now called the unified atomic mass unit (u), defined as 1/12 of the mass of a carbon atom of mass number 12. 1(u) = 1.66×10^{-27} kg.)

Atomising. A process in which liquid is induced or injected into an air-stream through a fine nozzle and mingling with the air forms an exceedingly fine spray.

Atrament process. A long immersion process for producing corrosion-resistant manganese phosphate films on steel and zinc products.

Attraction force (in atoms). A free atom is stable, with electrostatic attraction between electrons and nucleus balanced by gravitational forces of the angular momentum of the electrons.

As two like atoms are brought together, there is no change in this balance of forces, until a point where the electrons of one atom begin to attract the nucleus of the other, in each case. This attraction increases as the separation distance decreases, and, in some types of combination, additional attraction forces may come into play.

The attraction cannot increase indefinitely, because the electrons from each atom are themselves now coming close together, and will begin to repel each other. A repulsion force therefore builds up in opposition to the attraction. The two nuclei would also repel each other, if the distance of approach is close enough, and so towards the end, the repulsion forces build up very rapidly indeed. Obviously, a compromise is reached, when attraction and repulsion forces are equal, and the equilibrium, or "closest distance of approach" results. The potential energy is now at a minimum, the resultant force is zero and the assembly is stable. The potential energy of attraction is equal to $-(a/r^m)$ where a is a constant, r the distance apart, and m an index dependent on the type of combination achieved, including the additional forces of attraction referred to.

Auger electron spectroscopy. When a primary beam of X-rays or electrons is used to excite the atoms of a specimen there are several possible results. One of these is the emission of a secondary "Auger" electron (instead of an X-ray photon) which occurs when the excited atom decays and an electron "hole" at a low energy level is filled by an outer electron. This process in fact occurs more frequently than X-ray emission. The energy of the Auger electron is the difference in energy of the two electron levels involved minus its own binding energy, and is therefore characteristic of the atom which emitted it. The technique can be carried out in a scanning electron microscope, suitably fitted to prevent stray magnetic fields (which would deflect the Auger electrons) and with a spectrometer taking into account the low energy of these electrons.

Augite. A silicate of calcium, magnesium, iron and aluminium. The composition, which is variable, may be expressed $Ca.Mg(SiO_3)_2$ with $(Mg.Fe)(Al.Fe)_2 SiO_6$.

Auriferous. Means "containing gold". The term is mostly used in respect of gold-bearing lodes and mineral deposits.

Austempering. A process for the heat-treatment of steel that has been previously heated above the transformation range and thereby converted into austenite. The steel is quenched into a bath held at a uniform temperature of the order of 250-450 °C — i.e. at a temperature at which the products, mainly bainite or nodular troostite, are formed. The resulting structure is very fine and is claimed to possess considerably enhanced ductility for a given strength, together with greater freedom from internal stresses. The quenching bath requires to be either a salt bath or a molten metal bath.

Austenite. Solid solutions in which γ-iron acts as the solvent. The maximum solubility of C in γ-iron decreases from 1.7% at 1,130 °C to 0.8% at the eutectoid temperature, 723 °C. In certain highly alloyed steels the austenitic structure can be retained by rapid quenching. Austenitic steels are non-magnetic and highly resistant to most forms of corrosion. Main applications: 18/8 chromium-nickel stainless steel. Hadfield's manganese steel, 10-14% Mn, 1-1.1% C, possesses high work-hardening capacity and high resistance to wear and abrasion.

Austenitic alloy steels. The manganese steel (10% Mn) discovered by Hadfield does not harden on quenching. This characteristic is associated with the property of an austenitic alloy: that the austenitic solid solution does not change into hard martensite when the temperature is decreased to about 200 °C, and the austenite remains stable at ordinary temperatures. This effect, produced by manganese, is likewise produced by cobalt, and especially by nickel.

However, heating or cold-working the austenitic alloy permits the martensitic change to occur. Cutting, filing, etc., of such steels causes intense hardness in the areas affected, and thus these steels are useful where resistance to wear in service is required.

In the case of chromium-nickel steel (18% Cr, 8% Ni) or stainless steel, which is unhardenable by quenching, heating in the range 400-850 °C causes reactions to occur, resulting in the precipitation of chromium carbide at grain boundaries. The chromium is thus removed from the grains and is no longer effective in preventing corrosion. The deterioration of stainless steel in this respect can be prevented by the addition of small amounts of titanium (usually 5 X C content) or niobium (usually 10 X C content), which are strong carbide-forming elements and inhibit the formation of chromium carbide.

Austenitic cast iron. Cast iron, with a structure consisting of flake graphite in an austenitic matrix, produced by using suitable alloying elements in sufficient quantity to depress the eutectoid transformation temperature below atmospheric temperatures. Nickel is the alloying element mostly used for this purpose, and its effect may be assisted or modified by copper, chromium or manganese. Austenitic cast irons possess good powers of resistance to corrosion and unusual magnetic properties.

Austrian cinnabar. A pigment used in paints. It consists of basic lead chromate.

Autoclave. A closed vessel for heating chemical substances or for sterilising. The vessel is usually of steel and suitably lined for the required purpose. The contained substances can be heated to high temperature under high pressures.

Autofrettage. A prestressing treatment used in the construction of guns and gun tubes. The residual stress induced by preliminary overstrain renders the total stress encountered in service more nearly uniform than would otherwise occur, and enables the tube to withstand a greater load before permanent deformation occurs.

Autogenous grinding. A process for reducing ores and rock material to a fine state of division in tumbling mills, in which the grinding media is the same as the material that is being ground. The mills are large in diameter compared with their length (e.g. Aerofall mill, diam. 6.7 m, length 1.5 m) and they are run near their critical speeds in order to obtain a high momentum for the falling load. The mill is air swept, and classifiers are employed to return the oversize material to the mill for regrinding.

Avogadro's number. The number of molecules in 1 mole of the substance. If the substance be elementary, it is the number of atoms in the gram-atomic weight of the element ($= 6.023 \times 10^{23}$) 1 mole being the atomic weight measured in grams.

Azurite. Blue carbonate of copper $2CuCO_3 . Cu(OH)_2$ crystallising in the monoclinic system, which is found in association with copper ores in the weathered zone of copper lodes. When present in sufficient quantity it is a valuable source of copper. Occurs in France, Siberia and the USA. Azurite is also known as chessylite (cf. **Malachite**).

B

Babbitt's metal. The name applied to a group of tin-rich tin-antimony-copper alloys used in bearings as anti-friction metals. The original patent, by Issac Babbitt, covered the method of manufacture using a soft metal lining on a backing shell of hard metal (*see* Table 8).

Baddeleyite. Zirconium oxide ZrO_2, first discovered in the gem gravels of Sri Lanka. The main occurrence is in Brazilian stream gravels. The mineral is a source of zirconia used in refractories.

Bainite. A decomposition product of martensite obtained by tempering fully hardened steel at about 200-350 °C, or by austempering at 500-360 °C. Also known as troosto-martensite, troostite-martensite.

Bakelite. A thermosetting polymer, formed by condensation reaction between phenol C_6H_5OH base, and formaldehyde (methanal), $CHOH$, with the elimination of water. Mechanically strong but brittle, electrically a very good insulator.

Balas ruby. A variety of spinel $MgO.Al_2O_3$ which occurs in igneous rocks (particularly basic), limestones, gneisses and alluvial deposits.

Ball clay. A fine-textured clay which on firing yields a white pottery. It is also known as potter's clay.

Balling. The act of spheroidisation, as of the cementite constituent of steels during prolonged annealing at sub-critical temperatures.

Ball mill. A mill for grinding materials (e.g. ores) into a fine powder. It usually consists of a revolving cylinder within which are heavy stone or iron balls.

Banded structure. A segregated structure of parallel or nearly parallel bands which run in the direction of working. It is frequently observed in rolled or forged steels as alternate bands of pearlite or ferrite, rich in phosphorus, as an uneven distribution of phosphorus causes the carbon to be concentrated in areas low in phosphorus.

Band structure. When atoms are brought together, to unite in combinations for practical use, the electron energy levels are modified by the process (*see* **Attraction force** and **Bonds**). The discrete energies set by the equilibrium in the free atom are now "smeared" into broader ranges called "energy bands", and the electrons involved are described as having a "band structure". The first electrons affected are obviously those in the outer (highest) energy levels, and the interatomic distance has to decrease much more to affect the lower energy electrons. The electrons closest to the nucleus are hardly ever affected, because the repulsion forces do not allow of such a close distance of approach.

When like atoms are involved, the energy bands may lead to complete overlap of energy for some electrons, so that it becomes impossible to describe them by the precise quantum numbers used for the free atom. They become "hybrid" energy levels, and

Fig. 7 shows hybridisation of 3s and 3p electron energy in solid magnesium. Hybridisation also occurs in covalent bonding, particularly carbon compounds.

Banka tin. Tin metal (purity $> 99.75\%$) produced from Indonesian ores. The main source is the Isle of Banka.

Banket. A term originally applied by early Dutch settlers to the gold-bearing conglomerate or pudding stone of Witwatersrand. It consists of rounded pebbles of white quartz cemented together by siliceous material. The gold occurs in the cementing material and is associated with iron pyrites. The name is now used generally for all similar conglomerates.

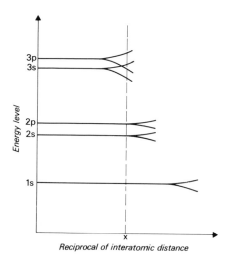

Fig. 7 *Energy levels of electron states in magnesium.*

Banox process. A phosphating process used as a preliminary treatment before drawing steel wire. The phosphate coating acts as a lubricant and provides a rust-resisting coat.

Bar. The name applied to any piece of solid metal having a simple cross-section (e.g. rectangular, hexagonal, circular), and of length considerably greater than either the width or the thickness. The name is also commonly used to denote a piece of metal fixed across the top of a moulding box to retain the sand.

Bar drawing. A process for making cold-drawn seamless tubes of narrow bore. A bar is inserted into a hollow cylinder of metal, and both bar and tube are pulled through the die together. Behind the die, the reduction in thickness and diameter of the tube causes a relative movement of the tube back along the bar. The tube is reeled off the bar by a cross rolling operation.

Barffing. A surface-finishing treatment applied to certain steel products, which renders the surface inert by producing a coating of magnetic oxide of iron Fe_3O_4. The articles to be treated are cleaned, placed in an air-tight chamber and heated to a dull-red heat. Super-heated steam at 400-700 kN m^{-2} pressure is then introduced. The coatings can be dyed and treated with oil or wax to improve resistance to moisture.

Barilla. A term applied to Bolivian tin concentrates.

Bar iron. Wrought iron in bar form, used by blacksmiths and millwrights.

Barite. *See* **Barytes.**

Barium. A metallic element belonging to the group of alkaline-earth metals. It occurs as barytes or heavy spar (barium sulphate) and also as witherite (barium carbonate). The metal is usually extracted by heating barium oxide with aluminium or ferro-silicon in a vacuum at about 1,200 °C and condensing the barium vapour in the cooler end of the apparatus. It may also be obtained by the electrolysis of barium chloride solution using a mercury cathode. Barium is isolated from the resultant amalgam by distilling off the mercury. Barium has a white metallic lustre and is very soft and malleable. It inflames spontaneously in moist air and burns readily to barium oxide BaO when heated in air.

The metal is used in the radio industry to remove the last traces of gas from vacuum tubes. Ternary alloys of lead with barium and calcium are used as bearing alloys. High-nickel alloys containing small amounts of barium are used in automatic ignition equipment and in electrical filaments.

Barium compounds find increasing use in industry, particularly in the glass and chemical industries. Barium sulphate exhibits phosphorescence. Barium is used largely in Very light signals, flares and fireworks to produce the green colour.

The thermionic emission of filaments is improved by coating with barium and strontium oxides. Barium carbonate is used in the refining of beet sugar and for pottery (*see* Table 2).

Barium titanate $BaTiO_3$. A typical ferro-electric material, with a crystal structure typified by perovskite, calcium titanate. This resembles three interpenetrating crystal lattices in one, and accounts for the reversibility of polarisation in an external electric field.

Bark. A skin or layer of decarburised material found just beneath the scale when steel has been heated in an oxidising atmosphere.

Barkhausen effect. The term used in reference to the phenomenon observed when iron or other ferromagnetic material is magnetised. As the magnetising current is continuously increased it is observed that the magnetisation does not also increase continuously, but rises in a series of minute jumps. The same effect is observable when the magnetising current is slowly decreased.

Barlow's rule. In any given molecule the volumes of space occupied by the various atoms are approximately proportional to the valencies, the lowest value of valency being the one applicable.

Bar mill. A rolling mill used for producing bars of metal.

Barn (unit). A unit (not SI) to describe the "target" area offered by nuclei to neutrons in radiation damage. The nucleus of an atom presents approximately 10^{-28} m^2 cross-sectional area, and this is the value of 1 barn. The best way to use the unit is to express the probability of collision, followed by absorption or deflection, when a neutron enters a matrix of atomic nuclei. (*See also* **Absorption cross-section**).

Barrel plating. An electroplating process in which the parts are plated whilst revolving in barrels made of hard wood or steel lined with suitable insulating material.

Barretter. This is a metallic resistor, which may or may not be enclosed in a gas-filled envelope. It has the characteristic of a positive temperature coefficient of resistivity. Barretters are used in control and measurement applications, because they can provide automatic adjustment of resistance over a certain voltage range. Also called ballast tubes.

Bartyes (heavy spar). A natural form of barium sulphate $BaSO_4$. It is very widely distributed and is often found as cawk in lead veins, where it is associated with galena, calcite, fluorspar, etc. It occurs as nodules (Bologna stone), resulting from the decay of barytes-bearing clays or limestones. It crystallises in the ortho-rhombic system and has a specific gravity of 4.5 and a hardness of between 2.5 and 3.5 on the Mohs scale. Barytes is insoluble in almost all liquids. Under the name of *blanc fixe* it is used as a pigment and as an extender for paints. It is also used as a loader for paper, textiles, soap, rubber and linoleum. Barytes is used extensively with zinc sulphide in the manufacture of the pigment lithopone.

Baryta. An oxide of barium BaO, whitish in colour, which is commonly obtained by heating the carbonate in the presence of carbon. A solution in water, called baryta water, was used in the estimation of carbon dioxide.

Barytocalcite. A double carbonate of calcium and barium occurring in lead veins.

Basalt. The name applied to basic lavas that are rich in ferrous and magnesian silicates (up to 52% Si). In structure they may range from glassy to finely crystalline rocks and the colour varies from grey to black. The constituent minerals are essentially plagioclase felspars, olivine and augite with accessory iron ores.

Base (transistor). One semiconductor component of a transistor, constituting a potential barrier between two other components, one on each side of the base, called emitter and collector respectively. (*See also* **Transistor.**)

Bash and Harsch test. An accelerated test for determining the life of resistance materials for electric furnaces. A straight length of wire hangs vertically under a fixed tension. The temperature of the wire is initially 1,066 °C and a constant voltage is maintained. The current flows intermittently, 2 minutes on and 2 minutes off, and the life is determined by an increase in resistance of 10% or failure.

Basic Bessemer process. A modification of the Bessemer steel-making process used for the treatment of basic (high-phosphorus-content) pig irons. The process differs somewhat from the ordinary Bessemer process (q.v.), as the converter is lined with basic refractory materials (e.g. calcined dolomite or magnesite) and lime is introduced into the charge. Instead of stopping the blast at the point at which the flame drops, the blow is continued for some minutes longer to oxidise the phosphorus, which then combines with the lime, forming calcium phosphate, and passes into the slag.

Basic lead carbonate (white lead). Hydroxycarbonate of lead $2PbCO_3.Pb(OH)_2$, used extensively in the manufacture of paints and for glazing pottery.

Basic lead chromate. The naturally occurring mineral, phoenicite $2PbCrO_4.Pb(OH)_2$, is found as hyacinth-red crystals (sp. gr. 5.75). The substance occurring in commerce, known as chrome red $PbCrO_4.Pb(OH)_2$, is obtained by digesting chrome yellow $PbCrO_4$ with caustic soda. It is used as a pigment.

Basic lining. Refractory materials used in the hearth and lining of a melting furnace in which silica is absent, or present in such small quantities that the basic properties of the basic oxides predominate. As a general rule, basic linings resist the action of metallic oxides, but are readily attacked by silica and siliceous materials at high temperature. The basic materials used in linings include lime, dolomite or magnesium limestone and magnesite.

Basic open-hearth process. A process for dealing with charges of pig iron and scrap containing only moderate amounts of phosphorus, the amount being too high to be dealt with successfully in a furnace with an acid lining, and too low to yield heat sufficient for Bessemerising in a converter. The basic lining is made of burnt dolomite or burnt magnesite, bound with hot tar. It is usually laid on a layer of firebrick. A lining of calcined magnesite cemented with magnesium chloride is also used.

Basic rock. Igneous rocks containing less than 52% Si and characterised by the minerals felspar, augite and olivine. Examples are basalt and dolerite.

Basic slag. A phosphatic manure which promotes the growth of clover on poor land. It is produced as a by-product in the manufacture of steel from phosphoric pig-irons. Average qualities contain 25-35% total phosphates, of which 20-90% may be soluble in a 2% solution of citric acid (Wagner's test).

Basic steel. Steel melted and produced in a furnace having a basic hearth and lining (q.v.) and under a slag which is predominantly basic.

Basil buff. A polishing mop made up from sheepskin discs used for finishing brass, copper, chromium-plated articles, etc.

Basis box. Tinplate sheets for manufacture into cans and boxes were traditionally packed into wooden boxes to protect the polished tin surface. Sheets cut to 35 cm × 50 cm (14 × 20 in) and of 0.3 mm thickness (30 SWG) weighed approx. 454 g (1 lb) and 112 were packed into a box, known as the "basis box". Variation in thickness of sheet cut to the same size would vary the weight, and larger sizes were made although these were then known as "tinned sheets". Modern production methods favour reeling into rolls for shipment, and protection with interleaving of paper or plastic where necessary.

Batch furnace. A furnace in which the material is treated in batches, as distinct from a continuous furnace. The charge is placed in the furnace, heated through the required cycle and subsequently discharged. The furnace itself may be heated and cooled with the charge or it may be maintained at a certain temperature.

Bath. A solution used for pickling, electroplating or galvanising.

Bath brick. A brick made from river silt, used for scouring and, when finely ground, in certain liquid metal polishes.

Batty's secondary nozzle. A device for controlling the pouring rate of steel ingots and castings by means of a secondary nozzle box secured to the bottom of the ladle. It enables the size of the casting nozzle to be changed, if necessary, between any two ingots or groups of ingots.

Baumann printing. A simple method of determining the points of sulphur concentration in metal specimens. An ordinary photographic paper is soaked with dilute sulphuric acid and the surplus acid removed by wiping. The emulsion side of the paper is carefully pressed against a clean (ground and degreased) surface of the specimen. After a short time the paper is removed. Dark stains, due to the reaction between the sulphuretted hydrogen and the silver salt in the paper, indicate the location of the sulphur-containing areas. The rest of the paper should remain comparatively white. The paper is fixed in "hypo" and washed and dried in the usual way.

Bauschinger effect. The phenomenon by which plastic deformation increases the yield strength of a metal in the direction of plastic flow and decreases it in other directions.

Bauschinger's theory refers to failure under repeated stresses. When bars are subjected to cyclic variations of stress, the elastic limits in tension and compression are changed, the range between the two limits depending on the material and on the stress at the lower limit of elasticity. Thus if the elastic limit in tension is raised by strain it is simultaneously lowered by compression, and for that condition of loading two new limits are set up. These are termed the natural elastic limits. The theory further shows that the range between these limits is the maximum range of stress (i.e. the limiting safe range) in fatigue.

Bauxite. A clayey, earthy mineral consisting mainly of hydrated alumina $Al_2O_3.2H_2O$ together with silica and usually some iron Fe_2O_3. It is the main source of the metal aluminium and derives its name from the town of Les Baux in the south of France. Considerable quantities of the mineral are used as abrasives, etc., in the manufacture of aluminium compounds and refractory materials for furnace linings and for bricks. The main sources are France, the USA, Hungary and the Guianas.

Bayer process. A process for purifying bauxite prior to the extraction of the metal aluminium. The bauxite is digested with caustic soda solution which dissolves the alumina. The silica and the insoluble metallic oxides present are filtered from the solution and the aluminium is then precipitated as hydroxide.

Bearing metals. These consist of all the alloys such as Babbitt's metal or bronzes which are used in bearings because of their anti-frictional properties (*see* Tables 5 and 8).

Bearing modulus. This is a quantity that may be used to compare the performances of bearings using the formula Zn/p. In this $Z =$ the absolute viscosity of the lubricant at the working temperature measured in poises: p is the pressure on the bearing of projected area, and n is the number of revolutions per minute of the journal.

Bears. Infusible masses of iron, often containing titanium, which may sometimes form in the hearth and lower part of blast furnaces. Bears are very seldom found in hot-blast furnaces.

Becking mill. A mill used in the production of railway tyres. It consists of a heavy rotating table through which projects a centre roll, and a massive outside roll which is mounted on to a diagonal shaft. The punched and flattened bloom is placed on the table so that it encircles the centre roll, and the outside roll is advanced. The bloom is decreased in thickness and increased in diameter by the squeezing action of the rolls. Rough flanges may also be formed by this process.

Beehive coke. A hard metallurgical coke produced by the old method of partial combustion of bituminous coal in a bee-hive shaped oven without external heating and with limited access of air.

Beilby layer. Beilby's experiments led him to conclude that the action of polishing a metal surface caused the surface layer to flow like a liquid and then to solidify without recrystallisation, forming an amorphous layer. In specimens prepared by the usual metallographic techniques, the thickness of the layer is approximately 3 nm. Beneath this layer are small crystals gradually increasing in size until they merge into the normal crystalline structure of the unaltered metal.

Bell. A cone-shaped device which seals the top of a blast furnace and which is lowered to admit the charge.

Bell metal. A copper-tin alloy containing 20-25% Sn which was originally used for bells but has since found more extensive application as a bearing bronze. A little zinc is usually added as a deoxidiser and to improve casting properties. Immediately after solidification the alloy consists of two constituents—α and β solid-solution phases—which may be retained by quenching from about 600 °C. At temperatures in excess of 600 °C, and in quenched condition, the alloys are tough, malleable and can be hammered or otherwise fabricated.

If slowly cooled, the β constituent undergoes transformation at 520 °C and the resultant alloy consists of particles of the very hard δ constituent in a comparatively plastic matrix of the α solid-solution phase. Slowly cooled (sand-cast) alloys are used for slide valves and heavily loaded bearings. The above transformation can be suppressed to a certain extent by quenching.

The tone of a bell is influenced considerably more by the design and by the soundness of the casting than by the composition of the alloy, but is improved by heat-treatment involving quenching from about 600 °C, which produces a tougher alloy.

Bell-metal ore. A sulphide of tin, copper and iron $Cu_2.FeS.SnS_2$ found in association with cassiterite, copper pyrites, lead and galena in Cornwall, Zinnwald Czechoslovakia, and Bolivia. It is also known as stannine and tin pyrites.

Belly-pipe (or blowpipe). A cast tube usually 1.2-1.5 m long and slightly tapered, joining the goose neck to the tuyères of the blast furnace.

Bend test. A transverse bending test carried out to determine the transverse strength of cast iron. The term is also used in connection with the deflection of a horizontal beam under load. As commonly used it is the standard test for malleable cast iron and certain sheet metals in which flat bars or test pieces are bent through 180° as a test of ductility.

Benefication. A term applied to all ore concentration processes.

Benet metal. An aluminium-magnesium alloy of French origin. It contains a small amount of tungsten to improve the ductility.

Bengough-Stuart process. An anodising process in which aluminium is treated with a 3% solution of chromic acid to produce a grey surface coating.

Bentonite. A clay similar in properties to fuller's earth, mainly a hydrated silicate of calcium and magnesium. It absorbs up to five times its own weight of water and is used to increase plasticity of ceramics and as a bonding clay for foundry moulding sand. Bentonites are wind-blown deposits, usually of volcanic origin, that are deposited in marine waters. They may contain fragments of volcanic glass, but are mainly colloidal silicates.

Benzene C_6H_6. An aromatic hydrocarbon, in which the carbon atoms are arranged in a hexagonal ring, one hydrogen atom being attached to each of the six carbon ones. Such cyclic compounds have different properties to the "chain" hydrocarbons, but both can form "branches" of similar or other atomic groups. Benzene is unsaturated, i.e. the carbon atoms have bonds which are readily broken. Highly flammable as a fuel, forms multitudinous compounds and derivatives of great importance, especially in dyestuffs, drugs, pharmaceuticals and polymers.

Berkelium. An element of atomic number 97 that is prepared artificially by subjecting the element americum to irradiation by helium nuclei (α particles) in a cyclotron.

Bertrandite. A secondary ore of beryllium that contains only a small percentage of the metal.

Beryl. A silicate of beryllium and aluminium $3BeO.Al_2O_3.6SiO_2$ crystallising in the hexagonal system. This mineral is the sole source of beryllium. It is opaque and may be emerald green, pale green, pale blue, yellow or white in colour. It has a hardness 7.5-8 (Mohs scale) and sp. gr. 2.7. Beryl occurs as an accessory mineral in acid igneous rocks, granites and pegmatites, particularly in Massachusetts, Brazil, India and Siberia. Gem varieties of the mineral which are lustrous and transparent are emerald (green) and aquamarine (pale blue).

Beryllia BeO. The oxide of beryllium metal, with a melting point around $2,570°C$. It possesses high electrical resistance, good resistance to thermal shock, but is toxic in finely divided form. As an oxide layer, it protects metals such as magnesium in the melt, and improves oxidation resistance in the solid.

Beryllium. A steely-white metal closely akin to magnesium, which has a very low specific gravity (1.85) and melts at $1\,278°C$. It has a great affinity for oxygen, and when in powder form burns with a brilliant light when ignited in air. At ordinary temperatures beryllium is very brittle, but it can be forged and hot rolled to a certain extent. The metal is extracted from beryl (q.v.), which occurs as an accessory mineral in acid igneous rocks, and is obtained as a by-product from felspar, mica and lithium ores. The separation processes are somewhat complex. Beryllium salts can be obtained by fusing beryl with sodium carbonate, digesting the mass with hydrochloric acid, evaporating to dryness, separating from silica by extracting the soluble salts with hydrochloric acid, then precipitating a mixture of aluminium and beryllium with ammonia; dissolving the precipitate in caustic potash and boiling the solution to precipitate the beryllium hydroxide. The precipitate is redissolved and reprecipitated several times to ensure complete elimination of aluminium. The beryllium oxide may then be reduced with charcoal.

The metal may also be produced by the electrolytic reduction of beryllium chloride. Pure beryllium chloride, obtained from beryl as indicated, is mixed with an equal quantity of sodium chloride and electrolysed in a heated nickel crucible, which serves as the anode, at about $350°C$. Dendritic flakes of beryllium are washed, briquetted and melted in a beryllium-oxide crucible in an atmosphere of hydrogen prior to final melting and casting in a vacuum of about 13 m bar ($1.3\ kNm^{-2}$).

Beryllium is used for the windows in X-ray tubes, for electrodes in neon lights and for targets in cyclotrons, but finds its main use as a deoxidiser for copper and as a constituent of various copper and nickel alloys. It has a remarkable strengthening effect on copper, particularly when the alloy has been heat-treated, and beryllium-copper is one of the strongest of non-ferrous alloys. Beryllium-nickel alloys possessing high strength/density ratio and good strength are used for highly-stressed structural components and for springs. Beryllium has been used for surface hardening steel, but the beryllium-aluminium alloys have not yet proved successful (*see* Table 2).

Beryllium-copper. A copper-base alloy with high thermal and electrical conductivity and good fatigue and corrosion resistance. It is used for springs, wire and non-magnetic components.

Bessemer converter. This is an egg-shaped vessel in which an oxidation process, called the Bessemer process (q.v.), is carried out. It is usually made of mild steel plates bolted together and lined with an infusible siliceous rock (ganister) which is ground, moistened with water and applied as a paste to the interior. The lining may also be made of basic refractory materials. The lower portion of the converter has an interchangeable bottom and the lining is pierced with about 250 small holes which connect the

inside of the converter with the wind-box and serve for the passage of the blast. This interchangeable bottom consists of a shallow lower section of the converter with wind-box and tuyères, together with the necessary arrangements for fixing these in their places. The bottom is attached to the converter in a manner such that the narrow space between the two may be rammed with plastic ganister by men working outside the converter. When a used bottom is withdrawn the new one can be inserted and the joint quickly made. This permits a considerable increase in the number of "blows" per day. The converter is supported on trunnions on which it can be made to rotate. One of the trunnions is hollow and serves as a wind duct through which a blast of air can be supplied to the wind-box. In operation, the converter is tilted into a horizontal position to take the charge of molten pig iron. The design of the converter is such that when in this position molten metal will not cover any of the tuyère holes and escape into the wind-box whilst the blast is turned off.

Bessemerising. An operation in the extraction of copper from sulphide ores. Air, blown through the molten matte (or mixture of sulphides), oxidises the sulphur, which is removed as sulphur dioxide, whilst iron is oxidised to ferrous oxide. The latter combines with silica to form a slag. The product is an impure copper known as "blister copper".

Bessemer process. This is a steel-making process in which air, blown through molten pig iron, oxidises the impurities silicon, manganese, carbon and phosphorus. The characteristics of this process are: (a) speed of operation, (b) no extraneous fuel is required and (c) the metal is not melted in the converting furnace. Molten pig iron is poured into the converter and a strong blast of air is forced through it as it is rotated back to the vertical position. In about 4 minutes or so the silicon and manganese have been oxidised and have formed a slag. During this stage the temperature rises rapidly and a flame, which slowly becomes larger and more luminous, is seen at the mouth of the converter. As the carbon is oxidised to carbon monoxide, the latter "boils" through the metal and burns with a long blue flame. When this flame shortens and "drops", the carbon has been reduced to the lowest practical limit and the operation is stopped. The required amounts of molten ferromanganese or spiegeleisen are added to deoxidise the melt and to produce the desired composition of steel and the molten steel, which is considerably hotter than the original pig iron, is then cast.

When the process is carried out in a furnace lined with siliceous materials the process is commonly known as the acid Bessemer process. In this latter process the metal used is grey hematite pig iron containing about 2-2.5% Si. Oxidation of this element provides the heat required to maintain the metal in the fluid state. The metal should be free from sulphur and phosphorus, as these elements are not removed in the acid process. Phosphoric irons are treated by the basic Bessemer process (q.v.).

Best cokes. A grade of tin-plate carrying heavier coatings of tin than coke grades but less than charcoal grades.

Beta iron. The original designation given to the body-centred cubic lattice of iron above the Curie point, $770°C$. Between this point and the transformation to the face-centred cubic lattice ($910°C$) the material behaves as a paramagnetic. Below $770°C$ it is strongly ferromagnetic. Beta iron was originally thought to be an allotrope, but is no longer thought to exist.

Beta rays. The name given to electrons and positrons emitted in the process of radioactive decay. They are deflected by magnetic and electric fields, but are less affected than cathode emitted electrons, since they have greater energy and therefore a higher velocity. They are more penetrating than alpha rays, have greater effect on photographic emulsions, but produce less ionisation. When alpha and beta rays are ejected in radioactive decay, a new substance remains as the remainder of the original atom. A succession of transformations is possible, e.g. in the case of radium, the final product is lead.

Bethanising. An electrolytic process for plating steel wire with zinc. The characteristics of the process are that insoluble anodes are used and the electrolyte is prepared by dissolving zinc ore or dross in sulphuric acid and purifying the resultant solution.

Betterton's process. A process for removing bismuth from lead, that consists of adding a calcium-magnesium alloy to the softened, desilvered and dezinced lead. In practice a calcium-lead alloy, containing 3% Ca, together with magnesium in the calculated proportion, is added. The molten metal is skimmed, the skimmings containing the bismuth. The lead is further treated to remove other dissolved elements by oxidation—e.g. by the use of chlorine.

Bett's process. An electrolytic process for refining lead, invented in 1901. The electrolyte consists of lead fluosilicate. The anodes are of cast, crude metal and the cathode of pure lead. In order to obtain solid deposits of lead the use of addition agents is necessary. In this process, gelatine is usually employed at a concentration of 1 in 5,000 parts of solution. The impurities remain in the anode slime and pure lead is obtained at the cathode.

B—H loop. In ferromagnetic materials, a magnetising field H produces an intensity of magnetisation I, and the ratio I/H is known as the "susceptibility" to magnetisation. The magnetic induction, or magnetic flux density, B, is made up of two components, the field H, plus the intensity of magnetisation. If there were no applied field, $B = \mu_0.I$, where μ_0 is the permeability of free space. When the field is present, $B = \mu_0 (H + I)$, but

$$\frac{I}{H} = \chi,$$

the susceptibility, so $B = \mu_0 (H + \chi H) = \mu_0 H (1 + \chi)$. The proportionality constant $(1 + \chi)$ is termed the relative permeability. When a magnetic field is applied to an unmagnetised ferromagnetic material, the flux density B follows a curve shown in Fig. 8. The value of B increases slowly at first, then very rapidly, and finally tails off. On decrease of H, the value of B does not follow the original curve in reverse, but falls very slowly, leaving residual magnetic flux when $H = 0$, known as the remanence B_R. If the applied field is now increased in the reverse direction, the value of B decreases to zero. Continuation of H in this reversed direction then remagnetises the material, in the opposite polarity to the initial magnetisation, and following a pattern which falls off in rate as H increases, reaching an identical point, but of opposite polarity, to the previous values of H and B at saturation. Decrease of H then repeats the previous pattern, first to decrease B to another value B_R, in the opposite direction, when $H = 0$ and then to a second value H_c when $B = 0$, if H is increased again in the original direction. Finally, the original saturation point is reached. The value H_c is called the coercive force.

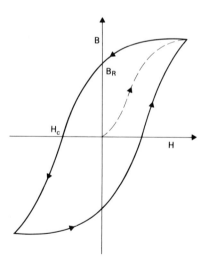

Fig. 8 *The B—H magnetisation curve.*

The whole cycle is called a **B—H** loop, or hysteresis loop, in view of the loss of energy represented by completing the cycle. In alternating current electromagnetic fields, the loop is traversed once per cycle, and the energy loss manifest as heat and/or sound and vibration.

Permanent magnets (hard) need a high remanence value and "fat" **B—H** curve. "Soft" magnetic materials, say, for electromagnet cores, transformers, require a low coercive force, a very small area hysteresis loop, with a high saturation value of **B**.

Bias. An electric potential difference applied to electronic devices such as p–n junctions in semiconductor systems, to modify the current flow. In the transistor this can be arranged to produce power amplification.

Billet. A semi-finished metal product of rectangular section with rounded corners, produced by rolling ingots.

Billet mill. A rolling-mill used to reduce ingots or blooms to billets. (*See also* **Rolling-mill**.)

Billon. A term used to describe alloys of gold or silver with copper in which the base metal largely predominates. It is also used to describe coins made from such alloys. The proportion of the precious metal is rarely more than about 25% of the alloy.

Bimetallic. Composed of two different metals.

Bimetallic fuse. A fuse element composed of two different metals, e.g. copper wire coated with lead or tin.

Bimetallic strip. A composite strip made of two metals having different coefficients of expansion, which bends or deflects when subjected to a change of temperature.

Binary alloy. All the alloy compositions, ranging from 0% to 100% of each component metal, which can be formed by two metals.

Binding energy. Applied to elemental atoms, this is a measure of the attraction force which holds together the individual constituents of the nucleus (the nucleons) and is of the order of several MeV per nucleon for most nuclei. In small nuclei, binding is reduced by the distribution of a small number of units. A release of nuclear energy occurs if these nuclei can be made to fuse together to form larger ones. This is the process in the release of energy in the sun, the fusion bomb, and the fusion power producing experiments. If the nuclei are very large, the protons repel each other strongly and become unstable. When the ratio of neutrons to protons increases above 1.50, the instability leads to spontaneous radioactivity, and when the ratio exceeds 1.55, the material is susceptible to nuclear fission when bombarded by neutrons.

The so-called nuclear energy is therefore derived from binding energy, either in building up large nuclei from smaller ones (fusion) or in fissioning large nuclei to form two smaller ones (*see* Fig. 9).

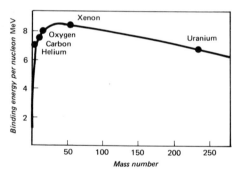

Fig. 9 *Binding energy per nucleon.*

When applied to electrons, binding energy refers to the mobility of electrons in the electric field of the nucleus, modified by the other electrons between the one in question, and the nucleus. Obviously, the larger the atom, with more electrons, the less tightly bound are the electrons in highest energy levels, because they are more shielded from the nucleus than those of lower energy level. Mobility of electrons is very important in electronics, since the relative ease with which electrons move through a solid determines the electrical conductivity.

Biotite. This is common mica consisting of silicates of magnesium, aluminium and potassium with varying amounts of iron salts $K_2 HAl_3 (SiO_4)_3 .(Mg.Fe)_6 .(SiO_4)_3$.

Birmingham wire gauge. A system of designating the diameters of rods and wires by numbers rather than by dimensional figures. The system is very generally used for iron and steel, and telephone and telegraph wires and sheet. (*See also* **Brown and Sharp gauge**.)

Bismite. An ore of bismuth consisting essentially of the hydrated oxide $Bi_2 O_3 .3H_2 O$. Also known as bismuth ochre.

Bismuth. A greyish-white lustrous metal (sp. gr. 9.8) which melts at 271 °C and boils at 1,560 °C. It is not a very abundant metal, but it is fairly widely distributed as native bismuth, which occurs in veins in granite along with tin, silver, copper and cobalt ores. It is also found as bismuthinite (Bi_2S_3), bismuth ochre and tetradymite (bismuth telluride), but these ores, which do not occur in quantity, are of minor importance. Bismuth may be prepared by heating the native metal till it melts and runs away from the matrix. It may also be prepared by roasting the sulphide and then reducing the oxide with carbon and a suitable flux (e.g. sodium carbonate, limestone, fluorspar). The main sources of supply are the native metal and the anode slimes from the electrolytic refining of copper and lead. Bismuth crystallises in the hexagonal system as rhombohedral crystals. It is brittle at normal temperatures and is a poor conductor of heat. Bismuth and bismuth-rich alloys expand on solidification and are able to produce sharply defined castings. The largest use of the element is for the production of medical compounds, but alloys of the metal with lead, tin, cadmium and some other metals find many applications as fusible alloys (e.g. Wood's metal, Rose's metal) in safety devices, low-melting point solders, for tempering steel, etc. The most important uses of the bismuth-containing fusible alloys are in the construction of aircraft assembly jigs and press tools for sheet-metal working. They are also extensively used in bending thin-walled tubes and other hollow sections (*see* Tables 2 and 8).

Bismuthinite. Bismuth sulphide Bi_2S_3 which crystallises in the orthorhombic system. It has a specific gravity of 6.4 and a hardness of 2 (Mohs scale). It occurs in veins associated with lead and copper minerals. Also called bismuth glance.

Bismuth ochre (Bismite). A yellow earthy alteration product of bismuth and bismuthine, corresponding to the oxide $Bi_2O_3.3H_2O$.

Bismutite. A basic carbonate of bismuth

$$3(BiO)_2.CO_3.2Bi(OH)_3.3H_2O.$$

Bitter spar. A ferruginous variety of dolomite $CaCO_3.MgCO_3$ which turns brown on exposure due to oxidation of the iron salts present.

Bitumen. A name applied to a number of natural flammable pitchy or oily substances, consisting of a mixture of hydrocarbons, which are believed to have originated from organic matter. Bitumens are sometimes classified into four groups: *(a)* solid, e.g. asphalt or mineral pitch; *(b)* semi-solid, e.g. maltha or mineral tar; *(c)* liquid, e.g. mineral naphtha, petroleum and Rangoon oil; *(d)* gaseous.

Bituminous asphalt. A viscous tarry fluid, originating in natural deposits (as in C. and S. America) or from petroleum or coal distillation. It is waterproof when consolidated and forms a good bond with aggregate material such as stone and gravel, hence its principal use as a binder, in road making and floor or roof covering. The material is probably composed of long chain hydrocarbons, with oxygen atoms attached to some chains. Change in orientation of these giant molecules, by adding a mineral filler, inhibits flow, and hence rapidly increases viscosity, so that the material, now a composite, behaves more like a solid than a fluid.

Bituminous coal. Coals under this name may vary considerably in character and composition. They all burn with a yellow smoky flame, and during combustion soften and swell in a manner resembling the fusion of bitumen, though there is no actual bitumen present. They are mostly black in colour, though some may be brown, and they have a bright pitchy lustre. The specific gravity varies from 1.14 to 1.41. The varieties of bituminous coal are frequently classified by the nature of the solid residue (coke) produced when the coal is heated in a closed retort, as caking or non-caking coals. Cannel coal is often classified as a bituminous coal.

Black annealing. A box-annealing process used in tin-plate manufacture and the tinning industry, which imparts a black colour to the surface of the untinned steel sheets.

Black anodising. A surface finishing process for aluminium products. The work to be treated is made the anode in a cell containing 13-15% solution of sulphuric acid, operated at 21-23 °C, and a current density of 1.1 A dm^{-2}. The coating produced in about 1 hour's treatment dye well in a solution containing 12 g nigrosine and 2 g Toronto yellow per litre at 71 °C and a pH of 4.5-4.8.

Black ash. Impure sodium carbonate.

Blackband iron ore. A carbonaceous variety of clay ironstone in which the iron is present as carbonate $FeCO_3$.

Black body. A body which can completely absorb the entire range of frequencies of radiation produced in a uniform temperature enclosure. Such a body appears black by whatever light it is viewed, does not affect the radiation stream in a heated enclosure, and so must restore to it by emission the radiation it takes from it by absorption, both in quantity and in quality (spectrum of frequencies).

Black copper. The product obtained when oxidised copper ores are smelted in a blast furnace.

Blackening. Stainless steel may be blackened by heating for about 15 minutes in a bath of sodium dichromate at 390-400 °C.

Blackheart. A defect sometimes observed in fireclay bricks. It is caused by vitrification of the surface before organic matter in the interior has been oxidised.

Blackheart malleable cast iron. A variety of cast iron which is characterised by toughness and malleability and by a sooty-black fracture. The castings are made in white iron and subsequently "annealed" for several days in an inert (non-decarburising) medium at about 850 °C. This malleableising treatment decomposes the cementite and produces a structure consisting of nodules of temper carbon in a ferritic matrix. For certain purposes a pearlite structure may be required in the matrix.

Blacking. Powdered graphite or other carbonaceous matter used by foundrymen as a powder or wash on the inner surfaces of a mould, to protect the sand and improve the finish of the casting.

Blacking holes. Casting defects which are usually revealed as irregular-shaped cavities containing carbonaceous matter when the casting is machined.

Black jack. A sulphide of zinc ZnS which crystallises in the cubic system, has a specific gravity of 3.9-4.2 and a hardness of 3.5-4 (Mohs scale). It is quite commonly found associated with galena in veins, and is then called false galena. Part of the zinc is usually replaced by iron, and small amounts of cadmium may also be present. Occurrence: Australia, North America, Great Britain and Westphalia. It is also known as blende and sphalerite.

Black lead. A natural form of graphite (carbon) used in the manufacture of pencils, crucibles and other articles of a refractory nature. It is used also as a lubricant where oil is not admissible or desirable, and for electrodes in batteries, for the coating of moulds and certain metals. A polished coating of black lead provides a conducting surface and enables non-metallic articles to be electroplated.

Black nickel. An electroplated finish used for staining a black colour on zinc and other metals. The process usually involves plating at low current density from a solution of a nickel salt (sulphate) containing zinc sulphate and sodium thiocyanate.

Black oxide of cobalt. Essentially a hydrated oxide of manganese containing a variable percentage (up to 40%) of cobalt oxide in mechanical mixture. Sulphide of cobalt and oxides of iron, copper and nickel may also be present. It occurs as an amorphous, earthy deposit with other ores of manganese and of cobalt in Missouri and New Caledonia. It is also known as asbolane and earthy-cobalt.

Black oxide finish. A surface-finishing process used for ferrous materials. It consists in heating, at about 300 °C, in fused sodium nitrate or a mixture of sodium and potassium nitrates. A black oxide finish may also be obtained by immersion in caustic alkali solutions containing an oxidising agent—usually sodium nitrate—with or without activating agents, such as cyanides, tartrates, etc.

Black sand. A mixture of sand and powdered coal used to form the floor of an iron foundry.

Black sands. Magnetite Fe_3O_4, which occurs in placer deposits, principally in the St. Lawrence and Columbia rivers as small sand-like particles.

Black tellurium. A telluride and sulphide of gold and lead crystallising in the orthorhombic system. Antimony is also present. The gold content varies from 6% to 12%. Hardness 1-1.5 (Mohs scale), sp. gr. 6.85-7.2.

Black thermit. The thermit mixture in which the iron is present as black oxide. The name is used to distinguish this mixture from red thermit which contains ferric oxide.

Black tin. Concentrates of cassiterite SnO_2 which have been dressed and are ready for smelting.

Blanking. Shearing out a piece of sheet metal in preparation for deep drawing.

Blast. Air (hot or cold) under pressure, usually blown into a furnace. Furnaces usually employ a hot blast. Cold blast is also used for cooling solid metals, for cleaning and for such purposes as shot-blasting and sand-blasting.

Blast furnace. A type of vertical smelting furnace, heated by solid fuel—usually coke—through which a blast of air is blown to increase the rate of combustion. Blast furnaces are most extensively used for producing pig iron from iron ores, but are also used for smelting certain non-ferrous metals, i.e. copper, lead, tin, antimony, etc.

The dimensions and constructional details vary considerably from the large blast furnace (14 m hearth diameter) used for producing pig iron, to the small furnaces used for smelting lead ores. The structure is roughly cylindrical, and is usually built of masonry and lined with a thick layer of firebrick or silica brick. It is surrounded by an outer shell of steel plates welded or riveted together.

Small furnaces may be truly cylindrical or may taper slightly towards the top, but as a general rule a furnace consists of three parts: an upper conical shaft; a middle part, or "bosh", which is of the form of an inverted truncated cone; and a bottom cylindrical part, or "hearth", in which the molten products of the furnace reactions accumulate.

The charge, consisting of ore, coke and flux, is admitted at the top of the shaft, usually by a cup-and-cone arrangement which prevents the escape of gases during the charging. An air blast, usually pre-heated, is supplied to the furnace through a series of nozzles or tuyères which are located at the top of the hearth. The chemical changes which take place during smelting are complex, as the materials undergoing change are in contact with solids and gaseous fuels throughout the process, and several reactions take place simultaneously in the different zones of the furnace. In the main, the changes consist of the reduction of the metallic oxides to the metallic condition by carbon and carbon monoxide.

Blast furnaces work continuously, a fresh charge being admitted as required, and the molten metal and slag are tapped from the hearth at intervals. The gaseous products of the blast furnace are combustible and are used for heating the incoming blast or for steam raising for other purposes. The rating of a blast furnace may be given in terms of its height in feet, or of the diameter of the hearth, or on the output in tons per day. As, however, this latter is more dependent on the materials used than on furnace design, the size of a furnace is best given in terms of the hearth diameter.

Blast-furnace gas. A by-product of the blast furnace, which is a gaseous fuel of low calorific value that is frequently used for heating the blast, steam raising, etc.

Blast-furnace slag. The term that refers to the layer of fused non-metallic material that forms during the process of iron ore reduction in blast furnaces. It is tapped from the slag hole, or cinder notch, in the furnace as a molten stream consisting of the oxides of calcium, aluminium, silicon, and sometimes magnesium depending upon the nature of the ore and flux. The proportions of the constituents are approximately SiO_2 32, CaO 47, Al_2O_3 14, MgO 2; the balance is phosphorus and small amounts of other elements.

Crushed slag is used in making concrete and as road ballast, and for making slag wool (q.v.). Slags resulting from the reduction of phosphatic ores give a basic phosphate in the slag which can be used as a fertiliser.

Blast roasting. A roasting process in which a blast of air passing over the heated charge accelerates the removal of sulphur.

Blazed iron. A pig iron, rich in silicon, produced in the blast furnace during the early stages of "blowing in".

Blende. *See* **Black jack.**

Blister. A defect caused by gas bubbles on or near the surface of metal.

Blister copper. An impure form of copper produced by blowing air through molten copper matte. During the conversion process, sulphur, iron and other impurities in the matte are oxidised. Volatile impurities escape, and iron oxide combines with silica to form a slag. As the molten metal (which contains about 98% Cu, and less than 1% Fe) cools, the sulphur dioxide dissolved in the metal is expelled and gives the metal a blistered appearance.

Blister steel. The product of the cementation process in which bars of wrought iron are packed in charcoal and heated at 650-700 °C for several days, to allow the carbon to diffuse into the metal. It was formerly used in the manufacture of crucible steel and shear steel.

Block copolymer. When different monomers are polymerised together, the result is a copolymer. If the units are randomly arranged it is called a random copolymer, and the properties will vary with the sample. If the units are arranged in more sequential form, the result is a block copolymer, with a behaviour much more like a metallic alloy, with more consistent and predictable properties. When the blocks are very small, almost consecutive monomer units, the material may be termed a regular copolymer.

Blocking. A drop-forging process in which the rough stock is formed to the approximate shape of the forging.

Bloom. A semi-finished metal product of rectangular cross-section with rounded corners, produced by rolling or forging ingots. The cross-sectional area of a bloom is greater than that of a billet and, in the case of steel, usually exceeds $0.025 \ m^2$.

Blooming mill. *See* **Cogging mill.**

Blow-hole. A cavity formed by the evolution of dissolved gas that fails to escape during the solidification of the metal. The holes are appreciable in size and the cavity walls are usually smooth and oxidised.

Blowing-in. The operation of starting up a blast furnace.

Blowing-out. The act of stopping down a blast furnace.

Blown casting. A casting which is defective because of the presence of blow-holes.

Blue annealing. A process of tempering that produces a blue-black surface on ferrous materials. It is used for certain types of rolled-steel sheet and for helical springs cold-coiled from wire. In the case of springs, it is used primarily for the purpose of relieving cooling strains, but the temper may also increase the elastic limit of the spring wire and increase the amount by which the spring may be deflected without taking a permanent set.

The colours developed depend on the composition of the steel as well as on the temperature, but vary from light straw-yellow at about 200 °C, through browns to purple at about 280 °C, full blue at about 310 °C and light blue at about 340 °C.

Blue billy. A mixture of iron oxides produced when iron pyrites (iron sulphide ore) is roasted in air to eliminate sulphur.

Blue brittleness. A form of embrittlement sometimes encountered in annealed or normalised low-carbon steels of very low alloy content, which have been strained at temperatures of about 260-270 °C. The impact strength of such steels at room temperatures is lower than that of similar steels to which the same amount of strain has been applied at room temperatures.

Blue carbonate of copper. A hydrated carbonate of copper $2CuCO_3.Cu(OH)_2$ found with other copper ores in the weathered zone of copper lodes. It is a valuable source of copper. Also known as azurite and chessylite.

Blue dip. A mercuric cyanide solution used for amalgamating.

Blue iron earth. A hydrated phosphate of iron $Fe_3(PO_4)_2.8H_2O$ found in association with iron, copper and tin ores, also in clay and in a peat, particularly with bog iron ores.

Blueing salts. A mixture of caustic soda and sodium nitrate used for producing a blue oxidised surface on steels. The cleaned and degreased metal is treated at 145-150 °C in an aqueous solution containing approximately 0.9 kg of caustic soda and 0.3 kg of sodium nitrate per litre. The metal must be thoroughly washed after treatment.

Blue john. A form of calcium fluoride CaF_2 crystallising in the cubic system, which is characterised by an attractive blue or purple colour. It is obtained from certain natural caves in Castleton in Derbyshire (UK). The best varieties are used as gemstones and ornaments, but inferior varieties (e.g. ordinary fluorspar) are used in enamelling, in the glass and chemical industries and as fluxes for metallurgical processes.

Blue lead. A term applied to metallic lead as distinct from lead products such as white lead, etc.

Blue metal. A powder consisting of zinc and zinc oxide, obtained as a by-product when zinc is vaporised.

Blue powder. *See* **Blue metal.**

Blue vitriol. Hydrated copper sulphate. The blue triclinic crystals contain five molecules of water of crystallisation per molecule of $CuSO_4$. Also called blue stone.

Blue-water gas. A gaseous fuel consisting of a mixture of carbon-monoxide and hydrogen, produced by passing steam over red-hot coke. The gas burns with a pale blue flame.

BNF jet test. A test for determining the thickness of metal coatings. A continuous jet of a specified reagent is allowed to impinge upon the surface of the deposit, under carefully standardised conditions, and the time to produce penetration is noted. Since the rate of perforation will depend on the nature and strength of the reagent and the temperature, the method must be calibrated against a number of deposits whose thicknesses are accurately known. Curves are then produced, and from these the thickness of the deposit under test can be computed. Suitable etching solutions are: for cadmium, $17.5\,g$ ammonium nitrate and $17.5\,cm^3$ hydrochloric acid in 1 litre of water; for zinc, $70\,g$ ammonium nitrate and $70\,cm^3$ hydrochloric acid in 1 litre of water; for nickel, cobalt and copper, $300\,g$ hydrated iron III (ferric) chloride and $100\,g$ hydrated copper sulphate in 1 litre of water.

Board of Trade Unit (United Kingdom). A legal unit (of energy) originally fixed by the Board of Trade (now the Department of Trade). It is equal to 1 kWh. Usually written BTU, it is equivalent to $3.6\,MJ$ or $1.34\,h.p.h^{-1}$. This unit is still in use in the electrical industry.

Bob. A wheel used for polishing electroplated wares. The wheel is usually a solid disc of leather, felt or similar material or a disc edged with one of these materials.

Bod. A ball of clay used to stop the tap-hole of a cupola or furnace.

Body-centred cubic (b.c.c.) lattice. The structure is obtained from super-imposition of two square two-dimensional lattices in ABAB stacking, with a 60° angle to the vertical from one such plane to the centre of corresponding atoms in the next. Alternatively, it can be regarded as two interpenetrating simple cubes, the body-centred atom of one unit cell being the corner atom of one of the related interpenetrating cubes. The cell unit is then that of a cube with an atom at each corner and one in the centre. The most closely packed planes are those which include the body-centred atom and the diagonally opposite edges of the basic cube, Miller indices (110) the dodecahedral planes, of which there are six per unit cell. The closest packed direction is that of index [111] and there are four such directions per unit.

These close-packed planes are more open than those of face-centred cubic lattices, and there are fewer close-packed directions. Only 68% of the available packing space is occupied, and the co-ordination number is 8. The closest neighbours are at a distance $\frac{1}{2}a(3)^{\frac{1}{2}} = 0.866a$ where a is the length of the cell side. There are also six next nearest neighbours at distance a. The near neighbours are thus only 15% further away than the nearest, which is much less than in the close-packed structures, where there is 40% difference. However, the close-packed systems have twelve nearest neighbours, as against eight for b.c.c. The (112) and (123) planes of b.c.c. are sufficiently close-packed for deformation to occur on them by the usual shear process, but the b.c.c structure, although possessing many slip systems (planes and directions), is rarely as ductile as f.c.c. structures.

Tetrahedral voids occur in the space between the two body-centred atoms and the two closest cube corners at the face common to two cells. Four such positions occur at each face each shared by two cells, so that 24 occur on all faces of one cube, and 12 of these can be attributed to any one unit cube. The tetrahedral holes have two sides of length a, and four of length $\frac{1}{2}a(3)^{\frac{1}{2}} = 0.866a$. A sphere of radius $0.291R$ can be contained in that space, which is larger than the tetrahedral void in f.c.c. but

smaller than the octahedral void in f.c.c.

Octahedral holes occur at the centre of the faces of two body-centred cubes, and at the centre of cube edges. They are irregular in shape, having two atoms at a distance of $a/2$ and four at $\frac{1}{2}a(2)^{\frac{1}{2}} = 0.71a$. A sphere of radius $0.154R$ can fit in this space, which is very small (*see* Fig. 10).

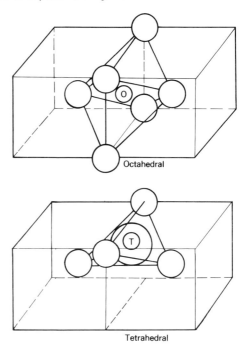

Fig. 10 *Voids in the body-centred cubic structure.*

Voids in body-centred cubic lattices are therefore smaller than the most favourable (octahedral) voids in face-centred cubic, which is one reason why carbon atoms in steel are more soluble in the higher temperature face-centred cubic iron structure (austenite) than in the lower temperature body-centred cubic one (ferrite). This gives rise to the multiplicity of structures available in the heat treatment of steel.

Many transition metals form a body-centred cubic lattice. Iron, molybdenum and tantalum are examples, with lattice parameters 2.86, 3.15, 3.3×10^{-10} m respectively. Sodium and potassium are more open b.c.c. structures, with $a = 4.29$ and 5.32×10^{-10} m respectively.

Boehmite. An aluminium ore, which consists of hydrated oxide of aluminium $Al_2O_3.H_2O$.

Boghead coal. A variety of cannel coal found in Scotland and New South Wales. It is believed to have been formed by the deposition of vegetable matter in lakes.

Bog iron ore. A loose earthy form of limonite (q.v.). It is a hydrated iron III (ferric) oxide $2Fe_2O_3.3H_2O$ found in low-lying or swampy ground.

Bohler. A high-carbon steel delivered in the form of ground rod with a smooth polished surface and with dimensional tolerances not greater than 0.012 mm from diameter required.

Bohr atom. The first model of the atom which explained the nucleus and the electrons moving in orbits around it. Bohr imagined quantum numbers to define the occupancy of discrete energy levels, and hence predicted the frequency of spectral lines obtained by excitation of atoms, in accordance with experiment. His model was oversimplified following de Broglie's proposals of the electron wave motion, but did pave the way for much of our modern understanding of the chemical behaviour and combination of atoms.

Bohr magneton. The unit of magnetic moment derived from the electric charge of the moving electron in an atom, equal to $eh/(4\pi mc)$ where e is the charge on the electron, m the mass of the electron, h Planck's constant, and c the velocity of light.

Boiler plate. Mild-steel plate usually produced by the open-hearth process, and used for making boiler shells and drums.

Boiling point. The point on a temperature scale where the vapour pressure above the surface of a liquid equals the external pressure. Molecules continually leave the liquid as gas at this point, the number of molecules leaving exceeding those which fall back. Bubbles of gas may nucleate at any location in the liquid above the boiling point, and may grow during their escape to the surface, being of lower density.

Bole. A hydrated silicate of aluminium (with appreciable amounts of iron oxide) occurring as a brownish-yellow or red clay. It breaks with a conchoidal fracture and falls to pieces with a crackling sound when placed in water.

Bolster. A block of steel used to support a drop-forging die. The bolster, which is usually dove-tailed, rests on the base-block of the hammer. The bottom die is keyed into it.

Boltzmann's constant. The constant derived by Boltzmann to interpret the kinetic behaviour of molecules of a gas, in a statistical sense. He stated the average energy of each degree of freedom of the motion of particles is the same, and equal to $\frac{1}{2}kT$, where k is Boltzmann's constant, and T the absolute temperature. Since a particle can normally move anywhere in space available to it, by referring to three co-ordinates in space, the average translational energy will be $3/2\,kT$. The ideal gas equation links pressure p, volume V, absolute temperature T, by the expression $pV = RT$, where R is the so-called gas constant for 1 mole of a gas. If the number of actual molecules in 1 mole of gas is N, called Avogadro's number, then $k = R/N$.

Bomb calorimeter. A strong iron vessel used for determining the calorific values of fuels (liquids or powdered solid fuels) by burning the fuel in oxygen under pressure.

Bondactor process. A method of applying refractory materials for patching or lining. The mixture is fed dry to a pressure gun, where it is moistened by an atomised spray.

Bond angles. The stereochemical arrangement of atoms, particularly in cases of covalent bonding, is largely governed by the repulsion forces between the electrons. Where electrons are bonded together (shared), they take up positions in space to minimise the repulsion forces between other bonded electrons (shared pairs) and any unbonded, or lone pairs. The angles between the directions of the electron bonds or lone pairs are the bond angles. In the tetrahedral arrangement of carbon in the diamond structure, the bond angles are 109.5°, and the structure forms one giant macromolecule, of great stability.

Bonded metals. Sheet metals that have the "finish" metal rolled on the basis metal. (*See also* **Cladding.**)

Bond energies. A measure of the attraction forces balancing the repulsion forces when atoms bond together, particularly in the covalent bond (shared electrons). When the bond energy is high, the material is likely to be strong and the structure stable. Bond energies are of the order of a few hundred kilojoules per mole, or a few electron volts per atomic bond.

Bonderising. A phosphating process used for the surface protection of ferrous materials. The steel is treated with phosphoric acid and an accelerator (catalyst). The process is mainly intended to produce a suitable basis for paint, enamel or lacquer.

Bonds. *(a) Atomic.* When two like atoms approach, two types of forces are generated, attraction of nuclei by electrons and repulsion between nucleus to nucleus and electron cloud to electron cloud. Initially, the attraction forces are the greater and as the interatomic distance decreases, the repulsion forces build up more slowly, but eventually at an increasing rate. When the two are equal, equilibrium is reached, and a "separation distance" is established, sometimes called "closest distance of approach". This is the atomic bond, a balance of forces between attraction and repulsion (*see* Fig. 11).

Inevitably, the electron clouds between the two atoms overlap, and this adds a further restriction affecting the balance of forces, since the electrons now overlapping, must still satisfy the Pauli exclusion principle, whilst sharing the same wave function. In practice, the energy levels of the outermost electrons in an atom become diffuse when overlapped, and a "band" of energy develops (*see* **Band structure**). Other aspects of electron behaviour may modify the behaviour of these higher energy level electrons, and are dealt with below.

The chemical inertness of the rare gases was soon recognised as a feature of the electronic configuration, since all have a completed sub group of the combinations of quantum numbers n and ℓ. Helium has two 1s electrons, neon two 2s and six 2p electrons, argon two 3s and six 3p, and so on. Apart from helium all have a "stable octet", of two s electrons and six p electrons. All have very low boiling points, and hence are gases over a wide range of temperature, and very few compounds of these gases have been recognised.

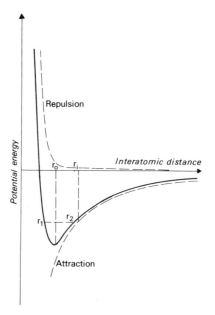

Fig. 11 *Potential energy of attraction and repulsion between atoms.*

This feature gave rise to the suggestion that combinations of other atoms might involve an increase in stability, if electrons could redistribute between the constituent atoms, to approach some resemblance to the stable octet of the inert gases. All atoms have more electrons than the inert gas preceeding them in the periodic system, and less than the succeeding one (*see* Table 13). By losing electrons, therefore, an atom could approach the structure of the inert gas preceding it, or by gaining electrons, approach the one succeeding it. Obviously, these processes are limited, and there will be preferences, on grounds of stability, which affect the type of bond.

(b) Covalent. In this type of bond, the stable octet is achieved by the sharing of electrons. The shared electrons do not "belong" to any one atom, they form part of a group of atoms. The atoms must then arrange themselves in a way to allow sharing of electrons in the simplest way and this leads to "shared pairs" of electrons, with strong directional properties. Hydrogen, as the simplest atom, has one s electron, but two hydrogen atoms can share the two electrons between them, and approach the two s electron state of helium

$$\text{Hydrogen (1s)}^1 + \text{Hydrogen (1s)}^1 \rightarrow \text{Hydrogen (1s)}^2.$$
$$\text{Hydrogen resembles Helium (1s)}^2.$$

The two shared electrons must then have opposite spin, to satisfy the Pauli exclusion principle, and the electron clouds will overlap. Two fluorine atoms, can similarly share one electron between them and approach the structure of neon, the succeeding rare gas.

$$\text{Fluorine (1s)}^2 \text{ (2s)}^2 \text{ (2p)}^5$$
$$\text{Neon} \quad \text{(1s)}^2 \text{ (2s)}^2 \text{ (2p)}^6$$

The arrangement is better seen diagrammatically in Fig. 12. As there is no polarity of electric charges, the covalent bond is also referred to as "homopolar".

More than one electron may have to be shared to produce a covalent bond, but these are always in shared pairs, of course.

Double covalent bonds will involve four shared electrons in two pairs of opposite spin. Triple bonds will obviously need six shared electrons in three pairs of opposite spin.

Oxygen atoms, two short of the stable octet of neon, form diatomic molecules by sharing four electrons between them, two from each atom. Nitrogen atoms have one less than oxygen, so need three electrons from each atom to be shared, in three pairs. Because of this requirement for pairing, the covalent bonds are directional and in the case of the compound ammonia NH_3, three hydrogen atoms supply their one electron to three of nitrogen, to form three shared pairs, and these are situated in three mutually perpendicular directions. Figure 12 shows the diagrammatic arrangement of oxygen, nitrogen, water and ammonia, but in two dimensions only.

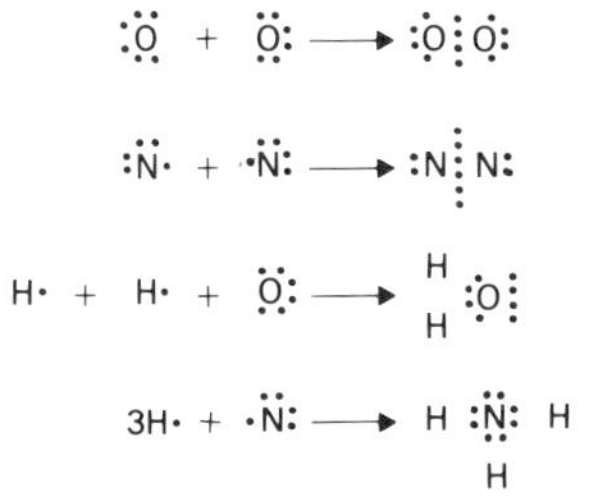

Fig. 12 *The covalent bond in oxygen, nitrogen, water and ammonia.*

The classical example of covalent bonding is that of carbon atoms. Carbon has four electrons short of a stable octet:

$$C\ (1s)^2\ (2s)^2\ (2p)^2.$$

To satisfy the Pauli principle, the electrons must be paired off in opposite spin pairs. If this happened only to the four electrons coming from other atoms to complete the stable octet, then these would form a double bond of two opposite spin pairs. The 2s electrons are already an opposite spin pair, but the original two 2p electrons are of parallel spin, and would have to change spin. In fact, the approach of other carbon atoms, to provide the four shared electrons, creates a new situation, the hybridisation of all the electrons involved. One of the two 2s electrons from each atom moves into the 2p band of energy, and the structure of the carbon atoms, now overlapping the outer electron clouds, becomes

$$C\ (1s)^2\ (2s)^1\ (2p)^3$$

and the electron spins of the 2s and three 2p electrons are parallel. They are then in position to pair off with four shared electrons of opposite spin, to form four shared pairs, and the Pauli principle is satisfied. The higher state of energy that one 2s electron acquired, in moving into the 2p band, is compensated by the lower binding energy of four shared pairs rather than two which would otherwise result. These hybrid sp^3 bonds are obviously highly directional and the largest possible angle which can exist between four bonds in spherical space is $109.5°$. This fixes the four neighbouring atoms at the corners of a regular tetrahedron. A large macromolecule can build up from this arrangement, that of the solid diamond. Diamond has a high melting point and high mechanical strength, indicating that covalent bonds are indeed very strong. Figure 13 shows the nature of the electron clouds in the diamond structure, with the four shared pairs.

(c) Ionic. The atom of sodium has one more electron than neon. The simplest way for sodium to reach the stable octet is to lose the outer 3s electron, and approach the state of neon, but the sodium atom will now have one positive charge on the nucleus more than the number of electrons. It will become positively charged or singly ionised, i.e. a "cation" (q.v.).

Neon $(1s)^2\ (2s)^2\ (2p)^6$
Sodium $(1s)^2\ (2s)^2\ (2p)^6\ (3s)^1$

Similarly, chlorine can simply gain one electron, become negatively charged in doing so, an "anion" (q.v.) and so approach the basic configuration of argon.

Chlorine $(1s)^2\ (2s)^2\ (2p)^6\ (3s)^2\ (3p)^5$
Argon $(1s)^2\ (2s)^2\ (2p)^6\ (3s)^2\ (3p)^6$

If the sodium atom gives its electron to chlorine, the two requirements are satisfied at the same time, but obviously there will be a strong attraction of the two oppositely charged ions, and hence a

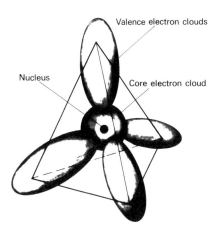

Fig. 13 *The electron clouds in the carbon tetrahedron (diamond).*

considerable bond strength results, whilst allowing both ions to retain the appropriate stable octet. This is the ionic bond, because ions are involved, and the extra potential energy of attraction is called "Coulomb" energy, given by the expression

$$\phi = \frac{-e\,e_2}{r}$$

where e, e_2 are the charges involved, and r is the interatomic distance.

The Pauli exclusion principle must still be satisfied, however, and in ionic bonds, this means that the overlapping energy levels of the individual ions are broadened into bands, and that the outer electrons must increase in energy, to allow the overlapped 2p bands for instance, to increase also. There is therefore an increased force of repulsion (repulsion forces are represented by positive, or increased energy) which counteracts to some extent the coulomb attraction. In the ionic bond, therefore, both attraction and repulsion forces are increased, the former by Coulomb forces, the second by exchange forces due to the energy level constraints.

The resultant bond is still a strong one, with high mechanical strength and melting point characteristic of materials with ionic bonds. Despite the requirement to balance electric charges, to maintain the ionic bond concept, it is relatively simple for ions involved in such bonds to form crystal lattices. The type of lattice is dependent on the relative sizes of the ions involved, and due to the high Coulomb attraction, a relatively close packing is

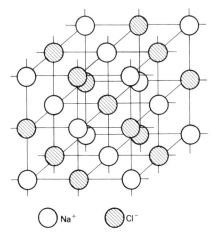

Na$^+$ Cl$^-$

Fig. 14 *The structure of sodium chloride.*

characteristic of ionic bonds, given the constraint of size difference. In sodium chloride, the relatively small sodium ions and the relatively large chlorine ions are arranged alternately along the three mutually perpendicular axes of a cubic system, thus preserving the balance of charge, and achieving the closest possible packing. Over 60% of the available space is filled, with a ratio of ionic diameters of 0.6. When atoms concerned in ionic bonds contain more than one above or below the stable octet, then more than one atom of the donor or acceptor type may be involved in achieving equalisation of charge on the ions. A metal such as magnesium, with two 3s electrons, can donate one each to two chlorine atoms, and so combine to form magnesium chloride $MgCl_2$. With oxygen, which can accept two electrons, one magnesium ion donates to one oxygen ion, in forming magnesium oxide MgO.

Figure 14 shows the structure of sodium chloride as equal sized points on the lattice. Figure 15 shows the actual size of the ions, affecting the packing density.

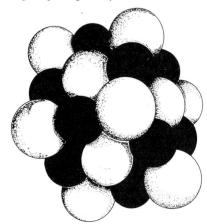

Fig. 15 *The ionic sizes in sodium chloride.*

(d) Metallic. The metallic atoms form crystal lattices readily, particularly the alkali and alkaline earth metals, with s electrons (e.g. sodium Na, magnesium Mg) and the transition metals having d or f electrons in the outer sub groups (e.g. iron Fe). The metals which are less strongly metallic and those which form oxides which may be acidic or basic (amphoteric) have p electrons, and frequently have a mixed bond arrangement. In the case of p electrons, there is a trend towards partial covalent bonds.

The metallic bond can be described as an arrangement of ions, whose higher energy electrons "float" through the whole space surrounding the ions, so that a continuous sharing takes place, sometimes described as a "free electron gas". The chemist's view of the metallic bond is to regard the individual ions as sharing electron(s) with nearest neighbours, and each of those neighbours sharing the same electron(s) with its own neighbours. By this means, the ions take up a position on a crystal lattice which provides the appropriate arrangement, e.g. in alkali metals, with one outer s electron, there are eight nearest neighbours so that one electron from each forms a shared bond with eight others, to achieve stability. The "bond" can then be considered to "resonate" from any one ion around the eight neighbours in turn, and the lattice is a body-centred cube.

In practice, many metallic ions do not share all the available outer (or valency) electrons amongst themselves, and a concept of positive ions arranged on a suitable lattice, with free electrons enclosing the whole, and permeating it completely, may be simpler to appreciate. To satisfy the Pauli principle, these electrons must assume higher energy, in a hybridised state, as occurs in the ionic bond. This is simplest with s electrons, more complex with transition metals and, as has already been stated, may involve partial covalent bonds of shared electrons, with p electron states.

The arrangement, whatever the detail, bestows special characteristic properties on metals. The ions can readily move relative to one another, giving rise to ductility, diffusion and the possibility of substitution by other metallic ions provided the dissimilarity is not too great (alloying). The electrons possess higher energy, partly due to the type of bond, and partly due to their essential mobility (Fermi energy). This allows them to be moved under the influence of applied thermal and electrical energy, as with the ions, and so metals have high thermal and electrical conductivity, in addition to ductility.

The balance of attraction and repulsion forces, exemplified by

the equation for potential energy of a stable arrangement of atoms $\phi_R = -a/r^m + b/r^n$ where $-a/r^m$ represents attraction, and $+b/r^n$ repulsion, requires in the metals, that $m = 1$, as it is for ionic bonds. For alkali metals, with the type of ideal metallic bond described, $n = 2$, and Fermi energy varies theoretically as $1/r^2$.

(e) Van der Waals'. Some arrangements of atoms into molecules cannot be envisaged in terms of ionic, covalent or metallic types of bond. The rare gases are good examples. They have atoms with stable octets, so cannot take part in the three types of bond discussed already, yet they do form molecules under certain conditions. The explanation is in polarisation, the position of the various electrical charges at individual points in time. It is reasonable to suppose that even electrons in stable octets are not always absolutely uniformly distributed the whole time. Figure 16 shows a sequence of events in which attraction and repulsion may alternate between electron clouds and nuclei of two like atoms. A net attraction force, of very low bond strength results, due to these polarisation, or van der Waals' forces. Where complex atoms or ions may be involved, polarisation may be increased, in which case the forces may be stronger, but never equal to that of other bonds.

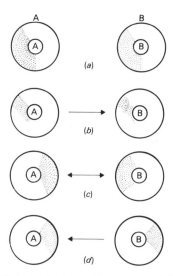

Fig. 16 *Van der Waals' attraction between like atoms:* →, ← *indicate attraction;* ↔ *repulsion.*

The attraction of Coulomb forces varies as $1/r$ in the ionic bond, the van der Waals' as $1/r^6$, where r is the interatomic distance. This serves to indicate the weaker nature of the polarisation forces. Occasionally, van der Waals' forces are present in addition to other types of bond, and will obviously be directional in nature. This gives rise to the properties of graphite for instance, where covalent bonds form in sheets of carbon atoms, with weak van der Waals' forces between the sheets, so that "sliding" of sheets over each other occurs, giving excellent lubricating properties.

Bond strength. The energy required to break the bond, which can be measured by the thermal energy involved in vaporisation, i.e. separating atoms from the main bulk, or by the energy involved in melting the material, when the thermal energy supplied overcomes the attraction forces and gives mobility to the atoms. The strongest bonds are the covalent and ionic type, the metallic bond is weaker, and the van der Waals' bond weakest of all. The bond strength should not be confused with the mechanical strength of a bulk of material, where crystal structure is also involved.

Bone black. A form of animal charcoal, used in the finely-ground condition as a pigment. It is also used to decolorise liquids, as a filtering medium and for the absorption of gases.

Borax. Sodium biborate $Na_2B_4O_7$. Found in Tibet, where it is called tincal. It is produced by neutralising the boric acid from Tuscan lakes with soda ash. It readily fuses to a transparent glass which can be coloured in different tints by oxides of metals. Borax is used as a detergent, and for many uses it replaces boric acid. It finds employment as a laundry glaze, as a flux in soldering and as a solvent for shellac and casein in glazing compounds.

Borchers-Schmidt process. A process for refining aluminium using mercury as a separating medium.

Bornite. A valuable source of copper. It consists of the sulphides of iron and copper. The ratio of the metal content varies, but an average composition may be represented by the formula Cu_3FeS_3. Hardness is 3 (Mohs scale) and sp. gr. 4.9-5.4. It is also known as erubescite (q.v.) and variegated copper ore. It occurs in some Cornish mines, where it is known as "horse-flesh" ore.

Boron. An element of atomic number 5, extracted from boric oxide or sometimes from potassium borofluoride. The oxide may be reduced to the metallic state by heating strongly with magnesium, sodium, potassium or aluminium or it may be dissolved in molten sodium fluoride and electrolysed. The borofluoride decomposes by heating with magnesium or aluminium. When obtained in any of the above ways, boron appears as a dark brown powder.

The element forms hard borides with carbon and silicon and these can be used as abrasives or as wear-resisting constituents of bearing alloys. In steel and in copper very small amounts of boron are useful as deoxidisers; but in the former case larger amounts (0.5-2%) give high strength steels whose brittleness increases with increasing amounts of boron. Amounts of the order of 0.003% are employed to facilitate the nitriding of steel. Melting point 2,300 °C, sp. gr. 2.34 (crystalline) and 2.37 (amorphous), at. wt. 10.81 (*see* Table 2).

Boron carbide. An abrasive material B_4C, used in cutting tools. It is obtained by heating boron oxide B_2O_3 with carbon at high temperatures—in the region of 2,500 °C. It is highly resistant to most chemical reagents.

Boron fibres. Filaments produced by chemical vapour deposition of boron on to a core of tungsten wire or a glass filament. They exhibit low density, high strength and very high elastic modulus, but are limited at present by high cost.

Boron nitride. A white, soft-feeling infusible powder having the formula BN. Boron is one of the few elements that combines directly with nitrogen. The nitride may also be formed by heating borax with sal-ammoniac or by heating boron in nitric oxide. It is not attacked by mineral acids, chlorine or solutions of alkalis, but it can be decomposed by fusion with potash or by steam.

When subjected to pressures of the order of 6 GN m^{-2} (0.06 Mb) at temperatures about 1,650 °C, boron nitride assumes a new physical state and develops new properties which are potentially useful. The crystals so formed bear a striking similarity to those of diamond, but they are harder than the latter and can withstand temperatures up to about 1,900 °C (diamond begins to burn at 850 °C). When lower pressures are employed, a material is produced that can be cut and shaped, and this shows some promise as a useful refractory. Also known as Borazon.

Bort. A compact variety of diamond, useless for gemstones, but extensively used for abrasive and cutting purposes.

Bosh. The tapering part of a blast furnace, lying between the top of the hearth and the bottom of the stack. The term is also used for a soft brush employed by foundrymen for moistening the edges of the sand moulds (i.e. around the pattern). This brush is also referred to as a swab.

Bossing. The operation of shaping malleable metals, such as sheet lead, by beating with mallets. It is also the name applied to the roughening of roll surfaces to improve their bite on the metal during rolling.

Boswell classification. A system of classifying moulding sands according to the proportion of grains occurring in the different size groups.

Bottom gate. A channel or in-gate leading from the runner into the bottom of a mould, through which the metal is poured.

Boundary. *(a) Domain.* In a ferromagnetic material, the periphery of a volume of material at which the direction of magnetisation changes as the boundary is crossed, often referred to as a domain wall. Domain boundaries move during magnetisation, and this movement is affected by imperfections present in or near the boundary. *(b) Grain.* The surface of a volume of crystalline material across which the atomic planes undergo a change in angular direction. Because of the finite spacing of atoms in a crystal, this change in direction inevitably involves a degree of mismatch

in the lattice, with the creation of vacant lattice sites or dislocations. *(c) Twin.* A plane in a crystal, which is characterised by the portions of crystal lattice on either side being each a mirror image of the other. This type of structure arises due to the deformation following mechanical shear, when atoms on successive planes move through a distance proportional to the distance from the twinning plane. Twin crystals can also arise during solidification, transformation or growth of crystals, with precisely the same features.

Bournonite. An ore containing the sulphides of copper, lead and antimony $3(Pb.Cu_2)S.Sb_2S_3$, which occurs with other ores of copper in Cornwall, Transylvania, Germany, Chile and Bolivia. It crystallises in the orthorhombic system, usually as modified prisms, often twinned, producing a cruciform or cog-wheel-like arrangement. This has given rise to the name "wheel ore" (radelez). It is also known as endellianite. Hardness 2.5-3 (Mohs scale); sp. gr. 5.7-5.9.

Bower-Barff process. A rust-proofing treatment applied to steel products. The steel objects are heated in a sealed retort to about 870 °C. Superheated steam is then injected to form a coating of iron oxides Fe_3O_4 and Fe_2O_3. Carbon monoxide is then injected to reduce the oxides. The cycle is repeated until the desired thickness of magnetic oxide coating has been obtained.

Box annealing. An annealing process in which the metal is enclosed in a suitable metal box or pot to protect the contents from oxidation. The process is also called close annealing or pot annealing.

Boyle's law. The volume of a given mass of gas is inversely proportional to the pressure at constant temperature. Expressed in terms of a perfect gas, i.e. one which does not show variations in normal kinetic behaviour, then $pV = N_0kT$ where p = pressure, V = volume, N_0 = Avogadro's number, k is Boltzmann's constant, and T = absolute temperature. It is usual to substitute $R = N_0k$, where R, the universal gas constant, is the constant for one mole of gas.

Bragg's law. In the reflection and diffraction of radiation by obstacles such as gratings, atoms on crystal lattice planes, etc., the condition for reinforcement of waves after reflection or diffraction is a function of path length traversed by individual waves. The path lengths must be equal, or differ only by an integral number of wavelengths.

If d is the spacing of the scattering units (slits in a grating or atoms in a crystal plane), θ the angle between the direction of the wave and plane involved, λ the wavelength, and n an integer, then $n\lambda = 2d \sin \theta$, which is Bragg's law. θ is often called the Bragg angle.

Bragite. A platinum ore, (Pd.Pt.Ni)S, found in the nickel deposits of Sudbury, Ontario.

Brake. A bar of solder. The name is also applied to the equipment used for bending sheet metal.

Brale. The diamond penetrator, conical in shape, used in the Rockwell hardness-testing machine.

Branched polymer. In the formation of a polymer, the individual monomer units may be situated on a straight chain, or one of the side groups may be replaced by a series of monomer units as a "branch". The properties of a branched polymer are generally modified towards higher strength and decreased plasticity, compared with a straight chain of similar molecular weight.

Brass. This is essentially an alloy of copper and zinc, but for special purposes small proportions of other metals are sometimes added to obtain increased strength, hardness or resistance to corrosion. The industrially important alloys of the zinc-copper system contain up to 50% Zn and are often designated by their copper-zinc content. The tensile strength of brass increases with increasing zinc content, rising sharply with the appearance of the β-phase at about 32% Zn and reaching a maximum when the α- and β-phases are present in approximately equal proportions. When the γ-phase is present (in excess of about 43.5% Zn) the strength falls rapidly and the alloy becomes extremely brittle. Alloys consisting entirely of the α-phase are malleable and ductile. They are particularly suitable for working cold, and may be rolled, drawn or otherwise fabricated hot or cold. Brasses containing both the α- and β-phases will withstand only a very limited amount of cold deformation without cracking, and may only be fabricated hot. The β-phase may be worked by forging, drawing or extrusion, hot, but alloys containing the γ-constituent are unsuitable for any form of mechanical working. Brasses are used as engineers' castings and, in the wrought form, as wire, sheet, tube and fabricated products.

The colour of brass changes with increasing zinc content, from a rich coppery red, through pale primrose yellow to white. The alloys containing 2.5% Zn (95% Cu) are usually known as zinc deoxidised copper. They will withstand very severe cold working and are used for small hollow-drawn parts.

Gilding metal containing 10-15% Zn is suitable for cold working by cupping, etc., and is extensively used for ornaments and cheap jewellery which is to be gilded, and for military badges and ornaments. Dutch metal, also known as Dutch brass or German brass, contains about 20% Zn and, because of its golden colour, is used in the manufacture of cheap ornaments. This alloy is also rolled or hammered into thin foils for use as a substitute for gold leaf.

Red brass, an alloy containing 30% Zn (70% Cu), gives the best combination of ductility and strength in this series of alloys and is used particularly for wire, for components fabricated from sheet by deep drawing and for tubing. When corrosion resistance is required, Admiralty brass is recommended. It contains 29% Zn and 1% Sn, the balance being copper.

The commonest form of brass (known as 60/40) contains 40% Zn and is known as yellow brass or Muntz metal. It consists of a mixture of the α- and β-phases, the relative proportions of which may be controlled to some extent by regulating the rate of cooling. δ-Metal contains approximately 41% Cu, 55% Zn and about 4% of other metals such as iron, lead and manganese. It possesses high strength and good resistance to corrosion. Naval brass contains 1-15% Sn with 61% Cu, the remainder being zinc.

Brazing solders contain equal proportions of zinc and copper and, when cooled sufficiently slowly through the solidifying range, consist of homogeneous β-solution (*see* Table 5).

Brass colouring. Brass components used in certain instruments were often coloured black by a dipping process in which the solution used was made by dissolving platinic chloride in an aqueous solution of a tin salt. With the development of less expensive methods, the practice of colouring small brass or copper-alloy products has been widely adopted. An extensive range of colours can be obtained by the use of various dipping solutions and by varying the conditions of immersion. Great care should be taken to ensure that the articles are free from grease before dipping in the colouring solution.

Typical Solutions

For brown shades, immerse in a solution of barium sulphide 12.5 kg m^{-3} (2 oz per gal) at 50 °C.

For brown, reddish-bronze or black shades, immerse in the following solution at 80 °C for a period sufficient to produce the colour desired: potassium sulphide, 84 g (3 oz); caustic soda, 84 g (3 oz); water, 4.5 × 10^{-3} m^3 (1 gal).

For black colour (oxide coating), use a copper sulphate solution 25 kg m^{-3} (4 oz to 1 gal of water) to which ammonia is added in sufficient quantity to just redissolve the green precipitate first formed. This solution is used at 60 °C and the black oxide coating which forms in a few seconds may be hardened in a caustic-soda solution 12.5 kg m^{-3} (2 oz per gal).

For steel-black colour (arsenic coating), use a solution of arsenic III oxide 560 g (20 oz) and ammonium chloride 336 g (12 oz) in 4.5 × 10^{-3} m^3 (1 gal) of hydrochloric acid. The solution is used cold and gives a strongly adhering coating of arsenic.

For bluish-black colour, use a solution of sodium thiosulphate 560 g (20 oz) and lead acetate 336 g (12 oz) in 4.5 × 10^{-3} m^3 (1 gal) of water. The solution is used at 80 °C. Various shades of colour ranging from dark gold to purple and finally bluish-black may be secured by varying the strength and temperature of the solution and the time of immersion.

Brassert acid-hardening process. A blast-furnace process in which the limestone content of the charge is restricted in order to produce a siliceous slag and a pig iron with a relatively high sulphur content. The metal is tapped into ladles containing sodium carbonate with which the ferrous sulphide in the metal reacts, forming sodium sulphide and ferrous oxide. These float on top of the molten iron and may easily be skimmed off. The metal may be cast into pigs, but is more usually transferred in the molten state to the steel plant.

Brass finishing. This is a comprehensive term which is applied to all processes used in preparing brass castings and wrought products for the market. In foundry work it usually refers to the grinding and fettling of castings, but in the case of wrought or heat-treated products it is generally applied to pickling or colouring processes. The removal of scale, etc., formed during annealing is generally performed in dilute sulphuric acid. The composition of this "pickling bath" need not be very accurately controlled, and a suitable solution may be made up by adding 2.25-4.5 × 10^{-3} m^3 (½-1 gal) of brown vitriol (BOV) to .045 m^3 (10 gal) of water. If more vigorous action is desired, approximately 5% nitric acid, chromic acid or potassium bichromate may be added to the bath.

For "bright dipping", which produces a bright golden colour, a much more vigorous pickling agent is required. The most common solutions used for this purpose are made up with nitric and sulphuric acids. The compositions are varied to suit individual needs, but a typical bath contains nitric acid (sp. gr. 1.56), 30 parts; water, 30 parts. The addition of a little common salt is usually considered advantageous. Products are immersed for only ½ minute or so in the acid bath and then thoroughly washed in clean water. A lighter yellow colour may be obtained by dipping for periods of several minutes in a solution of brown sulphuric acid 2.25 × 10^{-3} m^3 (½ gal) and potassium bichromate 2.27 kg (5 lb) in 0.045 m^3 (10 gal) of water.

Brass plating. Electroplated brass coatings are usually applied for decorative purposes. Typical details are as follows. The bath should contain copper cyanide, 100 g (3.6 oz); zinc cyanide, 34 g (1.2 oz); sodium cyanide, 210 g (7.5 oz); sodium carbonate, 112 g (4 oz); water, 4.5 × 10^{-3} m^3 (1 gal). The temperature should be maintained between 32 °C and 50 °C and a current density of 0.2-0.45 A m^{-2} will be found satisfactory.

Brass polishes. These are pastes or liquids used for polishing brass and some other metals. The various types of polish are produced to suit the severity of the operations in which they are to be used. The fine abrasive constituent varies from comparatively soft rouge (iron oxide) to hard alumina. In pastes, tripoli or pumice is mixed with oxalic acid and paraffin. Saponifiable greases are sometimes used. Liquid polishes often contain fine pumice or diatomaceous earth or an alkali with an organic solvent such as naptha.

Bravais lattice. The fourteen types of unit cell which can be shown to form extended lattices in crystals. Since the degree of symmetry of these indicates considerable basic similarity in some, the number of basic systems can be further reduced to six.

Brazing. One of the methods used to join metals and alloys together. It involves the use of an alloy more fusible than the metals to be joined. As distinct from soft soldering with lead-tin alloys, brazing usually refers to joining with alloys of copper and zinc. The materials to be joined are brought into as close contact as possible and the alloy melted so as to run between them. The heat may be applied electrically or by a blow-lamp. Brazing is preferred to soldering in those cases where a strong joint, which is heat-resistant, is required. The process is usually carried out on a hearth. Brazing solders (q.v.) or spelter solders, often referred to as hard solders, are employed.

Brazing solders. The common brazing or spelter solders contain approximately equal parts of copper and zinc and melt at temperatures from 870 °C to 900 °C.

The addition of silver to binary copper-zinc alloys lowers the melting point, and a series of ternary alloys containing 10-80% Ag, 16-50% Cu and 4-38% Zn, commonly called silver solders, are used for brazing alloys. Silver solders melt at 675 °C-820 °C, the lower-melting-point alloys having relatively high silver and low zinc content. They are used for joining silver to silver or to other metals. Gold may be added to silver solders for joining gold articles, the amount of gold and copper present being controlled by the carat of the gold to be joined. Nickel silver (q.v.) is also used as a brazing alloy. For special purposes tin, cadmium, nickel, chromium and manganese may be added to the silver brazing alloys. The silver-copper-phosphorus eutectic, which melts at 643 °C and contains up to 7% P, is frequently used.

Breaking stress. The stress, measured in MN m^{-2} (or smaller multiples), required to fracture a material in tension, compression or shear. (*See* **Ultimate tensile strength**.)

Break-out. A puncture in a mould or furnace hearth that permits molten metal to escape.

Breast. The working face of a mineral deposit. The term is also applied to the sloping sides of a furnace between the doors and the hearth or working bed.

Breeze. Very fine particles of coke.

Brick. Traditionally a block of material in the shape of a rectangular prism. Usually applied to an artefact of this shape made from clay, dried, then fired in a kiln to remove chemically combined water. Bricks for building purposes often use clay obtained locally, and containing small quantities of impurities, usually iron oxide, imparting yellow, brown or red colour to the finished product. The clay is rendered amorphous, or vitrified in firing, and though porous, does not absorb water chemically on a measurable scale. Previously fired brick (grog) may be crushed and added to clay in manufacture, to decrease the shrinkage which takes place on drying and firing of clay. This is very important in the manufacture of furnace or "fire" bricks. The amount of impurity affects the melting and softening temperature of fire clay used for high temperature operations, and hence the clay must be purer than for building brick. Different kinds of clay, hard marl and ground shale produce bricks of different texture and strength. Additions that may be made to the clay, e.g. sand, lime, etc., may increase the crushing strength from 10-20 MN m^{-2}. Marl bricks have an average strength of 85.5 MN m^{-2} and the upper limit may be as high as 96.5 MN m^{-2}.

Silica bricks are made by mixing crushed siliceous (quartz) rock or ganister with cream of lime, moulding, drying and kiln firing. Magnesia bricks are composed of burnt ground magnesia, bound with tar, and kiln fired. Bauxite bricks are composed of crushed bauxite with plastic clay binding. Other materials used are Dinas rock, flint and lime. Fire-bricks are made from any of the refractory materials above. For special purposes carbon and carborundum are employed.

Bridgman crystal growth. To control the growth of crystals from a melt the liquid may be held in a mould or crucible and subjected to a temperature gradient which will induce solidification beginning at the base. This can be done by lowering the mould through a furnace with a marked temperature gradient; alternatively, the mould can remain stationary and the furnace moved upwards. A third method is to keep the mould and furnace stationary but lower the temperature of the furnace by means of a controlled variable transformer. The latter method produces minimum vibration of the melt.

Bright annealing. An annealing process, usually carried out in an inert or reducing atmosphere, in which the surface oxidation is reduced to a minimum, so that the annealed product may have a relatively bright surface.

Brighteners. Materials used in electroplating baths to produce bright deposits. Substances such as cadmium salts, zinc salts, aldehydes, glue, gelatine, dextrine, naphthalene, sulphonates, polysulphonic acids and aryl sulphonamides are among those used for this purpose.

Brinell hardness test. A test in which the hardness of a metal is determined by applying a standard load to a standard-size steel ball placed upon the prepared surface of the metal, and measuring the diameter of the impression produced.

The surfaces under test should be flat and reasonably free from scratches and all tests should be made at a sufficient depth to be representative of the material. The specimens or objects under test should be firmly supported in such a way that the load is applied in a direction normal to the test surface. In the usual tests for steels the diameter of the ball is 10 mm ± 0.0025 mm, and a load of 3,000 kg is applied for 30 s. For softer metals such as brass, a load of 500 kg is applied for 60 s. The diameter of the impression should be measured in two directions at right angles to each other, using an instrument whose reading error does not exceed 0.02 mm. The average of the two readings is used in calculating the hardness from the following formula

$$H = \frac{P}{\frac{\pi D}{2}(D - \sqrt{D^2 - d^2})}$$

where H = Brinell hardness number; P = load applied (kg); D = diameter of the ball (mm) and d = diameter of the impression

made by the ball (mm).

The Brinell test is unsuitable for use on thin samples or on pieces which are so small that the metal would flow at the edges

TABLE 14. *Brinell Hardness Numbers and Conversion Table*

Diameter of ball Impression in mm	Brinell Number (10 mm ball 3,000 kg)	ROCKWELL			Vickers Pyramid No. VPN*	Tensile Strength (Steel)	
		C (150 kg)	D (100 kg)	A (60 kg)		Tons/ sq in	MN m^{-2}
2.00	946	–	–	–	–	–	–
2.05	898	–	–	–	–	–	–
2.10	857	–	–	–	–	–	–
2.15	817	–	–	–	–	180	2 780
2.20	782	72	78	84	1,224	171	2 640
2.25	744	70	77	83	1,116	162	2 500
2.30	713	68	75	82	1,022	155	2 400
2.35	683	66	74	81	941	149	2 300
2.40	652	64	72	80	868	142	2 200
2.45	627	62	71	79	804	136	2 100
2.50	600	60	70	78	746	131	2 020
2.55	578	58	68	78	694	126	1 950
2.60	555	56	67	77	650	121	1 860
2.65	532	54	65	76	606	116	1 800
2.70	512	52	64	75	587	112	1 730
2.75	495	50	63	74	551	108	1 670
2.80	477	49	62	74	534	104	1 610
2.85	460	48	61	73	502	100	1 540
2.90	444	46	60	73	474	97	1 500
2.95	430	45	59	72	460	94	1 450
3.00	418	43	58	72	435	91	1 400
3.05	402	42	57	71	423	88	1 360
3.10	387	41	56	71	401	84	1 300
3.15	375	40	56	70	390	82	1 260
3.20	364	39	55	70	380	79	1 220
3.25	351	38	54	69	361	76	1 170
3.30	340	36	53	68	344	74	1 140
3.35	332	35	52	67	334	72	1 110
3.40	321	33	50	67	320	70	1 080
3.45	311	32	50	66	311	68	1 050
3.50	302	31	49	66	303	66	1 020
3.55	293	30	48	65	292	64	990
3.60	286	29	47	65	285	62	950
3.65	277	28	47	64	278	60	930
3.70	269	27	46	64	270	59	908
3.75	262	26	45	63	261	57	880
3.80	255	25	45	63	255	55	850
3.85	248	24	44	62	249	54	830
3.90	241	23	43	62	240	52	800
3.95	235	21	42	61	235	51	780
4.00	228	20	41	61	228	50	770
4.05	223	19	40	60	222	49	755
4.10	217	17	39	60	217	47	725
4.15	212	15	38	59	213	46	710
4.20	207	14	37	59	208	45	695
4.25	202	13	37	58	201	44	680
4.30	196	12	36	58	197	43	662
4.35	192	11	35	57	192	42	648
4.40	187	9	34	57	186	41	632
4.45	183	8	34	56	183	40	618
4.50	179	7	33	56	178	39.5	610
4.55	174	6	33	55	174	39	602
4.60	170	4	32	55	171	38.5	596
4.65	166	3	32	54	166	38	587
4.70	163	2	31	53	162	37.5	576
4.75	159	1	31	53	159	36.5	565
4.80	156	0	30	52	155	36	557
4.85	153	B ($\frac{1}{16}$ in.)	E ($\frac{1}{8}$ in.)	30T	152	35	540
4.90	149	82	103	71	149	34	525
4.95	146	81	103	71	146	33.5	518
5.00	143	80	102	70	143	33	510
5.05	140	79	102	69	140	32	495
5.10	137	78	101	68	138	31.5	485
5.15	134	77	101	68	134	31	478
5.20	131	76	100	67	131	30	464
5.25	128	75	100	67	129	29.5	455
5.30	126	74	99	66	127	29	448
5.35	124	73	99	66	123	28.5	440
5.40	121	72	98	65	121	28	432
5.45	118	71	98	64	118	27	416
5.50	116	70	97	63	116	26.5	410
5.55	114	68	97	63	115	26	402
5.60	112	67	96	62	113	25.5	395
5.65	109	65	95	60	109	25	388
5.70	107	64	94	59	108	24.5	380
5.75	105	63	93	58	107	24	372
5.80	103	–	–	–	–	23.6	365
5.85	101	–	–	–	–	23	354
5.90	99	–	–	–	–	22.7	350
5.95	97	–	–	–	–	22.3	345
6.00	96	–	–	–	–	22	340

* Also referred to as DPH (Diamond Pyramid Hardness). DPN (Diamond Pyramid Number).

Rockwell B is the diameter of the ball indenter $\frac{1}{16}$ in = 1.6 mm.
Rockwell E is the diameter of the ball indenter $\frac{1}{8}$ in = 3.2 mm.

as a result of the ball impression.

A table of Brinell hardness numbers which eliminates the necessity for the calculation is given (Table 14). From the Brinell hardness number (BHN) an approximate value for the tensile strength of steels may be obtained. These values are included in the table. It should be clearly understood that the relationship between hardness and tensile strength depends to a large extent on the condition of the steel, i.e. hardened, tempered, annealed, normalised, etc.

British thermal unit. Originally defined as the amount of heat required to raise the temperature of 1 lb of water through 1° Fahrenheit, usually in the temperature range 60°F to 61°F. It was utilised as the basis of tables for calculation of heat or power derived from steam, and the conversion to metric units required new tables on the properties of steam. The International Standards Organisation agreed on a conversion, 1 B.T.U. is equivalent to 1.055 kJ, in 1956.

Brittle fracture. Theoretically, the separation of atomic planes in almost the entire absence of plastic deformation. Practically, the sudden and catastrophic failure of engineering components without prior plastic deformation. Brittle failures are the result of transition from elastic deformation to complete fracture, by propagation of fine cracks, the leading edge of which must be of very small radius, of the order of one or two atomic diameters. Such a sharp angle leads to severe triaxial stress concentration in the atomic bonds immediately ahead of the crack tip, with inevitable separation of atoms during propagation of the crack, often at extremely high velocity. The separation of the faces of the cracks absorbs some of the energy of the stressing system involved, but such a process simply serves to increase the stress ahead of the crack tip, hence the catastrophic failure which often results. Where the energy can be absorbed by plastic deformation, the crack tip is less likely to propagate, so that brittle failure tends to be associated with high strength materials possessing fewer slip systems (close packed directions and close packed planes) than more plastic materials. Body-centred cubic metals, hexagonal crystal lattices, and complex compounds of mixed bond type, all tend towards brittle behaviour, yet often show very high strength. Transitions at low or high temperature, and the effects of oxidation or chemical corrosion, can also transform a relatively ductile to a relatively brittle material. Crack-forming mechanisms add to the problem, and residual stresses due to manufacture, can be added to operational stresses to swing the balance in some cases. The rate of strain is important, high strain rates tending to exaggerate crack propagation velocity. Numerous tests devised for the assessment of brittle fracture, are based on artificial notching of specimens to induce stress concentrations, high strain rates such as those induced by a striking pendulum or a falling weight, and low temperatures of test. (*See* **Crack.**)

Brittle silver ore. Also called stephanite, this is a sulphide of silver and antimony $5Ag_2S.Sb_2S_3$ containing about 68% Ag. It occurs with other ores of silver in vein deposits in Europe, California and South America. It crystallises in the orthorhombic system, commonly in flat tubular prisms. Hardness 2-2.5 (Mohs scale), sp. gr. 6.26.

Broaching. A process for enlarging or finishing off holes in steel and other metals. The tool, called a broach, has teeth which increase in height towards one end. The first few teeth are the lowest to permit the tool to pass into the hole, the intermediate teeth remove most of the metal and the last few teeth finish the hole to size. The process is frequently used to produce holes other than circular section from cylindrical holes.

roggerite. A variety of pitchblende. Its composition is variable, but it may be regarded as uranium oxide U_3O_4 together with thorium, zinc and lead.

Bromlite. *See* **Alstonite.**

Bronze. Strictly speaking, this is an alloy of copper and tin, but the title has during recent years been conferred on a number of copper-base alloys that contain little, if any, tin. Some of these latter would be more aptly described as high-tensile brasses.

Tin bronzes are the oldest alloys known to man and were in use in Britain nearly 2,000 years before the Roman invasion. They still find many important applications as castings and

wrought products where high strength and resistance to corrosion are required.

Wrought bronzes are commonly known as phosphor-bronze and usually contain 4-6% Sn and less than 0.3% P, but the tin content may be as high as 14% in special cases. Wrought bronzes are used for turbine blading, springs, diaphragms, condenser tubes, spindles and electrical contacting mechanisms. Coinage bronze contains 95% Cu, 4% Sn and 1% Zn, but since 1940 the metal used for "copper" coinage in Britain contains 0.5% Sn and 2.5% Zn.

Binary tin bronzes containing about 20% Sn are used as castings for bells and the heavily loaded bearings of swing bridges and railway turn-tables, but most commercial bronzes usually contain additional alloying elements, such as zinc, lead and phosphorus. In casting qualities of phosphor-bronze which are often used for bearings and gear blanks, the tin content varies between 5% and 13%, but is usually around 10%. The phosphorus content (0.3-1%) is considerably higher than in wrought bronzes and is added to obtain increased hardness. Lead may be added to improve machinability, to ensure the production of pressure-tight castings or to improve anti-friction properties.

Since lead is practically insoluble in bronze, and appears in the structure as particles of practically pure lead, this metal should not be present in bronzes liable to become heated to temperatures in excess of about 300 °C, or the lead will exude. Lead up to about 2% improves the machinability without seriously impairing the mechanical properties. Bearing bronzes may contain up to 30% Pb, but the strength of the alloy decreases with increasing lead content and high-duty bearings are seldom leaded. Nickel may be used to improve casting properties and to refine the grain-size of the metal.

Zinc, like phosphorus, acts as a deoxidiser and improves the fluidity and casting properties of bronze, but the two elements are never added together, and when phosphorus is used to deoxidise a zinc-containing bronze the amount of phosphorus retained in the melt should be extremely small.

When zinc is used to replace tin the strength and corrosion resistance of the alloy are impaired. Bronzes containing 2% or more of zinc are termed gunmetals.

Bronze castings are used for bearing bushes, slide valves, air-compression machinery, marine fittings, ornamental work and statuary.

Aluminium bronzes are copper-aluminium alloys which often contain additional elements such as iron, nickel and manganese. These alloys have an attractive golden colour and possess good resistance to tarnishing owing to the formation of an extremely thin but very adherent surface film of alumina.

Wrought or α-aluminium bronzes usually contain about 4-7% Al. In the fully annealed condition they are soft (Diamond Pyramid Hardness 85) and ductile, but can be hardened by cold work to a DPH of about 250. The duplex aluminium bronzes used for castings, hot forgings, etc., usually contain between 9% and 10% Al and their structure consists of a mixture of the α and β solid solutions. The alloy containing 10% Al with 5% Fe and 5% Ni combines good strength, 370 MN m^{-2}, with resistance to oxidation and creep at moderately elevated temperatures (in the region of 400 °C). At room temperatures its tensile strength exceeds 620 MN m^{-2}.

The "manganese bronzes", now known as high-tensile brasses, are based on the 60% Cu-40% Zn (α + β) brasses modified by additional alloying elements such as tin (0.4-1.5%), aluminium (0.4-3.3%), iron (0.4-1.6%), manganese (0.6-2%), nickel (0.3-3%) and lead (0-0.9%). They find many industrial applications as sand cast, chill cast and hot worked components. Nickel bronzes and antimonial bronzes are formed by adding the appropriate metal, as well as tin, to copper (*see* Table 5).

Bronze welding. A jointing process in which filler rods consisting of brass containing small amounts of silicon are used. The melting points of these materials are below that of the metal to be joined.

Bronzing. The process of imparting to metals the colour of bronze. Brass may be "bronzed" by treatment with a dilute solution of hydrochloric acid containing potassium permanganate and green vitriol. Variations of the colour can be induced by subsequent heating. Other solutions used are: (*a*) a dilute solution of equal parts of lead acetate and hypo (sodium thiosulphate), (*b*) dilute sulphuric acid containing hypo. Zinc may be treated with a 3% solution of ammonium chloride, containing 1% potassium oxalate, in acetic acid.

Bronzing effect may also be produced by coating the metal with lacquers in which finely divided metal is suspended. The bronzing or "browning" of gun-barrels was produced by treatment with a strong solution of antimony chloride. The attractive patina—green or blue—in bronze occurs naturally after long exposure in the air, but can be induced artificially by treatment with ammonium chloride in a very dilute solution of acetic acid.

Brown coal. A carbonaceous fuel, intermediate between peat and true coals, which has a high moisture content. It is widely distributed in Europe, the principal deposits being in Czechoslovakia. The only British deposits are found at Bovey Tracy in Devon.

Brown hematite. Hydrated oxide of iron $2Fe_2O_3.3H_2O$ is more correctly known as limonite, since hematite is not combined with water. It is amorphous and is considered to be the result of alteration of other iron minerals.

Brownian motion. The irregular motion which occurs when the kinetic energy of gas and liquid molecules is translated by collision with secondary particles, such as dust in the atmosphere, or specks of solid suspended in a liquid.

Browning. A surface-finishing treatment applied to certain steel products for the purpose of improving appearance and resistance to atmospheric corrosion. There are numerous variations of the process, chiefly in the composition of the browning mixture ("witch's brew"). This latter is usually a solution containing iron III (ferric) chloride, alcohol and nitric acid, with or without additions of the chlorides of copper, mercury, antimony and bismuth, copper sulphate, ammonium chloride and hydrochloric acid.

The articles are thoroughly cleaned by boiling in a soda solution, dried and then treated with the browning solution at 50 °C. After drying for about half an hour, the articles are again treated with the solution and allowed to dry before placing in a warming cabinet at about 70 °C, whence they are transferred to a rusting chamber maintained at 80 °C with humidity up to saturation point. After about 90 minutes the rusted articles are dried in a chamber maintained at 60 °C to "set" the rust, and then placed in clean boiling water for 15 minutes. After draining, the articles are wire brushed to remove any loose particles of oxide. The whole series of operations (except the degreasing) may be repeated three or four times and the browned surface is finally coated with oil.

Brown ochre. Limonite (q.v.), used as a pigment.

Brown and Sharp gauge. A system of designating by numbers the diameters of wires. It is used throughout the USA for bare wires in brass, copper (except telephone wire), phosphor-bronze, nickel silver, aluminium and zinc, and for rods in non-ferrous metals. (See Table 15.)

Brucite. A natural hydrated oxide of magnesium $MgO.H_2O$ found in Ontario. Hardness 2.5 (Mohs scale), sp. gr. 2.39. Used in furnace linings.

Brunofix. A proprietary process for producing a black oxide film on steel by treatment in a strongly alkaline solution containing an oxidising agent.

Brytal process. An electrolytic brightening process used in the production of aluminium reflectors. The parts are anodically etched in an alkaline bath containing sodium carbonate and sodium phosphate, and subsequently anodised in a 25% solution of sodium bisulphite.

Buck plates. Iron or steel plates tied together by transverse rods and often strengthened by iron bands used to support the brickwork of a furnace.

Buddle. A shallow circular pit used for concentrating and classifying pulverised ores. In appearance it resembles an aerating tank seen on sewage farms.

Buff. A disc made of layers of cloth or leather and charged with abrasive. It is revolved at moderate speeds and is used in polishing metals.

Buffer reagent. A substance or a mixture of substances added to an electrolyte to prevent too rapid variations in the concentration of a particular ion.

TABLE 15. *Dimensions of Standard Wire Gauges*

No. of Wire Gauge	Diameter, in				Diameter, mm
	Brown & Sharp (B. & S.)	British Imperial (I.S.W.G.)	Birmingham Wire (B.W.G.)	Paris Gauge (P.G.)	Paris Gauge
7/0	—	0.5000	—	—	—
6/0	0.5800	0.4640	—	—	—
5/0	0.5165	0.4320	—	0.0122(5p)	0.31
4/0	0.4600	0.4000	0.454	0.0133(4p)	0.34
3/0	0.4096	0.3720	0.425	0.0150(3p)	0.38
2/0	0.3648	0.3480	0.380	0.0165(2p)	0.42
0	0.3249	0.3240	0.340	0.0196(p)	0.50
1	0.2893	0.3000	0.300	0.0236	0.60
2	0.2576	0.2760	0.284	0.0275	0.70
3	0.2294	0.2520	0.259	0.0315	0.80
4	0.2043	0.2320	0.238	0.0354	0.90
5	0.1819	0.2120	0.220	0.0394	1.00
6	0.1620	0.1920	0.203	0.0433	1.10
7	0.1443	0.1760	0.180	0.0472	1.20
8	0.1285	0.1600	0.165	0.0512	1.30
9	0.1144	0.1440	0.148	0.0551	1.40
10	0.1019	0.1280	0.134	0.0590	1.50
11	0.0907	0.1160	0.120	0.0630	1.60
12	0.0808	0.1040	0.109	0.0708	1.80
13	0.0720	0.0920	0.095	0.0787	2.00
14	0.0641	0.0800	0.083	0.0866	2.20
15	0.0571	0.0720	0.072	0.0945	2.40
16	0.0508	0.0640	0.065	0.1063	2.70
17	0.0453	0.0560	0.058	0.1181	3.00
18	0.0403	0.0480	0.049	0.1338	3.40
19	0.0359	0.0400	0.042	0.1535	3.90
20	0.0320	0.0360	0.035	0.1732	4.40
21	0.0285	0.0320	0.032	0.1929	4.90
22	0.0253	0.0280	0.028	0.2126	5.40
23	0.0226	0.0240	0.025	0.2323	5.90
24	0.0201	0.0220	0.022	0.2520	6.40
25	0.0179	0.0200	0.020	0.2756	7.00
26	0.0159	0.0180	0.018	0.2992	7.60
27	0.0142	0.0164	0.016	0.3228	8.20
28	0.0126	0.0148	0.014	0.3464	8.80
29	0.0113	0.0136	0.013	0.3700	9.40
30	0.0100	0.0124	0.012	0.3937	10.00
31	0.0089	0.0116	0.010	—	—
32	0.0080	0.0108	0.009	—	—
33	0.0071	0.0100	0.008	—	—
34	0.0063	0.0092	0.007	—	—
35	0.0056	0.0084	0.005	—	—
36	0.0050	0.0076	0.004	—	—
37	0.0045	0.0068	—	—	—
38	0.0040	0.0060	—	—	—
39	0.0035	0.0052	—	—	—
40	0.0031	0.0048	—	—	—
41	0.0028	0.0044	—	—	—
42	0.0025	0.0040	—	—	—
43	0.0022	0.0036	—	—	—

Building up. The deposition, either electrolytically, by metal spraying, or by welding techniques, of a comparatively heavy coating of metal on worn or undersized components, to bring them up to desired dimensions.

Bulk density. The term used in reference to the density of a loose powder. Also termed apparent density. It is important in powder metallurgy, as it affects the weight of powder that fills a die under constant-volume conditions.

Bulk modulus. The ratio of isotropic pressure to the relative volume change produced by that pressure within the elastic region. Isotropic pressure is equivalent to three equal and normal tensile or compression stresses acting in three mutually perpendicular directions. Bulk modulus is generally indicated by the symbol K

$$K = -\frac{p}{v/V}$$

where p is pressure, V the volume, v the change in volume. For metallic materials, K is approximately equal to E, Young's modulus of elasticity. For polymeric materials, K is much greater than E, and for covalent bonded and amorphous structures, about $60\% E$.

Bullard Dunn process. An electrolytic descaling process used for pickling iron and steel. It removes scale quickly from heat-treated parts without affecting the basis metal. The steel is first degreased electrolytically in special alkaline cleaners and then made the cathode in a solution of sulphuric acid containing 1 g l^{-1} of tin. The bath is operated at about 65 °C and a current density of 5.6-7A m^{-2}.

As the scale is removed tin is deposited as a uniform coating, which prevents further attack by the acid. The tin may be removed electrolytically or left on the steel to afford protection against corrosion.

Bulldog. Calcined tap cinder (oxide of iron) produced in the puddling process for making wrought iron, which is re-used for making the hearths of puddling furnaces.

Bullion. Gold or silver in bulk form.

Bumper. A machine used in foundries for compacting sand in a mould by repeated jolting.

Bunch. A pocket of ore. Many ore deposits follow the general lie of the rocks in which they occur, and such deposits are known as beds. When the ore occurs irregularly and is accumulated at certain points the deposit is called a "bunch" or "pocket".

Burden. The material charged into a smelting furnace, i.e. coke, ore and flux.

Burgers' vector. The magnitude and direction of slip associated with the movement of a dislocation, and given the symbol b. The direction of the vector is the direction of slip produced as the dislocation moves, taken as positive when directed towards the unslipped portion of crystal.

If the shear stress on a material is τ, then the force F_d per unit length of dislocation is given by $F_d = \tau\, b\, dl$ acting on length dl.

Burning. Permanent damage caused when a metal is exposed to unduly high temperatures. The damage, which cannot be removed by subsequent heat-treatment, may be due to the melting of the more fusible constituents, segregation of certain elements or reactions that produce penetrating gases.

Burning on. A process used in foundries for repairing or adding new pieces to castings, by making a mould around the points of jointing and pouring molten metal into the cavity.

Burnishing. A polishing treatment for imparting to metal a smooth, bright and glossy surface by flattening out the irregularities by the application of pressure.

Burnt deposits. A term applied to powdery electrodeposits produced when excessive rates of deposition occur.

Burnt limestone. Lime or calcium oxide CaO, a white amorphous powder produced by heating limestone $CaCO_3$ at a temperature in excess of 800 °C under such conditions that the carbon dioxide produced is continually removed. Also called quicklime.

Burnt magnesite. Magnesium oxide MgO. A white powder produced by calcining the carbonate $MgCO_3$ or the hydroxide $Mg(OH)_2$. It is used in medicine under the name of calcined magnesia, but finds its principal applications as a basic refractory material in bricks, crucibles, linings, etc.

Burnt pyrites. Consists essentially of iron oxide Fe_2O_3 and is a by-product of a process for making sulphuric acid by roasting iron pyrites FeS_2 in a current of air.

Burnt sand. A term applied by foundrymen to sands that have lost their powers of adhesion. Such sands may be revived by sieving through a very fine mesh, to remove the fine particles of dehydrated clay, and adding clay wash or an organic revivifier such as treacle.

Burr. A rough or sharp edge left on metal by cutting tools, etc.

Bustle pipe. The large iron pipe, refractory lined, that surrounds the blast furnace above the level of the hearth. It acts as a hot blast main, feeding the hot blast to the tuyères via the swan neck (*see* Fig. 17).

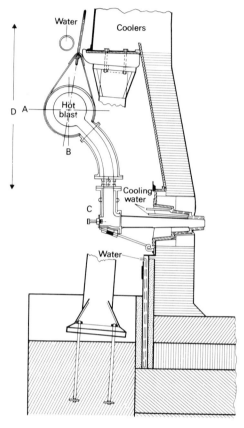

Fig. 17 *Cross-section of a blast furnace showing bustle pipe (A), swan-neck (B), belly-pipe (C), and bosh (D).*

Butt-welding. A process for joining two metal plates by placing them edge to edge, holding them together under pressure and welding along the line of contact.

C

Cadmium. A bluish-white metal, which takes a high polish, discovered in 1817. It occurs as a constituent of zinc ores and as sulphide in the somewhat rare mineral greenockite. The latter is found in Scotland, Czechoslovakia and Pennsylvania. The amount of cadmium found in the zinc ores, blende and calamine, varies from 1.5 to 3%. The metal is usually obtained as a by-product in the treatment of zinc ores, since cadmium is more volatile than zinc. The metal is easily rolled into foil or drawn into wire, but at temperatures above 80 °C it becomes brittle. Cadmium is mainly used as a plating metal and in the production of pigments, but it is also employed as a constituent of solders, low-melting-point alloys, bearing metals and in cadmium copper for conductor wires (*see* Table 2).

Cadmium selenide. A semiconductor material, with an energy gap around 1.75 eV, used as a detector of visible or near visible light.

Cadmium sulphide CdS. A semiconductor material, with an energy gap of approximately 2.5 eV, used as a detector of visible light.

Caesium. This was the first metal to be discovered by spectrum analysis (1860). It is silvery-white in colour, and at room temperatures is soft enough to be cut with a knife. It is the most electropositive of all the metals. Caesium takes fire on exposure to air and decomposes water at ordinary temperatures so that the metal is usually stored beneath a surface of oil. It is obtained chiefly from pollucite — a hydrated aluminium-caesium-sodium-iron silicate found in Germany and the USA. Caesium has only limited uses and is at present mainly employed in photo-electric cells (*see* Table 2).

Caesium chloride structure. A characteristic arrangement of an ionic bonded crystal lattice, in which the Cs cations (positive) are arranged on a simple cubic lattice, and the Cl anions (negative) are similarly arranged, the two cells being interpenetrating. The slightly smaller caesium ion just fits between eight of the larger chlorine ions, hence the co-ordination number is eight. The ratio of the ionic radii is 0.93, characteristic of this type of structure.

This structure becomes unstable when the radius ratio falls below 0.732, because the cation would not be "touching" the anions. (The ion is here regarded as a solid sphere, which it is not.) Over 60% of the available space volume is filled by the ions in the caesium chloride structure.

Caking coal. A type of bituminous coal that softens and appears to fuse on heating. The particles become aggregated and form a hard, compact cake in which there is no trace of the original coal particles. When lumps of such coal are heated in a retort they fuse together and yield a solid coherent mass of coke.

Calamine. This is zinc carbonate $ZnCO_3$, a very important zinc ore that occurs in beds and veins and is usually associated with blende ZnS and galena PbS and iron and copper ores in limestone areas. The main sources are Belgium, Germany, Spain and North America, as well as the Mendip Hills in Somerset and the Peak district of Derbyshire in England. The zinc is often replaced, in part, by iron and manganese. The colour of the ore varies from dirty white, deepening into yellow, grey, green or brown. It has a pearly lustrous appearance and is sometimes called zinc spar. Mohs hardness 5, sp. gr. 4-4.5.

The name calamine is applied commercially to include zinc silicate $Zn_2 SiO_4 .H_2 O$, which is known also as siliceous or electric calamine or hemimorphite.

Calaverite. A telluride of gold $AuTe_2$ found in Colorado and often associated with silver. Mohs hardness 2.5, sp. gr. 9.

Calcareous. Coated with or containing limestone or other calcium compounds.

Calcination. The operation of heating an ore or a refractory material at fairly high temperatures in order to drive off moisture, carbon dioxide, sulphur or other volatile constituents. The process differs from roasting because air is not supplied to the charge during the heating.

Calcite. A natural variety of calcium carbonate $CaCO_3$ which crystallises in the hexagonal system. When perfectly transparent, it is known as Iceland spar, which on account of its property of strong double refraction is used for polarising light in optical instruments. Mohs hardness 3, sp. gr. 2.7.

Calcium. A silvery white metal of sp. gr. 1.54 which tarnishes rapidly in air with the formation of a bluish-grey film of oxide which protects it against further attack. Unlike sodium, calcium metal can be handled without danger to the skin. It is so soft that it can be cut with a knife, and drawn, pressed, hammered or extruded.

Calcium occurs chiefly as carbonate in limestone, as sulphate in gypsum, as fluoride in fluorspar and as phosphate. The metal is produced by the electrolysis of fused calcium chloride $CaCl_2$ at about 800 °C, or by the thermal reduction of calcium oxide with aluminium *in vacuo*. Calcium is used as a deoxidising agent for a variety of metals and alloys, particularly copper, aluminium, magnesium, nickel and nickel-chrome alloys. Lead-calcium alloys containing up to 0.04% Ca are used for cable sheathing and those with 0.05-0.1% Ca for the grids and plates of storage cells. Calcium is used in certain types of magnesium-base casting alloys for improving surface finish. Alloys of aluminium containing up to 8% Ca have been patented.

Calcium oxide (lime) is a valuable slag-forming material. Calcium fluoride or fluorspar CaF_2 is used to improve the fluidity and sulphur-removing properties of slags. Calcium chloride is also used for the removal of sulphur from molten iron and steel (Saniter process) (*see* Table 2).

Calcium oxide CaO. An ionic compound, of very high melting point, crystallising with the "rock salt" structure. (*See* **Lime**.)

Calcium silicates. The oxides of calcium CaO and silicon SiO_2 combine to form several compounds, which also have hydrated forms. The compounds dicalcium silicate $2CaO.SiO_2$ and tricalcium silicate $3CaO.SiO_2$ are important in the manufacture of cements, since they combine with water to form a hard mass which can bind other rock materials together. The hydration process is slow, but more rapid than in the case of dehydrated aluminium silicates (vitrified clay). Nevertheless, the complete hardening of concrete takes several years.

Calcium sulphate $CaSO_4$. The natural mineral is a hydrated form. (*See* **Gypsum**.)

Calc spar. A form of calcium carbonate that crystallises in the hexagonal system. Sp. gr. 2.70-2.75. A form of calcite (q.v.).

Calgon. Sodium hexametaphosphate, used in alkaline metal-cleaning solutions, particularly in districts where the water is hard. It prevents the precipitation of insoluble soaps.

Caliche. A clayey or sandy material containing mainly sodium nitrate and sodium chloride, found in the rainless districts of Tarapaca Peru, North Chile and Bolivia. It is an important source of sodium nitrate (Chile saltpetre). The material may contain borates and up to 1% of sodium iodate. These form a source of boron and iodine respectively.

Calk. A variety of barytes $BaSO_4$ found in Derbyshire lead mines.

Calomel. Mercurous chloride Hg_2Cl_2. It occurs as the mineral horn quicksilver, crystallising in rhombic prisms. It usually appears grey or brown in colour and is often associated with the mineral cinnabar. Mohs hardness 1-2, sp. gr. 6.48.

Calorie. The amount of heat required to raise the temperature of 1 g of water at 15 °C by 1 °C. There are various definitions of the calorie dependent on the interval of temperature and the medium of measurement. The Internation Steam Table Conference in 1929 defined it in terms of the International Watt-hour, and this is known as the calorie IT equal to 4.1868 J. Although many tables may continue to quote heat energy in calories, the unit is now obsolete and has been replaced by the joule.

Calorific value. The number of heat units obtained by the complete combustion of unit mass of a fuel. The heat units were expressed either in centigrade-heat units (Chu), that is in pound-calories, or in gram-calories. The best method of finding the calorific value of a fuel is by direct experiment, using, for example, a bomb calorimeter (q.v.). An approximate value may be found from the analysis of the fuel. If, for example, 1 lb of fuel contains C lb of carbon, H lb of hydrogen and O lb of oxygen the approximate calorific value is given by

$$8,080\,C + 34,500\,(H - O/8).$$

In most cases the steam produced by the burning of the hydrogen escapes with the flue gases, so that its latent heat is lost, and the above formula gives a value that is too high. The "lower calorific value" is calculated from:

$$8,080\,C + 29,650\,(H - O/8).$$

Approximate calorific values for common combustible substances are given below.

Combustible	Calorific Value (Chu per cu ft)
Coal gas	260
Mond gas	85
Water gas	193
Natural gas	495
	(Chu per lb)
Hydrogen	34,500
Petroleum (average)	11,200
Fuel oil (average)	11,000
Carbon (burned to CO_2)	8,080
Carbon (burned to CO)	2,420
Sulphur	2,220
Anthracite coal	8,700
Bituminous (Welsh coal)	8,400
Cannel coal	8,000
Lignite	7,200
Coke	7,400
Dry wood	3,000

Calorimeter. A vessel in which heat quantities are measured. The apparatus may take several forms, since the quantities of heat are measured by observing the several effects they produce — e.g. a change in temperature, a transformation of heat into some other form of energy (mechanical or electrical) or vice versa.

Calorising. A trade name of a process for protecting the surfaces of steel from oxidation at elevated temperatures, by aluminising the metal. The coating of aluminium may be applied by dipping, by a chemical process of replacement or by spraying.

Campylite. A variety of mimetite $3Pb_3As_2O_8.PbCl_2$ that occurs in the form of barrel-shaped crystals of a brown or yellowish colour. It is used as a source of lead.

Can. A term used in the nuclear energy industry to denote the outer protective casing around the fuel element in a nuclear reactor. The metal used for the can depends upon the type of reactor, and such materials as zirconium, austenitic steel and special magnesium and aluminium alloys have been used for this purpose.

Cannel coal. A grey or dull black, easily ignited, amorphous type of bituminous coal, that breaks along joints or with a sub-conchoidal fracture. It burns with a long, luminous or smoky flame, whence the name "candle coal". These coals are close and compact in texture, and some of them, when heated, decrepitate, producing a crackling sound. On distillation they yield a large quantity of gas of good illuminating properties. There is also a good yield of tar oils.

Capacitance. The capacity of a body to store electric charge. Coulomb's law of capacitance states that $Q = CV$ where $+Q$ is the charge on one surface of the body, $-Q$ the charge on the opposite surface, V the potential drop between the surfaces, and C the capacitance. The charge is measured in coulombs, the potential drop in volts, so that

$$\text{capacitance} = \frac{\text{coulombs}}{\text{volts}}$$

and the unit of capacitance is the farad, after Michael Faraday. Capacitance is an intrinsic property of the material forming the body (capacitor) and the geometry of the system.

Capacitor. An arrangement to collect and retain electric charge, most frequently as two parallel conductor plates, separated by a gap, which gap may be filled by a dielectric (insulator).

If the plates are of area A square metres, the permittivity of the dielectric is ϵ, and the separation distance D metres,

$$C = \frac{\epsilon A}{D}$$

where C is the capacitance.

Cape ruby. A red garnet — pyrope or magnesium aluminium silicate $3MgO.Al_2O_3.3SiO_2$ — found in the Kimberley diamond mines. It is a typical garnet used in jewellery. Mohs hardness 7, sp. gr. 3.7, refractive index 1.71.

Capillary pyrites. Nickel sulphide NiS. It usually occurs in veins with other nickel and cobalt minerals as capillary crystals. Mohs hardness 3-3.5, sp. gr. 4.6-5.6.

Capture cross-section (nuclear). *See* **Absorption cross-section**.

Carat. A term used to express the degree of purity or "fineness" of gold. Pure gold is 24 carat or "1,000 fine". The fineness of alloyed gold is expressed by stating the number of parts of gold that are contained in 24 parts of the alloy. Thus 18 carat gold contains 18/24 gold, and this may be expressed also as 75% gold, or "750 fine".

A carat, equal to 200 mg, is used for weighing precious stones.

Carbide. A compound of an element (other than hydrogen, oxygen, nitrogen or chlorine) with carbon. Of the metals in Group I of the periodic table, only lithium combines directly with carbon, forming Li_2C_2. In Group II, calcium is the only one of the alkaline-earth metals that forms a carbide by direct union with carbon. The calcium carbide CaC_2 when treated with water yields ethyne. Beryllium is the only other metal in this group that combines directly with carbon to form a carbide Be_2C, but the remaining metals can be induced to form carbides if their oxides are heated with carbon in the electric furnace.

Aluminium is the one metal in Group III that forms a carbide Al_4C_3 by direct union, though the remaining metals will combine with carbon if their oxides are strongly heated with it. Silicon carbide SiC, a valuable abrasive, is produced by heating sand SiO_2 with coke and other carbonaceous matter in an electric furnace. It is marketed as carborundum (q.v.).

Of the remaining metals in the table the following form carbides: iron Fe_3C, chromium Cr_3C_2 and Cr_4C, tungsten W_2C and WC, molybdenum MoC and Mo_2C, manganese Mn_3C and uranium U_2C_3. All other metals may dissolve carbon but do not combine with it.

Alkali and alkaline-earth carbides react with water to form ethyne. Beryllium and aluminium carbides produce methane; manganese carbide forms methane and hydrogen; uranium carbide yields a mixture of hydrocarbons. All the rest are not attacked by water.

Carbon. A non-metallic element, which in its various forms finds many uses in metallurgical processes, and is an important alloying element in all ferrous metals. It is an obvious main constituent of a wide range of polymeric materials. Elementary carbon is found naturally in the crystalline forms of diamond and graphite (q.v.) and in the amorphous form as anthracite (which is about 95% carbon). In combination with other elements, it is found in mineral oil and is widely distributed in the carbonates of calcium and magnesium. Of the prepared amorphous varieties, wood charcoal and animal charcoal are obtained by the destructive distillation of vegetable and animal matter in the absence of air. Lampblack and soot result from the burning of carbonaceous fuels such as coal, oil, waxes, etc., with insufficient air-supply. Coke and gas carbon are the residues from the dry distillation of coal (see Table 2).

Coke, coal and mineral oil are used as fuels, while the two former are also used in the reduction of the ores of iron, copper, zinc, lead, etc. Gas carbon is used for electrodes and charcoal in the carburising of steel. The approximate percentage of carbon in commercial solid fuels is: lignite 45, dry peat 57, coking coals 79-83; anthracite: South Wales 90, Pennsylvania 93.

Carbonaceous. Pertaining to, or containing, carbon.

Carbonado. Coarse black diamonds, unsuitable as gemstones, used industrially for drills or for polishing. Also known as bort.

Carbon black. A form of elemental carbon obtained by condensation of the vapours produced by decomposition of certain carbon compounds, including hydrocarbons. The particle size is extremely small, giving rise to the suggestion that the structure is amorphous in type, although this is not universally accepted.

The material is used in paints, and as a filler in rubber, and other polymeric materials. In rubber, the carbon acts as an antioxidant to reduce ageing.

Carbon dioxide. A stable oxide of carbon CO_2 produced by the complete combustion of carbon or carbonaceous matter in air or oxygen. The heat of combustion of carbon to carbon dioxide is 406 kJ kg^{-1}.

Carbon fibres. The strong binding forces existing from covalent bonds have always attracted attention, but since such materials are not particularly ductile, the production of bulk artefacts is often not feasible. If the material can be used as a dispersed reinforcement in a more ductile matrix, the spread of brittle cracks is limited to local points, and a composite of this kind may be very strong but sufficiently ductile for practical use. Glass and silicate mineral fibres have been used extensively in this way, but carbon fibres were difficult to produce originally. As a result of extensive work on graphite for nuclear purposes, the achievement of preferred orientation in blocks was demonstrated, and finally, the means of producing graphitised fibres from a number of sources. In general, the first stage is to produce a fibre, drawing from a viscous "melt", such as tar; by forming a carbonised material such as coke into an extrudable material with tar as a binder; or by degrading a hydrocarbon or similar polymeric material. The simplest technique appears to be degradation of polymer, since this can be produced as a fibre very easily in the first instance, and all that is required is to pass the fibre continuously through a series of furnaces with carefully controlled heating zones. All such processes produce a "raw" graphite-carbon mixture, which then requires a high temperature graphitising treatment to complete graphitisation and to orient the hexagonal layer lattices along the fibre axis.

Acrylonitrile fibres are perhaps the most widely used. The fibre turns yellow after 30 minutes at 150 °C, and blackens in 5 minutes at 250 °C, due mainly to oxidation. At 1,000 °C in an inert atmosphere, the material is completely carbonised, and 1,500-3,000 °C is used for the final graphitising temperature. The carbon fibres produced have very high tensile strength (2-3 GN m^{-2}) and elastic modulus (350 GN m^{-2}). The incorporation of these fibres into polymeric resins has produced high strength/weight ratio components for aircraft and such sporting equipment as fishing rods, vaulting poles, rowing oars, golf club shafts and tennis racquets.

The fibres are also electrical conductors, and the more perfect the orientation the lower the resistivity. Treatment by nitric acid and alkali metals, for example, produces "intercalation" compounds between the oriented layers and improves the conductivity further. A whole series of semiconductor materials is feasible by this means. (See **Graphite, Fibres, Composite materials**.)

Carboniferous. Paleozoic rocks lying between the Devonian old red sandstone and the Permian formations, often containing coal. The carboniferous system comprises three main divisions: Carboniferous limestone is at the base, followed by the millstone grit, which in turn is followed by the coal measures.

This system is of great economic value and provides coal, iron ores, fireclays and lead ores, as well as limestone, building stone and road metal.

Carbonisation. The destructive distillation of carbonaceous matter that results in the formation of carbon, accompanied by the escape of volatile compounds.

Carbo-nitriding. *See* **Cyanide hardening.**

Carbon monoxide CO. A product of the incomplete combustion of carbon. It is produced when carbon burns in a limited supply of air or when carbon dioxide is passed over heated carbon. It is one of the constituents of water-gas (q.v.) which is produced when steam passes over red-hot carbon. The monoxide is a valuable fuel and reducing agent.

It is absorbed by a solution of cuprous chloride in hydrochloric acid, and a white crystalline compound $CuCl.CO.2H_2O$ is formed. This solution is used for the removal and estimation of the gas.

Carbon monoxide combines readily with such metals as nickel, cobalt, iron and molybdenum, forming carbonyls e.g. $Ni(CO)_4$. The carbonyls decompose on heating, leaving the pure metal, and this reaction is used in the refining of nickel (q.v.). As a result of its action on iron, the gas may eventually be able to escape through iron flues of stoves burning with an insufficient supply of air.

The gas occurs in coal-gas, water-gas, producer-gas, etc. It is very poisonous.

Carbon steel. Steel that owes its properties chiefly to the carbon content of the metal rather than to the presence of other alloying elements, which are seldom present in appreciable amounts. The carbon content is usually greater than 0.2%. The manganese content should not be greater than 1% or the silicon content greater than 0.75%. Also known as "straight carbon steel" or "ordinary steel".

Carbon tetrachloride (tetrachloromethane) (CCl_4). A colourless liquid that boils at 76 °C and which is used as a solvent for oils and fats in degreasing equipment. On account of its strong antiburning properties it is also used in certain fire extinguishers. Articles to be degreased are put on trays in a large container. The carbon tetrachloride is vaporised by hot water circulating through pipes in the base of the container, and allowed to condense on the articles to be degreased. Escape of the vapour is prevented by the presence of cold-water pipes below the lid of the container, where the vapour is condensed and drips back for recirculation in the degreasing cycle.

Carbonyl. The name applied to compounds formed by the combination of carbon monoxide and certain metals (e.g. nickel, cobalt, iron, molybdenum and ruthenium). Nickel, for example, combines with carbon monoxide at about 50 °C, to form the volatile $Ni(CO)_4$. At 180 °C this nickel carbonyl decomposes to give pure nickel and carbon monoxide. The corresponding cobalt compound is only formed at 150 °C and under a pressure of about $3MN\ m^{-2}$ (3MPa).

Carbonyl powder. The name given to particles of metal powder that are produced by the thermal decomposition of a metal carbonyl.

Carbonyl process. A process employed in the extraction of metals from treated ore, by reaction with carbon monoxide, thereby forming a metallic compound known as carbonyl (q.v.). The separation of the metal from this compound is accomplished by vaporising the carbonyl and obtaining the metallic powder as microscopic spherical particles of high purity.

Carborundum. Silicon carbide SiC. It is made commercially by passing an electric current through a mixture of sand and crushed coke (5 to 3) in an electric furnace. A little salt and sawdust are usually added and the temperature employed is above 1,550 °C. The product is a black crystalline material, used in place of emery as an abrasive. Since it does not readily fuse it is sometimes used in furnace linings. The pure substance, formed by heating silicon and carbon in an electric furnace, is nearly colourless and has a specific gravity of 3.1. Fused caustic soda reacts with carborundum to form sodium silicate.

Carburisation. The process of increasing the carbon content of the surface layers of a metal (usually steel) by heating it below its melting point with carbonaceous matter (e.g. charcoal, cyanides, carbon monoxide or other carbon-bearing media which may be solid, liquid or gaseous). If the steel has been deoxidised (e.g. by addition of aluminium), then the steel core structure will be fine grained, and quenching from the carburising temperature is satisfactory. If the steel is coarse grained, one or more heat-treatments may be necessary.

Carnallite. A hydrated chloride of magnesium and potassium $KCl.MgCl_2.6H_2O$ found as a saline residue at Stassfurt in Germany. The double salt fuses at about 176 °C, and deposits practically all the potassium salt, leaving the molten hexahydrate of magnesium chloride. It is an important source of potassium salts. It is also used as a fertiliser. Mohs hardness 1, sp. gr. 1.60.

Carnotite. A vanadate of uranium and potassium containing about 40% U, found in the sandstones of Colorado. It is a source of radium.

Caron's cement. A mixture of powdered charcoal and barium carbonate in the proportion of 3 to 2, used in the cementation process of iron and steel. Also called hardenite.

Case. The surface region of a metallic component in which the composition and properties have been modified, as for example by carburising, nitriding, etc.

Case hardening. A process consisting of one or more heat-treatments for producing a hard surface layer on metals, as by carburising, etc. In the case of steel the carbon content of the surface of a low-carbon nickel steel is increased by heating in some suitable medium containing the carbon in available form. Subsequent heat-treatment is required to produce the properties desired both in the core and in the case.

Caspersson's method. A method of pouring steel, suggested by Caspersson, in which a vessel with a number of small perforations is used. The steel is thus divided into a number of fine streams, the object of which is to afford an opportunity for the escape of contained gases.

Casseopeium. Synonymous with lutecium, a rare-earth metal of atomic weight 174.

Cassiterite. Tin oxide SnO_2. It is the principal source of tin and occurs in pneumatolytic veins associated with certain acid rocks, e.g. granites, pegmatites, and also in placer deposits resulting from prolonged weathering of the tin veins. Cassiterite is usually black or brown and crystallises in the tetragonal system as tetragonal prisms terminated by tetragonal pyramids. It breaks with a subconchoidal or uneven brittle fracture. More than half the world's supply of tin is obtained from alluvial cassiterite deposits in and around the Malay Peninsula, Indonesia, China, Zaire and Nigeria. Primary deposits of cassiterite and of other more complex tin ores yield from 35,000 to 40,000 tonnes of tin per year. Other sources include Cornwall, Spain, Portugal, Germany and the South American republics. Mohs hardness 6-7, sp. gr. 6.4-7.1. Also known as tinstone.

Casting. The operation of pouring molten metal into sand, metal or other moulds and allowing it to solidify.

A metallic object which has been made by casting the metal into the shape required without any working other than machining.

Casting ladle. A steel vessel lined with refractory material in which molten metal is transported to the casting shop for pouring or tapping into moulds.

Casting pit. An area in which molten metal is cast into ingots. The term generally applies to steelmaking, and the pit lies several feet below the level of the melting platform and extends for a considerable distance behind the furnaces. Travelling cranes span the pit and carry ladles of molten metal from the furnaces to the ingot moulds.

Casting strains. Strains, accompanied by internal stresses which result from the cooling of a casting where the rates of cooling of different parts are unequal. The stresses may be relieved by suitable heat-treatment.

Cast iron. An iron-carbon alloy produced by remelting pig iron and scrap, with or without additions of alloying elements, in a cupola or other furnace and casting the metal into sand or metal moulds. Cast irons usually contain 2-4% C and in the "as-cast" condition are not malleable. They are usually regarded as falling into three main groups.

(a) Grey iron, in which carbon occurs as flakes or graphite. This material has excellent casting properties and is readily machinable. Alloying elements such as nickel, chromium and molybdenum may be used to modify and enhance certain desirable properties. Grey iron is available in a wide range of strengths and is widely used where high shock resistance is not required. Many modern alloy cast irons combine strength with improved shock resistance due to a modified graphite structure. The so-called "high-duty" cast irons owe their properties to the use of special melting and casting techniques, the addition of alloying elements such as nickel, chromium, molybdenum, etc., and in certain cases to heat-treatment. Nickel, like silicon, has a graphitising effect and tends to produce grey iron, chromium forms stable carbides and molybdenum tends to produce an acicular structure.

Inoculated irons have relatively low carbon and silicon contents, and are graphitised in the ladle by adding compounds such as calcium silicide to the melt. The martensitic irons, used for resisting abrasion, contain 4-5% Ni with other alloying elements. Austenitic irons such as Ni-Resist have good resistance to heat and corrosion. They are non-magnetic and contain 14-25% Ni.

Spheroidal graphite cast irons combine high strength with ductility and are produced by treating suitable cast irons with cerium or magnesium. Spheroidal graphite cast irons develop optimum ductility on annealing in the range 850-700 °C for a few hours.

(b) White or *chilled irons* contain practically all their carbon in combined form as cementite Fe_3C and as pearlite, and are usually hard and brittle. Low silicon content is usually required and stabilising elements, such as chromium, may be added when white iron is to be used in the "as-cast" condition. White irons are frequently alloyed, e.g. with 4-5% Ni and 1.5% Cr (Ni-Hard), and are used for wear-resisting applications and particularly for metal working rolls.

(c) Malleable cast irons. This type of casting is obtained by casting white iron of suitable composition into the desired form, and subsequently converting the combined carbon into free carbon or temper carbon by suitable heat-treatment. In the "Whiteheart" malleablising process a certain amount of the carbon is eliminated from the surface of the casting by oxidation.

Castolin. A trade name of synthetic cast-iron alloys used for solder welding of cast iron. The weld metal has a lower melting point than the cast iron, and the joint is formed at a temperature below the cast-iron melting point, e.g. Castolin 14 will weld at 760 °C.

Castomatic process. An automatic process that uses a machine for casting bars of tin-lead solder and other white-metal alloys. In the process metal is removed from the bottom of the melting kettle and cast into enclosed moulds. Arrangements are made to offset shrinkage by feeding molten metal to the mould during solidification.

Cast steel. Any object made by pouring molten steel into moulds.

Catalan furnace. A primitive type of furnace designed to produce wrought iron direct from the ore. The bottom and back of the open hearth were made of sandstone lined with carbon. A blast was supplied by a trompe worked by water from a mountain stream. The strength of the blàst rarely exceeded 200 N m^{-2}.

Catalan process. A direct process for producing wrought iron by smelting specially selected ores with charcoal under reducing conditions. The ore generally used was brown hematite. The glowing mass of iron, after removal from the fire, was hammered while still hot, to expel the slag. The cinder, which contains upwards of 30% Fe, was a waste product as far as this method was concerned.

Catalysis. A process in which the speed of a reaction is altered by the presence of an added substance which remains unchanged at the end of the reaction. There appear to be several types of catalytic process: *(a)* the catalyst may cause the reaction to follow a course or mechanism different from that which occurs when the catalyst is absent, or *(b)* the catalyst may cause a change in the medium or of the concentration of the reacting substances, as a result of which the reaction velocity changes. This conception of catalysis does not exclude those cases in which the catalyst takes part in the reaction but is regenerated in the cycle of changes. Many metals of variable valency act as carriers, while some metals in a finely divided form act as "contact" catalysts. Hydrogen dissolved in, or adsorbed by, a metal is of different concentration and in a different medium from gaseous hydrogen. Nickel is a good hydrogenating agent, thorium dioxide is a useful oxygenating agent, as is finely-divided platinum.

Catalysts are usually employed to accelerate reactions and are called "positive" catalysts, but in some cases "negative" or retarding catalysts are important, e.g. glycerol, phenol and mannite retard the oxidation of sodium sulphite.

Certain substances, such as traces of arsenic, destroy the activity of catalysts such as platinum. They are known as catalytic poisons.

Catalytic agents. Substances which when added to a reaction mixture change the velocity of a reaction. They remain unchanged at the end of the reaction, but may or may not undergo a cycle of changes. In most cases the amount of catalyst employed is small and bears no relationship to the quantity of the substances involved in the reaction. But in reactions in which the hydrogen ion acts catalytically it is found that the reaction velocity does increase in proportion to the hydrogen-ion concentration. Catalysts have no influence on the final equilibrium, but only hasten the attainment of the equilibrium of the reacting substances (see also **Catalysis**).

Cathode. In electrolysis the cathode is the negative electrode or the plate by which the current leaves the electrolyte. In the electrolysis of salts the metals and hydrogen are liberated at the cathode. In the following electrochemical series any element in the list will be positive to one that comes after it, if they are placed in a conducting solution. The e.m.f. generated will be greater, the greater the difference in position in the list. The element that is earlier in the list will be the one that is consumed under these conditions. It should be noted, however, that the order of the elements may be slightly changed by varying the nature, strength and temperature of the solution used as the electrolyte. The electrochemical series is: magnesium, aluminium, manganese, zinc, iron, cadmium, nickel, lead, tin, copper, mercury, silver, platinum and gold.

Cathode copper. An electrolytically refined copper which is remelted and cast into bars or billets for fabrication into the required form. It is the raw material used for producing electrolytic high-conductivity copper, both tough-pitch and oxygen-free. It is also used for high-conductivity copper castings and for making alloys such as cadmium-copper, silver-copper, chromium-copper where low specific resistance is important.

Cathode rays (electrons). The name given to the rays that are emitted from the cathode during the passage of electricity between the metal terminals in an exhausted tube. The rays travel in straight lines and normal to the cathode, and they cause luminescence when they fall upon various materials, and in some cases change the colour of the substance exposed to them. When the rays are concentrated by using a cathode in the shape of a hollow cylinder or a portion of a sphere they produce heating effects at the focus. The mechanical effects of the rays when incident upon a light vane are due mainly to a secondary result of the thermal effect. The path of the rays may be deflected by imposing an external magnetic or electric field. Cathode rays can pass from the exhausted tube into the air through thin metal windows.

Cathode-ray tube. A funnel-shaped vacuum tube in which is generated a beam of electrons. An "electron gun" produces and projects the electrons while two pairs of parallel plates deflect the beam. A screen is provided on the wide end of the tube on which a luminous image is formed. In Fig. 18, K is the indirectly heated cathode. This latter is coated with certain oxides which readily emit electrons and these pass through the small opening in the negatively charged grid G. The anodes P_1 and P_2 at different potentials (with respect to the cathode) focus the electrons on the screen S. The electric impulse under test is applied to the deflecting plates D_1 and D_2 and changes in it are shown by the luminous trace marked out by the electron beam on the fluorescent screen. Cathode-ray oscillographs are used with directors of the electrical type to detect noise, phase relations, wave shapes and frequency. The abbreviation C.R.T. is often used.

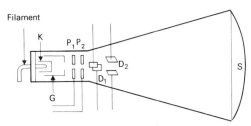

Fig. 18 *Principle of the cathode ray tube.*

Cathodic protection. A method of protecting ferrous metal structures by contact with "sacrificial" anodes, which form the positive of a natural electrolytic cell and are themselves corroded.

Cation. The positively charged ion in an electrolyte. It migrates towards the cathode under the influence of the potential difference between the electrodes.

Cat's eye. A peculiar opalescent form of quartz with a minutely fibrous structure due to the presence of asbestos. It has usually a greenish-yellow hue and when cut with a convex face the asbestos produces in certain lights a silky flash resembling the gleam of a cat's eye.

Caulk weld. A weld made for the purpose of sealing a joint.

Caulking. A process of burring down by means of a special tool a strip of metal on one edge of a plate on to an adjacent plate. Rivetted joints do not ensure steam tightness, and caulking of the joints is usually done both internally and externally, in the case of boilers. The method of sealing the joint may be done by packing it with suitable material (caulking rings or caulking strips, q.v.) and hammering in. In the case of overlapping plates, the edges are usually hammered into intimate contact.

Caulking rings. Rings of mild steel placed between the adjoining flanges of boiler plates to facilitate the caulking of the seams.

Caulking strips. Strips of mild steel used in caulking rivetted joints. They are usually employed in the larger type of work to ensure steam or water tightness.

Caulking tool. A chisel-like tool having a blunt edge suitable for burring down the edges of plates in sealing joints and for inserting rings or strips without cutting the metal.

Causal metal. An austenitic grey cast iron possessing good resistance to various types of corrosive media. Its resistance to corrosion is comparable with that of nickel.

Caustic. Generally used to describe any substance which corrodes or burns organic matter. Caustic soda NaOH and caustic potash KOH are the commonest substances of this class. They saponify (i.e. they form soaps with) most animal and vegetable fats and oils, and the former is employed widely as a degreasing agent. Other caustic substances are silver nitrate, mercuric nitrate, zinc choride and phenol.

Caustic embrittlement. Embrittlement resulting from immersion of metal in caustic alkali solutions.

Cawk. The name given by miners to the mineral heavy spar occurring in veins in lead mines, where it is associated with galena, calcite, fluorite and quartz. Sometimes spelt "calk".

Celestine. Strontium sulphate $SrSO_4$, occurring in beds of rock salt, gypsum and clay, in sulphur deposits in Sicily and also in certain limestones. It crystallises in the orthorhombic system. Mohs hardness 3-3.5, sp. gr. 3.96.

Cell (crystallographic). Most often referred to in terms of the "unit cell", it is a means of delineating the basic structural unit which can be repeated infinitely along orthogonal, or nearly orthogonal, three-dimensional crystal lattices, by simple translation.

Every unit cell pattern has identical surroundings in the one individual space lattice. If, in two dimensions, atoms are placed at the four corners of a square, the square becomes the unit of the two-dimensional lattice. If the third dimension is added by sitting a similar square lattice on top and below the atoms in the original, then this becomes three dimensional, the atoms are situated at the corners of a cube, and a simple cube becomes the unit cell. If, however, the two-dimensional layers above and below are placed in the centre of the squares of the original, and the third layer is placed directly above the first, the lattice cell is a "body-centred" cube (*see* Fig. 19).

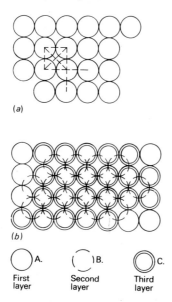

(a)

(b)

A. First layer   B. Second layer   C. Third layer

Fig. 19 (a) A square-packed plane of spheres. (b) Three layers of square-packed planes, ABAB... stacking (b.c.c.).

If the original two-dimensional layer is in a hexagonal array, with the second row and subsequent rows fitting in the space between atoms in the previous row, then several possibilities occur in three dimensions, with much closer packing than the square lattice.

The unit cell is not always obvious when building up a crystal lattice, but the characteristics obviously stem from the way in which atoms are packed together. The co-ordination number is also characteristic, since it indicates the number of near neighbours. (*See* **Crystal classification, Bravais lattice, Miller indices.**)

Cell constant. Used in connection with electroplating baths; it is the ratio of the specific conductance to the actual conductance of the electrolyte in a given cell. Numerically it is equal to the mean distance between the electrodes divided by the mean cross-sectional area of the current path.

Cellular solidification. When constitutional supercooling (q.v.) occurs, the liquid/solid interface cannot remain planar; it is unstable. Normal statistical variations must lead here and there to

small bulges being formed in the plane of the interface. This projects solid into the more supercooled liquid, and so acts as a nucleus and more rapid freezing occurs at the tip of a bulge. The bulge therefore grows still more and, as the effect must diminish to either side of the original bulge, rejection of solute occurs at the sides, so producing a high concentration of solute in the liquid. This increases the constitutional supercooling around the base of the bulge, and the liquid here remains unfrozen as the bulge grows. Because of the new concentration gradients to each side of the bulge, further bulges must form, and the whole interface tends to advance with the production of distinct columnar distribution of solute as the last liquid eventually freezes, hence the honeycomb or cellular type of structure commonly observed.

Cellular structure. A type of structure in which one constituent of an alloy forms a continuous network around the grain boundaries.

Cellulose. A carbon compound related to the sugars, since it can be regarded as an "anhydroglucose", in that glucose rings $C_6H_{12}O_6$ are linked by terminal hydroxyl (—OH) groups, with the elimination of water (*see* Fig. 20).

The hydroxyl (—OH) groups are highly polarised, and provide a strong bond between individual cellulose molecules, which tend to become oriented along a fibre axis. In wood, where cellulose is the main constituent, the fibre axis is termed the "grain" of the wood, and it is along this axis that the greatest strength is developed.

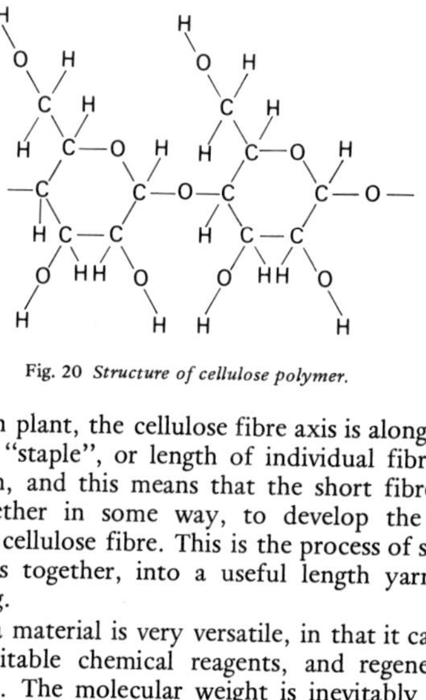

Fig. 20 *Structure of cellulose polymer.*

In the cotton plant, the cellulose fibre axis is along the cotton fibres, but the "staple", or length of individual fibre, is rather short in cotton, and this means that the short fibres must be connected together in some way, to develop the maximum strength of the cellulose fibre. This is the process of spinning, to twist the fibres together, into a useful length yarn, for subsequent weaving.

Cellulose as a material is very versatile, in that it can easily be dissolved in suitable chemical reagents, and regenerated into modified forms. The molecular weight is inevitably altered by these processes (*see* **Regeneration**).

Celsius scale. An international scale of temperature similar to the centigrade scale (i.e. the interval between the fixed points as determined by the freezing point of water and the boiling point of water at standard pressure is divided into 100 degrees). There are, however, an additional number of basic fixed points as the foundation of the scale as well as a range of secondary fixed points. The instruments available for interpolation make it convenient to divide the whole scale into four divisions.

(a) From $-190\,°C$ to $100\,°C$. The basic fixed points are the boiling point of oxygen under 101 kPa (760 mm Hg) pressure, the freezing point of water and the boiling point of water under 101 kPa pressure. Temperatures in this range are deduced from the resistance of a standard platinum resistance thermometer.

(b) From $0\,°C$ to $660\,°C$. The basic fixed points are the melting point of ice, the boiling point of water under 101 kPa pressure and the freezing point of sulphur. Temperatures in this range are deduced from the resistance of a standard platinum resistance thermometer.

(c) From $660\,°C$ to $1,063\,°C$. The basic fixed points are the freezing point of antimony ($630\,°C$), the freezing point of silver ($960.5\,°C$) and the freezing point of gold ($1,063\,°C$). Temperatures in this range are deduced from the e.m.f. of a thermocouple, one junction of which is maintained at $0\,°C$. The thermocouple is platinum-platinum rhodium, the latter alloy containing 10% Rh.

(d) Above 1,063 °C. The temperatures in this range are deduced from the ratio of the intensity of visible radiation (of given wavelength) to the intensity of radiation of the same wavelength emitted by a black body at the standard gold point (1,063 °C).

Celtium. Synonymous with hafnium. A rare-earth metal obtained from gadolinite earths. It is often found in association with zirconium.

Cement. Portland cement is made by heating argillaceous limestone or by calcining a mixture of suitable clay with the requisite quantity of limestone. Lime and clay are ground together and heated in a rotary kiln with coke. The product contains 55-59% CaO; 23-24% SiO_2; 7-8% Al_2O_3; 3.5-5.5% Fe_2O_3. When water is added to the calcined product, hydration of calcium silicate and calcium aluminate occurs, forming a mesh of interlaced crystals. The hydration takes from hours to months, depending on the hydrate involved, so that hardening proceeds over a lengthy period. Hydraulic mortar is made from dead burned lime, to which 5% clay has been added. Such a material hardens under water. Other cements are made from high alumina-lime mixtures (ciment fondu) or magnesium oxychloride.

Cementation. The process by which wrought-iron bars are converted into steel by heating in a carburising medium. The latter consists of powdered charcoal which is placed over and in the spaces between the stacked bars. The mass is finally covered with grinder's waste, luted and heated at a bright-red heat from 7 to 10 days. Carbon gradually diffuses into the metal, the amount of carbon and the depth of penetration depending on the temperature, the time of cementation and the material used. Other cementation processes include case-hardening in fused cyanides or gaseous atmospheres.

Cement copper. Impure copper produced by precipitating the metal from leaching solutions with iron.

Cemented carbides. These are prepared by powder metallurgy methods and are used for cutting tools, wire-drawing dies and parts subject to heavy abrasive wear. The principal carbides used in the manufacture of cemented carbides or "hard metals" are tungsten carbide WC, titanium carbide TiC. Tantalum carbide TaC is used occasionally and molybdenum carbide Mo_2C and vanadium carbide VC have also been used. The carbides are mixed with cobalt metal powder and milled in ball mills prior to pressing into the required shape. They are then sintered at temperatures between 1,350 °C and 1,550 °C. (See Table 9.)

Cementite. A hard, brittle compound of iron and carbon containing about 6.6% C, which corresponds to the composition Fe_3C, and crystallises in the orthorhombic system. It plays an important part in ferrous alloys and in reality alloys of the iron-carbon system are iron-cementite alloys. It is completely soluble in molten iron, but solubility in austenite (γ-iron) decreases from about 1.8% C at the eutectic temperature (1,130 °C) to about 0.9% at the eutectoid temperature (about 723 °C). At the eutectoid temperature (Ar_1) slowly cooled austenite decomposes, yielding ferrite and pearlite, which consists of laminated cementite and ferrite. (Note that in the presence of alloying elements the position of the change points may be altered.)

In high-carbon steels or in cases of carburised steels a network of cementite has an embrittling effect. The term cementite is applied to embrace all the carbides in cast iron and steel containing manganese, chromium, etc. Cementite, which is the hardest constituent in ferrous metals, may exist as fine plates, spheroids or as comparatively hard masses. (See also **Pearlite**.)

Cementite carbide. Carbide of iron, Fe_3C, synonymous with cementite.

Centigrade. A term denoting the thermometric scale associated with the name of Celsius, in which the temperature of melting ice is the zero of the scale and the boiling point of water (at normal pressure) is marked 100. The scale between the two fixed points is divided into 100 degrees. Temperatures on this scale are usually written °C. To convert centigrade temperatures into Fahrenheit temperatures multiply the former by 9/5 and add 32. Now discarded in SI units, use Celsius (q.v.) instead.

Centre of gravity. It is that point in a body through which the line of action of its weight always passes, no matter what the position of the body. Hence if the centre of gravity be supported, the body will rest in equilibrium. If the point of support be below the centre of gravity, the body will usually be unstable. Also called the centre of mass.

Centre spinning. A special casting technique that consists essentially in pouring molten metal into a rapidly rotating mould in such a way that the metal is directed by centrifugal force to take up the shape of the mould into which it is cast.

Centrifugal casting. A method for producing metal castings (usually cylindrical) by pouring the molten metal into a mould which is rotating rapidly about its longitudinal axis. This process, which produces sound, dense castings, is widely used for making cast-iron pipes, piston rings, etc., and for non-ferrous components such as gear blanks.

Centrifugal force. When a body moves in a curved path there is a tendency for it to move off "at a tangent". A force is therefore necessary to keep the body in its curvilinear path. The force exerted by the body in its tendency to move outwards from the centre of the curve is called centrifugal force and is measured by the quantity Mv^2/r, where M = mass of body, v = velocity of body and r = radius of the curved path at the instant under consideration.

Centrifuge. A device for separating solids from liquids or liquids from other liquids of different specific gravities by centrifugal force set up by rotation at high speeds. Used for the treatment of cleaning liquors and for finishing small hot-tinned wares.

Ceramics. A generic term for a special class of materials. Derived from the Greek *keramos*, which can be translated as "burnt earth", originally applied to potter's clay and pottery, but now extended to any material subjected to high temperatures during processing, but usually without actual melting. The high temperature is significant because it indicates a material of relatively high melting point, it is the effect of that temperature which is important. The processing results very often in the formation of a "vitrified" or glassy substance, relatively non-crystalline, or amorphous. Because of the high temperature used, the reversible reaction requires a very long period of time, and so "devitrification" is not common. This means that relatively plastic mixtures, such as clay and water, can be readily shaped and worked, but fixed permanently by the vitrification on firing in the kiln. When the process is applied to less plastic materials, the techniques are still similar, in that powdered material can be compacted, often with the aid of a lubricant, to form the shape, and then fired to sinter the particles together, with or without vitrification. Additions can be made to provide a small quantity of relatively low melting point material, which coats the particles and aids the sintering, without detracting too much from the physical properties of the matrix material. Such a material becomes less "refractory", because of the lower melting point material, and is avoided in the applications of refractory ceramics in furnaces and similar high temperature artefacts.

The amorphous structure produced by firing is of short range order, and therefore very "open", leaving "holes" for impurity or addition atoms, which again may modify the structure in terms of resultant physical and chemical properties. Glasses, for example, can be produced soft (of low softening point) as in decorative glassware or window glass, or hard, as in higher softening point laboratory glassware. The refractive index or the colour can be varied for optical instrument applications, and glass can be made electrically conducting, for electronic uses.

Ceramic materials can be made with great strength, because they deform plastically only at high temperature, when the viscosity is lowered. They suffer from brittle behaviour at lower temperatures, and are extremely notch sensitive (see **Notch sensitivity** and **Brittle fracture**), but provided these aspects can be taken into account offer the greatest scope for the future for high temperature high strength structures. The ceramic materials are frequently characterised by covalent bonds, or mixed ionic and covalent bonds. Although not yet of commercial significance, even metals have been treated to give amorphous bonds, with characteristics resembling ceramic materials of similar melting point.

One of the most common ceramic materials is silica, SiO_2, which can exist in three allotropic forms, is the main single constituent of the earth's crust, and can be processed in many different forms, including the complete amorphous state after melting. Clay contains a high proportion of silica, as do cement, common glass, refractory minerals, and many abrasives and semi-precious stones. Variations in the degree of vitrification allow the ceramist to produce "tailor-made" ceramics for various applications, involving both chemical and physical specifications. Alumina, Al_2O_3, is another common ceramic material, another constituent of clay and cement, and many other compounds.

The commonest types of ceramic are prepared from:

(a) silicates, e.g. aluminium silicates (mullite, china clay, zircon and sillimanite); or from magnesium silicates (steatite, forsterite); or from mixed silicates (cordierite);

(b) aluminates, e.g. spinels;

(c) oxides, e.g. zirconia, alumina, rutile, thoria;

(d) nitrides, e.g. boron nitride, silicon nitride;

(e) titanates, e.g. barium titanate;

(f) borides, e.g. chromium boride, titanium boride.

Most ceramics are hard or very hard and may be used as abrasives. They have a low coefficient of thermal expansion and are good insulators. Some of them, however, are unable to withstand large and sudden changes of temperature and thus fail by thermal shock. The upper limit of the working range is generally determined by their melting point, but it must be noted that in a few cases the particular ceramic may undergo some physical or chemical change (e.g. sublimation or decomposition) below the melting point (see Table 3).

Ceramics are usually chemically inert, and may thus be used in tanks and vats in chemical processes, and as containers for molten metals and molten slags and glasses. One of their main uses has been as electric insulators, but the newer types are also used as dielectrics, thermistors, transistors and transducers.

Ceramics may also be sprayed on metals to provide protection against oxidation, corrosion and erosion. Only very thin coatings are necessary. (See also **Cermets**.)

Cerargyrite. Silver chloride AgCl which crystallises in the cubic system and occurs with native silver and oxidised silver compounds in the weathered zone of silver sulphide lodes. Found in Saxony, New South Wales, Chile and California. Mohs hardness 1-1.5, sp. gr. 5.5. Also known as kerargyrite and horn silver, on account of its brown waxy appearance.

Cerite. A hydrated silicate of calcium and the cerium group metals. It usually contains 24% cerium oxide and about 35% of the oxides of lanthanum, praseodymium, neodymium and samarium taken together. Found in Scandinavia, Siberia, Greenland, North America and Brazil.

Cerium. The commonest metallic element belonging to the rare-earth group, it was discovered by Klaproth in 1803. It occurs most abundantly in monazite, cerite and allanite (together with other members of the group). The metal is obtained as a by-product in the extraction of thorium from monazite sands, by the electrolysis of the chlorides. Obtained in this way the cerium is mixed with lanthanum, etc., and is then known as Mischmetall. The pure metal is obtained by a complicated process of fractional precipitation and crystallisation. The metal is iron-grey in colour and is about as soft as lead, its BHN being about 28. It can readily be rolled and hammered and at moderate temperatures can be drawn into wire. In the form of wire it burns brilliantly when heated in air. Cerium III compounds, in alkaline solution, are strong reducing agents, precipitating gold and silver from their salts. The specific gravity of the metal is about 7 (see Table 2).

When alloyed with iron, nickel, tungsten, manganese or cadmium the products are hard and brittle and yield sparks when struck with a file. By contrast, the cerium-antimony alloy is soft. The cerium-iron alloy (containing about 40% Fe) is used for flints in automatic lighters. As a constituent of certain aluminium alloys cerium improves the strength and ductility of the alloy. The properties of magnesium alloys are likewise improved at high temperatures when cerium is present in quantities up to 6%. In the steel industry cerium is used as a deoxidiser, to refine grain size and to improve the mechanical quantities of the steel. In the electrical industry the metal is used as a getter in electronic valves and neon tubes, etc. (see Table 9).

Cermets. Pressed or sintered ceramic-metal mixtures used in jet engines and other high-temperature applications. The materials are usually combinations of inorganic substances with metallic constituents and the bond may be mechanical or chemical.

Examples are silicon-silicon carbide, chromium-alumina, nickel-titanium-carbide, beryllium-beryllium oxide.

Cerussite. A carbonate of lead $PbCO_3$ crystallising in the orthorhombic system. It is a valuable lead ore, white or grey in colour, and brittle. It is usually associated with galena, and occurs in Cornwall, South Scotland (Leadhills) (UK), many parts of Europe and in the Mississippi valley. Mohs hardness 3-3.5, sp. gr. 6.47.

Cervantite. An antimony ore $Sb_2O_3.Sb_2O_5$. Also known as antimony ochre.

Chabazite. A hydrated silicate of aluminium, calcium and potassium $(Ca.K_2)O.Al_2O_3.4SiO_4.6H_2O$. It is a member of the zeolite group of minerals and crystallises in the hexagonal system.

Chain structure. A series of linkages between atoms, or groups of atoms, which extends in one dimension predominantly, thereby resembling a chain or ladderlike structure. Chain structures may be regular, or formed of mixed groups, including side branches. Such structures are characteristic of polymers, and give rise to particular properties. (See **Polymer**.)

Chalcanthite. A hydrated sulphate of copper that occurs in the zone of weathering of copper lodes $CuSO_4.5H_2O$. It crystallises in the triclinic system. Also known as blue vitriol or copper vitriol. Mohs hardness 2.5, sp. gr. 2.12-2.30.

Chalcedony. A variety of quartz composed of amorphous and crystalline silica which is translucent and yellow in colour. Included in the same species are the gem and ornamental stones carnelian (red), sard (brownish-red), onyx (red), sardonyx (red) and chrysoprase (apple-green) and the agates.

Chalcocite. A sulphide ore of copper Cu_2S. It is dark grey in colour and crystallises in the orthorhombic system. Also called copper glance, redruthite.

Chalcopyrite. One of the commonest ores of copper, which is an iron and copper sulphide $CuFeS_2$. It occurs frequently as yellow masses having an iridescent surface tarnish. It crystallises in the tetragonal system. Also known as copper pyrite, peacock copper.

Chalcosphere. One of the geographical concentric zones of the earth's crust. In this zone the sulphides and oxides of the heavy metals, especially iron, are concentrated.

Chalk. A naturally occurring form of calcium carbonate $CaCO_3$ that has been formed from the shells of minute marine organisms, deposited in sea-water.

Chalkos. The early Greek term for brass and bronze.

Chalybite. An important iron ore consisting mainly of the carbonate $FeCO_3$. Also called brown spar.

Chamosite. A silicate of iron and aluminium.

Channel induction furnaces. One of the earliest designs of electric melting furnace incorporated the use of mains frequency, the furnace consisting essentially of a simple step down transformer. The secondary circuit was a simple loop or ring, of metal to be melted (the channel), the primary a coil whose winding embraced part of the channel containing the metal. Whereas a power transformer is designed to minimise heat losses in the secondary, the channel induction furnace endeavours to increase it.

The earliest furnaces placed the metal in a cylindrical shell lined with refractory, but this proved inefficient as the volume of metal was much smaller than the volume of refractory holding it. The surface area of exposed metal was also large in proportion to total volume and a covering slag had to derive its heat from the metal by conduction and convection. Refining processes involving slag-metal reactions were therefore ineffective.

The Ajax-Wyatt furnace employed a cylindrical bath of metal with a connection "vee" channel below the pool, the legs of which formed the secondary circuit. The primary coil surrounded the "vee" and so formed a figure of eight. The metal experiences a repulsion effect at the apex of the vee (i.e. at the base of the furnace) and a "pinch" effect at the top of each leg, causing a rapid circulation of molten metal into the main pool above.

These furnaces operate on any commercial mains frequency from 25 to 60 Hz, and are made in various sizes from 50 to 500 kW. They not only form useful melting furnaces, but can be used as "holding" furnaces supplied by other types, to act as reservoirs in a metal foundry. Modern designs incorporate gas seals to enable vacuum degassing or pressurised pouring to be carried out, and are used for vacuum treatment of alloy steels, nodular cast irons, and alloys which may be subject to oxidation.

Chaplets. A term used in foundry work in reference to a number of devices that are employed for holding or supporting a core or a section of a mould. They are generally, but not essentially, made of sheet iron and are tin-coated when made of metal. Galvanised chaplets are unsuitable for use where castings of steel or cast iron are concerned, because zinc vaporises at a temperature of a little over 900 °C, which is much below the temperature of molten steel or iron. (*See also* **Pipe chaplet.**)

Characteristic temperature (Debye). The input of thermal energy to a solid structure results in increasing vibration of the constituent atoms. The vibration is very complex, particularly in a long range order crystalline material, and is measured by two main parameters. The capacity to absorb the thermal energy is termed specific heat, and the increased mobility of the atoms is manifest as thermal expansion. As the temperature rises above absolute zero, the vibrations change in frequency as well as amplitude, and hence the value of specific heat changes with temperature. Einstein originally attempted a quantum mechanical calculation of specific heat, assuming that atoms could vibrate independently of each other. The results were adequate at higher temperature, but not towards absolute zero. Debye performed more sophisticated calculations, assuming the atomic vibrations were not independent, and that the frequency distribution was therefore unlimited. He assumed a cut-off point N_D at which the number of possible frequencies was equal to three times N the number of atoms involved, (there being three degrees of freedom for the vibrations). He then adopted Einstein's technique of defining a characteristic temperature, related to the characteristic frequency at the cut-off, N_D, by a simple quantum relationship,

$$\Theta_D = \frac{h N_D}{k}$$

where Θ_D is the Debye characteristic temperature, h Planck's constant, and k Boltzmann's constant.

The specific heats of solids are roughly equal at temperatures that are the same fraction of their Debye characteristic temperature, and a plot of specific heat against T/Θ_D, where T is the actual temperature, shows the same curve for almost all solids. Figure 21 shows the variations of specific heat with temperature for three different materials, which all follow the same curve when corrected for characteristic temperature.

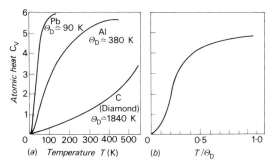

Fig. 21 *(a) Variation of atomic heat with temperature. (b) Variation of atomic heat with ratio of absolute temperature to Debye characteristic temperature.*

Charcoal. This is the black residue, rich in carbon, that remains when vegetable substances such as wood or sugar are heated in the absence of air. The destructive distillation of wood yields wood charcoal, pyroligneous liquor and a combustible gas. Wood charcoal is dull black, amorphous and friable. Its specific gravity varies from 1.4 to 1.9, according to the nature of the wood used, but because it is porous it floats on water. When freshly made or when freshly activated (by heating) it can absorb various gases and vapours, showing a preferential absorption for vapours of liquids that are volatile at ordinary temperatures. One volume of carbon can absorb 170 vols. of ammonia or 148 vols. of alcohol vapour, as compared with about 16 vols. of oxygen or nitrogen.

Wood charcoal rarely contains above 85% C. It is used as a fuel, in the making of gunpowder, in the carburising of steel (q.v.) and sometimes in the reduction of the oxides of metals.

When animal matter is heated out of contact with air, animal charcoal is produced. This is used in clarifying liquids. When the crude animal charcoal is treated with hydrochloric acid, ivory black is the result.

Charcoal pig iron. The product of the smelting of ore with charcoal, rather than with coke. The lower temperature of the charcoal furnace results in a much lower absorption of silicon by the iron, and since charcoal is free from sulphur, the pig iron produced is also sulphur-free. Charcoal pig iron has a more uniform grain than coke or anthracite irons.

Charge (electrostatic, electronic). The accumulation of elementary particles, atoms or groups of atoms with assymetric distribution of such particles, on a body, giving rise to an electrical potential difference between the body and its surroundings. If the elementary particles in excess are electrons, then the charge is said to be negative. If the elementary particles are deficient in electron balance, then the charge is positive.

Charles' law. The volume of a gas, maintained at constant pressure, varies directly as its absolute temperature. The coefficient of increase of volume is 1/273 of the volume at 0 °C and hence $V_t/V_0 = (273 + t)^{\circ}/273$, where V_0 = volume at 273 K, V_t = volume at $(273 + t)$ K.

Charlestone phosphate. A "soft" phosphate readily decomposed by sulphuric acid and found in river-beds in South Carolina. It contains 27% of phosphorus pentoxide.

Charpy test. A notched-bar or impact test in which a notched specimen, fixed at both ends, is struck behind the notch by a striker mounted at the lower end of a bar that can swing as a pendulum. The energy that is absorbed in fracture is calculated from the height to which the striker would have risen had there been no specimen and the height to which it actually rises after fracture of the specimen. The test is similar to the Izod test (q.v.) except that a "key-hole" notch is commonly used instead of a V-notch and the specimen is mounted horizontally. As the Izod notch is easier to make it is frequently used in beam tests of the Charpy type.

Chert. A rock consisting chiefly of fine-grained silica that is closely allied to flint. Cherts of volcanic origin may be rich in iron, and it was from such deposits that the iron ore of Lake Superior was obtained.

Chessylite. A naturally occurring basic carbonate of copper $2CuCO_3.Cu(OH)_2$ found in the Urals. It is deep blue in colour and is used in works of art. Also called azurite.

Chevreul's salt. This is copper II, copper III sulphite

$$CuSO_3.Cu_2SO_3.H_2O.$$

It is used in electroplating baths.

Chile saltpetre. Crude sodium nitrate $NaNO_3$ found in the rainless districts of Chile. It occurs in a distinct layer of earth known as caliche. This contains about 25-50% nitre. The nitrate is extracted by crushing the earth, dissolving in water and crystallising. The resultant product is about 95% pure and is used mainly as a source of nitric acid and as a fertiliser.

Chili bar. A crude form of copper containing about 1% S in addition to the usual impurities.

Chilian mill. A device, similar to an ordinary mortar mill, used for crushing ores.

Chill cast. Metal cast into metal moulds or chills.

Chill castings. Cast iron that has been set in metal moulds or castings that have been allowed to set in moulds in which metal pieces are placed in the walls of the mould to accelerate cooling.

Chilled iron. A type of cast iron used for mill rolls, ploughshares, wheels and other applications where high resistance to abrasion is required. One part of the cross-section is white iron, which merges into mottled or grey iron. This type of casting is produced by adjusting the composition to suit the depth of chill required and casting against metal "chills".

Chill mould. An iron mould used to accelerate the rate of cooling of cast iron so that the surface of the iron is hardened by retaining much of the carbon in the combined state.

China clay. The common or commercial name for kaolin (q.v.).

Chinese white. Zinc oxide prepared by burning zinc in air and condensing the oxide. It is used as a pigment in place of white lead, it does not darken when exposed to industrial atmospheres. Also called flowers of zinc.

Chipping. A method of removing surface defects from ingots with hammer and chisel or gouge, so that the defects will not be rolled or forged into the finished product. The method is also used for removing flash and other unwanted metal from castings, forgings, etc.

Chloanthite. A nickel arsenide, $(Ni.Co.Fe)As_2$, often found with kupfer nickel. It occurs in Europe and America. Also known as white nickel.

Chlorapatite. A hard acid-insoluble mineral that consists of the phosphate and chloride of calcium $3Ca_3(PO_4)_2.CaCl_2$. It is a source of phosphorus.

Chlorargyrite. One of the important ores of silver. It consists mainly of silver chloride. (*See also* **Cerargyrite**.)

Chlorides. Compounds formed between the halogen element, chlorine, and metals or other groups of atoms capable of donating electrons to provide the bond with chlorine ions. Chlorides are normally ionic compounds, highly crystalline, and of reasonable high melting point. In aqueous solution, they are readily electrolysed and chlorine is released at the anode. Some chlorides have special optical properties, e.g. rock salt NaCl and silver chloride is used in photography.

Chlorine. A halogen element, which boils at $-34.6\,°C$ and hence is gaseous at normal temperatures. Highly reactive, because the atom can readily accept an electron, to become ionised with a negative charge, and combine with metals in particular to form an ionic bond. Important in chemical processing, including purification of heavy metals, and in the manufacture of polymeric materials such as polyvinyl chloride, P.V.C.

Chlorite. A mineral group consisting of the hydrated silicates of magnesium, iron and aluminium and generally dark green in colour. They crystallise in the monoclinic system.

The salts of chlorous acid are termed chlorites, and the alkaline chlorites are bleaching agents. Lead chlorite, $Pb(ClO_2)_2$, is used in making detonators.

Chromate coating. A method of inhibiting or reducing corrosion of a metal surface, by immersion of the metal in a chromate solution to form a coating of chromate of the basis metal. The process is particularly applied to magnesium and zinc alloys.

Chromatic aberration. Since the focal length of a simple lens is shorter for blue rays of light than for red rays, a point source of light never gives rise to an image at a single point. The complete image, in fact, consists of a small linear spectrum lying along the axis. This divergence from the simple laws of refraction with the introduction of coloured images is termed chromatic aberration. It is measured by the product of the mean focal length and the dispersive power of the material of the lens.

Chromatic aberration is corrected by using a combination of lenses in which the refractive indices and the focal lengths are different, but the total dispersion is zero. Most microscope objectives have compound lenses of crown and flint glass, the former being converging and the latter diverging.

Chrome alum. A double sulphate of chromium and potassium $K_2SO_4.Cr_2(SO_4)_3.24H_2O$. It can be made by the reduction of potassium bichromate, and is also obtained in large quantities as a by-product in the manufacture of artificial alizarin from anthracene. The crystals are dark purple in colour and are isomorphous with those of the other alums. It is used in dyeing (as a mordant), in calico printing and in tanning leather.

Chrome amalgam. When a solution of chromium trichloride in strong hydrochloric acid is electrolysed using a mercury cathode, a solid amalgam Hg_3Cr is formed which under high pressure loses mercury and becomes HgCr. If this latter is heated *in vacuo* at about $300\,°C$, the amalgam loses the whole of its mercury and the chromium is left as a fine pyrophoric powder.

Chrome ironstone. The chief ore of chromium $FeO.Cr_2O_3$. The ore containing a minimum of 40% chromium oxide is used as a refractory in open-hearth furnaces. It is found in Asia Minor, Zimbabwe, North America, India, New Caledonia and the USSR. Also called chromite, chrome iron and chrome iron ore. It is usually dark brown to black in colour and crystallises in the cubic system.

Chromel-alumel couple. The most widely used of all thermocouples. The approximate compositions of the two wires are:

(a) Ni 90%, Cr 10% (chromel), and *(b)* Ni 98%, Al 2% (alumel). The couple has a fairly straight calibration curve and possesses good resistance to oxidation. It can be used in continuous service up to $1,100\,°C$ and intermittently up to about $1,300\,°C$. Figure 22 gives the e.m.f. produced by various temperature differences.

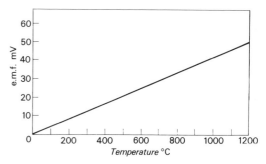

Fig. 22 *E.m.f. v. temperature. Chromel-alumel thermocouple, cold junction $0\,°C$.*

Chrome ochre. A naturally occurring form of chromium oxide Cr_2O_3. It is also produced artificially from the hydroxide or from the dichromate. It is green in colour and is used in colouring glass and porcelain and in enamels and paints. Specific gravity 5.01-5.20.

Chrome-tin-pink. An intimate mixture of the metallic oxides produced when chromic acid is calcined with stannic acid. It is used as a ceramic colour.

Chromising. A process for improving the hard-wearing properties and corrosion resistance of ferrous products by forming a layer of chromium-rich material on the surface. It is also a process of chromium impregnation. The material to be treated is packed in a mixture of powdered alumina and powdered chromium and heated to $1,300\text{-}1,400\,°C$ in hydrogen for 3-4 hours. The composition of the mixture is variable, but usually averages Al 45% and Cr 55%. The time of treatment is governed by the degree of penetration required and the chromium concentration. The process may also be carried out by heating in gaseous chromium chloride or by diffusing electrodeposited chromium into the steel surfaces.

Chromite. *See* **Chrome ironstone**.

Chromitite. A somewhat rare mineral of chromium which consists of ferric chromate $Fe_2O_3.Cr_2O_3$.

Chromium. A metal obtained from the mineral chromite (q.v.). Ores that can be worked economically contain not less than 48% chromium oxide, with only small percentages of silica, sulphur and phosphorus. The metal is obtained by the reduction of the oxide in various ways. In the thermit process aluminium or magnesium are used, usually the former. The reaction evolves so much heat that the alumina produced is fused and crystallises to form corubin. The product contains about 95.5% Cr. Reduction can also be achieved by heating the oxide in an electric furnace with carbon or silicon, the excess carbon and the carbides formed in the former case being removed by treatment with lime and water and acid, respectively. When heated with carbon and lime and fluorspar, the mineral yields an iron-chromium alloy called ferrochrome which contains 60-70% Cr and is used in the making of chrome steels. At a sufficiently high temperature hydrogen will reduce the oxide of chromium. Pure chromium is obtained by the electrolysis of chromium chloride $CrCl_3$ using a mercury cathode. (*See also* **Chrome amalgam**.)

Chromium is a steel-grey metal. (For mechanical and physical properties, *see* Table 2.) The importance of the metal is associated with its high tensile strength, its resistance to atmospheric corrosion at temperatures up to $1,000\,°C$ and with its ability to resist the corrosive action of certain acids and alkalis at ordinary temperatures. Alloyed with steel, it confers on the alloy a high degree of hardness and resistance to penetration that make it suitable for armour plate, high-speed tools, dies for drawing, etc. Stainless steel and iron contain 12-15% Cr; austenitic stainless steels of the Staybrite type that are more resistant to corrosion and are capable of being worked contain 18% Cr and 8% Ni.

Chromium finds application in alloys with nickel for use as electric heating elements, and in magnet steels. Ductile metal can be obtained by powder metallurgy techniques. Chromium can be deposited electrolytically from solution as a hard deposit which is decorative as well as important in providing a wear-resisting coating.

Chromium copper. A commercial alloy containing 0.5% Cr, which is added to improve the mechanical properties of the copper. The addition of the chromium, however, reduces the electrical and thermal conductivities. The optimum combination of mechanical properties with good electrical and thermal conductivity is obtained by heat-treatment. Most of the 0.5% Cr present in this alloy dissolves in the copper at 1,000 °C, and if this alloy is quenched from this temperature, the material becomes soft and ductile, but has poor electrical properties. Heat-treating the quenched material at 500 °C for upwards of 2 hours greatly improves the conductivity, strength and hardness, but reduces ductility. Chromium copper is useful for cylinder heads of internal-combustion engines and resistance welding electrodes and for any application where the working temperatures will not exceed about 450 °C. Typical properties of solution heat-treated (*left*) and precipitation hardened (*right*) chromium copper are as follows:

Tensile strength (MN m^{-2})	230	460
Elongation %	60	22
Hardness (DPH)	65	155
Electrical conductivity (% IACS)	45	80

Chromium oxides. As a transition metal, chromium demonstrates the ability to form a number of oxides, CrO; Cr_2O_3; CrO_2; CrO_3. The sesquioxide Cr_2O_3 is of high melting point, and used as an abrasive. It forms the surface coating on certain alloys with iron and other metals, in conjunction with other oxides, and particularly in the formation of spinel oxides, which are tough and protective against further oxidation.

The dioxide is ferrimagnetic, and used in applications involving recording of magnetic impulses on plastic tape coated with the oxide, and in switching devices.

The trioxide dissolves in water to form chromic acid H_2CrO_4, which forms the basis of plating baths for chromium deposition, and as an oxidising agent for cleaning laboratory glassware and similar articles.

Chromium plating. A process in which a thin layer of chromium is formed on the surface of another metal by electrodeposition. Articles to be plated with chromium must be thoroughly cleaned and are preferably given an undercoating of nickel. This under-coat is usually 10 to 100 times thicker than the chromium coating, the latter being about 0.25 μm in decorative plating. One chromium bath contains 250-450 g of chromic acid together with one-hundredth of the quantity of sulphuric acid per litre. The containing tank should be of glass or antimonial lead. This latter alloy is also used for the anodes. The working temperature is between 44 °C and 48 °C and the current density should be from 11 to 18.5 A m^{-2}. All parts to be plated should be highly polished before dipping into the bath. This is most important in the case of bearings, otherwise the advantage of the low coefficient of friction of the chromium will be lost. Aluminium cannot be plated directly in such a bath, but must first be given a zinc coating (from a solution of sodium zincate), on to which a thin coating of copper is deposited, before the chromium plating can be successfully accomplished.

Chrysoberyl. Beryllium aluminate $BeO.Al_2O_3$ found in Iceland, Sri Lanka and the Urals. It occurs in varying shades of green. Used as a gemstone. The most valuable form is alexandrite, which is reddish by transmitted light. It crystallises in the orthorhombic system. Mohs hardness 8.5, sp. gr. 3.5-3.8.

Chrysotile. A fibrous variety of serpentine $3MgO.2SiO_2.2H_2O$. It is a variety of asbestos and is sometimes known as Canadian asbestos.

Chvorinov's theory. This refers to the solidification of metals, and states that freezing time of a casting is proportional to $(V/A)^2$, where V = volume and A = area of the casting. Schwartz and Bock found that for white cast iron solidifying in green sand moulds the time is roughly proportional to $(V/A)^{3/2}$.

Cinnabar. Sulphide of mercury HgS. A red or black mineral which is the most important ore of mercury. It is found in Spain, Italy,

California, Bavaria (Germany), China and Japan. Roasting this ore in an excess of air removes the sulphur as sulphur dioxide, and the mercury, as a vapour, is collected and condensed.

Cire perdue. Lost wax process. A method of casting formerly much used by sculptors of the Middle and Early Ages. The object to be cast is first modelled in clay and then coated with wax. The wax surface is then given the exact form and the final finish of the object to be cast. A thin smooth paste of finely ground clay, ashes and brickdust is applied to the wax surface until a sufficiently thick covering is formed. Vent holes are left open and the inner and outer clay secured by driving in metal rods. The mould is then completed and reinforced and finally baked, during which time the wax escapes, leaving the space into which the molten metal is poured when the mould has cooled down. This process is still used, and the modern investment casting process has developed from the cire perdue method. An interesting description of the early Florentine method is given in the autobiography of Benvenuto Cellini (1500-71).

cis-**isoprene 1, 3 Butadiene, 2 methyl.** The monomer of isoprene is a simple hydrocarbon structure, but is asymmetrical (*see* Fig. 23(*a*)). When polymerised, the double bonds at each end break open to join the mers together, and a double carbon-carbon bond is left in the centre of each mer (Fig. 23(*b*)). The two ends are now symmetrical, but not the centre portion, and two forms of polymer can exist. In one, the carbon-carbon double bonds have a methyl (CH_3) group and a hydrogen atom on the same side. This gives flexibility in the coiling up of long polymer chains, without the interference of hydrogen atoms one with another, along with the chain. This is the *cis*-form, found in natural rubber. When the methyl group and hydrogen atom are on opposite sides, flexibility is reduced, and this is the *trans*-form, found in gutta-percha.

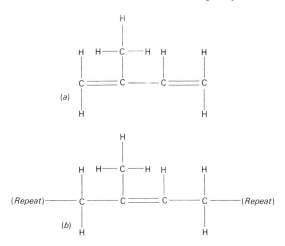

Fig. 23 (a) Isoprene. (b) Mer to form cis-isoprene polymer.

Cladding. A process for covering one metal with another. It is designed to combine the strength or the cheapness of the base (clad) metal with the attractive appearance or superior corrosion resistance of the cladding metal. For these purposes brass and other base metals are often clad with gold, silver and other precious metals (e.g. rolled gold); lead is clad with tin as in tin-coated lead; steel is clad with nickel, nickel alloys or stainless steel; aluminium alloys are clad with pure aluminium (e.g. Alclad).

The term cladding is not usually applied to electrodeposition, chemical replacement or dipping processes, but refers to pressure welding or pressure rolling processes carried out on relatively thick slabs or sheets of metals which are rolled together as a sandwich under high pressure so that the metals are firmly bonded into a composite sheet.

Thorough cleaning and intimate contact between the surfaces are essential, and large reductions in thickness are usually required before bonding can take place. The process is usually operated at high temperatures.

In one cladding process, casting methods are used. Two sheets of stainless steel are thoroughly cleaned and placed back to back in a mould with separating material (e.g. paper) between them.

Molten steel is poured into the mould so that it is cast against the clean surfaces of the stainless-steel sheets. A certain amount of alloying and inter-diffusion occurs and the steel adheres to the stainless steel. The sheets are then rolled at high temperatures and under heavy pressure and the component materials become firmly bonded into a composite sheet.

Clapeyron's equation. An equation from which the latent heat of fusion of a substance can be determined from the lowering of the melting point under the influence of high pressure, it being assumed that the substance melts without molecular change.

$$L = T(V_1 - V_s)\frac{dp}{dT} \times 0.0982 \text{ J}$$

where L = latent heat of fusion; $(V_1 - V_s)$ = change in volume from liquid to solid. If T_1 and T_2 are the temperatures of melting at pressures p_1 and p_2, then dp/dT may be taken as approximately

$$\frac{p_1 - p_2}{T_1 - T_2} \text{ and } T = \frac{T_1 + T_2}{2}.$$

Clarain. One of four constituents of banded coal which shows distinct behaviour on coking. When heated out of contact with air this constituent fuses and swells, giving rise to a brown coherent coke.

Claudetite. Arsenious oxide As_2O_3. It crystallises in the monoclinic system. Specific gravity 3.85.

Clausius-Clapeyron equation. An equation from which the latent heat of evaporation of a substance can be determined from the vapour pressures at known temperatures.

$$L = 19.16 \, T_1.T_2 \frac{\log p_1 - \log p_2}{T_1 - T_2} \text{ J}$$

where L = latent heat of fusion, p_1 and p_2 are the vapour pressures at absolute temperatures T_1 and T_2.

Clausthalite. Lead selenide PbSe occurring with the selenides of copper and silver at Clausthal in the Hartz district of Germany.

Clay. Clay is usually considered as a very soft fine-grained substance and is defined by its properties of being more or less coherent when dry, retentive of water and plastic when wet. The chemical composition varies according to the finely divided mineral matter present. The principal constituent minerals are hydrated oxides of aluminium and iron, hydrated silicates of aluminium, micas, chlorites, etc.

Clay occurs in nearly every geological formation either as massive clay substance or as a constituent of conglomerates and sandstones. It results from the weathering of older rocks, and the most important supplies of clay are obtained from decomposed felspar-bearing rocks. When admixed with sand, clays form mud or loam; with lime or dolomite they yield marl or calcareous clay schists. Few clays are suitable for fireclays and other refractory materials and these are essentially hydrated aluminium silicates such as kaolinite or kaolin $Al_2O_3.2SiO_2.2H_2O$.

A satisfactory fireclay must have the following properties, and be suitable for shaping into brick form or crucible: (a) it must be infusible at high temperatures; (b) it must clinker or become reasonably hard, and possess adequate crushing strength when burned; (c) it must also burn to a dense and reasonably porous brick; (d) it must resist chemical attack; and (e) it must not show any appreciable change of volume at medium or high temperatures or suffer damage from changes of temperature.

Clays suitable for making bricks and tiles are argillaceous in nature, the colour and texture depending on the impurities. For making bricks a proportion of sand or silt is added, to prevent too great shrinkage on firing.

China clay, which is a white and friable earth, results from the decomposition of granite. It is purified by repeated levigation and settlement in tanks. The constituent particles of clay are commonly 0.002 mm in diameter, the upper limit of size being about 0.005 mm.

Clay gun. A piece of equipment used in foundries, and in blast furnace or other furnace work for sealing the tap-holes of furnaces. A ball or fireclay is projected at high speed into the orifice to stop the flow of slag or metal and to seal the tap-hole.

Clay ironstone. Ferrous carbonate $FeCO_3$ mixed with clay is known as clay ironstone and is an important source of iron.

Cleavage. A tendency to split or fracture along definite crystallographic planes.

Cleavage planes. A crystallographic plane on which fracture may occur. (See also **Crystal planes.**)

Many metals and minerals show a tendency for fracture to take place along certain definite planes which are closely related to the crystalline form of the material. In certain cases fracture may take place on more than one crystalline plane, but in general, splitting or cleavage occurs most easily across the direction of least cohesion, i.e. along gliding planes where the packing of the molecules is least close.

Rock salt, calcite, fluorspar and mica provide good examples of cleavage planes in minerals. Certain other rocks, such as slates, sandstones and millstone grits which split along their planes of bedding, are also said to possess cleavage planes, but this type of cleavage is due to the method in which these rocks were formed under pressure and is quite distinct from the true cleavage planes of crystalline substances.

Cleveite. A variety of pitchblende. When heated *in vacuo* or with sulphuric acid it yields a gas containing 20% N, the remainder being a mixture of helium and argon.

Climb in dislocations. The normal movement of dislocations is along a slip plane, and screw dislocations can only move by this method, because they are oriented parallel to the slip plane. Edge dislocations however, having the extra atoms perpendicular to the slip plane, may move in the direction of the extra half plane itself. To do so, atoms must migrate to or from the edge-half plane, by diffusion. Atoms cannot readily move away but vacant lattice sites can diffuse towards the dislocation, which is the equivalent. A row of vacancies will move the dislocation edge away from the original slip plane, a row of atoms diffusing in will move it away also, but in the opposite direction. This phenomenon is called "climb" (*see* Fig. 24). Since diffusion is involved, temperature has a marked effect on climb.

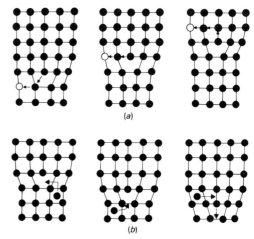

Fig. 24. *Dislocation climb by diffusion mechanisms. (a) Vacancy diffusion. (b) Interstitial diffusion.*

Clinker. The name applied to the coal residues that are formed in furnaces, etc. It is usually formed in lumps, partly fused. It is used in making the poorer qualities of Portland cement.

Clinking. A defect sometimes encountered in high carbon or alloy steels. Internal cracks sometimes known as "snow-flakes" or "hair-line cracks" are formed, often with a characteristic clinking sound during cooling under unfavourable conditions after hot working. During further working or heat-treatment the cracks may propagate and failure occur. Recent work has confirmed that the presence of hydrogen can be responsible for clinking.

Clipping tools. A collection of tools used for removing "flash" or excess metal from drop forgings or die-castings. When a metal is being forged or cast in dies, a certain amount of excess metal may be forced between the die-faces. Clipping tools, shaped to fit the forging or die-castings at the parting line of the dies, are used to trim the excess metal and burr from the articles and leave a good smooth surface. This operation is usually performed in a mechanical press or under the forging hammer.

Close annealing. The heating of metal in a sealed container — with or without packing materials — or in a special atmosphere so as to anneal it and preserve a bright surface free from oxidation. The charge is usually heated to, and held at, a temperature below the transformation range, and then slowly cooled. Also called pack, pot or box annealing.

Closed electron shells. An expression used to indicate the completion of the variations in the four quantum numbers for one value of the principal quantum number n, within the Pauli principle. The so-called inert gases are the best examples of closed electron shells, but every atom after hydrogen has one or more closed shells (*see* **Atom**).

Close-packed structures. The electron clouds of individual atoms are so distributed that they can be regarded as roughly spherical. The closest packing of identical spheres in a two-dimensional plane is hexagonal, i.e. each has six close neighbours. When two such planes are placed one above the other, the closest packing results when each sphere in the upper layer rests in the space between three spheres in the lower plane. The axis joining individual spheres in the two layers is then at an angle of 60° to the planes themselves. When a third layer of spheres is placed on the two already aligned, there are two possibilities.

(*a*) The third layer could be placed with the centre of its spheres directly above those of the first layer. A fourth layer would then be placed with centres vertically above those of the second, producing a continuous distribution expressed as ABABAB . . . packing (*see* Fig. 25(*a*)).

(*b*) In the second method of placing the third layer, the spaces between three spheres in the second layer are used, but again displaced with respect to the vertical axis, to an angle of 60° with the planes. A fourth layer would then follow the same pattern, but now each sphere would be vertically above those in the first layer, and so this becomes ABCABC . . . stacking (*see* Fig. 25(*b*)).

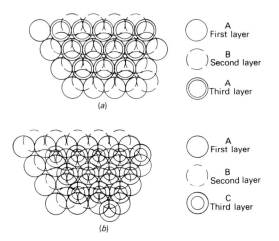

Fig. 25 *Three layers of close-packed planes. (a) ABAB. . . stacking. (b) ABCABC. . . stacking.*

A
First layer

B
Second layer

A
Third layer

(*a*)

A
First layer

B
Second layer

C
Third layer

(*b*)

Both these systems are close packed, each sphere having twelve close neighbours in three dimensions, and 74% of the total volume available is occupied by the spheres.

The ABAB . . . stacking is designated close-packed hexagonal, since the cell structure which results resembles a hexagonal prism. The ABC . . . stacking is called face-centred cubic, since the cell is represented by an atom at each corner of a cube, with one atom in the centre of each face of the cube. The planes of the close-packed hexagonal system become the basal planes and intermediate ones, of the hexagonal prism. The planes of the face-centred cubic system are octahedral planes of the unit cell, envisaged by lines drawn across the diagonal of an upper face, and down from the corners across two adjacent faces to the lower corner (*see* Fig. 26).

Since these two structures are so similar, it is possible for transformations to occur from one to the other in an element, when assisted by increased thermal energy. Cobalt for example, is close-packed hexagonal at low temperatures and face-centred cubic at higher temperatures.

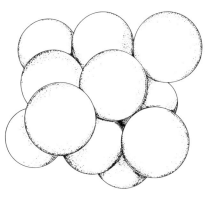

Fig. 26 *Octohedral plane of face-centred cubic structure.*

Cloudbursting. Peening the surface of a metal by the impingement of steel or white iron, glass-hard balls. It may be carried out for the purpose of testing for soft spots in case-hardened surfaces. Hard steel shot is then used and allowed to cascade on to the metal surface. The soft spots show up as dull (matt) areas in the brighter hard surroundings.

Cluster mill. A rolling mill in which each of the two working rolls is supported by two or more back-up rolls.

Coal. This black amorphous substance is the outcome of a process whereby most of the oxygen and hydrogen of vegetable matter, under the influence of heat, pressure and biochemical reactions, is eliminated, leaving a residue having a high percentage of carbon. Strictly speaking, coal is not a mineral and, having a very variable composition, is not readily definable. Coals are classified in different ways, taking into account some or all of the following properties: type of flame when burning, composition, calorific value, fuel ratio (i.e. weight of fixed carbon/weight of volatile matter), split volatile ratio (weight of fixed carbon plus volatile combustible/hygroscopic moisture plus half of the volatile combustible) and caking properties.

In the process of the formation of coal the substance peat probably represents the first stage in the conversion of the vegetable matter. The next stage may be represented by lignite, which is more compact than peat, but still shows distinct signs of its vegetable origin. Common coal — i.e. bituminous coal — represents the next stage. There are many varieties of bituminous coal, but they can be classified metallurgically into two groups: caking and non-caking. Bituminous coal is used in making gas, for coking and for house and steam fuel. Anthracite represents a still later stage in coal formation. It contains a high percentage of carbon. It has a higher ignition temperature than other types of coal, but it gives considerable heat and little flame when burning. Graphite probably represents the final stage in this process of decomposition.

Coal is used metallurgically as a fuel and as a reducing agent in the production of metals from their ores.

The satisfactory classification of coals offers numerous difficulties. The Seyler system (Table 16) is based on defined limits to the carbon and hydrogen content. Below 84% C (lignitious coals) the genus is subdivided entirely on the carbon, and above this figure the five types are distinguished by the hydrogen content.

TABLE 16. *Classification of Coals Using Seyler System*

Genus	Hydrogen, %	Approx. Volatile Matter, %	Typical Coals
Anthracite	Less than 4	Less than 10	Anthracite
Carbonaceous	4.0-4.5	10-16	Smokeless steam
Semi-bituminous	4.5-5.0	16-24	Navigation (bunker)
Bituminous	5.0-5.8	Over 24	Gas, steam, etc.
Per-bituminous	Over 5.8	—	Long flame steam

Coalesced copper. A form of oxygen-free copper, obtained by briquetting small particles of cathode copper at ordinary temperatures under a pressure of about 155 mN m⁻² heating the briquettes in a reducing atmosphere at 870-910 °C and extruding the heated briquettes in a controlled atmosphere.

Coalescence of cementite. The balling-up of the carbide lamellae in pearlite during a spheroidising anneal near but below the transformation range of a carbon steel. The same structure is achieved by long, high-temperature tempering of hardened steel.

Coal gas. The gas produced by the destructive distillation of coal, usually bituminous coal, with or without admixture with cannel coal. The heating is carried out in closed iron retorts and usually producer gas, sometimes coal, is used for heating. The gases evolved are first passed into a hydraulic main which also acts as a water-seal. Here the first separation occurs. Tar and ammonia accumulate in the main, and the crude gas passes to the scrubbers, where the remaining tar and ammonia are removed. Purifiers containing lime remove the carbon dioxide and some of the sulphuretted hydrogen. Iron III (ferric) oxide purifiers remove the rest of the sulphuretted hydrogen. Coal gas, as delivered to the consumer, should not contain more than one part of sulphuretted hydrogen in ten million parts of the gas.

The valuable cyanides are removed before final purification by passing the crude gas through ammoniacal liquor containing ammonium sulphide; ammonium thiocyanate is thus formed.

Coal gas, so produced, consists of three main groups of gases, viz, the combustible non-luminous gases (hydrogen 50%, methane 30%, carbon monoxide 8% approximately); the illuminants (unsaturated hydrocarbons and benzene — about 5%); and the inert constituents (nitrogen, carbon dioxide — about 7%). The calorific value of coal gas is about $44 \, MJ \, kg^{-1}$ (or approximately $22 \, MJ \, m^{-3}$). The hydrogen present is probably derived from the decomposition of hydrocarbons on contact with hot walls of the retort. Carbon, in a very pure state, is thus deposited in the retort, and is known as gas carbon. It is used in arc lamps and in electric batteries.

In most cases water gas (q.v.) is mixed with coal gas to improve its calorific value. The proportions of hydrogen and carbon monoxide are thus increased, while the proportions of the illuminants decreases.

Coarse metal. In the smelting of sulphide copper ores, the ore is first roasted in a flat furnace to oxidise part of the iron and copper. The ore is then fused in a reverberatory furnace with silica. This latter combines with the oxide of iron to form a fusible slag, while the copper and iron sulphide form a layer below the slag. This lower layer is referred to as coarse metal.

Coating. A form of protective surface, or decorative finish. Protective surfaces, usually against corrosion or abrasion, may be applied by electrodeposition, chemical deposition, dipping, welding, spraying or thermal evaporation in vacuo or inert gas atmospheres. Coatings are not effective unless some bonding occurs with the substrate, either mechanical or by diffusion. Decorative finishes may be applied by brushing, pressing, rolling, spraying. Electrical potentials may be applied in spray processes, and adhesives in rolling or pressing, to increase the bond. (*See* **Cladding**.)

Cobalt. The chief minerals from which cobalt is extracted are smaltite or cobalt speiss [a cobalt arsenide $(Fe.Co.Ni)As_2$], cobaltite or cobalt glance [a sulpharsenide $(Co.Fe)S.As$] and cobalt bloom or erythrite $Co_3(AsO_4)_2.8H_2O$. But the most important sources at present are the silver ores of Cobalt City, Ontario, and the manganese ores of New Caledonia. The direct reduction of these ores is difficult and laborious.

Cobalt ores are roasted to get rid of arsenic and sulphur. The product comes into commerce as zaffre, which is heated in a blast furnace with limestone and sand. Any iron passes into the slag, and the impure cobalt oxide (speiss) settles out and is extracted with hydrochloric acid, treated with sulphuretted hydrogen and oxidised with bleaching-powder (to remove iron). The cobalt and nickel so obtained are further treated with bleaching-powder so that the cobalt is oxidised to Co_2O_3. This oxide is then reduced by hydrogen, carbon monoxide or magnesium.

In Canada, roasted ores are leached with sulphuric acid, and from the solution the oxide is obtained and reduced as above. Commonly, however, the metal is obtained by the electrolysis of cobalt sulphate in the presence of ammonia and ammonium sulphate.

Cobalt is tenacious, silver-white, readily polished and takes a high lustre. It slowly oxidises when heated in air. In many of its properties it resembles iron, but it is harder, and is magnetic up to $1,115 \, °C$. Some of its physical properties are given in Table 2.

The metal is used alone only as electrodeposits, e.g. on mirrors. But its most important uses are as an alloying element in both ferrous and non-ferrous alloys. Tool steels containing from 8% to 10% Co have improved cutting qualities over plain carbon steels, but as cobalt carbide is unstable, cobalt is used only in conjunction with chromium. Cobalt is frequently used in conjunction with tungsten, chromium, molybdenum and vanadium in high-speed tool steels. Stellite, an alloy containing a high proportion of cobalt (up to 75%), is used for steam valves and medical implants. It is also used for high-speed cutting tools and as a hard facing for articles, such as turbine blades, that are required to be resistant to wear and corrosion. In the latter case the proportion of cobalt varies from 45% to 50%, together with about 25-30% Cr. Alloyed with aluminium (about 12% Co) the product is twice as hard as pure aluminium and has nearly twice the tensile strength, but the corrosion resistance is reduced. A series of alloys containing aluminium, cobalt and chromium and known as Kanthal is used for resistance heating elements of electric furnaces, operating at temperatures up to about $1,300 \, °C$. Cobalt is also an important constituent of alloys used in making permanent magnets of high coercive force — the other constituents being nickel, aluminium and sometimes tungsten. Magnetic steels contain some 30-40% Co and are used in the hardened state. Beryllium copper containing 2-3% Co gives an alloy of high electrical conductivity. Konel, an alloy whose approximate composition is Co 17%, Ni 73%, ferrotitanium 10%, has the special characteristic of maintaining high tensile strength at elevated temperatures, i.e. at $600 \, °C$, a tensile strength of approximately $450 \, MN \, m^{-2}$ is retained. Vitallium is a hard, heat-resisting alloy, used in making blades for aircraft turbines. Its approximate composition is Co 60%, Cr 35%, Mo 5%.

Salts of cobalt are used in porcelain, glass and also for vitreous enamelling to produce the characteristic and attractive blue colour. Cobalt compounds produce a blue compound (Thenard's blue) when heated with aluminium; a green compound (Rinman's green) with zinc; a pink compound with magnesium. Phosphates also give a blue compound. These colorations are used in blowpipe analysis for the detection of aluminium, zinc, etc.

Cobalt bloom. A pinkish coating on the surface of cobalt-bearing minerals, such as smaltite, which indicates the presence of cobalt. (*See also* **Cobalt**.)

Cobaltite. A sulpharsenide ore of cobalt and iron $(Co.Fe)S.As$. Also known as cobalt glance. Mohs hardness 5.5, sp. gr. 6-6.3.

Coercive force. The de-magnetising force that has to be employed to remove the residual magnetism of a metal or alloy.

Cogging. A process used for reducing the cross-section and increasing the length of ingots, which may be from ¾ to 4 Mg in weight. The rolls vary in size according to the ingots to be passed through them. Usually the ingots are brought direct from the soaking pit or the reheating furnace and passed backwards and forwards between the rolls until the required section is obtained. When the bar is of square section and more than 15 cm on the side, it is called a bloom. Hence the terms blooming and blooming mill.

Cogging mill. A machine for producing the semi-finished products — billets, blooms and sheet bars. It usually consists of one stand of rolls, either two or three high, about 76-100 cm in diameter. The rolls are divided by collars into passes of different widths. The ingots may weigh up to 3.6 Mg or more. Also called billet mill or blooming mill.

Cohenite. A term applied to cementite Fe_3C when present in meteorites.

Coherent precipitate. When solute atoms form a precipitate within a structure, and a definite orientation relationship exists across slip planes or in slip directions between the precipitate and the parent phase, the precipitate is said to be coherent. Such a phase produces the maximum strain and impedance to dislocation movement, and so increases strength.

Coinage gold. The fineness of coinage gold is usually expressed in parts per 1,000 (*see* **Carat**). Roman gold coins were originally made of the purest gold then obtainable and were 990-997 fine.

The British standard was adopted in 1526 and was 22 carat (11/12), or 916.6 fine, the balance being mainly copper. There is usually from 0.1 to 0.2% of silver in coinage gold. The Austrian ducat was minted 986¼ fine and the Dutch 983 fine, though these coins were not used for domestic trading. In France the standard of 900 fine was adopted in 1794. Specific gravity, British coin, 17.46.

Coining. A compression process for embossing the contours of a die or set of dies on plane metallic surfaces. It is essentially a cold-forming process in which the articles are produced to very close dimensional tolerance in dies. A good appearance and accurate reproduction of the impressions are required, and the operation is usually carried out in highly-finished alloy steel die blocks.

The coining equipment depends largely on the hardness of the material to be coined, and ranges from drop hammers to coining presses, fly presses, screw presses and hydraulic presses.

The term is also used in reference to the working or densifying of sintered powder metallurgical products. The working may be done hot or cold, but the former produces more satisfactory properties, since powder materials are usually cold short (q.v.).

Coke. The residue obtained by heating coal out of contact with air, until all the volatile constituents are driven off. It consists mainly of carbon, called "fixed carbon", together with the incombustible substances in ash derived from the coal. Coke also contains from 2% to 3% of gases, oxygen, hydrogen and nitrogen, and variable amounts of moisture. The nature of the coke depends on the type of coal used, on the temperature at which the coking takes place and on the time of heating. Low-temperature coke is dull black and has a sponge-like texture. High-temperature, long heating-time coke is hard and dense, is grey in colour and can only be burned in a furnace with an adequate (forced) draught. The nature of the coke, i.e. its strength and cohesive properties, depends on whether the coal used is caking or non-caking. The former gives a compact coke and the latter a fragmentary coke.

Coke is produced in closed vessels called ovens and these are differentiated according to the method of heating the charge of coal, e.g. direct heating (beehive ovens); flue heating (retort ovens); and indirect (by-product ovens). In beehive ovens the gas produced by heating the coal was allowed to burn in the oven, the required air being supplied and controlled through holes in the oven door. The process usually occupied two to three days. Retort ovens vary considerably in design due to the different arrangements of the heating flues. The gas produced from the coal is drawn into a by-product extraction plant where tar, ammonia and benzene are removed, and then returned to the oven. At the base of the flue, the gas mixes with heated air and combustion occurs. In the by-product oven, the evolved gas passes into a complete gas purification plant and the gas returned to the indirect heating of the oven therefore consists mainly of hydrogen, methane and some carbon monoxide.

High-temperature carbonisation in retorts yields from 62 to 80% coke, according to the origin of the coal. Though the product is not so dense as that produced in the beehive oven, there is a saving of about 3% tar and 1% benzine that would otherwise be wasted.

About one-third of the coke produced in the United Kingdom is used for metallurgical purposes.

Coke oven. This is usually a tall rectangular brick container, built in batteries, for the carbonisation of coal to produce metallurgical coke. The size of a single oven varies, but modern ovens are 2-4 m high, 30-60 cm wide and 9 m or more long. Such an oven would have a capacity of 9-10 Mg. Assuming a carbonising period of 10-12 hours, the oven would produce above 15 Mg of coke per day, or about half this quantity if the charge were wet compressed coal. In this latter case the heating period is longer, about 12-18 hours according to the temperature employed. The heating is done by blast-furnace gas or waste heat, supplemented by coke-oven gas or occasionally by oil. But the coke-oven gas, which is a product of the carbonisation, is a valuable by-product and is preferably used for purposes where blast-furnace gas would be unsuitable. Coke-oven gas is treated for the extraction of tar, ammonia and benzine, etc.

Coke-oven gas. The volatile product obtained in the preparation of metallurgical coke by the carbonisation of coal. (See also Coke oven, Coke.)

Cokes. The standard grades of hot-dipped tinplates used for packaging, canning and general purposes. They carry about 0.56 kg of tin per basis box (i.e. lighter coating than the charcoal grades) and consist of mild-steel sheet coated with tin.

Coking tests. A group of tests designed to give a coking index for various coals. The Campredon test represents the amount of fine white sand which 1 g of the coal is capable of just binding into a coherent mass. The Porter and Durley test is carried out on 22.7 kg of the coal packed into a rectangular box and treated along with an ordinary charge in a coke oven.

Colclad. The trade name of a series of clad steels used in the construction of oil refineries, chemical plants, etc. It consists of two different materials, either nickel or Monel or stainless steel and constructional steel, firmly bonded together so as to form a composite plate, which during pressing, spinning or drawing behaves as a solid plate. Six grades of stainless clad steel are in use, including 18-8-Ti; 18-8-Nb; 18-10-Mo-Nb and 14% Cr.

Colcothar. A variety of iron III (ferric) oxide obtained by igniting iron II (ferrous) sulphate in air. It is used as a paint and as a polishing powder.

Cold bend. A rough-and-ready engineers' test of the ductility of a metal. It consists in bending a bar through a specified angle at atmospheric temperature. The bar should bend without cracking.

Coldbond. A colloidal clay bonding material used for improving the mechanical strength of moulding sands.

Cold cracking. A defect evidenced by the presence of unoxidised cracks in cold or nearly cold metal. It is caused by excessive internal stress due to contraction on cooling.

Cold galvanising. A method of applying protective coatings of zinc. The coating consists of finely divided zinc powder and a binder taken up in a solvent, such as polystyrene, isomerised rubber or chlorinated rubber. The solvent evaporates leaving a film containing about 95% Zn.

Cold laps. Wrinkled markings on the surface of an ingot that represents incipient freezing of the surface of the metal. The defect is due to too low a casting temperature. Oxide may occur at the base of the wrinkles and in subsequent hot working may lead to cracks.

Cold quenching. A process for eliminating ageing of aluminium alloys during transfer from a water-quenching tank into a refrigerated store. The cold quench tank contains paraffin, white spirit or methylated spirit, and is maintained at a temperature slightly below the low storage temperature. Components are cold quenched immediately after water quenching, and from the cold-quench tank the work passes directly to the refrigerators for storage.

Cold short. Lack of ductility when a metal or alloy is worked in the cold. The metal shows brittleness at comparatively low temperatures.

Cold shut. A casting defect that arises from the freezing over of the surface of the ingot or casting before the mould has been filled. This embraces splashings in the mould, interrupted pouring, separations by dirty films in sluggish metal that fails to fill the mould completely, and the cooling, and consequent failure to unite, of metal entering a mould by different gates or runners.

Cold working. The plastic deformation of a metal at a temperature low enough to cause permanent strain-hardening. The treatment usually consists in rolling, hammering or drawing at ordinary temperatures, when the hardness and tensile strength are progressively increased with the amount of cold work, but the ductility and impact strength are reduced. The term cold working does not, however, refer only to deformation at ambient temperatures, but to deformation at temperatures below the recrystallisation temperature. The ductility can be restored and the hardness reduced by annealing at various stages between and after the cold-working operations. The process is used to improve the tensile strength of metals and alloys. When metals are deformed at temperatures below the equicohesive temperature (q.v.), they become harder and more brittle as the crystals are broken up and deformed by working. Thus an ordinary bar of pure tin or tinman's solder can be bent quite easily, but considerable force is required to straighten the bent bar. The work-hardening effect is removed by annealing at a temperature just above the recrystallisation temperature, when new equiaxed grains are formed and the metal becomes soft and suitable for future cold work.

Colemanite. A hydrated calcium borate $2CaO.3B_2O_3.5H_2O$ used as a source of borax. It is found in Asia Minor and North America. Mohs hardness 4-4.5, sp. gr. 2.42.

Collargol. A soluble variety of silver produced from colloidal silver. When silver nitrate is treated with sodium protalbite or sodium lysalbate, a yellow solution of colloidal silver is produced. If this solution is dialysed and then carefully evaporated, brownish-black collargol is formed.

Collaring. A term used in rolling-mills to describe the tendency of the metal being processed to encircle one of the rolls.

Collector. A term used in ore-dressing to denote a reagent which will increase the tendency of the desired mineral particles to attach themselves to the air-bubbles during froth flotation. The collector may be an inert oil such as paraffin or creosote, or an organic compound such as thiocarbamate or a xanthate.

Collector (transistor). A component of a semiconductor device, forming a potential barrier with the central unit, called the base, the other unit being the emitter (see **Transistor**).

Collision (electron-lattice). Neither the electron nor the atoms in a crystal lattice can be regarded strictly as "solid particles". An electron possessing kinetic energy can move through an array of atoms in a crystal lattice as if it were a small sphere, and there would obviously be a probability of collision between electron and atom. Since the atom is largely empty space, there is a low probability of direct contact with an atomic nucleus, which could be "head-on", leading to absorption of the electron into the nucleus, or at a "glancing" angle, leading to deflection of the electron path.

The electron behaves also as a wave motion, and in this context its interaction with nuclei or atomic electrons becomes important in the electrical conductivity of materials, which is simply the transfer of charges. Obviously, electrons moving through an electric field of other electrons, and positively charged nuclei, must be affected by the size of the atoms giving rise to these fields, their spacing, and the kinetic energy of both the moving electrons and the vibrating atoms. (See **Conduction**; for neutron-lattice collisions see **Absorption** and **Scattering cross-section**.)

Colloid. A substance (usually, but not always non-crystalline) which when dispersed in a liquid produces a clear suspension that cannot be separated by ordinary filtration. The substance assumes a state of sub-division intermediate between true solution and ordinary dispersion or suspension. The size of the particle in the dispersed phase (i.e. the colloidal substance) usually lies between 0.1 and 1 μm and the colloidal solution shows the Lyndall phenomenon, i.e. the scattering of light, made visible by the use of the ultramicroscope. Colloidal particles in suspension carry electrical charges and exhibit the Brownian movement. Precipitation can usually be effected by adding an electrolyte such as common salt or hydrochloric acid. Colloids have high molecular weights.

Colour centres (F-centres). The electron energy states in an insulator are such that an energy gap occurs between the occupied levels (valence band) and those available for electron mobility (conduction band). Transparent insulators are sometimes affected by visible light, with the appearance of colour in the crystals. Colour implies absorption of selected wavelengths, and hence must be due to energy changes too small to produce electrical conductivity (i.e. less than the energy gap). Photons of energy too low to bridge the energy gap are believed to be absorbed by energy levels associated with impurities in the crystals, these energy levels being situated between the valence and conduction band. This is a similar phenomenon to the effect of impurities in semiconductors, but in this case producing only the colour centres, not affecting conductivity. Alkali halide crystals are substances which show these effects, and

non-stoichiometry, i.e. an excess of metal ions in the crystal, produces colour centres. Excess potassium ions in KCl produces a blue colour, excess sodium in common salt a yellow. Many gemstones are coloured through the same agency.

In some cases, the proximity of the impurity energy level close to the conduction band means that the photons may bridge this gap, and the material become photoconductive. Photographic processes depend on similar effects to colour centre formations initially, although chemical precipitation is an added phenomenon.

Colour (of materials). When radiation falls on to a material surface, there may be reflection, transmission, or absorption. Reflection is never total, but when little selective absorption occurs, the reflection is said to be "specular", i.e. like a mirror. Polished metals show considerable specular reflection, but in some cases this rapidly deteriorates with formation of oxide films or corrosion (e.g. lead, steel). Transmission is also never total, but when there is little selective absorption, the material is said to be transparent. If the incident radiation is diffused, by scattering, but there is still little selective absorption, the material is translucent. If no transmission occurs, i.e. complete absorption, the material is opaque.

Selective absorption with considerable transmission gives rise to coloured transmission, i.e. the radiation leaving the material is modified by the absorption. Reflection can also involve some selective absorption, and opaque materials do involve some reflection. Hence, many materials appear coloured in visible light, or even with infra-red or ultra-violet incident radiation.

In the visible spectrum, a full range of frequencies is termed "white light", since this describes the sensation on the retina of the observer. When white light falls on a material, the radiation is first affected by the surface, and the electron clouds of those atoms forming that surface. Where the electron cloud is fixed by covalent bonding, there is only a small possible change in the energy of the electrons as a result, and often little absorption occurs. Many such materials are transparent or translucent, particularly if they have an amorphous structure. In metals however, the energy of the "free" electrons may be easily raised by absorption of energy from the incident radiation, and the radiation does not penetrate very far. As the electron energy in the conduction band is rapidly dispersed, metals are only transparent in extremely thin films, e.g. deposition from the vapour phase on to a transparent substrate. The reflected light from metals will be affected by the selectivity of absorption at the surface, and the "colour" of a metal will be the complementary colour to that which is absorbed or transmitted. Sodium metal films transmit visible light when less than 1 μm thick, but will absorb a wide range of frequencies, even in the infra-red. Freshly cut sodium reflects visible light with a dull greyish colour, but soon oxidises.

Copper absorbs visible light in the "blue" end of the spectrum, including ultra-violet, and so reflects (from a polished surface) the remainder, namely the "red" end of the spectrum. In very thin films, of 100 nm thick, copper appears blue-green, because the red colours have already been reflected away. Gold absorbs blue light also, but more yellow remains than in the case of copper, and so the reflected light is yellow rather than red. Transmission through gold films 100 nm thick is greenish blue. Silver absorbs some blue light, but much less than copper or gold, so that its reflection is largely white, but slightly yellow, more so than say chromium. This is why chromium plating looks more harsh (blue) than silver does. Most metals vary slightly in colour of reflected light, but the human eye cannot detect the differences. When metals are alloyed, the absorption changes, and hence red copper becomes yellow when alloyed with 5% zinc (brass), and the colour disappears entirely at 50% zinc. Tin also affects the colour of copper above 5% (bronze) but the reflected light is more reddish than with zinc, and the colour disappears at 30% tin. One alloy of copper and antimony is a beautiful purple.

Semiconductors are more or less transparent in the infra-red, but as the frequency increases there is often a sudden increase in absorption, at a frequency of photons corresponding to the energy gap between valence and conduction bands (with germanium about 1.5×10^{14} Hz). As energy is now absorbed, the material becomes opaque. Further increase in frequency continues the absorption, the electrons being excited into higher states in the conduction band. Some electrons fall back into lower energy states, giving rise to emission, and thus the material becomes reflective like a metal. Rise in temperature, which supplies energy in addition, to bridge the energy gap, automatically affects the frequency at which light is absorbed.

Insulators, having large energy gaps between valence and conduction bands, are transparent because the photons cannot supply sufficient energy to bridge the gap. They should, however, become opaque with radiation of sufficiently high energy, i.e. with very high frequencies. This does in fact happen in the ultra-violet, but with a more complex mechanism than in the case of semiconductors, involving interaction between electrons and positive holes, forming pairs known as excitons. They absorb energy in their formation and migration, but since they are electrically neutral, insulators producing them are not photoconductors as semiconductors are.

Columbite. A naturally occurring niobate and tantalate of iron and manganese $(Fe.Mn)(Nb,Ta)_2O_6$. When the niobium content exceeds the niobium content the mineral is known as tantalite. It is found in Colorado, South Dakota and Central Africa as a black or brownish-black opaque mass sometimes showing iridescence. The mineral is an important source of both niobium and tantalum. It is brittle, having a Mohs hardness of 6 and specific gravity between 5.3 and 5.9.

Columbium. The name used in the USA for niobium (q.v.).

Columnar structure. A type of crystalline structure produced when solidification begins at the walls of a mould and progresses towards the centre of the ingot or casting. Crystallites growing inwards from the surface often acquire considerable length before their growth is impeded by meeting other crystallites. In a square ingot crystals growing in different directions may meet and give rise to planes of weakness. This effect is reduced by rounding off the angle.

Combarloy. The trade name of a high conductivity copper intended primarily for use in commutator bars.

Combined carbon. The carbon that is present in steels and cast iron as metallic carbides as distinct from that in the free state as graphite or temper carbon.

Combustion. The combination of substances with oxygen when accompanied by light and heat, and usually referred to as burning. Slow combustion occurs when certain oxidisable matter is in contact with air or oxygen. The heat produced raises the temperature of the material, so that if precautions are not taken, the ignition temperature is reached when flame (light) and heat result, and the slow combustion changes to rapid combustion. Explosion is exceedingly rapid burning.

Comminution. Pulverising or reducing to a fine state of division ores or other materials by crushing or grinding. The term is usually applied to the treatment (by crushing, etc.) of ores prior to concentration.

Compact. The product resulting from the compression of metal powders, with or without the addition of non-metallic bonding agents.

Compactability. This is generally defined as the ratio of the green density of the compact to the bulk density of the powder. A high compactability is desirable, for, among other things, to ensure that the compact can be handled without undue risk of fracture, and that the necessary particle-to-particle contact is present which will favour the subsequent sintering process. Powders that are not readily compactable can be improved by adding a small percentage of lubricants to improve the flow of the powder. Such lubricants are stearic acid, paraffin wax and lithium stearate.

Compatibility (of materials). Generally, the degree of interaction between two or more dissimilar materials in interfacial contact, and the resultant effects of that interaction. It involves gas/solid or liquid/solid interaction, as in corrosion, oxidation or diffusion reactions; gas/liquid interactions as in solubility of gases in liquids, and solid/solid reactions in diffusion phenomena including alloying, bonding and joining processes.

At low temperature, the problems are generally minimal, because all the mechanisms involved are temperature dependent.

Many industrial and engineering applications involve high temperatures, high pressures (or stresses) and long times of exposure, and here there is an increasing tendency for interaction.

Solid state reactions are of paramount importance in moving machinery, and compatibility of materials was an immense problem from the earliest days of the Industrial Revolution. The seizing of pistons in cylinders, or crankshafts in bearings is still a problem of high speed reciprocating engines. Friction represents a waste of energy, and is now sufficiently important to warrant a special line of study. (*See* **Tribology**; the detailed processes involved are discussed under **Diffusion** and **Solid solution**.)

Complex salt. When aqueous solutions of two simple salts are mixed and the resulting solution evaporated, it frequently happens that the crystals obtained contain the two salts combined in molecular proportions. Thus when potassium sulphate and aluminium sulphate are so mixed the resulting crystals are potash alum. In solution this substance produces the same ions as the simple salts from which the alum was produced. Potash alum is, therefore, called a double salt, and this type of salt should be carefully distinguished from the product obtained when potassium chloride is added to platinic chloride. In this case a complex salt K_2PtCl_6 named potassium platinichloride is formed and this on ionisation yields an ion $(PtCl_6)^{2+}$ different from the ions of the simple salts. Similarly, potassium cyanide, when added to a deep blue ammoniacal solution of copper sulphate, yields colourless potassium copper II cyanide, which gives rise to the complex ion $[Cu(CN)_4]^{2+}$. As there are no free copper ions in such a solution, it does not give the characteristic reactions of copper (e.g. no black precipitate with hydrogen sulphide). The formation of complex salts is made use of in analysis in the separation or identification of cadmium in presence of copper, cobalt in the presence of nickel, etc.

Compliance. In the analysis of elastic deformation in solids, the relationships between strains (ϵ, γ) and stresses (σ, τ) are called elastic compliance constants, whereas the relationships between stresses and strains are called moduli. The compliance is not strictly the reciprocal of the modulus, since displacements are affected by preceeding displacement, whereas stress is not so affected (*see* **Deformation (Elastic)**).

Composite materials. The use of two or more dissimilar materials, in conjunction, to produce a material superior in one or more respects to any individual one. Wood is typical of a natural composite, having cells of highly crystalline cellulose in fibres (tracheids), bonded together with amorphous lignone. Concrete is a composite containing stone of various particle sizes, bonded together by Portland cement. Vehicle tyres contain reinforcing bands of steel or textile material, embedded in rubber. Glass reinforced plastic (GRP) consists of glass fibres woven into strip or ribbon, loosely intertwined as yarn, or randomly held by soluble adhesives into a mesh (mat), bonded together by polyester, epoxy or similar synthetic polymeric materials.

In principle, one component of a composite is generally load bearing, but if used alone possesses other undesirable properties, often that of brittleness, or, at the other extreme, too great a rigidity for the purpose of the application. The composite principle allows for more uniform distribution of load in the load bearing constituent, avoiding the concentration of stress so undesirable for brittle fracture, or preventing excessive flow or deformation, in the matrix of softer bonding material. Unless the volume fraction of load bearing particles is high, the flow of matrix will control the deformation, and the simplest way to achieve a high volume fraction is with fibres, oriented parallel to the direction of tensile stress, with a uniform and continuous bond of matrix material around and along the length of each fibre. When the stress is compressive, the simplest form is the aggregate, as in concrete, with the particulate sizes graded to pack together with the closest possible packing, so that little space is available for flow, and a continuous cement bond exists around each particle.

If cracks form in the fibre composite, they are limited to the cross-section of the fibre, and so are very short. They do not propagate into further fibres, unless the matrix is not uniform, and the stress is redistributed when the cracks form, by flow of the matrix, and frictional force between the matrix and remaining fibres. The surface properties of materials such as glass give rise to the starting point of such cracks, and the matrix resin in GRP protects that surface.

Obviously, all composites depend on good technology in their manufacture, and this includes the basic raw materials as well as the process of assembling them together. Composites are best designed for specific duties, with a manufacturing process which lends itself to the appropriate regulation of orientation, volume fractions, and absence of unwanted extraneous defects.

Compound semiconductors. A compound formed between materials which shows a characteristic energy gap between the electrons in the so-called valence band (producing the bond strength of the physical compound itself) and the higher energy level of the conduction band, which represents the acceptable conditions for those electrons of higher energy. These compounds are often intermetallic, and may contain one element from Group III of the periodic table with one from Group V, or one from Group II with one from Group VI. The net distribution of available electrons is obviously similar to that of the Group IV elements of the table, silicon and germanium. Gallium arsenide (Ga from Group III, As from V) has an energy gap of 1.45 eV, indium antimonide (In Group III, Sb Group V) a gap of 0.18 eV; cadmium sulphide (Cd Group II, S group VI) with a gap of 2.45 eV is obviously employed for a different purpose than the other two. By experimenting with these compounds a wide range of properties becomes available, varying the energy gap, and therefore the energy required to become conductive, but also showing variations in charge carrying mobility, whether electrons or positive holes.

Compounds (magnetic). Compounds formed between ferromagnetic materials and other elements might be expected to show magnetic properties of interest, and this is so, but many metals, themselves not strongly magnetic, form compounds which are. This is particularly true of the transition metals, and particularly those closely resembling the ferromagnetic metals Fe, Co, Ni. Manganese and chromium, with fewer electrons in the 3d level than Fe, Co, Ni, are antiferromagnetic, but when combined with other elements the interatomic distances are changed, electrons may align with parallel spin in the 3d level, and the compounds become strongly magnetic.

Compression test. A test for assessing the ability of a metal to withstand compressive loads. The test piece is usually a cylinder whose length is twice the diameter. Plastic metals do not show any well-defined breakdown point under compression, and it is usual to record compressive strength at a given amount of deformation. Such values are, of course, only comparative. With brittle metals the ultimate stress can be recorded with precision.

In engineering design the ultimate compressive strength (S) of columns or struts is given by Rankine's formula

$$S = p \left(1 + \frac{l^2}{Cr^2} \right)$$

where p = ultimate load in lb./sq. in.; l = length of column or strut in inches; r = least radius of gyration

$$\text{i.e. } r = \left(\frac{\text{moment of inertia}}{\text{area of section}} \right)^{½}$$

and C = 25,000 for steel, 5,000 for cast iron, 35,000 for wrought iron. (For conversion to non-Imperial quantities, *see* Table 1.)

Compton rule. This rule, which is not based on any fundamental relationships (and is classed as empirical), states that the atomic weight of an element (W) is approximately given by

$$W = \frac{2T}{L}$$

where T = the melting point of the element in degrees absolute, and L = latent heat of fusion.

Concentration. A process for the separation of the metalliferous ore from worthless gangue. In the simplest case the ore, as taken from the mine, is broken small and subjected to a separation process in which the difference between the specific gravity of the useful and unwanted material is utilised.

Concrete. A composite containing cement, sand and aggregate. The cement is generally Portland cement, the aggregate may be stone, gravel, furnace ash or clinker, or broken brick and the sand should not contain much clay or other binding material, which may interfere with the set. Concrete has high compression strength but is brittle in tension. It may be reinforced with steel bars, expanded metal mesh, or corrugated bars, to produce ferroconcrete, the metal component taking the tensile loads. When the bars of metal are prestressed in tension during the concrete set, the concrete is put under compressive load when the prestress load is removed. This helps to ensure that the concrete is not immediately in tension when tensile stress is applied. The volume mix of concrete for ferroconcrete is 15% cement, 30% sand and 55% aggregate. The aggregate and sand should be graded to give the closest packing of spheres. 10% cement, 30% sand and 60% stone of size greater than 2 cm but less than 8 cm, gives moderate strength for bulk concrete mixing. 25% cement, 25% sand and 50% stone less than 2 cm size, gives high density for reinforcement work. The use of chopped glass fibre to reinforce cement produces a sprayed concrete of considerable strength for shell structures or mouldings.

Condensation of gases. The gaseous state is a condition of high entropy, that is, disorder, with all the atoms or molecules vibrating within their containing vessel. The impact on these walls is the pressure p, the volume of the container V, and, at temperature T, the equation of state for a perfect gas is given by $pV = RT$ and R is the universal gas constant, RT representing the energy of one mole at temperature T.

If we refer to a single molecule, the equation becomes $pV = kT$ where k is Boltzmann's constant. In a monatomic gas, the atoms have three degrees of vibrational freedom, so the average energy is $3/2\ kT$, per atom. If there are N_o atoms in a mole, the energy for a mass of gas is $N_o (3/2\ kT) = 3/2\ RT$, since by definition $N_o k = R$. N_o is called Avogadro's number.

If we now compress the gas, by decreasing V, the pressure p will rise, but the mean free path of the atoms will also decrease, and the atoms will be forced closer together. The close proximity of the atoms, and the increasing likelihood of collisions between them, gives rise to an increase in the van der Waals' forces of attraction, and eventually, when the pressure is sufficiently high, the tendency to fly apart again will have disappeared. At this stage the entropy has considerably decreased and the material is said to be condensed. Further increase in pressure will not produce the same change in volume as before, and the material will not expand to fill the containing vessel. Its viscosity has increased, and it is now described as liquid.

The atoms or molecules still vibrate within the liquid, but in a much more restricted way. They are influenced by stronger attraction forces than before, and hence a greater force is required to move them relative to each other (viscosity). At the surface the stronger attraction from the interior than above the surface produces a net inward directed attraction, the surface tension. (*See* **Critical constants.**)

Condensation of liquids. To decrease the vibration of atoms or molecules in a liquid, it is necessary to produce even stronger attraction forces to bring them closer together, and decrease the total energy in the system. Both these objectives can be achieved by a decrease in temperature, pressure having much less effect. As temperature decreases, the vibration is reduced because the kinetic energy is reduced, and at this stage, new forces of attraction may appear, because the unit atoms or molecules are becoming much closer in their approach, the entropy is further reduced, and a greater state of order exists. In fact, the atoms may align themselves in a regular array, a crystal lattice, but if not at least they are brought together in a less random pattern. Noticeably, the viscosity falls sharply, and by several orders of magnitude to the solid state, whereas in the condensation of gases, the change is only of two or three orders of magnitude. Even if crystallisation does not occur, the material nevertheless becomes so viscous that it behaves as a rigid fluid, and it is said to be a glass, a supercooled liquid. Such materials do not exhibit a true melting point, they simply become less viscous on heating again. Many natural materials are glassy in nature, and many synthetic polymers exhibit a "glass point", sometimes in addition to a melting point.

Liquids then, condense into viscous fluids, crystalline solids, amorphous glasses, or a mixture of all. The attraction forces are now very high between individual atoms, but are limited by the strong repulsion forces built up when atoms approach very closely. There is therefore a limit to the compressibility of solids, and disruption of the crystal lattice or amorphous structure takes place under high compressive stress. (*See* **Viscosity**.)

Condensation polymerisation. In combining individual "mer" units to form a polymer, dissimilar units frequently combine first by chemical reaction, to produce the mer, before the units form the polymer. The elimination of some parts of the individual chemical units may be involved in mer formation, and hence this type of process is known as condensation polymerisation. The early polymeric materials were often formed this way, the prime example being phenol-formaldehyde condensation (Bakelite). (*See* Fig. 27.)

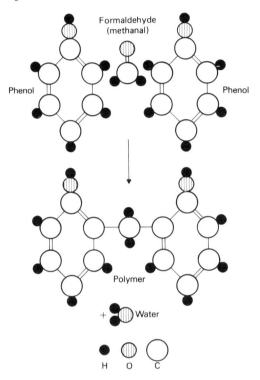

Fig. 27 *Condensation to form phenol formaldehyde (methanal).*

Conductance. Is defined as that property which a body has for conducting electricity. A body of resistance R ohms has a conductance of $1/R$ mho; a mho originally being the conductance of a column of mercury 106.3 cm long and 1 mm in cross-section, and of mass 14.4521 g at the temperature of melting ice (mho is pronounced moe). This unit is also called the siemens.

Conductance (equivalent). The electrolytic conductance of 1 g equivalent of a substance under standard conditions of a potential difference of 1 volt between electrodes 1 cm apart (i.e. conductivity $\times$ volume (in cm^3)).

Conduction band. That range of energy levels available to the electrons of highest energy in a solid material, which is not completely occupied by electrons of each available energy level in the range, and so allows for increased occupation by electrons at the top end of the range, when the material is excited by increasing temperature or in an applied electric field. The conduction band can be said to be partially filled, which refers to the density of states of electron energy for that particular material, complying with the principles of Wave Mechanical behaviour, and the Pauli Exclusion Principle (*see* **Density of electron states**).

In a metal, there is always a conduction band available to the electrons of highest energy, either by virtue of fewer electrons per atom than the number of states available, or the overlapping of another state of energy into which the electrons of highest

energy can expand. In an insulator, where most of the outer electrons, of highest energy, are involved in the process of bonding (covalent), there is no higher state of energy available in the valence band (involved in bonding) and the next higher state of available energy does not overlap the valence band, but a considerable energy gap exists between valence and conduction band. In such cases, extremely high temperatures, or very high electrical fields, are necessary to boost the electron energy so that this energy gap can be crossed into the conduction band. Under such extreme conditions, insulators break down and become conductors, although this frequently implies destruction of the original structure.

In semiconductors, the same energy gap exists as in insulators, but is very much smaller in energy terms. Increase in temperature then elevates the electron energy sufficiently to bridge the gap, and the material becomes a conductor. At lower temperatures, it remains an insulator unless subjected to very high electrical fields, as with the insulator.

Conduction by electrons. In an applied electric field, the electrons in a material are accelerated in a direction opposite to that of the applied field. Only those electrons not tightly bound to each nucleus (i.e. not involved in bonding or close to the nucleus) are affected, and the electron cloud is said to move with a drift velocity v_d. If there are n electrons per unit volume, and e is the charge on each electron, then current density $= n e v_d$. The value v_d is quite small, and is additive to the existing velocities of the electrons involved, due to their kinetic (Fermi) energy.

The electrons are bound to undergo collision with the atoms in the lattice through which they move, and will therefore be attenuated in energy, and therefore velocity, with each collision. This explains why the acceleration of the applied electric field does not give rise to an increasing drift velocity. The mean free path, the average distance between collisions, is related to the lattice spacing in crystalline materials, so that on average one collision occurs in passing roughly 100 to 200 atoms. From the value of drift velocity, this enables an estimate of the mean free time elapsed between collisions, and the conductivity of the material expressed as $n e^2 \tau/m$ where τ is the mean free time, m the mass of the electron. The value $e\tau/m$ is called the mobility of the electron, and is the drift velocity in unit electric field.

The atoms are continually vibrating, and vibration amplitude increases with rise in temperature, so that the mean free time will be inversely proportional to temperature, the number of collisions increasing with rise in temperature. Since the conductivity is directly proportional to mean free time, this implies that the conductivity should be infinite at absolute zero. This is not so however, because structural defects ensure that there is a finite limit to conductivity even when temperature is lowered. These defects include "impurity" atoms in substitutional or interstitial positions, vacant lattice sites, and dislocations in the structure, including grain boundaries.

Modification of the defect structures causes a change in conductivity, e.g. the work hardening of a metal by rolling or wire drawing will increase the density of dislocations. Quenching from a high temperature increases the density of vacancies.

In an applied thermal field, any electrons available with excess Fermi energy will also be carriers of thermal energy, in addition to their potential electrical conductivity. Thus, a metal bar heated at one end, will transfer the energy to the cold end, at least in part due to electrons moving in the direction of the thermal gradient. The mean velocity of the electrons is very high compared with the velocity of phonons, the elastic waves produced by atomic vibrations, and so in metals, electrons contribute largely to thermal conductivity, and metals are good conductors, both of heat and electric charge.

In materials where electrons are not available, as in ionic and covalent bonded solids, no contribution to thermal conductivity by electrons is possible.

Conduction by holes. In covalent bonded materials, where the valence electrons are all involved in the bonding process, no free electrons are available. There is, however, in some materials (semiconductors) a discrete but not excessively large energy gap between the valence and conduction band, and thermal vibration by rise in temperature may be sufficient to excite some electrons into the conduction band. At room temperature, the number of electrons made available by this process, in silicon, is about one in 10^{13} of the total number of valence electrons involved. As temperature rises, the number of available charge carriers increases exponentially in semiconductors, and so another process must be involved.

When an electron jumps out of its position in a bond to become a charge carrier, it inevitably leaves an incomplete bond behind, and when an electric field is applied, there is an added potential gradient to assist any electron in an adjacent bond to jump into the incomplete bond, and so complete it. This, however, leaves another incomplete bond, and the process can be repeated, so long as the potential gradient of the applied field favours the motion of such electrons. The incomplete bond can be described as a positive hole, meaning minus one electron, and the net effect will be that such holes appear to move in the opposite direction to that of the electrons, in other words to act as if positive charge carriers. This represents a further conductivity contribution, limited by the same influences as those on electrons, i.e. collisions.

Since the electron contribution to thermal conductivity is a major contribution in metals, so it must play a part in semiconductors, and so must the positive holes.

Conduction by ions. Ionic lattices (or ions in solution) display the same thermal vibration as other materials, and collisions between ions, impurity atoms (and solvent molecules in a solution) take place continuously. These movements are random, and increase with rise in temperature. If an external electric field is applied, there is a potential gradient which must automatically attract the ions towards the poles of opposite sign to themselves. The frequent collisions which must occur make this movement very slow, and the relatively large size of the ions also lowers the velocity. Although the mobility of the ions will increase with rising temperature, inevitably the value of ionic electrical conductivity is much less than electronic conductivity.

Impurities, flaws and defects in the lattice occur just as frequently in ionic materials as in others, and ionic conduction is very dependent on purity. As impurities will themselves be moving in the lattice, in an applied electric field, there must be a gradual removal of impurities, leading to decreasing ionic conductivity as a contribution of the pure ionic material alone becomes involved.

In the molten state, the ionic mobility is quite high, since collisions are now governed by a higher entropy state than in the solid. Similarly, in solution, ionic mobility is very high, by virtue of reduced concentration of ions with which to collide, and increased diffusion rate.

In the gaseous state, the individual atoms or molecules are separated or at least associated only in very simple groupings by van der Waals' forces. They are also of very high mobility. No normal charge carriers exist, and it is necessary to form ions by removing electrons from atoms or molecules. The potentials required to ionise, i.e. withdraw an electron to infinity, are high, of the order of 10-20 volts. Furthermore, the high mobility of gaseous atoms means a high probability of a released electron being recaptured before it has moved very far. At normal pressures, most gases are insulators, since the mean free path is very short, less than $1\,\mu m$, the kinetic energy acquired by the electron is low, and a very high potential is needed to accelerate the released electrons, and to increase the number of ionisations taking place. In fact, a spark discharge is necessary, breaking down the insulating properties of the gas. This occurs, of course, in lightning flashes, and breakdown of electrical systems in air.

An alternative approach is to increase the mean free path of the electron, decreasing the number of collisions and so increasing mobility. A lower value of the applied field will then be feasible. The only way to reduce collisions is to reduce the number of gas atoms or molecules, i.e. to lower the pressure. Too low a gas pressure, say, $0.1\,mN\,m^{-2}\,(10^{-6}\,mm\,Hg)$, increases the mean free path to 0.5 km, and few collisions take place again. A mean free path of the order of 1 cm is ideal, and this occurs at a pressure of about $100\,N\,m^{-2}\,(1\,mm\,Hg)$. An applied voltage of 50 to 500 V then gives a reasonable number of collisions between electrodes a few centimetres apart.

As current flows in such a gas discharge tube immediately the voltage is applied, current carriers must be present initially. These arise from stray ionisation due to ultra-violet radiation, cosmic rays, and other stray radiations. When the field is applied, electrons drift to the positive electrode, and the positively charged ions to the negative. The current rises to the point where the number of stray ionisations now equals the rate of drift of carriers to the electrodes, and a saturation level is reached, current being about $10^{-20}\,A\,cm^{-3}$ of gas. Further increase in voltage gives rise to ionisation by collision, the acceleration of electrons now helping to "knock off" electrons from gas atoms and creating more ions. The current rises again, and this is called the Townsend discharge. A chain reaction proceeds, as the number of collisions increases, and eventually a glow discharge begins. At this point, the positive ions hitting the negative electrode (cathode) knock out electrons from the cathode, a secondary emission of electrons. The chain reaction now proceeds with renewed vigour, and electrode material may begin to vaporise, leading to transfer of ions as a plasma — an electric arc discharge is set up. The breakdown voltage depends on the product of gas pressure and electrode spacing, and reaches a minimum when pressure is of the order of $130\,N\,m^{-2}$ and electrode space about 1 cm.

Conduction by the lattice. The atoms arranged in a crystal lattice, whether ionic, metallic, or covalent bonded, vibrate when supplied with thermal energy, and so are in motion at all temperatures above absolute zero. The atoms must interact with each other, and hence produce elastic waves, called phonons, by which the energy is dissipated through the mass when thermal energy is applied. At constant temperatures, the phonons will be moving in all possible directions, and a state of equilibrium exists, with as many moving in one direction as in the opposite one, at any given point. If a thermal gradient is created, by applying more energy at one point, then the equilibrium is disturbed, and more phonons move in the direction of the applied thermal gradient than in the opposite one. There is therefore a transference of thermal energy, and such transmission, the thermal conductivity, is dependent on the value of the phonon velocities. Phonons will be deflected by collision with structural defects, exactly as electrons are, and so the purity and perfection of a structure affects the thermal conductivity. Variations in thermal conductivity of materials depend on the mass of the atoms involved (inertia), the packing of the atoms (lattice constants), and the directionality of bonds where these exist. The phonon velocities are obviously temperature dependent, the defects and impurities are not. Thermal conductivity decreases in almost inverse proportion to thermal energy in crystalline material, as the mean free path of the phonons decreases. In amorphous materials, there is less regularity of atom positions, and conductivity is proportional to thermal energy. Substances which can exist in both amorphous and crystalline forms illustrate this variation. (*See* Fig. 28.)

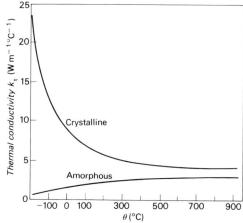

Fig. 28 *Variation of thermal conductivity between crystalline and amorphous forms of silica, SiO_2.*

The elastic waves (phonons) can be assumed to move with the velocity of sound (s) in the material, and if the mean free path is assumed roughly equal to the lattice parameter (d), then thermal conductivity is approximately $3\,ksd^{-2}$ where k is Boltzmann's constant. This means that the thermal conductivity of an insulator should be inversely proportional to the square of lattice parameter, and directly proportional to the velocity of sound.

In crystals, where electrons may contribute to thermal conductivity, an added, and much greater contribution is involved, as described under Conduction by electrons. Since electrical conductivity is also dependent on electrons, a relationship exists between the two. The Wiedemann-Franz law states that

$$k_t/k_{el} = LT$$

where k_t and k_{el} are thermal and electrical conductivity respectively, T is absolute temperature, and L is the Lorenz constant, whose theoretical value is 2.45×10^{-8} V^2 K^{-2}. Experimentally, this is found to be reasonably close for metals, but variations arise between the lattice and electron contribution to thermal conductivity in some cases, varying with temperature. Measurement of thermal conductivity at low temperatures has been used to separate the effects due to dislocations (proportional to T^2) from those due to impurities (proportional to T). As the electron contribution is much decreased at low temperatures, the lattice contribution is more noticeable, and the technique still reasonably sensitive.

The contribution of the lattice vibrations to electrical conductivity is very small, but the atomic vibrations have already been referred to, and there can be contributions if the electron density is affected by other processes, such as the formation of positive holes.

Conductivity (electrical). The conductance of a conductor is the reciprocal of the resistivity. The unit is the reciprocal ohm, written mho.

Conductivity (thermal). The thermal conductivity or the conducting power of a substance is the number of joules of heat which is transmitted in 1 second through a plate 1 m thick and area of face $1\,m^2$ when the temperature difference between the faces is 1K. It is sometimes called coefficient of conductivity and is measured by: $k = Hd/A(t_1 - t_2)$ where H = heat transmitted per second; d = thickness of plate; A = area of one face of the plate; t_1 and t_2 = temperatures of the two faces of the plate. In this formula $t_1 - t_2/d$ is called the temperature gradient. The SI unit for k is $Wm^{-1}\,K^{-1}$.

Conform. A new extrusion process in which the normal friction between billet and container is utilised to provide the pressure for extrusion through the die rather than a loss of energy input. The stock (up to 40 mm diameter rod with aluminium) is fed into a grooved wheel (up to 400 mm diameter) and then into a stationary block of metal (the shoe) which carries the extrusion die. The shoe forms the closing side of the extrusion chamber, against the groove in the wheel carrying the feed over part of its circumference. The frictional forces generated as the wheel rotates increase the pressure on the stock and raise the temperature. At the end of the shoe is the abutment to which the die is fixed and through which the stock is forced, as in normal extrusion processes. The die can be radial to the wheel, or at a tangent, the plasticity of the metal being sufficiently high to extrude into simple sections or bridge dies for tubing. Pressures are as high as 10 kbar (70 t. in.$^{-2}$) at the die face, and temperatures may reach 500 °C.

The process is licensed worldwide by the United Kingdom Atomic Energy Authority, who developed it, and is the first successful continuous extrusion process.

Conglomerated swarf. The product of a process for preparing swarf for charging into a cupola or furnace. The machine for the process consists of a bowl-shaped container that is caused to rotate while tilted at an angle of about 45°. The container holds the swarf on to which molten iron is poured. The resultant material consists of balls of solidified cast iron in which swarf is embedded and is suitable for including in furnace charges.

Conjugate solutions. Two solutions which exist together in equilibrium at a given temperature, and such that a variation of the temperature results in a change in the compositions and relative proportions of the solutions. The term is usually applied to two immiscible liquids but is also applicable to two immiscible solid solutions, e.g. the separate constituents of a eutectic.

Constantan. A nickel-copper alloy containing 40% Ni and 60% Cu used for resistance coils. The presence of the nickel greatly increases the resistance of the alloy, which has a low temperature coefficient and the highest thermal e.m.f. against platinum of any of the copper-nickel alloys. Resistivity, 49.1 $\mu\Omega$ cm^{-1}; sp. gr. 8.88; average temperature coefficient = -0.00001. Eureka is practically identical with constantan. Also used to a limited extent for making thermocouples, the maximum temperature for use being about 900 °C.

Constitutional diagram. A diagram which shows graphically the compositions at which the various solid solutions and compounds of metals are formed, and the temperatures at which the alloys of a system undergo changes of state or of structure, i.e. a method of expressing the phase rule (q.v.). Such diagrams are usually known as equilibrium diagrams.

When two metals are melted together a homogeneous liquid is usually formed, neither metal being visible separately. If the amount of one metal is quite small compared with the other, it may happen that on cooling a homogeneous solid is produced. Such a solid, having no fixed composition, for it can be varied over a certain range, is called a solid solution. Solid solutions are common constituents of useful industrial alloys. Examples of solid solutions are 5% tin in copper (bronze), 3% antimony in tin, nickel in copper. Solid solutions occur only in certain groups of alloys.

In contrast to those molten metals which are completely soluble one in the other, there are a few cases where the metals are mutually soluble over a very restricted range of proportions. Outside these proportions the immiscible liquids separate. More frequently, however, it is found that a uniform solution on solidification forms an intermetallic compound of fixed composition. Such compounds can often be given a chemical formula. They are usually hard and brittle and are therefore only useful as constituents of alloys.

In a large number of cases binary alloys do not consist entirely of a homogeneous liquid, or of a solid solution or of a compound, but contain two or more of these and the physically distinct substances that are present in an alloy are known as phases. In binary alloys, three phases may co-exist only at a limited number of combinations of temperature and composition, but two phases may co-exist over a much wider range, and the properties of the alloy will depend on the nature of the phases.

Most of the common alloys undergo structural changes with change of temperature and these transformations in the solid state are usually associated with changes in mechanical properties, tensile strength, hardness and ductility. The constitutional diagram provides a guide to the metallurgist both in melting and in heat-treating alloys of known composition. The usual thermal equilibrium diagram of a binary (two metal) alloy system shows the complete range of both constituent metals from 0 to 100%, temperature and composition being the co-ordinates. Such diagrams are divided into a number of areas or fields, and the boundary lines which have been determined experimentally indicate the temperatures at which alloys of any given composition undergo a change of state. For metallurgical purposes the vapour phase can usually be neglected, and the points indicated in Fig. 29 show:

(a) the liquidus curve, i.e. the temperature at which freezing begins on cooling, or at which an alloy becomes completely molten on heating;

(b) the solidus curve, or the melting-point curve, which lies below the liquidus curve and shows the temperature at which solidification is complete on cooling, or at which melting begins on heating;

(c) lines of liquid solubility, in certain alloy systems, indicating the boundaries of immiscible liquid phases;

(d) horizontal lines, extending over a range of compositions, show changes which occur at constant temperature over the indicated composition range. Such horizontals include eutectic lines (in which case they form part of the solidus curve) and lines representing eutectoid changes, peritectics and changes in crystal structure;

(e) sloping curves or lines below the solidus indicate the limits of composition within which the solid phases of the system

are in equilibrium. Intermetallic compounds, which in theory have a fixed composition, would be expected to show as a single vertical line, but in practice there is usually a certain amount of solid solubility of the component metals in the compound. Peritectic reactions, such as that shown by the iron-carbon system at 0.2%C, 1,494 °C, where two phases δ and liquid react to form a single phase γ, are represented by a point at which the three phases may exist in equilibrium.

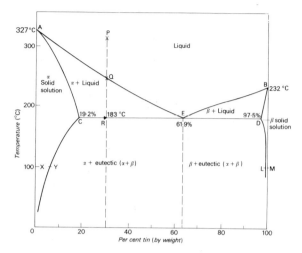

Fig. 29 *Constitutional diagram of lead-tin alloys.*

In ternary alloys, which can only be fully represented by three-dimensional models, volumes replace the areas or fields in binary diagrams, surfaces take the place of lines, etc. It is, however, common practice to make use of projections on to horizontal planes, or of vertical sections at various compositions through a model in order to study the system more easily.

Equilibrium diagrams have many important advantages, but it is essential to remember that they have also certain limitations. They express only equilibrium conditions produced in favourable circumstances which presuppose stability. Industrial conditions seldom reproduce experimental conditions, and it is necessary to make allowances and adjustments for metastable phases and for transformations and diffusions in the solid state that may proceed far more slowly than production conditions would permit to proceed to equilibrium.

In the constitutional diagrams of tin-lead alloys, the line *AEB* is called the liquidus (*see* Fig. 29). Any point on the diagram above the liquidus represents an alloy in the molten condition, the temperature and the composition being dependent on the position of the point. Similarly, any point below the line *ACEDB* represents completely solid alloy and this line is called the solidus. Between the two lines are areas in which solid and liquid alloy may co-exist as a more or less pasty mass.

The point *P* represents a molten alloy containing 30% Sn at a temperature of 350 °C. On cooling to 255 °C (point *Q*) solidification begins, crystals of lead containing a little tin in solid solution being deposited. As the solidifying temperature falls, the liquid becomes richer in tin, until at *R* (183 °C) the remaining liquid suddenly solidifies.

In the above case there is a freezing range of over 70 °C. The point *E* represents the melting point of an alloy containing 62% Sn that melts sharply at 183 °C. This is the eutectic alloy, and it is characterised by having a lower melting point than any other alloy of tin and lead. All the alloys of these two metals, containing from about 20 to 97% Sn, consist of eutectic together with a proportion of other higher-melting-point constituents. Thus at *R* the alloy consists of the lead-rich solid solution of composition *C* and eutectic alloy *E*, the ratio of these being represented by the ratio of the lengths *RE* to *RC*.

The tin-lead alloys represent a type of alloy system in which the components are completely soluble in one another in the liquid state and are partially soluble in the solid state. The solid solution of lead in tin is named β, and the solid solution of tin in lead is α, and the solubilities at 100 °C are given by *LM* and *XY*. The compositions of α and β at the eutectic point (183 °C) are

given by *C* and *D* and the solid that separates out at that temperature, the eutectic alloy, is not the pure metals, but a mixture of the solid solutions α and β. The proportions of α and β in the just solid eutectic alloy are given by *DE* : *CE*.

Constitutional supercooling. In the solidification of alloys, the initial solid formed is of higher melting point than the liquid remaining, and so the liquid becomes enriched in the solute which leads to a variation in composition of the liquid dependent on the the distance from the liquid/solid interface. As the freezing point of the solution varies with the proportion of solute the freezing point of the liquid must also vary with the distance from the liquid/solid interface. Rejection of solute occurs at the interface only, and hence a concentration gradient of solute must be set up such that at a given distance from the interface, the liquid is of original concentration, with a higher freezing point than the actual temperature. There will be a layer of liquid stretching out from the interface therefore, where the actual temperature is below the freezing point of the solution, i.e. the liquid is supercooled. This phenomenon, dependent on the constitution of the alloy and the changes which occur as solidification proceeds, is therfore known as constitutional supercooling. It leads to the formation of cellular structure (q.v.).

Contact bimetal. A combination of two metals, one of which is usually a precious metal employed for its suitability as an electrical contact, while the other is a material suitable for supporting the precious metal and acting as a current-carrying member.

Contact potential. In general, when two different metals are permitted to touch they acquire opposite charges and maintain a difference of potential while touching. This difference of potential is characteristic of the metals used. When measured *in vacuo* the p.d. is referred to as the intrinsic potential difference, and for two metals X and Y the potential difference (intrinsic) $V_{xy} = b \ (\nu_x - \nu_y)/e$ where ν_x and ν_y are the critical frequencies of photoelectric emission; b = Planck's contant and e = electronic charge. And if W_x and W_y are the electronic work functions of the metals: $V_{xy} = (W_x - W_y)/e$.

Contact tin plating. A process for depositing dense and adherent coatings of tin on most metals, including cast iron and copper alloys, by simple immersion, in contact with aluminium, in hot alkaline sodium stannate solutions such as are normally used for electrotinning. Deposition does not cease when the base metal is covered. Fairly thick deposits can be obtained in reasonable time.

Contamination. A term used to denote the inhibitory effects of small amounts of certain substances (e.g. arsenic, sulphur, phosphorus, mercury, etc.) on the activity of catalysts. These contaminating substances are also referred to as catalyst poisons. The term is also used in reference to the change in e.m.f. of thermocouples during use, due to the physical and chemical action of the medium in which the hot junction is placed.

Continuous casting. A process in which ingots, bars or tubes are cast continuously. Solidification takes place axially as the metal passes through the casting die, and the length of the casting is not restricted by mould dimensions. The method is satisfactory in respect of the elimination of gas cavities and shrinkage porosity.

Contraction cavities. Holes formed when the supply of molten metal is insufficient to compensate for the shrinkage that occurs on solidification and cooling. These cavities are usually concentrated in feeder heads and risers, but if the molten metal is unable to feed the whole of the casting, cavities may be formed in parts of the casting itself.

Controlled atmosphere. The term used in reference to the particular gaseous atmosphere in which a metal may be enveloped during thermal treatments. The object of using controlled atmosphere (e.g. coke-oven gas, natural gas, nitrogen, ammonia, inert gases and hydrocarbons such as butane, propane) may be to inhibit oxidation and so reduce undesirable surface actions, or to produce desirable effects (e.g. as in nitriding or carbonitriding). The atmosphere may be produced in the furnace itself by regulating the combustion of the fuel, or it may be produced external to the furnace.

Converter. A refractory-lined container in which molten metal is treated with a blast of air under pressure. The Bessemer converter (q.v.) is the general type used in the production of steel from cast iron. Other metals, e.g. copper, may be treated in a converter of the Peirce-Smith type having a basic (magnesite) lining. The copper matte (about 40% Cu) is converted into blister copper (70-75% Cu) prior to refining.

Coolant. These are liquids, usually mixtures of oils, or emulsions that are sprayed on metal cutting tools to carry away the heat developed in separating the chip from the work. The choice of a coolant will depend on the metal to be cut and the type of operation to be carried out (e.g. grinding, milling, tapping, etc.); on its lubricating qualities (reduction of friction, and in the case of steel, rust prevention) and on its ability to flush out the small chips and improve the finish of the work. Lard oil and mineral oils are used in varying proportions. Quite commonly mineral oils are used with water to produce emulsions having good body. Cheap coolants are often made up with sodium carbonate (0.5 kg), hard oil and soft soap (0.01 m^3) and water (0.05 m^3). For use of coolants in rolling-mill practice see **Rolling.**

Cooling curve. Graphs showing the rate of cooling of a body, obtained by plotting temperature against time for a metal cooling under constant conditions. The transformation ranges are indicated by the intervals between horizontal portions (or sudden change in direction) of the curve (time plotted horizontally). All pure metals freeze at a constant temperature, but most alloys freeze over a range of temperatures. Freezing commences at a definite temperature and the alloy becomes completely solid at a lower but equally definite temperature.

Inverse-rate cooling curves are often used in the thermal analysis of alloy systems. In this method, small intervals of temperature are used and the temperature is plotted against the time intervals required to cool through equal intervals of temperature change, i.e. $\Delta t/\Delta \theta$ is plotted against θ. In direct rate cooling curves the temperature drop for a definite interval of time (i.e. cooling rate) is plotted against the temperature. (See also **Arrest-points.**)

Cooperite. A platinum ore containing platinum sulphide PtS found in the copper-nickel deposits of Sudbury, Ontario.

Co-ordination number. The number of nearest neighbours to an atom in a structure. The closest packing of like atoms possible (face-centred cubic and close-packed hexagonal) gives a co-ordination number of 12. The next in order is the body-centred cube, with 8, then the simple cube, co-ordination number 6, and the tetrahedral or diamond type, with 4.

When dealing with compounds, it is customary to refer to the co-ordination numbers of the individual atoms or ions separately. The co-ordination numbers refer to the number of ions of opposite sign surrounding each of them. In CsCl, the ions are of similar size, and each Cs$^+$ ion lies at the centre of eight Cl$^-$ ions, situated at the corners of a cube. The co-ordination number for Cs is therefore 8, and for Cl it is also 8. In NaCl, where the ions alternate along each three-dimensional axis, the co-ordination is 6 in each case, whereas in ZnS (zinc blende) it falls to 4 in each case.

When two ions are involved, but twice as many of one as the other (AB$_2$ compounds) the numbers may differ. In CaF$_2$ the fluorine ions occupy tetrahedral spaces in a face-centred cubic lattice of calcium ions, and the co-ordination numbers are therefore 4 for fluorine, and 8 for calcium. The figure for calcium is less than that for a normal f.c.c. lattice, because the fluorine ions are closer than the normal positions in such a lattice.

Cope. The name given to the upper segment of a sand moulding box. (See also **Drag.**)

Copolymer. When more than one monomer is utilised to form a polymer, the result is called a copolymer. The number of ways in which two monomers, X and Y, may be joined together, gives rise to a nomenclature indicating the arrangement.

Alternation of the two monomers, XYXYXY . . . is termed a regular copolymer. Random distribution XXYXYYXYXXY . . . is a random copolymer. Alternation of polymeric units of X and Y, XYYYYXXXXXYYYYXX. . . is a block copolymer.

The properties of copolymers vary according to X and Y, and the particular arrangement as shown.

Copper. Copper is one of the few metals that occur native in workable quantities. It is found in masses and in veins in sandstone

in the Urals, in Sweden and on the shores of Lake Superior. As oxide it occurs in small amounts as cuprite Cu$_2$O and as melaconite CuO. In the form of carbonate and hydroxide, copper is found in malachite and azurite (q.v.), but the most important sources are copper pyrites CuFeS$_2$ and erubescite Cu$_3$FeS$_3$.

Copper is the only metal other than gold that possesses a natural colour. When pure copper appears a salmon-pink colour, the deeper reddish tint often noticed is probably due to a small amount of oxide film. The pure metal possesses high thermal and electrical conductivity, is malleable, has good mechanical strength and fair resistance to corrosion. All cold-working operations harden the metal and increase its tensile strength; at the same time ductility is decreased and the conductivity reduced by a few per cent. Annealing restores the softness of the metal and permits further cold working.

Copper is usually sold by refiners in the form of wire bars, cakes, billets, ingot bars, ingots and cathodes, and may be one of several qualities, the assessment of which is based on the conductivity and on the composition (percentage copper), viz, cathode copper (high purity), BS 1035; electrolytic, tough pitch HC copper (0.02-0.04% O) BS 1036; fire-refined tough pitch HC copper (conductivity 100% IACS) BS 1037; tough pitch copper (99.85, 99.75 and 99.5%, conductivity <100% IACS) BS 1038, BS 1039 and BS 1040 respectively; non-arsenical, phosphorus deoxidised copper (0.02-0.08% P) BS 1172; arsenical, tough pitch copper (0.3-0.5% As) BS 1173; arsenical, phosphorus deoxidised copper (low conductivity) BS 1174; oxygen free, high conductivity (99.95% Cu) BS 1861.

With zinc, copper forms a wide range of alloys called brasses (q.v.) containing up to 50% Zn with or without additions of small amounts of other elements (e.g. tin, lead, iron, nickel, manganese, aluminium and silicon). The properties vary from those of nearly pure copper to the high-tensile brasses with strengths up to 775 MN m^{-2}. Included in this series are the gilding metals and Muntz metal.

With tin, copper forms a series of alloys called bronzes (q.v.). Additional elements are often present, e.g. zinc, phosphorus, lead and nickel. Bronzes containing zinc are known as gun-metals. Cupro-nickel alloys are essentially alloys of nickel and copper with a wide range of compositions. They find wide application where strength, ductility and resistance to corrosion are required (e.g. bullet envelopes, coinage, etc.). The addition of 5-10% Fe greatly improves the resistance to corrosion by sea-water (cf. Cunife, Cunico). Those alloys containing a high percentage of nickel, e.g. Monel (68% Ni), are hard, of good tensile strength and resistant to corrosion and are used for valves, etc.

Nickel silvers are copper alloys containing zinc and nickel and are so called because of their appearance and ability to take a high polish. They have good mechanical strength and are resistant to corrosion, and find application for decorative purposes and for electrical contacts.

Silicon enhances the strength of copper, improves its resistance to acid corrosion and also its weldability.

Commercial copper begins to soften at about 200 °C, but in general the presence of small quantities of alloying elements raises the annealing temperature, increases the hardness and tensile strength and decreases the ductility and the electrical conductivity. (See also Tables 2, 5.)

Copper-clad steel. A product obtained by depositing electrolytically a coating of copper on steel, heating to the required rolling temperature and then squeezing the metals into intimate contact in a rolling-mill.

Copper-constantan. A thermocouple combination, consisting of pure copper coupled to Constantan. It has a limited application, being used in cases where the hot junction does not exceed about 315 °C. It can be used for measuring temperatures below 0 °C.

Copper glance. A sulphide of copper Cu$_2$S also known as chalcocite or vitreous copper.

Copper matte. A more or less impure mixture of copper I (cuprous) and iron II (ferrous) sulphides, containing 45-75% Cu. It is obtained as an intermediate product during the extraction of copper.

Copper nickel. Nickel arsenide (NiAs), one of the principal sources of nickel. Usually antimony and traces of cobalt, iron and sulphur are present. It is often associated in veins with cobalt, silver and copper ores. The principal sources are Ontario, Cornwall, Saxony and the Harz district of Germany. Mohs hardness

5-5.5, sp. gr. 7.3-7.6. Also known as kupfernickel and niccolite, the latter being the preferred name for the ore.

Copper oxides. Copper I (cuprous) oxide Cu_2O is a red powder insoluble in water. It oxidises slowly in air, to cupric oxide, and is reduced by heating in hydrogen to metallic copper. It imparts a deep red colour to silicate glass. Copper II (cupric) oxide CuO is black, insoluble in water but hygroscopic. It colours glass green or blue depending on other ions present.

A layer of Cu_2O on copper shows electrical rectification effects, but this use is now obsolete with improvements in semiconductors.

Copper powder. This is a fine red powder consisting of copper and prepared in several ways. The rather expensive method of reducing copper oxide (by heating in a stream of reducing gas) is not able to compete with the leaching process that starts with the copper ore. This is treated with water or dilute sulphuric acid until all the copper has been dissolved. The solution is filtered and scrap iron is added. The iron displaces the copper and goes into solution, while the copper is deposited as a fine powder. After washing and drying the powder is ready to be used in powder metallurgy operations.

Copper pyrites. A sulphide of copper and iron $CuFeS_2$ and an important source of copper. Also known as chalcopyrite.

Coprolite. A soft phosphate, i.e. one easily soluble in dilute sulphuric acid. It is found in many sedimentary rocks and consists mainly of calcium orthophosphate. It is of animal origin.

Coracite. A uranium ore, originally found on the eastern shores of Lake Superior. It is now known as pitchblende.

Core. The inner and unaffected part of a case-hardened steel which is always considerably softer than the surface layer called the case. The term is also applied to the inner portion of rolled rimming steel. In art founding, the term is applied to the solid mass of sand or loam that forms part of the mould. In general founding, a core is a shaped piece of sand placed in a mould in such a way that it will be encased by liquid metal, and when removed will leave a hollow in the resultant casting. In electrical work the core is part of a magnetic circuit. The material is usually iron — solid or laminated — and on this the coil windings are placed.

Core binders. Materials added to foundry sands to produce a good "green" bond during the moulding process, and which will produce a hard but not friable core on drying. Wheat flour, maize flour, linseed oil, etc., are commonly used, but substitutes may have to be employed due to non-availability of the traditional constituents. Such substitutes include potato starch, fish and mineral oils and synthetic resins. A satisfactory core binder should confer on the sand mixture the following properties: (a) good green bond; (b) flowability; (c) absence of stickiness; (d) good drying qualities; (e) absence of fumes on heating; (f) good dry strength.

Core-blowing. A mechanical device for rapid production of cores. The sand is mixed and blown into the mould at about $0.69\,MN\,m^{-2}$ air pressure, the time occupied being of the order of 3-5 seconds. The method requires an adequate volume of air free from moisture.

Core dressing. A solution applied to dried or baked cores for the purpose of providing protection against the action of the liquid metal and to assist in the formation of a smooth clear skin on the casting. The dressing may consist of ethyl silicate in a very dilute solution of hydrochloric acid in water or in a mixture of water and alcohol. The mixture is agitated until an emulsion is formed, and painted or sprayed on the core. The effect is to produce a hard but non-friable surface layer. If spirit is used the excess must be burned off.

Corhart. The trade name of fusion-cast refractories made from natural bauxites.

Coring. The microscopic segregation developed by the progressive freezing of zones successively richer in one metal when a liquid solution of two metals is solidifying to form a solid solution. When an alloy of any composition, say X (see Fig. 30), reaches the temperature t, freezing begins. The first solid formed is not of composition X, however, but of composition Y, as shown on the solidus curve at temperature t. Since the first crystallites are relatively rich in metal B, the melt will be correspondingly rich in metal A.

During rapid cooling, which prevents the attainment of equilibrium, each layer of solid metal deposited on the original crystal-

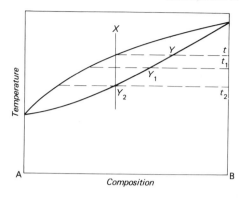

Fig. 30 *Constitutional diagram of alloys which show complete mutual solid solubility.*

lite differs in composition from the preceding layer and the alloy ultimately consists of a number of cored crystals, having the average composition X. In each crystal, however, the primary skeleton is rich in one component metal and the outer parts of each crystal are rich in the other metal.

Coring is a type of structure encountered in many alloys, e.g. cupro-nickels and tin-bronzes, which solidify as solid solutions over a fairly wide temperature range.

Cornish crucible. A type of crucible having a wide-open top originally developed in Cornwall for copper assaying. The material usually consists of about 72% silica, 25% alumina, 1% iron III (ferric) oxide, 1.1% potash and 0.4% lime. For more refractory vessels a little china clay is added.

Corronising. A patented process for producing protective coatings. A uniform layer of nickel is deposited on the basis metal, and a coating of tin superimposed on the nickel. The coated material is then heated to a temperature above $176\,°C$ (but below $230\,°C$, which is the melting point of tin) for a period sufficiently long to cause diffusion of part of the tin into the outer layers of the nickel coating.

Corrosion. Most metals are slowly attacked and gradually eaten away by liquids and gases to which they are exposed, and the general use of a metal is largely determined by its ability to resist attack by the environments in which it is likely to be used. Corrosion is quite distinct from erosion or abrasion, in which the metal is worn away mechanically, but in many cases a metal may be subjected to both corrosive and erosive influences.

The commonest example of corrosion is the rusting of iron which takes place in the presence of moist gases (of the atmosphere). It is a form of electrochemical attack. It is estimated that the protection of iron and steel against rusting costs over £1000-million a year in the United Kingdom alone.

The rate of corrosion is affected to a considerable extent by the conditions of exposure, the type of corrosive medium, the nature of the products of corrosion and by the presence of certain bacteria or marine growths. On exposure to air certain metals acquire a thin surface film, often only a few molecules thick, which protects the underlying metal from further attack. Such films may be "self-healing", i.e. they are able to repair any damage they may sustain. They may also be so thin as to be invisible to the naked eye.

Thicker films, such as rust on iron, are not self-healing, and because of their ability to retain moisture, afford little or no protection against further attack.

Atmospheric "rusting" is not the only kind of corrosive influence metals may have to resist. Painting, plating, anodising and many patented methods have been employed to reduce the influence of marine atmospheres and the various products of combustion in the air of towns. A number of corrosion-resistant alloys have been developed for use in a wide range of applications, in which ability to resist stress and corrosive influences at high temperatures for considerable periods of time is required. Certain metals, such as titanium and zirconium, are strongly resistant to mineral acids, and their presence as alloying elements often confers on the basis metal a greater capacity to resist corrosion.

Corrosion fatigue. Failure in service of fabricated parts that are subjected simultaneously to alternating stresses and to corrosion.

The endurance of a metal is considerably reduced when chemical attack is taking place during alternating cycles of stress. Failure will occur in time, at stresses below the fatigue limit of the material under normal (non-corroding) test conditions. The phenomenon is not due to general weakening by corrosion, but may be attributed to breakdown of the protective oxide film and to penetration of the corrosive agent into ultra-microscopic cracks in the surface. In such circumstances the usual relationships between ultimate tensile strength and endurance limit do not apply. There is no marked endurance limit and the useful application of the metal is governed mainly by its resistance to corrosion.

Corrosive sublimate. The very poisonous volatile chloride of mercury $HgCl_2$. Used in dilute solution for preserving skins and as a bactericide.

Corundum. A naturally occurring oxide of aluminium Al_2O_3 which forms rhombohedral crystals nearly as hard as diamond. Emery is an impure, fine-grained variety used as an abrasive (q.v.). When transparent, corundum forms a number of gems: oriental topaz (yellow), sapphire (blue, due to presence of oxides of cobalt, chromium or titanium), ruby (red, due to the presence of manganese) and oriental emerald (green). Mohs hardness 9, sp. gr. 3.9-4.1. Artificial corundum is made by fusing bauxite in an electric furnace, and marketed under the names Alundum, Aloxite.

Corundum in minute crystals often occurs in metals and alloys produced by aluminothermic processes.

Coslettising. A process for rust-proofing iron and steel. The articles are boiled for some hours in a solution of phosphoric acid saturated with iron II (ferrous) phosphate, when a protective coating of iron phosphate is formed on the surface of the metal. This coating forms a good base for painting.

Cottrell process. A method of precipitating the solid matter from the smoke or fume of smelting furnaces, blast furnaces, etc. The fume is passed into a chamber in which hang insulated conductors. These latter are charged to about 70,000 volts above the potential of the floor of the chamber. The solid matter is deposited on the walls and floor of the chamber and is removed from the former by tapping with a hammer. Lead conductors, suitably charged, are used in the electrostatic precipitation of acid fumes in the concentration of sulphuric acid. The acid droplets run off, from the conductors, into collecting tanks.

Cotunnite. Native lead chloride $PbCl_2$. A mineral found in the craters of certain volcanoes.

Coulomb. The practical unit of quantity of electricity. The c.g.s. electrostatic unit of quantity was such that if placed 1 cm from an equal and like quantity in air, it repelled it with a force of 1 dyne. This unit had no name, but 3×10^9 e.s. units = 1 coulomb. In current electricity the electromagnetic unit is employed. It was based on the electromagnetic unit of current, defined as that current, which in a circular wire, 1 cm radius, produced on a unit pole situated at the centre, a force of 2π dynes. The ampere was one-tenth of this unit. Thus 1 coulomb = 10^{-1} e.m. units = 3×10^9 e.s. units = quantity of electricity conveyed by 1 ampere in 1 second. (*See* Table 1 and SI units.)

Couverture. A term applied to the specially prepared slag used as a cover during the refining of antimony. It consists principally of antimony tetroxide, antimony trioxide, potassium carbonate, soda ash and charcoal.

Covalent bond. The sharing of electrons between like or unlike atoms, to achieve closed electron shell type structures (*see* **Bonds, Homopolar bond**).

Covelline. A naturally occurring sulphide of copper CuS. It is sometimes found in hexagonal crystals, but occurs more frequently in the massive state, as at Mansfeld Germany, in Vesuvian lavas and in Chile. It has a semi-metallic lustre and an indigo blue colour. Specific gravity 4.6. Also called covellite and indigo copper.

Covellite. *See* **Covelline.**

Covering power. A term used in electrodeposition to describe the extent to which the electrodeposited metal covers the surface of the cathode (cf. throwing power, which refers to the uniformity of distribution of the electrodeposit over the cathode surface). Also used to describe the spreading power of paint, and particularly the area which can be rendered opaque by a standard weight of basic pigment, e.g. titanium dioxide as against zinc oxide.

Cowper stove. A tall iron cylinder lined with firebrick and packed inside with chequer brickwork. The hot gas from the blast furnace, which contains about 28% combustible constituents, together with sufficient air to burn it, passes through the stove until the brickwork is heated to redness. The gas and air are then diverted to a second stove, where the same thing is allowed to occur. Meanwhile the air-blast to the tuyères is passed through the hot brickwork of the first stove, until the brickwork has cooled. The air-blast is thus pre-heated to a temperature of from $700\,^\circ C$ to $800\,^\circ C$. Cowper stoves work in pairs and are alternatively used to absorb and to emit heat.

Crack. The rupture of bonds between the atoms of a material, and the finite separation of the two surfaces created. Cracks may exist on the atomic scale, or grow to microscopic or macroscopic size. The area of the surfaces of separation obviously increases when macroscopic cracks are involved, until complete fracture results.

Fatigue cracks are those produced by the application of cyclic stressing below the level of the normal elastic limit. The cyclic stressing may be tension-compression (push-pull) or shear (bending or torsion). The initial requirement is for the setting up of high stress concentrations, which grow into microscopic and ultimately macroscopic cracks. These cracks can be detected by X- or gamma-ray radiography, by ultrasonic testing, magnetic testing and similar non-destructive methods, once they reach macroscopic size. Detection of microscopic cracks is not impossible, but dependent on particular conditions.

At fracture, the development of fatigue cracks can be seen by the smooth appearance of the surfaces which formed slowly, and the propagating crack front may be visible as roughly concentric or other symmetrical markings. The final fracture may be ductile, in which case the final portion shows marked deformation, or brittle, when a rough faceted type of fracture surface is visible. Without stress concentrations there can be no fatigue failure, but these can be produced either as part of the normal deformation process (dislocation movements) or by the presence of defects already in the material, such as non-metallic inclusions in metals, notches or pits due to bad machining, rough handling or corrosion.

Brittle cracks, or more strictly cracks producing brittle fracture, are also initiated as stress concentrations. The transition from elastic stress conditions directly to failure, without appreciable intervening deformation, is the characteristic of brittle failure. The stress concentration front, at which atomic bonds first separate, can be of such sharp radius, that further bonds separate in front of it, rather than deformation occur in the surrounding atoms, to relieve the stress concentration and thereby prevent propagation. In certain conditions or materials, the stress required to cause yielding, i.e. plastic deformation, may be higher than that to extend the crack front, and so brittle fracture may occur. The crack must propagate at such a speed, and under such a stress loading, that deformation is not possible, and to extend throughout the mass of material, the force required to separate the two surfaces must also be comparatively low. All these conditions require careful consideration. Stress concentrations can exist in the material from the beginning, even small cracks, as Griffiths first showed with the brittle behaviour of glass. Stress concentrations can result from particular dislocation movements, as in metallic oxides and ceramics. The rate of stressing has a marked effect, slow loading generally being more likely to give deformation. Temperature has an even greater effect, so that many materials which are ductile at room temperature (e.g. mild steel) become brittle at sub-zero temperatures. Equally, materials brittle at room temperature (ceramics) can be very ductile at extremely high temperatures. It is customary to speak therefore, of a transition temperature, below which brittle behaviour occurs, and above which the material is ductile.

It is not strictly correct to speak of ductile cracks, since high stress concentrations at an advancing crack front are not involved. What occurs is separation of two surfaces, after considerable plastic deformation. It is the change in shape of the surrounding material which is involved, and which gradually forms a cavity, by coalescence of vacant lattice sites created by the plastic deformation. These cavities grow, strictly by addition, and so increase the stress on the remaining structure. When the material is very pure, the plastic deformation produces localised reduction in area, which then causes further "necking" until the material finally parts at a point. This occurs in high purity copper and aluminium. Where a cavity forms in the centre of a mass, the growing cavity may meet the inward moving plastic deformation at the neck. In a cylindrical test piece, this produces a combination of mechanisms, the characteristic cup and cone fracture of mild steel for example. Some polymers fail in a ductile manner by a similar mechanism.

Griffith cracks are so called from their discoverer, who demonstrated the existence on the surface of glass of minute cracks which could propagate under quite low load. From this work has stemmed much of the present understanding of brittle fracture in all materials.

Craze or crazing. Minute hair-cracks, usually in a criss-cross pattern, which sometimes appear on enamelled ware and on badly painted or electroplated surfaces.

Crazing is also a common cause of failure of ingot moulds, when it describes the progressive deterioration of the working face, owing to the formation of a network of minor cracks. It is believed to be initiated by oxidation and growth beginning in the graphite cavities.

Creep. When a constant stress is applied to a ductile body over a period of time, plastic flow continues at a rate dependent on stress, time and temperature, and the process is described generally as creep. Application of load under creep conditions causes an instantaneous deformation significant to the design engineer in allowances for initial clearances. The first stage of creep is one of fairly rapid deformation, but at a rate decreasing with time, and hence known as transient, primary, or first stage creep.

At low temperatures and low stresses, the transient creep may follow a logarithmic law, $\epsilon = A \log t$ where ϵ is strain, t time, and A a constant. Pure f.c.c. metals and some c.p.h. metals behave thus at low temperature, as do some polymers and fibres such as glass.

At higher temperatures and stresses, the form can also be

$$\epsilon = Bt^{\frac{1}{3}}$$

where B is another constant. This was first demonstrated by Andrade.

After the transient stage, the flow settles into a steady-state, quasi-viscous, secondary or second stage creep, which is often of long duration. $\epsilon = Kt$ where K is a third constant. The rate of creep $d\epsilon/dt$ varies markedly with temperature and/or stress, and can be represented by $d\epsilon/dt = \text{constant} \exp(-Q/k\theta)$ for constant stress and $d\epsilon/dt = \text{constant} \exp(-\sigma/k\theta)$ for constant temperature where Q is an activation energy for steady-state creep, k is Boltzmann's constant, and θ is operating temperature.

Finally, at the end of the steady state, a rapid rise occurs in the rate of deformation, leading to fracture. This is the tertiary or third stage creep.

Creep can be explained in terms of dislocation movements, and the barriers to that movement imposed by solute atoms, precipitates, grain boundaries, etc., in other words, the detailed structure. The transient stage establishes the pattern, dislocations easily overcome the barriers to movement, but in doing so, create a tangle of dislocations themselves, and modify the barriers. When the thermal activation of the temperature conditions causes thermal recovery of these complex tangles, and diffusion of solute atoms etc. away from the dislocations, the dislocations can move again, and the steady state is an equilibrium between this thermal recovery and the recreation of new barriers to movement. Dislocation climb is thought to be one rate controlling process, cross slip another. Grain boundary sliding occurs in metals, and subsequently leads to vacancy formation and the creation of cavities. Coarse grained material is preferable to fine grain, because of the decreased surface area of boundaries involved. Diffusion of atoms has a marked influence on creep, and precipitation is desirable so long as coherent. In the tertiary stage, microcracks in the grain boundaries, probably joining the cavities created slowly in the steady state, begin to form, and creep accelerates again.

Polymeric materials deform with time, but generally behave as complex viscous fluids, particularly the thermoplastics. Cross-linked polymers show behaviour more like metals under stress, but have different mechanisms of deformation. Ceramics and some high temperature glasses creep less than metals at a given ratio of temperature to melting point, and if their brittle

behaviour at room temperature can be overcome, must replace metals for many high temperature applications.

Creep limit. The maximum stress which a metal can withstand at any given temperature without producing any measurable deformation. Alternately, the maximum elongation allowable may be specified, e.g. 10^{-4} mm per hour on a gauge length of 25 mm.

Creep tests. Methods of determining the extension of metals under a given load at a given temperature. The determination usually involves the plotting of time-elongation curves under constant load and the results are often expressed as the elongation (in mm) per hour on a given gauge length (e.g. 25 mm).

Short-time tensile tests do not provide reliable data for the design of structures which are to be used at elevated temperatures and strength values determined by long-time creep tests are preferable. These tests are most frequently made on bars under constant tensile load at constant temperature and may require from a few days to several years (*see* Fig. 31).

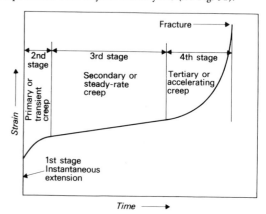

Fig. 31 *Time-strain graph showing stages of creep.*

Cristobalite. One of the allotropic forms of silica SiO_2, formed by transformation from quartz when heated to temperatures of the order of 1,250 °C or upwards. Cristobalite is covalent bonded, and hence shows high hardness and melting point. The structure is one of silica tetrahedra, the small silicon ion sitting in the centre of a tetrahedron of four large oxygen atoms, but each oxygen atom is shared by two silicons, i.e. forms part of two tetrahedra. This builds up an interconnected network, and in cristobalite the arrangement is cubic, the oxygen atoms being aligned on octahedral planes. This form of silica is of lower density than others, but is the form stable up to the melting point of 1,710 °C.

Critical air blast. The minimum rate of blast which will maintain combustion in an ignited bed of coke. For metallurgical cokes the test has been standardised and refers to coke of 10-14 BS mesh contained in a glass or quartz tube. Usual abbreviation CAB.

Critical constants. The values of certain physical properties of substance above or below which certain physical changes will not occur. The term has a wide application in many branches of science where the word critical is used in the sense of limiting value.

In the case of magnetic materials, the temperature at which the material loses its magnetic properties is often referred to as the critical temperature; the context usually suffices to indicate the change of property that occurs at the critical temperature. Critical angle in reference to prisms and lenses usually means the limiting angle of incidence at which refraction into a less dense medium will just not occur. In mechanics the same term refers to the minimum angle of a plane that will just allow a body to slide. In an oscillating system the "critical damping" is the minimum amount of damping that will cause the value of its potential energy to fall to zero without further oscillation.

The ideal gas equation, $pV = RT$, refers to a perfect gas, where the increase of pressure is assumed not to cause van der Waals' attraction as the mean free path decreases, and the actual volume of the gas molecules is disregarded as the total volume decreases. This is clearly not valid in practice.

Plots of pressure against volume, at constant temperature, produce a series of isothermal curves. At high temperatures, the perfect gas law is obeyed, but at lower temperatures, the attraction forces produce some condensation to the liquid state as the molecules are forced closer together and become a significant portion of the total volume. Figure 32 is based on the classic measurements of Andrews on CO_2 in 1863. At 13.1 °C, starting at point A, increase in pressure is seen to decrease the volume of gas along the line AB, but at point B, with the pressure 5.05 MN m^{-2} (5 046 kPa), liquid droplets appear. As the volume decreases, the pressure remains constant, the amount of liquid continuing to increase until all the vapour has been condensed (point C). The liquid is virtually incompressible, and hence as a very large increase in pressure is needed to decrease the volume, the isothermal has a very steep gradient. At higher temperatures, the isothermals look very similar, but the pressure at which liquefaction commences is increased, and the horizontal portion of the curve decreased. At a particular temperature, there is no sudden liquefaction, and at even higher temperatures there is a gradual approach to the perfect gas behaviour. The point at which the isothermal becomes continuous is called the critical point, the pressure, volume and temperature are also given the prefix "critical". Below the critical point, gas and liquid exist together, and the gas under these conditions is called a vapour, because of coexistence with the liquid, below the critical temperature. The value of pressure on the horizontals below the critical temperature, establishes the vapour pressure of the liquid at the particular temperature. The boiling point of the liquid is defined as that temperature when the vapour pressure reaches 0.1013 MN m^{-2} (101 kPa). Above the critical point, a gas cannot be liquified by the application of pressure alone, and below it, within the dotted curve, gas and liquid coexist.

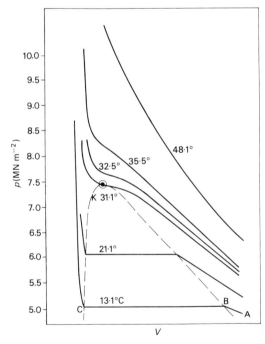

Fig. 32 *p-V diagram for CO₂.*

Critical cooling rate. The rate of cooling required to suppress phase changes. The term is most usually applied in ferrous metallurgy to the rate of quenching required to produce a martensitic structure. If the quenching is sufficiently rapid the austenite-pearlite transformation can be avoided, and at the lower temperatures the steel transforms to martensite and is hard. The addition of certain alloying elements, such as manganese or chromium, permits a considerable reduction in the critical cooling rate.

Critical point. In metallurgy this generally refers to a temperature at which some chemical or physical change takes place. These transformations cause evolution of heat on cooling or absorption of heat on heating, and appear as discontinuities or arrest points in the heating and cooling curves.

Critical range. The range of temperature within which reversible phase changes take place.

Critical shear stress. In a material under load, that value of shear stress acting on a given slip plane, and in a given slip direction, at which slip first takes place. It is a constant for the material. When a tensile load is applied to a cylindrical bar of material, the bonds between the atoms do not part, to form a crack perpendicular to the tensile force, except under special circumstances, and in brittle materials. The packing of atoms results in favourable planes and directions in those planes, along and across which the atoms can slip, or glide. This is a simpler mechanism for relieving the stress than the separation of atomic bonds, and hence is the common occurrence. To translate a normal stress into its components on particular planes, Fig. 33 shows the geometry of the slip plane involved, and the slip process can take place when $\tau = \sigma \cos \phi \cos \lambda$ where τ is the critical shear stress, σ the applied normal stress, ϕ the angle between the direction of applied stress and the normal to the slip plane, and λ the angle between σ and the slip direction.

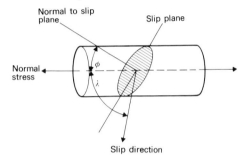

Fig. 33 *Resolved shear stress in a cylindrical specimen.*

The resolved shear stress has a maximum value when the slip plane is at 45° to the tensile axis. At a smaller angle, the atomic bonds would tend to be separated, since the plane would be approaching the perpendicular to the tensile axis. At a greater angle, the slip plane would be almost parallel to the tensile axis, and slip is extremely unlikely to result.

Individual crystals in a polycrystalline material will be oriented in different directions, and hence slip will commence first in those nearest to 45° to the tensile axis, where the shear stress is a maximum. Once slip commences, it is held up at grain boundaries, where the angle of the slip planes in the adjacent grains is different. Rotation of that or other grains then occurs, until another slip plane is favourably oriented, and slip occurs there. Such processes continue through a polycrystal, the distortion of the grains being clearly visible when viewed under a microscope. The smaller the grain size of the polycrystal, the greater the strength, as the more frequently must grain rotation occur, to bring slip planes into favourable orientation for slip.

Crocus. A reddish powder consisting of iron III (ferric) oxide Fe_2O_3 obtained as the solid residue after distilling green vitriol $FeSO_4$. The least calcined portions — scarlet in colour — are used as jewellers' rouge, while the coarser, more calcined portions, which have a bluish tint, are termed crocus. It is used as an abrasive for buffing steel cutlery and non-ferrous metals.

Cronak. A process for protecting zinc and zinc-base alloys against corrosion. It involves immersion in a solution consisting mainly of acid dichromate.

Croning process. A method of making moulds with a mixture of sand and a thermo-setting resin.

Crookesite. A selenide of copper and thallium $(Cu,Tl,Ag)_2 Se$ found at Skrikerum in Sweden. It contains about 17% Tl.

Crop. The ends of an ingot, bloom or billet which are cut off and discarded. These ends of ingots contain the pipe and other defects.

Cross-linking. The formation of bonds between individual polymer chains, to unite them and so give added resistance to deformation. Cross-linking can be deliberate, as in the production of thermosetting polymers, and in the stiffening of elastomers (the vulcanisation of natural rubber) or it can be accidental, as in the production of highly branched linear polymers (high density polyethene), or undesirable, as in the oxidation of natural rubber or other thermoplastic materials.

Cross-linking considerably modifies the properties of the basic material to which it is applied. Linking of branched polymers gives rise to a network, which is much more rigid, whilst not destroying completely the characteristic properties, but oxidation, when oxygen atoms, being divalent, form covalent bonds linking two individual chains, can convert the material to a very brittle one, since the number of oxygen bonds can continually increase under appropriate conditions. The vulcanisation of rubber, using sulphur, relies on very short cross-links, retaining the high elasticity of natural rubber, but slowing down plastic deformation. Excessive vulcanisation leads to an excess of cross-links, with almost complete removal of the elasticity. This was the basis of an early polymeric electrical insulator, ebonite. Figure 34 shows the cross-links in oxidised rubber. The vulcanised rubber has sulphur atoms in place of oxygen, and fewer of them.

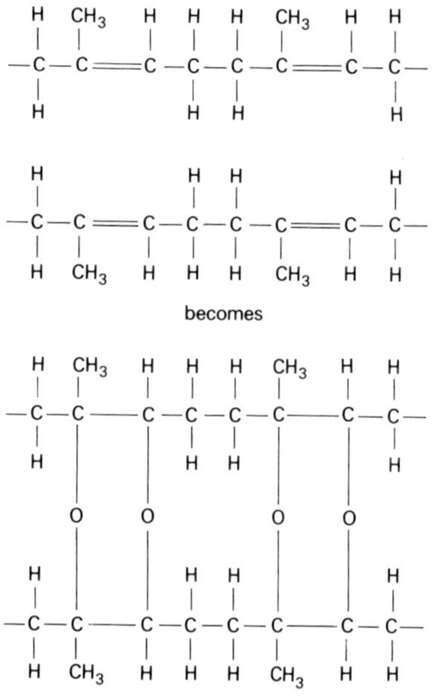

Fig. 34 *Cross-linking in oxidation of rubber.*

Cross rolling. A term applied to the transverse rolling of semi-finished material.

Cross slip. The motion of a screw dislocation into another slip plane having a common slip direction to the original one. It is a thermally activated process, and a means of restoring the equilibrium of a distorted structure or enabling further deformation to proceed after work hardening has created tangles of dislocations and slowed down their movement. It occurs in the thermal recovery process during creep, in annealing of metals, and is believed to be involved in the formation of fatigue cracks in the early stages.

Crucible. An open-topped refractory vessel in which material may be heated in furnaces and of such size that it may be lifted by means of tongs. A crucible should be able to resist the temperatures and variations in temperature likely to be encountered in service, and to resist the attack by the materials likely to be treated therein. It should have sufficient strength to withstand handling or grasping by tongs. Crucibles are made of a variety of refractory materials or of graphite. Small types are made of porcelain, platinum or nickel and are used in laboratories for igniting precipitated substances for analysis and for fusing alkalis, ores, etc.

Crucible furnace. A furnace in which metals contained in crucibles are melted. Such furnaces may be gas, oil, coke or coal fired.

Crucible steel. A steel of high quality made by melting wrought iron or blister bar with appropriate additions of ferro-alloys and carbon, in crucibles. A crucible holds a charge of about 45 kg

(100 lb). The process is now used only for special tool steels, having been largely superseded by electric furnace melting.

Crush. A casting defect evidenced by a deformation on the casting due to the displacement of sand when the mould was closed.

Cryolite. A fluoride of sodium and aluminium $AlF_3.3NaF$ which occurs as a pegmatic vein in Greenland. It is used in the manufacture of aluminium as the molten mass in which the purified alumina is dissolved. Also used in the making of sodium and aluminium salts and in the manufacture of white porcelain glass.

Cryptocrystalline. Consisting of very minute or indiscernible crystals.

Crystal. Many solids exist naturally in crystalline form. The definition of a crystal is that the atoms are arranged in a regular pattern, the regularity or order extending over distances many times the effective diameter of the atoms, a long range order. Because of the regularity of the atomic arrangement, crystals exhibit some characteristic physical properties. Since regular planes of atoms can be recognised, radiation and elementary particles are scattered by the planes, and can be arranged to produce diffraction patterns which identify the atomic arrangement from scattering angles and intensities.

Mineral crystals frequently exhibit external symmetry as a result of well defined planes of growth. The angles between corresponding faces are constant, the crystal can often be cleaved along these planes, and axes of symmetry can be identified by rotation of such crystals.

Polycrystalline materials are those in which crystallisation commenced from a number of common points, so that a large number of individual crystallites together make up the whole mass. Since the crystallites are identical in structure, but vary only in size and the orientation of their principal planes of symmetry, they act as if averaging all the properties which vary with orientation, and any anisotropy in the crystal will not show in a polycrystal of the same material. The boundaries at which the orientation of the individual crystallites change, are known as grain boundaries or crystal boundaries. In a two-dimensional cross-section, the individual units are referred to as grains, and the boundaries are seen as lines, not necessarily of a regular pattern, but generally approaching a polygonal outline.

The regular pattern of crystals is normally referred to a space lattice. The points of a space lattice need not be atomic sites, but in many cases it is more convenient to make the two coincide, particularly when like atoms are involved, and especially when only a single type of atom is involved, as with the elements which form crystals (usually metals). The classification of space lattices, the arrangement of the atoms, and the relationship between types, forms the basis of crystallography.

Crystal classification. A space lattice is regular along three-dimensional axes. The unit cell is the smallest grouping of atoms in the lattice which is repeated indefinitely along the three axes, to form the complete space lattice, and is therefore recognisable

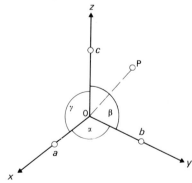

Fig. 35 *Crystallographic axes.*

as characteristic of the crystal system. There are other descriptions characteristic of each system, but which are not necessarily unit cells, but are often used in place of the correct definition.

The consideration of unit cell types serves to classify structures into groups, and to correlate similar patterns of behaviour and intrinsic properties in these groups.

The crystallographic axes to which the unit cell is referred, are designated by x, y, and z, and are related by the angles α, β, γ between them (*see* Fig. 35). The point P in the lattice of unit cell of sides a, b, c can be described by the vector sum

$$OP = u_a + v_b + w_c$$

(where u, v, w are the steps along the x, y, z axes). Any direction in such a lattice can thus be defined by the general expression $[uvw]$. The square brackets are used to indicate a direction, and u, v, and w are reduced to the smallest possible integers for convenience.

When a plane is to be described, the notation used is that of *Miller indices*. The plane is extended to cut the x, y, and z axes, and the intercepts in steps of the unit cell dimensions, a, b, c are written down. The reciprocals of these intercepts are next obtained, and fractions cleared to give the smallest integers again. Figure 36 shows three such planes in the cubic system where $a = b = c$, *(i)* parallel to the y axis and the z axis, *(ii)* parallel only to the z axis, and *(iii)* intercepting all three axes. In turn, the intercepts on plane *(i)* are 2 on the x axis, infinity on the y and z axes.

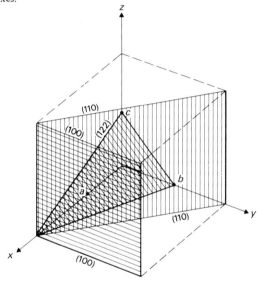

Fig. 36 *Miller indices of some planes in the cubic system.*

The intercepts are written 2; ∞; ∞
   reciprocals taken: ½; 0; 0
   remove fractions: 1 0 0
This plane is therefore known as the (100) plane, the parentheses indicating a plane.
   Intercepts in plane *(ii)* are 2; 2; ∞
   reciprocals taken: ½; ½; 0
   remove fractions: 1 1 0
Plane *(ii)* is therefore a (110) plane.
For plane *(iii)*
   Intercepts   2; 1; 1
   reciprocals taken: ½; 1; 1
   remove fractions: 1 2 2
Plane *(iii)* is therefore a (122) plane. (Round brackets are used for plane indices.)

The basic types of crystal system are few. There are six main forms although in the past others were recognised. The rhombohedral system, for instance, is a special form of the hexagonal, and was once classified separately. The essential values of the unit cell sides, a, b, c, and the angles α, β, γ are shown below.

Crystal systems

System	Unit cell	
Isometric (cubic)	$a = b = c$	$\alpha = \beta = \gamma = 90°$
Hexagonal	$a = b \neq c$	$\alpha = \beta = 90°\ \gamma = 120°$
Tetragonal	$a = b \neq c$	$\alpha = \beta = \gamma = 90°$
Orthorhombic	$a \neq b \neq c$	$\alpha = \beta = \gamma = 90°$
Monoclinic	$a \neq b \neq c$	$\alpha = \beta = 90° \neq \gamma$
Triclinic	$a \neq b \neq c$	$\alpha \neq \beta \neq \gamma$

In the hexagonal crystal system, it is often more convenient to use four axes than three, to emphasize the hexagonal symmetry. The z axis (side c of the unit cell) is retained, but three axes u, v, w, at 120° to each other replace the x and y axes, and so define the side a of the unit cell.

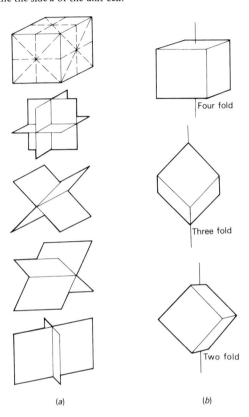

Fig. 37 *Diagrams showing planes of symmetry (a) and axes of symmetry (b) of a cube.*

A plane of symmetry divides a crystal into two similar and similarly placed halves, one half being the mirror image or reflection of the other. In the simple case of a cube nine planes of symmetry can be displayed (Fig. 37 (*a*)). If a crystal, when rotated through one complete turn, comes to occupy the same position in space more than once, the line or axis about which rotation has taken place is an axis of symmetry. In the case of a cube there are axes of four-fold symmetry, of three-fold symmetry and of two-fold symmetry (*see* Fig. 37 (*b*)). A crystal has a centre of symmetry when like faces are arranged in pairs in corresponding positions and on opposite sides of a central point. The main features of the six systems are illustrated in Fig. 38.

Crystal defect. Any departure from the perfect regularity of a crystal lattice, which locally destroys the long range order. Defects can be point (vacant lattice sites, interstitial atoms, substitutional atoms or impurity atoms, greater or smaller in size than the host), line or plane (dislocations, grain boundaries, twin boundaries).

Crystal directions. Miller indices can be ascribed to directions in the crystal, by a similar notation to that used for planes. The only directions of real interest are the close-packed ones, since these are concerned in slip systems (along with the close-packed slip planes) and influence some other physical properties.

To determine the Miller index of a direction, draw an imaginary line through the origin of the three crystallographic axes involved, parallel to the direction required. Take intercepts on each axis in turn, to establish the co-ordinates. If a cubic lattice is involved, sides a, b and c are equal, and from the origin of the x, y, z axes, imagine ua steps along the x axis; vb steps along the y axis; wc steps along the z axis. Hence, if u, v, w are the smallest integers (rationalise to make this so) then the direction index is $[uvw]$. (Square brackets are used for directions, to distinguish from planes.)

The diagonal of a cube face becomes [110], since it intercepts two axes at equal distance, and never intercepts the third. A cube edge, intercepting only one axis, becomes [100], and the diagonal from one corner through the body centre of a cube to the opposite upper or lower corner, [111] because it intercepts three axes equally. Since the [110] direction also has corresponding directions [101], [011], [$\bar{1}$10], [1$\bar{1}$0] etc., the family of related directions is represented by $< 110 >$.

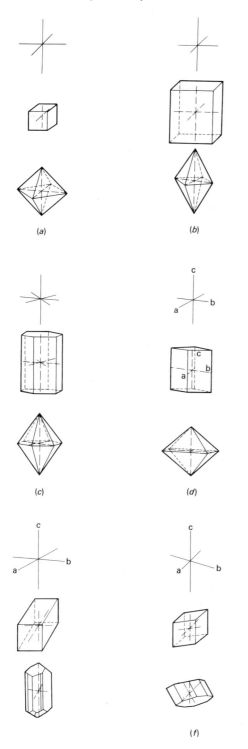

(a)

(b)

(c)

(d)

(f)

Fig. 38 *Crystal structure of the six crystallographic systems. (a) Cubic. (b) Tetragonal. (c) Hexagonal. (d) Orthorhombic. (e) Monoclinic. (f) Triclinic.*

Crystallisation. The production of crystals of long range order, by condensation of liquids (solidification), by sublimation from the vapour phase, or by transport of ions as in vacuum evaporation or electrolysis.

The process is thermodynamically favourable, in that lower free energy results, a reduction in entropy. A sharp transition occurs on freezing some liquids, with release of excess energy, as latent heat. The mechanism first manifests itself as an increase in viscosity of the liquid, as temperature is lowered, by almost an order of magnitude as the first crystals begin to form. Nucleation is required however, to cause solidification, and this can be homogeneous (from the material itself) or heterogeneous (from the impurities present, or added, or the walls of the container). Nuclei may always be present, but they increase in number as temperature is lowered, and often markedly so when the temperature is reduced below the freezing point (supercooling). Until the nuclei reach a critical size however, they are equally likely to become liquid again, a continual equilibrium existing between liquid and solid. The rate of nucleation therefore, and its dependence on temperature, is equally important with the rate of growth, and the relationship between the two determines not only the rate of crystallisation, but also the size of individual crystallites. The greater the number of nuclei, the more centres for growth, and the quicker the growing crystallites interfere with each other and form grain boundaries. Slow rates of nucleation may produce large crystallites, unless the rate of growth is equally slow.

Crystal growth takes place by addition on to nuclei, i.e. condensation of liquid at specific points, atom by atom. Since crystals have close-packed planes and directions, the specific points of growth are precisely those which build up the basic structure of the crystal, and any departure from this becomes a crystal defect. Experimental work has shown spiral growth by screw dislocation mechanisms to be a characteristic of several substances, even when formed under different conditions such as solidification and evaporation. The preferred crystallographic directions may become aligned with temperature gradients during solidification, e.g. heat rapidly abstracted from the walls of the container causes growth perpendicular to the walls. When this occurs, there may be a predominant axis of growth, and the crystals are called columnar. If there is no marked gradient the crystals are more uniform, and termed equi-axed. The first layer of crystals formed on a mould surface may give a very fine crystallite size, due to the rapid chill, and these are called chill crystals.

As the nuclei grow, both with columnar and equi-axed growth, they follow roughly the crystal axes or related directions, as described, and an infilling of solid between the growth points may take time. The result is a tree- or fern-like type of growth called dendritic, and the resultant crystal is a dendrite, from the Greek word for tree. The crystallisation produces a reduction in volume in almost all cases, and it is possible for holes to result, if liquid cannot reach the fine growth points between dendritic arms. Vacant lattice sites are very common in freshly crystallised materials, and if they coalesce, become small cavities. Porosity is therefore common in such materials.

If supercooling occurs without nucleation, then the attainment of long-range order is not possible, and the order prevalent in the liquid state may be partially retained. Such structures show little regularity, except over a few atomic distances, and the short-range order which remains is referred to as a "glassy state" amorphous, i.e. that of a supercooled liquid. The viscosity of such glasses may be lower than that of crystalline solids, and no sharp melting point can be observed. On reheating, these viscous supercooled fluids simply become less viscous. Some materials, like polymers, may partially crystallise, and yet become so supercooled in the structure that amorphous material is left between crystallites. These show a melting point, but also a "glass point".

The crystallisation of polymers is more complex, in that polymerised molecules are present in the liquid, as distinct from single atoms or much simpler molecules, in the case of metals and simple compounds. Supercooling is readily achieved with polymer melts, because of the difficulty in aligning long chains or complex shaped molecules, in what is the polymer crystal. The glass point, T_g, is the temperature at which the supercooling has so far increased the viscosity, that the material behaves to all intents and purposes as a solid. If crystallisation does occur, to greater or lesser degree, the release of latent heat at the melting point T_m can be detected, and cooling curves on polymer melts can demonstrate the production of some crystallisation by this means. In polymers, there is always a small degree of supercooling, but technical processing can be used to vary the degree of crystallinity and amorphous material, and so vary the properties. Carbon-based polymers melt at comparatively low temperatures, often between 100 °C and 250 °C, and the glass point may be below room temperature in many cases.

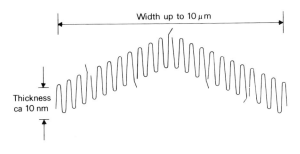

Fig. 39 *A folded-chain single crystal (polyethene).*

The crystallisation of polymers is not always dendritic, as with metals and similar simple systems. Some form tangled chains, in which there exists some parallelism between adjacent chains along a certain length, and the parallel portions are the crystalline volume. If all the chains can be grown parallel, then the material forms a lamellar single crystal. Slow growth from solutions can often achieve this. Normally, long chains fold up at regular intervals to produce these crystals, so that the degree of crystallinity cannot be absolutely perfect, at the bends in the folding for example. Polyethene forms folded chains, with a herring bone structure in cross-section, so that two kinds of symmetry result, the parallelism of the chain folds, and the pyramidal lamellae of the mass. The thickness of the crystals is of the order of 10 nm, and the chains may be of the order of 10 µm in length, so that 500 folds or so must occur in each chain (*see* Fig. 39). The most common crystalline feature in polymers is the spherulite (*see* Fig. 40), which develops from a nucleus, and from

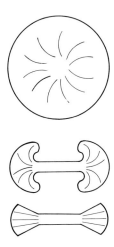

Fig. 40 *The development of spherulites in polymer crystallisation.*

which grow fibrils, or radial arms each formed from chains folded perpendicular to the axis of growth. The normal considerations of nucleation and growth will then apply, in determining whether a large number of small spherulites, or a small number of large ones, will be produced from the melt. If nuclei exist on a surface, parallel growth of fibrils can result, in place of spherulites. In natural cellulose, the chain folds are also parallel to the fibril axis and there is no tendency to form spherulites. This, however, is a biological formation, not related to growth from a melt.

Crystallography. The science of crystal structure.

Crystal planes. The two-dimensional figures intercepting the atoms in a crystal, and represented by Miller indices obtained by rational intercepts on the crystallographic axes. (*See* **Crystal classification.**) Generally, the close-packed planes, those with the greatest density of atoms in the system, are the important ones. The symmetry of crystal structures means that the Miller index of a plane represents only one of a whole family of related planes. To distinguish the whole family, the indices are written (111) for an individual plane, and {111} to indicate the family. This example, {111}, represents the family of octahedral planes, the close packed ones of the face-centred cubic system.

Cubic system. One of the six main classes into which all crystals may be divided. Crystals of this system are referable to three mutually perpendicular axes with equal parameters.

Cupel. A thick-bottomed, shallow vessel used in assaying laboratories and in the cupellation process for refining silver. The lining may be made of finely powdered bone-ash, marl, barytes or cement.

Cupellation. A process for refining silver by alloying it with lead and then removing the lead by oxidation. The oxide of lead and of any other base metals present are removed from the surface, while the silver remains unoxidised.

Cup fracture. When the exterior portions of a tensile-test specimen are extended beyond the interior part, so that the fracture resembles a cup, this type of fracture is referred to as a cup fracture.

Cupola. A shaft furnace used for melting metals, as distinct from a blast furnace in which ore is smelted. Metal, coke and flux are charged from the top of the furnace on to a bed of hot coke through which air is blown.

Cupping test. A mechanical test used to determine the ductility and drawing properties of sheet metal. It consists in measuring the maximum depth of bulge which can be formed before fracture. The test — e.g. Erichsen's test (q.v.) — is commonly carried out by drawing the test piece into a circular die by means of a punch with a hemispherical end.

Cuppy wire. A term applied to wire which, though apparently sound, shows on a longitudinal section a series of well-developed internal fractures which open to the surface under bending stresses. It may be due to the effects of axial segregation and pipe or to the use of badly designed dies and excessive reduction of area.

Cupralum. The trade name of a lead-clad copper. It is used in the chemical industries as an acid-resistant material and for the anodes in the electrodeposition of chromium.

Cuprite. Oxide of copper Cu_2O containing about 89% Cu. It occurs in the zone of weathering of copper lodes in Cornwall (UK), France, Spain, the USA, Chile, Peru and South Australia. It crystallises in the cubic system. Mohs hardness 3.5-4, sp. gr. 5.8-6.15.

Cupro-nickel. A series of copper-nickel alloys containing from 15% to 70% Ni. Copper and nickel form a continuous series of solid solutions and all the alloys are characterised by high ductility and good resistance to corrosion. The commonest alloys contain 20-30% Ni and are mainly used in condenser tubes. The 25% nickel alloy is used for the nickel coinage of the USA, and also in the new "silver" coinage of the United Kingdom.

Alloys containing more than about 15% Ni are white in colour and find many applications in ornamental castings. The 80% Cu, 20% Ni alloy is readily workable and is very suitable for cold pressings and stampings. On account of their high electrical resistance and low-temperature coefficient of resistance, alloys containing 40-50% Ni find many useful applications in the electrical industries. The higher nickel alloys containing about 70% Ni have high resistance to scaling and oxidation (*see* Tables 5 and 7).

Cure — of thermosetting polymers. Cross-linking in some polymers is carried out as a separate process, after preparation of polymeric material, by addition of a hardener, with or without a small quantity of a catalyst. The cross-linking is usually speeded up by the catalyst, and/or the application of heat. The process of heating, or the addition of the hardening agents, or both where they are used, is termed curing.

The word is also used in the vulcanisation of rubber, using sulphur compounds as cross-linking agents, and heat to mould the product and so cure and form a shape at the same time.

Curie. This is the old unit of radioactivity (not SI) and was defined as the quantity of any radioactive nuclide in which the number of disintegrations is 3.700×10^{10} per second.

Curie point. A transformation point below which iron, cobalt, nickel and alloys containing these metals are ferromagnetic and above which they become paramagnetic. In the case of pure iron the change occurs at about 770 °C.

Curie-Weiss law. Magnetic susceptibility is dependent on temperature. Curie defined paramagnetic susceptibility as $\chi_p = C/T$ where χ_p is the paramagnetic susceptibility, C the Curie constant and T the absolute temperature. Certain materials retain their magnetism permanently after removal of an externally applied field, but only below a critical temperature, the Curie point or Curie temperature. In these ferromagnetic materials, the susceptibility is defined by the Curie-Weiss law,

$$\chi_F = \frac{C}{T - \theta_c}$$

where θ_c is the Curie temperature.

Ferroelectric materials, in which polarisation of electric charges replaces the magnetic moments of the constituents of ferromagnetic materials, obey a similar law. Antiferromagnetics behave in the opposite way, being antiferromagnetic below a critical (Néel) temperature, and paramagnetic above it. In these cases the negative in the denominator becomes positive.

Curium. A metallic element, symbol Cm, at. no. 96 and at. wt. 242. This very rare metal has been isolated in extremely small quantities from neutron-irradiated Americium during work on atomic-energy projects. The metal was prepared by reduction of the fluoride CmF_2 with barium metal vapour at 1,275 °C.

Current (alternating). An electric current which grows to a maximum value, decreases again, changes its direction (with respect to a datum line), reaches a maximum value in the new direction, and then returns to the original value. This cycle is repeated indefinitely so long as the current flows. The interval between attainment of a given value of current on two successive cycles is termed the period, and the number of cycles per second, the frequency measured in units of Hertz (Hz). The amplitude is the deviation of the maximum, or of the minimum, from the datum line.

The graphical representation of current against time, for normal types of alternator, indicates the a.c. cycle to be of sine wave form. The current lags behind the potential difference producing it. The angle of lag is called the phase difference. (*See* **Angle of loss** for mechanical analogy.)

Current (direct). An unidirectional electric current, involving no periodic variations in polarity of the potential difference producing it.

Current density. A term used in electrodeposition to denote the average current flowing in the electrolyte, expressed as amperes per square decimetre of cathode or, more occasionally, of anode surface.

The term generally refers to an average figure, since the current density may vary considerably from place to place on the surface of an irregularly shaped electrode. High concentrations of current may occur at points or sharp corners and on areas where the cathode and the anode are in relatively close proximity. In electrodeposition processes where uniformity of current distribution over the whole surface of the work is most desirable, the disposition of the anodes in relation to the work is very important.

Current efficiency. The ratio of the mass of a substance liberated

or chemically changed in an electrochemical process to that which would be anticipated from Faraday's Law.

Cutting wheel. A thin composition disc impregnated with carborundum, diamond dust or Aloxite which can be rotated at high speeds. It is used for cutting hardened steel specimens, etc. Considerable care is required to prevent the specimens from becoming heated, and their structure consequently modified, during cutting. Continuous cooling with water or cutting fluids is essential. Also called slitting wheel and cut-off wheel.

Cuttlefish process. A process in the production of jewellery in which gold or platinum is cast in moulds made of the porous shells of cuttlefish.

Cyanide hardening. A process for increasing the carbon content of the surface layers of low-carbon steel components by immersing them in a bath of molten sodium cyanide. When low-carbon steels are treated in this way they absorb both carbon and nitrogen from the fused salt, so that when they are quenched direct from the cyaniding operation or given other suitable heat-treatment they develop a hard case.

Cyaniding. The term is applied to a process for extracting gold from pulverised ores, which depends on the property of cyanide solutions of dissolving gold. (*See also* **Cyanide hardening.**)

Cyanite. An aluminium silicate $Al_2O_3.SiO_2$. It crystallises in the triclinic system and usually occurs in the form of long, blade-like crystals in gneiss and schists. It is used, like andalusite and sillimanite, as a refractory material. Also known as disthene and kyanite.

Cyanosite. Hydrated sulphate of copper $CuSO_4.5H_2O$. Also known as chalcanthite and blue vitriol.

Cycling (mechanical). Variations in quality and/or quantity of mechanical stress, often produced by regular and systematic changes in a machine, building or structure. The most common is an alternating stress, as in reciprocating machinery and structures such as bridges. When tension-compression are involved, there may be problems of fatigue, and the same effect can be produced by vibrations in structures, leading to shear stresses of varying signs and magnitude, e.g. aircraft wings and control surfaces. The design of structures subjected to cycling is complex, if unexpected failures are to be avoided. (*See* **Crack.**)

Cycling (thermal). Variation in temperature which leads to expansion and contraction in the materials of the structure, and which inevitably gives rise to some mechanical stress, especially in a rigid framework or complex structure such as rotating and reciprocating heat engines. Thermal cycling is all too often ignored in design, but changes in ambient temperature in some climates can be considerable and stresses very high. Thermal cycling inevitably leads to alternating stress, and therefore can contribute to fatigue failure. The greater the specific heat of a material, and the higher the temperature coefficient of expansion, the greater the effect of thermal cycling in a structure. In some scientific instruments, the thermal stress is not high, but the effect of thermal cycling can upset the accuracy of readings, e.g. the balance wheel mechanism of time recording instruments. The fatigue of bearing metals is tied to thermal cycling, and anisotropic metals such as tin fail by thermal fatigue as a direct result.

Cylindrite. A complex sulphide of lead, tin and antimony found in Bolivia, usually in stanniferous veins and associated with other metallic ores.

Czochralski crystal growth. A technique in which a single crystal can be formed by extracting a solidification "tip" in the form of a wire or rod, from the surface of a melt held at constant temperature just above the freezing point. Solidification occurs on the tip, which is extracted at a rate equal to the rate of solidification. Also known as "crystal-pulling".

D

Dabbers. Projections cast into the surface of a loam plate to hold and reinforce the mould. The loam plate is a foundation plate for a core or mould and is always separately cast in iron to suit each job; in making the mould the face is "dabbed" with a sharp rod to make projections on the plate.

The term is also used to describe a kind of rammer used by foundrymen for ramming the sand between the mould and the pattern.

Dalton's laws. Dalton's laws of chemical combination are two in number: *(a)* the law of Definite Proportions states that when two or more elements enter into chemical combination to form a particular compound, they do so in fixed proportions by weight; *(b)* the Law of Multiple Proportions states that when two elements combine to form more than one compound, the amounts of one element that unite to a fixed weight of the other element bear a simple ratio to one another. Dalton also enunciated the Law of Partial Pressures, which states: "In a gaseous mixture each gas exerts a pressure equal to the pressure it would exert if it were confined alone in the space occupied by the gaseous mixture".

Damascening. An old process of ornamenting a metal surface with a pattern. In the early Middle Ages swords so adorned were made in Damascus (whence the name Damascening) by repeatedly welding, drawing out and doubling up a bar composed of a mixture of steel and iron, the surface of which was afterwards treated with an acid. The surface of the iron retains its bright metallic lustre under the action of the acid, but the steel surface is darkened with a pearlitic coating of carbonaceous matter. In the East the process of inlaying metal on metal is widely carried on, particularly in Iran and parts of India, where it is known as Kuft work.

Damper. An iron plate or shutter fitted across a boiler flue which may be adjusted to regulate the draught.

Damping. Under angle of loss, the lag between the driving force (stress) and the displacement in a material (strain) is described. Mention is made of the damping capacity of the material, the ability to damp out vibrations imposed on it. If the stress σ is plotted against the strain ϵ where such an alternating driving force is involved, an elliptical shaped figure results. The area of the ellipse is the energy dissipation per cycle, and if $\bar{\epsilon}$ is the amplitude of vibration, the relative decrease in amplitude $d\bar{\epsilon}/\bar{\epsilon}$ is equal to $\pi\delta$ where δ is the angle of loss.

The value $d\bar{\epsilon}/\bar{\epsilon}$ is called the logarithmic decrement, since

$$\bar{\epsilon} = \bar{\epsilon}_m \exp. \frac{(-\pi\delta t)}{T}$$

where $\bar{\epsilon}_m$ is maximum amplitude, t the actual time, T the cycle time. The time required for the amplitude to diminish to $1/e$ of its original value is called the relaxation time t_r, and the distance l over which it occurs is called the relaxation distance.

The damping of vibrations, in the absence of any external forces, is dependent solely on the internal friction of the material. For metals, δ is of the order of 10^{-2} to 10^{-4}, and for quartz, about the same. Brick and concrete show angle of loss of about 10^{-1}, wood about 1, cork 3 to 4, and rubber 5 to 6. The presence of impurities, internal stresses and grain boundaries affect the value markedly. Impurities increase δ, as do internal stresses from cold work or heat treatment. The fewer grain boundaries, the larger the angle of loss. A more complex crystal lattice decreases the value of δ, as does rise in temperature. Increasing the amplitude of vibrations tends to increase δ, and magnetisation of a ferromagnetic material decreases damping.

Damping is related to the after-effect of elastic stressing (*see* **Elastic after effect**). Components whose damping is required to be small, e.g. cymbals, bells, gongs, are made of materials of low damping capacity, of which copper alloys are good examples. High damping is required for anti-vibration mountings, sound insulation, and shock absorbers, hence rubber, cork, foamed polymers, are used. Wood is a suitable alternative, but less effective unless very porous, as with balsa.

Damping capacity. The percentage of the internal energy possessed by a metal at the beginning of an oscillatory cycle which is absorbed by internal friction during the cycle.

The amount of energy expended during each complete cycle of stress is indicated by the area enclosed by the hysteresis loop in the stress-strain curve. Metals with a high damping capacity usually have a low notch sensitivity, but the relationship to ductility is not very definite. The hard, highly notch-sensitive steels have low damping capacity; the softer less notch-sensitive steels have a higher damping capacity, while the cast irons, highly notch-insensitive, have very high damping capacity. Stainless steels of the 18:8 type are very notch-insensitive and have a high damping capacity, but so have the magnesium alloys, which have a high notch sensitivity. Alclad materials have high damping capacities, so they tend to inhibit the building up of resonant vibrations.

Damping down. The temporary stopping of a furnace by reducing the air supply or cutting off the blast, thereby lowering the temperature.

Damping test. A test to provide an indication of the rate at which energy is absorbed when a metal is subjected to alternating stresses below the elastic limit. The specimen swings through a very small arc as a pendulum, and the rate at which the amplitude of the pendulum decreases is noted.

Danburite. A borosilicate of calcium $CaO.B_2O_3.2SiO_2$ which melts readily to form a clear glass. It is found in pegmatites.

Darkfield illuminator. A piece of equipment incorporated into some large inverted metallurgical microscopes, which illuminates the specimen obliquely from all sides. The advantage of this type of illumination is that smooth surfaces appear dark, and grain boundaries and constituents slightly in relief appear light.

Datolite. Calcium borosilicate. Borosilicates in glass confer upon the product a high resistance to the action of water and ability to withstand higher temperatures without melting. Borosilicate lime-glass has a refractive index of about 1.49 and a dispersion C–F of 0.007.

Dead burnt. A term used in reference to the calcining of magnesite and limestone. The presence of clay in the minerals may cause them to vitrify if too strongly heated during the calcining process. A hard crust of silicates may here form on the surface. In the case of limestone this causes the lime to slake very slowly, and it is then referred to as "dead burnt", in contradistinction to the pure lime, which slakes readily and is known as a "fat" lime.

Dead head. A projection on a casting formed by metal from the runners or risers.

Dead roasting. A roasting process in which the sulphur in an ore is, as far as possible, eliminated. The process is carried out with a free supply of air and should be distinguished from sulphate roasting.

Dealuminising. A type of corrosion sometimes encountered in aluminium bronze in which the aluminium content of the alloy appears to be preferentially attacked. The phenomenon is analogous to dezincification as met with in brasses and manganese bronzes.

de Broglie. First postulated the duality of electron behaviour, i.e. as a particle and as a wave motion, in his Ph.D. thesis. From this concept of the electron as a wave

$$\lambda = \frac{h}{mv}$$

where λ is wavelength, h Planck's constant, m the mass of the electron and v the velocity of the electron, stemmed the basis of electron diffraction, electron microscopy and wave mechanics.

Debye-Scherrer-Hall powder method. This is a useful technique applied to powder metallurgy. A beam of X-rays is directed upon a sample of the powder and the diffracted beam is received upon a photographic plate. The particular effect produced is a result of the crystals lying in all orientations, and thus showing on the plate a series of concentric rings. The pattern is characteristic of the material and may be used to identify it. The method may also be used to obtain close estimates of the lattice dimensions.

Decalescence. A term used in reference to the absorption of heat without a corresponding increase in temperature when a metal is heated through the critical points.

Decarburisation. The loss or removal of carbon from the surface of steel during hot working heat-treatment or reheating or whenever the metal is heated in an oxidising or other decarburising atmosphere.

Decay. Decomposition by corrosion. The term usually refers to welds. (*See also* **Weld decay.**)

Dechenite. One of the principal vanadium minerals. It is a vanadate of lead $(Pb,Zn)(VO_3)_2$.

Decomposition value. The least voltage that will just cause a continuous electric current to flow through a molar solution of an electrolyte. (*See also* **Decomposition voltage.**)

Decomposition voltage. From Faraday's law it is shown that 96 516 coulombs of current will deposit the equivalent weight of an element from the electrolyte through which the current passes. Since energy = electric charge × potential, and the voltage required to cause the applied current to flow through the electrolyte varies according to the value of the back e.m.f. generated by the electrode products, there is a minimum voltage that must be applied before electrolysis can proceed. This is called the decomposition voltage. The back or opposing e.m.f. is that which would be generated in a cell composed of the electrode products in the same electrolyte, and this value must be exceeded before rapid electrolysis can begin. When the current is plotted against the voltage there is a sudden break in the curve where the decomposition voltage is reached.

Deep drawing. A cold-working process, involving considerable plastic deformation of metals in which sheet or strip metal is drawn into the desired shape by means of dies.

Deep-drawing steel. A high-quality steel, soft and ductile, which is capable of severe cold, visible, plastic deformation in the form of sheet or strip.

Deep-etch test. The treating of the surface of a metal with an appropriate acid solution in order to reveal defects.

Deformation (elastic). An atomic bond between two like atoms, separation distance between centres a, is subjected to a balanced force f (acting on each of the two atoms, so as to separate them and stretch the bond). A balanced force has no resultant and no couple, so the body containing the bond is not set in motion.

When $f = 0$, the equilibrium value of $a = a_0$. When f is finite, but small, the new equilibrium spacing is a, and the applied forces are balanced. Displacement $u = a - a_0$. Equilibrium requires that

$$f = \frac{d\phi(u)}{du}$$

where $\phi(u)$ is the bond energy at displacement u. When f is tensile, $a > a_0$; when compressive $a < a_0$.

$\phi(u)$ varies continuously with u, and a Taylor series can be used to express this.

$$\phi(u) = \phi_0 + \left(\frac{d\phi}{du}\right)_0 u + \tfrac{1}{2}\left(\frac{d^2\phi}{du^2}\right)_0 u^2 + \text{higher terms.}$$

ϕ_0 is the bond energy at $u = 0$, and all differential coefficients are measured at $u = 0$. $d\phi/du = 0$ at $u = 0$ (since $f = 0$ at $u = 0$), hence the second term can be eliminated and, because u is very much smaller than a_0, higher terms are eliminated (since they contain even higher powers of u, a very small quantity).

Hence $\phi(u)$ is approximately equal to

$$\phi_0 + \tfrac{1}{2}\left(\frac{d^2\phi}{du^2}\right)_0 u^2.$$

Differentiate: $\dfrac{d\phi(u)}{du} = \left(\dfrac{d^2\phi}{du^2}\right)_0 u$ (approximately),

and since $f = \left(\dfrac{d\phi}{du}\right) u$, then f is approximately equal to $\left(\dfrac{d^2\phi}{du^2}\right)_0 u$.

$\left(\dfrac{d^2\phi}{du^2}\right)_0$ is thus the curvature of the ϕ, u curve at the minimum point and, being independent of u, is therefore a constant. Hence f is approximately equal to a constant × u. Force is therefore proportional to displacement when the latter is small, to satisfy the conditions applied above.

This was first demonstrated experimentally by Hooke, and known as Hooke's law. Deformation which obeys this law, i.e. displacements of atomic bonds which are directly dependent on the force applied, is termed elastic, and is completely reversible when the force is removed. Solids in which this occurs are called elastic or Hookeian solids, and the deformation is called linear or Hookeian.

If the force is expressed in terms of the unit area of a mass of solid, then the force per unit area is the stress, and if displacement is expressed in terms of a unit length of that area of solid subjected to the force, displacement per unit length is the strain. (*See* **Young's modulus, Bulk modulus, Elasticity.**)

Deformation (plastic). The point at which reversible elastic strain ceases, is the point where plastic deformation, which is irreversible, begins. Due to difficulties in testing procedures between different manufacturers, the measurement of stress to produce a given value of permanent set is often used in place of elastic limit. A stress strain curve is plotted in tension, beyond the elastic limit and then a line is drawn on the strain axis, parallel to the slope of the elastic portion for a given small value of strain. Where this cuts the stress strain curve, is the stress required to produce that portion of permanent plastic strain which was the intercept on the strain axis used to draw the elastic line. The values of permanent set usually used, are 0.1, 0.2 or 0.5% strain. The stress is called a proof stress, to produce the given value of permanent set, hence 0.1% proof stress would be quoted if 0.1% was the value used in the test. (*See* Fig. 58.)

Ductile materials produce considerable plastic deformation beyond the proof stress levels, since the definition of ductility is the potential for plastic deformation. Materials which are not ductile however, pass from the elastic stage through a very small amount of plastic deformation before they fracture — these are the brittle materials. Brittle behaviour does not imply weakness, as is so often assumed, and ductility most certainly does not imply strength. Brittleness implies that the stress to break atomic bonds and separate the surfaces produced (often concentrated by a defect) is lower than the stress to cause deformation, whereas ductility implies a lower stress to cause plastic deformation rather than separate bonds.

A similar type of argument as that used to demonstrate Hooke's law theoretically can be used to estimate the strength of ductile materials in plastic deformation. The atomic planes are assumed to move along the close-packed planes, and in close-packed directions, such that each atom in the plane "climbs" up the hill of the atom in the plane below (or above) and so has to surmount the peak, or top dead centre, before rushing down the other side, then having achieved a displacement of one atomic spacing. Up to the top dead centre (half an atomic spacing) is assumed to be the elastic portion of the deformation, the atoms being able to slide back down again if the load is removed. The type of deformation in which whole lattice spacing movement is achieved, step by step, was known as slip, and the atomic planes were assumed to slide over each other like a pack of cards.

A crystal of lattice spacing "a" between planes, and "b" between atoms in the planes, is subjected to a shear stress, to cause slip. The shear stress will vary periodically with the displacement u, and the period will be b, the lattice spacing in the planes. τ, the shear stress, will be 0 at $u = 0$, b, $2b$, $3b$, etc., and also 0 at midway positions, when $u = b/2$, $3b/2$, $5b/2$, etc. τ will be greater than 0 when $b/2 > u > 0$, but less than 0 when $b > u > b/2$. The variation of τ with u will be roughly a sine curve, and the initial slope in each period will be equal to the shear modulus, this being the elastic portion.

Assuming the sine curve pattern, $\tau = \tau_{max} \sin (2\pi u/b)$ but G, the shear modulus is defined by $\tau = G\gamma$ and $\gamma = u/a$, so $\tau = Gu/a$. When γ is small $\sin (2\pi u/b) = (2\pi u/b)$. Hence $\tau = Gu/a = \tau_{max} (2\pi u/b)$, so that $\tau_{max} = (Gu/a)(b/2\pi u) = Gb/2\pi a$. As a most often equals b, τ_{max} is approximately equal to $G/2\pi = 0.167G$. A number of assumptions made can be refined, but τ_{max} is approximately between $0.1 G$ and $0.03 G$. For metals this gives a tensile strength of the order of $0.1 G$. The measured values of tensile strength (or shear stress) are less than this, by two or three orders of magnitude, and so the theoretical picture is incorrect.

The deformation process is in fact even simpler than that described, but it is the imperfections in the lattice which make this so. Not only do these imperfections illustrate the reasons for the deformation process, they explain many of the modes of behavior of crystals in terms of stress, thermal and electrical fields, with respect to time and temperature. The defects are of two main kinds, point and line defects, and reference to the section on crystal defects and dislocations will explain them in detail. Plastic deformation ceases when the potential for further deformation is exhausted, i.e. fracture results. For ductile failure *see* **Crack**, and the contrast with brittle behaviour, under **Fracture brittle**. (*See also* **Ductility, Necking, Reduction in area.**)

Degassing. The removal of dissolved gases from a molten metal prior to casting. The process usually involves the use of deoxidising agents, such as phosphorus, silicon or aluminium. Oxidising slags or fluxes may be used to remove hydrogen. In the case of some of the rarer metals heating to a high temperature *in vacuo* is effective in the removal of hydrogen. Chlorine, and a number of proprietary compounds that decompose to yield chlorine, are used for degassing aluminium.

Degradation (of polymers). The irreversible destruction of chemical bonds, which in polymers destroys the large molecules by separation of cross-links in a thermosetting or network polymer, or breaks up a long chain polymer by rupture of carbon-carbon bonds. The destruction of bonds is most readily achieved by rise in temperature, but can also be effected by radiation, and bombardment by particles such as electrons and neutrons.

Degreasing. The removal of mineral and vegetable oils and greases from the surface of metal, in particular the oily film left on metals after machining operations. The process is an essential preliminary to electroplating, enamelling, tinning and galvanising. The substances used are either alkaline solutions or organic solvents (e.g. carbon tetrachloride (tetrachloromethane), trichlorethylene (trichloroethene), etc.).

Degrees of freedom. A term used in the interpretation of the phase rule. For the attainment of equilibrium in a system it is necessary

that the energy intensity factor shall be the same for all the different phases in which the components of the system may co-exist. This energy intensity factor is called the "chemical potential", and for a component of a system in any phase (i.e. in any of the physically distinct and mechanically separable states in which it may exist) the potential depends on the temperature and pressure as well as on the composition of the phase. In a system of C components existing in P phases, to fix the composition of unit mass of each phase it would be necessary to know the masses of $(C - 1)$ components in each of the P phases. Hence, considering the composition of the whole system, there are $P(C - 1)$ variables, in addition to the two other variables, temperature and pressure. To define the system completely there must be the same number of equations as there are variables, i.e. $P(C - 1) + 2$.

If there are fewer equations, the difference between the above number and the actual number of equations will give "degrees of freedom" of the system.

The equations referred to are obtained from the relationship between the chemical potential of a component and the composition of the phase, the pressure and the temperature. For equilibrium the potential of each component must be the same in the different phases in which it is present. If one of the phases in which all the components occur is chosen as standard, then for each phase in equilibrium with the standard phase there must be a definite equation of state for each component. Since there are P phases for each component, the number of equations obtainable will be $C(P - 1)$. The degrees of freedom, F, will therefore be $P(C - 1) + 2 - C(P - 1)$ or $F = C + 2 - P$.

Degree of polymerisation (D.P.). In the build-up of a polymer molecule, the number of mer units in the molecule is known as the degree of polymerisation. The D.P. value also affects the molecular weight of the polymer, which is equal to the D.P. multiplied by the molecular weight of the monomer. There is virtually no limit to the D.P. value, but in practice, processing considerations and cost do in fact impose a maximum. Nevertheless, enormous lengths of polymer chains have been measured, to four or five orders of magnitude, with molecular weights sometimes over 10^6.

Degree of superheat. The amount by which the pouring temperature of a molten metal exceeds the liquidus temperature.

Delf. A thin seam of coal or iron ore.

Deliquescence. A term that refers to the liquefaction of solids when exposed to moist air. It should be noted that substances that absorb moisture without liquefaction are termed hygroscopic. All deliquescent substances are soluble in water. When a little moisture is present on the surface of a deliquescent salt a small amount of saturated solution is formed. The vapour pressure of this concentrated solution is less than the partial pressure of the aqueous vapour in the atmosphere, and so moisture is attracted by the salt, which gradually liquefies completely to a saturated solution.

Delta iron. The allotropic form of iron which exists between about 1,400 °C and the melting point of the metal. Delta iron has a crystalline structure similar to α-iron, i.e. the body-centred cubic form.

Demagnetisation. A process in which a magnetised body is caused to lose its magnetism. This may be carried out by heating the body above the Curie point and allowing it to cool. This method, however, is not always convenient and it is more usual to place the material in a strong magnetic field which can be repeatedly reversed. At the same time the field is gradually diminished to zero, causing the material to become demagnetised.

Dendrite. A tree-like crystal formation (from the Greek). A branched crystal formation that occurs when solidification takes place over part or all of the equilibrium melting range ΔT of an alloy. At a sufficiently low solidification rate R, in the presence of a temperature gradient T_g, the diffusion concentration gradient in the melt leads to local thermodynamic stability at all points in the liquid phase and plane front solidification occurs at the solidus temperature. Below a critical value of the ratio

$$\frac{T_g}{R} = \frac{\Delta T}{D},$$

where D is the self diffusivity of the melt, constitutional supercooling (q.v.) occurs in some parts of the melt, and solidification extends over a range of temperatures and compositions, forming cellular structures (q.v.) and dendrites.

The term is also used in reference to stones or minerals in or on which appear markings of an arborescent form. They occur in the joints or cleavage planes of close-grained rocks.

Densener. A piece of metal used as a chill in foundries. The metal is inserted in the face of a sand-mould to promote rapid solidification in that section and to ensure that freezing takes place progressively towards the risers.

Densifier. A term sometimes used in reference to the additions made to molten metals for the purpose of refining the grain size and securing a fine-grain homogeneous structure.

Density. The density of a substance is the mass of any volume of it divided by the volume. The units in which density is expressed depend on the units of mass and volume employed. For solids the units are $kg\,m^{-3}$ or $g\,cm^{-3}$. In the case of liquids the density is expressed in $g\,cm^{-3}$. The densities of gases are usually referred to the density of hydrogen, which is taken as 1. On this basis the density of a gas is one-half its molecular weight $(g\,l^{-1})$.

The specific volume of a substance is the reciprocal of the density. (For details of densities *see* Tables 2 and 3.)

The mass of a substance can be determined to a considerable degree of accuracy using a balance, and the accurate determination of density therefore depends very largely on the accurate measurement of volume (e.g. by the pyknometer method). In the case of gases, density may also be determined from the rate of diffusion of a gas into another gas of known density. (*See also* **Diffusion.**)

As most substances change in volume when heated or cooled, it is important, especially in the case of liquids and gases, to state the temperature to which the stated density refers. If d_1 is the density of a substance at $t_1\,°C$, and d_2 the density at $t_2\,°C$, then $d_1/d_2 = (1 + ct_2)/(1 + ct_1)$, where c is the coefficient of cubic expansion.

The importance of accurate values may be illustrated by naming a few of the important applications. The variation in the density of an electrolyte at the electrodes is measured and used in the determination of transport numbers and the migration constants of ions in solution. Changes in density have been measured in the study of the effects of cold working on metals. A knowledge of the density of crystals is important in the X-ray analysis of crystal structure.

The density formula, which may be used in verifying the lattice parameters is:

$$density = \frac{M.n}{m.p^3}$$

where m is Avogadro's number, M is molar mass, n the number of atoms per unit cell and p is lattice parameter, in cm. For body-centred cubic iron, density = $7.874\,g\,cm^{-3}$, molar mass = $55.85\,kg$, $n = 2$.

$$p^3 = \frac{55.85 \times 2}{6.022 \times 10^{23} \times 7.874}$$ therefore $p = 2.865 \times 10^{-8}\,cm$.

The term "bulk density" is often used for porous materials such as coke and for minerals such as ores, limestone, etc. The particle size and shape, as well as the true density of the material, influence the bulk density. Thus the bulk density of charcoal varies with the nature of the wood from which it was produced from about 0.135 to 0.203. Typical bulk densities, in $g\,cm^{-3}$:

Bricks	2.096	Coke	0.53
Cement	1.2	Limestone	1.5
Clay (dry)	1.57	Sand (dry)	1.3
Coal	0.83	Sand (moist)	1.76

(*See also* **Specific gravity.**)

Density of alloys. Since the mass of the two or more metal atoms making up an alloy cannot be the same, and the packing of the atoms depends on several factors, the density will vary according to the densities of the constituents, and the type of alloy structure produced. Vegard's law applies to some solid solutions, and assumes that the atomic concentration of solute produces a linear change in the size of the lattice. For solutes which occupy interstitial positions, $\rho_I = (n_A A + n_B B)m/V$ where ρ is density, n_A the number of atoms of metal A, A the mass of atom A, n_B the number of B atoms, and B the mass of each B atom, m the mass of the hydrogen atom, V the volume of the unit cell.

For substitutional solid solutions $\rho_S = nM_{AB}m/V$ where n is the number of atoms, and M_{AB} the average mass of A and B atoms.

Density of electron states. In considering the energy and distribution of electrons at the highest energy levels, a number of considerations have to be taken into account. The restrictions of the Pauli exclusion principle, the Heisenberg uncertainty principle, and the duality behaviour of the electron that stems from wave mechanics, all play a part. One of the more simplified approaches is to consider the electrons in the outer energy levels as "free", i.e. restricted by the various principles involved, but not affected unduly by the nucleus and the more tightly bound electrons close to it. When a large number of bound atoms are involved, the distribution of energy is considered spread over a range, but with limits which impose restriction on further electron occupations, or the movements of electrons with Fermi energy, i.e. extra kinetic energy over that involved in bonding and movement within the normal confines of the position in the atom.

A density of states curve can be derived from the Fermi-Dirac statistical treatment of energy levels. The density of energy states is plotted as a quantity $N(E)$ against E, where $N(E)dE$ is the number of energy states per unit volume, with energy ranging from E to $E + dE$. The number of electrons of particular energy is derived from the calculation of the number of possible states $[Z(E)]$ and the probability of their occupation $[F(E)]$.

$$N(E)dE = Z(E),\ F(E)dE.$$

$[Z(E)]$ is a function of the system, and can be calculated for the elements in the periodic system to a reasonable degree of accuracy. The value of Fermi energy E_F can be obtained from the number of electrons in the system. At absolute zero temperature, addition of electrons would fill the lowest states of energy first, and the energy available for any additional electron will depend on the states already filled with electrons. When few energy states are available at each energy level, the last electrons added will have to be of high energy. If many states are available, the rise in energy with additional electrons will be slower, and hence more flexibility is possible.

The total number of electrons, N, is given by

$$N = \int_0^{E_F} N(E)dE = \int_0^{E_F} Z(E)F(E)dE$$

From the free electron model, calculation of $Z(E)dE$ indicates that an enormous number of electrons can occupy a volume of $1\,cm^3$ before the energy exceeds, say $1\,eV$. Such a number is high enough for all the outer (valence) electrons to be accommodated without the value of Fermi energy having to rise to more than a few eV in every case.

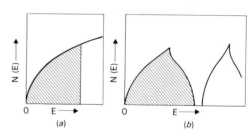

Fig. 41 *Density of states (schematic). (a) Sodium. (b) Carbon (diamond).*

Figure 41 shows a schematic representation of the $N(E)$ curve for sodium, and carbon in the form of diamond. The shaded areas represent the occupation of energy states, and in sodium there is an unshaded portion under the curve, indicating that not all energy states are occupied. Any extra energy, such as that supplied by an electric field, allows the electrons in sodium to move freely into the higher states of energy, and sodium is therefore an electrical conductor. In the case of diamond however, the area under the curve is entirely occupied, and a sharp boundary exists in the curve itself. This indicates one of the restrictions to the wave mechanical behaviour of the electron, and an energy gap is shown on the $N(E)$ curve, before other available electron energy states, at a much higher level of energy. Since all the available states are filled in the first instance, additional energy cannot allow the electrons to enter the higher states available to the electrons in diamond, without crossing this energy gap, i.e. a very high level of energy is necessary. Diamond is therefore an electrical insulator; it does not conduct electrons when an electric potential is applied.

Obviously, the shape of these density of states curves, and in particular, the size of the forbidden energy gaps, will affect the behaviour of the materials under various conditions. If the forbidden energy gap is small, then even the extra energy of a rise in temperature may be sufficient to allow electrons to jump across, and in that case, an electric potential will now cause conductivity, since there is a large availability of energy states in the higher zone. Once this has occurred, there is also an availability of energy states in the first zone, some electrons having crossed into the higher energy one. The net result is an increase in conductivity as temperature rises, a negative temperature coefficient of resistivity. Such materials are known as intrinsic semiconductors, because they are insulators below a given temperature, and conductors above it. Silicon and germanium are examples of these semiconductors.

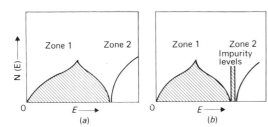

Fig. 42 *Density of states. (a) Intrinsic semiconductor. (b) Extrinsic (impurity) semi-conductor.*

The presence of a small amount of certain impurity atoms can also affect the availability of energy states. Figure 42 shows an intrinsic semiconductor density of states curve, and one in which the curve for the impurity atoms is a narrow band between zones for the basic material, within the forbidden energy gap. Impurity atoms must occupy positions in the lattice in place of the atoms of the pure material, or within interstitial positions. (In fact, the relevant atoms are substitutional.) If the impurity atom has one more electron per atom than the basic one, its density of states curve lies close to zone 2 on the diagram. Such an atom now has a free electron which has only to jump a very small energy gap to enter zone 2, and when an electric field is applied, this does occur, and the material becomes a conductor. Such a material is an extrinsic semiconductor, because it depends on impurity atoms. Since the latter provide the electrons to cause the conductivity, they are called donor, or n-type semiconductors, n being an abbreviation for negative, the sign of the electron. Impurity atoms which give rise to this effect in germanium are phosphorus, arsenic, antimony. The addition of 10^{20} atoms per cubic metre of arsenic decreases the resistivity of germanium at room temperature, from $0.5 \,\Omega\,m$ to $0.2 \,\Omega\,m$. Ten times this concentration of arsenic causes a further fall to $0.01 \,\Omega\,m$.

If the impurity atoms have one less electron per atom than the base, the density of states curve lies close to zone 1 of Fig. 42(b), and not close to zone 2. However, it is now favourable for an electron from one of the base atoms, to enter temporarily the impurity level, which is not completely filled. There is now a lack of occupancy of zone 1 of the base atom, and so it becomes a conductor, albeit by a different mechanism. Because the impurity atom is now an acceptor, there is virtually the creation of "positive holes" to accept the electrons of the host atoms, and the material is known as a p-type semiconductor. Boron, gallium and indium behave in this way in germanium. The resistivity of germanium at room temperature falls to $0.03 \,\Omega\,m$ with 10^{20} atoms m^{-3} of gallium, and to $0.02 \,\Omega\,m$ with a concentration of 10^{21} atoms m^{-3}.

The density of states curves are important therefore, in explaining the behaviour of semiconductors, and, in particular, extrinsic semiconductors. There is no necessity to plot these curves for every element however, the important levels are the upper level of the valence band, which houses the electrons binding the material together, and the lower level of the conduction band, the difference of which give the value of the forbidden energy gap. The

levels of electron energy of impurity atoms can then be plotted in relation to this energy gap, and indicate likely contributions of n- or p-type semiconduction.

The impurity atoms are present in small concentration, because the constitution and structure of the material would change if they were present in quantity and this would radically affect the density of states curves. There is no conduction between the impurity atoms themselves, because they are far apart. It is their contribution to the host atoms which is important. Electron scattering (*see* **Conduction by electrons** and **Conduction by holes**) increases resistivity, and the high purity of semiconductor material is necessary for this reason also.

Density of powders. In powder technology, the problem of handling loosely packed powder particles has led to the measurement of density by several techniques. The "pour" density is that achieved by pouring loose powder into a graduated vessel, to a given volume, with no attempt to compact the material. This gives an indication of the volume to be handled in filling dies. The "tap" density measures the result of lightly compacting the pour density specimen, say by vibrating the base of the cylinder. This gives an indication of the easily abstracted air volume when a filled die is vibrated.

Dental alloys. A comprehensive term embracing all the alloys used in dentistry, from the fusible alloys used for patterns to the silver-tin amalgams used for fillings and the alloys used for wire and dental plates. Dental gold covers a wide range of cast and wrought alloys consisting of gold alloyed with varying quantities of silver, copper, palladium and platinum with minor additions of other elements. Some of the alloys are in fact gold-free and consist mainly of silver, palladium and copper. Austenitic steels and cobalt alloys are also used.

Deoxidation. The elimination of oxygen from molten metal. The process is usually effected by adding elements having a high oxygen affinity to form fusible oxides or oxides of low specific gravity, which will float to the surface of the melt and thus be removed in the slag. For iron and steel the deoxidisers used are aluminium, silicon and manganese, which are added in various forms as silicomanganese (about 15-20% Si, 55-70% Mn) or ferrosilicon (40-90% Si, 60-10% Fe), calcium silicide or ferro-silico-aluminium (35-40% Si, 45-50% Al, 12-15% Fe).

For Babbitt metals the deoxidiser is commonly phosphor-tin (5-10% P), and for copper alloys phosphor-copper (10-15% P). Magnesium may also be used as a deoxidiser.

Deoxidised copper. Copper from which the oxygen has been removed by the addition of a deoxidising agent such as phosphorus. The electrical conductivity of deoxidised copper is lower than that of the tough-pitch grades, but the product is better suited for cold-working operations.

Deoxo process. A process whereby oxygen is catalytically removed from nitrogen, hydrogen and other industrial gases. It is used when high purity atmospheres are required, as in the production of powder metals and for such metallurgical operations as bright annealing and brazing.

Depletion layer. In a semiconductor device, when a section of p-type material is joined to a section of n-type, making a p-n junction, there is a migration of electrons into the p-type section, and of positive holes into the n-type. This creates a zone where no current carriers are available, one neutralising the other. In this depletion layer, about 10^{-4} cm thick, there is a heavy accumulation of charge however, positive in the n-type section and negative in the p-type. This acts as a capacitor, and has the effect of impeding the flow of charge carriers, and so modifying the behaviour of the p-n junction.

Depolarisation. During electrolysis of aqueous solutions a gas, e.g. hydrogen, may collect at an electrode. This weakens the current in two ways: *(a)* the gas itself has a large resistance; *(b)* a back e.m.f. is set up. Polarisation is thus defined as the increase of potential between an electrode and a solution, above that which would occur with the electrode at rest. Depolarisation is the process used for the reduction of the counter-e.m.f. and to ensure even dissolution of the anode.

As used in electroplating practice, depolarisation may be regarded as the reduction in the counter-e.m.f. and the ensuring of an even dissolution of the anode. In nickel-plating, for example depolarised anodes are prepared by carefully incorporating a small amount of nickel oxide in the rolled nickel anodes. This prevents the formation of passive films and permits working at high current densities.

The rate at which oxygen is supplied in the gaseous form (e.g. from aeration by air) or in the form of an oxidising agent will influence the rate of corrosion. In this latter connection it should be noted that polarisation may occur at both anode and cathode.

Depolymerisation. The reverse of polymerisation, known industrially as "cracking". It produces a similar result to degradation, but is usually applied in the presence of catalysts or other substances, to control the degree of depolymerisation and the products. In the petroleum industry, the process forms the feed stock for hydrocarbon fuel and petrochemical production.

Deposit. Any mass of material laid down as a bed or accumulated in veins or seams by a natural agency. The term is also applied to the metallic coating produced on any material by electrolytic means.

Depth of chill. A term used in reference to grey-iron castings such as rolls, wheels, etc., where one or more surfaces must be hard and resistant to abrasion, while the casting as a whole must be strong and not brittle.

Increasing the rate of cooling (i.e. chilling) is the means adopted to regulate the depth of chill and the depth to which white iron is produced. Other factors that influence the depth of chill are composition of the metal, the casting temperature, the temperature of the chill and of the mould and the size and location of the chill.

Derbyshire spar. Calcium fluoride or fluorspar CaF_2. It occurs as a common gangue with lead and tin lodes. High-purity fluorspar is used in the enamelling and glass industries. Inferior grades are used as a flux in steel-making and in foundry work. Mohs hardness 4, sp. gr. 3-3.25.

Descaling. The removal of scale and metallic oxides from the surface of a metal by mechanical or chemical means. The former includes the use of steam, scale-breakers and chipping tools; the latter method includes pickling in acid solutions.

The term is also used in reference to the removal of calcareous scale from the water-tubes or the interior shells of boilers. Modern methods of descaling include the use of ultra-sonic vibrations communicated to the feed water of boilers, and the use of specially designed oxy-gas burners.

Descloizite. A vanadate of lead $Pb_2V_2O_7$, one of the principal ores of vanadium.

Deseaming. The removal of surface defects from ingots, blooms, etc., by chipping or scarfing.

Desilverisation. The elimination of precious metals, gold and silver, from lead by processes such as Parke's process or Pattinson's process (q.v.).

Desulphurisation. The removal of sulphur from a metal or an ore. It is usually applied to the process for removing sulphur from iron. Soda-ash (sodium carbonate) is made to react with the molten iron, when sodium sulphide is formed and carbon dioxide evolved.

Detergents. Cleaning or cleansing agents, mild alkalis or solvents, used for removing dirt, grease, paint, etc., from objects. For most metals a dilute solution, about 3% mild alkali (e.g. sodium carbonate), to which may be added 2% caustic soda and 2% sodium phosphate, can be used hot. For aluminium a solution of 3% sodium carbonate and 1% sodium silicate is recommended. Organic solvents usually employed are tetrachloromethane (carbon tetrachloride) and trichloroethene (trichlorethylene).

Water is not an efficient cleansing agent, due largely to its surface tension, which prevents intimate contact with both the surface of the material to be cleaned, and the dirt to be removed from that surface.

Detergents lower the surface tension of water, and so allow spreading over a greater surface area, then coat the dirt or grease particles, float them away from the surface to be cleaned, and keep them apart during rinsing, so that recontamination does not occur. Some inorganic salts lower the surface tension of water, and many affect the compounds causing "hardness" in water so that the dirt particles cannot be so easily separated. The addition

of trisodium phosphate or sodium hexametaphosphate, will do this, and so help to "wet" the articles to be cleaned.

Soaps, which are compounds formed between alkalis (usually caustic soda) and fatty acids, were the original detergents. The process of saponification (soap making) is one of heating together the prime constituents of alkali, animal or vegetable oils or fats, and with the elimination of glycerol. The soapless detergents are synthetic, with hydrocarbon chains combined with sulphate or sulphonate, or long chain alcohols reacted with ethylene oxide.

Both soapy and soapless detergents rely on the dual structure of a hydrophilic (water-loving) head end, combined with a hydrophobic (water-hating) tail end. When added to water, the hydrophobic tails push between the water molecules at the surface (which give rise to the surface tension) and so break down the tension. A droplet of water collapses to an oblate ellipsoid if detergent is added, and finally spreads completely over the surface as a very thin film. As the water droplet spreads, it releases more detergent molecules, thereby breaking down the surface tension even further.

Solid particles of dirt attract the detergent molecules, usually the hydrophilic heads. These cover the particle, and then the hydrophobic tails attract the hydrophobic tails of other molecules, so that a second layer now surrounds the dirt particles—with hydrophilic heads sticking outwards. These are then detached by agitation, and so separated from the surface of contamination. Grease particles attract the hydrophobic tails of the detergent molecules, and so are surrounded by a layer of hydrophilic heads. This separates them from the surface, and disperses them in water by agitation.

Soap and most synthetic detergents are anionic, i.e. the head carries a negative charge. In soaps the hydrophilic head is usually a carboxylate —COONa— but the tails may vary according to the raw materials used, e.g. palm oil, stearates, olive oil. In synthetics, the sulphonate head is common, with a benzene ring incorporated, and the tail is similar to that of soaps, a hydrocarbon chain. Non-ionic synthetic detergents are more readily soluble in water, and carry a hydroxyl group —OH at the head. Biodegradable detergents have been evolved to prevent the build up of foam on rivers and sewage ponds, and enzyme detergents to react on protein material forming part of the "dirt". Cleaning and degreasing of metallic components in industry utilises cheaper materials, usually mild alkalis or solvents.

Detinning. A method of recovering tin from scrap tin-plate. The material is washed with an alkali detergent to remove grease, etc., rinsed and dried in air. It is then heated to melt any solder present, which latter is run off. The metal is then treated in iron cylinders, which are kept cool, with chlorine gas to form stannic chloride. The chloride, which is volatile, is led away and condensed, while the remaining iron (containing less than 0.1% Sn) is suitable for charging as scrap in the open-hearth process.

Detrital deposits. Mineral grains found in alluvial or placer deposits. They result from the natural weathering and wearing down of lode deposits, and have been carried along from the parent rock by stream or river action.

Detritus. The product of the natural process of wearing down (detrition) and weathering of rocks. It generally consists of sand, stone and mineral particles.

Detroit rocking arc furnace. An electric furnace in which the charge is heated indirectly by an arc struck between two graphite electrodes which enter the cylindrical melting chamber at opposite ends. The furnace, which can be effectively closed and operates with a reducing atmosphere, is rocked on its long axis by an electric motor.

Deuterium. An isotope of hydrogen having an atomic weight of 2. It was first prepared by the distillation of liquid hydrogen at a temperature of 13.9 K, the deuterium remaining in the concentrate after distillation. It is now prepared by the electrolysis of water, making use of the fact that the rate of discharge of hydrogen is more rapid than that for deuterium, resulting in a concentration of deuterium in the electrolyte. The chemical properties of hydrogen and deuterium are qualitatively similar, but there are marked differences quantitatively, e.g. the latter reacts with chlorine much more slowly than hydrogen.

Deuterium is known as heavy hydrogen, and its nucleus, called the deuteron, can be used in transmutation reactions and as a tracer element (in heavy water).

Devitrification. The completion of the delayed action of supercooling on a glassy state solid, by crystallisation. In most glassy or vitrified solids, the amorphous structure is so open, that rearrangement of bonds into a more crystalline long-range order is almost impossible, and in any case would take a very long time.

Transparent silicate glasses, which are in a state of extreme residual stress, and have not undergone an annealing treatment to remove much of this stress, can devitrify by innoculation with another piece of glass dropped on to the surface. The crystallisation then produces a shattering effect, as the residual stress is channelled into moving the atoms from an open structure to a more long range order. Vitrified clay, as brick, takes centuries to devitrify, but examples of ancient glass and ceramics have been found to be crystalline in parts.

Dewar flask. A special type of flask associated with the name of its inventor, Sir James Dewar. It is a double-walled vessel having the inside surface of the outer wall and the outside surface of the inner wall both silvered to reduce radiation. The space between the walls is evacuated of air. Since there is very little heat exchange through the walls of the flask, it is used to maintain liquids at steady temperatures. In the form of the well-known vacuum flask it is used to maintain the temperature of hot or cold liquids over periods of several hours.

Dew point. The temperature at which the moisture present in the air at any time would be just sufficient to saturate it. The dew point is determined as a preliminary to calculating the relative humidity of the air, and the instruments used in making the determination are called hygrometers. If the temperature of the air is $t_1\,^{\circ}C$ and of the dew point $t_2\,^{\circ}C$, and if the maximum vapour pressures at these two temperatures are F and f, then relative humidity $= f/F$. Usually the relative humidity is found by the aid of wet-and-dry bulb thermometers. The dew point is not calculated in this case (the calculation is somewhat involved), but the humidity is read off directly from tables such as Table 17.

TABLE 17. *Relative Humidity (%) (revised for SI units).*

Dry bulb Temp., °C	Difference between readings of dry and wet bulb, °C												
	0.5	0.8	1.0	1.5	2.0	2.5	3.0	3.5	4.0	4.5	5.0	5.5	6.0
−10	84	74	67	51	35	19							
−5	88	81	76	64	52	41	29	18	7				
0	91	85	81	72.5	64	55	46	38	29	21	13	5	
5	92.5	88	86	78.5	72	65.5	58	52	45	39	33	26	20
10	94	90	88	82	77	71	66	60	55	50	44	39	34
15	95	92	90	85	80	75	71	66	61	57	53	48	44
20	96		91	87	83	78	74	70	66	63	59	55	51
25	96		92	88	84	81	77	74	70	67	63	60	57
30	96		93	89	86	83	79	76	73	70	67	64	61
35	97		94	90	87	84	81	78	75	72	69	67	64

Dezincification. A type of corrosion sometimes encountered in brass. The corrosion products consist of a porous deposit of metallic copper associated with zinc compounds, such as basic zinc chloride. It was originally believed that zinc had been corroded away, but it has since been shown that the copper is re-deposited metal formed by reactions between copper corrosion products and brass. Brasses which contain more than 0.01% As do not show dezincification. Manganese and iron accentuate the tendency to dezincification, but tin, nickel, aluminium and lead act as inhibitors.

Dialogite. Manganese carbonate $MnCO_3$, found as a vein-stone in lead and silver-lead ore veins, and also in certain limestones. In the pure state it occurs as rose-red crystals, which, like calc-spar, form rhombohedral crystals. Also known as manganese spar or rhodochrosite. Mohs hardness 3.5-4.5, sp. gr. 3.45-3.6.

Dialysis. A process for separating colloidal suspensions, such as glue, gelatin and albumen, from soluble salts by allowing solutions of the latter to diffuse through a parchment membrane. (*See also* **Diffusion.**)

Diamagnetism. Atoms having p electrons in the outer energy levels, but not completing the p state (i.e. other than inert gases) show a resultant magnetic moment, from the angular momentum of

the electrons. When placed in a magnetic field, a bar of such material sets up an induced magnetic moment to oppose the applied field, and the bar aligns itself perpendicular to the applied field. Such substances are diamagnetic.

Removal of the magnetic field removes the induced opposing moment, and there are no residual effects. The ratio of intensity of magnetism to the applied field is called the susceptibility χ, defined as $\chi = I/H$ where I is intensity of magnetisation and H the applied field.

For diamagnetics, χ is negative. An increase in atomic number, which increases the equivalent outer radius of the effective outer electrons, also increases the diamagnetic susceptibility. The fluorine ion has a value of -9.4×10^{-6}, but chlorine ions, with higher atomic number, -24.2×10^{-6}. Anisotropic materials show anisotropy in diamagnetic susceptibility, but the value of χ is small for diamagnetics, the highest being antimony and graphite, around 10^{-3}. Many superconductors are diamagnetic, and when an applied field is present, the small induced electric currents which result from the induced field of the electrons are rapidly conducted through the material because of its low electrical conductivity. When the external field is removed, these small currents still flow, whereas in normal diamagnetics they would be lost. For this reason, such material can be used in the memory storage of computers.

Diamantine. Trade name of a variety of fused bauxite used as a refractory material. The fused alumina is crushed and mixed with a plastic clay and with felspar to assist in moulding the material and to provide the necessary binding power in the fired product.

The name is also applied to a very finely powdered form of alumina, widely used as a metallographic polishing medium. This material may be prepared on a laboratory scale by heating ammonium alum in a porcelain or silica dish until the water of crystallisation is driven off and the compound decomposes. The residue is heated in a muffle furnace at 1,000 °C for several hours. In the early stages of heating a gentle current of air is passed through the furnace to remove the fumes. Any crust forming on the heated mass is broken up at intervals. As the material cools it crumbles to a very fine powder which can be stored in bottles for use as required, without further preparation.

Diamond. A very pure form of carbon, crystallising in the cubic system. Found as yellow rounded pebbles in India, Brazil, New South Wales, Arkansas, South Africa and the USSR. The Indian diamonds occur in river gravels and in alluvial deposits and are separated by washing. Kimberley diamonds occur in the original "blue ground". This is a form of olivine, which after being blasted out is allowed to weather. After weathering the lighter earthy material is washed away and the heavier material remaining runs over beds of grease, which retain the diamonds.

Diamond is the hardest mineral (Mohs hardness 10). It is 140 times harder than corundum and is probably approached in hardness only by boron carbide B_6C. Black diamonds, which are valueless as gemstones, are known as carbonado and bort and are used as abrasive material and as a rock-cutting medium. Also used as indentor for hardness testing. Refractive index 2.42, sp. gr. 3-3.55. Diamond is transparent to X-rays. (*See* **Graphite, Lamp black.**)

Diamond cubic. A term which describes atomic arrangements which are similar to the diamond in having two face-centred cubic arrangements of atom centres, one of which is displaced with respect to the other by a quarter of the diagonal of the unit cube.

Diamond pyramid hardness test. A test to determine hardness by pressing a square-based diamond pyramid (having an angle of 136° between opposite faces) under a standard load into the surface of the material to be tested and measuring the diagonal of the indentation produced. The diamond pyramid hardness (DPH) is obtained from the relationship DPH = load (in kg)/pyramidal area of the impression (in mm²).

The abbreviations DPH, DPN (diamond pyramid hardness number) and VPN (Vickers pyramid number) are synonymous. (*See also* **Brinell hardness test.**)

Diamond substitutes. These are substances of great hardness possessing considerable resistance to wear and abrasion. They are usually carbides of tungsten, tantalum, titanium and boron.

Diaspore. A hydrated aluminium oxide $Al_2O_3.H_2O$ found with corundum and emery and in some bauxites. It is used as a refractory material. Melting point 2,050 °C, Mohs hardness 7, sp. gr. 3.5. Also called diasporite.

Diatomaceous earth. A deposit consisting of the skeletons of siliceous organisms such as diatoms. It is a hydrated silica and is in a very finely divided form. It is employed as a polishing medium, as a refractory material, as an insulating material and as an absorbent in the manufacture of explosives and as a filling medium under the name kieselguhr.

Diatoms. A group of unicellular algae characterised by heavily silicified cell walls. They inhabit both fresh and salt water.

Dichroism. A form of pleochroism (different colours when viewed in different directions) found in certain uniaxial crystals. This phenomenon is due to unequal absorption of light rays by different crystal planes so that the crystals display different colours when viewed in different directions by transmitted light. Crystals exhibiting two colours are said to exhibit dichroism, those showing three colours are trichroic, while those showing more than three colours are said to exhibit pleochroism.

Dickite. A hydrated silicate of aluminium $Al_2O_3.2SiO_2.2H_2O$ similar in general respects to kaolin.

Die. A metal or sintered carbide block used in metal shaping and forming operations. Perforated dies are used in wire drawing.

Die-casting. A process by which metal castings are produced in permanent moulds. The metal may be fed by gravity or under pressure.

Die-casting alloys. Alloys suitable for use in die-casting processes. Dimensional accuracy and absence of damage to the mould are prime requisites. Aluminium, copper, tin, zinc and lead base alloys are most generally used for die-casting.

Dielectrics. The section on Density of electron states describes the forbidden energy gap, which prevents an insulator from becoming a conductor. When an electric field is applied to an insulator, there can be no charge carriers available, but obviously some effects must be suffered by the material, especially if the potential applied is high. What occurs is the slight displacement of some of the positive and negative charge distribution present in the material. There is no transfer of charge, but some separation which did not previously exist. Removal of the field would allow removal of the displacements. These slight displacements cause the formation of local dipoles, and therefore some polarisation of the material. If an insulator already contains dipoles, these will be orientated in the direction of the applied field. Thermal agitation destroys this orderliness, and so polarisation is temperature dependent. There is a reasonable analogy between the effect of an applied electric field on polarisation, and that of a magnetic field on the magnetic moments in a material. The permeability of magnetisation is analogous to the permittivity of a dielectric, and the dielectric constant is the ratio of permittivity of the material to that of a vacuum.

Coulomb's law of force, governing the force F between charges q_1 and q_2 distance r apart, states that $F = \mathbf{a}.q_1.q_2/4\pi\epsilon r^2$ where ϵ is the permittivity and $\mathbf{a}$, a unit vector along the line between the two charges. When q_1 and q_2 are of the same sign, the force is one of repulsion.

The total dipole moment induced in unit volume of a material is the polarisation density. This can arise from three possible processes: *(a)* relative displacement of electrons and associated nuclei (electronic); *(b)* relative displacement of positive and negative ions (ionic); and *(c)* displacement of permanent dipole moments in complex ions or molecules.

Die-welding. A process used to join overlapping metal sheets, using a strictly localised pressure between heated dies that are arranged to hold the sheets together during the welding.

Differential cooling curve. A type of curve sometimes used in thermal analysis. Temperature is plotted as the ordinate, and the difference in temperature between the specimen having a transformation point and a neutral body having no transformation points is the abscissa. (See Fig. 44; see also **Differential heating curves.**)

Differential flotation. A process for the separation of an ore from the associated gangue, which consists in stirring up the finely divided ore with water to which latter certain additions have been made. When a small quantity of an oily substance—e.g. eucalyptus oil—is added to the wet pulp (finely divided ore and water) it causes certain changes to occur in the surface of those mineral particles which are metallic or are of a semi-metallic nature, and they become less easily wetted by the water than the gangue. Air is then blown in or the mixture strongly agitated so that a froth is formed. The "oiled" mineral particles adhere strongly to the air bubbles and are carried to the surface, where a stiff froth is formed. This mineralised froth may be separated. The success of the process depends on the difference in behaviour of the ore and the gangue towards froth bubbles, and it may be necessary to increase this difference, by adding oil to make the ore particles less readily wetted, and by adding acids or alkalis to make the gangue more readily wetted.

The process is particularly suitable for separating sulphide ores from each other and from the gangue associated with them.

Differential heating. Any method of heating so controlled as to produce a desired non-uniform distribution of temperature in one article.

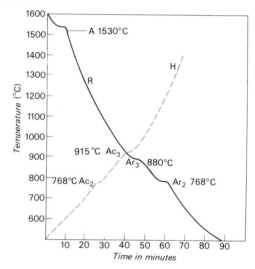

Fig. 43 *Typical heating and cooling curves of a plain carbon steel. The broken line indicates the heating curve and the solid line indicates the cooling curve.*

Differential heating curves. Ordinary heating and cooling curves are obtained by noting the temperature and the time when a mass of material is allowed to cool or is heated slowly. The form of such curves is shown in the accompanying figure (Fig. 43), in which the arrest points are indicated. To define the critical points more clearly, curves are drawn in which the difference in temperature between the specimen under examination and a pure material that does not undergo transformation in the range considered, is plotted against the temperature. Thus if θ is the temperature of the specimen and θ_1 that of the material that undergoes no change, then $\theta - \theta_1$ is plotted against θ, giving a difference curve. The differential curve is derived from this by plotting $d(\theta - \theta_1)/d\theta$. The form of these curves is shown in Fig. 44. The interpretation of differential heating and cooling curves to identify phase changes is known as Differential Thermal Analysis (DTA).

Differential quenching. A process for the selective quenching by immersion or spraying, etc., of different portions of one article. The process is often associated with differential heating.

Diffraction. The wave nature of radiation can account for the time lag of wave packets arriving at an object, having left the same source, by virtue of the difference in distance travelled. The phase differences produce interference, with complete extinction if that difference is half a wavelength or an odd multiple of this. When radiation is directed through a narrow slit, bands are produced, reinforcement from phase equivalence, and extinction from the half wave difference. With visible light, a series of light and dark bands can be produced on a screen, or recorded on a photographic plate. The phenomenon requires the provision of a source, a restricting slit, and a collimating system to produce sharp bands. If a continuous set of fine slits is used, as in the diffraction grating, the components of white light can be dispersed selectively, to produce spectra on each side of a central spot. The

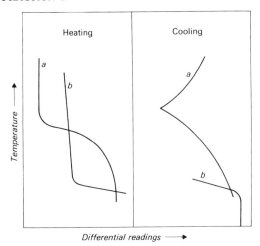

Fig. 44 *Differential heating and cooling curves: (a) transformations involving no latent heat; and (b) phase changes with latent-heat transformation.*

finer the slits in the grating, the greater the angle of dispersion for each wavelength, and the more spread out the spectrum.

Diffraction results from all types of radiation, but the gratings required outside the visible light region are obviously different, in view of the changed wavelengths. The atomic planes of a crystal lattice, are spaced apart roughly in the order 10^{-10} m and therefore can act as a diffraction grating, by reflecting radiation whose wavelength is of a similar order, and this is true of X-rays, the radiation emitted by electron bombardment of atoms. (*See* **X-ray diffraction.**) Electrons themselves will also be diffracted, but their penetration is low, and hence confined to surface layers, unless the specimen is a very thin foil. Neutrons can similarly be used. (*See* **Electron diffraction** and **Neutron diffraction.**)

Diffraction grating. The first instruments made by Fraunhofer were gratings made of fine wires. More accurate and more useful instruments are now made by ruling, with a diamond, on a glass plate or a polished metallic mirror, an immense number of parallel lines. Copies of these gratings are made by photography and replicas can be made by direct impression upon a celluloid or other plate. The lines on the grating act as opaque barriers with transparent "slits" between each adjacent pair. Gratings with 6,000 lines per centimetre are not uncommon. The spectra obtained by the aid of a grating are not as bright as would be obtained with the same source using glass or quartz prisms.

If a grating is ruled on a polished concave mirror, one perfect-diffraction spectrum can be formed without the aid of lenses.

Diffraction pattern. A characteristic spacing of lines produced when radiation (of known wavelength) is diffracted by a crystal.

Diffusion. The migration of a constituent, in a solid, liquid or gaseous system, from one location to another in the same system. The driving force for such migration may be concentration gradients, temperature or pressure gradients, and the movement is dependent on collisions with other atoms or molecules (as in a gas), the mean free path, or on the jump length (as in liquids and solids), the distance a particle is able to move when possessed of the appropriate activation energy.

In gases, Graham's law of diffusion states that the rate of diffusion is dependent on the inverse of the square root of density. Because of this, gases can be separated by compression and diffusion through a porous membrane, the lighter gas diffusing more rapidly. As the square root of density is involved, the effect is small, and many repetitions of the process may be needed for densities which are similar. Nevertheless, the technique has been successfully used to separate isotopes, notably the separation of ^{235}U from ^{238}U in the nuclear energy industry, the gases used being the hexafluorides.

In aqueous solutions, Fick's first law expresses the quantity passing through a unit area in a given time as, $dN/dt = -D \, dC/dx$ where N is the number of atoms per unit area in time t, C the concentration of one atom in distance x, and D is a constant called the diffusion coefficient. The higher the temperature and the concentration gradient, the more rapid the diffusion, as D depends on both.

Chemical compounds which are crystalline in the solid state diffuse relatively rapidly in solution, but amorphous substances do not. The latter tend to form colloids and related viscous solutions, and, like gases, can be separated from crystalline material in solution by diffusion through a porous membrane—the phenomenon of dialysis, so important in medicine.

Metals diffuse readily in solid solution. Those which are interstitial, i.e. occupy the "voids" in crystal lattices, can readily move via interstitial channels in the lattice, but this applies to very few atoms, hydrogen, carbon, nitrogen and sometimes boron and oxygen. Substitutional atoms, (the majority of metallic alloys involve these) require an appropriate activation energy, which is very high for interstitial channels, although feasible. The simplest mechanism, and the lowest activation energy, arises from vacancy transport, i.e. the relative interchange between vacant lattice sites and the migrating atoms (*see* Fig. 45). There is evidence that a vacancy-solute atom pair is formed, and the two move in conjunction. This can only assist diffusion, the solute atom having its vacancy to assist the transfer of positions. Diffusion can occur between like atoms in a pure metal, and vacancy diffusion is still the mechanism, but less complex since no vacancy-solute pair exists.

Diffusion occurs more rapidly in the vicinity of grain boundaries and dislocations owing to the irregular nature of the lattice and the obvious association with vacancies. Portions of a crystal lattice under strain assist the diffusion process, external pressure has little effect except where contact between dissimilar materials is involved. Close-packed lattices decrease the diffusion rate, open ones increase it.

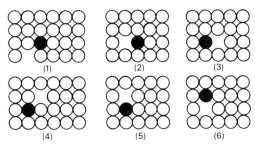

Fig. 45 *Solute atom-vacancy movements in diffusion.*

Two dissimilar metals A and B do not diffuse at the same rate into each other, a fact demonstrated by Kirkendall. There are two rates of diffusion because there are two diffusion coefficients, D_A and D_B, and these are rarely equal. The atom diffusing faster is not replaced by an equal number of the slower moving ones, and a concentration of vacancies builds up on that side of the interface from which the faster moving one originated. Porosity then results if these vacancies coalesce. Such an occurrence is manifest in the dezincification of brass, where zinc is corroded away at the surface, diffuses more rapidly from the interior, and leaves behind a porous skeleton of copper.

When a solid metal A is in contact with a liquid metal B, the interface gives rise to diffusion as with solid/solid contact, but there is increased energy in the liquid, and as the temperature will be elevated (since B is molten) diffusion is enhanced. The grain boundaries of A will be vulnerable, as rapid diffusion sites, and penetration of B into A may cause separation of whole grains from the surface.

Diffusion bonding. A method of joining metals by forming a bond between two similar or dissimilar surfaces without the presence of a liquid phase at the interface. The surfaces to be bonded are thoroughly cleaned and freed from oxide by chemical means or by scratch brushing, and then brought into intimate contact under high pressure. This pressure, which may be applied statically or by rolling, breaks down any residual film of surface oxide, and bonding takes place by diffusion across the faces. Preheating may be required. The process is also known as cladding.

Diffusion column. A vertical column used to measure specific gravity of solid particles, or a means of separating isotopes, in liquid or gaseous form, by thermal diffusion. A temperature gradient in a mixture of fluids will give rise to a concentration gradient, and when diffusion equilibrium is established, the steady state will have separated the two fluids along the equilibrium concentration gradient. The result is similar to gaseous diffusion through a membrane, and has been used to separate isotopes of uranium and boron.

In the instrument for determining the specific gravities of fine mineral particles, two completely miscible liquids of different specific gravity are placed in a graduated tube. The heavier liquid is placed at the bottom and the light liquid above, so that there is no mixing. The tube is then allowed to stand for a day or more until diffusion has occurred, and the resultant column of liquid varies regularly in specific gravity from top to bottom. Small fragments of minerals of known specific gravities, which are introduced into the liquid, and which come to rest at particular points, serve as indexes. The pulverised sample for testing is then introduced, and its several constituents are separated into bands according to their different specific gravities.

Diffusion creep. At low stresses and high temperatures, the migration of vacancies through the lattice or along grain boundaries, can be the rate-controlling mechanism in creep deformation. The rate of deformation can be expressed as

$$\dot{\epsilon} = \frac{B_1}{L^2} \cdot \frac{\Omega\sigma}{kT} \cdot D_1 \qquad \text{for lattice-diffusion creep}$$

and

$$\dot{\epsilon} = \frac{B_2}{L^3} \cdot \frac{\Omega\sigma}{kT} \cdot w.D_{\text{g.b}} \qquad \text{for grain-boundary diffusion creep}$$

where B_1 and B_2 are constants, L is grain-size, Ω the atomic volume, w the grain boundary width, k Boltzmann's constant, T the temperature, σ the stress and D_1 and $D_{\text{g.b}}$ the respective diffusion coefficients.

Diffusion layer. A term used in electrolysis to denote the thin layer of liquid surrounding an electrode across which the concentration of the electrolyte changes.

Diffusivity. The rate of propagation of temperature along a body. The rate is dependent on the thermal conductivity (K) of the material, on the density (ρ) and on the specific heat (s), and is expressed by $K/\rho s$. If the material is surrounded by non-conducting material to reduce heat loss from the surface the equation of propagation becomes:

$$\frac{K}{\rho s} \cdot \frac{d^2\theta}{dx^2} = \frac{d\theta}{dt}.$$

Also the name used for atomic diffusion.

Dilatometer. An instrument for magnifying and measuring the expansion and contraction of a solid during heating and subsequent cooling. Dilatometers are often used in the determination of changes of state which may occur with change of temperature, and have been adopted for practical use in the heat-treatment of steels. A typical dilatometric curve is shown in Fig. 46; for a heating and cooling cycle for a steel it shows: A to B, normal

expansion during heating; B transformation begins; C, transformation ends; C to D, further normal expansion; D, heating discontinued; D to E, normal contraction during slow cooling; E, transformation on cooling begins; F, transformation complete; F to G normal contraction during slow cooling to room temperature.

Dilution law. The variation of the degree of electrolytic dissociation with the concentration of the solution, is calculated on the assumption that the law of mass action applies equally to dissolved molecules and to the ions produced from them. For a substance yielding two ions in solution the dissociation constant K is given by

$$K = \frac{x^2}{(1-x)V},$$

where x is the fraction dissociated and V is the molar volume of solution. This equation is known as Ostwald's dilution law, and may be used to calculate the percentage dissociation at one dilution if the amount of dissociation at another given concentration is known. The law applies only to weak electrolytes, and for such the approximate equation $x^2/V = K$ may be used. This then leads to further simplification that $x \propto \sqrt{V}$ or the percentage dissociation is proportional to the square root of the dilution.

Dimorphism. Substances which are able to crystallise in two distinct systems are said to exhibit dimorphism. Calcium carbonate crystallises in the rhombohedral system as calcite or calc spar, while as aragonite it crystallises in the orthorhombic system. Substances capable of existing in three distinct systems are said to exhibit polymorphism.

Dinas bricks. Refractory bricks made almost entirely from silica sand. (*See also* **Brick**.)

Dioctyl phthalate (1, 2-Benzene dicarboxylic acid mono (2 octyl) ester). A plasticiser used in association with polymeric materials to modify the properties. It suppresses the glass transition temperature and so assists the proportion of crystalline material formed. In cross-linked materials, it modifies the inter-molecular forces between the network linked units.

Diode. An electrical device consisting of two electrodes, arranged in circuit so that current can flow in only one direction, thereby acting as a rectifier in modifying alternating current to direct.

In the vacuum valve version, electrons are ejected from a heated filament, and collected on a plate or cylindrical anode. They are unreliable in operation and the characteristics change with age or leakage of gas into the tube.

In the semiconductor version, a predominantly electron carrier material (n-type) is joined to a positive hole or acceptor material (p-type) to form a p-n junction. When a potential difference is applied, the predominant current is increased in a forward direction (n→p). With reversed bias, i.e. a reversed potential, the current is very small, one component (n→p) being reduced, the other (p→n) being unchanged. At higher temperatures, these effects are reduced, and the reverse current increased. Any heating of the device as a result of heavy current flow, therefore, varies the characteristics, and actually leads to destruction of the device. Apart from this, semiconductor devices are more reliable as rectifiers, although rectification can never be 100% (*see* Fig. 47).

When high reverse bias is applied to a p-n junction, collisions by carriers lead to ionisation, and an avalanche breakdown can occur. If the power is not excessive, no harm arises, but the voltage remains fairly constant over a wide range of current. Such a voltage controller is known as a Zener diode.

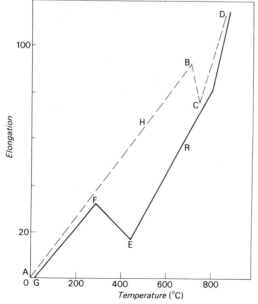

Fig. 46 *Dilatation curve typical of low-alloy steel. H indicates heating and R indicates cooling.*

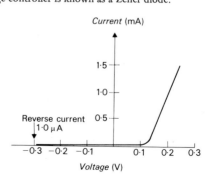

Fig. 47 *The voltage-current relationship for a silicon p-n junction.*

Dioptase. A hydrated silicate of copper $CuO.SiO_2.H_2O$ occasionally found in the zone of weathering of copper lodes. Also known as emerald copper.

Diorite. A sub-basic plutonic igneous rock containing plagioclase felspar with some orthoclase and hornblende.

Dioxan. An organic solvent used in certain cleaning mixtures which are employed in preparing the surface of a metal prior to electroplating, polishing, etc.

Dip-brazing. A process for joining metals by brazing, in which the metal pieces are immersed in a bath of the molten filler metal. The bath must be covered by a layer of a suitable molten flux.

Dipolar material (electrostatic). A material which demonstrates orientational polarisation, and usually a dipolar molecule, though not necessarily so. The most common example is water, but materials with -O-H groups and -C=O are dipolar, and often impart dipolar properties to larger molecules in which they occur. The charges must be separated by a distance, and so introduce a potential force, a dipole moment.

Dipole (magnetic). Similar to the electrostatic dipolar material, but separation of magnetic polarisation is involved, e.g. North and South poles in a bar magnet. In an applied field, a bar magnet freely suspended will orient itself with the attraction forces of the applied field, e.g. the magnetised compass needle in the Earth's magnet field. The magnetic moment is the measure of the dipole moment of that particular magnetic system or unit.

Dipping. A method for producing metallic coating by immersing a base metal in a molten metal (e.g. tin or zinc).

A pickling process involving brief immersion in acid solutions.

The immersion of a product in paint, lacquer or other coating material.

Direct arc furnace. An electric arc furnace in which the arc is struck between an electrode and the charge in the furnace.

Directional bonds. In ionic bonds, the charges can be assumed roughly spherical, and therefore more uniformly distributed. In the covalent or homopolar bond, there is a requirement for shared electron pairs, to satisfy the Pauli principle, in forming the bonds. Covalent bonds therefore tend to be directional, in order to direct the electrons shared between adjacent atoms in the bond. In nitrogen, with three electrons to be shared, the three shared pairs are in three mutually perpendicular directions. In carbon, the four close neighbours of the diamond structure give rise to the tetrahedral type of arrangement, with the bonds directed to the corners of each tetrahedron. The electrons are best envisaged as an electron cloud, but the directionality is still evident, in the major axes of the figures which most closely represent the cloud volume. Van der Waals' bonds, involving polarisation, must also be directional.

Directional properties. Anisotropy of structure automatically gives rise to anisotropy of properties. The atoms in an anisotropic crystal have some near neighbours, and some reasonably near but not quite so close. Any property which involves atomic vibrations, or the movement of electrons, must therefore involve variations along specific directions. Electrical conductivity, thermal conductivity, and expansion and deformation, are obvious examples. Directional properties also result from mixed bonds, e.g. in graphite, where the hexagonal layer lattice is covalent, but bonds between the layers are van der Waals' bonds. Metals with p electrons frequently show mixed bonding, e.g. antimony, bismuth, arsenic, tin. In this case metallic bonding is predominant, and so the anisotropy of properties is less marked than with graphite.

A normal ingot is usually isotropic in its mechanical properties, since it may be regarded as a large mass of equi-axed crystals. But as a result of forging, rolling, etc., the metal develops a structure in which the properties vary with the direction of the grain. Directional properties are developed, for example, by rolling in one direction only. The properties (e.g. ductility, resistance to shock, etc.) are superior in directions parallel to the direction of rolling than in directions at right angles to this. The directional properties may be reduced by rolling in two directions at right angles to each other.

Directional solidification. A method of casting or solidifying a metal to constrain solidification in a single direction to produce an anisotropic cast structure. The method is used commercially to produce columnar and single crystal components such as gas turbine blades from nickel-base superalloys. For f.c.c. alloys the naturally preferred crystal growth direction is <100>.

Direct oxidation. A reaction between metal and dry oxygen or oxidising gases leading to the formation of oxides on the metallic surface as primary products.

Direct process. A process for producing wrought iron from ore by direct smelting without passing through the intermediate pig-iron stage. Many processes which make use of the fact that iron oxides can be reduced by carbon or by carbon monoxide at dull red heat have been devised.

A process for producing copper from ores of special purity.

Direct rate curve. A type of cooling curve used in the thermal analysis of alloys. Temperature is plotted as the ordinate, and the temperature drop for a given interval of time as the abscissa.

Discard. That portion of an ingot or bar which is cut off to remove the pipe and other defects. (*See also* **Crop**.)

Discharge in gases. *See* **Conduction in gases**.

Disco process. A method for low-temperature carburisation of small-size coal. The coke which is produced in rotary retorts is suitable for metallurgical purposes. In the Ore-Disco process up to 60% of fine iron-ore concentrates are incorporated in the charge and the agglomerate used for smelting in low shaft cupolas or electric furnaces. The process is also used for heating ores of non-ferrous metal.

Discrete. A term used in reference to units or particles which can remain separate or distinct.

Dished. A termed meaning pressed, forged or beaten into a dish-like shape.

Dislocation. Under deformation, the theoretical strength of crystals is shown to be two to three orders of magnitude higher than experimental determinations indicate. When considering the crystallisation of a solid from a liquid, the fact that upwards of 10^{20} atoms per cm^3 are involved, indicates the impossibility of perfect alignment through the whole mass of a large crystal. Any imperfections, whether vacant lattice sites or atoms in the wrong places, will place the lattice under local strains, and the application of an external load must automatically affect the balance of forces internally.

To explain these variations, G.I. Taylor and others proposed in 1934 that some crystal defects took the form of extra half planes in the lattice, and that a lower stress than normal, along a slip plane, would cause movement, since only a small distance of slip was needed at each step, much less than one whole atomic spacing, and not involving shear across the whole of a slip plane at each movement. Figure 48 shows the edge dislocation which he proposed, and Fig. 49 the analogy of moving a carpet by creating a small fold and propagating that fold by steps, rather than movement of the whole. The movement of individual atoms is seen to be very small, but the total deformation will obviously depend on the number of dislocations present, and their distribution. Slip as a result of edge dislocations takes place along a unique slip plane, and the slip plane and slip direction are characteristic of the lattice involved.

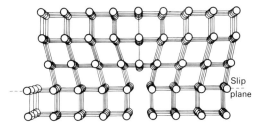

Fig. 48 *An edge dislocation.*

Burgers proposed another form of dislocation, in which two planes perpendicular to each other might be involved, with movement taking place parallel to the dislocation, which formed a helical path, and hence was termed a screw dislocation (*see* Fig. 50). In practice, not only do both edge and screw dislocations occur in the crystal, there are combinations of both. When elements of both are present, distinct areas of dislocated structure result, in the form of loops, such that edge dislocations will be perpendicular to the slip direction, and screw dislocations parallel to it. A screw dislocation can move into another slip plane having a common slip direction (*see* **Cross slip**) and an edge dislocation can move up or down at the extremity of the extra half plane (*see* **Climb**).

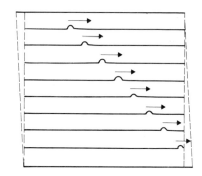

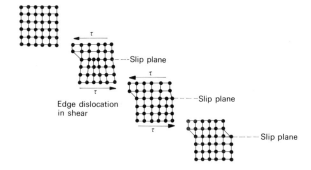

Fig. 49 *Movement of an edge dislocation in shear, with analogy of a carpet. Note the successive positions of the fold and carpet as the fold moves. The carpet corresponds to the slip plane only.*

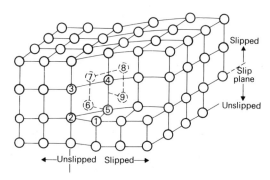

Fig. 50 *A screw dislocation in a cubic lattice. (The numbered atoms indicate the helical part of the screw.)*

The displacement of the dislocation when slip occurs is the Burger's vector, **b**, and a shear stress τ acting on length dl of dislocation gives rise to a force $F_d = \tau \, \mathbf{b} dl$ tending to move the dislocation, directed towards the unslipped portion of the crystal. The density of dislocations in a soft, well-annealed strain-free metal is of the order of 10^5 to 10^8 cm^{-2}. Heavy deformation multiplies the number, to values of 10^{11} to 10^{12} cm^{-2}.

Dispersion (of light). The separation of heterogeneous radiation (in this case white light) into separate components of wavelengths varying from roughly 3.5×10^{-7} m (violet) to 7.5×10^{-7} m (red) as a continuous band, known as a spectrum. Light can be dispersed by prisms, because the velocity varies with wavelength, and by gratings, or fine lines drawn on a transparent base, where the separate wavelengths are diffracted to various angles.

Dispersion hardening. *See* **Precipitation hardening.**

Displacement series. The series obtained when the elements are arranged in the order of their standard electrode potentials. Two standard electrodes are commonly used: *(a)* the normal calomel electrode covered by a molar solution of potassium chloride saturated with mercurous chloride, *(b)* the molal hydrogen electrode, consisting of hydrogen gas under a pressure of 101.3 kN m^{-2},

passing over a platinum plate and in contact with an acid of such concentration that the hydrogen ion is of molar concentration (e.g. 2 molar acid practically fulfils this condition). A cell is made by combining the standard electrode with the one whose electrode potential is required. This latter value is then the difference between the measured e.m.f. of the cell and the standard electrode potential. Table 18 gives the electrode potentials at 25 °C referred to the potential of the hydrogen electrode with hydrogen ion at unit activity as zero. This sequence of metals is also called the electromotive series.

TABLE 18. *Electrode Potentials at 25 °C, or Electromotive Series (volts).*

	E_0		E_0
K/K^+	-2.92	Ni/Ni^{2+}	-0.23
Ba/Ba^{2+}	-2.90	Sn/Sn^{2+}	-0.136
Sr/Sr^{2+}	-2.89	Pb/Pb^{2+}	-0.126
Ca/Ca^{2+}	-2.76	Bi/Bi^{2+}	$+0.30$
Na/Na^+	-2.71	Cu/Cu^+	$+0.345$
Mg/Mg^{2+}	-2.38	$2Hg/Hg^{2+}$	$+0.79$
Al/Al^{3+}	-1.71	Ag/Ag^+	$+0.80$
Mn/Mn^{2+}	-1.03	Hg/Hg^{2+}	$+0.86$
Zn/Zn^{2+}	-0.76	Pd/Pd^{2+}	$+0.83$
Fe/Fe^{2+}	-0.41	Pt/Pt^{2+}	$+1.2$
Cd/Cd^{2+}	-0.40	Au/Au^{3+}	$+1.42$
Co/Co^{2+}	-0.28		

Displacement spike. The path of an atom through a crystal lattice, when displaced by collision with a neutron, from its normal lattice position. The displaced atom collides with other atoms, which in turn may displace further atoms, until the energy acquired by the original collision is finally dissipated. The damage to the lattice is known as "knock-on" or radiation damage (*see* Fig. 51). The energy transmitted to an atom causing a displacement spike may be of the order of 10^5 eV (10^{-14} J), and since on average 25 eV (4×10^{-18} J) is sufficient to displace an atom, a number of such collisions can be involved in one displacement spike. Some of the damage may be annealed by the inevitable heat produced, some may remain for a considerable time.

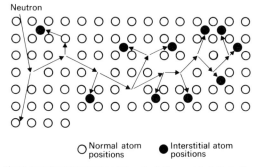

Fig. 51 *Two-dimensional representation of a displacement (radiation) spike.*

Dissociation. A term originally restricted to reversible decompositions in which one or more of the products were gaseous, and which were endothermic. The term is now applied to any reversible chemical change in which *(a)* complex molecules are resolved into simpler ones, *(b)* compounds are resolved into their free elements and *(c)* in the case of electrolytes, salts, bases, and acids that are resolved into their ions.

When the dissociation is brought about by the application of heat, it is referred to as thermal dissociation. Solid substances that are vaporised as well as dissociated by heat have vapour densities less than half the accepted molecular weights of the substances. The ratio of the true to the abnormal vapour densities is taken as the degree of dissociation.

Dissociation constant. According to the theory of electrolytic dissociation, the law of mass action is applicable to the equilibrium between an electrolyte and its ions. In the case of a weak acid, $HA \rightleftharpoons H^+ + A^-$, and if a molar weight of it is dissolved in V litres, the degree of ionisation (m) is given by

$$K = \frac{m^2}{(1-m)V}.$$

This relationship is known as Ostwald's dilution law and K is called the dissociation constant.

Dissociation pressure. According to the phase rule, a system of two components and three phases must be univariant. If one of the phases is gaseous, then the pressure of the gaseous phase is determined by the temperature. The maximum gas pressure at each temperature is known as the dissociation pressure, and such a system behaves like the one-component liquid-vapour system. Deville's law summarises the facts in the statement "The pressure of the gas is a function of the temperature only and is independent of the quantities". This law holds so long as all the chemical substances involved in the reversible reaction are in contact.

Dissolution. The disposition of a gaseous or solid material into a liquid (or a liquid into a liquid) to provide an apparently homogeneous material, the solution. The structure of liquids is such that attraction forces retain the individual atoms or molecules in reasonable order, albeit of rather short range. When the molecules of a solid can fit between the network of the liquid molecules, but not be attracted to each other, except again over a very short range, the properties tend to remain those of the liquid, modified by the presence of the solid in solution. If the solid particles grow too large, they can no longer be held in solution, and this occurs if the temperature of a solution is lowered, a fall in solubility occurs. If more solid is added, there is a finite upper limit to the solubility at a given temperature, and the solid molecules can no longer separate and disperse through the liquid. They fall to the base of the container, and the solution is said to be supersaturated. A rise in temperature is then necessary to take more of the solid into solution.

Similar arguments apply to gases and liquids dissolved in liquids, but in the case of gases, any excess over the saturation of solubility will simply escape, and with liquids, would separate out as an immiscible layer.

Dissolved carbon. Carbon in solution in molten or in solid iron. Pure iron (δ-ferrite) melts at 1,533 °C and the addition of carbon lowers the melting point as indicated by the liquidus curve XY (*see* Fig. 52) while the solidus curve XZ represents the lowering of the melting point due to the solution of carbon in solid iron.

Fused cast iron usually contains 2.2-4.5% C, and when cooled the carbon is wholly or partially retained in combination by the metal (as carbide), while the rest separates out in the form of graphite. The actual behaviour of the carbon is determined by the composition of the mass, the conditions of manufacture and the mode of cooling. Chromium and manganese favour the retention of a high proportion of the carbon in combination with the iron. Silicon and aluminium have the opposite effect.

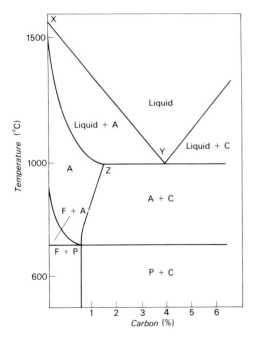

Fig. 52 *Diagram showing solubility of carbon in iron. A is austenite; C, cementite; F, ferrite; and P, Pearlite.*

Slow cooling promotes the crystallisation of graphite and favours the production of grey iron. Sudden cooling favours the retention of the carbon as carbide and the production of the more homogeneous white iron. All carbon steels that are slowly cooled from above the arrest point Ar_1 (about 700 °C) contain the greater proportion of the carbon in the form of carbide Fe_3C. Steel which is suddenly cooled from a higher temperature contains little free carbide. Austenite is a solid solution of carbon in γ-iron.

Divorced cementite. The cementite which occurs, as small discrete, more or less spheroidised particles, in certain slowly cooled steels or in steels after sub-critical annealing, particularly in fine-grained aluminium-killed steel.

Divorced pearlite. Pearlite of which the cementite has been spheroidised by annealing. Also known as spheroidised carbide.

DKT (Dipotassium tartrate). A piezo-electric material similar to Rochelle salt, but in which potassium has replaced the sodium ion.

Doctor. A device for electrodepositing metal on parts of articles or on imperfectly plated areas. It consists of an anode of the plating metal which is covered with a pad of absorbent material, saturated with the electrolyte.

Dog tooth spar. A form of calcite $CaCO_3$ found in Derbyshire lead mines and caverns. It is so called because of its scalenohedron habit.

Dolerite. A basic igneous rock. The name was formerly used to describe coarse-grained basalts, but is now restricted to basic igneous intrusive rocks. The essential mineral content is basic plagioclase felspar. Dolerites occur in central Scotland, Northern Ireland and Northumberland (UK), and very abundantly in South Africa and Moravia (Czechoslovakia). All dolerites are dark coloured and dense. The stone, being heavy and tough, is of considerable economic value for road-making and paving setts.

Dolomite. A calcium, maganesium carbonate $Ca.Mg(CO_3)_2$ in which the carbonates are present in equimolecular proportions, i.e. it is a double salt. It can be distinguished from calcite because it is not acted upon by dilute acids. It occurs as rhombohedral crystals in Switzerland, Piedmont (Italy), Cumberland (UK), Missouri Valley, Austria, South-East France and the Dolomite Alps. Calcined dolomite is used as a basic refractory material for withstanding high temperatures and the action of basic slags. Mohs hardness 3.5-4, sp. gr. 2.85. Also called bitter spar.

Dolomite limestone. A sedimentary rock containing calcite as well as dolomite.

Domain. When a material property is dependent on ordering of the distribution of some electronic phenomenon, whether charge, momentum or spin, then there may be regions of parallelism set up, so that in any one region all the effects act in the same direction, but direction differs between regions. Such regions are called domains.

In ferroelectrics, the polarisation of the material can be reversed by the application of an external electric field. In the "unpolarised" state, domains showing every possible direction of polarisation exist, so that the overall effect is neutral. When an external field is applied, domains whose direction of polarisation is closest to that of the applied field, will readily change to parallel direction with the field. As the field is increased, the more difficult domain directions will change also, the effect finally flattening out as the few most unfavourable directions become involved. Once oriented parallel to the field, domain size can grow, to absorb parts of other domains.

This phenomenon is analogous to the production of the **B-H** curve in ferromagnetism where the magnetic dipoles are arranged in a domain structure. When the electric field is decreased, and then reversed, a similar hysteresis loop to the **B-H** curve is demonstrated by ferroelectrics.

In the ferroelectric, any one domain shows parallel direction of polarisation, in the ferromagnetic, it is the alignment of unpaired electron spin. Domains are not coincident with crystal boundaries, and are often very much smaller than the crystallite grain size, because the minimum free energy arises from an increase in the number of domains. The transition from one domain to the next is not a sharp demarcation, but a gradual transition about 300 atoms thick, known as a "Bloch wall".

Figure 53 shows the analysis of the magnetisation curve (*see* Fig. 8) in terms of the domain concept. Domains of order are also recognised in the formation of superlattices. (*See* **Superlattice**.)

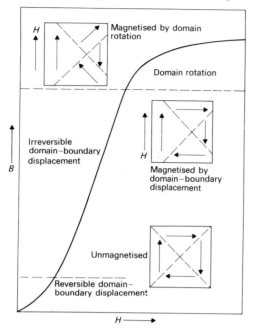

Fig. 53 *Magnetisation by domain growth and rotation.*

Donor (semiconductor). An atom present in an intrinsic semiconductor material (i.e. one with a narrow energy gap between the valence and conduction band) which has one more electron per atom than the host material, and an outer electron energy level close to the conduction band of the host atoms. The extra electron (not involved in the covalent bond) is able to break away from its own nucleus attraction, and move through the lattice when an electric field is applied. It is called a donor because of its contribution of an electron charge carrier.

Dore silver. Silver that contains some gold, i.e. silver bullion.

Dot-welding. A welding process for reclaiming castings and for joining metals. The electrode is fused by a low-temperature arc and is fed through a special pistol. Trigger controls operate the electrode feed and a needle valve in the pistol body controls the air supply used for cooling the welds.

Double-double iron. Blackplate produced by the pack-rolling process. In this process pairs of sheets are rolled, parted and doubled to form "fours". These are reheated and rolled to slightly more than the required length of sheet, trimmed to approximate size, doubled, reheated and rolled to length as double-double packs, or a pack of eight sheets which can be trimmed to size. The sheets are then separated.

Double refining. A duplex refining process in which a fire-refined metal is further purified by electrolytic refining. The term is also used in the production of wrought iron to indicate the processes in which merchant bar (single rolled iron) is cut into short length, piled or "faggoted", reheated and rerolled or forged to bar. During this double refining a certain amount of slag is expelled from the metal and the metal bar is made more uniform in composition. The slag fibres are considerably elongated and reduced in cross-section. This treatment gives a decided improvement in strength and ductility.

Double sintering. A term applied to a method of producing closer tolerances in components made by powder metallurgy methods. The powder is first compacted cold and then subjected to a low-temperature sintering. It is thereafter subjected to further compacting, usually in a closed die, and given a final sintering at a higher temperature.

Double skin. A defect in semi-finished products in which a surface layer is separated from the main body of the metal by a thin film of oxide. This defect is most frequently found near the bottom parts of an ingot and is often caused when an original splash

solidifies and is subsequently partly remelted but not completely welded into the mass.

Double treated. A term sometimes used in the USA to differentiate steels which have been hardened and toughened by quenching and tempering from those which have been single treated or oil-quenched.

Dowel. A slightly tapered pin or spigot used for locating the various parts of foundry patterns. Each pin is securely fixed in one part of the pattern and fits snugly into the corresponding hole of the adjoining part so that the complete pattern can be reassembled accurately whenever required.

Dowel plate. A steel plate drilled with a number of holes which serves to locate the pins of a divided pattern.

Downcomer. A tube that conveys the gas from a blast furnace. Usually there are several downcomers to one furnace. They are brick-lined and they join up into one main duct before entering the dustcatcher.

Dowson gas. A gaseous fuel obtained by passing steam and air over heated coal or coke. It may be regarded as a mixture of producer gas and water-gas. Since the reaction of steam with red-hot coke is endothermic and the reaction of air with red-hot coke exothermic, it is possible to average the proportions of air and steam so that the mass of fuel is kept at a fairly constant temperature and the process made continuous. The gas produced by such a process is known as Dowson gas, and it has a calorific value of 5 500-6 000 kJ m^{-3}.

Dowson gas plants use either coke or anthracite coal because the tar from bituminous coal gives rise to difficulties in working. The following table gives the comparative analyses of Siemen's (producer) gas (q.v.), Dowson gas and Mond gas (q.v.).

	Siemen's Gas	Dowson Gas	Mond Gas
Carbon dioxide, %	5.2	6.1	15.0
Carbon monoxide, %	24.4	26.0	12.0
Hydrogen, %	8.6	18.5	28.0
Methane, %	2.4	0.3	2.0
Nitrogen, %	59.4	49.1	43.0

Dozzle. A refractory-lined box fitted to the top of a mould to counteract the tendency to "piping". It is heated up before the hot metal is poured in, and provides a reservoir of liquid metal which can fill up the pipe as it forms. Also called feeder-head and hot-top.

Draft. The taper given to the sides of a foundry pattern to permit easy withdrawal from the sand mould, or of a die casting from the dies.

The term is also used in rolling to indicate the amount of reduction in cross-section per pass. This reduction is usually expressed as a percentage of the original cross-sectional area.

Drag. The bottom half of a foundry moulding box.

Drag-over mill. The classification of rolling-mills is based on the manner of arranging the rolls in their housings. When the rolling is in one direction only on 2-high mills, and the piece is returned over the top of the rolls to be re-rolled in the next pass, the mill is known as a drag-over mill. This type of mill was formerly used mainly for the production of light sheets and tin-plate. It is still used by merchant mills for rolling tool and high-alloy steels. Also known as "pull-over mill".

Draper washer. A device for washing coal in which water is introduced into a cone so that it has a low current velocity at the apex, where the coal enters.

Draught. The flow of air through a furnace.

Draw. An internal fissure caused by inadequate feeding during the solidification of a casting.

Drawbench. The stand that holds the die and drawhead used in drawing wire, rod and tubing.

Drawing. (a) Contraction during solidification leading to the formation of cavities and spongy areas; (b) the process of forming wire, tubes or rods by pulling through metal dies; (c) a forming process for producing hollow vessels or tubes from metal blanks; (d) the removal of a pattern from a mould; (e) an American term for tempering.

Drawing back. *See* **Tempering**.

Drawing down. The operation of reducing the cross-section of a bar by forging.

Draws. Contraction cavities or spongy areas found in castings near abrupt changes of section, due to lack of feeding and unequal contraction.

Dresser. (a) A mallet used for shaping sheet metal; (b) an iron block used in forging angular metal work; (c) a diamond-tipped tool used for truing the surfaces of abrasive wheels.

Dressing. (a) The separation of ore concentrates from associated impurities; (b) the removal of runners, risers, flashes, etc., from metal castings; (c) the solution or liquid painted on the mould or core to give protection against the action of liquid metal and help to produce a smooth, clean skin on the casting.

Mould dressings are commonly of two types: (a) refractory material such as bone ash, lamp-black, chalk, etc., either applied dry or more usually mixed to a thin paste with water and sprayed or brushed on the mould. If ethyl silicate is used it may be in either aqueous or spirit solution and is usually brushed on. When the spirit solution is used the excess spirit is burned off; (b) flaming, dressing of oil or resin often thickened with graphite, tar or similar material. This latter type is the one most commonly used.

Drift. A tapered tool of circular section (truncated cone) used for expanding the tubes and as a punch.

Drift test. A workshop test in which the ductility of a metal is assessed by driving a conical drift into a hole drilled in a plate. A hole of a given diameter is drilled into a plate near to an edge. The drift usually has a taper of one in ten, and is driven into the hole until a specified increase in diameter is obtained or until cracking occurs.

The name is also applied to a test applied to tubes, in which the diameter is increased at one end by forcing into the bore a conical drift of specified taper. This latter test is also referred to as a flaring test.

Drop forging. A process for the production in quantity of articles in metal by means of a falling weight forcing heated metal into a die. The process is essentially a moulding operation, the metal, at a sufficient temperature to bring it to a plastic condition but not to the molten state, being worked with the aid of machines. Two dies are required, one of which is fixed to the anvil and the other is fixed to the drop hammer or "tup". The heated metal is placed in the die in the anvil and the hammer allowed to fall, being controlled by guide-blocks. Several blows are usually required in quick succession. The small amount of metal extruded between the dies is called flash or fin and is removed by trimming tools. (*See* Fig. 54.)

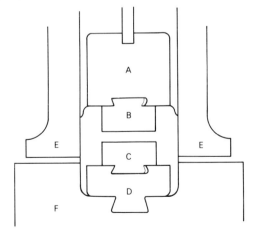

Fig. 54 *Cross-section of a simple drop hammer. A, is the tup; B, the top die; C, the bottom die; D, the bolster; E, the guides; and F, the hammer base.*

Drop forgings are made from bars or billets in which the "grain" or "fibre" runs longitudinally. The grain can therefore be worked into a position parallel to the direction of the principal stresses that are likely to occur when the finished product is in service. This represents one of the main advantages of drop forgings over castings. Crankshafts, connecting rods, flywheels, gear blanks, surgical instruments, hand tools and axles are examples of drop forgings.

Drop gate. A pouring gate or runner that leads directly into the top of a mould.

Drop stamping. A drop forging process (q.v.) or a component produced by such a process.

Drop test. An impact test in which a specified weight is dropped from a specified height on to the test piece and the deformation (or cracking) measured.

In the case of railway tyres, the latter are allowed to drop freely, in a running position from a specified height on to a steel rail attached to a heavy block. The tyre is then turned circumferentially through 90 degrees and the test repeated. The height of the fall depends on the diameter of the tyre.

Dross. Metallic oxides which float to or form on the surface of molten metal. The term is also applied to small coal of inferior quality.

Drossing. A process for the removing of impurities, consisting of oxides, etc., which form on the surface of molten metal. The dross is usually skimmed from the surface of the melt by long-handled scraper tools.

Dry cleaning. A term applied to coal cleaning processes. The fuel is freed from dust and graded according to size before it is discharged on to a reciprocating table equipped with riffles and air inlets. Under the influence of the moving table and the action of continuous or pulsating jets of air, the coal and refuse are stratified, and these materials carried to separate discharge points. The term also applies to the removal of oil from hot-dipped tin-plate with bran, middlings or other absorbent material.

Dry copper. Copper containing appreciable amounts of oxygen by reason of which it is liable to exhibit brittleness in working.

Dry cyaniding. A process for case-hardening steel components in a gas-carburising atmosphere containing a small, carefully controlled amount of ammonia.

During treatment both carbon and nitrogen are absorbed by the surfaces of the metal, and diffuse slowly towards the core. The rates of absorption and of diffusion are governed by the composition of the atmosphere and by the temperature. The process, which may be carried out at temperatures between about 780 °C and 930 °C, is also known as carbonitriding and gas cyaniding.

Dry drawing. A wire-drawing process in which the cleaned and pickled wire is drawn dry through a lubricant, as in the drawing of rods. By this process a highly polished surface is produced, and in steel wires the surface finish is known as lime-bright or bright.

The distinction between dry drawing and wet drawing is that in the latter process steel wire is immersed in a dilute solution of copper sulphate or stannous sulphate or in a mixture of these salts, and then in fermented rye liquor. From this latter bath the wire is drawn in the wet state. Wet drawn steel wire may have a coppery surface produced by chemical replacement of copper by iron in the sulphate solution.

Dry dross. A term sometimes applied to the sandy granular material which can be removed from a galvanising bath. It usually contains 5% or more of iron.

Drying stove. A stove used in foundries for drying cores and moulds. It is also called a core oven.

Dry moulding. The production of foundry moulds in dry sand, as distinct from greensand or loam moulding. Dry sand moulds are usually produced in greensand and subsequently dried in ovens.

Dry quenching. A process of cooling a hot substance without bringing it into direct contact with liquid. It is primarily used for quenching coke from coke ovens. The coke on discharge is usually at a temperature of about 1,000 °C. If water is sprayed directly on the hot mass, the heat in the coke is lost, the vapours produced have a corrosive action on steel and concrete in the vicinity of the plant, fine dust is carried away with the vapours, and the coke absorbs moisture, so reducing its heating value.

Dry quenching is accomplished by circulating inert gases through the coke. The inert gas is obtained by allowing a small amount of the coke to burn, producing a mixture of carbon dioxide and nitrogen. This is circulated through the mass of the coke by means of fans, and then led through tubes in a steam boiler where the heat passes into the water for steam raising. The cooled gases are re-circulated through the coke.

Several hours are required to cool the coke to a temperature suitable for handling and loading. The heat extracted from 450 kg of coke may be calculated as sufficient to produce more than 180 kg of steam.

DTA (Differential Thermal Analysis). *See* **Differential heating curves.**

Duct. A tube or passage in stone, concrete or metal used for conveying air or liquid.

Ductile-brittle transition. A characteristic of brittle materials, which show little plastic deformation under stress at ambient temperatures, is the gradual transition to more plastic behaviour with rise in temperature and/or slower rates of stressing.

All materials should theoretically show more brittle behaviour as temperature is lowered, or the rate of stressing increased, and this does occur. The requirement for plasticity is movement of dislocations, and any condition which impedes this must lead to lower plasticity. Materials which are to be used at low temperatures, especially if likely to be subjected to high velocity stresses (impact) require to be tested for the transition. Tests are usually high speed impact by a striker pendulum, and often use a notched bar to provide a high local stress concentration. Reduction in temperature of the test pieces enables a graph of energy to fracture against temperature of test to be constructed. Statistically, there is no immediate step transition, but a range of ductile-brittle behaviour, and the mid-point of the steepest slope of the curve is normally taken as the transition temperature. With a standardised test, all materials can then be placed in order of merit of transition temperature. (*See* **Crack** and **Impact tests.**) Figure 55 shows the results of typical tests, and the influence of radiation on the transition.

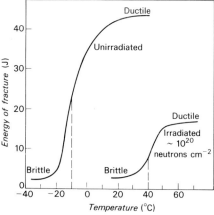

Fig. 55 *Change in ductile-brittle transition temperature with irradiation (En 2 steel).*

Ductile cast iron. A type of cast iron possessing superior strength and impact resistance, in which spheroidisation of the graphite is produced by treating the molten metal with magnesium. Maximum ductility is obtained when the matrix is entirely ferritic, and though such a structure can be produced in the as cast condition, heat-treatment may be required to develop optimum properties. The normal treatment recommended is to hold in the austenitic zone (at least 5 hours at 930 °C) to break down any cementite that may be present, then cool to below Ar_1 and hold at about 680 °C for a sufficient time to give highest ductility. The term ductile cast iron is used mainly in the USA for spheroidal graphite cast iron (q.v.).

Ductile fracture. *See* **Crack.**

Ductility. The property that enables solid substances, particularly metals, to undergo cold, visible, plastic deformation. The metal thus becomes permanently extended (e.g. by drawing), with corresponding reduction in cross-sectional area without actual fracture or separation.

The tests most commonly used for determining ductility are cupping tests (q.v.), such as the Erichsen and Olsen tests. Bend tests and elongation in tensile tests provide a good indication of ductility.

Duff. Fine coal that is separated out by the screening plant.

Dull finish. For certain purposes the smooth, bright surface of cold-reduced steel sheet and strip is unsuitable and a dull finish is required. Finishes of this type are produced by etching the polished roll surfaces with sulphuric acid or other pickling liquors, or by shot blasting the rolls. Otherwise the finished sheets may be "blued" by oxidising the surface.

Dulong and Petit's law. In the condensation of gases and liquids, the vibration of atoms can be demonstrated to account for the energy considerations involved. In a crystal of N atoms, there will be $3N$ degrees of freedom (the three mutually perpendicular directions of crystallography) and therefore $3N$ normal modes of vibration. If the atoms are envisaged as three harmonic oscillators, with $3N$ possible frequencies, then the average kinetic energy of each oscillator is $\frac{1}{2}kT$, and the potential energy is also $\frac{1}{2}kT$, where k is Boltzmann's constant and T the absolute temperature. The total average energy is then kT per degree of freedom. The average energy per mole must be $3N_0 kT$ (N_0 is Avogadro's number) and hence the capacity of the material to absorb energy (the specific heat at constant volume, C_v) given by

$$C_v = \frac{d}{dT} \ (3N_0 kT) = 3N_0 k = 3R$$

since $N_0 k = R$, the universal gas constant.

This is the law of Dulong and Petit, that the specific heat of substances tends to a constant value of $3R$, $1.48\,\text{kJ}^{-1}\,\text{K}^{-1}$ (about 6.2 calories per degree). The law applies well at elevated temperatures, but at lower temperatures the specific heat of crystals tends towards zero.

The quantum mechanical approach, first attempted by Einstein and refined by Debye, gives a more complex solution, to account for the variations in vibrational frequencies with temperature, and fits the experimental data well (*see* **Characteristic temperature**). The only exceptions are Be, B and C, which have high values of elastic modulus and a Debye characteristic temperature above room temperature.

Dumet. A composite material used in the form of wire. It consists of a core of low expansion alloy of nickel and iron (about 42% nickel) sheathed in copper. The amount of this latter metal may vary from 20 to 25% of the total weight. It is mainly used to replace platinum as the "seal-in" wire in lamps and vacuum tubes, the copper assisting in the production of a gas-tight seal. Owing to the very different radial and longitudinal characteristics, the maximum practical diameter of the wire is restricted to about 1 mm.

Dummying. A preliminary forging operation used for rough shaping heated metal before it is drop forged between dies.

Dump test. A test to determine the suitability of billets, bars, wire, etc., for hot or cold forging. As a hot-forging test, a sample length is heated to the forging temperature and hammered on end until it is reduced by a specified amount. As a cold-forging test, the specimen is similarly treated, but without heating. When used as a mechanical test for rivets, for example, short lengths of rivet bar equal to twice their diameter are taken. These should withstand being compressed cold to half their length. This test is also known as jump, knock down, slug up-ending or upsetting test.

Duplexing. The combination of two processes for melting or refining a metal. Steel may be melted in an open-hearth furnace and refined in an electric furnace. Iron may be melted in a cupola and subsequently passed to an electric furnace for further treatment.

Duplex structure. A microstructure composed of two phases.

Durain. A separable constituent of dull coal. It has a hard, firm, granular structure and frequently contains an appreciable number of spores. Other separable constituents of coal are known as clarain, fusain and vitrain (q.v.).

Duralplat. A composite material consisting of a sheet of aluminium alloy of the duralumin type coated on both sides with an aluminium-magnesium alloy. In Alclad (q.v.) the coating is pure aluminium.

Durex bearings. Copper-nickel powder impregnated with lead-base alloy, sintered on steel.

Durionising. An electrolytic process for depositing coatings of hard chromium.

Duroskop. A pocket hardness tester in which a diamond-tipped pendulum hammer is released from a fixed height and allowed to rebound freely from the surface of the specimen. The instrument is normally held against a vertical test surface, but it can also be used against horizontal specimens provided that, in the latter case, the rebound readings are converted to "standard" values.

Durville process. A process by which castings can be produced without intermingling of the molten metal with slag and at the same time avoiding contact of the metal with air during the casting process. The equipment consists of the mould and reservoir or ladle, which are rigidly attached to each other and connected by a passage having similar sectional dimensions to the mould. The ladle and mould are mounted in line and so arranged that the whole assembly can be tilted through 180 degrees (*see* Fig. 56).

The molten metal is first fed into the reservoir, where it may be treated if required by the requisite additions. Casting is effected by tilting the assembly so that the metal flows into the mould without turbulence. The surface of the liquid metal remains horizontal and continuous throughout, and there is no real pouring of the mass of metal.

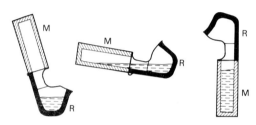

Fig. 56 *Three stages of the Durville process. M, is the mould; and R, the reservoir.*

The process is particularly applicable to alloys containing aluminium which are ordinarily prone to suffer from entrapped alumina films. The surface quality of ingots produced by this process is excellent and much superior to that obtained with other methods of casting. No mould dressing is required, and as the transfer of the liquid is a bulk transfer, low casting temperatures can be used.

Dust catcher. A device in which dust is extracted from furnace gases. The devices used to remove solid particles from gases may be classified according to the principle on which they operate: (*a*) sudden changes in the direction of the gas flow caused by baffles, etc.; (*b*) mechanical collector imparting a circular motion to the gas, causing dust to be thrown out of the gas stream by centrifugal force; (*c*) washer or wet scrubber, which depends on high velocities to atomise the water and wet the dust particles; (*d*) sonic and ultrasonic vibrations which cause the particles to collide and agglomerate; (*e*) electrical precipitator (*see* **Cottrell process**); (*f*) pebble filter which has a bed of granular material moving transversely across the path of the gases.

Dutch gilding. A transparent yellow lacquer usually applied to tin foil.

Dymaz. A black finish for zinc-base die casting alloy Mazak, known in the USA and on the continent of Europe as Zamak.

Dynamic strength. Resistance to suddenly applied or to rapidly changing loads, e.g. the impact of a falling weight. The effect of alternating stresses that vary between fixed limits is dealt with under **Fatigue** (q.v.).

Dynamic tests. These are impact tests that give a qualitative measure of one type of toughness, rather than an absolute figure. The toughness so measured is not directly related to fatigue. The tests are usually of the single-blow notched-bar impact type. The dimensions of the test piece and of the notch are defined in such tests as the Izod, Charpy and Frémont tests (q.v.). The resistance to impact is generally stated in terms of the energy absorbed (in joules or foot-pounds) in breaking the test piece.

Dystectic mixture. The term used to describe a mixture of two or more substances, in such proportions as to produce a maximum melting point. Thus on altering these proportions in any way, the melting point of the mixture will be lowered.

E

Ear. A wavy projection formed during the deep drawing process, in consequence of directional properties in the sheet metal.

Earthy cobalt. An amorphous deposit of variable composition consisting largely of hydrated oxide of manganese containing up to 40% cobalt oxide mechanically mixed with it. Sulphide of cobalt and oxides of iron, copper and nickel are sometimes present.

Ebonite. The product of high sulphur additions to natural rubber, followed by heating. The extensively cross-linked isoprene is hard and rigid, with high electrical insulation properties. It was extensively used in the early days of radio transmission and reception, but cheaper insulators now exist.

Eccentric converter. A bottom-blown steel-converting furnace having the mouth located at the top and placed to one side so that the opening lies in a plane inclined to the base of the vessel.

Eclair. An undercooling effect sometimes observed in the operation of cupelling gold. When the globule of gold is perfectly clear and free from vibration, crystallisation may not begin until the temperature has fallen below the freezing point. Then the development of heat (due to the evolution of latent heat) causes a brilliant light or "flashing" known as "eclair" at the moment of solidification.

Eddy current. In the armature of a dynamo the core is made of iron, to cause the total flux due to the field magnet to be concentrated where it is cut by the coils wound on the core. The rotation of a mass of iron, in a magnetic field tends to produce currents in the iron, and these, circulating locally, are termed eddy currents. They flow at right angles to the magnetic field and parallel to the rotating shaft. They necessitate a large expenditure of energy and produce a large amount of heat. Such currents cannot be entirely eliminated, but they are kept down to reasonable proportions by building up the core of thin sheets lightly insulated from one another.

The sheets are usually about 0.6 mm thick, and the insulation consists of resin, thin paper or an oxide coating produced by pickling. The insulation offers considerable resistance to the current flow parallel to the shaft, but practically no resistance to the magnetic flux. The armature is usually ventilated to keep down the temperature.

In the case of transformers there is an alternating flux which causes eddy current losses in the core, and since the losses are proportional to the square of the thickness of the core strips, these must be made as thin as practicable.

The high-frequency currents used in induction furnaces produce in the metal in the bath eddy currents that may generate sufficient heat to melt the charge. The maximum frequency that is necessary to melt a charge in a given bath depends on the resistivity of the metal to be melted and on the diameter of the crucible.

Edge dislocation. *See* **Dislocation**.

Edger. The part of a drop-forging die used to distribute the metal so that it will fill the die impression.

Edge rolls. Some rolling-mill lines are equipped with grooved rolls that control the width of the strip and produce the required contours on the edges. Edging rolls are usually inverted vertically immediately ahead of the horizontal work rolls.

Edison gauge. An American wire gauge devised by Edison to simplify electrical calculations. It is based on the circular mil (i.e. the area of a circle 1 mil (0.001 in) in diameter) and is sometimes known as the circular mil gauge.

Efflorescence. Crystalline substances which lose water on simple exposure to air are said to exhibit efflorescence, and the occurrence of the phenomenon is indicated by the presence of a fine powdery deposit on the surface of a crystalline rock mineral, or brick.

Effluents. A term used to indicate the various spent liquors that are allowed to flow away as waste from plating shops, pickling tanks, etc. The main interest in effluents is the rendering of them harmless and innocuous before disposal. Effluent treatment usually involves (a) the neutralisation of acid liquids — usually with lime; (b) the precipitation of the salts of heavy minerals as hydroxides by treatment with lime; (c) the treatment of cyanides, either by removal as hydrocyanic acid or by conversion into cyanates or into prussian blue. The former is accomplished by acidification and subsequent aeration, the latter by treating with chlorine in the presence of alkali or by treatment with iron II (ferrous) sulphate and lime; (d) chromates are usually reduced with iron II sulphate and precipitated with lime.

Effusion. The study of the kinetic theory of gases leads to the conclusion that the molecular kinetic energy of all gases at the same temperature is constant and equal to 3.41 kJ. The molecular speeds of gases are therefore inversely proportional to the square roots of their molecular weights. When gases are forced by pressure through small apertures into (say) air, the actual speeds are reduced by collision with the gaseous molecules in the air, but the relative rates are still inversely proportional to the square roots of their molecular weights. This phenomenon is called effusion. By measuring the rates of effusion of gases their molecular weights can be compared. The usual apparatus employed is of the type devised by Bunsen or some modification of this.

Egg sleeker. A tool used by foundrymen for making rounded corners in a mould, one end of the tool being spoon-shaped.

Eight minus N rule (8 − N). If N is the number of the group in the periodic table into which an element fits, then (8 − N) gives the principal valency of the element. The group number, N, is in fact the number of electrons in the outermost energy levels, i.e. s plus p electrons, and (8 − N) is therefore the discrepancy between the nearest inert gas structure following the element (with 8 outer electrons) and the element itself. In ionic or covalent compounds, (8 − N) indicates the number of electrons to be obtained from the other compounding element(s). Note that in the case of metals with only s outer electrons, (8 − N) will be 7 or 6, and hence it becomes simpler to lose these outer electrons to combine with ions whose value of (8 − N) is lower.

Elastic after-effect. When a load is applied to a body, an instantaneous strain results, which increases slowly over a period of time without further increase in load. The rate of increase diminishes, and finally vanishes. On removal of load, the reverse pattern occurs, an instantaneous contraction, and then further contraction diminishing with time. This is the elastic after-effect, not normally measured in mechanical testing. The lag arises from heat evolved slowly as a result of the stress, the effect of this heat energy on the vibration of the atoms, effects on the diffusion of impurity atoms, and mechanical hysteresis arising from the varying orientation of individual grains. Impurity atoms may diffuse into positions offering less strain when the material is stressed, first observed by Snoek. This type of influence would obviously be temperature dependent, for at low temperatures diffusion is slow, and at high temperatures it can occur too quickly to slow down the strain.

Measurements of internal friction can reveal these effects more readily than straight tensile tests, and much can be derived from these tests about the detailed mechanisms of deformation and fine structure. (*See also* **Visco-elasticity.**)

Elasticity. The property which enables deformation resulting from loading of a body to be completely reversible, so that none remains on removal of the original load.

A theoretical treatment indicates the behaviour discovered experimentally by Hooke, that for small strains, stress and strain are proportional. Elasticity is therefore confined to small strains, and a finite limit of stress can be observed, above which deformation is no longer reversible.

Hooke's law can then be written: stress = constant × strain. This constant is a fundamental property of the material, dependent on the attraction and repulsion forces in the atomic bond. It is called Young's modulus of elasticity, after the first person to recognise its significance. Young's modulus is independent of the sign of u, the displacement, and so holds for tension and compression. It is usually given the symbol E, stress is σ and strain ϵ. Then $\sigma = E\epsilon$.

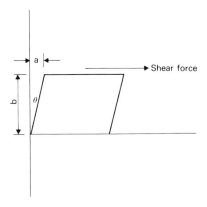

Fig. 57 *Shear strain.*

If a stress is applied tangentially through a plane of atoms in the material (*see* Fig. 57) a lateral strain results by displacement of this plane relative to other planes parallel to it. Tangential stresses are referred to as shear stresses, and the strains as shear strains. As with tension and compression, the stress and strain are proportional provided the strain is small.

$\tau = G\gamma$ where τ = shear stress, G is the shear modulus, another elastic constant, and γ is shear strain. Shear strain can then be measured in terms of angular displacement of the planes, then $\tan \theta = a/b = \gamma$ where a is the linear displacement in the planes, and b the distance between the planes.

Analysis of a body under stress can be interpreted for all the crystal systems, to evaluate elastic constants. In engineering terms, this is simplified by the fact that only systems of high symmetry are normally suitable for mechanical loading, and a new term, Poisson's ratio μ is introduced, the ratio of deformation in one direction to that in the perpendicular direction. For cubic crystals of high symmetry only three constants are required in such analysis, E, G and μ. If the body is perfectly isotropic, only E and μ are needed. The relationship between the normal stress and shear stress condition is then:

$$G = \frac{E}{2(\mu + 1)}.$$

Isotropic pressure is equivalent to three equal and normal tensile or compressive stresses, acting in three mutually perpendicular directions. A bulk modulus K can be used in this consideration.

$$K = \frac{2G}{3} \frac{(1+\mu)}{(1-2\mu)} \quad \text{or} \quad K = \frac{E}{3(1-2\mu)}$$

For the three main classes of materials, approximate figures can be given for the three important constants.

(a) Covalent and amorphous structures $\mu \approx 0.25$; $G \approx 0.4E$; $K \approx 0.6E$; *(b)* metallic lattices, crystals, alloys $\mu \approx 0.30$; $G \approx 0.38E$; $K \approx E$; *(c)* long chain molecules (polymers) $\mu \approx 0.50$; $G \approx 0.3E$; $K \gg E$.

When the higher terms of the Taylor series in the force-displacement equation for stretching a bond involve measurable values of u^n then Hooke's law no longer applies. As u, the displacement, increases, a non-linear relationship between stress and strain is involved, and a gradual change takes place in the stress strain curve, such that a much smaller increment of stress is needed to produce quite a measurable displacement. The curve becomes convex in the stress direction, and removal of the load does not now return the body to its original dimensions, as only the elastic strain is reversible. The permanent increase in length which must follow any application of stress beyond the so-called elastic limit, after the load is removed, is called a permanent set, or more correctly, plastic deformation (*see* Fig. 58).

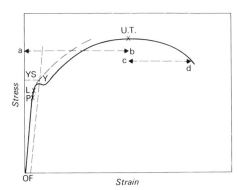

Fig. 58 *Typical stress-strain diagram of a metal under tension. OF, 0.2% set; ab general extension; cd, local extension; P, limit of proportionality; L, elastic limit; Y, yield point; YS, yield strength; UT, ultimate stress.*

Due to hysteresis, the inevitable lag in time of the deformation, it is possible to recognise two similar points on this curve, around the point where non-linearity becomes evident. The limit of proportionality is strictly the stress at which Hooke's law just ceases to be obeyed. The elastic limit is the stress beyond which a permanent set is measurable on removal of load. One or other of these points is an important parameter in engineering design for each material used.

Elastic limit. The maximum stress that can be applied to a body, and after the removal of which the body is able to regain its original dimensions, is called the elastic limit for that body. It is, in fact, the maximum stress that has no after-effects as far as can be measured. In some cases the elastic limit and the limit of proportionality are practically the same.

Elastomer. A polymer with a considerable potential for elastic extension, greater than all other solids. The structure of elastomers is generally one of long chains (often formed by addition polymerisation) with rotation of carbon-carbon bonds to give a coiled or zig-zag molecule, whose overall length is shorter than the actual chain length. Application of tensile stress first extends the contorted chain, pulling it straight, and only when this process is complete begins to stretch the actual bonds themselves. By performing a small amount of cross-linking between the chains, the strain can be transferred from one chain to another, so that no one chain becomes subjected to bond stretching before the majority of chains have been stretched from the convoluted form. No plastic deformation can result in this case, until almost 100% of chains have contributed to the elastic deformation. Elastic strains of three orders of magnitude become possible by this means.

Not all polymers with long chains are elastomers, however, because the bond rotation may be difficult with certain groupings and stretching of convolutions is limited in many cases by large groups of atoms along the chain, which interfere both with bond rotation and with sliding. The variation in properties between polyethene, a straight hydrocarbon, and polyvinyl chloride, in which one in four of the hydrogen atoms of polyethene has been replaced by the much larger chlorine atom, is an example. The addition of even larger groups, like the benzene ring in polystyrene, serves to emphasise the point, polystyrene being more rigid than PVC.

Elastomers can be modified, not only by the amount of cross-linking, but by the inclusion of plasticisers. PVC can be a rigid plastic with tangled chains, or an elastomer (but not so elastic as natural rubber) with the addition of a plasticiser. Natural rubber is an elastomer, whose full properties require some cross-linking (vulcanisation) but contains the *cis* form of isoprene, whose bonds can rotate. The alternative *trans* form of the isoprene chain cannot rotate so readily, and so is not elastomeric, even if other materials are added to it.

Electrical conductivity. The property of a substance by virtue of which electricity is transferred through it when an electric potential is applied at two points in the substance. The unit of measurement is the reciprocal ohm (mho). A substance in which unit quantity of electricity is transferred through unit cross-section under the influence of unit potential gradient is said to have unit conductivity. This unit is also called the "siemens". Good electrical conductors generally possess good thermal conductivity and there is a relationship between these quantities. (*See* **Conductivity**.)

Electrical resistance thermometers or pyrometers. All metallic substances, with the exception of a few alloys, show increase in electrical resistance with rise in temperature, and the rate of increase for pure metals is approximately linear. The increase in resistance is thought to depend on the internal energy of the metal, so that over any considerable range of temperature the resistance R is more conveniently represented by

$$R_t = R_o(1 + \alpha t + \beta t^2),$$

as suggested by Callendar.

Electrical resistance thermometers are almost invariably made of platinum. Such a thermometer consists of a pure thin platinum wire wound upon a mica frame and placed in a hard glass or porcelain tube. For use in furnaces this latter tube is usually protected by an outer steel tube, which may be from 150–300 cm long. The leads may be made of copper, but are preferably of platinum, and compensation leads are included so that the readings are independent of the length of tube immersed.

The platinum coil is used in conjunction with a deflection galvanometer of the moving-coil type, with or without a potentiometer to give rapid readings. But for more accurate readings it is usual to balance the resistance or the e.m.f. against known values. The fundamental interval of each individual thermometer must be determined by observations in ice and in steam.

The standard melting points of a number of pure elements may also be used, e.g. lead, sulphur, silver, etc.

The reductions of the thermometer readings to the thermodynamical scale, i.e. the scale correction, may be effected by the use of the difference formula

$$t - t_p = d.t. \frac{(t - 100)}{10\,000}\,;$$

where $d = 1.5$. This value of d is fairly constant provided the platinum coil does not become contaminated with furnace gas or refractory substances.

Electrical resistivity. This is the reciprocal of the electrical conductivity (q.v.). The unit of measurement, ohm, is defined as the resistance between two points of a conductor when a constant difference of potential of 1 volt, applied between these two points, produces in this conductor a current of 1 ampere, the conductor not being the source of any electromotive force.

The resistivity or the specific resistance (ρ) of a metal is the resistance of a unit cube of it measured across opposite faces. This is sometimes called the volume-resistivity as opposed to the mass-resistivity. This latter is defined as $\rho d \times 10^4$, where d = the density of the metal, and it is sometimes used in considering the weight of a conductor that is capable of carrying a given current at a given voltage.

At absolute zero, an ideal metal should have zero electrical resistivity, but since structural imperfections occur, a residual resistance remains (*see* **Conductivity**). As the temperature rises above absolute zero, the resistance changes proportional to the fourth or fifth power of temperature, up to a level of about one tenth of the Debye characteristic temperature. Above this, the contribution of thermal energy approaches its optimum, and the scattering effect on electrons becomes directly proportional to absolute temperature between $0.5\,\Theta_D$ and Θ_D. This was demonstrated experimentally by Mathiessen, whose law states that

$$\rho_e = T.\frac{d\rho_o}{dT} - \rho_o + \rho_{str}$$

where ρ_e is electrical resistivity, ρ_o the displacement from direct dependence on T, ρ_{str} the resistivity due to structural imperfections and impurities, and T the absolute temperature. The value of ρ_{str} is increased by quenching, cold-working, and the amount of impurities. Measurement of resistivity at low temperatures thus becomes an important tool in structural studies (*see* Fig. 59).

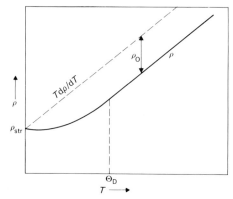

Fig. 59 *Mathiessen's law.*

When pressure is applied to a metal, the resistivity decreases, since the number of electrons per unit volume increases. This fact is used as the basis of electrical resistance strain gauges, whereby changes in resistivity can be correlated with elongation.

The electrical resistivity-temperature graphs of certain transition metals show a pronounced curvature at temperatures round about Θ_D. This is believed to result from the hydridisation of s and d electrons, and the effect of this on the electron cloud lattice perturbations. These effects disappear at higher temperatures. Near the melting point, an abnormally large increase in resistivity occurs with some metals, which indicates that vacant lattice sites are being created in profusion, as the long range order of crystallinity begins to change to short-range order.

A number of metals and their compounds lose their resistivity near absolute zero, and become superconductors.

Electrical steels. A name given to a group of steels used as sheet for the generators, motors and transformers of electrical equipment. They all have a low or very low carbon content, and contain from 0.5 to 5% Si, according to the grade and its intended application.

Electric arc welding. A process of welding by fusion in which the heat is developed by an arc struck across a gap between the electrode and the work. It is usual to distinguish the different types of arc welding according to (a) the nature of the electrode (e.g. carbon or metal), (b) the coating on the electrode, i.e. whether the electrode is bare or coated with flux material, and (c) the presence or otherwise of an inert gas for shielding the weld. Shielding is generally accomplished by arranging for a layer of molten flux to cover the molten metal or for an atmosphere of inert gas to surround the arc.

In carbon-arc welding direct current is invariably used, and the carbon rod or graphite electrode is the negative pole of the arc, while the work itself is the positive pole. This arrangement makes for a steadier arc and obviates carbon being carried into the weld.

A filler rod is not always required, but when used may or may not be part of the electrical circuit. The arc may be shielded or unshielded, in the former case usually with non-metallic substances. When the process is automatic the arc has a magnetic field superimposed to keep it steady. (See also **Metallic arc welding, Welding.**)

Electric flux density. When a dielectric material is placed in an electric field which already exists in air or a vacuum, the field is changed, the extent being a measure of the relative permittivity of the material referred to air or vacuum.

If D is the electric flux density, Q a point charge, **a** the unit vector of Coulomb's law (reflecting the direction of force on the charge), and r the distance of the measuring point from the point charge,

$$D = \frac{Q}{4\pi r^2} \mathbf{a}.$$

The flux density is therefore independent of the medium. If the point charge is assumed to be at the centre of a sphere of radius r, the flux density is identical at all points on the surface of the sphere. The total flux will be the flux density multiplied by the area of the sphere, i.e. the value of the charge, Q. This is the basis of Gauss's law.

TABLE 19. *Approximate Fusing Currents for Common Fuse Wire.*

S.w.g.	Fusing currents (amperes)				
	Cu	Standard alloy	Pb	Sn	Al
12	344	–	47	55	260
14	232	–	32	37	170
16	166	–	23	27	120
18	110	–	15	17	80
20	70	–	10	11	50
21	–	8.0	–	–	–
22	50	–	7.0	8.0	35
23	–	5.0	–	–	–
24	35	–	5.0	5.0	25
26	25	–	4.0	4.0	18
27	–	3.0	–	–	–
28	17	–	3.0	3.0	14
30	15	–	2.0	2.0	10
32	12	–	1.5	2.0	8.0
34	9.0	–	1.0	1.5	7.5
36	7.0	–	1.0	1.0	5.0
38	5.0	–	0.75	0.75	4.0
40	3.0	–	0.5	0.5	2.5

Electric fuse metals. To protect premises from fire and instruments and circuits from injury should an excessive current flow, it is usual to introduce fuse wires into the electrical circuits. These wires are made of suitable materials and of such dimensions that they will melt and break the circuit when the current exceeds a certain predetermined value. Of the materials used for fuses, copper wire or thin copper strip is often employed for very large currents; zinc strip is satisfactory where a fuse with a fair time-lag is required, i.e. one that will not quickly melt on the smallest overload; tin is useful for fuse wires for all ordinary

circuits and low pressures. Standard alloy contains 37% lead, 63% tin.

For the protection of electrical apparatus, the diameter of a wire that will fuse at a known current is obtained from the expression $C = Ad^n$, where C = current, A is a constant calculated from the required temperature of the wire (T), its specific resistance ρ and the emissivity (α), whence $A = \sqrt{T\alpha}/0.097\rho$. ($A$ has the value for tin = 12.8, for lead 10.8 and for copper 80.) The diameter of the wire is d and the value of n varies from about 1.2 to 1.5, having the lower value for small currents of 10 amperes or less.

The melting point of the material is not, however, the sole criterion of suitability for service, since in some applications the formation of surface corrosion products may delay or inhibit the operation of the fuse.

For currents up to 10 amperes the fuse wire should be at least 4 cm long. The approximate fusing currents for common fuse wires of various gauges are given in Table 19.

Electro-analysis. A variation of the usual procedure of gravimetric analysis in which the metal is deposited electrolytically on a weighed electrode. The method is often simpler and capable of giving more accurate results than the usual methods. The success of electro-analysis depends on being able to obtain coherent deposits of metal on the electrode, and in the case of separations, in being able to adjust the conditions for the complete deposition of the desired metal while the rest remain in solution. In all cases the potential difference across the electrodes must be carefully adjusted, while the temperature, in most cases, must be maintained within certain limits. Table 20 gives details of the conditions for a few types of analysis, but the method can also be used to advantage in the case of salts of cobalt and indium. The current density as stated is that for stationary electrodes. If the electrolyte is agitated by using a rotating stirrer or rotating electrodes, much higher current density can be used, e.g. of the order of ten to twenty times the values in the table.

TABLE 20. *Details for Electro-analysis of Metals.*

Metal	Electrolyte	Potential difference, (volts)	Current density (A/dm²)	Temp. (°C)
Copper	Add nitric acid, 2-3%	2.2	0.5-2.0	50-60
Lead	Add nitric acid, 10%	2.3-2.7	1.0-2.0	50-60
Nickel	Add ammonium sulphate and ammonia	2.8-3.5	0.5-1.5	15-20
Nickel and	Add ammonium sulphate, ammonia	–	Ni 0.2	90-92
Zinc	Add sodium sulphite	–	Zn 0.6-1.0	15-20
Silver and	Slightly acid (HNO₃)	Ag 1.36	–	55
Copper	Add alcohol	Cu 2.2	–	55
Cadmium	Add sodium cyanide	–	2.0-3.0	35
Brass	Add sulphuric acid	–	–	–

Electro brighteners. Materials added to the electroplating bath to improve the quality of the deposit. A wide variety of substances is available, including glue, gelatine and many organic substances. In the case of copper, arsenious oxide As_2O_3 in solution in potassium cyanide, or hypo have been recommended. For zinc and nickel, molybdic oxide has been used, while cobalt salts are favoured for certain other metals.

Electrochemical corrosion. *See* **Electrolytic corrosion.**

Electrochemical equivalent. Since the same amount of current, when passed through solutions of different electrolytes, discharges weights of ions that are in the ratio of their chemical equivalents, it may be concluded that a definite amount of electricity is associated with the discharge or dissolution of 1 gram of any ion. This amount is called 1 faraday and is equal to 96 500 coulombs. The weight of an ion deposited by 1 coulomb of electricity is known as the electrochemical equivalent. The weight (m) of any ion deposited from an electrolyte by a current of A amperes in t seconds is given by $m = At\epsilon$, where ϵ is the electrochemical equivalent. The value of the electrochemical equivalent of any ion is the chemical equivalent in g × 0.00001036.

Electro-colour process. An electrodeposition process for applying thin films of metallic oxides which vary in colour according to the conditions of deposition. Such films may provide a very attractive finish, but they are not durable and must be lacquered if long service is desired. The process usually involves the deposition of copper from a copper lactate solution using a voltage of about 0.5 volt and a current density of 0.027-0.27 A m^{-2}.

Electrode. In the decomposition of a liquid by the passage of an electric current through it, the conductors that are placed in the liquid, and by which the current enters and leaves the bath, are called the electrodes. The one by which the current enters the liquid is the positive electrode or anode. The other electrode is called the cathode.

The term is also used in reference to the carbon or metal rods that convey current, used in welding.

Electrode efficiency. The ratio between the amount of metal deposited at the cathode of a plating cell and the amount which should theoretically be deposited in accordance with Faraday's law.

Electrodeposition. The processes for depositing a layer of metal on to the surfaces of metallic or non-metallic conductors by immersing the articles in an electrolyte containing a salt of the metal to be deposited, and making them the cathodes of the electrolytic cell. Also referred to as electroplating.

Electrode potential. When a crystal of a soluble substance is placed in water, there is a tendency for molecules to pass from the crystal to the surrounding liquid. This tendency, because of its similarity to the vapour-pressure phenomena manifested by liquids, is called solution pressure. The molecules of the salt diffuse about in the solvent and some may recrystallise on the original crystal. When this reverse diffusion pressure is just equal to the solution pressure, the solution is said to be saturated. The salt and its ions are in equilibrium.

This description is applicable, with certain important differences, to the case of a piece of zinc introduced into a solution of (say) zinc sulphate. There is no appreciable solution pressure of the metal itself (the metal being practically insoluble in water). There are, however, ions of the metal produced which pass into the water, each atom in the process releasing two electrons which are left on the metal. Thus the metal becomes negatively charged and the surrounding liquid is positively charged. Opposing the tendency to produce more ions are (a) the diffusion pressure of the ions and (b) the electrostatic attraction between the ions and the metal. With stronger solutions of the zinc salt, the dissolving tendency will become less, since the diffusion pressure will have increased.

When the dissolving tendency is just balanced by the reverse diffusion tendency alone, there will be true equilibrium between the metal and its ions and there will be no potential developed between the metal and the solution. In that case $Zn \rightleftharpoons Zn^{2+} + 2\epsilon$ where Zn^{2+} is the ion of zinc and ϵ is the electron.

For equilibrium $[Zn^{2+}]/[Zn] = K_1$, or, since the concentration of zinc metal is constant, $[Zn^{2+}] = K_1[Zn] = K_{Zn, Zn^{2+}}$. The value of this latter constant is found to be of the order of 10^{41}, from which it would appear that zinc could be in true equilibrium with a solution only if the latter contained 10^{41} mols, or about 66×10^{38} kg of zinc per litre. Since this could never be realised, it is obvious that there must always be a potential whenever zinc is placed in a solution of one of its salts, though this will vary with the strength of the solution. This potential established between a metal and a solution is called the electrode potential of the metal.

TABLE 21. *Electrode Potential at 25°C of Common Metals.*

Metal	E_0, volts	Metal	E_0, volts
K/K$^+$	−2.92	Pb/Pb^{2+}	−0.12
Na/Na$^+$	−2.71	H$_2$/2H$^+$	0.00
Zn/Zn^{2+}	−0.76	Bi/Bi^{3+}	0.30
Fe/Fe^{2+}	−0.41	Cu/Cu^{2+}	0.34
Cd/Cd^{2+}	−0.41	Ag/Ag$^+$	0.80
Co/Co^{2+}	−0.28	Hg/Hg^{2+}	0.86
Ni/Ni^{2+}	−0.23	Pt/Pt^{2+}	1.20
Sn/Sn^{2+}	−0.13	Au/Au^{3+}	1.42

The measurement of the potential between a metal and a solution is made by combining the electrode under test with a standard electrode and measuring the e.m.f. of the voltaic cell so produced. Two standard electrodes are the calomel electrode and the hydrogen electrode. The potential of the latter is taken as zero and of the former is taken as +0.281 V at 25°C. The

electrode potentials of the commoner metals are given in Table 21.

When an electrolyte has been decomposed by a current there is set up an opposing or back e.m.f. which is considered the result of the potential energy of separated ions. This back e.m.f. has the value $E = HJZ/10^8$ volts, where Z is the electrochemical equivalent of the ion deposited; $J = 4.2 \times 10^7$; H = heat of combination of 1 g of the ion in question with the other ion produced in the electrolysis. Obviously the electric supply used for the continued electrolysis must have a greater voltage than the value for E.

Electro-extraction. The extraction of a metal from its ore by electrolysis.

Electro-forming. The production of metallic articles by electrodeposition of the required metal upon a pattern of the desired shape. It is largely used for making seamless hollow containers and moulds for plastic materials. The pattern may be metallic or non-metallic. In the latter case a very thin coating of silver is used as a conductor. For hollow vessels and intricate patterns, fusible alloy patterns, which can be removed readily by melting in hot water, possess many advantages.

Electro-granodising. A phosphating process applied to steel in which the steel is immersed in a phosphate solution and the chemical action accelerated by electrolysis (q.v.).

Electrography. A means of showing porosity in metallic or non-metallic coatings on metals. Paper, impregnated with a suitable electrolyte, is pressed against the coating and a counter-electrode is placed on the other side of the paper. When an e.m.f. is applied between the tested specimen and the counter-electrode, porosity is indicated by stains upon the paper. The technique is developed to reveal composition and heterogeneity of mineral samples.

Electroless plating. A process usually concerned with the deposition of a nickel-phosphorus alloy on metallic surfaces from an ammoniacal solution of a nickel salt with sodium hypophosphite, by chemical reduction in the presence of a catalytic metal such as iron or nickel. The method has been extended to the deposition of cobalt and of nickel-cobalt alloys from ammoniacal solution, and to the tinning of small mass articles by chemical replacement. The technique requires no externally applied current and all the surfaces of any article immersed in the plating solution are coated uniformly.

Electrolysis. Substances that undergo chemical decomposition by electrolysis include the solutions of the great majority of metallic salts, acids and bases, as well as a number of fused salts, etc. In the electrolysis of salts, the metal is always deposited at the cathode, while the anode may in certain circumstances dissolve in the solution.

The process of electrolysis is employed in electroplating and electro-refining, usually in aqueous solutions, and in electro-extraction, where fused salts may be used.

Electrolytic corrosion. When two dissimilar metals or alloys are in electrical contact with each other in the presence of an electrolyte they form the electrodes of a galvanic cell. One of the metals − the anodic electrode − is corroded more rapidly than if there were no contact between them. This type of corrosion takes place only when the electrical circuit through the metals and the solution is complete. The rate of corrosion is governed by the e.m.f. developed in the cell and is highest when the component metals are far apart in the electrochemical series scale, and when the metal ions go readily into solution.

Electrolytic dissociation. The molecules of acids, bases and salts, when dissolved in a suitable solvent, dissociate into particles (ions) that are charged electrically. The number of electrically positive charges is equal to the number of negative charges, so that the solution is on the whole neutral. The process of dissociation is regarded as a true chemical change and is reversible. The extent to which dissociation proceeds in any case depends on the following. (a) The nature of the solvent. It is established that the process of ionisation goes hand in hand with a chemical interaction between the solute and the solvent. Liquids that exhibit a high degree of association are the best ionising media. Additionally solvents such as water that have a high dielectric constant have high ionising power. *(b)* The nature of the solute. The dissociation of an electrolyte depends to a large extent on the valency of its ions. Atoms (and ions) of high valency tend not to dissociate to the same extent as ions of low valency. *(c)* The concentration of the solution. Generally the percentage dissociation increases with the dilution.

Electrolytic polishing. An anodic polishing process in which the object to be polished is made the anode in a suitable electrolyte. Solution of the anode occurs under the influence of the current and takes place in such a manner that its surface becomes progressively smoother and brighter. The process is characterised by selective attack on the surface of the anode whereby the upstanding roughnesses are preferentially dissolved. It therefore differs from mechanical polishing in which the prominences are caused to flow, producing an amorphous layer (*see* **Beilby layer**). The success of the process depends on the composition and temperature of the electrolyte, the current density and the time of polishing.

Table 22 lists some of the solutions that have been recommended for use with various metals. Of these, the electrolytes containing perchloric acid have been shown to be dangerous, especially in the presence of small amounts of organic matter.

Electrolytic polishing has already found useful application in the preparation of specimens for microscopical examination. When so polished, the surface of the material is free from the effects of cold working and from included abrasives, which defects are likely to appear as a result of mechanical methods.

TABLE 22. *Solutions for Electrolytic Polishing of Metals.*

Electrolyte	Weight, %	Conditions			Suggested Application
		Current density, A m^{-2} $\times 10^{-2}$	Temp., °C	Time, min	
1. Perchloric acid Acetic anhydride Water	19 76 5	0.04 −0.26	30	4-10	Most metals
2. Phosphoric acid Sulphuric acid Chromic acid Water (with or without addition agents, e.g. sodium dichromate, acetic acid, aniline, hydrofluoric acid)	67 20 2 11	0.11	50	60	Non-ferrous metals, stainless steel, brasses
3. Phosphoric acid Sulphuric acid Water (with or without addition agents, e.g. lactic acid, dextrose)	42 15 20	0.23	90		Carbon steels
4. Phosphoric acid Glycerol Water	42 47 —	0.0065 −0.04	120	8-15	Stainless steel and most metals
5. Phosphoric acid	100	0.065	65	5	Stainless steel, copper, brass
6. Sulphuric acid Water (additional phosphoric acid, chromium and aluminium sulphate)	73 27	0.11	30	0.5-2.5	Nickel, aluminium (no additional)

Electrolytic reduction. Ordinarily reduction is defined as the removal of oxygen (or the addition of hydrogen) from a substance, or in more general terms as the decreasing of the active valency of a positive element or increasing that of a negative element. In other words, reduction consists simply and always in the addition of electrons to an atom. When this process is produced by electrical means directly it is referred to as electrolytic reduction.

In a two-compartment cell containing as the negative pole a platinum plate charged with hydrogen and immersed in dilute acid, and as the positive pole a platinum plate immersed in iron III (ferric) chloride solution, the following action occurs. The hydrogen dissolved in the negative plate throws off hydrogen ions into the solution, leaving the plate charged negatively. This negative charge passes via the external circuit to the positive plate. Any iron III ions touching the plate gain an electron and are discharged to iron II ions, the change being represented: $Fe^{3+} + \epsilon \rightarrow Fe^{2+}$.

All reactions that take place in a voltaic cell are oxidation-reduction reactions. In the case of the Daniell cell the zinc loses electrons (oxidation) and the copper gains electrons (reduction), and the change is represented: $Zn + Cu^{2+} \rightarrow Zn^{2+} + Cu$. Any spontaneous oxidation-reduction reaction occurring between two substances can be utilised to produce electrical energy. Substances with a high positive electrode potential (q.v.) show a pronounced tendency to be oxidised, that is, they are reducing agents.

When platinised electrodes — one saturated with oxygen and the other with hydrogen — are placed, the former in an oxidising solution and the other in a series of reducing solutions, the values of the e.m.f.s produced can be used as a means of arranging the various solutions in the order of their reducing power.

Electrolytic refining. The purification of a metal by an electrolytic process. Ingots of the impure metal form the anodes of the electrolytic cells and the refined metal is deposited on the cathode. The impurities collect as a sludge at the bottom of the cell. The electrolyte is a solution of a salt of the metal that is being refined. The quantity of electricity required to produce 1 g of a metal from such a cell is $96\,500/n$ coulombs, where n = chemical equivalent of the metal.

In general if a current of C ampères is passed through the cell the applied voltage must be E_1, given by the equation $E_1 = E_0 + CR + p$ where E_0 is the decomposition voltage of the electrolyte, R is the resistance of the column of electrolyte between the electrodes and p is the polarisation of the cell. In refining, where metal is transferred from one electrode to the other, so that the total chemical change is none, the value of E_0 is zero.

Conditions are so arranged that the impurities in the crude anode do not deposit with the metal on the cathode and the voltage required is merely that required to overcome resistance and polarisation effects. (*See also* **Decomposition potential.**)

Electrolytic solution pressure. Every metal, in contact with a liquid, possesses a certain tendency whereby the metal tends to pass into solution in the form of ions. As these ions carry a positive charge, the metal electrode must receive a corresponding negative charge. An electrical double layer is thus formed, the negatively charged metal attracting the positively charged ions so that a state of equilibrium is finally reached when the electrostatic attraction in the double layer just balances the tendency of the metal to pass into solution as ions. This latter is called the electrolytic solution pressure.

If ions of the metal are already present in the solution, they will by virtue of their osmotic pressure oppose the solution tendency of the metal. Then the potential difference set up between the metal and the electrolyte will just equal the difference between the solution pressure and the osmotic pressure. If the former is greater than the latter (as in the case of zinc), the metal will be negatively charged with respect to the electrolyte. If the reverse is the case (as in the case of copper), the metal will be charged positively with respect to the electrolyte.

The value of the potential (in volts) that is set up may be calculated from the work done by the ions changing from the electrolytic solution pressure (P_1) to the osmotic pressure (P_2), since

$$\text{work} = \int_{P_2}^{P_1} v\,dp,$$

where $v = RT/p$. But work = $n.e \times 96540$, n being the valency of the metal and e being the voltage; and $R = 1.99/0.239$ in electrical units. Therefore $96540\,ne = RT \log_e P_1/P_2$

$$e = \frac{8.316\,T}{0.4343 \times 96540\,n} \log \frac{P_1}{P_2} \text{ volts.}$$

For a univalent substance (e.g. silver) at 18 °C (291 K)

$$e = 0.058 \log P_1/P_2.$$

Electromagnetic separation. An ore-dressing process used for separating magnetic minerals from other ores. The crushed material is carried on an endless band over powerful electromagnets. By varying the intensity of the magnetic field minerals of different magnetic properties can also be separated from one another.

The process is also used for separating ferrous metals from non-ferrous scrap.

Electromagnets. Electromagnets are made by winding many turns of insulated wire on iron cores of high permeability and passing direct current through the coils. They are extensively used where it is desired to apply and control forces at a distance. The smallest electromagnets are used in relays, while the large ones find useful applications in magnetic separators and in lifting devices. Circular magnets are usually employed for lifting and transporting cold steel ingots, breaker balls and pig iron. Since this type of magnet has the greatest depth of magnetic field, it is the most suitable for handling scrap.

Rectangular magnets are commonly applied to handling — lifting and transporting, steel plates, bars, sections and tubes. Bipolar magnets are useful for most of the applications for which rectangular magnets are suitable, but because the design permits the poles to project well beyond the coils, this type can be used for lifting hot billets and ingots. Of course, in such applications the coils must have proper arrangements for heat deflection and insulation.

The coil of any type of magnet must usually be designed so as to give the value of B (flux density) corresponding to the maximum value of μ (permeability). The lifting power of an electromagnet is given by $F = B^2 A/8\pi \times 10^{-5}$ N, where A = area of face of magnet in cm^2.

Electrometric titration. The use of indicators in titration is sometimes rendered difficult or even impossible by the presence of coloured substances or precipitates. However, determinations of acidity can be made accurately by hydrogen electrode measurements, since the hydrogen-ion concentration changes very rapidly with addition of alkali in the neighbourhood of the equivalent point. Each tenfold reduction in the concentration of hydrogen ion raises its electrode potential by 0.059 volt, and thus there results a sharp break in the voltage curve when the equivalent point is reached.

In titrating electrometrically, base is added to the unknown acid solution containing a hydrogen electrode (q.v.) and joined to a calomel electrode. After each addition of base the e.m.f. of the cell is measured and the results plotted directly. The end-point of the titration is taken as the mid-point of the steep part of the curve.

A similar method may often be employed to determine the end-point of oxidation-reduction reactions, e.g. the oxidation of an iron II (ferrous) salt with potassium dichromate. If a platinum electrode is placed in a solution of iron II salt it will come to equilibrium with the iron II ions. When the dichromate solution is added, oxidation occurs and the end-point is indicated by the sudden change in the electrode potential (*see* Fig. 60).

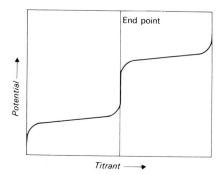

Fig. 60 *Electrometric titration curve of an oxidation-reduction reaction.*

Electromotive force. The energy of an electric current is measured by the quantity of electricity (in coulombs) transported by the current and the electrical pressure (in volts) that drives the electrons composing the current. This electrical pressure is called the electromotive force or voltage. The unit is the volt which is defined as the potential difference between two points of a conducting wire carrying a constant current of one ampere,

when the power dissipated between these points is one watt. Volts $\times$ coulombs = joules. (One joule per second is 1 watt; volts $\times$ amperes = watts.)

In the case of reactions involving electrolytes, the affinity of a reaction or the decrease of free energy may be obtained from the values of the e.m.f. of suitable voltaic cells.

The e.m.f. is usually measured by a voltmeter, but for electrochemical purposes a method of comparing the unknown e.m.f. with that of a standard cell whose e.m.f. is accurately known is frequently adopted. The apparatus used is called a potentiometer. Two standard cells are: (*a*) Weston standard cell of e.m.f. 1.0183 volts at 20 °C, and (*b*) Latimer-Clark cell of e.m.f. 1.433 volts at 15 °C.

Electromotive series. Single electrode potentials are determined using a cell which is made up such that a standard hydrogen electrode forms one half, while the other half consists of the unknown electrode dipping into a solution of one of its salts (of concentration 1 mole). The two halves are joined by a salt bridge and the e.m.f. is measured. This is known as the electrode potential. Taking the potential of the hydrogen electrode as zero, the metals can be arranged in order of their electrode potentials to give what is called the electromotive series or sometimes "the reduction-potential table".

Lithium	Zinc	Lead
Rubidium	Chromium (II)	Iron (III)
Potassium	Iron (II)	Hydrogen
Barium	Cadmium	Copper
Sodium	Thallium	Mercury (I)
Magnesium	Cobalt (II)	Silver
Aluminium	Nickel (II)	Platinum
Manganese (II)	Tin (II)	Gold (I)

In this series any metal in the list is electronegative to any other metal below it. The further apart the two elements are in the list the greater is the e.m.f. they can give in a voltaic pair. The actual e.m.f. is the difference between the e.m.f. of the metal lower in the list and the one that is higher. Some electrode potentials are given in Tables 18 and 21. (*See also* **Electrode potentials.**)

Electron cloud. As the electron cannot be described entirely by particulate or corpuscular behaviour, nor entirely by wave motion, it is generally assumed that the most appropriate condition is used for specific argument. In the case of extra-nuclear electrons in atoms, the old Bohr theory of orbital motion could not satisfy all the electronic behaviour patterns, and electrons are now assumed to exhibit a probability of spatial position, which is limited by the uncertainty principle, from precise definition. The electron cloud concept is sufficient to indicate the bounds within which the electrons may be assumed to function, and this is of great assistance in explaining most phenomena.

Electron diffraction. Electrons are rapidly absorbed at a surface on which they impinge so that penetration is small. If however, the conditions of diffraction can be met by reflection from atomic planes in the surface, a glancing angle diffraction pattern can be obtained. The pattern shows almost complete semi-circular arcs, and if d is the interplanar spacing of the crystals, L the effective camera length, λ the effective wavelength, and r the ring radius of the pattern, $rd = L\lambda$. $L\lambda$ is often called the camera constant, obtained by using a standard substance in calibration of the instrument. Values of d can thus be obtained from measurements of the ring radius.

The angles used for the beam are only a few degrees from the horizontal at the surface, and a depth of penetration of 5×10^{-8} m is then effectively 10 to 30×10^{-10} m normal to the surface. Very thin surface films formed by corrosion or oxidation can thus be studied, and surface finish, so-called amorphous layers, and contaminating phases on a surface, have all been studied by the technique.

Transmission patterns can be obtained from powders mounted on a copper grid, or particles precipitated on a polycrystalline foil, as well as very thin foils such as those used in the electron microscope.

Electronegative valency effect. When the difference in the electrochemical characteristics of the two components of a binary alloy is sufficiently great, there is a tendency towards the formation of very stable intermetallic compounds with limited homogeneity ranges. This effect restricts the formation of the primary solid solution, but as the solid solution tends to show a marked increase with rise in temperature, may indicate age-hardening properties.

Electronegativity. The degree to which atoms tend to attract electrons to themselves. This property normally increases across the periodic table from Group I to VIII, in other words, it increases with the buildup of closed electron shells. It varies within a group, decreasing with increasing atomic number, except for the transition elements where buildup of d and f electrons masks other effects.

Similar electronegativity tends to formation of metallic or covalent bonds, depending on release or acceptance of electrons. When electronegativities differ, the bond is partially ionic, the ionic character increasing as the difference in electronegativity increases.

Electron microscope *(a) Transmission electron microscope (TEM).* The resolution, or resolving power, of an optical system determines the ability to distinguish two images, whereas magnification simply enlarges whatever image is visible. If δ is the distance apart of two objects to be resolved, λ the wavelength of the illumination, NA the numerical aperture of the objective, μ the refractive index of the medium, and α one-half of the angular aperture (the apex angle of the reflected cone of light entering the objective) then:

$$\delta = \frac{\lambda}{2\,NA} = \frac{\lambda}{2\,\mu\sin\alpha}.$$

Resolution can be improved therefore, by high angular aperture of the objective, and an immersion medium of high refractive index, and this is done for the light microscope by using an oil immersion lens of high angular aperture. There are limits to this however, and a decrease in the wavelength λ is the only course left open.

The maximum resolving power of the best light microscopes is about $0.1\,\mu m$. The electron microscope beam corresponds to a wavelength given by

$$\lambda = \left(\frac{1.5}{P}\right)^{1/2}\,nm$$

where P is the potential drop, corresponding to electron energy. For $P = 100\,kV$, δ becomes 10^{-10} m, and this must be the eventual limit of such instruments, already giving over three orders of magnitude improvement on the light microscope, and with magnifications of $1\text{-}2 \times 10^5$ times.

The electron microscope generates a beam of electrons in an "electron gun", using thermionic emission combined with an accelerating potential. The beam is collimated by magnetic or electric fields in the shape of coils, so that it can be made parallel or convergent to a fine focus. The specimen is held in a vacuum chamber which can readily be sealed off from the remainder so that changing specimens does not involve pumping down the whole system each time. The transmission of electrons through the specimen is then displayed on a fluorescent screen, or allowed to activate a photographic plate.

Specimens of metal must be thinned to allow transmission, and special electrochemical thinning techniques are used. Alternatively, deposition of thin films of carbon or polymeric materials can be used to obtain a replica of the surface of the specimen, and the replica itself (with better transmission characteristics) used for the examination. Larger potentials can be used in the instruments, up to 10^6 V, so that better resolution is obtained, and thicker specimens can be penetrated. The transmission microscope can also be used for diffraction experiments, and variants of the instrument may be fitted with other devices, e.g. for microanalysis (EMMA).

(b) Scanning electron microscope (SEM). If a beam of electrons is scanned across the face of a specimen, and the secondary electrons, of low energy, are detected and displayed on a CRT the image can be magnified by this means, and the resolution will depend on the diameter of the electron beam. Using lower energy electrons than the transmission microscope (2-50 kV potentials) the beam can be focused to 5-10 nm at the specimen surface, and the resolution is then better than 10 nm. Although this is inferior to the TEM, there is great depth of field, due to the small angular aperture, and the SEM has been invaluable in examining surface topography on a fine scale. In medicine and biology, the lower energy of the electrons inflicts less damage, and the images throw new light on many aspects of tissue development, cell structure and membranes.

A combination of transmission and scanning techniques has been developed, with the secondary electron detector of the SEM situated below the specimen, not to one side. This gives better detection than the conventional TEM, which can utilise only electrons of high energy, and needs a very thin specimen as a result. The STEM (scanning transmission electron microscope) uses specimens up to three times the thickness of the TEM, but there is still the limitation of resolution — governed by the beam diameter.

Diffraction patterns can be obtained from a scanning microscope, but the STEM instrument is superior again for this purpose.

Electron probe micro-analysis (EPMA). Using an electron gun similar to that in an electron microscope, the specimen can be bombarded by the electron beam focused on to a small area (about $1\,\mu m$) and the X-radiation emitted by the elements present, analysed to identify them quantitatively from the X-ray spectra. Although coarser than the microscope, the analyser can display an electron micrograph of the area under observation, so that precision analysis is possible, and also scanning across a grain boundary or non-metallic inclusion for example, to identify constituents.

The beam penetrates a distance of about $1\,\mu m$ into the specimen, and so a volume of material is actually emitting X-radiation. The scattering of this radiation by surrounding material is also a difficulty. Complex mathematical methods of interpretation of results have been studied, to reduce errors, and refine the techniques. The instrument is a valuable tool in the armoury of electron optics.

Electrons. The negatively charged elementary particles which occupy the space outside the nucleus in every atom. The mass of an electron is very small, 1/1840 that of the hydrogen atom, about 9.1×10^{-28} g, and the radius about 10^{-12} mm. Electrons move with a velocity around 10^7 mm s^{-1}, and can be accelerated in specific directions by an electric field or deflected by a magnetic one. The charge on the electron is 1.6×10^{-19} C.

The number of extra-nuclear electrons in an atom is the atomic number, equal to the number of net positive charges on the nucleus.

Electrons may be produced by the action of ultra-violet light on metals (*see* **Photocell**) by raising the temperature of metals to incandescence (*see* **Thermionic effect**) and as a product of radioactivity. In the latter case the β-rays emitted consist of negatively charged electrons which move with a velocity approaching that of light and have considerable penetrating power. (*See also* **X-ray spectra, Isotopes.**)

Electron spectroscopy for chemical analysis (ESCA). Also known as X-ray photoelectron spectroscopy (q.v.) because this describes the technique better. The method employs a monochromatic beam of X-rays (normally Al $K\alpha$ or Mg $K\alpha$) to excite electrons at low energy levels in the atoms of the specimen. The kinetic energy of the ejected photoelectrons is given by the energy of the X-ray photons minus the electron binding energy, and is therefore characteristic of the atoms emitting them. Not only can an emitting atom be identified, but also its state of chemical combination, since there will be a detectable difference in the electron energy. The electrons arise from a very thin layer on the surface of the specimen.

Electron volt. A measure of energy, formerly used in expressing energies of electrons in their atomic behaviour, and when subjected to radiation or other impulses. 1 electron volt equals 1.6×10^{-19} joules. The thermal energy of a vibrating atom is a few electron volts. The energy liberated in the fission of ^{235}U is of the order of 200 MeV.

Electro-osmosis. When a solid, e.g. very finely divided clay, is placed in a liquid of different dielectric constant the solid becomes electrically charged. If two electrodes, connected to the opposite terminals of a battery, are put into the suspension, it is found that the particles move towards the positive electrode.

This phenomenon, which is known as cataphoresis, has led to the separation of colloids into two classes, electropositive and electronegative. Clay, alumina and ferric oxide are electropositive, while colloidal gold, silver, platinum and silicic acid are electronegative.

If the water between two charged electrodes is separated by a clay diaphragm, water is forced by the current towards the cathode, the relative motion of the clay and water being the same as in cataphoresis. The phenomenon is known as electro-osmosis and the direction of flow of the water depends on the diaphragm and on the electrolytes present. There may be a reversal of flow by the presence of strongly adsorbed ions on the diaphragm. Acid solutions cause the diaphragm to become positively charged and alkaline solutions negatively charged.

Electro-percussion welding. A recent development of the spot-welding technique in which an intense discharge of electrical energy is applied to the area of the proposed weld for a very short time. The electrical energy may be stored either in condensers or in capacity storage machines. Where the latter are used, the current from a special type of accumulator (which is not adversely affected by heavy current discharge) may be suddenly discharged into the primary of a transformer, so that a high voltage current is momentarily produced in the secondary windings to which the electrodes are connected. Alternatively a direct current may be fed into the primary of a transformer and the high voltage discharge in secondary windings effected by suddenly interrupting the primary current.

The process is especially suitable for spot welding metals that have a low electrical resistance and a high thermal conductivity, e.g. aluminium, aluminium alloys, copper, etc. The actual welding, however, must be rapidly followed up by a high pressure of the upper electrode upon the work, and this must be maintained during the cooling period. In order to prevent porosity or cracks in the weld it is necessary to apply the pressure rapidly and automatically after the weld is made. The pressure is usually provided by compressed air acting on the upper electrode.

Electrophoresis. One of the reasons why colloidal particles remain in suspension is the fact that they carry electric charges. Colloidal particles of arsenic trisulphide are negatively charged, while those of colloidal iron III (ferric) hydroxide are positively charged. When a current is passed through a colloidal solution, the charged particles migrate to one electrode or the other, according to the charge they carry, where they form coarse aggregates. The phenomenon is known as electrophoresis.

Electroplating. The deposition of metals from solutions of their salts by means of an electric current. The article to be plated is made the cathode of the electrolytic cell. Before the plating can be successfully carried out the article should be chemically clean. Grease and dirt may be removed by organic solvents or by immersion in a hot alkaline bath; rust or oxide film may be removed by wire-brushing, sand-blasting or by pickling in an acid bath. Electrolytic cleaning may also be used. The nature of the deposit on the cathode is influenced by the composition and temperature of the electrolyte, by the concentration and circulation of the electrolyte and by the current density. Metals such as iron, nickel and cobalt are normally deposited from solutions of simple salts, while in the case of copper, zinc and gold the metal is liberated from a complex ion. A small amount of a suitable organic colloid or metallic salts may be added to the bath to improve the quality of the deposit. (*See also* **Brighteners.**)

By using an electrolyte made up of mixtures of salts, and employing two or more anodes, alloys having attractive colours or distinctive properties can be deposited, e.g. tin-copper (speculum), tin-nickel, brass, bronze, etc.

Electroplating is used for (*a*) decorative effect; (*b*) corrosion resistance; (*c*) abrasion resistance; (*d*) building up worn or undersized components.

Electro-polishing. A method of producing a very smooth surface on metals by a process which is the reverse of electroplating. The piece to be polished is made the anode, and thus tends to be dissolved in the electrolyte. The electrolyte and the current density are controlled so that the oxygen is liberated at the anode, and this reacts with the metal surface, preferentially oxidising the high points. The oxidised metal dissolves in the electrolyte, and the result is a finish at least as good as that produced by mechanical buffing, but the surface layer is strain-free. The method is useful for polishing parts of irregular contour, where the require-

ment is smoothness of finish rather than a high lustre. (*See also* **Electrolytic polishing.**)

Electro-slag refining. The use of a submerged electric arc, beneath a pool of molten slag was developed in the USSR for the welding of very thick steel sections. The same technique can be employed in a refining process, with metal in a refractory container rather than sections of metal to be joined. The energy of the arc is available at the slag/metal interface, to speed up chemical reactions.

Electrostriction. The resemblance between ferroelectrics and ferromagnetics is described under **Domains, Ferroelectricity** and **Ferromagnetism.** Just as ferromagnetics show distortion when a magnetic field is applied, depending on the elastic properties of the crystal and the easy magnetisation directions of that crystal, so ferroelectrics distort in an electric field, due entirely to the crystal structure. Small displacements occur in the negative and positive charges, which give induced dipole moments. The field interacts with these dipole moments, and the distortion is proportional to the square of field strength. This is one difference with magnetostriction, where the direction of the field determines magnetostriction, whereas in electrostriction the distortion is always in the same sense; whatever the field direction.

Electrothermic. A term used in reference to the utilisation of electrical energy for the production of heat in chemical or metallurgical processes. The high temperatures attainable are employed for melting metals, and particularly in the electrical smelting of iron ore for the production of steel. The electric arc may be between electrodes, the latter being arranged so that the arc is deflected downwards on to the charge (radiation arc furnace). Alternatively the arc may be struck between carbon electrodes and the surface of the charge (direct arc furnace). In a third type of furnace (induction furnace) that is mainly employed for melting metals, the metal under treatment forms the single-turn secondary of a transformer, and an iron core or yoke of the transformer passes through and interlinks the primary and secondary coils. (*See also* **Furnaces, Eddy currents.**)

Electro-tinplate. Tin-plate produced by depositing tin electrolytically on to mild steel sheet. Though tin coatings of any desired thickness can be produced by electrodeposition, electro-tinplate usually carries coatings considerably thinner than those obtained by hot dipping. The usual amounts of tin per basis box may be 110, 225 or 335 g (4, 8, or 12 oz), compared with 540 g (19.2 oz) for hot dipped plate. Electro-tinplate of these coating weights is suitable for packaging certain dried goods, but not for canning processed foodstuffs.

Electrum. An alloy of gold and silver prepared and used by the ancients. Pliny states that the alloy contained 20% Ag and 80% Au. The colour is pale amber and the name is derived from the Greek word for amber. At various times the name appears to have been applied to alloys having a similar colour but containing no precious metals, e.g. copper-zinc alloys with or without additions of nickel or tin.

Native argentiferous gold ores containing 20% or more of silver are also called electrum or electron.

Elluvial deposits. These are deposits formed by the decomposition, with or without solution, of metalliferous rocks and lodes, but without being mechanically transported from the area of their formation. Such types of detrital deposit overlie the rocks or the ore from which they were derived. When the deposits are transported and redeposited by running water they become alluvial.

Elmore process. An ore-dressing process in which sulphides are separated from gangue by agitation in a mixture of oil and water. The oil is ultimately allowed to accumulate in a layer above the water, carrying with it the sulphides in suspension.

Elongation. A term used in mechanical testing to describe the amount of extension of a test piece when stressed. The amount of permanent extension in the vicinity of the fracture of a test piece broken in a tensile test is usually expressed as a percentage of the original gauge length, e.g. x% in 5 cm.

Elongation may also refer to the increase in length at any stage of a process which continuously increases the length of a metal object—as in rolling.

Eloxal. An anodic oxidation process applied to aluminium and its alloys.

Elutriation. A term used in reference to a method of separation of particles of various sizes by means of a moving column of fluid. Generally, water is used and it is caused to flow upwards at such a rate that gravitational forces that would cause the particles to descend are equal to, or less than, the forces produced by the upward motion of the fluid. By using different rates of flow, the sample can be separated into fractions each having a small but different size-range. Using air as the moving fluid with dry powders, separation of bulk material can be effected in sub-sieve sizes.

Embolite. A mineral found in Chile and Mexico. It consists of silver chloride and silver bromide in varying proportions.

Embrittlement. This is a condition in metals whereby they lose their impact toughness. This may be brought about in several ways.

(a) Low-temperature brittleness occurs at sub-normal temperatures in the case of mild steels, wrought iron and zinc alloys and magnesium alloys. The effect is temporary and disappears when the temperature is raised to normal.

(b) Hydrogen embrittlement, which is caused by the absorption of hydrogen during such processes as electroplating or acid pickling. This embrittlement can sometimes be removed by ageing for a long or short time at slightly elevated temperatures or at higher temperatures respectively.

(c) Alloys of zinc and copper containing a high percentage of zinc may suffer from a form of embrittlement known as season cracking (q.v.), or segregation of impurities (e.g. S, P, Bi).

(d) Embrittlement of steel plate may result from any of those heat treatments that result in oxide or carbide deposition at grain boundaries.

(e) High temperature embrittlement due to the formation of cavities at grain boundaries by diffusion to give a low fracture strain.

(f) Caustic embrittlement occurs most commonly in boiler plate, and is caused by the presence of caustic soda in the water. Cracks appear at the surface and follow a line along intercrystalline boundaries. Reduction of the risk of boiler explosion due to this type of embrittlement is secured by reducing the causticity of the boiler feed water. This is done by adding either sodium phosphate or the cheaper sodium sulphate to the water. The amount of the latter to be added is about three times the carbonate content of the water. The caustic in the water arises from the hydrolysis of the carbonate solution in the boiler.

$$Na_2CO_3 + H_2O = 2NaOH + CO_2$$

Emerald. A bright green variety of beryl $3BeO.Al_2O_3.6SiO_2$ prized as a gemstone. It crystallises in the hexagonal system and the green colour is due to the presence of chromium. Found in Colombia and other parts of South America and in the Urals. Mohs hardness 7.5, sp. gr. 2.67. Oriental emeralds are a variety of crystalline alumina.

Emerald copper. A silicate of copper $CuO.SiO_2.H_2O$ also called dioptase, found in limestone in the Kirghese Steppes and in Siberian gold-washings. It occurs in emerald green crystals of specific gravity of 3.3.

Emerald nickel. A hydrated basic carbonate of nickel

$$NiCO_3.2Ni(OH)_2.4H_2O$$

that occurs as an incrustation on other minerals. Also called Texasite.

Emery. A greyish-black variety of corundum Al_2O_3 containing much admixed magnetite or hematite. Pulverised emery is used as an abrasive for polishing hard surfaces. Crystallised alumina is excelled in hardness only by diamond and silicon carbide (carborundum). Its specific gravity varies with its composition from 3.7 to 4.3, Mohs hardness 8. A very finely divided form known as flour of emery is used by lapidaries and plate-glass manufacturers.

Emery wheels are made by consolidating the powdered mineral with media such as shellac, silicate of soda, vulcanised rubber, etc.

Emission (acoustic). During the magnetisation of a ferromagnetic material, the domain movements which occur can be identified by an induction coil surrounding the specimen, connected to an amplifier. Audible clicks are heard (Barkhausen effect) as the domains move. Improved detection techniques, using a wider range of frequencies, have revealed acoustic emission during many types of transformation in solids. (*See* **Non-destructive testing**.)

Emission (photoelectric). Radiation in the range of visible and near visible light can cause emission of electrons from materials. The shorter wavelengths are more effective, since the photon energy is greater. For a given frequency, the rate of emission is proportional to the intensity of the radiation. For constant intensity, the emission varies with the frequency, and a limiting frequency or threshold, below which no electrons are produced, can be measured for all materials which exhibit the effect. The minimum photon energy required is the photoelectric work function for the material.

Emissive power. When radiations of a given wavelength fall on a body, the absorptive power A of the body for that wavelength is defined as the ratio of the radiant energy absorbed to the total incident radiant energy. A perfectly black body absorbs all the radiations which fall upon it, and the value of A for such a body is unity for all wavelengths. For all actual substances, however, A is a proper fraction, its value depending on the nature of the body, its temperature and the wavelength of the incident radiation.

The radiant energy of given wavelength emitted per second by unit surface of a body is called its emissive power. The Stewart-Kirchhoff second law states that the ratio of the emissive power of a body to its absorptive power is a function of temperature only and is the same for all bodies emitting temperature radiation and is equal to the emissive power of a perfectly black body.

The absolute emissive power of a body is the emissive power of the body at 1K or the energy radiated per second by unit surface at 1K to a surrounding enclosure at absolute zero.

Emissivity. Also called surface emissivity or surface conductivity, it is defined as the quantity of heat which a body loses per unit surface area per unit time, under given conditions. The emissivity of a black body is proportional to the radiation density, and hence the amount of energy radiated is proportional to the fourth power of the absolute temperature (Stefan's law). (*See also* **Emissive power**.)

Emitter (transistor). The third component (with base and collector) of a transistor (*see* **Transistor**). In a device consisting of n-p-n junction material, the emitter is the main source of hole charge carriers, which pass into the base (middle) region when a forward bias is applied.

Emulsion. A mixture of two or more liquids in which one is present in the form of minute droplets of microscopic or ultra-microscopic size. The emulsion may be formed spontaneously on mixing the liquids or by mechanical agitation of liquids that have little or no mutual solubility. Unstable emulsions separate into distinct layers on standing. They can be made stable by the addition of emulsifying agents which may either form films at the surface of the droplets or may impart mechanical stability to the droplets. Soap is an example of the former and colloidal carbon is an example of the latter type of agent.

Emulsions may be destroyed either by destroying the emulsifying agent chemically or by heating or freezing the solution. Emulsions are important in the separation of low-grade ores by flotation processes (q.v.).

Emulsoid. A type of colloidal sol in which water or other liquid is the dispersion medium, and which has a higher viscosity and a lower surface tension than the dispersion medium. The dispersion medium is taken up or adsorbed by the colloid, and the greater the adsorption the more will the properties of the colloidal sol depend on the medium and the less on the electric charge on the colloid. Emulsoids are distinguished from suspensoids because the former are reversible colloids, that is, they can pass back into the sol state after being evaporated to dryness. Suspensoids are non-reversible colloids.

Enamelling. The process of coating metal, pottery or glass with a hard glazed finish, properly a fused vitreous finish (*see* **Enamelling (vitreous)**) for protection and decoration. Now also loosely applied to the application of certain types of paint having a high gloss and a reasonably tough surface.

Enamelling (vitreous). The application of a fused vitreous coating to metal surfaces. Vitreous enamels are typically a glass or frit consisting of complex mixtures of silicates, borates and aluminates of the alkali metals. These are fritted at temperatures around 1,200 °C and milled with certain additions to give desired properties of colour, opacity and coefficient of expansion. Most enamel is applied by dipping or spraying on an aqueous slip or suspension of milled enamel produced by the addition of small amounts of clay or other suspension agent at the mill. Tin oxide, antimony oxide, barium sulphate and titanium oxide, are all used as opacifying agents, and a wide range of colours can be produced by the inclusion of the oxides of such metals as copper, cobalt, manganese, chromium, etc. Vitreous-enamelled coatings are fired in a furnace at 700-800 °C to give a finish which is highly resistant to abrasion and atmospheric corrosion.

Enamelling sheets. Steel sheets for vitreous enamelling should ideally have very low carbon, sulphur, silicon and phosphorus content. Where components have to be deep drawn prior to enamelling, carbon contents up to 0.04% are tolerable.

Enamel plating. A patented finish for the decoration and protection of aluminium and aluminium alloys. The process is an electrolytic one involving a heavy coating of alumina on the metal surface, produced by anodising in a suitable solution.

Enargite. A copper arsenic sulphide $3Cu_2S.As_2S_5$ and an important copper ore. It is found at Butte (Montana), where it frequently contains antimony. It crystallises in the orthorhombic system. Mohs hardness 3, sp. gr. 4.43-4.45.

End-centred. Orthorhombic space-lattices having equivalent points at the corners of a unit cell and at the centres of a pair of opposite faces parallel to the c-axis. This is the same as side-centred with a different choice of axes.

Endellionite. A sulphide of copper, lead and antimony

$$3(Pb,Cu)S.Sb_2S_3.$$

Also called bournonite. It crystallises in the orthorhombic system.

Endoscope. An instrument providing a built-in source of illumination together with a light collecting system to transmit back an image of the illuminated area. Usually of small diameter, to enable inspection of inaccessible places, and can incorporate fibre optical devices to add even greater flexibility and accessibility. A camera can be incorporated at the viewing end, to enable recording of the image as the endoscope is inserted into a suitable orifice. Its most extensive use is in medicine, for examination of organs and passages via natural orifices.

Endothermic reaction. A chemical reaction during which heat is absorbed. Endothermic compounds are often unstable. Certain explosives which are endothermic compounds decompose with the liberation of a large amount of heat. In the case of combustible compounds, the heat of combustion of the compound is greater than the sum of the heats of combustion of the constituent elements. (*See also* **Energy.**)

End quenching. A process for hardening the ends of steel rails. As the rails are removed from the furnace at a temperature above the critical range, their heads are hardened at the ends by quenching with jets of cold compressed air. Heat from the rest of the rail suffices to temper the quenched portion. Also known as dry quenching. (*See also* **Jominy test.**)

End sizing. An operation used in the production of high-grade welded pipes, etc., to close-dimensional tolerances. The operation, which is carried out by forcing a mandrel into the bore of the tube or by forcing a die over the outside of the pipe, is performed in a large hydraulic press.

Endurance. The ability of a material or a structure to withstand stresses set up by fluctuating or repeated loadings. The term fatigue is usually applied to failures under the fluctuating or alternating stresses.

Endurance limit. The limiting stress below which a metal will withstand an indefinitely large number of cycles of stress without failure by fatigue fracture. Above this limit failure by fatigue occurs. If the term is used without qualification, the cycles of stress refer to reversals of flexural stress. The term may also refer to cycles within a specified stress range. The stress is taken at one-half of the maximum stress range. Steels have true endurance limits, but most non-ferrous metals do not show true endurance limits. In the latter cases the term endurance strength is used. It is defined as the repeated stress at which failure will not occur before a stated number of stress cycles. Also termed fatigue limit.

Endurance range. The maximum range of stress which a material can withstand for a specified large number of applications without fracture. When the mean value of the stress is zero, the endurance range is twice the endurance limit. Also termed fatigue range.

Endurance ratio. The ratio of the endurance limit for cycles of reversed flexural stress, to the tensile strength of the material.

Endurance tests. The object of endurance or fatigue tests is to determine for a definite stress cycle, applied at a definite frequency, the number of cycles that a material can withstand without fracture. The number of cycles (N) is called the "endurance". If the total stress range is R, the endurance limit is $R/2$, and this is denoted by S. The relationship between S and N is plotted as a curve known as the S-N curve, the values of N being usually from 0 to 10^7 at least. This curve may be of the form shown (*see* Fig. 61), where the curve appears to become parallel to the N-axis. In such cases there would appear to be a limiting range of stress below which fracture will not occur for any number of reversals, however large. The commoner types of test are known as (*a*) Haigh type—a tension/compression test, (*b*) Wöhler type—rotating cantilever test, (*c*) rotating beam test. Also referred to as fatigue tests.

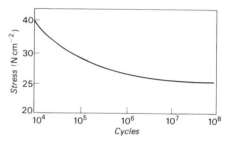

Fig. 61 *Typical S-N or endurance test curve.*

Energisers. Chemicals which are added to carburising media in the pack carburising process for case-hardening steels. The energiser is usually barium oxide BaO; barium, calcium or sodium carbonate or sodium cyanide. The object of using an energising agent is to provide a source of carbon dioxide which reacts with the carbon to increase the concentration of carbon monoxide.

Energy. The energy of a body is defined as its capacity to do work. A moving body possesses energy by virtue of its mass and its velocity, and the energy is measured by the quantity $\frac{1}{2}mv^2$. This is known as the kinetic energy of the body, and work must be done in reducing the velocity from the value v to zero. In general, kinetic energy = $\int Pds$, where P is the force required and s the distance through which the force acts. The potential energy of a body at rest is equal to the work the body can do in changing its position. It is measured by the quantity mgs, where m = mass of the body and s the distance moved in the direction of the force mg.

Different forms of energy are often readily interchangeable, e.g. heat energy as steam may be converted to electrical energy, when the turbine drives a generator. Subsequently that electrical energy can be converted back to heat some distance away, in the form of a resistance heater. The SI unit of energy is the joule, which can be expressed as a force times a distance (newton metre), so introducing mechanical energy. The unit of power, the watt, is also expressed as joules per second, so that an electrical energy unit, the watt hour, is in fact 3 600 joules. (The old heat unit of the c.g.s. system, the calorie, is equal to 4.182 joules.)

The total (or internal) energy of a substance or group of substances depends on the mass, the chemical nature and the physical conditions of the system, and is always an unknown quantity. Only changes in energy can be measured. When a system undergoes a change such that there is an increase of internal energy (E), then the increase must be equivalent to the heat absorbed (Q) less the external work (A) done on the system. That is, $E = Q - A$.

Where the work done is represented by an increase of volume at constant pressure, then: $\Delta E = Q - p(v_2 - v_1) = E_2 - E_1$ or $(E_2 - pv_2) - (E_1 - pv_1) = Q$. Since $(E - pv)$ represents the heat content of the system (H), $Q = H_2 - H_1$ and $\Delta H = \Delta E \times p \Delta v$. ΔH is the heat of the reaction at constant pressure and the heat of reaction measures the change in total energy only when $p \Delta v = 0$, that is, when no external work is done.

The heat of a reaction depends on the temperature at which it occurs, and from Kirchhoff's law $\Delta Q/\Delta t = \Sigma C_1 - \Sigma C_2$, which means that the increase of the heat of a reaction for $1°$ rise of temperature is equal to the difference between the specific heats of the reactants (C_1) and the resultants (C_2). The above considerations are based upon the first law of thermodynamics, which, in effect, states that energy can be neither created nor destroyed. The second law of thermodynamics deals with the conditions under which transformation of heat energy into other forms of energy takes place. The condition under which the maximum amount of external work may be obtained from a given process is that the process must be carried out reversibly. For 1 mole of a perfect gas expanding from volume v_1 to volume v_2

$$\text{work} = \int_{v_1}^{v_2} p\,dv = \int_{v_1}^{v_2} \frac{RT}{v}\,dv = RT \log_e \frac{v_2}{v_1}$$

When a quantity of a perfect gas is passed through a cycle of reversible operations (Carnot cycle) it can be shown that

$$A = Q\frac{(T_2 - T_1)}{T_2},$$

where A = external work, Q = heat absorbed by the gas, and T_1 and T_2 are the absolute temperatures between which the cycle of operations is carried out. The efficiency of the process is thus $A/Q = (T_2 - T_1)/T_2$. Where a physical or chemical process takes place isothermally $dA/Q = dT/T$.

For 1 g of liquid whose latent heat is l, vaporising at temperature T under a pressure p, the work done *by* the system is $(p + dp)$ $(v_2 - v_1)$, where v_1 is the volume of the liquid and v_2 the volume of the vapour. The work done *on* the system in the reversible change is $p(v_2 - v_1)$. The net gain of work is the difference between these two quantities, viz. $dp(v_2 - v_1)$, and the heat taken up is l.

Hence,

$$\frac{dp(v_2 - v_1)}{l} = \frac{dT}{T}.$$

This is known as the Clausius-Clapeyron equation, and is applicable to the process of fusion as well as to vaporisation. Van't Hoff pointed out that a measure of chemical affinity—on the basis of the second law of thermodynamics—is to be found in the maximum external work which can be obtained when a chemical reaction is carried out reversibly and isothermally. When a reaction is carried out at constant temperature and constant pressure, the total external work is the sum of the electrical or other energy available for use, and the work done by expansion against the pressure. The work available for use is called the free energy (G) and thus $-\Delta A = -\Delta G + pdv$, which means that the maximum work done is equal to the difference in the free energy plus the work done by expansion against the pressure.

Combining the algebraic expressions for the two laws of thermodynamics,

$$A - E = T\frac{dA}{dT},$$

which is the Gibbs-Helmholtz equation. If ΔA = change in total energy, ΔH = change in heat content:

$$\Delta A - \Delta E = T\left(\frac{\delta(\Delta A)}{\delta T}\right)_v; \quad \Delta G - \Delta H = T\left(\frac{\delta(\Delta G)}{\delta T}\right)_p$$

since $G = A + pv$ and $H = E + pv$.

The quantity $T(dA/dT)$ is the energy unavailable for external work and is called latent energy. In an endothermic reaction there is a decrease of free energy (or external work is done), but an absorption of heat takes place because more energy becomes latent energy than is given out in external work. If the quantity $T(dA/dT)$ is positive and greater than A (which can be the case, especially if the temperature is high), then $A - TdA/dT$ is negative and thus E is negative which corresponds to an absorption of heat when the reaction takes place.

The methods generally used for estimating relative chemical activity are: *(a)* by observing the speed of chemical action, i.e. the amount of matter undergoing chemical transformation in a suitable unit of time; *(b)* by measuring the quantity of heat that accompanies chemical change, using a calorimeter; *(c)* by measuring the e.m.f. of the current when the materials are arranged in the form of a battery cell.

For useful comparisons by any of these methods account must be taken of the temperature of the reaction and the concentration of the reacting substances.

In accordance with Planck's theory of energy quanta, the radiation energy which is taken by an electron is not absorbed continuously but in definite amounts called quanta. The quantum of energy of radiation is not constant, but depends on the frequency of the radiation, and is in fact equal to $h\nu$, where $h = 6.625 \times 10^{-34}$ Js and is known as Planck's constant, and ν is the frequency of the radiation.

When an electron falls back from energy level E_2 to a lower energy level E_1, the frequency of the radiation—spectral line—emitted is obtainable from the relationship $E_2 - E_1 = h\nu$.

Einstein put forward the hypothesis that when a substance undergoes photochemical reaction, each molecule absorbs a a quantum of radiation energy $h\nu$. Each quantum absorbed brings about the decomposition of one molecule. The law of photochemical equivalence may thus be expressed:

$$E = Nh\nu = Nh\frac{c}{\lambda},$$

where N = Avogadro's number, c = velocity of light, λ = wavelength of radiation.

Energy of dislocations. The energy can be derived from a cylindrical element, with either an edge dislocation or a screw. Deformation of a thin annulus, of length l, thickness dr, at radius r, is of the magnitude b, the Burgers vector, so that for a screw, shear strain $\gamma = b/2\pi r$. If Hooke's law is obeyed, $\tau = G\gamma$, and $\tau_{average} = Gb/2\pi r$. Strain energy

$$U = \tfrac{1}{2}G\gamma^2 = \frac{G}{2}\left(\frac{b}{2\pi r}\right)^2,$$

and the unit volume of the annulus is $2\pi r.dr.l$. The total strain energy can thus be obtained by integrating from r_0 to r on this unit volume:

$$\int_{r_0}^{r} \frac{G}{2}\cdot\left(\frac{b}{2\pi r}\right)^2 . 2\pi r.dr.l = \frac{Gb^2 l}{4\pi} \int_{r_0}^{r} \frac{dr}{r}$$

$$= \frac{Gb^2 l}{4\pi} \log_n\left(\frac{r}{r_0}\right).$$

The energy per unit length for a screw dislocation is therefore:

$$\frac{Gb^2}{4\pi} . \log_n\left(\frac{r}{r_0}\right).$$

Similarly, for an edge dislocation, the energy per unit length =

$$\frac{Gb^2}{4\pi(1-\mu)} . \log_n\left(\frac{r}{r_0}\right).$$

where μ is Poissons' ratio. The energy of a screw dislocation is therefore less than that of an edge in the same crystal, but of the order of several eV per atomic spacing. This means that a dislocation of reasonable length will tend to move, since the strain energy is of the order of the thermal vibrations.

Enthalpy. May be defined as the heat content of a substance per unit mass. It is equal to $E + pV$, where E is the intrinsic energy.

Entropy. The degree of disorder which exists in a system. The entropy energy increases with increasing disorder, and this energy is not available for doing work. The entropy is given the symbol S, and the entropy energy is conventionally the product of absolute temperature T, and the entropy, or TS. The free energy of a system is then the total energy (enthalpy) minus the entropy energy, $G = E + pV - TS$ where G is the free energy (Gibbs), E the internal energy of the system, and p the pressure, V the volume. It is this free energy which is available for doing work.

Epitaxy. Growth of a crystalline substance on a substrate crystal, in which the substrate determines the crystal structure adopted. Since crystal structures vary in lattice parameter and crystal type, quite apart from variations in atomic radius, it is obvious that epitaxial growth must be restricted, and that considerable stresses may be generated even when it occurs. In general, the two lattices involved (substrate and deposit) should be reasonably commensurate, the binding energy should not be too dissimilar, and in ionic substances, the arrangement of positive and negative ions should be capable of similar alignment.

For example, NaCl can be deposited on PbS. Both have the "rock salt" type of cubic crystal lattice, and the difference in lattice spacing is only 5%. On the other hand, NaCl and silver metal are mutually compatible in this respect, even though the lattice spacing varies by 38%. This is because the (100) planes of both coincide when NaCl is deposited on silver, with the [100] direction of the salt being aligned with the [110] direction of the metal. NH₄Br and CsCl are also orientated by metallic silver. Once deposition has been achieved by epitaxy, the deposited layer can act as a nucleating surface for further deposition, and any stresses tend to be relieved considerably in the outer layer of a thick deposit.

Electroplating of metal often produces epitaxial growth, and where stresses are involved, these produce considerable hardness in the deposited layer. Chromium is a typical example, and hard surface treatments frequently use chromium plating for wear resistance. (The effect here is made more complex by the adsorption of hydrogen from the electrolysis.)

The modern techniques of vapour phase deposition involve epitaxial growth, particularly when complex but thin layers are involved, as in many engineering and electronic techniques. (*See* **Crystal classification, Electroplating** and **Vacuum coating.**)

Epsomite. Hydrated magnesium sulphate $MgSO_4.7H_2O$. Also called Epsom salts.

Epstein test. A method of testing core losses of silicon-steel sheets intended for use in transformers, etc. The standard size strips, 3×50 cm, are arranged in the form of a square, and the total weight of strip used in building up the arrangement is 10 kg. The primary is connected to a supply of normal frequency and voltage through a wattmeter. The wattmeter reading (corrected where necessary for low power factor) gives the total core losses. Since, however, the dimensions of the core are known and the constants of the magnetising coils can be found, it is possible to separate the hysteresis and eddy current losses. To do this, the total losses are plotted against values of $B_{max.}$ for different values of the e.m.f. The quantity $B_{max.}$ is calculated from the usual e.m.f. equation:

$$B_{max.} = \frac{E \times 10^8}{4.44 \times AN_1 f_1}$$

where E = e.m.f. applied; A = magnetic cross-section of the core; N_1 = number of turns of the primary; f_1 = frequency of the a.c. A second curve is then plotted of losses against $B_{max.}$ for different values of E, using frequency f_2. Now, for any given value of $B_{max.}$ the hysteresis losses are proportional to f; and the eddy current losses are proportional to f^2. From ordinate PQR on the graph (Fig. 62) it will be clear that $L_1 = af_1 + bf_1{}^2$, $L_2 = af_2 + bf_2{}^2$, where a and b are constants. The solution of these equations gives the values of a and b, and hence the separate values of the hysteresis and eddy current losses.

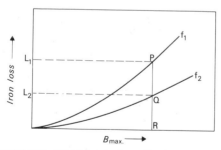

Fig. 62 *Typical graph of Epstein test showing total iron losses plotted against $B_{max.}$ with variation in alternating current.*

Equation of state. *See* **Condensation of gases, Critical point.**

Equiaxed. A term applied to crystals that are approximately the same size in three axial directions. Equiaxed structures are formed in cast metals when solidification begins at numerous points throughout the cooling metal, and results in the formation of a large number of grains of small average grain size. This type of structure should be contrasted with the columnar crystals found on the outside of a chill casting or ingot.

Equicohesive temperature. An original concept of comparison between the resistance to deformation of crystalline grains and their boundaries. It assumed that the material in grain boundaries

was stronger than that within the grains, but the boundary material became less resistant at higher temperatures, and the equicohesive temperature was the point at which the strengths became equal.

The modern interpretation is that grain boundary sliding becomes more rapid above the equicohesive temperature, so that below it ductile failure is almost universal, above it, the failure mode becomes centred on grain boundaries. The temperature is of the order of $0.5\,T_m$ where T_m is the melting point in absolute units. The rate of loading has a marked effect, and the duration of testing, so that it becomes important in creep testing.

Equilibrium. A state of equal balance. Mechanically a body is said to be in equilibrium if the forces acting upon it produce no change in its state of rest or of uniform (unaccelerated) motion.

The term also applies: to the coexistence for an indefinite time of a solid and its liquid and vapour; to the coexistence of a solution with its solute and vapour; to reversible chemical actions in which the reactants and the resultants coexist. When two opposite processes, such as evaporation and condensation, the ionisation of an electrolyte in solution, or the dissociation of a compound on heating, are occurring simultaneously and to the same extent, a state of equilibrium (kinetic) is established.

The characteristics of a state of equilibrium are: (*a*) there are always two opposing tendencies or actions which exactly counteract each other; (*b*) these tendencies are in full operation at equilibrium, i.e. the equilibrium is kinetic and not static; (*c*) a slight change in the conditions produces a corresponding change in the state of the system; (*d*) the reaction mixture will always come to the same composition (under the same final conditions), however the equilibrium point is reached. The fundamental relation that underlies all the various chemical equilibrium principles is the equation:

$$\frac{C_1{}^a.C_2{}^b.C_3{}^c \ldots}{C_4{}^d.C_5{}^e.C_6{}^f \ldots} = \text{a constant,}$$

where C represents the molar concentrations of the substances taking part in the reactions, or the partial pressures if the substances are gases, and a, b, c, etc. represent the number of molecules of the respective substances.

In the case of an alloy, equilibrium is said to have been attained when the various constituent phases are stable at the particular temperature and compositions under consideration.

Equilibrium diagram. A graphical representation in which the axes are composition and temperature showing the limits of composition and temperature within which the various constituents or phases of an alloy are stable. From such a diagram information concerning the structure and composition of the constituents of an alloy at a stated temperature may be deduced. (*See also* **Constitutional diagram.**)

Equipartition of energy. A principle enunciated by Maxwell which states that each degree of freedom of a body (or an atom) entails an equal increase in energy when, for instance, the body is heated.

Equivalence factor. A term used to indicate the relative effects of various alloying elements on the structure and properties of the basis metal. In this sense values may be assigned, as percentages, to the amounts of aluminium, iron or manganese that can be regarded as equivalent in effect to a certain percentage of zinc in brass.

Equivalent weight. That weight of an element or radical which combines with or replaces 1 part by weight of hydrogen or 8 parts by weight of oxygen. The atomic weight of an atom is always a small multiple of the equivalent weight. Also known as combining weight because the equivalent weights of elements provide the data for knowing in what proportions substances react.

Erbium. The existence of an earth which was described by Mosander and called by him erbia was at first discounted but later was confirmed. This earth was then renamed terbia because in the meanwhile the designation erbia had been applied to a rose-coloured earth. The terbia originally isolated from crude yttria by Mosander consists of at least four earths: erbia, holmia, thulia and dysprosia. The metal has a specific gravity of 9.05 and its atomic weight is 167.2. (Crystal structure cph, lattice constants $a = 3.532 \times 10^{-10}$ m, $c = 5.589 \times 10^{-10}$ m.) It forms a rose-coloured oxide and salts of the same colour. (*See* Table 2.)

Erg. The c.g.s. unit of energy or work. It is the work done by a force of 1 dyne moving through a distance of 1 cm. The unit is very small and for practical purposes in engineering three other units are employed:

$$1 \text{ joule} = 10^7 \text{ ergs}$$
$$1 \text{ watt-hour} = 3\ 600 \text{ joules}$$
$$1 \text{ kilowatt-hour} = 3.6 \times 10^6 \text{ joules}$$

Erichsen's test. A cupping test used for assessing the ductility of sheet metals. The method consists in forcing a conical or hemi-spherical-ended plunger into the specimen and measuring the depth of the impression at fracture.

Erinofort. A nonflammable cellulose acetate thermoplastic.

Erinoid. A thermoplastic made by heating casein with formaldehyde.

Erosion. The process of the wearing away of the land surface by natural agencies, e.g. by weathering, corrosion and transportation under the influence of gravity, wind and running water.

In the case of metals the term usually implies the wearing away of the surface by corrosion and flowing water, steam or gases which may or may not contain mineral particles in suspension.

Erubescite. A sulphide ore of copper, sometimes called purple copper, Cu_3FeS_3. The composition is variable. It is reddish-brown in colour, but rapidly darkens on exposure to air. Crystal system cubic. Also known as bornite (q.v.) and variegated copper ore.

Etchant. A solution of chemical reagents applied to prepared or polished surfaces of metals or alloys, that will differentially attack or colour selectively areas of different chemical composition to reveal structural features or components. When a pure metal is treated with a suitable etchant, the polished surface remains bright, but a network of fine lines becomes visible. These lines are the grain boundaries. A certain amount of segregation is inevitable during solidification, and etching reveals minute differences in chemical composition even in single-phase alloys. The various grains of an otherwise homogeneous alloy show different degrees of brightness after etching because the orientation of the crystal axes within the grains determines the rate at which the crystal is attacked and its polished surfaces roughened by the reagent.

Where the action of the etchant is to dissolve a component, the parts affected may assume a slightly roughened appearance and so produce contrast by causing a change in the degree of brightness. Colouring may be the result of a chemical change or of the deposition of products resulting from the action between the etchant and a particular component. Selective colouring may be due to the different rates at which such reactions occur.

The etching reagents are usually aqueous solutions or solutions in ethanol (ethyl alcohol). Less commonly methanol (methyl alcohol); pentanol (amyl alcohol); 1, 2, 3, propanetriol (glycerine); or propanone (acetone) are used as solvents. The etching effect is a function of the solvent, as the solution is not saturated with respect to all the solutes; it is also dependent on the extent to which the solution is dissociated and on the viscosity. Aqueous solutions wet metal surfaces less evenly than alcoholic solutions, and this may be associated with the fact that the latter are often good grease solvents.

For micro-examination the time of etching is relatively short, since general attack, which destroys structural features and impairs the reflectivity of the polished surface, is undesirable. For macro-examination a high degree of polish is not usually required, and reagents which act quickly to reveal heterogeneity, unsoundness or porosity are generally used.

In many cases the action of the etchant on the metal can be put into one or other of the following classes: (a) the solvent effect of an acid, e.g. hydrofluoric, hydrochloric, sulphuric or nitric acids; (b) the oxidising effect of an acid, e.g. nitric, sulphuric (when hot and strong), chromic and picric acids; (c) the oxidising effect of oxidising agents, e.g. hydrogen peroxide, ammonium persulphate, etc.; (d) the solvent action of alkalis, e.g. caustic soda and potash. Ammonia in the presence of an oxidising agent dissolves copper; (e) the chemical replacement of one metal by another, e.g. iron III (ferric) chloride or copper II (cupric) chloride; (f) two or more reactions occurring simultaneously, e.g. the liberation of hydrogen sulphide from a metallic sulphide by the action of an acid and the reaction of the hydrogen sulphide with a soluble salt, e.g. mercury II (mercuric) chloride

to produce a distinctively coloured product.

There are a number of etchants that are associated with the names of investigators, and, they are assembled in Table 11.

The number of etching solutions that have been proposed is very large. Some of them are quite commonly used, others are used very infrequently or only in special cases. Details of the commonly used etchants are included in Table 11.

Etchells furnace. An electric furnace using three-phase current. Two phases are introduced through vertical electrodes and the third through a bottom electrode embedded in the refractory lining. The resistance of the hot refractory bottom generates the heat.

Etch figures. When the surface of an alloy, consisting of one phase only, is etched with a suitable reagent, the different grains often show variations in brightness. When this is due to the formation of pits bounded by surfaces that approximate to planes with simple crystallographic indices, the facets are usually arranged in parallel groups in each crystal giving rise to etch figures.

Etching. The revealing of the structure of a metal or alloy by attacking a polished surface with a reagent that has a differential effect on the various constituents or on different crystals. (*See also* **Etchants.**)

Etch pits. Small cavities formed on the highly polished surface of a metal during etching.

Etch tests. A series of tests in which various reagents are used to bring out the macrostructure of a metal, usually to reveal porosity, flow lines or slag streaks. The McQuaid-Ehn test for abnormality and grain size in steel may be regarded as an etch test, though the primary purpose is to determine the hardening characteristics and the austentic grain-size characteristics of steels.

Ethane C_2H_6. Second in the series of saturated unbranched acyclic hydrocarbons. Boiling point $-88.6\ °C$.

Ethyl alcohol, ethanol C_2H_5OH. Liquid at room temperature, boils $78.5\ °C$. Suitable solvent for resins, hydrocarbons, fatty acids and many mineral salts. Readily oxidised to acetaldehyde and acetic acid. Source material for plasticisers, paints, fine chemicals and pharmaceuticals.

Ethylene, ethene C_2H_4. First in the series of unsaturated unbranched acyclic hydrocarbons. Boiling point $-103.7\ °C$. Burns with bright luminous flame, and forms explosive mixtures with oxygen. Decomposes on heating to yield carbon, methane and ethyne (acetylene). Derived from catalytic decomposition of crude petroleum and fractional distillation. Source material for wide range of petrochemical products, including petrol additives, paints, plasticisers, synthetic rubbers, antifreeze, detergents, brake fluid, lubricants, weed killers, degreasing agents, PVC, polyethenyl benzene (polystyrene), and polyethene (polyethylene) polymers.

Ettinghausen effect. The name applied to the temperature difference produced on the two edges of an elongated metal plate through which a current is flowing, by the application of a magnetic field at right angles to it. This effect is analogous to the Hall effect (q.v.), except that temperature difference is observed in place of potential difference.

Europium. A rare metal of the rare-earth group. It is found in small quantities (0.02%) in monazite. Its atomic weight is 152 and its density at $20\ °C$ is $5.25\ g/cm^3$. Its crystal structure is body-centred cubic. The oxide Eu_2O_3 and its salts have a pale pink colour. Lattice constants $a = 4.573 \times 10^{-10}$ m, $c = 3.960 \times 10^{-10}$ m. (*See* Table 2.)

Eutectic. In Fig. 63, AE represents the liquidus showing the point of first solidification of tin-rich solid, and BE the liquidus for separation of zinc-rich solid. The point of intersection E, at which composition equilibrium exists between liquid E, solid of composition C and solid of composition D, is the binary eutectic, and the temperature the eutectic temperature. An alloy of composition E solidifies at the eutectic temperature as an intimate mixture of solids C and D (in the case of the simplified Fig. 63, tin and zinc). In practice there would be some mutual solubility of each metal in the other to solidify as an intimate mixture of two solid solutions, α and β. The solidus, at which all compositions are completely solid, is the straight line CED, and occurs at the eutectic temperature.

The eutectic alloy E is the most fusible of all alloys in a given

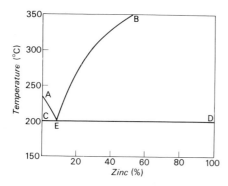

Fig. 63 *A simple binary equilibrium diagram for tin-zinc alloys. AE shows the separation of tin; BE the separation of zinc; E the eutectic; and CED the range of series in which eutectic mixtures will be found.*

series, or in a given range of that series. More than one eutectic may occur in a series of alloys, and in these cases each corresponds to a different set of constituents. A eutectic, though of constant composition, is not a chemical compound, but an intimate mixture and usually consists of laminated plates, spheroids in a matrix or combinations of these geometrical arrangements, and may be formed between pure metals, solid solutions, intermetallic compounds, or combinations of these.

Directional solidification of eutectic alloys has been used to produce composite materials consisting of fibres or lamellae of on phase (say α) in a matrix of the other (β). They have been termed "in situ" composites.

Eutectic alloy. An alloy of eutectic composition, i.e. one in which liquid of the fixed eutectic ratio of two or more solid phases can exist in equilibrium with them, and finally solidify at a constant temperature to give an intimate mixture of those phases. (*See* **Eutectic.**)

Eutectic point. The point in an equilibrium diagram at which the inclined branches of the liquidus curve meet the horizontal "eutectic line". The temperature corresponding with this horizontal line which indicates the completion of solidification, or the beginning of melting, is termed the eutectic temperature. The eutectic point indicates also the *composition* of the eutectic. (In the case of ternary diagrams the eutectic point is the point of intersection of three descending liquidus surfaces.)

Eutectic solder. A tin-lead alloy containing 62% Sn, 38% Pb. Its melting point is 183 °C. It is the fastest-working solder, used for high-speed machine soldering and fine work.

Eutectic structure. A characteristic, often laminated, duplex structure produced as the constituents of a eutectic alloy simultaneously solidify at constant temperature. It frequently consists of alternate lamellae of the constituents.

Eutectic temperature. *See* **Eutectic point.**

Eutectoid. A eutectoid is similar to a eutectic, except that it involves the simultaneous formation of two or more constituents from another solid constituent instead of from a melt.

Eutectoid steel. A steel of the eutectoid composition which in pure iron-carbon steel alloys is 0.89% C. Variations from this composition are found in commercial steels, particularly in alloy steels in which the eutectoid carbon content is usually lower. The structure of a eutectoid steel in the annealed condition is 100% pearlite.

Evaporation. A process in which a liquid is slowly converted into vapour, at temperatures below the boiling point of the liquid. The rate of evaporation is determined by the temperature and by the pressure of vapour at the liquid surface.

Evaporation process. A method of depositing a very thin film of metal on metallic or non-metallic articles. The process is conducted in a vacuum chamber. The metal to be deposited is put on a wire, which is heated electrically, and the article to be coated is slowly rotated before the wire. Many metals having relatively low melting points can be used, but aluminium is the most successful as far as colour and brilliance are concerned. It is recommended that the article, after coating, be given a further coating of transparent lacquer.

Exchange energy. In the approach of two or more atoms to form a bond, the interaction energy involved depends on a number of factors, not least of which is the overlap of wave functions. The energy may be decreased if the electron spin of the unpaired electrons assumes a parallel orientation. In other cases, antiparallel orientation may be more favourable.

Whenever such an adjustment is made to satisfy the various conditions applicable, the net balance of energy must include energy required to make adjustments, the exchange energy. In forming ionic bonds for example, the exchange of electrons involves some overlapping of similar energy levels in the resultant molecule, and although there will be broadening of levels into bands, the levels available in those bands will be filled. A net increase in electron energy is therefore involved inevitably, to satisfy the Pauli principle and the wave functions of the electrons. Any increase in energy is positive (repulsive), and hence this amount of energy decreases somewhat the available negative energy of attraction. The exchange repulsion therefore lowers the overall energy obtained from the ionic bond, but the net attraction still provides a strong bond, as is evident from the properties of the molecule.

In ferromagnetism, a similar effect arises from the alignment of electron spins. When one atom unites with its neighbours, there will be energy overlap, and broad bands of electron energy will be involved, but if parallel spins are favoured for say the 3d electrons, this can only mean more of one half band of energy containing more electrons than the other, and hence being filled to a higher energy level. The exchange has therefore improved the situation on the one hand, by an alignment of spins, but has involved an increase of energy because of the Pauli principle and wave mechanical considerations. The broadening of the 3d band is in fact such that the net spin moment is achieved with a net decrease in energy, and so in ferromagnetics this is what occurs. However, the influence of interatomic spacing on the 3d band broadening is such that only iron, nickel and cobalt have favourable parallel orientation of spins. Manganese can be made ferromagnetic by alloying with elements which increase the atomic separation distance, although pure manganese is antiferromagnetic.

Exclusion principle (Pauli). Not more than one electron can occupy a given state of energy as defined by all four quantum numbers. As the number of electrons increases, and the principal quantum number n varies 1,2,3,4 . . . the number of electrons that can be contained is $2n^2$ for each value of n (2,8,18,32 . . .). For a given value of n the maximum number of electrons when $\ell = 0$, the s state, is 2; for $\ell = 1$, the p state, 6; for $\ell = 2$, the d state, 10; and for $\ell = 3$, the f state, 14.

Exfoliation. The flaking off of the outer layer of a specimen, e.g. the case of a case-hardened component. Also referred to as spalling or peeling.

Exothermic reaction. A reaction or transformation during which heat is liberated. In a reaction from one definite state of a system to another the decrement of total energy is equal to the work expended on external bodies together with the heat emitted. If the process is reversible, taking any given system, it can be undone at a different temperature. Thus it admits of a Carnot's cycle, giving after reduction $A - E = T(dA/dT)$ where A = available energy, E = total energy, and T = temperature.

According to this principle, transformation at constant temperature can proceed spontaneously only in the direction in which the available energy (A) diminishes.

Berthelot's rule states that chemical change accomplished without the addition of external energy tends towards the formation of that compound or that system the production of which is attended with the maximum heat evolution. The rule, however, serves only as a rough-and-ready guide to the probable course of a chemical change. It is exact only when $A = E$. This tends to be the case at low temperatures of reaction and is exactly true when $T = 0$.

Expanded metal. An open-meshed metal network used for reinforcing concrete and plaster-work and for various other purposes, e.g. roofing. Produced by stamping or perforating sheet metal and stretching to form the required pattern.

Expanded polymers. Polymeric melts which are caused to "foam" during the forming operation, so that after cooling a large number of air or gas cells are retained as closed pores, leaving a lower density but reasonably strong material. The foaming action may be part of the polymerisation process, or be produced by added agents which gasify at a controlled rate.

Expanding metals. Certain alloys of bismuth which expand during or after solidification. This is a property possessed by some but not all the fusible alloys.

Explosive antimony. When an acid solution of antimony trichloride (sp. gr. 1.35), or a solution of antimony trioxide in hydrochloric acid, is electrolysed, using antimony as the anode and copper or platinum as the cathode, a grey lustrous deposit is obtained on the cathode. The metal has an amorphous fracture and a specific gravity of 5.78. It contains from 4.8 to 7.9% antimony chloride, together with some free hydrochloric acid. When scratched with a needle or touched with a red-hot wire it decomposes with liberation of heat and production of antimony trichloride. If it is heated to 200 °C it disintegrates into powder with a loud explosion.

Similar products are obtained by the electrolysis of the acid solutions of antimony bromide and iodide.

Explosive forming. A high strain rate process of forming metals, usually in sheet form. An explosive charge is detonated, to create a shock wave transmitted in air, water or other fluid, to the workpiece. The explosive charge is smaller with fluid transmission than it is with air. The explosive is not in contact with the workpiece, but situated at a stand off distance; this distance and the weight of charge determine the extent of deformation achieved. The blank to be formed may be clamped to a die, although this is not a prerequisite for forming. Only simple shapes are tackled on the large scale, such as circular diaphragms or end caps. The advantages are stated to be:

(a) die costs. Only a female die is needed, and this can be in cheap easily formed materials, hardwood, concrete, polymeric resins, metals, unless mass production is envisaged;

(b) absence of a punch. Apart from the cost saving, frictional problems disappear;

(c) re-entrant shapes. Less difficulty with this method;

(d) large size. Where installation of new press equipment or die tooling is not justified, the explosive method is a perfect solution;

(e) interstage annealing. The high rate of strain can produce greater total strain per forming operation, thereby reducing the number of annealing stages in tough alloys. This is particularly true of alloys with only a few slip systems capable of contributing to ductile behaviour;

(f) time. Large components require more handling and setting up for conventional processes, and hot forging materials with a narrow working range can be awkward in these circumstances. The speed and simplicity of explosive forming can overcome these objections.

Explosive forming can be used for closures in final assembly. Rivets have been closed by this means, the charge being contained in the head and detonated either by percussion or by heating. Firebox stays in boilers and similar installations have been upset in position by the same method. Powdered metals have been pressed by explosive charges replacing the conventional press.

Explosive welding. Similar to explosive forming, but here the charge not only provides the force for pressure welding, the shock wave also serves to remove oxide and other contaminants at the moment of impact and actual joining. The techniques are more difficult to apply than with forming, e.g. in joining two sheets together the pressure wave must be progressive across or along the area to be connected. A series of charges may be necessary in advance of the detonation. Nevertheless the method is lower in cost and can be applied to large areas when in the form of simple plate or sheet.

Extensometer. An instrument used for measuring small changes in length over a given gauge length, particularly in tensile tests. Since for steels the elongation within the elastic limit of an ordinary test piece is less than $25\,\mu$m, in most cases it is desirable that the instrument should be capable of reading to about $0.5\,\mu$m. The displacements are therefore magnified either mechanically, optically or electrically.

Extraction. A term used in relation to all those processes that are used in obtaining metals from their ores. Broadly, these processes involve the breaking down of the ore both mechanically (crushing) and chemically (decomposition) and the separation of the metal from the associated gangue. Extraction metallurgy may be conveniently divided into the following groups.

(a) Ore dressing. This includes all those processes for the removal of the unwanted material (gangue) and the concentration of the ore. In general, the ore must first be crushed, then screened and sized, preparatory to passing over or through a concentrator. Where mechanical or magnetic separation is not used, the process most generally employed is that known as the wet gravity concentration process. In this the laws governing the falling of bodies through water are made use of to increase the efficiency. When water alone is used the gangue is carried away in the stream of water. More recently the oil process of flotation-concentration has been used. (*See also* **Concentration**.)

(b) Pyrometallurgy. After concentration the ore is treated according to the nature of the compound in which the metal occurs. This treatment may include calcining, e.g. to reduce carbonates; roasting—a process of heating in the presence of air, e.g. to eliminate sulphur by converting sulphides into oxides, etc.; smelting—a process for the reduction of ores in the presence of a suitable flux in a blast or reverberatory furnace. Oxides of metals may also be reduced by heating with other metals, such as sodium, potassium, magnesium, aluminium or with hydrogen.

(c) Hydrometallurgy. This consists in the separation of the metal or its compounds through the medium of aqueous solutions prepared from metallic ores. The metal is dissolved, by addition of suitable chemical reagents, from the ore and is recovered from the solution by some suitable precipitating agent (or sometimes by electrolysis of the solution). Hydrometallurgical processes, as a rule, are applied to ores that are too low in metallic content to permit of their being treated economically by a smelting process, or that cannot be successfully concentrated into a product that will permit of being treated economically by smelting.

(d) Electrometallurgy. Under this heading are included all those processes that employ electrical energy either for the decomposition of a fused mass or a solution of a salt (electrolysis) or as a source of heat (electrothermic process). Electric furnaces are now used for smelting and refining in many extraction processes.

Extra solution. The name given to the leaching solution used in the Russell process for the extraction of silver from ores containing appreciable amounts of lead and zinc sulphides which are unsuitable for treatment by the ordinary "hypo" process. The usual thiosulphate leaching treatment is supplemented by leaching with the "extra solution" consisting of the double thiosulphate of copper and sodium $2Na_2S_2O_3.3CuS_2O_3$. As this salt decomposes readily on exposure to air, sodium thiosulphate is passed through a perforated box containing copper sulphate which is immersed in a closed-in leaching vat, just above the ore. The action of the extra solution, which is not rapid, is to dissolve the silver sulphide as the double salt $2Na_2S_2O_3.3Ag_2S_2O_3$.

Extrinsic semiconductor. A semiconductor, i.e. a material with a narrow forbidden energy gap between the valence and conduction bands, to which has been added a small quantity of material whose outer electron energy levels lie within that gap. If the "impurity" level is close to the conduction band of the semiconductor, the impurity atom must contain one more electron per atom than the host atom, or be capable of providing free electrons by other means. If the energy levels are closer to the valence band, then one less electron per atom is required.

The first case is called an n-type or donor impurity, the second a p-type or acceptor. Combinations of two different types of atom may be made to function in a similar way, provided always that electrons are available and there is a suitable acceptance route. Certain combinations of elements from groups III and V of the periodic table, or between groups II and VI, can achieve this.

Extrusion. A fabrication process for producing rods, tubes and various shapes of solid and hollow sections from the billet or blank of solid metal. There are basically three types of extrusion, known respectively as direct, inverse (or indirect) and impact. In the direct process the billet is placed in a strong cylinder, which can be heated if required, and a follow plate placed behind it. Sufficient pressure is then applied by means of a hydraulic press to cause the metal to flow plastically through the restricted orifice of the die. In the indirect process the follow plate and the ram of the direct process are replaced by a closure plate. The die is caused to move towards the billet so that the extruded part passes through the hollow die stand. Since the billet does not move bodily, there is no friction between the cylinder wall and the body of the metal as in the direct extrusion (*see* Fig. 64).

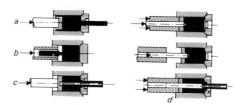

Fig. 64 *Diagrams showing different extrusion methods. (a) Direct; (b) indirect; (c) extrusion of tube from hollow billet; (d) extrusion of tube from solid billet.*

In the impact process an untreated slug of metal is placed in a shallow die and subjected to a percussive blow by a punch or former. The metal is caused to flow up over the punch through the annular orifice between it and the sides of the die. This method is used for making cup-shaped blanks, collapsible tubes, etc. (*see* Fig. 65).

Direct or, less frequently, indirect extrusion is commonly applied to numerous ferrous and non-ferrous metals and alloys, but it is most widely used for brass, lead and tin, copper, aluminium and magnesium alloys. Copper and brass tubing, cable sheathing, lead pipe and cored solder are usually produced by extrusion. Many attempts have been made to provide continuous extrusion processes. For one example, *see* **Conform**.

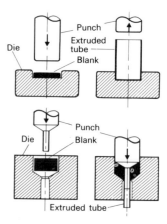

Fig. 65 *Diagram showing progressive stages in impact extrusion.*

Exudation. The phenomenon in which the liquid produced by partial or complete melting of a solid is liberated and escapes. It is commonly applied to the liberation of liquid metal from the solid, as in the case of the production of molten lead from a leaded brass on heating at temperatures in excess of the melting point of lead.

Face-centred cubic (f.c.c.). One of the two most close-packed structures in crystallography. The ABCABC stacking of close-packed hexagonal two-dimensional sheets leads to the two-dimensional planes becoming the octahedral planes of a structure which can be represented by atoms at each corner of a cube and atoms at the centre of each face of that cube (*see* Figs. 25, 26). The direction of the closest packed rows of atoms is [110], and octahedral close-packed planes are (111). There are four sets of (111) planes, and twelve close-packed directions. The coordination number is 12, and if a is the length of the side of the unit cell ($a = b = c$) the nearest neighbours are at a distance $\frac{1}{2}a(2)^{1/2} = 0.707a$. There are also six next nearest neighbours to each atom, at a distance a.

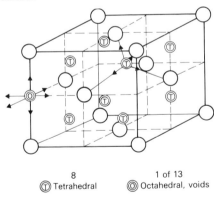

8	1 of 13
ⓣ Tetrahedral	ⓞ Octahedral, voids

Fig. 66 *Voids in face-centre cubic structure. Arrows indicate atoms forming a void boundary.*

Voids in this structure occur in tetrahedral spaces between the octahedral planes (i.e. near each corner atom of the cube) and can each contain a sphere of radius $0.225R$ where R is the atomic radius of the primary spheres of the lattice. There are eight tetrahedral voids per unit cube. There are also octahedral voids situated between the atoms at each corner of the cube, and therefore involving four face-centred atoms and two corner ones on each edge of the cube. There is also an octahedral void at the centre of each unit cube, involving all the face-centred atoms. Each octahedral void on cube edges is shared by four unit cubes, and there are twelve such positions on each cube. There are therefore four octahedral voids per unit cube, three attributable to each unit cube from the cube edge positions, and one at each centre. The voids will each contain a sphere of radius $0.414R$, but neither octahedral nor tetrahedral void spaces are large enough to accommodate any elemental atoms other than very small ones, such as

hydrogen and nitrogen, and so distortion of the lattice occurs if impurities occupy these interstitial positions. This is not surprising, considering that 74% of the available space is taken up by the primary spheres of the lattice (*see* Fig. 66).

Typical examples of f.c.c. lattices are copper, lead and aluminium, with lattice parameters 3.62, 4.95 and 4.05 $\times 10^{-10}$ m respectively.

Facet. A term used in reference to the flat surfaces of a crystal, and particularly to denote the bright crystal faces that can be observed in fractured surfaces.

Facing sand. A fine-grain sand used by foundrymen. It is used next to the pattern and on the surfaces of the mould so that it will be in direct contact with the casting. The sand is often mixed with clay, flour or molasses to improve its binding properties and to impart a smooth finish to the mould, but the material must be sufficiently porous for gases to escape and refractory enough to permit easy extraction of the casting from the mould.

Fadgenising. A mechanical finishing process for polishing zinc-base die-castings preparatory to plating. The castings are mounted on studs or pins attached to cylindrical racks which are rotated in an abrasive slurry consisting of flour silica, wood flour and maize meal.

Faggot. A stack of flat wrought-iron bars which is heated to welding heat and subsequently forged or rolled to form a solid bar.

Faggoting. A term used in reference to the various methods of piling bar iron or merchant bar for reheating in the production of refined wrought irons.

Fahl ore. Grey copper sulphide ore (tetrahedrite) which is worked for its copper and silver content.

Fahralloy. A trade name applied to a group of heat-resisting ferrous alloys of the iron-chromium-nickel-aluminium type.

Fahrite. A trade name of a series of heat- and corrosion-resisting chromium-nickel-iron alloys made in the USA.

Fajans, Russell and Soddy's law. A law that relates to the position of radio-elements in the periodic system. It states that in all cases in an α-ray change, i.e. a transformation in which an α-particle is expelled from the atom, the product so generated falls into a group of the periodic system two places lower than that to which the parent substance belongs. In a β-ray change, i.e. one in which an electron is expelled from the atom, the product falls into a group one place higher than that of the parent substance.

Falconbridge process. A process for recovering nickel from ores containing nickel and copper. The ore is smelted to obtain a high metal matte, which is then crushed and roasted to remove most of the sulphur, agitated with an acid solution to remove copper, and then filtered. The filtrate is then passed to electrolytic tanks, where copper is plated out. The filter cake containing the nickel is melted and cast into anodes which are refined electrolytically.

Falling-weight test. A simple form of impact test still used for testing rails, car wheels and other components. In the case of steel tyres the heavy weight may be dropped on to the tyre held in an upright position, or the tyre may be dropped from a specified height on to a steel rail. Also called the drop test.

Fanning. A term applied by blast-furnace men to the idling period between the blowing periods when the blast pressure is reduced to a minimum. During this period no charging or tapping of slag or metal is carried out.

Fan steel process. A process for the production of metallic tungsten. Wolframite $FeWO_4$ is heated with soda ash at about 800 °C to produce sodium tungstate $Na_{10}W_{12}O_{41}.28H_2O$. Any manganese present is converted into the dioxide. The furnace product is leached and the clear sodium tungstate solution filtered out and treated with calcium chloride to precipitate the tungsten as calcium tungstate. This latter is washed and boiled with hydrochloric acid, when tungstic acid is formed. This is then converted into ammonium paratungstate, and the relatively pure tungstic acid is formed by adding nitric acid. The tungstic acid is converted into oxide and the latter is reduced to metal by heating it in a current of hydrogen at 1,200 °C.

Fantail. A short tapering flue at the end of the slag pockets of an open-earth furnace. It leads to the top of the regenerator chamber.

Farad. The unit of capacitance. A conductor is said to have a capacitance of 1 farad if a charge of 1 coulomb raises its potential by 1 volt. A condenser has a capacitance of 1 farad if a charge of 1 coulomb produces a potential difference of 1 volt between its coatings.

Faraday lines. A convention in electrical theory, where it is assumed that Q Faraday lines emanate from an electric charge Q, and hence one Faraday line from a unit charge. This should be compared with $4\pi/k$ (Maxwell) lines of force and 4π lines of induction emanating from unit charge. Faraday lines are most useful in electrostatic theory.

Faraday's laws. These laws relate to the electrolysis of solutions and fused salts. The first law states: "The amount of chemical decomposition which takes place in a cell is proportional to the total quantity of electricity which passes". This may be expressed in the form of an equation: $w = ZIt$, where w = mass of a particular ion deposited, Z = electrochemical equivalent of the liberated substance, I is the current in amperes and t is the time during which the current passes (in seconds).

The second law states: "If the same current flows through several electrolytes the masses of the ions liberated are proportional to their chemical equivalents".

In the above equation $Z = Ba/v$, where a = atomic weight of the element, v = valency of the element and B is a constant which is 0.00001038. Thus a/v = equivalent weight and $0.00001038 \times a/v$ = equivalent or the number of grams of a substance deposited by 1 coulomb.

Fatigue. A phenomenon that gives rise to failure under conditions involving repeated flexure or fluctuating stress below the ultimate stress of the material. Fatigue failure generally occurs at loads which applied statically would produce little perceptible effect. The fracture is usually progressive, and in most cases there is evidence of the way in which it has occurred. Susceptibility to progressive failure is called fatigue. Resistance to progressive failure is called endurance (q.v.).

In the loading of a metal part, the maximum stress is usually located at the surface, so the performance in service under repeated stresses is directly dependent on the stress range at the surface and the properties of the surface. Failure often starts as a small—even a microscopic—crack which spreads at each repetition of the stress. The faces of a crack rub together, producing a sort of burnished appearance. When the service conditions are such that the stress varies, being sometimes of a low value and sometimes of a higher value, the rate of growth of the crack will vary. This gives rise to zones that can be distinguished because they have been subjected to different degrees of rubbing and the fatigue fracture will have an oyster-shell appearance.

Local stress concentration produced by nicks, notches, key-

ways, oil holes, screw threads, etc., are nuclei for the progressive or fatigue failure. Other stress raisers are scratches, tool marks, rough surface, quenching and grinding cracks, sharp changes in section, poor fillets, as well as inclusions in the metal, corrosion pits, etc.

For a given maximum and minimum stress the number of repetitions (stress cycles) before fracture occurs increases as the range of stress is diminished, and there is a range of stress known as the limiting range at which the number of stress cycles is infinite. The limiting range of stress becomes less as the maximum stress increases.

Fatigue test results are usually plotted as S-N curves, i.e. stress range (S) against number of stress cycles (N) (see Fig. 67). These curves show the endurance of all broken test pieces and can be used to determine the limiting safe range. (See **Thermal fatigue.**)

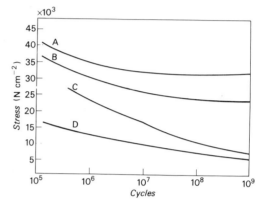

Fig. 67 *Typical S-N curves. Curve A for fine-grained quenched steel; B for fine-grained annealed steel; C for heat-treated aluminium alloy; and D for annealed aluminium alloy.*

Fatigue limit. Fatigue tests, which are extensively used for design purposes, have shown that the endurance (N) of a metal part—i.e. the number of stress cycles that can be endured before fracture—depends on the range of the applied stress (S) rather than on the maximum stress. The fatigue limit is the highest stress which, regardless of the number of times it is repeated, will not cause fracture. (See also **Endurance limit.**)

Fatigue range. The range of stress which a metal will withstand indefinitely. When the maximum stress in tension is equal to the stress in compression, the fatigue range is twice the fatigue limit. When comparing the fatigue (or endurance) results on rotating-beam specimens determined under completely reversed stress, with those of other types of test where the stresses are not thus balanced, it should be remembered that the former are generally given as half the total range of stress, while others are given as the total range, stating the extremes of the range.

Fatigue tests. See **Endurance tests.**

Fat sand. A term used in describing moulding sand that contains a large amount of clay or alumina.

Fayalite. An iron silicate $2FeO.SiO_2$ that crystallises in the orthorhombic system. It occurs at Fayal in the Azores and in the Mourne Mountains of Ireland and in slags. It has a melting point of 990 °C. Specific gravity 4·4.2.

F-centres. See **Colour centres.**

Feather ore. A sulphide of lead and antimony $2PbS.Sb_2S_3$ that crystallises in the orthorhombic system as feathery acicular crystals. It occurs with antimonite in Cornwall, the Harz Mountains in Germany.

Federal hearth. An ore-roasting or reduction hearth fitted with mechanical handling and rabbling equipment.

Feeder. The runner or riser of a mould which maintains an adequate supply of molten metal to compensate for shrinkage as the casting solidifies.

Feeder head. The upper part of a mould, usually made of refractory materials so arranged that the metal poured into it is maintained in the liquid state and can thus feed the contraction cavity which is formed as the metal solidifies in the mould. (See also **Feeder** and **Dozzle.**)

Feeding rod. A metal bar or rod which is inserted into the feeder head of a mould and worked up and down to maintain a free passage for molten metal during the solidification of a casting.

Fehling's solution. This solution is made from equal parts of two solutions containing (1) 69 g of pure copper sulphate dissolved in 1 litre of water, with the addition of 1 drop of sulphuric acid, and (2) 350 g of sodium potassium tartrate and 100 g caustic soda in 1 litre of water. Reducing agents, such as glucose, reduce the deep blue solution of the cupric salt to red copper I (cuprous) oxide.

Felspar. A group of minerals consisting essentially of silicate of aluminium with varying amounts of silicates of potassium, sodium, calcium and barium, which enter into the composition of igneous rock, e.g. granite. By the disintegration of these rocks soluble alkali salts and insoluble hydrated aluminium silicates (clay) pass into the soil.

Feran. Aluminium-coated steel strip produced by cladding extra deep-drawing quality steel with aluminium. Subsequent heat-treatment at 535-550 °C for 10-15 hours ensures maximum ductility of the steel with the minimum formation of compound $FeAl_3$ at the interface.

Ferberite. A tungstate of iron $FeWO_4$. It is usually associated with wolfram, so that its formula is often given as $2(FeMn)WO_4$. $(FeMn)O$.

Fergusonite. A niobate of yttrium $YNbO_4$. The Australian fergusonite is a tantalate of yttrium $YTaO_4$. The former contains about 44% Nb reckoned as Nb_2O_5.

Fermi level. The difficulty of assigning particular energy levels to each and every atom in a polycrystalline solid, together with the restrictions of the Pauli exclusion principle, led Fermi to envisage a probability for the occupation of a particular energy level in terms of the wave functions involved. The probability can be expressed $F(E) = e^{-(E - E_F)/kT} + 1$ where $F(E)$ is called the Fermi function, E represents the energy of the electron level involved, E_F is the Fermi energy and a constant for the system, k is Boltzmann's constant and T the absolute temperature.

The Fermi energy E_F is the extra kinetic energy, which is in excess of the energy needed for bonding, and allows for electron movements in the lattice. If it did not exist, then all the electrons would be tightly bound to the nucleus. Obviously those close to the nucleus are tightly bound, and the concern is with the outer electron energy levels only.

At 0K, the function can be represented graphically (see Fig. 68(a)). The function $F(E)$ has a value of 1 for $E < E_F$ and zero for $E > E_F$. At 0K, energy states with $E > E_F$ will be completely occupied, those above will be completely empty. E_F is thus the maximum energy of filled states at 0K for those materials whose electrons have a continuously available range of energy. This applies therefore to most metals, but not to covalent bonded or partially bonded materials, such as insulators and semiconductors.

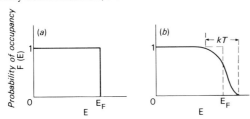

Fig. 68 *Fermi distribution curves at (a) T = 0 K; (b) T > 0 K.*

Above 0K, for a finite temperature, $F(E) = \frac{1}{2}$ when $E = E_F$. To define the effect of temperature, a Fermi temperature T_F can be introduced $kT_F = E_F$. The function for a finite temperature is shown in Fig. 68(b). The drop from $F(E) = 1$ to 0 is now not sharp, but the effect is noticeable only at very high temperatures. $F(E) = 0.2$ would give $(E - E_F) = 1.39 kT$, i.e. of the order of kT, whereas the function as a whole is of the order of kT_F. T_F for sodium is 5×10^4 K, for copper 8×10^4 K. Hence the effect of temperature is small until T approaches T_F. The top of the distribution function gives $E - E_F \gg kT$, and hence, $F(E)$ is approximately $e^{-(E - E_F)/kT}$ which is identical with the Boltzmann distribution for molecules of a gas, in classical mechanics. This simplifies

calculations for higher temperatures.

The Fermi function gives only a probability for occupation of an energy state by a single electron. To deal with a mass of material, the number of possible states must be multiplied by the probability of occupation, to arrive at the actual number of electrons involved. If $N(E)$ is the number of electrons with energy E and $Z(E)$ the number of possible states of that energy, then $N(E)dE$ is the number of states whose energy lies between E and $E + dE$, the population of electrons, and $N(E)dE = Z(E).F(E) dE$.

Above 0K, electrons can absorb energy by transfer of thermal energy as the atoms vibrate, but this energy can only be absorbed if there is a vacant state of that level. The curve of $N(E)$ against E is a parabola. At 0K, the cut-off in occupied states will be sharp, in accordance with the Fermi function. At higher temperatures, the tail of the Fermi function will operate, and the density of states curve will fall to zero with a similar shape (see **Density of electron states**). Because E_F represents the maximum energy state which can be occupied, it is called the Fermi level.

Ferobestos. The trade name of a group of laminated plastic materials used for gears, bearings, etc. The material is essentially laminated asbestos bonded with thermo-setting resins or, if a high-temperature resistance is required, with silicone.

Ferric ion (iron (III) ion). The triple charged ion of iron, Fe^{3+}, which occurs when three electrons are lost, to form compounds with electronegative elements that obey the $(8 - N)$ rule. Iron (III) (ferric) oxide, Fe_2O_3, is an example, as is $FeCl_3$, iron (III) chloride.

The iron (III) ion is important in the formation of complex oxides, many of which have important electrical and magnetic properties. In particular, it can occur side by side with the ferrous (iron (II)) ion, as in magnetite, which is ferrimagnetic.

Ferric oxide (iron (III) oxide). An oxide of iron (Fe_2O_3) which is one of the most important ores of iron. It occurs as red hematite and as specular iron, crystallising in rhombohedra and scalenohedra and possessing a steel-grey colour. The oxide can be prepared artificially in various ways, e.g. by igniting the hydroxide or any iron (III) salt containing a volatile acid, and is then produced as a steel-grey powder which, like all forms of the oxide, gives a brownish-red powder when finely triturated. Its specific gravity is 5.17, while that of the natural forms varies from 5.19 to 5.25.

The natural iron (III) oxide is only slightly paramagnetic and dissolves only slowly in acids. The best solvent is a boiling mixture of 8 parts of sulphuric acid to 3 of water.

Iron rust always contains iron (III) oxide with considerable quantities of iron (II) hydroxide and carbonate.

The residue left in the process of distilling fuming sulphuric acid from green vitriol $FeSO_4$ is termed colcothar and is largely used as a pigment in oil-paints. The least calcined portions, which are scarlet in colour, are used as jeweller's rouge, whilst the more calcined portions, which have a bluish tint and are termed crocus, are employed for polishing metals.

Brown hematite or limonite has the formula $2Fe_2O_3.H_2O$, goethite is $Fe_2O_3.H_2O$ and hydrohematite is $2Fe_2O_3.H_2O$. All three are crystalline.

Ferrielectricity. The electrical analogue of ferrimagnetism. In ferrielectric materials, there are antiparallel electric dipoles in the structure, but the moments are unequal (see **Ferroelectricity**).

Ferrimagnetism. The alignment of electron spin between the constituents, usually a metal ion and oxygen ions, is largely antiparallel, but not exactly so, and the net imbalance of spin direction imparts magnetic moments, which would be very much higher but for the antiparallel spin of most of the ions. Many such substances show a characteristic **B-H** curve of rectangular shape, with almost instantaneous saturation in the direction of the applied field. The great difference between ferrimagnetic and ferromagnetic materials, is that the former are insulators at room temperature, with resistivity of the order of 10^{10} times that of metallic ferromagnets. As eddy currents induced in them cannot circulate, these materials can be used in magnetic switching and storage systems in computer technology.

Ferrite. An allotropic modification of iron. There are three allotropes of iron, viz. (a) α-iron (b.c.c.), which is the chief constituent of wrought iron and which is stable below 760 °C. It is soft and magnetic but is a poor solvent for cementite (Fe_3C). The maximum solid solubility of carbon in β-iron is approximately 0.035% at 723 °C, but this decreases with falling temperature, until at room temperature the solubility has decreased to less than 0.01%. (b) γ-iron (f.c.c.) is stable above 910 °C. It is non-magnetic, but is capable of forming solid solutions with iron carbide. γ-iron containing iron carbide is known as austenite. (c) δ-iron (b.c.c) is stable above 1,400 °C and may contain up to about 0.1% C. Its melting point is above 1,500 °C. The changes that pure iron undergoes with temperature change are

$$\alpha\text{-iron} \rightleftarrows \gamma\text{-iron} \leftrightarrows \delta\text{-iron}$$
$$\quad 910\,°C \qquad 1,400\,°C$$

Ferrite is the term applied to substantially pure α-iron occurring in iron-carbon alloys. It is precipitated during the cooling of steels containing less than 0.85% C, and is so called to distinguish it from the iron of the eutectoid. Etched pure iron shows under the microscope as many-sided crystals. Certain alloying elements such Mn, Ni, Si, Cr, dissolve in solid α-iron to give solid solutions that have structures very like that of ferrite. For this reason ferrite is now used to include any solid solution based on α-iron as distinct from those solid solutions in γ-iron, such as austenite.

Ferrite ghost. A faint band of ferrite sometimes observed when freshly machined steel surfaces are viewed at a suitable angle. It is due to an uneven distribution of phosphorus and carbon (see also **Phosphorus**). The "ghosts" appear dark when etched.

Ferrites. Materials closely related chemically to magnetite, but having superior magnetic properties. These properties are associated with the fact that ferrites have a very low electrical conductivity, and when used as magnetic materials, the tendency to give rise to eddy currents is small. Magnetic ferrites have the general formula $Fe^{3+}(X^{2+}Fe^{3+})O_4{}^{2-}$ where X stands for any suitable element, e.g. manganese. The zinc and cadmium ferrites are non-magnetic and have different structures from the above.

Ferrites find applications in certain types of radio aerials, electronic computers and in equipment transmitting and receiving electromagnetic radiation at frequencies of the order of 10 hertz.

Naturally occurring ferrites are: (a) magnoferrite $Fe_2O_3.MgO$, which crystallises in black octahedra, and (b) franklinite Fe_2O_3. ZnO, which also forms black octahedral crystals. Other ferrites are prepared by igniting ferric oxide with concentrated caustic alkalis (alkali ferrites) or with lime (calcium ferrite). The crystals of the latter have a dark metallic lustre and are magnetic.

Ferritic stainless steel. A corrosion-resistant steel of which the following is the typical composition: C 0.12% (max.), Mn 1% (max.), P 0.04% (max.), S 0.03% (max.), Si 1% (max.), Cr 14-18%. This steel, which can be cold formed, possesses fair abrasion resistance, has good corrosion resistance and is used for tanks, cars and screens. The chief mechanical properties are given in Table 23.

TABLE 23. *Properties of Annealed and Cold-worked Ferritic Stainless Steels.*

	Annealed	Cold-worked
Tensile strength, MN m^{-2}	480-620	620-780
Yield strength, MN m^{-2}	250-380	550-720
Elongation, %	35-20	25-8
Hardness (BHN)	145-185	
Scaling temperature	760 °C	

Ferro-alloy. An alloy of iron with one or more of the elements, chromium, cobalt, niobium, manganese, molybdenum, nickel, selenium, silicon, tantalum, titanium, tungsten, aluminium, vanadium or zirconium. The amount of the last-named element is sufficient in quantity to be useful for adding that element to molten steel. The chief constituents of the commoner ferro-alloys are given in Table 24.

The effects of adding alloying elements to steels (and cast irons) depend on the amount of the alloying element; on the presence of other elements (added intentionally or otherwise); on the subsequent treatment. The chief advantages to be gained by the above additions are given in general terms below.

(a) *Chromium.* Retards the rate of decomposition of austenite and offers resistance to tempering. It increases the hardenability and the tensile strength without appreciably affecting the ductility.

It improves the wear resistance of high-speed cutting tools, and the presence of chromium carbides results in a high uniform hardness. Steels containing up to 1% give enhanced resistance to atmospheric corrosion. In amounts up to 6% it enhances the magnetic and creep properties of steel and in amounts of 14% produces stainless steels for cutlery and stainless engineering steels. Corrosion-resisting steel of the Staybrite type contains 18% Cr and 8% Ni.

TABLE 24. *Chief Constituents of the More Common Ferro-alloys.*

Ferro-cobalt	75-78% Co
Ferro-chromium	60-75% Cr; 1-8% C (also 1-7% Al, when prepared by electrothermic method)
Ferro-manganese	76-78% Mn; 0.6-7% C
Ferro-manganese-silicon	20-25% Mn; 47-54% Si
Ferro-molybdenum	70-75% Mo; 0.1-2% Si; 0.6-3.6% C
Ferro-molybdenum-tungsten	43% Mo; 10% W; 1.8% C
Ferro-niobium	50-60% Nb; 6% Ta
Ferro-nickel	25-45% Ni
Ferro-selenium	52% Se; 0.9% C
Ferro-silico-aluminium	45-50% Al; 34-40% Si; 2-3% Ti
Ferro-silicon	25-95% Si (low S and P)
Ferro-tantalum-niobium	70-80% Ta and Nb taken together
Ferro-titanium	25-42% Ti; 0.5-8% Al; 4-20% Si
Ferro-tungsten	75-85% W (low carbon, silicon and phosphorus)
Ferro-vanadium	50-60% V; 1.2-3.5% C
Ferro-zirconium	14-40% Zr; 39-52% Si

(b) *Cobalt.* Raises the tempering temperature of steels and improves wear resistance of high-speed cutting tools. High proportions of cobalt raise the coercive force of magnetic steels.

(c) *Manganese.* Is a deoxidiser and desulphidiser, and is therefore almost always present in steels. It is a useful carbide stabiliser, and in quantities in excess of that required to combine with oxygen and sulphur produces increased hardness and reduces ductility. High manganese austenitic steels are exceedingly resistant to wear and abrasion. Manganese austenite steels have a great work-hardening capacity.

(d) *Molybdenum.* The effects of molybdenum are very similar to those of chromium in producing increased hardness, but the percentages required are smaller than with chromium. It improves creep resistance in steels, reduces temper brittleness and enhances the resistance of stainless steels to pitting corrosion.

(e) *Nickel.* The presence of nickel increases hardenability and lowers the critical temperature. In low-carbon steel quite high (up to 10%) percentages of nickel may be added with advantage, the resulting steel having a higher tensile strength and greater toughness at low temperatures. Improves the resistance to corrosion and is an important alloying element (with chromium) in stainless steels.

(f) *Niobium.* Improves the nitriding of steels and diminishes weld decay in austenitic stainless steels. Improves the heat resistance of high-temperature alloys.

(g) *Selenium.* With aluminium, selenium reduces the porosity of steel and with zirconium is used for the production of free cutting steel. It improves both the ductility and the machining qualities of stainless steels.

(h) *Silicon.* In the absence of carbon, iron-silicon alloys (4% Si) have a high electrical resistance but a low hysteresis loss and are useful for transformers. It is a strong deoxidiser and dissolves in ferrite, raising the critical temperatures. High silicon-iron alloys are acid resistant, and the presence of the silicon improves the fluidity of the molten metal. Up to 2% Si in steel improves the tensile strength and the heat-resisting qualities. Silicon-killed steels tend to resist graphitisation at temperatures over 500 °C.

(i) *Titanium.* Is a good deoxidiser and denitrogeniser. It may form two carbides which are very stable. Its presence prevents the formation of martensite and it is most useful in preventing weld decay. It is therefore used as a carbide stabiliser in stainless steels.

(j) *Tungsten.* It increases the tensile strength of steels at all temperatures. Its carbide is exceedingly hard. It is a most valuable alloying element in high-speed tool steels because of the hardness of the carbide and the effect it produces on secondary hardening. It is used to improve the magnetic properties of carbon steels.

(k) *Vanadium.* It is a strong deoxidiser and a good grain-refining element. It forms a stable carbide and increases the hardenability. It resists tempering, reducing the softening effect. It is used in high-speed tool steels and its effect in secondary hardening is similar to that of tungsten. It is a useful addition to cast irons, in which it acts as a carbide stabiliser. In conjunction with molybdenum it improves creep resistance. (*See also* Table 6.)

Ferrochrome. When the ore of chromium, chromite $FeCr_2O_4$ is reduced by carbon, in the presence of lime and fluorspar in the electric furnace, iron with 60-70% Cr is produced. This is termed ferrochrome and is used in the making of chromium steels.

Ferroelectricity. The electrical analogue of ferromagnetism. Materials which are ferroelectric contain dipoles, and domains exist within which the direction of polarisation is parallel, i.e. each domain is spontaneously polarised in a given direction. The direction of polarisation varies between domains however, so that the net polarisation is zero (very similar to the domain structure in ferromagnetics).

When an electric field is applied, those domains whose direction of polarisation is most nearly parallel to the field will have a lower energy, and domains will grow to lower the energy by aligning with the direction of the applied field. Polarisation will now be changed in many domains, and net polarisation will not be zero. The growth of domains of favourable polarisation direction, and the movement of boundaries, will continue, and a hysteresis loop will be demonstrated, similar to ferromagnetics. To reverse the polarisation direction, the electric field will have to be reversed in direction, and coercive and remanence values can be measured, as expected.

The mechanical distortion which accompanies these effects is known as electrostriction, but it is the crystal structure itself which determines polarisation and therefore electrostriction effects, whereas in magnetostriction it is the direction of the applied field acting on specific directions in the structure.

Thermal energy aids randomising of the directions of polarisation, and a Curie temperature exists, above which the permanent polarisation is reduced, the dielectric constant falls in value and the hysteresis loop shrinks to a line. Above the Curie temperature the material has been described as paraelectric, to show the resemblance to the magnetic case.

Ferroelectricity is demonstrated by a number of compounds, but chiefly phosphates and tartrates of univalent metals, sulphates, borates, oxides and alums. The polarisation may be uniaxial, as in tartrates, or multiaxial as in amphoteric oxides of metals such as barium titanate, and lead niobate.

Ferrograph. An instrument for presenting the hysteresis curves—either normal or differential—of small specimens on the screen of an oscilloscope.

Ferrolum. A lead-clad steel used in the chemical industry to combine the corrosion resistance of lead with the strength of steel. Pure lead and various lead alloys have been used for the cladding metal.

Ferromagnetism. The diamagnetism produced by the angular momentum of electrons is very weak (*see* **Diamagnetism**). When electron spin in unpaired electrons contributes a further magnetic moment, the susceptibility becomes positive, and paramagnetism results. The metals with one s electron (alkali metals) are weakly paramagnetic, because the electron spin imbalance is limited to one electron only. The alkaline earth metals, with two s electrons, develop an antiparallel pair, and so are diamagnetic, as are the ions of alkali metals, since the s electron has been lost. The elements with one or two p electrons, if they remain parallel spin, are paramagnetic.

The transition metals, with d electrons, build up to a considerable imbalance of spin, and are therefore strongly paramagnetic. In the solid state, the interaction between atoms in the lattice affects the filling of electron energy levels, and the requirements of the Pauli principle mean that electron spin may be a variable to allow lower energy states than would otherwise be necessary, (*see* **Fermi level** and **Exchange energy**). The width of the 3d band of energy is important in the first long period transition elements, since the proximity of the atoms must affect this width, and therefore the range of energy levels available for occupation. In the three elements iron, cobalt, nickel, a state is reached where parallel spin is favourable, because the exchange energy is positive, and the magnetic moments due to spin are therefore enhanced. This condition applies only when the lattice spacing is of particular size, in fact the ratio of one-half the interatomic distance to

the average radius of the 3d level must be greater than 1.5. The ratio for iron is 1.63, cobalt 1.82, and nickel 1.98. The elements manganese and chromium, preceding iron, do not satisfy this condition, although manganese is close to it.

Materials which exhibit the parallel spin orientation and satisfy all other conditions exhibit spontaneous magnetisation, even in the absence of an externally applied field. When an applied field is present, the domains of magnetic dipoles, all in the same direction in any one domain, can rotate, and grow by domain boundary movement, so that orientation of dipoles becomes gradually uniaxial throughout the material. When the field is removed, the magnetisation remains, and can be removed by an applied field which reverses the directions. Such materials are termed ferromagnetic.

Thermal energy, however, can destroy the conditions necessary for ferromagnetism, as the dipoles become more disordered and therefore random. Above their Curie temperature, the ferromagnetics become paramagnetic (*see* **Curie Weiss law**).

Elements such as manganese and chromium, which are paramagnetic above the Neel temperature, become antiferromagnetic below it, when the magnetic moments cancel out, as many being in one direction as in the opposite. If the ratio of atomic distance to the radius of the 3d level can be varied however, conversion to ferromagnetism can occur, and in the case of manganese, this is fairly simple, with the alloying of copper and aluminium, Heusler alloys. (For the hysteresis effects and magnetisation details, *see* *B-H* loop.)

Ferromanganese. White iron in which manganese is present as an essential constituent. It contains carbon in amounts which can vary from 3.5 to 6%, practically all of which is in the combined form. It is called spiegel or specular iron and is hard and brittle. Its specific gravity is approximately 7.6. When this alloy contains more than about 25% Mn it attains a granular structure and is termed ferromanganese. Both ferromanganese and spiegel are largely used in the manufacture of steel by the Bessemer and open-hearth processes.

Ferro-molybdenum. A molybdenum-rich iron alloy used in the manufacture of special steels. It is prepared by reducing molybdenite MoS_2 with iron and carbon in an electric furnace. It usually contains 65-75% Mo. It may contain up to 6% C.

Ferro-nickel. A nickel-rich iron alloy sometimes used in the manufacture of nickel steel. The percentage of nickel varies from 25 to 45%, the balance being iron.

Ferro-silicon. A silicon-rich alloy used as a deoxidiser in making castings of steel, copper or bronze, and as a constituent of certain acid-resisting cast irons such as Duriron. It is made by heating a mixture of oxide of iron, silica and carbon in an electric furnace, when both oxides suffer reduction. The composition varies widely, from 25 to 95% Si, but usually those containing more than 50% Si are the most useful.

Ferroso ferric oxide (iron II iron III oxide). The name applied to the oxide of iron having the formula Fe_3O_4. It occurs naturally as the important ore, magnetite, so called because certain varieties, e.g. lodestone, are permanently magnetic. It occurs in Lapland, Sweden, Siberia, Germany and North America. It contains about 72% Fe and is the richest ore of the metal.

Since it is analogous and isomorphous with magnesium aluminate (spinel), it is sometimes given the formula $Fe(FeO_2)_2$, ferrous ferrite, or $FeO.Fe_2O_3$. It is the chief constituent of scale.

Ferrostan. Electrolytic tin-plate produced by the electrodeposition of tin from an acid sulphate bath on a moving (cathode) strip of continuously cold-reduced steel sheet. The coating weight is usually 110-335 g (4-12 oz) per basis box. The material may be used with the tin coating in "matte" as deposited condition or satin-finished by scratch-brushing, but is more usually "flow brightened" to produce the bright surface characteristic of tinplate. After flow brightening, the coating is usually "filmed" to improve its corrosion resistance.

Ferro-steel. Grey cast iron made from a mixture of pig iron, cast iron, scrap and steel, with or without small amounts of alloying additions.

Ferro-titanium. A titanium-rich alloy that may be added to molten steel to improve the tensile strength of the resulting cast metal.

It also exerts a "refining" effect. Titanium steel is used for bridge construction and on railways. (*See also* **Ferro-alloy**.)

Ferro-tungsten. A tungsten-rich iron alloy used in the making of special magnet and high-speed tool steels. It is prepared by reducing the trioxide with carbon in the presence of iron in the electric furnace. It usually contains 75-85% W.

Ferrous ion (iron (II) ion). The double-charged ion of iron, Fe^{2+}, when two electrons are lost, to form compounds with electronegative elements or groups. Iron II oxide, FeO, is an example, and FeS, iron II sulphide. The iron II ion can interchange with the iron III ion in complex compounds, and contributes to the electrical and magnetic properties of inverse spinels.

Ferrox cube. A group of sintered materials in which low hysteresis loss is combined with high resistivity. Also known as cubic ferrites and used as core materials for high-frequency transformers and chokes.

Ferroxyl indicator. A reagent for indicating ferrous corrosion. It is prepared by mixing dilute aqueous solutions of phenolphthalein and potassium ferricyanide. It becomes pink in the presence of an excess of hydroxyl ions or gives a deep blue precipitate in the presence of iron II ions. When a permanent record of the state of an iron surface is required, the solution is thickened with gelatine or agar.

Ferroxyl test. A test to reveal porosity in the coatings of tinned, galvanised or painted steel. (*See also* **Ferroxyl indicator**.)

Ferruginous deposits. Sedimentary deposits of iron ore which usually occur as carbonates, oxides or silicates.

Féry pyrometer. A radiation pyrometer consisting of a millivolt meter and a small telescope incorporating a concave reflector and a thermocouple. When the telescope is directed towards a radiant surface, from a certain distance, a portion of the radiant heat falls upon the mirror and is concentrated upon the couple. The e.m.f. generated can be measured by the millivolt meter and the reading converted into degrees Celsius. In this instrument and other radiation pyrometers the laws of heat radiation apply—the energy emitted by a heated black body is proportional to the fourth power of its absolute temperature.

Radiation pyrometers tend to give low-temperature readings, as radiation from furnaces is affected by smoke and dust.

Fettle. In general, this term means "to put right" and is usually applied to the cleaning and dressing operations carried out on castings and forgings. The term is also used in reference to the preparatory dressing or patching of a furnace lining or a furnace hearth.

Fiberfrax. A synthetic refractory fibre produced by melting alumina and sand in an electric furnace at about 1,820 °C and blowing a stream of the molten material into a fluffy mass of very fine fibres by means of controlled jets of compressed air. It withstands temperatures up to 1,260 °C without loss of properties and is used for high-temperature insulation.

Fibre. The arrangement of the constituents of a metal or the elongation of the crystals which indicates the direction of working. The term is also applied to the elongation of the inclusions in hot-worked metal. All wrought metals have in a marked degree "grain" or fibre, and in forging care is taken to ensure that the grain is worked into a position parallel to the principal stresses likely to occur in the article to be forged. One of the advantages of forgings (q.v.) over castings is the opportunity the former present to control the disposition of the grain.

Fibres. Material artefacts occurring naturally or processed, to give a high aspect ratio, i.e. great length with small diameter. Due to the method of formation, there is directionality of crystal orientation where crystalline, with resultant anisotropy of properties.

Glass fibres may be obtained from the melt by spinning or drawing, and develop considerable strength. Due to the presence of Griffith cracks on the surface, however, the utilisation of glass fibre alone is difficult, since bending stress causes propagation of the cracks through the diameter. If the fibres can be sealed at the surface by a viscous fluid or a crystalline solid which does not attack the glass chemically, then the reinforcement by the glass fibre results in a very strong composite material. Glass fibre can be woven into tape or ribbon, bundled into strands, or loosely held by a soluble adhesive in a randomly oriented mesh or mat, for the purpose of reinforcement. When chopped into lengths of

approximately 1 to 5 mm it can be used as a filler in reinforcing polymers used in mouldings or rods, or in cement for spray moulding.

The strength of glass fibres is of the order of 2 GN m^{-2} in tension, and the tensile strength of a glass fibre reinforced polyester about 70 MN m^{-2}. The tensile strength of the unreinforced polymer would be about 25 MN m^{-2}. To achieve these results, the fibre must be clean and uncontaminated, and the resin must perfectly wet the glass. The composition of the glass is important. Type A (A for alkali) contains a high proportion of sodium and potassium ions, in a straight silicate glass. Type E (E for electrical) is a borosilicate of very low alkali content, but high in alumina and lime. Type A should not be used in the presence of alkali, as in reinforcement of cement, where Type E is essential. The fibres are strongest when the diameter is small, and 0.01 to 0.005 mm diameter is common.

Glass fibre is also woven with textile fibres, or used alone as woven fabric, for high temperature electrical insulation (cloth, tape, braid) or fireproof fabrics such as curtains. The coarse fibre is used as acoustic or thermal insulation and for filtration.

Mineral fibres are generally those occurring naturally, of which asbestos is by far the most common. It occurs as one of the serpentine or amphibole minerals, hydrated metal silicates, and is naturally fibrous, with very short staple (about 50 mm maximum length) but extremely fine in diameter (ultimately about 0.5 pm). One of its main problems, as a material, is the very fine short fibre, because this can damage the human lung when inhaled, and is extremely difficult to filter out.

The main source is chrysotile or white asbestos (serpentine), a hydrated magnesium silicate, which is attacked by strong acids. The coarser material, less refractory but more acid resistant, is crocidolite, or blue asbestos (amphibole), a sodium magnesium iron silicate, which unfortunately has a greater effect on the human lung. A third form, amosite, is a hydrated magnesium aluminium silicate.

The fibres are very strong, up to 2 GN m^{-2}, and often stronger than glass fibre of the same dimensions. They are used for polymer reinforcement in the same way as glass, for acoustic and thermal insulation, and for filling polymers required to resist wear, such as floor tiles. Cement and bitumen can also be reinforced, the latter to produce roofing felt. The most extensive use, however, is in heat resistance applications, for electrical machines, brake and clutch linings, packing for steam valves, pumps and glands. Legislation now controls the use of asbestos in many countries, with severe restrictions on blue asbestos.

Polymeric fibre constitutes a major use of such materials, and brought synthetic polymeric materials into prominence. Long before the synthetic materials however, the natural vegetable fibres based on cellulose (cotton, flax, jute, kapok, coir, sisal, ramie) had been employed in clothing, buildings, and other equipment, from very early times. The disadvantages of natural vegetable fibres are variations in the length of the staple, and consignment variations in properties when processed, and the sometimes difficult conditions for cultivation and fabrication. Flax has been in use for nearly six thousand years in the form of woven linen, and hemp was used by the Chinese perhaps 20 centuries ago. Both were known to the Greeks, and the maritime exploits of the classical world have been attributed to their use. Cotton was known in India in 3,000 BC and in Mexico 7,000 years ago. Native cellulose fibres show a degree of polymerisation between 2-6,000, and tensile strength ranging from 500 MN m^{-2} for cotton, to 800 MN m^{-2} for flax. (Regenerated cellulose has less strength than both on average.)

These fibres are spun into yarn, by twisting the short staples together, and can then be woven into fabrics. Cotton, flax, jute, hemp and ramie are most commonly woven. Kapok, coir are often used as packing, in upholstery, mattresses, cushions. The uses are too numerous to list, and well enough known. (*See* **Cellulose** and **Crystallisation** for structural details.)

Animal protein fibres, are less numerous, silk and wool being the two important ones. Natural silk contains fibroin protein, and wool contains keratin. Silk is largely a linear polymer, but wool keratin is cross-linked by sulphur linkages. The degree of polymerisation is less than that of natural cellulose, but greater than that of regenerated cellulose. The tensile strength of natural silk is of the order of 400-500 MN m^{-2} and of wool, 200-300

$MN\,m^{-2}$, lower than natural cellulose fibres. Variations occur but the filament secreted by the silk worm caterpillar has a length up to 2 500 m, with a diameter of 0.02-0.025 mm. Wool staple length varies with the breed of sheep, but is usually 25 to 200 mm, the diameter is too variable to justify a quotation figure. Silk is used only for the finest fabrics, thermal and electrical insulation, and has a high affinity for dyes. Wool is traditionally the best thermal insulator in clothing, and is used for felting and carpets of great durability. The longer wool fibres are combed out and spun into worsted (smooth) yarn for dress fabrics, the shorter fibres are used for weft yarns.

Natural cellulose is capable of regeneration to produce more controlled diameter, length and properties, and the sources are mainly wood-pulp and cotton. Rayon was the earliest regenerated product, made by dissolution of the purified cellulose in cuprammonium solutions followed by spinning into a water bath, or by treating an alkaline solution of cellulose with carbon bisulphide to form sodium cellulose xanthate, which was then spun into a bath of dilute acid. Rayon has now given way to a wider range of regenerated products, cellulose compounds with acetic, propanoic, butanoic and nitric acids (acetate, propionate, butyrate and nitrate) and copolymers of these. The purely synthetic fibres are discussed under separate headings (see **Polyamide, Polyester, Acrylic resins**).

Fibre stress. A term sometimes applied to the magnitude of the stress at any point or along a line on a cross-section over which the stress is not uniform, e.g. the cross-section of a spring or the cross-section of a beam under a bending load.

Fibro-ferrite. A basic iron III sulphate

$$2Fe_2(SO_4)_2.Fe_2SO_4(OH)_4.24H_2O.$$

It forms a pale yellow or nearly white pearly or silky mass. The mineral is formed by the oxidation of the sulphides of iron in the presence of water.

Fibrolite. A silicate of aluminium $Al_2O_3.SiO_2$, also known as sillimanite (q.v.).

Fibrous fracture. A grey and amorphous fracture that results when a metal is sufficiently ductile for the crystals to elongate before fracture occurs. When a fibrous fracture is obtained in an impact test it may be regarded as definite evidence of toughness of the metal.

Fibrox. The carborundum formed in the electric furnace is usually surrounded by a layer of a substance, siloxicon, the fibrous variety of which is known as fibrox and is used as a heat insulator in place of asbestos. The non-fibrous variety is used as a refractory. Siloxicon has been assigned the formula Si_2OC_2, but it may be a solid solution of silica SiO_2 in silicon carbide SiC.

Fick's law. If diffusion occurs across a section between two parallel planes distance x apart, and the concentrations at the two planes are c_1 and c_2 (where $c_1 > c_2$), the amount of substance diffusing in time t is

$$Qt = K(c_1 - c_2)\frac{A}{x}\,t$$

where A = area of cross-section and K is a constant. More accurately

$$Q = K\frac{dc}{dx}\,At$$

where dc/dx = fall in concentration. As applied to the diffusion of gases in metals the usual form of this law is given in somewhat simplified form as

$$\frac{\delta c}{\delta t} = K\frac{\delta^2 c}{\delta x^2}$$

where K is the diffusion constant, c the concentration and x the distance of a given element from a fixed boundary.

Field (electric). The imposition of a potential difference between two points in a material causes a perturbation in the electronic equilibrium. In the presence of electron carriers, a drift of electrons will commence in the opposite direction to that of the applied field (assumed to be from positive to negative electrodes), the phenomenon of conductivity.

Field (magnetic). The imposition of a magnetic flux distribution near a material, either from a permanent magnet, or from a coil through which a current flows to produce such a flux. The field induces magnetic response in the material dependent on its type, the phenomenon of magnetic induction.

Field Ion microscopy. A large potential difference applied between the tip of a finely pointed cold specimen rod and a flourescent screen, with low gas pressure in the intervening space, causes ionisation of the gas atoms in the very strong field thus produced around the specimen tip. The ions formed are attracted to the screen, producing specks of light. Ionisation occurs preferentially just above atoms in the specimen, so that the final image on the screen produces a pattern of bright spots corresponding to individual atoms in the specimen. The method is naturally restricted to electrical conductors, and the radius of the specimen tip must be less than 1 μm in order for atoms to be observed. Although the region involved is minute in size, it has proven possible to study vacant lattice sites, grain boundary and dislocation structures.

File hardness test. A simple workshop method of testing the hardness of steel surfaces, for estimating the hardness of quenched or quenched and tempered steels, or for the control of hardened components such as bearings, etc., in production. The handle of the file is grasped in the hand with the index finger extended along the file, and the surface under test is rubbed slowly but firmly with the sharp teeth. The file is removed just as soon as it is apparent whether or not the teeth will bite. Files used in this test should be of standard shape, size and hardness, and the slower the movement the more accurate the test.

The hardness of a hardened steel file is usually of the order of 820-900 Vickers (62-65 Rockwell C), and it is often convenient to temper a set of files at different temperatures and so cover the desired range of hardness.

Filiform. A term denoting "wire-like". It is applied to minerals and certain native metals (e.g. copper) that may occur as thin wires or as twisted strands.

Filigree. Ornamental work, usually on works of art and on jewellery, formed with metal wires—usually precious metals.

Filing board. A device for removing the burrs from the edges of strip (usually narrow strip) and for shaping its edges. It consists of a number of small files mounted on a narrow plate, and is installed immediately behind the slitting cutters. The files are kept in contact with the edges of the strip by means of springs set at different angles according to the edge contours required.

Filled electron shells. See **Closed electron shells.**

Fillers. Materials added to provide bulk in a material and thereby reduce costs, or to improve the properties by reinforcement, or modify them by separating constituents. In polymeric materials, diluent fillers must be cheap, inert, and wetted by the polymer during processing. They include powdered clay, gypsum, sawdust, powdered slate, sand, and similar byproducts of industrial processes. They provide no reinforcement but may contribute to colour and opacity.

Reinforcing fillers may be fibrous, such as chopped glass, cotton flock, asbestos or other mineral fibres, or particulate, such as metal powder, carbon black, silicates, carbonates, oxides. Fibrous fillers generally increase tensile strength, particle fillers improve compression strength and abrasion resistance, and modify colour and opacity.

Filler sand. The ordinary foundry sand used by foundrymen. It should be coarse and have good venting properties, but need not be of such high quality as facing sand (q.v.) which comes into direct contact with the casting.

Fillet. The term used in reference to the concave section which is used to reinforce the re-entrant angle formed at the intersection of two plane surfaces. Sharp corners are stress raisers: lines of stress radiate from the corners during the cooling of castings. Arrangements for introducing a "radius" or fillet are made to reduce the stresses. In a fillet weld the weld metal is introduced in the angle between the plates that are welded.

Filter. A device for separating liquids from suspended solids. In many industrial processes it is advantageous to allow the insoluble matter to settle before filtration is attempted. As much as possible of the clear liquid is then decanted so as to minimise the bulk of the liquid to be filtered. The chief considerations in the design of industrial filters are: (a) the filters shall present a maximum available surface in as small a space as possible; (b) they shall be able to withstand the required pressures (pressure filtering); and (c) the filter shall not tend to clog and shall be easily cleaned.

The choice of materials for the filter medium depends on the fineness of the material to be filtered, the chemical nature (corrosive or otherwise) of the liquid and the convenience of collecting the solid material after filtering. The main types of filter material now in use include: *(a)* porous paper, woven cotton, wool or linen, felt and woven metal; *(b)* unwoven fibrous material such as cotton wool, linen and asbestos fibre, cellulose pulp and unwoven metal fibres; *(c)* granular materials such as gravel, sand, coke, powdered cork and sawdust; *(d)* plates of porous stone, unglazed porcelain, carbon, silica and fritted glass.

The filtration process may procede slowly by allowing the liquid to seep through the filter-bed and flow away. Pressure filters accelerate the process, pressure being exerted on the liquid surface. Vacuum (or suction) filters accelerate the filtration by reducing the pressure below the filter-bed.

Filtration. A process for the separation of solids and liquids when they occur together in the same system. Usually a porous medium is interposed which allows of the passage of the liquid but not the solid particles. This medium may be unglazed paper, cloth, wire mesh, porous or unglazed earthenware or a plastic membrane, and the process may proceed at ordinary pressures, at higher than normal pressures or *in vacuo*. In the case of pressure filters, the initial pressure is low, but builds up as the thickness of the filter cake increases. On the commercial scale, where very large volumes are filtered, beds of fine sand lying over layers of gravel of increasing coarseness are used. Gelatinous materials tend to clog the pores of a filter and slow down the process. In such cases, pressure filtration rarely succeeds, and the alternative is to use what is referred to as a filter aid. This is a finely divided material (e.g. diatomaceous earth) which is added to the liquid to be filtered. The gelatinous substances adhere to the solid and are then less liable to clog the filter.

Fin. A thin section of metal protruding from the surface of a casting or a forging, where a small portion of the metal has been forced from the mould or between the faces of the dies during production.

Fineness. A term used in relation to the particle size of pulverised ore. It is usually expressed as the percentages remaining on each of a series of ten sieves. The term is also used to express the degree of purity of gold or silver alloy, the fineness being given as the parts of gold or silver per 1,000.

Finery. A furnace in which metal is refined. The term is often used in reference to the hearth in which Swedish pig iron is puddled to produce wrought iron in the two-stage process used for relatively low-grade materials. In the single-stage process the metal, which gradually becomes pasty as carbon is removed, is puddled, i.e. worked, into a ball. This latter is removed from the furnace and hammered hot. This first hearth is known as the refinery.

If the metal from the first hearth is allowed to flow directly into the second hearth for further treatment, this second hearth is known as the "finery" or as the "charcoal finery". In America the term "knobbling fires" was used.

Finished steel. Steel that is ready for marketing without further work or treatment.

Finishing temperature. The temperature at which mechanical deformation of metal is completed in hot-working processes.

Fink process. A method of alloying the surface of carbon or alloy steel with aluminium. The process involves treatment of the steel in a hydrogen atmosphere immediately followed by immersion in molten aluminium. The method is chiefly used for treating strip and wire which are passed continuously through a hydrogen tunnel.

Fire-bars. The cast iron or alloy cast-iron bars that form a fire-grate in boiler furnaces, etc.

Fire-box steel. A high-quality carbon steel that can be bent cold to sharp angles without cracking. It is usually a low- or medium-carbon open hearth steel. One of its applications is implied in the name. Also called flange steel.

Fire-brick. A type of brick that can withstand high temperatures in presence of flame or hot gases. The important characteristics of a good fire-brick are: *(a)* resistance to melting or softening at high temperatures; *(b)* resistance to erosion by the action of molten slag in furnace linings; *(c)* resistance to erosion by hot ash-laden gases in chimneys; *(d)* resistance to spalling when subjected to sudden change of temperature; *(e)* mechanical strength should not decrease with rise of temperature. These properties are usually compatible with a good thermal conductivity, and thus fire-bricks often need to have the addition of good thermal insulation.

Fireclay. An acidic refractory material containing essentially kaolin $Al_2O_3.2SiO_2.2H_2O$, which occurs as a decomposition product of various rocks, chiefly felspar. Other substances frequently present, such as SiO_2, Fe_2O_3, CaO, MgO, TiO_2 and alkalis, have an appreciable effect on the refractoriness and on the amount of shrinkage on drying and firing.

The purity of fireclay affects the softening point, and therefore the temperature of service. The plasticity is also affected by purity, but more by the degree of weathering which the original rocks received, and the rate of sedimentation when the deposits were laid down, i.e. on the thickness of the sheets in the layer lattice structure, and any intermediate layers. Kaolinite with $2SiO_2$ molecules in the structure is different from pyrophillite with $4SiO_2$ molecules. The same layers of hydrated silica and alumina are present, but the order of the layer lattices in the pile is different, and the water added for plasticity gives rise to variation in plasticity. Clay is best left in contact with water for a long period, to achieve maximum plasticity; "souring" was known in ancient China. Some minerals may be leached out by this process — particularly iron compounds and alkali.

The heating of fireclay during firing of components is all important, to avoid distortion, cracks and internal fissures. Clays containing montmorillonite, with $4SiO_2$ molecules, lose water of hydration more readily than kaolinite, and care is needed to allow this to occur slowly. Simple dehydration is not regarded as the only chemical change on firing, dissociation of the alumina and silica is thought to take place. As the water is removed, porosity increases, but at a later stage sintering of the mass assists in closing some of the porosity. Permeability is not directly related to porosity, since some porosity is sealed from the surface, but in furnace applications, permeability may be very important, e.g. in the lining of the blast furnace.

To counteract shrinkage and porosity, fireclay is prepared by mixing in a proportion of "grog" or previously fired brick, finely ground. 10-15% is the maximum addition normally, but very plastic clay (British ball clay) can still be worked with over 50% grog. Fireclays may be mixed to vary the properties or the processing variables. Bentonite is a useful plasticiser, containing montmorillonite, and silica is added for countering the shrinkage, since silica expands at the temperature where clay shrinks (500-600 °C). Grading of particle size is necessary when quartzite rock is added to ensure close packing.

Insulating bricks can be made from fireclay, by incorporating a frothing agent to give porosity, or adding combustible material which gasifies on heating during the firing stage.

Fire-cracking. The term refers to the failure of a stressed metal at elevated temperatures. The term is also used in reference to the formation of longitudinal or circumferential cracks when deep-drawn metals, e.g. copper alloys, are heated rapidly to above the recrystallisation temperature.

Fire-cracks. Surface cracks found on metals that have been subjected to continuous alternating heating and cooling, as for example the rolls used in hot-rolling mills. The term is also used in reference to the surface irregularities sometimes found on wrought metals that have been rolled through defective rolls. The irregularities correspond to surface cracks on the rolls, and are usually caused by the alternate heating and cooling or by the overheating of the rolls in service.

Firedamp. Methane CH_4, the simplest of the hydrocarbons of the paraffin series, sometimes known as "marsh gas". It is a gas which occurs in coal-mines, and is a constituent of coal gas. It is formed when organic matter decays in the presence of moisture.

Fired ceramics. Ceramic materials which have been dried to remove surplus water added to make the material plastic for fabrication, then heated to a high temperature, often close to the fusion point, to remove water of hydration or to cause chemical or physical changes in the constituents. Ceramics should always be fired at temperatures higher than the service temperature if possible. Firing causes vitrification in many cases, the production of a glassy state in the main material or in any binding agent added.

Fire-gilt. The term refers to a method of covering metallic articles with gold. Gold dissolved in mercury is applied to the article to be coated and the mercury is burnt out by firing. The term is mostly used in the jewellery trade.

Firth hardometer. A device for measuring the hardness of metals using a diamond pyramid indentor. This latter is a square pyramid with angles between opposite faces equal to 140°. The machine is similar to, but simpler than, the Vickers hardness tester, which also uses a square pyramid indentor, the angle of which is 136°. The Firth hardometer also uses hardened steel balls of 1, 2 and 4 mm diameter. The load is applied through a calibrated spring, and the results are expressed as Brinell numbers.

Fish eyes. A term used in reference to microfissures occurring in steel. These microfissures occur as a result of the simultaneous crystallisation (or recrystallisation) while the metal is still under stress due to heat gradients. There is normally a shrinkage when a liquid freezes to a solid, but there is additional shrinkage also due to polymorphic changes. The shrinkage cavities or fissures appear as minute unbonded areas within the metal. They have the effect of reducing both tensile strength and ductility.

Fish plate. A simple bar used for joining rails in railway tracks. One side of the bar is formed to fit tightly into the space between the head and base of the rail (the fishing of the rail). The fish plate is bolted to each rail. Also known as splice bar and rail-joint bar.

Fish scale. A defect sometimes encountered in vitreous enamel coatings. It has the characteristic appearance that the name implies and is often associated with flaking of minute areas of enamel.

Fish tailing. A defect that occurs at the trailing end of rolled sheet, producing an effect that the name implies.

Fission. *See* **Nuclear fission.**

Fission products. The products of nuclear disintegration of a fissile material. Some of the binding energy of the fissile nuclei is released, together with elementary particles, radiation, and two fission products per nucleus of the original material.

The fission products themselves are highly radioactive, but may decay into more stable products within a reasonable time. A cooling period is therefore allowed for nuclear reactor fuel elements to approach this greater stability.

Some 60 primary products have been detected when a heavy nucleus splits and the range of their mass numbers is from 72 (an isotope of zinc) to 158 (an isotope of europium). The products fall mainly into two groups—light and heavy products. The former have mass numbers between 85 and 104 and the latter between 130 and 149. Together these account for about 97% of the total. The products which appear in largest amounts are the isotopes of strontium and xenon.

Since each direct fission product may contain from 1 to 3 neutrons too many in its nucleus they undergo decay into stable products in 1 to 3 stages, emitting beta radiation. This accounts for the large number of species found in the fission products.

Fissure veins. Ore deposits that have filled pre-existing fissures. The term has no special significance concerning the depth of the vein or the value of the deposit.

Fissuring. A surface defect in the form of a dense network of fine lines, seen particularly in certain cases of rolled aluminium. The defect may be due to: (*a*) hot cracking, inverse segregation or massive segregation of a second phase during solidification, or (*b*) mechanical damage during hot or cold rolling. Also known as resillage.

Flakes. Minute transverse internal fissures formed during the cooling of forged or wrought alloy steels and sometimes in carbon steels. They occur most frequently in large forgings and appear on fractured surfaces as small white specks resembling snowflakes. They can be eliminated by retarding the cooling of the forgings down to temperatures below 100 °C. It is believed they may be due to dissolved hydrogen. Also known as hair-line cracks, snowflakes and chrome checks. In America, where the defects have been found in the heads of rails, they are referred to as shatter cracks.

Flame cleaning. A method of preparing a steel surface for painting. It is usually applied to structural steel and other rolled-steel sections and consists in removing the millscale, etc., with an oxy-acetylene flame.

Flame cutting. A method of cutting ferrous metals, using a stream of oxygen in a torch. The metal is heated to about 820 °C by the torch, and oxidation of the hot metal to magnetic oxide of iron occurs. The heat of the reaction $3Fe + 2O_2 = Fe_3O_4$ contributes a large proportion of the heat for the continuation of the process. In practice, it happens that nearly one-third of the metal removed is actually washed away by the stream of gas, without oxidation. Machine-operated flame cutters can be made to work to close tolerances, and the finished edge is smooth.

Flame hardening. A process in which the surface of steel or cast iron is heated rapidly to the hardening temperature by means of high-intensity burners, and cooled at a rate demanded by the characteristics of the material. Plain carbon steels containing 0.35–0.6% C and cast irons containing about the same range of combined carbon are most commonly treated, but materials with higher carbon contents or alloy steels may be treated if exceptional wear resistance or core strength is required. The oxy-acetylene (ethyne) flame is frequently used as the heat source, but other oxy-gas or air-gas mixtures may be used.

The process is particularly suitable where only certain areas are required to be hardened, and where it is essential to eliminate distortion and consequent final grinding so often necessary after case-hardening. Plant is now available that can give automatic control of temperature and depth of hardening and the process is equally applicable to large and small components. In the former case a combined burner and quenching head move progressively across the surface to be hardened. Also known as shortering.

Flame plating. A method of depositing hard, wear-resistant coatings on the surfaces of metal components. The materials to be deposited, e.g. tungsten carbide, chromium carbide or refractory oxides, are suspended in a mixture of oxygen and acetylene (ethyne) and the mixture is detonated in a strong tube. The temperature of the detonation is high enough to raise the carbides or oxides to the plastic state, while the velocity of the resulting gas stream blasts them on to the target, where they become welded to the metal of the component. The process must be carried out in a purpose-designed chamber and the operation is by remote control.

Flame scaling. A zinc coating process normally used for steel wire. After annealing, the wire is allowed to cool and then passed through a hydrochloric acid pickling bath, a fluxing solution and finally a bath of molten zinc. The coated wire is carried through wipers which regulate the coating thickness and finally through a flame scaling unit which consolidates the coating.

Flame spectra. Most of the metals are non-volatile at the temperature of the Bunsen flame. However, the salts of the metals of the alkalis and of the alkaline earths are slightly volatile, and when introduced in the flame volatilise sufficiently to cause the flame to assume a characteristic colour, and this flame when examined by the spectroscope exhibits the peculiar spectrum of the given substance. The spectrum of every element is characteristic and may be used for the identification of the element. The colours imparted to the flame by the slightly volatile salts of the metals are: sodium salts—yellow; potassium salts—purple; lithium salts—crimson; thallium salts—green; barium salts—apple-green; strontium salts—red; calcium salts—orange-red.

Flame test. When certain volatile substances are vaporised in the Bunsen flame their vapours become incandescent and impart to the flame characteristic colours which may serve for their identification. Since chlorides are among the most volatile compounds, it is advisable, before applying the flame test, to moisten the salt with concentrated hydrochloric acid. A little of the moistened mass is then taken up on a loop of platinum wire and introduced into the flame.

In addition to the colorations given under flame spectra, the chlorides and some other compounds colour the flame as follows: boron (boric acid)—bright green; copper—bluish-green; tin—greyish-blue; lead—bluish-grey; arsenic—grey; antimony—greyish-green; zinc (sulphate)—green; manganese (chloride)—green.

If several metals occur together their flame colorations may mask one another—sodium especially, owing to its brilliant yellow colour. If, however, the colorations are viewed through deep blue glass, the colours are modified as shown in Table 25.

When a spectroscopic examination is made of the colours of the flame produced by compounds of copper, tin, lead, arsenic and antimony, no definite lines or bands are observed. The reason for this is that the temperature of the Bunsen flame is not sufficient to dissociate the compounds into simple atoms.

TABLE 25

Metal	Flame Coloration	Colour as Seen through Blue Glass
Sodium	Bright yellow	Nil
Potassium	Purple lilac	Crimson
Calcium	Orange red	Light green
Strontium	Red	Red
Barium	Apple green	Green

Flange steel. *See* **Fire-box steel.**

Flash. A thin fin of metal formed at the sides of a forging or casting when some of the metal may be forced between the faces of the die or between the joints of the mould. When two plates are spot-welded a fin of metal may appear between the plates and concentric with the weld, and is referred to as flash.

Flash-butt welding. A resistance welding process in which an arc is struck and maintained between the joint members until welding heat is attained. The current is then shut off and the weld made by forcing the parts together.

Flash roasting. A process for removing sulphur from ores by blowing the pulverised concentrates through a combustion chamber.

Flat back. A foundry pattern which has a flat surface at the joint of the mould, so that the whole of the pattern lies within the bottom half of the box.

Flat lead. A term applied to sheet lead.

Flats. Light-gauge narrow steel-mill products ranging from about 2.5-6 mm in thickness and up to about 600 mm in width.

Flex-tester. A device for testing the drawing quality of sheet metals. It is based on the old workshop test, in which the press operator turns back a corner of the sheet, but this instrument measures the resistance to bending. It is adaptable to sheets of different gauges.

The instrument may be used in conjunction with a spherometer, which measures the diameter of the bend produced by the flex-tester. Material that can be bent around a very small radius is not likely to develop "stretcher strains".

Flexure. A term used in reference to the curved state of a loaded beam. When a horizontal elastic beam bends under the action of vertical loads the deflection of the beam sets up stresses in the material which supports the load. If it be accepted that the deformation of any part of the beam is proportional to its distance from the neutral axis, and that the modulus of elasticity of the material is the same in tension as in compression, then $f = My/I$ where f = stress at a distance y from the neutral axis, M is the bending moment and I = moment of inertia.

I/y is called the modulus of the section and is usually represented by Z. $f = M/Z$.

Flint. A crypto-crystalline form of silica which breaks with a well-defined conchoidal fracture. It varies in colour from greyish-black to brown, and sometimes shows a vitreous lustre. Ground flint is used as the source of silica in glazes and pottery, and as an abrasive.

Flintshire process. A lead-smelting process in which galena PbS ore is roasted in the hearth of a reverberatory furnace to convert a part of the sulphide into oxide and sulphate. The temperature is then increased and the oxide, sulphide and sulphate react, producing metallic lead and evolving sulphur dioxide. The unreduced ore is calcined with lime.

Float. A granite trough used as a reservoir or settling-tank in tin smelting. The term is also applied to very finely pulverised ores which float at the top of washing waters.

Floatstone. A porous form of silica which floats on water.

Flocculation. The coalescence of a finely divided precipitate. Colloidal clay, silica, alumina, etc., carry an electronegative charge, and this is dissipated by the action of electrolytes or by electropositive colloids, e.g. iron III (ferric) oxide. The particles of clay, etc., therefore flocculate together and settle as larger grains and with greater rapidity.

Flong. Papier mache, made from layered sheets of tissue paper and absorbent paper. It is used in the form of sheets for making moulds in the casting of stereo-printing plates and other low-melting-point alloys.

Floor moulding. The making of foundry moulds in the bed of sand which constitutes the floor of the shop.

Flotation. An ore-dressing (concentration) process in which finely pulverised ore is agitated in a mixture of oil and water. The various processes depend on surface tension and on differences in specific gravity. The constituent minerals of the ore are separated from one another by virtue of their respective "wettabilities". Minerals that are hydrophobic (resist wetting) tend to retain an air film and so float. Wettable materials tend to sink.

Certain reagents which influence the ability of the particles to float are usually added in small quantity to the circulating water. Agitation is usually required to assist the separation.

Flow. A term for plastic deformation, usually applied to fluids, but also to crystalline solids. When pure viscous flow is involved, the characteristic pattern of deformation follows predictable laws of behaviour, but when combined with elastic behaviour, many variants are possible (*see* **Viscous flow**).

Flowers. A spangled appearance, resembling a frosted window-pane, found on the surface of galvanised-iron sheets. It is produced by the crystallisation of the zinc (spelter) coating, and is sometimes known as flakes, crystals or spangles.

Flow index. A term used in static indentation (hardness) tests to denote the rate at which a ball indent enlarges with the time of application of the load. In Hargreave's tests, using balls of various diameters, he found that the diameter (d) of separate impressions obtained at a constant loading ratio (L/D^2) on various metals was related to the time (t) for which the load was applied in accordance with the equation $d = Ct^2$. The rate at which the indent enlarges with time is usually denoted by S. When S is zero there is no viscous flow. In viscous materials like pitch S is high. Cold working of metals usually increases the value of S. The flow index gives a good indication of creep-resisting properties.

Flow lines. Lines which appear on the polished and macro-etched surface of a metal and indicate the direction in which plastic flow has taken place during fabrication. In certain cases flow lines can be seen on the surfaces of metal (iron and steel) components which have been locally stressed beyond the yield point. The elongation of the crystals by cold working produces local electrolytic differences of potential, and the flow lines indicate the principal directions in which movement of the material has taken place.

Flow off. A channel that permits molten metal to over-flow when a certain pre-determined level has been reached.

Flow sheet. A diagrammatic representation of the sequence of operations performed in a process or series of processes.

Flow structure. The alignment of the constituents or crystals of a metal that is produced as a result of plastic deformation.

Fluidised bed. When a gas is forced through a bed of particles, the volume of the bed expands when the gas pressure overcomes the gravitational force on the particles. The system then behaves as a viscous fluid (*see* **Viscosity**). The technique is used in chemical engineering to increase rates of reaction, in the transport of solids and in heat transfer. In the latter case it can be employed in metallurgy for heat treatment.

Fluidity. The reciprocal of dynamical viscosity. Flow in fluids is an interchange of atomic or molecular neighbours, a shear process, as in solids. Time is involved in the shear process, and necessitates the introduction of rate of shear rather than shear strain itself. A dot above a symbol is used to indicate a rate.

$\dot{\gamma} = f(\tau)$ where $\dot{\gamma}$ is shear rate, and τ is shear stress. When $\tau = 0$, $\dot{\gamma} = 0$, assuming that no internal stress or spontaneous changes exist. The function $f(\tau)$ can be expanded into series, as for Hooke's law with solids. $f(\tau) = \alpha\tau + \beta\tau^2 +$ further terms. When small shear strains are involved, the higher terms can be neglected. $\dot{\gamma} = \alpha\tau$ the constant α is the fluidity of the material. Expressed in terms of its reciprocal, vicosity (η) $\tau = \eta\dot{\gamma}$ which is Newton's law of viscous flow. Obviously there is great similarity between this law and that of Hooke, and hence solids can be regarded as a particular class of fluids, in which η is very large.

The word is also used in metallurgy to indicate the ability of a metal to flow freely and evenly into a mould, before freezing prevents further flow. It is often referred to as "castability". The fluidity or fluid life of a metal is influenced by factors such as degree of purity of the metal, the shape and nature of the mould, the degree of superheat and the rate of pouring, as well as upon surface tension and viscosity.

Fluids. Materials whose viscosity under the conditions of ambient temperature is by convention less that 10^{17} centipoise. Such a material cannot resist indefinitely the action of a shearing force, and so deforms by viscous flow. The viscosity may be very low, as with gases, which flow easily, or only a little higher, as with liquids like water. When the viscosity reaches 10^4-10^5 centipoise the flow becomes much slower, but nevertheless inevitable. (*See* **Viscosity**.)

Fluon. A trade name for polytetra-fluorethylene (polytetra-fluoroethene). (*See* Table 10.)

Fluorescence. Incident radiation falling on a material may produce electron transitions to higher energy states. The electrons may later return to the original states, with the re-emission of radiation, but not at the same wavelength. Such materials are said to be luminescent, and when the re-emission takes place less than 10^{-8} s after absorption, they are called fluorescent. The emitted light is of longer wavelength than the absorbed, and so no further re-absorption occurs. Due to thermal vibration of atoms during the re-emission, a band of wavelengths is emitted, not single ones. Traces of impurities can exercise a marked influence on the behaviour, and some impurities are called activators because they trigger off a series of possible absorption-emission processes. Typical examples of fluorescent materials are eosin and fluorescein.

Fluorite lattice. Typical of AB_2 compounds, calcium fluoride CaF_2 (fluorite) crystallises in a cubic lattice, with the calcium ions in positions of a face-centred cube, and fluorine ions in the tetrahedral voids of that lattice. The calcium ions have a co-ordination number 8, and fluorine 4. This is a close-packed lattice, with the diameters of the two ions very similar.

Fluorspar. Calcium fluoride CaF_2, a mineral that crystallises in the cubic system and occurs as a common gangue of lead and tin ores, associated with galena, barytes, zinc blende and calcite. The first grades are used in enamelling and in the chemical and glass industries. Inferior grades are extensively used as a flux in smelting operations and other metallurgical processes. The name fluorspar is in fact derived from this particular application. It is found in Derbyshire, where the blue variety is known as Blue John, in Saxony and in many other countries.

The property which fluorspar possesses of becoming luminous when heated gave rise to the term fluorescence.

Fluting. A kinking or series of flat faces formed when metal sheet or strip is bent round a radius, so small in relation to the thickness that the elastic limit is exceeded in the outer surface layers of the material.

Fluviatile. A term used in describing deposits that occur in rivers or streams.

Flux. A term used generally to describe a material which lowers the fusion point or the viscosity. The term therefore refers to substances added to a furnace charge to facilitate extraction by combining with the gangue and forming a more fusible slag. In metal jointing and hot-dip coating processes the term flux is applied to the substance used to remove oxide films from the joint members of the basis metal, to protect them from further oxidation during the joining or coating operations.

In assaying, the term is used in reference to the readily fusible substances that are used for the purpose of obtaining more rapid liquefaction than can be obtained by heat alone.

Flux density. The flux density at any point in a magnetic field is the field (or flux or lines of force) passing through an area of 1 m^2 at right angles to the direction of the field at that point. The unit is the tesla.

The force that a unit magnetic pole would experience when placed d m from a pole of strength m, in a medium of permeability μ is $m/\mu d^2$. This is called the field intensity and is usually denoted by H. The magnetic induction (B) is defined by the equation $\mu H = B$. If a piece of iron or steel, of cross-section A, be placed with its length parallel to the direction of the field whose strength is H lines of force per square metre (measured in webers), then magnetic poles will appear at the ends of the bar and an extra number of magnetic lines will pass through the bar. Those due to the effect of the field on the bar will equal $4\pi m$ (where m = strength of the poles) and the total number of lines = $AH + 4\pi m$, which is the total magnetic flux. Hence the flux density $B = (AH + 4\pi m)/A = H + 4\pi I$ where I = intensity of magnetisation.

Flux-neutron. A measure of the number of neutrons crossing unit area in any direction per unit time. It is defined as the product of neutron density and mean velocity.

Fly ash. The finely divided solid ash from fossil fuels, etc., sometimes found in metals melted in open-hearth furnaces, but regularly in emission from pulverised coal combustion.

Fly-away ingots. A term applied to cutlery steel made by melting wrought iron with charcoal, and casting without chilling as soon as it is molten. A stream of sparks is emitted during teeming and the resulting ingots contain many small blow-holes.

Flying shears. A device incorporated into rolling-mill lines to cut billets, bars and sheet while in motion.

Foam. The term applied to a suspension of minute bubbles of gas in a liquid. The effect may be produced by chemical means (e.g. fermentation, etc.) or, more usually, by shaking a gas with a liquid of low surface tension (e.g. soap solution). Aqueous foams may be separated by adding alcohol or by centrifuging. Because of the existence of surface tension, the concentration of solute molecules is not uniform throughout a solution. Substances that lower the surface tension become concentrated in the surface layer (and vice versa). If the excess of solute in the surface layer above that in the body of the solution is S, and if C is the concentration of the solute, then for dilute solutions: $-S = C/RT \cdot d\gamma/dC$ where R is the gas constant, T the absolute temperature and $d\gamma/dC$ the change of surface tension with concentration. It is evident that though the production of foam is associated with a decrease in surface tension, it is also dependent on an increase in viscosity of the liquid, brought about by the solute. Hence foam production is prevented by substances that diminish the viscosity, even though they also diminish surface tension.

Foaming agent. Substance used in pickling baths to prevent pollution of the atmosphere by acid fumes, etc. The function of the foaming agent is to cover the surface of the solution with a thick layer of suds which blankets the acid spray. A material which produces gas at a controlled rate under the action of heat or light, by chemical reaction or decomposition. Used to impart porosity to plastic materials, including clay-water mixtures, polymers, or rubber latex.

Fog. When molten caustic soda (Castner process) or molten sodium chloride (Downs process) are electrolysed to give metallic sodium, some of the metal tends to dissolve in the electrolyte and produce metallic fog. The effect can be reduced by keeping the temperature as low as possible.

Fog quenching. A method of quenching in which a vapour or fine mist is used as the quenching medium.

Foil. Very thin sheet metal, usually in such materials as tin, aluminium, copper, gold, silver and sometimes lead. It is used for wrapping, decoration and in the manufacture of electrical condensers. There is no standard thickness of foil, but it is generally regarded as being intermediate in thickness between "leaf" and sheet.

Folding test. A type of "forge test" that consists in bending a metal specimen, hot or cold, through 180° and noting if the material exhibits signs of cracking. This is a "gradually applied" rather than a "shock" test, designed to test ductility, and is applicable to both ferrous and non-ferrous materials to ascertain their behaviour should they be required to be forged and worked.

Foliated. A term used to indicate that the material referred to consists of thin, separable lamellae or leaves.

Footner process. A rust-proofing process for steels. After pickling in 5% sulphuric acid at 65 °C for 15-20 minutes, to remove scale, the work is thoroughly rinsed in hot water and then immersed for 3-5 minutes in a solution containing 2% free phosphoric acid and 0.3-0.5% Fe at 85 °C. Whilst still warm the work is painted with the primer.

Forced draught. A term used in reference to the air supply to furnaces. The air is driven or induced by fans or injected into the furnace through special openings near the base. The purpose is to accelerate the process of combustion.

Fore plates. A form of shelf incorporated into rolling-mill stands to support the work as it enters and leaves the mill. In many mills stationary stands are also provided.

Forge. A machine, or plant, in which metal, usually hot, is formed into the desired shape by hammering or pressing. A forge may take the form of a gravity drop hammer, the impact force of which now may be augmented by pneumatic or steam pressure, or presses.

More generally a forge is the name given to the whole bay or workshop in an industrial plant in which the forging operation is done.

Forge iron. Hematite pig irons used for making wrought iron and for mixing with other pig irons in foundries. Approximate composition is graphitic carbon 2-2.8%, combined carbon 1.4-1%, Si 1-1.5%, S 0.05-0.11%, P 0.02-0.05%, Mn 0.1-0.3%.

Forge pig iron. Pig iron suitable for conversion into wrought iron by the puddling process. Sometimes referred to as forge pigs and puddling iron. It is a grey iron containing less silicon than other grey irons and has a low sulphur and phosphorus content.

Forge scale. An oxide coating formed when iron and steel are exposed to atmospheric oxidation at forging temperature. The scale is friable and usually falls from the article during forging.

Forging. The forming of steel or non-ferrous metals into the required shapes by hammering or pressing. For the former, steam and pneumatic hammers, drop hammers and horizontally operating machines are used. For pressing, hydraulic presses are used, and these give a squeeze to the metal instead of a blow, producing a considerably greater reduction in one stroke than is possible in one stroke of a hammer. Machine forgings, which may be of various shapes and sizes, are formed between dies, one being moveable and the other stationary. For relatively simple shapes open dies can be used, but for more complex shapes conforming to the contours and dimensions of the finished product, two or more sets of dies may be required, and are known as closed-dies. The process of forging is usually confined to shapes that cannot be produced by rolling.

The more accurate dimensioned finished forgings are usually drop forgings.

In hand forging, the tools used are simple and are mainly counterparts of the shapes desired, and the moulding is done by impact. They are also of a "general purpose" type—i.e. they can be used on many different forgings rather than being designed for a particular piece of work. The blows may be delivered either by hand hammer or by power hammer. Hand forging may involve the following operations: (a) swaging or fullering—the reducing or drawing down from a larger to a smaller section; (b) upsetting—the enlargement of a part from a smaller to a larger section; (c) punching—the formation of holes; (d) bending; (e) welding; and (f) cutting off.

Forging strains. Elastic strains resulting from forging or from cooling from forging temperatures.

Formaldehyde (methanal) HCHO. The first in the series of saturated unbranched acyclic hydrocarbon aldehydes, i.e. carrying the group —CHO. A gas, with boiling point −21 °C, but usually handled as an aqueous solution, or in polymerised hydrated form as paraformaldehyde $(HCHO)_n . H_2O$, which is solid and contains 95% formaldehyde. Formaldehyde solutions contain about 40% by weight of formaldehyde, with 10% methanol added to inhibit polymerisation. Extensively used in condensation polymerisation of thermosetting plastics, and in the preparation of pharmaceuticals, dyes, explosives, insecticides, plasticisers. It provides a powerful disinfectant in plant propagation and medicine. The paper and textile industry consumes large quantities for modification of the physical properties of cellulose and related materials.

Former. A term applied to a shaping tool used in metal-spinning operations. It is also the name applied to a mandrel used in the production of hollow ware.

The name is also applied to a ridged portion of a roll which forms the fourth side of the rolling pass. As the work passes through the rolls, three sides are embedded in the groove formed by the bottom roll; the former on the top roll closes the fourth side of the pass. Also known as tongue.

Forsterite. A member of the isomorphous group of minerals in which olivine $(Mg.Fe)_2 SiO_4$ is the intermediate member. It is the magnesium-rich member and is usually assigned the formula $Mg_2 SiO_4$. Melting point 1,910 °C.

Forsterite bricks are made from crushed olivine mixed with a small quantity of dead-burned magnesite. These are used as refractories.

Foundry. A workshop in which molten metals are cast into the required shapes, usually in sand, loam or chill metal moulds. The name, foundry, also refers to any workrooms in which any of the following processes are carried on as incidental or supplemental processes in connection with casting operations, e.g. the preparation and mixing of materials used in the foundry process, preparation of moulds or cores, knockout operations, heat-treatment or welding of castings, mould-dressing operations and fettling operations.

Fountain. The runner through which molten steel is introduced into a bottom-poured ingot mould.

Four-high mill. A type of rolling mill used for rolling flat material such as sheets and plates. The rolls are arranged one above the other in the housings, and may be regarded as a special type of two-high rolling mill in which the smaller working rolls are reinforced by backing-up rolls which minimise bending and distortion of the work rolls and variations in thickness across the section of the work.

Fracture. The irregular surface produced when a piece of metal is broken. The fracture of metals is an important characteristic intimately connected with the crystalline form, and in many instances it is useful in giving an indication of the purity or otherwise of the metal. The older iron and steel workers judged the quality of their product by the appearance of the fracture. The following varieties of fracture are generally recognised: (a) faceted fracture as seen in zinc, spiegel iron, bismuth, antimony, etc.; (b) granular fracture—as seen in grey forge pig iron; (c) fibrous fracture—as seen in bar and wrought iron when partly broken in bending; (d) silky fracture—as seen in tough copper; (e) columnar fracture—as seen in commercial tin; (f) conchoidal fracture—as seen in certain alloys, e.g. brass (33% Cu); (g) intergranular fracture where the failure passes around discrete grains — as in impure materials and in high temperature (creep) fracture. Fatigue fracture occurs without perceptible stretching. The failure often shows a spot or nucleus at the surface from which a crack started and progressed until the reduced cross-section became too small to carry the load. The fracture appears "brittle", while the rest of the section is smoothed by the rubbing of the faces of the crack.

Frequently the propagation of the crack proceeds at different rates, resulting in the appearance of zones differentiated by different degrees of rubbing. This is referred to as "oyster-shell" appearance. (*See also* **Fatigue.**)

Fracture mechanics. Study of the theoretical and experimental aspects of failure by fracture, in particular of brittle behaviour in materials. (*See* **Crack, Fracture, Fracture toughness.**)

Fracture test. Breaking a piece of metal for the purpose of examining the fractured surface in order to determine the structure or composition of the metal or the presence of internal defects. This is one of the oldest tests for metals. The material is nicked and a section broken off, so that the fracture can be examined to observe grain-size, pipe, pipe segregation, etc.

More refined versions of this test include the Jernkontoret test for tool-steel fractures and the Shepherd test in which ten steels of varying grain sizes are used as standard.

Fracture toughness. Since the stress required for fracture measured in conventional mechanical testing does not provide a reliable guide to the behaviour of cracks in real engineering structures, attempts are continually being made to estimate those stresses which actually can cause cracks to propagate at values below the yield stress measured in tensile tests. There is no single criterion for all materials, the characteristics of each type of material are unique in crack propagation. Furthermore, the structural design leads to a complex stress pattern which may affect the crack shape in addition to the stress distribution around it.

A new material property was devised from plane strain measurements on the propagation of cracks, and called plane strain fracture toughness k_{1c}. Since the conditions of plane strain are most favourable for propagation of a crack, it seems reasonable to measure crack extension under those conditions, but in practice plane strain conditions do not apply, and plastic deformation around the crack tip may occur in addition. The argument is that k_{1c} is a conservative estimate, since it indicated the worst conditions, and therefore is a "safe" criterion. There remains one more difficulty, that of devising a test which in fact measures crack propagation under conditions of plane strain. In fact, more ductile materials are difficult to test, and conditions of plane strain may not be reached with them in a test devised for less ductile materials.

The test uses either three point loading in bending, or tension. The bend test is the simpler, using a rectangular cross-section test piece in which a notch is machined of root radius 0.08 mm or less. The notch extends from one edge through half the depth of the specimen and the specimen width is half the depth, so that the cross-section embracing the notch at mid length, is approximately equal to the section ahead of the notch. The length of the specimen is 4.2 times the depth, and the test span in bending is 4 times depth. Knife edges at the open end of the notch are used to fix a displacement gauge, which measures crack propagation by the opening of the open end. The specimen is first subjected to fatigue loading to cause a crack to form at the root of the notch, and loading is continued until the crack has propagated 1.3 mm or 5% of the overall notch length plus the fatigue crack length, whichever is the greater. The crack length can be observed with a travelling microscope, and measured accurately after test. The specimen is then placed in the three point bending jig, and load applied is plotted against displacement of the open end of the notch. The load P_Q at which the crack has extended 2%, with limited plastic flow, is obtained from the load-displacement curve by geometrical construction. If an upper limit occurs by yielding, before the set point, that upper limit is regarded as P_Q unless there has been excessive plastic flow, in which case the test is invalid.

The value of k_{Ic} is obtained by calculation from the formula

$$k_{Ic} = 0.0348 \frac{P_Q \cdot L}{WD^{3/2}} \left[2.9\left(\frac{d}{D}\right)^{1/2} - 4.6\left(\frac{d}{D}\right)^{3/2} + 21.8\left(\frac{d}{D}\right)^{5/2} - 37.6\left(\frac{d}{D}\right)^{7/2} + 38.7\left(\frac{d}{D}\right)^{9/2} \right]$$

where P_Q is the load as described, L the span of the bend test, W the width of the specimen, D the depth of the specimen, d the crack length measured from the notch edge of the specimen. All lengths in mm and P_Q in newtons, gives k_{Ic} units of N mm^{-2} mm$^{1/2}$.

If the specimen is insufficiently large, plane strain conditions may not be reached at the crack front, and the k_{Ic} value is too high. The specimen width W, and therefore the crack length d, must be at least $2.5(k_{Ic}/\sigma_y)^2$ where σ_y is the 0.2% proof stress.

The test was based on the reasoning of fracture mechanics in which a constant k_I, known as stress intensity factor, is involved for conditions at the leading edge of a crack. The value k_{Ic} is simply the value of k_I which is critical for a particular material in extension of a crack. (*See also* **Crack**.)

Francium. A metallic element, symbol Fa and at. no. 87, at. wt. 223. (*See* Table 2.)

Franckeite. A mineral consisting of tin sulphide, associated with silver.

Franklinite. A zinc ferrite containing manganese that occurs in New Jersey (Fe, Mn)$_2$O$_3$(Fe, Zn)O. It normally contains about 67% iron oxide, 16% manganese oxide and 17% zinc oxide. It is worked first for zinc, and the residue is used as an iron ore for the production of spiegeleisen.

Frank-Read source. One of the problems of dislocation theory in interpreting the deformation modes of crystals, has been the mechanism of continuous generation of dislocations during deformation as distinct from their origin as a result of thermal treatment, radiation bombardment or electric fields. Dislocations can be observed in the electron microscope, and their propagation through a crystal followed, but generation has not been convincingly demonstrated experimentally. F.C. Frank and W.T. Read envisaged the generation of dislocations by the deformation process itself, involving the "pinning" of parts of a dislocation system, which automatically leads to a sweep round of the dislocation line to reform itself. A dislocation line pinned at one end (by a precipitate or other obstacle in the lattice) can be shown to sweep round the pinning points as a pivot, causing further slip as it does so, and returning to its point of origin, whereupon it repeats the process. When the dislocation line is pinned at both ends, the mechanism is perhaps easier to follow. Figure 69 shows the line in position (1); pinned at both ends and

bowing out under the applied shear stress (2); sweeping round to expand (3); returning to the starting point (4); the recombination to a new line, as segments of opposite sign annihilate, and leave a loop and the original link to expand further (5). Electron microscope evidence of such observations has been reported, in oxide crystals.

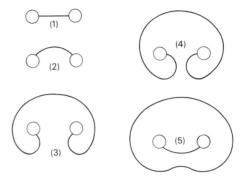

Fig. 69 Stages involved in the operation of the Frank-Read source.

Fraunhofer lines. These are dark lines which appear in the solar spectrum and which always maintain the same relative positions in such spectra. The most important of these lines are designated by the letters of the alphabet. Incandescent bodies produce a continuous spectrum without any dark lines. Vapours are able to absorb the same kind of rays as they emit. If an incandescent body is surrounded by the vapour of an element such as sodium, the vapour may absorb more of the particular rays than it emits. This will cause a weakening in the intensity of the light in the part of the spectrum corresponding to sodium. Hence by contrast the lines will appear dark.

Free cementite. The carbides of iron (or alloying elements) that occur in iron and steel as a separate constituent, i.e. not in association with ferrite as pearlite.

Free-cutting. A term applied to metals, most usually to steel alloys and copper alloys that have good machinability. This latter is not easy to define. It relates either to the resistance to cutting offered by a metal or to its effect on the cutting capacity of a given tool, and depends on the hardness of the metal, on the cutting speed and on some other factors. The elements lead, sulphur and some others are commonly added to metals to improve their machinability. The additions tend to produce short brittle chips rather than long spirals of metal during machining.

Free electron model. The interpretation of electron behaviour in the bonding of atoms is rendered difficult by the restrictions of the Pauli exclusion principle and the Heisenberg uncertainty principle. The latter, in particular, involves assumptions about the electron because the precise position and movement of the electron cannot be simultaneously declared. In the consideration of metallic bonding in particular, where high electrical conductivity is associated with ease of electron movement, a number of difficulties can be overcome by assuming that ions of metal atoms form the crystal lattice, with "valency" electrons freely distributed through the whole lattice, as a free electron gas. This serves to hold the lattice together, since positive ions will naturally tend to repel each other, but no one electron of the "free" type can be directly associated with any one atom.

From this assumption, calculations of the energy states available to the electrons leads to density of states curves, which can explain the conductivity of metals, and the insulating properties of insulators, even though electrons there are tied up in covalent bonds. The free electron model is not necessarily the true picture of electron behaviour, but it serves well to account for many observed facts. (*See* **Wave mechanical behaviour of the electron**.)

Free energy. The energy available in a system for doing work. This is always less than the total energy, because some of the total is bound up with the entropy of the system, the state of disorder. The free energy is at a minimum when the system is in equilibrium, so that all systems tend to come to equilibrium by lowering the free energy so far as the conditions allow. This automatically means that the entropy will reach a maximum at equilibrium, for the specified conditions. Gibbs (who conceived much of these general principles) expressed the free energy thus: $G = E + pV - TS$ where G is the Gibbs free energy; E the total internal energy of the system; p and V are pressure and volume respectively; T absolute temperature; S the entropy. Since in solid and liquid systems, pV is not particularly significant, the equation can be simplified to $G = E - TS$. (See also **Energy**.)

Free ferrite. Ferrite formed from austenite during cooling without simultaneous separation of carbide. The ferrite is present as such, and not associated with cementite to form pearlite.

Frémont test. A dynamic hardness test in which a weight with a pointed tip is allowed to fall freely upon the test surface. More usually the name applies to a type of impact test in which a falling weight is dropped on a test piece to cause fracture. The standard test piece is 3 cm long, 0.8 cm deep and 1 cm wide. It has a notch on the under side 0.1 cm deep and 0.1 cm wide. It is supported horizontally over a gap 2.1 cm wide. The tup falls upon the test piece, and the rounded knife-edge strikes it centrally, breaking it through the notch. The tup then strikes a copper cylinder, which it compresses, the amount of the compression being a measure of the energy remaining in the tup after the test piece has been fractured. The tup weighs about 5 kg and falls through a height of several metres. The striking velocity is thus higher than in the Izod or in the Charpy test.

French chalk. A form of soapstone $3MgO.4SiO_2.H_2O$ used as an absorbent for grease. The ease with which this mineral can be worked, together with its heat-resisting properties, have led to its use in stoves as firebricks. When finely ground it can be used for covering steam-pipes and as a lubricant. It can be fused and drawn into fine threads. Also known as talc, steatite.

Frenkel defect. The migration of an atom in a crystal lattice from its regular position to an interstitial position, leaving behind a vacant lattice site. Because of the lattice strains involved, these defects are regarded as forming an interstitial-vacancy pair, and may diffuse through the lattice in association. This is one contributing mechanism to diffusion, and affects deformation mechanisms because of the lattice strain involved.

Fretting corrosion. A form of corrosion that occurs on the contacting surfaces between heavily loaded metals which are subjected to slight relative movement. Clamping stresses under a collar or in a press-fit are severe stress raisers, and an additional factor in this is the roughening or galling of the surfaces in such collars or fits. There is always a very small amount of play between such surfaces which gives rise to the galling and tearing off of tiny metallic particles which sift out and quickly become oxidised. Particles of steel, therefore, appear as iron rust, and this has given rise to the colloquial terms "cocoa" or "bleeding". In the case of aluminium or magnesium and their alloys, the oxides usually appear as a fine black powder.

This type of corrosion occurs rapidly, but the rate depends on such factors as hardness, surface finish, atmosphere, amount of movement, lubrication, etc. Temporary avoidance may be obtained by oiling the surfaces, but to prevent galling and to distribute the contact stresses, a fibre, leather or paper bushing may be used in certain cases. Electroplating with a soft metal or a soft metal skin is sometimes found useful. Prevention of oxidation may be secured by the use of asphaltum varnish or by employing thin skins or gaskets of materials impregnated with phenolic resins.

Friction sawing. A band-sawing operation that utilises the heat generated by friction to supplement the cutting action of the saw teeth. A band saw moving at high speed (i.e. 15-25 ms^{-1}) is used without coolant or lubricant so that the metal immediately ahead of the saw is softened and more readily cut. Conduction of heat from the cutting zone restricts the use of the process to sections of not more than about 3 cm thickness. Also referred to as fusion cutting.

Friction welding. The joining of two artefacts by generating heat at the interface by means of friction, i.e. vibrating the artefacts or rubbing them together by rotation and pressure. The method serves to remove oxide and contaminating films before reaching the joining temperature, and can be regarded as a combination of pressure and fusion welding. Although it is not essential that actual melting occurs, the frictional forces at asperities must be such as to produce local melting, albeit in very small volumes.

The control of the process is capable of automation, and thus the amount of deformation which occurs just prior to joining and just after, before the machine comes to a halt, can be minimised. The junction exhibits severe plastic flow, but in view of the temperature involved, self annealing can take place to remove much of the strain.

The method is excellent for junctions between dissimilar materials, and where heat-treated structures may be involved, since the minimum thermal energy is involved and the heat-affected zone is small. Many machines resemble two lathes face to face, with chucks to hold the components. Clutches are fitted to one side or the other to slip when joining has been achieved.

Friedel-crafts catalysts. Strong electron acceptors such as aluminium chloride ($AlCl_3$), titanium tetrachloride ($TiCl_4$) and boron tri-fluoride (BF_3) used as catalysts in the cationic polymerisation of poly formaldehyde, polyisobutylene and butyl rubber.

Frieslebenite. A sulphide of silver, lead and antimony $5(Pb.Ag)_2 S.2Sb_2 S_3$ that contains about 22% Ag and is a valuable ore for the extraction of this metal. It is found associated with other silver ores and galena, etc., in Spain, Saxony, etc. Mohs hardness 2-2.5, sp. gr. 6-6.4.

Frit. Fused small friable pieces of enamel glass produced by quenching molten enamel in water. It is used as a basis for the vitreous enamelling process.

Fritting. A term used in powder metallurgy in relation to the production of solid compacts. If no liquid phase accompanies the production of the compact (formed by heating only), the process is known as fritting. The temperature to which the compact is heated is known as the fritting temperature. Fritting may be regarded as a heat-treatment process and may follow the application of pressure to form a synthetic compact. When pressure and heat are applied simultaneously, the process is called hot pressing of the solid compact.

Frosting. A metal finishing process in which a fine matt surface is imparted by acid-etching, scratch-brushing or sand-blasting.

Froth. In metallurgy the term is synonymous with foam (q.v.).

Frother. A substance—usually an organic compound or oil—which is added to the circulating water of a froth flotation plant. When agitated with air it forms a copious supply of bubbles and retains the mineral in the froth.

Froth flotation. A method of separating the constituents of finely pulverised minerals by causing some to sink and others to float in a froth. By this means the ore may be concentrated. The materials used must be soluble in water and must reduce the surface tension of the solution and raise the viscosity. Organic oils (pine, eucalyptus) and other substances are used as the frother.

Frue vanner. A device used for concentrating finely pulverised and classified ores. It consists of a wide endless rubber belt passing over pulleys and arranged as an inclined table with the belt moving slowly upwards. A rapid sideways shaking movement is imparted to the belt. A stream of water flowing over the belt separates minerals of different specific gravity; the lighter gangue material is washed off the table, whilst the heavier mineral travels upwards until it is finally discharged.

Fuel. A term used to describe materials that are employed to produce heat by combustion in air. Most of the natural fuels (wood, peat, coal, oil, natural gas, etc.) are chiefly compounds of carbon, hydrogen and oxygen with small proportions of other elements, e.g. nitrogen and sulphur. Solid fuels are divisible into two broad classes: (a) naturally occurring solid fuels and (b) manufactured solid fuels. The former include coal, wood and peat, while the latter include coke and charcoal. Pulverised solid fuel (coal dust) in a stream of air has been used in cement-drying kilns, in metallurgical furnaces and steam boilers. Pulverised coal in colloidal suspension in oil has also been used for firing steam boilers.

Natural liquid fuel (crude petroleum) forms the basis of practically all industrial fuels. It is a mixture of hydrocarbons, and separation of the constituents is effected by fractional distillation. Artificial liquid fuels are produced (a) by the destructive distillation of coal, lignite or shale at low temperatures; (b) by the hydrogenation of coal; (c) by the recombination of the constituents of water gas in the presence of a suitable catalyst.

Gaseous fuels include (a) natural gas and (b) manufactured gases. The latter are produced in a variety of ways, in some of which the gas is the "prime" product and in others it is a by-product, e.g. coke-oven and blast-furnace gas. The major combustible constituents in all these gases are either carbon monoxide, hydrogen, hydrocarbons or a mixture of these. The compositions may vary according to the nature of the materials used and the temperature at which the reactions occur. Typical compositions are given in Table 26.

TABLE 26. *Typical Compositions of Some Gaseous Fuels.*

Fuel	CO, %	H_2, %	Hydrocarbon, %
Coal gas	8.0	50.0	3.0
Water gas	39.0	52.0	0.8
Siemens gas	24.0	9.0	2.5
Dowson gas	26.0	19.0	0.3
Mond gas	12.0	28.0	2.0
Blast-furnace gas	28.0	3.0	0.2

Fugacity. The escaping tendency of a gas confined in an enclosed space. It is identical with the active mass of the gas and is measured by the pressure. For a perfect gas the fugacity is proportional to the concentration of the gas and, as all gases approximate the perfect state at low temperatures, fugacity will then be proportional to the density.

Full annealing. A heat-treatment process in which metals are heated to temperatures above the critical temperature range, and are held at these temperatures for sufficient time to allow transformations to proceed to completion. The heating is followed by slow cooling to below the critical range, usually in the furnace or in a soaking pit.

In the case of steels the annealing temperature is usually 55 °C above the upper limit of the critical temperature range, and the time of holding at that temperature is not less than 1 hour for each 2.5 cm of the maximum section being treated.

Fuller. A forging die (two are used) employed to reduce a portion of the stock to a smaller cross-section.

Fullering. A hand-forging process that consists in roughing down material to reduce the cross-section, prior to swaging. The tools used are called fullers, of which two are needed. The bottom fuller is usually set in an anvil and the bar placed on this bottom fuller. The top fuller is placed on the bar directly above the bottom fuller and blows are delivered on it. The resulting bar therefore shows slight corrugations corresponding to the shape of the heads of the tools. The corrugations are removed and the final drawing-down is done between swages.

Fuller-Lehigh mill. A ball mill used for pulverising coal and other minerals. The grinding operations are performed by one or more rings of balls rotating between grinding rings, which latter may also rotate or they may be stationary. Springs are used to maintain a uniform pressure that can be controlled by adjusting rods.

Fuller's earth. A clay-like mineral consisting mainly of hydrated aluminium silicate. It varies in colour from green to greenish-grey. It is so named because of its use by fullers as an absorbent for the grease and oil of cloth. This use, however, has declined, but the material is still widely used in the filtration of mineral oils and for the decolorising of certain vegetable oils. Its absorbent properties are due to the physical rather than the chemical nature of the earth. It is mined or dug in the United Kingdom, in Saxony and in many parts of North America. Specific gravity varies from 1.7 to 2.4.

Fulminate of mercury. A highly sensitive compound produced by the action of strong nitric acid and alcohol, on mercury Hg $(ONC)_2$. It readily detonates by friction or percussion even when lightly confined. Heating to 150 °C or the addition of a little concentrated sulphuric acid will cause violent detonation. Mixed with other substances, e.g. powdered glass, potassium chlorate, etc., it is used in detonators.

Fulminating gold. A dirty olive-green powder produced by the action of concentrated ammonia on auric chloride. It has the formula $2AuN_2H_3.3H_2O$. It is very sensitive to percussion and to heat. Hydrochloric acid decomposes it to form harmless gold (III) and ammonium chlorides.

Fulminating platinum. The action of caustic potash on ammonium platinichloride gives a series of compounds that are all explosive. The yellow precipitate obtained in such a reaction is said to have the formula $PtClNH_6O_2$. The dry compounds are all sensitive to friction and percussion, but may be rendered harmless by treating with hydrochloric acid.

Fulminating silver. When the oxide of silver is dissolved in ammonia and the solution exposed to air, a black precipitate of silver nitride Ag_3N is produced. This is called fulminating silver, since when it is dry it is very explosive.

Fume. A term used to describe the fine particles of a solid or liquid which may be suspended in a gas. The size of the particle is generally taken as $1 \mu m$ or less, and the suspension is really a colloidal system produced from industrial processes, e.g. combustion, distillation, calcination and sublimation.

Fume precipitation. A process in which the solid particles in fume or smoke are electrified and then discharged on to an electrode of opposite polarity. In some cases the metallic walls of the treating chamber are used as the negative electrode, while the positive electrode is a rod or wire placed axially in the chamber. In other cases, plates are hung in the chamber with positive electrode wires running vertically between them. High voltage, direct, pulsating current is required and the voltage should be such that a glow discharge can be maintained without arcing. For efficient working it is usually considered good practice to move the fume through the treating chamber at not more than $5 \, m \, s^{-1}$, and to proportion the chamber so as to allow the fume to travel at this speed for some 12 m between the electrodes. A usual efficiency of 98% is generally obtainable.

Fume precipitation can be applied to the removal of dust from, and the cleaning of, gases, as well as the recovery of suspended acid in fumes, etc., the material of the electrode being governed to a large extent by the nature of the material to be separated.

Furfural. A colourless liquid that turns brown when exposed to air and light. It is prepared from oat hulls, cotton-seed hulls, rice hulls and corn cobs. Formula $C_4H_3O.CHO$, sp. gr. 1.6 at 20 °C, b. pt. 161.7 °C. It is somewhat soluble in water and mixes freely with alcohol.

It is used in the production of synthetic rubber and in the refining of lubrication oils. It can be resinified by condensation with phenols or by acids in the presence of suitable catalysts. The resin is used for impregnating porous materials and as a wetting agent in the moulding of grinding wheels.

Furnace linings. Since furnace linings are exposed to high temperatures and to the scouring action of slags, they must be constructed of materials capable of withstanding these destructive actions. Refractory materials may be divided into three classes:

(a) acid refractories. These contain a high proportion of silica, e.g. flint, ganister, sand, Dinas rock and fire clay. They combine readily with basic oxides;

(b) neutral refractories. These are refractories that do not exhibit acidic or basic properties. Examples are graphite and chromite.

(c) basic refractories. In these the oxides of metals such as calcium, magnesium or aluminium predominate and there is an almost complete absence of silica. Examples are lime (not extensively used), dolomite, magnesite and bauxite.

Furnaces. A metallurgical furnace is a plant or contrivance in which metallurgical operations are carried out with the aid of heat. The heat may be produced by combustion (usually but not always of fuel) or from the heating effect of an electric current. The size and shape of furnaces vary considerably, as do the internal arrangements. These variations are associated with the kind of operation to be carried out in the furnace, with the method of heating, and sometimes with the nature of the materials available. Furnaces may be divided into the following groups:

(a) Those in which the charge to be heated and the fuel are in contact. These include:

(1) Shaft furnaces. These are usually either cylindrical or in the shape of a slightly truncated cone, though elliptical and rectangular section furnaces are still used. This type of furnace may be sub-divided into:

(i) Blast (or forced-draught) furnaces, in which the air blast is usually heated and is supplied under pressure. High temperatures are attainable so that the slag and the metal are in a molten state and can be tapped at intervals. Examples are found in the blast furnaces used for smelting iron, copper, zinc and lead. A smaller

type of blast furnace, used for melting iron in foundries, is called a cupola. (*See* Fig. 70.)

(*ii*) Natural-draught furnaces, which are sometimes referred to as kilns. They do not operate at such high temperatures as (*i*) above, since the products are not required in the molten state. Examples are seen in lime kilns and in such types as Gjers kiln (for calcining iron ores).

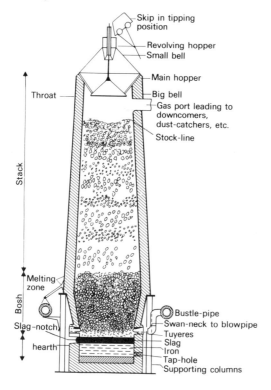

Fig. 70 *Cross-section of a blast furnace producing pig iron.*

(2) *Hearth furnaces.* This type usually consists of a shallow chamber, the height of which is not much greater than the width. The chamber is lined with suitable refractory material, and the blast is introduced through tuyères that direct the air blast either on to, or under, the charge. Examples can be observed in such types as the Scotch hearth, used for the reduction of galena, the Walloon hearth, used for the production of wrought iron, and the Siemen's open-hearth type furnaces used in steelmaking.

(*b*) Reverberatory furnaces.

In this type of furnace the charge is heated by, and is in contact with, the products of combustion, but is not in contact with the fuel itself. This type of furnace has the roof so arched that the flames and hot gases are deflected on to the charge. There are two types:

(1) *Low-temperature operation furnaces.* These are employed when the metallurgical operation is to be conducted at relatively low temperatures. They are often referred to as roasting or calcining furnaces, indicative of the operations carried out in them. The charge is not melted at any stage, but it may have to be turned over during roasting, and is finally raked out from the hearth in the solid state. As these operations are frequently done by hand, the hearth is made shallow to facilitate this. In the MacDougall roasting furnace there are several superimposed horizontal hearths and the ore is raked automatically from hearth to hearth.

(2) *Melting furnaces.* These are used to melt the metal, so that the hearth is basin-shaped to facilitate the tapping of the molten mass. A fuel giving a long, luminous flame is desirable for this kind of furnace. The higher the temperature desired, the shorter must be the hearth—as a general rule. On the other hand, the reverberatory furnace used in the preparation of copper may be as much as 40 m long. The open-hearth furnace for steel production is an example of this type. (*See* Fig. 71.)

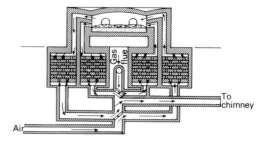

Fig. 71 *Cross-section of an open-hearth reverberatory furnace for steel-making.*

(*c*) Furnaces in which neither the fuel nor the products of combustion come into direct contact with the charge. These can be considered under three headings:

(1) *Muffle furnaces.* In this type the material being treated is to be kept separated from the products of combustion and from ash that may be carried over by the draught. The furnace usually consists of a brickwork or fireclay chamber heated by a fire located below it, or else heated by the products of combustion that are directed to circulate round it. Small muffle furnaces are often gas-fired. An example of this type of furnace is that used for cupellation, a silver refining process.

(2) *Crucible furnaces.* In these types of furnace the material is put into a closed refractory crucible, which is placed in the fire itself. It is frequently possible to obtain higher temperatures in this manner than in muffle furnaces. The crucible is removed from the fire when the charge is sufficiently heated or ready for pouring. Such furnaces are more generally fired by coke, but modern versions are often fired by oil or gas burners. They are referred to as "pit" furnaces when sunk in the ground.

(3) *Retort furnaces.* These are used when the material has to be vaporised and later condensed. The retorts are made of fire-clay, as also are the condensers, and they are usually placed in rows above an arched chamber which contains the fire. A good example is the Belgian furnace for the extraction of zinc.

(*d*) Converters.

This kind of furnace is operated by means of a forced draught, but unlike the blast-furnace type, they receive their charge in the molten condition. They also differ from most other types of furnace in that they can be rotated about a horizontal axis. The heat required for the operation is not obtained from external fuel, but arises from the oxidation of the unwanted impurities in the charge by means of the air blast. This latter may be introduced into the converter either through the base, as in the Bessemer type, through the side, as in the Tropenas, or from above, as in modern oxygen steelmaking. (*See* Fig. 72.)

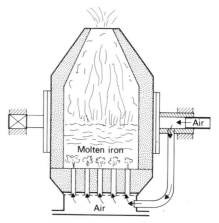

Fig. 72 *Section through a steel converter.*

(*e*) Electrical furnaces.

In all these types electrical energy is converted into heat energy, the amount depending on the flow of current and the resistance offered to the passage of the current. There are three main types:

(1) *Resistance-type furnaces.* In these furnaces the electrical resistance of a conductor is so arranged that in the passage of a

given current the conversion to heat of the requisite amount of energy shall take place.

(i) Metallic conductors. The material for these conductors is usually a heat- and corrosion-resisting alloy such as the nickel-chromium alloys, tungsten, molybdenum or platinum. The nickel-chromium alloys are generally employed for temperatures up to about 1,000 °C and the other elements for temperatures above 1,100 °C, but since tungsten and molybdenum react with oxygen at high temperatures, the furnace atmospheres must be controlled. Hydrogen atmosphere is allowable, or the furnace may be evacuated of gas. The conductors are frequently wound on solid cylinders of refractory material or around the refractory furnace chamber, and good electrical and thermal insulation is required. At temperatures above about 1,500 °C most of the refractory materials available react with these metals. Hence, for furnaces employing temperatures of this order, a molybdenum tube is used as the heating element and the furnace chamber. It is surrounded by concentric tubes of the same metal, and the annular spaces are evacuated of gas.

(ii) Non-metallic conductors. In these cases substances such as graphite or carborundum, the former being used preferably in the absence of air, are employed. One of the disadvantages of carborundum is the fact that it increases its resistance with usage.

(2) *Arc furnaces.* There are three types of arc furnace:

(i) Indirect arc furnaces, in which the arc is struck above the charge in the furnace, the charge being heated by radiation (in some cases the arc is deflected downwards magnetically).

(ii) Direct arc furnaces, in which the arc is struck between an electrode and the charge, the current passing through the charge and returning via a second arc to the electrode(s).

(iii) Bottom electrode furnaces, in which electrodes are built into the hearth and arc through the metal to electrodes above the bath, an example being the Greaves-Etchells furnace.

(3) *Induction furnaces.* In this type there is no direct connection between the metal and the current supply, but heat is produced by electrical induction in either:

(i) Cored induction furnaces, in which the bath holding the material is ring-shaped and acts as a single secondary winding of a transformer, the core of this latter being a closed iron circuit on which is wound the primary winding (an example is the Ajax-Wyatt furnace); or

(ii) Coreless high-frequency furnaces, in which the charge is placed in a crucible and this latter is surrounded by a water-cooled tube which functions as part of the electric circuit. Eddy currents set up in the metal in the crucible generate heat in amounts sufficient to melt the charge. In such furnaces it can be arranged that the molten metal circulates within the crucible and the latter can be arranged to tilt for pouring. An example is the Ajax-Northrup furnace.

Fusain. One of the important constituents of coal. It consists essentially of the remains of decomposed plants from which the volatile matter has been eliminated. It has a higher percentage of fixed carbon than the other separable constituents of bituminous coal (clarain, durain, vitrain). It does not cake, but may improve the caking properties of bituminous coal, of which it usually forms from 0.1 to 10%. In appearance it is dull and friable.

Fused silica. The vitreous form of SiO_2, achieved by melting and cooling to prevent complete crystallisation. The material is then impervious, acid resistant and tough, excellent for laboratory and industrial use, and with a very low coefficient of thermal expansion. (*See* Fig. 3, **Amorphous material.**)

Fusible alloys. A group of non-ferrous alloys with low melting points. They approximate to ternary, quaternary and quinary eutectics and are based on the low-melting-point metals (bismuth 271.3 °C, lead 327.4 °C, tin 231.9 °C, cadmium 320.9 °C, etc.) They are best known through their use in safety devices such as fire-alarm sprinklers, boiler plugs and electrical fuses, but have recently acquired greater significance through their use as tube-bending alloys, low-temperature solders, and in foundry patterns, assembly jigs, dies, etc. Some of the alloys expand slightly on solidification and continue to do so for several days afterwards. Others are free from dimensional changes, while some show slight contraction. A list of the commoner fusible alloys, together with their composition and chief properties, is given in Table 8.

Fusible plug. A safety device used in boilers, pressure cookers, etc. It consists of a perforated plug sealed with a low-melting-point alloy which melts when the safe operating temperature is exceeded.

Fusion welding. A welding process in which the joint is made between metals in a state of fusion without hammering or pressure. The welding operation is usually carried out by means of an oxy-ethyne gas flame or an arc-welding process such as a carbon arc, metal arc or inert gas shielded arc.

G

Gabbro. A basic plutonic igneous rock composed chiefly of felspars, labradorite and diallage (silicates of sodium, potassium, calcium, aluminium, iron and magnesium). They are usually coarse grained and are very widespread, though not occurring in great masses. Gabbros are used locally as road metal or for building stone. In Scandinavia, segregations of iron ore are found in association with gabbro and may contain up to 1.7% titanium oxide and about 14% iron oxide. In Norway and the Urals masses of iron pyrites, often copper bearing, are found with gabbro-type rocks, while in Ontario the norites are rich in nickel.

Gadolinite. A mineral in which occur the rare earths, yttrium, cerbium, erbium, etc. Silicates of iron and beryllium are also usually present $(Fe.Be)_2 Y_2 Si_2 O_{10}$. It is one of the most important ores of the rare earths.

Gadolinium. A metallic element belonging to the rare-earth group. It occurs in yttrspar, gadolinite, samarskite and monazite. It is separated from other members of this group by a process of fractional decomposition of the double magnesium nitrates. Symbol Gd., at. no. 64, at. wt. 157.2, density (20℃) 7.9 g cm^{-3}, crystal structure—close-packed hexagonal, lattice constants $a = 3.629 \times 10^{-10}$ m, $c = 5.76 \times 10^{-10}$ m.

Gagger. A hook-shaped bar used by foundrymen to support the sand in a cope mould. The upper end of the gagger is hooked into a box bar of the cope. Gaggers reinforce the sand and hold it together when the mould is moved or the pattern withdrawn.

Gagging. A semi-mechanical process for straightening bars, rods and tubing by means of a press.

Gahnite. A somewhat rare mineral found in New Jersey. It consists chiefly of zinc spinel $ZnO.Al_2O_3$ and is usually associated with other zinc ores. (*See also* **Franklinite.**)

Galena. A sulphide of lead PbS and the most important source of lead. It is frequently associated with blende, pyrites, calcite, fluorspar, quartz, barytes, etc., as gangue materials. It occurs in beds, and in veins traversing crystalline rocks, clay-slates and limestones in the UK, the Harz district, the USA, and Australia. The mineral nearly always contains silver, argentiferous lead being an important source of this metal; the amount of silver is usually about 0.85 g per kg of mineral. Galena crystallises in the cubic system and the colour of the mineral is lead-grey. Mohs hardness 2.5, sp. gr. 7.2-7.7.

Galling. The removal of particles from localised areas of one or both the metallic surfaces that are in contact during sliding friction. Even small amounts of movement under high pressure result in the tearing off of tiny metallic particles. These eventually appear at the free surface as oxidised metal. (*See also* **Fretting corrosion.**)

Gallium. A rare metal that is widely distributed in nature, though found in only tiny quantities in any of the minerals that contain it. It is present in zinc ores, in bauxite and in certain iron ores. Middlesbrough iron contains about 1 part of gallium in 33,000 and was at one time the richest source of the element known. The first commercial source was the sludge from American zinc ores. More recently it has been produced from bauxite in America and from flue dust in the UK. The metal is still expensive, on account of the very small content in any of its sources. 1,000 kg of bauxite yields about 28 g of gallium and the extraction is a long and tedious process.

Pure gallium is bluish-white in colour and has the remarkable property for a metal of melting at 30 ℃. The liquid metal retains its silvery colour and can remain liquid for a long time, even though the temperature is reduced well below the melting point. When solid, the metal is tough, though it can be cut with a knife. On solidifying it expands, a property that introduces a small complication in storage, since the liquid must be kept in flexible and unbreakable containers. The sprouting of the metal when suddenly solidified by rapid cooling is associated with the large increase in volume that accompanies the change in state.

Since the metal is liquid at 30 ℃ and does not boil below 1,900 ℃, it would seem to be a very suitable thermometric liquid. Its use in thermometers for measuring high temperatures has been recommended and tried successfully.

Liquid gallium will "wet" glass, and thus may be used for the masking of mirrors. An alloy of gallium and aluminium can be similarly used.

A whole range of useful intermetallic compounds of gallium has been investigated for semiconductor applications and there appears to be a field of application of gallium alloys for making precision castings and in dentistry. Most of the alloys expand on solidification. Gallium-aluminium alloy can be used in metal vapour lamps (in place of mercury vapour), and gallium-cadmium alloys in such lamps gives a very pure red light that has applications in photography and optical researches. Liquid gallium attacks many metals causing embrittlement (*see* Table 2).

Gallium arsenide. An intermetallic compound between gallium Ga, with two 4s electrons and one 4p available for bonding, and arsenic As, with two 4s and three 4p electrons. The combination produces a semiconductor with an energy gap of 1.35 eV, slightly higher than that of pure silicon, the covalent bonded structure having a diamond type of lattice, with the two atoms alternating.

Diodes of the material emit infra red radiation when a forward bias is applied, the radiation corresponding in energy to the energy gap of the semiconductor. The material is also used in tunnel diodes, at temperatures up to 350 ℃.

Galmei. A hydrated silicate of zinc $2ZnO.SiO_2.H_2O$. It is a valuable zinc ore, known also as calamine hemimorphite.

Galvanic cell. An important electrochemical phenomenon, first demonstrated by Galvani as a source of e.m.f., but involved in the mechanism of corrosion when dissimilar metals are present in an electrolyte.

Two metals, such as zinc and copper (used by Galvani), immersed in dilute sulphuric acid, will set up electrode potentials by the following reactions.

$$Cu \rightleftharpoons Cu^{2+} + 2e^-, \text{ electrode potential } - 0.34 \text{ V}$$
$$Zn \rightleftharpoons Zn^{2+} + 2e^-, \text{ electrode potential } + 0.76 \text{ V}$$

The zinc has the higher electrode potential (referred to hydrogen) and if the two metal samples are connected together by an external wire, electrons will flow from the zinc to the copper, via the external wire. The copper now is charged with more electrons than those produced by its normal electrode potential, and so some of the copper ions which entered the acid solution on immersion, will return to the copper cathode to neutralise the electron excess, and restore equilibrium at the cathode. This will continue to occur, until all copper ions have come out of solution and deposited on the copper cathode. The zinc will continue to cause a flow of electrons however, and now these must be reduced at the copper by another mechanism. The water of the electrolyte is ionised into H^+ and OH^- ions, and the acid into $2H^+$ and SO_4^{2-} ions, so positive hydrogen ions are available to neutralise the excess electrons at the copper. When they do so, they cease to be ions and become hydrogen atoms, which escape as bubbles and combine into H_2 molecules. The zinc will continue to send electrons along the external wire, and zinc ions will continue to enter the solution (in fact this will continue until all the zinc anode is dissolved, or the water level depleted by the escape of hydrogen). Note that the anode dissolves, the metal with the highest electrode potential. The greater the difference in electrode potential, the greater the current flow (electrons in the external circuit) and the more rapidly the anode is dissolved. The higher the temperature, the more rapid are all the reactions, and the more hydrogen ions in the electrolyte, the more rapidly they travel to the cathode. This is electrochemical or galvanic corrosion, which is intentional when producing current, but disastrous when involving attack of engineering components.

Galvanised iron. Iron (or mild steel) in the form of plate, steel or wire that has been coated with zinc to protect the surface from atmospheric corrosion. The zinc dissolves before the iron in presence of oxygen and moisture, since it is more electropositive than the iron. The zinc therefore affords sacrificial protection. (*See also* **Galvanising.**)

Galvanised wire. Wire on which a coating of zinc has been deposited to afford protection against corrosion. The galvanising is accomplished by a continuous process in which the wire passes through an annealing furnace, acid pickling bath, washing and fluxing tanks, and the galvanising bath. The last-named consists of a tank of molten zinc, the surface of which may be covered by a flux.

For most purposes the zinc coating must be flexible, i.e. it should not tend to break away when the wire is bent. This is ensured by regulating the thickness and physical qualities of the coating by passing the wire through wipers as it leaves the galvanising bath.

Galvanised wire is generally used in the making of wire cloth, wire mattresses and wire rope. Netting made from black annealed wire is galvanised after making.

Galvanising. The process of applying a coating of zinc to the surface of iron or steel to provide a corrosion-resistant surface. There are several methods of applying the coating:

(a) *The hot-dip process.* This oldest method consists in dipping the prepared article into molten zinc and draining off the excess metal. In the case of sheet or wire, the metal is cleaned by pickling in an acid bath (usually hydrochloric acid), washed, fluxed and dipped into molten zinc (temperature 430-450 °C), the surface of which is covered with flux. The thickness of the coating is regulated by passing the metal between rollers for sheet, or through wipers for wire. The amount of zinc taken up on the surface of sheet metal varies from 0.36-0.96 kg m^{-2} (1.2 to 3.2 oz ft^{-2}).

(b) *The zinc-plating process* is an electrolytic one which can be carried out at room temperature. It is therefore very suitable for all those cases where the temperature of the zinc bath would be harmful in causing buckling of flat surfaces or modification of the temper of steel springs. The thickness of the coating deposition can be carefully controlled, but the electrolytic deposit is more "spongy" than that obtained by dipping.

(c) *Sherardising* — *see* **Sherardising.**

(d) *Metal spraying* is a process in which a very fine spray of molten metal is projected on to the material to be galvanised. Previous cleaning of the metal is required, and the zinc is deposited by the aid of a special pistol, using an oxygen-gas flame. The method is especially suitable for treating large vessels after construction, for girders erected and generally for most articles after fabrication.

Galvannealing. A process sometimes used in the production of galvanised (zinc-coated) steel sheet or wire. It is designed to improve the adhesion and remove the porosity of the zinc coating. As the sheets emerge from the molten zinc galvanising plant they are passed without removal of the excess zinc through a chamber heated to a temperature of above 450 °C, so that the zinc coating remains molten long enough to produce a coating consisting almost entirely of zinc-iron alloy. This coating has a very smooth finish, which on bending tends to produce very fine cracks but does not flake away.

Galvanometer. An instrument used for measuring electric current. For laboratory purposes galvanometers are chiefly used for the measurement of very small currents and are also used as a means for "null methods" of measurement, i.e. they are used as a means of detecting when no current is flowing in a particular branch of a circuit.

The types used for accurate work are invariably of the mirror type. Attached to the suspension wire of the moving system is a small concave mirror upon which a beam of light from a lamp is focused. A fine wire is stretched across the lens of the lamp. The light is reflected from the galvanometer mirror on to a graduated scale, the image of the wire serving as the indicator.

The type most commonly used is the "moving-coil" or D'Arsonval type. It is little affected by stray magnetic fields. The moving coil is usually rectangular in shape and is free to rotate between the poles of a permanent magnet. The instrument is dead-beat, the damping being usually due to eddy currents generated in the coil frame. In some forms the coil is mounted in a long silver tube which, by rotating with the movement, gives efficient damping.

The moving magnet or Kelvin type consists of a fixed coil of many turns, at the centre of which three or four small magnets attached to the back of a mirror are suspended. The sensitivity of the movement can be controlled by a permanent magnet placed above the instrument, and usually such instruments are capable of indicating a current of the order of one-millionth of an ampere. The mirror acts as an air-type damping vane. The instrument is, however, sensitive to stray fields.

Ballistic galvanometers may be either moving coil or moving-magnet type instruments, constructed for work with currents of short duration, e.g. induced currents, condenser discharges, etc. "Damping" must be as small as possible, but, in any case, a correction must be made for it. The moving system must have a large moment of inertia so that practically the total discharge passes before the system begins to move.

The Duddell thermogalvanometer is primarily intended for the detection and measurement of very small alternating or variable currents, such as are met with in the receiving aerials of wireless telegraphy and in telephone circuits. The moving system consists of a loop of silver wire suspended between the poles of a permanent magnet. The lower ends of the loop are attached to pieces of bismuth and antimony, respectively, these being connected with the junction situated just above a wire filament called the heater. When the current to be measured passes through the heater the bimetal junction is heated and the thermoelectric current generated is measured by the deflection of the loop.

In the String galvanometer a powerful electromagnet replaces the permanent magnet and the moving system is a glass fibre on which a thin film of silver is deposited. It can be used to detect the very small currents due to small differences of potential. The movement of the fibre can be observed through a microscope.

Gamma brass. A solid solution of zinc and copper containing 60-68% Zn and which begins to appear when the percentage of zinc is greater than about 45%. As the percentage of zinc approaches 45% the strength of the alloy falls rapidly, and it becomes brittle and unsuitable for mechanical working. Gamma brass is the hard, brittle constituent.

The unworkable gamma brasses are known as white brasses and are not of great industrial importance. The 50% zinc alloy can be used (in granulated form) in brazing, and those alloys having a higher percentage can be cast or used as the basis of certain useful drop-forging alloys.

Gamma iron. An allotropic modification of iron, crystallising in the face-centred cubic system. It is stable within the range of 910-1,400 °C, and results from a transformation, in the case of pure iron, of the body-centred alpha iron at 910 °C. Gamma iron is capable, at 1,110 °C, of holding carbon in solid solution up to 1.7% by weight. When heating is continued beyond 1,403 °C the gamma iron changes to delta iron. Solid solutions of carbon in gamma iron are known as austenite, and this phase is often referred to in carbon steels as the gamma phase. Gamma iron is non-magnetic. The range of stability of the gamma phase is varied by taking alloy elements, e.g. nickel or chromium, into solid solution.

Gamma rays. The penetrating radiations that frequently accompany radioactive changes or are associated with nuclear reactions. The wavelengths lie within the range 3 to 0.005×10^{-10} m. They are electromagnetic radiations related to light rays and X-rays, but are of shorter wavelength, and hence of higher energy. The longer γ rays are identical with the shorter X-rays.

The emission of gamma rays is an important method by which a nucleus in a state of high internal energy can lose its excess energy. Because of their greater penetrating power they may be used instead of X-rays in radiography. There are certain advantages in the use of sources of gamma radiation, since they are compact and readily portable. Many of the artificially produced radioactive isotopes have a high specific activity with a reasonably long half life. From Aston's curve showing the packing effect, it is clear that for elements of atomic weight less than 80, energy can only be given out when the nuclei are built up by the addition of protons, and not when they break down. The heavier elements give out energy—with gamma rays as a secondary phenomenon—on disintegrating. From the β-ray spectra it may be concluded that gamma rays are characteristic radiations emitted from the radioactive nucleus in its rearrangement after the ejection of a β-particle. These radiations from the nucleus, in their passage through the atom, excite the characteristic radiation of the external electronic system. The total gamma radiation of a radioactive atom consists not only of gamma rays from the nucleus, but also of the characteristic X-rays of the atom.

Gamma rays may produce one or more of three main effects on the material through which they pass. (a) The Compton scattering effect that occurs when a gamma-ray photon is scattered by an elastic collision with an electron present in an atom. In this interaction a beam of gamma rays will lose energy, but the number of photons will remain unchanged. This effect is proportional to the number of electrons in the atom, and is thus greater for atoms of "heavy" elements than for atoms of low atomic weight. It is also inversely proportional to the energy of the radiation. (b) The photo-electrical effect occurs when a gamma photon, having energy greater than the binding energy of an electron in an atom, transfers all its energy to an electron. This latter is therefore ejected from the atom. The excess of the transferred energy over and above what is required to detach the electron appears in the latter as kinetic energy. (c) When a gamma-ray photon with energy in excess of 6.25×10^{-12} J (1.02 MeV) passes close to the nucleus of an atom, the photon may be annihilated, with the formation of a positive and a negative electron. Energy in excess of 6.25×10^{-12} J (1.02 MeV) passes into the kinetic energy of the "electron pair".

The electron pair and the photoelectric effect represent interaction between an atomic nucleus or an atom and a gamma-ray photon, with the disappearance of the photon. The Compton effect involves a photon and an individual electron. The effects on the reduction of the gamma radiation are summarised thus:

	Atomic Number of Element (N)	Intensity of Gamma Radiation (I)
Compton effect Photoelectric effect	Increases as N increases Increases as $(N)^5$	Decreases as I increases Decreases rapidly as I increases

	Atomic Number of Element (N)	Intensity of Gamma Radiation (I)
Electron pair effect	Increases as $(N)^2$	Increases with excess of energy over 6×10^{-12} J

The absorption coefficient for gamma rays passing through a metal is roughly proportional to the density of the material. Common gamma-ray sources for industrial radiography include cobalt 60, tantalum 182 and iridium 192, as well as radium and radon. To give useful service a source of gamma radiation must (a) have a high specific activity, (b) preferably be of small dimensions, (c) have a reasonably long half-life period—say a year or more—and (d) must emit radiation of energy suitable to the particular application. As a general guide it is usually accepted that radium, radon, cobalt 60 and tantalum 182 are suitable for penetration of steel 5-15 cm thick, and that iridium 192 is suitable for penetration of steel 1-6 cm thick.

Radioactive materials are usually supplied in small containers in the form of pellets, or wires or evaporated or deposited upon tin foil.

The original (non SI) unit of activity (strength) of a radioactive source was the curie. One gram of pure radium undergoes 3.7×10^{10} disintegrations s^{-1}, which is conventionally taken as a suitable unit for expressing decay rates. In general, a curie is the weight of any radioactive material that is undergoing 3.7×10^{10} disintegrations s^{-1}. The SI unit is s^{-1}, 1 disentegration per second is unit activity, called the "Bequerel" (Bq).

An electron volt is the energy acquired by a unit positive or negative charge when it is accelerated by a potential of 1 volt. It is represented by symbol eV. The practical unit is one million eV (MeV), which is equivalent to 1.6×10^{-13} joules.

The röntgen is the original (non SI) unit of radiation exposure dosages. The physiological damage caused by gamma radiation is due to the ionisation brought about by the high energy electrons that are ejected or produced in passing through matter. The energy absorbed by tissue is related to the ionisation produced in air by gamma rays over a wide range of energies.

The conventional unit of measurement of the amount of energy deposited in living tissue is the radiation absorbed dose, or *rad*, equivalent to 10^{-2} J kg^{-1}. This unit is being replaced by the SI unit the gray (Gy), equal to 100 *rad*, equivalent to 1 J kg^{-1}.

Radiation is of several basic types, however; alpha, beta and neutron particles, in addition to gamma and X-rays. Since these vary widely in their ability to penetrate tissue and alter molecular structures as a result of ionisation, the *rad* is weighted for the increased biological effectiveness of particles. One *rad* of exposure to alpha particles produces greater biological damage than 1 *rad* of X-rays. The weighted measure of exposure is the dose equivalent, the amount of radiation absorbed by the organism, called the *rem*, and therefore independent of the type of radiation involved. The *rem* is being replaced by the SI unit, the sievert (Sv) equal to 100 *rem*, but 1 *rem* is 10^{-2} J kg^{-1}, so that 1 sievert is 1 J kg^{-1}.

As an example of low level radiation exposure, the contribution of natural background radiation to the population of the UK is on average 37×10^{-3} *rem*, with a range 26-46 *mrem* per head. In Aberdeen, Scotland, a city built on granite, the average is 61 *mrem* per head. On average, medical irradiation brings the total to just over 50 *mrem* in the UK, there is a further 0.54 *mrem* additional from other man-made sources, and 0.08 *mrem* from the nuclear industry. The natural background therefore accounts for 67% of the total.

Gangue. The minerals of no particular economic value that are associated with metallic ores in primary and secondary deposits. The processes in which concentration is carried out, whereby valuable minerals in an ore are separated from worthless impurities or gangue are referred to as ore dressing.

Ganister. A highly siliceous, even-grained sandstone used in the ground condition as a refractory material for furnace linings, etc. It is also made into bricks for the same purpose. It generally occurs in the coal measures and usually contains a certain amount of clayey matter, and is designated as argillaceous sandstone. The corresponding clay-bonded material is known as quartzite in America and as dinas in Germany. Ganister bricks have good mechanical strength.

Garnet. A group of minerals found in limestone, dolomite, granites and metamorphic rocks. They are essentially orthosilicates of various divalent and trivalent metals such as calcium, magnesium, iron, aluminium and sometimes manganese and chromium. Lime-alumina garnet is called grossular and has the formula $Ca_3 Al_2 (SiO_4)_3$; iron-alumina garnet is almandine

$$(FeMg)_3 Al_2 (SiO_4)_3.$$

Noble garnet has the same formula, the transparent crystals varying in colour from pale yellow to dark red, according to the amount of iron present. Lime-iron garnet or andradite is $Ca_3 Fe_2 (SiO_4)_3$. They all crystallise in the cubic system—usually as rhombic dodecahedra. Mohs hardness 6.5-7.5, sp. gr. 3.5-4.2.

Garnierite. An important source of nickel found in New Caledonia, Urals and the USA. It is a hydrated silicate of nickel, $2(NiMg)_5 Si_4 O_{13}.3H_2 O$, containing about 24% Ni, but the composition is variable. Specific gravity 3.2-2.8.

Gas analysis. The quantitative determination of the nature and amount of gases in a mixture by absorption successively in various reagents. The reagents generally used are potassium hydrate solution (for CO_2 and acid gases generally); alkaline pyrogallate solution (for oxygen); copper II (cuprous) chloride solution—either acid or ammoniacal (for CO); alcohol (for hydrocarbons); iodine solutions (for SO_2); potassium dichromate solution (for SO_2); iron II (ferrous) sulphate (for NO); acidified potassium permanganate solution (for NO); bromine water (for olefine hydrocarbons).

Modern techniques of mass spectrometry and gas chromatography are more accurate, reliable and rapid.

Gas carburising. A process for increasing the carbon content of the surfaces of ferrous components by heating in a carbon-rich gas or gases, usually CO or hydrocarbons, such as methane, butane or propane or a mixture of these gases. The carbon-bearing gases are mixed with air or manufactured gas. Iron or low-carbon steel when so treated develops a case or shell containing $Fe_3 C$, and this case gradually grows in thickness as the heating is continued. The presence of nickel retards the carburising, but chromium and molybdenum promote it. The surfaces to be carburised are usually machined, while the portions that are not to be hardened are protected by copper plating, enamelling, etc. The process is successful with small articles having complicated shapes, using rotary type furnaces, but for larger articles that would be damaged by tumbling, a vertical furnace is used and the gases are circulated within it.

Gas coal. A term sometimes applied to types of coal that, on distillation, give a good yield of gas of good illuminating power. Bituminous coals are divisible for commercial purposes into five classes, of which gas coal is one.

Gas coke. The solid residue produced by carbonising bituminous coal in a retort at about 1,000 °C in the process of gas making. It contains 83-90% fixed carbon, but is not so dense or hard as metallurgical coke. When produced by low-temperature carbonisation (400-800 °C) the coke may contain 6-12% volatile hydrocarbons, and is thus a useful household fuel. Coke made by the high-temperature carbonisation process (about 1,000 °C) is difficult to burn and is chiefly usable for metallurgical processes.

Gas constant. Designated by the letter R, the gas constant is the value of the quantity pv/T, where p = pressure of the gas, v = volume of gas and T is the absolute temperature. The value of R is 8.314 J mol^{-1} K^{-1}.

Gas holes. Small cavities, usually spherical in shape and with bright walls, caused by the release of dissolved gases during the solidification of a casting.

Gasket cement. In the workshop and laboratory cements for gaskets and similar applications usually contain asbestos and a bonding cement. This latter frequently consists of silicate of soda mixed with slaked lime, magnesia, lead oxide, barytes, fine sand or fireclay.

Gassing. A term that is applied to the absorption of gas by a metal. It is also applied to the evolution of gas by a metal on cooling and to the passing of gas through a molten metal.

Gas welding. A fusion-welding process in which neither pressure nor hammering is applied. The heat is produced by a fuel gas—oxygen flame, and the joint is formed either by the deposition of a filler metal on the joint or simply by fusing the welding surfaces together. The melting point of the filler metal must be very near to that of the parent metal, and usually a flux is required. The excess of this latter must be carefully removed after welding, using dilute acid and hot rinsing. In many cases the plates to be joined must have their edges chamfered. When a weld made with plates having bevelled edges cools, the surface of the deposit shrinks. This causes the slight raising of the plates on either side to an extent dependent on the angle of the V. Fusion welding using high-temperature oxygas flame in which neither additional filler metal nor flux is used is known as autogenous welding.

Gate. A term used in foundry work to indicate that part of the mould which serves as an inlet for molten metal into the mould or for the channels through which molten metal is led into the cavity of a foundry mould. The terms "gate" and "sprue" are synonymous.

Gate stick. A wooden or metallic rod used in foundries during the preparation of a sand mould. It is rammed up in the sand, and on removal leaves a channel through which metal can be poured into the mould.

Gathering. A defect sometimes encountered in packed-rolled steel sheet. Small areas of metal tend to stick to the roll, and to cause patchy markings at irregular intervals on succeeding sheets. The defect may be found on any metals which tend to weld or adhere to the rolls.

Gauss. The old unit of magnetic field strength now replaced by the tesla, T (10^{-4} gauss). A magnetic field is said to be unit strength at any point if a unit pole placed there is acted upon with a force of one dyne. If the medium has a permeability μ, then the field strength at a distance d from a N pole strength m is $H = m/\mu d^2$ gauss.

The intensity of a magnetic field is represented numerically by the number of lines of force per unit area (perpendicular to their direction). A field of strength H has H lines of force per unit area. In air 4π lines of force radiate from a unit pole. (See Table 1.)

Gay-Lussac's laws.—*First Law*—"When the pressure is maintained constant, all gases expand or contract by 0.00366 of their volume at 0 °C for each degree rise or fall of temperature respectively." By adopting the absolute scale of temperature, Gay-Lussac's law can be simplified as follows:"At constant pressure, the volume of a given mass of gas is proportional to the absolute temperature." This is known also as Charles' law, and is expressed simply as v/T = constant.

Second law.— "When gases react they do so in proportions by volume that bear a simple ratio to one another and to the product, if it be a gas."

Gaylussite. A hydrated sodium calcium carbonate, $Na_2 CO_3.CaCO_3.5H_2 O$, occurring as a saline residue. Mohs hardness 2-3, sp. gr. 1.94.

Geisdorffite. A sulphide-arsenide of nickel NiAsS. Was originally obtained from a cobalt mine in Halsingland, Sweden, and used by Cronstedt in 1751 in experiments that led to the isolation of nickel.

Gel. A weak solid, formed by a continuous network of material, with an interstitial filling of liquid. The viscosity is high, and gels must be on the borderline of definition between highly viscous fluids and weak solids. They have low values of Young's modulus (10% gelatine in water has $E = 24$ kNm^{-2}). Many gels show thixotropic behaviour, they do not resist a shearing force, and break up into sols if agitated sufficiently, with considerable decrease in viscosity. The same occurs on heating, but on cooling again the gel structure is restored, as it is by resting at constant temperature after shear thinning.

Some inorganic precipitates (formed in aqueous media) gel, but these do not liquefy on warming, and considerable heating is necessary to cause them to lose the retained water. Silicic acid and some metal hydroxides are in this category. When the water is evaporated away, they form xerogels or aerogels, and can absorb water again, acting as drying agents. This absorbed water can then be driven off more readily than the original. (See **Sols.**)

Geodes. Cavities, usually in igneous rocks, into which well-formed minerals project.

Geolin. The trade name of a polish, made in Germany, suitable for polishing aluminium and aluminium alloys.

Gerber's law. A mathematical expression of the relationship between mean stress and endurance range in fatigue testing. If M = mean stress; R = endurance range for a mean stress M; R_1 = endurance range for a mean stress O; and f = the ultimate tensile strength, then

$$R = R_1 \left[1 - \left(\frac{M}{f} \right)^2 \right]$$

Germanative conditions. A term applied to the condition of a metal which has undergone a certain critical amount of cold work that causes grain growth and the development of a very coarse grain structure when metal is heated. Sometimes also referred to as critical strain conditions.

Germanite. A germanium ore found associated with copper in South Africa. It contains about 5-10% Ge together with about 0.5% Ga.

Germanium. A comparatively rare metal analogous to tin in many of its physical and chemical properties. The mineral sources of the metal are argyrodite $GeS_2.4Ag_2S$, a rare mineral containing 5-7% of the metal; euxenite, complex ore containing mainly niobium and tin in which about 0.04-0.1% Ge is found; germanite, a comparatively new source of germanium in which the metal is found in amounts varying from 5 to 10%. It is often associated with zinc ores and is extracted from the zinc sludges. Flue dust from certain types of coal may contain up to 2% Ge.

The general mechanical and physical properties are given in Table 2.

The main interest in germanium lies in its electrical properties and in its alloys. In contradistinction to most metallic substances, the resistivity of germanium continually increases with decrease of temperature. The exploitation of this property for the making of metallic resistance thermometers for measuring low temperatures has not yet been made. Quite small amounts of impurities alter the resistivity considerably. Germanium is a semiconductor, and possesses unusual properties as a rectifier and a tin-germanium alloy is used in radar apparatus. Germanium crystals have found applications as modulators, voltage regulators, low-frequency oscillators and polarising devices. The metal is also well suited to the working of film resistors. Very thin films can be deposited on silver or directly upon glass, and a range of resistances from 10^3 to 10^6 ohms, depending on the size and thickness of the film, has been produced.

Germanium forms an alloy with gold (12% Ge) which has good colour and is harder than pure gold. It melts at 937 °C, and is therefore suitable for hot-dip processes of gold plating. A dental alloy (10% Ge) is hard and corrosion resistant. It casts well and expands on cooling. When added to aluminium it produces a better hardening effect than silicon, and the rolling properties are improved. Copper-germanium alloys do not show improved properties.

Crystal structure is diamond cubic. Lattice constant a = 5.647 $\times 10^{-10}$ m.

Getter. A material used in low pressure systems to remove the last traces of gases, especially oxygen, nitrogen and hydrogen. In sealed tubes such as electric filament lamps, thermionic valves, cathode ray tubes, the getter is placed in the envelope in the form of wire or ribbon, and a current is passed to volatilise it, after the envelope has been evacuated and sealed. The getter reacts with the gases to form a non-volatile and innocuous compound, or provides a film on the envelope or the components which can absorb the gases. Where transparency must be retained, there is greater difficulty, as few materials would react efficiently without impairing transmission as in electric filament lamps. Gas filled bulbs became more popular than vacuum bulbs partly for that reason.

In continuously evacuated systems, or those in which the vacuum is broken and restored at intervals, getters must be renewed, or provided with a separate circuit. The materials used are those with high free energy for formation of the appropriate compounds, such as magnesium, sodium, barium, niobium, tantalum, titanium, zirconium, rare earths.

Ghosh's law. Good electrolytes do not obey Ostwald's law (q.v.), which holds well for dilute solutions of weak acids and bases.

With good electrolytes if v is the dilution of the solution and m is the degree of ionisation, Ghosh's law states $\sqrt[3]{v} \log m$ = constant.

Ghost. The characteristic banded appearance seen in carbon steels that is due to the particular distribution of the ferrite. This distribution is greatly affected by the presence of impurities and inclusions. It appears as a faint, light-coloured streak that is visible only at certain angles of viewing and under suitable lighting conditions. It is caused by long local segregation of nearly pure iron (ferrite) in the steel, with a high proportion of phosphorus in solution as impurity and also sulphide inclusions. The appearance is visible on freshly machined surfaces and may be removed by further machining. However, a similar effect may appear at lower layers, so that machining may not completely eliminate the ghost lines. From the nature of these lines it is clear that they cannot be got rid of by any form of heat-treatment.

Gibbs free energy. The general equations expressing free energy at constant volume are usually attributed to Helmholtz. The one attributed to Gibbs is that referring to reactions in which pressure and temperature are constant, $G = E + pV - TS$. For spontaneous changes at constant pressure and temperature, the Gibbs free energy always decreases, and reaches its lowest value when the system comes to equilibrium. (See **Free energy**.)

Gibbsite. A hydrated oxide of aluminium $Al_2O_3.3H_2O$. It is crystalline in form and occurs along with amorphous bauxite. Mohs hardness 3, sp. gr. 2.35. Also called hydrargillite.

Gilding. A process of covering surfaces with gold either by mechanical or chemical means. In mechanical gilding the gold in the form (usually) of gold leaf is mechanically attached to surfaces using adhesives (gold size) or heat or mercury.

In the chemical processes gold chloride in ether solution is painted on the metal, which is heated till the metal remains and can then be polished. When gold amalgam is used the surface of the metal is treated either with mercury or mercuric nitrate before the amalgam is applied. The mercury is then evaporated and the surface burnished.

The gilding of pottery is accomplished by applying a mixture of gold chloride, bismuth oxide and borax in gum arabic and subsequently firing. Gilding by immersion is carried out by immersing the cleaned article in a solution of gold chloride and potassium bicarbonate.

The above methods have been superseded by the process of electrodeposition for small metallic articles.

Gilt. Silver or base metal having gold deposited thereon by chemical or electrodeposition process.

Gjer's kiln. A kiln designed especially for heating spathic iron ore. The kilns are built in rows and each has heavy cast-iron hearthplates to facilitate the movement of the charge. The charge consists of spathic ore $FeCO_3$, limestone and small-size coal or coke breeze. The latter is placed in thin layers between each charge or ore and limestone. The heating process is continuous and a charge is calcined in approximately four days.

Glass. The generic term for material which has been supercooled from a melt, and has not crystallised by a normal freezing mechanism. The material is therefore correctly described as a supercooled liquid, and its structure is amorphous, i.e. with short range order only. Accordingly, a glassy substance does not melt, but merely softens with rise in temperature, the decrease in viscosity being gradual and not sudden as with the melting of a true crystalline solid.

Glasses occur in nature, particularly in volcanic rock or rocks which have been subjected to high pressure or strain, and/or high temperature, so that a melt was produced, but subsequently no crystallisation. Obsidian is a natural silicate glass containing quartz, with felspar, and basalt, with no quartz content, can be partly glassy.

Obviously, compounds which are likely to form glasses are those in which it is fairly simple to induce supercooling. The silicates are typical of such compounds, but borates, phosphates and aluminates can also respond in this way. Even metals and metallic oxides have been produced with a high proportion of amorphous structure.

The synthetic glasses are in the main, silicates. The silica tetrahedra which form the basis of the pure silica structures, quartz, tridymite and cristobalite, arise from the association of a small silicon ion with two oxygen ions of much greater size, to form the basic SiO_2 composition. In forming tetrahedra each oxygen ion is shared by two separate silicon ions in two separate tetrahedra, so that every silicon is in fact associated with four oxygen ions, and $(SiO_4)^{4-}$ becomes the basic tetrahedral unit. (Since the oxygen ions are shared the basic composition is still one Si to $\frac{1}{2}O_4$, i.e. SiO_2). The structure is extremely open, and so other small ions can fit into the void spaces, and appropriate ions could substitute for silicon ions provided the size factor was acceptable, and electrical neutrality still preserved. Substances in which one or both of these mechanisms occur, form the silicates, a very wide range, many of which occur also in natural rocks.

The size of the metal ions involved determines the composition of a silicate and therefore the properties. Beryllium ions will easily replace silicon ions, being even smaller. Aluminium ions are slightly larger than silicon, but only a little distortion will accommodate them. However, aluminium ions will easily fit into the octahedral spaces in the close packed silica tetrahedra of tridymite or cristobalite, and the same applies to iron, magnesium and titanium. These metals therefore, form silicate structures with little distortion, and simply improve the close packing of the structure by filling in more spaces. The larger metal ions, sodium, potassium and calcium, can only enter the silica lattice with distortion, and can only occupy octahedral voids. Silicates containing them are therefore more open than pure silica lattices, or those containing Al, Fe, Ti or Mg.

The silica lattice is a hybrid, mainly covalent but involving some ionic forces too, in view of the great discrepancy in size of the two ions. More strongly ionised metal atoms, substituted for silicon, will alter the balance of forces, particularly when entering octahedral voids. Of the natural minerals, orthoclase, a felspar, has aluminium and potassium ions replacing silicon. The aluminium ion has a treble charge, the potassium single, so one of each in close proximity, the aluminium in the silicon position and the potassium in an octahedral void, will replace one quadruple charged silicon. In the mica group of minerals, aluminium again replaces silicon, but magnesium or iron fit into the octahedral spaces and, as they are double charged, one of them must be associated with two aluminium ions, or two of them must replace one silicon, to preserve neutrality. In garnet, calcium, magnesium or iron, all doubly charged, may be combined with aluminium to replace the silicon ions.

The introduction of larger ions, or the combination of two smaller ones (but greater in size than silicon) creates sufficient distortion in the lattice to hinder the process of crystallisation. The large ions of sodium and potassium and the slightly smaller than these, of lead, all encourage supercooling in silicates, and hence much of the commercial glass uses just these additions. (The difference in binding energy between the ions also affects supercooling.) Sodium ions are smaller than potassium, and soda ash (Na_2O) is a cheaper chemical than potash (K_2O), so soda glass is the cheap and softer glass used for rolling into window glass. Lead ions are smaller than calcium, but either can be added to soda glass for producing soft and easily worked material. The glass used for laboratory apparatus, often requiring blowing and much more fabrication, contains more lead than calcium. Hard glass for optical instruments, and for taking a fine polish, uses potassium. Sodium and potassium ions can actually migrate through the amorphous structure, by diffusion, and so a small degree of electrical conductivity can be detected in these glasses.

When the small ions are used, such as boron (B^{3+}) and phosphorus (P^{5+}) they are capable of replacing silicon in the network of tetrahedra, and so are called network-forming ions.

In the other cases, the larger ions like sodium and potassium distort the lattice, even in voids, and so are called network modifying ions. Those of intermediate size, Al^{3+}, Be^{2+}, Zn^{2+}, Fe^{2+} may be network forming or modifying, and calcium, magnesium, lead and titanium may act similarly. As the size of ions increases, or the electric charge decreases, the softening temperature is lowered because the bond strengths within the structure are reduced. The mechanical strength is lower for the same reason, and the resistance to chemical attack reduced. If halogen ions are substituted for oxygen, as in fluor-silicate glass, the lower charge on the fluorine ions also reduces strength and softening temperature.

The colour of glass is affected by metallic ions, even when very small quantities are present. Cobalt produces blue or pink colouration, depending on whether it acts as a network former or modi-

fier respectively. By varying the composition of glasses therefore, control of properties can be exercised sufficiently to produce tailor made materials.

Glass reinforced polymer (GRP). (Variously expressed in terms of polymer or more specifically as polyester). The material obtained by bonding fibres of glass, with a polymeric matrix which can wet the glass and serve as a barrier to crack propagation from one fibre to the next. The material is therefore a composite of high strength and often chemically inert to many environmental conditions. (*See* **Composite material.**)

Glass structure. *See* **Crystallisation** and **Condensation of liquids.**

Glass transition. Refers to polymeric materials which are not highly crystalline, but differs from the glassy state of silicates and similar materials, in that transition from liquid to supercooled state is *not* involved. In polymers, the glass transition applies to materials which are inhibited from crystallisation by virtue of difficulty in achieving a regular arrangement of polymer chains, either due to strong intermolecular forces at localised points, or the geometry of the bulky side groups on the chain, or both.

When such materials are above their glass transition temperature, the polymer chains are readily moved relative to one another, or can be stretched elastically when coiled or folded. Below the glass transition, the elasticity is drastically reduced, and the material becomes more like a rigid solid. The forces which operate between the molecules have obviously increased, and the capacity for rotation, uncoiling and other elastomeric behaviour, has almost disappeared. The transition is completely reversible, which confirms that no chemical or drastic physical change has taken place, simply the viscosity of the material has been enormously increased.

Glass wool. A resilient, white, fleece-like product consisting of fine fibres of glass, diameters of which vary from $9\,\mu m$-$25\,\mu m$. Molten glass is allowed to pass through platinum alloy bushings in the melting tank. These bushings contain very fine holes, and as the glass issues in a fine stream it is caught by jets of high-pressure steam or air. The fineness or coarseness of the fibre produced depends on the size of the orifice, the pressure exerted by the jets and the temperature. (*See* **Fibres.**)

Glassy polymers. *See* **Crystallisation** and **Condensation of liquids.**

Glauconite. A hydrated silicate of potassium and iron. It is essentially the same material as the green iron silicate that is the chief constituent of many iron ores of marine origin. Mohs hardness 2, sp. gr. 2.2-2.4.

Glazing. A process of polishing iron and steel by means of a rapidly rotating, well-worn, emery bob, the abrasive action of which has been reduced to a minimum. Also refers to a process for producing a smooth finish on wiped (soldered) joints by passing a red hot iron over the surface. The process of fixing a glassy coating on pottery, by fusion of metallic oxides with silica or similar compounds.

Glazy pig. A grey pig iron containing 10-12% Si.

Glide plane. In the plastic yielding of crystalline solids under stress the regular orientation of the particles in a crystal causes slipping of the particles over one another more readily in certain directions than in others. These preferred directions are the glide planes, along which slipping may occur without fracture. The translatory motion may take place along only one plane or along a number of planes parallel to one another. A crystal may therefore possess several distinct systems of glide planes. The successive small slips along glide planes of the same system give rise to slip bands.

Globar. Silicon carbide used in the form of rods for resistors in furnaces working at temperatures up to 1,250 °C.

Globular cementite. The globular condition of iron carbide Fe_3C. (*See also* **Spheroidal cementite.**)

Glo-crack. A method of crack detection by fluorescence.

Glost oven. A furnace in which refractory materials, pottery or enamel ware are fired.

Glutamate flux. A type of flux, which at soldering temperatures yields a neutral salt. It is much more active than resin and less corrosive than zinc chloride. It consists of 400 g urea in 1,500 cm³ of water, added to a solution of 700 g of glutamic acid hydrochloride in 3,500 cm³ of water. The mixed solutions must be heated to about 80 °C and then stored.

As the stored solution is slightly acid and does not, in fact, become completely neutral until at the soldering temperature, it should not be allowed to flow on to the parts that will not be heated. Both the flux and its products on heating are water soluble. It is effective for soldering brass, copper, silver, steel and many non-ferrous alloys, and for electrocoatings of zinc, nickel, tin, cadmium and silver.

Gluten. A sticky nitrogenous substance, obtained from potatoes, flour or other similar substance rich in carbohydrates. It is used as a "core-gum" in sand moulding, since the gluten acts as a binder without increasing the fusibility of the sand.

Ordinary flour contains about 10-12% gluten, which may be separated by kneading the flour, contained in a fine muslin bag, with water. When stored for any length of time, gluten decomposes and becomes disagreeable in appearance and odour.

Glyptal resins. Synthetic resins that are formed by condensation of polyhydric alcohols and polybasic acids. It is found that the products of condensation of dibasic acids and dihydric alcohols are usually thermoplastic, while if a trihydric alcohol is substituted for a dihydric alcohol (e.g. glycol) the product will be thermo-hardening. These latter substances are widely used in various metal-coating compositions, since they dry and harden on stoving, without oxidation.

Gneiss. An important type of metamorphic rock, characterised by its holocrystalline nature, and as a rule by a coarse foliation and abundance of felspar. The term, which is of Slavonic origin, meaning "decomposed" in reference to the altered character of the rock in the vicinity of ore veins, is used generically to signify a large and varied series of rocks having a banded structure. Layers of minerals having a granular structure alternate with thinner layers of fibrous mineral or sometimes the layers may consist of granular minerals of different compositions. The minerals of the granular bands may consist of quartz or felspar or both. The fibrous or lamellar bands usually consist of chlorite, mica, graphite, amphibole, sillimanite, etc.

Goethite. A crystalline hydrated oxide of iron $Fe_2O_3.H_2O$ which is often associated with limonite and hematite in several deposits, e.g. Westphalia in Germany and in Cornwall. It varies in colour from yellow to dark brown, but the colour by transmitted light is often blood-red. It crystallises in the orthorhombic system. Mohs hardness 5-5.5, sp. gr. 4-4.4. It is identical with lepedocrocite and pyrosiderite.

Gold. A widely distributed precious metal of pleasing yellow colour that is most frequently found in the native condition, i.e. as metal, but invariably alloyed with silver or copper and occasionally with bismuth, mercury and other metals. Natural gold has been known to contain as much as 99.8% Au, but as a general rule the gold content ranges from 85% to 95%, the balance being for the most part silver. Gold also occurs in vein deposits as tellurides. The principal sources of gold are South Africa (Transvaal), Zimbabwe, the USA, Australia, Canada, Russia, Mexico and India.

Native gold occurs as nuggets or as grains in alluvial sand. It is recovered from the alluvial deposits by some form of water concentration, followed by amalgamation. In vein mining the gold-bearing quartz obtained by blasting is crushed to fine powder in stamps or mills, and the gold is extracted by amalgamation, by leaching with potassium cyanide (0.25-1%) solution—a process devised by McArthur and Forrest. For complex ores in which gold is intimately associated with base metal sulphide minerals, gravity concentration or flotation is used as a preliminary treatment. The concentration can then be roasted or cyanided.

Gold is refined by chlorine treatment, electrolysis or by cupellation. In the compact form gold has a characteristic yellow colour, but in the finely divided state it may appear red or purple.

When pure it is the most malleable and ductile of all metals and can be beaten into foil 120 nm thick or drawn into extremely fine wire. Deposits on gold lace are only about 2 nm thick. Very thin films, not more than 0.1 μm thick, transmit green light. On heating gold leaf the gold recrystallises and the film then transmits red light.

Colloidal gold is formed by reducing solutions of gold chloride. The reducing agents used are iron II sulphate, phosphorus, methanal, hydrazine, etc., and according to the size of colloidal particle the solutions vary widely in colour. The solutions containing the larger particles are blue, and as the size decreases the colour becomes successively purple, ruby-red and finally yellow. If the reducing agent is a mixture of tin II (stannous) chloride with tin IV (stannic) chloride, a purple powder called "purple of Cassius" is formed. This is used in the making of a very fine ruby glass.

The electrical conductivity of gold is about 67% that of silver, and the thermal conductivity is 72% that of silver. The other physical properties of the metal are given in Table 2.

Gold was formerly used for coinage, but is now chiefly used for jewellery, and is usually alloyed with silver or copper to increase hardness and improve wearing properties. "Fineness" is expressed in parts per 1,000. Standard British coinage was 916.6 parts of gold and 83.4 parts of copper (22 carat gold). USA coinage is 900 fine. The legal standard for jewellery is the carat and in UK 22, 18 or 9 parts per 24 parts are usual. In other countries different standards exist which include 12 and 14 parts per 24 parts in addition to the above.

The metal is used in electroplating and as "liquid gold". The latter is a suspension of finely divided gold in essential oils and is used in the decoration of pottery.

Gold is permanent in air and water under all conditions of temperature. It is unattacked by any single common acid, but readily dissolves in *aqua regia* (a mixture of nitric and hydrochloric acids). In presence of air the metal dissolves in potassium cyanide solution to form a double cyanide $K.Au(CN)_2$. The yellow colour of pure gold is considerably modified by the presence of alloying elements. Small amounts of silver produce a pale yellow alloy, while amounts of the order of 35% Ag yield an alloy having a distinct greenish tint. Copper, on the other hand, produces a red colour. Platinum in amounts of 20-25% and nickel give a white alloy known as "white gold", while about 12% Pd produces an even whiter alloy. With aluminium (about 20%) a purple alloy is produced which, however, has not found useful applications on account of its brittleness.

Gold amalgam. Gold readily combines with mercury to form amalgams. Those in which mercury is in large excess are liquid, but when 15% or more of gold is present the alloy is solid.

Crystalline amalgams of varying composition can be prepared artificially, while others occur naturally. The latter may contain gold up to approximately 40% and silver is also usually present. Natural gold amalgam is found in California, Columbia, the Urals and Victoria (Australia). It is often associated with platinum.

Gold-beaters' skin. The fine membrane used to make the leaves of the shoder and the mould for the production of gold leaf. It is prepared from the outer coating of the blind gut of oxen. The degreased skins are strongly stretched before being glued together in pairs. They are then treated with camphor and egg albumen and cut to size, about 15 cm square.

Gold-beating. The process for producing gold leaf by beating sheet gold, first between sheets of vellum or very tough paper and then between sheets of gold beaters' skin. The sheet gold is rolled down until it is of such thickness that 10 g will yield about 25 sheets each 4 cm square and about 0.1 mm thick. The final beating yields about 1,200 sheets roughly 15 cm square. These are trimmed down to about 10 cm square and are made available in books of 25 leaves. The gold in such a book weighs, on the average, 0.3 g or 4.5 grains.

Gold bullion. Gold regarded as an article of merchandise, and not as coin. It is the metal after refining and after being brought to a certain standard of purity, but has not been converted into coin or jewellery.

Gold cased. Silver or base metal having on its surface an electrodeposit of gold of such density that neither nitric acid of specific gravity 1.4 nor silver nitrate applied for 2 minutes will discolour or stain it.

Gold coinage. Gold is too soft for use alone in jewellery or coinage, and it is always alloyed for these purposes with copper and/or silver. The reddish colour of the English gold coin was due to alloying with copper, while the Australian sovereign, which had a pale yellow colour, contained silver. The English gold coin (22 carat) was 916.67 fine (i.e. contained 916.67 parts of gold per 1,000 parts of the alloy). American, German and Italian standard for gold coins was 900 fine (i.e. 21.7 carat gold).

Gold filled. A sheet of gold of a standard not lower than 9 carats, sweated or soldered to each side of a bar of silver or base metal, the whole being rolled down together. In the case of wire the core is of silver or base metal which is completely surrounded with gold of a standard not lower than 9 carats. The gold is sweated or soldered on, the whole being drawn down together in such a way that on annealing and pickling a surface of gold will be shown.

Gold front. A term used in the jewellery trade, descriptive of those articles that are made from a sheet of gold, of standard not lower than 9 carats, which is made in a separate portion and afterwards attached to a silver or base-metal back. The attachment is such that the sheet of gold can be removed from the silver or base metal.

Gold leaf. Thin sheet of gold chiefly used in gilding. It is prepared by first casting the metal in small ingots, rolling down the ingots into a ribbon each 3 m long and 4 cm wide per ounce of metal; cutting the ribbon into small square pieces; piling the squares between sheets of special paper, about 150 at a time, and beating with a heavy hammer till each piece is about 10 cm square; cutting each piece into four; piling and beating again between sheets of gold-beaters' skin; cutting again into four; piling again and beating till the pieces are 10 cm square before trimming. The thin leaves of metal, which are transparent to green light, are about 74 μm thick.

Gold number. The weight in milligrams of an emulsoid which just fails to prevent the change from red to violet when 1 cm^3 of a 10% solution of sodium chloride is added to 10 cm^3 of a gold solution of concentration 0.0055%. Emulsoids in general exert a protective action on suspensoids in regard to their precipitation by electrolytes. It is supposed that the emulsoid particles surround the solid suspensoid particles and prevent their coalescence. A small quantity of gelatine, for instance, will prevent the coagulation and precipitation of gold in colloidal solution. The degree to which this protection is afforded to a colloid is given in terms of the gold number. For gelatine the gold number is 0.005.

Gold paint. Paints made with bronze powders and transparent varnishes or amyl acetate. They vary in colour from yellow to brown, and do not contain any gold or gold alloy.

Gold plating. The electrodeposition of gold from solutions of gold cyanide in potassium cyanide. The process is carried out in much the same way as silver plating, but the requisite amounts of silver and copper salts are added so that these metals are deposited as an alloy with gold if a suitable voltage is used.

Goldschmidt's process. A process in which the great evolution of resulting from the combination of aluminium with oxygen for reducing metallic oxides, e.g. Fe$_2$O$_3$, Cr$_2$O$_3$, MnO$_2$. of molten steel in welding broken articles, e.g. ors, *in situ*, a mixture of aluminium powder thy scales) is placed in a crucible above the are is ignited by a magnesium wire and a Molten iron covered with a layer of and is tapped from the bottom of the art to be welded. by the reduction of the sesquioxide the chromium forms a fused mass a purity of about 99.5%. A very pure ase of manganese if the oxide Mn$_3$O$_4$ The process is also used for the prod- c alloys. ferred to as the thermit process and

d of a standard not lower than 9 carats, the case being of such thickness that i in acid the gold shell is left intact.

AuTe$_2$ is found associated with silver s petzite (Ag,Au)$_2$Te, sylvanite (Au,Ag)Te$_2$. ciated with lead and antimony in nagyagite 17. These, together with krennerite, are australia, Colorado and Dakota.

strument for measuring the interfacial angles of tact instrument is mainly used to obtain an approximate measurement of the angles of large crystals. The reflecting goniometer is always used for the accurate measurement of small crystals. The bright smooth surface of a crystal will reflect a sharply defined image of a bright object. When the crystal is turned about an axis parallel to the edge between two faces the image reflected from the second face can be brought into the same position as that formerly occupied by the image reflected in the first face. The angle through which the crystal has been rotated is the supplement of the angle between the faces of the crystal.

Gooch crucible. A crucible of platinum or porcelain having a perforated base and used in rapid filtering. The crucible is fitted to a small funnel by means of a broad piece of rubber tubing and the bottom of the crucible is provided with a layer of asbestos-felt. A filter pump is used in conjunction with this crucible.

Goose-neck. A bent air-duct or penstock that is fire-brick lined and joins the bustle-pipe to the belly pipe of a blast furnace. It conveys the hot blast to the tuyères.

Goslarite. A hydrated sulphate of zinc ZnSO$_4$.7H$_2$O that results from the decomposition and oxidation of blende. It is found at Goslar (in the Harz, Germany) and in certain other zinc mines. It is isomorphous with Epsom salts.

G-P (Guinier-Preston) zones. In the precipitation of aluminium-rich compounds with copper, from duralumin type alloys (aluminium containing about 4% copper) the X-ray structures studied by Guinier and Preston showed reflections from the lattice along $<100>_{Al}$ directions. This was attributed to copper rich platelets on $\{100\}$ planes in the aluminium matrix.

Further study showed several stages in the precipitation, commencing with the above platelets, now called GP[1] followed by ordered zones of GP[2], and then a phase known as θ'. This, together with the equilibrium θ, or CuAl$_2$ precipitate, leads to incoherency with the matrix, and softening of the structure, whereas the first two stages, the GP[1] and GP[2] zones are coherent, and give rise to the strength of correctly aged material.

Similar GP zones are formed in aluminium alloys containing magnesium and silicon, and in those containing magnesium and copper. The platelets of GP[1] are clusters of copper atoms segregated on to the $\{100\}$ planes of the aluminium matrix, and only a few atomic planes thick, but about 10 nm long. The GP[2] zones are really a coherent intermediate precipitate, of maximum thickness 10 nm and up to 150 nm in diameter. The structure is tetragonal, fitting perfectly in the a and b directions with aluminium, but not in the c.

Graham's law. Graham's law of diffusion states that the velocity of diffusion of gas is inversely proportional to the square root of its density.

In his experiments on liquid diffusion Graham showed that acids and salts diffuse rapidly, whereas such substances as glue, albumin and starch diffuse only very slowly. The rapidly diffusing substances (except acids) were all crystalline in the solid state and were named, by Graham, crystalloids. Since gum, albumin etc. form amorphous solid masses, these were called colloids. Crystalloids can be separated using Graham's dialyser.

Grain. A term which, when used in reference to pure metals and homogeneous solid solutions, is synonymous with crystal. In duplex alloys the term is more usually applied to type or size of structure. In fractures the term refers to size of facets.

Grain-boundary attack. In the etching of the polished surface of a metal to develop the crystal facets, it is often found that the etchant has a specific attack at the junctions of crystals, developing dark lines round the grains. This is referred to as grain-boundary attack corrosion, and is due to the presence of a trace of impurity or variations in structure. In the case of a metal containing impurities, which has been rolled, there may result layers of different compositions. Grain-boundary attack then proceeds at different rates and may cause what is termed zonal corrosion, the evidence of which is seen in the flaking and the swelling of the metal. Generally speaking, corrosion is slower in coarse-grained material than in fine-grained material, but the reverse is sometimes true, e.g. pure tin, tin bronzes and pure iron.

Grain growth. An increase in the average grain size resulting from certain crystals absorbing their neighbours. The fundamental principle in connection with grain growth is the same as that governing the production of a single crystal. Large grains result from recrystallisation proceeding gradually from one point to another. Small grains are produced when recrystallisation starts from many points simultaneously. One of the chief factors in the starting of the recrystallisation is the presence of impurities— a pure metal or pure alloy is a prerequisite for the production of large grains. Tiny particles of substances such as oxides, carbides or other precipitated substances constitute nuclei around which crystallisation proceeds in all directions and small grains result.

The existence of allotropic transformations makes the growing of large grains difficult. If the metal is heated above the transformation temperature, it will of course recrystallise on cooling, and as it is not possible to restrain the transformation starting simultaneously at a number of points, the result is always grain refining.

Grain growth takes place at elevated temperatures, particularly after severe cold working. Grain size can be refined by suitable heat-treatment consisting of heating to a temperature slightly in excess of the temperature of recrystallisation. Prolonged heating at such a temperature, or heating to higher temperatures favours further grain growth. But grain growth does not take place to any appreciable extent *during* the heat-treatment of metal castings (or cast metals that have not been plastically deformed after casting) except in the case of iron and ferrous alloys which can recrystallise on heating above Ac_1.

After plastic deformation, hot or cold, metals exhibit grain growth at temperatures above the temperature of recrystallisation. It proceeds by the absorption of grains—usually the smaller ones by the larger ones—and a decrease in grain boundary area results.

Straining a polycrystalline material above its elastic limit followed by annealing at elevated temperatures, but not above its transformation temperature (if any), gives large grains. The straining beyond the elastic limit produces lattice distortions within the grains and on the surface and hence a tendency towards crystallisation. There probably exists a definite relationship between the cold deformation and grain growth. For unalloyed iron the greatest grain growth seems to occur above about 2.5% cold deformation and by subsequent annealing at 880 °C for some 60 hours. In the case of silicon irons containing more than 2% Si, and having no transformation, the annealing can be done at any temperature. Large grains are readily produced in 4% Si-Fe sheets by a small amount of cold rolling followed by annealing at 1,000 °C.

Graining. A process for the production of metal powder by violently agitating the molten metal during solidification. The product almost always requires grading.

Grain refining. A process in which the grain size of a metal or alloy is markedly reduced. The refining may be brought about by suitable heat-treatment, mechanical treatment or any addition of small quantities of alloying elements. In the case of steel, grain refining or normalising is carried out by heating the metal to, and holding it at, a suitable temperature (e.g. about 50 °C above the transformation range), followed by moderately rapid cooling in still air at room temperature. Not only is the grain size thus refined, but internal stresses are relieved and mechanical properties improved. Such treatment increases the tensile strength and raises the yield point, while the reduction of area and elongation are slightly reduced. Vanadium is a very active grain refiner for steel; titanium and niobium are also used in the light alloys as well as in steel. Boron is an excellent grain refiner for aluminium and aluminium alloys.

Grain rolls. Sand cast iron rolls, used for roughing and intermediate rolling. The surface is often densed by casting into chills or denseners faced with an appreciable thickness of sand or other refractory material. They may contain small amounts of alloying elements such as nickel, chromium, molybdenum and vanadium, but are usually plain cast iron with about 2.7-3.7% C, 0.7-1.2% Si and 0.5-1.5% Mn. The sulphur and phosphorus content should be low.

Grain size. When a metal is cooled slowly and undisturbed from a high temperature, it will, in general, show a coarsely granular or crystalline structure. The size of the grain is a function of the temperature and the time during which the metal is held at the maximum temperature, and the rate at which it is cooled. In large masses the structure tends to be coarser at the centre than

at the surface owing to the difference in the rate of cooling. In order to overcome this difference and produce a homogeneous uniform material, the metal may be worked during the period in which grain growth would normally take place. Steel that has been hot worked down to Ar_1 will show a fine grain and this will be accompanied by greater strength and higher ductility than the unworked metal. Steel that has been worked below the Ar_1 range, i.e. cold worked, will show considerable grain distortion and may even become hardened. Cold rolling develops a weak laminated structure. Grain-size control is also obtained in the treatment of steel, e.g. by the addition of aluminium up to 0.05%, in order to control the rate at which austenite grains grow when heated above the critical range. Vanadium and molybdenum both reduce the rate of grain growth on heating, as compared with straight carbon steels.

The A.S.T.M. has issued a standard classification of grain size and Fig. 73 gives the idealised form for austenitic steels. These are magnified 100 times, and the actual grain size is given in the following table. The manner of establishing (carburising or heat-treating) and the methods of revealing the grain boundaries are included in the specification.

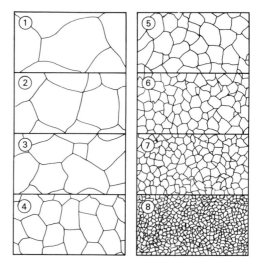

Grain Size No. (A.S.T.M.)	No of Grains per sq. in. (Actual) as Viewed at × 100 Diameters	Diameter of Grain (Actual) Regarded as Spherical	
		mm.	in.
1	¾— 1½	0.287	0.0113
2	1½— 3	0.203	0.0080
3	3 — 6	0.144	0.00566
4	6 — 12	0.101	0.0040·
5	12 — 24	0.0718	0.00283
6	24 — 48	0.0507	0.0020
7	48 — 96	0.0359	0.00142
8	96 —192	0.0254	0.00100

Fig. 73 *A.S.T.M. standard classified grain size (× 100 approx.).*

Granite. An igneous rock in which the essential constituents are quartz (more than 66%) and felspar. Typical granites may contain in addition, muscovite and biotite or green hornblende and sometimes plagioclase felspars. The different varieties are named biotite granite, hornblende granite, tourmaline granite, etc., according to the particular non-essential constituent present. Topaz-granite is called greisen.

Granodising. A process for producing a layer of zinc phosphate on zinc or on zinc-coated steel. It is usually effected by dipping into a hot phosphate bath. The coating can also be produced electrochemically and the dark deposit so obtained affords a very good basis for painting.

Granosealing. A phosphate treatment for reducing wear and scruffing, particularly between piston rings and cylinder walls of internal combustion engines.

Granular ash. A form of sodium carbonate Na_2CO_3 for use in foundries.

Granular cementite. *See* **Granular pearlite.**

Granular pearlite. A structure produced when pearlite steels are annealed for prolonged periods at temperatures below, but approximating to, the lower critical point, so that the cementite lamellae are transformed into spheroids of cementite in a ferritic matrix. Also known as "divorced pearlite" and sometimes as "globular pearlite". It is preferably described as granular cementite or spheroidised cementite.

Granulated metal. Small irregular pellets of metal produced by pouring fine streams of molten metal through a wire mesh into water or by dispersing through a rotating disc.

Granulation. A process whereby the columnar crystalline dendrites of a freshly solidified metal are recrystallised to produce more or less equiaxed crystals at temperatures near the solidus.

Granulite. A rock in which the minerals are in rounded, closely packed grains. This probably results from thermal metamorphism or the grinding down by movement of an igneous rock.

Graphic tellurium. Telluride of gold and silver $(Au.Ag)Te_2$ containing about 24.5% Au and about 62% Te. Antimony and lead are sometimes present. Crystal system—monoclinic; Mohs hardness 1.5-2, sp. gr. 5.73-8.28.

Graphite. An allotropic condition of carbon that crystallises in the trigonal system as grey shining hexagonal plates. The bonding in the hexagonal sheets of the layer lattice is covalent, an alternating single and double bond creating the linked network (*see* Fig. 74). Between the layers, van der Waals' attraction results from the polarised nature of the carbon rings. The double bonds (three per hexagon) resonate around the ring, so that all four bonds of any one carbon atom will be on average double bonds for one quarter of the time scale. Because the van der Waals' forces are much weaker than the covalent bonds, the layers are readily separated, hence the use of graphite as a lubricant. The structure is electrically conducting within the layers, because of the bond resonance, and in a solid lump, there will be artefacts at all possible angles, so that anisotropy of conduction may be hardly noticeable. When extruded into special shapes however, with a large preferred orientation of the layers in the extrusion direction, then anisotropy of electrical conduction, thermal expansion and conduction, and mechanical strength, will become marked.

Graphite is found naturally in Cumberland (UK), Siberia, Sri Lanka, Bohemia (Czechoslovakia) and California. It is made artificially by the Acheson process, sand and powdered coke are heated in an electric furnace and the silicon carbide at first formed decomposes into silicon and graphite.

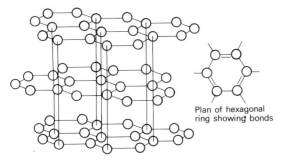

Plan of hexagonal ring showing bonds

Fig. 74 The graphite lattice and double bonds.

Graphite has a Mohs hardness of 1-2 and a specific gravity of 2.1-2.6. It is a good conductor of heat and electricity. It burns only at a high temperature. On account of these properties it is used in the cores of arc-carbons and as anodes in electrolytic cells, as the conducting deposit that covers the wax moulds on which copper is deposited in electrotyping, and when mixed with sand and clay it is used for making crucibles. A granular mixture of graphite, carborundum and clay is used as a resistance in electric furnaces under the name Kryptol. Graphite as a "dag" ("Aquadag" and "Oildag" being proprietary names) with water or oil is used as a mould dressing or for lubrication in metal-working and other processes.

Metallurgically, graphite is free elementary carbon that occurs in iron and steel. It is brittle and is often partially removed by polishing operations, leaving cavities. In unetched specimens, graphite areas appear black. Etched specimens often retain some of the etchant in the graphitic areas and cavities, if not thoroughly washed. This residual etchant creeps over the specimen and causes staining.

Graphite occurs in pig iron and in grey cast iron as flakes, which separate from the molten metal during and immediately after solification. When separating from molten cast iron, graphite appears as thin plates. On the other hand, temper graphite (or temper carbon) resulting from the decomposition of cementite during annealing appears as small rounded particles surrounded by ferrite.

Graphite softening. A corrosion effect often noted on cast-iron pipes and generally associated with bacterial corrosion. When chemical action is set up by the bacteria in the vicinity of the pipe it is found that the graphite and certain other constituents in the metal remain unaffected. The surface layers are found to consist of graphite, etc. embedded in an iron sulphide which is very soft and can readily be cut. The effect is probably due to sulphate-reducing organisms that liberate sulphuretted hydrogen and this latter sets up a chemical corrosive action.

Graphitic carbon. *See* **Graphite.**

Graphitisation. The term is sometimes applied to an annealing treatment, whereby some or all of the carbon of a cast iron is converted to free graphitic carbon.

Gravitational segregation. A form of macrosegregation in which constituents having different densities separate under the influence of gravity, or of centrifugal force in the case of centrifugal castings, during the solidification of the alloy.

Gravity cell. A cell in which the force of gravity (or centrifugal force) can be converted into electrical energy. In a gravity cell, barium chloride (concentration 17%) with the electrodes 3.77 m apart, an e.m.f. gradient of 0.64 $\mu V\ cm^{-1}$ is obtained. Higher voltages are obtained in tubes laid along the radii of a horizontal disc and rotated at high speed. The e.m.f. produced in gravity cells is too small to be of practical use, but the phenomenon is taken as striking evidence of the existence of free ions in an electrolyte.

Gravity die casting. A process in which castings of various metals and alloys are made by pouring metal in permanent moulds. It is so called in contra-distinction to *pressure* die casting in which the molten metal is injected into moulds under pressure. The metal is sometimes poured by hand and the dies are made of steel, cast iron or bronze. The method gives closer dimensional accuracy and a smoother finish than sand casting and the dies can be used repeatedly.

Grease marks. An undesirable marking sometimes found on charcoal grades of tin-plate and on other heavily tin-coated sheet. The defect is associated with the palm-oil used for fluxing, and can usually be removed by polishing.

Greaves-Etchell's furnace. A direct-arc furnace of the "bottom electrode" type. The current, after arcing between carbon electrodes and the charge, passes through the melt to a copper conductor embedded in the furnace hearth.

Green gold. (*a*) Electrodeposited gold containing small amounts of cadmium or silver which are added to produce the desired colour. (*b*) the addition of silver to gold reduces the reflectivity in the extreme red of the spectrum. An 18-carat gold containing 25% Ag has a good green colour. 12% Cd or 4.3% Cd together with 9.7% Cu and 11.5% Ag also gives a good green colour in 18-carat gold. With gold of fineness less than 750 the addition of copper and silver in varying quantities can be used to develop the greenish colour.

Green rot. A type of failure due to corrosion attack met with in high temperature nickel-chromium or nickel-chromium-iron alloys. It is characterised by a greenish fracture and is generally attributed to intergranular oxidation and formation of oxide within the grain accompanied by general swelling or growth of the section. Failures of this type occur when components such as heating elements are exposed to high temperatures in certain types of atmospheres. Those that appear to be most corrosive are those containing CO_2, H_2O and H_2, such as occur in carburising atmospheres or when coal gas is only partially burnt. The green rot, which may be due to carburisation of the chromium followed by oxidation of precipitated carbide, can be substantially reduced by (a) increasing considerably the percentage of iron and (b) the addition of about 2% Si to the alloy.

Greensand. A term used in geology to indicate a sedimentary sandstone having abundant grains of glauconite. In foundry work the term is used to described the sand used for moulds that are to be used without drying. It consists of an intimate mixture of ground sand and coke dust that is made moist and moulded but is not subsequently baked. Sometimes the *surface* of the mould is dried by means of a torch after the pattern has been removed. More frequently a flammable wash is applied by brushing, to the mould surface. When this is ignited, a hard surface is produced. Greensand is most frequently used in steel foundries. The main desirable properties of a good greensand are (a) it should be a sharp bank sand; (b) it should contain the minimum amount of clay that is sufficient to bind the particles together; (c) it should have sufficient porosity to allow the escape of gases when the mould is being filled with molten metal.

Green vitriol. A crystalline hydrated sulphate of iron $FeSO_4.7H_2O$ that is green in colour. It crystallises in the monoclinic system, has a Mohs hardness of 2 and specific gravity of 1.83. It results from the decomposition of pyrites and is found wherever pyritic iron ores occur. It is used by tanners and dyers and in the manufacture of inks, since it yields a good black colour with tannic acid. When treated with potassium iron III cyanide it yields Prussian blue.

A solution of green vitriol, $158\,g\,l^{-1}$, and a solution of anhydrous sodium carbonate, $60\,g\,l^{-1}$, when mixed in equal quantities has been recommended as an antidote for cyanide poisoning in metallic plating shops. The crystals tend to oxidise in a moist atmosphere, producing iron III (ferric) hydroxide.

Greisen. A modified granite containing quartz and white mica but no felspar or biotite. The accessory minerals are usually topaz, tourmaline, fluorspar and oxides of iron. It has been formed from granite by the action of fluids or vapours containing fluorine, boron, etc., rising through fissures in the granite.

The rock frequently carries small percentages of tin oxide, the amount of which is economically workable (e.g. Saxony, Tasmania), and often wolfram is found as veins in greisen bands.

Grey antimony. A sulphide of antimony Sb_2S_3. Also known as stibnite and antimonite.

Grey forge pig. A fine grain hematite pig iron containing little or very little sulphur and phosphorus, and rather less silicon than other grey irons.

Grey iron. A pig or cast iron which has a dark grey fracture. In cast irons, if the silicon or phosphorus content is high the carbon content is usually low. Manganiferous irons usually have a high carbon content. The colour of the fracture of pig iron is dependent on the state in which the carbon is present. When all, or nearly all, the carbon is present as iron carbide (cementite), the fracture is white. If a high percentage of the carbon is in the form of graphite, the fracture is grey. When the carbon present is intermediate between these two extremes, the irons are termed mottled. Grey irons are relatively soft and are easily filed, chipped or machined.

On polished specimens the graphite may be seen as long black threads (due to graphite separation from liquid) or as short threads, or as rounded particles. As a general rule cementite and pearlite are both present, in addition to graphite and ferrite. Picric acid as an etchant reveals the pearlite, while alkaline picrate reveals the cementite.

Grey slag. Slag produced when lead ores are smelted on an ore hearth. It contains a considerable quantity of lead and is usually subjected to retreatment.

Griffith crack. *See* **Crack.**

Grinding. A process for removing small amounts of metal from a metallic surface, usually by rapidly rotating abrasive wheels in contact with the surface. The wheels consist of an abrasive substance embedded in a bonding material, the tenacity with which this latter holds the cutting particles being the factor that decides the "grade" of a grinding wheel. The "grain" of a wheel depends on the coarseness of the cutting particles as determined by standard screens or sieves.

Grinding cracks. A defect that may occur during the removal of metal by abrasion. Since grinding is usually carried out on hardened components, the material is in a highly stressed condition, and uneven removal of excess stock tends to redistribute stresses and cause warping or cracking. Overheating during grinding, rapid grinding, loaded wheels or inadequate cooling lead to cracking. Grinding cracks usually show a typical fine criss-cross or crazy check pattern. The cracks are most likely to occur with high carbon or case-hardened material.

Grinding wheel. A wheel composed of a cutting material embedded in a bonding material used for cutting, grinding or finishing metals. Formerly emery and corundum were the only abrasives available, but many artificial products, e.g. carborundum, alundum, aloxite, etc., are now used. The bonding material may be vitrified clay, silicate of soda, shellac or vulcanised rubber.

Grit. Angular particles of sand, siliceous matter or metal used as an abrasive or cleaning agent, particularly in grit blasting.

Grizzly. A device used for coarse screening of ores. It consists essentially of a number of parallel bars over which the crushed ore is passed.

Grog. Non-plastic material, such as old burnt and crushed fireclay, used in the manufacture of refractories. It is added to refractory batches to reduce shrinkage and increase resistance to thermal shock

Gross calorific value. The heat produced by the combustion of unit quantity of fuel in an oxygen atmosphere at constant volume in a bomb calorimeter. The general conditions of the determination include an initial pressure of oxygen of 0.2-0.4 MN m^{-2}, and a final temperature in the range of 20-35 °C, so that the final products are ash, liquid water and gaseous carbon dioxide, and sometimes sulphur dioxide and nitrogen.

Groundmass. Metallurgically the term is used in reference to the matrix in which primary crystals of any alloy with a duplex structure are embedded. In geology the term refers to finer-grained portions of igneous rocks in which the larger crystals are embedded.

Ground state. Refers to the energy level of electrons. In the normal equilibrium state, the energy level of all electrons determined by the principles of atomic structure and the bonding of atoms, is that of the ground state. Electrons near the nucleus seldom change that state unless violently energised by electrical or thermal gradients, but the highest energy electrons frequently leave their ground state when energised. In many cases, the electron returns to the ground state with re-emission of energy, but in the extreme case, it can be lost to the atom—the state of ionisation. Ground states express the datum line for electron transitions involved in many mechanisms of thermal or electrical nature.

Group (periodic table). The elements arranged in the periodic table fall into vertical columns of similarity of chemical and physical behaviour, which originally led to the periodic classification. The electronic structure in the group is identical in type in the outer energy levels, but the energy increases naturally as the atomic number increases. The valency is therefore constant in a group, and the major changes in members of a group are due to increasing atomic number, size, and mass of the atoms involved. (*See* **Atom** and **Periodic law.**)

Growth. A term applied to certain cast irons in reference to the permanent increase in volume which sometimes occurs when grey iron castings are subjected to repeated cycles of heating and cooling. Also occurs with crystals of anisotropic structure.

Grozing iron. A tool used by plumbers for smoothing wiped joints in lead pipes.

Grummet. A jointing material used by plumbers for making watertight joints in pipes, etc. It consists of a mixture of red lead, putty and hemp.

Grunerite. A ferrous silicate often found in blast-furnace slags $FeO.SiO_2$.

Gruner's theorem. The statement of a scheme for the ideal working of a blast furnace. The theorem states that all carbon burnt in the furnace should reach the tuyères and be oxidised there to carbon monoxide, and all reduction of oxides above the tuyères should be effected by this carbon monoxide, with carbon dioxide as the stable product of interaction. This "ideal" does not in practice always lead to the highest fuel economy and, since for economic commercial operation there must be heavy burdening, fast driving and low fuel consumption, the theorem has been restated in the following sense: in all cases, for a given set of conditions, maximum fuel economy will be attained when the maximum use, for reduction purposes, is made of the carbon monoxide generated at the tuyères.

Guanidine aluminium sulphate hexa hydrate. (Often abbreviated to GASH by electronic engineers.) $CN_3H_6.Al(SO_4)_2.6H_2O$. A uniaxial ferroelectric material. (*See* **Ferroelectricity**.)

Guide mill. A rolling-mill fitted with guides to ensure that the stock enters the roll correctly.

Guillery test. An impact test carried out on a type of impact machine in which the specimen to be tested is fixed in position close to a flywheel that carries a knife edge or wedge. The flywheel is rotated by means of a motor until a definite speed is reached. The knife is carried in a recess in the flywheel and is released into the operating position at the correct moment by a spring. The fracture of the test piece slows up the flywheel and the reduction in speed serves as a means of measuring the amount of energy required to fracture the test piece. Guillery tests also include (*a*) a cupping test for sheet and strip metal; (*b*) an indentation hardness test closely resembling the Brinell test, but with the load applied through springs.

Gummite. An oxidised uranium mineral found in dolomites.

Gun metal. A series of copper-tin-zinc alloys containing 5-10% Sn and 2-5% Zn. The alloys possess good tensile strength and good resistance to corrosion and are used for bearings, steampipe fittings and marine castings. Lead and nickel may also be present. (*See* Table 5.)

Gutta-percha. The *trans* form of isoprene. (*See cis*-isoprene.)

Gutter. A small channel in a die. It is connected to the main impression by the flash gap and is intended to relieve excessive pressure on the metal in the die by providing a small channel into which excess metal may be expressed, so reducing the resistance to flow.

Gypsum. Hydrated calcium sulphate that crystallises in the monoclinic system $CaSO_4.2H_2O$, also known as selenite. Natural fibrous gypsum is known as satin-spar, and alabaster is the crystalline variety of specific gravity of 2.32. Gypsum occurs in saline residues, in dolomitic limestone and in many clay formations. When heated to a temperature of 120-130 °C for some time the hemihydrate $2CaSO_4.H_2O$, known also as plaster of Paris, is formed. Estrich gypsum is produced by heating the natural crystalline substance to a temperature above 130 °C, but below the temperature of "dead-burnt" sulphate. Its usefulness depends on the fact that it hardens slowly. Gypsum is also used as a fertiliser.

Gyratory crusher. A machine for breaking and for coarse-crushing of ores. The material to be crushed is fed in at the top and is crushed between a gyratory head and a fixed cone until it is small enough to drop through and out of the bottom.

H

Haematite. *See* **Hematite**.

Hafnium. A metallic element that always occurs in close association with zirconium, from which it is difficult to separate. It has a high melting point (2,227 °C) and high electronic emissivity. Crystalline structure—close-packed hexagonal. Density 13.3 $\times$ 10^3 kg m^{-3} at 20 °C. Lattice constants $a = 3.200 \times 10^{-10}$ m, $c = 5.077 \times 10^{-10}$ m. (*See* Table 2.)

Haigh machine. A fatigue-testing machine in which the specimen is subjected to alternating direct stress by means of a powerful electro-magnet excited by an alternating current. Since the stress is tension-compression, the test is referred to as in "push-pull" mode.

Hair-line cracks. Minute internal fissures sometimes found in wrought alloy steel products. The phenomenon is noticed when steels exposed to hydrogen at high temperatures are cooled down rapidly through the Ar_1 point. It is not generally accepted that the hydrogen is the main factor in causing hair-line cracking, but it is agreed that it does intensify the effect. The addition of titanium to steel prevents the occurrence of hair-line cracking, and it is thought that the presence of this element (which fixes the carbon as titanium carbide) indicates that carbon plays an important part in the phenomenon by preventing the diffusion of the hydrogen.

Half cell. An electrode of an electrolytic cell and the electrolyte with which it is in contact. When, for instance, a copper electrode is in contact with even a dilute solution of copper sulphate, the metal becomes positively charged with respect to the solution and the copper and its sulphate solution constitute a half cell. Similarly, the calomel electrode is a half cell. (*See also* **Electrode potential**.)

Half element. *See* **Half cell**.

Half-life. A term used to indicate the time required for one-half of a quantity of a substance to undergo a change or disintegration. For radioactive substances the half-life is given by t where

$$t = \frac{\log 2}{\lambda} = 0.693l$$

and λ is the disintegration constant and l is the mean life (*see* Table 27).

TABLE 27. *Radioactive Isotopes — Half-life*

Material	Half-life		Beta energy, MeV	$J \times 10^{-13}$	Gamma energy, MeV	$J \times 10^{-13}$
Cobalt 60	5.3	years	0.3	0.48	1.1-2.2	1.76-3.52
Tantalum 182	115	days	0.43	0.68	0.03-1.4	0.04-2.24
Iridium 192	70	days	0.68	1.09	0.13-1.06	0.21-1.70
Strontium 89	53	days	1.46	2.34	0.91	1.46
Radium	1590	years			0.19	0.30
Radon	3.83	days			0.50	0.80

Halfplane. The extra rows of atoms forming an edge dislocation, extending to the slip plane, and producing distortion in the lattice.

Halides. Salts in which the acid radical is one of the halogen elements, i.e. fluorine, chlorine, bromide and iodine.

Halite. Rock salt, NaCl.

Hall effect. When a current flows along a rectangular conductor that is subjected to a transverse magnetic field there is produced a difference in potential between the sides of the conductor. This phenomenon is known as the Hall effect and the Hall coefficient (r) is given by

$$r = \frac{Ed}{IH}$$

where E is the transverse potential difference, d = thickness of plate, H = the intensity of the magnetic field and I = the current.

Hall-Petch effect. When dislocations move along slip planes and arrive at grain boundaries they cannot easily continue to glide, in view of the change in orientation. The "pile-up" which results causes an increase in stress in the adjacent grains, and this eventually causes glide in those grains. The yielding of adjacent grains can only occur when the stress reaches a value σ_d and by analogy with the stress concentrations at crack tips, the stress at distance r ahead of the pile-up is $(\sigma_y - \sigma_i)(d/r)^{1/2}$, where σ_y is the yield stress, σ_i the frictional stress opposing dislocation glide, and d is the grain size. Writing $k_y = \sigma_d r^{1/2}$, gives the Petch equation for the lower yield stress $\sigma_y = \sigma_i + k_y d^{-1/2}$.

This shows that fine grained metal is stronger than coarse grained, and has been proved experimentally for mild steel and other alloys which behave in a similar manner.

Hall process. The process of producing aluminium by the electrolysis of fused cryolite (or cryolite and fluorspar) containing 20-25% purified alumina. Carbon electrodes are used and the current of 7 V has a density of 60-70 A m^{-2} of cathode. The passage of the current suffices to melt the mixture of cryolite and alumina and to maintain it at about 900 °C. With a current efficiency of 95% it requires about 6 kW h kg^{-1} (21 MJ kg^{-1}) of aluminium produced. (*See also* **Heroult process**.)

Hall's process. A puddling process introduced in 1830 by Joseph Hall for the production of wrought iron. The old sand hearth, which produced considerable amounts of slag, was replaced by old slag and scale, thus permitting the process to be adapted for any grade of iron, and at the same time increasing the yield of iron and decreasing the fuel consumption and the time for each heat. Because of the small amount of slag formed, the process was also known as "dry puddling".

Haloes. *See* **Fish eyes**.

Hamburg white. A mixture of 1 part of white lead (basic lead carbonate, $2PbCO_3.Pb(OH)_2$) and 2 parts of barium sulphate.

Hamilton's principle. This principle establishes a formal similarity between certain optical and dynamical principles, in particular between the path of a particle in a mechanical system and a supposed optical system in which the refractive index is correlated with the potential or kinetic energy of the mechanical system. If a particle of kinetic energy K moves from A to B, the path described will depend on the attractions and repulsions in the field between A and B. The principle of least action states that

$$\int_A^B 2K\,dt$$

will have a minimum value for the actual mechanical path.

If the passage of a beam of light from A to B be considered, this will depend on the refractive index of the space between A and B, and Fermat's principle of least time leads to the statement that the actual optical path is characterised by the minimum value of

$$\int_A^B \frac{ds}{v}\,,$$

where v is the velocity of light and depends on the value of μ, the refractive index.

The importance of the analogy lies in the interpretation of it given by de Broglie and the establishment of the existence of the de Broglie wavelength:

$$\lambda = h/(2m(E - V))^{\frac{1}{2}}$$
$$= h/mv$$

where h = Planck's constant, E = total energy and V = potential energy.

Hand forging. The art of forming or shaping metal by heating and hammering. The metal is heated to such temperature that it becomes plastic but not molten, and the method of working is moulding by impact, usually employing hand tools. Power-driven tools or machines to deliver blows are also used.

Hand ladle. A small foundry ladle which can be carried and used by one (or two) men.

Hand moulding. A method of producing firebricks, etc., of the best qualities for use in furnaces. The material used is a stiff plastic mass made from marls and shales and containing about 12-14% water. When a lump of this material is rolled in sand and thrown into a sanded mould it readily flows into the shape of the latter and yields a strong well-formed brick.

Hand picking. An operation carried out in ore-dressing that is sometimes used to remove either rich material or waste material, by hand, from a stream of crushed ore on a conveyor belt or moving table. The method is not considered economical for material finer than about 5 cm. Cobbing is a hand-picking process in which the material is broken by hammers to assist in separating the waste material from the more valuable ore.

Hand shank. A foundry ladle supported in a ring at the centre of a long iron rod, one or both ends of which are formed into a pair of handles. It can be carried by two men, and is used in foundries for transporting and pouring molten metal.

Hanger crack. A crack sometimes found in ingots which have been cast in moulds with badly fitting feeder heads. The cracks are due to restriction during cooling. They may also be found in ingots when overfilling of the mould has taken place.

Hanging. A term used in reference to a blast furnace to indicate the accumulation of material (mainly carbonaceous) on the walls. This frequently occurs when there is insufficient batter in the stack walls and when the blast velocity is too high. Usually an increase in the quantity of limestone of large size and a reduction in the blast velocity will effect an improvement.

Hansgirg process. A method for the production of magnesium by the reduction of its oxide with carbon.

Hanson-Van Winkle-Munning process. An electrolytic process for descaling and producing a bright silvery surface on iron and steel by two successive steps. Scale is removed by making the work cathodic in a warm dilute sulphuric acid bath (10-20% sulphuric acid) at a current density of 9-$14\,A\,m^{-2}$ for several minutes. The work is then rinsed and treated anodically in cold sulphuric acid (40-50% sulphuric acid) for 2-3 minutes at a current density of 9.4-$14\,A\,m^{-2}$.

Hard drawn. A temper produced by cold-drawing wire, rod or tube.

Hard-drawn steel wire. Wire drawn direct from the rod or from an annealed and patented bar, with a reduction of area greater than about 10%.

Hardenability. A property of metals which determines the depth and distribution of hardness induced by quenching a ferrous alloy. The standard test for hardenability is the Jominy test (q.v.).

Hardener. An alloy containing a high percentage of one or more alloying elements which is used in making alloys of these latter. Hardeners permit closer control of composition than is attainable using pure metals. Sometimes also called master alloys.

Hardener (polymeric). A material added to a polymer to cause cross-linking and therefore increase the hardness. Peroxides are efficient hardeners for polyester resins and similar materials, being strong oxidising agents. Hardening by cross-linking can take place naturally over a period of time. (*See* **Ageing of thermoplastic polymers.**)

Hardening. A heating and quenching process applicable to certain iron-base alloys. These are heated and quickly quenched from a temperature within, or above, the critical temperature range for the purpose of producing a hardness greater than that obtainable when the alloy is not quenched. It is usually restricted to the formation of martensite.

Change of properties as a result of heat-treatment is not confined to ferrous alloys. Hardening can occur with any alloy in which a change in solubility of a constituent in the basis metal can take place. Alloys of the duralumin type when quenched from about 500 °C will slowly harden if allowed to stand at ordinary temperatures for several days. This is known as age-hardening. The process can be accelerated by heating up to about 160 °C and this is known as artificial age-hardening. Both processes are included under the title "precipitation hardening" (q.v.). Non-heat-treated alloys may develop increased hardness by cold-working.

Hard-facing. A process in which the surfaces of a metal component that are liable to heavy wear are coated with a layer of some metal or alloy possessing better resistance to wear and abrasion. The harder metal or alloy may be welded to, or sprayed on, the softer metal. The process is most frequently used on metal working dies, well-drilling tools and excavating equipment. (*See also* **Hard-facing alloys.**)

Hard-facing alloys. Hard materials that are applied to the surface of mild steel, etc., in order to obtain a wear-resistant face. The usual materials, which are frequently welded to working faces, include: (a) iron-base alloys containing up to 20% of the alloying elements: chromium, manganese and silicon, together with small amounts of tungsten, molybdenum and nickel; (b) iron-base alloys containing from 20 to 50% of the following elements: chromium, tungsten, molybdenum, nickel, cobalt, manganese, silicon, vanadium and zirconium; (c) stellites or chromium, cobalt, tungsten alloys; (d) composite materials consisting of tungsten carbide particles in a matrix of steel; (e) austenitic steel; (f) hard bronze; nickel base alloys containing silicon and boron.

Hardhead. A mixture of impure tin-iron compounds obtained as an infusible residue during the refining of crude smelted tin by liquation.

Hard lead. A term applied to an impure form of lead that lacks the characteristic malleability of pure lead.

The term is also commonly used to describe an antimonial lead containing 1-12% Sb that is used for chemical stopcocks in the sulphuric-acid industry. For storage batteries the composition is usually Sb 7-12%, and about Sn 0.25%, the balance being lead. For building construction a hard lead containing 6% Sb is usually employed. Shrapnel bullets were made of hard lead.

Hard metals. As the name implies, metals and alloys (including metallic compounds) used for their qualities of hardness and wear and abrasion resistance. Materials used for cutting tools and surface coatings for wear resistance are included. (*See* Table 9, **Hard-facing alloys.**)

Hardness. A property of a material by virtue of which it resists deformation by external forces. Hardness is related to (a) the tensile strength of metals in the case of steels, (b) anti-friction properties, (c) degree of temper as a result of heat-treatment, and (d) the homogeneity of the metal.

Exact methods of measuring hardness are thus necessary, not only to identify constituents, but also to determine a standard of quality. The common methods of measuring hardness are: (a) the scratch test proposed by Mohs, who devised the Mohs scale (q.v.). This test is most frequently used for minerals; (b) the more exact scratch test devised by Martens, in which hardness is measured by the width of scratch produced on the surface by a diamond point under a definite load; (c) indentation tests, such as the Brinell test (q.v.), that use a hardened-steel sphere which is pressed into the metal under a certain load for a restricted time. The Rockwell and Vickers tests use a diamond point in place of a sphere; (d) the rebound test, in which a scleroscope is used to measure the rebound of a diamond-pointed hammer from a horizontal surface; and (e) the recoil method, using the recoil of a loaded pendulum after impact on a vertical surface of the metal to be tested. (See also **Herbert hardness tester, Microhardness**.)

Hard plating. A term applied to a fairly thick coating of electro-deposited chromium on other metals.

Hard water. Water that contains dissolved salts, such as the bicarbonate of calcium or the chlorides or sulphates of magnesium and calcium, which prevent the ready formation of a lather with soap solution. Hard water may be "softened" by boiling (temporary hardness removed) if the dissolved salt is calcium bicarbonate. The addition of the right quantity of milk of lime can be used to soften large volumes of temporary hard water. "Permanent" hardness is due to the presence of soluble calcium and magnesium salts. These can be removed either by the addition of a solution of sodium carbonate or by passing the water over zeolites (sodium aluminium silicates).

Hard zinc. A dross which collects in the bottom of galvanising pots or kettles during prolonged operation. It consists of an iron-zinc alloy of variable composition, but usually contains about 2.5-4% Fe.

Harris process. A process for softening lead by the use of salts of sodium, e.g. $NaCl$, $NaOH$, $NaNO_3$, and an oxidising agent such as litharge PbO. The arsenic, antimony and tin present as impurities are first oxidised and then converted into sodium arsenate, antimonate and stannate respectively. The process consists in allowing the molten impure lead to fall through a layer of the fused salts. The spent salts may be worked for the recovery of antimony, arsenic and tin.

Hartmann lines. A macroscopic pattern of lines, etc., sometimes found on the polished surfaces of metals that have been plastically deformed.

When iron and mild steel are stretched by gradually increasing loads a point in the stress-strain diagram, termed the yield-point, is reached soon after the material has ceased to be elastic, and further extension of the material occurs without the application of an additional load. When this occurs it is often noticed that scale attached to the surface of the metal begins to flake off in lines inclined at approximately 45 degrees to the direction of the stress. If the surface is polished, lines appear having approximately the same direction. These lines can often be recognised by touch, as ridges or steps on the smooth polished surface. Also known as Lüder's lines and as stretcher-strains.

Hausmanite. An ore of manganese Mn_3O_4 that occurs as sedimentary or residual deposits. It crystallises in tetragonal pyramids, and one of its best localities is Ilmenau in Thuringia (Germany). Mohs hardness 5-5.5, sp. gr. 4.72-4.85.

Hausner process. A process for hard chromium plating in which a high-frequency alternating current is imposed on the plating bath.

Hazelette process. A continuous casting process in which the metal is poured into a mould pocket formed between two cylindrical rolls or two water-cooled steel belts. The metal passes between the mould-rolls and over successive pairs of guide-rolls until it finally emerges as a continuous solid strip.

Head. The "hydrostatic" pressure of the metal in the riser of a mould due to its elevation above the uppermost parts of a casting.

Header. A foundryman's term for the feeder-head which is used to compensate for contraction during solidification, and to provide a reserve of molten metal which will not solidify until the casting itself is solid.

The term is also used to describe a device for removing excess zinc from galvanised wire. It consists of a split box packed with sand, fine cinders or charcoal, and is so placed that the wires pass through the header as they emerge from the galvanising tank. A hollow cone of zinc forms round the wire and acts as the wiper.

Head metal. The term applied to the metal contained in the riser or feeder head of a mould.

Heap leaching. A process used for the recovery of copper from weathered ore and from the material from mine-dumps. The material is laid in beds alternately fine and coarse until the thickness is roughly 6 m. It is treated with water or the spent liquor from a previous operation. Intervals are allowed between watering to allow oxidation to occur, and the beds are provided with ventilating flues to assist the oxidation of the sulphides to sulphate. The liquor seeping through the beds is led to tanks, where it is treated with scrap iron to precipitate the copper from solution.

Hearth. That part of a furnace which contains the molten metal.

Heat. A batch-smelting operation, a melting or a heat-treatment cycle.

Heat affected zone (HAZ). Welding involves the heating of the parent metal or alloy up to and beyond the melting point, in order to secure adequate penetration in making the joint. The conduction of heat into the parent metal creates a steep temperature gradient and, in the case of an alloy such as steel, brings about all possible phase transformations somewhere in this heat affected zone. On cooling, the same will happen in reverse, so that almost every type of heat treatment will have occured somewhere. The cooling will be more rapid than with many heat treatments due to the mass of colder material in the artefact being welded, and will almost be equal to a mild quench in some cases, and at the best similar to a normalising treatment.

Apart from generating micro and even macro-stresses, this effect may cause cracking due to the restraint offered, when a phase transformation may occur too quickly to establish equilibrium. In some alloy steels this can be difficult to avoid, and pre-heating or even post-heating of the joint may be necessary to decrease cooling rates.

Even in low carbon steels, the grain size will be much increased close to the weld joint, where the temperature of the parent metal is highest and the cooling still fairly rapid. Some coarsening of the structure will occur wherever the steel reaches a temperature much above Ac_3, but the areas at Ac_3 or just above will be refined. The transition zone between Ac_1 and Ac_3 will consist of a partially recrystallised area, of mixed grain size, and this merges into the completely unaffected original structure.

Above 0.3% carbon in plain carbon steels, and above 0.15% carbon in low alloy steels, there is always a danger of forming brittle martensite. Time-temperature-transformation diagrams assist in assessing the problems with alloy steels.

Heating curves. Curves obtained by plotting time against temperature, when a metal is heated under constant conditions. The curves show the absorption of heat during melting and during any transformations which may take place in the solid state.

Heating elements. Furnace heating elements are made from alloys having good resistance to corrosion at high temperatures. Nickel-chromium alloys are very commonly used, but are unsuitable at temperatures above 1,050 °C. For higher ranges, globar or silit are useful up to about 1,450 °C. Tungsten and molybdenum are suitable in those cases where oxygen can be rigorously excluded, because the oxides of both these metals are volatile at high temperatures. Graphite can be used in furnaces at temperatures up to 1,800 °C, but it reacts with all the common refractory materials. In all cases it is advisable to operate graphite resistance furnaces in an inert atmosphere, e.g. nitrogen. Platinum is used in some small furnaces and is suitable for temperatures up to 1,600 °C.

Heat of hydration. Energy released when water molecules become chemically bound to a compound to form a new stable hydrated compound. Typical of such evolution is the hydration of gypsum (plaster of Paris) and Portland cement. Unless the evolution is controlled, it can produce cracking or weak structures.

Heat capacity. Another term for specific heat, usually at constant volume.

Heat of vaporisation. The quantity of thermal energy required to vaporise a solid, i.e. transform the bonded atoms into individual free atoms.

Heat-resisting alloys. The use of higher temperatures in the efforts to improve the thermal efficiency of heat engines has led to the demand for metals or alloys that can withstand oxidation, have good tensile and other mechanical properties at elevated temperatures and have a high resistance to creep. Such materials are referred to as "heat-resisting", since they resist to a marked degree changes in those properties, chemical and mechanical, that are essential to good service, but which tend to deteriorate with rise in temperature. Heat-resisting steels contain relatively large proportions of nickel and chromium, together with niobium, molybdenum or tungsten, the presence of one or more of the last three being essential for good creep resistance. Non-ferrous heat-resisting alloys contain mainly nickel and chromium.

Heat-resisting steels. These are usually high-alloy steels especially used in situations requiring resistance to oxidation, creep and fatigue. They are normally stainless steel types, i.e. they always contain chromium with or without nickel or silicon, with enhanced contents of chromium and/or nickel, and of carbon, together with, as a rule, carbide-forming elements, such as titanium, tungsten or molybdenum. For most uses the alloying elements tungsten and molybdenum are essential, but in some cases titanium, vanadium or niobium can be used as substitutes. Heat-resisting steels were developed from iron-based austenitic steels by greatly increasing the proportions of the alloying elements. These latter include chromium (up to 25%), nickel (up to 30%) and cobalt (up to 20%). The carbon content is usually up to 0.5% and the rarer elements added—all usually less than 5%—are molybdenum, tungsten and niobium. (See Table 6.)

Heat tinting. A method of distinguishing and of identifying the micro-constituents of a polished surface of a metallographic specimen. The method is based on the fact that temper colours or heat tints, due to the interference of light in the thin oxide film, appear when oxidation begins on a polished surface that is being heated. The polished specimen, which must be quite dry, may be heated on an iron plate, but if better control of temperature is desired it may be heated in a thermostatically-controlled oven or by floating it on the surface of molten tin. The progress of the colouring is carefully observed and is stopped when this is sufficient by quenching in mercury. The colour attained depends on the nature of the metal and the time and temperature of heating.

When a specimen of alloy contains several constituents that have different susceptibilities to oxidation, and therefore oxidise at different rates, they can often be differentiated by heat tinting. The method is often used for the detection of phosphide in cast iron and steel, since phosphide colours more slowly than cementite. In certain cases the colour contrasts may be enhanced by giving the specimen a light etch with a 2% solution of phosphoric acid before heat tinting.

The temper colours corresponding to the various temperatures for polished steel are given as:

220-230 °C	Pale yellow
240 °C	Dark yellow
255 °C	Yellowish-brown
265 °C	Brownish-red
275 °C	Purple
285 °C	Violet
295 °C	Cornflower blue
315 °C	Pale blue

The actual colour obtained should not be regarded as a reliable indication of the temperature, since prolonged heating at, say, 230 °C will result in a specimen slowly passing through the range of colour from pale yellow to blue.

For metals that do not oxidise at ordinary temperatures the method of heat tinting at elevated temperatures may be used to produce decorative effects, as in the case of rhodium metal or rhodium-rich alloys.

Heat-treatment. An operation or combination of operations in which metal in the solid state is taken through cycles of heating and cooling for the purpose of obtaining certain desired conditions or properties. The various temperature cycles are designated and described as follows.

(a) Annealing. A heating to and holding at a suitable temperature, followed by relatively slow cooling operation. The purpose of such treatment may be (1) to induce softness and improve machinability; (2) to remove stresses; (3) to refine the grain structure, and (4) to obtain other desirable properties in the material.

(b) Full annealing. This operation consists in heating the material to about 50 °C above the critical temperature range, holding at this temperature for a sufficient time, depending on the material and its size, and cooling slowly either in a furnace or under ashes through the transformation range.

(c) Sub-critical annealing. This consists in heating a steel to a temperature below transformation range, i.e. below the carbon change point, holding at this temperature and then cooling at a suitable rate. The effect is to produce a resolved or spheroidised structure.

(d) Blank carburising. In this the carburising procedure is followed without using the carburising medium.

(e) Blank nitriding. The nitriding procedure is carried out without the nitriding medium.

(f) Case hardening or carburising. A process in which steel is heated in contact with carbon or a carbon-containing medium to a temperature above the critical range. A high-carbon case which is hard and wear-resistant is thus formed. It is usual to quench after hardening at two different temperatures: a higher one (about 900 °C) to refine the relatively soft core, and a lower one (about 760 °C) to refine and harden the outer case.

(g) Nitriding. A surface-hardening process for producing a hard case by the absorption of nitrogen on certain types of steel by heating in an atmosphere of ammonia or other suitable nitrogen-containing medium. The temperature should be below the critical range, and there is no subsequent quenching. Very high values of hardness are attainable. (See also **Nitriding.**) Carbo-nitriding is the process of case-hardening iron-base alloys by the simultaneous absorption of carbon and nitrogen (see(f)) by heating in a suitable atmosphere, followed by quenching or slow cooling as required.

(h) Black annealing. A process of box-annealing iron-base alloys in sheet form. The sheets are hot rolled (in which condition they are black), pickled and annealed.

(i) Blue annealing. A heat-treatment process applicable to iron-base alloys. Hot-rolled sheet is heated in an open furnace to a temperature within the critical range and subsequently cooled in air. The oxide that forms on the surface has a bluish colour.

(j) Box annealing. A process of heating batches of components in a sealed metal container with or without packing material. The material is heated slowly to a temperature below (sometimes within) the critical range. After holding at this temperature for a suitable time, the container is slowly cooled. Also known as pot or pack annealing, and sometimes as close annealing.

(k) Bright annealing. A process of annealing in which the heating is carried out in a controlled atmosphere, so that oxidation is reduced to a minimum and the surface remains relatively bright.

(l) Flame annealing and hardening. The process in which an iron alloy is treated with a high-temperature flame. The heat is applied locally, and results in local change of hardness.

(m) Inverse annealing. This is a process applicable to cast iron that contains appreciable amounts of austenite. The treatment results in an increase of hardness and in strength of the metal.

(n) Process annealing. A treatment applicable to iron alloy sheet and wire, in which the metal is heated to a temperature just below the critical range. On cooling, the material is softened, and can then be further "processed" or cold worked.

(o) Cyaniding. A process for the hardening of the surface of steel by the absorption of nitrogen and carbon, applicable to low-carbon steels. The steel is usually heated to a suitable temperature in contact with sodium cyanide and subsequently quenched.

(p) Hardening. The process of heating a metal to a temperature above the critical range and cooling rapidly by quenching.

(q) Homogenising. A heat-treatment applicable to ingots, sections of ingots and compacts of powdered metals that is carried out with the object of producing as fine a microstructure as possible and eliminating segregation or other forms of uneven composition. This method of ensuring a composition uniform throughout the sample is carried out at high temperature—usually about 20 °C below the solidus—and the material must be held at this temperature for 2 hours or more and subsequently quenched.

(r) Isothermal annealing. In this process the steel is heated to a temperature above the transformation range and held at this temperature for some time. The material is then cooled to a lower temperature, at which the transformation from austenite to pearlite will proceed to completion. The steel is held at this lower temperature for some time before quenching.

(s) Normalising. The material is heated as in full annealing and then allowed to cool in still air at room temperature. This permits

a moderately rapid rate of cooling, the object of which is the relieving of internal stresses, refining of the grain size, producing a more uniform structure and generally improving the mechanical properties.

(t) Quenching. A term applied to the rapid cooling of a hot metal by quickly immersing it in a bath of cold water, oil, a solution of salt, a fused salt or molten metal, etc.

(u) Surface hardening. A term applied primarily to the production of a hard surface layer by heat-treatment alone. The usual methods are by the use of a high-temperature flame, for local hardening, or by the use of induction heating.

(v) Tempering. A process of reheating hardened steel to some temperature below the critical range, followed by a suitable rate of cooling. The object of the treatment is to modify the hardness and increase the ductility. The American term for tempering is "drawing".

Heavy burden. A term used in reference to the charge or burden of a blast furnace, when the charge contains a higher proportion of ore and flux than the normal charge. The term burden is applied to the total charge consisting of ore, flux and coke.

Heavy case. A term used in reference to case-hardened steels to indicate a carburised zone more than 1.5 mm thick. Such cases are usually obtained by pack carburising.

Heavy hydrogen. An isotope of hydrogen in which the nucleus consists of one proton and one neutron, as compared with the single proton characteristic of ordinary hydrogen. The ratio of heavy hydrogen to ordinary hydrogen in the gas prepared is about 1 to 5,000. It is obtained by the distillation of liquid hydrogen at the triple point (13.9 K), and it may be produced by the electrolysis of cold water. Atomic weight 2.0142.

This isotope is also called deuterium, and its ions are called deuterons. These latter, when moving with high energy, are very effective in producing nuclear transmutations, probably because of the loosely bound neutrons they contain.

Heavy iron. A very thick heavy coating sometimes found on galvanised iron. It is produced when the sheets are hot-dipped in a zinc bath which is so badly contaminated with iron that the zinc cannot alloy with the steel sheet. Intermetallic compounds of zinc and iron are carried from the galvanising tank to form a coating of "heavy iron".

Heavy metal. An intermetallic compound of iron and tin $FeSn_2$ formed in tinning pots which have become badly contaminated with iron. The compound tends to settle to the bottom of the pot as solid crystals, and can be removed with a perforated ladle.

Heavy spar. Another name for barytes $BaSO_4$, a common vein stone in lead mines, where it is associated with galena, calcite and quartz and is known by miners as cawk.

Heavy water. The oxide of the hydrogen isotope deuterium, and hence called deuterium oxide. When cold water is electrolysed till about 1 part in 100,000 remains, the liquid so produced is practically pure heavy water. It can be used as a moderator in nuclear reactors.

Hegeler furnace. A muffle-type roasting furnace used in roasting zinc ores. It consists of several (usually seven) hearths arranged one above the other. The charge enters at the top and is slowly worked by mechanical rabbles and then drawn on to the hearth next below. After leaving the lowest hearth the charge ordinarily contains not more than 1% S.

Heisenberg uncertainty principle. It is impossible to determine accurately the position and momentum of a particle (like the electron) simultaneously. The reasons are related to the behaviour of the electron as a wave, with an equivalent wavelength $\lambda = h/mv$ where h is Planck's constant and mv the momentum of the electron (de Broglie). For electrons accelerated by 15 kV, $\lambda = 10^{-11}$ m, Bohr postulated the use of a hypothetical γ-ray microscope to attempt to measure the simultaneous position and velocity of an electron, as an illustration of the uncertainty principle. The precision with which a microscope can distinguish changes of position, the resolving power, is limited by the wavelength of the light employed, and the aperture of the objective. Since the electron may have an equivalent wavelength similar to that required by the hypothetical microscope used to observe it, an interaction must result (first described by Compton) whereby the photons of observation will produce a recoil in the observed electrons as

they interact. Thus, the very means used to observe will affect the particle being observed, and vice versa. The position of the electron cannot be estimated without affecting the momentum, and if the momentum is affected the position will change. This is the basis of the uncertainty principle, that accurate estimation of one parameter must affect the other, and the two cannot simultaneously be measured. If Δx is the uncertainty of position and Δp the uncertainty of momentum, then Heisenberg expressed the relation as $\Delta x.\Delta p \approx h$, Planck's constant.

By similar arguments, a similar relationship can be stated for the uncertainty in energy ΔE (corresponding to momentum) and time Δt (corresponding to position) $\Delta E.\Delta t \approx h$.

By using probabilities rather than accurate statements of values for these parameters, a form of mechanics can be evolved which replaces the classical mechanics of Newton, where uncertainties of this kind are not involved.

Helium (He). An inert gas first noticed in the atmosphere of the sun and later found in traces in the terrestrial atmosphere. It is formed by the decay of radium and radon—1 g of radium produces 34×10^9 atoms of helium per second. It is found occluded in a number of minerals, and was probably produced as a result of radioactive changes at a remote period. Natural gas may contain up to 1% He, and from this natural source, especially from the gases evolved at Kansas (USA) and Medicine Hat (Canada), helium is now prepared in quantity. It is used as an inert or shielding atmosphere in certain melting processes and welding techniques and for heat transfer.

The gas liquefies at $-268.9\,°C$ and by rapid boiling of the liquid a temperature as low as 0.82 K has been attained. At this temperature it was shown that metals have practically no electrical resistance (i.e. a current initiated in a metal ring at this temperature will continue to circulate for days). Atomic weight 4.003, critical temperature $-267.8\,°C$, critical pressure 0.229 MN m^{-2}, density 0.1786 g l^{-1}. (*See* Table 2.)

Helve. A simple form of hammer, of ancient origin, used to a limited extent in forges. In all the various forms of helve a mass of iron is raised by means of a cam attached to a revolving wheel and then allowed to fall by its own weight on to the metal to be hammered.

Hematite. A mineral consisting mainly of iron oxide Fe_2O_3 corresponding to an iron content of about 70%. It is a valuable iron ore free from phosphorus and when crystallised is often black. It crystallises in the hexagonal system. Mohs hardness 5.5-6.5, sp. gr. 4.5-5.3. Also spelt haematite.

Hemimorphite. A hydrated silicate of zinc $2ZnO.SiO_2.H_2O$ that contains about 54% Zn and is regarded as a valuable ore of this metal. It occurs both in veins and in beds along with sulphides of zinc, iron and lead and also with calamine. It is found in several locations in Great Britain, in Belgium, Nerchinsk, Siberia and Montana (USA). It crystallises in the rhombic system, but the crystals are differently terminated at the ends. This is connected with the phenomenon that when the crystals are subjected to change of temperature they become positively charged at one end and negatively charged at the other, i.e. pyroelectric. Mohs hardness 4.5-5, sp. gr. 3.16-3.49.

Henry. The practical unit of inductance of a circuit. A circuit has an inductance of 1 henry when a current increasing at the rate of 1 A s^{-1} produces an opposing e.m.f. of 1 volt.

Henry's law. The amount of any gas which dissolves in a liquid with which it does not react chemically, is dependent on the nature of the gas, its temperature and pressure. Henry's law refers only to the effect of pressure and states that the amount of the gas absorbed by the liquid is proportional to the pressure. It should be noted that the volume of the gas is reduced by the pressure (Boyle's law) so that, in effect, a given quantity of liquid dissolves the same *volume* of gas at all pressures.

Hepatic pyrites. A variety of iron pyrites in which the mineral crystallises in the orthorhombic instead of in the cubic system. The crystals are isomorphous with mispickel, but are rarely well developed. The fractured surface of the pure ore is tin-white, and the lustre is metallic. The mineral oxidises on exposure to moist air. Also called marcasite. Mohs hardness 6-6.5, sp. gr. 4.8-4.9.

Herbert hardness tester. An apparatus for the measurement of hardness that consists of a weight system resting on a steel ball which constitutes a compound pendulum. The pendulum is placed on the surface to be tested, on which the ball makes an impression, and is then allowed to swing through a small angle. The time of swing is taken as an indication of the hardness of the material.

Hercynite. An iron spinel that is black in colour $FeO.Al_2O_3$. It is found as an alteration product in emery (corundum).

Héroult furnace. A furnace heated by direct arc and consisting of a steel casing lined with refractory material. Current at low voltage is supplied by a step-down transformer with various voltage tappings for different stages of melting or refining processes. The furnace can be tilted to facilitate pouring. It is used for melting steel and other metals and the arc is struck between the electrodes and the charge.

Héroult process. An electrolytic process for extracting aluminium from bauxite. The purified bauxite Al_2O_3 is dissolved in molten cryolite, and the electrolysis of the oxide results in molten aluminium being liberated at the cathode (the carbon lining of the cell) and the carbon electrodes gradually burn away. Low-voltage direct current is used and the heat developed in the burning of the electrodes to CO, and in overcoming the resistance of the cell suffices to maintain the contents of the bath molten and effect the decomposition. (*See* **Hall process.**)

Hertz. The unit of frequency, equal to number of cycles per second, and abbreviated to Hz for convenience.

Hessite. A tellurium mineral consisting essentially of silver telluride. It occurs with other silver ores in many parts of the world, notably at Savodinski (Siberia).

Hess's law. This is the law of "constant heat summation", and states that the amount of heat produced or absorbed in a chemical reaction is the same whether the reaction takes place in one stage or in several stages. Thus all chemical reactions that start with the same reacting substances and end with the same resultant substances will liberate or absorb the same amount of heat per mole of substances involved, irrespective of the steps by which the final state is reached.

Heterogeneity. Literally, showing different origins. In materials science, a non-uniform material, varying in chemical composition, physical structure or properties. A composite material is heterogeneous by intent, but any heterogeneity in a semiconductor may ruin its performance. Metals, produced mainly by high temperature processes, are rarely absolutely pure in chemical composition, but in general heterogeneity refers to microscopic or macroscopic impurities or artefacts introduced as a result of industrial processes. Polymeric materials are heterogeneous in polymer chain length and molecular weight, even when perfectly pure. Ceramics are frequently heterogeneous for the same reason as metals, i.e. the difficulty of achieving perfect control over high temperature fabrication processes.

The fact that heterogeneity is almost universal in industrial materials limits some applications of those materials, or requires heavier section structures in design than would otherwise be necessary — part of the design engineer's factor of safety. The cost of minimising heterogeneity commercially limits its application to very special projects, where the higher cost leads to a worthwhile gain in performance, reliability or safety, as in aerospace, nuclear power reactors, racing vehicles, or defence installations.

For examples of purity in materials *see* **Intrinsic** and **Extrinsic semiconductors.** For physical heterogeneity, *see* **Anisotropy, Crystallisation, Dislocations, Preferred orientation.**

Hexagonal close-packed structures (c.p.h.). The general arrangement is dealt with under close-packed structures, indicating that ABAB . . . stacking of hexagonal close-packed two-dimensional layers produces the hexagonal prism cell characteristic of this type. The close-packed layers of A and B are identical, but in this case they are only the close-packed planes, as compared to the octahedral planes of the f.c.c. structure. Since close-packed planes are likely to be the sites of slip planes in plastic deformation, c.p.h. structures show such deformation mainly in the basal plane, the close-packed A and B layers. The close-packed directions will also lie in these layers, and the Miller-Bravais notation is the most suitable to describe them. In this notation, there are three a axes of equal length at $120°$ to each other in the hexagonal plane, and a

fourth c axis at right angles to the other three. The basal plane is then given the index (0001) since this plane is parallel to all three a axes and intercepts the c axis only. In the Miller-Bravais notation the sum of the rationalised intercepts on the a axes (*see* **Miller indices**) must be zero, a special feature of this system.

One of the three close-packed directions in the basal plane is [1120], the others being [2110] and [1210]. These of course, line up three atoms across the centre of a single hexagon with one atom in the centre. A less closely packed plane in the c.p.h. lattice is that joining two A layers in a vertical plane, the "prismatic" face of the basic cell. There are six such prismatic faces in one cell, and all will have the last index in the Bravais notation equal to 0, since they are parallel to the c axis. Each of the six faces is also parallel to one of the three a axes, and intercepts the other two, one with a positive and one a negative value. There are therefore three sets of parallel prismatic faces in the basic cell, $(01\bar{1}0)$ and $(0\bar{1}10)$; $(10\bar{1}0)$ and $(\bar{1}010)$; $(1\bar{1}00)$ and $(\bar{1}100)$. (*See* Fig. 75.)

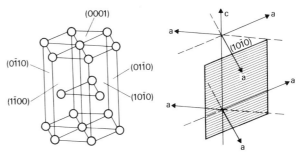

Fig. 75 *Close-packed basal and prismatic planes, close-packed hexagonal system.*

Reference to Fig. 75 and to Fig. 25(*a*) will indicate that the atoms in the basal plane layers A, are much closer together than the A layers are themselves, hence the prismatic planes are rectangular in outline. Deformation on these planes will involve larger movements, and also there will be anisotropy, i.e. variations in properties along different crystallographic axes, because of the "stratification" of the closely packed basal planes and the more open packing in the prismatic ones.

In the c.p.h. system, each atom in a basal plane has six neighbours at distance a, and neighbours in the planes above and below at distance $(\frac{1}{3}a^2 + \frac{1}{4}c^2)^{\frac{1}{2}}$ where c is the hexagonal separation. When $c/a = (\frac{8}{3})^{\frac{1}{2}} = 1.633$, the separation of atoms above and below a basal plane $= a$, and all nearest neighbours (twelve) are at the same distance. The next nearest neighbours will be at distance $a2^{\frac{1}{2}} = 1.414a$. The third closest are those in the prismatic planes, distance c, or $1.633a$ minimum.

Voids in close-packed structures are of the same two types, whether the structure be ABAB or ABCABC stacking. In the c.p.h. lattice, the octahedral holes are situated between and below two atoms in the basal plane (along an a axis) and between and above the two nearest atoms in the plane below (B layer). There are three such positions above layer B and three below in each unit (*see* Fig. 76). Similarly, the tetrahedral holes are directly above an atom in the B layer, and equidistant from three atoms in the basal plane, i.e. not along an a axis. There are again three such positions above and three below each B layer (*see* Fig. 76). For close-packed structures, the sizes of the voids are always in the same ratio to the atomic radius R, whether f.c.c. or c.p.h. Octahedral voids can accommodate a sphere of radius $0.414R$, and tetrahedral voids one of radius $0.225R$.

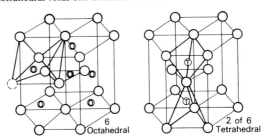

Fig. 76 *Voids in the close-packed hexagonal system.*

From the argument on close neighbours, it is obvious that not all hexagonal structures are close-packed. Because of the electron cloud overlap, any repulsive forces generated by atoms coming into close proximity in a crystal can be accommodated by a ratio of c to a greater than 1.633. When this occurs, the neighbours in basal planes will not all be nearest neighbours, and the prismatic plane will be even more rectangular. Beryllium is a metal which packs the atoms even closer at high temperature, and c/a is 1.63. Zinc and cadmium are not close-packed, $c/a = 1.86$ and 1.89 respectively. The space occupied in these non close-packed hexagonal lattices is obviously less than 74%.

Because of the similarity between f.c.c. and c.p.h. structures, transition from one to the other often takes place when thermal energy favours the rearrangement. Cobalt is c.p.h. at low temperatures, and f.c.c. above 400 °C. The structure becomes more close-packed at higher temperature, having $c/a = 1.624$ in the hexagonal form. Strontium is face-centred cubic up to 235 °C, changes to c.p.h. of $c/a = 1.63$ at that temperature, and to body-centred cubic at 540 °C.

A partial transition between the two close-packed structures can also occur over a restricted number of planes. An A layer of the c.p.h. system has only to move a small distance to become a C layer of the f.c.c. system, and if these occur during dislocation movement, it is known as a "stacking fault".

High alumina cement. As the name implies, made by fusing bauxite, alumina Al_2O_3 with chalk or limestone, calcium carbonate $CaCO_3$ to a melt, and then grinding to fine powder. The product is largely tricalcium aluminate $3CaO.Al_2O_3$ but contains other aluminates of varying proportions of CaO to Al_2O_3. The aluminate hydrates rapidly on mixing with water, with considerable evolution of heat. It develops high strength, over twice that of Portland cement for the same cement/sand or aggregate mixture, and is resistant to sulphates, weak acids and sugar. In furnace applications, it can be an excellent bond for heat resistant aggregates.

Unfortunately, changes can occur in the structure of the aluminates, particularly above 30 °C and in the presence of moisture. Poor control of mixing the original material causes this reversion of structure to be exaggerated, as does hastening of setting time. High alumina cement must not be mixed with Portland cement, as a "flash" set results, with loss of strength.

High density polyethylene (polyethene). The original processes for polymerisation of ethene used high temperatures and pressures, which caused branching of the polymer chains, and a wide variation in molecular weight. The plant was expensive to run and maintain, and the density of the product was low, about 0.92.

Low pressure polymerisation gives high density polyethene, more crystalline or more cross-linked, depending on the process. The Ziegler process uses a metal halide catalyst, and a temperature of 50-75 °C at about atmospheric pressure. The density achieved is 0.94-0.95. In the Phillips process, a metal oxide catalyst is used at 100-175 °C at a pressure of 2.8-3.5 MN m^{-2}. The Standard Oil process employs a metal catalyst at 100 °C and 3.5 MN m^{-2} pressure. The density of Phillips and Standard Oil material may be 0.96-0.97.

High-frequency furnace. A high-frequency induction furnace is a melting furnace in which currents at a frequency above 500 Hz are used to induce eddy currents in the charge, which in turn generate heat in the material sufficient to melt it. The principle on which the furnace is designed is essentially similar to that for an air transformer. The primary of the transformer is usually a spiral tube, water cooled, and the secondary is the metal charge that is being melted. The heating effect is proportional to the voltage used and to the square of the product of the frequency and the flux for a given furnace.

High-speed steel. An alloy steel, usually containing 0.6-0.7% C, 12-18% W, 3-4% Cr and small amounts of other alloying elements, e.g. vanadium, molybdenum. The alloys of this type are hardened by heating to 1,300 °C and cooling in air. As the tempering temperature is raised from 300 °C to 650 °C the steels become harder as austenite is converted into martensite. Heating up to about 600 °C constitutes a secondary hardening. These alloys retain their hardness up to about 700 °C (i.e. at dull red heat), and thus can be used for metal-cutting tools operated at high speeds. The tungsten is the element mainly responsible for the special pro-

perties of high-speed steel. Other compositions are given in Table 6.

Boron and zirconium have been used with chromium, molybdenum, cobalt and vanadium to conserve tungsten. Mushet's special steel (1865) (q.v.) was one of this self-hardening type. Molybdenum may be used to partially replace tungsten on the basis of 1 part molybdenum for each 2 parts of tungsten.

High-tensile brass. A group of alloys of the 60% copper-40% zinc type, in which small amounts of iron, manganese and aluminium are used to increase hardness and strength — generally at the expense of ductility.

Hilgenstock process. A process for reducing free sulphur in molten iron or steel by adding manganese and permitting the melt to stand so that the manganese sulphide formed may rise to the surface.

Hilger spectrometer. A spectrometer in which a prism of constant deviation is used. The prism is optically equivalent to a 60° prism of the same kind of glass. Though it is actually one prism, it can be considered as made up of two prisms each having a refracting angle of 30°. If such a prism is rotated about a vertical axis, different wavelengths may be observed in a fixed direction at right angles to the direction of the incident ray. The measurements of wavelength are made on a large drum, which is actually the head of a fine adjustment screw by means of which the prism is rotated. The divisions on the drum give direct indications of wavelength.

Hoesch process. A type of duplex process (q.v.) for the production of steel. Ordinarily in such processes the raw materials for the production of steel are treated in one type of furnace and the liquid product is transferred to a second type of furnace for finishing. The Hoesch process differs from the usual procedure in that the partly purified metal, after separation from the slag, is charged back into the furnace from which it has just been tapped.

Hogging. A flame-machining process used for cutting superfluous metal risers, sprues, etc., from steel castings.

Holding furnace. A furnace used as a reservoir for molten metal from other melting furnaces. Adjustments in composition may be made whilst the correct casting temperature is awaited.

Hole, electron conduction. Sometimes called a positive hole, or positive carrier. The concept is an analogy, to explain the movement of electrons from atoms covalent bonded, but containing impurity atoms with fewer electrons per atom, which can, when substituted in the structure, accept an electron from the matrix material, whose valence band is completely filled, and where a forbidden energy gap exists between the valence and conduction band. The energy requirements favour electron movement of this kind in preference to jumping the energy gap, and thus the impurity atoms act as sinks, or "holes" for electron migration. (See **Extrinsic semiconductor, Conduction by electrons.**)

Holfos process. A patented method of producing high-quality castings in bronze, gunmetal, etc. It employs the basic principles of centrifugal casting combined with the use of a rotating chill core which is progressively withdrawn as the casting process proceeds.

Hollow forging. A process for making steel tubes or hollow forgings by expanding a hole trepanned or punched in a solid forging. The expansion is carried out hot on a mandrel under a forging press.

Holmium. A trivalent metal of the yttrium sub-group of rare-earth metals. Atomic weight 164.94; at. no. 67. The oxide has the formula Ho_2O_3 and the salts of the metal are generally yellow in colour. (See Table 2.)

Homoeomorphous. A term used to describe substances that are chemically quite distinct, but exhibit similarity of crystalline form. For example, greenockite CdS and zincite ZnO are practically identical in crystalline form. Other examples are calcite $CaCO_3$ and Chile saltpetre; celestine $SrSO_4$ and marcasite FeS_2.

Homogeneous. A term used to describe a material that consists of one substance only, or of a single phase. An alloy in which any small portion has the same chemical composition and the same physical structure as any other portion. For elements and compounds the condition of homogeneity applies if they occur in the system in a single state; the same refers to solid, liquid and gaseous solutions which are the same in all parts and show such a molecular interpenetration that mechanical separation of their constituents is impossible. (See **Heterogeneity.**)

Homogenising. A heat-treatment involving holding at high temperatures for sufficient time to permit a uniform composition and structure to be attained by diffusion. The process is intended to eliminate or decrease chemical segregation by diffusion.

Homopolar (covalent) bond. Since electrons are shared uniformly amongst all the atoms in such a bond, there are no ions created, and therefore no polarisation. Homopolar simply means equal polarity and implies no electrical conduction by ion movement, and no polarisation properties in an electric field. (*See also* **Bond, Covalent bond.**)

Hone. A smooth stone used either dry or lubricated with oil or water for sharpening the edges of cutting tools or for grinding micro-specimens or other smooth surfaces.

Honing. A process for improving the surface finish and dimensional accuracy of cylinder bores, etc., by abrasion with smooth stone slips.

Hood and Fender stock. An American term for the highly polished, hard-rolled steel sheet used for automobile hoods and bumper bars.

Hooke's law. Within certain limits (called the limit of proportionality) stress is proportional to strain. (*See* **Deformation** and **Elasticity.**)

Hoop iron. Thin mild (low carbon) steel strip used for securing barrels, etc. and in various building operations. It is usually made by the basic open-hearth or Bessemer processes.

Hopeite. Hydrated zinc phosphate $Zn_3P_2O_8.4H_2O$. It is an important zinc ore worked in Zambia.

Hopper. A large circular section trough or funnel through which the burden is charged into a blast furnace. It is usually used in conjunction with a bell which can be raised or lowered at will to close the top of the furnace, or to allow the charge to fall into the furnace. In all modern furnaces a double bell and hopper are used to conserve the large volume of gas which formerly escaped at each lowering of the bell. The arrangement consists of a small bell and hopper, below which is a larger pair, the space between the two pairs (of bell and hopper) being large enough to accommodate the charge. The volume of this chamber represents the small amount of gas lost at each charging. The term is also applied to any inverted cone or pyramid used to feed fuel or pulverised material to a conveyor. (*See* Fig. 70, p. 114.)

Hornblende. An important rock-forming mineral of complex and variable composition, but consisting essentially of silicates of calcium, magnesium, iron and aluminium with smaller amounts of sodium and potassium silicates also present. Approximate composition: CaO 10%, MgO 13%, FeO 14%, Fe_2O_3 7%, Al_2O_3 12%, SiO_2 40%. The Na_2O and K_2O usually amount to about 4%. The colour is grey, green, brown or black, according to the composition. Basaltic hornblende may contain significant amounts of titanium. Crystal system, monoclinic; hardness 5-6 Mohs scale, sp. gr. 3-3.4.

Horngate. The name applied to the tapering, horn-shaped channels radiating from the bottom of a runner used to supply molten metal to a number of moulds contained in the same moulding box.

Horn lead. The double chloride and carbonate of lead $PbCl_2$. $PbCO_3$ found in Derbyshire and Cornwall (UK).

Horns. The name given to the long, fishtail-shaped projections formed at the trailing end of sheet metal during rolling when the sides of the sheet are elongated more than the centre.

Horn silver. A silver chloride ore AgCl containing approximately 75% Ag. It occurs with native silver and oxidised silver compounds in the zone of weathering of silver sulphide lodes in Europe, Australia, California and South America. It crystallises in the cubic system as greenish-grey cubes. Mohs hardness 1-1.5, sp. gr. 5.5.

Horse-power. The power of an agent or machine is measured by the rate at which it can supply work, i.e. power = work/time (or, what is equivalent, power = force $\times$ velocity). James Watt is said to have made some tests on the power of some of the heavy dray horses belonging to Barclay and Perkins' Brewery, London. The horses were set to raise a weight of 100 lb (45.36 kg) from the bottom of a deep well by means of a rope passing over a pulley. In doing this he found the horses could walk at 2½ miles/hr (4.02 km hr^{-1}). Allowing 50% extra for work wasted in friction, he arrived at the estimate of 33,000 ft lb/min (0.746 MJ s^{-1}). This is rather more than the average horse is capable of, but it has been adopted by engineers and is known as a horse-power. 1 h.p. = 746 watts (1 kilowatt = 1.34 h.p., 1 h.p. = 746 J s^{-1}).

Hot blast. The term used to describe the heated air forced, under pressure, into the base of a furnace used in smelting practice. The temperature of the air varies from about 340-820 °C. It enables larger quantities of ore to be smelted than would otherwise be the case and there results some economy in fuel consumption.

Hot blast furnace. *See* **Blast furnace.**

Hot-dip coating. A process for coating or covering the surfaces of iron or mild steel with a thin layer of zinc by immersion of the metal in a bath of molten spelter. The metal to be coated is cleaned or pickled in dilute hydrochloric or sulphuric acid and then washed before passing into the molten metal. The surface of the latter is usually covered with ammonium chloride. If the metal is used "wet" it is first passed through a flux box containing ammonium chloride. The thickness of the coating in the case of steel sheets is regulated by passing them between rollers as they emerge from the bath. The weight of zinc taken up on the surface of sheet metal usually varies from 0.03-0.08 g m^{-2}.

The process is also known as hot-dip galvanising. Hot-dipped galvanising produces a brittle alloy layer between the zinc surface and the steel. It is found that hot dipping reduces the endurance of the steel, since the cracks that form in the iron-zinc alloy are propagated through the steel. The reduction varies from about 5% for 90 BHN irons to about 40% for hard steels of BHN above 300.

Hot-dip process. *See* **Hot-dip coating.**

Hot-galvanising. *See* **Hot-dip coating.**

Hot-metal process. A term that usually refers to a steel-making process in which molten metal is charged into an electric or open-hearth furnace and treated under a suitable slag for further refining and finishing.

Hot-short. Steel or wrought iron which has a tendency to display embrittlement at elevated temperatures. The cracking and formation of minute flaws is generally due to a high sulphur content.

Hot tearing. A term used in reference to the rupture of a casting due to overstressing of the metal during cooling and contraction. This type of failure occurs at or near the solidus temperature, when the last small percentage of the metal is freezing, and it is attributed to undue restraint or to unsuitable pouring temperature.

Hot-top. A refractory-lined container used on top of an ingot mould to maintain a reservoir of molten metal as long as possible while the ingot is solidifying. It is a special type of feeder head, which draws the shrinkage cavity or pipe and the highly segregated metal towards the top and into the hot top portion if possible. In this manner a maximum amount of sound steel is produced. Hot-topping may be aided by electrical or Thermit-type heating. (*See also* **Dozzle.**)

Hot working. The operations of rolling, or forging or extruding metals at elevated temperatures, the temperatures employed varying with the composition of the material. The plastic deformations take place at temperatures and at rates that do not cause strain hardening. The lower limit of temperature is the crystallisation temperature.

Huey test. A corrosion test used for stainless steels to determine their resistance to intergranular corrosion. A test sample is boiled in 60% nitric acid for five periods each of 48 hours' duration. The loss in weight after each period is determined and the average of the last three periods is usually reported in some standard unit such as "inches (or mm) penetration per month or per year".

Hull cell. A device for testing the properties of electroplating solutions or for controlling electroplating baths by means of plating tests. These tests are used as supplementary to the usual chemical tests and to the judgment of practical operators.

Humidity. The relationship—usually expressed as a percentage—between the amount of water vapour actually present in the air at any given temperature and the amount that would be required to saturate the air at the same temperature. Since the vapour obeys (approximately) Boyle's law up to the point of saturation, the weight of it contained in a given volume is simply proportional to the pressure, and hence:

$$humidity = \frac{\text{pressure of vapour at } t\,°C}{\text{pressure of vapour in saturated air at } t\,°C}$$

There are thus two general methods of determining the above rates. One by actually determining the *weight* of vapour in a measured volume of air (w_1) and then consulting Regnault's (or similar) tables to find the weight of moisture (W) that would saturate the same volume at the same temperature. This is termed the chemical method and here humidity = w_1/W. The second method consists in determining the *pressure* (f) of the vapour in the air and then ascertaining the maximum pressure at the same temperature F, from vapour pressure tables. This is the method practised in all "dewpoint" instruments, and here humidity = f/F.

Wet- and dry-bulb thermometers are frequently used in industry. The dry-bulb temperature is noted, and also the difference between the reading of the dry and the wet bulb. By consulting hygrometric tables the humidity percentage is obtained. (*See* Dew point, Table 17.)

Hund's rule. In evaluating the energy of electrons in terms of the four quantum numbers, Hund deduced (from spectroscopic measurements) that electrons arrange to give the maximum possible total spin momentum consistent with the Pauli exclusion principle. Thus, when l is greater than 1, and m_l can have the values $+l$, 0 and $-l$, it is the value $m_l = -l$ which is filled first, then $m_l = -(l - 1)$, and so on through 0 to $+l$. When the shell is half full, the only possible change is in the electron spin, and as this merely changes sign, all the values of m_l are then repeated. (*See* **Atom.**)

Hunter process. A process for the production of metallic zirconium. It is based on the reduction *in vacuo* of zirconium tetrachloride by sodium.

Huntsman process. The original crucible steelmaking process introduced into England in 1742 by Benjamin Huntsman. The charge may be either blister steel, iron bar and carbon, or wrought iron and white cast iron. The crucibles may be made of either clay or plumbago—the latter mainly for castings and the former for tool steels. The charge is heated in a furnace chamber in closed crucibles with coke piled up round them. Usually ferro-manganese is added before casting (sometimes aluminium is used) to ensure gas-free products. The original natural draught furnaces accommodated two crucibles each containing about 45 kg (100 lb) charge. This capacity can be greatly increased by using gas-fired furnaces. Huntsman steel finds its chief application in the manufacture of high-grade cutlery and tool steel, but has been almost superseded by electric furnace steel.

Hyacinth. A transparent red variety of zircon $ZrO_2 .SiO_2$ used as a gemstone. It occurs, with other zircons, in the gem gravels of Sri Lanka and in New South Wales. Mohs hardness 7.5, sp. gr. 4-4.5.

Hydrargillite. A hydrated aluminium oxide, $Al_2 O_3 .3H_2 O$. Also known as Gibbsite.

Hydratogenesis. A process of ore formation in which the ore body has been deposited from solution. It represents the final stage in the intrusion of an igneous magma and consists of the evolution of heated waters of high chemical activity which are capable of dissolving most of the metals of economic importance, and transporting them for considerable distances from the parent rock. By cooling, or by chemical change, the dissolved metal is deposited in cavities or fissures (lodes) or between the grains of loose sediments (impregnations).

Hydraulicking. A mining process in which alluvial or placer deposits are broken down by powerful jets of water under high pressure directed against the working faces.

Hydraulic test. An acceptance test for pressure tightness applied to castings, boilers, etc. Water is slowly pumped into the cavity or vessel until the pressure exceeds the working pressures likely to be attained in service by a specified amount.

Hydrides. A term used to describe a compound of an element with hydrogen. Though this term could be applied to compounds of hydrogen and the non-metals, it is mostly used to describe the binary compounds of metals and hydrogen. All the elements in groups IV, V, VI, and VII of the periodic table that combine with hydrogen form gaseous hydrides. Boron also forms a hydride. The hydrides of the alkali metals, of the metals of the alkaline earths and the rare-earth metals are all solids. The hydrides of

thorium, copper, chromium, cobalt, nickel and iron are also solids. LiH, NaH, KH and CaH_2 are well known and their composition is established. The absorption of hydrogen by palladium and by platinum was at one time thought to give rise to a hydride or to an alloy of the palladium and a volatile metal that was called hydrogenium. It is now considered that "occlusion" (q.v.) is a more complicated process than was at first thought and that at least five different phenomena may be involved.

Hydrik process. A process that was devised for the production of hydrogen on an industrial scale. With the military use of hydrogen for inflating balloons, methods were devised for producing hydrogen in the field using materials that were relatively light and cheap and did not involve transport difficulties. One of these methods, known as the Hydrik process, depends on the reaction of aluminium metal with caustic soda.

Hydroblast. A method of cleaning castings by means of powerful jets of water. The water is circulated in a closed system and any foundry sand removed can be recovered.

Hydrocarbon. The generic term for compounds between carbon and hydrogen. Because carbon requires four electrons to satisfy the closed shell arrangement, it can combine in covalent bonding with four hydrogen atoms, but this is not essential, if bonds can also form between carbon atoms themselves, as they do. This leads to saturated hydrocarbons, in which all carbon bonds are with hydrogen atoms, and unsaturated, where carbon-carbon bonds are involved. Unsaturated can have double bonds, with two shared pairs of electrons between two carbon atoms, or triple bonds, with three shared pairs.

The nomenclature used for saturated hydrocarbons linked with carbon bonds of a linear rather than a cyclic nature, is a generic term "alkanes", the prefix indicating the number of carbon atoms involved, e.g. methane CH_4, ethane $C_2 H_6$. The generic formula is thus $C_n H_{2n + 2}$.

Of the unsaturated acyclic hydrocarbons, "alkenes" contain one double bond, with the formula $C_n H_{2n}$, and "alkynes" contain one triple bond of formula $C_n H_{2n-2}$. There are other names for mixtures of double and triple bonds and multiples of either in one compound.

Cyclic compounds, in which the bonds form a closed ring, can be saturated or unsaturated too. The prefix "cyclo" is used with the name of the corresponding acyclic hydrocarbon with the same number of C atoms, e.g. cyclohexane is $C_6 H_{12}$, because hexane $(C_6 H_{14})$ has the same number of carbon atoms, but being acyclic, has more hydrogen atoms (one at each end of a straight chain). The corresponding cyclic unsaturated hydrocarbons follow a similar rule. The Council of the International Union of Pure and Applied Chemistry agreed this nomenclature in 1957.

Hydroceramic. The name applied to porous unglazed pottery used for filters, etc.

Hydrofluoric acid. An extremely corrosive, strongly fuming liquid HF. The aqueous solution rapidly attacks glass and most metals. In the former case it removes the silica as silicon fluoride and in the latter case fluorides of the metals are formed. The solution gives a clear etch on glass, but the gas itself gives an opaque etch. The noble metals (gold, platinum) are not attacked, and the dilute aqueous acid may be kept in *gutta-percha* (q.v.) containers. It is used in electroplating, electrolytic polishing, bright polishing and pickling.

Hydroforming. A process for producing sheet-metal parts for aircraft. etc. A hydraulic press is used in the drawing and forming operations.

Hydrogel. The term applied to the colloidal system in which the solid or jelly-like form has water as one of its components. It is the insoluble gelatinous form in which many colloids may exist. Hydrosols are frequently transformed into hydrogels by contact with small quantities of an electrolyte. It would appear as though the gelatinous colloid consists of the enormous colloid molecules piled together so that numerous gaps or interstices are left. These are filled with water. This honeycomb arrangement of water dissolves the electrolytes, and sufficient freedom of movement is still left to the latter compounds; and since their molecular particles are very much smaller, their usual reactions proceed unimpeded. Since hydrogels have a remarkable capacity for dissolving electrolytes to form "dry" solutions, this property has been exploited in numerous technical applications, e.g. "dry" photographic plates, electric "dry" batteries, etc., as well as in tanning and dyeing.

Hydrogen. A gas that has neither colour nor odour and has the least density of any known element. It is a constituent of all acids and of water, and may be obtained from these either by electrolysis or by the action on a suitable metal at a suitable temperature. Certain metals such as aluminium and zinc evolve hydrogen with aqueous solutions of alkalis. The gas burns with a hot flame in air, and with oxygen produces the oxy-hydrogen flame in which a temperature of about 2,800 °C is attained. It is a good reducing agent and readily reduces the oxides of copper, lead and iron (but not the oxides of zinc and aluminium).

It is used industrially in the air-hydrogen blowpipe in autogenous welding. The oxy-hydrogen blowpipe is used for fusing silica in making silica apparatus.

The gas combines with many metals at high temperatures—or is absorbed by them. On cooling, some of the gas is evolved, creating problems of soundness in castings. At the temperatures common in steelmaking, hydrogen probably exists only in the diatomic form (H_2), but the dissociation into atoms may occur in the metal.

Hydrogen atoms are small enough to occupy interstitial positions in many metallic lattices, and may associate again in voids, since this lowers the free energy. If such a process continues in the solid metal, large pressures may build up, creating hair-line cracks or other fissures.

In aluminium alloys, hydrogen solubility in the liquid state is high, and drops radically on transformation to the solid state. Unsoundness in castings of these alloys can be avoided by suitable fluxing and degassing operations, especially prior to pouring the metal into moulds.

Some metals readily form hydrides, but as these usually decompose at high temperatures, they are not often a problem in metallurgy. Hydrogen produced by electrolytic action however, is in the atomic form at the cathode, and can enter the structure of the solid metal before associating into molecules and rising as bubbles of gas. Any situation where hydrogen embrittlement of a metal or alloy can occur requires remedial action when such electrolytic action provides a possible source of the hydrogen.

The electronic structure of the hydrogen atom has been intensively studied theoretically, since the element has the simplest atom of all. The concepts of electron bonding, in forming molecules, and the application of wave mechanics, frequently deal with the hydrogen model for the same reason.

Hydrogen is manufactured by the following processes:

(a) Lane Process, in which steam is passed over reduced iron at a temperature of about 650-850 °C: $3Fe + 4H_2O \rightleftharpoons Fe_3O_4 + 4H_2$ (the iron oxide can be reduced by water-gas and the product used again).

(b) Bergius Process, in which water is heated with iron in a closed bomb. A temperature of 300 °C is required and the pressure of the steam within the bomb rises to about 100 atm: $Fe + H_2O \rightleftharpoons FeO + H_2$.

(c) Silicol Process. In this powdered silicon or an alloy of silicon and iron is heated with strong caustic soda solution:

$$Si + 2NaOH + H_2O = Na_2SiO_3 + 2H_2.$$

(d) The electrolysis of caustic soda solution using iron or nickel electrodes. (*See* Table 2.)

Hydrogen annealing. An annealing process conducted in an atmosphere of hydrogen. The process is restricted by economic considerations to a few special items, but when used for steel sheet, iron and manganese oxides are reduced and carbon and sulphur can be almost completely eliminated.

Hydrogenation. A term applied to those processes in which a substance combines directly with hydrogen. It is now almost always restricted to those cases in which an "unsaturated" organic substance is treated with hydrogen in presence of a catalyst. The latter may be finely divided nickel, cobalt, iron, copper or metals of the platinum group. Hydrogenation is usually associated with the well-known processes for the hardening of oils to produce fats, but the process is also used for the preparation of organic solvents and, under suitable conditions of temperature, for the preparation of alcohol.

Hydrogen electrode. *See* **Electrode potential.**

Hydrogen embrittlement. A condition of low ductility resulting from the absorption of hydrogen gas by a metal. The absorption of hydrogen may arise during pickling or cleaning in acid or alkaline solutions or during electrolytic processes in which water is the solvent medium. Hydrogen may also be absorbed from gases (including steam) at high temperatures. In ethylene production plant, hydrogen may diffuse into certain steels and react with carbides to form methane bubbles at grain boundaries, causing premature failure.

Hydrogen-ion concentration. Since an acid can be defined as a compound that produces hydrogen ions in aqueous solution, the extent to which this dissociation takes place in solutions of different acids but of equivalent concentrations is a measure of the hydrogen-ion concentration, and hence of the strengths of the acids. A rough measure of the hydrogen-ion concentration is given by *(a)* the specific conductivity of the solution, since the high velocity of the hydrogen ion compared with that of any other ion ensures that practically all the current will be carried by the hydrogen ion; *(b)* if the ionisation of an acid be represented $HA \rightleftharpoons H^+ + A^-$.

Hence
$$\frac{[A^-]}{[HA]} = \frac{K}{[H^+]}$$

but $[HA] = [\text{total acid}] - [A^-]$

$$\frac{[A^-]}{[\text{acid}] - [A^-]} = \frac{K}{[H^+]} \text{ or } \frac{[A^-]}{[\text{acid}]} = \frac{K}{K + [H^+]} = \alpha$$

Hence $\dfrac{1}{[H^+]} = \dfrac{1}{K} \cdot \dfrac{\alpha}{1 - \alpha}$

and $\log \dfrac{1}{[H^+]} = \log \dfrac{1}{K} + \log \dfrac{\alpha}{1 - \alpha}$

which is an expression from which the hydrogen-ion concentration can be plotted against the degree of ionisation, α being the fraction ionised. The value of the hydrolysis constant in the case of a weak acid (K_b) is given by

$$K_b = \frac{\alpha^2}{(1 - \alpha)V}$$

where V = volume of solution containing 1 g equivalent of acid. This generally becomes $K_b = \alpha 10^{-n}$, where n is a whole number.

Hydrogen-ion concentrations are now expressed in terms of what is called the hydrogen-ion exponent—a number obtained by giving a positive value to the negative power of 10 in the above expression. For pure water the hydrogen-ion concentration is 10^{-7} and this is written as pH = 7, which represents water neutrality. All acid solutions which contain a higher concentration of hydrogen ions will have pH values less than 7. And solutions more alkaline than water will have pH values between 7 and 14, since $[H^+] \times [OH^-] = 1 \times 10^{-14}$. It is useful to remember that

$$pH = \log \frac{1}{[H^+]}$$

Hydrogenite. The name applied to a mixture of 25 parts of silicon, 60 parts of caustic soda and 20 parts of slaked lime. When ignited the substance burns, evolving about 270-370 litres of hydrogen per kilogram. The residue consists of sodium and calcium silicates. The process is one of the industrial processes for the production of hydrogen gas.

Hydrogen telluride. A combustible gas obtained by treating zinc telluride with acids (e.g. HCl). It may also be obtained by the electrolysis of a 50% solution of sulphuric or phosphoric acids using a tellurium cathode. The temperature must be kept in the neighbourhood of −20 °C. It is a gas at ordinary temperatures, but liquefies at −1.8 °C. An aqueous solution when exposed to air is slowly oxidised, giving a wine-red solution of collodial tellurium.

Hydrogen welding. A type of shielded arc-welding process in which oxidation and nitriding are prevented by enveloping the weld by an atmosphere of hydrogen gas.

Hydrohematite. A crystalline mineral consisting of hydrated iron III (ferric) oxide $2Fe_2O_3 . H_2O$.

Hydrol. The name given to the simplest molecular condition in which liquid water exists. The anomalous expansion of water below $4\,°C$ and the abnormal vapour density of steam just about the boiling point have led to the supposition that liquid water is associated and that the more complex molecules are dihydrol $(H_2O)_2$ and trihydrol $(H_2O)_3$ but the existence of these latter is hypothetical.

Hydrolith. The hydride of calcium CaH_2 that is formed when hydrogen is passed over the heated metal. In contact with water this compound produces hydrogen and leaves slaked lime. This is the basis of an industrial method for producing hydrogen.

Hydrolysis. A process that takes place in aqueous solution, and which consists in the production of free acid and free alkali by the interaction of the ions of water with the ions of the salt. The fact that water is ionised into H^+ and OH^- gives rise to the possibility of the interaction of these ions and those of the salt in solution.

Substances such as sodium carbonate, sodium borate, potassium acetate and potassium cyanide, which are formed by the neutralisation of a weak acid by a strong base, do not produce a neutral solution in water. The "acid" ions combine with the H^+ to form unionised acid, thus leaving free OH^- ions to which the alkalinity of the solution is due. Similarly, solutions of salts formed from a weak base and a strong acid show an acidic reaction.

Hydrometers. Instruments used for comparing liquid densities. A common type of hydrometer consists of a glass (sometimes metal) bulb to the upper end of which is attached a graduated tube, and to the lower end is attached a small bulb, weighted so that the whole instrument will float upright in the liquid whose density is read off on the graduated stem, and from this reading the density of the liquid is obtained either directly or after consulting tables.

Hydrone. An alloy of lead and sodium containing 35% Na. It can be used in place of sodium amalgam for liberating hydrogen from water. This is the basis of an industrial process for making hydrogen for use in balloons.

Hydropaste. A non-inflammable paste consisting of flake aluminium and a binder and used as an ingot mould dressing.

Hydrostatic forces. Originally defined as the forces due to immersion in deep water, but gradually related to uniform isotropic pressure, equivalent to three equal and normal tensile or compression forces acting in three mutually perpendicular directions. The bulk modulus K is defined as the ratio of isotropic pressure to the relative volume change produced by that pressure.

Hydroxyl group (ion). The unit of one oxygen and one hydrogen ion, leaving a net negative charge, which allows the hydroxyl ion to attach itself to positive ions or groups. In water, the hydroxyl ion is combined with a hydrogen ion, easily dissociated, so that the two ions can migrate to appropriately polarised groups on solids, or dissolve in the liquid. Hydroxyl ions take part in many chemical reactions, and are associated with chemical bases, such as alkalis, whereas the corresponding hydrogen ion is associated with acids. To indicate the bond involved, the ion can be written $-O-H$, but more often simply as $(OH)^-$.

Hypereutectoid alloy. An alloy containing more than the eutectoid percentage of alloying element.

Hypereutectoid steel. A steel containing more than the eutectoid proportion of carbon (i.e. with more carbon than can be contained in pearlite). During the process of cooling of steels from a red heat at a moderate rate, the cementite forms a mechanical mixture with a definite amount of ferrite so that the resultant mixture contains approximately 0.9% C. This constituent is called pearlite, and consists of interstratified layers or bands of ferrite and cementite. A eutectoid steel contains about 0.9% C and consists entirely of pearlite. In the hypereutectoid steel pearlite is present with excess of free cementite.

Hyperm. Magnetic materials of high permeability. They include pure iron, silicon-iron and nickel-iron alloys and were introduced by Krupps as high-permeability alloys for high field strengths. The Ni/Fe 50/50 alloy resembles the American alloy Hipernik.

Hypoeutectoid steel. Steels containing less than the eutectoid ratio of carbon that consist of a definite amount of pearlite, varying according to the carbon content, the remainder being free ferrite.

Hypoferrite. Iron, which is not appreciably attacked by dilute alkali, since it quickly becomes passive, but which is attacked by hot, very concentrated alkali to produce a compound of the type Na_2FeO_2, which is called sodium hypoferrite. Also called ferroate.

Hysteresis. A term used to indicate the phenomenon of the lagging of physical effects behind their cause. For magnetic hysteresis, *see* **B - H loop**. For mechanical effects, *see* **Elastic after-effect** and **Angle of loss**. For electrical effects, *see* **Angle of loss** and **Current, alternating**.

I

Ice. The solid form of H_2O, characterised by a lower density (approximately 0.92 g cm^{-3}) and crystallisation in the hexagonal system.

The crystal structure is made up by each oxygen atom being at the centre of a tetrahedron of four other oxygen atoms, with a hydrogen atom situated between each adjacent pair of oxygen atoms, so that each oxygen is close to four hydrogen and each hydrogen close to two oxygen atoms. This, repeated in three dimensions, leads to hexagonal rings, and resembles the silica structure in tridymite. Such a structure is very open, hence the low density, but Bridgman has applied high pressure techniques to produce forms of ice much denser than water, the result of pressure being to close the distance of approach of the atoms.

Deformation of ice by pressure causes melting, and as this relieves the stress, the liquid freezes again, without stress. This explains the movement of glaciers over obstructions or rough ground.

The freezing of water to form ice takes place by gradual cooling to a temperature of 4 °C, when the maximum density of water is attained. The dense water falls through the depth of liquid until the temperature reaches 4 °C, this effect assisting the cooling by heat transfer to the surface. At 4 °C, when the maximum density is attained, the surface falls in temperature below 4 °C, and the surface layers are now lower in density than those below (at 4° C). The surface layers freeze, and heat transfer to lower depths is now inhibited, so that the ice layer grows only slowly. If this were not so, water ponds would freeze from the bottom up, and marine life would cease to exist in cold climates.

Iceland spar. A very pure form of calcite $CaCO_3$ (q.v.). It readily cleaves into perfect rhombohedra and exhibits double refraction in a marked degree. It is used in making nicol prisms.

Ideal diameter. The diameter of an infinitely long cylinder which, in an ideal quench, will just transform to a given specific microstructure. The term is used mainly in reference to consideration of hardenability.

Ideal quench. A quench in which the rate of heat transfer at the surface is high, i.e. one in which the surface of the metal instantly attains the temperature of the quenching medium.

Igneous magma. The molten fluids and gases generated within the earth and from which igneous rocks at or near the surface have been formed. These magmas are also regarded as the original source of the concentrations of the metals. There are many agents by which the metallic constituents of the magma are segregated or extracted from the rocks and finally concentrated. In the case of molten magmas the process is termed differentiation, and in the case of gases and vapours it is known as pneumatolysis.

Igneous rocks. Rocks that have been formed from certain materials once in a molten condition. They are usually sub-divided into three classes:

(a) Effusive rocks. The lavas that have been poured out from volcanoes to form sheets on the surface.

(b) Intrusive or bypabyssal rocks. Those that have been formed by molten material being forced into veins, fissures, etc., beneath masses of other rocks.

(c) Plutonic rocks. Those that have been solidified under great pressure and at great depths.

Igneous rocks show a wide variation in chemical composition, but consist mainly of silica, alumina, iron oxide, lime, magnesia, potash and soda, with small amounts of certain elements, phosphorus, chlorine and sulphur.

Magmas sometimes contain a considerable amount of water, and are then in a state referred to as aqueo-igneous fusion. In such cases coarse crystalline rocks (pegmatites) often result and minerals of various types may be formed as well-developed crystals. Those minerals that are present in large amounts in igneous rocks are referred to as the essential constituents. These are quartz, felspars, pyroxenes, amphiboles, micas, etc. Other minerals that occur in smaller amounts are known as accessory minerals, and in these ore deposits are sometimes found.

Ignition. This term is used to define the commencement of combustion, in which the substances involved react to give out light and heat. The ignition temperature is the lowest temperature to which the substance (or a mixture of substances) must be raised in order that flame may be initiated and combustion spread throughout the mixture (*see* Table 28). Flame is produced only when a gas or vapour burns. A solid burning with a visible flame must therefore be heated to such temperature that some of the material shall be converted to vapour.

TABLE 28. *Ignition Temperatures*

Substance	Ignition temperature, °C	Substance	Ignition temperature, °C
Carbon	400	Methane	660
Carbon monoxide	675	Heptane	260
Hydrogen	585	Kerosene	295
Sulphur	235	Cylinder oil	420
Phosphorus	34	Alcohol	558

Ihrigising. The process for the impregnation of iron or steel with silicon. The process consists in heating the metal at 930-1,010 °C in silicon carbide or ferro-silicon and chlorine so as to form a siliconised case containing about 14% Si, which is very resistant to corrosion, heat and wear at temperatures up to 870 °C.

Illam. A gravel occurring in Sri Lanka which is worked for corundum, spinels and zircons.

Illinium. A rare-earth metal discovered in 1926. Atomic number 61. Also sometimes called florentium. Now called promethium.

Illite. A band clay which usually occurs in association with shales.

Ilmenite. A titaniferous iron ore consisting mainly of the oxides of iron and titanium $FeO.TiO_2$. It is very variable in composition, especially in the ratio of iron to titanium. Some magnesia is often present. It is of widespread occurrence as an accessory constituent in the non-basic igneous rocks and in certain sands. The titanium content varies between about 18 and 24% or even more. The separation of ilmenite from the associated heavy minerals and siliceous gangue is effected either by froth flotation, or by gravity or magnetic separation. The latter is the most expensive. In dealing with beach sands the method of separation consists in grading—since the grain size of ilmenite is more restricted than that of felspar or quartz—followed by gravity separation carried out with jigs, tables or spiral concentrators.

Immersion coating. A process of producing a metallic coating on a metal by immersion in a solution of some suitable salts of the coating metal. A more electropositive metal may be deposited on a less electropositive metal, the later dissolving in the solution and precipitating the coating metal.

Impact extrusion. A method used for the production of collapsible and rigid containers up to about 115 mm diameter. The blank or slug is placed in a die cavity, and when struck by the punch, the metal is extruded backwards along the outside of the punch, so forming a thin shell around it. When the punch is raised, the extruded metal is held back by a stripper plate, and thus is able to fall away. The process is usually automatic, so that high rates of production are possible. The metals chiefly used are tin alloys and pure aluminium. (*See* Fig. 77.)

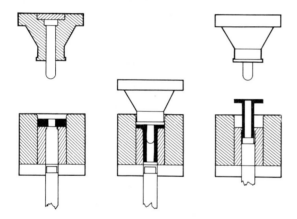

Fig. 77 *Impact extrusion of short tubular product.*

Impact tests. Tests devised to determine the energy absorbed in fracturing a test piece at high velocity, as distinct from static tests. They may be carried out in tension, bending or torsion, and the test-bar may be notched or unnotched. Under practical conditions of service, materials may be fractured by shock, i.e. by high-intensity stresses maintained for a short period of time. Resistance to shock is referred to as "toughness", and impact tests are used to investigate this property. All the tests involve the breaking of a test piece of standardised form. Commonly the test piece is notched in order to secure a concentration of stress at one section. The fracture is produced by a falling weight, and the energy absorbed is taken as a figure of merit for the material, but the results obtained by different methods are not always easy to compare or interpret.

A number of tests have been devised, but those most commonly used are the Charpy, Fremont and Izod tests (q.v.). Table 29 gives some data distinguishing the various types of impact test.

TABLE 29. *Details of Various Types of Impact Test*

Type of Test	Test Piece	Notch	Weight	Fall
Charpy	60 × 10 × 10 mm	Hole 1.35 mm dia.	Pendulum	—
Frémont	30 × 10 × 8 mm	1 mm sq.	5 kg	4 m
Izod	75 × 10 × 10 mm	V 2 mm	Pendulum	—

Impact toughness. A term introduced to differentiate high velocity impact tests from the usual notched-bar tests. In these latter the

notch tends to restrict the normal plastic deformation and the failure of the test piece is more akin to brittle fracture and does not take full account of the velocity effect. Impact toughness thus refers to the results of high velocity rather than the comparatively slow impact of pendulums or falling weights. Low-temperature brittleness is detectable by these impact tests but not by the usual tensile tests.

Imperfect crystals. Very few crystalline materials can be produced free from defects. The nearest approach to date is that of "whiskers", fibre-shaped artefacts of very high aspect ratio. These have a lower defect density than normal materials. (*See* **Crystal defects**.)

Imperfect dislocation. The Burgers vector defines the displacement of atoms resulting from dislocation glide or slip. When this is equal to a whole number multiple of the lattice spacing in the crystal, the slipped regions are restored to perfect matching of the crystal planes after the dislocation has passed, and such dislocations are described as perfect. It is not essential, however, that a whole number multiple of lattice spacings should be involved, in fact to achieve this may prove an obstacle. If the dislocation splits into two portions which move together as a unit, but in which each portion slips by only a fraction of lattice spacing, it is described as imperfect, or a partial dislocation. Partial dislocations may also be involved in twinning, where many atomic movements involve a fraction of lattice spacing. (*See* **Twinning**.) The crystal lattice existing between the two partial dislocations must be disordered, since movements of less than a whole lattice spacing are involved. The disordered portion of the lattice is called a

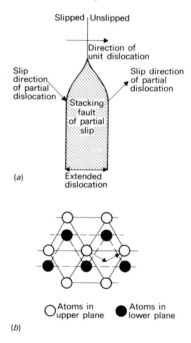

Fig. 78 *(a) An imperfect dislocation; (b) one possible partial atom movement involved.*

stacking fault. Once the second partial has moved through the disordered lattice, order is restored, and the stacking fault no longer exists there. The width of the stacking fault, i.e. the distance between the two partial dislocations, depends on the energy involved for that particular crystal. The higher the stacking fault energy, the narrower the strip of disordered lattice. Copper, silver and gold have low values of stacking fault energy, and so readily form partial dislocations with wide ribbons of stacking fault. On the other hand, aluminium and magnesium have high values, and so areas of stacking fault are more limited in those metals. It is significant that twinned crystals rarely form in aluminium. When zinc is alloyed with copper, the stacking fault energy is further reduced below that for copper, and stacking faults and twinned crystals are common in alpha brass which has been deformed.

Figure 78 illustrates the ribbon of stacking fault, and a possible atom movement involved. Here it is the possible change from f.c.c. (ABC stacking) to a small area of c.p.h. (AB stacking) which may occur. In copper, the stacking fault may be up to 10 atoms wide, in aluminium, only 2.

Impervite. A refractory material used as protection tubes for thermocouples, etc. Its composition approximates to that of silli-manite, and it is suitable for use at temperatures above 1,100 °C.

Impingement attack. A form of corrosion caused by impingement of moving water which results in erosion of the protective film on the metal, followed by corrosion of the exposed areas. The severity of the attack is increased by the entrained air and is intensified with rising water speeds.

Impingement tests are often used to assess the resistance of condenser tube materials to moving sea-water.

Imprest process. A process for embossing aluminium sheet by cold rolling in contact with materials such as embossed fabrics, lace, wire-gauze or paper cut to pattern.

Improving. A process for refining lead by oxidising impurities with furnace gases and the addition of lead oxide. A dross is formed which can be skimmed from the surface of the molten lead.

Impurity semiconductor. *See* **Extrinsic semiconductor.**

Incandescence. The phenomenon of the emission of light from the surface of a substance raised to a sufficiently high temperature. The temperature of flame usually exceeds that at which bodies become incandescent. Thus, substances can be maintained in a state of incandescence by allowing gaseous mixtures to burn, without flame, at the surfaces of refractory substances. The process has considerable advantages, e.g. combustion is accelerated by the incandescent surface, and high temperatures can be realised. The heat is transmitted from the incandescent surface by radiation, and the rate of transmission to the adjacent surfaces required to be heated is rapid.

Metals can be raised to incandescence by the passage of an electric current through them. The temperature attained will depend on the current that passes and the resistance of the metal. But for such employments as lamp filaments the permissible temperature must be well below the melting point of the metal. Other factors have also to be considered, viz. the effect of the temperature in reducing the mechanical strength of the metal and the tendency to vaporise. This latter can be suppressed by the presence in the lamp bulb of an inert gas. Most of the metals of the platinum group have high electrical resistances, particularly platinum and osmium. Both these metals have been used for making filaments, but they have been superseded by tungsten and tantalum on grounds of cost.

Inchrome process. A process for the treatment of small articles to render them rustproof by the diffusion of chromium into the outer layers of steel articles. Low-carbon steel is used and the process produces an outer zone consisting of about 30% Cr, decreasing to zero. The amount of chromium normally used is 200 g m^{-2} of surface, and it is claimed that scaling or stripping does not occur, and that the finished products are corrosion-resistant to industrial atmospheres and sea-water and can withstand temperatures up to 850 °C.

Incision. A cut made by a sharp tool.

Inclusions. The name applied to the impurities which have been trapped in metal during solidification. In steel, the inclusions most frequently encountered are manganese sulphide and silicate, slag and alumina. In brasses the inclusions are chiefly dross (such as oxides or silicates) and charcoal. Inclusions are usually non-metallic and may arise from the slag or from the refractory lining of the furnace, but more commonly they are the resultant materials of reactions occurring in the metal itself during finishing or oxidation processes or during pouring and solidification.

The inclusions are sometimes segregated in stringers, plates or laminations and are mostly objectionable in steels that are to be subjected to impact or to repeated stresses (e.g. tool steels).

Incoherent precipitate. A precipitate (usually in metallic alloys) whose crystal structure is not aligned with the structure of the matrix in which it is produced, so that dislocation movement cannot proceed along related slip planes. (*See* **Coherent precipitate.**)

Incongruent melting point. A compound is said to have an incongruent melting point if the solid having the composition of this compound is incapable of co-existence with the liquid.

Indentation hardness. A term applied to the resistance of a metal to indentation as in the usual hardness tests. The indenter may be spherical or diamond-shaped, and is pressed into the surface of the metal under a specified load for a given time. (*See also* **Hardness.**)

Indentation test. A hardness test in which a steady load is applied to the prepared surface of a metal for a specified time. The load is applied through a small, hardened-steel sphere or a pointed diamond indenter, and the dimensions or the depth of indentation are measured. When the test is used for soft metals, a smaller load is employed, and it is important to maintain standard temperatures, since variation of temperature markedly affects the results. (*See also* **Hardness.**)

Indicators. These are mainly organic compounds that are either weak acids or weak bases and which readily dissociate in aqueous solution. The colour of the undissociated atom is different from that of the negative ion, so that the indicator changes colour according to the presence of excess of hydrogen or hydroxyl ions.

The ionisation of a weak acid is strictly in accordance with the law of mass action. In the case where an indicator is ionised to the extent of 50%, $k = [H^+]$, i.e. k is equal to the concentration of hydrogen ions in an indicator ionised to the extent of 50%. A knowledge of the value of k enables the experimenter to select the best indicator for any particular titration.

Universal indicators are prepared by mixing certain indicators so that the colour changes can be made to extend over a considerable portion of the pH range. They are unsuitable for accurate determinations. Typical composition in grams/500 cm^3 absolute alcohol is: phenolphthalein 0.006, methyl red 0.013, methyl yellow 0.019, bromothymol blue 0.026, thymol blue 0.065. NaOH is added till the colour is just yellow. The colour changes are:

pH	Colour
2	Red
4	Orange
6	Yellow
8	Green
10	Blue

Indirect extrusion. An extrusion process in which the metal is forced back inside a hollow ram that pushes the die forward into the metal. Also called inverted extrusion.

Indium. A greyish-white metallic element which bears certain relationships with aluminium and gallium in Group III of the periodic classification. It occurs in small quantities in many varieties of zinc blende, and is also obtained from the flue dust of the zinc plants of Arizona. Some tin and lead ores also contain small amounts of the sulphide. The metal is soft and crystalline and is stable in air at ordinary temperatures. When heated above its melting point (156 °C) it burns with a violet flame. Its chief applications are in electroplating and in the production of fusible metals and dental alloys. It has also been used for coating the surfaces of aircraft engine bearings, and as the "weld" metal in joining copper sheets. Its alloy with palladium resembles red gold in colour. Mohs hardness 1.2. (*See* Table 2.)

Indium antimonide. One of the compound semiconductor materials produced by combining a Group III with a Group V element. The energy gap between valence and conduction band is small, 0.18 eV, and infrared radiation can bridge this gap. As a result, the material finds uses in fire detectors and magnetic field detectors. (*See* **Extrinsic semiconductor** and compare **Gallium arsenide.**)

Induced draught. A term applied to the actual air current or to the methods by which an increased air supply to a furnace or cupola is obtained. The simplest method consists in providing a tall chimney so that a difference in pressure is maintained between the hot air in the chimney and the cooler air outside. Steam jets suitably located near the top of a cupola are effective in producing an induced draught. In most cases, however, the required air supply is obtained by the use of a fan.

Induced polarisation. The production of dipole moments in a material, by proximity to an already polarised material. This effect increases van der Waals' attraction forces, and the more complex the dipole moments of the two constituents, the more complex the attraction.

Induced polarisation affects the crystallinity of polymeric materials, leading to alignment of polymer chains and increase in strength, due to the electrostatic attraction between parallel or related chains or portions of them.

Inductance. The property of an electrical conductor by virtue of which an e.m.f. is produced in the circuit due to varying currents in that circuit or in a neighbouring circuit. When a current is passed through a coil a magnetic field is set up. If the strength of the current varies in any way, so also does the magnetic field and, as this latter changes, an e.m.f. is induced in the coil. This induced e.m.f. acts always in opposition to the applied e.m.f.

Flux through coil $\propto$ current;

induced e.m.f. = rate of change of flux
$$= -L \times \text{rate of change of current}$$
$$= -L \frac{dI}{dt}$$

where L = the coefficient of induction (or inductance).

In a circuit of negligible resistance,

$$\text{current} = \frac{\text{applied voltage} \times \text{time}}{\text{inductance}}$$

$$\text{and } L = \frac{S\phi \times 10^{-8}}{I}$$

where S = number of turns in the circuit, ϕ = total flux, I = current in amperes. In this equation L is measured in henrys.

Induction. The production of an e.m.f. in a circuit by variations in the current in the circuit or in a neighbouring circuit. Induction may result from the motion of a conductor in a magnetic field (dynamically induced e.m.f.) or by a change in the flux (statically induced e.m.f.). The coefficient of mutual induction of two coils is proportional to the product of their numbers of coils, whereas the coefficient of self-induction of a single coil is proportional to the square of its number of turns.

Induction coil. An apparatus in which the principles of mutual induction are applied for the purpose of transforming a low potential difference between the terminals of a primary coil into a high potential difference between the terminals of a secondary coil. The primary coil consists of a few turns of low-resistance wire and is in series with a contact breaker. Across the contact breaker is placed in parallel a small condenser which reduces the tendency of a spark at the break in the current and intensifies the current in the secondary coil at the break in the primary current. If I_1 = primary current at the instant of "break", I_2 = secondary current, L = inductance of the secondary coil and M = the mutual inductance, then

$$\frac{I_2}{I_1} = \frac{2M}{L}$$

The usual make and break in the primary circuit is of the vibrating hammer type, but this may be replaced by the Wehnelt electrolytic interrupter or the motor mercury interrupter. The latter admits of control of the rate of make and break and also the time during which the current is made and broken. The apparatus is used for producing silent discharge through gases and high-voltage discharge through gases at low pressure.

Induction hardening. A hardening process which depends on the generation of heat in a conducting metal by means of induced electric currents. It was originally developed for producing local surface hardening on the bearing surfaces of crankshafts which had already been heat-treated to obtain the required mechanical properties throughout the mass.

The applications of the process have been extended into many other fields as new techniques and new equipment have been developed. High-frequency current is passed through an inductor which surrounds, but does not touch, the area to be hardened. The current produces a magnetic field which cuts the surface of the object being treated and causes heating of the surface by currents induced in the surface layers as well as by the hysteresis losses in the area subjected to the magnetic field. These combined effects rapidly heat the surface under treatment to the Ac$_2$ point, at which temperature the area is quenched by water or oil sprayed

through suitable orifices. The rate of heating must be extremely rapid in order to prevent undesirable heating of adjacent areas or of underlying metal by conduction.

Induction heating. A method of heating that is applied to melting and other furnaces in which there is no direct electrical connection between the electrical supply and the metal in the bath of the furnace, though there is an indirect connection by electrical induction. The main advantages of this type of heating are (a) freedom of metal in the bath from contamination from outside sources, (b) elimination of heat losses due to flues, etc., and shortening of heating time, (c) loss of metal by evaporation is reduced since the heat is developed in the charge itself, (d) accurate control of temperature and (e) a good circulation of the molten metal in the bath is ensured by the operation of electromagnetic forces. Cored furnaces are usually of the low-frequency type, while coreless furnaces are of the high-frequency type.

Induction period. A term used in reference to the time that elapses between the introduction of a metallic substance into a corrosive medium and the commencement of vigorous corrosive action. The term is also used in reference to a type of chemical reaction or metallurgical transformation in which there is a delay period before the main reaction or transformation commences. In processes which involve successive reactions, the speed of the whole process is determined by the slowest component reaction. The delay in the appearance of the final product is called the period of induction.

When the stimulus for the reaction is photochemical, the period of induction is called the Draper effect.

Inductor. A term used in electrical science to describe a component that has a low electrical resistance and a high capacity for inductance. Chemically, an inductor is a substance which accelerates a chemical action by reacting itself with one of the substances taking part in the reaction.

Indurated. A term used to describe the hardening of a mineral or other chemical substance.

Indurated talc. An impure slaty variety of talc 3MgO.4SiO$_2$.H$_2$O that is somewhat harder than French chalk.

Inert-gas shielded-arc process. A welding process in which the arc is surrounded by an inert gas such as argon or helium. No flux is required.

Infection. A term used in reference to the increase in the probability of corrosive attack in the neighbourhood of a point where corrosion has already started. The effect is associated with the behaviour of the products of corrosion which may act with the corrosive medium and so prevent its accumulation over the area, or deposit a protective coating of insoluble salts on the area affected.

Infiltration. A method of increasing the density and strength of powdered metal products by filling the pores with a fluid, lower-melting point metal instead of compacting the powder. Numerous methods have been devised, the more important being:

(a) *Capillary dip.* This consists in immersing a small portion of the permeable sponge body in a bath of molten metal.

(b) *Full dip infiltration.* This consists in the complete immersion of the (sponge) skeleton body in molten metal.

(c) *Contact infiltration.* Here the infiltration is brought about by the penetration of the molten metal which is originally in contact with the skeleton as solid metal.

The process is widely used for refractory metal contacts and electrodes which are permeated with a conductor metal such as copper or silver, and for jet engine compressor blades.

Infra-red. Originally termed "heat waves", the infra-red radiation is that portion of the electromagnetic radiation spectrum of longer wavelength than the visible spectrum, extending continuously into the range of radio waves. The lower limit of the range of wavelengths attributed to the infra-red is about 0.75 to 0.8 μm, and the maximum about 50 μm. The photons involved possess energy of the order of 1.5 eV down to 25 meV within that range (2.4 to 40 $\times$ 10^{-19} J).

Infra-red radiation is generated by any body undergoing thermal vibrations, and can be detected by thermocouples and bolometers, both of which have sensors which rise in temperature due to the absorption of the infra-red, and then measure that rise in temperature. Photographic emulsions can be made to react to this radiation, and the phenomena of reflection and refraction demonstrated, the latter by coarse gratings.

Infra-red radiation can be employed in drying processes, particularly with resinous materials such as coatings, paints and varnishes. Silicate glasses do not transmit infra-red, except the shorter wavelengths near to the visible spectrum range. Quartz transmits the longer wavelengths, and many transparent plastics the shorter ones. The longer wavelengths are also transmitted by rock salt and fluorite crystals.

Ingate. The channel (or channels) by which molten metal is led from the runner into the interior of a foundry mould.

Ingot. A term applied to metal that has solidified from the molten state into forms that are suitable for, or are intended for, subsequent mechanical working, e.g. rolling, pressing, forging, extruding and drawing. The term is also applied to the products obtained from a refinery or smelting furnace which are to be used for remelting or further refining.

Ingot bleeding. A term used in reference to the break down of the crust of a partly solidified ingot. If an ingot is stripped too soon after pouring, a portion of the crust or shell may give way under pressure from the still molten metal within. This oozing or flow of molten metal is known as "bleeding". The resulting ingots are unsatisfactory and the bleeding may be a source of danger.

Ingot iron. A term applied to pure iron (wrought iron) produced in an open-hearth furnace.

Ingotism. The term applied to a defect, due to the formation of large dendritic crystals, in ingots and castings which have been cast at too high a temperature, or cooled too slowly through the solidification range. Such structure lacks strength or toughness. In the case of ingots the defect can be reduced by hot working, provided care is taken during the early stages. In steel castings the defect is not easily rectified, but annealing at a suitable temperature may have a beneficial effect.

Ingot structure. The general arrangement of the crystals in an ingot usually consists of long columnar crystals more or less normal to the chill surface of the ingot mould, and an interior core of equi-axed crystals. The proportions of the different types of crystal depend on the temperature difference between molten metal and the mould and on the rate of cooling through the freezing range. Under favourable conditions one type of crystal only may be formed.

Ingot tipper. A mechanical device used in rolling-mills for manipulating ingots into any required position on the rolling tables, and for guiding the billets through the rolls.

Inhibitor. A substance that prevents or retards a reaction. The term is applied to the organic materials used in pickling baths to minimise acid attack on the basis metal. Flour, sulphonated oils, etc., are frequently used in pickling baths and fluxing solutions. Substances such as fluorides, boric acid, etc., are used to prevent the oxidation of magnesium alloys and these are also referred to as inhibitors.

In electrochemistry inhibitors are substances that diminish the corrosion rate of a metal without increasing the intensity of the attack. They are usually divided into the following classes: (a) anodic inhibitors, which include most of the sodium or potassium salts whose anions form sparingly soluble salts with the metal in question; (b) cathodic inhibitors, e.g. soluble salts of metals that form sparingly soluble hydroxides; and (c) absorption inhibitors, which are usually organic substances of high molecular weight—many of them that are colloidal form a protective film on the metal. Cathodic inhibitors are not very efficient, but they do not cause intensification of attack. Most anodic inhibitors are fairly efficient, but may cause an increase in intensity of the attack.

Initiator. A term used to describe a material which stimulates polymerisation of monomer.

Although often called catalysts, the term initiator is more correct. In radical polymerisation, the initiator itself requires activation, either by radiation or another compound, often an amine, mercaptan, or metal naphthenate. The initiator, often a peroxide, is activated to produce free radicals, which then combine with monomer molecules, and these then grow by addition of further monomer molecules at the growing end, where activity is retained.

In ionic polymerisation, either a strong acid (cationic) or alkali metal alkyl, amide or alkoxide (anionic) causes initiation by conversion of monomer to carbon "ions" by the subtraction (cationic) or addition (anionic) of electrons. Condensation polymerisation does not always require an initiator, but acids or bases are often added to improve the speed of reaction or yield.

Injection moulding. The production of components to precise shape and dimensions, by the use of precision-made metal moulds and the injection into them, under pressure, of plastic material (and sometimes low melting point metallic alloys). The injected material is generally molten, or at least in a state approaching the viscosity of the liquid, and heating in a preliminary chamber is usually used. The heating chamber is fed with new material from a hopper on each stroke, and the energy of the pressure stroke assists in lowering the viscosity to the correct level, as the material is being injected into the mould. The moulds are usually split, with internal cores, to facilitate extraction of the finished component and allow mass production on a continuous basis.

Injector. A device to securing an induced draught to a furnace or cupola that consists of a pipe having its end widened. Steam is forced in, carrying with it a stream of air, the quantity of which can be regulated as required. Injectors are also used for supplying the correct proportions of steam and air to various types of gas producer.

Inner electrons. Those electrons close to the nucleus, of lower energy level and therefore tightly bound to the regions of the nucleus. They play little part in physical and chemical reactions, except for the production of spectra and in radiation damage by bombarding particles.

Inoculated cast iron. High-quality castings produced by adding to the molten metal some material which will modify the structure and properties of the iron. The additions are usually made in the ladle and frequently consist of ferro-silicon, calcium silicon, ferro-manganese silicon, zirconium-silicon or other graphitising agents which are sold under various trade names. When these inoculants are used the iron is melted with a lower silicon content than is desired in the casting and the remainder of the silicon is added in the ladle.

Carbide stabilisers containing chromium are also used as inoculants. Magnesium and cerium are used as graphitisers.

Inoculation. The addition to molten metal of substances that are intended to form nuclei for crystallisation. The inoculation processes are most frequently used in the production of high-grade cast irons. (*See also* **Inoculated cast iron.**)

Inspissate. A term meaning to thicken by evaporation that is usually applied in reference to aqueous fluxes.

Insulator. Electrical insulators have no available states of electron energy for conduction by electrons or positive holes, their outer electrons being involved in covalent bonds, or unable to bridge an energy gap between the valence state and the next highest (conduction) band of feasible energy levels. Insulators usually show a forbidden energy gap of the order of several electron volts in magnitude, and resistivity of the order of 10^{12} or greater. They generally disintegrate when subjected to a high electrical field or high temperature and so cease to exist in their original form when any attempt is made to make them conducting. Naturally enough, they find use in electrical circuits and machines, to eliminate discharges, control switching, and improve safety. (*See* **Conduction by electrons.**)

Thermal insulators have no available electron energy levels to assist in the transmission of phonons, and the vibration of atoms is entirely responsible for the transmission. The thermal conductivity of an insulator is inversely proportional to the square of the lattice parameter, if crystalline, and if amorphous, proportional to the thermal energy content. The effectiveness of a thermal insulator is also dependent on its thermal capacity, since a good insulator does not necessarily remain cold. If the thermal capacity is low, the material soon heats up, is saturated, and does not go on absorbing heat from a source indefinitely. The parameter which best quantifies this is the thermal diffusivity a, the ratio of thermal conductivity to thermal capacity.

$$a = \frac{k_t}{\rho c_p}$$

where k_t is thermal conductivity, ρ the density and c_p the specific heat at constant pressure.

Metals have a high value of a because the conductivity k_t is high. Many good insulating materials are porous, as both conductivity and heat capacity are low, although the value of diffusivity may not be very different from that of a more dense material. The propagation of temperature must therefore consider both conductivity and capacity.

The rate of abstraction of heat when a cold and hot surface are placed in contact, is important industrially and domestically. The contact coefficient b is defined by the product of conductivity and capacity, $b = k_t . \rho . c_p$. When b is small, the temperature difference between the interior and surface of a body is high, whether heat is being added or subtracted. Good insulating materials need a low value of b, so that less heat is gained or lost by transmission, i.e. both k_t and $\rho.c_p$ are low, as demonstrated in considering diffusivity. Materials for floor coverings, upholstery, handles, are required to have a low value of b for this reason. Cotton is a textile with a high value of b and hence cotton sheets feel cold when in contact with a warm body. Cotton has a high value of k_t, whereas wool has a low one. Metals feel cold to a warmer body, since they have both a high k_t and frequently a high $\rho.c_p$ also.

During the night, an unheated room falls in temperature, and water in the atmosphere is absorbed by textiles, because their low b value also means a rapid cooling at the surface, when the ambient temperature falls. Water thus condenses on textiles, as it does out of doors on wooden fences or bridges over streams, on a summer evening, whereas stone, with a higher b value, remains dry. Next day, the increase in ambient temperature causes textiles to heat up again, driving off the water, which will condense on windows or metals, if they remain colder, which they do, because of their value of b, which means they heat up more slowly. The condensation of water on glasses or metal objects brought from a cold to a heated room is another example. (*See* **Conduction by the lattice**.)

Intensifier. A term applied to some boron alloys used in steel making for increasing the hardenability of steel. The amount of boron added for this purpose seldom exceeds a few thousandths of 1%. The intensifiers are available in various grades containing 0.20-24% B, together with other elements that reduce the oxidation of boron in the steel.

The term is also used in reference to those substances that increase the intensity of corrosion, whether or not the corrosion rate is increased. The intensity of corrosion attack is defined as the corrosion per unit time per unit area, calculated only on the portion of the specimen affected.

Intensity of magnetisation. The magnetic moment per unit volume of a magnetised body. The ratio of this intensity to the applied magnetising field is the susceptibility.

Interatomic distance. As the name implies, the distance between like atoms in a crystal lattice, or in a compound, the distance between constituent atoms. When atoms exist in a "free" state, i.e. as vapour, there will be an equilibrium between electrical forces and the momentum of the electrons, retaining an approximately spherical distribution of the electron clouds in the highest energy levels. Presumably, the radius of the outermost extremities of this assumed spherical shape can be regarded as the atomic radius of that particular material. When atoms are brought together however, in molecules, in a lattice, or in compounds, the electron clouds overlap, and, depending on the type of bonding developed, will assume a new atomic radius, which is indicative of the true interatomic distances in those circumstances. Like atoms will presumably be symmetrical, i.e. the atomic radius is half the interatomic distance. Ions which form in ionic bonding however, react differently according to whether they are positive or negative. Metals, which lose electrons to become positively charged, will contract in atomic radius, because the now unequal positive charge on the nucleus will attract the outer electrons closer to the nucleus. Non-metals which form anions, negatively charged, will expand, because the attraction forces are decreased. It is necessary therefore to distinguish an ionic radius quite separately from the atomic radius of the free atom.

When like atoms form a crystal lattice, there will be greater overlap of electron clouds with close packing (co-ordination number 12) than with more open packing (say co-ordination

number 8). In fact, this difference is of the order of 3%. With crystal lattices of like atoms however, the co-ordination number is fixed by the crystal structure, and so the atomic radius will be one half of the interatomic distance between nearest neighbours. When atoms substitute for each other, as in metallic alloys, there will be differences in atomic radius used for each, unless both or all have close packing in their elemental crystal lattice, which is clearly not so. Atomic radii are therefore calculated on a basis of co-ordination number 12, whatever the actual packing in the elemental lattice, as this gives at least a consistent basis for argument in consideration of metallic alloys.

The combination of ions leads to further complications when the ions of one element may be doubly charged, as against an element which is singly ionised. An empirical basis is again used, calculating a "univalent" ionic radius although this does not truly represent the limit of the electron cloud in free space.

Similarly, values can be calculated for covalent atomic radii, where covalent bonds are formed. Pauling found that, in carbon, the interatomic distances decreased as the bond became more unsaturated, i.e. in double and triple bonds. He evolved an equation $R(1) - R(n) = 0.3 \log n$ where $R(1)$ is the covalent radius for the normal single covalent bond, and $R(n)$ is the radius in a structure where v single bonds resonate among N positions, and $n = v/N$. Pauling later used this equation to calculate single bond radii for some metallic elements, to interpret their crystal lattice formations. A satisfactory theory of covalent bond lengths has developed from this concept, but modification is needed when appreciable differences occur between electronegativity of the elements in combination.

There are therefore three possible values for interatomic distances, dependent on the element, the structure and the nature of the bond. Care is needed not to confuse these, especially the large differences between atomic and ionic radii.

Intercrystalline cracks. Cracks or fractures that follow the grain boundaries. They are frequently the result of repeated stress caused by vibration. The best method of detecting cracks at the surface in magnetic materials is by magnetising the metal and applying a magnetic powder. The latter gathers about the crack and renders it more readily visible. The method, known as Magnaflux, has been developed commercially. For non-magnetic materials the use of suitable etchants is generally adopted. (*See* **Non-destructive testing**.)

Interface. A term used to indicate the surface of separation between two components. These may consist of immiscible liquids or solids. The term is also used for the contact between basis metal and a coating upon it.

Intergranular corrosion. A form of corrosion in which the attack of the corrosive medium takes place preferentially at, and is concentrated on, the grain boundaries. This type of corrosion leads to disintegration before the bulk of the metallic mass has been attacked to any considerable extent. Available evidence seems to indicate that specific attack at grain boundaries is frequently due to the presence of a trace of impurity which can accumulate as a eutectic or otherwise. On the other hand, very pure aluminium and magnesium often reveal intergranular attack. The penetration of the corrosive attack along grain boundaries is dependent on the nature of the metal and of the corrosive medium. The formation of an insoluble corrosion product which may be deposited between grains tends to reduce the action, while the formation of complex ions which prevents the formation of simple metallic cations will often favour the intergranular attack.

Intermediate constituent. A metallic phase formed when two metals in an alloy combine in definite proportions to form a new constituent which has a different crystal structure from that of the component metals, i.e. an intermetallic compound.

Intermediate metals. When a circuit is made by joining several pieces of different metals and all the junctions are maintained at the same temperature there will be no current flow in the circuit. If the temperature of one of the junctions is raised, a current flows, the value of which is the same as that which would be obtained in a circuit of the same resistance and composed only of the two metals at the heated junction. If the contact potential between three metals in such a circuit is V_{AB}, V_{BC}, and V_{AC}, then $V_{AB} = V_{AC} - V_{BC}$. This equation is the mathematical statement of the law of intermediate metals used in the theory of thermocouples.

Intermetallic compound. When two or more metals are alloyed together, the possible phases which form are determined by three factors; the relative atomic sizes of the constituents, and the relative valency of the metals. If the range of homogeneity of a phase is comparatively small, it is often termed an intermetallic compound, even though it does not necessarily obey any valency laws, or have a fixed composition. Three types of "compound" are recognised; electrochemical, size-factor, and electron compounds.

When one element is strongly electropositive and the other strongly electronegative, there is a strong tendency to form compounds which do resemble normal chemical compounds like inorganic salts. These electrochemical intermetallic compounds satisfy the valency laws, have a small range of solubility outside the fixed composition, and usually have high melting points, similar to ionic compounds. The crystal structures are also very similar if not identical, with the classic ionic compound lattices such as rock salt (NaCl) and fluorite (CaF$_2$). Magnesium, a strongly electropositive element, forms one series of such compounds Mg$_2$X, which have a structure anti-isomorphous with fluorite, the magnesium atoms being in the place of the fluorine, and X in place of calcium (*see* **Fluorite**). The element X in this series may be from group IV of the periodic table, and the electronegativity increases in the order lead, tin, silicon, so that the compounds increase in melting point in the order Mg$_2$Pb, Mg$_2$Sn and Mg$_2$Si. Similar effects occur with group V elements, although to satisfy the valency laws the general composition is Mg$_3$Y$_2$ and the melting point increases in order when Y = Bi, Sb or As.

When the atomic diameters differ appreciably, compounds may be formed in which the atoms take up appropriate positions in the lattice, either interstitial or substitutional. When the radius of an atom is more than 0.41 times that of the basic metal, it cannot fit into an interstitial position without distortion, but if the radius is between 0.41 and 0.59 times that of the parent, interstitial compounds are formed, of which hydrides, borides, carbides and nitrides of the transition metals are examples. These compounds are cubic or hexagonal in structure, the non-metal being in an interstitial position in the lattice, and the phases exist over a narrow range of general composition M$_2$X or MX. The rock salt structure is adopted by the carbides and nitrides of Ti, Zr, Hf, V, Nb and Ta. The metal atoms have obviously altered their structure to f.c.c., so that the interstitial void is as large as possible, and the non-metal atom has six close neighbours of metal atoms.

When the ratio of atomic radii exceeds 0.59, the interstitial positions would create excessive distortion, and more complex structures usually result. The carbide of iron, cementite Fe$_3$C is an example, for which the radius ratio is 0.63. For size differences between metallic atoms whose radius ratio is 1.2 to 1.3, a special structure, the Laves phases, may accommodate the atoms in an efficient packing. Each atom A has twelve B neighbours and four A neighbours, whilst each B atom has twelve neighbours six of which are A and six B. The general formula is AB$_2$, and the average co-ordination number is higher than that of close-packed lattices of equal sized atoms. Three structural types can be recognised, with the small atoms arranged on a tetrahedral spacing, the final arrangement being cubic or hexagonal, and represented by the structures of the compounds MgCu$_2$, MgNi$_2$ and MgZn$_2$. Because of the complex geometry, such structures have a narrow range of composition, and are most common in the transition metals.

The third type of intermetallic compound occurs when the ratio of electrons to atoms is a particular quantity, namely 3:2; 21:13; and 7:4. These structures were originally studied in the alloys between copper, silver and gold, but do apply to others to a limited extent. In copper-aluminium, three valency electrons from aluminium and one from copper combine in Cu$_3$Al to give 6 electrons from 4 atoms, the 3:2 ratio, with a body-centred cubic structure typical of β brass in the copper-zinc system. The β brass occurs as CuZn, two electrons from zinc and one from copper, again a 3:2 ratio. In copper-tin, Cu$_5$Sn, 9 electrons from 6 atoms, also gives a β brass b.c.c. structure. The phase γ brass, Cu$_5$Zn$_8$, with a complex cubic structure of 52 atoms to the unit cell, has an electron/atom ratio of 21:13, the ϵ phase of brass, CuZn$_3$, a close-packed hexagonal lattice of 7:4 ratio.

In applying these rules to transition metals, they are credited with zero valency, which is reasonable when considering the incomplete complement of d electrons and the filled outermost s electron level. Further consideration of these electron compounds indicates that, although the electron concentration is by far the most important, the atomic size, valency, and electrochemical influences also play some part.

The intermetallic compounds are therefore a fruitful source of study of many facets of alloy theory, but by no means a set of fixed rules applicable across the periodic table.

Internal chills. Castings are regarded as chilled when they have been poured under such conditions that the whole or some part of the metal has been caused to solidify much more rapidly than it would have done with the ordinary methods of sand moulding. The cooling effect is secured by employing a metal face in the mould wherever the change is to be produced. If it is an interior that is being chilled, as in cases where mandrels are used instead of cores, the internal chilling will cause the metal to bind on the core former.

Internal energy. The internal energy of a system is made up of the internal energies of its components. The internal energy of a separate body is dependent on its mass, its chemical nature and its physical condition. For given conditions it is proportional to mass and the internal energy of a given substance may be written $U = \Sigma n\mu$ where μ = molar internal energy and n = number of moles. The change in the energy of a substance is given by $\Delta U = A - Q$ where A = external work done and Q = heat absorbed.

If heat is added to a body while the volume remains constant (no external work), then the heat serves only to increase the internal energy. If C_V is the amount of heat required to raise the temperature of 1 mole by 1 °C (= molar heat), the change in internal energy

$$U = \int_{T_1}^{T_2} C_V \, dT \text{ or } \frac{dU}{dT} = C_V$$

This provides a definition of internal (molar) energy, viz. the differential coefficient of the molar internal energy is the true molar heat at constant volume.

It should be noted that the absolute value of the internal energy of an atom is unknown. In most thermodynamical calculations only changes in the internal energy are involved.

Internal friction. The effect of mechanical vibrations on the physical equilibrium of crystals already undergoing thermal vibrations. The interaction produces a damping of the mechanical vibrations, which depends on the inherent properties of the crystalline material, the physical structure, and the impurities or crystal defects present. The energy of the mechanical vibrations is generally absorbed as increased thermal vibration, i.e. a rise in temperature occurs. (*See* **Damping**.)

Internal stresses. Stresses which arise due to thermal gradients, mechanical stress gradients, or other cause, leading to strain in the material, sometimes referred to as "stored energy". The casting of metals and alloys into fairly rigid moulds, e.g. sand, leads to thermal gradients and hindering of natural thermal contraction on cooling. Local plastic deformation may occur, particularly at higher temperatures when the yield stress is low, and further cooling may then reverse the stresss in local areas, now remaining as elastic stress, although often close to the yield point. Friction between mould surfaces and a casting may also cause internal stress.

During welding, or other joining processes involving heat, thermal expansion and contraction may be hindered by the restraint imposed on some structural members, e.g. the welding of a square or circular plate on to another larger plate, automatically imposes restraint on the small plate when it cools down. The larger plate does not reach the same temperature, does not expand so much, and the weld joint prevents contraction of the small plate.

Thermal gradients may be produced in casting, welding and heat treatment, and often in service, when different cross-sections of material heat up or cool down to different levels or at different rates. Any localised plastic deformation which results to accommodate this is not necessarily reversible, particularly on a cooling cycle, and hence internal stress occurs. Transformations in the solid state produce micro stresses of a similar nature, but in reasonably large sections, these micro stresses can become macro stress by the same process. Variations in composition produce stress, as in carburised steel bars, where the outer layers are higher in carbon, and the transformation stresses therefore much greater. Friction may lead to transformations and subsequent internal stress on removal, as in the surface layers of railway tracks or brake drums.

The stress can readily be demonstrated by destructive methods. A fine saw cut in a member under tension, will lead to the opening of a larger gap than the width of a saw cut. Cold rolled sheets of metal are under internal stress unless annealed, and cutting with shears leads to twisting and warping unless performed uniformly across the sheet.

The methods of alleviation are thermal or mechanical, and heat treatment is by far the most effective. Heating is carried out reasonably slowly, to a point where the yield stress is sufficiently low for plastic deformation to take place, removing the stress. Time must be allowed for adequate creep to take place, and then very slow cooling to prevent reintroduction of stress during the cooling cycle. This type of treatment cannot be used where transformations are involved, as the properties of the material may be radically changed. In such cases, mechanical methods rely on applying compression stress to those parts of the structure in tension, by peening with a hammer or shot. The surface is affected, but this may be less objectionable than the residual stress.

Claims have also been made for lowering temperatures, and for vibration of structures. No doubt some stress is relieved by these treatments, but to a restricted level. In cases where very large welded structures are involved, and a brittle transition may occur during construction or service, thermal stress relief is almost essential. Pressure vessels in the chemical, petroleum and nuclear industries are examples.

International annealed copper standard. A standard according to which the specific resistance of pure annealed copper at 20 °C is 1.7241 $\mu\Omega$ cm^{-2} cm^{-1}. The electrical conductivity is therefore 0.58001 reciprocal $\mu\Omega$ cm^{-2} cm^{-1}. The conductivity of other metals is usually defined as a percentage of this value. Abbreviation is I.A.C.S.

Interrupted ageing. The process of ageing an alloy at two or more different temperatures. The ageing is done in stages, cooling the material to room temperature between the stages.

Interrupted quenching. A quenching process in which the metal is removed from the quenching medium while it is still at a substantially higher temperature than that of the medium.

Interstices. Another name for voids, the interstitial spaces between atoms in a crystal lattice.

Interstitial atoms. Atoms occupying void spaces. This gives rise to two possible effects — the diffusion of such atoms through the lattice, or the creation of an interstitial solid solution of one atom in another matrix.

To occupy an interstitial position, the atoms must be very much smaller than those of the matrix, usually of atomic radius less than half that of the host atoms. In metals, only hydrogen, boron, nitrogen and carbon are sufficiently small to merit consideration in this context.

When the interstitial atom is identical with the matrix, this is a lattice defect, since there is no difference in atomic size involved. Such an atom imposes great strain on the surrounding lattice, and is frequently associated with a vacant lattice site close by. Diffusion is then a matter of the pair, the interstitial atom and the vacancy, migrating together. (*See* **Frenkel defect.**)

Intertype. A type-metal casting machine that is capable of using the harder kinds of printing alloys. In principle it resembles the Linotype, but there are important modifications in the size and shape of the moulds.

Intracrystalline fracture. The type of fracture that occurs across, i.e. through, the crystals of a metal.

Intrinsic energy. The term used in reference to the total amount of energy in a body. The actual value of this quantity is unknown, but since the elements do not undergo change into one another, the absolute value is of no consequence when dealing only with changes of energy. The intrinsic energy of an uncombined atom is usually taken as zero and thus, since the intrinsic energy of a compound differs from the intrinsic energy of its constituent atoms by the amount of the heat of formation, the intrinsic energy of a compound becomes equal to its heat of formation, but with reversed sign.

Intrinsic semiconductor. A material having an energy gap between electrons in the valence band, and the conduction band of higher energy. When the gap is very large, the material is an insulator, when small enough to be bridged by a boost of energy such as bombardment by electromagnetic radiation, or by thermal energy, the electrons jump the energy gap and the material becomes conducting, and such a material is an intrinsic semiconductor. (*See* **Conduction by electrons.**)

Intrusions. An igneous rock formation produced by a flow of molten rock magma into the cavities of existing rocks.

Inverse chill. A term in common use in connection with the appearance of the fracture of a casting—but especially in reference to malleable cast iron.

The fracture shows a white core and a dark outer shell. The name appears to have arisen because this type of fracture is the reverse of the appearance of the fracture of grey cast iron when cast against a chill. The term is not particularly appropriate and could well be discontinued.

Inverse cooling curve. A type of cooling curve used in thermal analysis in which temperature is plotted as the ordinate and the time required to cool through equal intervals of temperature as the abcissa. Usually the times are taken for a fall in temperature of 2 °C, and these are plotted against the temperatures at the intervals concerned. These curves are rather more spectacular, but are not more accurate than direct cooling curves. (*See also* **Cooling curve.**)

Inverse segregation. A concentration of constituents having relatively low melting points near the outer surfaces of an alloy ingot or casting. It occurs most generally in non-ferrous alloys where a constituent having a lower melting point than that of the primary crystals tends to be concentrated near the cooling surfaces of the casting.

Investment casting. A process based on the old "lost-wax" process and used for producing precision castings. The process is also called "precision casting", but the degree of precision obtainable, though high when compared with the tolerances characteristic of other hot-metal-forming methods, is not comparable with precision machined parts. The process, however, shows advantages over other methods by virtue of the superior finish, close tolerance, better physical properties and flexibility of design of finished product. There are limitations on the size of casting for economical production, and the advantages are most readily achieved if the total weight of casting does not exceed 2 to 3 kg.

A master pattern, preferably in steel, brass or bronze, is made, and from this dies are produced, usually in soft metal. An alloy of tin and bismuth is frequently employed. Patterns are produced by injection into the metal dies by means of a "grease gun", of either wax or some plastic material that can be completely removed at a later stage by heating. The wax pattern is coated with a slurry of fine refractory powder which serves as a mould-facing material, and then transferred to a moulding-box, where refractory material is packed around the pattern. After the mould has been dried and cured, the wax pattern is melted out, a fairly high temperature being used to ensure that all the wax is removed from the mould. The molten metal is usually introduced into the mould under (air) pressure. (*See* **Cire perdue.**)

Involute. A term meaning having the edges rolled inwards.

Iodine. A halogen element of atomic number 53 and atomic weight 126.9. It is a solid which gives a brown solution in potassium iodide and a violet solution in chloroform and other organic solvents. The main sources of the element are: (*a*) Chile saltpetre deposits and (*b*) oil-well brines and seaweed. In solution or in the vapour form it attacks metals, and this action has been utilised in studying the phenomenon of film formation and the measurement of the film thickness on metals. The latter has been accomplished by gauging the film thickness by colour against standard colours and by determining the coulombs required for the cathodic reduction of the deposit. The action of the element on metals has further been used in stripping off the oxide film that forms on iron, a 10% solution in potassium iodide being used, which preferentially attacks the metal, causing the oxide film to flake off.

The action of iodine on sodium thiosulphate, using starch as the indicator, is used in a number of quantitative determinations in metallurgical analysis, e.g. determination of copper in copper-brass- and bronze-plating solutions; gold and lead in plating solutions and certain organic brighteners. (*See* Table 2.)

Ion. An atom or a radical that has gained (or lost) one or more electrons and thus has more (or fewer) electrons than are normally associated with the electrically neutral atom or radical. Ions are charged positively or negatively, according to whether they lose or gain electrons. The soluble salts of all metals dissociate more or less on going into solution, forming negatively charged ions (acid radical) and positively charged ions (metal). Under the influence of an applied potential difference the positively charged ions move towards the cathode, and are therefore named cations. The negatively charged ions under the same influence move towards the anode, and are therefore termed anions.

Ionic bond. The term applied to a type of bonding or linkage in the formation of molecules and chemical compounds in which the essential feature is the transfer of electrons from one atom to another. An atom of an element in Group II of the periodic table, such as magnesium, having two electrons in its outermost energy level, can loose these electrons to form a bi-positive ion Mg^{2+}. An atom of oxygen has six electrons in its outer shell and can accept two electrons to complete the stable octet. In the same way two atoms of chlorine, each having seven electrons in their outer shells, can each accept one electron. The result is the formation of a stable compound MgO or $MgCl_2$, and this stability, especially towards heat, is associated with the attainment of the "inert gas" structure. Also known as electrovalent, electronic, polar and heteropolar bond.

Ionic mobility. Because of the relationship of ion conductances to the velocities at which the ions migrate under the influence of an applied potential difference, they are often referred to as ionic mobilities. The mobility is measured as the velocity attained by an ion under the influence of a potential gradient of 1 V cm^{-1} in dilute solution. The values of the ionic mobilities for a number of common ions are as follow:

Ion	cm s^{-1}
Potassium	0.00067
Silver	0.00057
Sodium	0.00045
Chlorine	0.00068
Hydrogen	0.00326

These remarkably small values are due to the enormous frictional resistance due to the viscosity of the solvent.

Ionic radius. *See* **Interatomic distance.**

Ionisation. The removal or addition of one or more electrons from an atom, to disturb the electrical neutrality, and leave an ion, positively or negatively charged. Ionisation may occur as a result of high energy input to the atom, whether thermal or electrical. (In the case of mechanical energy, e.g. an explosion, it is difficult to distinguish mechanical from the accompanying thermal energy which results.)

The word is also used for the dissociation of an ionic bonded material into its constituent ions, when dissolved in an ionising solvent. The degree of ionisation is often referred to, but is strictly the degree of dissociation, the material is already ionised in the solid state, but held in an electrically neutral configuration by its bond and structure.

Ionisation potential. This is defined as the energy per unit charge required to remove an electron from an atom and is numerically equal to the work done, expressed in electron volts, in removing the electron from the atom. Since the ionisation potential may vary somewhat according to the particular configuration of an atom, it is more precisely defined as the energy corresponding to the passage from the most stable state of the atom to the most stable state of the ion, and then the first ionisation potential corresponds to the energy required to remove the most loosely bound electron from the atom.

Iridescent. The production of fine colours on a surface, due to interference of light reflected from front and back of a very thin film.

Iridium. A steely-grey metallic element that is obtained as a by-product, of 99.7% purity, in the extraction of platinum from electrolytic copper and nickel refinery residues. Iridium is insoluble in all acids, including *aqua regia,* and this property combined with high hardness makes it suitable for several special applications, despite the relatively high cost. The metal is brittle, but working properties are improved by adding 10% or more of rhodium. Grains of iridium metal, used for tipping fountain-pen nibs, are produced by adding approximately 7% Pt to a hot iridium sponge. The chief uses are as a hardener for platinum used in pen-points, watch and compass bearings and electrical contacts. It has also been employed for the extrusion dies for high-melting glasses and for small crucibles for high-temperature experiments.

Iridium is not wetted by mercury, and this phenomenon has been utilised in some electrical devices. Platinum-iridium alloys are used where resistance to attack by acids and corrosive solutions is desired and in jewellery. The standard metre of Paris is constructed in an alloy of 90% Pt, 10% Ir, and this same alloy together with platinum is used for thermocouples measuring high temperatures. Since, however, iridium volatilises above 1,000 °C, the platinum-rhodium alloy is used at higher temperatures. Tensile strength (annealed) 1.03 GN m^{-2}; crystal structure f.c.c.; lattice constant (20 °C) $a = 3.831 \times 10^{-10}$ m; neutron capture cross-section for thermal neutrons 430 barns; hardness (annealed) 220 D.P.N. (*See* Table 2.)

Iridosmine. An alloy of iridium and osmium in variable proportions that occurs usually in small flattened grains together with iridium and platinum in gold washing in the Urals, New South Wales, South America and Canada. It is used for tipping fountain-pen nibs.

Iris. A form of quartz showing the chromatic reflections of light from fractures, often produced artificially by suddenly cooling a heated crystal. Also known as rainbow quartz.

Iron. A heavy whitish metal and one of the most abundant and widely distributed of the metallic elements. It constitutes about 4.6% of the earth's crust and is the most important of all metals. It is the basis of the many types of steels and cast irons and provides essential materials for the many and varied needs of aeronautical, agricultural, automobile, civil, electrical, marine and mechanical engineering, and is used extensively in sanitary and waterworks engineering. It has important applications in the packaging and food-canning industries.

Iron is occasionally found in the free state as meteorites, but the most important sources are the oxide ores (*see* **Iron ores**). These are smelted in blast furnaces with coke and limestone to yield pig iron. This latter is an impure form of iron containing 3-3.75% C, 0.1-2% Si, 0.3 to about 1% Mn and varying amounts of sulphur and phosphorus. The actual composition of the iron depends on the purity of the raw materials and on the conditions under which the furnace is operated. The pig iron may be transferred as molten metal to steelmaking furnaces or cast into open sand or chill moulds for subsequent processes—the cold metal ingots are then referred to as pigs.

Wrought iron is produced by refining pig iron under an oxidising slag in reverberatory or puddling furnaces at temperatures just below the melting point of pure iron. The pasty mass of practically pure iron, together with adherent slag, is collected into a spongy ball, which is consolidated by forging and rolling. During these processes a considerable amount of slag is squeezed out of the metal, but though the wrought iron is practically free from metallic impurities, about 1% of slag remains, and this is present as elongated stringers.

The mechanical properties of pure iron are too low for many important applications, but there are several commercial grades of iron such as Armco iron, which may be used in the form of sheet for deep-drawn components and pressed work.

The presence of carbon, even in small percentages, has an appreciable effect on the properties of iron and the metal is most commonly used in the form of iron-carbon alloys such as steel (q.v.) and cast iron. The principal difference between iron and steel is the carbon content. Steels may contain from about 0.07 to about 1.5% C together with other alloying elements, but deleterious elements such as sulphur and phosphorus are restricted to 0.06% or less. (*See* Tables 2 and 6.)

Cast irons contain over 1.5% C (usually from 2.5 to 3.5%) and

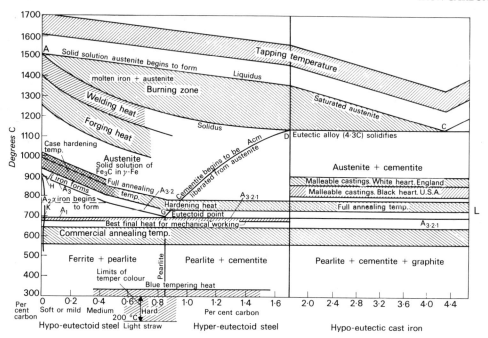

Fig. 79 *Iron-carbon equilibrium diagram showing ranges of temperature for various operations.*

silicon and manganese are also present. Other elements may be added for special purposes, and the sulphur and phosphorus content are not so closely controlled as in steels. In ordinary cast irons the elements carbon, silicon, manganese, phosphorus and sulphur together frequently exceed 5%.

Irons may be classified broadly according to whether they are produced by cold-blast or hot-blast furnaces, and each of these may be sub-divided into grey, foundry, forge and white irons:

(a) Cold-blast pig irons are produced in furnaces where the temperature of the blast does not exceed about 100 °C. Usually hematite iron ores are used. The product is used for high-duty castings.

(b) Hot-blast pig irons are produced in furnaces where the air blast is preheated to a much higher temperature than in *(a)* and basic pig irons are usually produced. The cast iron produced contains a relatively high percentage of phosphorus, and the product is used mainly for general iron casting or for making basic steel. If magnetite or red hematite be used as the raw material the product has a relatively low phosphorus content.

The form in which the carbon is present, either as graphite or as iron carbide, has a marked effect on the appearance of the metal when fractured and on its physical properties. When the greater part of the carbon is present as graphite, the fracture is grey and the metal is known as grey iron. When most of the carbon is present as iron carbide Fe_3C, the fracture is white and the metal is known as white iron. Between these two are several intermediate conditions together classed as mottled irons. A comparison of grey and white irons is given in Table 30.

TABLE 30. *Comparison of Grey and White Irons*

Type	Grey iron	White iron
Production from molten iron	Slowly cooled	Rapidly cooled
Carbon present as	Graphite	Iron carbide
Silicon content	High, promoting the formation of graphite	Low
Manganese content	High	Low, promoting the retention of Fe_3C
Grain	Open (coarse)	Close (fine)
Properties	Soft. Tough. Readily machinable	Hard. Brittle. Not machined

Malleable cast iron is produced from ordinary white pig iron by annealing, in boxes containing red hematite ore, at 700-800 °C. The white fracture becomes grey, due to the decomposition of the iron carbide. In America the oxide of iron is frequently omitted and the temperature of annealing is lower, so that the iron carbide near the surface is decomposed by heat alone. The terms black heart and white heart—which are self-explanatory—are used in reference to the two types of castings.

Iron Carbide. A compound of iron and carbon Fe_3C that is a constituent of steels, its presence being associated with the changes that occur with the hardening of steel. When ordinary carbon steels containing varying amounts of carbon are allowed to cool from a high temperature it is found that there are two or more temperatures at which the regular rate of cooling is interrupted. The phenomenon is referred to as recalescence, and the temperatures at which retardation during cooling takes place are indicated by Ar_1, Ar_2, Ar_3 (*see* **Arrest points**). The Ar_3 change takes place in steels containing not more than 0.83% C at temperatures varying from 900 °C in pure iron down to 770 °C in steel of this composition. The Ar_2 change takes place at about 770 °C and the Ar_1 at about 700 °C. The Ar_1 change is due to the formation of iron carbide. All steels that are slowly cooled from above this temperature contain the greater part of their carbon as Fe_3C, while those that are cooled suddenly from above 700 °C contain scarcely any of this compound. Pearlite is a definite mixture of iron and iron carbide forming the eutectoid. Sorbite is a form of pearlite in which the iron carbide has not had time to segregate. It is formed by rapidly cooling the steel immediately after the Ar_1 change.

Iron-carbon diagram. A diagram showing the equilibrium of the solid and liquid phases of the pure metal and its alloys with carbon as a function of the temperature and composition. Pure iron melts at 1,535 °C, but the presence of carbon depresses the

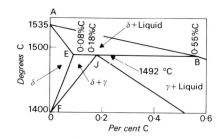

Fig. 80 *Formation of δ phase in the iron-carbon system.*

melting point; the larger the percentage of carbon, the lower is the melting point. This is plotted on the line $A-C$, which is the liquidus. At all temperatures and compositions above and to the right of AC the alloy is molten (see Fig. 79). At 1,130 °C hot solid iron will dissolve 1.7% C (point D), and rather less than this amount as the temperature increases. This relationship is plotted on the line $A-D$, which is the solidus. At all temperatures below and to the left of $A-D$ the alloys are solid. The region between the liquidus and solidus shows the temperatures and compositions at which the alloy is in a pasty state—part solid, part liquid.

The area $ADGH$ is labelled austenite—a solid solution of iron carbide in γ-iron. The limit of the solubility is indicated by the line DG. Austenite is an aggregate of crystals of γ-iron, the iron atoms being arranged in the face-centred cubic system with the carbon atoms in voids within the space lattice. The atoms, though not free to move at random, do have a degree of diffusion which increases with temperature and, under the influence of heat, considerable rearrangement of atoms can take place, both by interchange of position and by moving in of other available atoms. These changes, which are associated with an absorption or evolution of heat, are evident on the time-temperature diagram (cooling curves) as arrest points. They are listed in Table 31.

Cooling curves of low-carbon steels show that the change occurring at 910 °C from γ- to α-iron is lowered as the carbon content is increased. When the percentage of carbon is a little below 0.9 the arrest points A_3 and A_1 coincide. These facts are summarised on the diagram by lines HGD (H being the point A_3) and KGL. At G the eutectoid steel contains 0.83% C. Low-carbon steels, on cooling, show that the constituent that is first precipitated is iron or ferrite. High-carbon steels first precipitate the excess of iron carbide (cementite). These facts are indicated in the area under the lines HG and GD, respectively.

A 0.83% C steel, after slowly cooling, has the appearance of small lamellar crystals of iron and iron carbide more or less parallel, and exhibits a pearly lustre. This is called pearlite.

In hypereutectoid steels, on slow cooling, the excess of cementite surrounds the pearlite masses.

Figure 80 shows detail of the formation of δ-iron omitted from Fig. 79 for clarity.

TABLE 31. *Arrest Points Associated with Iron-Carbon Alloys*

Temperature, °C	Arrest	Crystal Structure	State
1,535	—	Body-centred cubic	Melting point of δ-iron
1,390	A_4	Face-centred cubic	Recrystallisation of γ-iron
910	A_3	Body-centred cubic	Recrystallisation of β-iron (non-magnetic α-iron)
770	A_2	Body-centred cubic	α-iron; no recrystallisation (magnetic point)
690	A_1	—	—

Iron-constantan couple. A base metal couple having a somewhat longer life than chromel-alumel in reducing atmospheres at temperatures up to 980 °C, but not recommended for continuous use in air at temperatures above 760 °C. Above about 200 °C the volts/temperature ratio is practically constant at approximately 6 mV/100 °C.

Ironing. A term applied to the process of smoothing (and possibly also reducing the wall thickness of) objects during deep drawing operations.

Iron meteorites. Meteorites that consist essentially of nickel-iron alloys.

Iron ores. The commonest compounds in which iron occurs in nature are the oxides, carbonates and sulphides. The most important ores are:

(a) Hematite (red) Fe_2O_3 containing when pure 70% Fe. This is found mainly in Sweden, Belgium, Alabama and the Lake Superior region of the USA. The iron content of the ore as mined may be as low as 25%.

(b) Limonite or brown hematite $2Fe_2O_3.3H_2O$ which contains 56.9% Fe when pure. This is found in Germany, France, Spain and the USA in Alabama, Virginia and Tennessee.

(c) Magnetite Fe_3O_4, a natural magnetic ore containing 72% Fe when pure, but 30-65% Fe may be considered more general. It is found in Sweden, Norway, the Urals and in the USA (New York State, Pennsylvania and New Jersey).

(d) The carbonate ores include siderite (chalybite) or spathic iron ore $FeCO_3$ and contain 48% Fe when pure. Other carbonate ores are clay ironstone (iron II (ferrous) carbonate mixed with clay) and blackband ironstone (iron II carbonate mixed with clay and coal). These ores are found in the UK.

All ores that are suitable for smelting must have low sulphur content. This means that the plentiful iron pyrites is not a commercial source of iron. However, a certain amount of phosphorus is permitted in ores that are to be used in cast iron or for the manufacture of steel by a basic process. Formerly only iron ores containing more than 30% Fe could be smelted profitably, but lower-grade ores can now be used, and the value of an ore deposit depends on geographical location, accessibility and the composition of the associated gangue as well as the actual iron content.

Iron pan. The name given to the hard layer found a short distance below the surface of sands and gravels. It is formed by the deposition of iron salts from percolating waters.

Iron pyrites. The disulphide of iron FeS_2. It crystallises in the regular system, often in cubes. It has the colour of yellow brass (fool's gold) and is very hard, striking sparks with steel. It often occurs in coal, and is the chief source of the sulphur dioxide occurring in coal gas. Mohs hardness 6-6.5, sp. gr. 5.19.

Another form of the disulphide is marcasite, which occurs in rhombic crystals often in the form of radiating nodules. It is grey to black in colour with a shiny surface and has a specific gravity of 4.68-4.85.

Ironstone. An impure form of iron carbonate $FeCO_3$ that occurs along with clay and carbonaceous matter. It is found as beds and nodules in many coal measures and in layers in many other formations. It is common in most of the British coalfields and in those of Pennsylvania.

Isoelectric point. Substances such as gelatin, albumin and casein (used in electroplating baths) are proteins that exhibit colloidal behaviour and the properties of amphoteric electrolytes. The isoelectric point is defined as that hydrogen-ion concentration at which the acidic and basic ionisation of the protein molecule are equal. In this unionised form it is electrically neutral and will not tend to move in either direction in an electric field. For gelatin the isoelectric point occurs at pH = 4.7. The complex ion is formed as a cation when the acidity is greater than corresponds to the isoelectric point. If the acidity is less, the complex ion occurs as an anion.

Isokawimeter. An instrument in which the electrical circuit is designed to suppress the effects of variation in conductivity, so indicating only changes in coupling between the test coil and the specimen. It is used for measuring the thickness of non-metallic coatings on non-ferrous metals in the form of flat sheets.

Isomers. The name given to those chemical substances that have the same molecular weight but have different structural formulae and possess different chemical properties.

Isometric. An alternative name for the regular (cubic) system of classification of crystalline forms.

Isomorphism. Substances that have analogous formulae and crystallise in the same form are isomorphous. As a result of the study of a large number of minerals, it became clear that ability to crystallise in the same form is not the criterion of isomorphism. In addition, the substances must be able to form (a) mixed crystals and (b) layer crystals. The mixed crystals are homogeneous crystals containing the isomorphous substances in variable proportions, and are therefore more appropriately called solid solutions. It is thus clear that isomorphous substances cannot be separated in a state of purity by crystallisation.

Since a certain amount of an element can be replaced in one of its compounds by an equivalent amount of an isomorphous element, the formula of such a compound when calculated from the analysis will not give a whole number of atoms of each of the isomorphous elements. In writing the formulae of such isomorphous mixtures the isomorphous elements are enclosed together in a bracket with a comma separating them. The common elements have been divided into the following series, the members of which form isomorphous compounds:

(1) K, Rb, Cs, Tl; Li, Na, Ag.
(2) Be, Mg, Zn, Cd, Mn, Fe, Co, Ni, Cu; Ca, Sr, Ba, Pb.

(3) La, Ce.
(4) Al, Fe, Cr, Co, Mn, Ir, Rh, Ca, In, Ti.
(5) Cu, Hg, Pb, Ag, Au.
(6) Si, Ti, Ge, Zr, Sn, Pb, Th, Mo, Mn, Pt, Fe.
(7) P, As, V, Sb, Bi.
(8) Cr, Mn, Mo, W, As, Sb, Fe.

As an example it may be quoted that spathic iron ore may have the iron partly replaced by magnesium, manganese, calcium and the formula is often written $(Fe,Mn,Ca,Mg)CO_3$.

There are many exceptions to the "law of isomorphism". The following examples indicate that the chemical as well as the crystallographic properties must be investigated before the materials can be regarded as isomorphous and before conclusions are drawn concerning appropriate methods of separation:

	System
Argentite (Ag_2S) and galena (PbS)	Cubic
Barytes ($BaSO_4$) and potassium borofluoride (KBF_4)	Orthorhombic
Scheelite ($CaWO_4$) and potassium periodate	Tetragonal

Some of the exceptions have been shown to be due to the existence of two or more varieties of a substance—polymorphism. Where each of two forms of a dimorphic substance are isomorphous with a form of another polymorphic substance the phenomenon is known as isodimorphism. Arsenic trioxide and antimony trioxide are isodimorphous; stannic oxide and titanium dioxide are isotrimorphous.

Polymorphism when exhibited by an element is termed allotropy.

Isoprene (1, 3-Butadiene, 2-methyl). The monomer of natural rubber. (*See Cis*-isoprene.)

Isosklers. Lines drawn on a surface connecting points of equal hardness.

Isothermal annealing. A heat-treatment process sometimes applied to steels. It involves heating to, and soaking at, some temperature above the transformation range and then cooling to, and holding at, a suitable temperature until the transformation from austenite to pearlite is complete. The final cooling is done quite freely.

Isothermal transformation diagram. A diagram giving a graphical representation of the rate of decomposition of austenite when held at constant sub-critical temperatures. From such a diagram can be obtained the time interval required for a steel to transform isothermally from austenite to ferrite and cementite. Within limits such diagrams provide a useful means of understanding hardenability and the effects produced in heat-treatment. They

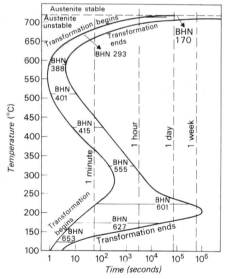

Fig. 81 *Isothermal transformation diagram of a hypoeutectoid steel.*

also contribute to the understanding of the controls that are necessary to improve heat-treatment (*see* Fig. 81). Also known as S-curves. (*See also* **Time-temperature-transformation diagrams.**)

Isotopes. The name applied to those instances where an element is known to exist in the form of atoms which have different atomic weights, but which have the same positive nuclear charge so that they occupy the same position in the periodic table and are chemically indistinguishable. An atom having an atomic number Z and atomic weight W contains Z protons and $W-Z$ neutrons. It is the atomic number, i.e. the number of electrons, which determines the chemical properties of an element. Atoms with nuclei containing the same number of protons, i.e. with the same atomic number and the same nuclear charge, but different numbers of neutrons (and hence different mass numbers or atomic weights), are essentially identical from a chemical standpoint, and are called isotopes. Most elements present in nature exist in two or more stable isotopic forms which are virtually indistinguishable except for some minor differences in physical properties. Hydrogen has two stable isotopes of mass numbers 1 and 2; oxygen has three isotopes and tin has ten.

In addition to the stable isotopes, and a few unstable ones which occur naturally, it is possible to produce, by nuclear reactions, unstable radioactive isotopes of all the known elements. More than 1,000 radioactive isotopes are now known. If a particular atom is to remain stable the ratio of the number of neutrons to protons in the nucleus must lie between certain limiting values.

When the numbers of neutrons and protons in a specific nucleus fall outside the neutron-proton limits for the given mass number, that isotope will be radioactive and will show beta activity.

As the physical properties of isotopes vary, it has become necessary to separate them for specific applications, e.g. boron 10, has a high capture cross-section for neutrons, and therefore can be used as a control or safety device in a nuclear reactor. The isotope ^{235}U is fissile by thermal neutrons, whereas ^{238}U is not. Separation has been achieved by diffusion processes, initially through membranes, later by centrifuging, or in some cases, by thermal columns. The process is very slow, taking advantage as it does of very small physical property differences, and consequently is very expensive. In medicine, the use of radioactive isotopes may justify the expense for specific therapeutic applications.

Isotropic. The term applied to a substance that possesses the same physical property in all directions. In cubic crystals, it applies to many properties but not to elastic ones.

If a transparent mineral is slowly rotated and remains black when viewed between crossed nicol prisms it is optically isotropic.

Itabarite. An iron ore composed of alternate layers or laminations of hematite and quartz. It is found in Brazil.

Itacolumite. A porous yellow sandstone found in Brazil. The stone is slightly flexible and gradually bends under its own weight when suitably supported. It is thought to be the source of the diamonds found in the district of Minas Geraes.

Italian asbestos. A calcium magnesium silicate $CaO.3MgO.4SiO_2$ that occurs as white or dark grey crystals. It is usually found in blade-shaped fibrous or compact columnar masses. Mohs hardness 5-6, sp. gr, 2.9-3.1.

Iwaarite. A form of andradite (q.v.) $3CaO.Fe_2O_3.SiO_2$ that contains a significant percentage of titanium. It crystallises in the cubic system, often in rhombic dodecahedra which are anisotropic, the cause of which is not known. Its occurrence is limited to intermediate and basic alkaline igneous rock. That found in Finland contains about 25% TiO_2. The mineral is dull and black with Mohs hardness 6.5-7.5.

Izod test. A type of impact test in which the notched specimen, mounted vertically, is subjected to a sudden blow delivered by the weight at the end of a pendulum arm. The energy required to break off the free end is a measure of the impact strength or toughness of the material.

J

Jacinth. A somewhat uncommon mineral consisting of zirconium silicate $ZrSiO_4$ frequently coloured with iron oxide. It occurs in crystalline rocks, especially in granular limestone, schist, gneiss, syenite and granite. The chief localities are the alluvial sands in Sri Lanka, the Urals, Greenland, Australia and in parts of North America. It crystallises in the tetragonal system. When pure the mineral is colourless, and this and the smoke-coloured varieties are called jargoon (q.v.). Many of the gemstones known in the trade as hyacinth are mere garnets (hessonite or cinnamon stones). Other types, e.g. Spanish compostella hyacinths, are coloured quartz. Also called hyacinth (q.v.).

Jack hammer. A portable type of pneumatic hammer used for drilling rock, brickwork or concrete.

Jacquets method. An electrolytic polishing (bright pickling) process for obtaining highly polished surfaces, free from mechanical deformation, for metallographic specimens. The specimens are first mechanically polished and then treated anodically in a suitable solution with careful control of current density and voltage. The polished surface obtained is superior to that obtained by the usual mechanical methods, and the surface structure shows important differences. The distortion produced by mechanical polishing is avoided, and there is an absence of the Beilby layer of disorganised metal.

Jacutinga. An iron ore composed of loose plates of hematite Fe_2O_3 and found at Minas Geraes, Brazil.

Jade. The name applied to two minerals, nephrite and jadeite, which are, respectively, the silicates of calcium and aluminium, and sodium and aluminium, containing insignificant amounts of titanium dioxide. The mineral is tough, but can be cut to intricate shapes. The colour varies from white to green and through shades of brown to black. The green colour is due to chromium, and the brown and black colours are due to the presence of iron. The material has been used so far in making highly decorated ornaments, amulets and axe-heads. It is found in many parts of Europe, China, Mexico and New Zealand. Mohs hardness 6-7, sp. gr. 2.9-3.3.

Jamesonite. A sulphide of lead and antimony $2PbS.Sb_2S_3$ that crystallises in the orthorhombic system and is usually found as acicular crystals, frequently with feather-like forms. It is dark grey in colour and usually contains about 20% sulphur. It often occurs in association with antimonite in Mexico and the western states of the USA. Mohs hardness 2-3, sp. gr. 5.5-6.

Japanese goldsize. A type of oil varnish which when applied thinly dries on exposure to air, and forms a hard film, giving a decorative or protective effect. Like other oil varnishes, it contains fossilised gum dissolved in oil and "thinned" by the addition of turpentine. Japanese goldsize is made for gilder's use and as a binder for colours. It contains 50 parts of gum to 4 parts of oil and 12 parts of turpentine. The gum should be pale in colour and selected from copal or manila gum and linseed oil is the basis.

Japanning. A process mainly used for coating metals with varnishes that are dried and hardened on, in hot stoves. The temperature employed varies somewhat with the metal or other base, but is usually about 90 °C. The process is so named because it is an imitation of the famous lacquering of Japan. The varnishes may be colourless or black, or may incorporate mineral paints. The black colour is due to the presence of asphaltum, which with gum animé is dissolved in linseed oil and thinned with turpentine. The transparent varnish is usually made from copal. The brilliant polished surface is much more durable and much more resistant to heat, moisture, etc., than ordinary painting or varnishing. The process is mainly applied to metals (motor-car bodies, cycle frames, domestic articles, etc.), but may also be used on wood, papier mache, slate (imitation black marble) and, when slightly modified, to leather (patent leather).

Jargoon. The name given to those varieties of zircons that are fine enough to be cut as gemstones but are not coloured orange or red like hyacinth. Certain jargoons exhibit a green colour, while others are brown or yellowish. The colourless jargoon may be obtained by heating certain of the coloured stones, and the change in colour is accompanied by an increase in brilliance. The "Matura" diamonds of Sri Lanka are in fact zircons that have been decolorised in this way. The hardness, however, is only about 7.5 (Mohs scale). (*See also* **Zircon.**)

Jasper. An impure variety of crypto-crystalline silica SiO_2 —quartz— that is hard and compact and is capable of taking a fine polish. It occurs in a variety of colours—dark green, brown, yellow, dark blue or black—and occasionally it is found banded with different colours. The term is now restricted to the opaque stone to distinguish it from chalcedony. It is in fact opal deposited in layers of different colours. Jasper occurs in Siberia (banded), Libya and Egypt (brown), Hessen, Lohlbach (Germany) (various) and Dakota (USA) (red).

Javelle water. A bleaching solution made by passing chlorine into a cold solution of potassium hydroxide. It is used in bleaching, stripping or discharging colours and is a powerful germicide. It is also used for cleaning, reducing and removing hypo from negatives.

Jaw breaker. A rock-crushing machine with one fixed vertical jaw and an inclined movable jaw that can swing inwards.

Jenolite. A proprietary liquid for removing scale, rust and dirt from ferrous metals. It is claimed that treatment with the liquid also provides a phosphate coating which protects against atmospheric corrosion for a considerable time.

Jerking tables. The name given to types of machine used in ore-dressing for separating ore. The table is a smooth rectangular surface that slopes downwards slightly. In one type riffles are tacked on the table parallel to one another at intervals. The riffles have a maximum height at one end and taper to nothing at the other. Ore and water are fed through a hopper at one end of the upper edge, while wash-water is fed along the rest of the upper edge. A differential reciprocating action moves the table back and forth and is so arranged that the table is moving at maximum velocity when it reverses direction at one end of the motion, and with minimum velocity when it reverses direction at the other end of the motion. The minerals on the table therefore move lengthwise and the heavier material settles into the riffles. The lighter materials are carried over the lower edge of the table into the tailings by the wash-water. Tables such as the Wilfley concentration table can deal with graded material from about 1.5 to 3 mm diameter. With clean minerals properly sized, the tables are fairly successful, except where specific gravities are very close together. Very fine material (i.e. less than 0.5 mm) is unsuitable for treatment on jerking tables, and vanners and slime tables are much less used than formerly, having been superseded by flotation processes. Also called bumping tables.

Jet. A resinous, hard, coal-black variety of lignite or anthracite. Under the microscope it exhibits the structure of partially decomposed vegetable matter (coniferous wood). It was formerly found on the seashore, especially near Whitby (UK), but is now mined in cliffs and in certain inland dales (Eskdale) (UK), where it is found embedded in hard shales. It is hard, compact and tough, and can be carved or turned in a lathe. The fine polish is obtained by means of rouge, rotten stone and lampblack.

Jet. A nozzle or orifice through which a stream or atomised spray of liquid can be directed in any desired direction. The term is likewise applied to a nozzle through which gas can be delivered, as, for example, a gas burner.

Jetal. A process for producing decorative black oxide films on steel. It involves treatment in a strongly alkaline solution containing an oxidising agent.

Jets. The term applied to the actual stream of fluid issuing from an orifice usually under high pressure. Jets of water are used in hydraulic mining. They are often 25 cm in diameter, and a head of 300 m is not uncommon. For such a case a thrust of about 275 kN would be obtained.

Jets have also been applied to cutting of paper, plastics, ceramics and wood.

If a jet has a cross-sectional area of A m^2 and the liquid of density d is issuing under a head of h m the reaction R is given by $R = n A h d$.

Theoretically $n = 2$, but in practice it rarely exceeds 1.2. The velocity v at which the liquid issues from an orifice under a head of h m is given by

$$v = \sqrt{(2gh)} \text{ m s}^{-1}.$$

The power of a jet acting on a flat surface perpendicular to the jet is:

$$P = \frac{Av^3 d}{g} \text{ W}.$$

Jet tapping. A method of opening furnace tapholes by using a small directional explosive charge. Such charges have a small cavity in one end, and this end faces the furnace when inserted for use.

Jet test. *See* **B.N.F. jet test.**

Jeweller's borax. Sodium pyroborate crystallising with 5 molecules of water of crystallisation $Na_2 B_4 O_7 . 5H_2 O$. It is also known as octahedral borax to distinguish it from common or prismatic borax which has 10 molecules of water of crystallisation. Both forms are used as fluxes, the octahedral being more convenient for soldering and brazing small parts. It is obtained by dissolving common borax in hot water and crystallising above 35 °C.

Jeweller's cement. A cement for general purpose use made of isinglass, mastic and gum ammoniac dissolved in spirit. It is also called armenian cement.

A mixture of pitch and resin together with a little tallow and thickened with brickdust is used by gold and silver chasers for keeping their work firm and in place.

For use in lathe work, for holding jewellery and lenses during polishing and grinding, a temporary cement that consists of resin and calcined whiting mixed with wax is used.

Jeweller's rouge. A polishing powder that consists of red oxide of iron $Fe_2 O_3$. It is prepared by calcining iron II (ferrous) sulphate. The residue left in the process of distilling fuming sulphuric acid from green vitriol is termed colcothar or caput mortuum vitrioli, and is largely used as an oil paint as well as a polishing medium. The least calcined portions, which are scarlet in colour, are known as jeweller's rouge and are used for polishing lenses, stones and metals. The more calcined portions, which have a bluish tint, are termed crocus and are employed for polishing brass and steel.

Jig. A device for use in the accurate machining of a casting or forging. The device holds the work firmly and also guides the tool exactly to position. The guides take the form of plates having holes of suitable size to receive the tools as they move towards the work.

A device for separating heavy minerals from associated gangue or from other minerals utilising the differences of specific gravity of the materials. The material to be separated is placed in a container having a screen for a bottom. This is suspended in water. By moving the container up and down in the water a separation of the mineral into layers is obtained. By plunging the container quickly in a downward direction water is forced upwards through the material. Due to the difference in specific gravity of the constituents of the mixed materials, a classification is obtained. The succeeding upward motion accentuates and improves the classification. Often the operations are carried out by machines that provide for the automatic discharge of the separated materials.

Jigger. A device used in mining for transporting material from the working face to the point of collection into tubs, etc. A long trough made up in sections so as to have the required length is mounted on rollers placed at suitable intervals. An engine is coupled to the trough and communicates to it a succession of jerks. These cause the material to travel along the trough. Sometimes also called shaker-conveyor.

Jigging. The operation of classifying (separating) mineral substances using a jig (q.v.).

Jim crow. A term used to describe a crowbar fitted with a claw end. The name is also applied to a tool used for bending rails.

Jog. The addition or subtraction of atoms from an edge dislocation need not be uniform across the half plane. When moving across a slip plane, the dislocation will then have parts on a different slip plane, one or more steps having been created along the length of the half plane. Such steps are known as jogs (*see* Fig. 82). Screw dislocations can also have jogs, but since the slip direction is that of the dislocation line in this case, the jog resembles a small piece of edge dislocation, and cannot move so easily. If the jog in a screw dislocation climbs into another slip plane, point defects are created, but the screw dislocation containing the jog can continue to slip. Otherwise the jog must move sideways along the screw dislocation and attach itself to an edge component of the dislocation line. These movements are more difficult, and are thought to be involved in the work hardening of metals.

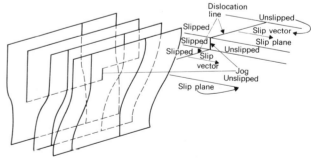

Fig. 82 *A dislocation jog.*

Johnnite. A uranium mineral formed by the gradual oxidation of the pitchblende found in Joachimsthal (Germany). It is a sulphate of the metal and is known also as uranium-vitriol.

Joint. In foundry work, a term used to describe the plane of partition in a sand mould. The adjoining faces are dusted with a thin covering of parting sand so that the cope and drag can be separated intact to permit the withdrawal of the foundry pattern.

Joints. A term used in reference to the binding or uniting of metal parts by riveting, brazing, welding, etc. When the metal parts, e.g. metal plates, are to be joined by rivets, overlap and are held together by one or more rows of rivets, a lap-joint is formed.

In a combination lap-joint there is a cover-plate, inside or outside the lap. In the single-riveted lap-joint with cover-plate there are in fact three rows of rivets, but only one row actually passes through both the plates to be jointed and the cover-plate. When there are two rows of rivets passing through the overlapping plates and the cover-plate the joint is known as a double-riveted lapjoint with cover-plate. If the plates are curved, the cover-plate is usually placed on the concave side.

The strength of riveted joints is dependent on a number of factors. They may fail by shearing the rivets, tearing the plate between the rivets, crushing the rivets or the plate, or by a combination of two or more of the above causes. To find the efficiency of a joint it is necessary to calculate the breaking strength for each of the possible ways of failure. If the least of these results is R and the tensile strength of the solid plate is S, then $100\,R/S$ is the efficiency expressed as a percentage.

Jolly. The name of the machine that is used for moulding hollow refractory materials.

Jolt rammer. A moulding machine used in foundries. The box containing the pattern is placed on the platform of the machine and roughly filled with moulding sand. The platform is repeatedly raised by air pressure and allowed to fall freely, so that the sand is packed and consolidated round the pattern as a result of the jolting.

Jolt squeeze. A type of moulding machine used in foundries. After preliminary ramming by jolting, the sand is more closely packed by a hydraulic ram which presses on the top of the moulding box.

Jominy test. An end-quench hardenability test for determining the hardenability of steel. The test specimen is shown in the diagram (see Fig. 83). Before machining to the required dimensions, the material is pretreated in a manner similar to that which the steel will normally undergo in practice before quenching and tempering. This is important, since the structure of the steel before quenching may materially alter the hardening characteristics.

The steel is usually held at the normalising temperature for about an hour and cooled off in still air. In cases where the normalised steel is still too hard to machine readily a short time temper at about 55 °C below the Ac_1 is recommended.

The specimen is heated to the austenitising (quenching) temperature, the selection of which varies according to composition. It is important that there be as little scaling or decarburisation as possible. To secure this the specimen is usually allowed to stand on a graphite disc and in a suitable container. The furnace atmosphere must be non-oxidising. The time to heat the specimen is 30-45 minutes, and about 20 minutes time of soaking at the austenitising temperature is allowed before quenching. The end-quenching is carried out by means of a jet of water acting on the specimen for a minimum of 10 minutes in still air. Details of the jet are: temperature of water 5-25 °C, free height 5-7.5 cm, diameter of orifice 12.5 mm, distance of specimen from orifice 12.5 mm.

After cooling, two flats 180° apart are ground to a depth of 0.5 mm ± 0.1 mm. Hardness tests on these flats are taken at convenient distances apart, and a curve constructed to show hardness values at increasing distances from the quenched end up to a minimum of 6 cm. A typical curve obtained is shown. (See Fig. 83.)

Josephenite. A nickel-iron compound $FeNi_3$ that is found in serpentine, particularly that obtained in Oregon (USA).

Joshi effect. This refers to the change (either increase or decrease) in an electric current passing through gases or vapours when they are irradiated by light which can ionise them.

Joule. The unit of energy, including work and quantity of heat. The work done when the point of application of a force of 1 newton is displaced through a distance of 1 metre in the direction of the force. It is defined by the equation $W = JH$, where W = work done and H = heat produced, so that J represents the number of units

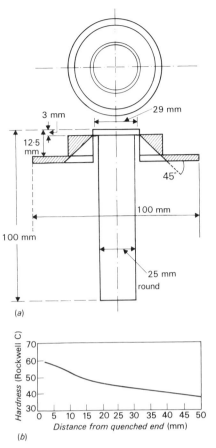

Fig. 83 *Jominy test. (a) Details of specimen. (b) Typical curve obtained.*

of work necessary to generate one unit of heat. The numerical value of J is thus dependent on the units in which W and H are measured:

$$
\begin{aligned}
1 \text{ joule} &= 10^7 \text{ ergs.} \\
&= 2.778 \times 10^{-4} \text{ watt-hours.} \\
&= 2.389 \times 10^{-4} \text{ kilogram-calories.} \\
&= 0.1020 \text{ kilogram-metres.} \\
&= 0.2391 \text{ calories.} \\
1 \text{ calorie} &= 4.182 \text{ joules.} \\
&= 42{,}670 \text{ gram centimetres.}
\end{aligned}
$$

Joule effect. The term used in reference to the heating effect caused by an electric current passing through a resistance. The heating effect is given by $I^2 Rt$, where I = current in amperes, R = resistance in ohms, and t = time in seconds.

Hence,

$$
\begin{aligned}
J_E &= I^2 Rt \\
&= 3\,600\, I^2 R \quad \text{watt-hours.}
\end{aligned}
$$

Joule-Kelvin effect. When a gas is compressed and then allowed to escape to a lower pressure through a fine nozzle, a cooling effect is noted. This applies to most of the common gases, but with hydrogen a slight *increase* in temperature is noted. The fall in temperature is due to the absorption of energy to do internal work in separating the molecules of the gas. This cooling effect is employed commercially in the liquefaction of gases. In the case of air the cooling effect in degrees C is given by

$$
\frac{A_2 - A_1}{4} \times \left(\frac{273}{T_1}\right)^2
$$

where A_1 and A_2 are the initial and final pressures of the gas in bars and T_1 is the absolute temperature of the air before expansion.

Joule's energy law. This law states that the internal energy of a given mass of gas depends only on its temperature. Thus if a gas increase in temperature from T_2 to T_1, internal energy change = $C_v(T_1 - T_2)$. If a gas is heated at constant pressure it will expand, thus increasing the internal energy as well as doing external work during expansion. Thus, $C_p(T_1 - T_2) = C_v(T_1 - T_2) + E$, when E = external work done.

Joule-Thomson effect. *See* **Joule-Kelvin effect.**

Jump. A term used in reference to the overlap in pack-rolled steel sheet, that is caused by uneven heating of the pack.

Jumping up. A term used to describe the forging operation in which the ends of bars or components are thickened, i.e. upset, by hammering.

Jump join. A butt joint in which the ends of the two members to be joined are thickened, i.e. upset, by forging before they are welded together.

Junction (n-p and p-n). In semiconductor technology, a junction is simply an electrical contact between two dissimilar materials. The contact is normally made as low in resistivity as possible, e.g. by diffusion, so that the true function of the junction materials is not complicated by bad contact, but care must be exercised not to overdo any diffusion, as otherwise the materials become radically changed in function. (*See* **p-n junction.**)

Junghans-Rossi process. A process for the continuous casting of non-ferrous metals. It is used chiefly for aluminium and its alloys, but may also be used for copper and copper alloys, especially brass. In the Junghans-Rossi machine for casting aluminium the mould is of copper, is often plated inside with chromium and is water cooled. In shape it is cylindrical, and the length is about 2½ times the diameter. The molten metal is poured continuously into the mould through a special feeding arrangement at the top.

The mould has a reciprocating movement, descending with the metal billet for a stroke of adjustable length 16 mm (approx. 5/8 in.) and then returning at a much faster speed. The mould is sprayed periodically at the top with oil to facilitate the movement of the mould relative to the billet. The hot billet issuing from the bottom of the mould is water cooled by spraying and is held by rolls which control the speed of the billet through the mould.

Junker's mould. A type of water-cooled mould that is used for casting billets or slabs of non-ferrous metals and their alloys. The mould is surrounded by a water-tight jacket, and in this are placed baffles to ensure that cooling water will operate effectively over the whole of the casting surface.

Jurassic. The term used to describe the rocks laid down in the geological period between the Triassic and the Cretaceous periods. The rocks contain many fossils. They were deposited in relatively shallow water and consist mainly of clays, sandstones and limestones. The Jurassic rocks are regarded as constituting two groups, *(a)* lias and *(b)* oolite, and these are represented separately on geological maps. The whole system is divided into seven important series; four are mainly clay and three are mainly limestone and these alternate.

This rock formation is important economically because in it occur coal, iron, oil shales and building stone. The calcareous beds are burnt for lime and the clays produce excellent bricks. The Jurassic limestones are comparatively hard, and below them lie softer shales. The main divisions with their important economic features are shown in Table 32.

TABLE 32. *Main Divisions of Jurassic Rocks*

Subdivision	Economic Importance
Purbeck	Mainly limestone
	Purbeck marble
Portland	Limestones and yardstones
	Freestone (building)
Kimmeridge	Dark grey or black shaley bituminous clay
	Yields oil: useful for brick making
Corallian	Hard limestone
Oxford	Clay useful for brick making
Great oolite	Sandstone, slates, fuller's earth
Inferior oolite	Sandstones
Lias	Clays, ferruginous limestones and sands

Jurin law. A law of particular interest in determining surface tension of a liquid by measuring the rise *(h)* of a liquid in a capillary tube of radius *(r)*. In the formula γ is the surface tension, α is the angle of a contact between the liquid and the tube and ρ is the density of the liquid,

$$h = \frac{2\gamma \cos \alpha}{\rho \, gr} \, .$$

K

Kainite. A white, hydrated magnesium potassium sulphate and chloride $MgSO_4 K_2 SO_4 MgCl_2 . 6H_2 O$ that occurs in the Stassfurt salt deposits of Germany. Magnesium salts considerably in excess of demand are obtained from these deposits, and consequently kainite is not worked to any large extent. It is sometimes used as a fertiliser.

Kalamein. A type of window and door construction in which sheet steel or other sheet metal is used to cover the wood core. Such doors, etc., are very nearly fireproof.

Kaldo process. A process in steelmaking in which a water-cooled oxygen lance is used to project a stream of oxygen over the surface of the molten slag that covers the molten metal in a converter. The lance is inclined at a small angle to the surface of the slag, and thus the reaction with the oxygen is diffuse rather than local. Intense local temperatures are avoided and fume formation is reduced. The converter is a rotary vessel suitably lined, but the motion promotes refractory wear, which is the main serious drawback to the process.

Kalunite. A naturally occurring mineral found in the USA and consisting mainly of alum (q.v.).

Kalunite process. A method of recovering aluminium and potassium from the extensive alunite deposits in Utah, California and Arizona USA. The crushed ore is dehydrated at 550 °C, cooled to 200 °C, and then leached counter-currently with a solution of potassium sulphate and sulphuric acid (10%). Potassium alum is filtered from the cold solution, redissolved in dilute potassium sulphate solution and heated in an autoclave at 200 °C. This results in the formation of a basic alum (together with potassium sulphate and sulphuric acid). The former, on calcining, yields alumina and $K_2 SO_4$ and this latter is removed by leaching.

Kalvan. A fine-particle-size variety of precipitated copper carbonate $CuCO_3$ that is used as a pigment. It is also employed to strengthen rubber, and especially to increase its resistance to tearing.

Kamacite. A nickel-iron alloy found in meteorites.

Kamenol. An aryl-alkyl sulphonate of the type $R.A.SO_3 Na$ that is used as a strong detergent.

Kamplyte. A mineral that was formerly found in good quantities in Cumberland UK. It consists of lead chloro-arsenate $(Pb.Cl).Pb_4 (AsO_4)_3$ in which a considerable part of the arsenic is replaced by phosphorus. The crystals are six-sided barrel shape and vary in colour from reddish-brown to orange-yellow. The mineral shows marked curvature of the crystal faces (hence its name). It is found in the upper parts of veins of lead ore where it has been formed by the oxidation of galena and mispickel. Only when found in large amounts is it considered an important source of lead, e.g. in Saxony (Germany) and SW Africa. Mohs hardness about 3.5, sp. gr. 7.1-7.25. Also called campylite.

Kaolin (kaolinite). A pure form of hydrated aluminium silicate $Al_2 O_3 . 2SiO_2 . 2H_2 O$, also known as china clay. It is formed by the decomposition of acid igneous rock such as granite, of which the felspar crystals, containing potassium oxide, alumina and silica, are attacked by carbon dioxide to remove the potassium. Impure kaolin may be contaminated with metallic oxides such as iron, and excess silica.

Kaolin crystallises as a layer lattice, thin platelets about 500 nm wide and 30 nm thick, of irregular hexagonal shape. The silica tetrahedra are linked in hexagonal rings, each tetrahedron sharing three of its oxygen atoms with three other tetrahedra. This creates "holes" in the centre of each ring, into which can fit hydroxyl ions having a hexagonal pattern matching that of the silica rings. The aluminium ions are also arranged in a hexagonal pattern, with a lower layer of hydroxyl ions, so that each aluminium ion rests between one face of a hydroxyl tetrahedron and the corner hydroxyl of an adjacent tetrahedron (*see* Fig. 84).

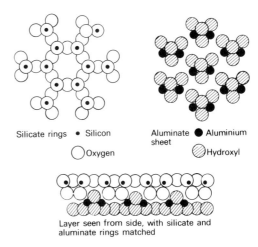

Silicate rings • Silicon ○ Oxygen

Aluminate sheet ● Aluminium ◢ Hydroxyl

Layer seen from side, with silicate and aluminate rings matched

Fig. 84 *Structure of kaolinite.*

The two layers fit to make a compound layer, with each aluminium ion having six close neighbours, oxygen and hydroxyl ions. The lower layer of hydroxyl ions can now attract hydrogen ions from water molecules, which then form dipoles, the hydrogen loosely attached to the kaolin hydroxyl ions, and the water hydroxyl ions hanging below. Those water hydroxyl ions may now attract further water dipoles, and a thick layer of water can be built up between the sheets of kaolinite. Thus, the kaolin may have varying amounts of water between its layer lattices, and its properties will vary accordingly, the sheets of kaolinite sliding over the water layers in the sandwich. With small amounts of water, the clay becomes plastic, then with increased water, sticky, and finally, a thin slip which behaves like a fluid.

When heated, moisture is first driven off and colloidal matter is coagulated. At 500 °C the kaolinite is decomposed into the anhydrous substance. At temperatures up to 800 °C the alumina begins to polymerise, and this is accompanied by shrinkage. Above 1,000 °C there occurs combination between silica and the alumina with the production of sillimanite, and this latter at temperatures in the region of 1,500 °C sinters into a hard stony mass which only begins to soften about 1,650 °C. It is used for making porcelain, china and refractory for bricks for furnace linings. The latter have a high resistance to spalling. It is also used for making electric insulators, as an abrasive and as a filler for plastics. When mixed with zinc oxide, the product is known as Chinese White. Specific gravity 2.6.

Kappa carbide. A complex iron-molybdenum carbide to which the formula $(FeMo)_{23}C_6$ is given. The molybdenum content may vary from about 3.5% to over 10%, though the upper limit is uncertain. It has a face-centered cubic structure, is more stable than cementite Fe_3C and is magnetic. It is formed by direct transformation of austenite or by tempering martensite within certain temperature limits.

Karbate. A carbon or graphite brick. The latter has a higher thermal conductivity, but is softer than the former. The binder for the finely powdered carbon is usually a bituminous substance, but for the graphite a furfural resin is used. On sintering, the bituminous material is carbonised, whereas the resin used with the graphite becomes polymerised in the pores of the powder. The bricks have a high crushing strength of about 69 MN m^{-2} and a density of 1.84×10^3 kg m^{-3}.

Kata thermometer. An instrument used to indicate the cooling and evaporative power of the air. It is a large bulbed alcohol thermometer of standard size, and the stem is graduated from 95° to 100° F (70.5-73.3 °C). In use the instrument is warmed up until the meniscus of the liquid is above the 100° mark. The bulb is dried and the rate of cooling from 100° to 95° is taken with a stop-watch. From a factor determined for each instrument the cooling power is deduced in millicalories. In the case of the wet Kata thermometer a muslin glove covers the bulb.

The instrument records not only the effect of the temperature of the surrounding air on the cooling of its surface when at the approximate temperature of the human body, but of any air movements, and also how quickly cooling takes place from a wet surface, as, for example, the surface of the skin when perspiration is occurring.

Katayama equation. An expression giving the change in molecular surface energy of a liquid with temperature. If one mole of a liquid be in the shape of sphere the surface of the sphere may be called the "molecular surface". For different liquids there will be the same number of molecules distributed on the surface. The product of the surface tension (γ) and the molecular surface is called the molecular surface energy and is equal to $\gamma(M/D)^{2/3}$, where M/D is the molecular volume and D the density of the liquid. The molecular surface energy decreases with temperature, the relationship being given by the Katayama equation which is $\gamma[M/(D_1 - D_2)]^{2/3} = a(T_c - T)$. In this equation D_1 and D_2 are are densities of the liquid and vapour at temperature T, T_c is the critical temperature and a is a constant for all normal liquids and is equal to 2.04.

Katharin. The German name for carbon tetrachloride (tetrachloromethane), CCl_4, a liquid used as a grease solvent and also as a fire-extinguisher. It is non-flammable, but when strongly heated decomposes giving poisonous phosgene. Now regarded as highly toxic.

Keep's hardness test. A hardness test in which a standard steel drill is caused to make a definite number of revolutions while it is pressed with a standard force against the material to be tested. The hardness is automatically recorded on a diagram on which a dead soft material gives a horizontal line, while a material as hard as the drill itself gives a vertical line. A substance of intermediate hardness is represented by the angle (between 0° and 90°) its line makes with the horizontal. There is a further type of hardness test in which a steel-testing tool having 100 small pyramidal points is used. One corner of the tool is held against the polished specimen and a 12.5 kg weight allowed to fall upon it from a height of 2.5 cm. The number of point markings is counted and reported as the hardness of the specimen.

Keep's shrinkage test. A test in which bars, 30 cm long and 1.3 cm square, are cast between jaws of cast-iron yokes. The shrinkage is measured by the aid of a graduated steel taper gauge.

Kelcaloy. A method of making special types of high-quality, high-cost alloy steels. The material is formed into a continuous tube by passing through a tube-mill. It is then passed to a water-cooled mould which functions as an electric furnace, the tube being one of the electrodes. The arc melts the tube, while a second tube delivers a predetermined amount of alloy and flux. These also melt in the arc and the latter forms a thin protective coating which floats on top of the melt.

Keller's spark test. A method of testing metals by observing the characteristics of form and colour of the sparks produced on grinding. It is a qualitative test that has been used for determining different types of steel. To obtain results that can be of value in discriminating between different types of steel it is essential to use a wheel of fine texture, and to make a careful observation of the complexity of the "sprig" arising from the swollen part of the spark. The test is based on the assumption that all highly carburised steels will produce bright sparks that tend to "explode". Pure iron gives sparks that are not so bright and not explosive. The brightness apparently depends on the presence of alloying elements such as nickel or tungsten that are resistant to oxidation and thus reduce the brightness—or to the presence of elements such as manganese or silicon which increase the brightness.

Kellog hot-top process. An ingot casting process in which heat is supplied to the top of the metal in the ingot mould by the discharge of a controlled current from an electrode which is submerged in a protective layer of flux.

Kellog's process. A method of producing composite steel slabs. Carbon steel electrodes are fused inside tubular electrodes through which the alloying materials are supplied to the surface of the carbon steel under cover of a protective flux.

Kelly converter. A type of converter similar in certain respects to the Bessemer converter, invented by William Kelly in America. Kelly's trials were made concurrently with Bessemer's, but the apparatus used by the former inventor never came into commercial use.

Kelly filter. A type of pressure filter that is employed for material that is difficult to wash or filter. In the treatment of slimes the removal of solution from the solids is conveniently effected by first thickening the pulp and then using some form of pressure filter. The Kelly type differs from a frame press, since it consists of filter leaves inserted in a cylindrical drum. The pulp is pumped into the drum, and when a cake has formed, wash-water is pumped in. After washing, the filter leaves are withdrawn and the solid residue is blown off.

Kelpchar. A trade name for an activated charcoal made from seaweed. It has a high capacity for absorbing gases.

Kelvin. A name sometimes applied to the Board of Trade (now Department of Trade) unit of energy. One kelvin = 1 kilowatt hour, or 1,000 watt-hours, or 3.6 MJ.

Kelvin electrometer. An instrument that can be used for measuring differences of electrical potential. The Kelvin types (attracted disc and quadrant electrometers) are more sensitive than the gold-leaf electroscope and can be adapted to measure large differences of potential.

In the attracted disc electrometer, the potential is given by

$$V = d \left(\frac{8\pi Mg}{s}\right)^{1/2}$$

where d = distance apart of the discs, s = area of attracted disc and Mg is the attractive force. Since all these quantities are dependent only on measurements of length and force the instrument is called an absolute electrometer.

In the case of the quadrant electrometer, if the needle and one pair of quadrants has a potential v_1 and the other pair of quadrants has a potential v_2, then $\theta \alpha (v_1 - v_2)^2$, where θ = angle of deflection of the needle. The Leppman electrometer is useful in measuring electromotive forces less than 1 volt.

Kelvin temperature scale. In the case of an ideal engine working between temperatures θ_1 and θ_2 taking in Q_1 units of heat and discharging Q_2 units of heat to the condenser, the efficiency

$$\eta = \frac{Q_1 - Q_2}{Q_1} = f(\theta_1 \theta_2),$$

where $f(\theta_1 \theta_2)$ is some function of θ_1 and θ_2. But $Q_1 = kf(\theta_1)$ and $Q_2 = kf(\theta_2)$.

$$\therefore \frac{Q_1}{Q_2} = \frac{f(\theta_1)}{f(\theta_2)} = \frac{t_1}{t_2} \text{ (say)}$$

$$\therefore \eta = \frac{Q_1 - Q_2}{Q_1} = \frac{t_1 - t_2}{t_1}$$

If the numbers expressing t_1 and t_2 are taken to represent the temperatures at which the heat is taken in and rejected, then the ratio of any two temperatures t_1 and t_2 on this scale $= Q_1/Q_2$. If $Q_2 = 0$, then t_2 will also be zero. All the heat (Q_1) taken in at t_1 is then converted into work, and since it is inconceivable that more heat can be converted into work than that which is drawn from the source, t_2 cannot be negative. This means that t_2 is the lowest possible temperature, and that the absolute zero on the Kelvin scale is the same as the zero on the perfect gas thermometer, i.e. $-273\,°C$. The kelvin is the SI unit of temperature.

Kemidol. The trade name for a finely powdered "lime" made from dolomite.

Kemrock. Sandstone that has been impregnated with black furfural resin and baked till hard. It is resistant to acids and alkalis and is used for bench and table tops in laboratories, etc.

Kenmore process. A process in which the desired metal (copper or nickel) is electrodeposited on a steel core, and the plated wire subsequently drawn to the required gauge.

Kennametal. A hard metal consisting of tungsten-titanium carbide sintered with cobalt (or other) binder. Tungsten, titanium and graphite in suitable proportions are heated to about $2,000\,°C$ in graphite crucibles. Nickel is added to the charge, which assists the ingredients to combine. On cooling the nickel is dissolved out in *aqua regia* and the excess graphite removed by washing. The material is then ground in ball mills with matrix forming metal (e.g. cobalt). When very fine it is pressed into the required shape and sintered.

Kennel coal. An American term for a coal that can be ignited with a match to burn with a bright flame. It is also known as "candle coal", and this latter name is probably the origin of the term "cannel coal".

Kerargyrite. Native silver chloride AgCl, also known as Horn Silver and as cerargyrite.

Kerf. The narrow channel formed by an oxygen jet when ferrous metals are flame cut. This channel should have uniformly smooth and parallel walls.

Kermesite. An antimony ore $2Sb_2S_3.Sb_2O_3$ resulting from the alteration of antimonite. It occurs as needle-shaped crystals or thin six-sided prisms which have a cherry-red colour and a metallic lustre. Also known as antimony blende.

Kernite. Sodium borate $Na_2B_4O_7.4H_2O$. An important source of borax, found in California in the Mojave Desert (USA). (*See also* **Colemanite.**)

Kern's process. A method of testing one of the physical characteristics of paint coats on metals. The surface to be tested is bombarded with an air blast containing carborundum powder. The weight of this latter, under stated conditions, that is required to wear through the coat of paint is taken as a measure of its abrasion resistance.

Kettle. An open-top vessel used in refining metals of relatively low melting point, e.g. lead, tin, etc.

Kevlar. The trade name for an aromatic polyamide (aramid) corresponding to polypara-phenylene terephthalamide, available as cord or fibres. Possesses high strength and stiffness.

Keyhole test piece. An impact test piece in which the notch is formed by a hole and slot. The notch resembles a keyhole and gives multi-axial stresses.

Kibble. A term used in reference to a large bucket that is used in mining.

Kick's rule of similarity. The area of the impression obtained in an indentation hardness test bears a constant relation to the applied load for any material provided that the angle included by the opposite edges and the bottom of the impression is independent of the depth.

Kidney ore. A reniform variety of hematite Fe_2O_3 containing, when pure, about 70% Fe. Clay and sandy impurities are frequently present.

Kieselguhr. A variety of tripoli or diatomaceous earth. It is a hydrous and opalescent form of silica containing 88% SiO_2, 0.1% Al_2O_3, 8.4% H_2O and the balance organic matter. It is not unlike chalk in appearance, but is very much lighter. It is used as an absorbing material, heat insulator, filler for paints and as a mild abrasive (often in metal polishes). It is obtained mainly from Germany.

A grey variety, known as randanite, which is found in central France, has a very similar composition and is used in making insulating bricks. The randanite is mixed with ground cork, using a binder. It is moulded and then dried. Finally, the cork is burnt out, leaving a very porous brick which can withstand temperatures up to $1,000\,°C$. Also known as infusorial earth.

Kieve. A form of gravity concentrator. It is used in the "wet" method of concentration, employing water as the medium, and is in effect a stirred-bed concentrator bearing some resemblance to the common type of cylinder washer used in the cleaning of coal. In the latter the centrifugal action is produced by high velocity injection.

Killed spirits. The name applied to zinc chloride $ZnCl_2$ obtained by adding excess zinc to hydrochloric acid. It is commonly used as a soldering flux.

Killed steel. The name given to steel that has been fully deoxidised before casting, as distinct from balanced and rimming steel. Silicon, manganese and aluminium are among the commonly used deoxidisers. There is little evolution of gas during solidification. The top surface of a killed steel ingot exhibits shrinkage with a central pipe below, but sound ingots are obtained.

Killing. A term used in reference to the application of a small amount of cold work to finally heat-treated steel strip. The process, which is usually performed by skin passing, is intended to prevent the development of kinks and stretcher strains during subsequent manipulation.

Killing pickle. A term used in industry that is applied to the process of working out a sulphuric acid pickling bath that has become heavily contaminated with iron II (ferrous) sulphate. When the iron content of the bath reaches some 10-15%, no further sulphuric acid is added, but pickling is continued until the bath is so nearly spent that it acts too slowly for economic service.

Kiln. A furnace in which heating operations that do not involve fusion are conducted. Kilns are most frequently used for calcining, and free access of air is usually permitted. The raw materials may be heated by the combustion of solid fuel with which they are mixed, but more usually they are heated by gas or the waste heat from other furnaces.

Kilowatt. A unit of power that is equal to 1,000 watts. It is the power in a circuit when the product of the potential difference (in volts) and current (in amperes) is 1,000. Since power is a rate of working, a kilowatt is a rate of working equivalent to 1,000/746 = 1.34 h.p.

Kilowatt-hour. A unit of energy, also known as the kelvin (*see* **Kelvin**).

Kimmeridge coal. A type of coal that is sometimes regarded as a bituminous shale, though it resembles the Sapropelic coals of carboniferous age, in appearance. It occurs in Dorset (UK) in beds about 60 cm thick. It has often taken fire spontaneously and has been worked, without much success, as a source of paraffin and even gas. It is unsuitable for metallurgical operations and is useless as a household fuel on account of its very unpleasant smell.

Kinematic viscosity. The ratio of the absolute viscosity of a liquid or gas to its density. The unit, in the SI system, is the Stoke, $10^{-4}\ m^2\,s^{-1}$ but centistoke may be used. Since viscosity changes rapidly with temperature, the kinematic viscosity must always be calculated at stated temperatures. The dimensions of kinematic viscosity are L^2/T and 100 centistoke = 1 cm^2 s^{-1}.

Kinematic viscosity is important in all problems connected with the flow of liquids or gases, since it is one of the factors on which the pressure drop depends.

Kinetic energy. The energy (i.e. the capacity to do work) of a body that is in motion, the energy being expended as the velocity is reduced. Mechanically energy may be expressed as $\int P ds$, where P = force and ds is the distance moved. $\int P ds = \frac{1}{2} mv^2$.

The theory of relativity leads to the above expression for small velocities, but as the velocity increases the kinetic energy increases more rapidly than this expression would indicate and tends to become infinitely great as the velocity approaches that of light. It can be shown that the increase in the mass of a body due to its speed is equal to the energy due to the motion divided by the square of the velocity of light.

Kinetic theory. The occurrence of diffusion, in opposition to gravity, that has been observed in gases and liquids, leads to the conclusion that the molecules of gases and liquids are in ceaseless motion. This conception of the motion of molecules is referred to as the kinetic theory. Though the molecules of a gas exert practically no force one on the other, they do in fact bombard the walls of the containing vessel, and the total effect constitutes the pressure of the gas. The pressure per unit area due to all the particles of a gas bombarding the walls of the containing vessel is given by $p = \frac{1}{3} M G^2/V$, where M = mass of molecules, V = volume and G = mean square speed. The mean speeds of certain gases at 0 °C are given in Table 33.

TABLE 33. *Speed of Gases at 0 °C*

Gas	Mean Speed, m s^{-1}	Speed of Sound in Gas, m s^{-1}
Hydrogen	1,695	1,290
Oxygen	425	316
Carbon dioxide	362	256
Helium	1,213	–
Nitrogen	455	305
Chlorine	288	–
Mercury vapour	170	–
Ammonia	628	428
Carbon monoxide	392	337

In the solid state it is assumed that the molecules are in a state of motion, performing small oscillations about a fixed or equilibrium position. On applying heat to the solid the amplitude of the oscillations increases, and eventually the movement becomes such that the molecules collide and break away from their original positions. At the point of fusion the molecules are able to move about amongst each other.

Kink. A bar when delivered from the finishing rolls may be wavy, either up and down or sideways. The former is known as a buckle, while the latter is a kink. Also used to describe an offsett for a dislocation confined to a glide plane.

Kirchhoff's law. (a) If the radiations (from a hot source) of a given wavelength fall on the surface of any body the absorptive power of the body for that wavelength is defined as the ratio of the radiant energy absorbed to the total incident radiant energy. Since a perfectly black body absorbs all the radiations incident upon it, the value of the ratio for such a body is unity. Kirchhoff's law states that, for radiations of the same wavelength and the same temperature, the emissive power divided by the absorptive power is the same for all bodies and is equal to the emissive power of a perfectly black body.

(b) Kirchhoff enunciated two laws that facilitate the calculation of the currents in a network or a complicated arrangement of resistances. The first of these is based upon the fact that electricity cannot accumulate at any point in a circuit and states that at any junction of resistances the total current flowing towards it is equal to the total current flowing away from it. The second of these laws is based on Ohm's law and states that in passing round any complete circuit the algebraic sum of the quantities "IR" for the separate parts is equal to the resultant e.m.f. in that circuit.

When stated in terms of vector quantities these laws are applicable to a.c. circuits.

Kirkendall effect. A diffusion phenomenon sometimes encountered in solid metals and alloys. When composite specimens made by electroplating a 70/30 brass with copper are heated to bring about interdiffusion, the original interface is displaced towards the brass. This preferential diffusion effect has also been observed with copper to copper-tin and copper-aluminium alloys, from copper to gold or nickel and from gold to silver.

Also known as Smigelskas effect. (*See* **Diffusion**.)

Kirsch test. A hardness test which is a measure based on the load required to press a steel punch, 5 mm diameter, into the specimen to a depth of 0.1 mm.

Kish. The free carbon which separates from molten cast iron during cooling down to the temperature of solidification. It often floats on top of the molten metal.

Kisser. A local patch of scale sometimes found on steel sheets when two sheets have remained in close contact during pickling.

Kiss process. A process that was used in the extraction of silver from its ores. The ores were first roasted with common salt to produce silver chloride. This latter was then dissolved out by a solution of calcium thiosulphate. Such a solution, however, has no effect on metallic silver or on silver sulphide. The process has been replaced by the cyanide process.

Kjellin furnace. An electric induction furnace used for melting ferrous metals. It is operated with low-frequency high-voltage current and with large furnaces special generators are required. The crucible containing the metal is annular in shape and rather narrow, so that the hearth is not readily accessible.

When large capacity is required the crucibles are built of masonry and are lined with basic materials similar to those used in the Bessemer converter. The furnace is usually covered during operation and is arranged to tilt for convenience in pouring. When pure raw materials are used the quality of the steel will compete with good crucible steel.

Knock-down test. A test to determine the suitability of billets, bars, wire, etc., for hot or cold forging. As a hot forging test, a sample of suitable length is heated to the forging temperature and hammered on end until it is reduced by a specified amount. In cold forging tests the sample must withstand a similar flattening without heating. The test is also called dump, jump, slug, upending and upsetting test.

Knock-on damage. Another term for damage to materials by the impact of elementary particles or radiation. (*See* **Displacement spike**.)

Knock out. A term used in the foundry to describe the operation of separating of sand-castings from the moulds and cores. It is also used to describe the device for removing work from dies after forming operations such as stamping, forging or deep drawing.

Knoop hardness test. A diamond indention hardness test used for determining the hardness of the individual micro-constituents of alloys. The test is similar to that devised by Brinell, but the indenter is rhomboid in shape.

Kobold. The copper-miners of the Harz district of Germany often obtained ore that resembled copper ore. But on roasting, the ore produced an unpleasant garlic-like smell, but yielded no copper. The occurrence of this "false-ore" was attributed to the work of an evil spirit, Kobold. From kobold is derived the name cobalt.

Koft-kari. The name given in the Punjab (India) for inlaid metal work—usually gold wire on steel. The main centres of this work are Gujrat and Sialkot.

Kohlrauch's law. This law is also known as the "law of independent migration of ions". It states that every ion at infinite dilution contributes a definite amount towards the equivalent conductance irrespective of the other ion with which it may be associated in the electrolyte. It is mathematically expressed as follows:
$\Lambda_0 = l_+ + l_-$ where Λ_0 is the limiting value of the equivalent conductance which is attained with decreasing concentration, and l_+ and l_- are the ion conductances at infinite dilution of the cation and anion, respectively.

Koldflo. The name given to an extrusion process carried out at room temperature.

Koldweld. A process for welding aluminium and other nonferrous metals by pressure alone at room temperature. In the process the metal is made to flow away from the welding point as the tools are brought together. The shape and size of the dies are critical. (*See also* **Pressure welding**.)

158

Kolthoff buffer solutions. Solutions whose pH values undergo little change on the addition of small quantities of acid or alkali. They are important because the success of the indicator method of determining the hydrogen-ion concentration of a solution depends on the availability of standard solutions, the pH values of which are definitely known and readily reproducible.

Buffer, or regulator, solutions contain only a small actual concentration of hydrogen ions. For this reason they are often said to possess a reserve acidity and reserve alkalinity. The variation of pH during the neutralisation of a weak acid by a strong alkali is given by the expression

$$pH = \log\frac{1}{K} + \log\frac{[salt]}{[acid]}$$

The action of buffer solutions can be understood from a consideration of the effect of adding sodium acetate (ethanoate) to acetic (ethanoic) acid. The first result, of course, is to reduce the $[H^+]$ concentration. If to such a solution some highly ionised HCl solution is added, the H^+ ions will combine with ethanoate ions and produce ethanoic acid which is practically un-ionised. Thus, the total increase in $[H^+]$ is very slight. The principal pairs of substances employed as buffers are: (a) ethanoic acid and sodium ethanoate, (b) boric acid and borax, (c) citric acid and sodium citrate, (d) sodium phosphate and hydrochloric acid, (e) sodium carbonate and sodium bicarbonate and (f) disodium phosphate and sodium dihydrogen phosphate.

An example of the practical application of buffer solutions is the removal of the phosphate radical in analysis (Group IIIa).

Though the pH value of a buffer solution may be calculated from the equation given, it is usually determined electrometrically.

Konimeter. An apparatus for measuring the amount of dust in a gas in the atmosphere of foundries, mines, etc. A measured volume of a sample of the air is caused to impinge on to a circular glass plate which is coated with a thin film of glycerine jelly. The dust retained in the glycerine film is examined under a microscope, and the number of particles per cm^3 is reported. Owing to the fact that this method does not reveal particles of diameter less than $0.8\mu m$, and since the dust in steel foundries is generally below this limit, the Konimeter is favoured for use in mines while Owen's jet test (q.v.) is mainly used in foundries.

Koppers oven. A widely used type of coke oven having vertical heating flues and regenerators below the ovens. In America the modified type of oven known as Koppers-Becker is favoured. The heating is by producer gas supplemented by coke-oven gas, and the arrangement of flues is said to ensure more uniform heating, permit heavier wall structures and allows the flues to be built up higher.

Kopp's law. This is an extension of the law of Dulong and Petit concerning atomic heats. It states that the molecular heat of a solid compound is approximately equal to the sum of the atomic heats of the constituents (also calculated in the solid state). In general, the molecular heat of a compound is a simple multiple of 6.3.

Kranz triplex process. A process used to make malleable iron. A charge, commonly made up of pig iron or a 50-50 mixture of pig iron and scrap, is melted in a cupola furnace. The metal is tapped into a large ladle and about one-sixth of this is then treated in a side-blown converter to remove most of the silicon, carbon and manganese, and after treatment is returned to the ladle. The ladle of mixed metal is then poured into a Heroult-type electric furnace for storage. It is tapped from the holding furnace, as needed, into small ladles and thence poured into hand ladles for casting.

Krause mill. A sheet rolling-mill in which a large reduction of thickness is effected in a single pass. A reciprocating action is used.

Kroll's process. A process used particularly for the production of titanium and zirconium metal. It consists in the reduction of the tetrachlorides of the metal with magnesium. The process must be carried out in the absence of air (usually argon at a slight positive pressure is used) and the reaction temperature is about 800-850 °C. The exothermic reaction may be represented by the equation: $TiCl_4 + 2Mg = Ti + 2MgCl_2$.

The crude product contains magnesium and magnesium chloride as impurities. The removal of these latter two substances is effected by heating the mixture to a temperature below 1,000 °C in a high vacuum. Most of the $MgCl_2$ can be removed as it melts, and the remainder, together with the magnesium, is removed by vaporisation. The coke-like sponge of titanium that results is 99.5% pure.

Krupp-Renn. Process for the production of iron and steel from medium-grade ores, such for instance as may contain 44-57% Fe and high silicon content. The process involves a continuous reduction and is carried out in a revolving tube furnace, which is designed in the first instance for the production of iron. The iron is reduced into a sponge and then converted into low-carbon metallic grains which are called "pellets".

Further applications of the technique are enrichment of a poor-quality ore, especially one with a high silica content, and production of non-ferrous metals. In the former process the pellets, which are practically slag-free, may be smelted again in the blast furnace with proportionately less coke. In the latter process the non-ferrous metals are collected within iron pellets.

Krupp's process. A method of purifying molten pig iron by the addition of iron and manganese oxides. These remove about 80-95% of the silicon and the phosphorus present, but have little effect on the carbon.

Kryptol. A mixture of graphite, carborundum and clay. It is used as a resistance material in electric furnaces.

Krypton. One of the five inactive gaseous elements found in very small quantity in the atmosphere. It can be separated from the other constituents of the air by absorption in charcoal at −100 °C. Its specific gravity (3.74) is more than twice that of argon and its solubility in water is likewise twice that of argon. Boiling point −152.3 °C. (See Table 2.)

Kuckersite. A dark, combustible material found in Silurian rocks, particularly those in the regions to the east of the Baltic. It consists of fossil algae and is of considerable interest, since many bituminous shales and cannel coals are of algal origin, as also are certain shales and coals of the Permian and Carboniferous periods.

Kulm. The natural slack of an anthracite coal, which is in the form of a powdery material and has been formed as a result of earth movements. It is sometimes used in black paints. Also called culm.

Kunzite. A transparent lilac-coloured variety of spodumene used as a gemstone. It is found in California (USA), and Madagascar. It is a lithium, aluminium silicate belonging to the pyroxene group $Li.Al(SiO_3)_2$. It crystallises in the monoclinic system. Mohs hardness 6-7, sp. gr. 3.15.

Kupfernickel. A nickel arsenide, NiAs, also known as niccolite. It is one of the principal sources of metallic nickel. It is usually associated with cobalt, silver and copper ores. Antimony and sulphur are sometimes present in small amounts. Mohs hardness 5-5.5, sp. gr. 7.3-7.6.

Kupfferite. A magnesium silicate $MgSiO_3$ belonging to the amphibole group of rock-forming minerals. It crystallises in the orthorhombic system. It always contains some water of crystallisation and often also some alkali.

Kutter's formula. An important formula for finding the flow of water in open channels: $v = c(rs)^{1/2}$ where r is the radius of the channel and s the slope of the water surface. The coefficient c is taken from tables.

kX units. The unit of length intended for use in crystallographic measurements was the Ångstrom (10^{-10} m). In X-ray crystallography, so called "crystal Ångstroms" were used, based on the Siegbahn scale of X-units, an X-unit being defined so that the (200) spacing of the calcite crystal at 18 °C is equal to 3029.45 X-units. This definition made an X-unit almost exactly equal to 10^{-3} Ångstrom units, based on the then known value for Avogadro's number. More accurate work showed up an error, and the crystallographic or kX unit is equal to 1.00202 Ångstrom units, or in SI terms, 100.202 pm. There is still controversy over the last digit of this number, and the units should be quoted in SI terms by making the correction and indicating which correction has been used.

Kyanite. An aluminium silicate $Al_2O_3.SiO_2$ used as a refractory, especially for lining furnaces for non-ferrous metals and glass. It is found in India, Kenya and in several parts of the USA. Synthetic kyanite is made from kaolin, alumina and felspar. Mohs hardness 6-7, sp. gr. 3.56-3.67. It crystallises in the triclinic system. Also called cyanite.

Kymograph. A machine for recording and measuring the velocities and characteristics of the strokes delivered by a punch or ram.

159

L

Labile. A term used synonymously with "unstable" in reference to systems of two phases that undergo spontaneous change as soon as the transition temperature is passed.

Labradorite. A plagioclase felspar consisting of sodium-calcium-aluminium silicate. The composition varies from

$$Na_2O.CaO.(Al_2O_3)_2.(SiO_2)_2 \text{ to } Na_2O.(Al_2O_3)_4.(CaO)_3.(SiO_2)_{12}.$$

It is an important constituent of many basic igneous rocks. The iridescence often noted in specimens of the mineral is due to the presence of fine lamellar inclusions of hematite or ilmenite.

Lacquer. A substance applied to metallic surfaces for decorative purposes or to provide protection against corrosion. It is usually a solution of film-forming substances such as shellac, synthetic or natural gums in a volatile solvent. Cellulose lacquers are solutions of nitrocellulose or acetylcellulose with admixtures of resins and plasticisers and, if desired, of dyestuffs or pigments.

In the canning industry lacquer on the inside of tin-plate cans is employed to prevent the discoloration of the fruit juice by tin or iron and to obviate "hydrogen-swells". It is used in aluminium cans for packaging beverages, including alcoholic ones, to prevent corrosion.

Lacustrine coal. Coal formed from deposits laid down in fresh water. It represents one of the sub-divisions into which coals are sometimes divided according to their supposed method of deposition. This sub-division includes sopropelic coal (cannel and boghead coal) and most of the humic coals. Anthracite is sometimes rather doubtfully included.

Ladder. The rigid frame on which the endless chain of digging buckets is carried by a mining dredger. The term is sometimes used of the chain itself. A dredger fitted with such a ladder is called a bucket-ladder dredger.

Ladle. A vessel lined with refractory material that can be used for holding molten metal after tapping or for transporting metal to the moulds in a foundry. Ladles may be designed either for lip or bottom pouring. The latter type is usually provided with a hole in the bottom, sealed by a vertical refractory stopper or plug. The plug is operated by hand. In order to avoid the defects that are traceable to inefficiently prepared ladles, the latter should be clean, dry and preheated before use.

Ladle drier. An apparatus designed to eliminate moisture and carbon particles (which may cause gassing of the metal) from the lining of a ladle. The apparatus may employ solid or liquid fuel. In the former case the unit is a type of forced-draught fire-box, the products of combustion passing into the inverted ladle. In the latter case, oil burns in compressed air, the burner being inserted into the ladle through a central hole in a refractory-lined lid.

Lag. A term used to indicate the time required for a slow diffusion process or a sluggish reaction or transformation to proceed to the desired degree of completion.

Lagging. A term applied to the heat-insulating material that is packed round furnaces, steam pipes, etc., to reduce heat loss. It may also be used in reference to the actual process of covering furnaces, etc., with a non-conducting material such as asbestos, cork, slag wool or a plaster made of asbestos and magnesia, to reduce transfer of heat.

The term may also be used to indicate "sluggishness" in certain physical and chemical changes.

Lake. The name given to the insoluble pigment obtained by precipitating solutions of organic colouring matters with alum or a tin salt.

Lake copper. Metallic copper produced from the high purity native ores found in the neighbourhood of Lake Superior. Before the development of modern refinery processes this was the purest form of copper.

Lakes. A term applied to a large segregated mass of impurities, e.g. lead, in a metallic structure.

Lalande cell. A primary cell of industrial importance in which amalgamated zinc is the anode, and the cathode consists of copper oxide depolariser either tamped into a perforated iron or copper container or pressed into a block. The electrolyte is sodium hydroxide, which should be protected against evaporation and ingress of CO_2 by a layer of paraffin oil. The open circuit voltage is 1-1.1 volts and the operating voltage about 0.7-0.85 volt. Additions of sulphur to the depolariser raise the voltage by about 0.1 volt. The cell is efficient, satisfactory for rugged construction and cheap fabrication and has a well-sustained operating voltage. The electrolyte can be made unspillable by gelatinising with starch.

Lamagal. A refractory material consisting of 40% alumina and 60% magnesia.

Lambda point. A term used in reference to the temperature at which the specific heat of a substance reaches its maximum value. Occasionally the specific heat falls when the temperature rises above the λ point. Such changes occur with ferromagnetic materials at the Curie point. They also occur in certain phase changes such as order-disorder transitions.

The term is also apecially applied to change of helium I to helium II which occurs under normal pressures at about $-270.81\,^{\circ}C$.

Lamellae. A term applied to thin plates or scales.

Lamellar. A term applied to a substance or material that is arranged in thin plates, layers or scales, e.g. pearlite. In the case of minerals such as tubular spar or wollastonite the material can often be separated into thin plates.

Lamellar pyrites. A mineral having the same composition FeS_2 as iron disulphide, but which crystallises in the orthorhombic system. It is known also as marcasite and hepatic pyrites. Mohs

160

hardness about 6, sp. gr. 4.85.

Lamina. A thin plate or a layer or coating of one material lying over another. The term is particularly applied to thin sheets of iron or steel used in electrical apparatus.

Laminated. A term applied to a structure that is composed of thin sheets or layers.

Laminated spring. A type of spring built up of a number of superimposed thin leaves, flat or curved, and so arranged that the leaves become progressively shorter. The spring may be in the form of a beam or a cantilever, and the arrangement of leaves is such that the spring is of uniform strength along its length. Such springs are used for carriages, automobiles, locomotives and railway rolling-stock.

Lamination. A term applied to separation into two or more layers which is sometimes found in rolled or forged products. In the case of steel it is usually due to some discontinuity, such as a layer of non-metallic inclusions in the metal.

Lampadite. A variety of wad, a black earthy mineral that contains up to 18% CuO. It is usually very soft, being a decomposition product of other minerals, and is often deposited in marshes or near springs.

Lampblack. A black pigment consisting of very finely divided carbon obtained by the imperfect combustion of highly carbonaceous substances, such as oils, fats, resins and pitch. The finest qualities are obtained by the combustion of coal-tar oils in a limited supply of air. It is used as a pigment, in printing ink, in vulcanised rubber and in dressing and lacquering of leather.

Lanarkite. A lead-bearing mineral that consists of basic lead sulphate $PbO.PbSO_4$.

Lane's process. A commercially important method of preparing hydrogen by passing steam over red-hot iron. The iron should preferably be in the form of "sponge" or reduced iron and must be heated to about 650 °C. The hydrogen is 98% pure. The residue of black magnetic oxide has a limited market.

Langbeinite. A potassium-magnesium ore that has the composition represented by $2MgSO_4.K_2SO_4$.

Langelier index. A figure which shows the difference in pH value of a sample of natural water before and after treatment with solid calcium carbonate (marble dust). The index has a positive value if the pH value falls, for this shows that the solution was already supersaturated with respect to calcium carbonate. In general, a positive index indicates that the water will not be highly corrosive. Waters that show a negative index if treated with lime will usually show a positive index, so that such treatment is likely to reduce the corrosive effects of the water on metals.

Lanolin. Wool fat (*adeps lanae*) obtained by the purification of "brown grease" or degras extracted from raw sheep's wool. It readily emulsifies with water and does not turn rancid. It is used as a lubricant and as a protection against corrosion. For ease of application it is often used with white spirit as a solvent and is sometimes thickened with linseed oil or rosin dissolved in kerosene. Additions of resin bitumen, zinc chromate or kaolin are beneficial in producing a good, fairly hard film. Beeswax and vaseline produce a softer film that gives useful protection for steel in storage. Inhibitive pigments, such as zinc dust, aluminium powder or zinc chromate, improve the protective qualities of the "grease".

Lanthanide elements. The group of transition metals in which the inner energy level of quantum number 4f begins to be occupied, in preference to proceeding with occupation of the 5d level. Lanthanum itself, following caesium and barium, has two 6s and one 5d electron in the outer levels. Cerium, which follows, is the first lanthanide element, with one 4f electron, and the series ends with lutetium, fourteen 4f electrons. These elements were formerly called "rare earths" because of their scarcity at the time, and the difficulty of separating them one from another by chemical means. This remains a difficulty, but all have been prepared in quantity and many have commercial applications, in additions to metallic alloys, in electronic applications such as phosphors for television colour screens, and in additions to glasses and ceramics for electronic devices.

Lanthanite. A mineral containing hydrated lanthanum carbonate $La_2(CO_3)_3.8H_2O$ in which varying quantities of cerium are present. It crystallises in greyish-white, pink or yellowish rhombic prisms and occurs at Bastnäs, in Sweden, as well as in Pennsylvania and other parts of the USA.

Lanthanum. The commonest and the most basic of the metals of the rare-earth group. It occurs along with other members of the group in cerite; allanite found in Brazil, Scandinavia and the USA; and monazite sands of India and Sri Lanka. It is obtained in fairly large quantities in the production of cerium and thorium from monazite sands, but so far has found only limited applications. The metal has been used in aluminium alloys, but the present main use is in the form of oxide for ceramic glazes, optical glass and as a "getter" in vacuum tubes. Solutions of lanthanum show no absorption spectrum. The main properties are: at. wt. 138.9; at. no. 57; crystal structure, hexagonal close-packed—$a = 3.754 \times 10^{-10}$ m; $c = 12.13 \times 10^{-10}$ m; at. vol. 22.6; density (20 °C) 6.17×10^3 kg m^{-3}; specific heat (20 °C) 0.197; melting point 920 °C. The metal is iron grey in colour and can take a high polish. It tarnishes in air and becomes steel blue in appearance. It can be beaten into foil and readily drawn into wire. With water and dilute acids it reacts to form the hydroxide or the corresponding salts. (*See* Table 2.)

Lanz pearlite process. A process in which low carbon, low silicon cast iron is cast in moulds heated to temperatures from 100 °C to 500 °C, depending on the iron and the type and size of casting. The object of the process is to retard the rate of cooling of high-duty metal of "border-line" composition, that is liable to produce white iron. If the composition is suitable there is produced a metal of improved quality which is associated with a high percentage of pearlite structure. The composition of the iron should be such that when poured into a cold mould or one of ambient temperature, a section of the casting will show a white (or a white to mottled) structure. On being poured into a heated mould (at a temperature dependent on the thinnest section), such a metal will then have a grey structure, will be readily machinable and will have enhanced physical and mechanical properties.

Lap. A surface defect produced when sharp corners or fins of hot metal are folded over the surface of hot metal, but are not welded to the metal during rolling or forging operations. It is sometimes found on wrought products by the folding of a surface against itself.

The name is also applied to the tool used in lapping operations. It generally consists of a soft cast iron, tin, copper, brass or lead base into which is charged the abrasive—usually by pressing it into the base or matrix by means of a steel block.

Lapis lazuli. A silicate of aluminium and sodium with some sulphur, found in Afghanistan, Siberia, Chile and Italy. It consists chiefly of lazurite (q.v.). It has a deep blue colour and takes a high polish, which, however, soon loses its lustre, since the mineral is moderately soft. The fine granular structure contains, in addition to the blue cubic minerals, some of the following substances which give the genuine stone its characteristic appearance: pyrite (gold-like specks), zircon, calcite, diopside, amphibole, felspar, mica, sphene and apatite. Mohs hardness 5.5, sp. gr. 2.38-2.45. When finely ground it produces the natural pigment ultramarine. This is now superseded by the prepared pigment.

Lap joint. A plate joint in which one piece of metal overlaps the other. The joint members are welded, glued or riveted together. The weld is made along the seam and the rivets are arranged in one, two or three rows (*see* **Joints**).

Lapping. A polishing and truing operation, the object of which is to give the final work a good finish and the highest possible accuracy. The tool used is called a lap (q.v.) and it is usually of the same outline as the work. The process is always used for finishing the bores of engine cylinders, gun barrels, etc., to fine limits.

Lap weld. A joint formed by welding together two overlapping pieces of metal. (*See also* **Joints** and **Welding**.)

Laser. The initial letters of "light amplification by stimulated emission of radiation", first discovered in 1960, in crystals of ruby. The ruby is essentially a form of alumina Al_2O_3, with chromium ions replacing some of the aluminium ones, hence the colour. When the ruby is irradiated with green light, the electrons in the chromium ions are stimulated into higher energy levels, where they are unstable, and fall back readily into an intermediate level where they are metastable, but can remain until stimulated again. The fall back to the ground state, from this metastable state, gives rise to an emission of fluorescent radiation, in the red end of the spectrum. The laser is simply a means of stimulating this emission from as many as possible or all the ions whose electrons are in the metastable state. To do this, one end of a ruby rod is fully silvered, and one half-silvered, so that the red light, when emitted at different angles, may be fed back into the rod to assist in stimulation of further ions. The application of red light commences the emission, and this spreads until a cascade of red light is finally beamed from the half-silvered end of the rod, with an intensity far greater than would normally occur from simple fluorescence. In the ruby laser there is a discontinuous system, because sufficient ions have to be stimulated to the higher electron energy levels before any emission can take place. The pulse is about 5×10^{-4} s in duration, with a power of about 10 kW from 1 cm² of cross-sectional area. Continuous laser systems have been developed, in which a number of transitions occur, or several steps, and power output has increased to the level required for melting and welding operations. In medicine the delicate nature of operations has provided a useful field for the laser in eye and neurological surgery. Figure 85 illustrates the principle of the ruby laser.

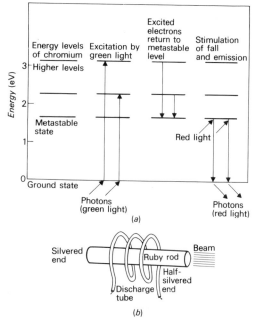

Fig. 85 *The ruby laser. (a) Excitation to higher states and return by stimulation. (b) Practical details.*

Lashing. A mining term used in reference to the broken rock after blasting. Probably originated in South Africa.

Latent heat. The amount of heat evolved (or absorbed) when 1 g of substance changes its state without change of temperature. The amount of heat absorbed in the transformation of unit mass of a solid at its melting point into liquid at the same temperature is termed the latent heat of fusion of the substance. Similarly the heat absorbed in changing unit mass of a liquid at its boiling point into vapour is the latent heat of vaporisation.

Laterite. A non-marine ferruginous clay formed by the tropical weathering of basic igneous rocks. The two characteristic constituents are limonitic matter Fe_2O_3 and H_2O and trihydrate of aluminium Al_2O_3 and H_2O. Typical laterite contains about equal proportions of limonitic matter and hydrated oxide of aluminium. It is resistant to atmospheric weathering and is used for building stone. If the ferruginous matter exceeds 50% the rock is limonitic laterite and can be used as iron ore. When the proportion of aluminous matter exceeds 50% the rock is known as bauxite and is used as an aluminium ore.

Latex. The viscous fluid of natural rubber as tapped from the tree.

Consisting largely of isoprene monomer, the material requires hardening by cross-linking, to become a useful elastomer. Many partially polymerised or monomer fluids are referred to as latex in the plastics industry, and the word has become synonymous with a viscous fluid entering a final process stage.

Lattens. A zinc-copper alloy, hot rolled and used for ornamental purposes, e.g. monumental brasses and effigies. The sheets may be of thickness from 0.4 to 0.5 mm, and there are three forms—black latten is unpolished, roll latten is polished both sides and shaven latten is the thinnest type of sheet.

The term is also applied to hot-rolled steel sheets of 25-27 gauge (B.G.).

Lattice. A regular arrangement of points in space is known as a space lattice. Many naturally occurring substances exhibit a near approach to perfection of external shape, and from this orderliness of geometric structure observed in crystals it is reasonable to infer an orderliness in the arrangement of atoms within a crystal, and this has been otherwise confirmed. In a plane parallel with a face of a crystal the atoms can be regarded as so arranged that if they could be joined by straight lines a definite self-repeating network would result. But the orderliness exists in space, so that if the corresponding atoms in different planes could be joined, a self-repeating crystal unit would result. This unit is called the unit cell and the repetition of this forms the lattice. The contents of each cell are identical. (*See* **Crystal classification.**)

Laue diagram. A diagram of the diffraction pattern produced by the Laue method of crystal analysis. The method was devised by Laue to determine indirectly the position in space of the atoms that build up a crystal. A fine pencil of X-rays is passed through a crystal, using the regular arrangement of atoms or atomic groups as a diffraction grating, and a record is taken on a photographic plate placed behind the crystal (or in front of the crystal if the X-rays are to be reflected from the surface of the crystal). Such a photograph shows a symmetrical arrangement of spots (Laue spots) round a central spot, the latter being due to the direct pencil of rays. The patterns obtained depend on the degree of symmetry of the crystals and they can be used in the determination of the dimensions of the space lattice.

When crystals of sufficient size are not available the X-rays may be passed through a quantity of crystalline powder and the measurements made by the method of Debye and Scherrer.

Laumonite. A hydrated silicate of aluminium and calcium, which on heating fuses to a white enamel $CaO.Al_2O_3.4SiO_2.4H_2O$. Mohs hardness 3.5-4, sp. gr. 2.2-2.3.

Launder. A trough or channel for conveying molten metal from a furnace spout to a ladle or from a container to a casting. The term is also applied to the troughs used for conveying liquids or a mixture of liquids and solids in coal-washers, ore-dressing plants, etc.

Laurionite. An oxychloride of lead $Pb(OH)Cl$ which occurs naturally and is also sometimes found in old slags. It was prepared artificially by treating galena with hydrochloric acid. The resulting $PbCl_2$ was then dissolved in water, and on the addition of lime-water a snow-white precipitate was produced. This was used as a paint in place of white lead.

Laurite. A sulphide of ruthenium and osmium found in association with platinum in placer deposits in Borneo, Oregon and Colombia. It usually occurs as small black octahedrons that are difficult to distinguish from magnetite.

Lauth mill. A three-high type of rolling-mill used for rolling steel plates. The top and bottom rolls are of the same diameter and are both power-driven. The middle is smaller than the other two and is friction driven, and can be brought into contact with either top or bottom roll. In making a bottom pass the plate passes through the gap between the centre and bottom rolls whilst the top roll acts as a backing roll. In the return pass the middle one is dropped into contact with bottom roll whilst the work passes through the upper roll gap.

Lava. Rock substance which has been ejected in molten form from volcanoes and subsequently solidified. It is distinguished, according to whether the silica content is high or low, into acidic and basic types. The latter is more fluid. Pumice is characteristic of lavas of rhyolitic composition and is extensively used as a polishing, smoothing and cleaning stone. When ground and mixed with soaps it forms a constituent of metal polishes.

Laxal process. A rust inhibiting process resembling the phosphating treatment for steel except that oxalic acid is used in place of phosphoric acid.

Layer lattice. The type of crystal lattice in which two-dimensional sheets with strong bonds are formed, but the bond in the third dimension is weaker than within the sheets. Usually the sheets or layers are covalent or ionic plus covalent bonded, with van der Waals' forces between the layers. In some semi-metallic lattices, corrugated sheets like layer lattices are formed, with predominantly metallic bonding, but covalent linkages between the sheets.

Graphite is typical of a covalent layer lattice, and kaolin is an example of ionic and covalent bonding. The characteristic of materials with a layer lattice, is the easy shear between layers, but strength within the layers, making them suitable lubricants (as with graphite and talc) or bestowing great plasticity when water can be absorbed, as with kaolin and the clay minerals generally. (*See* Figs. 74 (Graphite) and 84 (Kaolinite).)

Lazurite. A rock-forming mineral isomorphous with sodalite. It is a sodium aluminium silicate containing some sulphide

$$Na_4 (NaS_3 .Al)Al_2 .(SiO_4)_3 .$$

It crystallises in rhombic dodecahedra, is usually white or blue in colour and has a vitreous lustre. It is an important constituent of lapis lazuli. Mohs hardness 5.5, sp. gr. 2.2-2.4.

Leaching. The extraction of one or more constituents from a heterogeneous material by dissolving in water, dilute acid or other chemical solution. The process is used for concentrating and purifying ores. Metals, etc., may be recovered from the leaching liquors as well as from the solid residues. Types of solution that are commonly used contain $H_2 SO_4$, HCl, $H_2 S$, SO_2, $(NH_4)_2 CO_3$ or NaCN.

Lead. The densest and softest of the common metals that posseses high resistance to corrosion but little mechanical strength. In the cold state it flows readily under pressure and can be beaten into any desired form, extruded into pipe or rod or rolled into thin sheet. Since it possesses little tenacity it cannot be drawn into fine wire.

Lead has been extensively used for water-pipes since pre-Roman times, and finds many important uses in building, chemical plant and for other industrial applications. Alloys containing 0.25% Cd or 0.5% Sb and 1.5% Sn have greater strength and are frequently used in preference to pure lead. Lead alloys include babbitt metals, type metals, soft solders, leaded bronzes, cable sheathing and copper-lead bearing alloys. Lead additions up to about 0.25% are used to improve the machinability of mild steels and copper base alloys. Lead-tin alloys are used as a coating material for terne plate. The principal source of lead is galena PbS and the numerous oxy-salts found in the upper parts of galena deposits. It frequently contains silver in workable amounts. Other sources are cerussite $PbCO_3$ and anglesite $PbSO_4$. The principal producers are the USA, Mexico, Australia, Canada, Peru, Spain, Argentina, France and Zimbabwe.

The extraction of the metal is now usually by smelting the ore in a blast furnace. The ore is first concentrated either by flotation or by a wet gravity method, and then calcined or sintered either in reverberatory furnaces or on a Dwight-Lloyd sintering plant. The calcined or sintered concentrate is charged into a smelting furnace together with coke, limestone and some iron. The impure lead so obtained is purified either by kettle refining or by an electrolytic process (known as Bett's process, q.v.). (*See* Table 2.)

Lead annealing. A method of annealing alloys by immersion in molten lead. It is most widely used for the heat-treatment of steel wire. The process protects the work from atmospheric oxidation during heat-treatment and affords a simple means of obtaining close control of temperatures.

Lead azide. An explosive compound having the formula PbN_6. When ammonia passes over sodium metal, sodamide is formed, and when this latter is treated with a stream of $N_2 O$ at about 200 °C, sodium azide is produced. This is not a sensitive compound, but the precipitate obtained by adding lead acetate solution to the sodium azide will readily detonate when struck. Lead azide is used in detonators.

Lead-bearing steel. A steel to which about 0.2% of lead has been added to improve its machinability.

Lead burning. A method of welding together two pieces of lead by fusion without the use of solder.

Leaded bronze. A copper base alloy containing 5-10% Sn with 5-30% Pb used in high-duty bearings, slide valves, etc.

Leader. A term sometimes used in rolling-mills for the first forming pass. It comes immediately after the "roughing" or breakdown passes.

Lead frit. A disilicate of lead obtained by fusing the lead oxide (litharge) with silica. It is used for making lead glazes.

Lead glance. An alternative name for galena PbS, q.v.

Lead glass. Ordinary glass consists of a mixture of silicates of calcium and the alkalis, though for special purposes the calcium and part of the alkali silicates may be replaced by other silicates, and phosphoric or boric oxides may partly replace silica. If the materials are slightly impure the glass may become coloured or opaque, e.g. iron II (ferrous) iron gives a deep yellowish-brown tint. To oxidise these impurities lead dioxide may be used, and when the sodium silicate is wholly replaced by lead silicate a potassium-lead glass having a high refractive index (1.70-1.78) and a specific gravity of 3-3.8 is produced. This is known as lead glass, flint glass or crystal glass. It is readily fusible and is easily worked, but the heating must be done in a non-reducing atmosphere, otherwise the glass becomes opaque, due to the precipitation of lead. Highcontent lead glass is used for optical purposes, but all lead glasses are readily attacked by aqueous solutions of many substances. They are therefore unsuitable for making chemical apparatus. A typical composition is SiO_2 40-50%, PbO 38-53%, $K_2 O$ 8-11% and the balance should be not more than 1% $Fe_2 O_3$ and $Al_2 O_3$.

Lead gruff. A by-product of cyanide manufacture. It contains 23-28% Pb and about 30% soluble lead salts. Usually about 3% Fe is also present.

Leadhillite. A greyish-white mineral occurring in laminae and consisting of sulphocarbonate of lead $PbSO_4 .3PbCO_3$.

Lead metaniobate. $Pb(NbO_3)_2$. A multiaxial ferroelectric, with a pyrochlore type of structure. The application of an electric field allows polarisation along several axes.

Lead monoxide. The oxide formed when lead is heated in air. The grey dross so produced contains a mixture of lead monoxide and metallic lead. If this is heated strongly in iron pans the mixture turns yellow and is known as massicot. If this is fused and powdered the reddish-yellow crystalline form known as litharge is obtained. Litharge is also produced in the refining of silver, and is largely used in making flint glass, in glazing pottery, in preparing lead salts and in the making of paints and varnishes. When added to linseed oil it acts as a "drier", since it accelerates catalytically the absorption of oxygen by the oil to form a solid compound called linoxyn (the basis of glazing putty). Olive oil — the oleic ester of glycerin — is saponified when boiled with water and litharge. Glycerin goes into solution and a sticky adhesive mass of lead oleate is formed. This latter is used in making lead-plaster.

Lead nail. A small copper alloy nail used for fastening lead flashing and lead roofing sheets.

Lead patenting. A process of cooling a metal at a controlled rate that consists in quenching in a bath of molten lead.

Lead tellurium. Lead containing from 0.02 to 0.085% Te. The addition of tellurium imparts hardness, toughness and increased corrosion resistance to the lead. It is used in underground cable sheathing.

Lead tetraethyl. A compound formed by the interaction of lead chloride $PbCl_2$ and magnesium ethyl iodide $Mg(C_2 H_5)I$. It has the formula $Pb(C_2 H_5)_4$ and acts as a negative catalyst when added to certain types of petrol, inhibiting the tendency for pre-ignition.

Lead vitriol. Lead sulphate or anglesite $PbSO_4$. The latter is found associated with galena and carbonates of lead at Leadhills (Scotland).

Lead zirconate titanate. $Pb(Ti,Zr)O_3$. (Also known as PZT to electronic engineers.) A piezo-electric material, used in earphones, hearing aids, microphones and underwater sonar transducers.

Leakage current. When an alternating e.m.f. is applied across a capacitor, the current is out of phase with the applied voltage by 90°. However, with a dielectric there is a small current directly in phase with the voltage, which is due to the resistivity of the dielectric not being infinitely high. The phase angle is less than 90° therefore, and the current through the dielectric is termed the leakage current, or resistive current.

Le Chatelier's principle. A rule by which it is possible to indicate qualitatively the direction in which equilibrium will be displaced by a variation in the conditions. It can be stated in the following terms: when a system in equilibrium is subjected to a change in the equilibrium conditions it behaves in such a way as to minimise (or oppose) the change. The principle is a general one and is not restricted to chemical reaction. The conditions mentioned in the rule refer to the effects of temperature, pressure and volume on the equilibrium state.

Leckie process. A direct process for the production of iron and steel. The ore is mixed with flux and carbon, the charge is heated on a reverberatory furnace, and the reduced product is melted on the same hearth.

Leclanche cell. A low-resistance cell especially useful for intermittent work. The low potential element consists of a rod of zinc placed in a solution of ammonium chloride. A rod of carbon forms the high-potential element, and this is placed in a porous pot and surrounded by broken carbon and black oxide of manganese, which acts as a depolariser. However, the hydrogen, during the flow of current, is produced faster than the MnO_2 can absorb it but, on standing, the hydrogen is oxidised and the cell regains its strength. The advantages of this cell are: (a) it is inexpensive and does not emit objectionable odours; (b) it can be used intermittently over many months with only a little attention, e.g. addition of water. The crystals of ammonium chloride are apt to "creep", and this can be prevented by coating the upper part of the inside of the jar with paraffin wax.

Nearly all forms of dry cell are modifications of the Leclanche principle, the electrolyte usually consists of a paste of plaster of Paris, flour, zinc chloride, ammonium chloride and water. Voltage is about 1.5 volts. Internal resistance of porous-pot type is about 5 ohms. In the dry type the internal resistance is usually less than 0.5 ohm.

Lecocq test. A sensitive test for molybdenum. An alcoholic solution of diphenylcarbazide is added to ammonium or sodium molybdate acidulated with HCl, when a violet or indigo colour is produced. The reaction does not occur in presence of excess alkali.

Ledeburite. The name, conferred by Wust on the cementite-austenite eutectic containing 4.3% C which freezes at 1,130 °C. During cooling the austenite in ledeburite may transform into a mixture of austenite and cementite. This iron-carbon phase is found in cast-irons and certain high alloy steels.

Leg pipe. A term sometimes applied in blast-furnace work to the tuyère pipes through which the hot blast is delivered to the tuyères.

Leidie process. A process due to Leidie and Quennessen which was formerly much used for the separation of the constituents of platinum ores. It is still used, with minor modifications, as a laboratory method. The material is powdered and gently heated in a nickel crucible with sodium peroxide. The product is treated with water and filtered. The filtrate contains sodium salts of osmium, ruthenium and palladium. If it is blue in colour iridium is present. If the solution is yellow it may contain osmium, ruthenium and palladium, in which case chlorine is passed into the solution which is then distilled. To the distillate ammonium sulphide is added and a black precipitate indicates osmium and/or ruthenium. If, on heating the solution with potassium nitrate, a violet product is formed, osmium is present. The solution will give a black precipitate on adding alcoholic potash if ruthenium is present.

If the solution is blue, as already indicated, the presence of iridium is confirmed by adding hydrochloric acid, ammonia and potassium chloride and evaporating. Black crystals will result.

The residue from the above distillate contains all the palladium. Its presence is confirmed by adding hydrochloric acid, ammonia and potassium chloride when red crystals are obtained

on evaporating.

The original residue contains the nickel, platinum and rhodium and is heated with hydrochloric acid, filtered and evaporated. Sodium nitrite and sodium carbonate are added and the solution boiled and filtered. The residue is nickel and the solution contains the platinum and rhodium. The latter is heated to dryness and then treated with ammonium chloride. Platinum is precipitated but the rhodium remains in solution as a reddish double chloride.

Lemon chrome. The insoluble yellow barium chromate $BaCrO_4$. It was used as a pigment but has been superseded by the lead chromates, which are brighter in colour and have more body. Specific gravity 3.9. Also known as yellow ultramarine.

Lenticular. Having the form of a double convex lens.

Lenz's law. A law of electromagnetic induction which states that the direction of the induced current (and e.m.f.) is such that it tends to oppose the motion or change producing it. As applied to self-induction of a circuit, it explains why self-induction behaves like inertia, choking back a current that is being set up and giving the "extra current" in a circuit where the current is being cut off. This extra current is the cause of the dangerous spark produced when circuits are broken by unsuitable switches.

Lepidocrocite. A hydrated oxide of iron $Fe_2O_3.H_2O$ associated with limonite in iron ores. It is the form of iron rust (gamma) that is produced by the *slow* oxidation of iron II(ferrous) hydroxide.

Lepidolite. A silicate of potassium, lithium and aluminium
$$(Li,K,Na)_2.Al_2(SiO_3)_3.(F,OH)_2.$$
It is a source of lithium, the mineral containing from 1.3 to 5.7% Li. It crystallises in the monoclinic system, but is often found in the form of scales. It is usually lilac to peach-blossom in colour, but sometimes occurs in greyish-white masses which show a pearly lustre on cleavage. Mohs hardness 2.5-3, sp. gr. 2.85. It is found in Germany, Czechoslovakia, California (USA), Canada, Sweden, the Urals and in Central Australia. Also called lithium mica.

Leucite. A silicate of aluminium and potassium $K_2O.Al_2O_3.4SiO_2$ that occurs as a primary constituent of volcanic rocks. In Italy, where it occurs in volcanic lavas, it is used as a fertiliser. The mineral is remarkable for its optical properties, which vary with the temperature. This is due to the fact that the crystals consist of a complex intergrowth of orthorhombic or monoclinic individuals which are optically biaxial and repeatedly twinned, giving rise to twin-lamellae and to striations on the faces. Boracite has a similar pseudo-cubic character. The crystals are white or grey in colour, often dull and opaque, but sometimes transparent and glassy. The latter are known as white garnet. Leucrite is a source of potash, but the extraction is not easy. Mohs hardness 5.5, sp. gr. 2.5.

Leuocophane. A sodium, calcium silicate $Na,Ca,Be,F(SiO_3)_2$ containing beryllium. It is one of the sources of beryllium.

Levelling solution. A term applied in electrodeposition to electrolytes that produce plated surfaces of greater smoothness than the original base. The process does not necessarily produce *bright* electroplate and the solutions operate satisfactorily only within a limited range of compositions and operating conditions. The substance most commonly added to the electrolytes to produce levelling is coumarin but aromatic amides, saccharin, etc., or a mixture of these are sometimes used.

Level set. A term applied to copper ingot castings which have a flat or slightly convex upper surface caused by the liberation of absorbed gases, during solidification.

Lever rule. An adaptation of the mechanical lever force calculations, applied to thermal equilibrium diagrams to obtain the proportion of constituents present for a given composition. If two constituents A and B, are separated by a distance x in the diagram, then any composition between them, say distance y from A, will have $x-y$ proportion of constituent A and y proportion of B.

Levigation. The process of grinding a mineral, etc., to a fine powder in water, followed by the fractional sedimentation and separation into grades of various particle sizes.

Lias. The lowest division of the Jurassic system. It consists chiefly of clays, sands and limestones often 300 m thick, and is divided into lower, middle and upper lias. The liassic beds are typically shales which are often grey or bluish-grey in colour. They contain some coal and abundant fossil remains. Typical lias coal (Hungary) contains 78% C, 4% H, 10.5% ash. The usual carbon-hydrogen ratio is about 20.

Liberian ore. Iron ore obtained from deposits near Monrovia (Liberia). The typical analysis is 68.5% Fe, 0.8% SiO_2, 0.8% Al_2O_3, 0.1% CaO, 0.07% Mg, 0.15% S, 0.1% Mn and P.

Libethenite. A naturally occurring basic phosphate of copper $Cu_3(PO_4)_2.CuO.H_2O$. It crystallises in dark olive-green rhombic prisms having a waxy lustre.

Liebigite. A secondary uranium mineral consisting mainly of the carbonate of the metal. It contains about 37% U_3O_8.

Lifter. A bar of cast or wrought iron used for supporting the sand in the cope of a foundry mould. The bar is usually S- or L- shaped, and the upper end is hooked on to a bar of the box.

The term is also used in the USA to indicate a "cleaner", a foundry tool used for lifting dirt out of the mould, for repairing broken mould faces and for finishing the bottom and sides of deep narrow openings.

Lifting plate. A small iron plate let into foundry patterns. A screw-threaded bolt is inserted into a hole drilled through the plate, to facilitate the removal of the pattern from the mould.

Light metals. The common metals that are characterised by comparatively low specific gravities, and which can be used alone or in alloy form for structural or engineering purposes. The term usually excludes the alkali metals, but includes aluminium, magnesium, beryllium and titanium.

Light metals of the platinum group. The six metals of the platinum group divide themselves naturally into two groups according to their specific gravities. The specific gravity of the heavy group averages 22 and for the light group 12. The three lighter metals are ruthenium, rhodium and palladium, and in some properties they resemble silver. Variable valency, however, is not a *characteristic* of silver as for metals of this group.

Light-weight refractories. The name applied to refractory material in the form of bricks and other shapes, that have a high porosity. They are generally used as insulators. They may be made from fireclay or silica and the high porosity is obtained by mixing with these a bulky material such as sawdust (which is later burnt out) or a volatile solid such as naphthalene (which is later sublimed), or by forcing air through the plastic batch.

Lignin. The organic material forming the binding constituent of woody fibres. It is the undissolved residue left after boiling wood in dilute alcohol. Also called xylogen.

Lignite. A coal in which the dominant constituent is wood, so little altered that the fibrous form and structure can be observed. It is probably formed from trees of coniferous type. It is an intermediate stage between wood and coal, and is more compact than peat and is lustrous. It is dried and pressed into blocks and used as fuel. Large beds of lignite occur near the surface in many parts of Germany, Hungary and in the Mississippi valley. It is sometimes called brown coal. Jet is a hard variety of lignite used for ornaments.

Limburgite. A volcanic rock consisting of augite, magnetite and olivine. It is dark in colour resembling basalt in appearance. The accessory minerals are titaniferous iron oxides and apatite. It occurs in Germany, Scotland, central France, Spain and Brazil.

Lime. The white oxide of calcium usually prepared by heating limestone $CaCO_3$ in a kiln. The process is a continuous one, the charge of coal and limestone (about 1:5) is fed at the top of the kiln. The quality of the lime depends on the nature of the limestone. If the latter is pure, the resulting lime slakes readily and is known as "fat" lime. When magnesia is present a "poor" lime that does not readily slake is formed, and if the amount of magnesia exceeds about 25% the lime is useless for building purposes. The presence of a considerable quantity of silica in the limestone causes the production of a lime-silica mixture which yields a mortar that hardens under water. If clay is present in the limestone the lime formed will slake very slowly as silicates are formed during the heating and these incrustations retard the slaking. Such a lime is referred to as "dead burnt".

Lime is used for drying certain gases (not chlorine or acidic gases) and for purifying coal gas. It is also used in the making of mortar which is a mixture of sand and lime. It hardens very slowly

in air by absorbing CO_2. It may slowly become converted into silicate. Slaked lime is formed by adding water to the dry anhydrous powdery oxide of calcium. It is used in making glass, bleaching powder, caustic soda, artificial building stone and in the purification of sugars. It is used also as a flux in many smelting operations. Specific gravity 3.30, melting point about 1,900 °C.

Limebag. A bag of powdered lime used in the foundry for testing the fit of mould joints.

Lime boil. A stage in the basic open-hearth steelmaking process which follows immediately after the "ore boil". During the "ore boil" there is a boiling action as carbon is removed by the action of the oxides of iron, forming CO and CO_2. As the carbon content of the bath decreases and the temperature rises, limestone decomposes, liberating oxides of carbon which bubble through the metal, causing a vigorous "boiling" agitation. During this stage lime rises to the surface, liberating iron and manganese from the slag and forming calcium silicates.

Lime burning. The term applied to the process of producing lime from limestone by heating in kilns. The lime and the necessary fuel (usually coal) are mixed and charged into the kiln. This is the usual procedure and the process can be continuous. The older (intermittent) batch production is carried out by locating the fire at the base of the kiln and piling the lumps of limestone on a grating above the fire. This method occupies about 48 hours per charge. A continuous draught must be maintained during the burning. The carbon dioxide formed as a by-product is lost.

Lime coating. A cleaning process used in the manufacture of steel rods and wire. It consists of immersing the metal in milk of lime (a suspension of lime in water) in order to produce a coating of the desired thickness, which will act as a lubricant during subsequent drawing operations.

Lime felspar. A holocrystalline igneous rock belonging to the gabbro group. It contains less than 52% SiO_2 and is basic in character. It consists of silicate of alumina with silicates of calcium and sodium and occurs in gabbro together with salts of iron and magnesium.

Limestone. A rock consisting mainly of calcium carbonate $CaCO_3$. It usually contains varying amounts of silica, alumina, carbonate or oxide of iron, calcium phosphate and magnesium carbonate. All forms of limestone (except dolomite) react readily with dilute acids. When pure, limestone is white, but in nature it is frequently discoloured by oxides of iron (yellow to brown) or by iron sulphide or iron carbonate or bituminous matter (grey to black). Crystalline limestone formed as a result of great heat and pressure is known as marble. This may assume a variety of colours due to the presence of other minerals. Most limestones are of organic origin.

Pure limestone is burnt to produce "fat" lime for use in chemical glass, soap-making and silicate industries. Argillaceous limestone is burnt for the production of hydraulic lime. Limestone is used for road material and some varieties for building stone. Metallurgically, limestone is important as a flux in the making of basic steel and in blast-furnace smelting. Mohs hardness 3, sp. gr. 2.6-2.8.

Lime uranite. Hydrated phosphate of calcium and uranium

$$CaO.2UO_2.P_2O_5.8H_2O.$$

It closely resembles the tetragonal torbernite in form, the latter being the corresponding copper compound. It crystallises in the orthorhombic system and is optically biaxial. In colour it is sulphur yellow and thus readily distinguishable from the emerald-green torbernite. Mohs hardness 2-2.5, sp. gr. 3.05-3.19. The mineral is usually found with pitchblende and other uranium ores. Also called autunite, uranium mica and calcouranite.

Lime water. The alkaline solution obtained by dissolving calcium hydroxide in water. It is "hard", but can be softened by the addition of soda ash (washing soda). The solution absorbs carbon dioxide and precipitates insoluble calcium carbonate. Hence it is used as a test for carbon dioxide, and for removing carbon dioxide from gaseous mixtures, but in the latter application it must be remembered that it also absorbs Cl_2 and SO_2 and NO_2. Its most useful application is thus in separating CO and CO_2.

Liming. A term used to describe a process which involves treatment with lime either to reduce acidity or to remove carbonic acid. It is also used to describe the application of a film of lime to pickled steel in the wire-drawing process. Acidic soils benefit by treatment with lime and the removal of the aggressive CO_2 from water. In the latter case an adherent deposit of calcium carbonate occurs on the interior of pipes, and this prevents the corrosion of the metal.

Limiting range of stress. The greatest range of alternating stress (the mean stress being zero) a metal can withstand for an infinitely large number of cycles without failure. Also called endurance range.

Limit of proportionality. The mathematical treatment of the relationship between stress and strain in a material leads to the statement of Hooke's law that "stress is proportional to strain", up to a definite limit for each material, which represents a considerable stress but usually a very small extension (about 0.5%). This limit is termed the "limit of proportionality" for the type of stress considered. After this limit has been passed strain *increases* at a much greater rate in relation to the stress, and the strain is not entirely removed when the load is removed. (*See* **Elasticity**).

Limonite. A naturally occurring hydrated oxide of iron

$$2Fe_2O_3 3H_2O$$

containing about 60% Fe. It is yellow to brown in colour and has been formed by the alteration of other iron minerals, e.g. by oxidation and or hydration. It is a common and important ore found in the USA and Europe. The substance is amorphous. Mohs hardness 5, sp. gr. 3.5-4.

Linarite. A hydrated sulphate and oxide of lead and copper found in the weathered zone of lodes.

Linear expansion. Most substances expand on heating. When the substance is in the form of a rod, wire, column, etc., in which the length is the important dimension, the change in length per unit length when the substance is heated through 1 °C is known as the linear expansion coefficient. The length (l_t) at t °C of a material whose length at 0 °C was (l_o) is given by $l_t = l_o (1 + \alpha t)$. The values of the coefficient α for the elements are given in Table 2. The values of α for certain alloys and common materials are given in Table 3.

Linear polymer. A polymer, usually formed by addition polymerisation, in which the monomer unit is repeated along a chain without, or almost without, any branching. Examples are polyethene (polyethylene) and PVC.

Line defects. *See* **Crystal defect.**

Line of segregation. The limit of the segregation of low-melting-point constituents of an alloy, near the centre and top of an ingot or casting. The segregation varies in amount with the composition of the different types of alloy and with the different rates of solidification.

Line spectra. Spectra are broadly divided into two classses—absorption spectra and emission spectra. Line spectra are characteristic of atomic emissions, while band spectra originate in molecules. No two substances produce the same spectrum and so it is possible to determine the chemical nature of a substance by an examination of its spectrum. The line spectrum really consists of a succession of images of the slit in the spectrometer. In identifying an element it is necessary to have a method of expressing the position of the lines in the spectrum, on a scale of wavelengths, but as the values are very large they are inconvenient, and are therefore replaced by wave numbers.

Lining. The refractory brickwork and slurry which forms the hearth, bosh, roof and walls of a furnace or a stove.

Linnaeite. A cobalt ore consisting of cobalt-nickel sulphide

$$(Co,Ni,Fe)_3S_4$$

which usually contains some iron and copper. It is steel grey or copper-red in colour and occurs in regular octahedra in Germany and the USA. Also called cobalt pyrites, and may be used as a source of nickel.

Linotype metal. A printing metal used in high-speed machines, such as Linotype, Intertype and Ludlow casters, in which slugs of metal each carrying a line of type are die-cast. (*See* Table 8 for compositions.)

Linz-Donawitz process. A steelmaking process in which a long water-cooled oxygen lance is used to direct a high-energy jet of oxygen downwards on to the surface of the slag covering the molten metal. The jet produces a highly turbulent motion in the slag and in the metal just below it. This results in a rapid reaction between oxygen and metal. Alternately, the oxygen can be allowed to react mainly with the slag and then the reaction between slag and metal is speeded up by rotating the converter to give the necessary mechanical agitation.

Because the combustion temperature is high there is produced a considerable quantity of iron oxide fume, and this tends to attack the refractory lining as well as pollute the atmosphere. Suitably shaping the interior of the converter may reduce fume turbulence, and hence the severity of the attack on the converter lining.

Desulphurisation of the metal has been achieved in a modification of the process, in which powdered lime is injected into the molten metal along with the oxygen (referred to as the L.D.A.C. process) (*see* **Oxygen steelmaking**).

Lionite. An artificially prepared variety of corundum, crystalline aluminium oxide. It is made by fusing bauxite in an electric furnace, and it is used as an abrasive. Mohs hardness 9.

Liquation. A term synonymous with segregation as applied to alloys. It also refers to a partial melting operation which permits the separation of certain alloys into two or more constituents. It is used in the refining of tin. Bars of impure tin are slowly melted in a reverberatory furnace in which the floor slopes downwards from the fire to the flue. The tin, with any impurities that are capable of forming easily fusible alloys, melts and runs out of the furnace. The less fusible impurities and alloys that remain are poled or aerated so that oxidisable impurities can be skimmed as a dross. The remaining tin is then cast.

Liquidated surface. The surface of an ingot or casting which exhibits protuberances or exudations in consequence of inverse segregation, e.g. tin sweat or lead exudation.

Liquid crystals. Materials which appear to share properties common to both liquids and crystals. In 1888 the organic compound cholesteryl benzoate was found to have two distinct melting points. The solid melted at 145 °C, to form a turbid liquid, which on heating to 179 °C underwent a second transition to a perfectly clear fluid. The turbid liquid is birefringent, a property typical of a crystal, although its viscosity is typical of a liquid. The phase existing between true solid and true liquid is called a mesophase, and mesogen is the name given to a compound capable of yielding a mesophase.

Liquid crystals are compounds which deviate significantly from spherical symmetry, usually rod-like molecules with a strong tendency to align their long axes parallel in the mesophase. On melting the solid, the long-range order of the solid crystal is destroyed, but the angular correlation of the rod-like molecules persists until the isotropic liquid is formed at the higher temperature. Three types of structure are recognised; the nematic mesophase with turbidity and low viscosity, the cholesteric phase with higher viscosity but optically active, and the smectic mesophase which more closely resembles a solid. More than one of these phases may exist in one substance, but the nematic one is associated with the highest temperatures, and the smectic with the lowest (nearer to the solid state). Liquid crystals are normally thermotropic, i.e. the changes take place with change in temperature, but there are mesophases which are formed by dissolution of a compound in an appropriate solvent. The lipid membranes of biology are in this category, and some soaps.

The nematic mesophase is the simplest, with long axes of the molecules parallel, but showing no other symmetry, e.g. the molecular centres have no correlation (*see* Fig. 86(*a*)). The parallelism is not perfect, the root mean square fluctuation in orientation may be 40°.

Cholesteric mesophases are associated with the compounds of cholesterol, although cholesterol itself is not a liquid crystal. This phase is related to the nematic, in that the main characteristic is that of parallel long axes to the molecules. However, there is a further state of order, in that the molecules are arranged in layers, with the long axes rotated through a small angle on passing from one layer to the next, although all parallel within the layer. As Fig. 86(*b*) shows, the transition is itself ordered, forming a helix with a given periodicity. Because of this, the cholesteric mesophase has the power to rotate the plane of polarisation of linearly polarised incident light, has higher viscosity, but otherwise

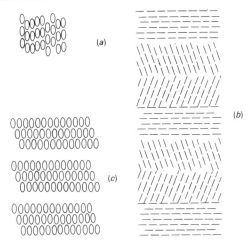

Fig. 86 Liquid crystal structures. (a) Nematic. (b) Cholesteric. (c) Smectic.

resembles the nematic phase. The twisted structure is believed to arise from the asymmetric centres of the molecules, affecting the intermolecular potential energy.

The smectic mesophase does not lose all spatial order when formed from the solid, unlike the other two. A characteristic lamellar structure results, with the individual layers able to slide or rotate over one another when subjected to mechanical shear. Smectic phases arise in strongly polarised molecules, and the interaction between polar groupings causes stabilisation of the layers. Several types of such layering have been observed, and smectic A to H phases reported. In A, the preferred orientation of the long axes is orthogonal to the layers, in B the molecular centres are regularly arranged, and in C the structure resembles A but there is a tilt between long axes and the layer plane. Smectic D phase is cubic in arrangement. Figure 86(c) shows the smectic B phase.

Nematic mesophases have found application in electro-optic devices such as digital read out systems. A voltage is applied to electrodes either side of a sheet of the material, one electrode being transparent. When voltage is zero, the mesophase is transparent, but application of a voltage causes light scattering, which can give read out by breaking up one electrode into segments and varying the voltage on the segments. Mechanical-optical transducers have also used nematic mesophases.

The cholesteric mesophase is sensitive to changes in temperature, pressure and vapour impurities, all of which change the pitch of the helix and therefore the light scattering. By mixing a number of compounds, temperature detectors which change colour quite dramatically with differences as small as 1 °C have been developed. They are used in engineering for detecting hot spots in machinery, but most important, in the medical application of thermography, detecting hot spots in the body associated with rheumatic disorders and cancer.

Liquid fuels. Apart from small amounts of animal and vegetable oil and some shale oil, practically all the industrial liquid fuels are derived from petroleum. Crude natural oil is a mixture of hydrocarbons, and it is roughly divisible into classes according to the

nature of the residue remaining after distillation, i.e. waxy or asphaltic. Crude petroleum is subjected to fractional distillation, the more volatile portions being used as motor spirit, while other portions are kerosene, fuel and lubricating oils. Artificial liquid fuels are also prepared from the tarry oil obtained in the dry distillation of wood and coal and lignite, and by the distillation of shale. Other methods include the recombination of the constituents of water-gas in presence of a catalyst, and the hydrogenation of coal. The properties of common liquid fuels are given in Table 34.

Liquid gold. A sulphoresinate of gold dissolved in essential oils. Sometimes bismuth and rhodium are present in small quantity. The liquid is applied to the surface of the glaze on pottery, etc., and after drying is fired at about 750 °C. A brilliant film of gold is left on the ceramic ware.

Liquid honing. A process for deburring and cleaning the surfaces of metals before electroplating or painting. A suspension of abrasive particles in a liquid is forced against the metals by compressed air.

Liquid-phase sintering. A term used in reference to a method of improving, by densifying, the properties of powder metal products, by sintering the material in such a manner that the metal of lower or of the lowest melting point in the alloy actually melts. The method is applicable to alloys or mixtures in which there is a large difference in melting point of the constituents (e.g. cobalt cemented tungsten carbide).

Liquid silver. A mixture of finely powdered silver, silver oxide or platinic chloride suspended in essential oils. The liquid is applied to ceramic ware by suitable means and after drying is heated to about 250 °C, when a bright film of metallic silver is deposited. The film is adherent to glass and can be increased in thickness if desired by electrodeposition. The film can be readily soldered.

Liquidus. A line on a binary constitutional diagram (or a surface on a ternary diagram) indicating the temperatures at which solidification begins during cooling and at which melting is completed during heating under equilibrium conditions.

Liquor finish. A reddish-bronze-coloured finish for steel wire, etc., produced by the wet drawing process. After pickling the wire is thoroughly cleaned before immersion in a dilute solution of mixed copper and tin sulphates, when a thin coating of copper and tin is deposited by chemical replacement. From this process the wire passes through a vat of fermented liquors, and while still wet to the drawing dies.

Lirconite. A hydrated arsenate of aluminium and copper, variable in composition. The probable composition can be represented by the formula $Cu_9 Al_4 (OH)_{15} (AsO_4)_5 .20H_2 O$. It is bright blue in colour. Mohs hardness 2.5, sp. gr. 2.95.

List edge. During the production of hot-dip, coated sheet-metals, such as tinplate, terneplate or galvanised iron, a certain amount of the coating metal tends to concentrate on the lower edge of the plate. This forms the list edge of the plate.

Litharge. If lead be heated in air the surface becomes covered with a greyish-yellow film. If the heating is continued the mixture of lead and lead oxide is converted entirely into the latter substance. The yellow oxide is called massicot, and if this is heated to fusion point litharge results which is reddish-yellow in colour.

Crystalline litharge occurs as a mineral in rhombic octahedra. Specific gravity 9.36. It is used in the manufacture of flint glass and as a glaze for earthenware, and also in the preparation of red lead, lead acetate, white lead and lead plaster and in drying oils.

Lithium. A light, silvery-white metal. It is obtained from a number of important minerals: triphylite (LiNa)(FeMn)PO_4 containing 1.6-3.7% Li; petalite LiAl(Si_2 O_5)_2 , containing 2.7-3.7% Li; lepidolite containing 1.3-5.7% Li; spodumene LiAl(SiO_3)_2 containing 3.8-5.6% Li. The principal sources are the USA, Australia, Brazil, Sweden and Germany.

Lithium is usually separated from the ore by wet processes involving precipitation as carbonate. The metal is finally extracted by the electrolysis of lithium chloride.

The metal is soft and tarnishes quickly in moist air. It is used as a hardener in some lead-base bearing alloys (Bahn metal) and in magnesium. Lithium-calcium alloys are used for deoxidising

TABLE 34. *Properties of Liquid Fuels*

	Absolute viscosity cp at 20 °C	Boiling point °C	Calorific value MJ kg^{-1}	Flash point °C	Specific gravity 15 °C
Methyl alcohol (methanol)	0.6	64.6	20		0.796
Ethyl alcohol (ethanol)	1.20	78.3	26.8		0.794
Benzole	0.63	86	40	−9	0.878
Aviation petrol	0.45	84	44		0.720
98 octane petrol	0.50	104	44		0.740
Tractor oil	1.10	166	43.5	37	0.780
Kerosene	1.27	196	43.5	40	0.793
Diesel oil	4.26	300	43	75	0.870
Light fuel oil	44.8	348	42	80	0.895
Heavy fuel oil	1139.0	360	41	110	0.949
Heptane	0.4	98	45		0.691
Methylated spirit	1.00	76	30	−8	0.832

copper and bronze and up to 0.25% Li has been used with success in silver solders. Lithium salts are used in glass and ceramic industries. Atomic weight 6.94, sp. gr. 0.534, melting point 180 °C (*see* Table 2).

Lithium alloys. They may be produced by the direct fusion of the constituents. Few of the alloys so far produced have found useful application. They are characterised by lightness and, if lithium is present in large proportion, by great activity. Light alloys suggested are (*a*) lithium-beryllium, lithium-magnesium, lithium-aluminium; (*b*) aluminium containing small quantities of lithium develops improved strength and elasticity; (*c*) lithium-magnesium alloys having a density of about 1.4 are not unduly subject to corrosion; (*d*) the 50/50 lithium-cadmium alloy in small quantities is very efficient in removing gases from copper. Silver solders and brazing alloys show increased strength, fluidity and wetting properties.

It is reported that lithium performs a useful function in ensuring an even distribution of carbon in cast iron; it also reduces the viscosity of stainless steels, facilitating the production of thin castings.

Lithium fluoride. LiF. An insulator with an energy gap of 11eV. LiF crystals can be etched to reveal edge and screw dislocations quite distinctly, and have been used to follow dislocation movement under stress, the growth of dislocation loops, and the form of such loops. Lithium fluoride exhibits a pronounced yield point, which becomes less marked as the number of free dislocations is increased by repeated straining.

Lithium selenite $LiH_3(SeO_3)_2$. A uniaxial ferroelectric material.

Lithomarge. A yellow-white or mottled clay found in Germany and Cornwall (UK). It is a hydrated silicate of aluminium.

Litho-oil. The product obtained by heating raw linseed oil to a little over 300 °C for 3 days. The properties of the oil are greatly modified if the oil is heated in closed vessels. It becomes more viscous and withstands the action of water better than raw or boiled oil. It is used in paints, etc. Also known as stand-oil.

Lithopone. A mixture of zinc sulphide and barium sulphate used for indoor painting. Barium sulphide is added to zinc sulphate and the double precipitate is heated. It contains zinc oxide as well as sulphide and has a better covering power than white lead. It does not darken on exposure to sulphuretted hydrogen but darkens on exposure to light.

Lithorizing. A phosphating process for zinc, and for galvanised or cadmium-plated materials. (*See also* **Lithoform.**)

Littoral deposits. Deposits lying on the sea-shore between high and low water-marks. The term refers particularly to the deposits brought down by rivers to their estuaries.

Liver of antimony. These are compounds of antimony trisulphide with the sulphides of other metals, and are properly known as thioantimonites. Those of the alkali metals are formed by fusing the constituents together. They are brown or black in colour, and their solubility in water depends on the proportion of antimony, decreasing as the proportion of this latter increases. Many thioantimonites occur naturally as minerals, e.g. berthierite $FeS.Sb_2S_3$, wolfsbergite or antimonial copper $Cu_2S.Sb_2S_3$, boulangerite $5PbS.Sb_2S_3$, jamesonite $2PbS.Sb_2S_3$, bournonite $2PbS.Cu_2S.Sb_2S_3$ and pyrargyrite or red silver ore $3Ag_2S.Sb_2S_3$. Livers of antimony are used in the vulcanising of rubber and in the making of matches.

Liver of sulphur. A mixture of various polysulphides of potassium with the sulphate and thiosulphate. It is prepared as a liver-coloured mass by gently heating potassium carbonate with sulphur in closed vessels. It is used in combating mildew and insect pests.

Lixiviation. A process of treating a mixture of soluble and insoluble mineral substances by a suitable solvent to effect a separation. It is a leaching process employed in the extraction of potash from wood ashes and in the extraction of certain metals from their ores, e.g. antimony, zinc, copper. The process is suitable for batch or continuous treatment. The latter, under the name of percolation, is used for the extraction of the active constituents of drugs and in the extraction of gold using cyanide solution.

Load. The name given to the weight that is supported by a structure or the force that is applied to a test piece.

Loading range. The limits or the range of weights or stresses within which a testing machine meets the specified limits of accuracy.

Loading rate. The rate at which the load or stress is applied to a test piece. The behaviour of a metal within the region of plastic deformation is influenced by the rate of loading because the full strain corresponding to a given load is reached only after a perceptible time.

Loam. Common clay is contaminated with limestone, quartz and oxide of iron. A mixture of clay and sand is known as loam and usually it contains a very small proportion of titanium oxide TiO_2. The amount of sand present must be sufficient to prevent the clay massing together. Loam containing straw is used for making casting moulds and bricks (*see also* **Moulding sands**).

Loam bricks. Blocks of foundry sand or building bricks coated with sand, used as supports for the walls of foundry moulds.

Loam mill. A mortar mill used for crushing and mixing foundry sand, clay, water or bonding liquids to form loam.

Loam-moulding. The type of moulding carried on in foundries, in which a mixture of clay and sand, mixed to the consistency of a stiff slurry, is used. The moulds are built of loam bricks (q.v.) covered with loam. Usually the loam is swept to the required contour, loose wooden patterns being used for forming irregular shapes. Complete or skeleton patterns are frequently used in loam-moulding. The moulds are partly dried before separating the sections or withdrawing the patterns.

Loam-plates. The name given to the steel or cast-iron plates on which loam-sand moulds are built up.

Local action. Ordinary commercial zinc used as electrodes in batteries contains a number of impurities, i.e. iron, lead, etc. These, together with the zinc and in the presence of the electrolyte (usually acid), give rise to a number of local currents all over the surface of the metal, the result being that the zinc is consumed without any advantage being derived therefrom. This is termed "local action", and its occurrence is prevented by coating the zinc surface with mercury. This latter dissolves the zinc and forms a uniformly soft amalgam which covers up the impurities. When the battery is not active the zinc is not readily dissolved, but when the cell is used the zinc is consumed in the normal way and the impurities fall to the bottom of the container.

The term is likewise used to describe the electrochemical corrosion due to microscopic or macroscopic variations in composition, or to physical differences (e.g. hardness) within the same alloy.

Lode. A general term for ore bodies occurring in forms which have considerable length and depth but comparatively small thickness, e.g. in fissures, cracks and as veins in the earth's crust.

Lodestone. A form of magnetite Fe_3O_4 which exhibits polarity and behaves as a magnet. Steel, when rubbed with this ore, acquires magnetic properties.

Logarithmic creep. *See* **Creep.**

Logarithmic curve. A fairly common type of curve met with in science. It is defined by the equation $y = me^{x/a}$. It consists of a single branch, and the *x*-axis is asymptotic. For all values of m the sub-tangent is constant and equal to a, and the area between the curve, the asymptote and any ordinate is equal to the area of a rectangle whose sides are equal, respectively, to the abscissa and the sub-tangent.

Logotype. Type-metal cast and made up into words or portions of words instead of letters.

Logwasher. An apparatus used in ore-dressing, the object of which is to disintegrate certain types of ores such as nodules of the iron oxide in a clay matrix. The ore is agitated with water, the iron nodules thus being freed from the clay, and the latter is washed away with the overflow at the lower end of the washer.

Long chain molecule. A linear polymer with an extended chain length, formed by addition polymerisation. By implication, the molecular weight may be high, and the amount of branching minimal.

Longitudinal direction. A term used to indicate the direction in which a wrought-metal product has been extended by rolling, drawing or extrusion. Longitudinal specimens, e.g. for test pieces, are cut with their principal axis parallel to the direction in which the metal has been extended.

Longmaid-Henderson process. A relatively cheap and efficient process for the recovery of copper from pyrites that has been burnt for the production of sulphur dioxide in the making of sulphuric acid. The burnt pyrites, or "purple" ore, as it is called, is crushed and mixed with about 10% of its weight of common salt or rocksalt. This mixture is roasted in a multiple hearth furnace having a slight down draught and working at temperatures between 500 °C and 600 °C. The gases evolved, which consist of sulphur dioxide and trioxide together with some hydrochloric acid, are absorbed in a suitable condensing tower, and the acid solution is used for leaching. After leaching the chloridised ore with the "tower" acid, the copper is precipitated by adding scrap iron. Since the increase in the manufacture of sulphuric acid from elementary sulphur, and other sulphides, the process has tended to become less important.

Long-range order. A state of repetition of structure over a considerable distance, usually in three dimensions, leading to relative stability, low entropy, and requiring considerable energy input to change the state of order. It is characteristic of the solid state, at its most refined in a crystal with low density of imperfections such as dislocations and vacant lattice sites. Although commercial crystalline materials may have a high density of imperfections, as in metals, nevertheless they represent materials of very long range order. Natural materials such as diamond and gemstones represent long range order of a similar level.

Long ternes. A type of terneplate (i.e. sheet steel coated with a tin-lead alloy). The term "long" refers to the linear dimensions which usually fall within the limits 0.6 × 1.5 m to 1.2 × 3.7 m. It is used to distinguish such plates from short ternes, which commonly measure 0.5 × 0.7 m.

Long ton. A term used to indicate a weight of 2,240 lb to distinguish between this weight, the "short" ton of 2,000 lb and the metric ton of 1,000 kg, i.e. 2,200 lb approx. (called the tonne).

Looking-glass ore. A lustrous, metallic and opaque variety of hematite Fe_2O_3. The brilliant metallic lustre of the steel-grey crystalline variety is known variously as looking-glass ore, specular iron ore or iron glance. It crystallises in the hexagonal system with rhombohedral symmetry and the occurrence is often in tabular form. Mohs hardness 6, sp. gr. 5.2.

Loomite. A trade name used to describe short-fibred talc (q.v.) products. It is mainly a magnesium silicate and is used as a mineral filler in many industries, e.g. paper-making, heat-insulating materials, etc. Also called Snofibre.

Loop (dislocation). A complex dislocation system which can be observed in the microstructure of etched materials, formed by elements of both edge and screw dislocation.

Loose pack rolling. A method of rolling automobile body steel and other wide sheets down to a thickness of 2.8-1.25 mm. After the bars have been rolled down to about half length, they are pickled and then immersed in a suspension of powdered charcoal in water (or other similar material). Sheets are then carefully matched to form packs of two, three or four before reheating and rolling to finished length.

Lorandite. A mineral rich in thallium, containing up to 59% Tl. It has the formula $TlAsS_2$, and occurs with realgar As_2S_2 in Macedonia (Greece).

Lorentz-Lorenz expression. An expression for the specific refractive power of a substance. It is independent of temperature and is characteristic for a given substance. The expression is generally represented by r_L, and is given in two forms:

$$(a) \qquad r_L = \frac{n^2 - 1}{n^2 + 2} \times \frac{1}{d}$$

$$(b) \qquad Mr_L = \frac{n^2 - 1}{n^2 + 2} \times \frac{M}{d}$$

where n = refractive index; d = density; and M = molecular weight. Formula (b) is known as the molecular refractive power. It is useful in studying the structural composition of different molecules.

Lost wax process. See **Investment casting** and **Cire perdue.**

Lota metalwork. Vessels for hand use to contain water. These are mainly for ceremonial use and are made of copper, brass or bell metal. They are made in India and Indonesia.

Low-alloy steel. A term applied somewhat loosely to steels that contain one or more of the usual alloying elements (e.g. chromium, nickel, vanadium, tungsten, molybdenum) which have been added to improve the properties of the product, but the amounts added are not in such proportions that the steel cannot be heat-treated (e.g. hardened or tempered) in somewhat similar manner as plain carbon steels.

Lowig process. A method of producing caustic soda from soda ash. This latter is mixed with ferric oxide and heated to a bright red heat in a revolving furnace, when sodium ferrite is formed. This product is broken up and thrown into hot water, when caustic soda goes into solution and ferric oxide is precipitated. This latter can be used again.

Low red heat. A term used to describe temperatures between about 550 °C and 700 °C. In the heat-treatment of steel, temperatures are sometimes judged by the colours on the surface of the metal, but the colour produced at a given temperature is affected by the composition of the steel. Table 35 gives the usual recognised colours and the approximate temperatures corresponding to these.

TABLE 35. *Colours Corresponding to Various Temperatures*

Colour	Temperature °C	
Red:		
Visible in dark	400	
Visible in twilight	475	
Visible in daylight	525	
Visible in sunlight	580	Low red heat
Dark red	700	
Dull cherry	800	
Cherry	900	Bright red heat
Bright cherry	1,000	
Orange:		
Orange-red	1,100	
Yellow	1,200	White heat
Yellow-white	1,300	Welding heat
White	1,400	
Brilliant	1,500	
Blue (dazzling)	1,600	

Low-temperature carbonisation. The process of making a semi-coke by heating coal in enclosed retorts at temperatures between 400 °C and 600 °C. The solid fuel resulting from this process is smokeless and has a high calorific value and can be used for domestic heating and for steam raising. This semi-coke may contain up to about 12% of the volatile hydrocarbons in the form of oils. Coke probably forms at about 420 °C, and at slightly above this temperature volatile hydrocarbons (but not tars, pitches, naphthalene, etc.) distil, so that the low-temperature process will yield a large percentage of paraffins, but small amounts of benzene and its homologues. The yield of gas is much smaller than that normally obtained at gasworks, but it has at least double the calorific value.

Lubricant. Solids, semi-solids or liquids applied to the surfaces of metallic or other substances that move, the one relative to the other. The usual solid lubricants are graphite and soapstone and MoS_2. Semi-solid lubricants are either fats or greases the latter being prepared from fats and oils mixed with soaps. Liquid lubricants are generally named oils, whether they are of animal, vegetable or mineral origin. The properties of a lubricant that determine its application are: (a) specific gravity; (b) flash point—the temperature at which vapours are produced that can be ignited to give a flash; (c) the fire-point, which is the temperature at which the oil will burn steadily in a supply of air; (d) the flow temperature, which is the temperature at which the liquid will not flow; (e) the fluidity (or viscosity); (f) acidity value, and certain other tests. Tests for all of these have been standardised.

The majority of the lubricants at present in use are of mineral origin and are distilled from crude petroleum. The selection of a lubricant is governed by the size of the bearing surfaces, the pressure (or load), the speed of the surfaces and the clearance necessary. Since viscosity is the governing physical property of lubricants, and as this depends on temperature, the temperature at which the determination is made must be stated. (See **Viscosity, Tribology.**)

Luce-Rozan process. A process of extracting silver from silver-bearing lead that is a modification of the older Pattinson process, which it has largely replaced. When molten low-grade silver-bearing lead is allowed to solidify, the lead that separates as crystals is considerably poorer in silver than the remaining liquid metal. The eutectic composition is Ag 2.5%, Pb 97.5%. Thus, when freezing occurs in the pot, the lead crystals are removed by a perforated ladle to a second pot, where melting, cooling and separation by ladle are repeated. This yields a high purity lead. To the remaining metal in the first pot is added more of the silver-bearing lead, and the processes are repeated until the enrichment of the lead in the first pot has proceeded to about 2% Ag when the latter is recovered by cupellation.

Lüder's lines. A surface defect sometimes encountered during the forming or deep-pressing of mild-steel sheet. It consists of a series of dull lines that appear on the surface of the metal and are due to non-uniform distortion. These appear on the metal as soon as deformation starts and are first noticeable on the parts that are least deformed. As the deformation proceeds, the lines appear in increasing numbers. They can be seen to be slight depressions or slight ridges, according to whether the metal has been subjected to tensile or compressive forces.

When large-grained materials are subjected to considerable plastic flow (elongation), wide bands cutting across many grains may destroy the surface flatness and spoil the appearance of the sheet. If the metal fractures, the lines coincide approximately with the direction of maximum shearing stress. The lines are visible under the microscope on polished surfaces of the metal, but if the specimen was coated with oxide, the latter is liable to flake off and the lines are revealed without polishing. These lines are also called flow lines.

The study of flow lines on etched specimens of steel has been used in the investigation of stress distribution in rolling, drawing, pressing, etc.

The appearance of the sheet may be improved by giving it a light roll of about 1-3% reduction in thickness.

Also called Stretcher Strains.

Ludwig test. A hardness test in which a 90-degree conical indentor is used. The hardness number is quoted as load divided by the conical contact area, i.e.

$$H = \frac{4L \sin \theta/2}{\pi d^2}.$$

Luffing. The operation of lifting or lowering the jib of a crane in the course of loading or unloading cargo of any materials (fuel, ore, etc.).

Luminescence. The absorption of electromagnetic radiation (usually in the visible or near visible spectrum) by electron transitions between different energy levels, followed by return of the electrons to lower energy levels or their ground state. If the emission takes place within 10 ns after absorption, the phenomenon is called fluorescence. If the time delay is greater than 10 ns, the material is phosphorescent, and the emission lasts for a longer period. The emitted light is always of longer wavelength (more red) than the incident light.

Luminous paint. There are three types of material in this category. The first is luminescent, usually phosphors, which emit their light after absorption of incident light, after a period greater than 10 ns. Many metallic sulphides are in this category; CaS, BaS, ZnS and PbS are examples. The intensity of emission fades with time, but can be revived by renewed exposure to incident light. The emission is believed to depend on impurities, called activators, often compounds related to the main material. Paint merely employs these phosphorescent compounds as the pigment. They are used to delineate safety hazards in the hours of darkness, as on highways and pavements, to illuminate the hands of clocks and watches, instrument dials, etc.

The second category is that of irradiation by radioactive substances, thereby eliminating the incident light required for luminescent materials. The radioactive material, in very small quantity, is mixed with a material which is luminescent as a result of bombardment with alpha or beta radiation. (In the early years of handling radioactive material, more dangerous materials were used, such as radium, leading to a health hazard for workers exposed to the material when applying it to articles. It was not unknown for workers to lick their brushes to a fine point, and contract cancer of the tongue.) The illumination of instrument dials in aircraft and at sea is the main application.

The third category is another dangerous one, the use of combustible material in finely divided form, which is prevented from burning on initial application by a layer of varnish or solvent. In time the varnish is permeable to oxygen, or the solvent evaporates, allowing combustion to take place. Typical of such materials is a solution of white phosphorus in carbon bisulphide. The combustion is accompanied by white smoke of phosphorus pentoxide. Such materials had application largely in theatrical effects or in rescue equipment at high altitude or at sea, but are now obsolete.

Lustre. A term used in describing minerals, and relates to the appearance of the surface—especially to the appearance of a crystal face as seen by reflected light. The term is usually modified by the addition of such adjectives as metallic, vitreous, resinous, greasy, pearly or adamantine.

Lute. A pasty mixture of fire-clay and other refractory materials used to seal the joints between annealing boxes and their covers or between crucibles and their lids, prior to heating.

Lutetium. A metal of the rare-earth group that is found along with ytterbium and erbium in the minerals gadolinite (q.v.), euxenite, etc. Symbol Lu, at. wt. 174.99, at. no. 71, density at 20 °C 9.84 × 10^3 kg m^{-3}, at. vol. 18 cm^3 mole^{-1}; crystal structure, close-packed hexagonal—a = 3.509 × 10^{-10}; c = 5.559 × 10^{-10}. (See Table 2.)

Lutetian. A state in the Eocene stratigraphy that marks the widest extent of the eocene marine invasion of Europe. It is a stage particularly well developed in the Paris Basin (France) and to a lesser extent in Bedfordshire and Hampshire (UK). The limestone which is characteristic is a useful flux as well as a building stone.

Lydian stone. A black stone used in former times to estimate the proportion of gold in alloys. The colour of the streak made by the metal under test on the stone was compared with that made by "needles" of metal of known composition. Three sets of "needles" containing various proportions of gold were used corresponding to the three alloys: (a) gold and copper, (b) gold and silver, and (c) gold, copper and silver.

Lye. The name by which is known the solution of alkali salts obtained by leaching or lixiviating wood ashes with water. The name is now used for any strong solution of the caustic alkalis NaOH and KOH, and for alkaline solutions used as detergents.

Lyophilic colloid. An emulsoid or reversible colloid which when evaporated to dryness will on contact with water pass back into the colloidal state.

Solutions of the type like albumin are not affected by the addition of small amounts of electrolytes. They usually require large concentrations of electrolytes to produce coagulation. But the coagulation is reversible if the salt is removed by dialysis. Emulsoids and lyophilic colloids are often used synonymously, though their connotation is not identical.

The behaviour of lyophilic sols is non-Newtonian, and often thixotropic. The colloidal particles assemble into densely linked networks when concentrated, and set to "gels" on resting at constant temperature. In this state they are easily broken up again, by shear forces, to become more fluid. Many varnishes, oil and plastic paints, aqueous solutions of starch and dextrin, are examples. The size of the colloid particles may vary from 1 to 500 nm in diameter.

Lyophobic colloids. These are suspensoid solutions (e.g. Prussian blue, arsenic trisulphide) that are irreversible. On evaporation to dryness they do not pass again into the colloidal state on the addition of water. They are all very sensitive to electrolytes, and there always exists a difference of potential between the surface of the particle and the surrounding liquid, and they thus exhibit cataphoresis (q.v.). All lyophobic colloids are very dilute concentrations; they are clear (or slightly opalescent) solutions; and they may vary widely in colour for a given substance. Gold solutions may be red, purple or blue, while silver solutions vary from blue or purple to red and yellow.

Lyotropic series. The series of ions that results from arranging them in the order of their precipitating powers on lyophobic colloids (q.v.). This order depends to a considerable extent on the valency of the cation or anion used, but as the following series shows it is not wholly dependent on valency. A typical series is aluminium> magnesium>potassium>sodium>lithium all added in the form of chloride in solution. A similar lyotropic series can be found for anions, e.g. citrate>tartrate>sulphate>acetate>chloride>nitrate >chlorate>iodide>thiocyanate.

M

Macarthur-Forrest process. A method of extraction of silver and gold from their ores. The process consists in digesting the finely crushed ore with a solution of sodium cyanide. The silver or gold in the metallic state dissolves in the presence of atmospheric oxygen to form a double cyanide. Compounds of silver form similar compounds without the aid of oxygen. The solution is then passed over zinc shavings when the metal (silver or gold) is precipitated. It is collected, melted and electrolytically refined. Also known as the cyanide process.

McDougall furnace. A furnace designed for the roasting of low-grade, finely crushed ores, particularly copper ores. Essentially, the furnace is a vertical cylinder with superimposed hearths and a central rotating shaft which is provided with arms for stirring the ores. The latter are fed at the top of the furnace and displaced by rabble arms until they fall over the edge of one hearth to the hearth below. The arms are arranged so that the charge falls alternately from the outer and the inner edge of (or through slots in) the hearths. The rabble arms and central shaft are cooled by compressed cold air, and the issuing hot air can be used in aiding combustion in the furnace.

Machinability. A term used in reference to the resistance of a metal to cutting. Other factors being constant, the machinability of a metal can be related to the tool life or the interval between successive grindings of a tool. But tool life depends not only on the hardness of a metal but also on the cutting speed, the coolant employed, the nature of the swarf produced and the physical properties of the tool itself. Thus the term may also be used in reference to the effect of a metal on the cutting capacity of a tool.

Machine-forging. The forming or shaping of metals by heating and hammering, the latter being performed by power-driven machines. In these latter, special tools or dies are frequently used. They are designed for the production of large numbers of pieces of a particular shape. The machines used are referred to as forging machines or forging presses and they include hammers operated by steam or compressed air, drop hammers, die forging machines and hydraulic presses. In all cases except the last-named type the work is done by a single blow or by a series of blows in quick succession. Hydraulic presses have a squeezing action that produces a fairly uniform texture in the metal. High-speed forging presses enable the operator to complete the forging before the metal has cooled to any marked extent.

Machine moulding. Any moulding process in which the manual labour involved in filling and ramming the moulds is wholly or partly replaced by machinery to do the same work more efficiently and much more quickly. Machine moulding may involve the filling of the mould by hand and the ramming by machines operated by electric, hydraulic or pneumatic power. This process involves the use of core blowers and jolt or squeeze rammers. Sand slingers (q.v.) may be used to fill the mould at rates of, say, $0.2\text{-}0.5 \times 10^{-2} \, \text{m}^3 \text{s}^{-1}$ and, in the act of so doing, also ram the sand.

Machine tools. Except in certain cases (e.g. precision casting), forgings and castings are usually not producible to the required size or finish, so that the product of a casting or forging process requires that at least some surfaces shall be finished smoothly and to true size in order that the parts may fit or work together. The correcting of these errors of contour, size or roughness that arise in cast, forged or rolled metal is accomplished by the aid of machine cutting. The chief types of metal cutting and finishing that are done by machine tools are listed as follows: *(a)* turning, *(b)* boring, *(c)* broaching, *(d)* drilling, *(e)* milling, *(f)* planing, *(g)* shaping, *(h)* slotting, *(i)* grinding, and *(j)* polishing. Many tools are now fitted with computerised control.

Mach number. The ratio of the speed of a body (usually supersonic) in any medium to the speed of sound in the same medium. It is usually represented by M.

McKee top. A device for charging an iron-making blast furnace. The charge is carried to the top of the furnace, where it is discharged into a large stationary hopper. From this, the charge falls by gravity to a rotating hopper, which can take up, in turn, several positions, and so distribute the successive skip-loads round the periphery of the large bell.

Mackenite metal. The name given to a series of heat-resisting alloys of nickel-chromium or nickel-chromium-iron.

Mackenzie's amalgam. An amalgam containing both lead and bismuth, usually made in two parts, one of which consists of a solid mercury-bismuth alloy and the other a solid lead-mercury alloy. When ground together in a mortar at room temperature, a liquid amalgam results.

Macle. A variety of andalusite $Al_2O_3.SiO_2$. The term is also used in reference to a twin crystal. In this latter use the term has been superseded by the term hemitrope, which is meant to refer to the fact that many twinned crystals may be explained by the rotation of one portion through two right angles (half turn).

McQuaid-Ehn test. A test originally developed during an investigation into the effect of melting-furnace practice on the hardenability of plain carbon case-hardening steel, but has since been extended to cover alloy steels and all types of plain carbon steel. It provides a means of connecting the properties inherent in a given heat of steel with the results to be expected in heat-treatment and in the finished product. Steel is carburised at 910-940 °C in a solid carburising medium until a definite hypereutectoid zone which can be studied at a magnification of $\times 100$ is developed (usually about 8 hours). A case of approximately 1.25 mm usually provides ample hypereutectoid zone for satisfactory study. The steel is allowed to cool very slowly from carburising temperatures in order to develop a pearlitic structure, with free cementite in the hypereutectoid zone. The interpretation of the test depends on a microscopic examination of the pearlite and the free cementite. In a normal steel, cementite exists as a fine but continuous envelope at the crystal boundaries, the structure within the grains being completely pearlitic. In an "abnormal" steel the pearlite, if present as such, appears in very coarse, irregular lamellar form, and at the grain boundaries it is broken down completely into massive cementite and free ferrite. Between these two extremes there are many intermediate degrees of abnormality, and in certain cases the cementite tends to become a series of thick discontinuous particles. The size of the grain in the hypereutectoid zone, which varies in different heats, serves as an indication of the hardenability of the steel and of the degree of distortion during quenching. Coarse-grained normal steels harden more deeply and tend to distort more than the finer-grained abnormal steels, but have usually much lower resistance to impact.

Steels can be selected and obtained commercially according to grain size. Standard charts which permit of classification according to grain size (q.v.) as revealed by this test, are available. (Grain size varies from No. 1, in which there are 0.23 grains cm^{-2}, to No. 8, in which there are 14.8 or more grains cm^{-2}.) (*See also* **Hardenability**.)

Macro. A prefix denoting large (or long), in contradistinction to "micro", which means small (or very small).

Macrography. The examination of the structure of metals and alloys using low-power magnification—say, up to 10 diameters.

Macro-molecule. A term applied to covalent bonded structures forming crystals, in which repetition of a unit structure extends over a very long range order. Examples are diamond, with its repetition of tetrahedra; silica with tetrahedra formed by silicon and oxygen.

Macroscopic. Visible with the naked eye or at magnifications up to about 10 diameters.

Macroscopic etching test. A test that consists in etching the surface of a metal with acids or other etchants to reveal the macrostructure, flow lines, defects, etc.

Macroscopic stress. The residual stress, involving relatively large areas or the whole of a specimen and producing measurable strain.

Macro-structure. The structure of metals and alloys as revealed to the naked eye or at magnifications up to about 10 diameters. The examination may be carried out using a smoothly machined surface which may or may not be etched. The term may also be used in reference to the general crystalline structure and the distribution of impurities as revealed on an etched surface when examined under a low-power lens.

Madsenell process. An electrolytic pickling process used for removing carbon smut and absorbed hydrogen from ferrous metals. It is mainly used for improving the surface condition of metal which is to be coated by electrodeposition or hot-dipping and is not primarily designed as a scale-removal process. The work is made anodic in a concentrated solution of sulphuric acid (over 85% H_2SO_4) and the potential difference about 12 volts.

Magistral. A name by which roasted copper pyrites is known. It contains both copper and iron sulphides.

Magma. The molten and gaseous products generated in the interior of the earth which have given rise to igneous rocks at or near the surface. Igneous magmas are the original source of metals now found in workable quantities (percentage of metal in ore), and this has been brought about by magmatic differentiation giving rise to ore bodies associated mainly with plutonic rocks.

Magmatic segregation. A mode of formation of ore deposits which depends on the concentration of minerals of economic importance in certain areas—usually the periphery of slowly cooling igneous rock magma. In deposits of this type the centre of the rock deposit is often siliceous.

Magnaflux. A magnetic method of detecting cracks in iron and steel. The specimen is magnetised and dusted with a dry magnetic powder, or immersed in low-viscosity oils containing finely divided magnetic particles in suspension. Surface cracks and some defects at small depth below the surface are revealed, since fine particles of the magnetic dust are attracted to and collect at regions where the strength of the magnetic field in air at the surface is increased by cracks or flaws. (*See* **Non-destructive testing**.)

Magnefer. The name given to a dead burned dolomite (magnesium limestone).

Magnesia. Magnesium oxide MgO obtained by calcining magnesite. It is extensively used as a refractory material. After strongly

heating, it becomes crystalline, forming periclase, a substance having a melting point of 2,800 °C; periclase occurs naturally, but is rare.

Magnesia alba. Commercial basic magnesium carbonate.

Magnesia alum. Hydrated sulphate of aluminium and magnesium, crystallising in the monoclinic system. It usually occurs in fibrous masses, and is formed by the weathering of pyrite-bearing schists. Sometimes known as pickeringite.

Magnesia cement. A cement made by heating hydrated magnesium chloride or carbonate to redness. When mixed with water it sets to a friable mass. When mixed with asbestos fibre it is used as an insulating material for covering steam pipes, etc. Sorel cement is composed of magnesium oxide and chloride. This latter is hard and strong and is used, as a filler, for floors.

Magnesia glass. Glass containing 3-4% of magnesium oxide. Electric-lamp bulbs have been mainly made from this type of glass since fully automatic methods of production were adopted.

Magnesia mortar. Mortars that contain an appreciable proportion of magnesia, the presence of which is not detrimental to the value of the mixture. Such mortars are strong and resist the action of sea-water better than lime-mortar.

Magnesian limestone. Crystalline granular dolomite $CaCO_3.MgCO_3$ that occurs naturally in the marine beds. It is used as a refractory material.

Magnesia usta. Commercial name for magnesium carbonate calcined at a low temperature for a long period.

Magnesioferrite. A typical member of the group of minerals known as spinel. Formula, $MgFe_2O_4$.

Magnesite. A mineral consisting mainly of non-crystalline magnesium carbonate that occurs mixed with magnesium silicate, in Greece, Southern India and California. It results from the alteration of rocks rich in magnesium silicate. It has the appearance of unglazed porcelain and is often white in colour, though some varieties are yellow, due to the presence of iron oxide.

It is used as a source of magnesium metal, but its main application is in the manufacture of basic refractory bricks, lightweight refractories and building stone. The dead burned magnesite prepared by heating above 1,500 °C is used, but the product has little resistance to sudden changes of temperature. A crystalline variety is known which has a Mohs hardness of 4 and a specific gravity of 3.1.

Magnesium. A brilliant white metallic element, and the lightest of all structural metals. It occurs in dolomite, magnesite and carnalite, as well as in certain salt brines and in sea-water; the latter containing about 0.5% $MgCl_2$. The commercial production was not attempted before the beginning of the present century. Before 1928 the usual method was the electrolysis of fused $MgCl_2$ or the electrolysis of MgO dissolved in molten fluorides. During the next eleven years production was usually carried on by the former method or by the reduction of MgO with carbon or ferrosilicon. The reduction with carbon yields magnesium dust, which is subsequently distilled (Hansgrig process), while reduction with ferrosilicon involves the condensation of the metal vapour (Pidgeon process). Since 1939 there has been a process for extracting magnesium from sea-water on a large scale.

When pure, the metal is very stable in dry air, but is vigorously attacked by air containing salt spray. Protective coatings are thus essential, and are usually obtained by anodising or by chemical treatment. Additions of manganese, calcium, cerium and certain other metals improve the corrosion resistance and the mechanical properties. The metal is resistant to hydrofluoric acid and gives satisfactory die-castings. In the pure state magnesium is used as a deoxidiser for brass, bronze, nickel silver and Monel metal. In powder form it is used in pyrotechnic mixtures. Magnesium-aluminium base light alloys are used as castings, die castings, wrought products, etc., for a wide range of applications. The combustible metal Elektron, used in incendiary bombs, is an alloy of magnesium and aluminium.

The physical properties of the metal are given in Table 2, heat of fusion, 374 $kJkg^{-1}$.

Magnet. Usually implies a permanent magnet, i.e. a piece of material which has retained its magnetism after magnetisation, but strictly refers to any material or device which has associated with it a magnetic flux capable of inducing a magnetic field in a susceptible material. Electromagnets are more flexible in application in industry, for lifting and securing loads, or for generating a magnetic field when required. Permanent magnets are most suitable for instrumentation, and for hand use when required. (See under specific headings for details.)

Magnetic ageing. A gradual spontaneous change in the physical properties of a metal at ordinary temperatures that manifests itself as an increase in hysteresis loss and a decrease in permeability with time. The use of silicon iron largely eliminates ageing, since it removes one of the ageing constituents (oxygen). Ageing is found in iron containing one or more of the impurities whose solubility curve for α-iron is similar to that for carbon (e.g. C> 0.001%; N_2>0.001%; O_2>0.005%).

Magnetic amplifier. A useful device for amplifying a small direct or a low-frequency alternating current. An alternating current flows in the primary winding of a transformer and the direct current in a third winding. The secondary winding carries the induced current which depends on both the primary alternating current and the direct current (signal) which is the subject of amplification.

Magnetic anneal. A term applied to the heat-treatment of iron-nickel alloys. The alloy is heat-treated above 600 °C and then maintained at about 500 °C, during which time it is subjected to a magnetic flux of several webers. After cooling to ordinary temperatures, it is found that the alloy has a much greater permeability and a lower coercive force is needed than if the metal is not so treated. The effect is most pronounced when the percentage of nickel in the alloy is about 65. The hysteresis loop is almost rectangular.

Magnetic change point. See **Curie point.**

Magnetic crack detection. See **Magnaflux** and **Non-destructive testing.**

Magnetic cycle. See **B-H loop.**

Magnetic dipole moment. All magnetic materials act as dipoles, i.e. there is an orientation of the field which causes an apparent concentration into two centres, the "poles", equivalent to the positive and negative poles of an electrical dipole. The analogy is such that a magnetic dipole behaves as if a circulating current I were traversing the cross-sectional area A, and the magnetic dipole moment is equal to IA. Since the material can be broken down into atoms, each atom must possess a dipole moment, and the moment for a block of material is the sum of all individual moments. The magnetic moment per unit volume is called the intensity of magnetisation, and the units ampere-metres squared, the product of current and area.

The atoms possess a magnetic moment from one or more of three possible sources of moment. The electron energy can be partially represented as angular momentum and a net angular momentum automatically implies a net magnetic dipole moment. Atoms with only s electrons in the outer energy levels will have an electron cloud distribution roughly represented by spherical symmetry, and therefore no net angular momentum. There can be no dipole moment therefore, from this source. Atoms with p electrons in the highest levels do not have a spherical cloud distribution, and so can display a magnetic moment, defined in terms of the Bohr magneton $eh/(4\pi mc)$ as the basic unit.

However, when all the p states are occupied, there will be once more a close approach to spherical symmetry, and so the rare gases, which conform to this rule, do not exhibit magnetic moments due to angular momentum. Similar arguments apply to the transition elements, where d and f electron energy levels are filled in preference to a normal sequence, even though these energy levels are not the highest in value. Consequently a magnetic dipole moment of some measurable value is to be expected from atoms with incompletely filled p, d or f states of electron energy. In the solid state however, the energy levels become smeared into bands, and the alignment of magnetic dipole moments may depend on other factors such as closest distance of approach and overlapping of energy levels in the bands; the magnetic moment is said to be quenched when there is no net moment. With transition metals, there is less quenching effect when the d or f levels are shielded in the solid state, by the higher levels, and in the case of f electrons in particular, some net moment may still occur in the solid due to angular momentum.

The electron spin however, also produces a magnetic moment,

unless balanced out in the atom or in the solid state by bonding. When the spin alignment is antiparallel, the spin moment is cancelled out, so that atoms with parallel spin electrons should display a moment, either in the free atom or in the solid state or both. There are also magnetic moments produced in the nuclei of atoms, due to the spin of protons among other reasons, but the nuclear magnetic moments are three orders of magnitude smaller than those due to electrons, and can be ignored from the standpoint of magnetic materials, although not for analytical purposes, as in nuclear magnetic resonance.

As described under diamagnetism, many materials produce an induced magnetic field to oppose any applied field, and those with closed electron shells, or a near approach to this state, show a higher value of diamagnetic susceptibility, which of course is negative in sign. Ions show similar behaviour, since they approach more spherical symmetry of electron cloud distribution, and the bigger the ion, the higher the value of the negative susceptibility.

The angular momentum effects are weak however, and overshadowed by the much greater effect of electron spin. In general, for unpaired electrons, parallel electron spin alignment leads to lower free energy, the spin moments assist the applied field, and susceptibility becomes positive, the material is said to be paramagnetic. The susceptibility is now affected by temperature (see **Curie-Weiss law**) but swamps the low values of moment due to angular momentum, or any nuclear effect.

Atoms with one s electron (alkali metals) therefore tend to be weakly paramagnetic having only one electron outside a closed shell to generate a dipole moment. With two s electrons (alkaline earth metals), the spins must be antiparallel, and hence cancel any moment from that source, and are weakly diamagnetic. When these s electron atoms lose electrons to become ions, they will change to become diamagnetic. Elements with one or two p electrons only (aluminium, gallium, indium) but which do not crystallise according to the (8-N) rule, tend to be weakly paramagnetic, because they exhibit parallel spin with these electrons, and do not have closed shells and therefore spherical symmetry.

Transition metals tend to be strongly paramagnetic, and those in odd numbered periodic groups to have the highest susceptibility. Those which adopt a particular relationship of atomic spacing in conjunction with the position of the d or f electrons are ferromagnetic, and retain their magnetism below a specific (Curie) temperature. Organic materials and the light elements (H_2, He) are diamagnetic, with low values of susceptibility. (See **Diamagnetism, Ferromagnetism, Paramagnetism, Curie-Weiss law.**)

Magnetic domains. See **Domain.**

Magnetic elements. These are defined as the three quantities— magnetic declination, magnetic dip and the horizontal intensity— which together completely define the earth's magnetic field at any point.

Magnetic field. The space in the neighbourhood of a permanent magnet or of a coil carrying a current, in which the magnetic force is appreciable. When a field of force exists at any point a magnet or a coiled wire conductor carrying a current placed there experiences a couple.

The fundamental units of the SI system include the ampere as the unit electric current. Current is linked with the other fundamental units by the force per unit length between parallel conductors separated by distance d (metres), the two conductors carrying currents I_1 and I_2 (amperes). Force (newtons) = $\mu_0 I_1 I_2 /2\pi d$ where μ_0 is the permeability of space, or magnetic constant, measured in units of henry per metre (value $4\pi \times 10^{-7}$ H m^{-1}). Magnetic field strength is measured in units of ampere per metre.

Magnetic flux. From the force between parallel conductors carrying currents, the definition of magnetic flux density can be derived. A conductor length l (metres) carrying a current I (amperes) experiences a force in a direction perpendicular to the flux density given by force (newtons) = $B.I.l$. where B is the flux density, measured in teslas, or webers per metre squared. The total magnetic flux over an area is the integral of the normal flux density, measured in webers.

The weber is defined as the magnetic flux which, linking a circuit of one turn, produces in it an electromotive force of one volt, as it is reduced to zero at a uniform rate in one second.

In the c.g.s. system, flux density was defined by $B = H + 4\pi I$ where H is the magnetic field strength and I the intensity of magnetisation. The ratio of B to H is known as permeability, and the ratio of I to H is known as susceptibility.

Magnetic gyration. When an atom is exposed to a magnetic field, it may be supposed to behave as though it were in rotation about an axis parallel to the direction of the field, while the speed of rotation is proportional to the strength of the field. If such an atom could be illuminated by circularly polarised light, the resulting effect will depend on whether the magnetic rotation is with or against the light vector, and the scattering power of the atom will then be capable of assuming two values. Its magnetic gyration is small for all usual magnetic fields, but it occurs in all transparent isotropic materials.

Magnetic hysteresis. The lagging of the magnetic flux produced behind the magnetising force producing it. (*See* **Hysteresis.**)

Magnetic induction. The product of applied field strength (H) and permeability of space, and hence the flux density B, measured in tesla.

Magnetic intensity. A definition in the c.g.s. system identical with field strength H, in SI units measured in amperes per metre.

Magnetic iron ore. Naturally occurring oxide of iron Fe_3O_4. When pure it contains 72.4% Fe and is one of the most valuable iron ores. Some of the iron may sometimes be replaced by small amounts of magnesium or titanium. The most famous deposits are found in north Sweden. Crystal system — cubic; Mohs hardness 5.5-6.5, sp. gr. 4.9-5.2. Also known as magnetite (q.v.).

Magnetic materials. All substances, when placed in a magnetic field of suitably large intensity, will develop magnetic properties. According to their behaviour when magnetised they can be divided into three classes: *(a)* diamagnetics; *(b)* paramagnetics; and *(c)* ferromagnetics. The term magnetic materials is generally applied to substances of class *(c)*. (*See also* Table 6.)

Magnetic oxide of iron. The oxide of iron Fe_3O_4 formed when iron is heated to redness in the presence of steam. It occurs naturally as magnetite.

Magnetic permeability. In a magnetic field of strength H, the introduction of a ferromagnetic medium will result in magnetic induction in that substance, so that the flux density becomes equal to B. The value of B is dependent on the property of the substance known as magnetic permeability, the ratio of B to H, measured in henry per metre, and given the symbol μ.

When μ is less than 1 the material is diamagnetic, greater than 1, paramagnetic. The value of μ for ferromagnetics varies with H, tending to a limiting saturation value.

Magnetic pyrites. A sulphide of iron, FeS_2, that often contains nickel, sometimes as much as 5% Ni, and is then useful as a source of this latter metal. The most important occurrence is at Sudbury (Ontario) and it is also found in the UK, Norway and the Harz district of Germany. Crystal system—hexagonal, Mohs hardness 3.5-4.5, sp. gr. 4.4-4.65.

Magnetic quantum number. A term sometimes used to describe the two quantum numbers m_l and m_s which refer to the electron behaviour in an applied magnetic field and with respect to electron spin, respectively.

Magnetic saturation. Strictly the point at which all magnetic domains are oriented so that their direction of magnetisation is parallel to the direction of the applied field. In practice, saturation is the point determined on the *B-H* loop, at which the rate of increase of B with increase in H, is negligible, although not all domains will then be absolutely parallel to the applied field.

Magnetic screens. Usually a thick-walled tube of iron or high permeability alloy which, when placed in a magnetic field, screens the space within from magnetic influences. The lines of force due to the field crowd into the iron, since this has a much greater permeability than air. Such screens are sometimes placed round galvanometers to protect them from stray magnetic fields. They are also used around electrical appliances such as furnaces if these give rise to electromagnetic interference in the vicinity.

Magnetic separation. A method for separating magnetic mineral or magnetic products from non-magnetic material by the use of a magnet or magnets. The material to be separated is usually carried on a conveyor belt and the magnets can be attached to the underside of the belt or suspended above it. Alternatively the tail pulley may be magnetised, causing a difference in the velocity of discharge of the magnetic and non-magnetic substances. This enables the materials to be collected in different hoppers.

The magnets above the belt can be so arranged to withdraw the magnetic constituents, and the magnets on the belt retain the magnetic material after the non-magnetic constituents have been discharged.

By utilising different magnetic strengths it is possible to separate magnetic minerals into several different concentrates.

There are mechanical objections to the use of magnets on the conveyor belt, but a large drum carrying magnets at the surface proves quite satisfactory. The charge in this case falls on the drum, and the magnetic materials are carried round, to be brushed off on the underside of the drum.

There are a number of varieties of magnetic separators or concentrators available.

The principle is also used to clean used foundry sand of any magnetic metallics before again putting it into circulation.

Magnetic susceptibility. The ratio of intensity of magnetisation (I) or magnetic polarisation to the magnetic field strength H. It is positive for paramagnetics and negative for diamagnetics, and varies inversely with absolute temperature for paramagnetics (Curie's law). Because of the relationships between H, I and B, there is also a relationship between susceptibility and permeability

$$\chi = \frac{\mu - 1}{4\pi}$$

where χ is susceptibility, μ the permeability.

Magnetic transformation point. Ferromagnetic substances can be magnetised strongly by magnetic fields much less strong than are required in the case of diamagnetic or paramagnetic substances. They also retain a large proportion of their magnetism when they are removed from the magnetising influence. But these peculiarities can occur only at temperatures below a certain critical value. This temperature is known as the magnetic transformation point, the Curie temperature or the magnetic change point. The critical temperatures for a few ferromagnetic substances are: Thermalloy, 10°-70 °C; nickel-iron alloy (30% Ni), 70 °C; nickel, 358 °C; permalloy (78% Ni), 550 °C; iron 770 °C; and cobalt, 1,120 °C. It should be noted that above the Curie temperature, ferromagnetic substances behave like paramagnetic substances.

Magnetite. Naturally occurring magnetic oxide of iron Fe_3O_4. It is black and opaque with a marked metallic lustre. Also known as magnetic iron ore (q.v.).

Magnetohydrodynamic accelerator. An equipment for producing electrical energy by passing a stream of ionised air at high velocity between the poles of a series of magnets. Speeds of gas of up to 90 m s^{-1} are used and temperatures approaching 3 500K are attained. For rocket propulsion, the propellant should be ionised by an electric arc before passing through the magnetic field.

Magnetomotive force. Magnetic flux can be regarded as an electric current in an electrical circuit, the flux proceeding round a magnetic circuit. The electromotive force which produces a current in a conductor, then has a parallel, the magnetomotive force, producing the magnetic flux. To carry the resemblance to a logical conclusion, a parallel exists with Ohm's law, in this case due to Rowland, and the ratio of magnetomotive force to magnetic flux is called the reluctance R, analogous to the resistance of the electrical circuit. In c.g.s. units, the magnetomotive force due to a current I flowing in a wire was given by $4\pi I$, and in a solenoid of S turns, $4\phi SI$, where ϕ is magnetic flux. In SI units, the force is measured in amperes and the reluctance as reciprocal henry. Permeance is the inverse of reluctance.

Magnetostriction. Investigation of the orientation within magnetic domains indicates that the crystal lattice behaves in an anisotropic manner with regard to magnetisation. The body-centred cubic lattice of iron shows the most ready magnetisation in the [100] direction, i.e. along a cube edge, and the most difficult is the [111] direction, along a cube diagonal. In nickel, it is the [111] direction which is easiest, and the [100] which is most difficult. Cobalt is a hexagonal lattice, and the easy direction of magnetisation is normal to the basal plane, the [0001] direction. These differences indicate anisotropy of magnetisation, and hence dimensional changes must occur in the process. Iron crystals expand along the direction of magnetisation and contract normal to it, whereas nickel crystals contract along the magnetisation direction and expand normal to it. The phenomenon is known as magnetostriction, and is obviously quantitatively dependent on the elastic properties of the crystal in addition to the magnetic ones. Imperfections in crystals affect domain growth, and in particular the domain rotation in the closing stages of magnetisation, and of course they affect the mechanical behaviour of crystals.

Depending on the mechanical treatment received by a ferromagnetic material, the material may reveal magnetostrictive behaviour by changes in the magnetisation resulting from internal mechanical strains, strain sensitivity and magnetostriction being related properties of anisotropy. High initial magnetic permeability is associated with low magnetostriction. In iron-nickel alloys, the saturation magnetostriction is zero at 81% nickel, and this is also the composition at which strain sensitivity is zero. Iron-cobalt alloys show the greatest dimensional changes on magnetisation, about 7×10^{-5} for 40-60% cobalt. Pure nickel decreases in length about 4×10^{-5} on magnetisation.

These properties are of value in transducers, translating mechanical movement into magnetic changes or vice versa.

Magnetron. An electronic device to generate high frequency oscillations. A combination of electric field control with a deflecting magnetic field, usually with a split anode, produces high frequency power at 1 to 30 GHz with efficiencies from 50 to 60%.

Magnet steels. The name given to carbon or alloy steels that are used for making permanent magnets. The early magnets were made of crucible steel, to which was added at a later date tungsten, chromium and manganese. The element molybdenum is often used in place of tungsten. They possess high remanence.

The amount of alloying element added varies considerably. Thus tungsten is usually present up to about 6%, molybdenum up to 22%, chromium up to 5%, cobalt (or cobalt and chromium together) 18-60%. With the more complex alloys, the amounts of each alloying element present would depend on the nature and amount of other elements present.

Annealing is done at 800°-850°C, and some of the materials can be aged at 100°C to improve the constancy of the magnetic field they can maintain.

Magno masse. A compact mass of oxides of calcium and magnesium obtained by calcining dolomite. When finely ground the product arranged as a thick bed may be used for filtering water.

Majolica. A type of ceramic ware the body of which is coated with enamel whitened with tin oxide. The enamel can also be applied to steel to produce a surface resembling marble.

Majority carriers. When more than one source contributes to the current carrying process of conduction, the predominant source is referred to as providing the majority carriers. In an n-type semiconductor for example, only a few electrons are thermally excited across the energy gap from the intrinsic semiconductor atoms, the vast majority are electrons from the donor atoms, and hence the electrons from the added impurity are majority carriers.

Malachite. A naturally occurring basic carbonate of copper

$$CuCO_3 . Cu(OH)_2$$

that is formed by the action of damp air on copper sulphides. It is thus found in all areas where copper occurs, and especially in the USSR, Arizona and Australia. It is light-green in colour and the monoclinic crystals are often found associated with azurite (q.v.). The massive form is frequently fibrous, and this is used for making vases, etc., since it takes a high polish.

Verdigris, or copper rust, has the same composition as malachite. Mohs hardness 3.5-4, sp. gr. 3.7-4.

Malachite green. The double salt formed by the combination of chloride of tetramethyldiaminotriphenyl carbinol and zinc chloride $3(C_{23}H_{25}N_2Cl).2(ZnCl_2).2H_2O$. It forms a deep green solution that is used as an indicator, the solution assuming a yellow colour at pH values below 0.5 and a green colour at values above 1.5.

Malacolite. A calcium-magnesium-iron silicate. It is a variety of pyroxene that does not contain alumina.

Malacon. A zirconium mineral in which the rare-earth element hafinium was discovered in 1923.

Malcomising. A process for the surface hardening of stainless steel that consists in heat-treating the chromium steel in an atmosphere of nitrogen.

Maldonite. A gold-bismuth ore Au_2Bi found in vein deposits.

Malinite. The name given to the halloysite found in Indiana (USA). It is very similar in composition to kaolinite (q.v.), but contains a higher percentage of alumina. It is used as a refractory material.

Malleability. The property of a metal that determines the ease of deformation, e.g. by hammering, rolling, etc., without fracture. It is exhibited by some metals, e.g. lead, tin, gold, at room temperatures, and by other metals at elevated temperatures. The property is related to ductility. Metals and alloys having a fine grain structure are, in general, more malleable than when the structure is coarse grained.

Many metals become hard and fracture readily when they contain dissolved or absorbed and combined oxygen, nitrogen, etc. When the gases are removed in the refining process the resulting metal has lost its tendency to brittle fracture and is then known as malleable metal.

Malleable brass. A 60/40 brass that is used for marine components because of its good resistance to corrosion by sea-water. Small additions of tin are made to increase the hardness and enhance the corrosion resistance in marine surroundings. The alloy is then called naval brass and contains 0.5-1.5% Sn, with total impurities 0.75% (max.). These latter are usually iron and lead. The annealed malleable brass has an elongation of about 50% with a tensile strength of 380 MN m^{-2}. Hot or cold rolling will usually increase the tensile strength up to about 620 MN m^{-2}.

Malleable cast iron. A form of cast iron possessing good resistance to impact. There are two different types, known respectively as whiteheart and blackheart. In both cases the castings are produced in white iron, in which the carbon is present in the form of cementite Fe_3C and, subsequently, annealed at 850°-950°C. In the whiteheart process a decarburising atmosphere removes some of the carbon and produces a structure consisting of carbon nodules in a ferritic matrix. This iron shows a steely fracture. In the blackheart process the castings are packed in an inert material, which inhibits decarburisation. The cementite is decomposed to produce either a pearlitic structure or carbon nodules in ferrite. The latter type has a sooty-black fracture. In America the usual practice is to remove the carbon near the surface, a method that requires lower temperatures than those given above and a shorter time of heating.

The annealing operation is carried out in a steel box called a "sagger" and the packing material may be "roll scale" (oxide of iron) or sand alone.

Malleable copper. Native copper, the name given by miners to virgin copper. The term is also used sometimes to distinguish between the relatively hard and easily-fractured "sett copper". This latter contains cuprous oxide, which is reduced by the forcing of green logs beneath the surface of the molten copper. This "poling" operation when correctly carried out produces soft, malleable copper known as "tough-pitch" copper.

Malleable iron. The term usually refers to malleable iron castings, but is also used in reference to wrought iron.

Malleable nickel. A high-purity nickel obtained by remelting and refining (usually deoxidising) electrolytic metal.

Malleablising. In the case of cast iron the term refers to the annealing process, in which the combined carbon (cementite) of white iron castings is wholly or partly converted into graphite or free carbon. In the whiteheart process (see **Malleable cast iron**) part of the carbon is eliminated by oxidation. In the case of nickel, for example, the term refers to the treatment required to produce adequate ductility for hot working and cold forming or processing operations. It involves the deoxidation and removal of CO, N_2, H_2 from the metal by small additions of magnesium, calcium, aluminium, phosphorus or boron.

Malmstone. A siliceous material used in building, but sometimes used for the hearths of furnaces. Also known as firestone.

Mammillated. An indeterminate form of aggregation, not necessarily dependent on crystal character, which displays large spheroidal surfaces. Stalactites and stalagmites of fibrous malachite (q.v.) often have a smooth, mammilated or botryoidal surface.

Manandonite. An aluminium silicate containing both boron and lithium $2LiO.2B_2O_3.7Al_2O_3.6SiO_2$. It is used as a source of lithium, of which it contains about 2.1-2.2%.

Manchester furnace. A coal-coke-fired furnace used for annealing. Special recuperative flues are built in alongside the fire-box for preheating the air. Temperatures up to 1,000 °C or 1,100 °C are readily attainable and can be maintained with close uniformity.

Mandrel. A metal bar round which other metal may be coiled, bent or shaped. A tapered mandrel is used in the production of seamless tubes to pierce the billet, while the solid metal flows over it.

Manganblende. A black sulphide of manganese MnS. Mohs hardness 4, sp. gr. 3.75.

Manganese. A hard, whitish-grey metallic element that does not occur in the uncombined state, but is very widely distributed as oxides, carbonate and silicate and sulphide. The most important minerals are the oxides: hausmannite Mn_3O_4, braunite Mn_2O_3, manganite $Mn_2O_3.H_2O$, pyrolusite MnO_2, polianite MnO_2, psilomelane (hydrated oxide with barium and potassium oxides), wad (similar to psilomelane) and absolane (cobaltiferrous wad). They occur as sedimentary or as residual deposits in the USSR, India, Ghana, S. Africa, Cuba, Brazil, Mexico, Sweden, Spain, Malaysia, etc. Nodules of oxide are found on the floors of oceans and inland seas and are becoming an important source.

Manganese is one of the most important of the strategic metals, but finds its main applications as an alloying element. Pure manganese and manganese-base alloys have very few, if any, uses, but developments may follow the production of high-purity (99.98%) manganese on a commercial scale.

Manganese containing carbon as an impurity, is obtained by reducing the oxide with carbon in an electric furnace. The Thermit process in which the reduction is effected by aluminium is also used. High-purity manganese is obtained by the electrolysis of a concentrated solution of manganese II chloride using a mercury cathode. The mercury is subsequently removed from the amalgam by distillation in vacuum at 250 °C (approx).

The chief uses of manganese are in the form of iron-manganese alloys such as ferro-manganese (80% Mn), spiegeleisen (or spiegel, 5-20% Mn) and silicon-spiegel (10% Si and 15-20% Mn) which are used extensively in the manufacture of steels. These are crude pig-metals containing about 4-7% C and they provide a convenient means of adding carbon and manganese, as well as of deoxidising the molten steel. Manganese combines readily with any sulphur present, producing small discrete particles of a duplex manganese-iron sulphide, in place of iron sulphide, which would otherwise form continuous films and so cause embrittlement. Any manganese in excess passes partly into solid solution in iron and partly combines with carbon forming Mn_3C, which occurs with Fe_3C in cementite.

Manganese, which is always present in commercial steels, reduces the amount of carbon in the eutectoid so that for any given carbon content the ratio of pearlite to ferrite is increased by an increase in manganese content. Manganese also modifies the properties of the ferrite and produces a more finely lamellar pearlite (i.e. sorbitic pearlite).

High-tensile structural steels used for bridges, ship plates, etc., usually contain 1.2-1.6% Mn with 0.25-0.3% C, and show tensile strength, particularly elastic limit and toughness, superior to those of plain carbon steels.

Manganese causes a marked lowering of the critical ranges Ar_3 and Ar_1 and, with manganese contents in excess of 1.6% and a carbon content of about 0.3%, gives rise to air-hardening properties. Pearlite steels containing 1.5-2% Mn have excellent properties when suitably heat-treated.

Austenitic manganese steels (containing 12-14% Mn and about 1.2% C—Hadfield's steel) are used in the water-quenched condition for railway points and crossings, rock-drills, stone crushers, armour, etc. Austenitic manganese chromium steels (containing 15-18% Mn, 14-17% Cr) possess high ductility and attract attention as stainless steels.

Manganese is also used in several nonferrous alloys; the so-called manganese bronze is a high-tensile brass containing up to 3.5% Mn. Other copper-base alloys containing 8-12% Mn and 4% Ni are used for electrical resistances.

Manganese enhances the mechanical properties of many aluminium alloys and improves the corrosion resistance of magnesium alloys.

The physical and mechanical properties of manganese are given in Table 2.

Crystal structure cubic (complex) $a = 8.894 \times 10^{-10}$ m.

Manganese alloys. Manganese appears as an alloying element (usually with nickel, iron, copper) in electrical resistance and magnetic alloys, and as a common constituent of all steels.

Manganese bronze. A copper-zinc alloy containing aluminium, manganese and iron, and occasionally nickel or tin. It is also, and more correctly, known as manganese brass, complex brass, high tensile brass or aluminium brass. When properly prepared, manganese bronze castings are strong and ductile, and possess good resistance to corrosion by sea-water and by marine or industrial atmospheres as well as by many other common corrosive agents. It is used for various parts of marine engine pumps, for propeller blades, hydraulic equipment, gears, worm-wheels as well as for general industrial and engineering castings.

Since manganese bronze has rather greater solidification shrinkage than most brass and bronze alloys, and since it also tends to form scum and dross rather readily, castings must be carefully designed and close attention given to the foundry techniques. Wrought manganese bronze is available in the usual forms of bars, shapes, sheet and strip. Forging is usually done in the 650°-750 °C range.

The usual type of manganese bronze is approximately 60/40 brass, where the manganese may vary from 1 to 3.5%, and the balance is made up with additions of aluminium, iron and tin. Tin imparts appreciable hardness and increases the tensile strength. It also enhances the resistance to corrosion by reducing dezincification, a type of attack to which such alloys are prone.

High-tensile manganese bronzes have a manganese content not less than 3.5% and often as high as 4.5%. The additional elements are aluminium and iron.

Manganese carbide. A soft compound formed by heating Mn_3O_4 with carbon at 1,500 °C. It belongs to the class of carbides known as methanides, since it evolves methane (and hydrogen) when treated with water. When manganese is added to steel in small amounts it preferentially combines with any sulphur present, and an excess of manganese then appears as Mn_3C. Specific gravity 6.89.

Manganese copper. An alloy of copper and manganese containing 3-5% of the latter metal. It is relatively resistant to oxidation and retains its strength at moderately elevated temperatures. The name is also used to designate a master alloy containing 25-30% Mn used for hardening brasses, bronzes and cupro-nickel. Copper and manganese form a continuous series of solid solutions with a minimum in the melting curve at about 30% Mn at approximately 880 °C. This composition, and grades containing 20-25% Mn, are most widely used, but most commercial products also contain 0.3-4% Fe, and a similar amount of silicon. These elements are naturally objectionable for high-tensile brasses and some cupro-nickels.

Manganese ferrite $Fe(MnFe)O_4$. An inverse spinel oxide material, in which 8 manganese ions occupy octahedral voids together with 8 trivalent iron (ferric) ions, and 8 iron III (ferric) ions occupy tetrahedral voids, in the cubic close-packed lattice of 32 oxygen ions. A similar inverse spinel exists with zinc ions in place of manganese, and a mixture of the two, with 48% $Fe(MnFe)O_4$ and 52% $Fe(ZnFe)O_4$ is commercially produced as manganese ferrite. It is a soft magnetic material of very high magnetic permeability ($\mu > 1500$) and negative magnetostriction. When magnesium or copper ions replace zinc, the material becomes magnetically hard.

Manganese glance. The naturally occurring sulphide of manganese MnS. It is a steel-grey crystalline substance having a green streak. It crystallises in the cubic system. Mohs hardness 3.5 4, sp. gr. 4.04. Also called manganese blende and alabandite.

Manganese green. A green pigment consisting of barium manganate $BaMnO_4$. It is insoluble in water and poisonous. Specific gravity 4.9. Also called Cassel's green.

Manganese spar. A carbonate ore of manganese $MnCO_3$. It is isomorphous with calcite and often occurs with iron partly replacing the manganese. The mineral is usually rose-red in colour, but may also be grey or brown with a vitreous lustre. It crystallises in the rhombohedral system and is found in veins (with silver, lead and copper) or in deposits of manganese ores. It is used in making ferromanganese and spiegeleisen. Mohs hardness 3.5-4.0, sp. gr. 3.5. Also known as rhodochrosite and dialogite.

Manganese steel. Steel to which manganese is intentionally added as a strengthener and toughener. In small percentages manganese deoxidises and desulphidises steels, rendering them capable of being hot rolled, forged, etc. without rupture, and improving machinability. In somewhat larger amounts (up to about 7%) the manganese is used to increase the hardenability of the steel at relatively low cost. In still larger percentages (up to about 15%) the manganese is used in high-carbon (1-1.45%) austenitic steels that work-harden rapidly, so giving rise to good wear and abrasion resistance. It is to these latter steels that the term manganese steel especially refers, since manganese in small quantities is a normal constituent of all steels. In relatively small quantities manganese toughens steel by raising the ductile-to-brittle transition temperature of the iron. In this connection it is important to remember that the carbon-manganese ratio has an important effect in determining the tendency to brittle fracture.

Manganite. An iron black mineral containing about 62% Mn, having the composition $Mn_2O_3.H_2O$. It is found in the UK, Germany and in the Lake Superior area of the USA. It is a source of manganese. Crystal structure, orthorhombic; Mohs hardness 4, sp. gr. 4.2-4.4.

Manganosite. Naturally occurring manganese oxide MnO. It is found in Sweden. This oxide occurs in steels and high manganese irons that have not been deoxidised by more efficient oxidising agents such as silicon or aluminium. Ordinarily it is a dark-grey or greyish-green powder which fuses at a white heat without loss of oxygen. When heated in an atmosphere of hydrogen crystalline octahedra are formed which are emerald-green by transmitted light and dark grey by reflected light and have an adamantine lustre. The presence of iron II (ferrous) oxide often modifies the colour from green to red or brown. Melting temperature 1,700 °C, Mohs hardness 5-6, sp. gr. 5.1.

Mangling. A flattening process in which metal plates, either hot or cold, are passed between suitable rollers.

Manipulation. A group of works acceptance tests carried out on tubes, etc. They usually consist of flattening, crushing, expanding and flanging tests on representative samples.

Mannesmann mill. A rolling-mill for producing seamless tubing.

Mannesmann process. A process used in the production of seamless tube. A tube billet is pierced by rotation between two heavy rolls mounted at an angle and forced over a stationary mandrel. (*See* Fig. 87.)

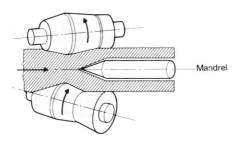

Fig. 87 *The Mannesman process for tube manufacture.*

Mannhes process. A method for treating copper matte by blowing air through the molten material to remove sulphur (as sulphur dioxide) and oxidise iron.

Mannitol. A hexahydric alcohol isomeric with sorbitol

$$(CH_2OH)_2.(CHOH)_4.$$

It is made by the hydrogenation of glucose. It is a white powder melting above 98% and is readily soluble in water. It is used as a humectant and in the production of synthetic resins, plasticizers and drying oils. It has been used as an addition agent in electroplating.

Mansfield process. A method devised for the smelting of copper sulphide ores in a blast furnace. The furnace is constructed of iron plates with a water-cooling jacket, and the lower portion is lined with firebrick. The charge consists of ore, anthracite or coke and a siliceous flux. Air is forced in at the base of the furnace, and eventually molten slag and matte collect at the base of the furnace. These are run off into a "fire-hearth", where the slag is run off at a higher level than the tap-hole for the matte. The molten matte is poured into a Bessemer-type converter and air again blown through the material. Sulphur burns and escapes as SO_2, iron is oxidised and passes into the slag as silicate, while any arsenic sublimes and is separated from the charge. Attempts are made to improve the efficiency of the process by using the heat in the escaping SO_2 to heat the air-blast.

Mantle. A term used to designate that part of a blast furnace which carries the weight of the stack. It conforms to the shape of the furnace immediately above the bosh and completely encircles it at that level. The mantle is constructed of heavy steel plates and sections, and is supported by fabricated steelwork or cast-iron pillars resting on the foundations of the furnace.

Maraging. The term used in reference to a process of improving the mechanical strength of certain nickel-iron alloys. These alloys contain 18-25% nickel, the ductile martensite formed in these alloys being "aged" at a moderate temperature without subsequent quenching. The range of tensile strength is given as 1.47-2.16 GN m⁻², and this is said to be combined with good ductility and good impact strength.

Marble. Crystalline limestone $CaCO_3$ occurring in large compact masses. When pure it is white in colour, but most of the marbles quarried are coloured or striped by a small quantity of metallic oxides. Good statuary marble has low porosity and takes a high polish. It is sometimes used for building stone in dry, non-industrial atmospheres. Marble dust is used as an abrasive in hand soap. Specific gravity 2.7.

Marcasite. A white form of iron pyrites FeS_2. It occurs in crystalline form and when brass yellow in colour is called fool's gold. Gem variety is white with Mohs hardness of 6-6.5 and sp. gr. 4.9-5.1.

Margarite. Crystalline aluminium silicate. It is an alteration product of natural corundum. It can be used as an abrasive.

It is sometimes known as calcium mica $CaO.Al_2O_3(SiO_2).H_2O$. It is usually yellow in colour, but other colours are not uncommon. It is useful as an insulator for heat and electricity, and when finely powdered is sometimes used as a filler for plastics.

Marialite. A double compound of sodium aluminium silicate and sodium chloride $2NaCl.3Na_2O.3Al_2O_3.18SiO_2$. It is a member of the scapolite series of rock-forming minerals, in some varieties of which sodium sulphate replaces sodium chloride. All the varieties crystallise in the tetragonal system, the crystals being prismatic hemihedral. Mohs hardness 5-6, sp. gr. 2.6.

It is fairly resistant to attack by acid, but is decomposed by heating to fusion point. The alteration products give rise to kaolin, mica and zeolites.

Marienglas. A term used to describe a form of gypsum $CaSO_4.2H_2O$ in the form of thin cleavage-plates. It was formerly used for glazing windows and in microscopic work to determine certain optical constants of minerals. Mohs hardness 2, sp. gr. 2.3.

Marl. A calcareous clay, containing up to 40% calcium carbonate, that is used in brick-making and is also used for sports pitches as a binder and fertiliser. The bricks must be fired at temperatures from 1,100° to 1,200 °C. The product is often yellow in colour.

Marles constituent. A micro-constituent sometimes found associated with cementite in rapidly cooled cast irons, particularly in high-silicon irons. It does not respond to heat tinting and may be regarded as an iron carbide containing silicon precipitated from the austenitic phase in the form of needles.

Marlstone. A calcareous shale used, with lime, for making Portland cement (q.v.).

Marmatite. A ferruginous variety of lead sulphide containing up to 20% Fe.

Marquenching. The name applied to a heat-treatment process for steel. The steel is usually quenched in hot mineral oil from about the transformation range and maintained at a temperature between 90 °C and 200 °C for a period long enough to permit of uniform heat penetration—or soaking. Thereafter the steel is allowed to cool in air down to room temperature. The effect of the quenching is to arrest the cooling at a temperature slightly above that at which martensite is formed, but below that at which austenite transforms to pearlite. The temperature of the oil bath is important, as it influences the steel microstructure. At the top of the range the structure will resemble that produced by martempering, while at the lower end of the range the steel will show a more or less uniform hardness and be less liable to exhibit distortion. This method is useful in treating steel sections of somewhat larger area than can be satisfactorily treated by martempering.

Martempering. A hardening process sometimes applied to steel. It involves quenching from a temperature above the transformation range to some temperature slightly above the upper limit of martensite formation, soaking at that temperature to permit equalisation of temperature throughout the mass without transformation of the austenite, followed by cooling—usually in air. The martensite structure may be tempered as desired.

In practice the steel is heated to such temperature that it consists wholly of austenite. It is then rapidly cooled to a temperature of some 100°-200 °C above that corresponding to the upper limit of martensite formation. The steel can be maintained at this temperature for a long time, since transformation begins very slowly. The time actually allowed is that which is considered sufficient to permit a considerable reduction in the temperature gradients within the mass. At the end of this time the steel is removed from the metal bath used for the first quench and either cooled in air or quenched in oil.

The process reduces the tendency of metal to crack or to distort and the resulting metal is reasonably ductile. Also known as "Interrupted quenching".

Martensite. A characteristic micro-constituent of hardened steels. It is a metastable transitional constituent indicating the first stage of the decomposition of austenite. It is the hardest of the transformation products of austenite, the hardness (and brittleness) depending on the percentage of carbon (up to 0.9%). It is produced by rapidly cooling from above Ac_3, such as by quenching in water. Less drastic quenching (e.g. in oil) often produces a mixed structure of martensite and troostite.

Martensite is produced by a diffusionless shear transformation, of body-centred tetragonal structure. The c/a ratio decreases linearly with carbon content from 1.1 to 1.7%, approaching 1 (b.c.c.) at carbon contents below 0.5%. Less than 10% of the possible interstitial sites are filled by carbon atoms, but the structure represents the closest approach to the body-centred cubic lattice of ferrite which can still retain carbon in supersaturated solid solution.

When martensite is tempered it gradually decomposes, iron carbide being thrown out of solid solution. The result is a structure consisting of ferrite in which the iron carbide (cementite) is dispersed as fine particles. This structure is known as troostite.

Martensite is magnetic. Acid etchants attack it more readily than they do ferrite or cementite. On polished surfaces, suitable etchants (q.v.) reveal the martensite as light-coloured acicular or needle-like markings.

Martensitic stainless steels. A group of hardenable stainless steels containing from 11 to 14% Cr and from about 0.15 to 0.45% C. A further steel of the martensitic type is one containing about 16-18% Cr and 1-3% Ni, and known as S80. They should contain not more than 1% each of silicon or manganese. Their hardness depends on the percentage of carbon, and they are magnetic at ordinary temperatures. Martensitic stainless steels harden readily on air-cooling from about 950 °C.

Mass absorption. A term used in reference to the absorption of gamma radiation in its passage through matter. If a beam of gamma photons is incident on a slab of matter of thickness x so that the intensity of the incident beam can be called I_0 and that of the emergent beam I, then $I = I_0 e^{-\mu x}$, where μ is the absorption coefficient. The value of μ divided by the density of the material is known as the mass absorption coefficient. Values of the total mass absorption coefficients for common materials are given in Table 36. In calculating these values the total absorption coefficient has been used. Also known as linear absorption coefficient (in cm) or molar coefficient (mole cm^{-2}).

TABLE 36. *Absorption Coefficients of Some Common Materials*

Gamma-ray Energy, MeV	Mass Absorption Coefficient (g cm^{-2})				
	Air, $\times 10^{-4}$	Water	Al	Fe	Pb
0.5	0.075	0.096	0.085	0.080	0.132
1.0	0.063	0.070	0.059	0.057	0.063
2.0	0.044	0.049	0.044	0.042	0.043
3.0	0.035	0.039	0.033	0.035	0.040
4.0	0.031	0.035	0.030	0.033	0.041
5.0	0.027	0.030	0.028	0.030	0.042
10.0	0.026	0.022	0.022	0.029	0.054

Mass action. A law enunciated by Guldberg and Waage which states that the velocity of a chemical reaction is proportional to the molecular concentration of the reacting substances. If in V litres of a reacting mixture there are a moles of substance A and b moles of reactant B, and if x moles of A react with y moles of B, then

$$v = k \frac{a}{V}^x \cdot \frac{b}{V}^y$$

The value of v is determined by chemical or electrochemical methods; and k is the velocity constant for the stated conditions of temperature (and in cases of gases, of pressure also).

Mass defect. In general, a nucleus has slightly lower mass than the sum of the individual masses of the nucleons which it contains, the difference being the mass defect. When a nucleus is formed, the nuclear forces of attraction bind the nucleons together, with release of energy — the binding energy, translated by Einstein's relation, into mass.

The helium nucleus, two protons and two neutrons, has a sum of nucleon masses equal to $(2 \times 1.672) + (2 \times 1.674) \times 10^{-27}$ kg $= 6.692 \times 10^{-27}$ kg. The mass of a helium nucleus is 6.644×10^{-27} kg, so that a mass defect of 0.048×10^{-27} kg is involved, equivalent to energy of 26.92 MeV, 4.31 pJ.

Mass effect. A term employed to signify the effect of size and shape in causing a deterioration in properties from the surface inwards, due to variations in the rate of cooling during heat-treatment. The term refers therefore to the rate of cooling rather than to the actual mass of the object, except in so far as a piece of metal bar 200 mm in diameter, for example, is more massive than a thinner bar of equal length.

Thus if a series of steel bars of identical composition, but of different diameters, are quenched from the same temperature under identical conditions, the physical and mechanical properties of the bars will vary in a more or less regular manner with the diameters of the bars. When the bar is thin or the surface large in comparison with the volume, extremely rapid cooling takes place. With increase in section the cooling rate is limited by the rate of conduction of heat from the interior of the bar to the quenched surface and it becomes impossible to quench the centre of the bar and to secure uniformity of properties across the section. Specific heat, thermal conductivity and the incidence of phase changes, as well as the actual "massiveness", of the section influence the rate of cooling.

Massener process. A method of reducing the sulphur content of molten pig irons by adding manganese to the melt and allowing it to stand. Manganese sulphide is formed and this rises to the surface, whence it may be removed.

Massicot. Lead monoxide PbO that may occur in association with galena. It is used as a pigment, in the making of glass and for fluxing earthenware. It finds additional uses as a filler in rubber and as a constituent of plumber's cement (a mixture of massicot and glycerine). Specific gravity 9.38.

Mass number. The atomic weight of an individual isotope expressed on a scale on which the most abundant isotope of oxygen has a weight exactly equal to 16. These weights are very nearly integral, the integer being referred to as the mass number of the isotope concerned. Also called isotopic weight.

Mass scattering. When a beam of X-rays falls upon any substance, part of the beam may be absorbed and transformed into radiations characteristic of the material absorbing the rays, another part may be scattered or dispersed in much the same way that the light is scattered by fog. The coefficient of mass scattering is the rate of decrease, per unit mass traversed, of the natural logarithm of the intensity of a parallel beam of X-rays in a substance, due to scattering alone. With materials composed of elements of low atomic weight the greater part of the X-radiation is scattered.

Mass spectrograph. An apparatus, due to Aston, for the determination of the exact mass of individual atoms, i.e. isotopic weights, by using the technique of positive ray analysis. The positive rays are, in fact, a stream of positively charged ions moving in a rarefied gas from the cathode in a direction away from the anode. The rays are deflected by an electrostatic and by a magnetic field, so that all rays having a specific value of e/m (charge/mass) are brought to a focus at a definite point on a photographic plate. There is a very nearly linear relationship between the mass of a particle and its position on the photographic plate. The method is capable of considerable accuracy.

Master alloy. The name given to an alloy or mixture of elements that is used for introducing desired elements into molten metals in the foundry. They usually contain a high percentage of the particular element that is to be introduced into the melt and are often used in the ladle to obtain good control over the composition of the final product.

Mastic. A pale yellow resin used in varnishes. It is obtained from the bark of a tree grown on Chios (an island off Asia Minor). The resin darkens slowly with age. A bituminous preparation is used for pointing joints in brickwork and for bedding floor-blocks.

Matching. A stage in the production of pack-rolled sheet. After the sheets have been parted they are reassembled to form a new pack that can be rolled to yield a product of the required gauge.

The pairing of two broken-down steel sheet bars for further rolling.

Match lines. Two lines, one along each of two right-angle edges of a die block, that serve in forging for base lines for measuring and alignment. Also called matched edges.

Match plate. A smooth solid piece of wood or sheet metal used in foundries, to form the base on which the bottom part of a pattern is laid—joint side down. A moulding box or flask is placed around the pattern on the match plate and the rest of the box is filled with moulding sand. The parts of the pattern, split along the parting line, are mounted back to back together with the gating system, so as to form an integral pattern.

Materials handling. The management and technology concerned with the economics and safety of the movement and storage of materials.

Mathesius metal. A lead-alkali metal alloy containing calcium or strontium which is of German origin. It contains up to 1% of the alkali metals, the latter forming metallic compounds of the type Pb_3Ca and Pb_3Sr, the crystals of which are distributed throughout the matrix. The tensile strength is about $170\,MN\,m^{-2}$, which is somewhat better than high-tin bearing metals.

Matheson joint. A leaded joint used in assembling pipes and other tubular products. An annular groove turned on the spigot provides for firm retention of the lead, and this lead combined with the special design of the bell forms a rigid joint.

Matlockite. An oxychloride of lead $PbO.PbCl_2$ that occurs as tabular tetragonal crystals. It is found in Derbyshire (UK).

Matrix. The ground mass of principal phase in which another constituent is embedded.

In electroforming, the form used as the cathode is known as the matrix.

Matrix brass. A brass of composition suitable for the making of matrices for linotype machines.

Matrix metal. A term used in powder metallurgy in reference to the metallic constituent of a powder mixture that has a lower melting point than the other constituents. On sintering this matrix metal fuses and acts as a cementing medium for the balance of the materials present.

Matte. An impure metallic sulphide product obtained during the smelting of metallic sulphide ores, e.g. copper, nickel, lead.

The term is also applied to a smooth but dull surface that has a low specular reflectivity. Also called matt.

Maximum tensile stress. Ultimate tensile strength (q.v.).

Maxwell-Boltzmann distribution. An assembly of fast moving particles will involve collisions between particles constantly, and a distribution of velocities will result. The Maxwell-Boltzmann distribution defines how N particles per unit volume, having velocities in the range c to $c + dc$, particle mass m, at temperature T K, will be given by $N(c) = 4\pi c^2 N(m/2\pi kT)^{3/2} \exp(-mc^2/2kT)$ where k is Boltzmann's constant. The distribution can be graphically illustrated in terms of temperature, and indicates that for hydrogen, the range of velocities spreads out over a wider distribution at high temperatures.

Mean fatigue stress. In most fatigue tests the evaluation is made under conditions of cyclic loading. The range of applied stress usually varies from a tension stress to a compression stress of equal magnitude. The stresses in such cases are usually represented by S_1 and S_2 and where $S_1 = S_2$ the mean fatigue stress is zero. In other cases the mean value may be a (+) stress or (−) compression.

Mean free path. The average distance traversed by a moving particle or body between successive collisions with other particles or bodies, or before involvement in some specified event (ionisation, capture, etc.) (see **Condensation, Conductivity**).

Mean square error. The square root of the mean of the squares of the deviations from the mean value of a quantity when a number of observations have been made and all known errors eliminated.

Mean stress. The average value of a range of stress in which tensile and compressive stresses are considered of opposite sign. If the maximum and minimum values of cyclic stresses are not constant, then the frequency with which the various values occur must be taken into account.

Mechanical equivalent of heat. Heat and mechanical work are mutually convertible, and in any operation involving such conversion 4.186 joule of mechanical work disappear (or appear) for each calorie of heat generated (or expended).

Mechanical properties. The term applied to those properties of metals which indicate their reaction to externally applied forces, e.g. those that involve the relationship between intensity of applied stress and the strain produced. The properties generally included under this heading are those that can be recorded in conventional mechanical tests. They include elastic limit, yield point, maximum stress, elongation, reduction in area, hardness, resistance to shock (impact), creep rate and fatigue range.

Mechanical testing. The methods by which the mechanical properties of a metal (q.v.) are determined. The tests include: (a) tensile; (b) bend; (c) hardness; (d) impact; (e) fatigue; (f) torsion; and (g) creep.

Mechanical twins. Twinned crystals produced by mechanical straining alone. If the twinned structure has been produced by straining and annealing, it is sometimes referred to as annealing twins. (See also **Neumann bands**.)

Mechanical working. The subjecting of metals to processes such as rolling, pressing, forging, etc., which change the form or modify the structure and so alter the mechanical and physical properties of the metals.

Medium alloy steel. A term somewhat loosely applied to include all steels containing one or more alloying elements such as nickel, chromium, molybdenum, vanadium, copper, etc., which are added to enhance certain desirable properties, whilst still retaining the pearlitic-ferritic structure of a plain carbon steel.

Steels containing manganese and silicon in the usual small amounts are not ordinarily classified as alloy steels.

Nickel (up to about 5%) is frequently used to strengthen and refine the grain-size of ferrite; chromium particularly in the presence of manganese, nickel, molybdenum and vanadium stabilises and modifies the carbide structure and increases hardenability, molybdenum strengthens and refines the ferrite of low carbon steels; small amounts of aluminium, vanadium and titanium are used to refine the grain.

Medium carbon steels. The name generally applied to steels that contain from 0.3 to 0.6% C. The tensile strength and the heat-treatment depend on the carbon content; the former usually lies between 400 and 600 MN m^{-2}.

Meehanite. The name given to various grades of cast iron produced under controlled conditions for the production of high-duty cast irons registered as Meehanite metals, whose composition is such that the molten metal is susceptible to improvement by inoculation in the ladle, usually by the addition of calcium silicide.

The Meehanite process depended on controlled additions based on a measurement of a chill cast wedge of metal leaving the cupola, and the addition of calcium silicide inoculant into the metal running down the spout, before it reached the ladle.

The cast iron may be heat-treated by oil quenching from 870 °C, and tempering at 450°-620 °C. Like other grey irons it can be stress-relieved at 450°-620 °C.

The process has shown that the coarse graphite and dendritic structure often associated with grey irons with low mechanical strength is not displayed. According to the composition and the process, the metal is heat, abrasion and/or corrosion resistant at will, and is used in castings for dies, hydraulic cylinders, brake drums, gear and pump accessories (*see* Table 3).

Meerschaum. Hydrated silicate of magnesium $2MgO.3SiO_2.2H_2O$. It is an amorphous rock-forming mineral. Specific gravity 2.

Melaconite. Native copper oxide, containing 79.8% Cu, that occurs mostly as a dull black powdery deposit in the zone of weathering of copper lodes. Also known as tenorite.

Melamine-formaldehyde resin. A condensation polymer formed between melamine (2, 4, 6,–triamino–1, 3, 5–triazine) and formaldehyde (methanal) (*see* Tables 3 and 10).

Melanterite. Naturally occurring iron sulphate $FeSO_4.7H_2O$. It is green in colour and is used as a mordant and in the production of pigments and ink. Specific gravity 1.89. Synonymous with copperas.

Mellosing. A metal-spraying process. Molten metal is fed into a jet of heated compressed air, is atomised and projected against the surface to be coated.

Melting furnace. A type of furnace in which the metal charge is to be brought into the molten condition. Various methods of heating may be used, including the use of solid, liquid and gaseous fuel, and electricity.

Melting point. The temperature at which a substance changes from solid to liquid or the temperature at which the solid and liquid can co-exist. For most pure substances the melting point is sharply determinable and is a specific property. The eutectic point is the lowest temperature at which the solidification of an alloy can begin.

Melting pots. Vessels of iron or steel, some having a refractory lining, used for melting lead, solder, tin and other metals of low melting point.

Melting range. The range of temperature through which a solid passes progressively as it changes from solid to liquid. The limits are set by the first appearance of liquid and the last disappearance of solid. Non-crystalline solids usually do not possess a definite melting point.

Menachite. The black magnetic sand found in Cornwall (UK) containing titanium oxide $Fe_2O_3.TiO_2$. It is important as the mineral in which titanium was discovered.

Mendozite. A soda alum $Na_2SO_4.Al_2(SO_4)_3.24H_2O$ found in the Andes.

Mer. The basic unit which, in addition polymerisation gives rise to the "polymer" by repetition along a molecular chain. The unit from which the mer is derived is the "monomer". With polyethene, the monomer is ethene C_2H_4, and the mer is $-CH_2-$, the polymer then becoming $(-CH_2-)_n$.

Mercast process. A precision casting process in which mercury is used in place of wax as a pattern material. The mould is first filled with acetone, which acts as a lubricant and is subsequently displaced with mercury. The filled mould is immersed in ethylene (ethene) at −60 °C until the mercury is frozen. The mercury pattern is then dipped in or coated with a crystobalitic slurry at −60 °C. The completed pattern is then allowed to stand at room temperature until the mercury has melted and drained away, and is then dried and fired in the normal manner.

Merchant iron. Wrought iron used for making chains, locks, etc. It is produced in bar form by repiling and rerolling bars of puddled iron. Also known as Crown Iron. It usually contains less than 0.12% C.

Merchant mill. A small rolling-mill that can be used for rolling a variety of products in a wide range of cross-sections and shapes. A general-purpose mill.

Merchant wire. The finished products made from wire, e.g. nails, staples, galvanised, barbed and woven fence wire.

Mercuric acid. The red oxide of mercury (HgO). It is unstable to heat. It combines with mercury II (mercuric) chloride in different proportions to yield oxychlorides. These are $2HgCl_2.HgO$ (red), $HgCl_2.2HgO$ (black) and $HgCl_2.3HgO$ (yellow). With warm ammonia solution it yields Millon's base. HgO is used in marine paints.

Mercury. A silvery-white metallic element which is liquid at ordinary temperatures. Native mercury occurs rather sparingly as fluid globules scattered through mercury sulphide (cinnabar, HgS), which latter is the principal source of the metal. Carniola in Yugoslavia, Almaden (Spain) and mines in California and Texas are the main sources, but appreciable amounts are also produced in Mexico, the USSR, Czechoslovakia, etc. The earliest recovery methods consisted in roasting cinnabar in an oxidising atmosphere to convert the sulphur into SO_2. The mixed vapours of SO_2 and mercury passed down long earthenware pipes in which the mercury condensed. It is now more usually obtained by distillation in retorts in the presence of iron, which removes the sulphur as FeS. The crude metal is purified by a further distillation.

The chief peace-time uses of mercury are in the making of drugs, pigments, insecticides, etc. Fulminate of mercury is used in percussion caps in ammunition primers. The oxide is used for military dry-cell batteries. Mercury finds employment also in making anti-fouling paints and for coating mirrors. The vapour of the metal is used in mercury-vapour lamps and in arc-rectifiers, power-control switches and certain scientific instruments. It dissolves most metals readily, forming amalgams, the chief interest in amalgams being in the extraction of gold and silver from their ores. A new development consists in its use in alloys used as soft solders (e.g. 4-8% Hg, 3-6% Sn, the balance being lead). Mercury vapour is highly toxic. (*See* Table 2.)

Merilising. A proprietary phosphating process that is applied to steel.

Merrill-Crowe process. A process for precipitating gold from deoxygenated cyanide solution by means of zinc dust. The occluded or dissolved air, which causes increased consumption of zinc, is removed by sucking the solution into a drum under a high vacuum. The solution is made to form a fine spray by passing through perforated trays, and escapes through a water-seal.

Merrill filter. A type of pressure filter used in the extraction of gold from ores. The crushed ore and cyanide solution are pumped into frames covered with filter cloth, which is stretched on corrugated plates. The cake formed in the press is washed and discharged, and the solution is treated for the recovery of gold.

Merrillite. A name given to the fine zinc dust of high purity that is used in the Merril-Crowe process for the precipitation of gold and silver in the cyanide-extraction process.

Mertone. A form of silica gel used as a coating on the surface of blue-print papers, in order to deepen the blue colour and obtain greater contrast in the print.

Mesitite. An impure variety of magnesite $MgCO_3$ that may contain from 30 to 50% iron carbonate.

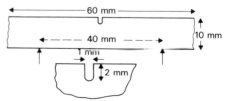

Fig. 88 *Dimensions of Mésnager testpiece.*

Mesnager test piece. A shallow notched impact test piece used in France and Italy. The form of the test piece is shown in Fig. 88 and the fracture is usually performed in a Charpy machine.

Mesons. Particles associated with cosmic rays. Positive, negative and neutral mesons are known to exist. Their existence and behaviour may throw light on the mechanism whereby the particles in a heavy nucleus are held together.

Messiter bedding system. In large plants for copper roasting or smelting, the ores and the concentrates that arrive for subsequent working must be suitably stored to permit of easy withdrawal when required. The Messiter bedding system consists of suitable belts carrying the ore, etc., and tripper belts cause the material to be dropped on to a concrete floor to form a bed. The system includes belts on which is scraped ore from the bed for removal to the smelter or roaster.

Metal. The following properties are exhibited by most metals, but they are not exclusively metallic: *(a)* metals are solid and opaque (mercury is liquid at ordinary temperatures); *(b)* when polished or freshly cut, metals show distinct "metallic lustre"; *(c)* metals are good conductors of heat (they are cold to the touch) and of electricity; *(d)* metals, when struck, produce a metallic ring or timbre (sodium, potassium, lead, tin, mercury, etc., do not ring); *(e)* metals are malleable and can be bent without fracture, and are ductile (cast iron is brittle); *(f)* metals usually have a high density (sodium, potassium, float on water; aluminium, magnesium are light metals); *(g)* metals can be melted and cast and exhibit a crystalline structure.

Chemically a metal is an element that: *(a)* can combine with oxygen to form a basic oxide which when dissolved in an acid produces a salt of the metal; *(b)* will form a chloride which is stable in the presence of water; and *(c)* the soluble salts will ionise in aqueous solution, giving an electropositive metallic ion, which can be plated out on to the cathode by an electric current.

Of over 100 known elements, 19 are definitely non-metallic, 5 are difficult to classify (*see* **Metalloids**), and the rest are generally regarded as metals. (*See also* **Bonds** *(d) metallic.*)

Metal infiltration. *See* **Infiltration**.

Metalite. Trade-name of a particular grade of alumina used as an abrasive.

Metallic arc welding. A welding process in which the rod or electrode is made of metal (as distinct from carbon). The arc is struck between the electrode, and the work and the high temperature (estimated at about 3,400 °C) obtained serves to melt the appropriate area of the work and also the electrode, the latter taking the place of a separate filler rod for supplying additional weld metal. The rods may be coated or uncoated, and in the former case the necessary flux, alloying elements and arc-stabilising material may be incorporated and a "shielded" arc obtained. Either d.c. or a.c. may be used, but in the former case the rod is made the negative electrode. The range of voltage is from about 20 to 50 volts, and the current varies according to the thickness or massiveness of the work.

Metallic cementation. A general term applicable to all surface absorption processes by which metals such as zinc, aluminium, etc. are impregnated into the surfaces of ferrous products. Examples are Sherardising, Calorising and Ihrigising (q.v.).

Metallic coatings. Thin coatings of metal applied to the surfaces of metal products for corrosion protection, decorative effect, improvement of electrical properties, etc. The processes include: *(a)* dipping in molten metal (galvinising, tinning); *(b)* metallising (spraying, etc.); *(c)* electroplating; *(d)* fire gilding (gold); *(e)* vacuum and gaseous deposition and *(f)* cladding.

Metallic paints. Paints in which the pigment consists of finely divided metal. In aluminium paint, flake powder is dispersed in linseed oil, varnish, gloss oil, laquer or a silicone resin. Finely divided mica or asbestos may be added to the aluminium powder. Lead paint contains colloidal lead in ordinary oil. Stainless steel paints consist of very finely powdered stainless steel in a synthetic resin. In making bronze or brass paints very finely powdered brass—containing 70-95% Cu and the balance zinc—is used. The colour may be varied by heating the powder. (*See also* **Gold paint**.)

Metalling jig. A jig for holding the mould which holds, for example, engine connecting-rods and bearings while being run up with white metal.

Metallisation. A process for spraying metal upon finished or semi-finished or worn products. It is applicable to large pieces of metal and to erected structures. The metal to be treated must be clean from scale, rust or oxide, and free from grease. In order to obtain a satisfactory "keying" for the sprayed metal the surface to be treated should be "undercut", a result that can be satisfactorily achieved by shot-blasting.

Aluminium, zinc, bronze, etc., can be sprayed on to metal to provide a protective coating. A special spray-gun is used which melts the metal (supplied in the form of wire) in an oxy-coal gas-flame and projects it in the form of droplets by a jet of compressed air. A continuous film of zinc or aluminium can easily be obtained which covers all the individual components of a structure so as to prevent ingress of electrolyte in the interstices.

Small articles can be heat-treated to improve the protection, while worn parts (e.g. bearings) can be built up to size using steel wire. The built-up portion is usually made oversize and then machined as required.

The word is also used for metallic coatings achieved by vacuum or gaseous deposition, particularly on polymeric material and glass. In the vacuum process, the articles to be coated are rotated under a heated filament of the material and streams of metallic ions deposit the material on to the substrate, with an applied potential between the two (only applies to advantage when a conducting coating has first been made on the substrate) or, far more generally, without any potential difference. As all vacuum processes are batch processes, metallisation of continuous plastic film uses rollers to transfer the film across the line of sight of the filament, and as large a roll of film as possible, and of the greatest feasible width, is used for economic conversion. The film-coating process is relatively cheap, and used for packaging material, decorative finishes, capacitor electrodes, reflecting and electrically conducting films. Glass lenses are coated, and many decorative artefacts of automobile trim, by the batch rotating on a turntable beneath the filament. Aluminium is very suitable for most purposes as the coating, since reflectivity or conductivity are essential properties both for technical and packaging purposes.

In gaseous deposition, a metallic compound is vaporised and decomposed in vacuo or in low pressure carrier gas streams, on the surface of the substrate. Many expensive metals with special properties, such as corrosion resistance, can be applied this way.

The plasma jet process is also available, particularly for building up worn parts of machinery, or applying special hard or corrosion resistant coatings. In this process, a plasma of ionised gas is generated in a separate arc chamber, and the coating material fed into the plasma leaving the jet under pressure. The effect on the substrate is much improved over normal spraying of molten metal, oxidation is reduced, and reactive metals can be successfully handled.

Metallography. That branch of metallurgy concerned with the preparation and examination of the surface of a metal. It is not practicable in many cases to prepare very thin sections of metal that could be examined by transmitted light. It is therefore necessary to carefully prepare a surface and look at it by the aid of a reflecting microscope. Metal surfaces are prepared by rubbing a flat specimen on successively finer grades of special polishing (emery) papers and removing the finest scratches by polishing on a cloth or leather disc that is fed with water and fine polishing powder such as alumina or magnesia. The effect of this preparation is to produce a very thin layer of "flowed" metal that has been dragged over the surface. Etchants (q.v.) are used to lightly attack this layer, and by selective action on various constituents of the metal reveal its structure under a high-power microscope.

In the macro-examination of a structure it is necessary only to polish the metal surface on fine grades of emery paper. A deep etch is usually made and the surface examined under a low-power objective. The pattern of the surface as thus seen is known as the macrostructure. Sulphur prints (q.v.) are also used in macroscopic examination of steels. (*See also* **X-ray diffraction** and Table 11).

Metalloids. Such elements as show some of the typical properties of metals and some of the more characteristic properties of non-metals are known as metalloids. Boron, silicon, germanium, arsenic, antimony, polonium and tellurium are metalloids.

In steelmaking, the elements carbon, silicon, phosphorus and sulphur which are present in small quantity are sometimes referred to as metalloids (incorrectly).

Metallurgical coke. A product of the process of high-temperature carbonisation of coal, during which the volatile hydrocarbons are removed by dry distillation. Good metallurgical coke can be prepared from coal that contains not more than 30% or less than 20% volatile matter, the determination being made under standard conditions. Since most coals contain some shale and other substances, washing is generally required so that the extraneous matter shall be removed either in gravity or flotation types of machine. The coal after washing and draining should not show more than 6% ash on carbonisation. The washed coal is then blended, to ensure a fairly uniform product and to permit use of a certain quantity of lower-grade coal. The blend is crushed so that the fineness is about 70% or 80% below 3 mm mesh.

Modern coke-ovens are made in silica brick and have the following dimensions: length 10-15 m, height up to 5 m and width 50 cm. The charge is about 17 Mg of crushed coal. The burning time varies from 18 to 24 hours (25% longer for foundry coke), and the flue temperature is between 1,150 °C and 1,200 °C. The regenerators that utilise the hot gases are situated under the ovens.

Metallurgy. A term of wide meaning, embracing the practice and science of extracting metals from their ores, the refining of crude metal, the production of alloys and the study of their constitution, structure and properties, and the relationship of physical and mechanical properties to thermal and mechanical treatment of metals and alloys. Hydrometallurgy is that branch of metallurgy concerned with the science and art of extraction of metals from their ores by processes involving solution in water. Electrometallurgy is concerned with the extraction and refining of metals using the electrothermal or electrolytic effects of a current.

Metal mixer. In the making of steel the molten metal from a blast furnace may be transferred to a converter or open-hearth via a large receptacle known as a "mixer". This is a refractory-lined storage vessel which is gas heated. From it the amount of liquid metal required for a single charge is withdrawn as needed. When the mixer has a suitable lining and the molten metal is covered with a suitable slag, a certain amount of impurity can be removed prior to "converting".

Metal polishes. These usually consist of mixtures of dolomite lime and a grease. The latter is usually present in amounts up to 25% and may be of animal or vegetable origin, and should be saponifiable. The lime may be wholly or partly replaced by other abrasives, such as emery flour, tripoli, pumice flour, silica or rouge. Paste polishes contain tripoli or pumice together with oxalic acid and paraffin. Modern liquid polishes contain fine abrasives such as very finely ground pumice or diatomaceous earth, together with a soap detergent, a little naphtha and fine oil or a caustic alkali. They are suitable for polishing specimens for metallographic examination.

Metal spraying. Synonymous with metallising or metallisation (q.v.). Titanium and zirconium can be applied as a metal spray by projection of the hydrides of these metals through an electric arc on to the target surface. The hydrides are decomposed on the target and the liberated hydrogen acts as a shield against oxidation during the short cooling interval.

Metamorphism. In its widest sense the term includes all the changes that a rock has undergone. It is, however, more usually employed in reference to the results upon a rock of earth movements (regional metamorphism) or of heat (contact or thermal metamorphism).

Metasomasis. A mode of formation of ore deposits by partial or complete replacement of a pre-existing rock by an ore deposit. Limestone is frequently the rock displaced, e.g. in Cleveland iron and some veins of zinc and lead ores.

Metastable. A term used in reference to metals indicating an "unnatural" condition or a state of false or apparent equilibrium corresponding to a local energy minimum but not the only local energy minimum or that of the lowest energy. The system concerned is stable in itself and shows no tendency for spontaneous change, but under certain conditions a change to the "natural" or other condition would occur. In the operation of tempering steel the metal has its tensile strength reduced, but its toughness increased. The structural changes induced consist in the partial or complete decomposition of the martensitic solid solution produced when the steel was hardened. At ordinary temperatures the natural condition of a plain carbon steel is not that of martensite. The martensitic solid solution was produced by the rapid cooling during quenching. Hence the quenched steel is in a metastable condition. In this sense metastable means departing from a state of physico-chemical equilibrium. Many alloys employed in industry are used in a more or less metastable state, as solid bodies offer very great internal resistance to molecular rearrangements, and excessively slow rates of cooling would (in most cases) be required in order to allow equilibrium conditions to be attained.

Metatectic. An American term for peritectoid (q.v.).

Metcolising. A process for depositing on cast iron a protective coating of aluminium. The aluminium and a special "sealer" are applied by spraying (see **Metallisation**) and the whole is heat-treated. The product can then be used to give useful service at higher temperatures than would otherwise be possible.

Meteoric iron. Native iron found in masses which have fallen to the surface of the earth from sources outside the atmosphere. It is usually alloyed with nickel, and small amounts of cobalt, manganese, tin, chromium, carbon, copper and phosphorus.

Methane CH_4. The lowest member of the saturated acyclic unbranched hydrocarbons, and part of the stream produced in petroleum distillation. It is the main constituent of natural gas emanating from oil fields, and can be converted readily into methanol, an excellent fuel for internal combustion engines, and starting point for solvent, pharmaceutical and plastics production.

Methane forms a starting point for methyl methacrylate, poly-tetra-fluoroethylene (tetra fluoroethene) and the condensation formed formaldehyde plastics with phenol, urea and melamine. Boiling point −164 °C.

Methanol CH_3OH. Also called carbinol and wood alcohol, the latter in view of its derivation by distillation of wood. Boiling point 65 °C. Used as a solvent, a fuel, and a starting point for manufacture of a wide range of products. Toxic, and controlled by adulteration in many countries, to limit abuse.

Methyl orange. The sodium salt of helianthin (dimethylaminoazobenzene sulphonic acid). It is a brilliant orange-yellow powder that dissolves readily in hot water, and the solution is yellow. It is seldom used as a dye because of its sensitivity to traces of acid. These latter turn it to a red colour, and hence its use as an indicator. Yellow, pH 4.3-red, pH 3.2.

Methyl violet. The chloride of pentamethyl *para*-rosaniline. It is usually sold in lumps that have a green metallic lustre and are soluble in hot water and in alcohol. The solution has a beautiful violet colour and is used for dyeing cotton, wool and silk. It is used as an indicator, since acids cause the violet colour to be discharged to pale yellow. Range: yellow, pH 0.3-violet, pH 1.6.

Metric ton (tonne). 1,000 kg or 2,204.6 lb. This is equal to 0.942 ton.

Meyer's constant. A constant representing the capacity of a metal for work-hardening. It is based on the Brinell hardness test. Meyer showed that the diameter of the Brinell ball impression varied with the applied load in accordance with the formula $L = ad^n$, where L = applied load; d = diameter of the impression; and "a" and "n" are constants. The constant "a" is a measure of the hardness level and is not related to the work-hardening capacity. The exponent "n" depends on the metallurgical condition of the specimen and thus indicates the capacity for cold work. A value of $n = 2.5$ indicates a dead soft anneal. This value falls as the material has been more or less cold worked.

If P_m = mean pressure obtained from the projected area of the indent, then Meyer's constant is obtained from $P_m = 4L/\pi d^2 = 4ad^{n-2}/\pi$.

Mica. A group of widely distributed rock-forming minerals, some of which are important commercially. They are distinguished by a perfect basal cleavage which causes them to split up into thin elastic plates, and by their pearly—sometimes dark metallic—lustre. They all crystallise in the monoclinic system.

Mica occurs as an essential constituent of igneous rocks and as an alteration product in some mineral silicates. As a group they are complex silicates—usually orthosilicates of aluminium, together with potassium, sodium, lithium, hydrogen and magnesium. Sometimes Fe^{2+} and Fe^{3+} are present and quite rarely rubidium, caesium, chromium, manganese and barium are found. The main use at present is as an electrical insulating material, filler in ceramic and plaster boards and finishes, and in the making of micanite cloth. The transparent varieties show a refractive index 1.58-1.60 and double refraction is strong. Mohs hardness 2-3, sp. gr. 2.7-3.1.

Micaceous hematite. A variety of specular iron ore Fe_2O_3 that is foliated or micaceous in form.

Micanite. Small sheets of mica cemented with shellac or other insulating cement on to cloth or paper and used as an insulating medium for a variety of purposes.

Micelle. In the formation of crystalline polymeric material, the micelle was regarded as a primitive orientation of polymer chains into parallel directions over all or part of the chain. Subsequent work has demonstrated that micelle formation is only part of the polymer crystallisation process, but may still be applied to materials where the degree of crystallisation is low. (*See* **Crystallisation.**)

Microanalysis. A special technique for the analysis of very small quantities of specimens or for the recognition and estimation of traces of impurities. The methods employed are often similar to those employed in ordinary analysis, and include the production of certain colorations or changes of colour, or of substances having distinctive odours. Though most of these methods are qualitative, they are usually very sensitive, and some of them are quantitative also. Special equipment, such as the microbalance and the spectroscope, is normally employed.

Micro-character. A hardness tester which consists of a very small, pointed diamond mounted under a load of 3 g. It is used for making minute scratches on a highly polished surface in order to determine the hardness of the different constitutents of an alloy. The width of the scratch at different parts of its length is measured in micrometres (0.001 mm) and converted into "microhardness" numbers by reference to a standard scale.

Microcline. A silicate of aluminium and potassium

$$K_2O.Al_2O_3.6SiO_2$$

that occurs in acid igneous rocks, especially granite. It crystallises in the triclinic system. Mohs hardness 6-6.5, sp. gr. 2.55.

Microcosmic salt. A white crystalline sodium ammonium hydrogen phosphate $NaNH_4.HPO_4.4H_2O$. It is used in place of borax in dry analysis. When heated in a loop of platinum wire it loses water and then ammonia and is converted into a transparent bead of $NaPO_3$. This substance, like sodium metaborate, will combine with metallic oxides (e.g. $CuO + NaPO_3 = CuNaPO_4$). The coloration caused by such oxides may be used for identification of minerals. Whilst, however, silica can displace boric oxide from sodium metaborate, combining with the alkali base to form a transparent glass, it does not react with the phosphate. Hence particles of silica may be seen floating in the bead after fusion. This reaction is employed as a method of detecting silica.

Microhardness. A term used in reference to the hardness of the individual constituents of an alloy—that is the hardness of localised areas of solid solutions or small particles of other phases that may be present. Testing is carried out under a microscope with a diamond indenter of Vickers pyramid shape, or a rhombus based pyramid. Load is in grams rather than kilograms.

Microlite. Essentially a pyrotantalate of calcium $Ca_4Ta_6O_{19}$. It frequently contains niobium combined with other bases. The percentage of tantalum (calculated as Ta_2O_5) is about 68 and the niobium present usually amounts to 7% or 8% calculated also as the pentoxide.

Micropegmatite. The name given to the fine intergrowth of quartz and felspar that fills up the irregular interspaces between crystals of other minerals. Because of its irregular form and indefinite shape, it is thought that this may have represented the remains of the mother liquor from which the other minerals crystallised. It is found in many igneous rocks that contain high percentages of silica. The structure is similar to that shown by certain pegmatites or coarse granitic veins on the large scale, but the finer grains can be detected only under the microscope.

Microporosity. The name given to the extremely fine porosity sometimes found in a metal. It is due to shrinkage or gas evolution during casting.

Microprocessor. An integrated circuit incorporating several types of electronic device, produced on one initial substrate of semiconductor material, usually silicon, by photolithographic or simi-

lar printing techniques followed by selective etching and "doping" to give n or p type areas. One small slice of substrate, say 4 cm square, can be so processed and manipulated under a microscope to produce as many as 1500 such "chips" each incorporating a complete circuit. Individual "chips" may be as small as 1 mm square. (*See* **Semiconductor, Transistor.**)

Microradiography. The name used in reference to that branch of radiography which deals with the examination of very small objects, the image of which must be enlarged. Since X-rays can only with difficulty be focused, the enlargement must usually be done photographically, and very fine grain emulsion is therefore required. In a complex alloy each separate phase shows a different absorbing power, and by trial or by calculation the appropriate wavelength to give maximum differentiation between phases can be arrived at. Thin metal sections are placed in contact with the photographic film and enlargements up to $\times 1000$ are usually prepared.

Microscopic stresses. Stresses which are balanced over small distances, of the order and magnitude of the grain size.

Microsegregation. The name used in reference to the variation in composition occurring within individual dendrites of a solid solution which results in the centre of the dendrite being richer in one constituent than in another. Also known as coring and dendritic segregation.

Microstructure. The structure of metals and alloys as revealed after polishing and etching a sample, by examination under a microscope at magnifications greater than 10 diameters.

Micro-throwing power. A term used in reference to plating baths to describe the ability of the bath to deposit metal in pores, i.e. its pore-filling capacity.

Middlings. A term applied to that portion of a pulverised ore which has a specific gravity intermediate between that of the concentrate and the tailings. Middlings obtained during a concentration process are recrushed to finer size and retreated to obtain a concentrate.

Mild-drawn wire. A term used in reference to steel wire drawn with a reduction of area of up to about 10%, from rod or annealed bar.

Mild steel. The name applied to steels that contain carbon in amounts ranging from 0.12 to 0.25%. Also known as mild carbon steel, low-carbon steel or soft steel. The classification of steels according to carbon content is purely a matter of convenience. A steel containing carbon at the maximum limit quoted would have a tensile strength of about 385 MN m^{-2}.

Milk of lime. A suspension of calcium hydroxide in water $Ca(OH)_2$.

Mill. A name applied to a single machine or to a complete plant for crushing and grinding ores and also for rolling metals. In the latter case the name is also applied to a variety of machines, e.g. billet mill, which is a blooming mill in which billets are formed; blooming or cogging mill in which ingots are reduced to blooms, billets, slabs or sheet bars by rolling or forging; continuous mills, in which the metal passes from one set of rolls to the next, being continuously reduced by successive passes to form the finished product; edging mill, for rolling sheet plate or strip, having special rolls that control the width of the product and give various edge finishes; sinking mill, in which tubular products are made by passing tubes through sizing rolls.

Miller indices. The indices used for crystal planes, which are the smallest integers proportional to the reciprocals of the intercepts, in terms of the parameters (written a, b and c) of the plane of the three crystal axes. The indices are written h, k and l.

For crystal planes in the hexagonal system the Miller-Bravais indices are used. These are similarly the smallest integers proportional to the reciprocals of the intercepts, in terms of the parameters of the plane of the four axes. These indices are written h, j, k and l. (*See* **Crystal classification.**)

Millerite. A naturally occurring nickel sulphide NiS that occurs with other nickel and cobalt minerals in Ontario, Cornwall (UK), Pennsylvania and Saxony (Germany). Traces of cobalt, copper and iron are usually present. It crystallises in the hexagonal system usually in very slender capillary crystals. Mohs hardness 3-3.5, sp. gr. 4.6-5.6.

Miller's process. A method of purifying and toughening gold. It involves treating the molten metal with mercury II chloride or passing gaseous chlorine through the melt. Bismuth, antimony and arsenic are eliminated as volatile chlorides. If silver is present, the silver chloride formed fuses and floats to the top.

Mill finish. The surface finish produced on wrought metals by rolling. The finish that is produced is dependent on the ground finish of the rolls in what is usually referred to as the finishing mill.

Milling. A term that refers to a method of grinding ores to fine particle size. It is usually a second stage crushing process using rotary methods such as in ball-mills, edge-mills and cone-crushers.

A system of open-cast mining used for coal, ironstone and other metalliferous ores in which the following sequence of operations occurs: (a) the removal of the overburden; (b) the working of a shaft or inclined runway to the bottom of the ore-bed; (c) the development of subsidiary underground tramways beneath the ore-body; (d) the raising of openings through the ore from underground drifts; and (e) the "milling" or shovelling the ore into the "raises", where it is loaded into haulage trucks for hoisting to the surface.

A machining operation for shaping metals. The milling machine has a revolving cutter and a sliding table to which the work can be fixed. By using cutters of different profile the number of machining operations can be reduced.

The term is also used in reference to the treatment of metal powder mixtures used in powder metallurgy. The object of milling in this latter case may be to affect the size or shape of the individual particles or to coat one constituent of the mixture with another constituent.

Milling is the term used to describe the forming of corrugations on the edges of coins and similar shaped articles. In the case of coins it was introduced to stop the "clipping" of coins. In the making of other types of disc, the milling is used as a "finish" or to provide a grip.

Mill pack. The product of pack-rolling.

Mill process. A process for producing iron direct from the ore by reduction with carbon monoxide in a rotary kiln or a kiln or a shaft furnace.

Mill scale. The layer of oxide which forms on the surface of iron or steel that has been worked at temperatures exceeding about 800 °C. It is black to brownish-black in colour. It is not compact enough to provide a protecting layer over the surface of the metal. Also called rolling scale.

Milner-Debye theory. A theory of interionic attraction of ionised salts. It is based on the idea that, owing to the electrical attraction between the positive and negative ions, there are, on the average, in the neighbourhood of any ion more ions of unlike sign than of like sign. Hence, when a solution is diluted, the separation of the ions involves the doing of some internal work against this electrical attraction and a corresponding increase in the energy content of the solution. The theory is mathematically derivable from Boltzmann's principle (kinetic theory) and Poisson's equation.

In the case where there are only two kinds of ions of equal valence v at the same molal concentration c, the change in energy attending the infinite dilution of that volume which contains 1 mol of each of the ions is

$$2\Delta U = \frac{2Av^2\sqrt{2cv^2}}{K^{1.5}\,T^{0.5}}$$

where A is a constant and K is the dielectric constant of the medium.

Mimetic twins. A term used in crystallography in reference to the production by twinning of forms that apparently display a higher degree of symmetry than that actually possessed by the substance. Also called pseudo-symmetric twins.

Mimetite. A chloro-arsenate of lead isomorphous with apatite and one of the pyromorphite set of minerals which occur associated with hydralogenic lead deposits in Cornwall, Derbyshire, Cumberland (UK), Saxony (Germany), Bohemia (Czechoslovakia), Mexico and parts of the USA. It crystallises in the hexagonal system. Mohs hardness 3.5, sp. gr. 7-7.25.

Mineral. A naturally occurring substance, the production of which in the earth's crust has not been related to decay or decomposition of organic matter, and which has certain definite and essential characteristics. These latter relate to (a) chemical composition, (b) crystalline form, (c) specific gravity and (d) crystallo-physical properties. These properties vary within quite narrow limits. The crystallo-physical properties referred to include optical characters, and magnetic, electrical and thermal characters.

Miner's pan. A shallow metal pan used in placer mining and prospecting. The matter to be tested or concentrated is placed in the pan and shaken with a swirling action with water. Any gold quickly settles at the bottom, and the rest of the sand is gradually washed away by dipping the pan in water and pouring it off slowly so that only heavy minerals remain.

Minervite. A type of spitite—an altered basaltic rock. It contains a high percentage of soda, is extensively altered and shows a "pillow" structure. The mineral is usually associated with areas that have experienced long-continued and slow subsidence. It usually contains up to 5% TiO_2.

Minium. Red lead Pb_3O_4, which is usually prepared by heating massicot (q.v.) on an open-hearth to a temperature not exceeding 400 °C. The brilliance of the colour depends on the slow oxidation at or below the temperature stated. When heated to 470 °C or above, the minium decomposes into massicot and evolves oxygen. The red oxide is used as a pigment and in the making of flint glass. In this latter application the minium should be free from iron, and, for making the brilliant glass, the oxide is prepared from white lead. The usual impurities in the commercial product are litharge and iron oxide. Specific gravity 8.6-9.1.

Miocene. The system of strata which forms the lower of the two divisions of the Neogene or newer tertiary period, and occurs between the Oligocene and the Pliocene. The Miocene period corresponds to a cycle of sedimentations; and the deposits are found—at least in Europe—in distinct basins that were directly related to the regression of the sea.

Misch metal. An alloy derived from a mixture of cerium earths obtained as a by-product in the chemical industry. The cerium earths are converted into chlorides by treating with hydrochloric acid. These chlorides are then intimately mixed with fused calcium chloride and packed into a graphite crucible. The mixture is submitted to a heavy current, when electrolytic decomposition occurs and the alloy—misch metal—remains. The composition is approximately Ce 50%, La 45% and balance the rare-earth metals.

When misch metal is alloyed (in vacuo) with iron or manganese or with both these metals a pyrophoric alloy is produced which is used for "flints" in petrol lighters.

The alloy containing 20% Ce is used as a "getter" in electronic tubes. Small additions (about 0.5%) to nickel-chromium resistance alloys reduces the tendency to "green rot". Quite small amounts (0.05-0.1%) are advantageous in the production of nodular cast iron. Cerium is a good desulphuriser as well as a good deoxidiser. In stainless steels and in aluminium and magnesium alloys it is used to stabilise carbides and nitrides and to refine the grain.

Miscibility gap. If the compositions of two conjugate solutions are plotted against the temperature, the typical curve obtained is convex upwards, the summit of the curve (known as the critical solution temperature) representing the temperature at which the compositions of the two solutions are identical. Only at points on the graph lying above the curve are the components completely miscible in one another in all proportions. The region under the curve is known as the miscibility gap.

Miscible. A term used of liquids, including molten metals, in reference to their ability to mix and form a homogeneous substance. Liquids that are not miscible separate into layers according to their specific gravity.

Mispickel. An important source of arsenic which is a sulphide of iron and arsenic FeAsS. It occurs in lead, silver, tin and copper veins and some varieties contain up to 9%, cobalt replacing some of the iron. It crystallises in the orthorhombic system. Mohs hardness 5.5-6, sp. gr. 6.3. Also known as arsenical pyrites.

Misrun. A casting, defective because of incomplete filling of the mould. This is usually due to the metal being poured too cold, so that it tends to solidify before the metal streams meet. The defects are usually found in thin sections, near the top, and at a distance from the runner. Sometimes they can be traced to runners that are too small or that have become choked.

Mitscherlich's law. The law of isomorphism stating that substances which are similar in crystalline form and in chemical composition can usually be represented by similar chemical formulae. The alums are typical examples of isomorphous substances whose formulae may be written $A_2SO_4.R_2(SO_4)_3.24H_2O$, where A may be sodium, potassium, NH_4, and R stands for aluminium, iron, chromium and other trivalent metals. It is also known that isomorphous substances can replace each other wholly or in part in the crystalline substances without altering the crystalline form.

Mixer. A vessel for storing hot metal from one (usually more than one) blast furnace. It conserves the heat from the liquid metal and levels out minor variations in composition so that a more uniform metal can be delivered to the steel furnaces. It is used principally in steelmaking.

Mixture rules. When artefacts are constructed by combining two incoherent materials, as in a composite or mixture bonded by adhesive or cement, it is advantageous to estimate the properties in terms of the proportions used and the properties of the individual materials. These factors can be considered by mixture rules, some of which take into account the type of dispersion adopted.

A simple rule can be applied to density, in that the contribution of each constituent is regarded as additive, and in proportion to the volume occupied by each. This is successful only where uniform distribution applies, since a layer structure, for example, would show anisotropy even of density. Spherical particles will behave differently from rod or disc shaped particles when embedded in a matrix, since spherical particles may not disrupt the coherency of the matrix if they are in small enough concentration, but discs or rods may be in contact over part of their surfaces.

A formula of the type $e^k = \Sigma_n \theta_n e_n^k$ where θ_n is the volume fraction occupied by the addition, e_n is the value of the parameter coefficient for the addition, e that for the mixture and k has a value from $+1$ to -1, can be used to express these factors. The value of k is dependent on distribution, and is $+1$ for density. When the particles are spherical, $k = 0$ when the constant for the matrix is lower than that for the dispersed phase, and $k = +\frac{1}{2}$ when the reverse is true. Lenticular or disc shaped dispersions show $k = -2/3$ when the matrix constant is lower and $k = -1/3$ in the reverse case. For layer structures $k = +1$ parallel to the laminations for conductivity measurements (resistance, -1) and $k = -1$ normal to the laminations. Metallic alloys of two or more phases may show random orientation, and hence some intermediate value of k will be necessary.

Maxwell derived more complex relationships, but related to spherical particles, though the results are usually found within the bounds of simple mixture rule for all values of k. A logarithmic rule is frequently successful, revising the simple equation to $\log e = \Sigma_n \theta_n \log e_n$. By plotting $\log e$ against θ_n, straight lines are obtained which fit well when there is a coherent matrix, or where there is a marked difference in properties between the constituents.

Composites made by fibre reinforcement are difficult to predict, since the fibres are rod-like, and therefore anisotropic, and moreover, the process may squeeze out matrix between some fibres and not others. Methods of manufacture have to be much more precise than usual, if wide variations in properties are to be avoided between one batch and the next.

Experimental work can markedly assist the derivation of an appropriate mixture rule, for development purposes, but theoretical assessment still falls far short of the standard required for prediction of properties of hitherto unexplored composite materials or mixtures.

Mizzonite. One of a group of rock-forming minerals known collectively as scapolites. These latter include a number of isomorphous mixtures and consist of (a) sodium or calcium silicate and (b) sodium or calcium carbonate or sulphate. Mohs hardness 5.5-6, sp. gr. 2.77. The colour, white to grey.

Mocha stone. A variety of chalcedony SiO_2 containing small dendrites of iron oxide or a ferruginous chlorite or oxide of manganese. These infiltrated oxides give the appearance of vegetable remains. The stone is found in the Deccan (India). Artificial varieties are produced at Oberstein (Germany).

Moderator. A material used in a nuclear reactor, to reduce the velocity and therefore the energy, of emitted neutrons, by collision between neutrons and moderator nuclei. The moderator material must possess low neutron capture cross-section, as the objective is to lower the energy, not reduce the neutron flux.

Modification. The term used in reference to minor adjustments in composition or production technique which alter the structure and/or the properties of an alloy. The term is most usually employed to describe the process for altering the structure and properties of aluminium-silicon alloys by small additions of an element such as sodium, or to the production of fine graphite in grey cast-iron by various treatments.

Modified Bauer-Vogel process. A process used to thicken the natural oxide film on aluminium surfaces. Articles to be treated are degreased before immersion in a solution of 7.5% washing soda Na_2CO_3 and 1.5% sodium or potassium chromate. This solution is used at a temperature of 90°-100°C and the immersion time is 3-5 minutes. The colour of the film produced is grey or dark grey, and this can be darkened by the addition to the bath of a solution containing 2.5% copper nitrate, 3% nitric acid and 0.5-1% potassium permanganate. If cobalt nitrate is substituted for copper nitrate, an excellent black film is obtained. The process is used to produce a surface protection and as a good key for painting.

Modulus of elasticity (Young's modulus). The ratio, within the limits of elasticity, of the stress to the strain produced by the stress. It may be regarded as a force which, acting parallel to the axis of a bar of unit cross-sectional area, would produce an elastic deformation equal to the original length of the bar. Such an occurrence is entirely imaginary. The stress is usually measured in $MN\ m^{-2}$ or $kN\ mm^{-2}$. The strain is the ratio of the linear extension to the original length (for Young's modulus) and is a dimensionless quantity.

$$E = \frac{Force}{Sectional\ area} \times \frac{Original\ length}{Extension}$$

When the strain is measured as the change in volume/unit volume, the modulus is known as the bulk modulus.

Modulus of rigidity. A term applied in torsion testing of metal specimens in which the strain is measured as the displacement per unit length caused by shear stress per unit area. The ratio of stress to strain, within the elastic limit, is called the shear or rigidity modulus.

Modulus of rupture. This is defined as the breaking load per unit of cross-sectional area required to rupture a specimen either in torsion or in bending. In the case of materials tested to rupture in tension, the modulus of rupture is then called the ultimate tensile strength. The modulus is usually represented by R and is regarded as the maximum stress in the outer fibres of a bar or beam tested to fracture. The computation is usually made by the empirical application of the flexural formulae to stresses above the transverse elastic limit. For a rectangular bar loaded at the centre

$$R = \frac{3}{2} \times \frac{load \times span}{sectional\ area \times depth}$$

Modulus of strain hardening. In the testing of metals the application of stress causes changes in shape and in volume. The calculation of the true stress involves taking into account the changes that occur in sectional area. If the true stress (σ) be plotted against the true strain (λ) in tensile testing, the slope of the curve, within the plastic range, is known as the modulus of strain and the ratio $d\sigma/d\lambda$ is the rate of strain hardening.

Moebius cell. The electrolytic cell used in the Moebius process for the refining of silver (q.v.). The unrefined silver is made into the form of plates about 12 mm thick, and these are placed alternately with the pure silver cathodes. The silver that is deposited on the cathode is scraped off continuously and accumulates in canvas trays at the bottom of the cell. The deposit can be made more compact and the process hastened if an addition agent (q.v.), such as glue, is present and a larger current used in conjunction with a rotating cathode. Any copper in the unrefined silver goes into solution as copper nitrate, and its deposition is prevented by adjusting the current and ensuring the presence of free nitric acid.

Moebius process. A process for the refining of silver—and incidentally for the recovery of any gold present. Dore silver, scrap silver and residues containing gold can be treated. The silver to be refined is cast into plates which are surrounded by cotton bags and made the anodes in the moebius cell (q.v.). The electrolyte is a dilute solution of silver nitrate containing free nitric acid. The cathodes are made of pure silver. The gold accumulates in the slimes of the anode bags.

Moffat process. A direct process for the production of steel, in which the iron ore was mixed with a reducing material and heated in a rotary kiln.

Mohs hardness scale. The degree of hardness of a mineral has long been determined by a method of scratching, and has been used for distinguishing one mineral from another. Mohs scale, devised in 1820, is still used as a scale of hardness for minerals. It is quite arbitrary and is merely comparative. It provides no absolute measure of degree of hardness. Measurements, under the microscope, of the depth of scratch produced by a diamond point under a certain load have been made, and the results have tended to show how arbitrary is the scale. It is found, for example, that the gap between 10 and 9 on the scale is greater than the gap between 9 and 1.

If a steel file will scratch felspar but will not scratch quartz, then its hardness is assessed as 6 to 7 or sometimes as 6½. The materials used as standards on the Mohs scale are placed in column II of Table 37. Other substances that could be used as substandards are placed separately in column III.

TABLE 37. *Mohs Hardness Scale*

Mohs Hardness Number		Mineral Used as "Standard"	Other Mineral Standards that Could be Employed
10		Diamond	
	9½		
9		Carborundum	Carborundum
8		Topaz	Sapphire
7		Quartz	
	6½		Garnet
			Steel file
6		Felspar	(Orthoclase)
	5½		Pumice. Common glass
5		Apatite	Opal
4		Fluorite	
3		Calcite	
2		Gypsum	(Rocksalt)
1		Talc	

Moirée Metallique. A beautiful crystalline pattern obtained when the surface of a tin ingot or a heavily tinned plate is treated with a mixture of nitric and sulphuric acids. It is often used as a metallic ornamentation and is then coated with a coloured varnish.

Moissanite. A silicide of carbon similar in composition to artificially prepared carborundum (q.v.) that is found in meteoric irons.

Molar solution. A term used in analytical and electrolytical work to denote a solution containing 1 mole of the solute in 1 litre of solution.

Mole (symbol:mol). Mass of substance containing the same number of elementary units as atoms in 0.012 kg of nuclide carbon 12. Elementary unit to be specified, e.g. atom, molecule, ion, electron, photon, etc.

Molecular conductivity. A term used to denote the conductivity of a solution containing 1 mole of electrolyte when placed between electrodes of indefinite size 1 cm apart. Strong electrolytes (e.g. most salts, mineral acids and strong bases) have large molecular conductivities which approach a constant limiting value when the dilution becomes very great. With weak electrolytes, the molecular conductivities are small, and show no signs of reaching a constant value as the dilution is increased.

Molecular polarisation. When a molecule forms from two separate constituent atoms, there will be a redistribution of electrons between them, and, if they are bound by ionic forces, a distinct separation of electric charges between the positive and negative ions. However they are bound, the electronic distribution is likely to create a dipole moment, except when two like atoms are involved.

If an electric field is now applied, the polarisation may be modified. Firstly, the relative position of the ions or atoms may change slightly — atomic polarisation; or there may be rotation of the molecule to align its axis with the field of direction — orientation polarisation.

Molecular susceptibility. A term used in reference to the magnetic susceptibility (q.v.) of compound substances. Most ionic and molecular compounds are diamagnetic and their susceptibilities are independent of temperature. From measurements on solid salts and on solutions it appears that there is, for each ion, an approximately constant value for the susceptibility (χ_A). It also appears that χ_A increases with the number of electrons.

In the case of water $\chi_M = 0.7218 \times 10^{-6}$ at 20 °C and increases by approximately 0.013% per degree celsius.

Molecular weight. The sum of the atomic weights of the constituents. The weight of one molecule compared to the weight of the oxygen molecule (=32). The least weight (expressed on the oxygen scale) of the substance which can have separate identifiable existence.

Mollerising. A process for impregnating the surface layers of steel with aluminium for the purpose of protecting the steel from corrosion. The steel parts are heated to 870°-1,095 °C in a salt bath, containing principally barium chloride. When the required temperature has been attained the parts are withdrawn through pure molten aluminium which floats on top of the fused salt. In this way a "hot-dipped" aluminium coating, bonded to the steel by an iron-aluminium alloy, is produced.

Molybdena. Synonymous with molybdine (q.v.).

Molybdenite. A naturally occurring ore of molybdenum MoS_2. It occurs in Cumberland (UK) and in Norway, Sweden, Bohemia (Czechoslovakia), Saxony (Germany) and in the Urals, as well as in several places in the USA, e.g. Connecticut, California, etc. It is commonly found in foliated masses or in scales, but sometimes occurs in tabular hexagonal prisms. It is lead-grey in colour and has a metallic lustre. It resembles graphite somewhat and leaves a grey trace on paper. It is oxidised readily by nitric acid and by strongly heating in air.

Molybdenum. A silvery-white metal that is not found native. It can be obtained from several ores by simple reduction. The most important ores are obtained from Climax (Colorado), Questa (New Mexico), Knaben (Norway) and Azigour (Morocco). During roasting in air a volatile oxide is formed, and this, on condensation, yields a crystalline oxide that is directly reducible by hydrogen. This reduced metal powder is pressed into bars, sintered and swaged.

The pure metal is malleable and becomes irridescent when heated to 600 °C in oxidising atmospheres. The oxide formed is volatile at high temperatures. The most important use of molybdenum is in steelmaking, where small percentages of the element are added to increase the strength and hardness and to improve the notched-bar impact performance. Molybdenum also causes a marked improvement in strength at elevated temperatures. Small percentages of molybdenum have beneficial effects on the properties of other alloy steels, improving creep strength as well as abrasion and corrosion resistance. Alloy steels containing molybdenum are now extensively used in aircraft, marine and automobile construction, particularly where large masses of metal are involved.

When present in steels, molybdenum is partly in solid solution in ferrite, and partly combined with carbon as complex carbides. It influences composition, mode of formation and distribution of the carbide, and the principal effects of molybdenum are to cause a marked lowering of the critical range on cooling and a sluggishness of the thermal changes, to raise grain-coarsening temperature of austenite, to increase depth of hardening. It prevents intercrystalline corrosion in stainless steels and reduces brittleness of the case of nitrided steels.

Molybdenum is used as an alloying element in manganese, molybdenum; molybdenum, nickel; molybdenum chromium, and nickel, chromium, molybdenum steels. The alloy containing 60% Mo, 10% Pt, 10% Ni and 20% cupro-nickel has been used in place of the very hard osmiridium for tipping pen-nibs.

Pure molybdenum is used for the making of grids in certain electronic equipment. When alloyed with silver there is obtained an ideal contact material for electrical switches, and when alloyed with tungsten (25% Mo, 75% W) the alloy may be used with pure tungsten as a thermocouple that has a useful range (up to 3,000 °C in inert atmospheres.

In electrodeposition processes a small quantity of molybdenum acts as a brightener for zinc plating. (*See* Table 2.)

Molybdenum carbides. Molybdenum combines directly with carbon to form two carbides. The usual methods of preparation are: (a) MoC, heat the metal with carbon in the presence of aluminium; and (b) Mo_2C, calcium carbide is heated with molybdenum dioxide. The Mo_2C, which has hexagonal structure and is non-magnetic, is known as theta carbide. It can dissolve up to at least 25% Fe by weight. There are in addition two substituted carbides. That known as kappa carbide is given the formula $(Fe,Mo)_{23}C_6$. It is formed by direct transformation of austenite or by tempering martensite within certain limits. The omega carbide $(Fe,Mo)_6C$ is probably formed from the melt by reaction of the metal with theta carbide under conditions of slow cooling. It is soluble in austenite.

Molybdenum steel. Molybdenum, when added to steel, produces much the same hardening effect as tungsten (see also **Tungsten steel**), and the alloy is used for very similar purposes, e.g. tools (0.3-1.35% Mo), permanent magnets (23% Mo) or about 1.5% Mo with or without additions of cobalt and chromium, etc. The amounts of molybdenum that are used to produce comparable effects are of the order of one-quarter the amounts of tungsten.

Molybdine. The trioxide of molybenum MoO_3 which occurs in molybdic ochre and as an oxidation product of molybdenite MoS_2. Also known as molybdite.

Molyte. A fused mixture of silicon, calcium and molybdenum oxides used in steelmaking.

Monazite. A mineral consisting mainly of the phosphates of the cerium (rare earth) metals together with variable amounts (up to about 10%) of thorium silicate. It occurs in granites and pegmatites, but is obtained only from sands where a certain amount of natural concentration has taken place. It is a source of thorium and of the cerium that is used in steelmaking and in the manufacture of "flints". The crystals belong to the monoclinic system. Mohs hardness 5.5, sp. gr. 5.1-5.2.

The deposits at present worked are the monazite-bearing sands of North Carolina, Travancore (India), Sri Lanka and Brazil. The monazite is separated from the other heavy minerals (zircon, ilmenite) by electromagnetic means.

Monazite is also obtained as a by-product in the final dressing of alluvial tin found in Nigeria and Malaysia.

Mond gas. A type of semi-water gas obtained by passing air which has been saturated with steam and then preheated to about 250 °C over strongly heated coke or coal slack. The large amount of steam keeps the temperature low (about 650 °C), and this allows of the recovery, as ammonia, of a larger proportion of the nitrogen of the coal than would be possible by heating the coal in retorts in the manner used in making coal-gas. The gas contains about 13% CO, 25% H_2 and 2.5% CH_4, the balance being non-combustible gases such as carbon dioxide and some nitrogen.

Mond process. A process for the extraction and refining of nickel. The product of the Orford process (q.v.) consists of about 72% Ni and 1 or 2% Cu and some combined sulphur. This is calcined to reduce still further the sulphur content, and the nickel oxide is then reduced by water gas at about 350 °C to the metallic state. This nickel is then treated with producer gas (rich in CO) at a temperature not exceeding 50 °C so that volatile nickel carbonyl is formed. This latter is then led into towers heated to 180 °C where $Ni(CO)_4$ is decomposed into metallic nickel. The carbon monoxide is recovered and used again.

When originally established the Mond process was designed to treat a nickel-copper matte from Canada, and consisted essentially of the following stages: (a) roasting to eliminate sulphur; (b) leaching with sulphuric acid to remove copper as copper sulphate; (c) reduction of the residual nickel oxide at 300°-400 °C to finely divided metal by means of hydrogen in water gas; (d) volatilisation of the nickel as carbonyl gas by reaction at 50°-60 °C with the carbon monoxide; and (e) the decomposition in towers, heated internally to about 180 °C, of the nickel carbonyl, the nickel being deposited on the surface of nickel particles, resulting in gradual growth into the familiar Mond pellets. As a result of changes in the processing in Canada, the stages (a) and (b) above have disappeared from the flow sheet of the Mond plant at Clydach, South Wales. (See also **Nickel carbonyl.**)

Monel. The name given to a "natural" nickel alloy containing mainly nickel and copper. The crushed nickel ores of Sudbury are concentrated (from flotation) and roasted in a reverberatory furnace to remove sulphur. The resulting matter is then blown in a basic-lined Bessemer-type converter. Most of the iron is removed, and the product is mainly nickel and copper sulphide, which is converted into the copper-nickel alloy known as Monel. It is one of the very few alloys smelted direct from mixed ores.

It is chiefly used for corrosion-resisting components such as condenser tubing, turbine blades, pump bodies, chemical plant and kitchen, laundry and marine fittings generally. It has good mechanical and corrosion-resisting properties up to about 400 °C.

Monell process. A process for making steel that may be regarded as a modification of the basic open-hearth process. In this process limestone and ore are charged into a basic open-hearth furnace and heated till the batch becomes pasty, when molten pig iron is added. The silicon, manganese and phosphorus become oxidised and pass into the slag. On further heating, the carbon oxidises to carbon dioxide and this escaping through the slag causes the latter to become foamy.

Monkey cooler. An annular casting which is inserted through the wall of a blast furnace to form the slag hole. It is usually conical and made of copper. Water circulates through the casting to minimise the fluxing attack of the slag on the brickwork and on the refractory materials.

Monochromatic. A term used in spectrographic work in reference to light that has a single wavelength.

Monoclinic. A crystal form having three unequal axes—one vertical (a), one at right angles to the vertical axis (b), and the third (c) making an oblique angle with the plane containing the other two. The crystallographic elements therefore are $\alpha = \gamma = 90°$; $\beta \neq 90°$; $a:b:c: = x:1:y$.

Monofrax. A group of fusion cast refractories. They contain a very small proportion of glassy matrix with crystalline constituents such as β-alumina or corundum and in some special cases iron and chromium spinels.

Monolithic lining. A lining formed by ramming or sintering into position a crushed refractory material. Such linings are free from joints.

Monomer. A basic unit substance, which, when an unsaturated bond is broken to form available bonds by catalytic or activator influence, may form the basic mer of a polymer. (See **Mer.**)

Monotectic. An alloy which solidifies by a liquid L_1 forming a mixture of a second liquid L_2 and a solid α: $L_1 \rightarrow L_2 + \alpha$.

Monotron test. An indentation hardness test. A diamond ball penetrator, 0.75 mm in diameter, is employed and the hardness reading corresponds to the load in kilograms required to penetrate the surface to the standard depth of 0.046 mm. When the depth-gauge shows this depth of penetration the hardness-reading is given directly on the pressure-gauge.

Monotropic. A substance which exists in one stable form, all the other forms in which it may exist being unstable in all circumstances, is said to be monotropic.

Monovalent. The property of an element, one atom of which will combine with or replace one atom of hydrogen. Elements like sodium whose atoms have a single electron in the outer shell that can be shared with another atom—or elements like chlorine whose atoms can accept an electron—to complete the required number of electrons in the outer shell and form a stable compound are monovalent.

Monox. The name by which the product of the reduction of silica is known commercially. It is a brown powder that was regarded as a mixture of silica and silicon, but it appears probable from the composition that it approximates to a monoxide of silicon.

Monthier's blue. A deep blue powder having an enamel blue reflex. It is iron III (ferric) ammonium iron II (ferro) cyanide

$$Fe(NH_4).Fe(CN)_6.H_2O.$$

It is water-soluble but is slowly decomposed by alkalis.

Montmorillonite. $Na_2O.2MgO.5Al_2O_3.4SiO_2\,xH_2O$. The major clay mineral in bentonite and fuller's earth. It contains alkaline oxides, those of sodium and magnesium, and the aluminate sheets of the structure are sandwiched between two silica ring layers. The multilayer spacing is wider than in kaolinite, and so larger quantities of water can be accommodated. (See **Kaolinite.**)

Moonstone. A bluish-white or white opalescent variety of orthoclase that is used as a gemstone, the colours being interference colours. It is found mainly in Sri Lanka.

Moorewood machine. A machine generally used for the production of hot-dipped tinplate and terneplate in the heavier coating grades.

Mop. A polishing wheel composed of layers of fabric supported on a central boss.

Morphotropy. A term used in reference to the general study of the variation of crystal form with a specific change of chemical composition, of which enantiomorphism and isomorphism are but particular cases. It is also sometimes used in reference to the resemblances between compounds that are not regarded as isomorphous because they do not have analogous compositions or because the degree of resemblance is small.

An example of a morphotropic series is ammonium iodide and the alkyl derivatives.

Moseley numbers. The number of protons in the nucleus of an atom or the number of planetary electrons that circulate round the nucleus. It is in fact the integer that determines the position of an element in the periodic classification. Solid elements when bombarded by electrons emit characteristic radiation that can be resolved into a spectrum, as Moseley showed by reflection from a crystal of potassium iron II (ferro) cyanide. There are four kinds of rays, designated K, L, M and N, which differ in wavelength. The square roots of the frequencies of corresponding lines in the spectra of successive elements taken in the order of their position in the periodic table, when plotted against the Moseley number, give practically a straight line. The atomic or Moseley number is now generally denoted by Z.

Moss agate. A form of fairly pure quartz which, because of its colour and opacity, is sometimes mistaken for jade. It is harder than jade, but has a lower specific gravity.

Mossotite. A form of calcium carbonate that may resemble jade in appearance. It is readily distinguished from the semi-precious stone by its softness, its density and its reaction with mineral acids.

Mossy zinc. The name by which the granulated zinc, obtained by pouring the molten metal into cold water, is known.

Mother of coal. A type of humic (bituminous) coal that is dull black in appearance and has a fibrous texture. Also called fusain and mineral coal.

Mottled cast iron. If the carbon contained in iron exists as iron carbide, which is hard and brittle, the iron itself possesses these same properties. It is difficult to work and breaks with a white or silvery fracture. It is known as white cast iron. If the carbon occurs in the free form as flakes of graphite, then the iron is softer and machinable. Such iron shows a dark fracture and is known as grey cast iron. The chief factors that determine whether the product shall be white or grey iron are (a) the silicon content, (b) the rate of cooling and (c) the manganese content. Mottled cast iron is intermediate between white and grey iron, and the fracture shows grey and white zones. Moderately slow cooling of iron containing small amounts of silicon produces mottled cast iron. It is difficult to work, but has good casting properties, and its lack of good mechanical properties in tension is partly compensated in some applications by its good compression strength.

Mottled iron is usually regarded as a medium silicon pig iron in which about half the total carbon is present as graphite, which exists in the form of star-shaped masses. These latter give rise to the characteristic mottled fractures. The material is hard, brittle and practically unmachinable.

Mottling. A term used to describe the appearance on metal surfaces, radiographic films, etc., which is due to a random occurrence of dark and light areas.

Mottramite. One of the more important vanadium minerals found in Cheshire (UK). It is a basic vanadate of lead and copper $(Pb,Cu)_3(VO_4)_2.2(Pb,Cu)(OH)_2$. The Keuper sandstone on which the mottramite is deposited as a film is treated with strong hydrochloric acid and the solution is evaporated in presence of excess of ammonium chloride when ammonium metavanadate is formed. When this is roasted the pentoxide V_2O_5 is obtained. When this is dissolved in fused ferrosilicon, or in a molten mixture of iron fluoride and calcium carbide, and electrolysed, ferrovanadium is obtained. About 90% of the vanadium extracted is used in this form.

Mould. A hollow container, which is made in a variety of metals and may or may not be water-cooled, or a cavity made in suitable moulding materials (e.g. sand) into which molten metal is poured to produce a casting of a desired shape.

Mould board. The smooth solid board used in mould-making on which the bottom part of the pattern is laid—joint side downwards. Also known as match plate (q.v.) and joint board.

Mould clamp. A device used to hold together the cope and the drag portions of a split pattern.

Mould cope. The name given to the upper part of a foundry mould.

Mould core. A distinct and separable part of a mould, the function of which is to enable a cavity of desired shape to be produced within a casting. In certain cases cores may be used to strengthen or improve the interior or exterior surfaces of a mould. Usually the core is made from the same kinds of sands that are used for making moulds, the bonding materials being either clay, organic substances such as cereal binders, or oil. As the core will be surrounded almost entirely by molten metal, it is usual to vent it to permit escape of gases.

Cores are most frequently used after suitable drying in core-drying ovens, provision being made for a circulation of hot air round the core.

Mould-core assembly. A method of moulding that is very suitable for intricate designs, since it gives good accuracy in dimensions. In this method every part of the mould is formed by a core.

Mould drag. In the majority of cases sand moulds are produced in two pieces, the pattern being split along a suitable plane. The drag is the name given to the lower part of the foundry mould. When the mould is in three parts, the middle one is called the cheek.

Mould dressings. These are frequently powders that are dusted on, or rubbed in by hand, on to the faces of green sand foundry moulds. The substances employed include flour, soapstone and talc for most of the non-ferrous metal castings and graphite (black lead) for cast iron. Mould paints (q.v.) are frequently used for steel castings.

Mould drying. Though it is the usual practice to dry or bake foundry cores, it is not the practice to treat foundry moulds in the same way except in the case of heavy castings. The treatment of moulds may be as follows: (a) air-dried moulds are made with the addition of a little dextrin, which confers upon the sand a property of air-drying at the surfaces; (b) skin-drying, i.e. drying to a predetermined depth of sand, is used for moulds made in the foundry floor, and for other medium-sized moulds. The heater is fired with gas, oil or coke and a small fan circulates the hot air and carries away the moisture evaporated. Infra-red lamps have also been used; (c) stove-drying is practised in foundries where the moulds cannot be poured within a short time of surface drying. Since the surface picks up moisture on cooling, such moulds are dried or stored in a stove, the temperature of which is slowly raised to about 200 °C. Since better results are obtained by pouring into hot rather than cold moulds, the latter are often stored overnight in the stove.

Moulder. A term used in reference to the product of rolling the first cogging stage when the slab or bar is sheared into suitable weights. The pieces are reheated for further rolling.

Moulder's rule. A steel rule marked off in units larger than the standard (marked) units in order to make allowance for the contraction of castings during solidification.

Moulding boxes. Rolled steel, open-top boxes made in sizes suitable to take the moulding sand and the pattern for making the mould. The size of the box exerts an important influence on the accuracy of the finished casting, since it determines the amount of sand that surrounds the pattern—and eventually the molten metal. It also frequently determines the position of the mould joint. Also called flasks.

Moulding machines. When large quantities of castings of the same kind are required, moulding machines are employed, since their use leads to uniformity in size, quality and finish of the casting. The patterns are mounted on pattern plates and the runners and risers are attached. Small patterns used in small snap flask-work or in squeeze machines may be double-sided, but large patterns are always mounted singly on the plate. The drags, but not the copes, are preferably made on roll-over machines, as this facilitates the removal of the pattern. The types of power-operated ramming moulding machines are: (a) squeeze machines (q.v.), which are useful for shallow patterns; (b) jolt machines, which produce a hard sand face in contact with the pattern, (c) jolt-squeeze machines; and (d) sand slingers (q.v.).

Moulding sands. These consist broadly of silica sand, clay and frequently a binding material. The kinds of sand available vary with the location so that analytical control is necessary. The kind of sand that can be used in a foundry depends on the metal or alloy that is to be cast in it. If the sand is not natural bonded (i.e. does not contain sufficient clay), binding material such as fireclay, fuller's earth or bentonite can be added. Organic binders such as molasses, dextrin, gelatinous starch, oils, pitch and resins—natural or synthetic—can be used for core-moulding sands. The green strength of moulding sands depends on grain size and the shape and distribution of the grains, the type and amount of clay or other binder, the moisture content and the milling time. The dry strength depends very largely on the baking temperature.

Mould paints. These are variously described as facings, washings, liquid dressings or liquid coatings and consist of finely ground refractory materials mixed with organic or inorganic binders and then diluted (usually with water, but volatile organic liquids have been used, the latter being subsequently burnt off). The dilution is made so that the paint can be conveniently applied to the mould face by brushing on, swabbing or spraying. The ingredients of some commonly used mould paints are listed in Table 38.

TABLE 38. *Commonly Used Mould Paints*

Metal	Mould Paint Constituents
Carbon steel	Silica flour; bentonite; oil
Manganese steel	Magnesite; bentonite; dextrin
Cast iron	Black lead; bentonite; dextrin
	Black lead; molasses
(large castings)	Silica flour; bentonite; dextrin and oil
Aluminium	Talc; whitening; molasses
Copper alloys	Talc; black lead; molasses
Phosphor bronze	China clay; black lead; molasses
Bronze (tin)	Sodium silicate

Mould reaction. When an alloy containing a readily oxidisable element is poured into a sand mould, a reaction between this element and the mould occurs which is referred to as mould reaction. The oxidisable element reacts with the steam that is produced when the hot material comes in contact with the mould material. A film of oxide is deposited on the metal surface, causing discoloration on the surface of the casting, while the liberated hydrogen dissolves in the molten material. On solidification the hydrogen is expelled, giving rise to gas porosity in the outer layers. The extent of these effects is dependent on the percentage of oxidisable element (e.g. magnesium, phosphorus) that is present, and on the time of solidification of the casting. They are therefore more severe in large castings than in small ones. In the case of magnesium-aluminium alloys it is found that prolonged drying of the mould even at 230°C does not prevent the reaction occurring, but the addition of about 0.004% Be to the molten metal, or about 2% of boric acid or of ammonium bifluoride to the moulding sand does reduce the amount of reaction. The use of both types of inhibitor appears to be desirable in the case of large castings.

Sulphur alone (up to about 5%) in moulding sands also inhibits mould reaction in the case of magnesium-aluminium alloys. In the case of stainless steel a little sulphur or tellurium or selenium added to the molten metal almost entirely prevents mould reaction and the accompanying porosity. For tin bronzes containing phosphorus, vanadium (0.1%) or chromium (0.5%) or aluminium (0.1%) act as inhibitors.

Mould reaction, when controlled in amount, has been found to exert a beneficial effect on the pressure tightness of lead gunmetal castings (valves).

Mould resistance. A term used to describe the effect of mould material in offering restraint to the contraction of castings as they cool in the mould, thus increasing the internal stresses. By using the correct type of facing sand (having a grain size distribution that permits close packing of the grains) at different parts of the mould, the residual stresses can be controlled.

Mouldrite. A proprietary synthetic resin used in foundry work as a core binder. There are two types: UF, which is urea formaldehyde (methanal); and PF, which is phenol formaldehyde (methanal).

Mould washes. Materials used as coatings on the interior surfaces of sand moulds in order to prevent surface defects on castings. Such washes include thin clay slurries, slaked lime, suspensions of aluminium, tar, etc., in water or alcohol or other organic liquid.

Muff. A term sometimes applied to the crude magnesium deposit produced by the Pidgeon ferrosilicon process.

Muffle furnace. A type of furnace in which the work chamber is surrounded by the heating chamber, but there is no communication between the two chambers. The work is thus protected from the fuel or the products of combustion, and the atmosphere in the chamber can be controlled. Muffle furnaces are used where heat-treatment is desired. For melting, a crucible furnace, which is another of this type of closed-vessel furnace, is used.

Muller. The name by which a type of roller-mixer is known. The rollers are raised slightly above the pan (16-30 mm) as required. A type of mill used for breaking up the clay in the preparation of moulding sands. The product of the mill is usually a flat cake which is passed through disintegrators. These latter contain rapidly revolving plates, teeth or pins which reduce the cake to a fine sand suitable for moulding or facing sand.

Mulling machine. Synonymous with Muller or muller-mixer. It commonly consists of a wide, shallow pan. Rollers, mounted on beams extending radially from a driven vertical shaft rising above the centre of the pan. The rollers rotate as they travel in continuous circles.

Mullite. A mineral of the composition $3Al_2O_3.2SiO_2$ which crystallises in the rhombic system. It is stable up to about 1,810°C, at which temperature it melts and corundum Al_2O_3 separates. It closely resembles sillimanite. It is prepared artificially in the USA and used as a refractory, but it is important that the proportions Al_2O_3/SiO_2 shall not depart appreciably from 72.5/27.5.

It is reported to be a good general-purpose refractory for melting copper-base alloys.

It is found in Mull (Scotland), whence the name is derived. Mohs hardness 6-8, sp. gr. 3.16.

Multiaxial ferroelectric. Ferroelectric materials are those polarisable materials in which the direction of their spontaneous polarisation can be reversed by the application of an electric field. In multiaxial ferroelectrics, this polarisation can take place along more than one crystal direction which are equivalent in the non-polar state.

Multi-hearth furnace. A type of roasting furnace used in the thermal method of ore preparation. The hearths are arranged vertically one above the other. The ore is fed at the top and revolving rakes or other devices are used to move it from one hearth to the next lower one. It is usual to provide for the regulation of the air supply at each hearth separately and for the regulation of heat supply.

Multiple process. The short title of the process for the refining of copper electrolytically using multiple electrodes. All the anodes are connected to another copper bar on the other side. The tanks are arranged in series. All the anodes of one cell and all the cathodes can be lifted out of the tank separately. The former are removed before they are spent and remelted, while the latter are sent to the refining furnace.

Muscovite. A rock-forming mineral of the mica type, at one time used in windows in Russia, whence its name. It is a potash mica having the formula $2H_2O.K_2O.3Al_2O_3.6SiO_2$. It crystallises in the monoclinic system. Mohs hardness 2-2.5, sp. gr. 2.8-2.9. It is widely distributed, but is found in large sheets of commercial value in India, the USA, East Africa and Brazil.

Mushet steel. A steel developed by R.F. Mushet in 1868 that was called self-hardening. It is a type in which the rate of decomposition of the austenitic solid solution is so slow that transformation to soft pearlitic products does not occur even when cooled in air. Mushet's steel is a high carbon-manganese-tungsten steel. It was used for machine tools. The approximate composition is C 1.8-2.15%, Mn 1.5-2.5%, W 5.5-9.2%. Modern air-hardening steels contain nickel or chromium in addition to manganese and tungsten, which they partly replace.

Music wire. A cold-drawn steel wire mainly used by spring manufacturers. It is of high carbon content (0.80-0.95%) and is usually made by the acid process. The high tensile properties are obtained by heavy cold reduction during repeated passes through drawing dies. During reduction the wire is toughened by a patenting process which involves passing through a furnace at a temperature of 980 °C and then quenching into molten lead, or air in order to produce a sorbitic structure.

Music wire gauge. A system of indicating the diameter of wire by means of a number, the number increasing as the diameter indicated increases, up to 44, which represents a diameter of 4.32 mm.

N

Nail-head spar. A variety of calcite $CaCO_3$.

Nailing. A slow-annealing process used in the manufacture of clay pots and crucibles. The articles are heated to just visible redness.

Naphtha. A product of the fractional distillation of petroleum. It is obtained from the portion distilling between $130\,°C$ and $170\,°C$ and consists chiefly of dimethyl benzene or xylene C_8H_{10} and *iso*propylbenzene or cumene C_9H_{12}. It is a useful liquid in which metals such as sodium and potassium can be stored.

Naples yellow. Lead antimoniate. It is a basic salt used in oil painting and is prepared by heating a mixture of 1 part of tartar emetic, 2 parts of lead nitrate and 4 parts of common salt at the temperature of the fusion point of the latter substance.

Nascent hydrogen. An atom being the smallest part of an element that can be produced in a chemical reaction, it is assumed that the hydrogen produced, for example, when an acid and a metal interact, is in the atomic state. The atoms either combine in pairs to form molecules of hydrogen or take part in any possible reaction. In the latter case, the hydrogen which is in the "just born" or nascent state is considered to be more reactive than the molecule hydrogen.

Natalite. A fuel that consists of 95% alcohol mixed with varying amounts of ether. The two liquids are completely miscible, and the mixture is used mainly as a motor fuel.

National coating. The name by which a process for the protection of iron and steel piping is known. The coating, which it is claimed is particularly suitable for underground steel, etc., consists of a mineral rubber compound (National Dip) reinforced with fabric.

Native alumina. The rhombohedral crystals of corundum Al_2O_3. The substance is nearly as hard as diamond. Emery is an impure form, while hydrated forms of alumina occur: diaspone $Al_2O_3.H_2O$, bauxite $Al_2O_3.2H_2O$ and hydrargillite $Al_2O_3.3H_2O$

Native amalgam. A naturally occurring mercury-silver alloy containing about 35% Ag. It occurs with cinnabar in certain deposits at Almaden (Spain) and Idria (Italy). It crystallises in the regular system.

Native antimony. The black mineral stibnite often contains native metal, sometimes in association with iron, silver and arsenic.

Native arsenic. The pure element occurs in veins with lead or silver ores in Saxony (Germany), Bohemia (Czechoslovakia),Siberia, Borneo and the USA. It usually occurs in kidney-shaped masses, but is occasionally found as distinct crystals.

Native bismuth. The metal bismuth occurs sparingly and usually in the native form together with arsenic and tellurium in Australia, Bolivia and Saxony. The bismuth metal is obtained from the native material by liquation. It is heated in sloping iron tubes, when the low-melting-point bismuth flows away.

Native copper. Metallic copper, sometimes containing a little silver and bismuth, occurs as a metasomatic deposit filling cracks and forming the cement of sandstone and conglomerate. Such deposits have been located in Keweenar (Lake Superior) and in Chile, Queensland and Zimbabwe. Native copper is also found in the upper working of copper mines, particularly in Australia. It crystallises in the cubic system.

Native gold. Though usually found in the native state, gold never occurs perfectly pure, being always alloyed with more or less silver, sometimes also with copper or iron or traces of the platinum-group metals. Native gold is generally found in quartz veins or reefs which intersect metamorphic rocks. The metal is frequently found crystalline—in octahedra or tetrahedra—but the crystals are sometimes acicular, or even in spongy form. It may be found in masses called nuggets, in thin laminae or flattened grains and in small masses in sand or gravel. The sands, gravels and clays that result from the disintegration of gold-bearing rocks form an important source of the metal. The amounts vary from 2 parts in 10^4 of deposit in a rich source to 1 part in 15×10^6 for a poor alluvial source.

Native iron. Iron seldom occurs in the metallic state in nature, though small amounts have been found in the basalt of the Giant's Causeway (Ireland) and in the older lavas of the Auvergne Massif (France). Native metal of extra-terrestrial origin has been found in large quantities—up to 20 000 kg (20 Mg). This meteoric iron always contains some nickel and frequently cobalt as well.

Native mercury. Mercury occurs in the native state, though the amount is small compared with its production from its ores. It is usually found in globules disseminated through the sulphide ore found in Hungary, California, Mexico and Peru. It also occurs as silver amalgam and as gold amalgam and is then referred to as native amalgam.

Native platinum. Except in the mineral sperrylite $PtAs_2$, platinum is always found native, though very seldom pure. It occurs with other members of the platinum group (palladium, rhodium, osmium, iridium, ruthenium) as well as with iron, copper and titanium. The purest native platinum containing some palladium comes from Brazil. Other sources are Colombia, the Urals, Borneo, Peru, California, India and Australia. The metal is also found in lead and silver ores. In the Urals platinum occurs along with chrome iron ore and in Brazil it is found with gold in syenite.

Native silver. Native silver, often in quite considerable quantity, is found together with gold, copper or mercury in the zone of weathering of silver sulphide lodes. It occurs in strings or veins in eruptive and in sedimentary rocks in Norway, Peru and the USA.

Native tellurium. Nearly pure tellurium, associated with a little gold and iron, occurs in small amounts in Transylvania (Romania). It is also found in several states in the USA, in Bolivia, Brazil and Western Australia.

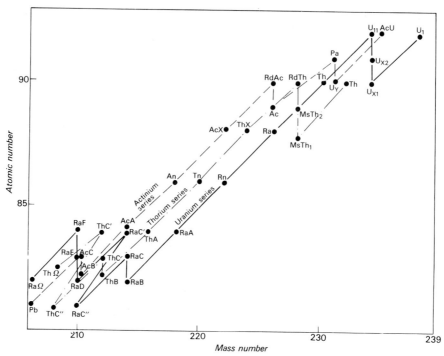

Fig. 89 *The natural radioactive series.*

Natrolite. A mineral of the zeolite group. It is hydrated sodium aluminium silicate $Na_2O.Al_2O_3.3SiO_2.2H_2O$. It is usually white, but sometimes yellowish-red in colour, and has a vitreous lustre. It crystallises in the rhombic system, but on account of the frequent occurrence of slender acicular crystals it is sometimes called Needlestone or needle zeolite. It readily fuses and is decomposed by hydrochloric acid. Mohs hardness 5.5, sp. gr. 2.2. It is used in water softening.

Natural ageing. The spontaneous precipitation of submicroscopic particles from a supersaturated solid solution at room temperature. The effect of rapid cooling in cases where the solubility decreases with fall in temperature is two-fold. Firstly it reduces the amount of precipitate that can occur, thus producing a supersaturated solution. The second effect is to refine the size of any precipitate that forms. When such a supersaturated phase is retained down to room temperature by rapid cooling some precipitation does in fact occur at the lower temperature, though on a submicroscopic scale, but the structure is still metastable. The effect of the precipitation is to produce considerable hardening of the alloy.

Also called age-hardening precipitation.

Natural bonded sands. The term used in reference to moulding sands which when quarried always contain a constituent (e.g. clay) which will serve to bind the sand in a coherent mass when moistened and pressed into shape in a moulding box.

Natural gas. The name by which natural flammable gas found in porous sub-surface formations is known. Such gas is found in the USA, South America, Canada, Romania, Poland, the USSR, India, Japan, Africa and the continental shelf of Europe. The gas usually consists of a mixture of gases of the alkane series. In a few cases the proportion of nitrogen or carbon dioxide is high enough to prevent the gas from igniting. Apart from their use for heating purposes, the gases are used as a source of petrol, carbon black, etc.

Natural radioactivity. The spontaneous disintegration of the unstable atomic nuclei of certain naturally occurring substances, to give more stable nuclei, is termed natural radioactivity. This is usually accompanied by the emission of alpha particles (helium nuclei) and/or beta particles (electrons) and gamma rays. These radiations can pass through substances opaque to ordinary light; they can affect a photographic plate; they ionise air or other gases through which they pass; they cause certain substances to fluoresce; and they generate heat. The process continues despite external influences.

Uranium is just one of about forty natural radioactive elements, practically all of which have atomic numbers lying between 18 and 92. They fall into three series: (a) the uranium-radium series; (b) the thorium series; and (c) the actinium series. In each of the series there are a number of distinct radioactive elements. Any one of these can be traced back to a parent element by transformations involving the emission of α or β particles.

It is not known if elements heavier than uranium have existed. If indeed they did exist they must have been characterised by comparatively short lives, and that would imply that uranium is an intermediate in a longer series which is now extinct. Among the light elements nine are known that are naturally radioactive. They are potassium, vanadium, rubidium indium, tellurium, lanthanum, cerium, neodymium and samarium, of atomic numbers 19, 23, 37, 49, 52, 57, 58, 60 and 62, respectively. It is possible that these represent the remnants of a more extensive series that is now extinct (*see* Fig. 89).

Natural rubber. *See* **Isoprene.**

Nature print. A print made specially to show the flow of metal during working. The surface of the metal must first be surfaceground (or moderately ground on emery). It is then deep-etched, using a suitable etchant. For steels, hot moderately strong HCl (1 in 1) is used; for light alloys a mixture of strong nitric and hydrofluoric acids is used; copper alloys are macro-etched in iron III (ferric) chloride solution made distinctly acid with strong HCl. The time of etching must be judged by the result. To obtain the print, suitable paper is pressed against the etched metal surface which has been treated with printer's ink, using a rubber roller.

Necking. The localised reduction in area of cross-section concentrated near the subsequent fracture that takes place when a ductile metal is tested in tension. Also called necking down.

Needled steel. The name by which certain boron-hardened steels are sometimes known. Carbon is the element which mainly determines the surface hardenability of steel, while other elements such as molybdenum, chromium or manganese affect the hardenability below the surface skin. Boron has a very marked effect on steel that has been previously completely deoxidised (or killed). It is used in amounts of 0.001-0.0015%, the effect so far as hardenability is concerned being several hundred times its own weight on any of the other elements commonly used. Hypoeutectoid steel containing 0.4-0.5% C, 1.5% Mn and 0.0002% B is known as needled steel.

Negative crystal. In the natural growth of crystals foreign matter may be accidentally caught up. When this results in cavities having the same shape as the surrounding crystals, then the cavities are referred to as negative crystals. They may be empty or contain liquid.

Negative electrode. *See* Cathode.

Negative hardening. A term sometimes used in reference to an annealing process applicable to certain types of steel. It is properly regarded as a "corrective" treatment for soft steel that has been made brittle by a previous treatment. Usually the steel will have been overheated or maintained at the treatment temperature for too long a time. The object of negative hardening is to improve the toughness and ductility without increasing the hardness. The method consists in raising the temperature of the steel to about 800 °C, soaking until the steel is uniform in temperature and then quenching in boiling water.

Negative replica. When only the surface of a polished metallic specimen is to be examined by the aid of the transmission electron microscope, it is necessary to first prepare a thin transparent replica of the surface. The negative replica is obtained by dipping the metal into a dilute solution of formvar or collodion in an organic solvent such as ether or dioxane. When dry, the film is carefully stripped off for examination. In some cases it may be necessary to reinforce this film before stripping, the reinforcement being subsequently dissolved away. Cellulose nitrate can be used as the backing or reinforcing layer. It is subsequently dissolved in amyl acetate. Figure 90(a) shows the surface of the original specimen, (b) indicates the formvar and cellulose deposited on specimen, (c) shows the final negative replica, and (d) the density distribution in the electron image.

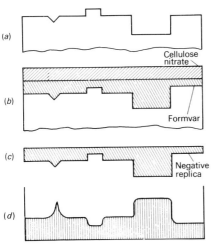

Fig. 90 *Diagram showing cross-sections of metal (a) and negative replica (c); (d) is a density distribution of the replica.*

Negative segregation. *See* Inverse segregation.

Neodymium. A rare metal which has a very faint yellow colour. It occurs in cerite and monazite in which it is present as oxide (neodymia). Crystal structure—close-packed hexagonal ($a = 3.650 \times 10^{-10}$ m, $c = 5.890 \times 10^{-10}$ m). Specific heat at 20 °C = 0.205 kJ kg^{-1}, m.p. 1,010 °C, resistivity 64.3 microhms cm, density 6.8 × 10^3 kg m^{-3}, at. vol. 20.5 cm^3 mole $^{-1}$, at. wt. 144.24, at. no. 60, symbol Nd. The oxide Nd_2O_3 is light blue in colour; it easily dissolves in hydrogen chloride and the metal is obtained by the electrolysis of the chloride.

Neogene. A term used in reference to the newer Tertiary period. The Neogene system is divided into the lower (or Miocene) and the upper (or Pliocene) divisions.

Neon. One of the rare, inert gases present in air to the extent of 16 parts per million. It is obtained from air by fractional distillation of liquid air or by heating monazite to 1,000 °C. This latter yields about 1 cm^3 g^{-1}. Atomic weight 20.18, at. no. 10, density 0.9004 g l^{-1}, b. pt. −245.9 °C. It is used in glow discharge tubes (red glow).

Neoprene $(C_4H_5Cl)_n$. A chlorinated synthetic rubber, with excep-

tional resistance to oxidation and to hydrocarbons such as petroleum products.

Neosyl. A trade name for an amorphous form of silica.

Nepheline. One of the uniaxial silicates containing aluminium, sodium and potassium $K_2O.3Na_2O.4Al_2O_3.9SiO_2$. Crystal system, hexagonal; Mohs hardness 5.5-6, sp. gr. 2.5-2.6. It readily dissolves in hydrochloric acid yielding gelatinous silica. Nepheline and alkali felspar—known as nepheline-syenite—is rather rare, but is notable for the fact that it is often rich in rare earths.

Nephelometric analysis. A method of quantitative analysis in which the concentration of solid matter in a liquid is determined by optical methods. In metallurgy the principal application is in the determination of the oxygen content of metals.

Nephrite. A compact form of actinolite $CaO.3(Mg.Fe)O.4SiO_2$. It is a form of jade.

Neptunium. A transuranic element that is not found in nature. It is produced in an atomic reactor when uranium 238, by radiative capture of moderately fast-moving neutrons, produces first an isotope of uranium, viz. ^{239}U: $^{238}U + n^1 \rightarrow ^{239}U + \gamma$. The ^{239}U has a half-life of 23 minutes. It decays by the emission of a negative beta particle yielding neptunium: $^{239}U \rightarrow \beta + ^{239}Np$. Neptunium has an atomic number 93; its half-life is 2.3 days. It exhibits negative beta activity, yielding plutonium: $^{239}Np \rightarrow \beta + ^{239}Pu$. (*See also* Plutonium and Table 2.)

Neritic zone. The name applied to that portion of the sea-bed which lies between low-water mark and the edge of the continental shelf, at about 200 m. The oscillation of the shelf over long periods, above and below sea-level, has taken place in many areas. In a number of instances the water-borne material has accumulated to provide useful ore deposits.

Nernst effect. When heat flows through a strip of metal, which is lying in a magnetic field perpendicular to its length, a difference of electrical potential is produced between opposite edges. The effect is taken as positive or negative according to the direction of potential drop. The positive effect is exhibited by bismuth and the negative effect by iron. (*See also* Hall effect.)

Nernst's distribution law. If a solute be dissolved in two liquids which are immiscible or only slightly miscible, then at any temperature the molar concentration of the substance dissolved in one liquid bears a fixed ratio to the molar concentration of the same substance dissolved in the other liquid. The actual ratio is known as the partition (or distribution) coefficient. The law is applicable to the chemical actions within and between slag and metal.

Nertalic process. An argon arc-welding process that uses a consumable electrode.

Nessler's solution. A reagent used as a sensitive test for the presence of ammonia, and as a colorimetric test for traces of ammonia in water, etc. The solution is made by adding a solution of potassium iodide to one of mercury II (mercuric) chloride. Enough of the former is added to just dissolve the red mercury II iodide that is first thrown down. The solution is then made alkaline with sodium hydroxide. The soluble complex iodide has the formula K_2HgI_4. The action of ammonia is to produce a yellowish-brown colour or a brown precipitate, according to the amount of ammonia. The precipitate has the composition NH_2Hg_2OI.

Nested electrode. A composite electrode consisting of two or more metal filler rods nested together, but insulated from each other, except at the end, where they are joined and gripped by the electrode holder. Through this latter the electrodes are connected to a source of current supply. They are used in welding.

Network polymer. *See* Cross-linking.

Network structure. Grain boundaries are the preferred locations for the appearance of new phases in alloys or other products. When the structure of an alloy is such that one constituent occurs partly or completely surrounding the crystals of another constituent, then the appearance of an etched section taken across the grains shows as a network structure. Also called cellular structure.

Neumann bands. Narrow, straight bands parallel to the crystallographic planes observed in the crystal structure of metals which have been deformed by impact. In some high-silicon irons, particularly at low temperatures, the bands are formed by less drastic deformation in ordinary cold-working processes. They are, in fact, narrow twin bands differently oriented within a grain of ferrite, and are most frequently observed in pure iron. Also called Neumann lamellae. They are an indication of mechanical twins.

Neutral axis. *See* **Neutral filament.**

Neutral filament. When a uniform rod or bar is bent by the application of couples applied at its ends, then the longitudinal filaments will take the form of arcs of circles, provided the radii of curvature of the filaments are large compared with a^2/b (where $2a$ = the width of the bar and $2b$ is the thickness of it). There will be an increase in the length of the filaments on the convex surface and a decrease in length on the concave surface. One of the filaments will be unchanged in length, and this is called the neutral filament. It passes through the centre of gravity of the strained section.

The trace of the neutral filament in the plane of bending is called the neutral axis.

Neutral flame. A term used in welding in reference to a gas flame which has neither oxidising nor reducing effect. It may also be used to describe the flame used in other operations where metals are heated by contact with flame. Hydrocarbons used as fuels must have the correct amount of air to just ensure the complete combustion of the gaseous fuel.

Neutral fluxes. A term sometimes applied to materials such as fluorspar, which render the slags formed during steelmaking more fusible and more fluid without appreciably altering the acidity or basicity. They have neither oxidising nor reducing properties.

Neutralisation. Acids owe their acidity to the presence of hydrogen ions in solution. Similarly bases owe their basic properties to the presence of hydroxyl ions. The H^+ and OH^- ions are those that possess the greatest mobility. When an acid is added to a base, the OH^- and H^+ ions combine to form nearly undissociated water. Thus the withdrawal of these ions causes a diminution in the conductivity of the solution. When the conductivity falls to a minimum value the acid and the base have been neutralised. The net result of neutralisation is therefore the union of H^+ and OH^- to form practically undissociated water. (*See also* **Indicators.**)

Neutral point. A term used in rolling-mills to denote the point at which the speed of the work is equal to the peripheral speed of the rolls. In analysis the term is sometimes used synonymously with "end point" of a titration.

Neutral refractory. A refractory material that is neither markedly acid nor basic in character. The most important of this class of refractory are graphite or plumbago and chromite. The latter is cheaper and very widely used. The naturally occurring ore is the double oxide of iron and chromium $FeO.Cr_2O_3$, and this when finely ground is bonded with clay, formed into the standard shape and fired to produce what are known as chromite bricks.

Neutral temperature. A term used in reference to thermocouples. It is the temperature of the hot junction at which the e.m.f. round the circuit has its highest value, while the rate of change of e.m.f. with temperature is a minimum.

Neutron. A constituent of the nuclei of all atoms except hydrogen. It is an electrically uncharged particle whose mass is slightly greater than that of the proton. It was originally discovered by Chadwick in 1932 by allowing alpha-particles from polonium to collide with beryllium. An atom of atomic number Z and atomic weight W contains Z protons and $(W-Z)$ neutrons. For elements of small atomic weight $W = 2Z$, but in the elements of high atomic weight the proportion of neutrons to protons increases. This increase in the relative proportions of $(W-Z)/Z$ enables a stable nucleus to be formed (*see also* **Isotopes**). Where the actual value of the ratio lies outside limits the isotope is radioactive. The mass of a neutron is taken as 1,850 times the mass of an electron, and since it is electrically neutral, the neutron can pass through matter.

Neutron diffraction. The neutron can be represented as a wave motion, in a similar way to the electron, but differing in that the neutron is electrically neutral and has a greater mass. The neutron mass is 1,850 times that of the electron, and therefore possesses considerable kinetic energy. Furthermore, neutrons are produced conveniently only by nuclear reactions, and since they are generated at very high energy and velocity, require to be moderated by

light elements before they can be easily utilised in diffraction. The kinetic energy of the neutrons is related to temperature by the equation $1/2(mv^2) = 3/2(kT)$ where m is mass, v velocity, k Boltzmann's constant and T absolute temperature. This kinetic energy then gives a formula for the equivalent wavelength λ for diffraction, as with the electron, $\lambda = h/(3mkT)^{1/2}$ where h is Planck's constant. At $0\,°C$, this corresponds to an equivalent wavelength of 0.155 nm, and at $100\,°C$, 0.133 nm. Such wavelengths are similar to those of X-rays, and therefore excellent for diffraction by crystal planes.

The formula above utilises the root mean square velocity, but a neutron beam leaving a reactor via a collimator will contain the equivalent of a wide band of wavelengths, and a crystal monochromator is needed to select a fixed wavelength. A calcium fluoride crystal, utilising the (111) plane, can satisfy the Bragg law $\lambda = 2d \sin \theta$ to give wavelengths in the collimated beam of the appropriate level. The wavelength selected is usually around $\lambda - 0.11$ nm, and the variation in the band of wavelengths is then ± 2.5 pm about the mean.

Neutron diffraction has been used for study of ordered structures in metallic alloys, particularly when the atoms involved are in close or adjacent groups of the periodic table. The magnetic structure can also be investigated, with respect to electron spins and type of ordering.

Newtonian fluid. A fluid which obeys Newton's law of viscous flow. (*See* **Fluidity, Viscosity.**)

Nibbling. A term used in reference to the cutting of metal stock by means of a nibbling machine. The metal in the form of sheet or plate is cut by the removal of small bites successively from the edges.

Ni-carbing. A process for case-hardening steel components in a gas carburising atmosphere containing ammonia gas. The composition of the gas is such that both carbon and nitrogen are absorbed simultaneously. The rate of cooling is adjusted to secure the desired properties. Also known as carbonitriding.

Nicaro process. A process for the recovery of nickel from low-grade ore. The pulverised ore is reduced by producer gas, and leached in an ammonia solution from which nickel is precipitated as carbonate. (Nicaro is in Cuba.)

Niccolite. One of the chief ores of nickel which occurs as the arsenide NiAs. Also called kupfernickel. It occurs in the Saxon mines in Austria (Styria), in Scotland (Leadhills) and the USA (Connecticut). It contains about 44% Ni. The colour is usually pale copper-red. Mohs hardness 5.5, sp. gr. 7.5.

Nick break test. A test for the determination of the quality of a weld. A cut or nick is made in the weld metal and the specimen fractured by bending. The fracture is examined for visible defects.

Nicked fracture test. A test in which a bar is nicked or sawn so that the centre of the bar is left uncut. The bar is broken by impact, and the resulting fracture is examined.

Nickel. A metallic element that occurs in numerous small deposits throughout the world (see **Nickel ores**). The only mines of commercial importance at present are those of New Caledonia, Cuba, W. Australia, Sudbury (Ontario) and Manitoba. By far the largest proportion is produced from the nickeliferous pyrrhotite deposits at Sudbury. There are three processes for treating this ore: the Orford (or top and bottom) process is now out-of-date. The other methods are the Mond and the Hybinette processes. They all start with a copper-nickel matte, which is produced by the same methods used in smelting copper ores. The preliminary treatment of this ore consists of crushing and grinding, followed by flotation to remove waste rock and to provide three types of concentrations; one consisting primarily of copper, a second ore containing the bulk of the nickel with more copper, and the third mainly of iron but with some nickel. The nickel concentrate is roasted in multiple hearth furnaces, the resulting calcine is then smelted in reverberatory furnaces, in the course of which a high proportion of the sulphur is removed. The matte produced from these operations is bessemerised in converters. This matte was formerly treated by the Orford process (q.v.) but is now put through an improved process known as matte flotation. The copper sulphide and nickel sulphide are separated by subjecting the matte to a combination of operations, including controlled special cooling, followed by grinding, magnetic separation and flotation. The copper sulphide is remelted, blown to blister copper, and transferred in the molten state to the nearby copper refinery.

The nickel sulphide is subjected to a sintering operation to produce a dense nodular nickel oxide sinter for refining by both the electrolytic process and the Mond carbonyl process (q.v.) or for use in the making of nickel alloys and alloy steels.

An alternative method for treating Canadian sulphide ores is known as the Hybinette process, in which the copper is not separated from the nickel during the smelting process. The final matte is roasted to remove most of the sulphur and then leached with sulphuric acid to remove most of the copper. The insoluble residue containing the nickel is melted and cast into anodes which are refined electrolytically.

The usual treatment of oxide ores such as those found in Cuba is, after reduction, to leach out the nickel with ammonia and ammonium carbonate. The basic nickel carbonate is calcined to give nickel oxide.

In the Orford process the copper-nickel matte was smelted with an alkali sulphide and cast into large pots, where it solidifies in two layers. The upper layer consists primarily of copper and alkali sulphides, while the lower layer is mainly nickel sulphide. In the fusion with sodium salt, the charge consists of coke, sodium sulphate and matte. The nickel-sulphide bottoms are crushed, leached, dried and sintered. They are mixed with low-ash coal and melted in a reverberatory furnace. The cast anodes are refined by an electrolytic process.

In the Mond process, crushed nickel-sulphide bottoms from the Orford process were calcined to remove sulphur and to convert into oxide. The solid residue was heated in reduction towers in which it meets an upward stream of water-gas at about 350 °C. The reduced nickel so obtained is then treated with plant or producer gas, rich in carbon monoxide, at a temperature of about 60 °C in volatilisers. This results in the formation of volatile nickel carbonyl (q.v.), which is led into decomposing towers, where it passes over nickel pellets at a temperature of about 180 °C. At this latter temperature the carbonyl decomposes, nickel being deposited on the pellets. The latter, which grow in size and are eventually separated by screening, are 99.8% pure nickel.

Malleable nickel is produced by casting the metal into ingot moulds after deoxidation. The metal, which is silvery-white in colour, is used in the food processing, radio and chemical industries; in coinage; in special accumulators (see **Ni-Fe accumulator**), and as a catalyst in the hydrogenation of unsaturated organic compounds. It is extensively used for electroplating, but the most important use is as an alloying element in ferrous and non-ferrous metals.

Nickel is one of the chief alloying elements in steelmaking, where it is used to increase hardness, strength and toughness. One of the effects of adding nickel to a plain carbon steel is to lower the temperature at which transformation proceeds. Nickel steel therefore hardens more readily than plain carbon steel, and drastic quenching is not essential. Most of the high tensile steels contain nickel in amounts ranging from about 0.5 to 5%. For example, nickel-chromium-molybdenum steels find many uses in automobiles, tractors and aircraft. A low-carbon high-nickel steel containing 8.5% Ni is used in oil-well engineering. Stainless steels containing 8-14% Ni with about 18% Cr find a wider range of uses in domestic equipment, chemical and food-processing equipment, and in the aircraft industry.

High nickel alloys such as Monel and Inconel are used in power and chemical plants. Inconel, which is able to withstand high temperatures, is used for sheathing electrical heating units. The Nimonic alloys (q.v.) are used for high temperature applications in gas turbines. Nickel-clad steel plate is used in the chemical industry, particularly for caustic alkali manufacture. Monel-clad steel is also produced and is used in petroleum refining, salt evaporators, etc.

In cast iron, nickel is used to refine grain structure, improve mechanical properties and increase resistance to abrasion, heat and corrosion. Ni-Hard, a martensitic cast iron, is used for pump bodies, and for components subject to abrasive conditions. Austenitic cast irons, known as Ni-Resist (q.v.), are resistant to corrosion, heat and wear.

A low-expansion wear-resistant cast iron containing 35% Ni is used in precision machine tools and instrument parts.

High-nickel irons, containing up to 80% Ni, are used as magnetic alloys in communications equipment transformers, etc. Wrought nickel silver alloys, containing up to 18% Ni are employed for making silver-plated ware.

The present British "silver" coinage is a copper-nickel alloy containing 75% Cu and 25% Ni. (*See also* Tables 2, 7.)

Nickel bloom. The name given to the hydrated oxidised nickel minerals formed on the exterior of nickel ores. They are all green in colour and include: *(a)* emerald nickel or zaratite, which is the hydrated basic carbonate $NiCO_3.2Ni(OH)_2.4H_2O$; *(b)* nickel vitriol or morenosite $NiSO_4.7H_2O$; *(c)* annabergite—a nickel arsenate $Ni_3As_2O_8.8H_2O$.

Nickel brass. The name used to designate a class of copper-zinc alloys that contain from 2 to 14% Ni and are useful for a wide range of applications (*see* Table 5).

Nickel bronze. A series of cast copper-tin-nickel alloys usually containing 5-8% each Ni and Sn, 1-2% Zn with a very small amount of phosphorus.

Small additions of from 0.5 to 2% Ni are used to improve the casting properties of tin bronzes, to refine the grain size and to improve the mechanical properties (particularly the yield strength) of the castings.

Some of the alloys are capable of being further hardened and strengthened by age-hardening treatment (*see* Table 5).

Nickel carbonyl. The volatile compound formed when carbon monoxide is passed over heated nickel. Combination takes place in excess of CO at about 50 °-60 °C. On heating the gas to 180 °C, decomposition occurs with the deposition of nickel and the regeneration of CO. These reactions are the basis of the Mond process for the purification of nickel (q.v.).

$$\overset{60\,°C}{\underset{180\,°C}{Ni + 4\,CO \rightleftarrows Ni(CO)_4}}$$

Nickel cast irons. Nickel is added to cast iron in order to refine the grain size, improve the mechanical properties (such as tensile and yield strength) and to improve the resistance to heat, abrasion or corrosion. Small amounts of nickel (2-5%) improve the resistance to abrasion (*see* Ni-hard). Moderate additions of nickel (15-36%) greatly increase the resistance to corrosion and high temperature. The nickel cast irons may be considered as falling into three grades: *(a)* the austenitic cast irons for corrosion and heat resistance; these contain upwards of 13% Ni together with chromium and frequently copper; *(b)* martensitic white irons contain about 2.5-5% Ni together with 1.5-2% Cr, and are used where good abrasion resistance is desired, e.g. in ball mills; *(c)* acicular irons contain 1-5% Ni together with 0.8-1% Mo, and are designated high-strength alloy irons. Spheroidal graphite cast iron (q.v.) commonly contains 1-2% Ni.

The addition of nickel to cast iron in amounts exceeding 10% gradually reduces the permeability until at 30% Ni the alloy is non-magnetic. Additions of nickel beyond 30% up to 85% give a range of nickel-iron alloys which, after suitable heat-treatment, show a very high permeability. They are also magnetically "soft", since they show a small hysteresis loop. When aluminium is alloyed with the nickel iron in suitable proportions, "hard" magnetic alloys, suitable for making permanent magnets, are obtained.

Nickel cast steel. The term usually applied to steels containing 2.25-3.5% Ni. A typical composition is C 0.15-0.25%, Mn 0.5-0.75% and Si 0.25-0.45%, the nickel lying within the above range. The effects of adding nickel to a plain carbon steel are: *(a)* to lower the temperature at which transformation from austenite to martensite occurs, *(b)* to produce a marked reduction in the rate at which transformation proceeds. Nickel steel therefore hardens more readily than plain carbon steel, and drastic quenching is not essential. A high-carbon, high-nickel steel is in fact an air-hardening steel.

Nickel clad. The term applied in reference to the bonding of a thin sheet of nickel to a thicker sheet or plate of low-carbon steel. The latter provides the strength, and the cladding gives the corrosion protection. Such a combination is economical, and in certain circumstances is quite serviceable for heavy structural applications.

Nickel ferrite $Fe(NiFe)O_4$. An inverse spinel in which 8 iron III (ferric) ions occupy the tetrahedral voids, and 8 iron III (ferric) and 8 nickel ions the octahedral voids, in the cubic close-packed lattice of 32 oxygen ions. If zinc is used in addition to nickel, so that zinc ions fill some of the octahedral voids in the same proportion to iron III ions as for nickel, then a magnetically soft material results, with a permeability of $\mu = 100$. Nickel ferrite alone has a lower permeability still ($\mu = 15$), but a high magnetostriction, which increases as the proportion of nickel or zinc atoms increases.

Nickel ores. Nickel is found in the following minerals:

(a) Sulphide ores:
   (1) Millerite NiS or nickel blende.
   (2) Pentlandite (Fe,Ni)S or NiS.2FeS found in Sudbury.
   (3) Horbachite (Fe,Ni)$_2$S$_3$.
   (4) Linnaeite (Ni,Co)$_3$S$_4$.
   (5) Gersdorffite NiAsS or nickel glance.
   (6) Ullmanite NiSbS or nickeliferous grey antimony.

(b) Arsenides and antimonides:
   (1) Niccolite NiAs—nickeline or kupfernickel.
   (2) Chloanthite NiAs$_2$.
   (3) Breithauptite NiSb.

(c) Silicates:
   (1) Garnierite 2(Ni,Mg)$_5$.Si$_4$O$_{13}$.3H$_2$O found in New Caledonia—a low grade ore.
   (2) Rewdanskite (Ni,Fe,Mg)$_3$Si$_2$O$_7$.2H$_2$O.

(d) Other ores:
   (1) Morenosite NiSO$_4$.7H$_2$O or nickel vitriol.
   (2) Annabergite Ni$_3$(AsO$_4$)$_2$.8H$_2$O or nickel ochre.
   (3) Zaratite NiCO$_3$.2Ni(OH)$_2$.4H$_2$O.

Nickel pellets. A commercial form of nickel of purity about 99.8% produced by the decomposition of nickel carbonyl.

Nickel plating. A process for the electrodeposition of metallic nickel from acid-sulphate baths. The tank must be rubber, wood, lead or plastic lined. The following solutions are commonly used. The approximate proportions are stated in kilograms per cubic metre:

(a) Crystalline nickel sulphate 124.7
   Ammonium chloride 37.4
   Boric acid 24.9
   pH=5.3, temperature 21-32 °C, current density 1-2A m^{-2}. Generally used as a barrel-plating solution.

(b) Crystalline nickel sulphate 199.5
   Nickel chloride 37.4
   pH=5.2, temperature 43-48 °C, current density 2.5-5A m^{-2}. Generally used for dull or normal finish.

(c) High-sulphate bath:
   Crystalline nickel sulphate 75
   Sodium sulphate 12.5
   Ammonium chloride 12.5
   Boric acid 12.5
   pH=5.5, temperature 21-26 °C, current density 2-3A m^{-2}.

(d) Bright plating bath:
   Crystalline nickel sulphate 240
   Cobalt sulphate 15.6
   Nickel chloride 45
   Sodium formate 34
   Boric acid 30
   pH=4.2, temperature 54-60 °C, current density 2-3A m^{-2}.

Organic brighteners are added to solutions of type *(b)*. Stirring or agitation of the bath is necessary and the electrolyte must be maintained free of impurities.

Nickel pyrites. Nickel sulphide, found in Pennsylvania, which contains about 5.6% Ni. A similar ore containing about 2.5% Ni together with an equal amount of copper sulphide is found in Ontario, where it is sometimes found associated with nickeliferous iron pyrites (Ni,Fe)S$_2$. Other ores have been found in Oregon.

Nickel silver. The name applied to a series of alloys of copper, nickel and zinc, possessing an attractive colour, good resistance to corrosion and good mechanical properties. It is used as a base for electroplated silver-ware, for architectural and ornamental metalwork, marine fittings and certain types of food-handling equipment. The commonly used alloys are given in Table 5.

Nickel substitutes. Alloys or bonded metals that have been suggested for use in order to economise in the use of nickel. They are to be used in sheet form: *(a)* aluminium-clad iron (P2 metal) for general use; *(b)* aluminium-iron-nickel (Sandwich metal PN); *(c)* nickel flashed steel for rust protection; *(d)* nickel clad steel (10-15% Ni).

Nicking. The cutting of a narrow groove in a metallic specimen in order to facilitate the fracture at the desired section.

Niclad. The name used in reference to a composite sheet, consisting of sheet steel sandwiched between, and bonded to, sheets of nickel. The composite sheet which is produced by rolling sheets of nickel and a sheet of steel possesses the strength of the steel used and the corrosion resistance of nickel.

Niclausse boiler. A type of boiler capable of meeting heavy demands for steam in which Field tubes are used. These latter consist essentially of double-walled tubes. The outer wall is exposed to the furnace gases. The water on being heated rises up the space between the inner and outer walls, and cooler water descends in the inner tube to maintain a rapid and efficient circulation. Such boilers are useful for driving turbines or supplying steam in such processes as the manufacture of water-gas, etc.

Nicolite. The ore from which Crondstedt first isolated nickel. It is the arsenide of nickel NiAs and is synonymous with Niccolite (q.v.).

Nicol's prism. Any type of polarised light can be turned into plane polarised light, and the instrument used in such investigations is called a Nicol's prism or simply a "nicol". In transmitting a beam of light, many transparent crystalline materials have the property of splitting the beam into two beams having different laws of refraction. For most directions of transmission a single ray of ordinary light separates into two, one of which is called the *ordinary* ray because it obeys the ordinary laws of refraction, and the other is called the *extraordinary* ray because it does not obey them. The mineral calcite (or Iceland spar) exhibits this doubly refracting property most strongly. It is therefore used in the making of a nicol. A rhomb of calcite, which is about three times as long as it is broad, is sawn across the line *AC*, and after the cut has been polished, the two halves are cemented together with Canada balsam. (*See* Fig. 91.) (Canada balsam has a refractive index intermediate between those of the ordinary and extraordinary waves.) The plane of the cut makes an angle of about 22° with *AB*. The two ends *AD* and *BC* are also cut to a slightly different angle (actually 68° instead of 72°).

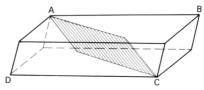

Fig. 91 *Schematic diagram showing plane of cut in a calcite crystal to prepare a nicol prism.*

A wave entering parallel to the prism is doubly refracted, and because of the slope of the face *AD*, the waves continue in somewhat different directions, the ordinary wave being most refracted. As the ordinary wave meets the Canada balsam surface very obliquely it becomes totally reflected, and so falls on the side *DC*, which is blackened to absorb it. (*See* Fig. 92.) The extraordinary

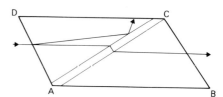

Fig. 92 *Cross-section of nicol prism showing passage of light ray.*

wave passes through the prism and emerges as plane polarised light. A nicol will polarise light over a range of angles of incidence of about 25°. When two nicols are used, the first produces polarised light by absorbing one component. If the second nicol is "crossed", being placed with its axis in the same line as the first but turned through 90°, the wave that was the extraordinary in the first prism becomes the ordinary wave in the second prism, and is thus absorbed. The use of polarised light in the microscopy of metals is illustrated in Fig. 93.

When used with polymers and ceramics, or any material which is transparent in thin sections, the light path from the source passes from beneath the specimen through the same train S-F-C_1-A_1-A_2-C_2P (see Fig. 93) and then through the specimen to the objective O. The simplest way to achieve this is to place the reflector V.R. beneath the specimen stage, and between A_1 and A_2 of the light train. The reflector V.R. can then be a silvered mirror, there is no necessity for any transmission, as with metallurgical microscopes.

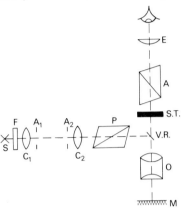

Fig. 93 *Schematic diagram of metal microscope using polarised light. S is the source; F the filter; C_1 the condenser; A_1 the lampiris; A_2 the fieldiris; C_2 the condenser; P the polariser; V.R. the vertical reflector; O the objective (refracting or reflecting); M the metal specimen; S.T. the sensitive tint; A the analyser; and E the eyepiece or photo-multiplier tube.*

Nife. According to Wegener's hypothesis, the visible part of the earth's crust, consisting chiefly of the lighter and more acid rocks (sial), has beneath it a layer of denser and more basic rocks (sima). The interior core, which is denser still, is referred to as nife.

Ni-Fe accumulator. A type of accumulator invented by Edison—and hence also called Edison accumulator—in which the positive plate consists of nickel hydroxide suitably enclosed, and the negative plate of iron oxide. The positive plate consists of $Ni(OH)_2$ enclosed in perforated nickeled-steel tubes. The hydroxide, being a poor conductor, is mixed with fine nickel flake, and the tubes are kept small in diameter (usually about 6 mm). They are strengthened by steel rings equi-distantly spaced and encircling the tubes. A double row of these tubes is mounted on a light nickeled-steel grid to form the positive plate.

The negative plate consists of a number of small perforated steel rectangular containers into each of which is pressed a mixture of iron oxide and a little mercury (to improve the conductivity). Cadmium may also be used. The electrolyte is caustic potash containing a little lithia.

The working voltage is 1.2 volts. The energy capacity varies from 17 to 30 watt-hour per kilogram of cell, according to the size of cell. Using large cells in the most efficient manner, an output of 1 h.p.-h can be obtained from a total weight of about 25 kg of cell.

Nigger heads. A term used in reference to the lumps of unfused matter that remain in a steel melt at the end of a lime boil. Fusion is accelerated and the fluidity of the slag is increased if a suitable flux (e.g. fluorspar, lime, ore or sand) be added in the correct proportions.

Nilo. A series of nickel-iron alloys containing different percentages of nickel, this difference being closely associated with the applications to which the alloys may be put. The alloy having 36% Ni (Nilo 36) has a low coefficient of expansion, and this property mainly determines its chief uses. As the percentage of nickel increases above 36%, the coefficient of thermal expansion increases, so that it is possible to prepare alloys having a range of coefficients from 1.5 to 12 × 10^{-6}. Compositions and applications are shown in Table 6.

Nilvar. An alloy similar in composition to Invar and containing 36% Ni. It has a low coefficient of expansion, being only 0.9 × 10^{-6} at temperatures up to 100 °C, but rising rapidly at temperatures above 250 °C.

Nimonic alloys. A series of alloys developed from the 80/20 nickel-chromic alloys to meet the need for materials having good resistance to oxidation and good creep resistance at high temperatures (see Table 7 and **Superalloys**).

Niobite. A mineral consisting of niobate and tantalate of iron and manganese. $(Fe,Mn)(Nb,Ta)_2O_6$. The iron and manganese contents vary greatly. When the iron is largely replaced by manganese, the mineral is termed manganiobite or manganotantalite. If the tantalum content is greater than the niobium content the mineral is called tantalite. Also called columbite.

Niobium. A steel grey lustrous metal. It is malleable and ductile and about as hard as wrought iron. It resists attack by nearly all chemical reagents. It is soluble with difficulty in *aqua regia,* and is attacked by concentrated sulphuric acid and by molten alkalis.

The metal has been prepared by aluminothermic reduction or reduction by carbon of the oxide, *in vacuo.* The modern processes use reduction by sodium of the halogen compounds, prepared by chlorination or fluorination of purified oxides, or reduction of pure oxide or chloride. Whatever process is used, the pure metal is contaminated with trace oxygen, which has to be removed by electron bombardment melting, or sintering in hydrogen, with or without small carbon additions to assist removal of oxygen as carbon monoxide. Niobium oxides are volatile above 2,000 °C, so that melting processes contribute to purification directly, provided the oxide is removed as it volatilises. Purity of 99.95% can be so achieved.

Niobium has been used as a getter in vacuum tubes, and extensively in steel making, due to its carbide forming tendency, the stability of the carbides, and their insolubility in iron. In austenitic steels it reduces the tendency to intergranular corrosion and weld decay, by preventing grain boundary precipitation of carbides. Added to nitriding steels containing chromium and aluminium, it increases the rate of thickening of the nitrided layer. Niobium was extensively explored in the nuclear energy programme of the UK, and was used as the fuel clad material of the Dounreay fast breeder reactor which operated successfully for eighteen years before final closure. (For physical properties, see Table 2.)

Nipple. The name used in reference to the short piece of iron (usually wrought iron) piping, screwed at each end and used for connecting two lengths of piping. The term is also used to describe the short hollow cylinder of metal used for securing a length of cable to an operating lever.

A small cylindrical orifice to admit oil or grease to a bearing or any other point requiring lubrication.

Ni-resist. A series of corrosion-resistant cast irons. The structure consists of graphite in an austenitic matrix. There are several grades of the alloy containing 14-22% Ni and 0-6% Cu. They are all much more resistant to corrosive action than cast iron and are used for pump bodies, paper-making machines, food-handling equipment (copper absent), sea-water strainers, etc. The types containing up to 22% Ni can be used with advantage for handling very dilute sulphuric acid or dilute inorganic non-oxidising acids at room temperatures. A more recent type containing 28-32% Ni can withstand the action of fuming sulphuric acid even at elevated temperatures. Most of the series are suitable for use where concentrated brine is encountered. Their heat- and wear-resisting properties are indicated by their use as cylinder liners, and valve guides in internal-combustion engines. (*See* Table 6.)

Nitralising. A method of preparing steel sheets for enamelling. After the usual degreasing and pickling treatments the metal is immersed in fused sodium nitrate at a temperature of about 500 °C. This treatment minimises the production of faulty coatings due to blister and copper-head formation.

Nitre. The commercial name for potassium nitrate KNO_3. Also called saltpetre.

Nitre bath. A fused mixture of sodium nitrate 50 parts, potassium nitrate 50 parts, manganese dioxide 1 part. It is used for blueing steel. Sodium nitrate may be used alone, if a cheaper medium is required, but higher operating temperatures are necessary.

Nitric acid. A colourless, corrosive liquid HNO_3 which, when concentrated, has a density of 1.52 (when 99.7% pure). The yellow fuming acid, used as a strong oxidising agent, contains oxides of nitrogen in solution which are produced by adding a little starch to the saltpetre before distilling with strong sulphuric acid. The pure acid boils at 86 °C. Nitric acid is prepared by distilling potassium nitrate with strong sulphuric acid, and on the manufacturing scale is produced by (a) the retort process (e.g. Guttermann or Valentiner), which involves heating Chile saltpetre ($NaNO_3$) with concentrated sulphuric acid in cast-iron retorts; (b) the arc process (e.g. Birkeland and Eyde furnace), involving the direct combination of nitrogen and oxygen of the air, and (c) the ammonia oxidation process, using a platinum-rhodium alloy catalyst. The acid reacts with all metals except platinum, gold, rhodium and iridium. With the following metals the oxide is produced: tin (white oxide), antimony (yellow), tungsten (yellow), molybdenum (pale yellow). In other cases a soluble nitrate is formed. Nitric acid is used in metallurgy as an etchant (see Table 11), as a constituent of pickling baths, as a constituent of solutions, used for evaluating the corrosion resistance of alloys, for inducing the "passive" state on metallic surfaces and in certain production processes, i.e. the purification of gold.

Nitric acid test. For the purposes of segregation or quick identification a useful test can be applied to certain alloy steels which involves putting a few drops of the concentrated acid on the metal and observing the effect. Such tests are referred to as spot tests (q.v.), and when nitric acid is the reagent used, as nitric acid (spot) tests. Nickel-chromium steels of high and low alloy content are readily differentiated, since the former are not soluble in the acid, while the latter readily dissolve. Monel (a copper-nickel alloy, q.v.) also dissolves in the acid, but, unlike straight chromium steels, is non-magnetic.

High-chromium stainless steels, Inconel and similar alloys, on the one hand, and other alloy steels and cast iron, on the other, are readily differentiated by the spot test, and in certain cases identification is possible. With stainless nickel-chromium steels and high chromium steels the concentrated acid does not react. In most other cases (e.g. low chromium, plain carbon, nickel, tungsten, molybdenum, vanadium steels and grey cast iron) a dark brown or black colour appears in the drop of acid. The presence of nickel may be indicated by a deep green coloration, while with tungsten a yellow sediment is produced.

Nitrides. These are binary compounds of metals and nitrogen produced when metals are heated in presence of nitrogen gas at high pressure, or in presence of ammonia at lower pressures, or in presence of solid nitrogenous material. When aluminium, magnesium, lithium, calcium and boron are strongly heated in nitrogen they burn and form nitrides of the following formulae: AlN, Mg_3N_2, Li_3N, Ca_3N_2, BN. Aluminium nitride is a yellow crystalline compound that decomposes on the addition of water, producing ammonia. Magnesium nitride is greenish-yellow and calcium nitride brownish-yellow, and both behave with water as does AlN. Boron nitride is white in colour and does not react with water, but is decomposed to produce ammonia when heated in a current of steam. It reduces many metallic oxides at the temperature of fusion, giving the metal together with a borate and oxides of nitrogen.

Other metals that form nitrides, e.g. manganese, chromium, iron, tungsten, do so when heated to a dull, red heat. The following interstitial compounds are known: Mn_2N, Mn_4N; Cr_2N, CrN; Fe_2N, Fe_4N; W_2N. They are all hard or very hard and brittle and they have high melting points.

The formation of iron nitride in the presence of ammonia at only moderately high temperatures, and in the absence of high pressures is explained as due to the breaking down of the ammonia, in contact with the metal, producing atomic nitrogen. The decomposition $2 NH_3 \rightleftarrows 2N + 3H_2$ is greatly favoured in the forward direction by rise of temperature. Iron-nitrogen alloys evolve nitrogen (except for about 0.05%) during solidification. At lower temperatures (400-500 °C) the decomposition of the nitride is very slow. This forms the basis of the nitriding process (q.v.) developed by Fry.

Aluminium nitride, AlN, is very stable, even at high temperatures, and hence may be used to denitride steel. Small additions of titanium have a similar effect, there being at least four known

stable nitrides. The solubility of nitrogen in manganese decreases rapidly with rise in temperature and hence the small amounts of manganese usually present in steel would be quite insufficient to denitride it.

Nitriding. A case-hardening process in which machined and heat-treated steel components are subjected to the action of a nitrogenous medium, commonly ammonia gas, at temperatures ranging from 480 °C to 520 °C. The alloy steels used for nitriding contain 0.85-1.20% C, and one or more of the elements aluminium, chromium and vanadium (see **Nitriding steels**). The parts to be nitrided are placed in a gas-tight box in such a way that the ammonia gas can circulate freely round them. Heating at about 500 °C is continued for a day or longer according to the thickness of nitrided case desired.

The term "case" in reference to nitriding is less appropriate than for carburising. The temperature at which the nitriding takes place causes the rate of diffusion of the nitrogen to be much slower and change in the nitrogen content from the outside layer to the core to be much more gradual. Peeling of the nitrided part cannot occur.

When the parts have been exposed to the action of the ammonia for the requisite time, the box is allowed to cool rapidly to about 360 °C, after which the parts can be removed as soon as it is convenient to handle them. No quenching is needed, and no further heat-treatment is required; the parts having been quenched and tempered and stress relieved where necessary before being submitted to the hardening process. Any portions of a component that are not to be hardened can be protected by applying a thin layer of tin or solder. If the tin coating does not exceed about 7 μm the tin will not run. Plating with nickel or copper is a suitable alternative, but the coatings must be thicker in order to give adequate protection.

During the nitriding process the components swell slightly so that in treating machined and finished parts an allowance of 4×10^{-4} per mm is usually made. The effect of nitriding on the appearance of the steel is to produce a thin greyish film and in most cases this is acceptable. The hardness produced is considerably greater than is obtained by any other case hardening process and the hardness is not affected if the parts are heated up to about 550 °C. The values normally obtained for hardness (D.P.H.) usually lie in the range 1,000-1,150.

The main advantages of nitriding over other processes for hardening the surface layers are: (a) low temperature at which the process is carried out diminishes the risk of distortion during treatment; (b) the process is cheaper than carburising; (c) quenching is eliminated and quenching cracks are avoided; (d) the structure of the steel is not altered during the treatment.

Nitriding steels. The name used in reference to those types of alloy steel that are susceptible to hardening by the nitriding process. The essential alloying elements are one or more of those that form stable nitrides: aluminium, chromium, vanadium. Molybdenum is also added because the nitriding process takes place in that range of temperature which favours the occurrence of temper brittleness in certain types of steel, viz. those that contain high proportions of manganese, chromium and phosphorus. The presence of molybdenum markedly reduces the tendency. Nitralloy is a special range of steels suitable for nitriding. (See Table 6.) Chromium-molybdenum-nickel and other medium carbon steels that do not contain aluminium are being nitrided, but the surface hardness obtained in such steels is much less than that obtained with steels containing aluminium—the values lying between 600 and 900 D.P.H. Nitrided parts can be softened by heating in a bath of fused salts containing equal parts of sodium and potassium chlorides at about 815 °C.

Nitrided steels show a great improvement in endurance and wear-resistance properties over other steels, and since the nitride case retains its hardness at elevated temperatures, they have marked advantages for use as liners in high-power aircraft engines, motor cars, etc. Leaded bronzes containing 80% Cu, 10% Sn and 10% Pb, and graphite bronzes are generally used as bushings against nitrided Nitralloy.

Nitrochalk. A mixture of calcium carbonate $CaCO_3$ and ammonium nitrate NH_4NO_3. It is used as a fertiliser.

Nitrogen. An element in the first short period of the periodic table, atomic number 7, and therefore in Group V of the table, between carbon and oxygen. Nitrogen forms diatomic molecules by covalent bonding, and in the solid and liquid state, large molecules by virtue of van der Waals' attraction. The boiling point, $-196\,°C$, indicates the weak forces involved. Nitrogen forms nitrides with many metals, and can be present in interstitial solid solution in metallic lattices. Nitrides tend to be harder than carbides when present in the surface layers of metals. (*See* **Nitriding.**) Nitrogen is a useful inert gas for protection of reactive metals at high temperature and as a carrier gas in chemical and material processing. It is a constituent of polyamide, nitrile and aminomethanal polymers.

Nixferrum. The name by which the oxide of antimony was known. It was produced by roasting the metal reduced by iron. The name was derived from the belief that iron was necessary for its formation.

Nobbing. The term used in reference to the process of consolidating and welding together the particles composing the balls of metal taken from the puddling furnace, and removing the slag therefrom. The operation, which is also known as shingling, was done by squeezing or hammering.

Noble gases. An alternative term for the elements of Group 0 of the periodic table, the rare or inert gases.

Noble metals. The name used in reference to the metals gold, platinum, palladium, etc., and silver, since they do not oxidise in the air. They can all be used as coating metals and are readily deposited on base metals. Their electrode potentials are positive and increase from 0.80 (for silver) to 1.68 for gold.

Nodular cast iron. *See* **Spheroidal graphite cast iron.**

Nodular fire clay. A sedimentary clay consisting mainly of kaolinite containing aluminous or ferrous nodules. It is used together with the more plastic variety in the making of fire-clay brick.

Nodule. A type of concretion frequently occurring in sedimentary rock, roughly spherical in shape. The term is also applied to the graphite in certain types of cast iron treated with cerium or magnesium in which the graphite occurs as spheroids instead of flakes.

Non-caking coal. Coals of the anthracite type which burn without forming into masses as a result of softening.

Non-destructive testing. Strictly, any test which can be conducted on a complete or partly complete artefact, leaving the specimen fit to perform the task for which it was made, or the next step in that process, after testing. The majority of such tests, however, are those to detect external and, more particularly, internal defects.

Perhaps the oldest of such tests was the "sounding" of metal components by striking. Centuries ago, the quality of musical instruments of the percussion type, gongs, bells and cymbals, were so tested, and it was natural for the industrial revolution to adapt the sonic test, of which railway wheel tapping is an example still within living memory. The change in damping when a flaw is present can be quite sensitive, but it cannot locate an internal defect, and is therefore a pass or failure test unless supplemented. Recent research into acoustic emission may revive this test in a more refined form in the future.

The magnetic "ink" test, whereby changes in magnetic flux around a defect are detected by a fine suspension of metal filings (usually iron), is limited to ferromagnetic material, and it was the discovery of X-rays which really brought non-destructive testing to the fore. However, an adaptation of the technique is available for non-conducting materials.

The detection of surface flaws such as pits, cracks, and seams, requires an increase in the visual contrast between the flaw and its surroundings, to highlight the imperfection. The oldest method has limited application, that of flooding the surface of a porous body with a suspension of coloured or fluorescent particles. The liquid is absorbed into areas where cracks or flaws occur, the material itself acting as a filter, and leaving a deposit of the suspended particles around the crack. Used for ceramics, concrete, sintered powder metallurgy compacts.

Magnetic ink has been referred to. A powerful magnetic field is induced in the surface of the object, and a suspension of finely divided magnetic particles applied to the surface. The magnetic induction may be carried out by insertion in a coil carrying a current (very suitable for small objects and repetitive tests) or by clamping contacts on to the object and passing the current

through the object. Alternating, direct or half wave direct currents are used, the half wave d.c. being suitable for subsurface flaws. Since cracks or flaws cause a break or leakage in the magnetic flux, a concentration of magnetic particles collects around the defect. The magnetising force should preferably be greatest in the direction which will give the best contrast, and this generally means passing a current via probes in a direction parallel to the direction of the defect. Where this cannot be deduced in advance, magnetisation in two perpendicular directions is usually applied. Used only on ferromagnetic materials.

Non-conducting material such as plastics, ceramics and glass, can be treated in a similar way, but the clean and dry object is first immersed in aqueous penetrant (low surface tension), so that penetrant remains in the cracks after draining. Particles of fine powder, such as powdered chalk $CaCO_3$ are blown on to the object, and become positively charged. The penetrant in the cracks supplies a source of electrons, which attract the positively charged powder. Powder collects around cracks in a manner similar to the magnetic ink on ferromagnetic material, and outlines the defect.

Liquid dye penetrants operate in a similar manner. The penetrant, of low surface tension, carries a deeply coloured dye, usually crimson, and is sprayed on to the degreased and clean surface of the object. When dry, the excess is removed by a second spray, which is not so penetrating, but serves to clean the surface. Penetrant within flaws still remains, and a fine white powder, capable of absorbing the dye, is sprayed on to the object. The dye is absorbed only from the defects where it still remains, and a sharp contrast is produced between flaw and normal surface. In this case, shape and size of the flaw are delineated more accurately, because dye only remains in the flaw proper. A fluorescent dye can be used in conjunction with ultra-violet light, to further enhance the contrast. Used on metals and a wide variety of materials provided that they are not absorbent, which reduces the contrast.

Radiography is a well-established technique for internal flaw detection, and is effectively a shadow picture obtained by the differential absorption of X-rays between flaw and sound material. Direct viewing on a fluorescent screen is possible for on line inspection, but safety precautions need to be stringently applied. Recording the shadow picture on a photographic film has the advantage of permanent reference when required. Since the wavelength of the radiation determines the penetration, and the accelerating voltage of the electrons in the X-ray generator determines the wavelength, there is every advantage in using the highest voltage commensurate with cost and material to be examined. Metallurgical work usually requires voltages from 100 kV to 2 MV, the radiation from a 200 kV tube being capable of penetrating up to 25 mm of steel. The maximum voltage for fluorescent screen viewing is about 50 kV, because of the effect on the eye, but with mirrors and added screening by lead glass, up to 150 kV can be handled in this way.

Radioactive isotopes, emitting gamma rays, can be used in place of X-rays, and have the advantage that they are more mobile. Although the amount of radioactive material is minute, the necessary screening for operators, and the heavy remote controlled shutter mechanism used to make the exposure, requires lifting and handling gear, but at least the source can be taken to a production site — impossible with all except very small X-ray generators, and little power (or none at all) is needed to operate the shutter. Gamma ray sources are available to cover a similar range to X-ray generators in penetrating power, but have some advantages. The spectrum of X-rays is a broad band of wavelengths, whereas gamma rays are produced as discrete wavelengths characteristic of the source isotope.

Radiation passing through an object is absorbed according to the equation $I_x = I_o \exp(-\mu x)$ where I_o is the intensity of the incident radiation, I_x the intensity of radiation which has travelled distance x through material of linear absorption μ for radiation of a given wavelength. In addition, there is considerable scattering effect, which serves only to confuse the shadow picture recorded after transmission, because it changes the effective wavelength of the radiation and therefore the value of μ. In general, an increase or decrease in absorption by the equivalent of 1% or more of the specimen thickness, is capable of interpretation on the radiograph. If excessive energy of radiation is used relevant to the sample thickness, the difference in μ is reduced by scattering in addition to the dependence on wavelength. This reduces the contrast recorded, and makes interpretation more difficult. Film type and processing can also influence contrast, and fine grain film always improves, although it lengthens exposure time. Intensifying screens, which fluoresce and attenuate the X-radiation to translate into wavelengths more appropriate for the film emulsion, also help. To remove scatter, intermediate screens may be used between specimen and film cassette, to filter part of the radiation transmitted. The size of the X-ray target is important, the smaller it is the sharper the shadow, and the greater the source to object distance, and the smaller the object-film distance, the better the result, but the longer the exposure time. Radiography is more highly skilled than tests for surface flaws as a result. The object may have to be radiographed more than once, to ensure that the orientation of a suspected defect is arranged to give maximum contrast on the film. Steels are often difficult, especially if coarse grained. Radiographic texts showing pictures of typical defects are available.

Ultrasonic testing developed from sonar instruments used in marine applications, whereby a pulse of high frequency sound waves is transmitted via the surface of the object, and reflections from the back wall of the object, or any defect which can give an echo, are detected by a suitable transducer and recorded on the fluorescent screen of an oscillograph. The frequency range 0.5 to 15 MHz is common, and the generation and detection of ultrasonic waves is usually by piezo-electric crystals of quartz, barium titanate, or lead zirconate. (*See* **Piezo-electric materials.**)

The propagation of the ultrasonic pulse can be represented by the equation $c = K(E/\rho)^{1/2}$ where c is velocity, K a constant, E Young's modulus, and ρ density. Longitudinal (compression) waves have the highest velocity, ranging from 300 m s^{-1} in air, to $1\,500 \text{ m s}^{-1}$ in water, $3\,000 \text{ m s}^{-1}$ in some polymers, $4\,700 \text{ m s}^{-1}$ in copper, $5\,900 \text{ m s}^{-1}$ in steel, to $6\,300 \text{ m s}^{-1}$ in aluminium. Transverse waves are also generated (usually referred to as shear waves) but travel more slowly, about half the speed of the longitudinal waves in metals, and slightly less than half in polymers. (These ratios are higher for lower frequencies and are related to Poisson's ratio.) Transverse waves are not generated in low viscosity fluids like air and water.

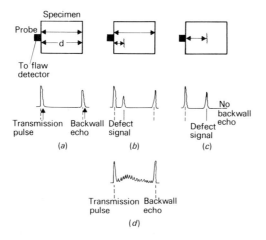

Fig. 94 *The technique of ultrasonic testing. (a) No defect. (b) Small defect. (c) Large defect, size greater than beam diameter. (d) Background noise (grass).*

The acoustic impedance of a material is the product of wave velocity and density, ρc. The greater this impedance, the greater the reflection of the ultrasonic pulse. For this reason, the very low impedance of air, and the comparatively high value for metals, requires a good coupling at the surface of the object, where the transducer is applied, as otherwise reflection occurs at the air/object interface. A liquid coupling of water, oil or grease is usually applied, but even then some reflection occurs at the interface. Any flaws within the object will be filled with air or gas usually, and hence such an interface gives an excellent reflection, due to the differential in acoustic impedance.

The oscillograph is adjusted to a time base to show each for-

ward pulse at time zero, and the back wall reflections at time $2d/c$ where d is the thickness of the material. About 500 pulses are needed to achieve a clearly visible display on the screen. By calibrating the screen in terms of distance travelled, defect echoes can be estimated in depth below the surface, and some indication of defect size obtained by comparing the amplitude of the defect echo signal, with those from a series of standard artificial defects in a similar thickness of the same material, say a series of drilled holes. Figure 94 shows the principle of ultrasonic testing. Used for a wide variety of materials, metals, ceramics, polymers and glass.

One of the best non-destructive testing methods for production lines is the use of eddy currents generated in the material, and detected by a search coil. The great advantage is speed, but the technique is limited to surface or near surface defects, and works best on standard sections or shapes. Alternating currents are electromagnetically induced by placing a coil close to the material under test, which must of course be an electrical conductor. The magnetic field induces alternating currents in the specimen which mirror the currents in the coil. The search coil picks up the magnetic field from these eddy currents and the apparent change in resistance R and/or inductance wL, depends on the electrical conductivity σ, the magnetic permeability μ, the dimensions of the sample, the frequency of the energising current ($= w/2\pi$) and upon the coil/sample geometry.

In non-ferrous materials, μ is constant, and in ferrous metals, additional d.c. saturating coils are used to make μ effectively constant (unless μ is under test).

A radial flaw in the metal tube, for example, severely affects the eddy currents, but a circumferential defect has less influence. When testing rod or tube on a continuous basis, the unwanted part of the signal is discarded by using a balanced bridge circuit, so that only signals connected with small changes in ΔR and ΔwL are recorded.

Eddy current testing can be used to measure σ and μ, and is suitable if identical specimens are being compared, as on a production line. The induced eddy currents have a limited penetration into metal, given by an approximate formula, for skin depth d, $d = 1/(\pi \nu \sigma \mu)^{1/2}$ where ν, frequency $= w/2\pi$. Defects can only be detected within the range of d, and at the extremity of skin depth, the currents are only one-third of those at the surface, hence the use on tubes in preference to rods. The frequencies adopted for eddy current testing are within the range 1 kHz to 2 MHz.

Among newer N.D.T. procedures is microwave examination, in which electromagnetic energy is applied via a wave guide, to the test object, and the transmitted or reflected energy analysed to determine composition, structure, density, moisture content or defects, in non-metallic materials such as plastics, ceramics, glass, wood and some mineral substances. Infra-red radiation emission is used to detect temperature variations of the order of 0.005 °C arising from irregularities in the material, processing variables, or the function of working parts. Thermography of this kind is already well established in medical diagnosis.

Holographic interferometry is used to detect internal strains, cracks and defects such as lack of bonding in composite materials. Stress-wave (acoustic) emission techniques monitor emissions above a pre-set amplitude level, when stored strain energy is released, as during deformation processes and shear transformations such as the production of martensite. The range 150 kHz to 1.4 MHz is common. Dislocation movements, growth of cracks, fracture of fibres in composites, debonding, delamination and failure of joints, have all been measured, and applied to a wide variety of materials and composites. Ultrasonic pulse echo techniques have been applied to materials sorting in terms of heat treatment, porosity, moisture content, and structural variations, including the estimate of non-metallic inclusions in metals, and moisture in moulding sands. Nuclear radiation gauges for levels in containers, density and moisture content, are in commercial use. They depend simply on attenuation by absorption or scatter of the radiation, and find application in metal, paper, mining, chemical and food industries.

Non-electrolyte. A substance which when dissolved in water yields a solution that does not dissociate, and does not conduct an electric current. All such substances have solutions which show normal osmotic pressure, and normal elevation of boiling point.

Non-ferrous alloy. An alloy that does not contain iron as a main constituent. Certain accepted non-ferrous alloys do however contain a little iron as impurity.

Non-magnetic steels. These are generally defined as steels having a permeability of 1.05 or less. Steels such as manganese steel and some of the austenitic chromium-nickel-iron alloys have a very low permeability. But not all the stainless steels are non-magnetic. The commonly used 18:8 grade tends to become magnetic after cold working or if the composition is modified to produce a duplex structure.

Non-polar compound. Covalent bonded material, in which the electron distribution is shared in such a way that polarisation of electrical charge, either positive or negative, does not arise. Homopolar means equal polarity, and hence homopolar or covalent bonds produce non-polar compounds.

Non-wettable surface. A term used in the flotation concentration of ores. The mineral substance is finely ground and fed into water that is agitated with air. Small amounts of frothing agents are added to produce a dry, strong and stable froth. Such substances as cresylic acid, pine oil, camphor oils or oil of eucalyptus are used. But most minerals require special treatment so that the surfaces acquire the properties that give them the ability to become attached to an air bubble. The important property is that of the non-wettable surface so that the water cannot displace the air bubble which will carry the mineral to the surface. The xanthates of sodium, potassium or ethyl, butyl or amyl xanthates (esters of ethoxydithioformic acid) are most commonly used. Quite small quantities are required to produce the non-wettable surface on finely divided minerals, there being no similar effect on the gangue, which is thus not collected in the froth.

Norbide. A compound of boron and carbon B_6C formed by heating coke and boric acid in an electric furnace. It is the hardest material produced for commercial use, being markedly harder than silicon carbide and second only to diamond. It is used in place of diamond dust for grinding and lapping dies and tools made of tungsten carbide or tantalum carbide. It is also very suitable for use as linings for nozzles used in sand-blasting because of its high abrasion resistance. It is used as a deoxidiser for steel. Also known as Norton boron carbide.

Nordhausen sulphuric acid. The acid originally produced at Nordhausen by distilling green vitriol (iron II sulphate, $FeSO_4$). The green vitriol was obtained by the oxidation of iron pyrites. It was roasted in air and evaporated down to form the basic iron III (ferric) sulphate, and this latter on strongly heating in fire-clay retorts yields SO_3. The latter is passed into water. The fuming acid is really a solution of SO_3 in sulphuric acid and has the formula $H_2S_2O_7$. It fumes in air, since SO_3 is readily evolved. It acts on many materials more vigorously than ordinary sulphuric acid and is much used in the coal-tar colour industry.

Norgine. A by-product extracted from seaweed. The latter is boiled with sodium carbonate and the residue after filtering is fairly pure cellulose. It was called algulose and used in making paper. The filtrate on acidifying yields a gelatinous substance which under the name algin was used for making jellies and sizing paper. The highly purified material, sodium alginate, is used extensively in the food industry. Norgine is used as an adhesive.

Noric. A sub-division of the Keuper division of the Triassic system. It contains typically limestones and dolomites with some red or variegated clays and sandstones.

Noric steel. A steel produced in Noricum—a district south of the Danube corresponding to part of Austria (Styria and Carinthia) and Bavaria. The mountainous country was rich in iron, and from it was made the famous Noric steel used by the Romans for weapons of war.

Norite. A hypersthene gabbro type of basic igneous rock, containing no quartz. It is unstratified and crystalline, having been formed at great depth below the earth's surface. As an intrusive rock in the Pre-Cambrian rocks of the Canadian shield, norite-micropegmatite occurs in the upper division or Keweenawan of Sudbury, Ontario. It contains copper and nickel norite-porphyrite, a hypabyssal type of igneous rock, has the same chemical and mineral composition as the norites, but shows porphyritic instead of granitic structure.

Normagal. A refractory material consisting of 40% alumina and 60% magnesia, and used for furnace linings.

Normalising. An annealing treatment that is aimed directly at raising the quality of the metal after it has undergone other mechanical or thermal treatments for special purposes. The metal will usually be in a strained condition after cooling from the casting operation or after forging, drop-forging, rolling, etc. If it has been heated to a high temperature and cooled without working, the crystal structure will be coarse. The purpose of normalising is therefore to refine the grain size, to render the structure more uniform and to remove the strains that have been induced in the metal while working. The treatment consists in heating the steel to a temperature only a little in excess of the the upper critical point (not more than 50 °C above Ac_3) and then cooling in still air.

Normal segregation. Alloying constituents or impurities are said to segregate at a particular place when their concentration at that place is greater than in the surrounding steel or other metal. In the usual process of solidification the metal solidifies progressively from the mould walls towards the centre. In alloys, the liquid solidifying near the centre and towards the top becomes enriched with alloying constituents or impurities having relatively low melting points. This concentration towards the centre of the casting is known as normal segregation.

Normal solution. A solution of an electrolyte which contains its equivalent weight in grams per litre of aqueous solution. For acids and bases this amount is the molecular weight divided by the basicity (acids) or the acidity (bases). Such solutions were much used in volumetric analysis. Normal solutions have been renamed molar solutions (q.v.).

Normal temperature and pressure. Since gases markedly expand on heating and contract under pressure, it is usual to state the volume of a gas under fixed conditions, now known as standard temperature and pressure, these being respectively 0 °C and 1.01325×10^5 N m^{-2}.

Northamptonshire ore. The brown earthy hematite $Fe_2O_3.2Fe(OH)_3$ found in the oolite and greensand in Northamptonshire (UK). It is high in phosphorus, containing an average of 1.12% P. The pig iron produced from these ores contains all the phosphorus, since this element is not removed in the blast furnace. Northamptonshire pig iron usually contains from 1.05 to 1.15% P. It is further treated in basic (dolomite) lined furnaces to remove the phosphorus.

Northrup furnace. An electric production furnace originally designed in 1916. The furnace is heated by electromagnetic induction and this is achieved by passing a high-frequency current through a conducting coil which surrounds the resistor (usually the metal charge) to be heated.

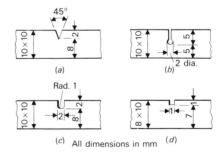

Fig. 95 *Types of notch used for impact testing. (a) Izod; (b) Charpy; (c) Mésnager; (d) Frémont.*

Notch. A type of weir gauge used for measuring the quantity of water flowing in an open channel. It is a simple and easy method of checking the volume of cooling water used in a condenser, etc. The weir may be V-shape or rectangular. In the former case if the angle of the notch is 90° the quantity of water discharged is given by $Q = 0.07 H^{\frac{5}{2}}$ m³ s⁻¹, where H is the vertical height of the water in the notch.

Also the slot or groove cut in a test piece that is to subjected to impact testing, to ensure that the specimen breaks at a desired section. The shape of the slot or notch varies in the different tests that are used to measure brittleness (or toughness, q.v.) (*see* Fig. 95).

Notch brittleness. The susceptibility of a metal to fracture at points of stress concentration caused by grooves, slots, notches, sharp fillets or scratches when subjected to a suddenly applied load. In notch brittle materials the notch or crack is propagated with great rapidity under sudden loading conditions and there is little or no plastic deformation. Austenitic steels are relatively free from this type of brittleness.

Notched-bar test. A device to obtain relative values of the brittleness of materials. In its usual form the test represents the ability of a steel to resist the formation or propagation of a crack through the metal. The various methods of carrying out the test are noted under Charpy, Frémont and Izod tests (q.v.). In each case a moving weight falls on or strikes against the test piece. The amount of energy absorbed by the material during fracture is the notched-bar value. (*See* Fig. 95.)

Notch sensitivity. The effect of a notch, whether in the form of a scratch, groove, or sharp corner, in a material, is to produce a large hydrostatic tensile stress near its root. In the direction longitudinal to the sharp end of the notch (y-axis) the stress rises to a maximum a little in front of the notch, but to a very high maximum perpendicular to this direction (x-axis) assuming the transverse axis of the notch to be the third perpendicular direction (z-axis). In the case of a scratch or groove, the z-axis is that along the scratch or groove, the y-axis that forward of the scratch, and the x-axis that in the direction above and below, or either side of the scratch. The high value of σx produces corresponding contraction strains along the y- and z-axes, but these strains are restrained by the metallic bonds. It is this concentration of stress which causes σx to be several times greater than the yield stress of the material, and the criterion for brittle fracture to be met under lower conditions of yield stress than would be met by plain tension. The temperature at which brittle crack propagation can proceed (transition temperature) is also raised by this hydrostatic stress system. The relief of stress can only be achieved by plastic deformation at the root of the notch, but as the volume of material involved is small, there is little sign of plastic deformation, and there is every likelihood of brittle crack propagation, since the condition continually repeats itself as the rupture of bonds occurs, often at high speed. There can also be notch strengthening, depending on the material, the notch geometry and the test conditions.

The sensitivity of material to notches is therefore a measure of its susceptibility to the stress system involved at the notch root, the value of the transition temperature and the effect of the notch on this, and the influence of the material structure in suppressing crack propagation once initiated. Obviously a material with high notch sensitivity is unsuitable for use in structures with high residual stress, and/or high operational stresses, if surface or internal notches could be produced (poor welding, bad machining, poor design) or the temperature may be low during construction or operation.

Notch toughness. A term sometimes used in reference to the high resistance of a metal to fracture by suddenly applied loads at a notch or other stress raiser.

Noumeite. A hydrated nickel magnesium silicate of variable composition found in serpentine in New Caledonia in veins associated with chromite and talc. A residual deposit, rich in nickel, is formed by the decay of the nickeliferous serpentine. It occurs also in Oregon, North Carolina and the Urals. The nickel content varies from 10 to 35%, and hence it is a source of nickel. Also known as garnierite.

Nowel. The lower or bottom section of a mould or foundry pattern. Synonymous with drag.

Nozzle. A general term applied to the terminal end of a pipe, as, for example, the removable end of a welding blowpipe. The name is also used in reference to the short refractory tube through which the flow of molten metal from a ladle is controlled during casting.

n-type semiconductor. An extrinsic semiconductor in which the number of electrons per atom of the impurity addition is greater than that of the base material (e.g. phosphorus in silicon) and the electron energy of the outermost level of the impurity is close to the conduction band energy of the base. A small increase in temperature is then capable of exciting the electrons from the impurity into the conduction band. As electrons are the carriers, the term n type, for negative type, was adopted. (*See* **Extrinsic semiconductor.**)

Nuclear charge. This is the positive charge on the nucleus of an atom, resulting from the number of protons it contains. The sum of the charges of the protons is $+ Ze$, where Z is the atomic number and $+ e$ is the charge on a proton.

Nuclear cross-section. The apparent cross-sectional area presented by a nucleus as a target for collision by a bombarding particle, e.g. neutron. The nucleus can "capture" or absorb the projectile, or merely cause a deflection. There are therefore two distinct cross-section values for such collisions. In the case of unstable nuclei, there is a third cross-section, relevant to nuclear fission processes which may occur on capture. (*See* **Absorption** and **Scattering cross-section.**)

Nuclear fission. The breaking up of a heavy atom, e.g. uranium, into two atoms of approximately equal mass, accompanied by the release of free neutrons and enormous amounts of energy. The disruption of the nucleus may be brought about by the bombardment of a heavy atom by neutrons.

Nuclear forces. A term which applies specifically to attractive forces between nucleons. Such forces exclude electromagnetic forces and are due to the exchange of π mesons between nucleons (protons or neutrons), the mesons being either positive, negative or neutral.

Nuclear reactor. A device utilising the fission of particular heavy elements by neutrons, to take advantage of the supernumary electrons to build up a controlled chain reaction, and to harness the thermal energy generated for power production or similar energy transfer.

Uranium was the first fissile material to be explored in this way. Natural uranium contains only 0.7% of the isotope ^{235}U, which is fissionable with slow and fast neutrons. The main constituent of the natural material, ^{238}U, is fissionable only with fast neutrons. The majority of nuclear reactors for power production use an enriched uranium in which the ^{235}U content has been increased by separation in a diffusion or centrifuge plant. ^{238}U captures neutrons, and decays to plutonium 239, which is also fissionable.

Since materials of construction capture neutrons, the main problem in nuclear reactors is to provide a containment of neutrons within the fissionable material with the minimum of loss by capture in other materials, whilst preserving the geometry of the system for control purposes, and providing adequate cooling to remove the heat. As uranium is a reactive metal, protection from the coolant is usually arranged by enclosing it in fuel element "cans", basically of aluminium, magnesium, zirconium or similar alloys. Coolant may be gas or liquid, air and carbon dioxide having been pioneered in the UK and water in the USA, for this purpose. To slow down neutron velocities, and allow more fission with ^{235}U atoms, a moderator, formed of light elements, permeates the structure. Graphite was first used, and remains the moderating material in UK power producers. Water can be used as a source of hydrogen, but is most effective if the hydrogen isotope deuterium is present, as in so-called heavy water. When pressurised, ordinary light water becomes more effective, the advantage of water moderation being that the moderator can become the coolant. The disadvantage with light water is that the whole system becomes a pressure vessel, and corrosion of materials of construction has to be guarded against. The disadvantage of graphite moderation is that the graphite stores energy from knock-on damage, and this has to be carefully released under controlled conditions at intervals. Furthermore, when CO_2 is used as the coolant, unusual species of compound may be produced by reaction between CO_2 and graphite, and these may affect corrosion of the heat-exchange system. (*See* Fig. 96 for the basic reactions taking place in a nuclear reactor.)

Control of the fission process in reactors utilising neutrons of moderated energy (thermal energy and therefore thermal reactors), is by insertion of neutron absorbing materials, which can be withdrawn under precise control as required. Safety control elements called shut-down rods are often fitted for rapid cessation of the nuclear reactions in emergency conditions.

Greater efficiency can be achieved by the use of fast neutrons without any moderator, but the heat flux is much higher, and the problem of cooling requires better heat transfer. To date, liquid metals have been used, but these are not the only feasible ones. Their use brings other problems, of corrosion, and of potential high reactivity in heat exchangers, since the low melting point

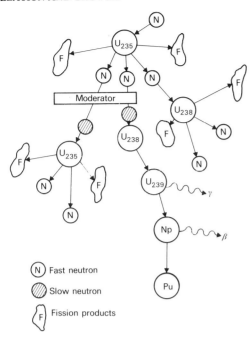

Fig. 96 *Diagrammatic representation of reactions in a nuclear reactor.*

(N) Fast neutron

(N) Slow neutron

(F) Fission products

alkali metals, sodium and potassium, are the ones with the desired combination of properties, low neutron capture cross-section, good heat transfer, and good pumping efficiency. In such reactors the excess neutrons can be utilised to produce plutonium or other materials in a special section outside the "core" of the reactor, to "breed" more fuel, hence the name fast fission breeder-reactor.

The use of nuclear reactions to generate power is more efficient in the utilisation of the earth's resources than any other proven system, and represents the greatest potential when cheap fossil fuel becomes exhausted towards the end of the twentieth century. The engineering and safety problems are well understood as a result of immense effort expended up to 1960 in the UK, the USA and the USSR in particular, so that the only problem outstanding is the disposal of radioactive waste from the irradiated material when reprocessed. Many solutions exist for this disposal problem, which involves a very small total bulk, and would still do so, even if all power were generated by nuclear means. The principal obstacle would appear to be an emotional one, in view of the deeply rooted fear that populations have of the results of the use of nuclear energy for military purposes.

Nucleation and growth. A change in "order" or entropy of a system requires a process of initiation and subsequent propagation. Nucleation and growth processes apply to all materials, and to many types of transformation.

The short range order gases, of high entropy, can condense into liquids by simple changes in temperature, pressure and volume (*see* **Condensation**) but also via heterogeneous nucleation. Condensation of gas on to small solid particles is not dissimilar to condensation on large areas of solid surface, and gases also condense to the solid phase, e.g. during evaporation in a low pressure system, and collection on a cooled metal surface. Liquids transforming to solids is a more common feature of materials science, but the principles remain the same.

In both gas and liquid, such short range ordered units as do exist are continually agglomerating into larger units by collision, and almost equally rapidly breaking up again. This reversible condensation process provides the mechanism for the principle of nucleation and growth. If a condensation unit can be made stable, it becomes a nucleus, on which further growth can take place, the reversible process is modified so that the rate of aggregation exceeds the rate of disintegration. A critical point can be described, in terms of size, and spontaneous aggregation of such units of longer range order than the mean, is described as homogeneous nucleation, because the material forms its own nuclei. The distribution of nuclei is usually random, and such nuclei

obey a first order dependence on time, $n = Nt$, where n is the number of nuclei per unit volume at time t, and N is a nucleation constant.

If however, nuclei are provided by other sites, e.g. impurities suspended in a gas or liquid or the surfaces of a containing vessel, there is zero order dependence on time once the temperature of crystallisation is reached, and growth centres become instantaneously active. This is heterogeneous nucleation, because the nuclei are different materials. In this case the number of nuclei available is important, because this affects the maximum size of growth units which are feasible. Growth has first order dependence on time, and if for convenience a spherical shape is assumed, $r = Gt$ where r is the radius of the spherical growth unit, t is time, and G the growth constant. The average volume of growth units on completion of the transformation is given by

$$\phi = \frac{V_F}{N^1 V_0}$$

where ϕ is average unit volume, V_0 the initial volume of short range order material, V_F the final volume, and N^1 the number of nuclei per unit volume. As V_0 and V_F are almost similar, ϕ is approximately the reciprocal of the number of nuclei N^1.

In homogeneous nucleation however, nuclei continue to appear as growth proceeds, and hence growth and nucleation are interlinked. Rapid growth does not allow the potential nuclei to grow to full size, because all growth occurs on existing nuclei, and the "grain" size is large. Slow growth however, continues as many more new nuclei appear, and a small grain size results. In this case the final average volume of growth units is more complex,

$$\phi = \frac{4\rho_L}{\rho_s \Gamma^{¼}} \times \left(\frac{\pi\rho_s}{\rho_L}\right)^{¼} \times \left(\frac{G}{N}\right)^{¾} \quad \text{which is approx.} \left(\frac{G}{N}\right)^{¾}$$

where N = nucleation rate per unit volume, Γ is a constant called the gamma function, ρ_L is the density of the liquid, ρ_s is the density of the solid, and G is the growth rate per unit volume. Varying sizes of growth unit must result, since some nuclei appear early in the process, and some very late.

High temperatures favour slow growth, because supercooling is limited, and generally coarse grain size results. In the case of heterogeneous nucleation, this would only be true if the number of nuclei per unit volume was lower at higher temperature. In many cases this is confirmed by experiment, the number of heterogeneous nuclei becoming operative seems to increase as temperature falls, although obviously there must be a finite limit even at the lowest temperatures. Nuclei added deliberately are of course under control.

Homogeneous nucleation is affected by temperature, because both G and N are affected, often both in the same direction. The ratio G/N is often temperature independent as a result. Turnbull and Fisher proposed a relation $N = N_0 . \exp(-E_D/kT - \Delta G/kT)$ where N = nucleation rate, E_D is the activation energy for transport across the surface of a nucleus, ΔG is the free energy of formation of the critical size nucleus, k is Boltzmann's constant, T is absolute temperature.

For a normal three-dimensional nucleus, e.g. a sphere, ΔG is proportional to $T_m^2/\Delta T^2$ where T_m is the melting point, ΔT the degree of supercool (i.e. T_m-T_z where T_z is the actual temperature). When there is moderate supercooling, an initial incubation period may occur, with no nucleation, followed by slow growth in the number of nuclei. With extensive supercooling, the incubation period has occurred already as temperature was falling, and so a rapid rate of nucleation occurs. Depending on the material, 5 °C to 20 °C supercooling is often necessary to overcome this incubation period. As the temperature of crystallisation falls, it is E_D/kT which is the predominating influence, not free energy. N therefore rises to a maximum, and then decreases with temperature. Figure 97 shows the change in free energy at different temperatures, plotted against radius of nuclei, and indicates that with radii less than the maximum point on each curve, nuclei redissolve or disperse. Above the maximum point, they grow. At the melting point, the maximum point on the curve is the critical size, usually less than 1 μm, but the size is still large, and growth may not occur, as there is little favourable change in free energy. As the degree of supercooling increases, the maximum point on

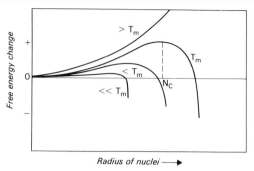

Fig. 97 *Change in free energy vs. radius of nuclei.* T_m *is the melting point; and* N_c *the critical size of nucleus for growth at melting point.*

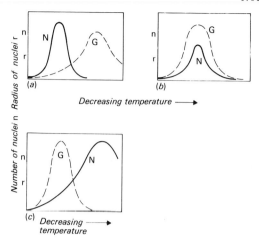

Fig. 99 *Nucleation and growth vs. temperature. (a) nucleation rate more rapid than growth; many nuclei, growth slow — fine grain size. (b) Growth rate similar to nucleation rate; few nuclei before maximum growth — fairly coarse, but mixed, grain size. (c) Growth rate more rapid than nucleation — coarse grain size.*

the curve moves to nucleus size well below the critical, and more favourable change in free energy, so that many more small nuclei result. This is observed in cooling curves with large supercooling, (taken under conditions to exclude heterogeneous nucleation) when a large rise in temperature occurs as the latent heat is evolved, a flood of nuclei form, and solidification proceeds rapidly.

Nucleation rate is obviously directly dependent on the creation of new nuclei, and therefore on temperature or the rate of abstraction of energy, since the latter leads to more supercooling and removes the latent heat to prevent temperature rising again. Nucleation is therefore slow in the centre of a volume of cooling liquid and more rapid nearer to the cooling surface of the container. The number of nuclei plotted against temperature is illustrated in Fig. 98, a Gaussian distribution.

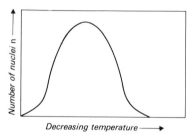

Fig. 98 *Number of nuclei vs. temperature.*

Growth rate follows a similar trend to nucleation rate, as already described, but the two are then in open competition. The shape of the two curves is similar, but the maximum may be reached at different temperature levels. Slow cooling will normally build up slowly to a maximum nucleation rate, but also to growth rate, and so a mixed grain size will result, from the early nuclei growing large, and later ones growing to a lesser size even though growth is now more rapid, because more nuclei are also produced, and solidification becomes complete before the later nuclei grow to very large size.

Rapid cooling will give a high rate of both nucleation and growth, but the relative maxima of each process will determine the final effect, whether coarse or fine. If the maximum of the nucleation curve is reached earlier than the growth one, then fine grain results, if the converse, then coarse grains. In practice, a mixed grain size often results in metal castings, very fine on the surface, coarse for the subsurface, and very fine in the interior. Figure 99 indicates some of the possible variants.

Heterogeneous nucleation requires identification of nucleating points. These can be irregularities on the mould surface — as with sand grains in a sand mould as against a metal mould, or impurities from furnace or ladle refractories, dust and dirt in the foundry, or intentional additions of inoculants. With non-ferrous metals, the lower melting points lead to less supercooling, and heterogeneous nucleation may be an excellent method of producing finer grain size, provided that any additions are not detrimental to desirable properties.

In polymers, additions are made as in metals, e.g. in polyamide production (Nylon) polyesters, phosphates and rutile crystals have been proved effective, reducing spherulite size from 50-60 μm at 150 °C, down to 5-15 μm. (*See* **Crystallisation** for structural considerations.)

Nucleus. The relatively small and compact structure at the centre of an atom. It carries a net positive charge equal to the negative charge on the electrons that move around it. The nucleus accounts for practically the whole mass of the atom and is inaccessible to ordinary laboratory agencies because of the strong electric field that surrounds it. Radioactivity arises in the nucleus of an atom and is in fact a nuclear process.

Nuevite. A somewhat rare mineral containing yttrium. It is found in California.

Nugget. The larger masses of a rounded or mammilated form in which native metals, particularly gold, are found.

Numerical aperture. The resolving power of an objective lens system of a microscope depends on the angle included between lines drawn from the focal point to the outer limits of the front lens (called the angle of aperture). It is calculated as the product of the sine of half this angle and the index of refraction of the medium between the objective and the specimen that is being examined. Thus $\alpha = \mu \sin n$, where $\mu = 1$ for air and 1.515 for cedar-wood oil. For dry lens the highest value of α is about 0.95. With a good oil-immersion lens the value may rise to 1.3 or 1.4.

The essential properties of a microscope such as illumination and resolving power depend on α.

Nylon. A polyamide, of which several types have been extensively used. (*See* Table 10 for nomenclature and details of processing.)

Objective. The lens system in a microscope that is nearest to the specimen being observed which forms the first real image.

Obsidian. An acid volcanic igneous rock formed as a result of the rapid cooling of an acid magma. It is a potassium sodium aluminium silicate containing about 70% silica. The vitreous substance on heating changes to pumice. The grey and black varieties are used in cheap jewellery, since they take a high polish. Smoky-coloured obsidian is sometimes known as marakanite.

The heat-expanded material when crushed yields a fine fluffy powder used as a light-weight aggregate in wall boards, and in heat and acoustic insulating tiles. Also called volcanic glass and basalt glass.

Occidental quartz. A variety of quartz cat's eye in which the luminous band effect, characteristic of chrysoberyl, is in fact due to inclusions of parallel fibres of asbestos. Though it originates in Sri Lanka it is called "occidental" to distinguish it from the finer "oriental" stone. Specific gravity 2.65.

Occlusion. The term used in reference to the retention of liquids or gases in an invisible form on the surface of solids. Platinum, iron and palladium become permeable to hydrogen when heated, while iron also occludes carbon monoxide. The amount of gas taken up appears to be proportional to the pressure of the gas and inversely proportional to the temperature. Various explanations have been offered to account for the large volume of hydrogen that can be taken up by palladium. These include: (a) the formation of a chemical compound; (b) simple solid solution; (c) surface condensation. The latter has found most favour, in view of the phenomena observed when the occluded hydrogen is gradually removed from the metal. Also called adsorption. Charcoal adsorbs gases, showing preferential adsorption for the gases that are more readily condensed.

Ochres. These are argillaceous earths often containing barium and calcium with varying quantities of iron III (ferric) oxide Fe_2O_3 which produce the characteristic red or brown colours. Red and yellow ochre are earthy forms of limonite $2Fe_2O_3.3H_2O$. They are usually mixed with oil and used as pigments, since they are inert and permanent in colour.

Octahedral voids. Spaces in crystal lattices formed by the presence of six close neighbouring atoms disposed in the form of a double rectangular pyramid, four atoms in a square, with one on each side equidistant from the centre of the square and normal to that surface. (*See* Fig. 100.)

Octahedrite. One of the three mineral forms of titanium dioxide TiO_2, which is always found as small, well-developed crystals. These are long, slender tetragonal pyramids. The mineral is usually deep blue to black in colour and has a steely lustre. It is commonly known as anatase. Mohs hardness 5.5, sp. gr. 3.9, refractive index 2.5.

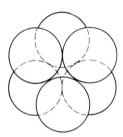

Fig. 100 *Octahedral void in two different forms.*

Octet structure. A term used in reference to the stable structure of 8 electrons in the outer shell of the atom of rare gases. The helium atom, which consists of the nucleus having 2 protons and 2 neutrons, and 2 extra-nuclear electrons is a very stable structure. In building up the atoms of complexity greater than helium, the additional electrons must be supposed to be arranged on an additional outer shell. When the number on this latter is 8 the stable gas neon results. This structure is repeated up to a further 8 electrons on another shell, when stable argon is produced and so on through the series.

Oddside. The term used in reference to the temporary cope used to support a full pattern which cannot be placed on a flat moulding-board. When single castings are required, the temporary oddsides are made in greensand, but if several are required strengthened sand containing linseed oil is used. For a large number of castings the oddside is made in wood, quick-setting plaster or metal. In making an oddside to receive the part of the pattern up to the mould joint, a top part box is filled with sand and strickled off level. The shape of the pattern is removed and the pattern bedded in to the joint line. The face of the temporary top part is smoothed, and then the drag is rammed up and the two boxes clamped together and turned over. The oddside is removed and subsequently the cope, to complete one mould, is rammed up.

Odontolite. A fossilised material known as bone-turquoise, since it consists of fossilised bones, teeth or ivory coloured by iron phosphate or stained by copper salts. Also known as occidental turquoise, to distinguish it from the true turquoise which is an aluminium phosphate containing variable proportions of iron and copper.

Offset method. A method of determining the proof stress, i.e. the stress corresponding to a specified permanent extension or "set". The apparatus required consists of an accurate extensometer in a testing (tensile) machine having an autographic recording device attached.

Ohm. The unit of electrical resistance. The resistance between two points of a conductor when a constant difference of potential of

1 volt, applied between these two points, produces in this conductor, a current of 1 ampere, the conductor not being the source of any electromotive force. The reciprocal of the resistance is known as the conductance and the unit is the reciprocal ohm (mho). This unit is also called the "siemens".

Ohmic resistance. The resistance offered to the flow of an electric current through a conductor, as distinguished from other types of resistance, e.g. reactance, inductance, etc.

Ohm's law. This law states that the ratio of the potential between the ends of a conductor and the current flowing in it is constant. If the potential difference E volts and the current is I amperes, then E/I is constant, and is called the resistance of the material. This resistance is measured in ohms (q.v.).

Oil bath. A bath containing an oil or a mixture of oils into which hot metals are plunged to produce rapid cooling. In the larger baths used for heavy pieces of metal, methods of circulating the oil to maintain the temperature at the desired value are necessary. Usually mixtures of mineral and vegetable oils are used, but the composition is of no vital importance, since the requirement is only the ability to extract heat rapidly.

Oil bonding. A term used in moulding and foundry work generally which refers to the addition of linseed oil to foundry sand. The sand is then referred to as oil sand (q.v.) and is specially useful in making cores and for intricate mould shapes.

Oil dag. The term used in reference to a mixture of finely divided graphite in oil. The graphite is sometimes mixed with tannin, and is then especially useful for the lubrication of internal-combustion engines. The essential constituent of a dag is finely divided or colloidal graphite, and this may be mixed with oil, glycerine, alcohol or even water.

Oil gas. A mixture of hydrocarbons and hydrogen that may contain up to 85% H_2 obtained by the action of heat on petroleum oil. In some cases the oil is sprayed into a hot retort, by which means the lower stable hydrocarbons (e.g. methane) are obtained in high proportion. It is frequently used in low-temperature flame-cutting torches and for similar applications. Also called carbo-hydrogen.

Oil hardening. A process in which alloy steels are hardened by heating to or above the critical range, and then cooled in oil. It is less drastic than water or spray hardening, and for nickel-chromium-molybdenum steels gives a product which after tempering has a tensile strength of 1.1-1.4 GN m^{-2}, provided the components are not too massive. Steels suitable for oil hardening may be found among those whose composition lies in the range: C 0.26-0.42%, Ni 0-3.5%, Cr 0-1.3%, Mo 0-0.5%. If chromium is absent, nickel *must* be present, and vice versa. The quenching temperature should not be below 820 °C and the temperature of subsequent tempering not below 550 °C.

Oil of turpentine. The oil obtained by the steam distillation of the oleo-resin that exudes from various coniferous trees. Its composition varies somewhat with the type of tree. Turpentine is a solution of various solids in the liquid called oil of turpentine, and when the former is distilled in steam the essential oil passes over, leaving a residue of colophony or violin resin. An oil similar to turpentine is obtained by distilling sawdust and waste wood. It rapidly absorbs oxygen from air, becomes darker and viscous and eventually produces a resin. With linseed oil it is able to transfer absorbed oxygen to the oil, thus producing a tough and durable film in paints. It is thus a drying oil. Oil of turpentine is used in making artificial camphor and artificial rubber.

Oil of vitriol. An oily, highly corrosive liquid, that consists of a solution of sulphur trioxide in sulphuric acid $H_2S_2O_7$. It is used for a wide variety of purposes, e.g. pickling baths, plating solution.

Oil quenching. The process of the rapid cooling of steel by plunging it, while at or above its critical temperature range, into oil. A wide variety of oils and mixtures of oils is available, though the nature of the oils is unimportant. Oil quenching is normally used with alloy steels in the process of hardening, the rate of cooling being intermediate between that of water quenching and cooling in air.

Oil-sand. All sands suitable for use in foundries must contain a

bonding agent. The natural bonding agent is clay, but in oil-sand moulding the silica sand mixed with fireclay, silica flour, dextrin, bentonite is bonded with molasses and a "core-oil" such as linseed oil. All oil-sand moulds are baked before use, usually by gas flame, while cores are dried in ovens. Since the oil bond disintegrates when heated to high temperatures, oil-sand used to surround the cast metal crumbles when subjected to pressure, and thus places no restraint on the cooling contraction of the casting. Oil-sand is always employed where a good surface finish of the casting is required, but in some cases it is economical to have only a surface finish of the mould in this type of sand.

Oil shale. A hard shale containing veins of a greasy solid known as kerogen which is oil mixed with organic matter. The crude oil, when extracted, is black and viscous and usually contains a high proportion of sulphur. It is cracked—i.e. heated out of contact with air and usually in presence of a catalyst—by heating above 400° C, yielding condensable oils and a solid residue. The yield varies according to the origin of the shale. Certain shales also yield waxes and resins.

Oilstone. A fine-grained naturally occurring slaty silica rock substance, used for sharpening steel tools. An artificial oilstone is made from compressed alumina.

Oldhamite. A meteoric stony substance which in addition to the usual native iron contains calcium sulphide. The importance of this latter consitutent lies in the fact that it does not, and indeed could not, exist as a mineral substance on the earth. The meteoric substance must have originated under conditions of lower oxidation and probably in dry atmospheres.

Oleum. "Oleum" is the name given to any mixture of sulphur trioxide and sulphuric acid. It is marketed on the basis of the percentage of SO_3 present. A 40% oleum means that the liquid contains 40 parts of SO_3 and 60 parts of H_2SO_4. The important uses of oleum in the order of tonnage are: (a) for making fertilisers; (b) chemical manufacture; (c) paints and pigments; (d) metallugical processes, which together account for 90% of acid produced.

Oligocene. The geological period that forms the second part of the Tertiary era. The system of rocks therefore includes those strata occurring between the Miocene (below) and the Eocene (above). They include sandstones, grits, marls and limestones and lignites. Oligocene beds are found in the Hampshire basin, including the Isle of Wight (UK). The sands and clays of this period are important in Belgium, from which country there was a good export trade in foundry sands. In Germany the brown coal occurs in the Oligocene deposits.

Oliver filter. A type of continuous rotating filter used for dewatering flotation concentrates.

Olivine. A rock-forming mineral composed of magnesium and iron II orthosilicate, $(Mg,Fe)_2SiO_4$. The purest kinds of olivine consist chiefly of forsterite, Mg_2SiO_4. The chief sources are found in the USSR (Ural Mountains), the USA and Norway. It has a high-melting temperature and can be used as a refractory and a heat-insulating material. In the former application it has a high resistance to crushing and a very high resistance to molten alloys. It is being used increasingly to replace wholly or in part ordinary quartz sand. Norwegian olivine, mixed with a small amount of magnesite, is used in the manufacture of refractory bricks. When the proportion of iron II silicate in the mineral is less than 15%, the mineral is olive-green in colour.

Olsen test. A cupping test for assessing the forming qualities of sheet and strip metals. In the test a steel ball or plunger is pressed into the surface of the material under test, this latter being held securely in clamps. The greatest depth of the cup that can be formed before the surface begins to tear is taken as an indication of the ductility of the material. If the ductility of the material varies in different directions, this may be assessed from the form and extent of the tear. A rough, orange-peel effect produced in the dome of the cup indicates that the material has an abnormally large grain size. The test is purely qualitative, but is regarded as providing valuable information in the hands of an experienced operator.

One-minute wire. A term used in reference to galvanised steel wire. It is applied to the copper sulphate test (Preece test) for wire, and means that the zinc coating will successfully withstand immersion in a standard neutral copper sulphate solution for 1 minute. Other grades of wire are known by the time in minutes that is required to dissolve the coating at its thinnest parts.

ONERA process. A method of applying a bright chromised surface to ferrous metals and alloys. A mixture of powdered chromium, alumina and moist ammonium fluoride and hydrofluoric acid is used, and the process is carried out in an atmosphere of hydrogen at temperatures in the region 1,050-1,100 °C. ONERA is from the initials of a French research body—Office National d'Etudes et de Récherches Aéronautiques.

Onofrite. A metallic selenide that occurs in Mexico. It has the formula HgSe.4HgS and is a source of the element selenium (q.v.).

Onyx. Amorphous silica occurs in nature in a variety of forms. Mixtures of amorphous silica with quartz occur as chalcedony, which is usually translucent and yellow, having a specific gravity of about 2.3. Onyx is a variety of chalcedony silica mineral differing from agate only in the straightness of the layers. Alternate bands of colour are usually black and white, or red and white. It can be artificially coloured by the use of mineral salts and acids. Argentine or Brazilian onyx is dark green to yellowish-green in colour and is used in decorative work. The red only is called sardonyx.

Onyx marble. Limestone having coloured impurities in band layers similar to onyx (q.v.).

Oolite. Calcium carbonate occurring as concentric deposits in layers around small nuclei.

Oolitic ironstone. An iron carbonate which has replaced the calcium carbonate of an oolitic limestone retaining the structure of the original rock, as in Cleveland iron ore.

Opal. A hydrated form of silica that contains 2-13% water

$$SiO_2.nH_2O.$$

Like other amorphous varieties of silica it has apparently been formed by the drying of colloidal silica. The gem opal shows brilliant colours by the interference of light in thin layers. Specific gravity 2.2, Mohs hardness 5.5-6, refractive index 1.44-1.46. Jasper is opal deposited in layers of different colours. Waxy opal is found in large quantity in Queensland.

Opalescence. A term used in reference to the play of colours seen in the opal. The coloration and iridescence are not due to ordinary absorption of light because the reflected and transmitted colours are complementary. The opalescence is probably due to the fact that as the silica gel (see **Opal**) dried it became riddled with fine cracks which became tilled with a gel ot slightly different composition. The heterogeneity so produced results in variations of refractive index. The range and brightness of the colours depend on the disposition of the cracks; the thinner and more uniform are the latter, the greater is the splendour of the colourings. Precious opals are rather porous and should not be immersed in water. The term is also used to describe any iridescence in a mineral substance.

Opalite. Fine white amorphous silica used as an abrasive and as a filler. Tripoli is a variety of silica similarly used. Opalite is employed also in oil-well drilling.

Opax. An opacifier used in ceramic glazes and vitreous enamels on metals. The non-transparency of the enamel is produced by the use of zirconium oxide to which a small percentage of silica is added.

Open die. Dies in which there is little restraint to lateral flow of material at the flash line. This generally results in excessive flash formation.

Open-end rolls. The term used in reference to rolls that are left open at one end to enable objects of ring form to be removed therefrom.

Open-hearth furnace. A reverberatory furnace, containing a basin-shaped hearth, for melting suitable types of pig iron, iron ore and scrap for the production of steel. The furnace is heated by means of gaseous fuel (often assisted by oil injection) fed in above the charge. Since it is desirable that the side walls should not bear the weight of the roof, the width of the furnace depends on design and is usually about 4.3-5 m for acid furnaces and up to 7.5 m for basic furnaces. The length must be such that it is convenient to work, and it must be arranged that the flames directed from one

end cannot reach the opposite end of the furnace. Furnaces arc of the order of 9-25 m long, the longer ones often being oil fired. The maintenance of the high temperatures needed in the steel-making process is ensured by the use of the Siemens regenerative system. (See Fig. 71.) The hearth, side walls and end blocks are lined with silica brick in the acid process, while for the basic process the furnace body and roof are lined with the mechanically strong silica brick dipped in tar, and the hearth is of magnesite bricks rammed with burnt dolomite, which is ground and mixed with dehydrated tar and applied in layers. In the acid process the bath of silica brick is lined with ganister and red sand (95/5) rammed in successive layers and a final layer of pure silica. In all open-hearth furnaces a large amount of dust from the ore and other materials and splashings from the slag are carried away by the waste gases. This is especially serious in the basic process. To prevent choking of the regenerators, and the fluxing and glazing of the chequer bricks, a supplementary chamber is provided in which slag and dust are collected. This is referred to as a slag-pocket or dust-catcher.

Before steelmaking can be started in the acid process, the silica bed must be saturated with slag from a preliminary heat, or by the fusing of cinder. Slag and pig iron in a small melt are then worked into the hearth by careful rabbling before a full charge is worked.

The open-hearth furnace is now obsolescent, although many are still in use. The oxygen steelmaking processes reduce the time for refining to 10% of that in the open hearth, and although oxygen lances can be used in the open hearth, furnaces designed for the purpose are more efficient, economical and easy to work. Estimates of world steel production in 1980 indicate open-hearth furnaces accounting for 14% or less.

Open-hearth process (acid). A process for making steel from pig iron, and scrap in an open basin having an acid (silica) lining, the heat being supplied by the burning of gaseous fuel within the furnace, the flames from which are directed upon the charge. The charge must have a low phosphorus and low sulphur content, the amount usually being restricted to a maximum of 0.04% of each, while the silicon content should be limited to 1%.

Since the atmosphere of the furnace is oxidising, the melt must be protected by a covering of slag, and during the working period this must be kept fluid. When the bath is molten, iron ore, to oxidise carbon, silicon and manganese, is added in controlled quantities, and the bath comes on the boil. This indicates the de-carburisation of the steel.

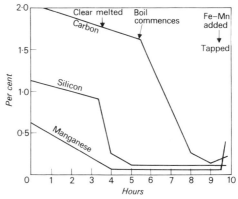

Fig. 101 *Removal of elements in open-hearth process.*

Open-hearth process (basic). A process for making steel from pig iron in which the phosphorus content may be as high as 2%, and thus not workable in the acid process (q.v.). The sulphur and the silicon should be low. After the open-hearth furnace (q.v.) has been prepared with a liberal coating of basic slag from iron ore and lime, the charge of pig iron is introduced, together with small amounts of lime and ore.

The phosphorus may not be completely eliminated before most of the carbon has been removed. Since the bath goes off the boil when the carbon is eliminated further removal of phosphorus would cease. It is therefore usual to add small quantities of low-phosphorus ore, by which means a little carbon is added and the boil recommences. The process was employed primarily for the production of mild or dead soft steel and was of economic importance in finding a use for the large amount of phosphorus-containing ore available.

Open joint. A form of joint used in certain foundry patterns, and so called because it is left slightly open with the object of avoiding the warping and the distortion of the pattern that is liable to occur when the pattern is subjected to the influence of damp sand for a long period. The joint should be made of hard wood, e.g. mahogany; pine is usually unsatisfactory.

Open sand moulding. A term applied in reference to moulds that are uncovered at the time of casting. It has the disadvantage in that it is suitable only for castings that have a level surface, and this fact limits its usefulness. Fluid metals at high temperature are in a state of agitation while in contact with air, which condition persists until the metal solidifies. Consequently the top surfaces of such castings are rough and are mainly suitable for bedding plates.

Ophorite. A mixture of magnesium powder and potassium perchlorate, used as an ignition mixture for incendiaries and initiating aluminothermic processes.

Optical properties. The incidence of electromagnetic radiation within the visible spectrum on a material produces either transmission, absorption or reflection, or a proportion of all three. The interaction involved is that between the electric field of the incident radiation and the electric field of the material, i.e. its charges. The visible spectrum wavelength ranges from approximately 0.4 to 0.7 μm, but the ultra-violet (0.1 μm) and the infra-red (50 μm) must be considered also, since many devices involve this range of wavelengths too.

The incidence of radiation, being electromagnetic, causes displacements in the material of the various charges, and results in the production of dipoles. Because the radiation is a wave motion, it resembles an alternating current, and the dipoles induced in the material oscillate about mean positions. This has the effect of slowing down the passage of the radiation, and so a transparent material involves a velocity of light below that of the velocity *in vacuo*.

The ratio of the two velocities is termed refractive index, μ. This is expressed as $\mu = c/v$ where c is the velocity of light *in vacuo*, and v the velocity in the medium. Since this index is brought about by polarisation, there must also be a relationship with the electrical permittivity.

The dielectric constant is defined as the ratio of permittivity of the material to that of a vacuum, and hence the velocity of light must vary. This is expressed as $v = c/Ke^{1/2}$ where Ke is the dielectric constant. Hence $\mu = Ke^{1/2}$, but this only holds where the material is transparent. For many solutions and gases, this is confined to the visible spectrum. Other frequencies affect the polarisation, and hence absorption may be involved, or reflection.

Although the induced dipoles can normally follow the field which induced them, they are governed also by the various attraction and repulsion forces involved in bonding. A natural resonance frequency is characteristic, dependent on mass, so that for electrons this frequency corresponds to ultra-violet radiation, and for ions, to the infra-red. Below the resonance frequency, dipoles will follow the field readily, and the refractive index is closer to 1, whereas near to it, the dipoles follow the field with difficulty, and the refractive index increases. At and just above it, some absorption results. The change in refractive index with frequency gives rise to dispersion, utilised in optical instruments for separation of wavelengths, and where different mechanisms for dipole formation exist in a material, there will be transitions of the type described over a band of frequencies.

Absorption of energy involves electrons moving into higher energy levels, and hence the energy gaps between available states must be directly related to the absorption which can result. In materials having bound electrons, such as insulators and semiconductors, a minimum photon energy exists, to excite the electron across the energy gap, and this photon energy can be equated to the wavelength of the radiation that will make it operative. The material will then act as an absorber to wavelengths short of that to give the critical energy, and be transparent to higher wavelengths. As the wavelength is further decreased, some of the energy of electrons in the higher state (conduction band)

will be re-emitted, and there will be reflectivity, with a more metallic appearance.

Insulators, having large energy gaps between the valence and conduction band, are generally transparent to visible light, because photon energies are insufficient to bridge the gap until the energy is very high, the wavelength very short. Insulators show dispersion, as described, with attendant absorption at specific and narrow bands of wavelength, due to the resonance phenomena described.

When free electrons are present, as with metals, the interaction with electromagnetic radiation is easy, and energy will be absorbed as any electron interacts with a photon. The more free electrons per atom, the more likely the interaction, and in general, metals absorb light readily in the infra-red. As wavelength decreases, energy changes can result in electrons being transposed into higher bands of energy, and as many such transitions are possible, there may be absorption across a wide range of the spectrum, but each transition is itself specific, and this gives rise to the colour of some metals when highly polished and reflective. (*See* **Colour of materials.**)

Ionic materials involve resonance processes as with insulators when the energy is low (infra-red) but electrons and band transitions can be responsible at higher energy (ultra-violet). The electrons in ionic bonds require considerable energy to move them to a higher energy level, so that a large range of transparency may exist between infra-red and ultra-violet. For this reason, many such materials find use in optical instruments (rocksalt, fluorite).

Impurities can play a very important part in the optical properties, by selective absorption, which may affect semiconductor materials and ionic ones, by localised transitions, or by producing colouring (*see* **Colour** or **F centres**). The production of excitons also affects absorption.

Reflection is strongest when absorption is greatest, as in metals, and least when absorption is low, as in glass. The colour of reflecting materials is an indication of the selective nature of absorption and re-emission, by whatever process.

Many of the applications of materials depend on optical properties, not only in optical instruments. The reflectivity of metals leads to obvious applications such as mirrors, but is also involved in decorative uses, in architecture, automobiles, furniture, packaging, and as an insulating material for buildings, in cooking, etc. Transparent materials like window glass serve also to transmit ultra-violet light from the sun, but absorb the infra-red given out when the inside of the chamber converts the ultra-violet to infra-red, as with the soil and plants in a hot-house, thereby conserving energy. Water vapour and carbon dioxide in the earth's upper atmosphere also absorb much of the ultra-violet from the sun, keeping the earth's highest surface temperatures to a more acceptable level. Concern is being expressed by the possibility of atmospheric pollution reacting to destroy some of this CO_2 layer. The absorption of light to give photoelectric emission may become an important source of alternative energy. (*See* **Photo-emission.**)

Optical pyrometer. An instrument for measuring high temperatures in which the temperature is estimated by the intensity of light emitted by the hot body. There are several different types: *(a)* the disappearing filament pyrometer. This is provided with a lamp, the filament of which is compared with the hot body, and when the lamp filament dissolves or disappears against the background of the hot body the estimation of temperature is made in terms of the reading of a milliammeter or a voltmeter. In an alternative type the measurement is made of the resistance of the lamp filament, using a "bridge"; *(b)* polarisation pyrometers of the Wanner type run the lamp at constant brightness, and the light from this and from the unknown source are polarised in planes at right angles. The field is then viewed through a rotatable Nicol's prism; *(c)* photoelectric pyrometers may be of the types *(a)* in which the human eye is replaced by a photoelectric or photoconductive cell.

Orange lead. A variety of red lead Pb_3O_4 that is produced by heating white lead. It is less dense than red lead produced by other methods. Although it is less satisfactory as a pigment in oil varnishes, it is valuable as a pigment for use in printing inks and in certain enamels.

Orange-peel effect. The term used in reference to the roughening of the surface of sheet-metal products that is encountered during forming from stock of coarse grain size. Also known as alligator skin and pebbles.

Orange red. The name used in reference to the pigment realgar or arsenic disulphide As_2S_2. It is used in paints, as a depilatory in tanning and pyrotechnics.

Orangite. A variety of thorite that is orange-yellow in colour. It is generally regarded as thorium silicate $ThSiO_4$, but usually contains some uranium and cerium. It contains up to 70% thoria, but the variable composition results in a wide range for the specific gravities of different specimens (sp. gr. 4.4-5.4). It is found in Norway and Sri Lanka.

Orbital. A term used in reference to the path of an electron round the nucleus of an atom. It is supposed that for any electron more than one orbital is possible, each being regarded as an "energy level". Since the angular momentum of an electron can be expressed as mvr, then $mvr = nh/2\pi$, where h = Planck's constant and n is the quantum number of the particular energy level. The possible orbitals are obtained by putting $n = 1, 2$, etc. (*See* **Atom.**)

Order. In materials science a term referring to regularity, of distribution in space, of energy level, or of rank in property value. (*See* **Entropy, Condensation, Crystallisation.**)

Ordering. A transformation in certain solid solutions in which a random arrangement of solvent and solute atoms is replaced by a regular or orderly arrangement of the different atoms on preferred lattice sites.

Ordinary ray. A term used in reference to the ray of light in a double refracting prism, which obeys the ordinary laws of refraction. When a ray of light is incident on such a material it is split into two rays, the ordinary and the extraordinary ray. The latter is plane polarised. (*See also* **Nicol's prism.**)

Ore boil. A term used in reference to the agitation of the molten steel in an open-hearth furnace that occurs after the clear melt and indicates the evolution of gas. At temperatures above 1,200 °C the iron oxide in the charge reacts with carbon, silicon and manganese. As the result of the reaction with carbon, considerable amounts of carbon monoxide are formed which, escaping through the slag, cause it to foam. This is known as the "ore boil". This should be distinguished from the agitation that occurs in the later stage of the basic O.H. process and which is known as "lime boil".

Ore down. The term used in reference to the additions of ore that are made to the charge during steelmaking in the open-hearth process. The iron ore is added in small amounts at intervals to assist in and hasten the removal of carbon. After each addition has had time to react the carbon content of the bath is determined.

Ore dressing. The concentration of a raw ore, the term used in reference to the separation of the desired mineral from the associated gangue and the conversion of the concentrates into products from which metals can be extracted on an economic basis.

The concentration processes may include: *(a)* grinding or crushing the former, yielding a finer product; *(b)* sizing by passing through suitable screens; *(c)* concentration using differences in specific gravity, e.g. classification, jigging, flotation; *(d)* magnetic separation; and *(e)* electrostatic separation.

Ores. Minerals which can be used economically as sources of one or more of the metals they contain. Earth and rocks that cannot be worked at a profit for the extraction of metals are not considered as ores. Ores may be oxides, sulphides, halides or oxygen salts, e.g. carbonate, silicate, etc. Ores must first be crushed and separated from gangue. This is a process of concentration and concentrates are either worked on the spot or sold on the basis of metal content.

Orford process. A process formerly used for separating mixed copper and nickel sulphides. The mixed copper-nickel matte from the Bessemer converters is melted with sodium sulphate and coke. The resulting product solidifies as two phases: nickel sulphide at the bottom and copper sodium sulphide at the top. The phases are readily separated mechanically, and by a repetition of the process the purity of the nickel bottoms can be increased to about 98% nickel sulphide. This material can then be passed to the refineries.

Organic material. Strictly material involved in life processes, but generally taken to concern materials based on carbon, whether derived from natural sources or not.

Orichalcum. An alloy prepared by the ancients by heating copper with a mineral known as cadmia together with charcoal. Cadmia is the old name for zinc so that the alloy was in fact a brass.

Oriental amethyst. A purple variety of corundum Al_2O_3, the colour being due to the presence of manganese. The crystals are rhombohedral.

Oriental emerald. A green transparent variety of corundum used as a gemstone and for watch-jewels. The light green colour is due to the presence of chromic oxide. Iron II (ferrous) oxide produces a bottle-green colour.

Oriental topaz. A light yellow variety of transparent corundum, the colour being due to the presence of iron III (ferric) oxide. The refractive index is about 1.77, Mohs hardness 8, sp. gr. 4.

Orientation. The relative direction of the axes in two or more crystals, or the relative position of the axes referred to a surface or a line such as the surface of a crystallographic plane.

Original mineral. A term used in reference to a mineral that was formed at the same time as the rock in which it occurs was formed.

Orpiment. A trisulphide of arsenic As_2S_3 containing 61% As. It is found in Peru, Utah, Central Europe. The ore is yellow in colour. The sulphide is prepared commercially by subliming white arsenic As_2O_3 with sulphur. It is extremely poisonous because of the excess of white arsenic present, and this artificially produced product was formerly much used under the name of King's yellow as a pigment. It is now entirely superseded in this use by the comparatively innocuous chrome yellow. Rusma is a mixture of orpiment, lime and water and is used in the East as a depilatory. Specific gravity 3.4, Mohs hardness 1.5-2.

Orthoclase. One of the group of abundant minerals used for vitreous enamels, pottery, tiles, glass and fertilisers. It is the potash felspar having the composition $K_2O.Al_2O_3.6SiO_2$ and occurs as an essential constituent of the more acid rocks such as granite and rhyolite. It is frequently green in colour due to oxide impurities. The ground mineral (used in glass-making) has a specific gravity of 2.44-2.51, and a melting point between 1,250 °C and 1,350 °C. Also called sunstone and microcline.

Orthophosphoric acid. A tribasic non-volatile acid, H_3PO_4. It can be made by oxidising phosphorus, but is prepared technically by heating bone ash with sulphuric acid. It melts about 39 °C. The crystalline trisodium salt $Na_3PO_4.12H_2O$ is not deliquescent, and is used under the name of tripsa for softening boiler-water. Only the alkali (except lithium) phosphates are soluble in water. The crystalline acid is used as a flux for soft soldering aluminium bronze and other copper-base alloys. The specific gravity should not be less than 1.75.

Orthorhombic. A crystal form in which the axes are at right angles, but are of unequal length. Common examples are found in the crystals of potassium sulphate and aragonite.

Orton cones. Standard pyrometric cones used in the U.S.A. The complete series includes 42 cones that soften and bend at temperatures from about 1,125 °C to 1,835 °C. They are reliable only under strictly standard conditions as to heating rate and atmosphere. If the heating be slow, or if a soaking at a steady temperature be involved, the cones are liable to fall at temperatures lower than those specified. Even under the best conditions, temperature measurement by this means can only be regarded as crude.

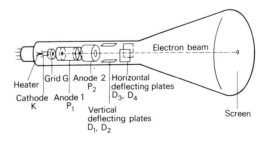

Fig. 102 *Schematic section of a cathode-ray oscillograph.*

Oscillograph. The most widely used is the cathode-ray oscillograph. This latter visualises electric and magnetic effects by causing them to deflect a narrow beam of electrons which produces a luminous image by impinging on a fluorescent screen. Since the electrons have practically no inertia and can be given high velocities, extremely rapid changes can be followed. In Fig. 102, K is the cathode, which on heating emits electrons. The grid G has a negative potential with respect to K to control the flow of electrons towards the screen. The anodes are maintained at a positive potential with respect to K, potential P_2 being higher than P_1. This arrangement constitutes the electron gun. The electric frequency or impulse under test is connected to the deflecting plates D_1, D_2, D_3 and D_4. The impulses on the deflecting plates change the direction of the electron beam from that established by the electron gun and the moving spot produces characteristic figures on a fluorescent screen.

Osmiridium. An alloy of osmium and iridium found in platinum ores. Small amounts of ruthenium and rhodium are also present. The compositions of the alloys found in different localities are given below.

Composition of Osmiridium, %				
	Urals	New Granada	California	Australia
Iridium	55.2	57.8	53.5	58.1
Osmium	27.2	35.1	43.4	33.5
Rhodium	1.5	0.6	2.6	3.0
Ruthenium	5.9	6.5	0.5	5.2
Platinum	10.2	—	—	—

The Californian metal, which contains less platinum and more osmium and iridium, is denser and harder than the Russian alloy and is preferred because of these properties for pen-nib alloys. It is exceedingly hard and was used for fountain-pen tips and gramophone needles. It is difficult to work and is usually fabricated by precision casting. It finds limited application in balance bearings, compasses and point bearings for delicate electrical instruments.

Osmite. A term applied to the nearly pure osmium (usually about 80%) alloy found in Borneo. It contains in addition about 10% Ir and 5% Rh. The term is also applied to iridosmine (*see* **Osmiridium**) that contains over 40% Os.

Osmium. One of the platinum group of heavy metals that occurs native in crude platinum and alloyed with iridium in osmiridium. It is usually extracted from this latter source. Its density calculated from the space lattice is 22.61×10^3 kg m^{-3}, but actual determinations have given values between 22.41 and 22.57×10^3 kg m^{-3}. Either this element or iridium is the densest of all metals, but available data do not permit a more definite statement. The metal is tin-white in colour and has a melting point about 3,000 °C, but when heated in air it oxidises fairly rapidly. The oxide has a low boiling point, so that it volatilises at moderate temperatures. In the absence of air osmium shows the lowest vapour pressure of all the platinum group metals, and on this account and also because of its high melting point, osmium filaments were used in incandescent lamps until displaced by tungsten filaments.

The metal is difficult to work and is brittle even at high temperatures. When pure it has a hardness of 350 V.H.N. It is not usually employed as a platinum alloy, though it has approximately double the hardening power that iridium has on platinum. It is probably somewhat less corrosion resistant than some of the other members of the group. The oxide has an unpleasant odour (hence the name osmium) and is highly poisonous.

The principal uses of osmium are in fact hard fountain-pen tipping and for hard electrical-contact alloys, some of the latter also containing tungsten carbide. It is a very efficient catalyst and is used in hydrogenation processes. Osmic acid has been used for the detection of finger-prints, since it is very readily reduced. (*See* Table 2.)

Osmium pen alloy. An alloy of osmium, rhodium and platinum in the proportions of 17:2:1 used for making the hard tips of fountain pens.

Osmosis. A term invented by Dutrochet, and derived from the Greek *osmos*, meaning a push, used in reference to the flow of a solvent liquid through a membrane or semi-permeable septum. If a quantity of a solution and pure solvent are brought together, diffusion will occur and eventually the concentration of solute molecules will be the same throughout the liquids. If, however, the solution and the solvent are separated by a membrane such as parchment, the rate of diffusion of the solvent into the solution will usually be greater than the rate of diffusion of the solute molecules into the pure solvent. There is thus a tendency for the solution to become more dilute and for the solvent to take up some of the solute, and there is an overall tendency for the solvent to pass to the stronger solution so that both become more nearly equal in concentration—this phenomenon is called osmosis. At one time the terms endosmose and exosmose were applied to the oppositely directed diffusion currents, but the single term osmosis is now used to denote the process of diffusion of a liquid through a membrane.

Osmotic pressure. This is defined as the equivalent of the hydrostatic pressure which is set up when a solvent and a solution in the same solvent are separated by a semi-permeable membrane. Since osmotic pressure results from a difference in the relative velocity of osmosis (q.v.) of solvent and solute molecules, it follows that as the membranes used become relatively less and less readily permeable to the solute, the osmotic pressure observed will increase. A membrane that is permeable to solvent molecules but not to solute molecules is termed semi-permeable and such a membrane allows osmosis to take place in one direction only. The osmotic pressure in such cases therefore increases until it is balanced by the hydrostatic pressure produced by the osmosis. It is to this equilibrium or maximum pressure that the term osmotic pressure is applied. Hence, osmotic pressure may alternatively be defined as the equivalent of the excess pressure which must be applied to a solution in order to prevent the passage into it of the solvent through a semi-permeable membrane.

Osmotic pressure is a measure of the difference which exists between the free energy (or the activity) of the solvent in the pure state and in the solution. It is a measure of the force producing osmosis, and osmosis itself is due to this difference in free energy. The difference also manifests itself in the lowering of the vapour pressure of the solution.

The osmotic pressure of the dilute solutions of non-electrolytes is proportional to (*a*) the concentration of the solution (or inversely as the volume of the solvent) and (*b*) the absolute temperature. The analogy between the behaviour of gases and dilute solutions of non-electrolytes is stated by van't Hoff's generalisation, viz. "the osmotic pressure of a solution is equal to the pressure that the solute would exert in the gaseous state if confined in a volume equal to the volume of the solution at the same temperature."

Osram. The first practical metal filament electric lamp contained osmium and the name Osram is a contraction from osmium and wolfram. However, osmium has not been used for about fifty years, but the name is still preserved as a trademark for certain tungsten filament lamps.

Ostwald-Planck dilution law. A mathematical expression of the relationship between the degree of association of a substance in solution and the concentration (or dilution) of the solution. Its application is confined to univalent molecules. It states that

$$K = C \frac{a^2}{1-a} = \frac{a^2}{v(1-a)}$$

where K = dissociation constant of the electrolyte, C = equivalent concentration of solute in solvent, $1/v = C$, a = degree of dissociation. Where a is small the equation may be simplified to give

$$(K/C)^{1/2}$$

Ostwald ripening. The absorption of small dispersed particles by larger ones, when a reduction in surface energy results. A process of diffusion is obviously involved, so that energy must be put into the system to effect the change.

Otto cycle. The cycle of operations that is the standard of comparison for internal-combustion engines in which the heat is received and rejected at constant volume. In Fig. 103 the line AB represents the induction of the charge and the thermal cycle commences at B. (*a*) The charge of fuel and air is compressed rapidly and adiabatically from B to C; combustion occurs and heat is liberated; (*b*) the volume remaining constant, the pressure rises from C to D so that both pressure and temperature have their maximum values at D; (*c*) the working stroke is represented by DE in which the gas expands adiabatically; (*d*) heat is removed at the end of the stroke at constant volume, and this is represented by EB. The temperatures, volumes and pressures at the beginning and end of the strokes of the 4-stroke cycle are indicated on the diagram. If γ = the ratio of the two specific heats of a gas (C_p/C_v), then efficiency

$$= 1 - \frac{T_4 - T_1}{T_3 - T_2} \qquad = \frac{T_2 - T_1}{T_2} \qquad = 1 - \left(\frac{1}{r}\right)^{\gamma - 1}$$

where r = compression ratio, $V_1/V_2 = V_4/V_3$. If air is taken as the working substance then $\gamma = 1.404$. The cycle is then called the "air standard cycle" and the standard efficiency is then

$$1 - \left(\frac{1}{r}\right)^{0.404}$$

Also called the constant volume cycle.

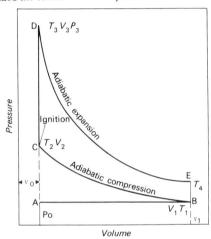

Fig. 103 *Four-stroke, or ideal Otto cycle.*

Outcrop. The area over which a particular rock or a mineral deposit comes to the surface of the ground.

Outer electrons. The highest energy levels reached by extra-nuclear electrons in an atom. They are the electrons involved in chemical reactions and many physical phenomena such as bonding, conductivity, ionisation, absorption and attenuation of radiation, and are least affected by the nuclear charge due to the shielding effect of the inner electrons.

Oven. The name generally applied to a variety of furnace in which materials are heated to moderate temperatures for purposes other than melting the charge. Low temperature ovens are used for such purposes as core drying, japanning or stoving. Moderately high temperatures are required for preheating, annealing, etc. The ovens are usually lined with fire brick. Solid fuels are less frequently used than gas or oil, and generally the recuperative principle (i.e. the utilisation of waste heat) is employed. In the muffle type, the source of heat and the products of combustion are external to the working chamber. The reverberatory type provides for the circulation of the products of combustion through the working chamber and the heat descends on the objects within the furnace.

Over-ageing. A phenomenon encountered in certain alloys which, when freshly quenched, in the form of a supersaturated solution, are soft. When a solution-treated alloy is hardened by heating at a predetermined temperature, the optimum properties are attained after a certain length of time. The alloy becomes harder when precipitation takes place, but after the peak properties have been reached the particle size of the precipitate becomes large and the material slowly begins to soften, i.e. it is said to become over-aged.

Overbending. A term used in reference to the allowance that is made for springback when bending metal to any prescribed angle or shape. The term also refers to the angle through which a metal component is bent, in excess of the specified angle for the completed bend.

Overblown. A term used in reference to the production of steel in the Bessemer process. In this process oxidation of impurities such as silicon, sulphur and phosphorus in the charged pig iron occurs first, followed by carbon. The carbon monoxide produced burns with a blue-edged flame at the mouth of the converter. After a short time the oxidation of the carbon is completed and the flame disappears. If the blast is allowed to continue after this,

oxidation of part of the iron occurs and the charge is then overblown.

Overdraft. A term used in reference to behaviour of sheet metal in passing through a rolling-mill. The metal tends to curve upwards on leaving the rolls due to the speed of the upper roll being less than the speed of the lower.

Overdrawing. A process in which an increase in tensile strength of a steel is obtained by excessive reduction of cross-section. This excessive cold-working also seriously reduces the ductility of the metal (as measured by the percentage elongation and reduction of cross-section) and increases the probability of breakdown by fracture.

Overheating. The exposure of a metal to unduly high temperature, whereby it acquires a coarse-grained structure, thus impairing the properties of the metal. Unlike a burnt structure, which represents permanent damage, overheated metal may be corrected by suitable heat-treatment or by mechanical working or by a combination of both.

In the case of (alloy) tool steels, overheating may cause melting to occur at boundaries. This in turn may result in the hard carbides associating with other constituents, and their separating out, or even their removal from the crystal boundaries. In either case their useful properties are lost.

Over-poled. A term used in reference to the refining operation of blister copper in which the metal contains oxygen in combination as copper I (cuprous) oxide. This latter renders the copper brittle, and thus it is "toughened" by the removal of the oxygen. This is effected by covering the surface of the molten metal with a thin layer of anthracite and plunging a green pole of wood into the molten mass. Large volumes of reducing gases are produced which cause the melt to "boil" and at the same time reduce the oxide. The reduction must not, however, proceed to completion, for then "tough-pitch" copper is not obtained. If the reduction has proceeded too far—which is judged by the appearance of the fracture of a solidified sample—the copper has been over-poled. It must then be exposed to the air for a short time to bring it back to "tough-pitch".

Over-reduced steel. In the production of steel for castings, the de-oxidation is a quite critical operation. Inadequate deoxidation results in unsoundness in castings. Over-reduced steel rapidly forms a solid film on the melter's test spoon, and because of this low fluidity often produces defective castings. A viscous slag, high in lime and manganese but with a low iron II oxide content may result in over-reducing the steel. Obviously careful control of the addition of deoxidants is required to avoid the low fluidity product.

Over-voltage. From a study of decomposition voltage (q.v.) it might appear at first glance to be impossible to deposit electrolytically from solution any metal that has an electrode potential more negative than that of hydrogen. This, however, is not the case, for zinc (electrode potential -0.7628) can be deposited from a weak acid solution. The explanation of this lies in the occurrence of an over-voltage when gases are evolved as a result of electrolysis.

Using a platinised platinum electrode, hydrogen is liberated at practically the reversible hydrogen potential if a dilute sulphuric acid solution is used. Using normal sulphuric acid, decomposition takes place when the voltage just exceeds 1.12 volts. If smooth platinum electrodes are used the decomposition potential is higher, at about 1.7 volts. With other metals the decomposition potential varies according to the nature of the metal. The voltage difference between the reversible hydrogen potential and that at which visible gas evolution occurs at a metal electrode in the same solution is termed the hydrogen over-voltage. This over-voltage is encountered not only in the case of hydrogen but also in the case of oxygen and other gases.

Owen jet test. A test of particulate pollution in the atmosphere, the dusty air being drawn through a slit as a fine ribbon-shaped jet and allowed to impinge on a coverglass placed 1 mm from the slit. The air is first passed through a damping chamber where it picks up sufficient moisture to produce condensation. The drop in pressure as the moist dusty air passes through the slit produces condensation due to the resulting fall in temperature. As the velocity falls off, pressure and temperature rise, the water evaporates from the coverglass, and the adhering dust particles can then be examined and counted under a microscope.

Oxidation. In its simplest terms oxidation means the combination of any substance with oxygen. Carbon burns in air, combining with oxygen to form carbon dioxide, so that this process is one of oxidation. On the other hand, when hydrogen is removed from a compound the process is likewise referred to as oxidation. When hydrogen sulphide burns in air under certain conditions it is oxidised to sulphur.

The definition is now extended to include all those cases where: (a) the oxygen content of a molecule is increased; (b) the valency of the electropositive part of a compound is increased and (c) the valency of the electronegative part of the compound is decreased. These latter very obviously correspond to an increase in the positive charge on the ion (i.e. the electropositive ion) or a decrease in the negative charge on the ion (the electronegative ion). Typical examples of these aspects of the definition are: (a) $H_2SO_3 \rightarrow H_2SO_4$; (b) $Fe^{2+} \rightarrow Fe^{3+}$; (c) $(MnO_4)^{2-} \rightarrow (MnO_4)^-$.

Oxidation-reduction indicator. Of two oxidation-reduction (O/R) systems, that which has the higher O/R potential (q.v.) will oxidise the other. The greater this difference in potential the more complete will be the oxidation. Figure 104 indicates the kind of potential change that occurs when an oxidising substance (the oxidant) is slowly added to the reductant. Assuming that the standard O/R potentials are markedly different (i.e. not less than 0.3 volt difference), the end-point of the oxidation will be marked by a sudden rise in the electrode potential as indicated at E (end point).

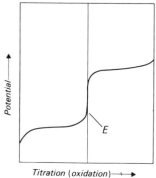

Fig. 104 *Curve showing potential change when oxidant is added to reducing agent.*

Substances which exhibit different colours in the oxidised and reduced states can be used as O/R indicators, but they must be so chosen that their standard O/R potentials lie between those of the oxidant and reductant, otherwise the end-point will not be sharply defined.

Oxidation-reduction potential. If, into a solution containing both oxidant and reductant ions, an electrode that is not attacked by them (e.g. platinum) is immersed, an electrode potential is established, the value of which will depend on the tendency of the ions to pass from one state of oxidation to the other. If the system has reducing properties, electrons will be given up to the electrode, which will then be negatively charged. The sign and magnitude of the electrode potential will provide a measure of the oxidising or reducing power of the system. This potential is termed the oxidation-reduction (O/R) potential. It can be expressed as

$$E = E_0 + 2.303 \frac{RT}{nF} \times \log \frac{C_1}{C_2}$$

where E_0 = potential when $C_1/C_2 = 1$; R = gas constant; T = absolute temperature; C_1/C_2 = the ratio of the concentration of oxidant to reductant ions; n = difference in the number of electrons in the oxidised and reduced states; F = Faraday constant.

Oxidising-reducing reactions are called "redox" reactions and the standard O/R potential obtained on a polished platinum electrode are defined as those at which the solutions are 50% oxidised.

Oxide. A binary compound of oxygen and another element. The compound in which the element combining with oxygen shows its normal valency is often called a normal oxide. Oxides of non-metals are usually acid, dissolving in alkalis to form salts, e.g. SO_2, CO_2, SiO_2, etc. The higher oxides of many metals and of all non-metals are also acidic. Oxides of metals are usually basic, dissolving in acids to form salts, e.g. CuO, MgO, FeO. Certain

oxides may behave as basic oxides towards strong acids and acidic towards strong bases. They are termed amphoteric oxides. Examples are found in the oxides of zinc, aluminium and lead. Thus zinc dissolves in sulphuric acid to form zinc sulphate $ZnSO_4$ and in caustic soda to form sodium zincate Na_2ZnO_2. Basic oxides such as Ag_2O, HgO evolve oxygen on heating and leave the metal. Peroxides contain a higher percentage of oxygen than the normal oxide but do not form corresponding salts as a rule, e.g. BaO_2, Na_2O_2, PbO_2. They readily evolve oxygen and in certain circumstances yield hydrogen peroxide. There are certain oxides that behave as if they were mixtures of two oxides. These are referred to as composite oxides. They are not mechanical mixtures and the constituent oxides are present in molecular proportions. Thus magnetic oxide of iron, iron II, iron III oxide, Fe_3O_4 behaves as $FeO + Fe_2O_3$, red lead Pb_3O_4 behaves as $PbO_2 + 2PbO$, trimanganic tetroxide Mn_3O_4 behaves as MnO_2 and 2MnO. Certain oxides, such as CO and NO, do not fall into any of the above classes and are appropriately known as neutral oxides.

Oxides form an important class of materials for industrial application. As many are ionic compounds, and the melting points high, they form useful ceramic materials for high temperature use as crucibles, furnace bricks, flame tubes and ducts. The oxides of silicon, aluminium, magnesium, calcium, beryllium, thorium, titanium, zirconium are in this category. Materials of high melting point also form good abrasives, particularly silica, alumina, magnesia, chromium and iron oxides.

Metals and alloys oxidise at the surface exposed to air, particularly when the temperature is raised. If the oxide coating is permeable, as it usually is to a greater or lesser degree, either oxygen can diffuse through it to attack the metal beneath, or metal ions diffuse outwards to oxidise on the surface. Depending on the specific volumes of metal and oxide, this may lead to disruption of the oxide film, causing cracking, and increased attack via the cracks. When oxide films are very coherent however, and the diffusion rates for both metal and oxygen ions are low, a resistant self-healing oxide film results in protection from attack. This occurs with aluminium and chromium, forming Al_2O_3 and Cr_2O_3 respectively, and in alloys containing them, particularly in austenitic steels containing more than 12% chromium.

On other alloys, oxides similar to the spinel structure may form. A normal spinel oxide contains 32 oxygen ions in a close-packed cubic unit cell, with 8 ions of metal A in tetrahedral voids, and 16 ions of metal B in the octahedral voids. Metal A must be divalent, and metal B trivalent, so that the formula is AB_2O_4. A combination of metal ions of valency 2 and 4 is also possible, provided that the total charge on the metal ions in the formula equals 8. When trivalent ions appear in the tetrahedral voids and a combination of trivalent and divalent in the octahedral voids then an inverse spinel results, general formula $B(AB)O_4$. Many compounds of both types exist, with not only heat resisting properties, but also remarkable magnetic and electrical properties. Nickel-chromium alloys form a pseudo-spinel oxide $NiCr_2O_4$ which is very resistant to diffusion, and this property alone assisted in the use of the Nimonic alloys to develop a successful gas turbine for jet propulsion. The use of similar alloys in electrical resistance heaters relies on this property too. Aluminium can replace chromium in the spinel structure, and one of the most common industrial thermocouples uses a combination of Chromel/Alumel, both alloys relying on spinel oxide coatings.

Magnetite, Fe_3O_4, is an inverse spinel whose formula can be written $Fe^{3+}(Fe^{3+}Fe^{2+})O_4$, and is ferrimagnetic with high resistivity (10^{10} times that of good conductors). The oxides which have antiparallel spin arrangements, leading to ferrimagnetism, are called ferrites, an unfortunate choice of name considering that metallurgists have used the name ferrite for the solution of carbon in alpha iron for more than a century.

As the arrangement of metal and oxygen ions in many oxides gives a net balance of electric charge in quantity, but the charges are separated by distances dependent on the crystal structure, dipoles may be formed, which will affect the behaviour in an electric field. Piezo, pyro and ferroelectricity can result in all these cases, and many oxides now form valuable sources of material for types of transducer; mechanical-electrical; thermal-electrical; electronic storage.

Oxide coatings. Oxide coatings on metallic surfaces are usually employed as a good basis for painting, though in certain cases the oxide itself forms a protective and even a decorative finish. Heating steel in contact with an oxygen-giving compound (potassium nitrate) produces the familiar "blueing". Anodic oxidation in the case of aluminium improves the oxide skin that is always present on the surface of aluminium and its alloys. The process is carried out by making the article the anode in an electrolytic cell which may contain chromic, oxalic or sulphuric acid. Aniline dyes may be added to the bath to produce attractive colours. Lead exposed to air usually develops an oxide-carbonate film.

Oxide replica. A type of replica of a metallic surface that is particularly applicable to aluminium. The etched surface is oxidised electrolytically and the oxide film subsequently removed by the action of a strong solution of mercuric chloride. Replicas of other metals can be obtained if the metal surface is pressed into soft aluminium foil. From the latter the oxide replica is prepared. The replicas are examined in the electron microscope.

Oxidising agent. A chemical substance that is able to supply oxygen directly or indirectly to the material to be oxidised. In more general terms an oxidising agent is a substance that can increase the proportion of the electropositive part of a compound, taking up one or more negative charges from the ion (*see also* **Oxidation**). The commonest oxidising agents are: (*a*) oxygen, peroxides, the oxides of the less reactive metals; (*b*) compounds containing a high percentage of oxygen, such as nitrates, chlorates, dichromates, persulphates and permanganates; (*c*) nitric acid, *aqua regia*, hot strong sulphuric acid; (*d*) chlorine, bromine, iodine (in potassium iodide solution).

Oxonium. The hydrated hydrogen ion H_3O^+. It may be formed when hydrochloric acid (gaseous) dissolves in water, the reaction being represented by $HCl + H_2O = H_3O^+ + Cl^-$. That the hydrogen ion exists in the hydrated form has been verified by Volmer's X-ray examination of perchloric acid. The components of the polar lattice are shown to be H_3O^+ and ClO_4^- when the crystals of the monohydrate $HClO_4 \cdot H_2O$ are examined.

Oxy-acetylene welding. When two pieces of metal are to be joined by holding them together and melting the interface, the process is known as fusion welding. If the heat is supplied by an oxy-acetylene (oxy-ethyne) torch, the process is known as oxy-acetylene welding. Extra metal is frequently added to the weld zone.

Oxygen. A gaseous substance at ordinary temperatures. It is the most abundant of all the elements, being contained in air (about 20% by volume), in water (about 88% by weight) and in many rocks of the earth's crust. For commercial uses, oxygen is usually prepared from the air by fractional distillation of liquid air. The gas combines with nearly all the elements except fluorine and the inert gases.

Oxygen is the active substance involved in roasting ores (whereby sulphur may be eliminated as SO_2); in the blast furnace oxygen in the preheated air first combines with carbon (in coke) producing carbon monoxide, which latter reduces the iron ores; oxygen-enriched air is used with advantage in both roasting and converter operations and in the smelting of certain metals, e.g. lead. Oxygen dissolves in certain metals, causing brittleness. The formation of oxide at ordinary temperatures on the surface of metals is referred to as rust (red rust on iron, white rust on zinc) or more generally as corrosion, and the prevention of corrosion usually involves coating the metal surface to exclude contact with air. (*See* Table 2.)

Oxygen-enriched blast. The use of oxygen in one form or another is an essential part of the steelmaking process. In smelting for the production of a basic pig iron it was early shown that the process can be made successful in a low shaft furnace. The use of an enriched blast will cause a saving of heat only (and thus a reduction in the rate of coke consumption) if either the top temperature is reduced or the burden increased, thereby raising the efficiency of the furnace. Usually the blast is enriched up to about 50% O_2. (Oxygen content of ordinary air is approximately 20% by volume.)

If the blast is enriched more than is necessary, so that the top temperature falls below about 120 °C, the coke combustion rate must be increased to provide for the loss of sensible heat entering the tuyères because of the reduction in the volume of the blast.

The advantages of enriched blast are: (*a*) a shorter time of blowing is required; (*b*) reduction in the amount of ferro-silicon required; and (*c*) hotter metal. The higher temperatures attained make it possible to manufacture ferro-alloys, e.g. ferro-chromium

(30% Cr) requires an enrichment up to about 55% and ferro-silicon-chrome (47% Cr, 13% Si) an enrichment up to about 80% O_2.

Oxygen-free, high-conductivity copper (O.F.H.C. copper). A type of copper containing neither oxygen nor deoxidising elements such as phosphorus, zinc, silicon, calcium, lithium, etc., which would reduce the electrical conductivity of the metal below that of tough-pitch copper (q.v.). It is produced from electrolytically refined copper by further refining and casting in a non-oxidising atmosphere. The electrolytic tough-pitch copper is deoxidised with carbon, since carbon is practically insoluble in copper. Casting is carried out in an atmosphere of nitrogen and carbon monoxide. More recently it has been found practicable to melt copper electrodes (cathodes from the electrolytic refining) in induction furnaces under a layer of carbon and in an inert atmosphere of nitrogen and carbon monoxide. Casting is done in the same atmosphere.

The absence of oxygen in the copper renders it immune from gassing (which causes embrittlement) when heated in a reducing atmosphere, and this fact makes it suitable for flame welding and brazing and for the preparation of glass-to-metal seals. O.F.H.C. copper is preferable to other grades of copper for cold impact extrusion or for any processes involving heavy degrees of cold work, such as abnormally severe deep-drawing or spinning operations. The metal can be heated to 800 °C in a reducing atmosphere without adverse effects.

Oxygen in steel. When low-alloy medium-carbon steels are exposed to the air at high temperatures, iron oxide inclusions may be formed beneath the scaled surface, so that dissolved iron oxides and oxygen may be present in addition to the observable iron oxide inclusions. The method of detecting the dissolved oxides, etc., consists in using an etchant consisting of alkaline chromate solution.

Oxygen lance. A steel tube—usually water-cooled—used for introducing oxygen under pressure to the bath of molten metal, in the production of "oxygen steel".

The lance may be used for injecting oxygen into the steel melted in the electric furnace, e.g. where stainless-steel scrap is used in the making of chromium steel.

Oxygen steelmaking. When atmospheric air is used in the blast in the usual type of converter, the inert nitrogen is picked up by the steel and applications of the latter may be limited by the effects of the nitrogen. On the other hand, the oxygen required in open-hearth practice is mainly obtained from the oxygen contained in the iron ore or scale in the charge. But the iron ore also introduces quantities of unwanted gangue material that must be slagged out. To overcome the limitations set by the above methods, and in order to produce a high-quality steel, low in residual alloys, pure oxygen may be directed on to the surface of the hot metal in a suitable container. A pressure of about 1 MN m^{-2} is normally employed.

The oxygen converts some of the iron into iron oxide, which immediately fluxes the lime portion of the charge. Any sulphur and phosphorus can then be taken up and eliminated by the molten slag. Any remaining oxide reacts with silicon, manganese or carbon, these latter actually acting as deoxidisers.

The action of the oxygen in producing carbon monoxide and the circulation of the latter before leaving the bath ensure a positive stirring action which is effective in the same way as the carbon boil in ordinary open-hearth practice. The process is continued for some 20 minutes or so, depending on the nature of the hot metal, by which time the silicon should have been removed and the other elements reduced to the approximate percentages given, viz. carbon, <0.1; manganese, 0.25; sulphur, 0.02; phosphorus, 0.015. The oxygen is introduced by means of the oxygen lance (q.v.).

The advent of bulk oxygen supplies (often referred to as tonnage oxygen) after 1945 caused a move to oxygen enrichment in Bessemer and similar converters, whereby all reactions were speeded up and carried further to completion. The amount of nitrogen in the steel is then reduced, partly because there is a lower quantity of nitrogen introduced in the blast, and partly because of the reduction in time of exposure.

The heavy manual nature and lengthy operation of open-hearth steelmaking led to considerations of combining the advantages of the converter as a furnace, with enriched oxygen, and the protective and refining slag advantages of the open-hearth, to improve the speed of steelmaking operations, reduce the manual labour, and increase the overall efficiency. Various attempts were made to allow for easy introduction and control of the oxygen lance, rotation of the furnace to give good slag/metal contact, and tipping of the furnace for charge and discharge.

The Kaldo process, developed in Sweden, used a vessel similar to a Bessemer converter but with a closed bottom, operated with the cylindrical axis at an angle to the horizontal, and introduced the oxygen lance at a steeper angle through the mouth blowing on to the metal surface (*see* Fig. 105). The slag was generated by oxidising the iron and the impurities via the oxygen lance, and the furnace rotated to assist even wear in the refractory lining. The removal of impurities follows the sequence of open-hearth practice, heat losses are small, and sufficient heat is generated by the oxidation to allow 40% of scrap metal to be charged. Carbon monoxide is generated by the lance, and burns largely within the vessel, the total heat time being 35 minutes.

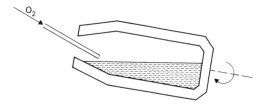

Fig. 105 *KALDO process.*

The Rotor process, in Germany, was similar, but the vessel more cylindrical in shape, supported horizontally, and the oxygen lance immersed below the surface. A large slag-metal area results from the shape, but considerable ejection of metal occurs above the surface, throwing hot metal on to the refractory lining above. The vessel is rotated to remove this metal by dissolution, and to even out hot spots on the refractories, but the Rotor certainly has high refractory wear. Good removal of sulphur and phosphorus occurs, which is due to the slag-metal interface area and the very high temperatures attained. Heat time about 30 minutes (*see* Fig. 106).

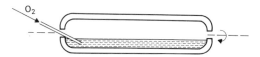

Fig. 106 *ROTOR process.*

The L.D. (Linz-Donawitz) process originated in Austria, using a vessel similar to a closed bottom Bessemer converter, but mounted vertically except for charge and discharge. The oxygen lance blows on to the surface from a vertical position, and the slag forms readily because of the deep penetration of the oxygen jet into the centre of the metal. There is no advantage in rotation of the vessel, so that very large vessels can be used, requiring only tipping like a Bessemer converter. The heat generated is conserved, as with Kaldo, but not as well as with Rotor. Nevertheless 25% of scrap can be used, and the heat time is approximately 40 minutes. On low phosphorus iron, the process was an outstanding success, but the larger amount of lime needed to form a fluid slag with high phosphorus iron increased the time involved. In Luxembourg and France, an adaption of the oxygen lance was made, to inject powdered lime with the oxygen (L.D.A.C. process) and this produces the slag at the point of highest temperature, and speeds the reaction to fix phosphorus early in the blow. The oxygen lance is also not impeded by the slag forming a thick layer on the surface, during the early stages before temperature has built up (*see* Fig. 107). The L.D. process is undoubtedly the best of the three, and has often replaced some of the earlier furnaces. It is now most frequently known as the basic oxygen furnace or BOF. By 1972, 400 tonne heats were being blown in L.D. converters in 12 minutes, using lances with 7 to 8 holes, and peak oxygen flow of 30 m³ s⁻¹. This is fifty times faster than the open-hearth operation, but confined to low phosphorus iron.

The use of BOF top blown converters for high phosphorus still requires removal of the first phosphorus slag, and hence increases heat time. The bottom blown Thomas converter (the adaptation of the Bessemer process to basic linings for high phosphorus iron) is still advantageous in this connection, and the obvious development is to modify the L.D. furnace for bottom blowing. The problem is the very high temperatures developed in front of the oxygen lance when immersed in metal, one of the problems of the Rotor. One solution is a concentric double tube tuyère, in which oxygen enters via the central tube, and a "heat shield" is formed by the outer tube, through which hydrocarbon fuel is injected. The O.B.M. process (oxygen bottom-blown Maxhuette)

Fig. 107 *L-D process.*

uses propane or butane, the L.W.S. (Loire, Wendel and Sprunk) uses liquid fuel oil. The hydrocarbon is "cracked" into carbon and hydrogen on entry, this process removing much heat around the tuyère and the refractories. Success is claimed for use of high phosphorus iron with powdered lime blown in with the oxygen, for converters up to 35 tonnes capacity. Large bottom-blown converters of this type (200 tonne O.B.M.) have been installed in the USA (1975) to prove the adaptation to larger heats (*see* Fig. 108).

Fig. 108 *OBM/Q-BOP process.*

Submerged injection processes (S.I.P.) are also under investigation, using annular tuyères fitted just below the slag/metal interface in existing tilting open-hearth furnaces, in Nova Scotia. Two tuyères in a 220 tonne furnace, with oxygen at 15 m³ s⁻¹ blowing rate decreased the tap time to 90 minutes starting with cold metal, and only 12 minutes was used for refining. This type of process could prolong the life of existing open-hearth furnaces, but bottom-blown converters also have the advantage of fitting into existing melting shops with minimal structural alteration. (L.D. converters require overhead bunkers and space for the vertically mounted oxygen lance.)

The trend in steelmaking, for new plant, or conversion of old, appears to be to top blown converters for basic structural steel, with possible modification to bottom-blown converters when proven. Electric arc furnaces are still likely to be used for alloy steels, but have been used in preference to converters for manufacture of basic steels.

In 1979, a world production of 700 million tonnes was made up by 14% open-hearth, 67% top-blown converter, 2% bottom-blown, 17% electric arc.

P

Pachuca tank. A cylindrical tank, often made of steel, which has a conical bottom. Down the centre of the tank is a wide tube, and into the lower open end is inserted the nozzle of an air-line. Air under pressure is forced into the air-line and is discharged into the wider tube known as the "air lift tube". The agitation of the liquid in the tank caused by the air raises material from the bottom of the cone to the top of the tank. Circulation is thus achieved, the material rising in the agitator or air-lift column and descending in the surrounding tank space. The circulation can be arranged to occur for a certain time and then the contents are discharged through an outlet near the base. When several tanks are used in series, a launder can be arranged to catch a proportion of the discharge from the top of the air lift and transfer it to the next tank. In this way continuous operation can be obtained. The tanks vary in height from 9 to 20 m and from 2.5 to 5 m diameter. They are used in the leaching, with sulphuric acid, of zinc ores and in the treatment of ores for the recovery of gold. The ores must be finely ground so that they may be kept in suspension.

Pack annealing. The process of continuous annealing of shallow packs of metal sheets.

Pack carburising. The carburising (q.v.) of metal parts by placing them in a heat-resistant alloy container packed with carburising material, and heating the closed container in a furnace. The closed box ensures that the carburising atmosphere is maintained, excludes air and conserves the carburising compound. Also called box carburising.

Packing. A term used in powder metallurgy in reference to the material employed to support the compacts before and during sintering. The term is also used in reference to the material used in annealing boxes to support the metal parts that are being treated. Such support reduces the tendency to warp during treatment.

Packing fraction. If a particular isotope whose atomic weight to the nearest whole number is N, the actual atomic weight will differ from this by an amount δ. Then δ/N is called the packing fraction. It is known that for most elements the actual mass of the nucleus is somewhat less than the sum of the actual protons and neutrons contained in the nucleus. This difference is known as the mass defect and represents the binding energy of the particles in the nucleus. The decrease in mass is usually defined by the packing fraction thus:

$$PF = \frac{(\text{Isotopic Mass}) - (\text{Mass Number})}{\text{Mass Number}} \times 10^4$$

The relationship of the packing fraction to the mass number is shown in Fig. 109, from which it can be seen that with the exception of the elements in the first short period of the periodic table, and the naturally occurring radioactive elements, the packing fraction is negative. The most stable elements are those indicated as lying on the lowest part of the curve. These are the elements in the first long period of the periodic table from chromium to zinc.

Also used to indicate the space occupied by atoms in a structure, assuming a spherical shape for the atoms, and reported as a percentage rather than a fraction. Closeness of packing gives some guide to void space available for interstitial atoms, possible ease of diffusion, availability for vibration, etc.

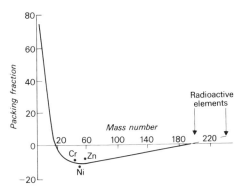

Fig. 109 *Relationship between packing fraction and mass number.*

Pack rolling. A rolling process in which two or more sheets of metal are passed through the rolls together. The process is used in the production of metal foils from sheets and in the older types of tinplate mills for producing basis sheet steel. This method gives more accurate control of the thickness of thin sheets than individual rolling.

Pad. The term used in reference to the portion of a casting, forging or stamping on which markings are to be made. It is also used as a term synonymous with boss to indicate a raised portion of a casting or forging that is subsequently intended to be drilled or tapped.

Paddle mixer. A type of mixing machine used particularly in the preparation of oil sands, for which the ordinary type of roll-mill is unsuitable. There are usually two (or more) sets of arms or paddles which revolve in opposite directions.

Paint. The term used in reference to liquids containing solid particles, which can be applied to metallic and other surfaces, and which on exposure to the air harden to produce a film that is opaque, and protective of the surface of the metal, etc. Protection by paint can be obtained in one or other or both of two ways. The film of paint may act by mechanically excluding the corrosive influences, or the film may contain a substance that inhibits the corrosion process.

Paints in common use were formerly based on drying oils, viz. linseed or tung oil. They contained a pigment, driers —frequently soaps of lead or manganese— and in many cases a thinner such as turpentine or white spirit. When the paint was applied to metal surfaces in a thin layer the thinner evaporated and the driers took up atmospheric oxygen, which was then passed on to oil base. The "drying" process consisted in both oxidation and polymerisation of the unsaturated constituents of the oil. These changes were usually accompanied by volume changes. These latter accounted for the expansion or contraction of the oil film, and since linseed oil finally showed a contraction, it was less water-proof than tung oil (which showed expansion). Pigments that required relatively small quantities of oil in the preparation gave better protective paints than pigments that required larger quantities of oil. Oxides of lead and of zinc acted as inhibitors, retarding the corrosion process at the metal surface. The prevention of corrosion by the presence of zinc dust in paints was cathodic, while chromates acted by depositing a film on the metal (steel) surface of ferric and chromic oxides.

Modern paints employ polymeric materials in place of drying oils, largely to decrease the time required for oxidation of a sufficiently thick layer of drying oil (all other aspects referred to above are still relevant). The monomer may be used with an inhibitor which breaks down due to the action of heat, ultra-violet light or oxygen, allowing polymerisation. Alternatively, polymerisation is arranged to occur in solution or in emulsions, the latter being particularly useful for water based paints. Acrylic and polyvinyl acetate resins are particularly suitable for this purpose, and polyesters used with drying oils (alkyd resins) to speed the drying of the protective film.

Acrylic emulsion paints are alkali resistant, and suitable for fresh plaster or concrete brick, in addition to very rapid drying. Polyvinyl acetate is used as a latex, a stabilised suspension in water, and dries simply by water evaporation. Droplets of resin coagulate on drying into a durable film, also resistant to alkali. Rubber based paints are also used as latex, usually a copolymer of styrene and butadiene, and dry by water evaporation, and again are alkali resistant. Silicone materials form heat resisting paints, either mixed with alkyd (polyester) resins, or incorporating aluminium powder as a pigment. Thixotropic paints are formed by the alkyd-drying oil combination. The unsaturated, carbon-carbon bonds of the drying oil are opened up to cross link with the polyester.

Pigments and corrosion inhibitors used with polymeric paints are the same as for oil paints.

Pairs. A term used in reference to the packs rolled in the hot-pack rolling-mill. In the multiple hot rolling of sheets, the sheet bars are rolled singly. These breakdowns are matched, reheated and rolled two in a pack. These latter packs are referred to as pairs.

Palladium. A silvery-white metallic element which occurs native in crude platinum, and in small quantities in cupriferous pyrites, particularly those containing nickel and iron sulphide (pyrrhotine). The chief source of supply is the copper-nickel ore deposits of Sudbury (Ontario), but the metal also occurs with platinum in Brazil, San Domingo, the Urals and South Africa.

Palladium is very ductile and can be worked hot or cold. Ingots are usually worked at about 800 °C to secure rapid reduction and finished by cold working. The metal withstands drastic cold working and can be beaten (in much the same way as gold) into leaf as thin as 0.1 μm. The metal is preferably annealed in nitrogen or carbon monoxide, since it oxidises in air below about 800 °C, and this range is preferred for annealing to obtain maximum ductility. Palladium absorbs up to 800 times its own volume of hydrogen at room temperature, and when molten readily absorbs oxygen, sulphur and carbon. The liberation of dissolved gases during solidification produces marked change (swelling) in volume. It is slightly harder and stronger than platinum and has a high resistance to tarnish in ordinary atmospheric conditions, but it is discoloured by sulphur dioxide.

Pure palladium is an excellent metal for electrical contacts, and large quantities are used in the telephone service. In dental alloys containing gold and platinum, palladium improves the strength and toughness and the response to heat-treatment.

When finely divided and suitably supported, palladium is a very effective catalyst in hydrogenation.

The thin foil is used for decorative purposes on ceramics and for ornamental metal-work on book bindings. Base metals clad

with palladium are used in the equipment for making fine chemicals. In the dental field alloys are used containing up to 50% Ag, 30% Pd with platinum and/or gold, and copper. The alloy containing 4.5% Ru is a very successful jewellery alloy. Other alloys (with copper, silver, chromium, iridium, etc.) are known, their main uses depending on their electrical resistivity and their resistance to corrosion. Palladium can be electrodeposited from a number of baths, particularly those containing the complex nitrites, and palladosamine chloride solutions. (*See* Table 2.)

Palladium-gold. There are a number of palladium-gold alloys which have important applications in pyrometry. The strongest and hardest alloys contain between 50% and 60% Au, the 55% Au-45% Pd alloy having maximum (cold-drawn) tensile strength of 570 MN m^{-2}. The same alloy possesses the maximum resistivity and minimum temperature coefficient of all the palladium-gold alloys. Alloys containing more than 10% Au are resistant to tarnishing in sulphur-bearing atmospheres, while those containing more than 20% Au are resistant also to nitric acid. All the alloys containing more than 15% Pd are white. The chief uses are: (*a*) in white gold for jewellery; (*b*) in gold-palladium/platinum thermocouples and (*c*) in solders for platinum. The respective maxima of gold percentage in these applications are approximately 20%, 60% and 75%, respectively.

Palladium oxide PdO. A covalent bonded oxide of co-ordination number 4, similar in structure to copper II oxide and platinum oxide.

Palladium plating. Palladium may be deposited directly upon silver or copper articles, but steel must first be copper or silver plated. The metal is usually deposited from a sodium-palladium nitrite solution in presence of sodium chloride, the usual composition being:

Na$_2$[Pd(NO$_2$)]$_4$	10 g
NaCl	30 g
Water	100 g

In this process the anodes must be made of hard-rolled palladium. An alternative process uses as the electrolyte a solution of tetramino-palladium II chloride [Pd(NH$_3$)$_4$]Cl$_2$ in a divided cell using a lead anode. Thin deposits only are satisfactory because thick deposits show a tendency to crack. Deposits of about 2.5 μm are bright and are capable of resisting tarnishing indefinitely.

Palladium-silver. Palladium and silver form simple solid solutions of varying composition but without any transformations in the solid state. All the alloys are very ductile and readily worked. The electrical resistance shows a maximum at a little under 40% Ag, while the temperature coefficient reaches a minimum at about 44%. Those alloys containing less than 50% Ag are tarnish resisting. When the palladium content is less than 50% the alloy may be ennobled by the addition of platinum, up to about 10%. Such alloys are used in jewellery. Palladium-silver alloys with copper and frequently gold and platinum are used for denture bases. Palladium-silver alloys hardened with copper have been used for pen-nibs.

The palladium-silver (56/44) alloy, because of its favourable electrical properties, may be used in the resistance-type strain gauges.

Parachor. A term used in reference to non-associated liquids and signifying a comparative volume. It represents the molecular volume of a liquid at such temperature that its surface tension is unity. Over a wide range of temperature it is found that the following relation holds:

$$\frac{\gamma^{1/4}}{D - d} = \text{constant } (C)$$

where γ = surface tension; D = density of the liquid; d = density of the vapour of the liquid at the same temperature at which γ is measured. If M = the molecular weight of the liquid, then

$$\frac{M\gamma^{1/4}}{D - d} = MC.$$

The constant MC is the parachor and since d is negligible compared with D,

$$\text{Parachor} = \frac{M\gamma^{1/4}}{D}$$

where M/D is the molecular volume.

Paraffin series C_nH_{2n+2}. The obsolete name for saturated un-branched acyclic alkanes, present in natural deposits of petroleum and separated principally by distillation processes. The source of material for many polymer derivatives.

Paraffin test. The term used in reference to a method of non-destructive testing. When used in the testing of a casting in which it is not convenient to make a pressure-tight joint, the casting is filled with paraffin and the outside is white-washed. The paraffin will find its way through any areas that are porous or unsound and discolour the whitewash. Castings that cannot be filled in this way may be immersed in *warm* paraffin for an hour or two. They are then removed and wiped dry, and the surface is dusted with finely powdered chalk. As the casting cools, any paraffin retained in a spongy or porous area or in any fine cracks is ex-pelled and causes discoloration of the chalk in the defective area. Fluorescent materials may be used in oil that is rubbed over the surface. After wiping dry the surface is viewed in ultra-violet light to reveal the presence of cracks.

Parallax. The term used in reference to the difference in direction or to the shift in the apparent position of a body, due to a small change in the position of the observer.

Parallel growths (epitaxy). A term used in reference to the property of a crystal of a compound forming a layer crystal in a solution of an isomorphous compound. The term isomorphism is now restricted to those substances whose similarity of form is unques-tionably due to a definite similarity of internal structure, so that crystals of isomorphous compounds are capable of forming regular overgrowths or parallel growths on each other.

Paramagnetism. When the unpaired electron spins in an atom are aligned parallel rather than antiparallel, the spin magnetic moments assist the applied field by aligning their moments with the field, and the susceptibility becomes positive, swamping any effects of angular momentum (diamagnetism). The substance is said to be paramagnetic. (*See* **Curie-Weiss law** and **Magnetic dipole moment.**)

Parameter. In mathematics and physics a parameter is a variable which is constant in the case under consideration but varies in different cases; a "measurable characteristic".

Parent metal. A term used in metallurgy to indicate the "most important" or "chief constituent". In the latter reference the parent metal is the main constituent in an alloy—the base of the alloy—after which the alloy is named. In the former cases the parent metal may be the metal which has to be treated in any way, e.g. the metal that is to be coated, clad or plated, or even the metal that is to be cut or welded.

Pargasite. A dark coloured hornblende (q.v.).

Paris blue. The name given to the finest variety of prussian blue. The commercial variety is a mixture of a number of iron III, iron II cyanides, and consequently varies in colour and appearance. The Paris blue is obtained by precipitation of potassium iron II cyanide with an iron III salt. Lumps of this product, especially when wet, show a splendid bronze colour, and by this the quality can be judged.

Paris green. A mixture of copper acetate (ethanoate) and copper arsenite $Cu(CH_3.CO_2)_2.3Cu(AsO_2)_2$. This was formerly used as a pigment, especially on wallpapers, but its use for this purpose has been discontinued, since the growth of certain moulds on the paper produced some very poisonous gaseous arsenic compounds. It is now chiefly used as an insecticide against plant pests.

Paris white. A fine grade of chalk that has been carefully ground and washed. It consists only of $CaCO_3$, and is used for paints, ink and making putty.

Parkerising. The name of a process that is used to produce on steel a sparingly soluble, protective anodic product. The bath used contains free phosphoric acid together with manganese dihydrogen phosphate. Copper salts, chlorates, nitrates and nitrites are some-times added to accelerate the process of formation of a fine-grain deposit of iron III phosphate. Aniline (aminobenzene) and tolu-idine (aminotoluene) are said to have similar effects. The bath is used hot and the time of immersion is short—of the order of a few minutes. After treatment the metal is rinsed in water or in a dilute solution of sodium chromate. After drying, the deposit is sealed with oil. The process usually produces a grey deposit, but this can be darkened by the addition of suitable dyes. The

process is used for treating small parts, such as nuts, bolts and small tools. It is said to be effective for intricate parts in auto-matic machines and in typewriters.

Parke's process. A process used for the desilverisation of lead. The process depends on the fact that molten lead and molten zinc do not mix (immiscible) and that silver is more soluble in the mol-ten zinc than in molten lead. The silver-bearing lead is melted in a furnace and about 1-2% Zn is added according to the silver present. The silver-zinc alloy floats on top of the molten lead and is skimmed off. The skimmings are heated with charcoal to reduce any oxides present, while the zinc distils off. The remaining silver-rich lead alloy is treated by cupellation to recover the silver.

Parrot coal. A dull black type of coal having a compact texture. It burns with a long luminous but smoky flame and has good quick-firing qualities. On dry distillation it yields a high proportion of illuminating gas. Low-temperature distillation produces a high yield of tarry oils—the residue is mainly ash. It is found in Scotland (where the name originated), and in other parts of Great Britain and in the USA. It is also known as cannel coal.

Partial dislocation. *See* **Imperfect dislocation.**

Partial pressures. If two or more gases that do not react with one another are confined in a given space, the pressure each exerts is that which it would exert if the same quantity were confined alone in the same space. This is called the partial pressure of the gas. In the case of several gases confined over a solvent, the amount of each gas which dissolves at any given temperature is proportional to the partial pressure. This is Dalton's extension of Henry's law.

Partial solid solubility. The behaviour of binary alloys—in which the two components are completely miscible in the molten state—may show three types of possibility on solidification. These are: (*a*) complete miscibility in the solid state as exemplified by the gold-silver alloy system; (*b*) complete immiscibility in the solid state as shown in the bismuth-cadmium alloy system; and (*c*) par-tial solid solubility. This latter is well shown in the tin-lead alloy system, the constitutional diagram of which is shown.(*See* Fig. 110.) Up to a certain small percentage, lead dissolves in tin, the amount depending on the temperature. The horizontal *CD* re-presents the solid solubility of lead in tin at 100°C. At the same temperature *EF* represents the solubility of tin in lead. Inter-mediate compositions between these two limits of solid solubility are composed of an aggregate of the two solid solutions (α and β).

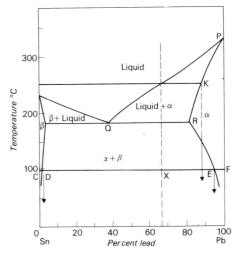

Fig. 110 *Lead-tin equilibrium diagram as an example of partial solubility in the solid state.*

In the solidification of an alloy containing 67% Pb the first solid that separates at 255 °C is β solid solution of composition denoted by *K*. The liquid then becomes richer in tin, and as the tempera-ture falls more solid is deposited, the composition of liquid and solid varying continuously along the lines *PQ* and *PR*, so that at 183 °C the solid has the composition represented by *R*. Just below this temperature the liquid solidifies as eutectic, and the alloy consists of β solid solution and eutectic in the proportion of

QX to *XR*. The β solid solution breaks down into β + α as the temperature falls so that at room temperature the solid is made up of the eutectic mixture of α and β and a further mixture of α and β produced from the continuous breakdown of β solid solution *R*. This description is applicable to binary alloys that show partial solid solubility in the solid state.

Particle size. A term used in reference to the linear dimension of a particle and normally reserved for the description of particle size in powder metallurgy, where the separation and analysis are accomplished by the use of a number of sieves or screens. The results are expressed by giving the particle size in terms of the sieve mesh or as the linear dimension in micrometres.

Particle-size distribution. In powder metallurgy, a method of specifying or describing a powder in terms of the percentage by weight of the various sizes of particles in the sample. The sample is passed through a series of screens and the weight of powder remaining on each, expressed as a percentage of the total weight, gives the distribution by agreed size ranges.

Parting. A method separating (or parting) silver, gold and platinum from one another. The old method of parting gold and silver in the bullion bead was carried out by dissolving the latter in nitric acid, and in order to make the separation successful the proportion of silver to gold was specified as 4 to 1, silver being added if necessary to achieve this ratio. The modern process is an electrolytic one, the anodes being obtained from the roasted slimes produced in the electro-refining of copper. The residues from the roasting having been freed from copper by leaching with sulphuric acid are pressed into Dore bullion. There are three processes: *(a)* the Balbach process in which the bullion is made the anodes in earthenware cells. The electrolyte consists of dilute nitric acid containing about 5% each of silver and copper. The cathode consists of graphite sheets placed at the bottom of the cell. The anodes rest on trays covered with canvas. On electrolysis the silver in the anode is deposited on the graphite cathodes in the form of fine crystals. The remaining precious metals are left in the black anode slimes on the canvas.

(b) In the Thum process the electrolyte is practically a neutral solution of silver nitrate and copper nitrate in the proportions of 60 to 3 l^{-1} of the respective salts. Vertical electrodes are used in the Moebius cell, the anodes being surrounded by cloth bags. In this cell the silver collected on the cathodes is scraped off periodically and drops to a tray in the bottom of the cell.

(c) In the Wohlwill process the black anode slimes from the Balbach process are melted and cast into anodes and placed in an electrolytic cell containing dilute hydrochloric acid in which about 6% or 7% of gold chloride is present. The gold is deposited on thin gold cathodes as well as some silver. By superimposing an alternating current upon the d.c. current, the silver is induced to fall from the cathodes. Any platinum present dissolves with the gold, but is not deposited on the cathode. When the concentration of platinum group metals reaches about 50 g l^{-1}, these metals are precipitated out by the addition of ammonium chloride.

Alternatively the black slimes may be treated with sulphuric acid and nitre, which dissolves silver and the platinum metals. The gold is left undissolved, while the platinum metals are worked up from the solution.

Parting compound. This usually consists of dried silica sand, sea sand or burnt sand which is sprinkled on the faces of patterns before moulding so that when the pattern is drawn the face of the mould impression is not damaged by adhesion of grains of sand to the pattern. Other parting compounds consist of pulverised limestone, dolomite or precipitated chalk, sillimanite or bone ash. Lycopodium powder is used for fine work.

A liquid parting material consists of paraffin with a small addition of colza oil. For low-melting-point metals French chalk or flour may be used. (*See also* **Mould washes.**)

Parting down. When moulding with two boxes the parting line or joint may have a step in it. The operation of making such a mould is known as parting down.

Parting limits. The maximum concentration of a metal of high (positive) electrode potential, i.e. more noble component, of an alloy above which selective corrosion, known as parting, does not occur in selected corrosive conditions.

Alloy	Limiting Percentage of Nobler Metal
Gold-silver	50 and 25-30
Gold-copper	50 and 22-25
Palladium-gold	25

The nobler metal is the first named in the different alloy systems.

Parting line. A term used in foundry work to indicate the line along which a pattern is divided for moulding. It is also the line along which the cope and drag separate. In forging the term refers to the dividing line between a pair of forging dies.

Parting sand. *See* **Parting compound.**

Parting tool. A name sometimes applied to the flat foundryman's trowel used for making joints or partings. The name is also used to describe a lathe tool for cutting pieces completely through, and thus separating or parting one piece from another.

Partition coefficient. When a solute, which is soluble in each of two solvents that do not react with one another, is placed in contact with a mixture of the solvents, it distributes itself between them in such a way that

$$\frac{\text{Concentration of solute in 1st solvent}}{\text{Concentration of solute in 2nd solvent}} = K$$

where *K* is a constant known as the partition coefficient.

Partition law. When a solute distributes itself between two immiscible solvents that have no mutual solubility, there exists for each molecular species, at a given temperature, a constant ratio of distribution between the two solvents, and this ratio (called the partition coefficient) is independent of any other molecular species which may be present.

Pass. A term used in reference to the passing of metal between the rolls of a rolling-mill. The term is also used to describe the movement of the work relative to the machine through a single working cycle in such operations as milling, grinding, etc. In welding, a pass is a single progressive operation of depositing filler metal along the whole length of the weld.

Passivating dip. A term used in reference to the acid sodium dichromate solution that is used to produce a protective film on zinc or zinc-plated articles. The result of this treatment should be to produce a greenish iridescence after the articles have been washed in cold water and dried in warm air.

Passivation. A term used in reference to the forming of a protective coating on the surface of metals. In certain cases the surface of a metal exposed to air becomes covered with a carbonate, e.g. lead, and these films are compact and continuous, so that protection of the metal below the film results. These are cases of natural passivation. Artificially produced films may also give surface protection to metals. Thus zinc-plated steel may be treated with a chromate solution to form a protective coating, the process being known as passivation or chromate passivation.

Passivator. A term sometimes used in reference to those reagents that produce passivation and so reduce corrosion. The term inhibitor is now more generally used, and inhibitors are of three types: *(a)* anodic inhibitors which restrain anodic reactions; *(b)* cathodic inhibitors which restrain cathodic reactions; and *(c)* adsorption inhibitors which are adsorbed generally over the surface of a metal, so preventing or retarding any form of corrosive attack. It is the type of inhibitor that diminishes the corrosion rate without increasing the intensity of attack that the older term mainly describes.

Passivity. A term used in reference to the abnormal behaviour of metals in an electrolytic cell, at elevated potentials. The term is sometimes used quite irrespective of whether the metal remains immune from attack or passes into solution at an abnormally high valency. It is preferable to reserve the reference to the passive state to those cases where the metal exhibits an "inert state" and resists attack, simulating a more noble metal and exhibiting a positive electrode potential.

Passivity occurs as the result of the accumulation of the anodic product on the anode. Thus relatively low potentials produce passivity in cases where the anode is horizontal and protected from the movement of circulating liquid. If the electrolyte in contact with the anode is agitated, high potentials are required to produce the effect.

Pass sequence. A term used in welding in reference to the number of passes or runs in the technique of depositing weld metal in fusion welding. In each pass the metal laid down is termed a bead, so that the term may be said to refer to the number of beads as well as to the size of the filler rod or electrode that is used.

Paste solder. This is a solder mixture used in paste form. The metallic solder in powder form is mixed with a flux, the tinning material and sometimes a cleaning agent. The convenience in using this type of solder results from the fact that the paste can be put into position and the work then heated to the required temperature.

Patented steel wire. The term used to describe carbon steel wire, hard drawn after patenting the rod or an intermediate size of wire.

Patenting. The term used in reference to a heat-treatment of steel, usually in wire or rod form, whereby it is heated well above the transformation temperature, e.g. up to 1,000 °C, and subsequently cooled in air, molten lead or a fused salt mixture held at about 500 °C. This latter temperature varies somewhat according to the carbon content of the steel and the properties required in the product. The treatment facilitates subsequent cold working and the achievement of suitable mechanical properties in the finished material.

Patent yellow. The hydrated oxychloride of lead, obtained by warming lead oxide with a solution of sodium chloride. Also known as Turner's yellow.

Patera process. The name used in reference to an old method of extracting silver from its ores. It was used in the second half of the nineteenth century, but has now been superseded by the cyanide process. The silver bearing ore was roasted with common salt in rotating cylindrical furnaces. The silver chloride so formed was washed out, using a solution of sodium thiosulphate (hypo), and from this latter silver was recovered.

Patina. The term originally used in reference to the artistic product of the corrosion of copper or copper-rich alloys that are exposed to moist air, or are buried in damp soil. It consists of basic sulphate of copper and is usually dark green in colour. It is formed by the action of sulphuric acid in air or soot and by the oxidation of copper sulphide. The deposit accumulates slowly, and when completely covering the surface offers resistance to further corrosion. The formation of the patina may be accelerated by directly inserting the metal sheets or other objects in an appropriate solution or by anodic treatment in a solution having the following composition:

Magnesium sulphate	10 parts
Magnesium hydroxide	2 parts
Potassium bromate	2 parts
Water	86 parts

The bath should have a temperature just below boiling point of water (approx. 95 °C), and the current density should not exceed 0.05 A m^{-2}, and the time of immersion is about 15 minutes.

Patio process. The name used in reference to the process of extracting silver from its ores by amalgamation. It was invented by Bartolomea de Medina about 1557. The ores, which may contain silver sulphide and chloride as well as metallic silver, are finely crushed in stamping mills. This wet mass is mixed with common salt and well trodden by mules on a paved floor or patio. Mercury is then added and a little roasted pyrites, and the treading continued for many days (up to about 7 weeks). Copper chloride which is first formed decomposes the silver salts to form silver chloride, and this latter dissolves in the brine and is then reduced by the mercury. The silver forms an amalgam with the mercury. After the excess of mercury has been squeezed from the amalgam the residue is distilled in iron retorts to recover the mercury, at the same time leaving the silver. Also called amalgamation process. It is replaced now by the cyanide process.

Patronite. An impure sulphide of vanadium V_2S_5, found in considerable quantities in Peru and used as a source of vanadium.

Pattern. A replica constructed in wood, metal, plaster or quick-setting compound, around which sand is moulded to produce a cavity into which a metal can be cast. Patterns are usually made in:

  (a) wood. These are cheap and easily made, but are susceptible to the influence of moisture in the sand which produces a tendency to warp. Wood patterns wear out rapidly in comparison with metal patterns, but they can be reinforced with metal insertions to resist sand abrasion. Mahogany is superior to pine, especially for repetition moulding;

  (b) metal. Cast iron is often used because of its strength and resistance to abrasion, though its brittleness is a disadvantage if carelessly handled. It can readily be filed and the smooth surface gives a satisfactory mould surface. Brass patterns are heavier than cast iron, but do not show tendency to rust. They can be readily altered or rectified by soldering and withstand sand abrasion. Aluminium alloy (Al 80%, Zn 14%, Cu 3%, Fe 1.5%, Si 1.5%), which is easy to cast and is light in weight, is often used for repetition work. It is easily machined, but is liable to damage by moulders' steel tools. White metal is not often used, though it is the best material that can be employed for lining stripping plates. It is too soft for general use;

  (c) quick-setting compounds and plaster of Paris are very good materials for patterns used for hand ramming or for squeeze moulding. They are too fragile for use in jolt machines.

In certain applications wax and frozen mercury may be used for patterns.

The usual types of pattern are: (a) loose full patterns; with these the moulder must cut his own runners and risers. When withdrawn by hand the cavity must be slightly enlarged to facilitate this. Greater accuracy of mould is ensured by mounting the pattern on a moulding machine and withdrawing it mechanically; (b) gated full patterns are satisfactory for short runs. They have the runners and feeding heads attached to the pattern; (c) match plate patterns (q.v.) are most useful for castings made on moulding machines. They are usually made in metal; (d) mounted split patterns, as the name implies, are made in "halves", split on a convenient joint line and mounted each on a separate plate; (e) skeleton patterns (q.v.) are used for very large castings; (f) strickles (q.v.) are employed for large castings which are wholly or partly circular in shape; (g) core assembly is a method used where the design is intricate in shape and where good accuracy is required.

Pattern metal. The name used in reference to a series of alloys used for making metal patterns for moulding. They are tough and strong, do not rust and take a better surface finish than cast iron patterns. Very thin sections can be cast and the metal is easily machinable. Small patterns are readily rectified, built up or fitted by soldering. (See Table 5.)

Pattison's process. A process for separating silver from lead, based on the fact that the two metals form a eutectic system with a eutectic point at 2.5% Ag (304 °C). During slow cooling, lead-rich crystals separate out at a temperature below the freezing point of pure lead, because of the depression of the freezing point by the dissolved silver. These crystals are withdrawn by perforated iron ladles. If this process were carried far enough, lead and silver would begin to separate out at the eutectic point. In practice, about seven-eighths of the original lead is removed. The process is carried out in a series of iron pots, desilvered lead accumulating at one end of the series and enriched lead at the other end. The cooling of the melt is brought about by careful sprinkling of the surface with cold water. The Luce-Rozan process (q.v.) is a modification of the above process.

Pattinson's white. A basic chloride of lead Pb(OH)Cl, produced when chloride of lead is boiled with milk of lime. It is used as a pigment.

Pauli's principle. This principle states that no two electrons in an atom can be in exactly the same state as described by the four quantum numbers n, l, m_l and m_s. (See **Atom.**)

Peacock copper. The name used in reference to copper pyrites and to which the formula $Cu_2S.Fe_2S_3$ is given.

Peacock ore. An important copper ore which is the double sulphide of iron and copper Cu_5FeS_4 and which contains 63% Cu. It occurs in large brown masses which on exposure become purple. The ore is fairly widely distributed and is mined in South America (Chile and Peru), Canada and the USA. Mohs hardness 3. Also known as bornite, horseflesh ore and variegated ore.

Pea iron ore. A variety of limonite having a pistolic structure. It is a hydrated oxide of iron $2Fe_2O_3.3H_2O$.

Pearl ash. The name used in reference to the white, alkaline granular powder that has the composition $K_2CO_3.H_2O$. It is produced from plant ashes and from natural deposits in Germany and Russia. Chemically it is potassium carbonate and is also named "potash".

Pearl filler. The name used in reference to the white anhydrous calcium sulphate $CaSO_4$ which is used as a filler for paper. The colour is not such a pure white as that of the hydrated sulphate. The latter is known as crown filler when used in the finer qualities of paper. The finely ground calcium sulphate when used as a paint pigment is called satin spar.

Pearlite. The lamellar aggregate of ferrite and cementite resulting from the direct eutectoid transformation of austenite at Ar_1 and in an iron-carbon alloy contains about 0.87% C (Fig. 79). The lamellae may be very thin, and resolvable only at high magnifications. The constituent usually shows a pearly lustre in white light. In cast iron it is a very desirable structure. The tensile strength, if it could be obtained throughout a specimen, would be about $690\,MN\,m^{-2}$. A good average in plain grey iron is $155\,MN\,m^{-2}$. When steel has been subjected to prolonged annealing the ductility improves at the expense of the mechanical strength, and the cementite lamellae in the pearlite tend to spherodise. This type of pearlite structure is called "divorced pearlite". (*See also* **Iron-carbon diagram.**)

Pearlitic malleable iron. When white iron is heated to a temperature between 800 °C and 900 °C some of the iron carbide is decomposed to yield graphite. When the process is controlled so that a significant amount of iron carbide still remains while the carbon forms in rosettes, the material is known as pearlitic malleable iron, and it is stronger, harder and more wear resistant than ordinary standard malleable iron (grey iron).

Pearl spar. The name used in reference to the fine-grained compact form of the double carbonate of magnesium and calcium $MgCO_3.CaCO_3$ or dolomite. It is usually white in colour and is used raw as a flux. For furnace linings it is first calcined to produce the oxides. Mohs hardness 3.5-4, sp. gr. 2.8-2.9.

Pearl white. The name of the white pigment, used in paints, that is known chemically as bismuth oxychloride $BiOCl$. It is a fine crystalline powder insoluble in water and having specific gravity of 7.7.

Peat. An earthy mass of vegetable origin produced by the accumulation of rapidly growing mosses which have partly decayed. It is generally found at or near the surface of the ground, the portions lying at the surface being brown while the lower layers are often black in colour. Usually it contains 70-80% moisture and must be dried before use as a fuel. Its calorific value is low (about $11.7\,MJ\,kg^{-1}$). After careful drying peat is often used as heat-insulating substance. When dry distilled it yields an illuminating gas and a tarry liquor. The wax extracted from peat has properties similar to those of montan wax and can be used in some applications as a substitute for it.

The fuel has been extensively and economically utilised for heating and for electric power generation in the Republic of Ireland. There the peat is harvested by cutting into rectangular blocks (sods) or by milling into granular texture. The drying process reduces the water content to less than half on site, and subsequent treatment depends on the type of boiler installation. Boilers were developed from those used for brown coal or lignite, in Germany, and require careful admission of air for complete combustion. About 25% of the demand for electric power in the Republic of Ireland relies on peat fuel, and in 1977 milled peat showed the lowest fuel cost in steam plant generating stations. The lowest figure was 0.79 pence per unit sent out total cost, of which 0.59 pence per unit was fuel cost, as against 1.6 pence per unit for coal, and 0.99 pence per unit for oil.

Pebble mill. A mill for grinding ore. It is similar to a rod mill except that the rods are replaced by flint pebbles so as to avoid contamination with iron.

Pebbles. The name used in reference to the surface defect which sometimes appears on drawn products. The roughening of the surface is caused by stretching beyond its elastic limit material that has a rather coarse-grained structure. Also called orange-peel effect.

Pedersen process. A process for the production of aluminium from low-grade bauxites by first producing a calcium aluminate slag and bleaching this with hot sodium carbonate.

Peeling. The stripping or detaching of an electrodeposited layer of plating metal from the base metal.

Peening. A method of cold working the surface of components by means of blows on the surface from a hammer or by other means. If a ball peen hammer is used the metal surface is worked locally. Bolt ends are often so treated and then the nut is prevented from running off. Peening improves the mechanical properties of the part treated and is frequently employed to increase the resistance of a component to fatigue. It is applied to automobile parts which normally experience vibrations in service. Peening sets up compressive stresses in the surface layers of a metal. This is an advantage, since both heat-treatment and machining of a component leave residual tensile stresses, and these are neutralised by the peening. The process may be carried out using iron shot in barrels or by air blasting.

Peening shot. Material in the form of shot, made by allowing molten iron to fall into water. The shot are graded according to their diameter, e.g. from 0.15 to 4 mm. Medium-size shot, e.g. Numbers 25-30, (0.6-0.75 mm) are frequently used for peening metal components to improve their fatigue resistance. Other sizes are used for cutting and grinding stones. The operation is often carried out in tumbling barrels.

Peen pin. A ramming tool having a wedge-shaped end, and used by foundrymen for finishing the corners and pockets in preparing moulds. Also called pin rammer.

Pegmatite. An economically important variety of acid igneous rock having an extremely coarse structure. It contains quartz orthoclase and mica as main constituents. Pegmatite represents the last portion of an igneous magma intruded in a very liquid state and occurs in dykes, strings and veins around the borders of granite masses. Other constituents may be zircon, cassiterite and garnet.

Peg rammer. A rammer used in the preparation of foundry sand moulds that has a flat or slightly wedge-shaped end. Such rammers are often worked by compressed air. Also called pin rammer.

Peirce amalgamator. An apparatus designed to bring the gold-bearing "pulp" into intimate contact with mercury to ensure that as much as possible of the finely divided gold amalgamates. Baffles are arranged so as to force the pulp into contact with the mercury which is contained in the bottom compartments. Finely divided gold may escape this first treatment and then the upper baffles may be coated with amalgam.

Peirce-Smith converter. A large barrel-shaped basic lined converter used for converting copper matte. The charge consists of matte and silica flux, the latter being added in several stages. The blast is led into the converter through the hollow trunnion at one end.

Pelleted pitch. Pulverised coal-tar pitch made from hard grades of pitch having a softening temperature in the range of 120-135 °C. It is used as an additive to foundry sand in the proportion of 1 to 3%. The particle size is important, and is controlled by the use of standard sieves. Ideally the size should be between 100 and 200 (standard) mesh, but a proportion (20%) may be less than 200 mesh and about 20% between 30 and 100 mesh. It is used to replace coal dust. It is claimed that the high volatile content gives good surface finish to casting, while the pitch as a binding agent gives increased strength to the moulds. High-melting point pitch (upwards of 1,400 °C) is the most satisfactory type to use. The pellets are spherical in shape and about 0.25 mm diameter on the average.

Pellin's test. A dynamic hardness test in which a ball attached to a weighted frame is allowed to fall freely upon the specimen, and the hardness is assessed from the diameter of the impression.

Peltier effect. A thermo-electric effect produced when a current flows round a circuit made up of two metals, in which heat is evolved at one junction and absorbed at the other. When the thermo-electric current passes round the circuit of a thermo-couple it absorbs heat at the hot junction and evolves heat at the cold junction, and the difference between the two quantities of heat is the source of the energy to which the current in the circuit is due. A thermo-electric couple thus acts like a small heat engine, taking in heat at the hot junction and rejecting heat at the cold junction as its condenser.

Pencil gates. A series of small inlets leading from the bottom of the pouring basin to the top of the casting to conduct the molten metal to the mould. They were originally devised for casting quite large castings with a minimum weight of head. The fine streams of metal fall to the bottom of the mould, where solidification rapidly occurs. Subsequent supply of metal is continuously breaking the surface of the metal in the mould, and so the slag is induced to float. There results a casting with a relatively small amount of top discard.

Penetrol black process. A process for forming a black oxide film on steel by treatment in an alkaline solution with an oxidising agent. The surface produced is a good base for subsequent painting, but if the black finish is required as a protection against corrosion, it is usual to give a finishing treatment with oil.

Pen stock. A short cast-iron pipe, lined with fire-brick, that is used to convey the air blast from the goose neck (q.v.) to the blow pipe in a blast furnace. Also called leg pipe.

Pentlandite. A nickel-iron mineral (Ni.Fe)S that occurs in the nickelferrous pyrrhotite deposits mined in Sudbury (Ontario). The nickel content is usually about 1-2%, and occasionally 5%. Copper is also present associated with the nickel to the extent of about 2.5%. These ores yield the metals of the platinum group as by-products.

Pepper blisters. During the cooling of metal from the molten state, bubbles of gas may fail to overcome the surface tension of the metal. They then give rise to blisters, which when they are very small are called pepper blisters or pin-head blisters.

Percussion welding. A development in electrical spot welding in which an intense discharge of electrical energy is applied to the area of the proposed weld. The duration of the discharge is very short, so that in the case of metals having low electrical resistance and high heat conductivity the heat loss is minimised. The electrical energy may be stored in condensers and then discharged at the required moment, or a large inductance may be used. The welding procedure is not identical with the two types of machine. With the condenser discharge machines the electrodes are arranged to maintain a high pressure at the moment of fusion and during the cooling down. The upper electrode automatically moves towards the weld at the appropriate moment. In the inductance machine high pressure is exerted by the electrode initially. This is then released somewhat to increase the contact resistance and increase the heat produced by the current impulse. Finally the pressure is increased again to promote the forging or upsetting.

Perfect gas. The term used in reference to the ideal gas that would exactly obey the gas laws as enunciated by Boyle and Charles. The dependence of the volume (V) of a perfect gas upon the pressure (p) and the absolute temperature (T) is given mathematically by the perfect gas equation $pV = nRT$, where R is the "gas" constant. All gases deviate from the above equation, and their behaviour is more nearly represented by van der Waals' equation (q.v.).

Perichlor. An acid-proof cement made by adding powdered ordinary cement to a strong solution of sodium silicate. It is used for lining and joining chemical tanks, ducts and drains.

Periclase, native magnesia MgO. Occurs as small grains disseminated in masses of limestone metamorphosed by volcanic lava. It originates from dolomite, the magnesium carbonate being dissociated by the high temperature. When hydrated the material is converted to brucite $MgO.H_2O$.

Periodic law. A law relating to the classification of the elements, and enunciated by Mendeleef, which states that "The properties of the elements are periodic functions of their atomic weights". It is now recognised that the atomic number is a more fundamental property of elements than the atomic weight and the modern tables show the arrangement of the elements quoting the atomic number. Table 39 shows the arrangements of elements according to Bohr.

Peritectic reaction. An isothermal reversible reaction between two phases which occurs at a constant temperature during the cooling of an alloy, as a result of which a new phase is produced. An alloy containing the two phases in a certain ratio consists after the reaction of an entirely new phase, but alloys containing an excess of one of the original phases will retain some of this phase after the reaction is complete. The peritectic reaction, which is a feature of many alloys, is illustrated in Fig. 111, which is a portion of the tin-copper diagram, *FAE*, being the liquidus and *GBCE* the

TABLE 39. *The Periodic Classification of the Elements (After Bohr)*

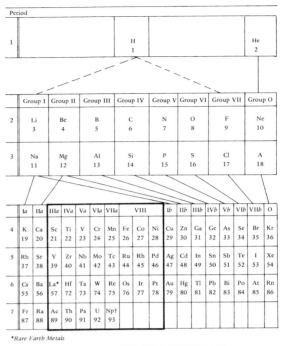

Rare Earth Metals

Ce	Pr	Nd	Pm	Sm	Eu	Gd	Tb	Dy	Ho	Er	Tm	Yb	Lu
58	59	60	61	62	63	64	65	66	67	68	69	70	71

Transuranic Elements

Np	Pu	Am	Cm	Bk	Cf	Es	Fm	Md	No	Lw
93	94	95	96	97	98	99	100	101	102	103

solidus. Here the α solid solution reacts at 790 °C with the liquid from which it has solidified to form a new solid solution β. The only alloys in which this reaction occurs are those containing between 74.5% and 91% Cu. At the peritectic point there is equilibrium between the liquid and the α and β phases.

An alloy containing 77.5% Cu on cooling to 790 °C (the peritectic temperature) would then consist of: (a) α solid solution containing 91% Cu; and (b) liquid containing 74.5% Cu, and they would be present in the ratio of *AB* to *BC*. The liquid and the solid solution then react to give the solid solution β, and the alloy when just solid consists entirely of this constituent. On further cooling β begins to split up into a mixture of β and α.

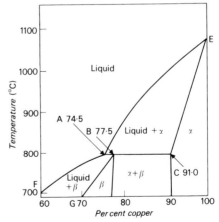

Fig. 111 *Peritectic reaction occurring in the copper-tin binary system.*

An alloy containing between 77.5% and 91% Cu would on cooling to 790 °C consist of the same two constituents, but in different proportions. They would react as before, but, due to the higher proportion of α solid solution, some of it will remain, and the result is a mixture of solid solutions α and β. In alloys containing between 74.5% and 77.5% Cu the peritectic reaction will occur on cooling to 790 °C, but in this case there is excess liquid which will remain. When the reaction is complete there will then remain alloys consisting of β solid solution and liquid.

Peritectoid reaction. An isothermal reversible reaction in which two solid phases of a binary alloy system react during cooling to form a new solid phase. Such reactions are similar to peritectics, except that the complete reaction occurs in the solid state, the additional solid constituent taking the place of the liquid phase. Figure 112 shows a part of the copper-tin diagram in which a reaction between the compound Cu_3Sn and the solid solution γ occurs, as a result of which another compound Cu_4Sn is produced.

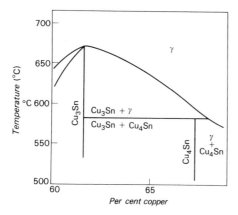

Fig. 112 *Peritectoid reaction in the copper-tin binary system.*

Perlit. The trade name of a high-strength pearlitic cast iron. It has a tensile strength of 460-490 MN m^{-2}.

Permabrasive. A type of blasting shot which is a treated malleable iron having a hard exterior. It is used mainly in metal cleaning.

Permaclad. A composite metal consisting of stainless steel bonded to one side of a plain carbon steel.

Permalloy. A series of nickel-iron alloys that have a much greater permeability than iron and are readily magnetised in weak fields. The composition is given in Table 7. They are used in magnetic cores for apparatus that operates on a feeble electric current. They are also employed in the loading of submarine cables, where the alloy in strip form is wrapped round the entire length of the cable. They show a very small hysteresis loss.

Permanent hardness. Water which does not give an immediate lather when shaken with soap solution is termed hard water. If the hardness cannot be removed by boiling it is termed permanent hardness, and is due to the presence of the sulphates (or chlorides) of calcium, magnesium or iron. Permanently hard water may be softened by the addition of sodium carbonate or by passing the water over zeolites (q.v.) when the calcium ions (for example) are exchanged for sodium ions.

Permanent magnet. A substance which possesses magnetic properties that are not dependent on the presence of an external magnetic field. Permanent magnets are produced by exposing suitable materials to the influence of a strong magnetic field. On removal of the latter the material retains a substantial proportion of magnetism induced in it. As a result of the irreversible nature of part of the magnetisation process, energy is stored in a magnetic material in the form of remanent induction. This remanent induction may be used to provide magnetic flux in a suitable circuit, and materials whose properties make them useful for this type of application are termed permanent magnet materials.

Permanent magnet alloys. Alloys suitable for use as permanent magnets which show a large remanent magnetisation and a large coercive force. The latter requirement arises because the magnet must maintain as large a magnetic field across an air gap as possible, and it must not be unduly sensitive to external effects such as stray magnetic fields and mechanical shock. This is usually expressed by saying that the energy product $(B \times H)_{max}$ should

be large. Plain carbon steels containing 1-1.5% C when quenched in water give good (martensitic) magnetic steels. Small percentages of molybdenum, tungsten or chromium improve the magnetic properties of steel. Cobalt in amounts from 3 to 35% increases the saturation induction and the coercive force. Nickel-iron-aluminium alloys when rapidly cooled become very hard, and because of the arrested precipitation usually have large internal stresses which are favourable to a large coercive force. Alloys containing aluminium, nickel and cobalt have greatly improved magnetic properties if heat-treatment is carried out in a strong magnetic field, but the effect is limited to a narrow range of 8-9% Al in the alloy. The composition of a number of common magnetic materials and their magnetic properties are given in Tables 6, 7 and 9.

Coarse-grained cast Alnico alloys with columnar growth during solidification show improved magnetic properties compared with random orientation of grains. Other cobalt alloys, with Fe + Mo, Fe + V, and Ni + Cu, have properties suitable to various applications, e.g. Co-Ni-Cu is relatively soft and ductile but with high coercivity. Cobalt-platinum alloys have been developed for miniature mechanisms in chronometers and satellite equipment.

The most recent developments are in the use of rare-earth elements with cobalt, mostly with R Co$_5$ hexagonal structures, R being one of the rare earths, usually yttrium, cerium, praseodymium, neodymium or samarium. The alloy is formed by melting in inert atmospheres, then milling to fine powder and compacting whilst aligned in an applied magnetic field.

Permanent mould. A mould, usually made of metal, which is used repeatedly for the production of a number of castings of the same form.

Permanent set. The deformation remaining after removal of the stresses which have exceeded the elastic limit.

Permanent white. A fine white powder which consists of barytes barium sulphate, $BaSO_4$. It is used as a pigment, but has a poor covering power. When barium sulphide and zinc sulphate are mixed, the precipitate consists of a mixture of barium sulphate and zinc sulphide. This is known as lithopone. It has then a better covering power than white lead. It does not darken in industrial atmospheres, but turns yellow on exposure to light.

Permeability. The term is used with reference to the flow of fluids under pressure gradient, through or into a porous material. Porosity measures the void size, permeability accounts for skin friction, connecting pores and so is dependent on physical parameters other than porosity, including the viscosity of the fluid. (*See also* **Magnetic permeability.**)

Permeability alloys. The general name for the nickel-iron alloys which are characterised by a greater susceptibility than ordinary iron. A number of these alloys are known by special names which are associated with special applications in electrical and electronic apparatus. Some of the better known ones are given in Table 7.

Permittivity. Coulomb's law of force between electrically charged bodies may be written

$$F = \frac{Q_1 \cdot Q_2}{4\pi\epsilon r^2} \times \frac{a}{\_}$$

where Q_1, Q_2, are the electrical charges (coulombs), r the distance apart (metres), ϵ the permittivity of the environment, and $\underline{a}$ a unit vector directed along the straight line between the two charges. The permittivity of a dielectric material is the product of the permittivity of a vacuum and the dielectric constant (dry air has a value roughly the same as that of a vacuum) and is characteristic of the material involved. Analogous to permeability in magnetic materials, as is evident from the Coulomb equation used to define magnetic field strength in SI units in terms of the ampere. (*See* **Magnetic field.**) Permittivity is a direct result of the change in polarisation produced by an electric field.

Permutit process. A process for softening any type of water introduced by Gans. Permutit is a double silicate of sodium and aluminium allied to the zeolites. When hard waters containing calcium and magnesium salts are percolated through permutit the water-hardening salts are removed and replaced by equivalents of sodium salts. The change is reversible and the spent permutit can be re-activitated by percolation by a solution of common salt.

Pernax. A substitute for *gutta-percha* supplied in sheet and strip form and used by electroplaters for "stopping off".

Perofskite. Calcium titanate $CaO.TiO_2$, an ore of titanium containing about 58% Ti.

Perrin process. A modification of the Bessemer method introduced about 1933, which makes possible the production of practically pure iron in a basic bessemer converter. The process was developed mainly to assist in the removal of phosphorus. Blown metal is separated from the first high-phosphorus slag and poured into or agitated with basic fluid slag consisting of a mixture of lime, alumina and fluorspar in which the former is present to the extent of about 80%.

Petzite. A tellurium mineral found in Hungary, Romania (Transylvania) and in many parts of the USA. It resembles sylvanite, having a massive form, showing a conchoidal fracture and a metallic lustre. It is a natural alloy of the three elements and corresponds to the formula $(Ag,Au)_2Te$. Mohs hardness 2.5, sp. gr. 9-9.4.

Pewter. The name of a tin-base alloy originally produced for making dishes and ornamental articles. Roman pewter contained 70% Sn and 30% Pb. Some samples of Tudor pewter have shown a 90% content of tin, and a harder, but still workable alloy contained 91% Sn and 9% Sb. The presence of much lead causes the alloy to be dark in colour and is undesirable for food-containing articles. When the percentage of tin is reduced, additional antimony must be added to make the alloy susceptible to polishing. High antimony content is also undesirable for any food-containing articles. Table 8 gives the approximate composition of various grades of pewter.

Because of the undesirability of lead in pewter, a number of white metal alloys have been introduced free from lead (*see* Table 8).

Pfanhauser's platinum bath. This is a solution used mainly for flash plating. The constituents of the bath are:

Diammonium phosphate	20 g
Disodium phosphate	100 g
Chloroplatinic acid	4 g
Water	1,000 cm³

The solution is prepared by boiling the above until a pale yellow colour is obtained, which indicates the formation of complex platinum ammino-phosphates. The solution is used at a temperature near to the boiling point at 3-4 volts and a current density of about 1 A dm⁻². Bright deposits are obtained in ½-2 minutes, using platinum anodes. The method is suitable for plating on lenses and telescope mirrors as well as on metals.

Pharmacolite. One of the chief minerals containing arsenic and from which the element may be extracted. It is a hydrated calcium arsenate, $2(CaHAsO_4).5H_2O$.

Phase. A homogeneous, physically distinct and mechanically separable portion of a mixture. A system which consists of ice, water and water vapour has three phases of the same chemical substance, but while a phase must be physically and chemically homogeneous it need not be chemically simple. Thus a gaseous mixture or a solution may each form a phase, but in a heterogeneous mixture of solid substances there are as many phases as there are distinct substances present.

The term phase is also used in reference to the position in its path of a harmonically vibrating point, and generally means the fraction of a whole period of its vibration which has elapsed since it has passed its mean position in the direction that is assumed positive. The difference in phase between two vibrating points may be defined by stating the difference between the times at which they cross their mean position in the positive direction, as a fraction of the whole period of either. It may also be stated as the corresponding angle of revolution of the uniformly revolving point, either in degrees or radian measure (*see* Fig. 113).

Phase diagram. When a molten alloy cools it will experience certain changes, e.g. change of state or change in composition in the solid state. When these changes occur, heat is evolved, and thus these changes can be indicated by arrest points on a cooling (time/temperature) curve.

The determination of cooling curves for a representative series of alloys is known as thermal analysis, and it is upon such an analysis that temperature/composition curves can be based. These latter graphs are known as phase diagrams. Since a phase diagram

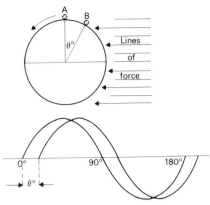

Fig. 113 *Phase difference. Conductors A and B rotate in a magnetic field. E.m.f. in B lags, that in A leads. θ is the phase difference, i.e. e.m.f. in B lags behind that in A by an angle θ. In the e.m.f. displacement curve θ is the phase difference.*

portrays equilibrium conditions (because it is based on an ideal rate of cooling so that equilibrium obtains at all temperatures), it is often called an equilibrium diagram. Also called constitutional diagram (q.v.).

Phase rule. When a substance exists in two different states or in two different phases of a system, equilibrium can occur only when the chemical potential of each component is the same in all the phases in which it occurs. This potential of the component in any phase depends on: *(a)* the composition of the phase; *(b)* the temperature; and *(c)* the pressure (or volume).

In a system of C components existing in P phases, it will be necessary to know the masses of $(C-1)$ components in each of the phases in order to fix the composition of unit mass of each phase. Hence, as regards composition, the system possesses $P(C-1)$ variables. To this we must add the other two variables—temperature and pressure—making $P(C-1) + 2$ variables in all.

If one of the phases in which all the components occur is taken as standard, then in any other phase in equilibrium with it the potential of each component will be the same as in the standard phase. Thus if there are P phases there will thus be possible $(P-1)$ equations for each component, and for all C components there will be $C(P-1)$ equations.

The difference between the number of variables and the number of equations to determine them gives the variability or degrees of freedom of the system. Hence, if F = degrees of freedom,

$$F = P(C-1) + 2 - C(P-1)$$
$$= C - P + 2$$

This equation is the mathematical statement of the phase rule. (*See* **Degrees of freedom.**)

Phenol C_6H_5OH. The aromatic alcohol of benzene, an unsaturated monocyclic hydrocarbon. Also known as carbolic acid (incorrectly) synonymous with benzenol, hydroxybenzene. A solid at room temperature, melting at 43 °C. The starting point of numerous reactions in the preparation of organic compounds, and a constituent of one of the oldest synthetic polymers, when condensed with formaldehyde (methanal).

Phenol-formaldehyde. (Hydroxybenzene-methanal). When phenol and formaldehyde are heated together, in molecular ratio 1:1.5, condensation polymerisation occurs, with the elimination of water. The reaction is exothermic, and can be catalysed by ammonia, sodium hydroxide or oxalic acid to speed up the rate of reaction. The products vary according to the basicity and can be linear polymers (acid catalyst) or compounds with methylol groups which on further heating cross link to form a network. The intermediate stages are known as resols or novolac, and it is these which are blended with fillers, pigments and plasticisers, to create a variety of resins for moulding into articles, when cross linking is completed, to form a strong network which no longer softens on heating. (*See* Table 10 for trade names and properties.)

Phlogopite. A magnesium mica $H_2KMg(SiO_4)_3$. It is often brown in colour and generally resembles muscovite in its physical properties. It is, however, somewhat softer than the latter, but has a greater heat resistance. It is used mainly in electrical work. The main sources are Madagascar, Canada and India. Also called amber mica.

pH measurement. The determination of the number used to express the *active* acidity or alkalinity of a liquid may be carried out in several ways, which are known as the "electrometric", the "colormetric" and the "capillator" methods, respectively. In the electrometric method is measured the difference in potential between the hydrogen electrode immersed in the solution and a standard calomel electrode. The results are absolute.

The colorimetric method depends on the use of indicators which are, in fact, organic colouring matters that change colour according to the pH value of the liquid to which they are added. A universal indicator has been prepared by mixing suitable indicators so that a number of colour changes may be obtained over a large pH range. Using measured amounts of the liquid and indicator, the colour of the solution obtained is compared with a series of solutions contained in colourless glass tubes. These solutions have known pH values.

The capillator method is a modification of the colorimetric method in which capillary tubes are used instead of test-tubes. This method, which uses only small amounts of liquids, can be used successfully with slightly coloured liquids, but it is unsuitable for determining the pH values of pure waters or aqueous solutions of pure substances having little or no buffer action.

Phonolite. One of the intermediate sub-acid igneous rocks. It is a variety of felspar of variable composition. Chemically it is an aluminium silicate in which the proportion of alumina may vary from 15 to 25%. It is used as a flux and in the making of glass and pottery. Those varieties having a high percentage of alumina have been used for the extraction of aluminium. A rather soft variety having a soapy feel is called agalmatolite, and is used by the Chinese for carving bas-reliefs. Also called clinkstone.

Phonon. A quantum of thermal energy, which leads to the concept of a thermal wave in the transmission of thermal energy along a temperature gradient in a material. The phonon is associated with specific frequencies of vibration of atoms when thermal energy is absorbed, in a similar way to the variation in electron energy leading to specific changes in energy level.

Phosphate beads. When microcosmic salt $NaNH_4.HPO_4.4H_2O$ is heated on a platinum wire it forms a colourless glass bead. If small quantities of metallic oxides are taken up on the bead and reheated, a coloured bead is obtained, the colour of the bead being characteristic of the metal. Copper and cobalt are readily detected in this way. Silica does not react with a phosphate bead as it does with a borax bead. Hence particles of silica can be seen floating in the former bead after fusion. This test is sometimes used to detect silica.

Phosphate cleaner. The white crystalline normal sodium phosphate $Na_3PO_4.12H_2O$. It is used in detergents, in soaps and in cleaning compounds for metal surfaces. It is a constituent of some boiler compounds used for water softening.

Phosphate coatings. As a pre-treatment of steel prior to painting, enamelling or treating with cellulose, the metal may be treated to form a layer of phosphate which, though porous, is strongly adherent. Most of the processes carry proprietary names. Coslettising is a process for producing phosphate coatings on iron and steel by immersion in a hot solution of phosphoric acid saturated with iron II phosphate. The Parkerising process uses phosphoric acid containing manganese salts as accelerators. The Bonderising process uses phosphoric acid together with copper salts as accelerators. After treatment in the phosphate bath the metal is washed and then usually given a short-time immersion in very dilute chromic acid. This improves the rust-proofing quality of the deposit (*See also* **Bonderising, Coslettising, Granodising and Parkerising, Phosphate of lime.**)

Phosphate of lime. The common name of the white crystalline compound which is chemically acid calcium phosphate

$$CaHPO_4.2H_2O.$$

It is sometimes used as a polishing medium and in the treatment of steel for phosphate coating. Also called calcium diphosphate.

Phosphatic bauxite. The creamy-coloured aluminium phosphate containing about 32% alumina. Silica is usually present up to about 7% and about 1.3% titanium oxide. The colour is mainly due to the presence of a little iron oxide. It is obtained from Brazil.

Phosphide banding. The formation of thin bands of iron phosphide Fe_3P occurring in a metal or alloy and revealed by etching. The bands generally run in the direction of rolling and the appearance is often referred to as "banded structure". (*See also* **Ferrite ghost.**)

Phosphide segregation (phosphide streak). A faint, light-coloured streak, discernible on the surface of freshly machined metal under suitable lighting conditions. It is due to local segregation of ferrite in wrought steel, where there is a high proportion of phosphorus in solution. The defect may be removed by machining, but heat-treatment is not effective.

Phosphorescence. The emission of light following absorption, with a time delay greater than 10^{-8} seconds, often some hours. The emitted light is usually of longer wavelength than that absorbed (*see* **Luminescence**).

Zinc sulphide as prepared in the laboratory is white in colour and does not glow after exposure to light. But if heated very strongly in the presence of traces of sulphides of copper or manganese the product is phosphorescent. Natural zinc sulphide exhibits this same effect without this treatment. The after-glow of zinc sulphide or of the double sulphide of zinc and cadmium when used on the screen of a cathode-ray tube is bright green, and it lasts long enough to be suitable for visual inspection. Cadmium sulphide after strong heating gives a phosphor having an excellent green after-glow. Zinc sulphate gives a red after-glow of long duration suitable for visual inspection of non-recurrent traces. Calcium tungstate gives a violet-blue trace suitable for photographing, as also does a mixture of zinc sulphide and copper phosphorogen (*see* **Phosphorogen**). Willemite, or natural zinc silicate, is quite commonly used, since the green trace persists long enough for making a photographic record.

Phosphorised copper. The name used in reference to commercial copper bars, wire, sheet, etc., that contain residual phosphorus over and above that required for deoxidising. The amount of phosphorus may vary, but in all cases the copper has higher strength, greater hardness and increased resistance to corrosion as compared with pure copper. The electrical conductivity, however, is lower.

Phosphorogens. The name used in reference to those materials that must be co-precipitated with zinc, cadmium and other sulphides that are used as phosphors. The metallic sulphides, generally only required in very small amounts, that are required to activate the phosphors are those of copper, silver or manganese. The presence of iron, nickel or cobalt is generally undesirable in phosphors, since they tend to kill the phosphorescence.

Phosphors. Substances that glow or "phosphoresce" for a longer or shorter time after being exposed to a strong source of illumination. (*See also* **Phosphorescence.**)

Phosphorus. A non-metallic element that occurs in nature chiefly as phosphates, such as apatite $3Ca_3(PO_4)_2.CaF_2$, or chlorapatite $3Ca_3(PO_4)_2.CaCl_2$. These are hard minerals which can be treated in the electric furnace, together with sand and coke, to produce phosphorus. The element combines readily with oxygen, sulphur and chlorine, and when strongly heated with lime forms calcium phosphide, Ca_3P_2. Phosphorus is an undesirable element in steel. It is usually present in foundry iron, in which it may be present as iron phosphide, Fe_3P, giving increased fluidity to the molten metal. In steel, phosphorus causes cold shortness, and the permitted content is frequently limited to 0.05% or less. Larger proportions, however, are present in free-machining steels. Phosphorus is removed during the steelmaking process either in the basic converter or the basic open-hearth processes. In the former, air is blown through the molten pig iron and the phosphorus present is converted into phosphoric oxide. Basic oxides, e.g. lime, are added and calcium phosphate is produced in the slag. The converter is usually lined with magnesite (also a basic oxide). Very similar changes, as far as the phosphorus is concerned, occur in the basic oxygen and open-hearth process. (*See also* **Perrin process.**)

Phosphorus tends to segregate in the steel, and the extent to which this occurs is an indication of the poor quality of the steel as well as of the indifferent method of steelmaking.

Phosphorus is used as a deoxidiser in certain non-ferrous metals, particularly copper and copper alloys. To ensure complete removal of oxide a slight excess is added. This reduces the electrical conductivity. (*See also* **Bronze** and **Phosphorised copper.**)

In rolled steel containing much phosphorus the appearance of a polished and etched surface shows distinct bands where the phosphorus occurs in the form of iron phosphide. (*See also* **Phosphorus banding** and Table 2.)

Phosphorus banding. The formation of thin bands of iron phosphide in steel due to segregation. These bands are made visible after polishing and etching the specimen.

(For reagents that are suitable for revealing phosphorus in steel *see* Table 11.)

Photocell. Synonymous with photoelectric cell.

Photochemical equivalent. When light is absorbed by an atom there is first formed an "excited" atom in which an electron from a low energy state is moved to one of greater energy. This is a primary process which occurs before any chemical change takes place. Each atom or molecule entering into the chemical reaction must first absorb one quantum of radiation [$h\nu$] known as the photochemical equivalent.

Photochemical induction. Certain substances can be caused to react under the influence of a photochemical catalyst (e.g. light). In certain cases it has been noted that there is an initial delay—a period of "hesitation" between the first exposure and the commencement of the reaction. This is known as the period of photochemical induction or the Draper effect.

Photo-elasticity. In the photo-elastic method a model of a component is cut from some isotropic, transparent material and is stressed in the same manner as the actual component would be in service. The stress distribution produced in the model is inferred from an examination of the interference figures that are produced when polarised light is passed through the stressed model. The materials used are glass, celluloid or transparent plastic substances, e.g. methyl methacrylate (2 methyl-propanoic acid, methyl ester). The whole model may be examined under polarised light, or sections cut from the model (when this is large and complex) and examined separately. When sections are used, the stress must be capable of being "frozen in" during stressing, so that a redistribution does not occur when cutting out the section. Careful selection of the material, and correct heat treatment during stressing and cutting, is necessary.

Photo-emission. The incidence of electromagnetic radiation on to the surface of materials is described in general under Optical properties of materials. If the energy of the photons is considerable, then the interaction with electrons in the material may be sufficient to overcome the attraction forces acting on the electron, and impart some additional kinetic energy to the electron so that it leaves the surface of the material and is emitted as a free electron. This is known as photo-emission. The wavelength of effective incident radiation and the number of electrons emitted will be characteristics for the material involved, and naturally are related to the electronic structure of that material. In metals, the Fermi level (the outer extremity of the electron energy in the conduction band) will determine the simplest release of an electron, and the photon energy $h\nu$ must exceed the work function ϕ before emission can occur. A critical wavelength exists therefore, and only radiation of wavelength less than this threshold can cause photoemission. The critical wavelength λ_0 can be defined by the equation

$$\lambda_0 = \frac{1.24}{\phi} \mu m$$

Most metals have work functions of the order of 2 eV upwards, which gives $\lambda_0 = 0.62 \mu m$, in the ultra-violet.

Semiconductors can become photo-conductive, when electromagnetic radiation is absorbed, since the photons may bring sufficient energy to bridge the gap between valence and conduction bands, but if the energy is even greater, they too may display photo-emission. As with metals, the energy required is such that ultra-violet or the blue end of the visible spectrum is the only radiation feasible.

The kinetic energy of the electron may be affected by phonon interaction, and since the atoms are always vibrating above 0K, the photo-emission in metals is limited to about 3 nm below the surface. This does not affect semiconductor materials and the only other possible interaction is with unexcited valence electrons,

so that the effective depth may be of the order of 25 nm. If the excited electrons have energy higher than the band gap, then electron-hole pairs may be produced as they escape, degrading their kinetic energy and preventing escape below 3 nm from the surface — like the metals.

Metals such as caesium, sodium, tungsten and silver show photo-emission in the blue end of the spectrum or ultra-violet, with low efficiency. Semiconductors like germanium, Na_3Sb, Cs_3Sb, are much more efficient, and antimonides containing Na and K or Na, K and Cs, even more so. A composite film of silver, oxide and caesium, is sensitive in the near infra-red, but with low efficiency.

Photogrid process. A method of studying the deformation of metals during cold forming, pressing, etc. A rectangular or polar grid recorded photographically on the surface of the metal before and after working indicates the direction and the degree of local distortion.

Photometer. An apparatus that may be used for comparing the brightness of surfaces or for measuring the illuminating power of a light source.

Photomicrography. The art and practice of the photographic reproduction of any object at magnifications in excess of about 10 diameters. A metallurgical microscope, when used for the production of a photomicrograph, produces the magnification in two stages. The objective usually gives a magnification of between 8 and 100, while the eye-piece increases this by a factor of 6-10. Further magnification is produced by projection.

Photon. The electromagnetic wave associated with electromagnetic radiation and expressed in terms of quanta, $h\nu$ where h is Planck's constant and ν is frequency. Compare with phonon which is the corresponding unit for thermal energy in the vibration of atoms.

Photo-sensitiser. The name used in reference to the photochemical process in which the substance which absorbs the radiant energy does not itself undergo change, but transfers the energy to another substance which will react. An example of this is found in the use of certain dyes (aurin, erythrosin, cyanin) that are used in the production of panchromatic photographic plates. Since the ordinary photographic plate is much less sensitive to red light than to the shorter wavelengths, a small addition of the dyestuff is made to the silver salts. This dye absorbs the light over the whole spectrum in a much more uniform manner. The energy is then transferred to the silver salts, and these are reduced just as effectively as if they themselves had absorbed the light in the same uniform manner.

Phototype. The name used in reference to a plate or block used in the production of printed diagrams or pictures. The blocks are prepared from a photograph by a process of electrodeposition.

pH values. A method of expressing the degree of acidity or alkalinity of solutions, i.e. the degree of dissociation. According to the theory of electrolytic dissociation, strong acids are largely dissociated into ions, but weak acids only slightly so. The latter therefore contain only a small number of H^+ ions. Similarly, strong bases yield a large number of $(OH)^-$ ions and weak bases a small number of hydroxyl ions. Purified water is only slightly dissociated into H^+ and $(OH)^-$ ions and has a small but definite conductivity. The law of mass action when applied to this case yields the equation

$$\frac{[H^+] \times [OH^-]}{[H_2O]} = constant$$

Since $[H_2O]$ is very large compared with $[H^+]$ and $[OH^-]$, it may be taken as constant, so that the product $[H^+] \times [OH^-]$ is assumed constant. It is equal to 10^{-14} mol l^{-1} at room temperature (18 °C). As pure water is taken as the standard of absolute neutrality the concentrations of hydrogen ions and of hydroxyl ions are equal and $[H^+] = 10^{-7}$. It is usual to consider only the concentration of hydrogen ions in a solution, and the index of the concentration, with its sign changed, is taken as the pH value. A decimolar solution of HCl has a hydrogen-ion concentration $[H^+] = 10^{-1}$ and its pH value is thus 1.

pH values **less than** 7 denote acidity.
pH values **greater** than 7 denote alkalinity.

The complete range of pH values extends from 0 to 14 and forms a graduated scale in which the concentrations represented by the numbers decrease in logarithmic order. Thus each whole number indicates one-tenth of the concentration of the preceding whole number:

$$
\begin{aligned}
\text{pH } 0 &= 1 \quad \text{mol l}^{-1} \quad \text{ionised hydrogen} \\
1 &= 0.1 \quad '' \qquad '' \qquad '' \\
2 &= 0.01 \quad '' \qquad '' \qquad '' \\
3 &= 0.001 \quad '' \qquad '' \qquad '' \\
4 &= 0.0001 \quad '' \qquad '' \qquad ''
\end{aligned}
$$

Intermediate (decimal) values of pH are used, and it should be noted that each *increase* of 0.1 in pH value corresponds to 20.6% *decrease* in hydrogen ions on the original pH value. That is:

$$
\text{When pH } 0 = 1 \text{ mol l}^{-1} \text{ ionised hydrogen}
$$

$$
0.1 = \frac{79.4}{100} \times 1.0 = 0.8 \text{ mol l}^{-1}
$$

$$
0.2 = \frac{79.4}{100} \times 0.8 = 0.63 \quad ''
$$

$$
0.3 = \frac{79.4}{100} \times 0.63 = 0.5 \quad ''
$$

It should be noted that the terms neutral, alkaline, etc., when used in conjunction with indicators have a slightly modified meaning. A solution that is described as neutral to methyl red is one that will not completely change the colour of that indicator. A liquid that is described as alkaline to methyl orange is one whose pH value is not less than 4 (the range of pH values of this indicator being 3.2-4.3). The liquid may actually be acid. (*See also* Indicators.)

Physical change. The name used in reference to the changes which may take place in matter that do not involve transformation of the matter into a new substance or substances having very different properties. Examples of physical change are: *(a)* change of shape or volume due to application of mechanical stress or heat; *(b)* change of state; *(c)* the acquisition or loss of a single particular property such as magnetism; *(d)* change of colour of a powder due to particle size; *(e)* change of crystalline form. As a general rule physical changes are not permanent.

Physical properties. The term used in reference to those familiar properties of matter, the determination of which does not involve the deformation or destruction of the specimen. Such properties are density, thermal expansion and conductivity, electrical conductivity and magnetic permeability. This definition is intended to exclude those properties which are more appropriately regarded as mechanical properties (q.v.). (*See also* Table 2, Table 3.)

Physical testing. The term used in reference to the whole range of physical properties and their determination. In addition to the density and the thermal, electrical and magnetic properties, which are usually determined, it may be required to determine or assess such simple fundamental physical properties as colour, crystalline form and melting point.

Piano wire. A high-grade carbon steel wire containing 0.8-0.9% C, 0.42% Mn and low sulphur content. Piano wire is usually made from acid open-hearth or electric steel. It is cold drawn with annealing between reductions. It should bend through 180° without cracking and permit of winding on to reels to form a close helix whose inside diameter is 1½ times the diameter of the wire. The wire is usually made in sizes ranging from 0.7 to 1.3 mm diameter. It is marketed in gauge sizes according to Washburn and Moen, American Steel and Wire Co., Roebling and the English music wire gauges.

Such wire is used in musical instruments and for making springs.

Picker. A tool in the form of a sharp-pointed wire used for removing small patterns from foundry moulds.

Pickle brittleness. A form of embrittlement sometimes encountered in strip or sheet metal after pickling. It is due to the absorption of nascent hydrogen during the removal of the scale by the pickling acid. The effect can be minimised by the addition of an inhibitor to the pickling bath, e.g. glue, or by heating the metal at about 100 °C for periods varying from a few minutes to several hours, according to the thickness of the sheet. Also known as hydrogen embrittlement.

Pickle inhibitors. These are substances added to pickling baths to restrain the action of the acid constituent on the metal without interfering with the action between the acid and the oxide scale. They are usually organic compounds such as glue, size, anthra-

cene, naphthalene, gelatine, agar, molasses, yeast. Other substances used are pyridine, (azine) and quinoline (benzazine). They suppress the evolution of hydrogen, and thus prevent roughness appearing on the surface of steel plates. They are also useful in reducing acid spray (also due to hydrogen) in pickling-rooms.

Pickle liquor. The name used in reference to the spent liquor that has been used in a pickling tank.

Pickling. The process of removing scale, oxide or foreign matter from the surface of metal by immersing it in a bath containing a suitable chemical reagent which will attack the oxide or scale, but will not appreciably act upon the metal during the period of pickling. (*See also* **Pickling solutions.**) Frequently it is necessary to immerse the metals in a detergent solution or to degrease them in a vapour degreaser before pickling.

Pickling solutions. The ideal pickling solution should be one that attacks the scale or oxide on the surface of a metal but leaves the metal itself unacted upon. Solutions vary in composition and concentration according to the nature of the metal to be treated, and in many cases according to type of finish, e.g. bright or matt, required. The solutions are often referred to as "dips"— a name which emphasises the fact that the period of immersion is of considerable importance and that it is usually short. The following dips are commonly used:

(a) Aluminium Dips. (1) A solution of trisodium phosphate in water. The strength is 38 g l^{-1} and the solution is used just below boiling point (about 90 °C). (2) A solution of caustic soda in water, strength 25 g l^{-1} and operating temperature about 90 °C. (3) A solution of strong nitric acid (70% acid, sp. gr. 1.42) and hydrofluoric acid (56% acid) in ratio of 3 to 1, and used at room temperature.

Solution (1) is used for sheet metal. It is mildly alkaline, but causes blackening (smut) if copper, iron, silicon, etc., are present in the aluminium. A short dip in dilute nitric acid and immediate washing generally removes smut. Solution (2) is used for more rapid action and to produce a deeper etch. Smut, if any, is removed by nitric acid. Solution (3) must be used for aluminium castings, since an alkali dip would always produce blackening.

(b) Dips for Copper and Copper-base Alloys. (1) Sulphuric acid solution containing 1 volume of strong acid to 4 of water. It is used hot (about 70 °C). (2) A similar solution to (1) above, but containing in addition 19 g of crystalline sodium bichromate to each litre of the solution. (3) A solution of strong sulphuric acid, strong nitric acid and water in the proportion by volume of 2:1:1. It is used at room temperature. (4) A mixture of strong sulphuric acid and strong nitric acid in equal parts by volume. To this is added zinc oxide—100 g l^{-1} of acid. This also is used at about 70 °C. (5) A solution containing 65 cm^3 of strong sulphuric acid per litre of water to which is added 20 g of crystalline sodium bichromate.

Solution (1) is the most commonly used dip for removing scale. Solution (2) acts more rapidly and produces a thin protective film. If staining develops the stains are removed by dipping in a dilute solution of sodium cyanide (25 g l^{-1}). Solution (3) produces a bright surface on the metal, while solution (4) is intended to produce a matt surface. In this solution increasing the proportion of nitric acid leads to a coarse finish, while decreasing it gives a finer matt surface. Solution (5) also gives a matt finish generally much finer than (4).

(c) Zinc and Zinc Alloy Dips. Solution (1) contains 250 g of chromic acid and 20 g of sodium sulphate dissolved in 1 litre of water. It is used at room temperature and the dipping time is very short—about ¼ minute. Solution (2) contains 150 g of chromic acid in 1 litre of hydrochloric acid of density 1.078. This is used at room temperature with dipping time up to 1 minute. Solution (3) contains 180 g of sodium bichromate in 1 litre of water, to which 7 cm^3 of strong sulphuric acid have been added.

Solution (1) gives a bright finish, but sometimes a yellow stain remains on the metal. This can be removed by a very short time dip in very dilute sulphuric acid (6 cm^3 of the strong acid per litre). Solution (2) always leaves the metal with a yellow stain which is removed by dipping in a 60% solution of chromic acid. Solution (3) is most useful for zinc die-castings and for articles on which a good coating of zinc has been deposited. There is formed on the article a greenish film which has the property of protecting the zinc against the effects of an acid environment.

(d) Dips for Iron and Steel. (1) Sulphuric acid of strength 10%, used at about 70 °C. (2) Hydrochloric acid solution of strength 15%. (3) A 2% solution of sulphuric acid containing 2.5 g of glue per litre. (4) A solution made by mixing one volume of strong sulphuric acid and two volumes of water. It is used cold. (5) A 75% solution of sulphuric acid in water.

Solutions (1) and (2) are those normally used for descaling iron and steel. Solution (3) is a very slow-acting solution, the dipping time being 4 hours or more. The glue acts as an inhibitor, preventing the acid from attacking the metal, but having no influence on the reaction between the acid and oxide scales. The solution (4) is used for steel which is to be plated, such steel having previously been treated with solutions (1) or (2). The etch surface (matt) is produced anodically, using a current density of 20 A dm^{-2}. By using solution (5) in place of (4) a polished surface is produced.

Picon test. Because of its high toxicity it is often required to detect small amounts of thallium. In the Picon test a drop of carbon disulphide and an excess of ammonia and ammonium sulphide are added to 1 or 2 cm^3 of the solution to be tested. The mixture is gently heated till it begins to boil. The thallium separates as black thallium I sulphide, which is transformed into a red complex by the carbon disulphide.

Picric acid (2, 4, 6 trinitro phenol). A nitration derivative of phenol having the formula $C_6H_2(NO_2)_3OH$. It is a yellow crystalline solid, which is readily soluble in water and in alcohol. The aqueous solution is used as a yellow dye and the alcoholic solution as an etchant called picral.

Picture-frame malleable cast iron. A type of malleable cast iron whose fracture shows a blackheart core surrounded by a thin white frame. This type of fracture is produced when a typical blackheart iron consisting of temper carbon nodules in ferritic matrix undergoes slight surface decarburisation. In picture-frame malleable irons the rim often has a pearlitic structure.

Pidgeon process. A process for the production of magnesium metal from the oxide by reduction with silicon. The magnesium oxide is used in the form of dolomite. Calcined dolomite CaO.MgO is heated in a vacuum furnace with ferrosilicon, when magnesium and calcium silicate are formed. The pressure is kept low continuously, a condition that favours the reaction, permitting the change to occur at a much lower temperature than would otherwise be required.

Piercing. In the production of tubes from solid bars a mandrel is used to pierce the metal. The mandrel is arranged in a circular die space and the extruded metal flows through the annular space so formed. Piercing and shaping often occur together when non-ferrous tubes are made by extrusion. In rotary piercing the hot billet is controlled by rolls that grip it and at the same time spin it. The hole formed in the centre of the billet by the plug thus tends to open out. In some cases the billet is pierced and subsequently shaped, the rough pierced blank being plug-rolled to produce the required shape. This operation is carried out in a Pilger mill (q.v.) or over a mandrel between grooved rolls.

Piezo-electric materials. Polarisation of the charges in a material can only occur when the electrons are displaced relative to the nucleus containing positive charges, and held there to produce a dipole moment; "free" electrons, as in a metal, are displaced by an electric field, but pass freely through the material and are returned at the negative contact, to constitute a current flowing through the circuit. With insulators however, the charges form part of the bonding mechanism, and so cannot "flow", but may still be displaced. When this occurs solely by mechanical pressure, the material is piezo-electric. Conversely, the application of an electric field causes mechanical deformation in the material by displacement of charges.

Materials which have crystalline form, but are symmetrical crystallographically will not demonstrate this effect, because the stress system will average out over the crystal with respect to displacement. Non-symmetrical crystals however, will produce displacement under pressure, and charges will become polarised. From the 32 basic symmetry crystal classes, 20 are piezo-electric because the symmetry is of a low order. If the electric field is alternating, a piezo-electric crystal will vibrate, and alternatively,

if vibration is transmitted to such a crystal, an alternating voltage will be set up in the crystal. Piezo-electric transducers therefore find application in sound reproduction, in microphones and gramophone cartridges, in ultrasonic transmitters and receivers for marine depth gauges and sonar installations, and in non-destructive testing for medical and industrial investigations. (*See* **Non-destructive testing.**) These crystals can also act as controls for electronic oscillators, by cutting the crystal to have a resonance frequency in the desired range. Quartz has a high piezo-electric activity, roughly 10^{-12} coulombs per newton or metres per volt. Ammonium di-hydrogen phosphate has an activity of the order of 10^{-11} coulombs per newton. New materials are continually being investigated, chiefly from phosphate, arsenate, selenite, sulphate, titanate and niobate compounds.

Pig. The name used in reference to a bar or block of pig iron. When the molten metal is tapped from the blast furnace, it flows along main open channels in the ground. These are called sows and the branch channels are called pigs. Pig is applied as a name to the branch channel as well as to the metal itself. The name also applies to the product of pig-casting machines.

Pig-and-ore process. The name sometimes used in reference to the Siemens open-hearth process of steelmaking. In this process iron ore Fe_2O_3 is the agent used for the controlled oxidation of the carbon and silicon in the pig iron. (*See* **Siemens process.**)

Pig-and-scrap process. The name sometimes used in reference to the Siemens-Martin open-hearth process of steelmaking. This process was developed from the Siemens process. Martin originated the idea of diluting the fluid metal in the furnace by dissolving therein selected wrought iron or steel scrap. Thus, an increase in the amount of metal in the bath was not accompanied by any appreciable increase in the amount of impurities and so the percentage of the latter was reduced. (*See also* **Siemens-Martin process.**)

Pig bed. A bed of sand formed into suitable channels into which molten iron from the blast furnace is run.

Pig boiling. The name used in reference to that stage in the puddling process of making wrought iron when the original pig iron is melted and thoroughly mixed with the oxidising substances. The temperature of the whole mass then rises and it becomes agitated by the escape of carbon monoxide giving the impression of "boiling".

Pigging back. A term used by steelmakers in reference to the addition of pig iron to an open-hearth charge which has melted out low in carbon. During the "boil" the carbon content drops steadily and the rate of fall is systematically checked by rapid analyses. If the carbon content is too low, rectification is made by the addition of a sufficient weight of pig to increase the carbon content above the desired amount. Boiling then continues until the bath has the required composition.

Pigging up. The name used in reference to the method of increasing the fluidity of molten steel. Pig iron is added to avoid over-oxidising the steel, as it helps to maintain the carbon content of the bath and raises the temperature by producing a small boil as the carbon, silicon, manganese, etc., are oxidised. (*See also* **Pigging back.**)

Pig iron. The product of the iron blast furnace. It results from the reduction of iron ore in which process coke and limestone are required, the former for the reduction and the latter to produce a slag in which most of the impurities are removed. The name is derived from the fact that the metal is cast into horizontal ingots called "pigs". It is used as the raw material for making steel and in iron founding. It usually contains 2.5-5% C, 1-3% Si, with varying amounts of manganese, sulphur and phosphorus which originate either in the ore itself or are picked up from the coke.

Pigments. Substances, mainly of mineral origin, that when added to oil or other paint medium give body and colour to the paint. Those substances which are called auxiliary pigments or inert pigments are really fillers that retain their opacity in the paint. They are better referred to as extenders, and common examples are barytes, blanc fixe, terra alba (gypsum), china clay, silica, Paris white (whiting) and magnesium silicate. Some of the common substances used as pigments are listed below, grouped according to colour:

(a) black pigments. Carbon-black, bone-black, magnetic oxide

of iron, ground slate. Bone-black is the one non-mineral pigment;

(b) blue pigments. Ultramarine, cobalt ultramarine, smalt (potassium, aluminium, cobalt, silicate), Prussian blue, Chinese blue, cobalt oxide, copper phthalocyanine (tetrabenzoporphyrazine). Of these, Prussian blue is not permanent on exposure to light, while smalt gives a pleasing permanent colour. Cobalt oxide, either alone or mixed with other oxides, is widely used in colouring china-ware;

(c) brown pigments. Umber, sienna, iron III oxide, vandyke. Terra-cotta is burnt clay. Mixtures of lead chromate, lead manganese and lead sulphate produce light brown to orange colours;

(d) green pigments. Chrome green, Plessy's green, Verona green, Paris green, Cassel's green (barium manganate) and chromium oxide. Chrome green is a mixture of chrome yellow and Prussian blue and is not permanent. Verona green consists of a potassium, iron III magnesium silicate containing some manganese;

(e) red pigments. Red ochre, iron III oxide, venetian red, Tuscan red, red lead and cadmium selenide. Red ochre is a natural iron oxide having a high content of iron III oxide. The colour may vary with the percentage, and with much clay it may become yellow or reddish-yellow. Mercury sulphide gives a fine vermilion;

(f) white pigments. White lead, zinc oxide, titanium oxide, zinc sulphide and lithopone. All these, with the exception of the first two, are permanent in the usual town atmospheres;

(g) yellow pigments. Lead chromate, zinc chrome, yellow ochre, sienna ferrite, cadmium sulphide, cobalt yellow (potassium cobaltinitrate). Of these, the lead chromate tends to darken in industrial atmospheres; cadmium sulphide is brilliant and permanent but is expensive; cobalt yellow is very stable.

All cadmium pigments are brilliant. A green pigment containing cadmium (e.g. CdS) has about the same reflectivity as ordinary grass. It is useful as a camouflage paint, since in aerial photography areas so painted are not distinguishable from grass. Fungicidal pigments often contain copper salts, while zinc-containing pigments generally have good corrosion and heat-resisting qualities. All pigments should be ground fine enough to pass through a 325-mesh sieve.

Pig-washing process. A term used in reference to those methods of refining molten pig iron by oxidising treatment at relatively low temperatures. In Bell's process the molten pig iron is mixed with iron oxide in order to remove the silicon and phosphorus without appreciable reduction in carbon. In Krupp's process iron oxide and manganese oxides are used. Talbot's process (q.v.) consists essentially in pouring the metal through a molten basic slag composed of lime and iron oxide. Air has also been used for the selective removal of phosphorus by oxidation in presence of a basic oxidising slag.

Pile. An obsolete name for a nuclear reactor.

Pile also refers to a column or sheeting sunk into the ground to support vertical loading or to resist lateral pressures.

Pilet plating baths. Palladium is plated in a bath of the following composition:

Palladium II chloride	3.7	g
Disodium phosphate	100.0	g
Diammonium phosphate	20.0	g
Benzoic acid	2.5	g
Water	1,000.0	cm³

The solution is boiled until the dark red colour changes to yellow. The palladium II chloride must be replenished from time to time and the solution boiled. Bath used at 50 °C and current density of $2\text{-}3 \times 10^{-3}$ A m^{-2}.

Pilger mill. A type of rolling-mill having specially designed rollers for tube rolling, using a mandrel. The process of making wrought-steel tubes from hot billets is completed in several stages, the rough pierced blank being improved and sized as the mandrel moves forwards and backwards. As the rolls rotate the pass also varies; the bloom can slide easily into the pass at its largest, and at its smallest, the pass is considerably smaller than the bloom. A pneumatic-operated piston drives forward the mandrel carrying the bloom into the largest pass—or the "gap" as it is called. The rolls continue to rotate and bring the pass to the "bellmouth" section, which bites into the bloom. The rolls rotate in such a direction that the gripped bloom is driven back again in the direction from which it has just come. The trapped wave of metal is then moved out by the next part or section of the rolls called the "parallel" section.

Thus the hollow bloom is converted into a tube whose internal diameter is that of the mandrel. During the latter period of the operation the piston is also forced back with the mandrel compressing the air in the cylinder, so that by the time the "gap" section of the rolls comes round again the piston is ready for its next forward stroke. The air cylinder is mounted on a carriage which is automatically moved forward after each stroke, to allow for the shortening of the bloom. The mandrel and bloom are also turned through 90 degrees each time they approach the rolls, as the roll pass is not completely circular.

The process is now known as the rotary forge process.

Piling. The term used in reference to the arranging of the cut lengths of puddled bars into piles prior to welding and subsequently rolling them.

The term is also used to signify the operation of sinking piles into the ground to support vertical loading.

Pillar. The term applied to a member of a structure designed to resist compressive loads which act in an axial direction. In architecture a pillar is an irregular column, round and insulate and deviating from the proportions of a just column. It is generally, but not always, load-bearing.

Pillaring. A troublesome phenomenon sometimes encountered in blast furnaces. It is attributable to segregation of the charge, so that the coarser materials collect around the walls whilst the fine particles tend to form a cold pillar in the centre. It is usually indicated by a cold hearth with low-quality iron, and eventually by the development of hot spots as the furnace reactions are concentrated near the brickwork.

Pillow block. The name used in reference to a shaft bearing, consisting principally of the base and the block, in one casting and the brasses and the cap. Also called plummer block.

Pimple metal. A term used in reference to the product of smelting roasted copper or coarse metal with coke and siliceous flux. Since the object of the smelting is to reduce the iron content, the product is graded as: *(a)* fine or white metal in which little or no iron is present; *(b)* blue metal in which more or less iron is present; and *(c)* pimple metal which is a grade intermediate between *(a)* and *(b)*.

Pimpling. A term used in reference to the very small (pin-head size) bulges that may occur on the surface of aluminium or aluminium-base alloys. The phenomenon is restricted to pressure die-castings in which air is entrapped during casting. When the casting is heated to the softening point of the alloy the expanding gas causes the surface to show bulging and distortion.

Pin bar. The name used in reference to the small diameter case-hardening steel used for making dowel-pins.

Pinch. The term used in reference to a longitudinal overlap in sheet metal. It is seen in hot-pack rolling of sheets and may be caused by variations in thickness of the sheet, or by cold spots.

Pinch-bar. Synonymous with crow-bar.

Pinch pass. The term used in reference to the passing of a sheet through the rolling-mill in which it is arranged that a very slight reduction in cross-sectional area occurs. (Also called skin pass.)

Pinhole. The term applied to the small or very small holes or voids in castings. They arise from the entrapment of gas in the molten metal and the evolution of the gas during solidification. Pinholes also occur in electrodeposited metallic coatings. When the coating metal is cathodic to the basis metal, it is clearly important that the deposit is continuous, i.e. it should not contain pinholes, since these greatly reduce the protective value of the coating. The usual method of detecting these pinholes is to use, on absorbent paper, some reagent that gives a colour at those points where the basis metal is exposed.

Pinion. The smaller of two toothed wheels in gear or the small tooth-wheel meshing with the teeth on a rack. In rolling-mills, the pinions, which are supported in housings similar to the roll housings, are gears, one of which is driven through a driving spindle in line with one of the rolls.

Pinite. A hydrous silicate of potassium and aluminium produced by the alteration of felspars. In the massive form it resembles steatite. It is used for kiln linings in cement plants, since it will bond alone like clay and has a low shrinkage. At 1,125 °C it is converted into mullite, a valuable refractory that does not soften below its melting point (1,819 °C).

Pin joint. A joint in which a cylindrical metal pin is used to connect together two parts or two rods. In such a joint rotational movement of one or both rods may be possible.

Pink meal. A mixture of bran, wheatings or hardened sawdust with a small quantity of calcium sulphate. It provides an absorbent material for cleaning tinplate. As the tin-coated plates leave the hot-tinning pots they carry a surface film of palm oil which is removed in cleaning machines charged with this material.

Pin-lift. A type of hand moulding machine in which the moulding box is pushed upwards away from the pattern by means of ejector pins pushed up through holes in the framework. The method is satisfactory for the withdrawal of the pattern except in those cases where there are deep cods of sand. These overhanging masses are liable to fall away and in such cases a roll-over machine (q.v.) is required.

Pinning (of dislocations). The dislocation represents a piece of mismatched lattice in a crystal, an imperfection or defect. When a moving dislocation reaches an obstacle such that greater force is required to continue that movement, the dislocation is said to be "pinned". An impurity atom, larger or smaller than the matrix, creates a stress field around it, as does a vacancy, but these would not create a great barrier for dislocation movement, although they affect the movements. A cluster of atoms in the form of a coherent precipitate, or a tangle of other interacting dislocations, can however, be a serious obstacle. When pinning occurs, other mechanisms of dislocation movement may be necessary to continue movement, or a change in temperature, as in heat-treatment processes for stress relief or softening.

Pinolin. A pale brown light spirit obtained by the dry distillation of resin. It is used for blending with turpentine, as a plasticiser for rubber and in the production of synthetic moulding resins. Also called oil of amber.

Pin rammer. A tool used by moulders for pressing sand into confined spaces about the pattern. Also called pegging rammer.

Piobert lines. The name used in reference to the surface defects which appear as lines or ridges during deep-drawing of steel sheets that have been incorrectly annealed. Also known as stretcher-strains, Luder's lines.

Pipe. The name used in reference to the cavity formed in the top of an ingot or the feeder head of a casting by contraction during solidification of the last portions of the liquid metal.

Pipe-chaplet. A specially shaped device used in moulding for holding the core in place. It usually has a large bearing surface and a long stem which will reach to the outside of the moulding box, where it is secured by wedges.

Pipe-mill. The name used in reference to a type of rolling-mill in which hot metal is plastically deformed to produce tube or pipe. (*See also* **Pilger mill, Plug rolling.**)

Pipe nails. An alternative name for pipe-chaplets which are in the form of broad-headed nails. The heads are curved and the nails are used to secure the cores in position.

Pipe-steel. A term usually applied to a very low carbon steel with alloying additions and particularly suitable for welding. It is made by both the Bessemer and open-hearth processes.

Pipe-stove. The name applied to a type of hot-blast stove in which the air for a blast furnace is heated by passing through pipes which are heated externally by blast furnace or other hot gases.

Piping. The name used in reference to the type of defect peculiar to extruded metals which consists of a ring of oxide running through the extruded bar or rod. The oxide originates on the outside of the billet due to reheating, and this may be further contaminated if the surface of the casting is dirty.

Pit. A sharp depression in the surface of a metal. The name is also applied to a type of furnace. (*See* **Soaking pit.**)

Pit casting. A method of casting in which the mould is set in a pit in the foundry floor. It is specially suitable for large castings, since the mould can be made very strong to take the weight of molten metal, and the pouring level can be arranged for greatest convenience.

Pitch. A term used to denote the distance from the centre of one roll to the centre of the other of a pair of rolls measured without clearance. It is also used to denote the distance from the centre of a bolt in one line to the centre of the next one in the same line, and the distance between adjacent threads of a machined screw.

When two toothed wheels are in gear they may be supposed to roll together on two imaginary circles called the pitch circles. The diameters of the circles are known as pitch diameters, and the distance from any point on one tooth of a wheel to the corresponding point on the next tooth, measured along the pitch circle, is called the circular pitch.

The name is also given to viscous dark coloured materials obtained naturally as asphalt or bitumen, and residues from the distillation of wood (the correct origin from the Greek and Latin for pine tree) or coal. The material can be softened by heat, and dissolves in hydrocarbons, particularly aromatics. It provides an excellent bond for stone, brick and sand, and is still occasionally used for these, apart from its obvious application in road-making material.

Pitch is a fluid which will pass slowly through an orifice which subjects it to a shearing force (viscosity about 10^{12} centipoise at room temperature).

Pitch binder. Coal-tar pitch used as a binding agent for sands for making cores.

Pitchblende. A complex oxide of uranium containing variable amounts of lead, iron, copper and bismuth, as well as the rarer elements radium, thorium, yttrium, helium and argon. It crystallises in the cubic system, but crystals are rare, the mineral occurring in the massive form. The content of uranium oxides varies from 64% to 89%. There are few commercially important deposits, the most important being those of Great Bear Lake, Zaire, the Joachimsthal district of Czechoslovakia and parts of Australia. Small isolated deposits are found in various parts of the USA. The variation in composition of the ore is commonly recognised by the different names, e.g. cleveite, nasturan, nivenite, broeggerite, by which it is known. These are crystalline varieties.

The mineral is used for the extraction of uranium, radium, etc. Though the names uraninite and pitchblende are often used synonymously, the latter is more appropriately applied to the amorphous and impure forms. Mohs hardness 5.5, specific gravity varies greatly with the purity from 5 to 9.6.

Pitch-on metal. The name given to mild-steel sheet that has been hot dipped in coal-tar pitch in order to produce an adherent coating. It is used as a substitute for galvanised iron and for copper sheeting in roofings and flashings.

Pit moulding. The term applied to the process of casting in an open pit made in the foundry floor.

Pit soaking. A process in which ingots are heated in a pit furnace so that their temperatures are equalised and adjusted prior to working.

Pitting. A form of corrosion associated with highly localised breakdown of a protective film on the surface of a metal. Pitting has been regarded as an intermediate stage between general corrosion and immunity. It involves less destruction of metal than general corrosion, and may be set up by the inefficient use of inhibitors as well as by impurities in the metal (inclusions) or in the corrosive medium.

Placer deposits. Deposits formed by the action of running streams and found where the velocity, and hence the carrying power, of the water has decreased. The deposits consist of gravelly or sandy material which becomes progressively finer the further the depositing waters are from their mountain source. Gold, platinum, tin, wolfram and magnetite are examples of metals and ores transported by water and found as placer deposits.

Planck's constant. In order to explain the energy distribution in the spectrum of black-body radiation, Planck proposed the quantum theory. This theory supposes that energy can be emitted or absorbed by a vibrating body only in multiples of a certain unit called a quantum. The size of this quantum depends on the frequency of the radiation and is equal to "$h\nu$", where ν is the frequency and h a universal constant known as Planck's constant. It is equal to 6.625×10^{-34} J s. In the case of elements that are solid at ordinary temperatures the value ν is the frequency of the light emitted by the incandescent vapour of the substance.

Planck's theory leads to the assumption of the atomic origins of radiation and agrees with the conception that radiation is due to the change of an electron from one energy level (electron shell) to another level.

Plane of symmetry. A plane that divides a crystal into two similar and similarly placed halves, each being the mirror image of the other.

Plane stress and strain. In an isotropic Hookeian solid, the generalised stress-strain relationships can sometimes be simplified. When a flat sheet or plate is considered, forces acting in the plane of the membrane give stress free surfaces normal to that plane, and the stress components directed out of that plane are zero. In a very thin sheet, the same applies through the whole thickness of the material, and only two dimensional stresses are involved. This is termed plane stress when the two dimensional stress components are constant through the thickness. The strains may not be zero through the thickness however, because of the effect of Poisson's ratio.

When the displacements produced are parallel to a plane, this is referred to as plane strain, and applies particularly to strains in thick plates. Considerations of plane stress and strain are used in solving problems in the fabrication of metals by rolling and pressing quite apart from the design of structures.

Plane surface. A perfectly flat surface free from elevations, depressions or irregularities.

Planimetric method. A method of metallographic grain counting. The field is divided into definite small areas and in these the grains or fractions of grains are counted.

Planishing. A process for improving the surface of wrought metals usually by the action of rolls. When a planishing mill is used it is located immediately preceding the finishing rolls. It is thus an intermediate operation designed to smooth the surface of rolled products. It is a cold-forming operation, and may be used to sharpen or flatten an edge of, or remove flash from a forged product. When a power hammer is used the intention of the process is to produce a smooth and polished surface and at the same time toughen the metal by delivering a rapid series of blows.

Planishing mill. A roll-stand in a rolling-mill, the object of which is to produce a smooth and polished surface by planishing. The material is passed between smooth rolls, the reduction in area usually being very slight.

Plasma. The name used in reference to a gas, usually at low pressure, which has been ionised by stripping one or more electrons from some or all of the atoms. It is therefore a mixture of neutral particles, positively charged ions and negative electrons, and is an excellent conductor of electricity.

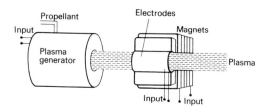

Fig. 114 *Schematic diagram of plasma jet.*

When heavy discharge from a bank of condensers passes through the gas, strong electromagnetic forces of repulsion violently accelerate the plasma away from the electrodes and give rise to a shock wave with resulting very high temperatures—of the order of 5×10^5 K. Using what is called a pinch effect and pre-ionisation of the gas, temperatures exceeding 10^6 K can be obtained.

The high temperature of the plasma jet may provide the key to controlled nuclear fusion. It is currently being used for spraying corrosion-resistant molten ceramics on to metals.

Caesium plasma is used in the plasma thermionic converter to convert nuclear or solar energy into electrical energy (Fig. 114).

In the direct generation of electricity from highly conductive plasma helium has been used seeded with pure metallic caesium, which enhances ionisation and permits generation to occur at

about 1800 °C. Argon with caesium has been used, and ionisation stimulated by a high-voltage electron beam enables electric generation to occur at temperatures in the region of 1200 °C.

Plastalloy. The name used in reference to a low-carbon steel having a fine-grain structure which will take a high polish and is suitable for making moulds. In the annealed condition the steel is readily machined and will take a deep imprint of a hot pattern. Such steels are case-hardened and mirror-polished before use.

Plaster. A term usually employed in reference to a mixture of plaster of Paris, sand and glue-water. Neat plaster is a mixture of plaster of Paris and lime-water or glue-water. Both these mixtures are used for repair work and for wall finishes. They are unsuitable for outside work, since the constituents are slightly soluble in water. If fuller's earth is used in place of plaster of Paris a more satisfactory material is obtained which is suitable for filling cracks and for subsequent painting.

Plaster of Paris. When gypsum $CaSO_4 \cdot 2H_2O$ is heated to about 130 °C it loses some of its water of crystallisation and forms the hemihydrate $2CaSO_4 \cdot H_2O$, which is plaster of Paris. This latter when mixed with water evolves heat, expands a little and quite quickly solidifies. If the hemihydrate is heated to about 200 °C it loses the whole of its water of crystallisation. The anhydrous salt readily takes up water again. But if the heating has been more intense the product takes up water only very slowly and sets slowly. It is then said to be dead-burnt. Slight decomposition occurs at about 400 °C and the product which sets only very slowly is known as Estrich-gips. Water solutions of synthetic resins may be used with plaster of Paris to produce strong and waterproof casts.

Plastic bearings. Bearings made of sheet or laminated plastics and used in special cases as substitutes for metals, since lubricating oil is not needed. Materials such as nylon can carry light loads at high speed without lubrication and are useful in certain types of textile machinery where insulation is required. Most plastics may be lubricated if required by water only or by water containing a little sodium oleate (soap).

Plastic bronze. A term sometimes used for bearing bronzes that contain a high proportion of lead, usually of the order of 30%. Such bearings are less hard and less strong than straight bronzes, but are rather superior in these properties to the Babbit metals. They are used in heavy bearings, e.g. railway coach journals. (*See* Table 5.)

Plastic deformation. The term used in reference to the permanent (inelastic) distortion of metals under applied stresses which strain the material beyond its elastic limit. The deformation is accompanied by changes in the internal state of the metal, involving distortion of the crystal structure. As distortion proceeds the resistance of the metal to deformation generally increases, while there is a corresponding reduction in the capacity to deform. The tendency of the metal is then to relieve the strain by recrystallisation, and this involves reorientation of the atomic arrangement. In some metals this tendency is noticeable at ordinary temperatures, while in others the effect is not noticeable until the metal is heated. Lead is an example of the former, while iron is an example of the latter, the temperature in this case being in excess of 450 °C.

Plasticiser. An ingredient added to polymeric material to lower the viscosity or change the rheological properties of the material in service. Generally, the action is one of penetration between polymer chains, lowering the glass transition temperature. Cross-linked polymers are hardly ever affected in this way, but amorphous linear polymers can be easily plasticised. Crystalline polymers can be affected, but the appropriate compounds are limited.

The word plasticiser is used only for additions to polymers, but similar effects can often be achieved by co-polymerisation with another polymer. In the latter case, the phenomenon is called internal plasticisation. Water can act as a plasticiser with hydrophilic polymers (e.g. cellulose) but alcohols such as glycol (1, 2-ethanediol) and glycerol (1, 2, 3-propanediol) are much better, because they are less volatile. Some indeed act as humectants, retaining moisture. Esters are often used, particularly phthalates, adipates and sebacates (esters of 1, 2-benzenedicarboxylic, hexanedioic and decanedioic acids respectively). Rubber is plasticised with mineral oil or similar petroleum derivatives.

TABLE 40. *Composition of Plating Baths*

Metal	Bath composition	g m^{-3}	Metal	Bath composition	g m^{-3}	Metal	Bath composition	g m^{-3}	Metal	Bath composition	g m^{-3}
Antimony	Antimony II oxide	1.2	Indium (cont.)	Boric acid	0.62	Platinum (cont.)	Diammonium hydrogen phosphate	0.9	Zinc (cont.)	Aluminium sulphate	0.4
	Hydrofluoric acid	2.3	Iron	Iron II chloride	12.4		Disodium phosphate	4.95		Sodium chloride	0.26
Bismuth	Bismuth oxide	0.8		Calcium chloride	13.55		Platinum II diaminonitrite	0.32		Boric acid	0.4
	Perchloric acid	3.0		Iron II ammonium chloride	6.2		Sodium nitrite	0.21		Zinc chloride	3.1
	Glue	0.006		Sulphuric acid	0.005		Ammonium nitrate	2.1		Aluminium chloride	0.52
Chromium	Chromium III oxide (CrO_3)	5.2	Lead	Basic lead carbonate	2.6		Ammonium hydroxide	1.03		Sodium chloride	5.2
	Sulphuric acid	0.05		Hydrofluoric acid	2.77	Rhenium	Rhenium sulphate	0.21		Zinc cyanide	1.16
	Oxalic acid	0.26		Boric acid	2.06		Potassium persulphate	0.58		Sodium cyanide	1.03
	Chromium III oxide (CrO_3)	9.3	Nickel	Nickel sulphate (cryst.)	2.06	Rhodium	Rhodium (as phosphate)	0.04		Sodium hydroxide	1.42
	Sulphuric acid	0.09		Nickel ammonium sulphate (cryst.)	0.58		Sulphuric acid	0.65	Brass	Sodium cyanide	1.3
	Oxalic acid	0.52		Ammonium chloride	0.39		Rhodium (as sulphate)	0.21		Copper cyanide	0.65
	Chromium III oxide (CrO_3)	6.8		Boric acid	0.32		Sulphuric acid	1.94		Zinc cyanide	0.19
	Caustic soda	0.97		Nickel sulphate (cryst.)	4.13	Ruthenium	Ruthenium nitroso-chloride	0.08		Sodium bicarbonate	0.26
	Sulphuric acid	.01		Nickel chloride (cryst.)	3.6		Sulphuric acid	0.77		Ammonia	0.26
Copper	Copper sulphate (cryst.)	4.13		Boric acid	0.84	Silver	Silver cyanide	0.06	Bronze (Cd)	Copper cyanide	0.45
	Sulphuric acid	1.03		Nickel chloride (cryst.)	6.2		Sodium cyanide	1.29		Cadmium oxide	0.06
	Copper cyanide	0.52		Boric acid	0.65		Silver cyanide	0.52		Sodium cyanide	0.71
	Potassium cyanide	0.74		Nickel fluoborate	4.64		Potassium cyanide	0.77		Sodium carbonate	0.32
	Sodium carbonate	0.13		Ammonium fluoborate	1.03		Potassium carbonate	0.52	Bronze (Sn)	Copper cyanide	0.74
	Copper cyanide	0.62		Nickel sulphate (cryst.)	5.2		Silver cyanide	0.77		Sodium stannate	1.03
	Potassium cyanide	0.78		Nickel chloride (cryst.)	0.78		Potassium cyanide	1.03		Sodium cyanide	0.55
	Rochelle salt	1.03		Boric acid	0.55		Potassium hydroxide	0.26	Bronze (Zn)	Copper cyanide	0.61
	Sodium carbonate	0.77		Sodium fluoride	0.26		Potassium carbonate	0.84		Zinc cyanide	0.05
	Copper tartrate	2.32	Palladium*	Palladium II chloride	0.078	Tin	Stannous sulphate	1.14		Sodium cyanide	0.77
	Potassium carbonate	0.71		Disodium phosphate	2.06		Sulphuric acid	1.24		Rochelle salt	0.32
	Copper sulphate	0.31		Diammonium phosphate	0.41		Cresol sulphonic acid (Methylphenol sulphonic acid)	2.06	Rhenium-Nickel	Rhenium sulphate	0.21
	Sodium oxalate	0.21		Sodium palladium nitrite	0.31		Sodium stannate	1.68		Potassium persulphate	0.58
Gold	Potassium gold cyanide	0.39		Sodium chloride	0.62		Sodium hydroxide	0.26		Nickel sulphate	0.23
	Potassium cyanide	0.26	Platinum	Platinum IV chloride	0.18		Sodium acetate	0.32	Speculum	Sodium stannate	2.1
	Caustic potash	0.26				Zinc	Zinc sulphate (cryst.)	6.2		Copper cyanide	0.26
Indium	Indium fluoborate	4.9								Sodium cyanide	0.52
	Ammonium fluoborate	0.97								Sodium hydroxide	0.32
									Tin-Nickel	Tin II chloride	1.03
										Nickel chloride	6.2
										Ammonium bifluoride	0.72
										Sodium fluoride	0.58
									Tin-Zinc	Sodium stannate	1.15
										Zinc cyanide	0.06
										Sodium hydroxide	0.09

*Palladium may also be deposited from palladosamine chloride or nitrate baths.

Plasticity. Deformation which is irreversible is termed plastic deformation, i.e. producing a permanent set. Such deformation behaviour often succeeds elastic behaviour, but in other cases deformation is part elastic part plastic from the beginning of stressing. (*See* **Elasticity**.)

A notable form of plastic behaviour is the phenomenon of yielding, which occurs in polymeric materials and in some metals and alloys, notably low carbon steel. Yielding is characterised by sudden and large amounts of deformation on reaching a critical stress level. Attempts have been made to express plasticity in mathematical form to aid stress analysis of structures, but with varying success in predicting the avoidance of failure by over-stressing.

Plastics. Used by the polymer industry to refer to a generic class of substances manufactured from polymeric material, but frequently containing fillers, plasticisers and sometimes gas pores. Frequently misused in the plural as an adjective, and in the singular as a noun.

Strictly, the word applies to materials which demonstrate plasticity when stressed, i.e. suffer deformation before rupture. However, any resinous material, synthetic or natural, thermo-plastic or thermo-setting, tends to be classified in this way, whether plasticity is a feature of its deformation or not.

Plastiron. A low-carbon iron used for making moulds and dies. It is hardened by carburising.

Plate. The product of rolling an ingot or slab in a plate-mill. The cross-section is rectangular and over 3 mm thick, and the width is very much greater than the thickness.

Plate-mill. Synonymous with rolling-mill (q.v.).

Plating. The term used in reference to the application of a thin film of metal to the surface of a material. The process may include dipping into molten metal, but it usually refers to the electro-deposition of an adherent coating.

Plating barrel. The barrel or drum in which small articles are put in the process of barrel plating. The method is specifically useful when large numbers of small articles have to be plated, e.g. washers, bolts, springs, nuts, etc. It is clearly important that the work must have direct or indirect contact with the cathode, and it is equally important to ensure that no item of the work shall make contact with the anode. In one apparatus the work is contained in a non-conducting, perforated barrel, within which is the cathode in the form of a star connected through the drum-shaft to the negative terminal of the battery. The anode is external to the barrel. In another the barrel is omitted and the work is put in the container, which itself revolves. Such a container may be tilted so that no part of the work shall remain shielded from the plating process. Since the anode must be within the bath, precautions must be taken to shield it from contact with the work. This is done by using a perforated cover.

Plating baths. Solutions of salts of the metals to be deposited with or without the addition of acids or alkalis or organic brighteners. The concentration of the bath, the temperature and the circulation of the solution have a marked effect on the nature of the deposit. Circulation of the liquid is frequently necessary to prevent marked decrease in concentration round the cathode.

Metals such as iron, nickel and cobalt can be deposited in a satisfactory form from simple salts, the solution being kept slightly acid to prevent the precipitation of basic salts. Copper can also be deposited from an acid solution (but not on iron or zinc), and zinc may be deposited from a simple salt in the presence of boric acid. Most metals are better deposited from complex ions. Brass, bronze, cadmium, copper, silver, gold, zinc and speculum are all deposited from a mixture of cyanides of which one is an alkaline cyanide and the other the cyanide of desired metal. Palladium, platinum, rhodium, tin and tin-nickel alloy are usually deposited from special baths in which the metal forms part of a complex ion. Bismuth is deposited from a perchloric acid bath (*see* Tables 40 and 41).

Plating rack. A frame from which articles are suspended in a plating bath and through which current passes to the articles which are the cathodes in the bath. The articles are moved occasionally to prevent wire marks.

Platinised asbestos. Very finely divided platinum deposited on asbestos. The process consists in first boiling asbestos fibres in

TABLE 41. *Plating Baths for the Deposition of Various Metals*

Metal or Alloy	Type of Plating Bath	Anode	Working Temp., °C.	Current Density, A m⁻²
Antimony	Fluoride	Sb	10	0.93
Bismuth	Perchlorate	Bi	10	2.8
Cadmium	Cyanide (alkaline)	Cd	18-30	0.9-1.9
Chromium	Chromium III oxide	7% Antimonial	40-50	11-19
	Sodium chromate	Lead	15-20	9-90
Cobalt	Sulphate (acid)	Co	25-35	8-14
Copper	" "	Cu	15-50	0.9-19
	Cyanide (alkaline)	"	45-60	0.5-0.9
	" (Rochelle salt)	"	50-70	1.9-5.6
	Tartrate (alkaline)	"	10-15	0.2-0.3
	Sulphate, oxalate (alkaline)	"	10-15	0.3-0.4
Gold	Cyanide (alkaline)	C, Au or Pt	50-80	0.2-0.56
Indium	Fluoborate	In	16-25	5-10
Iron	Chloride	Fe	60-70	0.9-11
	Double sulphate (acid)	"	20-30	1.9-3.0
Lead	Fluoborate (acid)	Pb	16-25	1.4-1.9
Nickel	Sulphate (acid)	Ni	16	0.5-1.0
	Sulphate-chloride (acid)	"	30-40	1.4-3.3
	Sulphate-chloride (strongly acid)	"	45	9.0
	Chloride (acid)	"	60	2.3-9.0
	Fluoborate (acid)	"	20	0.9-1.9
Palladium	Double nitrite	Pd	40-50	0.01
	Phosphate	"	50	0.2-0.3
Platinum	"	Pt	90-100	0.9
	Diaminonitrite	"	95	5.6
Rhenium	Persulphate	Re	25-45	8.4-9
Rhodium	Nitrite (acid)	Pt	40	0.5
Ruthenium	Chloride (acid)	Ru	40	2.0
Silver	Cyanide (alkaline)	Ag, steel	10	1.4-1.9
	" (alkaline, carbonate)	Ag	20-25	0.3-0.4
	" (strongly alkaline)	"	30-40	1.9-7.4
Tin	Stannate (acid)	Sn	20	0.2-0.9
	" (alkaline)	"	75	0.9
Zinc	Sulphate (acid)	Zn	16	0.9-9.0
	Cyanide (alkaline)	"	30-40	0.9-1.9
	" (alkaline)	"	60	1.4
	Chloride	"	20-40	1.4-14
Brass	Cyanides (alkalines)	70/30 brass	35-40	0.9
Bronze (cadmium)	" "	Cu	10	0.2-0.5
(tin)	Cyanides-stannate (alkaline)	Sn and Cu separate	65	0.9-5.0
(zinc)	Cyanides (Rochelle salt)	92/8 bronze	10	0.2-0.4
Rhenium-Nickel	Sulphate (acid)	Pt	25-45	5.0
Speculum	Stannate-cyanide	Cu and Sn	65	1.4-2.3
Tin-Nickel	Chloride-fluoboride	Sn and Ni	65	2.3
Tin-Zinc	Stannate-cyanide	Sn-Zn 78/22	65	0.9-2.8

strong hydrochloric acid. They are then soaked in platinum IV chloride solution, dried, and then heated with a little ammonium chloride. Alternatively the platinum salt may be reduced with sodium formate. It is used as a catalyst, catalysing such reactions as the oxidation of SO_2 to SO_3 for the making of sulphuric acid by the contact process.

Platinising. The process of coating other materials with platinum.

Platinum. The most important and the most abundant of the platinum group metals. It occurs native in alluvial deposits in Russia and Colombia and in association with other minerals in Canada, the USA, South America and South Africa. The chief sources for the Western world are the Canadian nickel-copper ores from which the platinum is recovered as a by-product. From the Russian alluvial deposits the metal is recovered by hydraulic concentration. The crude metal is dissolved in *aqua regia*. Ammonium chloride is added and the immediate precipitate is ammonium chloroplatinate. When this is calcined on a block of lime or magnesia crucible the platinum is left as a sponge, while the ammonia and hydrochloric acid volatilise. The refining processes required to effect the separation from associated metals are long and involved, but finally separate the platinum from palladium, rhodium and iridium and osmiridium. Platinum is a white metal, slightly greyer and somewhat less lustrous than silver. It is soft and is readily malleable and ductile, so that ingots can be rolled in the cold, but the metal is usually worked at about 800 °C when rapid reduction is possible. It is drawn into wire and rolled into sheets and the latter are formed by spinning and stamping.

Its resistance to common acids and most forms of chemical corrosion renders it particularly suitable for use in laboratory ware, dentistry and in the electrical and fine chemical industries. In normal times platinum metals are widely used by jewellers.

Platinum dies are used for making glass fibres, used in filter cloths, etc., and for making the glass insulators in electric lamp bases. The coefficient of expansion of platinum is 8.9×10^{-6} while that of soda glass is $8.5-8.7 \times 10^{-6}$, and hence the metal makes a good glass seal. It has, however, been partly replaced in this employment by less expensive metal alloys. Platinum and platinum alloys are largely used as catalysts in various manufacturing processes, in pollution control and in the construction of high-temperature thermocouples. (*See* Table 2.)

Platinum alloys. Platinum alloys—usually with other platinum group metals or with other precious metals—find very wide diversity of uses because of their resistance to corrosion, their pleasing colour, their high melting point and their favourable electrical resistance characteristics. Some of the commoner alloys with their main uses are given in Table 9.

Platinum black. A very finely divided powder consisting of metallic platinum obtained by reducing a solution of chloroplatinic acid with zinc or with sodium formate. It is very active catalytically, oxidising alcohol to aldehyde and causing a mixture of hydrogen and oxygen to explode.

Platinum catalysts. Platinum metal, in the form of wire, or gauze or very finely divided metal powder on an inert medium or as a colloid will materially increase the speed of a chemical reaction and so act as a catalyst. Each catalyst (and each form of it) will be more or less specific for particular reactions. Since the metal is usually employed to catalyse reactions between gases or reactions in liquids, it is termed a heterogeneous catalyst, and its action has been explained on the basis of a theory of surface adsorption. Such catalysts do not undergo any permanent chemical change, but they may change their physical state. Smooth platinum foil becomes roughened when used to accelerate the oxidation of ammonia to nitric acid; platinum wire or foil becomes coated with a very fine grey deposit of platinum during the catalysing of the combination of hydrogen and oxygen.

When deposited on asbestos, platinum readily causes the combination of sulphur dioxide and oxygen. When supported on suitable refractory materials it will catalyse the production of hydrocarbons, and is generally useful for the hydrogenation of certain organic compounds, and in the synthesis of hydrocarbons of high molecular weights. Platinum is frequently the catalyst used in flameless and in automatic lighters.

Platter. A term used in forging in reference to the whole mass of metal that is being worked in a complete operation.

Plessy's green. A bluish-green powder used as a pigment. It is chromium phosphate and is insoluble in water.

Plough steel wire. A patented type of steel wire made from steel free from segregations, and manufactured perfectly circular in section and drawn to have an ultimate tensile strength of 1.55-1.7 GN m⁻². It is used in making high-grade steel rope and as a reinforcing cord in automobile tyres.

Plug. A term having a variety of meanings such as a rod or mandrel used in tube drawing; a stopper used for closing the tap-hole of a furnace or a pouring ladle; a false bottom in a die; the projecting part of a forging die designed to produce a cavity in the forging.

Plug gauge. A term used in reference to a cylindrical male gauge used in inspection procedure for measuring the size and contour of a hole.

Plugging. A term used in reference to the rectification of castings, where the method is used to correct the appearance or stop a leak. It is most reliable when applied to non-machined surfaces.

Plug rolling. A method of making seamless tubes by rolling pierced billets over a mandrel in a two-high mill that is equipped with a movable trough and pusher on the entry side and with a stationary guide and mandrel-bar supports on the delivery side.

Plug-welding. A term used in reference to the welding of two plates, the upper member of which has a hole exposing a certain area of the lower member. The weld metal usually fills the hole. The method is applicable to welding the two members forming a lap-joint.

Plumbago. A naturally occurring variety of carbon. It has a metallic greyish-black appearance and is greasy to the touch. The amorphous kind is used in making lead pencils and as a pigment in certain paints. It is also used in foundry mould facings (e.g. combination lead). Most natural varieties contain a little silica, so that for lubricants the artificially produced kind, made in the electric furnace, is preferable. The material is usually specified and marketed according to purity, particle size and electrical resistance. Extremely fine particles of graphite in colloidal suspension in oils or water are marketed as dags. The foliated variety is used for making crucibles. Plumbago occurs in veins in Sri Lanka, Cumberland (UK) or in bedded masses in east Canada. It is used in making crucibles and other refractory articles, and in giving a black polish to metals and in electroplating to provide a conducting surface. Mohs hardness 0.9-2, sp. gr. 2-2.5. Also called graphite.

Plumbsol. A soft silver solder developed for use in installations where freedom from the toxic effects of lead is essential. It is a silver-tin alloy containing only a small percentage of the former. It melts in the temperature range 220-225 °C. In soldering, its behaviour resembles that of the 50-50 tin-lead solder, but it possesses higher strength at elevated temperatures.

Pluramelt. A composite steel with different types of stainless steel bonded or welded to a depth of one-fifth of the composite plate. The bonding is obtained by a process of inter-melting and rolling. The corrosion resisting steel is usually bonded to a steel of lower cost. The result is a cheaper product and a saving in the use of chromium and nickel.

Plutonic rocks. Igneous rocks which have cooled slowly at a great depth. They possess coarsely holocrystalline structure and occur in great masses.

Plutonium. A radio-active element that can be produced from uranium. When uranium ^{238}U is bombarded by slow-moving neutrons, neutron capture results and the isotope ^{239}U is produced. This ^{239}U is radioactive, emitting β particles (electrons) and becoming converted into neptunium (at. no. 93). Neptunium decays by β-particle emission, and plutonium results. The half lives of ^{239}U and neptunium are 23 minutes and 2.3 days respectively. Plutonium is comparatively stable. It has a half life of about 24,000 years and decays to ^{235}U. It is fissionable. (See Table 2.)

Plyer. A device used in tube drawing. It consists of a carriage equipped with jaws which grip the tube, and is usually drawn along rails by chains or cable from the die head to the limiting extent of the draw-bench.

Plymax. A composite material consisting of plywood to which thin sheets of aluminium are attached.

Plymetals. Sheet metals consisting of bonded layers of dissimilar metals. Also called clad metals. They are used in thermostats and in temperature-actuated devices. Also used to economise in cladding metal.

Plymetl. A composite material consisting of plywood core to which galvanised steel sheets are cemented.

Plywood. Thin sheets of wood cut from the trunk by turning, and then glued together under pressure with alternate sheets having the grain at an angle to each other, usually 90°. Water-repellent resins produce composites having better resistance to weather and marine conditions, and more resistant to shrinkage and warping. The mechanical properties approach those of an isotropic material, particularly when the sheets are thin and many are used to make the ply.

Pneumatic hammer. A hammer—usually a forging hammer—that utilises compressed air to produce the impacting blow.

p-n junction. When a piece of p-type semiconductor material is joined in electrical contact with a similar piece of n-type semiconductor material, this forms a p-n jucntion with very important properties, and constituting the basis of transistor devices. The two materials form potential sources of electron carriers (n-type) and positive hole carriers (p-type) and at the junction between the two, normal diffusion processes will ensure that some atoms will diffuse across, but many more electron or positive hole carriers, which will ensure a neutral zone as the two cancel out. This is the depletion layer, about 1 micrometre thick. Within this zone, the atoms providing electrons (n-type) are now positively charged, since their electron carriers have been neutralised after moving

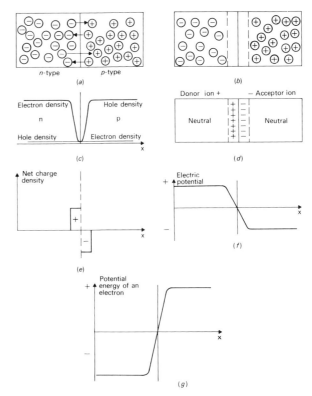

Fig. 115 *p-n junction and depletion layer. (a) Diffusion at junction. (b) Depletion layer. (c) Distribution of carrier densities. (d) Distribution of charges in depleted layer. (e) Charge density. (f) Potential gradient. (g) Potential energy of an electron across the junction.*

into the p-type material across the junction. Similarly, the atoms providing positive holes will be negatively charged on the other side of the junction. The depletion layer is thus a charged zone, with two layers of opposite charge across the junction, rather similar to a charged capacitor. A potential gradient will therefore exist across the junction, and is about 1 volt in silicon and 0.6 volts in germanium. An equilibrium must then result, with the potential gradient opposing the net movement of carriers (*see* Fig. 115) and no net current flows across the junction.

When a bias potential is applied to the junction, the dynamic equilibrium due to carrier diffusion must be upset. If the n-type material is made positive with respect to the p-type, reverse bias, the first effect is to increase the potential difference across the junction. This reduces the flow of positive carriers into the n-type and of electrons into the p-type, but does not affect the reverse flow very much, because there are only a few positive holes in the n-type depletion layer and only a few free electrons on the p-side of the layer. A small net current can now flow from p to n, as electrons move back to the n side and holes to the p-side (*see* Fig. 116). If, however, a forward bias is applied, this reduces the potential gradient due to the diffusion into the depletion layer. Again, the flow of carriers from n and p regions is scarcely affected, but the flow in the other direction is very substantially increased, because the barrier created by the diffusion (the potential gradient) has been partially neutralised. The probability of an electron crossing the barrier is multiplied by a factor exp. (eV/kT), and when the bias voltage is 0.1 volts, this factor is 52X.

Figure 47 (*see* **Diode**) shows the voltage current relationship which results. The reverse bias (voltage negative) shows that little or no current can pass, the junction becomes a rectifier. The exponential character of the current change when forward voltage is increased is obvious, leading to the amplification effect described. The only damage that can arise is that of overheating, because kT increases, the amplification becomes even greater, the current much higher, and self-destruction is the end point.

Point defect. Defects in a crystal lattice in which single atoms or atom positions are involved, in contrast with line defects such as dislocations. Point defects include vacant lattice sites and inter-

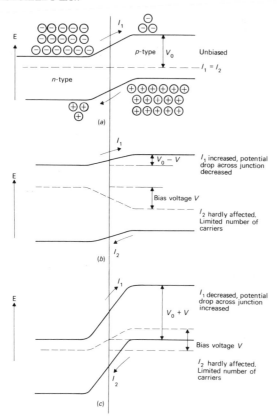

Fig. 116 *Energy bands in p-n juction with bias voltage. (a) Equilibrium. (b) Forward bias. (c) Reverse bias.*

stitial atoms of the parent lattice in interstitial positions (but not interstitial atoms of another material in interstitial positions in the host lattice). When interstitial atoms occur by migration within a material, they are called Frenkel defects. When vacant lattice sites occur by migration of an atom to another position, not necessarily interstitial, or to the surface, they are called Schottky defects. Point defects are important in diffusion.

Poiseuille's law. The rate of flow of fluid through a tube is inversely proportional to the fourth power of the radius of the tube.

Poiseuille assumed that no change in volume occurred in the fluid under the small stress involved, that there was no slip at the tube wall, and that inertial forces are small compared to viscous forces (laminar flow).

A tube of length l (metres), radius r (metres), with pressure difference $p(Nm^{-2})$ will deliver fluid of viscosity η (centipoise) at the rate of v (cubic metres) in one second according to the formula

$$v = \frac{\pi p r^4}{8l\eta}$$

Poisson's ratio. When a rod or bar is subjected to direct loading, it is extended or compressed longitudinally and contracts or dilates transversely. The ratio of the transverse change (in a line at right angles to the length) per unit length, to the change in length, longitudinally, per unit length is known as Poisson's ratio. This is constant for given materials over a small range of elongations. It is not an independent constant, and in the case of an isotropic substance whose bulk modulus is K and whose rigidity modulus is G, Poisson's ratio (μ) is related to K and G:

$$\mu = \frac{3K - 2G}{6K + 2G} = \frac{E}{2G} - 1$$

where E = modulus of elasticity. If μ is greater than ½ then either K or G must be negative. Hence the greatest value of μ for an isotropic material is ½. Except in the case of hardened and hard-drawn wire the usual values lie between 0.23 and 0.33.

Polarisation (electrolytic). In the electrolytic decomposition of a solution a certain minimum voltage must be applied to the elec-

trodes before continuous electrolysis will occur. This minimum voltage varies from electrolyte to electrolyte, and depends also on the nature of the electrodes. The cause of this phenomenon, known as polarisation, is the reverse e.m.f. produced by the products of electrolysis in contact with the electrodes. For a given concentration of electrolyte the polarisation e.m.f. for a given type of cell is constant. The polarising effected may be caused by: (a) bubbles of gas accumulating on the electrode; (b) change in the concentration of the electrolyte in the neighbourhood of the electrodes. (*See also* **Decomposition voltage.**)

In general an electrode has a characteristic reversible electrode potential corresponding to its condition of equilibrium. If the potential is raised above this figure the electrode is anodically polarised, while if the potential is lowered below the equilibrium or static value it is cathodically polarised. (*See also* **Molecular polarisation, Ferro-electricity, Piezo and Pyro-electricity, van der Waals' bond.**)

Polarisation (light). Ordinary light may be considered to be a transverse wave-motion in which the oscillations are at right angles to the direction of propagation, but are not confined to any one plane. If the oscillations are restricted to a particular plane, the light is then referred to as plane polarised. When light is incident on most polished surfaces it is partially polarised after reflection. The most favourable angle to produce this effect is called the angle of polarisation, and is such that the tangent of this angle is equal to the index of refraction of the reflecting medium. The polarised beam is rather faint after reflection at a single surface such as glass, but the effect can be greatly increased by using a pile of parallel-sided thin sheets of glass.

Light may also be polarised by passing through a crystal of tourmaline or calcite, the direction of displacement associated with the wave-motion being at right angles to the plane of polarisation. Similar effects are produced by some polymeric films (*see* **Polaroid**).

Biaxial crystals, e.g. quartz, have the power to rotate the plane of polarised light. Many liquids and vapours also possess this same property, which can be used to distinguish between optical isomers. Polarised light is used in the microscopic examination of minerals. It is also used in microscopy, particularly in studying the surfaces of anisotropic metals, since the crystals of these latter are able to cause the rotation of the plane of polarisation. The surfaces of cubic metals can be rendered reactive to polarised light either by suitable etching or by electrolytic treatment as a result of which an anisotropic film is formed on the surface. Anisotropic metals are reactive to polarised light after careful polishing. Using this technique it is frequently found possible to differentiate between inter-metallic phases.

Polar molecule. The term used in reference to a molecule, the configuration of charge in which constitutes a permanent dipole, the latter being two equal point charges of opposite sign separated by a small distance.

Polaroid. A trade name of the Polaroid Corporation of America, relating to polarising filters. Polymeric films such as cellulose hydrate and polyvinyl alcohol have the ability to absorb one set of linearly polarised waves produced by the double refraction of unpolarised incident light. The films are stretched and impregnated with an organic dye, and placed between protective optical glass flats for protection against scratches or indentation. They allow large aperture for the incident light, are compact, and much simpler to use than Nicol prisms. Modern polarising microscopes use them almost exclusively, and the design of microscopes has improved as a result of their use. Used in some photographic processes. (*See* **Nicol prism.**)

Pole. A term used to express a centre or source at which an imaginary concentration of electrical or magnetic charge or flux can be said to operate. Poles of opposite sign, separated by a distance, constitute a dipole moment or magnetic moment, from which a number of phenomena can be described in mathematical terms as direct effects of that moment.

Pole figure. A term used in X-ray examinations of crystal aggregates and applied to a stereoscopic projection showing the preferred orientation of the normals to planes of a given form.

Polianite. A naturally occurring ore consisting mainly of manganese dioxide, but containing also other oxides of the metal. Also called pollux.

Poling. In the fire-refining of copper the melt is treated with air under pressure when some of the impurities are preferentially oxidised, and with the flux form a slag. This latter is skimmed off at intervals. A small proportion of the copper is also oxidised, and when the amount corresponds to approximately 0.9%, the bath is skimmed and the surface covered with a layer of coke. Then green young tree-trunks are pushed below the surface. This operation is called "poling". The reducing gases and the water vapour evolved assist in the reduction of the copper. The melt is poured when the oxygen content is about 0.04%, and this produces what is termed "tough-pitch" copper. If the operation of poling is carried too far, the hydrogen content of the metal becomes excessive, and the copper is then said to be "over-poled".

Polish etch. A process in which an etchant produces a polished as well as an etched surface on a metal. In a few cases this double effect is obtained by immersing the specimen in nital (10% nitric acid in alcohol), as occurs with certain magnesium base alloys. A more generally applicable process is that in which occurs the anodic dissolution of the surface of the specimen. For the precious metals and certain of the rarer metals a polish-etch is secured by impregnating the polishing cloth containing the abrasive with a dilute etching reagent.

Polishing. A process in which the surface of a metal is made smooth by rubbing with abrasive materials, particles of which are embedded in a soft matrix or attached to the surface of a wheel or belt. If the abrasive is loosely applied in the operation, the latter is usually known as buffing (q.v.). The basic materials that can be used in polishing are many, including leather and canvas. The former is often used cemented to wooden wheels, while both leather and canvas may be cut into the shape of discs and sewn or glued together. Emery—in grades from coarse powder to emery flour—is the commonest abrasive for all types of polishing, but diamond dust reduces polishing time and produces an excellent surface finish. It is usually used in the form of a paste in which the diamond dust is dispersed in a mixture of stearic acid and a diol.

In metallographic work specimens are polished prior to microscopic examination. They are first ground to produce a flat surface and to remove the disturbed layer resulting from cutting the sample. This operation is carried out in two steps. Firstly a coarse grind is obtained using carborundum on a moving belt or a rotating disc. This is followed by rubbing the coarsely ground surface on flat sheets of successively finer grades of emery paper. Soft materials are liable to pick up particles of loose emery which become embedded in the polished surface. The use of lubricants such as white spirit or paraffin wax dissolved in a liquid hydrocarbon will greatly reduce this effect. A tin-lead alloy, containing equal parts of the two metals, in which emery is embedded is sometimes used as a grinding disc. The operation of finely polishing the ground surface, to remove the fine scratches, consists of hand polishing the specimen on a rotating disc covered with cloth. This latter is impregnated with alumina, magnesia or chromic oxide made into a paste with distilled water. Sometimes a "metal polish" containing either diatomaceous earth or fine pumice powder in paraffin is used before final polishing with magnesia. Soap is usually added to the polishing medium when polishing zinc-, magnesium- and lead-base alloys. Electrolytic polishing is now used as a commercial process as well as for metallographic applications. In the latter the ground specimen is made the anode in an electrolytic cell, the electrodes being only 3-5 cm apart. A suitable electrolyte is selected for each metal. In electrolytic polishing the ridges left after grinding are preferentially dissolved, and thus a very smooth (but very slightly undulating) surface results free from scratches and from the deformation produced by mechanical polishing. (*See* **Electrolytic polishing.**)

Polishing powders. These are usually abrasive materials in a very fine state of division that may be used dry (e.g. on disc, wheels, etc.), or wet (e.g. on a rotating horizontal disc) with water, spirit or soap solution. The most important of these powders are emery flour, alumina, magnesia, chromium II oxide, rouge, crocus, tripoli, diatomaceous earth, pumice. (*See also* **Abrasive.**) Some of the powders have trade or common names, e.g. Gamal is extremely fine alumina, mild polish is stannic oxide, green rouge

is chromium II (chromic) oxide. Whiting, kaolin and fuller's earth are also used.

Polishing wheels. The name used in reference to the rotatable discs that are made of materials of different abrasive properties and are used in smoothing and polishing metals. Discs made of coarse abrasive material bonded by suitable resins or cements are usually employed in grinding operations. For polishing the following materials are used: leather, canvas, cloth, wool, felt, paper and impregnated rubber. Leather is usually employed in two distinct ways. It may be glued to a wooden wheel and impregnated with emery powder or other abrasive, or it may be in the form of discs cemented together. This latter form is also generally suitable for most other materials, but as an alternative canvas, cloth, etc., are often used in the form of discs sewn together. Such discs generally need side metal plates for support. Wool and felt wheels are generally used for fine polishing, but paper and strawboard—cemented and compressed—are hard and suitable only for rough work. Walrus leather is very effective for fine polishing, but on account of its high cost it is less frequently used than compressed canvas, which is as effective.

Polonium. An intensely radioactive hexavalent substance discovered in pitchblende by M. Curie. Its general properties are closely allied to those of tellurium. It is a decay product of uranium and has a half life of approximately 140 days, giving off alpha particles (helium nuclei) and gamma radiation. Atomic weight 210, at. no. 84. It is sometimes plated on strip metal and used as a static dissipator. (*See* Table 2.)

Poluethov test. A test for the element germanium in which a 0.01% solution of hydroxynaphthalenequinonesulphonic acid in concentrated sulphuric acid is added to the solution to be tested. A bright pink colour observed when the solution is viewed in blue (or ultra-violet) light indicates the presence of germanium.

Polyacrylics. Acrylic (Propenoic) acid $CH_2 = CH.COOH$ and methylacrylic (2 methyl-propenoic) acid $CH_2 = C(CH_3)COOH$, can be readily polymerised, using radical initiators, to yield polymeric acids. These are hard, brittle, water soluble solids, used for thickening paints, coating textiles, rubber latex, aggregation of soils and drilling muds.

Esters of these acids, yield acrylate and methacrylate units,

$$-CH_2-CH- \qquad CH_3$$
$$| \qquad\qquad |$$
$$COOR \qquad -CH_2-C-$$
$$|$$
$$COOR$$

Acrylate   Methacrylate

which can polymerise by radical polymerisation initiated by heat, ultra-violet light, or in the presence of oxygen and peroxides. Sheets and blocks of the hard transparent polymer can be cast, moulded or extruded for optical use, dyed for decorative displays, bathroom fittings, telephone cases, illuminated signs. Solvent solutions are used for varnishes and printing inks, aqueous solutions and dispersions in size, paints, adhesives, screen printing binders, textile finishes.

Polyacrylate rubbers are formed from halogen monomers, and reactive copolymers can be obtained which cross-link under heat to produce excellent lacquer and paint finishes, adhesives to bond fabrics and coating of textiles. The most recent use is in cyanoacrylate and anaerobic adhesives. (*See* Table 10 for trade names and details.)

Polyamides. Obtainable in two main forms. Fibrous materials are obtained by condensation of amino acids or by polymerisation of their cyclic anhydrous derivatives, called lactams, and by condensation of diamine compounds with oxalic (ethanedioic) dicarboxylic acids. The generic name nylon is usually followed by a number, depicting the number of carbon atoms in the molecule of the initial monomer. When two molecules are involved, there will be two numbers.

Nylon 6, formula $C_6H_{11}ON$, is prepared from caprolactam, (6-amino, hexanoic acid, lactam), itself derived from phenol (hydroxy benzene). Nylon 11, $C_{11}H_{21}ON$, is derived from vegetable sources such as castor oil. Nylon 6.6, $C_{12}H_{22}O_2N_2$, is obtained from hexamethylene diamine (1-6, -diamino hexane) and adipic (hexanedioic) acid, both of which have six carbon atoms in the molecule.

Non-fibrous polyamides are condensations of alkylene polyamines with polymerised fatty acids, and are used in surface coatings.

Nylon fibres are extensively used because of their high mechanical strength, even when wet, their resistance to abrasion and toughness. Nylons 6 and 6.6 are mainly used in wearing apparel, industrial fabrics and carpets, often blended with other natural or synthetic fibres. Nylons 11 and 6.10 are used in bristles, surgical sutures, fishing line, ropes and cords, netting etc. (*See* Table 10 for trade names and details.)

Polyimides, related to polyamides, have great potential in resisting high temperatures, being serviceable to 250 °C plus, considerably higher than most polymers.

Polybasite. A complex sulphide of silver, copper with antimony and arsenic. It contains about 70% Ag and may be represented by the formula $9(Ag_2Cu)_2S.(Sb.As)_2S$. It occurs with other silver ores in Mexico, California and Chile. It is an important silver ore, but it cannot be treated by direct amalgamation or cyaniding.

Polybutadiene. Butadiene can be easily obtained from the C_4 fraction of petroleum distillation, and polymerises in the liquid state in the presence of an alkali metal such as Li, Na or K. Dissolved in a hydrocarbon solvent, metal compounds act as catalysts, and Li, Co, Ti and Ni have been used, often as organic compounds. The polymerised material forms an elastomer which can be cross-linked like natural rubber, and yields a material of greater abrasion resistance than natural rubber, but with some inferior tensile properties. The coefficient of friction is lower than that of natural rubber, but on ice and snow it gives better grip. Extensively used as a substitute rubber, in tyres, footwear, belting, hosepipe, floor covering and injection moulding.

The fundamental unit is obtained in *cis, trans* or vinyl form, the *cis* form producing the elastomeric properties, and the other two a thermoplastic material. The method of manufacture varies the proportions of all three, and the stereospecific catalysts employing cobalt, titanium and nickel, yield over 90% *cis* form in the product. The *trans* and vinyl forms behave more like *gutta-percha*.

Butadiene is also extensively used in copolymers with styrene, (ethenyl benzene), to increase the moulding potential of styrene, and in ABS, with acrylonitrile and styrene, to give better impact strength, lower brittleness. (*See* Table 10.)

Polychloroprene (Poly 2 chloro-1, 3-butadiene) $(-CH_2 - CCl = CH - CH_2 -)_n$. Like butadiene and isoprene, (2 methyl-1, 3-butadiene), forms of *cis, trans* and other units are possible, and the properties vary accordingly. It is prepared from acetylene (ethyne) by catalytic processes in aqueous solution, but will polymerise spontaneously to give an elastomeric material extremely resistant to acyclic hydrocarbons such as mineral oil, commercial fuels and greases. The original product, trade name Neoprene, was largely used for its solvent resistance, in seals and gaskets, in the motor vehicle, aircraft and marine engine fields. Uses have extended to impact adhesives, golf-ball covers, gloves, foams, additions to concrete and asphalt, and treatment of paper, leather and textiles. It bonds to metals, but the bond is improved by a priming coat of chlorinated natural rubber. (*See* Table 10.)

Polycrystalline. Containing many crystals. The crystallisation of materials proceeds by a process of nucleation and growth, so that polycrystallinity is the norm, and mono crystals require special production methods. Polycrystallinity inevitably averages out orientation and anisotropic effects to a marked degree, but involves grain boundaries, which have properties of disordered rather than crystalline nature, which can modify the properties yet again. (*See* **Crystallisation, Nucleation and growth.**)

Polydimethylsiloxane. The fundamental unit of silicones is the association of silicon and oxygen, as in the inorganic silica tetrahedron, but with organic radicals in place of one, two or three of the oxygen atoms. The radical is a methyl group in dimethylsiloxane, and the unit is then

$$
\begin{array}{c}
CH_3 \\
| \\
-Si-O \\
| \\
CH_3
\end{array}
$$

The units are always linked to share each oxygen atom between two adjacent silicon atoms, and in silicone rubber, there are units similar to dimethylsiloxane in a linear chain, with an extra radical at each end to round off the chain. In dimethylsiloxane the rubber has excellent properties over a very wide range of temperature, for fabric coating, electrical insulation, coating of containers for foodstuffs, beverages and medicines, and for seals, casting moulds for polyesters and white metal, potting of electronic devices. When 15% or so of the radical is changed from methyl to phenyl, C_6H_5, the low temperature properties are improved. The use of 5% or so of vinyl groups gives better cross-linking, and for cross-linking at room temperature, hydrogen, hydroxyl or alkoxy groups are involved.

Silicone resins use groups with three radicals to form cross-links, and methyl and phenyl groups are usually mixed to achieve the best properties. Such resins are used in protective coatings and paints, for impregnation of fabrics, laminated fibres, electrical insulation, release and non-stick coatings, room temperature sealing and encapsulation.

Silicone esters such as ethyl orthosilicate hydrolyse to form silica, and have been used to bond ceramics, to provide water repellent finishes on brick, concrete and stone, and as paint additives and anti-slip finishes on textiles. (*See* Table 10.)

Polyesters. The condensation of a polyhydric alcohol and a poly functional acid, produces a simple polyester, to which the name alkyd (alcohol and acid) was given. Depending on the number of reacting groups per molecule (functionality) the polyester may be linear, or branched and cross-linked. In the latter case, the incorporation of unsaturated fatty acids derived from vegetable oils (linseed, hemp, castor) modifies the product to increase solubility and flexibility. At the same time, the polyester modifies the oil to increase the gloss and drying rate, and the material is therefore ideal for paints.

When ethylene glycol (1, 2-ethane diol) and terephthalic (1, 4-benzenedicarboxylic) acid are condensed as a polyester, the linear polymer has a repeating unit of

$$
-OCH_2.CH_2O.C \underset{\overset{\|}{O}}{\bigcirc} C \underset{\overset{\|}{O}}{} ---
$$

polyethyleneterephthalate, readily fabricated into strong, tough, fibres and sheets. (*See* Table 10 for trade names such as Terylene, Dacron, Melinex and Mylar). Fabric manufactured from the fibre is quick drying, can be stretched and retains pleating when blended with wool for example. Industrial uses include filter and sieve cloth, cords in tyres, sailcloth and fishing nets. The film is an excellent packaging material, electrical insulator (in capacitors) and used for magnetic recording tape, film base for movies, typewriter ribbon, conveyor belts. The film is readily metallised (*see* **Metallisation**) for mirrors, packaging and decorative yarn (Lurex).

Polycarbonates are polyesters of simple structure, the most common based on diphenylolpropane (2, 2-bis (4 hydroxyphenyl) propane), whose unit is

$$
\begin{array}{c}
CH_3 \\
| \\
\bigcirc - C - \bigcirc - OCO --- \\
| \qquad\qquad\quad \| \\
CH_3 \qquad\qquad\quad O
\end{array}
$$

The material is thermoplastic, with good electrical and mechanical properties, and transparent. It can be manufactured into fibres, films and mouldings, suitable for electrical applications, gears and bearings, and as photographic film base.

Unsaturated polyesters are similar to the type described earlier, but with double bonds in one or other of the radicals, which then co-polymerise with a vinyl (ethenyl) monomer to yield a network structure of cross-linked material. Glycol (1, 2-ethane-diol) and maleic (butenedioic) acid form an ester which will co-polymerise with styrene (ethenylbenzene). The polyester resin is inhibited before blending with reactive monomer, and polymerisation and cross-linking attained by peroxides such as benzoyl and methylethyl ketone (butanone) peroxide, with acceleration by cobalt naphthenate or dimethylaniline.

The main use of the unsaturated polyester is in reinforced structures utilising glass fibre, for building panels, vehicle bodies, boats, ship superstructures, chemical vessels, aircraft and machinery components. The resins can be dyed, pigmented or filled, for casting purposes, sealants or mould making, decorative medallions in automobile trim, and can be metallised. (*See* Table 10 for details.)

Polyether. Ethenyl alkyl ethers polymerise to form soft, tacky substances used in adhesives, and can be copolymerised with alkyl halides or anhydrides such as that of butenedioic acid (maleic acid).

Intermediates in the production of epoxy resins are of polyether type, derived from reactions similar to those in the production of polycarbonates. New interest has arisen in a high performance material, polyetheretherketone, with a maximum service temperature of 200 °C plus.

Polyethylene (Polyethene). The polymer of unit $- CH_2 -$ derived from ethene. Synthetic material chemically similar to paraffin wax, and a simple linear polymer, although often highly crystalline. (*See* Table 10 for details and *see also* **High density polyethylene**.)

Polyisoprene (Poly 2-methyl 1, 3-butadiene). The polymer of natural rubber. (*See cis*-**isoprene**.)

Polymer. A material formed by linking units together in covalent bonds, the units being simple additions of a "mer" in the case of linear polymers, or more complex condensations of more than one organic material. They may also be cross-linked. (When there is no cross-link, van der Waals' forces of polarisation will provide some bond between chains, although a weak one.)

The polymer is a bonded structure, and so stronger than the same number of units held together only by van der Waals' forces. The larger the giant molecule, or the more branching, the higher the melting point and the greater the mechanical strength, as the molecules become intertwined. The more symmetrical the atoms on either side of the chain, the lower the mechanical strength, but larger atoms (chlorine as against hydrogen) or groups (styrene as against chlorine) increase strength because of greater difficulty in separating them. A change in molecular weight from 7,500 to 15,000 may increase tensile strength from 270 MN m^{-2} to nearly 700 MN m^{-2}. During polymerisation, not all molecules will form to the same size, and the properties will represent a mean value of the size distribution, and be affected by the degree of crystallinity developed. Molecules may vary from 1,000 to 10,000 mers per molecule, and molecular weights exceed 500,000. (*See* **Copolymer** for polymerisation of two or more monomers, and **Polymerisation** for the processes by which polymers may be formed. *See also* Tables 3, 10.)

Polymerisation. The process of converting monomer into polymer. Two main processes are involved, addition and condensation.

Addition polymerisation occurs, as the name implies, by simple addition of molecules, but the monomer first requires to be activated for this to happen. A saturated monomer would require rupture of a carbon-carbon bond or some specific chemical attack on one end of the molecule, but an unsaturated material requires opening out of the double or triple carbon bonds to give a free bond for the addition. The second molecule then becomes the activated one, and adds a third, which in turn becomes activated —a chain reaction. Cessation of polymerisation then depends on the activating agent becoming neutralised, being transferred elsewhere, or transport inhibiting the movement of sufficient monomer to the active end of the polymer for further addition to take place. Since many chains are being built up simultaneously, not all will be expected to reach the same size.

The activation process can occur by means of free radicals, and is then known as radical polymerisation. Free radicals are produced spontaneously by the action of electromagnetic radiation (particularly U.V.) or by thermal energy, and the monomer may supply its own activation by this means (styrene). Commercially, the speeding up of radical formation is necessary, and a catalyst or initiator is used, often a peroxide. To activate the catalyst, heat or light are necessary, or an accelerator, which may be an amine, mercaptan compound, or a heavy metal naphthenate (such as cobalt). Chemical reactions such as those involved in oxidation-reduction systems also yield free radicals.

By whatever process is used, the free radical is produced, and then combines with a monomer molecule, the combination now being active, with an unpaired electron. This combination adds on a second monomer molecule, and the activity is transferred to the end where addition takes place — the growing end. The termination of activity is by transfer or loss of the activity, as described,

and the activity then goes with the transfer radical or molecule, or is used up in a chemical reaction such as combination or disproportionation.

Initiator	$R - R$	$\rightarrow 2R^\circ$	free radicals, $^\circ$ is the unpaired electron.
Combination	$R^\circ + M$	$\rightarrow RM^\circ$	
Growth	$RM^\circ + M$	$\rightarrow RMM^\circ$	
	$RMM^\circ + M$	$\rightarrow RMMM^\circ$ etc.	

The rate of polymerisation is proportional to the square root of initiator concentration, and the rate rises exponentially with temperature. This is to be expected from a process such as that described.

When carried out in liquid monomer, the term bulk polymerisation is used, and tends to low molecular weight because the viscosity rises as polymerisation proceeds, and this causes difficulty in heat transfer.

When the monomer is dissolved in an inert solvent, the term solution polymerisation is used, and suffers from the same problem as the bulk method. Monomer suspended or emulsified in water (suspension or emulsion polymerisation) yields high molecular weight, but in the latter case contamination is possible since other components are present. The suspension technique has the advantage of producing minute spheres of powder, the emulsion method a "latex" which can be utilised as a spread coating or coagulated.

Initiation of polymerisation may also be effected by means of ionic catalysts. Cationic polymerisation proceeds by conversion of monomer to a carbonium ion C^+, through the medium of a proton donation catalyst such as a strong acid (H_2SO_4, $HClO_4$) or an acid forming compound ($AlCl_3$, BF_3, $SnCl_4$). The converse, anionic polymerisation, converts the monomer to a carbanion C^-, with an electron donating or proton abstracting substance. Organic compounds with metals, such as n-butyl lithium C_4H_9Li, sodium ethoxide $NaO.C_2H_5.2C_2H_5OH$ and potassium amide KNH_2, are used for this purpose.

An inert non-aqueous medium is used for ionic polymerisation, such as a hydrocarbon, or a polar liquid. (Ionisable solvents, such as water, prevent the process starting, or terminate it, by supplying ions of the opposite sign.) The catalyst forms the polymer ion, and itself becomes a counter ion, moving with the polymer ion as an ion-pair. In a non-polar solvent, the ion-pair is tightly bound, in a polar solvent less so, and dissociation leads to the termination already mentioned. No activation is needed by this means, and so elevated temperatures are unnecessary. Indeed, low temperatures are preferred, because termination depends on a proton transfer, itself increasing with temperature rise. Very high molecular weights can thus be obtained often at very low temperatures (isobutene at -100 °C).

Condensation polymerisation, involving chemical reaction which actually eliminates part of the monomer(s), gives a repeat unit in the polymer which is lower in molecular weight than the monomer(s). An initiating catalyst is not always essential and the polymer is not built up by growth on the end of a chain, two main points of difference with addition polymerisation. The growth of the polymer is therefore more random over a number of sites, and high molecular weight builds up only towards the end of reaction time. In this respect, the process is more close to the nucleation and growth processes involved in inorganic melts, e.g. metals and alloys. The degree of condensation is capable of control, and can be operated as a stepwise process to reach intermediate products, or if necessary complete the condensation into large molecules only at final fabrication stage.

Condensation can be obtained by self condensation in a single monomer or between two different monomers. In the latter case, the reactive groups must be present in equivalent quantities to build up to high molecular weight.

Special forms of polymerisation have been discovered in recent years, perhaps the most notable being that of control of symmetry of the polymer chain. Any asymmetrical monomer can produce two possible ways of addition, usually described as head to tail and head to head. Each repeat unit can also exist in enantiomorphic configurations or mirror images, (as in the dextro (*d*-)

and laevo rotatory (*l*-) forms of the sugar molecule). With certain catalysts however, a high molecular weight polymer of uniform configuration in space can occur, giving rise to regular structures which are highly crystalline. Such catalysts are described as stereospecific. If the configuration produced is identical along the chain, the catalyst is described as isotactic, if it produces a regular alternating *d* and *l* repeat unit, syndiotactic, and if completely random, atactic.

Such catalysts are formed from metal alkyl compounds dissolved in a hydrocarbon solvent and reacted with a halide of a transition metal (e.g. $Al(C_2H_5)_3$ with $TiCl_4$). Polyvalent metals or metal oxides supported on carbon or silica substrates are also used.

Other types of special polymerisation refer to interfacial reactions, from which the product can be withdrawn as fast as it forms at an interface between two streams or supplies of material, and ring compounds which can undergo scission to form monomers capable of condensation type reactions. (*See also* **Copolymer.**)

Polymorphism. A property possessed by certain chemical substances by virtue of which identical materials may form two or more minerals of different crystalline structure and physical properties, e.g. density. The phenomenon is due to differences in the number of atoms contained in a molecule and thus polymorphs will have differing energy contents.

Polystyrene (polyethenylbenzene) $(C_6H_5CH = CH_2)_n$. Ethenyl benzene (styrene) is obtained commercially by reaction of ethene, a product of petroleum distillation or natural gas, with benzene derived most frequently from coal but also from petroleum. Styrene readily polymerises, on heating or exposure to U.V. radiation, and in the presence of free radicals. Inhibitors are added to styrene to prevent spontaneous polymerisation. The polymer is tough, and transparent, but with variable molecular weight and physical properties, depending on the method of processing.

Polystyrene is a main stream thermoplastic used in packaging, small domestic containers, casings and liners for appliances, toys and electronic equipment. The expanded form is a good insulator for buildings and vehicles, and as acoustic tiles and shock absorbing packaging of glassware and electronic circuitry.

The incorporation of an elastomer such as butadiene into polystyrene improves the physical properties, and copolymers with butadiene or acrylonitrile (propenoic nitrile), or both, have produced extremely useful moulding materials for engineering applications.

Polytetrafluoroethylene (polytetrafluoroethene) (P.T.F.E.) $(C_2F_4)_n$. The polymer can be regarded as polyethene into which fluorine has been introduced to replace the hydrogen atoms. The tremendous chemical resistance which results is the most valuable of its properties. It is soluble in no known liquids, and relatively unaffected by concentrated acids and alkalis. Only alkali metals and their organic complexes have any effect, but prolonged exposure to fluorine causes decomposition. The material has a high molecular weight and is highly crystalline at normal temperatures.

The main uses for P.T.F.E. are electrical, due to its high temperature dielectric properties; chemical, because of the inertness in hot and corrosive conditions; and mechanical, due to the low coefficient of friction, in bearings and wear surfaces. Dispersions can be used for spray coating and impregnation.

Polyurethane. Condensation polymers produced by reacting aromatic di-isocyanates with polyhydric alcohols. Elastomers, with excellent resistance to abrasion, impact, oils and greases, ozone and radiation. Generally harder than natural rubber.

Polyvinyl (polyethenyl) **alcohol** $(C_2H_4O)_n$ (P.V.A.). An amorphous or polycrystalline material, depending on the mechanical treatment received. In fibre form, stretching causes a high degree of orientation of the chains, with up to 60% crystallinity. Since hydrogen and hydroxyl groups are relatively easily accommodated on the chains, stereospecificity does not produce much difference in properties.

The material is used in sheet or tube form in industry, withstanding many oils, fats and waxes. In aqueous solution it can be employed as an adhesive, a binder for pigments, and in sizing of fabrics and paper. The fibre forms are used in clothing, carpet fabrics, netting and filters, and as a substitute for leather and cotton in driving belts, filter cloths and cordage.

Polyvinyl chloride $(C_2H_3Cl)_n$. Virtually the polymer chain of polyethylene, with one in four hydrogen atoms replaced by chlorine, hydrogen and chlorine alternating along the chain. The material is mainly amorphous, of reasonably high molecular weight, and excellent in resistance to weather. It can be plasticised to produce sheet, tube and coatings, the properties depending on the plasticiser used.

The main uses for the rigid material are in the building industry, for external pipes, guttering, roofing, wall cladding and windows and in industry for chemical tank liners. The plasticised form is made into flexible hose, curtains, tarpaulins, aprons and raincoats. Floor and wall coverings can be processed with a wide variety of pigments, patterns and textures, as can footwear.

Pony-roughing mill. The roll standing in a rolling-mill that is designed to produce a section approximately that of the finished section. It is used for rolling sheet, plate, bar or other types of section and is located intermediate between the roughing stand and the planishing stand.

Pop-gate. A term used in foundry-work to describe a top-gate that consists of a series of small holes leading from the bottom of the pouring basin to the top of the casting. It is particularly useful in casting liners and the small holes are effective in preventing the slag from entering the mould. Also called pencil gate (*see* Fig. 117).

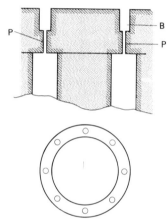

Fig. 117 *Cross-section and plan of a pop-gate. B is the pouring basin; and P the pop-gate.*

Pop-ups. The name used in reference to the small cavities made in the top box of a mould to prevent the formation of slight defects in the top surface of a casting. Gases escape into these cavities, which are usually made by the aid of a large vent wire or a moulder's cleaner.

Porcelain enamel. A vitreous coating usually applied to iron and steel to prevent corrosion. The enamel consists of a borosilicate glass, finely powdered and referred to as a frit. This is sprayed on the surface of the metal (or the part may be dipped) and the treated part is then fired at 820°-870° C, when the coating adheres firmly to the metal. Several coatings may be applied according to the proposed service of the part, the coatings on cast iron being, in general, thicker than those on steel. The thicker coatings are fired at the lower end of the temperature range. The brittleness of this enamel coating limits its applications. It is more generally used on domestic ware and sometimes for wall signs.

Porcelain jasper. A clay or shale that has been altered by contact with hot igneous rock.

Porcupines. The name by which are known the special cooling racks used in the production of galvanised-iron sheet. The racks, which may hold twenty or more sheets, rotate slowly so that the sheets can be inspected as they are turned over.

Pore. The term used in reference to a minute hole or cavity in a coating or in a powder compact.

Porosity. A term used in foundry-work to describe the presence of small gas-holes beneath the skin of castings. The holes more usually appear in the heavy sections, and they are approximately spherical, and the surface is bright and free from oxide tint. In most non-ferrous melting the gas may be sulphur dioxide (absorbed from furnace atmospheres or from oil in the charge) or it may be hydrogen (absorbed from water). Damp moulds and too high a pouring temperature increase the tendency to produce porous castings. With ferrous castings hydrogen may be the agency producing the porosity, but more usually the defect may be due to insufficiently de-oxidised metal, and thus the presence of oxygen causes the gas-holes to appear.

Porosity when referring to a refractory material is the ratio of the volume of the pores to the total volume. This is usually estimated by determining the volume of water absorbed.

The term is also used in reference to the large number of pores in any compacted material.

Porter. A metal-working tool in the form of a bar which is attached to an ingot or a forging. The purpose of the tool is to support and manipulate the metal during the forging.

Porthole die. A die made in two or several sections for extruding metals. Each die produces a specially shaped section and the various sections are produced simultaneously.

Portland cement. The reaction between clay substance (aluminium silicate) and limestone (calcium carbonate) at high temperatures produces a number of compounds, mainly tricalcium silicate, but containing also dicalcium silicate, tricalcium aluminate, and others depending on impurities in the clay itself. A typical composition would be:

Composition		% by weight
Tricalcium silicate	$3CaO.SiO_2$	46
Dicalcium silicate	$2CaO.SiO_2$	28
Tricalcium aluminate	$3CaO.Al_2O_3$	11
Calciumalumino ferrite	$4CaO.Al_2O_3.Fe_2O_3$	8
Gypsum	$CaSO_4$	3
Magnesia	MgO	3
Lime	CaO	0.5
Alkali metal oxides	K_2O,Na_2O	0.5

When hydrated, the various constituents react with water at very different rates, and with varying amounts of heat evolved. The aluminate reacts most rapidly and evolves most heat, but this reaction is slowed by gypsum, which also reacts rapidly. The tricalcium silicate hydrates 70% complete within one month, but the dicalcium silicate takes a year or more to complete the reaction. The impurity oxides all affect the reactions to some degree.

Positive flash mould. A type of steel mould used in moulding plastic articles, that resembles the positive mould (q.v.), except that a small allowance is made for flash in a vertical spew channel. This type of mould produces articles having close uniformity in dimensions, since a constant pressure can be applied and maintained during curing (*see* Fig. 118).

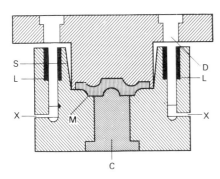

Fig. 118 *Cross-section of a positive flash mould. C is the core plug; D the dowel; L the liner; M the moulding; S the spew channel; and X the air outlet.*

Positive mould. A type of steel mould used in the production of plastic components, where the moulded product must have absence of flash or flash line and have a uniform density. The parts of the mould must have very close accuracy of fit within the rather massive chase ring or bolster. The plastic charge is usually in pellet form to ensure close control of weight and volume (*see* Fig. 119).

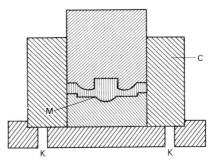

Fig. 119 *Cross-section of a positive mould. C is the chase ring; K the knock-out holes; and M the moulding.*

Positive-ray analysis. The positive ions produced in a tube containing an elementary substance at low pressure can be separated into a spectrum using a magnetic field and an electric field at right angles to it. The deflection is a function of the ratio of the charge to the mass of the particle. Such a spectrum can be photographed, and it is found that there are a number of lines each corresponding to a definite ratio of mass to charge. In this way isotopes were first discovered.

Positive rays. The name given to the stream of ions that passes from the anode to the cathode in a discharge tube when a difference of potential is applied to the electrodes. Since these rays were not discovered until a perforated cathode was used, through which they were able to pass, they are also called canal rays. These rays are deflected by a magnetic field, but the deflection is smaller than, and opposite to, that which would be experienced by cathode rays (electrons). The particles were thus known to be heavier than electrons and to carry a positive charge. A study of the positive rays emitted from elements led to the discovery of isotopes of many of the common elements.

Positive replica. A copy or replica of a metallic surface made in suitable material, in which the high and low points correspond directly with those on the original surface, such replicas being used in the electron microscope for the examination of metallic surfaces.

The double transfer process may be carried out: (*a*) a strong and comparatively thick negative replica is made by moulding polystyrene on the metal surface. A thin film of silica is condensed on the contoured surface of the polystyrene. Sometimes the silica film is strengthened by a thin film of transparent hard polymer. On drying, the polystyrene is dissolved away in ethyl bromide; (*b*) in another method silver is evaporated (in a vacuum) onto the metallic surface and this thin film is reinforced by a layer of electrolytically deposited silver. The positive is made in cellulose tetranitrate dissolved in acetic acid pentyl ester and the silver dissolved in nitric acid.

A process, devised to avoid as far as possible any mechanical straining of the replica, consists in making a strong negative replica from a solution of collodion in amyl acetate. This is dry-stripped and the contoured side treated with a solution of formvar in chloroform. The collodion film is then dissolved away in acetic acid pentyl ester.

Positron. An elementary charged particle having the same mass as the electron and having a positive charge equal numerically to that on the electron. Positrons are emitted by a number of artificial radioactive elements. They are produced by the bombardment of matter by hard gamma rays and by neutrons. The mass is 1.04 ± 0.14 (electron = 1). When high energy electrons from outer space pass into the earth's atmosphere they rapidly lose their energy, e.g. on passing into the electric field of an atomic nucleus, where the electron is absorbed after emitting a photon. In the process, known as collision radiation, both a positron and an electron are produced.

Post heating. A heat-treatment process applied as soon as possible after welding for the purpose of relieving stresses and of tempering the weld metal and the heat affected zone to avoid a hard or brittle structure.

Pot. A refractory vessel for holding molten metal. The name is also applied to the metal container in which castings are packed during malleablising. The metal container used to protect components from the furnace atmosphere during heat-treatment is also called a pot, as is the reduction cell used in certain electrolytic extraction processes (e.g. aluminium).

Pot annealing. A term synonymous with box annealing (q.v.).

Potash alum. The double sulphate of potassium and aluminium $K_2SO_4.Al_2(SO_4)_3.24H_2O$. It is an excellent mordant.

Potassium. A silvery white, soft and very reactive metal that resembles sodium in most of its chemical properties. It occurs very widely distributed in nature, mainly as felspars and related minerals. These potassium aluminium silicates on weathering form potassium carbonate. Potassium salts, e.g. carnallite, are also found in large deposits resulting from the evaporation of inland seas. The metal is usually produced by the electrolysis of fused potassium chloride to which small additions of the chlorides and fluorides of the alkaline earth have been made in order to lower the melting point. The metal itself has few commercial applications, but many of its salts are of importance, e.g. potassium hydride is used on the cathode of photocells to increase the light sensitivity; potassium cyanide is used in electroplating processes and in cyaniding steel; potassium carbonate is used in the manufacture of hard glass and soft soap. (*See* Table 2.)

Potassium dihydrogen phosphate KH_2PO_4. A uniaxial ferroelectric material of tetragonal structure, with a Curie point at 123 K.

Potential. A word that in its various applications implies latent potency. In electrical applications all practical purposes are served by regarding electric potential as pressure. The earth is taken as the practical zero of electrical potential, since none of our usual experiments are likely to cause any change in it. Thus the electrical potential of a body is its electrical pressure above or below that of the earth, and if two bodies are joined by a conducting material it decides which way current will flow. Mathematically the zero of potential is at an infinite distance, but the difference in potential of two bodies is the same whether the practical or the mathematical zero is adopted.

The unit of electric potential is the volt, the difference of electric potential between two points of a conducting wire carrying a constant current of 1 ampere, when the power dissipated between these points equals 1 watt. The ampere is defined as the constant current which, if maintained in two straight parallel conductors of infinite length of negligible circular cross-section and placed at a distance of 1 metre apart in a vacuum, will produce a force between them of 2×10^{-7} newtons per metre length. Hence the relationship between electric potential and mechanical force can be established.

The potential energy of a body in mechanical terms is its capacity to do work by virtue of its position or configuration. The energy is measured by the amount of work the body can do in passing from its energy position to a standard or normal position in which the potential energy is considered zero. The unit of mechanical energy is the joule, the work done when the point of application of a force of 1 newton is displaced through a distance of 1 metre in the direction of the force (note the relationship with the definition of ampere in terms of force). (*See also* **Electrode potential, Oxidation-reduction potential.**)

Potential barrier (electrons). In finding solutions of the wave mechanical behaviour of electrons, it is necessary to explore the possible ways in which an electron will react to potential discontinuities along its path. A potential barrier represents potential energy in excess of the ground state within which the electron is travelling when it meets that barrier. Such states exist in reality,

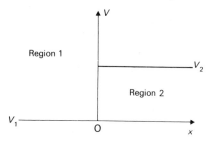

Fig. 120 *Electron moving through a discontinuity of potential.*

at the surface of a heated metal, for example, or at a junction between two dissimilar electrical conductors, or wherever an applied electrical field exists.

Figure 120 represents the potentials between region 1 in which the electron is travelling along the x-axis with constant potential energy V_1. From the diagram, $V_1 = 0$ when $x < 0$. A discontinuous change (the potential barrier) occurs at $x = 0$, and the potential rises to V_2 when $x > 0$. The total energy of the electron is E. Classical mechanics would indicate that if $E < V_2$ the electron would be reflected, but if $E > V_2$ the electron would pass through the barrier. (An example in classical mechanics would be a ball thrown at a window, in which V_2 represented the fracture strength of the window and E the kinetic energy of the ball.)

The time-independent Schrödinger equation (q.v.) for region 1 for the electron behaving as a wave is written

$$\frac{\delta^2 \psi}{\delta x^2} + \frac{8\pi^2 m}{h^2} E\psi = 0$$

Since

$$\frac{8\pi^2 mE}{h^2} \qquad \text{is constant, substitute } \alpha^2$$

Then

$$\frac{\delta^2 \psi}{\delta x^2} + \alpha^2 \psi = 0$$

This is a simple harmonic variation of ψ with x. The differential equation is insensitive to the direction of propagation of such a wave, so the integral may be written

Wave function $\Psi = e^{i\alpha x} + Ae^{-i\alpha x}$ (with reference to x, i.e. space).

The constant A depends on the boundary conditions, and the complete wave function to include time, is

$$\Psi = (e^{i\alpha x} + Ae^{-i\alpha x})e^{-\left(\frac{2\pi iE}{h}\right)t}$$

Each term is of the form $e^{i(kx - wt)}$ so that the whole solution is a travelling wave form.

For region 2, beyond the potential barrier, the Schrödinger equation is written

$$\frac{\partial^2 \psi}{\partial x^2} + \frac{8\pi^2 m}{h^2}(E - V_2)\psi = 0$$

If we write

$$\frac{8\pi^2 m}{h^2}(E - V_2)$$

as β^2, to correspond with the equation for region 1, the complete wave function solution is

$$\Psi = Be^{i\beta x} e^{-\left(\frac{2\pi iE}{h}\right)t}$$

for region 2 which is again a travelling wave form, and the constant B depends on the boundary conditions. ($e^{-i\beta x}$ would represent a wave travelling in the opposite direction.)

The boundary conditions for A and B can be shown to be ψ continuous at $x = 0$; $\partial \psi/\partial x$ continuous at $x = 0$. When $E > V_2$, $\beta^2 > 0$, the wave is real in region 2, and so transmission occurs at the discontinuity along with some reflection back into region 1. The amplitude of the transmitted and reflected waves will depend on boundary conditions. Wave mechanics depict changes in amplitude, classical mechanics would indicate uniform transmission when $E > V_2$.

When $E < V_2$, $\beta^2 < 0$ and the solution becomes

$$\Psi = Be^{-\gamma x} e^{-\left(\frac{2\pi iE}{h}\right)t}$$

where $\gamma = -\beta^2$. Since i in the exponent of x has disappeared, this is not a travelling wave, but an oscillation in time with amplitude decreasing as shown by the term $e^{-\gamma x}$. This is a so-called standing wave, with amplitude decaying exponentially from $x = 0$, and the rate of decay dependent on the value γ. When $E \ll V_2$, the decay is steep, when E approaches V_2, the decay is much slower.

This argument can be extended to a third region similar to 1, in which the potential $V_3 = 0$, and then region 2 becomes a potential barrier of width d. The solution of the Schrödinger equation is then

tinuities along its path. A potential barrier represents potential energy in excess of the ground state within which the electron is travelling when it meets that barrier. Such states exist in reality,

$$\Psi = C^{i\alpha x} \, e^{- \left(\frac{2\pi i E}{b} \right) t}$$

for region 3 which is again a travelling wave of amplitude C. The value of C now depends on the rate of decay across distance d of the value $Be^{-\gamma x}$. There now exists the possibility of transmission of an electron through such a potential barrier, even when $E < V_2$, a solution impossible in classical mechanics.

A probability can then be expressed in terms of transmission,

$$P_T = \frac{\text{Number of electrons penetrating barrier}}{\text{Number of electrons arriving at barrier}}$$

$$= \frac{4}{4 \cosh^2 \gamma d + \left(\frac{\gamma}{\alpha} - \frac{\alpha}{\gamma} \right) \sinh^2 \gamma d}$$

This is known as the electron tunnel effect, of great importance in electronic devices, thermionic emission, at differential metallic junctions, and in theories of metallic alloying. It explains for example, why two aluminium wires twisted together will conduct electricity even when oxide films are present on the surface of each.

Example. An electron of energy 5 eV, incident on an insulating barrier 5×10^{-10} m thick, with potential barrier level 6 eV (say two different metals) would give $\gamma = 5 \times 10^9$ and P_T about 1.5%. Each electron would have a 1.5% chance of transmission. Barriers thicker than about 10^{-9} can be neglected, as too great for transmission. Potential barriers can also affect other phenomena on the atomic scale, e.g. diffusion, nucleation and electrokinetics.

Potentiometer. An instrument used for measuring direct current e.m.f. or potential differences. It has been used for the accurate measurement of current and resistance and for the verification and calibration of ammeters, voltmeters, etc., as well as for the comparison of e.m.f.s. The use of the potentiometer in the above applications has three obvious advantages: (a) it is a "zero" method, i.e. one in which no deflection of the galvanometer needle has to be observed; (b) in testing batteries no current is taken from them, and thus the true e.m.f., not affected by polarisation, is obtained; (c) the resistance of the galvanometer and the battery tested are not taken into account in the calculations, since no current passes through them. In its simplest form it consists of a length of manganin wire (AB) of length L (platinum-iridium or other constant-resistance alloys are also used) of uniform section, connected to a source of e.m.f. An ammeter is placed in this circuit to detect any current variations during the experiment. A sliding contact C is connected in series with a sensitive galvanometer to one terminal of the e.m.f. to be measured. The other terminal is connected to A so that the e.m.f.s. across AC and AB are in opposition through the galvanometer. If E is the standard e.m.f., that under test is given by El_1/L, where l_1 is the length of resistance wire (AC) for zero current through the galvanometer (see Fig. 121).

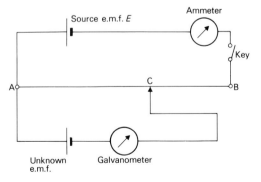

Fig. 121 *A potentiometer.*

Potentiometric titration. A method of determining the end-point of the reaction between an acid and an alkali in solution, by measuring the change in the e.m.f. using a potentiometer. The changes that occur in the values of the pH during the neutralisation of an acid by a base may easily be shown graphically, and it is clear that at the neutralisation point there is a sudden change in the pH of the solution. If a hydrogen electrode is placed in such a solution there would be a sudden change in the e.m.f. at the points of inflexion in the curves. A potentiometer is used to measure the e.m.f., and this constitutes an accurate method of finding the end point. This method is especially useful in titrating coloured indicators where ordinary indicators would be ineffective. For acid-alkali titrations the hydrogen or the quinhydrone electrode is used. The method is now applied extensively for the investigation of precipitation reactions.

Pot metal. The name used in reference to the glass that is tinted by adding metallic oxides to the molten glass in a melting pot. It is mainly used in making stained-glass windows. The term is also applied to a low-grade brass or copper-lead alloys containing both zinc and tin.

Pot quenching. The term used in reference to the operation of direct quenching from the carburising operation.

Potstone. A magnesium silicate otherwise known as talc. It has a slaty structure and breaks into thin semi-transparent flakes. It is used for laboratory table-tops. Found in northern Scotland, Cornwall (in the UK), Ireland and many parts of the USA.

Pottery. Literally, artefacts made from fired clay. Within this definition, a whole range of techniques of fabrication and composition of the basic clay body, can be described, including the term whiteware for the finest and lightest coloured pottery. Three main classifications can be distinguished, chinaware, stoneware and porcelain, and earthenware.

"Chinaware" should be reserved for true "bone china" but is also used for translucent porcelain. True bone china is almost exclusive to England, using china clay, china stone, and a proportion of calcined animal bone. The bone is essentially calcium phosphate, and imparts a delicate translucency. The fired material is dense, impermeable and vitreous in structure. The finishing is usually by glazing and decoration with elaborate designs and colours, including gold.

Stoneware and porcelain contain no bone, but aim at a similar texture and impermeability to that of chinaware. Translucency is almost entirely absent. Stoneware uses plastic clay with fluxing material, usually naturally incorporated in the clays selected, to promote vitrification on firing between 1,000 °C and 1,250 °C. Domestic ovenware and table ware is frequently made from fireclay and ballclay, and sometimes from synthetic mixtures of various clays, to vary the amount of flux present.

Porcelain used for table ware and laboratory purposes is generally prepared from a mixture of china clay, quartz and felspar. Porcelain artefacts for the electrical industry, as insulators, contain ballclay also. The requirements of both electrical and chemical porcelain involve very high impermeability, achieved by higher firing temperatures, approaching 1,400 °C. Maximum vitrification can be achieved by this means, without loss of shape.

Earthenware covers a very wide range of composition and appearance, varying in colour from red and brown to cream and white, according to the impurities in the clays used. True earthenware is essentially porous, and is then coated with an impervious glaze or engobe. The unglazed material is often referred to by the Italian name "terra cotta", the glazed varieties may have names such as "Delftware", "faience" or "majolica".

Earthenware is fired at a lower temperature than other pottery, usually 1,000 °C to 1,200 °C, and hence is less vitrified and more porous. Some finer grades of earthenware are made from ball and china clays mixed with flint and felspar or added flux, to give a white colour after firing, at higher temperatures.

Fireclay with low iron content is used to make buff coloured tiles and kitchen sinks, although the quantity is decreasing. Clay which fires to a red colour tends to be favoured for horticultural use, since colour is less important. Sanitary vitreous ware uses a more impervious body, with felspar flux to ensure vitrification, and clay which has little colour after firing.

The processing of materials for high quality pottery and whiteware has to be carried out to a high degree of fineness and purity. The final mixture prepared from individual raw materials is known as the "body". Raw clay is treated with water in a "blunger", a drum-like container with rotating blades, which reduces the clay to a thick "slip". The slip is sieved to remove coarse particles, and magnetic separators abstract any fragments of ferrous material, which would cause subsequent discolouration. The non-plastic raw materials, china stone, flint, felspar, are crushed and then ground in rod or ball mills to form a second slip, which is also sieved and passed through the magnetic separator. The mixing of clay and other material takes place in a chamber called the "ark". The mixed slips are filtered in a press, to remove a large excess of water, and the "cake" removed from the press contains only 25% water. When plastic clay is to be used, the clay body is passed through a "pug-mill", in which helical screws masticate the clay and remove air bubbles. The clay is extruded from the mill as a plastic consolidated body.

Plastic clay may be processed by hand throwing on the familiar potter's wheel, or shaped by a steel tool inside a prepared mould. To make cup shapes, a ball of clay is placed in the revolving hollow mould, and a profile tool shapes the inside, the mould shaping the exterior surface. This is known as "jollying". For disc shapes, such as plates, a disc or "batt" of clay is formed on its inside surface by a revolving mould, and the profile tool forms the back. This is known as "jiggering". Modern automatic or semi-automatic machines form more accurate shapes, by pressing finely ground material whose moisture content has been reduced to 10%.

Articles which do not lend themselves to any of the above processes are cast from slip. The slip is made up by adding water again, and a deflocculating agent to cause the particles of slip to settle in the mould. The mould is made from plaster (gypsum) and when it receives the slip, immediate removal of water occurs at the surface, the surplus slip being held internally. The excess slip is poured off after a time interval sufficient to produce an adequate thickness of clay at the mould surface, and reused. The mould is carefully dried, and the shrinkage of the clay causes it to separate from the plaster. The mould is opened and the cast removed. Where separate parts of an intricate shape have to be joined (e.g. spouts and handles) each is slip cast, and on removal from the mould, jointed by gentle pressure and smoothing, using a little liquid slip as a cement, to give a homogeneous bond.

The shaped ware is then dried slowly to avoid cracking or excessive shrinkage, and fired to convert the "green" clay to hard and strong "biscuit" ware. The biscuit is then coated with glaze slip, and fired again, to give a smooth impervious glaze. Decoration is then applied to the glazed material (sometimes to the biscuit) and a third firing to fix the decoration as "enamel". Elaborate designs may require more than one such enamel firing. Metallic compounds are used in these decorative glazes, to give a wide range of colours, mixed with silica in the form of ground quartz or quartz sand. Some whiteware and electrical porcelain is given a single firing, with the glaze applied to the "green" body. In these cases, no decoration is required.

The firing temperatures used are within the following ranges:

	Biscuit °C	Glaze Firing °C	Enamel Firing °C
Chinaware (bone china)	1,200-1,250	1,050-1,100	700-800
Porcelain (domestic)	900-1,000	1,350-1,400	700-800
(electrical)	nil	1,200-1,250	nil
(laboratory)	nil	1,350-1,400	nil
Stoneware	900-1,000	1,050-1,100	700-800
Earthenware	1,100-1,200	1,050-1,100	700-800
Redclay	1,000-1,050	1,050-1,100	700-800
Sanitary fireclay	nil	1,200	nil

Pouring. The term used in reference to the transfer of molten metal from the ladle to the ingot or other types of moulds. The operation is carried out in such a way as to reduce or eliminate gas unsoundness. In many cases top-poured ingots are poured relatively slowly to avoid turbulence. The use of a tun dish is also frequently beneficial. The pouring of molten steel is often referred to as teeming.

Pouring basin. A term used in foundry practice in reference to the reservoir at the top of gate into which the molten metal is poured.

Powder burning. A method of cutting metals and of removing the runners and risers from castings by using an oxyacetylene (oxy-ethyne) flame into which metallic powder is injected. The principle is also used in powder scarfing of steel billets.

Powder diffraction. An X-ray spectroscopic technique in which a large number of crystals arranged at random replace the single

crystal in the Laue method of examining crystals. The method has advantages, since it is possible to study crystals that are difficult (or impossible) to grow to large size, whereas all solids can be produced as a fine powder.

The fine crystals are placed in a fine bore silica tube or formed into a cylindrical rod by an adhesive and a central wire, then the rod or tube is suspended at the centre of a circular camera. The adhesive method is common at room temperature, the silica tube is used for high temperature exposures. One advantage of the powder method is that furnaces can be built to fit above and below the specimen but insulated from the film and the remainder of the camera. Only a very small gap is required for the X-ray beam and the diffracted beams, so that quite high temperatures can be attained. The main problem is measurement of temperature of the specimen.

Since there are a very large number of fine crystals present there will be some that are in a position to reflect from every system of atomic planes whose spacing is more than a certain minimum (dependent on the wavelength of the X-rays). The beam of X-rays is passed through slits and through a hole in the film, on to the rod. The rays are diffracted, giving a series of cones having their apices at the point of incidence of the rays on the rod (see Fig. 122). The curved sections of the cones, where they meet the photographic film, produce the powder diffraction photograph.

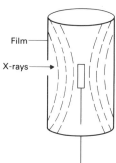

Fig. 122 X-ray diffraction photographs of powders.

Powder metallurgy. The branch of metallurgy that is concerned with the production of metal powders, their pressing or rolling into compacts and the sintering of the latter to produce shaped components or stock. The method was first introduced in connection with the metallurgy of tungsten and is especially useful also in the case of tantalum, zirconium, titanium, niobium and molybdenum, which have very high melting points. The metal powder is produced in a variety of ways, e.g. crushing brittle metals, reducing finely powdered oxides in hydrogen or carbon monoxide, by electrolysis and by atomising a stream of molten metal. The metal powders are graded and then strongly compressed in steel dies to the required shape, making due allowance for shrinkage. Hot pressing may be employed, in which case lower pressures are allowable. The resulting compact varies in density according to the grade of powder used, the pressure of the die, the presence of a (volatile) bonding agent. The compacts are sintered in a non-oxidising atmosphere or in vacuum. The temperatures employed are generally about three-quarters of the melting temperature of the metals concerned, or in the case of alloys of the major constituent. Thus in the case of alloys or mixed powders, one of the constituents may actually melt, an occurrence which gives increased density to the final product.

Powder metallurgical techniques are usually, but not necessarily, applied to the mass production of small components. The metals used are iron, steel, bronze and brass. Hard-metal cutting, punching and pressing tools are made by sintering the high-melting point (and brittle) carbides with cobalt. In the process the cobalt actually fuses. Porous metals suitable for self-lubricating bearings and for filters are made by the powder metallurgy processes, selecting in the first place the correct grade of powder and pressing to the correct density before sintering. Oilite bronze is an example of this type of application. De-icing strips on aeroplane wings may be made of porous metal.

Metals that do not mix in the liquid state, e.g. tungsten-copper, molybdenum-silver, etc., can be satisfactorily alloyed either by sintering a mixture of the metal powders or by immersing a porous sinter of one constituent in a molten bath of the other. The second metal is then absorbed in the pores of the first of the pair. Refractory oxides, e.g. Al_2O_3, may be mixed with metal powders or metal oxide powders to produce materials that have good high-temperature properties. Many of the magnet alloys now used can be produced by compacting the components with or without a binder and sintering.

Powder of algaroth. A white powder which is mainly antimony III chloride formed when the trichloride of antimony dissolved in HCl is poured into water. If little water is used SbOCl is formed, but with excess of water the compound $2SbOCl.Sb_2O_3$ is formed.

Powder welding. Just as an injection of metal powder into a high temperature gas flame can be used for cutting through metals, so can metal parts be joined by a similar technique. Metals or metallic alloys in powder form are deposited through the flame of a gas torch or plasma jet, the surfaces of the parent metal being heated to a temperature at which fusion of the deposited metal will occur, so that a bond is created between parent and deposit. The powdered metals are self-fluxing, and only in special cases are additional chemical fluxes added.

The process differs from metal spraying, where the base metal is relatively cold, by virtue of the fusion of the powdered metal or alloy, which is usually selected to have a melting point lower than the base metal. The base metal does not melt in bulk, but at the interface where the powder is applied, the molten powder particles diffuse rapidly into the base metal, lower its melting point, and ensure a good bond by fusion or partial fusion. The deposits are much more dense and mechanically better bonded than sprayed coatings.

The main advantages are the fact that the base metal is not fused in bulk, so that brittle materials can be welded without cracking. Arc welding can produce severe undercutting of the parent metal, the powder method avoids this. The disadvantages arise with highly active alloying elements in the base or the powder, and the production of oxide, nitride or hydride films which may prevent good bonding, e.g. with titanium or chromium bearing steels or their alloys.

Used for hard facings on automobile components for wear resistance; high speed electrical rotors; wear resisting components in rolling mills; shaft bushing, tools and dies in plastics fabrication; repair of foundry defects.

Powellite. A mineral containing both calcium tungstate and calcium molybdate $Cu(Mo.W)O_4$. It contains up to 48% Mo and about 10% tungsten VI oxide, and is regarded as an alteration product of scheelite. It is dull yellow in colour and is identified by its dull-yellow fluorescence.

Power factor. In many electrical machines using alternating current the current lags behind the voltage. The power is never equal to the voltage × amperage because they do not have their maximum values at the same instant. In fact, the actual power is $V \times A \cos\theta$, where V = voltage, A = amperage and θ is the angle of lag (or lead). Low-power factor means that additional current is required to generate a given power. Squirrel-cage motors have a low power factor at starting (they require four to six times the normal current for starting), but the factor improves until full speed is reached. Induction motors have a power factor that varies with the load up to a certain maximum value. Such motors are uneconomic when under or over loaded. It follows therefore that it is uneconomic to instal a motor that is too large for the job. In a three-phase induction motor the line current per phase is:

$$C = \frac{W \times 746}{\sqrt{3} \times \xi \times E \cos\theta}$$

where W = power of the motor in watts; ξ = efficiency; E = line voltage; θ = angle of lag.

Praseodymium. A metallic element derived from didymia. In addition to samarium, it contains both praseodymium and neodymium. These latter resemble one another closely, but the salts of praseodymium are green, while those of neodymium are rose-coloured. The metal is yellowish in colour, is not affected by atmospheric air, the oxide being usually obtained from the nitrate.

Melting point is 931 °C and the specific gravity is 6.78. Praseodymium and neodymium are used together under the name of didymium. They give a neutral grey colour to glass. It occurs in monazite sand. Atomic number 59, at. wt. 140.9. (*See* Table 2.)

Precious metal. A term used in reference to the costly and rare metals used in coinage, jewellery and ornaments. The name is usually restricted to silver, gold and the platinum metals, since these have both intrinsic value and permanence. Neither radium nor the rest of the platinum metals are precious, though they are rare. Gold and the platinum group metals are also referred to as noble metals, since they are highly resistant to acids and to corrosion.

Precipitate. The name used in reference to the solid that is thrown out of solution (either in a liquid or in a solid) as a result of the decrease in solubility usually due to a fall in temperature. In aqueous solutions, when the ionic product is greater than the solubility product, precipitation occurs. Chemical analysis depends in a large measure on the formation of precipitates by double decomposition. Two clear solutions may each contain one radical of an insoluble compound, and on mixing the insoluble compound is precipitated.

Precipitation hardening. The constitution of certain alloys changes if the temperature is raised or lowered, and this fact makes heat-treatment of metals possible. When a solid solution of limited solubility cools slowly some of the solute is precipitated and may be concentrated about the grain boundaries of the solvent metal crystals. On cooling rapidly the amount of solute precipitated is greatly decreased, so that supersaturation occurs. The structure is thus metastable, and at room temperature some precipitation on a very small scale may occur, resulting in considerable hardening of the alloy. It was first noticed in aluminium alloys of the Duralumin type which contain essentially copper, magnesium and silicon. The phenomenon is the basis of the industrial process known as precipitation hardening. The treatment involves fast cooling from an elevated temperature (450-550 °C) and then precipitation at room temperature. The hardening can be accelerated by heating to about 100-200 °C. Also known as dispersion hardening, age-hardening. Precipitation hardening occurs in certain types of aluminium alloys and in copper-beryllium alloys.

Precision casting. The name applied to a number of processes that have been devised for the production of castings to very close dimensional tolerances. The principal processes are: (*a*) the investment or lost-wax process (q.v.) in which a pattern in wax or other fusible material is enclosed in a refractory slurry which sets to form a mould. The pattern is removed by heating which also fixes and strengthens the refractory material; (*b*) the shell-moulding process (q.v.), of which there are several variants, uses metal patterns and a fine silica sand to make the thin mould or shell. In the Croning process a hot pattern is used, and the shell is formed over it by gravity flow of a sand containing a resin binder. The Dietert process uses a cold pattern and the shell is formed using a core-blower; (*c*) in the carbon-mould process a special permeable carbon mould is machined for the pattern impression, and the mould surface is coated for each casting with a silica-base wash; (*d*) a plaster-mould process in which the moulding material is plaster of Paris, employs metal patterns (*see* **Pressure-cast patterns**). It is useful for light-alloy castings; (*e*) a process that employs high-pressure moulding and uses a resin binder with silica sand and bentonite. Metal patterns are preferable, and the mould has an excellent surface finish.

Precoated sand. A term used in foundry-work to designate a sand that has been treated before use with an oil or resin. The sand, in the latter case, is heated to a temperature at which the resin flows freely so that blending can be as complete as possible and the grains of sand coated with the resin. Sometimes electrically heated patterns are used in moulding with such sand, and there results a fine varnish finish on the mould face.

Preece test. A test to determine the uniformity and thickness of zinc coatings on galvanised iron and steel. It consists in noting the number of one-minute dips in a 1.27M solution of copper sulphate at 18 °C that are needed to produce through-penetration. This is indicated when a bright *adherent* deposit of copper occurs. The specimen is washed in running water and lightly rubbed to remove loose copper, between each dip. The test shows up places where the coating is locally thin.

Preferential deformation. In the mechanical working of metals it must happen that the material is deformed or extended in one direction more than in any other direction. This is referred to as preferential deformation, and it gives rise to asymmetry in the mechanical properties of any worked specimen.

Preferred orientation. A term used in reference to the departure—in the case of a polycrystalline metal—from a random crystalline structure. Where the regular arrangement of the atoms of a metal is continuous through the whole mass, the metal is referred to as a single crystal. However, it often occurs that the pattern is discontinuous and there are changes, at numerous points, of the inclination of the pattern. This latter is referred to as the orientation. In cast metals the conditions under which a component is produced may favour similarity of orientation, and this is termed preferred orientation. Its importance lies in the fact that it gives rise to directional properties in the metal. Preferred orientation may be produced by mechanical working (rolling, drawing), since there is a tendency, with certain crystals, to force the slip planes of each grain towards the direction in which the metal flows. This should be taken into account when deciding where test pieces are to be cut from a sample of worked metal.

Preheating. A term applied to any heating operation, at a moderate temperature, that is carried out as a preliminary to further heating at a substantially higher temperature. Tool steels are often slowly heated to some intermediate temperature until the whole mass is uniformly hot before being subjected to a higher temperature. In the case of steels of high hardenability the operation must be carried out prior to welding in order to avoid hard spots and to reduce the tendency to cracking, and it is preferable to anneal after welding. Preheating temperatures for low carbon steels may be from 120 °C to 150 °C, but for high carbon-steels temperatures up to 450 °C may be necessary.

Presetting. A mechanical operation in which a spring or machine part is loaded in the same direction as the load on a component in service so as to produce permanent deformation or set. When the load is removed the surface fibres of the part in which the yield stress of the material was exceeded, will be in a state of stress in the opposite sense to that applied on those particular zones during presetting.

Presintering. A preliminary heating operation carried out on a powder compact. The temperature is usually well below that used in sintering, and the operation is intended to improve the handling of the compact as well as to remove any lubricant or binder that was used in its formation.

Press. A general name for the various types of machine in which pressure is employed for performing the work required. The press is usually named according to the operation it performs, e.g. flanging press, extrusion press, etc.

Press forging. A metal part that has been worked to a desired shape by a process involving pressure on the metal, produced in a special type of press called a forging press. The operation may be carried out on either hot or cold metal and dies may or may not be required. Since metals are more plastic when hot it is common practice to perform a rough shaping on the hot metal and then follow with cold working to the desired shape. The final cold working gives good dimensional accuracy and improved surface finish, while the strength of the metal is generally increased.

It is important to contrast this method of producing components with other types of forging, e.g. drop-forging, where a sudden hammer blow or blows is substituted for the comparatively slowly applied pressure.

Pressing. The term used in reference to the working of metals, and may include several metal forming processes known under particular names.

(a) Cutting processes. These may employ shears for cutting metal sheet; and blanking processes in which a punch of any shape and a die of corresponding shape are used to cut out shapes from sheet metal.

(b) Bending operations are performed on sheet metal and wire by pressing the material into cavities or slots by means of a punch. As the name implies, bending usually involves a two-dimensional change in shape e.g. folding.

(c) Drawing is an operation for achieving a three-dimensional change of shape. A flat sheet can be pressed into a die to produce a cup-shaped product. Deep drawing generally requires several operations or redrawings, because the final product would frequently suffer fracture if an attempt were made to produce the final shape in one drawing. So heat treatment after one or two draws—depending upon the severity of the deformation—may be necessary. If the major dimension across the open end of the "cup" is D two or more drawings are needed if the *depth* of the cup is greater than D/2.

(d) Reduction is a process for reducing or "necking" the open end of a cup-shaped component produced by drawing.

(e) If the final shape of the "cup" that is being drawn is not straight-sided but bulges or tapers upwards an operation known as "bulging" is resorted to. A hydraulic die, with oil within the blank, is compressed to the shape of the die by means of a ram.

(f) Curling is a special pressing operation in which the edge of the open end of the "cup" is rolled or flanged. If wire is to be enclosed within the flange the process is referred to as "wiring".

(g) Sizing by compressing or squeezing a component into a die involves the flow of metal, and hence is not used for cold-working cold-short metals and alloys, and in particular is unsuitable for cast iron and hardened steels. Coining is a particular application of this method, and can be used for making powder compacts in powder metallurgy. High pressures are used and close tolerances can be obtained; if the flow of metal is in a particular direction (usually but not always perpendicular to the direction of pressure exerted by the die) the process is referred to as extrusion.

(h) Metal spinning may be regarded as a pressing operation, and is used when the component to be formed from sheet metal has a surface of revolution. The metal is spun in a lathe and the wood or metal spinning tool is held against the spinning sheet. Holding and guiding the tool may be done manually or mechanically.

Pressure. In a gas, the force per unit area exerted on the walls of the container by the gas molecules, that force being the rate of change of momentum when the molecules collide with the container walls due to their kinetic energy.

The vapour pressure of a liquid or solid is the force exerted by the molecules of liquid or solid above the surface of that liquid or solid. In an enclosed space, this pressure reaches a maximum dependent on the substance involved and the temperature, the vapour is said to be saturated, and the pressure is then the saturated vapour pressure of the substance involved. Solids have very low vapour pressure, volatile liquids have very high vapour pressures.

Pressure cast patterns. A term used in reference to a method of producing aluminium patterns which give very good definition of the master pattern and have very good surface finish. The master wood pattern is moulded in plaster of Paris and the mould is stoved. Pouring is done with the aid of a container placed above the mould and sealed from the latter by an asbestos disc. Metal is poured into the container, which is then closed and attached to a low-pressure air-line. Sufficient pressure is used to break the seal and so admit metal to the mould under pressure.

Pressure die-casting. A process for making castings in which the metal is forced into the mould or die under high pressure. The purpose of the method is to obtain castings of high dimensional accuracy and good surface finish at a high rate of production. The success of the process depends on the temperature at which the metal is admitted to the mould, the rate at which it enters the cold mould and the pressure used. In modern practice, e.g. for aluminium, magnesium, zinc and copper-base alloys, pressures of the order of 35-42 MN m^{-2} are used, the alloy being fed at a temperature not much in excess of its melting temperature and forced into the die by means of an injection plunger. Pressure die-cast components usually require little or no machining, but the process is limited to small parts of, say, not more than 5.5 kg (12 lb) weight each. Die-casting machines are of two types—the hot chamber (used for alloys of zinc, tin and lead) and the cold-chamber type (used for aluminium and brass). The chief difference between the two types lies in the robustness of the latter, which must be designed to take much higher pressures than the former. The figure shows in outline the principle of operation of a cold-chamber die-casting machine. (*See* Fig. 123.)

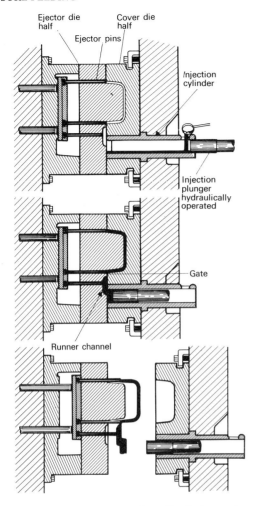

Ejector die half

Cover die half

Ejector pins

Injection cylinder

Injection plunger hydraulically operated

Gate

Runner channel

Fig. 123 *Cross-section of a pressure die-casting machine, showing operation.*

Pressure-feeding. A term used to describe the methods of applying pressure to the molten metal in sand-moulds for the making of steel or other castings, the object of which is to improve the quality of the casting. The usual methods employed are: *(a)* placing a sand-core in the top of the riser; *(b)* placing graphite rods in the riser; *(c)* using a blind riser, usually cylindrical in shape and having an hemispherical dome. In this latter case the pressure can be further increased by external means through the top of the riser, or in a completely sealed riser the additional pressure can be produced by inserting, before the mould is closed, a compound that gasifies slowly, to ensure that the pressure does not begin to act until the metal in the riser has had time to solidify. High pressures are unsuitable in such cases, since the effect may be to enlarge the casting or cause the metal to penetrate into the sand of the mould.

Pressure filter. A type of filter used in the treatment of slimes and of solutions that are otherwise difficult to filter. They are used either to remove the solution from the solids or for clarifying a solution prior to precipitation of a desired solute. In all cases the solution is driven out of the pulpy mass by the pressure, the solids collecting on the filter cloth which covers the filter frames. Pressure filters can be worked at quite high pressures (up to 0.45-0.55 MN m^{-2}) and therefore have considerable advantages over vacuum filters. They are used in collecting precipitated gold in the extraction process of this metal and in clarifying electroplating solutions.

Pressure leaching. A process used in hydrometallurgical extraction in which the slimes are treated in pressure filters to clarify the solutions before subsequent treatment. (*See* **Pressure filter.**)

Pressureless sintering. A process used in powder metallurgy in which the powder is compacted solely by vibration. It can be applied usefully to produce blanks of simple shape, and since pressure is not introduced, the compact may have a more uniform pore size and shape. It may be suitable for producing filter material, but precautions must be taken to avoid lamination, fissuring and warping. Also called loose sintering.

Pressure of fluidity. The term used in reference to the stress at which a loaded rigid punch continuously penetrates a block of ductile metal. The tests are usually carried out at room temperature. The value is approximately equal to the minimum pressure required to extrude a metal through a die under certain conditions. The stress is roughly equal to twice the Brinell hardness number of the annealed metal.

Pressure testing. A type of test that has been widely used both for inspection and acceptance of the different types of casting to which it can be applied. The parts to be tested are clamped firmly over the ends of a length of steel tube and pressure applied by pumping in air (or more usually) water. The test pieces are examined for leakage. The test has its limitations, since the penetration of air or water through the mass of the specimen will depend on there being interconnection between the pores or voids in the mass of the metal. Though it is not a very reliable test for soundness of castings it is used with advantage to test riveted joints, etc. (*See also* **Pressure tightness.**)

Pressure tightness. A term used in foundry-work in reference to a casting that is free from interconnecting pores and similar defects that would cause leakage when subjected to pressure testing (q.v.). The term is also used in reference to containers made from plates that are riveted together.

Pressure welding. The term applied to the processes of welding in which both heat and pressure are applied to the abutting faces, the temperature being below the solidus (or melting point in the case of elementary metallic substances). The process involves the forcing together of the mating surfaces into such close contact that atomic bonding is able to occur across the interface. Since rise of temperature increases the mobility of the atoms, pressure welds are usually more readily made at the higher temperatures. Though pressure welding is easier when the temperature is above the recrystallisation temperature of cold worked metal or above the critical point at which any phase change occurs, this is not essential to the process. Forge-welding and fire-welding as practised by the blacksmith are forms of pressure welding in which the mating parts are forced together by hammer-blows.

Primary carbide. The term used in reference to the carbon in steels in excess of the eutectic or eutectoid compositions.

Primary crystals. The first crystals formed during the cooling of an alloy through the freezing range.

Primary mineral. A mineral which was formed at the same time as the rock in which it is found.

Primary pipe. A term used in reference to the depression that normally forms in the central top portion of an ingot. In an ingot the solidification is directional, since it proceeds inwards from the sides and bottom, and compensation for the shrinkage occurs at the advancing solid surface. Unless, therefore, extra metal is supplied there will result a depression at the centre where the solidification has not yet proceeded to completion.

Primer. A coating which protects the underlying substrate as a preliminary to application of other finishing coatings, or provides a better bond with the substrate than the finishing coatings, so limiting the tendency to stripping, peeling or cracking of a coating subsequently. Also called priming coat.

Primes. A term used in reference to metal products such as sheet or plate of the highest quality. In coated sheets, e.g. tin-plate, etc., it usually means free from visible surface defects.

Priorite. A rare-earth mineral that contains oxides of niobium, tantalum and titanium.

Proactinium. A radioactive element of the actinium series resulting from the decay of uranium Y. It gives out alpha particles and is converted into the element actinium. Symbol Pa, at. no. 91, half-life period 3.2×10^4 years, at. wt. 231. (*See* Table 2.)

Probability distribution. Referred to the electron, the use of probability avoids the problem of defining the electron as a wave motion by describing such aspects as amplitude. Born proposed that the wave should represent the probability of finding the electron at a given point in space and time.

The wave motion may be represented by two functions, f and g, such that the electron density is proportional to $(f^2 + g^2)$ at a particular position in space. The assumption is made that time is independent, so that f and g must be one-quarter of a period apart. (If the functions were in phase, intensity would fluctuate with time.) If x is the spatial position, n the frequency, a the amplitude, λ the wavelength and t the time, then

$$f = a \cos 2\pi \left(nt - \frac{x}{\lambda}\right) \quad \text{and} \quad g = a \sin 2\pi \left(nt - \frac{x}{\lambda}\right)$$

represent the wave motion. As time is independent, $(f^2 + g^2) = a^2$.

The real functions of f and g are replaced by a single complex function $\Psi = f + ig$ where $i = (-1)^{1/2}$. The conjugate complex is $\Psi^* = f - ig$, and the electron density is proportional to

$$f^2 + g^2 = (f + ig)(f - ig) = \Psi\Psi^*.$$

This is a mathematical device only, to unite two vibrations one-quarter of a period apart, and so make them independent of time. The number of electrons at a particular position in a volume ΔV is proportional to $\Delta V \Psi \Psi^*$; when a quantum of energy hv is involved, $\Psi(x,y,z,t) = \psi(x,y,z)e^{2\pi int}$ where $\psi(x,y,z)$ is the space factor in co-ordinates x,y,z, often called the amplitude factor, and $e^{2\pi int}$ is the time factor. $e^{2\pi int}$ has a scalar value of unity, so Ψ and ψ can be interchangeable. (Some textbooks have confused these two symbols, and referred to them as "amplitude".) The symbol Ψ has come to represent the "waves" associated with electrons, and the probability of finding an electron at a given point in space and time is $\Psi\Psi^*$ or $|\Psi|^2$. From these concepts Schrödinger evolved the time free or time independent equation

$$\Delta^2 \Psi + \frac{8\pi^2 m}{h^2}(E - W)\Psi = 0$$

where W = potential energy of the electron, E the total energy, m the mass and h Planck's constant.

This equation represents a family of surfaces of wave form corresponding to one value of E. The solutions to such equations can have many values, and the problems in wave mechanics are to find a solution which appears physically possible. There are conditions involved, that ψ is finite and has a single value, and that

$$\psi \text{ and } \frac{d\psi}{dx} \text{ shall be continuous.}$$

When the solution represents a stationary state (standing wave) the conditions imply that the equivalent wavelength λ is related to some distance x which is involved in the problem. An electron in a unidimensional box of length L gives acceptable solutions only when

$$\lambda = \frac{2L}{N}$$

where N is a whole number. This means that only certain values of energy can be involved, the quantum numbers of quantum theory. Thus, although wave mechanics deals in continuous quantities, the acceptable solutions involve whole numbers, as required by quantum mechanics. There is also the analogy of a stretched string, a stationary state of vibration, where λ must be definite fractions of the string length. This is not to imply that the electron is a physical wave like a vibrating string, but simply that the same mathematical relationship governs the acceptable solutions in both cases.

These considerations give rise to the concept of the electron "cloud", which represents the probability of the bounds within which the electron may be found in specific cases.

Probability is a specialised topic in statistics, and must be considered in any topic or phenomenon involving energy changes and random events, e.g. nucleation and growth.

Process annealing. A term applied in reference to processes used in sheet and wire mills in which ferrous metals are heated to a temperature just below the lower limit of the critical range, and allowed to cool. The purpose of this heat-treatment is to soften the metal preparatory to further cold working.

Process wire. The term applied to wire that has been drawn down from bar and is to be heat-treated and further worked before finishing.

Producer gas. A gaseous mixture of carbon monoxide and nitrogen used as a fuel for heating purposes. It is made by passing air through a thick bed of incandescent coke. The producer used in the making of this gas is a closed fire-grate and is often sealed below by a water-seal. The primary air is then forced through the fuel by means of a fan. In metallurgical work the gas is often burned without cooling, though some heat is always lost in conveying the gas from the producer to the ovens or furnaces where it is burnt.

Profilometer. An instrument for measuring surface finish. The vertical movements of a stylus which traverses the surface are amplified electromagnetically and recorded (or indicated) as the r.m.s. of the irregularities. The machine is similar to the Talysurf instrument. Instruments are also available to measure the surface condition of wire-drawing dies.

Progressive ageing. The name of the process in which a metal is subjected to a rising temperature during ageing. The temperature may be increased slowly and progressively or may increase in stages between which long or short intervals are allowed.

Projection welding. A process developed from spot-welding in which one of the plates or parts to be welded is treated in a pressing operation so as to produce a number of depressions or "pimples". The method of welding is very similar to that used in spot-welding, the resistance heating being confined to the projections on the one part that make contact with the other part. The method involves the use of larger electrodes, higher contact pressure and greater electrical input.

Promethium. A rare element found in monazite sand. It has been prepared by the fission of uranium and by the irradiation with neutrons of neodymium. The atomic number is 61 and the chemical symbol Pm. (*See* Table 2.)

Promoters. The name used in reference to those substances which are not in themselves catalysts but which when mixed with these latter enhance their activity. In the Haber process for the synthesis of ammonia, finely divided iron or molybdenum may be used as catalysts, but a mixture containing iron, alumina and potassium oxide is particularly active. Platinum is the catalyst normally employed in the oxidation of ammonia to nitric oxide. Promoter action is secured by using iron oxide and bismuth oxide. A mixture of zinc oxide and chromium oxide can be used as promoter in the synthesis of methanol from hydrogen and carbon monoxide.

Proof bar. A bar of material, similar to that being treated, that is included in the work during surface-hardening processes. The bar is withdrawn at intervals during the operation to provide evidence of the progress of the treatment.

Proof-bend test. A test often applied to tubes in which the latter are suitably supported and then loaded to give a specified bending moment. The load required is reported.

Proof stress. A term used in mechanical testing to indicate that stress which will produce a specified amount of permanent deformation in a tension test. The stress is measured as the load per the original area of cross-section and the extension is usually specified as 0.1%, 0.2% or 0.5% on the gauge length (*see* Fig. 124 and **Tensile test piece**).

Proplatinum. A jewellery alloy used as a substitute or to imitate platinum. It is an alloy of nickel, bismuth and silver.

Proportional limit. The greatest stress which a material is capable of developing or withstanding without deviation from Hooke's law of proportionality. It is the highest stress at which the stress-strain curve is a straight line. For many metals the proportional limit and the elastic limit are taken as approximately equal.

Protal process. A chemical process for the protection of aluminium and for preparing this metal for painting. It involves immersion in, or spraying with, a solution containing titanium and chromium salts, together with alkali fluorides, in order to produce an insoluble surface layer.

Protection. A term generally used in reference to the various methods in use for coating the surface of a metal to enhance its resistance to corrosive influences. Among the methods used or recommended are the following:

(a) *Anodic treatment.* In the case of aluminium there are a number of processes available, and the final colour of the coating is dependent on the electrolyte used: (1) Bengough-Stewart process uses a 3% solution of chromic acid and is suitable for all alloys of aluminium except those which contain copper. The film

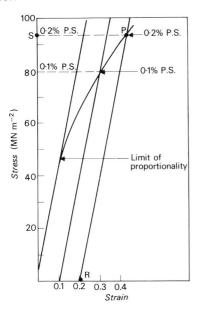

Fig. 124 *Stress-strain relationship showing 0.1% and 0.2% proof stress.*

is light grey in colour and is opaque. (2) Aluminite process in which the electrolyte is sulphuric acid containing small amounts of aluminium sulphate or glycerine as addition agents. (3) Eloxal process uses oxalic acid as the electrolyte, and small amounts of boric, chromic or sulphuric acid are added. There is an alternating current superimposed on the usual direct current used for the anodic oxidation. The deposit is opaque and yellow to brown in colour. (4) Sheppard process uses sulphuric acid as the electrolyte and glycerine is the addition agent. (5) Other processes which use phosphoric acid or ammonia or ammonium sulphide, and the deposits are cream to bluish-grey in colour. (*See also* **Protal process**).

(*b*) *Tar coatings.* A process usually restricted to iron pipes which are to be located under ground. Soaking in light tar oil, followed by dipping in coal-tar pitch forms the coating, which, when thoroughly dry, is wrapped in hessian also soaked in tar. In severe conditions of corrosion where temperatures may be high enough to soften the tar, the latter is stiffened with fine slate or sand dust and may be given an outer wash of lime. Asphalt-bitumen containing silica or slate dust or other filler may be used for the interior of iron or steel pipes.

(*c*) *Electrical (cathodic) methods.* These are based on a method of maintaining the metallic pipeline, cable etc., cathodic with respect to the surrounding earth, so that stray currents shall not produce anodic regions in the metal. This latter condition would, of course, lead to extensive corrosion. The methods of reducing the rate of corrosion are (1) by inserting numerous insulating gaps in the conductor, (2) by joining the pipeline or other conductor to the negative feeder of the source of current of the neighbouring cable system, and (3) by joining the underground system to be protected to a number of sacrificial anodes.

(*d*) *Cement coatings.* Such coatings are sometimes used in soils where sulphates are absent, but the thickness of the coating as usually required makes the method fairly expensive.

(*e*) *Metallic coatings.* On steel pipes, metallic coatings are sometimes successful. Lead has been found not to be useful in this employment, but zinc provides good protection in neutral or nearly neutral soils and in soils containing sulphate-reducing bacteria.

(*f*) *Non-metallic coatings.* Some protection can be obtained against corrosion by using vitreous enamels, paint or coatings of grease containing inhibitors such as finely divided aluminium or zinc dust.

(*g*) *Electrochemical protection.* This may be provided for metal surfaces in contact with water (e.g. boiler plates, ships' propellers and rudders, etc.), by using zinc blocks. The member to be protected is put in metallic contact with a suitably placed ring or block of zinc, the latter being renewed periodically. The method is used also in pipelines, storage tanks, etc. Zinc may also be used to protect aluminium alloys (e.g. Duralumin). Pure aluminium is employed as a protection for the mechanically stronger aluminium alloys.

Protoactinium. Synonymous with proactinium, (q.v.).

Protogenic. A term used in reference to solvents that are capable of giving up protons to the solute. The conception that an acid is a substance that produces hydrogen ions in water and a base is a substance that will give hydroxyl ions in similar conditions does not apply to non-aqueous solutions, and an extension of the definition of an acid and a base is needed. The modern definition describes an acid as a substance that has a tendency to lose a proton and a base as a substance that has a tendency to gain a proton. Dissociation of an acid cannot occur unless the solvent can take up protons given up by the acid. The definition is illustrated by the equations

$$Acid \rightleftharpoons proton + base\ conjugate$$

and

$$Base + proton \rightleftharpoons acid\ conjugate$$

Protogenic solvents (e.g. ethanoic acid) hinder the dissociation of even strong acids and assist the dissociation of bases. The strength of an acid is measured by the degree of dissociation, which is the same as measuring the tendency to lose protons. The proton-accepting capacity of water is such that all strong acids can be completely ionised in water, and hence their "strengths" cannot be distinguished. From the above explanation it is clear that the strengths must be measured in a protogenic solvent (e.g. ethanoic acid) and not in an amphiprotic solvent (e.g. water). When the dissociation constants of acids are measured in ethanoic acid solutions, the strongest acid is found to be perchloric acid and the order of decreasing strength is $HClO_4 > HBr > H_2SO_4 > HCl > HNO_3$. The strengths of strong bases can be best determined using a protophilic solvent.

Protolac process. Method of producing a protective film on aluminium by dipping into a solution of an alkali salt of a metal that will deposit an insoluble oxide on the aluminium. It is thus essential that the metal used must have a higher oxide that is soluble in alkaline solution and a lower oxide that is insoluble.

Proton. The name given to the relatively heavy fundamental nuclear particle that carries unit positive charge. It is identical with the nucleus of the hydrogen atom. The stream of positive rays (q.v.) produced by passing a discharge through gas at very low pressure consists of protons. The mass of a proton relative to that of the electron is 1840. The number of positive charges (protons) in the nucleus of an atom is the same as the atomic number.

Protophilic. A term used in reference to solvents that are able to accept protons from the solute. They reduce the ionisation of strong bases and assist the ionisation of acids. Liquid ammonia is an example of a protophilic solvent, and in this a weak acid such as ethanoic is highly ionised.

Proustite. A silver ore which is a sulpharsenite of silver Ag_3AsS_3 and contains about 65% Ag. On account of its colour it is sometimes called light ruby silver. Dark ruby silver is pyrargyrite (q.v.). It occurs in veins in Peru, Chile, Mexico and some southern states of the USA. Mohs hardness 2.5, sp. gr. 5.6.

Prussian blue. A deep-blue compound formed by the reaction of an iron III salt with potassium iron II cyanide which has the formula $K.Fe[Fe(CN)_6]$. A similar compound is formed with sodium or ammonium iron II cyanides and, in fact, the name Prussian blue is taken to include all the iron III iron II cyanides. Other methods of producing Prussian blue include: (*a*) interaction of an iron II salt with potassium iron II cyanide; (*b*) reduction of iron III iron II cyanides; (*c*) oxidation of iron II iron II cyanides. The substance formed under (*a*) was known as Turnbull's blue, but is probably the same as Prussian blue. The following varieties have been described: (1) α-soluble Prussian blue prepared from an iron III salt and potassium iron II cyanide. It is soluble in water, giving a deep-blue solution, and a bronze reflex when viewed as a dry powder. It also dissolves readily in oxalic (ethanedioic) acid and is precipitated by the addition of salts. Ammonia decomposes it. (2) β-soluble Prussian blue, obtained as under (*c*) in neutral solution. The oxidation occurs with great rapidity with atmospheric oxygen. The product is soluble in water but insoluble in ethanedioic acid. (3) γ-soluble Prussian blue, prepared as in (*c*) above in cold acid solution. It is water soluble and much more stable towards alkalis than (1) or (2) above. (4) Williamson's violet—obtained as under *(c)* using nitric acid as the oxidising agent. It is insoluble in ethanedioic acid and in dilute mineral acids, and appears as a violet-blue powder which is green in thin layers by transmitted light. (5) Monthier's blue is obtained as under *(c)* using an ammoniacal solution. (6) An insoluble Prussian blue is obtained by adding an excess of iron III salt to soluble Prussian blue. All these substances are colloidal except (6). When dried they retain some water and are very difficult to obtain in a state of purity. The commercial material is a mixture of the above types, and is generally produced as under *(c)* using bleaching-powder as the oxidising agent. The finest variety—which is known as Paris blue—is made by adding an iron III salt to an iron II cyanide. The dry powder shows a pleasing bronze lustre. It is used as a dye or in ethanedioic acid solution as an ink, but for both these employments has been partly replaced by aniline dyes.

Psilomelane. An amorphous, hydrated oxide of manganese often containing both barium and potassium. It occurs in sedimentary deposits along with other manganese minerals. It is black in colour and contains MnO_2.

p-type semiconductor. An extrinsic semiconductor in which the number of electrons per atom of the impurity addition is less than that of the base material (e.g. gallium in germanium) and the electron energy of the outermost level of the impurity is close to the valence band of the base. A small increase in temperature is then capable of exciting electrons from the valence band of the base atoms into the impurity atoms, creating the concept of "positive holes" in the base, into which electrons can also move from other base atoms. The carriers are the positive holes, leading to a net movement of electrons, and hence p-type, for positive, was the name adopted. (*See* **Extrinsic semiconductor** and **Acceptor.**)

Puckering. A term used in pressing and drawing operations to indicate defects associated with these processes. The defects referred to are usually corrugations on that part of the metal that has been formed. The defects usually indicate faulty tool design or sometimes unsatisfactory conditions of pressing.

Puddling. An obsolete process for producing wrought iron from pig iron. The latter is melted in a puddling furnace and roll scale or ore is added. The operation is carried out in an oxidising atmosphere, and during the early stages of melting the silicon, manganese and sulphur and part of the phosphorus are oxidised and removed in the slag. The temperature is raised to promote interaction between the iron oxide and the carbon, and carbon monoxide is evolved causing violent "boiling" of the charge. A good deal of the slag or puddle cinder is removed through the working door. The charge is then worked with a hand or mechanical rabble, and as the iron becomes purer it can no longer remain liquid at this working temperature. Patches of solid metal appear, causing the entire mass to become pasty. When the reaction ceases, the mass is separated into portions which are worked into balls. These balls are taken (in tongs) to the hammer and worked into blooms, the operation being designed to squeeze out occluded slag. The blooms are then rolled into puddle bars, during which process more slag is removed.

The first (rough) rolling produces an irregular "muck" bar. They are cut to suitable lengths, bound together and heated to welding temperature and rolled. The product is known as refined iron.

Pull crack. An irregular transverse crack sometimes found on the surface of an ingot or casting when free contraction during cooling has been prevented.

Pull-over mill. Synonymous with drag-over mill (q.v.).

Pulls. Surface defects sometimes encountered on ingots and rolled products. They are usually forked or Y-shaped and originate from transverse cracks which have been elongated in the direction of rolling. Deep isolated cracks starting from longitudinal ingot defects are known as splits.

In foundry-work, pulls or hot tears are cracks having jagged edges and are defects attributable to sand conditions. They may be caused by excessive mould hardness due to very hard ramming.

Pulsator jig. A name used in reference to a type of jig for concentration of ores in which a plunger operated by an eccentric cam causes the water to act upon the ore on a submerged screen, in surges. Usually the downward stroke of the plunger is referred to as the pulsion stroke, this producing an upward current of water

through the screen in the neighbouring compartment (*see* Fig. 125). Other methods of obtaining surges of water through the sieve include the rotating valve of the Richards type and the moving screen of Hancock-type jigs.

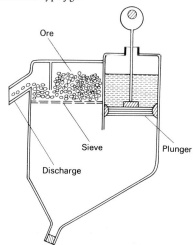

Fig. 125 *Schematic cross-section of a pulsator-type jig.*

Pulverised fuel. A term used in reference to finely divided combustible matter—either anthracite or bituminous coal—used for firing furnaces of various types. The air supply for the combustion of the fuel must be preheated for the best results. In the case of anthracite it is essential that the air be highly preheated.

Pulverised silica. Very finely crushed quartz, sometimes used to replace tripoli as an abrasive.

Pumice. A porous volcanic glass generally used in powder form as an abrasive in fine polishing of metal surfaces. It is a constituent of metal polishes and scouring compounds. In colour it is commonly greyish-white and contains 65-75% SiO_2 and 12-15% Al_2O_3, the balance being accounted for by sodium or potassium oxides.

Pumicite. A volcanic ash similar in colour and composition to pumice. It is graded to pass through 325-mesh sieve and is used as an abrasive.

Punch. The term usually refers to the tool that is used for producing holes in metal plates by pressing out a slug. In forging it is also used to denote the upper member of a die or tool set that is movable, producing the final forming of the upper side of the forging.

Punching. A forging operation in which the punch bears against the forging stock so that it forms a more or less deep impression or removes a slug and so forms a hole. This latter is usually circular in plan, but it may be of any desired shape. Punching operation causes severe stress in the material, so that a reduction of mechanical strength results. Because of this, holes in boiler-plates and plates for pressure vessels are drilled.

In punching thin sheet the punch should be a close sliding fit in the die. For thicker sheets or plates a certain clearance should be allowed. The clearance affects the pressure required for punching, but soft metal requires more clearance than hard metal. An approximate rule for determining the pressure to perforate metal sheet is:

$$\text{Pressure (in N)} = \frac{P}{3} \times tC$$

where P = perimeter of the hole in mm; t = thickness of sheet in mm; and C = a constant. The value of C is 5.2 for steel, 4.2 for brass. In the case of a circular hole $P/3$ can be taken as the diameter. Tin plate does not usually require a lubricant when punching or deep drawing. Aluminium should never be worked without a lubricant (kerosene, petroleum jelly or lard oil). With most other metals lard, sperm oil, flaked graphite or tallow may be used. White lead in fish oil or soap solution may be used, respectively, for steel and copper.

Punching bear. A small portable machine for punching and shearing sheet-metal. It is often actuated by a pump which supplies the water to the actuating cylinder.

Punching machines. Machines for punching sheet-metal. When thin sheets are to be worked, crank-driven power presses are used. Hydraulically operated machines are necessary for heavier work. A punching machine is commonly combined with a shearing machine.

Pure metal. Theoretically a metallic specimen is pure if it contains no detectable impurities. Practically the estimation of purity depends on and is limited `by the accuracy of the method of measurement employed. Pure metals have constant melting point and crystalline structure.

Commercial metals that are available in quantity have the following percentage purity: Zn 99.99, Pb 99.98, Cu 99.97, Sn 99.95, Ni 99.8 and Al 99.5.

Puron. A very pure highly refined iron that is used as a spectroscopic standard.

Purple copper ore. A copper pyrites having the composition Cu_3FeS_3 and containing about 55% Cu. Also called bornite and peacock ore (q.v.).

Purple of cassius. A purple powder obtained by precipitating gold chloride with a mixture of stannous (tin II) and stannic (tin III) chlorides, used in making ruby glass.

Push bench. An arrangement of reducing ring dies used in making hot-rolled steel tubes. A punch is forced into a square-rolled billet placed in a cylindrical die. This produces the rough tube blank. A mandrel is then pushed into the blank and the two are forced together through the series of reducing dies. During the final reeling the mandrel is removed leaving a tube with a closed end (*see* Fig. 126).

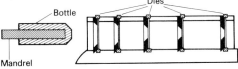

Fig. 126 *Section showing the principle of operation of a push-bench.*

Putty. A plastic material made from whiting (powdered chalk) and linseed oil. It hardens on exposure to air.

Putty powder. An abrasive material consisting of impure tin IV oxide.

Pyknometer. An apparatus that may be used for determining the density (and the coefficient of expansion) of liquids. It consists of a glass vessel graduated to hold a definite volume of liquid at a given temperature. By weighing it full of liquid at different temperatures the variations in density, and hence the apparent expansion, can be measured. For the determination of the specific gravity of minerals the pyknometer can be used when a weighable amount of pure material can be selected.

Pylumin process. An immersion process for the treatment of aluminium alloys prior to painting. It is similar to the modified Bauer Vogel process, in which a thin oxide film is formed on the metal by immersion in a solution containing chromates.

Pyramidal system. In crystallography, the system in which all the axes are at right angles, two having equal lengths while the third is of a different length. This latter is referred to as the principal or morphological axis. The parameters are therefore $a:a:c$. Also called the tetragonal or quadratic system.

Pyramid hardness number. The term used in reference to the Vickers hardness number (V.P.N.). In the hardness-testing machine a diamond pyramid is used as the indenter. In shape it is a square pyramid, the angle between opposite faces being $136°$. A range of loads up to 3,000 kg can be used according to the hardness of the material and the hardness value is computed by dividing the load in kilograms by the area of the impression in mm^2. The Brinell and Vickers scales are comparable up to about 200.

The Vickers test usually occupies ½-¾ minutes. The indentation is small, and thus a good surface finish is required. Though the V.P.N. is theoretically independent of load used, small variations do occur, and hence it is usual to report the load used.

Pyrargyrite. The sulphantimonite of silver ore Ag_3SbS_3. It is isomorphous with the sulpharsenite known as proustite (q.v.). It is usually dark grey in colour, but thin sections appear red by transmitted light, giving rise to the name "dark red" silver ore. It occurs in veins along with calcite, silver and lead ores. It is found mainly in Saxony (Germany), Mexico and many parts of the USA. The silver content is about 59%. Mohs hardness 2.5, sp. gr. 5.8.

Pyrex glass. A trade name for glass of low coefficient of thermal expansion, containing boron oxide and high in silica, and hence classed as a borosilicate glass. The chemical resistance is high, the resistance to thermal shock improved, and softening does not occur until about 800 °C (viscosity 10^{10} cP). The material is excellent for fabrication of laboratory glassware and domestic cooking utensils, and in the engineering industry for containers and electrical applications. Composition (typical): 80% SiO_2, 12% B_2O_3, 3.5% Na_2O, 0.5% K_2O, 2% Al_2O_3.

Pyrites. The term used in reference to the cubic form of the di-sulphide of iron FeS_2 which is dimorphous. (The orthorhombic form is usually termed marcasite.) The mineral is usually pale yellow in colour, and is widely distributed and commonly found in veins. In coal it often occurs as bands or nodules known as "brasses". It is worked mainly for the sulphur it contains, the latter being used in the manufacture of sulphuric acid. Pyrites is mined in Spain, France, Portugal and the USA. Many specimens contain copper, nickel, arsenic and cobalt. Traces of gold and thallium are also found. Mohs hardness 6, sp. gr. 4.9-5.1. Also called "fool's gold". Alteration products of pyrites depend on the conditions of oxidation. In the presence of pure water ferrous (iron II) sulphate, ferric (iron III) sulphate and basic iron III sulphates are formed. In water and in the presence of calcium carbonate the iron appears as hydroxide, while the sulphur appears as calcium sulphate (gypsum).

Pyritic smelting. The term used in reference to the oxidation of pyritic ores. In this process the oxidation and formation of the slag produce the heat required to carry on the operation. If the ore is deficient in sulphur, carbonaceous fuel must be added. The blast furnace used for the treatment of copper ores is smaller than that used for iron. The ore poor in sulphur yields an impure form of copper and a slag containing very little copper. The high sulphur content ore is roasted to burn away the sulphur to produce a matte containing only a small proportion of sulphur.

Pyritohedral. A term used in reference to a sub-division of the cubic system in crystallography. The crystals of this class have three cubic planes of symmetry but no dodecahedral planes. The pentagonal dodecahedron—a solid bounded by 12 pentagonal faces, which are, however, not regular pentagons—belongs to this class, and pyrites represents a commonly occurring example.

Pyrobitumen. A type of bitumen which is jet black in colour and has a brilliant lustre. It breaks with a conchoidal fracture. It occurs in veins and is a derivative of petroleum. Also called albertite.

Pyroceram. A glass which has been seeded with a nucleating agent, of which titanium dioxide is one, and which is fabricated by conventional glass working methods. After completion of fabrication, the artefacts are carefully reheated, become partially devitrified, and crystallise with high density to form a ceramic product of great strength and resistance to thermal shock. The mechanism is likely to be one of two liquid phases which gradually separate into vitrified and crystalline regions, the proportions depending upon the composition and the heat treatment applied.

Pyrochlore compounds. The mineral pyrochlore is a complex niobate, often given a formula $Ca_2Nb_2O_7$, in which some calcium may be replaced by sodium, niobium by tantalum, and oxygen by fluorine. The general formula is undoubtedly $A_2B_2O_7$ where A and B are metals, and the structure is built upon octahedra of formula NbO_6. In this respect, the structure bears some resemblance to the titanates based on perovskite $CaTiO_3$.

These compounds are ferroelectric, exhibiting spontaneous polarisation in the absence of an electric field, but which may be switched by application of an external electric field. This phenomenon arises from the lack of symmetry in the crystal structure. They are also pyroelectric and piezoelectric.

Pyroelectricity. A pyroelectric material exhibits spontaneous polarisation in the absence of an electric field, but changes its polarisation on heating. If the change in polarisation is ΔP on raising temperature by an interval ΔT, $\Delta P = \lambda \Delta T$ and λ is the pyroelectric coefficient. All pyroelectric materials are also piezoelectric, but the converse is not true. The ferroelectrics are a special class of pyroelectric materials, in which the direction of polarisation can also be changed by application of an external electric field.

Of the 32 classes of crystal symmetry, 11 are centrosymmetric and 21 non-centrosymmetric. Of this 21, 20 are piezoelectric, and of that 20, 10 are pyroelectric, the other 11 being non-pyroelectric. Tourmaline and quartz are pyroelectric.

Pyrolusite. The most important ore of manganese, consisting mainly of the dioxide MnO_2. It occurs in rhombic crystals and in crystalline and amorphous masses. It possesses a metallic lustre, a black or dark-grey colour and a black streak. It is found in Thuringia (Germany), Bohemia, Moravia (Czechoslovakia) and in the Caucasus. Other localities where small but important deposits occur are Vermont and San Francisco in the USA, in Nova Scotia (Canada) and India and, more recently, in South Africa. The mineral often results from the dehydration and oxidation of manganite $Mn_2O_3.H_2O$, and therefore frequently contains some water. It is always an alteration product of other manganese ores—mangenite, rhodochrosite, rhondonite, etc.

It is extensively used in the manufacture of spiegeleisen and ferromanganese and of master alloys such as manganese bronze. It is used in the manufacture of chlorine and permanganate (disinfectants) for discharging the green or brown colour of glass containing ferrous iron (II), as a colouring matter in glass and ceramics, in dyeing and in the making of green and violet paints and in dry batteries.

The dioxide possesses feebly basic and feebly acid properties in presence of acids and bases respectively—sp. gr. 4.82.

Pyrolysis. A term used in reference to the chemical decomposition of a substance by heat.

Pyrometallurgy. A term used in reference to those processes and techniques by which metals are extracted from their ores by the application and action of heat. Such processes include smelting, roasting, distilling, etc.

Pyrometric cones. A series of cones made of clay and of graduated melting points. Each is fusible within a certain small range of temperature estimated by observing which of the cones have softened and bent over so that the apex touches the base. They are used mainly in connection with the firing of ceramic materials. (*See* Seger cones.)

Pyrometry. The term used in reference to the measurement of high temperature. Various types of instrument have been proposed. The methods include: (a) measurement of temperature by observing the expansion of a metal rod; (b) the estimation of temperature using a series of ingots of alloys of different compositions and different melting points. They were placed in the furnace and the temperature was estimated by observing which of the alloys had melted. Silver-gold and gold-platinum alloys have been used; (c) temperature estimated by noting which of a series of clay cones softened when placed in a furnace. This method was much used in connection with the manufacture of ceramics. The cones were made of clay of different compositions and were known as Seger cones (q.v.); (d) radiation pyrometers are of various types. In the earliest practical form of total radiation instrument due to Féry the radiation is focused by a mirror on to a sensitive thermocouple and the e.m.f. generated is indicated on a galvanometer. In a later type a bimetal strip replaced the thermocouple and a direct reading obtained from indications of the needle attached to the end of the strip that uncoils on heating. The disappearing filament (optical pyrometer, q.v.) is also a radiation type. More recent types measure the brightness of the hot source (which is related to the temperature of the source) by the aid of photo-electric cells, which are chosen for their sensitivity to a range of several colours. The current produced is measured on a sensitive instrument, and this is related to the temperature of the source; (e) resistance-type pyrometers utilise the regular increase of electrical resistance of a metal with rise in the temperature; (f) the thermo-electric effect is also used to determine temperature, the dissimilar metals being usually alloys of high or very high melting point. Some thermocouples in common use are given in Table 42; the useful range of temperature is indicated.

The general principles on which high-temperature measurements are based are similar whatever the method employed. The change with temperature of some physical property of a substance or substances is measured within the range of temperature for which the hydrogen thermometer is standardised. This change is then represented by some mathematical formula, and on the assumption

TABLE 42.
The Maximum Working Temperature and e.m.f. Produced by Various Thermocouple Elements

Thermocouple Elements	Maximum Temperature, °C	e.m.f. Produced, Millivolts	
		At 100 °C	At. Max. Temp.
Copper-Constantan	400	4.0	20.0
Iron-Constantan	800	5.0	45.0
Chromel-Alumel	1,400	4.0	56.0
Pt-Pt/13% Rh	1,700	0.6	20.0
Pt-Pt/10% Rh	1,700	0.6	18.0
Pt-Pd/50% Au	1,000	3.0	39.0
Graphite-Silicon carbide	1,700	—	—

that this formula will hold good at temperatures above the range of the gas thermometer, the higher temperatures are estimated. It is usual, however, to check the pyrometer against some established temperature such as the melting point of sulphur and of gold.

Pyromorphite. A mineral consisting mainly of lead chlorophosphate and sometimes used as a source of lead $(PbCl)Pb_4(PO_4)_3$. It is an alteration product of galena and is frequently found in the upper levels of lead mines. It has usually a green colour and the crystals are hexagonal prixms. Yellow or yellowish-brown specimens are not uncommon and the mineral has a resinous lustre. Also called green lead ore. Mohs hardness 3.5, sp. gr. 6.6-7.1. It is isomorphous with mimetite.

Pyrope. A member of the well-defined group of orthosilicates known as garnet. It is a magnesium aluminium silicate

$$3MgO.Al_2O_3.3SiO_2$$

which crystallises with cubic symmetry—usually as dodecahedra. The colour of the mineral is due to the presence of iron, manganese or chromium. It is found in Saxony (Germany), Bohemia (Czechoslovakia), the USA and South Africa. The name refers to the deep-red colour (fiery eye) and a pyrope of typical blood-red colour is the common garnet of jewellery. It is distinguished from red almandine since it has a lower density and a lower refractive index ($\mu = 1.705$ for pyrope and 1.80 for almandine). Specific gravity 3.5. Also known as Bohemian garnet, Cape ruby and rhodolite.

Pyrophillite. A rather soft mineral consisting of hydrated aluminium silicate, $Al_2O_3.4SiO_2.H_2O$. Both a crystalline and an amorphous variety exist. The former has a pearly lustre and is somewhat flexible. Both varieties are white or pale green or even yellow. The amorphous variety is greasy to the touch and was used to make "slate pencils". It is still used for "tailor's chalk" or French chalk. Mohs hardness 1-2, sp. gr. 2.8. It occurs in Switzerland, the Urals and in North America. It also occurs in the East, since it was used by the Chinese for making small images and ornaments. It is easily machined.

Pyrophoric alloys. The name given to those alloys which when filed or scratched readily emit bright sparks.

Pyroxene. An important group of rock-forming minerals which resemble the amphiboles in general characteristics and chemical composition, but are distinguishable by the crystalline form and cleavage angle. The members of the group are conveniently divided into three series according to the symmetry of the crystals. They are all metasilicates: (a) the rhombic series (e.g. enstatite, bronzite and hypersthene), which are magnesium or magnesium-iron silicates; (b) the monoclinic series. A large number of minerals are included in this series, of which diopside is typical

$$Ca.Mg(SiO_3)_2.$$

(c) the triclinic series of which rhodonite is typical are silicates containing calcium, iron, manganese, rarely magnesium and occasionally zirconium.

Pyrrhotite. A mineral which consists of iron sulphide, though analysis suggests that it may be a solid solution of sulphur in iron II sulphide. Nickel, cobalt and sometimes copper occur in the mineral. The nickeliferous pyrrhotite is the most important source of nickel. It is found in Sudbury (Ontario). Precious metals may also be found in small amounts. The mineral is magnetic and is hence often referred to as magnetic pyrites. Mohs hardness 4, sp. gr. 4.6-4.65.

Quadrant plate. The name applied to the plate that is fixed to the fast-headstock end of a lathe, the purpose of which is to carry the change wheels.

Qualitative analysis. In chemical analysis this generally refers to the determination (and confirmation) of the nature of the constituents of a compound or mixture, without inquiring into the actual amounts that may be present.

Quality factor. A term synonymous with merit number.

Quantitative analysis. The term used in chemical analysis that refers to the determination of the amounts of the constituents present in a mixture or compound. The chemical methods employed may be divided broadly into two main classes—gravimetric and volumetric. In those cases where the amount of a particular constituent present in a mixture or compound, or where the total sample available is small, special techniques are employed, e.g. spectroscopic analysis, micro-analysis. Wet and dry analysis are the terms used to describe assaying of minerals and ores.

Quantometer. An instrument used in the spectrographic analysis of metals. (See **Spectrographic analysis.**)

Quantum. A definite amount of energy associated with electromagnetic waves. These latter include light, X-rays and γ-rays. The quantum is dependent on the frequency of the electromagnetic radiation and is equal to the product $h\nu$ where h = Planck's constant and ν is the frequency of the radiation. The energy emitted in the form of radiation must be an integral number of quanta, the amount being inversely proportional to the wavelength. The quantum of X-rays (which have very short wavelength) is greater than for visible light and in fact a quantum is not a fixed amount but varies with the type of radiation. The size of the quanta for radiations of different wavelength is given in Table 43.

TABLE 43.
Energy Quanta Corresponding to Different Wavelengths

Wavelength, λ (max.), $\times 10^{-10}$ m	Energy $h\nu \times 10^{-19}$ J	Colour
7,500	2.62	Red
6,500	3.26	Orange
5,900	3.33	Yellow
5,750	3.42	Green
4,900	4.00	Blue
4,550	4.32	Violet
3,950	4.98	Ultra-violet

Quantum efficiency. In a photochemical reaction the primary process in all cases is the absorption of light to form an excited atom or particle. Einstein assumed, in reference to primary chemical changes, that each molecule entering into reaction must be excited by the absorption of one quantum of radiation. (This is the law of the photochemical equivalent.) The number of molecules actually decomposed per quantum of radiation absorbed is known as the quantum efficiency. It is usually represented by γ. Also called the photochemical yield.

Quantum level. A term used to denote the so-called ground state of an electron.

When an atom is subjected to an electrical discharge it absorbs energy, and electrons are moved from an inner energy or quantum level to outer levels of higher energy. In returning to the stable configuration again, radiation, represented by $h\nu$, where ν = the frequency of the radiation, is emitted. It is monochromatic. This is regarded as the explanation of the line spectra of atoms.

Quantum numbers. *See* **Atom.**

Quantum theory. An hypothesis put forward by Planck which assumed that radiation is emitted or absorbed by oscillating or vibrating electric particles in a discontinuous manner. The energy lost or gained by a particle is proportional to the frequency of vibration of the particle. It is not, however, emitted or absorbed in a continuously varying manner, but only in whole-number multiples of a certain minimum quantity called a quantum, the ratio between frequency and energy being expressed by $E = h\nu$, where ν = frequency and h is a constant known as Planck's constant ($h = 6.625 \times 10^{-34}$ Js). This theory was originally applied to the explanation of heat-radiation, but was extended from the oscillating motion due to heat to include explanations of the particles in the atom.

Quarrying. The process of obtaining rock material from the earth's crust—usually in open working. For metallurgical purposes material that is to be used for fluxing is usually drilled, blasted and then crushed to the required size.

Quartation. The operation of combining three parts of silver with one part of gold as a preliminary to separating and purifying gold by means of nitric acid. Though silver alone dissolves readily in nitric acid it does not do so when alloyed with much gold, for the whole of the silver is insoluble when the quantity of it in the alloy is slightly more than the quantity of gold. It was formerly believed that in order to separate the whole of the silver, not more than one quarter of the alloy must consist of gold. This latter is the origin of the term quartation and the process is still used.

Quarter-wave plate. A plate of doubly refracting crystal cut parallel to the optic axis and of a certain minimum thickness. Such a plate produces no separation of a ray that is incident normally. But if the incident ray is polarised, the amplitudes of the normal and the extraordinary ray as well as the velocities of travel will be different. The rays will not be separated, but each particle in the common path will have two simple harmonic motions and in general the actual path of the particle will be an ellipse. If the thickness of the plate be such that the phases of the ordinary and extraordinary ray differ by 90° the light will emerge as circularly polarised light. The quarter-wave plate is used in the analysis of polarised light—particularly to distinguish unpolarised light from circularly polarised light.

Quartz. A very common and very widely distributed mineral which consists of silica SiO_2. The material is hard and is put to a large number of diverse uses. Several varieties are used as gemstones.

Quartz is one of the three crystalline forms of silica.

The silica tetrahedra, in which one small silicon atom sits at the centre of four oxygen atoms, each oxygen atom being shared by two tetrahedra, are arranged in a helical form in beta quartz. The oxygen atoms in the fourth plane are situated vertically above those in the first, oxygen ions in the fifth plane are directly above those in the second, and oxygen ions in the sixth are vertically above those in the third plane. The structure can be regarded as a distorted form of the related high temperature forms of silica, cristobalite (cubic) and tridymite (hexagonal). (Some authorities describe the quartz structure as a hexagonal form.) Covalent bonds imply high hardness and high melting point, and all the forms of silica are invaluable as constituents of high temperature ceramic materials. Quartz is the naturally occurring form, and therefore the more abundant form of silica.

The colourless crystals (rock crystal) are used in optical instruments. Opaque and coloured masses of quartz are quite common and the purple variety is cut as the semi-precious stone amethyst. The crystals are enantiomorphous. Fused quartz is used for making lenses and laboratory ware, and for making the elastic and strong quartz fibres used in physical apparatus.

Finely powdered quartz sand is very useful abrasive. Crystals of quartz exhibit pyro-electric properties (q.v.) and also piezo-electric characteristics. When suitably cut the crystal plate is circularly polarising (see **Quarter-wave plate**). Mohs hardness 7, sp. gr. 2.65.

Quartzite. Quartz being a mineral highly resistant to weathering, it forms the bulk of sandstones. When the sand grains are cemented together by a later deposit of secondary quartz the resulting rock formation is known as quartzite. The rock is compact and free from pores and gives a smooth fracture. Though the rock is hard and white it is not generally used in building because it is very splintery and readily breaks down under the action of frost.

Quartz lens. Lens made from quartz or rock crystal for use in photomicrography employing ultra-violet light. The crystalline structure of rock crystal must be reduced by fusing it, otherwise double refraction would occur. Since the light used in this work is monochromatic, it is not necessary to correct such lenses for chromatism. The use of ultra-violet light has now been greatly restricted because it is somewhat troublesome and also because much better resolution is obtained with the electron microscope.

Quartz-porphyry. The name used in reference to a group of hemicrystalline rocks containing quartz crystals in a fine felsitic matrix. They are acid in nature and may vary in colour from grey to green. Some specimens are reddish and may contain small but usually nearly perfect zircons.

Quasi-viscous curve. The curve showing the relationship between load and mean creep rate during the third stage of the creep curve (see Fig. 31).

Quasi-viscous flow. A term used to describe the K component of the creep curve. The latter can be resolved into two components, B and K components, and they are related in the equation due to Andrade

$$L = L_0 (1 + \beta t^{\frac{1}{3}}) e^{Kt}$$

K being the measure of the final flow that proceeds viscously. Here the strain rate is not proportional to the stress. L_0 is the length of the test piece immediately on loading and L is the length after time t.

Quaternary alloy. An alloy containing four principal elements apart from accidental impurities

Quaternary steels. Alloy steels that contain two alloying elements in addition to the carbon in the steel.

Quench ageing. A term used in reference to the changes in the physical and mechanical properties of an alloy which occur as a result of the passage of time. These changes may occur gradually at atmospheric temperatures and more rapidly at higher temperatures following upon rapid cooling. The term quench ageing in this connection refers to the method by which the metal is hardened. The hardening depends on the fact that the basis metal can, at an elevated temperature, hold in solid solution a larger amount of the alloying element than it can hold at room temperature. The operation of quenching produces, at room temperatures, a highly supersaturated solution. This latter is not hard so long as it remains in the supersaturated condition but becomes hard when precipitation begins. (See also **Precipitation hardening.**)

Quench hardening. The process by which alloys (usually ferrous alloys) are heated to within or above the critical temperature range and then rapidly cooled by quenching. A plain carbon steel at high temperatures (over 900 °C) consists of γ-iron holding carbon in solid solution. On slow cooling through the transformation range (down to about 750 °C) the carbon separates out as Fe_3C as the γ-iron is transformed to α-iron. The rapid quenching suppresses this change, and the presence of such alloying elements in the steel as chromium, nickel or manganese facilitates the retardation of the transformation. When the transformation has been entirely suppressed the product is soft, but when an intermediate condition is obtained in which the transformation has been partly suppressed, the product is hardened.

Quenching. A process of rapidly cooling a metal from an elevated temperature by contact with liquids, gases or solids. As heat is removed from solids by dissipation from the surface, rapid cooling is possible only when a cooling medium having a high specific heat and good thermal conductivity is kept in contact with the surface of the metal.

Liquids are most generally used, and their circulation in the cooling tank prevents overheating of the liquid and helps to remove gas bubbles from the metal surface, which is advantageous. Water and certain aqueous solutions are used for drastic quenching. Oils are generally more satisfactory when freedom from distortion and residual stresses is important.

Quenching crack. The term used in reference to a crack or fissure that results from stresses induced by rapid cooling or quenching. It is usually the result of volume changes during cooling.

Quenching intensity. A factor that is used to represent the quantity of heat transferred per unit area of the surface per unit time, per degree difference of temperature between the body and its surroundings, divided by the thermal conductivity of the material of the body. It is represented by the letter b,

$$b = \frac{Q}{A \times t \times (\theta - T) \times k}$$

where Q = quantity of heat transferred, A = surface area, θ = temperature of body, T = temperature of surroundings, t = time and k = thermal conductivity of body. For air cooling $b = 0.03$ and for oil $b = 0.8$.

TABLE 44.
Relative Cooling Rates of Quenching Media

Medium	Cooling Rate (Still water at 20 °C = 1)
10% Caustic soda solution	1.20
Brine	1.10
Iced water	1.07
Running water at 20 °C	1.05
Still water at 20 °C	1.00
Running water at 60 °C	0.60
Mineral base oils	0.40
Cotton seed oil	0.35
Fish oil (whale)	0.30
Machine oil	0.20

Quenching media. The term used in reference to the liquids or gases that are used as the vehicle for the removal of heat from hot metal. The selection of the correct quenching medium is of considerable importance, particularly in the case of steels, since the rate at which the heat is extracted as the metal cools controls the hardness. Apart from the actual amount of the medium used and its initial temperature, the amount of heat extracted from metal during quenching depends on the following physical properties of the medium: (a) specific heat; (b) thermal conductivity;

(c) latent heat of vaporisation; (d) viscosity; (e) volatility. The progressive order of severity due to the use of different media for quenching is (1) still air, (2) air blast, (3) oil, (4) water at room temperature, (5) iced water, (6) brine, all of which would reduce the temperature of the metal to about room temperature. For some purposes the quenching is required to terminate at higher temperatures than room temperature and in such cases molten metal or molten salt baths are used. The relative cooling rates of a number of quenching media are given in Table 44.

Water is the usual quenching medium and is frequently used at about 20 °C. Brine gives good results for files, while such articles as springs and saws that have to be both elastic and tough are quenched in lard or whale oil. Mercury is sometimes used for small articles that are required to be extremely hard.

Quenching oils. Mixtures of animal, vegetable or mineral oils used for the rapid cooling of heated metals. Oils are less drastic than water or aqueous solutions for quenching. (*See* **Quenching media.**)

Quenching stresses. The internal stresses set up in metals as a result of rapid cooling or quenching. A low-temperature tempering treatment usually referred to as a "stress-relieving" treatment is normally prescribed after hardening.

Quick immersion pyrometer. A type of pyrometer which can be plunged into liquid steel for just long enough to give a reading of the temperature and then withdrawn intact. Such a pyrometer must be lightly sheathed so as to assume the temperature of the bath quickly and the time of immersion must be short—of the order of a fraction of a minute—if the instrument is to be withdrawn undamaged. Platinum/platinum-rhodium thermocouples are available up to about 1,700 °C.

The quick immersion pyrometer cannot be successful with heavy refractory sheaths because of the excessive lag and the errors arising through conduction losses, so that accurate readings are not obtainable in the short time available. When, however, the platinum/platinum-rhodium thermocouple is lightly sheathed in a thin-walled silica tube a reading can be obtained in about 20 s. The silica is very resistant to thermal shock and has a very short lag, but it can be relied upon to remain intact for the period of the reading, even though it may soften and devitrify or even react with the slag and molten steel.

Quicking. A process sometimes used in connection with the plating of silver on copper or copper alloys. When copper or a copper-base alloy is placed in a solution of silver nitrate or cyanide copper tends to replace the silver and a rather poorly adherent deposit of silver collects on the copper. To prevent this such articles are dipped in a solution made by dissolving mercury II oxide in sodium cyanide. This dipping—which is called quicking—produces a thin film of mercury on the copper so that when subsequently placed in the silver bath, no copper replacement occurs.

Quicklime. The name used in reference to the oxide of lime obtained by heating limestone $CaCO_3$ in kilns, using coal in the proportion of about 1 part of coal to 5 parts of limestone. In modern kilns the process is continuous. The quality of the quicklime depends on the nature of the limestone from which it is made. The lime from pure limestone slakes readily and is known as "fat" lime. When the amount of magnesia present exceeds about 25% the "lime" forms a thin mixture with water which is useless for building purposes. If silica is present in considerable amount, the lime yields a mortar that hardens under water and is known as hydraulic cement. Limestone containing clay may yield a lime that slakes very slowly—due to a crust of silicate. It is then referred to as dead burnt.

Lime is used in making mortar and cement; in the glass industry, for making caustic soda and, when slaked, for making bleaching powder. It dissolves slightly in water and the alkaline solution — called lime-water — is used in analysis. The pure substance is white, with specific gravity of 3.3. It melts about 1,900 °C.

Quicksilver. A common name for the metal mercury (q.v.). It was known to the ancient philosophers as "liquid silver" and was obtained from cinnabar. It was used for the extraction of gold. (*See also* **Amalgamation.**)

Quinhydrone half-cell. The name of an apparatus—also called the quinhydrone electrode—for use in determining electrode potentials. It is less generally useful than the hydrogen electrode, for it cannot be relied upon in the presence of certain neutral salts and it can only be used in solutions in which pH is less than 7. It is, however, very easy to instal and, of course, it does not require a hydrogen supply. The electrode depends on the use of the organic substance quinhydrone, which is a compound of one molecule of quinone and one of hydroquinone. This substance changes from quinone to hydroquinone quantitatively according to the pH of the solution into which it is put. Since the first of these substances is oxidising while the other is reducing, the potential set up on a platinum or gold electrode will vary with the proportion of each substance in solution.

A small piece of gold or platinum foil is inserted into the solution to be tested and about 5 cm^3 of saturated quinhydrone solution added. A calomel electrode is put into the solution and the potential between the two is measured.

Because this quinhydrone must be added to the solution under test, this method is not suitable for continuous measurements.

Quinine sulphate periodide (herapathite). Fine crystalline needles, which, aggregated with specific orientation, and embedded in a plastic base, polarise light by absorption effects. The transmitted light is partly coloured, but more modern polarising materials have rendered the material obsolescent.

R

Rabbling. The operation of raking or stirring a bath of molten metal or a charge of heated ore in the furnace hearth to secure adequate mixing. The iron rod used in the operation is termed a rabble.

Rack. The straight member that is provided with teeth in the combination of a toothed wheel—referred to as the pinion—and a rack. The combination is used for a number of purposes, e.g. the moving of the tables of planing machines to and fro. (*See also* **Rack feed, Racking motion.**)

Rackarock. An explosive that was used in mining and quarrying consisting of a mixture of potassium chlorate and nitrobenzene.

Rack feed. A feed for a tool that is produced through the operation of a rack and pinion mechanism.

Racking motion. The movement of a body by means of a rack and pinion.

Radethein process. A process for the production of magnesium by the reduction of magnesia with carbon. The reaction takes place in an enclosed electric-arc furnace having a lining of carbon blocks. A temperature of about 2,000 °C is employed and the magnesium is drawn off by pumps, in the form of vapour which is suddenly cooled. Also called the Hansgirg process.

Radial stress. The stress (measured at right angles to the curved surface—that is, along a radius) in material in the shape of a cylinder that is subjected to internal or external pressure. For thin-walled vessels the radial pressure may be taken as equal to the applied pressure. Where the pressure is applied by means of a piston within the cylinder, the radial stress may be taken as the applied load per unit of cross-sectional area of the piston.

Radial test. Synonymous with transverse bend test (q.v.).

Radiated pyrites. Another name for marcasite or white pyrites (q.v.).

Radiation. A term widely used to include the emission of energy by means that do not depend on the presence of matter. It is, however, convenient to distinguish between (*a*) the radiation which consists in the propagation of a wave motion through a medium or a vacuum, and (*b*) the radiation which consists of a stream of particles or corpuscles moving in the same direction, but not associated with any form of matter. The electromagnetic radiations (*a*) travel with the same velocity as ordinary light and are usually described in terms of their wavelengths (*see* Table 45). All these radiations exhibit the characteristic phenomena associated with transverse wave motion (e.g. reflection, refraction, interference, polarisation). The radiations of shortest wavelength—the gamma rays—are generated when atomic nuclei disintegrate: those of somewhat longer wavelength (from 1.4 to about 136×10^{-10} m) are emitted when fast-moving electrons are incident upon and stopped by a heavy metal target. These are the X-rays. From about 140 to $4{,}000 \times 10^{-10}$ m are included in the ultra-violet rays emitted by incandescent or very hot bodies and by ionised gases. The visible spectrum has a short range from wavelength 4,000 to $8{,}000 \times 10^{-10}$ m. Beyond this up to about $4 \times 10^6 \times 10^{-10}$ m is the infra-red region—the radiation resulting from molecular motion and emitted by hot bodies. Solar radiation, as normally experienced at ground level, has a range overlapping both the ultra-violet and infra-red as indicated. The Hertzian waves resulting from oscillations in metal conductors extend from about wavelength 1×10^6 to $3 \times 10^{14} \times 10^{-10}$ m and include the relatively short waves produced by radio emitters and spark-gap discharge. The longest waves are generated when a coil of metal is rapidly rotated in a magnetic field.

TABLE 45.
Wavelengths and Frequencies of Various Radiations

Approx. Frequency, Hz	Wavelength		Description
	cm	Angstroms	
5×10^{19}	6×10^{-10}	0.06	GAMMA RAYS
6×10^{18}	5×10^{-9}	0.5	
2.14×10^{18}	1.4×10^{-8}	1.4	
			X-RAYS
2.2×10^{16}	1.36×10^{-6}	136	
			ULTRA-VIOLET RAYS
2.94×10^{15}	1.02×10^{-5}	1,020	
		4,000	
			VISIBLE RAYS
3.75×10^{14}	8×10^{-5}	8,000	SOLAR RADIATION
			INFRA-RED RAYS
3.0×10^{12}	1.0×10^{-2}	1×10^6	
7.5×10^{11}	4.0×10^{-2}	4×10^6	
			SHORT HERTZIAN WAVES
3.0×10^7	1×10^3	1×10^{11}	
			LONG HERTZIAN WAVES (RADIO)
1.0×10^4	3×10^6	3×10^{14}	
			VERY LONG WAVES

The corpuscular radiations include (*a*) radiation that consists of a stream of electrons, e.g. cathode rays, and β-rays emitted by radioactive substances; (*b*) streams of positively charged atoms, e.g. canal or positive rays, and α-rays or streams of helium nuclei emitted by radioactive substances; (*c*) streams of uncharged particles or atoms.

The term radiation is also used in reference to a mode of transmission of heat from a hot body to other bodies in its environment which are not directly in contact with it. It is a fundamental principle that the radiation within a uniformly heated enclosure depends only on the temperature of the enclosure. From this is derived the concept of a perfectly black body, i.e. one which absorbs all radiations that fall upon it whatever their wavelength. For *any* body the absorptive power is the ratio of radiant energy absorbed to the total incident energy. It is denoted by *a*, and for a perfectly black body its value is unity. For radiations of the same wavelength and the same temperature the ratio of the emissive power to the absorptive power is the same for all bodies and is equal to the emissive power of a perfectly black body (Kirchhoff's law). According to Stefan's law the total radiation of any body is proportional to the fourth power of the absolute temperature. This is strictly true only for black-body radiation.

The exposure of elementary substances to the products of disintegration of uranium (nuclear fission) for the purpose of inducing artificial radioactivity in those substances is referred to as "irradiation".

Radiation damage. Electromagnetic radiation and electrons may cause physical or chemical changes in substances such as polymers. Metals and ceramics, being more strongly bonded, can absorb most electromagnetic radiations with only minor effects, but can be affected by bombardment with particles such as neutrons, particularly high energy neutrons. Radiation damage implies the more permanent effects which result in changes in properties or behaviour as a result of such bombardment. The shorter the wavelength of radiation, the more penetrating, and hence the greater possibility of damage, but mass effects will obviously be involved when heavy particles are involved at high energy.

Ultra-violet light causes changes in some polymers, leading to degradation, and a rise in temperature in all materials. Similar effects apply to X-rays, γ-rays and electron beams (β-rays) and radio waves, the penetration being greatest in biological tissue. Helium nuclei (α particles) despite their mass, have only shallow penetration, comparable with electrons, but all affect polymeric materials to some degree, and living tissue particularly. Even short radio waves are known to produce deleterious effects in human tissue. Engineering materials are used to shield biological materials from these effects, because of their absorption power. Heavy metals absorb better than light, and all metals are better than wood, brick or earth. Concrete containing barium ions rather than calcium has better absorption, hence barytes concrete is used in hospital and nuclear installations for shielding against X- and γ-rays, electrons and β particles.

It is the neutron which produces the most drastic effects on all materials, but electrons can cause degradation of polymers into lower molecular weight units, or cross-linking in others, to form a network structure of increased hardness. The charge on the electron prevents it from doing more damage, but the neutron, electrically neutral, is a projectile of much greater mass (1,840 ×) which cannot be slowed down except by collision. (*See* **Nuclear cross-section**.) The mean free path of neutrons in most solids is at least a few centimetres, but in that space it can cause considerable damage. Collisions at shallow angles produce slight displacements and the dispersed energy appears as heat, but angles which result in considerable change in momentum cause displacement of atoms into non-equilibrium positions, leaving interstitial atoms and vacant lattice sites. The displaced atom may itself have sufficient kinetic energy to itself displace other atoms, and these ongoing effects may result in a displacement or radiation spike, of so-called "knock-on" damage. (*See* **Displacement spike** and Fig. 51.) It is the loss of momentum by such collisions which causes the neutron to lose its energy, and in the final stages, it may be captured by an atom favourably oriented, becoming part of the nucleus of that atom. Resonance effects can give rise to easier absorption of neutrons in certain atoms.

Nuclear reactors based on neutrons of thermal energy require to slow down the neutrons produced in fission (of very high energy) and employ a "moderator", a material which readily suffers neutron collisions but does not give rise to excessive capture. Light elements such as hydrogen, beryllium and carbon are excellent for this purpose. The energy transmitted to an atom by a single collision with a high energy neutron may be 10^2 keV, and on average there is one collision per 500 atoms through which

the neutron moves. The amount of radiation damage is therefore considerable, and a cluster of defects extends over a distance of 1 to 100 μm in each direction, from the passage of a single neutron. Fortunately, much of the energy involved is released as heat, as the atoms are forced through the lattice, and the rise in temperature which results causes some self annealing effects, allowing the defects to be neutralised as atoms return to available vacant lattice sites by diffusion or other collisions. Aluminium (f.c.c.), for example, will anneal at −50 °C, but iron (b.c.c.) requires a temperature of 300 °C or more. Unless the material can rise to such a temperature, ferrous materials will undoubtedly suffer hardening effects due to the displaced atoms, and become more brittle.

More complex lattices, oxides and silicates for example, do not anneal so readily, and anisotropic materials such as hexagonal metals, and graphite, suffer damage which is non-uniform in direction, and so largely irreversible. Quartz is converted to an amorphous type of structure, and silicates such as concrete tend to behave similarly although over a longer period of time. Concrete therefore tends to disintegrate, which is unfortunate since it is a cheap form of biological shielding around a reactor. In practice, a layer of metal such as steel is used to slow down the neutrons before entering the concrete and the steel is able to carry this absorption better than the concrete, still avoiding the need for excessive amounts of metal. The decreased ductility which arises in a metal such as steel as a result of damage, can be a problem, particularly in steel pressure vessels used to contain reactor components and subjected to neutron damage. If the steel is maintained at its self-annealing temperature, it may suffer plastic deformation due to creep, but if not, it may be embrittled and so suffer the possibility of brittle failure later. The yield strength may rise by 155 MN m^{-2} in a thermal reactor operated for 10 years whose average fast neutron flux is 10^{12} neutrons cm^{-2} s^{-1}. In mild steel, this would lead to a 30 °C rise in the brittle transition temperature and so shut down for maintenance in winter temperatures could run the risk of brittle failure. If more highly alloyed steels are used, the creep strength improves for higher temperature operation, but the brittle transition temperature also rises. As radiation also increases creep by radiation induced mechanisms, much effort has been expended in improvement of the brittle transition characteristics in such steels. (*See* Fig. 55.)

Graphite is the cheapest form of moderator for nuclear reactors, and due to its anisotropy, suffers considerable radiation damage. The energy tends to be stored in the lattice, and increase in temperature is needed to allow annealing to take place. The temperature of the reactor may in fact be above the self-annealing temperature, but the rate of damage in graphite usually exceeds the rate of annealing, and a regular energy release programme is essential. If the release is not controlled, catastrophic rise in temperature may result, as the release per unit mass may exceed the specific heat of the material leading to a runaway condition, as occurred at Windscale in 1957.

Chemical reactions are affected by radiation, as has long been known for infra-red and ultra-violet light. In the presence of neutron bombardment, active species of chemical compounds have been detected. The reactants may change, the nature and mechanisms of reaction may be modified, and the kinetics of reaction violently accelerated. Reactions between carbon dioxide and graphite are examples, leading to deposition of graphite in other parts of the reactor coolant circuit where CO_2 is employed as a coolant. Metal oxidation by CO_2, H_2O or even atmospheric air, can be accelerated, or the stability of the metal oxide affected.

Radiation pyrometers. Instruments used in measuring high temperatures whose method of operation depends on the fact that the intensity of the radiation of the hot body is related to its temperature. The various instruments in use may be classified roughly according as they employ part of or all the radiation emitted. The optical pyrometer (q.v.) matches the brightness of a lamp filament, that can be varied by adjusting the current, against the brightness of the hot body. The photoelectric effect is used in the various types of photo-electric cell pyrometers. The cells used are sensitive to and respond to a range of wavelengths, while the electrical effects are proportional to the intensity of light radiation.

In the total radiation pyrometer, the proper functioning of which depends on the conditions of black-body radiation (*see* **Radiation**) existing, the radiation is focused by a mirror upon a small black disc attached to the thermocouple junction. Approximate black-body conditions may be obtained either by sighting the instrument on a small opening in the furnace or, better still, by sighting down a tube inserted into the furnace and where applicable having the closed end immersed in the molten metal. If a lens is used either to focus the radiation or to protect the mirror, only a portion of the total radiation will be incident on the black disc, for glass and fused silica absorb some part of the longer infra-red rays. Corrections are thus necessary both for the deviation from ideal conditions and for the opacity of the instrument to certain wavelengths. The calibration is made experimentally by comparison with standard instruments. (*See also* **Féry pyrometer.**)

Since the pyrometer is sensitive to a wide band of frequencies, it is liable to be affected by gases intervening between the source viewed and the instrument, particularly if these gases have strong absorption bands in that portion of the spectrum in which the larger proportion of the radiated energy is concentrated. Products of combustion such as water vapour, carbon monoxide and carbon dioxide in the temperature range 600-1,600 °C can produce effects that considerably affect the pyrometer indications, generally giving low readings.

Radioactive decay. A term used to describe the fact that the activity of a simple radioactive product decreases with time. In fact, the activity diminishes (*see* Fig. 127) in a descending geometrical progression with lapse of time. The law governing this rate of transformation states that the rate is proportional to the number of atoms that are left unchanged. Thus if N is the number of atoms unchanged at any particular moment

$$\frac{-dN}{dt} = \lambda N,$$

where λ is a factor known as the radioactive decay constant. Since by integration $N_t = N_0 e^{-\lambda t}$, then

$$\log_e \frac{N_t}{N_0} = \lambda t$$

Since the integration gives an exponential function, all radioactive substances will require infinite time to become completely disintegrated. It is therefore usual to characterise all radioactive substances by a half-life period, i.e. the time required for the activity to be reduced by one-half, so that $N_0 = 2N_t$ and thus

$$\lambda t = \log_e 2$$

$$\text{or } t = \frac{0.693}{\lambda}$$

(For half-life periods of some common radioactive elements *see* Table 46.)

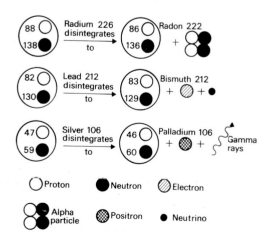

Fig. 127 *Types of radioactive decay.*

The decay constant is connected with the range (R) of alpha particles emitted during an α-radioactive transformation. The law (Geiger-Nuttal relationship) states that:

$$\log \lambda = A + 53.9 \log R$$

where A is a constant for any particular series, but varies from one series to another.

Radioactive isotopes. When a radioactive element loses an α particle its atomic number is reduced by 2 and a new element is formed. If this new element then loses two β-particles, the resulting element will have an atomic number the same as the original element but its atomic weight will be 4 units less. Thus two elements exist which would occupy the same place in the periodic table. (*See* Table 39.) Elements that have the same chemical properties and differ only in atomic weight are isotopes. In their atomic structure, isotopes have the same number of protons, but different numbers of neutrons.

The first isotopes to be discovered were among the radioactive elements (actually 230 thorium (ionium) and 232 thorium), but later the use of positive-ray analysis (q.v.) revealed that the majority of elements are mixtures of isotopes.

Artificial radioactive isotopes can be prepared by two general methods. The first consists in bombarding the element with fast-moving nuclear particles, and the second method is by the fission of atoms of certain elements, such as uranium and thorium. Under the first general method we include all those isotopes that are produced in a nuclear reactor. The inactive element (or one of its compounds) is placed in a suitable container—before insertion into the hole in the reactor. It should be ensured that other elements (or impurities) do not have a high neutron absorption. The principle of the technique is to bombard the element with the appropriate kind of particle. Sometimes the latter may enter the nucleus, producing a heavier nucleus, sometimes it may cause the displacement of another kind of particle from the nucleus. In the former case the nucleus gives out γ-radiation. The particles that are used are: (*a*) neutrons obtained from a nuclear reactor; (*b*) protons; (*c*) deuterons; or (*d*) α-particles which have been accelerated to high energies in a machine such as a cyclotron.

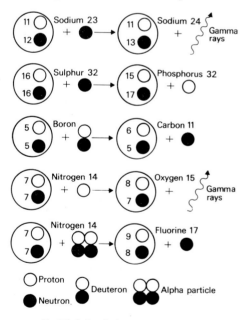

Fig. 128 *Radioactive isotopes: types of reaction.*

The reactions that occur can be listed as follows and are illustrated in Fig. 128: (*a*) the most generally applicable type is known as radiative capture (n, γ) type in which a neutron is absorbed, and γ-radiation emitted. e.g. the production of ^{24}Na from ^{23}Na; (*b*) the (n, p) type in which a neutron is absorbed and a proton displaced from the nucleus, e.g. the production of ^{32}P from ^{32}S. Since the product is different chemically from the parent substance, separation may be effected and the radioactive isotope may be obtained "carrier free"; (*c*) here the bombarding particle

TABLE 46.
Some Useful Radioactive Isotopes

Isotope	Half-life			Radiation (in MeV), approx. (1.6 × 10⁻¹³ J)		
				α	β	γ
Ra 222	Very long	1,590	years	4.79	—	0.19
Sr 90	Long	21.5	"	—	0.53	—
Co 60		5.3	"	—	0.31	1.17
Tl 204		2.7	"	—	0.77	—
Ca 45	100-200 days	152	days	—	0.25	—
Po 210		138	"	5.3	—	0.8
Ta 182		120	"	—	0.5	1.13
Sn 113		105	'	—	—	0.085
S 35	50-100 days	87	"	—	1.68	—
W 185		76	"	—	0.7	0.133
Ti 51		72	"	—	0.36	1.0
Ir 192		70	"	—	0.59	0.2
Zr 95		65	"	—	1.3	0.9
Sr 89		53	"	—	1.46	—
Fe 59	10-50 days	47	"	—	0.46	1.3
Hg 203		43	"	—	0.21	0.28
Ce 141		30	"	—	0.66	0.21
Cr 51		26.5	"	—	—	0.32
Rb 86		20	"	—	1.8	1.08
P 32		14	"	—	1.69	—
Ba 131		12	"	—	—	1.2
Ag 111	5-10 days	7.5	"	—	1.0	—
Bi 210		5.0	"	4.7	1.17	—
Mo 99	1-5 days	2.8	"	—	1.3	0.75
Au 198		2.7	"	—	0.96	0.41
Rh 105		1.5	"	—	0.78	0.33
Pt 197	Less than 1 day	18	hours	—	0.65	—
Na 24		15	"	—	1.38	2.7
Pd 109		13	"	—	1.1	—
Cu 64		12.75	"	—	0.57	1.2
K 42		12.5	"	—	3.5	1.5
Ru 105		4	"	—	1.5	0.75
Si 31		2.75	"	—	1.6	—

is a deuteron (consisting of one proton and one neutron)—the nucleus of the heavy isotope of hydrogen. The proton is absorbed into the nucleus and a neutron is liberated. This is the basis of the formation of ^{11}C from ^{10}B; *(d)* the (n, α) type in which a neutron is absorbed and an α-particle ejected, e.g. when ^{6}Li is converted into tritium, the isotope of hydrogen having an atomic weight of 3; *(e)* the (p, γ) type in which a proton is absorbed and γ-rays are emitted as in the production of ^{15}O from ^{14}N; *(f)* the (α, n) type in which an α-particle is absorbed and a neutron emitted, e.g. the production of ^{17}F from ^{14}N; *(g)* the (n, γ) type of reaction that is followed by β decay, e.g. the formation of ^{131}Te from ^{130}Te and the subsequent production of ^{131}I; *(h)* the Szilard-Chambers process in which when a (n, γ) reaction occurs, the atom emitting the γ-radiation is torn from its place in a chemical compound and may then assume a different chemical form, e.g. using potassium iron II cyanide as the target material, iron appears as iron III ^{59}Fe.

In the second general method the radioactive isotopes are secondary products in the fission process. In the case of uranium the nucleus may split in thirty to forty different ways, and more than sixty primary products have been detected, their mass numbers ranging from 72 to 158. About 97% of the ^{235}U nucleus yields products that fall into two groups; *(a)* a light group with mass numbers from 85 to 104,; and *(b)* a heavy group with mass numbers from 130 to 149. Arsenic and iodine are typical products, but both have short half-lives. Among the products with long half-lives are strontium 90 (half-life 21½ years), ^{144}Ce + ^{144}Pr (1½ years) and ^{137}Cs (33 years). (*See* Table 46.)

Radioactive tracers. A radioactive isotope used in a chemical reaction for the purpose of identifying and following the element through a series of chemical changes, e.g. the passage of gases through a blast furnace producing pig iron. A radioactive isotope has the same chemical properties as the corresponding non-radioactive element. No known process is available to accelerate or retard the radioactivity, and since extremely small amounts of the tracer element can be detected the technique has many useful chemical, biological and industrial applications. The detection of the radioactivity is made using equipment based on either a Geiger-Müller counter or a scintillation counter.

Radioactive waste. The main waste from nuclear reactors consists of fission products arising from the fissile material itself, but the handling of all radioactive materials automatically generates some secondary radioactivity, and any material which has passed through a nuclear reactor core or been in close proximity may be described as such. Low activity, i.e. a level of radioactive decay which does not give rise to high levels of radiation, can be disposed of fairly simply, and without danger. Solid material may be dumped in mines, quarries, or on the sea bed, provided that it is sealed against possible escape or corrosion by water or other solvents, or is situated out of reach of water supply which might otherwise be contaminated. Liquid activity could be similarly contained and dumped, or pumped into the sea at sufficient distance to disperse rapidly with a dilution factor so high that the levels of radiation are hardly above normal. Naturally, tidal and current effects have to be taken into account so that return to the shore before adequate dilution does not occur. Marine life, and to a smaller extent animal life in the case of terrestrial disposal, may digest some radioactive waste before adequate dilution. Some isotopes will be cumulative in the organism, rather than be excreted, and hence a secondary problem arises from the consumption of such organisms by other forms of life, leading to contamination of a biological chain which will nullify the effect of dilution by sea or other waters. Checks have therefore to be made of fish, crustaceans, seaweeds, and sea birds, for any possible accumulation of radioactivity in the vicinity of a nuclear waste dumping ground. On land, the problem was highlighted by the Windscale reactor accident of 1957, when radioactive fall-out contaminated grassland and was accumulated by grazing cows from whence it was passed into their milk. One cow covers a large area of grass in one day, so the very low level of waste falling on the grass became significantly larger in the day's milk yield, and consumption of that milk had to be discontinued for a period. Such effects are well understood, and rigorous tests are continually applied, so that the record of contamination measurement and accident prevention is better than for any other industry where comparable danger exists.

Medium-activity waste may have to be stored for a time until the level of activity has fallen to that which can be regarded as low level, and then disposed of as described. The time for storage will depend on the particular isotopes involved. High-level activity waste is that which has a very long half-life in decay, and from which the radiation is emitted at dangerous levels. This material has not only to be stored for very long periods, it has to be cooled, due to the heat energy given out during decay. This is mostly carried out by storage in corrosion resistant tanks, carefully welded and radiographically checked during commissioning, which in turn are buried in concrete of sufficient thickness to lower the level of any radiation reaching the outside to acceptable levels.

The most likely method of disposal in the future may well be the evaporation of liquid from chemically processed wastes, and the fixation of the solid remaining in a fused glass structure, which can resist corrosion and cannot be broken up into easily dispersed small particles again. The bulk will be enormously reduced by this process, and storage is envisaged in deep mines or drilled rock holes in geographically suitable areas, with no geological faults and good screening properties in terms of transmission of radiation. To give a correct perspective to the problem of the quantity of nuclear waste, if all the electricity generated in Great Britain were generated by nuclear power, the nuclear waste thus produced would amount in one year to a quantity equal to half an aspirin tablet per head of population. After 10 years' storage 99.9% of all radioactivity will have decayed, so the concentration of criticism of nuclear reactor problems is confined to 0.1% of the aspirin tablet per capita per annum.

Radioactivity. A property possessed by all the heavy elements (of atomic number greater than 83) of breaking down spontaneously into other elements by the emission of α, β or γ radiation. The process of emission of radiation is known as radioactive decay, and the rate of decay is characteristic and fixed for each radioactive element.

The nucleus of an atom consists of a number of protons and a number of neutrons, but the exact nature of the forces holding these together is not known. It is certain, however, that the nuclear particles (the nucleons) are in constant motion, and it may happen that one of the nucleons may escape providing it is possessed of sufficiently high energy to do so. For most of the chemical elements such an occurrence is virtually impossible and the nucleus is therefore said to be stable. In the case of the heavier atoms, however, the nuclei are unstable, and every so often the nuclear

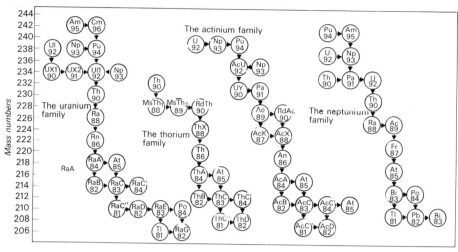

Fig. 129 *Stages in the radioactive decay of the uranium, thorium, actinium and neptunium families.*

forces are so disposed that the nucleus disintegrates. Such spontaneous disintegration is observed among elements of atomic number greater than 83 and the phenomenon is known as radioactivity.

The α-radiation consists of positively charged particles that are identified as the nuclei of helium atoms having a mass 4 and charge +2. β-rays consists of electrons or positrons. The γ-rays are electromagnetic radiation of the same nature as hard (short) X-rays. When a radioactive nucleus disintegrates with the emission of an α-particle, the nucleus loses 4 units of mass and 2 of electric charge and the atomic number therefore decreases by 2. When decay occurs by the emission of a β-particle, the loss of a negative charge (which is equivalent to a gain of a positive charge) results in an increase in the atomic number of one. Figure 129 shows the four radioactive families and the stages in the process of radioactive decay are indicated.

Radioautograph. The photograph of a mineral containing radioactive substance obtained by placing the mineral in contact with a photographic film and subsequently developing the latter in the usual way. In the finished photograph the halation produced tends to exaggerate the size of the radioactive particles in the mineral. The method can be used in appraising an ore containing, say, uranite. It is also used to discover the location of tracer elements in biological research. The method is also used in the case of metals and alloys to record the distribution of, say, phosphorus. Better termed autoradiograph.

Radiographic quality. For an X-ray tube this is defined as the load-carrying capacity per second, divided by the effective focus (w/f $\sin \theta$), where w = the load-carrying capacity per second (in kW), f = the actual area of focus in mm² and θ is the angle between the beam and the focal surface. The quality determines the sharpness of definition attained in a given time.

Radiography. The term used in reference to the use of radiant energy in the form of X-rays or γ-rays for the examination of opaque objects and the production of "photographs" for permanent record. It is used in non-destructive testing for the detection of flaws or unsoundness in metals.

Radio waves. Electromagnetic radiation of longer wavelength than visible light or infra-red radiation, of frequency about 10^3 Hz to 10^{12} Hz, often divided into general groups: short, medium and long. The advent of radar led to increased exploration of very short wavelengths, and the distinction between "microwaves" and infra-red radiation becomes largely irrelevant. The use of radio waves in communications is well known, but microwaves can also be used to transmit energy in localised volumes, as in heating devices for cookery, and this emphasises the continuous nature of the electromagnetic spectrum merging from radio waves into infra-red radiation.

Radium. A brilliant white metal usually obtained from pitchblende which latter is found in Canada (NW Territory), Austria, Zaire and elsewhere. It is also obtained from carnotite (a vanadate of uranium

and potassium) which is found in the USA (Colorado and Utah) and in the USSR. The radium, uranium and other metals which may be present are leached from the pulverised pitchblende with hot hydrochloric acid. A mixture of barium chloride and sodium sulphate is added to the solution to precipitate barium sulphate and lead chloride, together with the radium which separates out slowly. The lead chloride is removed by washing with brine and the barium and radium are separated by repeated crystallisation.

To obtain radium from carnotite the latter is heated with 80% sulphuric acid, filtered through glass wool and allowed to cool when barium and radium sulphates are deposited. These are soluble in concentrated sulphuric acid, and this property is used in separating them from other sulphates. The barium and radium sulphates are heated with carbon to form the sulphides, and these latter are dissolved in hydrobromic acid. Radium is now separated from barium bromide by repeated crystallisation.

Radium metal is obtained by electrolysis of the bromide in aqueous solution using a mercury cathode, and the radium is obtained by distilling the amalgam. The metal resembles barium and its salts resemble the corresponding barium salts. The dry salts will ozonise air, and they phosphoresce in the dark. Radium darkens on exposure to air owing to the formation of nitrides. Like the alkali metals, it rapidly decomposes water and it attacks glass and quartz. All the pure salts are white in colour and they evolve oxygen and hydrogen continuously in solution.

Radium (or its salts) are widely used in radiotherapy, and was formerly used in luminous paints for the dials of watches and scientific instruments, and in place of X-rays for the examination of metals.

Atomic weight 226.03, at. no. 88, sp. gr. 5, m. pt. 700 °C. The radium transformations are shown in Fig. 130. The half-life period

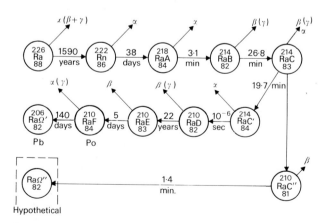

Fig. 130 *Radium decay.*

of Ra is 1,590 years and of RaD is 22 years. In the other cases the half-lives are short (or very short).

Radium sulphate mixed with finely divided beryllium powder and contained in platinum cylinders is normally used as a source of neutrons. (*See* Table 2.)

Radius ratio. Another term for the ratio of ionic or atomic sizes in determining the packing of units in formation of compounds of particular structure. The critical radius ratios are those concerned with closest distances of approach of anions and cations in forming typical cells such as the caesium chloride, rock-salt and fluorite structures. Figure 131 shows the ratios for compounds of AB type, and Fig. 132 for compounds of AB_2 composition. Note that as the discrepancy in size becomes more noticeable, the bonding becomes more covalent than ionic, and the structures become less close-packed. The hardness and melting points tend to rise as the compounds become more covalent than ionic, and they become more insoluble in water also.

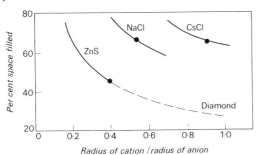

Fig. 131 *Radius ratio for compounds of AB type.*

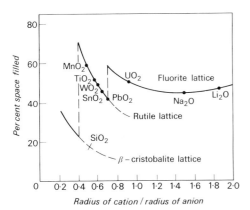

Fig. 132 *Radius ratio of compounds of AB_2 type.*

Radon. A gas evolved from radium as it slowly decays. The gas is colourless, and resembles the inert gases. It is distinctly soluble in water. Atomic weight 222.4, b. pt. $-62\,°C$, m. pt. $-71\,°C$. It is used in radiographic work, being suitable for use in the radiography of steel from 5 to 15 cm (2 to 6 in) thick. The gas is absorbed on a small carbon granule and the latter is sealed in a glass capsule. Mean energy 0.7 MeV. Half-life is only 3.8 days. (*See* Table 2.)

Rag. A term used in reference to forging defects that are the result of loose material having been driven into metal surface during the forging operation.

Ragged roll. *See* **Ragging.**

Ragging. The term used in reference to the series of well-rounded grooves cut into the surface of rolls to increase the limiting angle at which the rolls will bite the piece at entry. Since these grooves leave ridged impressions on the worked metal, their use is restricted to preliminary stages of rolling, e.g. blooming mills, billet mills, roughing stands, etc.

Rail mill. A rolling-mill used in the production of steel rails.

Rail test. A test formerly applied to steel rails, which consists in dropping a heavy weight from a specified height on to a rail supported at two points. The load and the height are specified and the maximum deflection is reported. Alternatively a dead load may be substituted for the live load.

Rake. A tool used for moving the fuel in a furnace or for removing ashes and clinker. It consists of a small rectangular plate attached to the end and at right angles to the length of the long bar or handle.

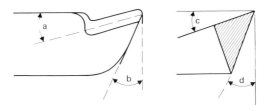

Fig. 133 *Rake on cutting tool.*

The term is also used in reference to certain angles to which certain faces of machine cutting tools are ground. In Fig. 133 *a* shows the top rake; *b* the bottom rake; *c* the top side rake; *d* the bottom side rake. The purpose of the rake is to enable the tool to take heavier cuts with minimum power consumption. To permit the tool to work without rubbing in the cut certain clearances are arranged. The design of the clearances must be carefully made because they must be adequate to permit the tool to cut freely. They should not be so large that the nose of the tool would vibrate in use. The cutting angle of the tool is the total angle made by the rake and the clearance angles. The higher the top rake used on a tool, the more readily it shears off the metal and the less power is consumed in the cutting. But a high top rake makes a weaker cutting angle and the tool is liable to break when used on high tensile materials. The weak angle is also less satisfactory in absorbing the heat produced by the cutting.

Rake-classifier. A type of mechanical classifier which consists of an inclined trough into which the ground ore is charged with water. The finer particles remain suspended and overflow at the lower end. The heavier and larger particles sink to the bottom of the trough and are conveyed by mechanical rakes to the top of the trough and are there discharged. It is frequently operated in close-circuit with a ball mill, to which the coarse particles are returned direct, the mill's products being fed to a point about two-thirds up the rake-classifier.

Raman spectra. When a beam of monochromatic light is passed through a transparent substance some of the light is scattered by the molecules in the substance, and appears as radiation of greater wavelength than the incident light. The frequency of the scattered light changes with that of the incident beam but the *difference* of the frequency between primary and scattered beam is independent of the frequency of the primary beam. If the wavelengths of the primary and scattered beams are v_1 and v_2, then the difference in energy is $h(v_1 - v_2)$, and this must be absorbed by a molecule in some quantum change. This difference corresponds to an infra-red frequency. The light, on being scattered by the molecules of the medium, loses (or picks up) one or more quanta of vibrational energy. If, for example, the light from mercury vapour is used there will be found, associated with each line of the primary beam, a line corresponding with an infra-red line characteristic of the spectra of the transparent material used. The incident radiation thus sacrifices part of its energy in an interaction with matter and reappears as a radiation of lower frequency. The phenomenon has been successfully used in the elucidation of the molecular structure of certain types of compound.

Rammers. Tools used by moulders in ramming or compressing sand round a pattern in a moulding-box. The work is carried out by the use of hand tools called "rammers", or power-ramming machines of the squeeze or jolt types may be used.

In sand-control tests the ramming must be standardised, and it is usual to employ a standard rammer, i.e. one having standard size and weight, which is dropped on the sand specimen (contained in a steel tube) a given number of times and from a given height.

Random copolymer. *See* **Copolymer.**

Randupson process. The name used in reference to a moulding process used in the production of bronze and ferrous castings, in which cement is employed as a bonding material for the moulding sands. A clean well-graded silica sand is mixed with about 10% of Portland cement and 4-6% water. The mixture is used as a facing sand about 2.5 cm (1 in) thick on a backing material of old crushed moulds and cores. One advantage is the elimination of the mould drying and stoving.

Raney nickel catalyst. A nickel catalyst prepared from Raney's alloy by the method detailed below. The catalyst is a highly active material and is used in the hydrogenation of a wide range of chemical compounds including benzene.

The Raney nickel catalyst may be prepared from the standard alloy (*see* Table 4) by treating this alloy with caustic soda at a certain temperature (25 °C or 50 °C, according to the process used) and then washing the catalyst—by decantation at 25 °C or by continuous washing at 50 °C. If the washing is carried out in the presence of hydrogen at 0.152 MN m^{-2} pressure, the catalyst is reputed to have a very high activity.

Ransburg process. The name used in reference to an electrostatic paint-spraying process used for metal finishing and metal protection. The "paint" particles acquire an electric charge as they pass through an ionised zone set up between two electrodes, and they are attracted towards the work. The process is said to give substantial reduction in the paint consumption and freedom from tears.

Raoult's law. A law discovered by Raoult (1887) which states that the vapour pressure of a solution is proportional to the mol-fraction of the solvent present in the solution. If N and n are the numbers of the molecules of the solvent and solute respectively, and if p_0 and p_s are the vapour pressures of the pure solvent and the solution, respectively, then the law is expressed:

$$\frac{p_0 - p_s}{p_0} = \frac{n}{N + n}$$

The law holds only for perfect solutions. Solid non-electrolytes in general obey this law, which can thus be used to determine their molecular weights; for if x g of solute of molecular weight: M_1 are dissolved in y g of solvent of molecular weight M_2, then

$$\frac{p_0 - p_s}{p_0} = \frac{x M_2}{y M_1}$$

(since x is always small compared with y).

Rapping. The process of loosening a pattern prior to its removal from the mould. This may be done by sticking a spiked rapping bar into the wooden pattern and knocking the bar from side to side with a striking bar. To prolong its life the pattern should be fitted with rapping and lifting plates. These should be firmly attached to the pattern, and should contain holes to receive the rapping bar.

Rapping bar. Usually a pointed iron bar that can be inserted into a pattern for the purpose of loosening it before withdrawal upon the mould. Usually a rapping plate in the pattern is provided to receive the bar.

Rapping hole. A hole formed in a foundry pattern or in a rapping plate attached thereto to receive the rapping bar. This latter is knocked from side to side with a hammer to loosen the pattern. Lifting screws are then fixed to the lifting plate and the pattern withdrawn vertically.

Rare earths. The title used in reference to the basic oxides of some fourteen uncommon metals included in Group III of the periodic table. The names and electronic structures of the metallic elements are given under rare-earth metals (q.v.).

The general formula for the oxides is R_2O_3, but in the case of the first member of the group, cerium, the most stable oxide is CeO_2. All the oxides are more basic than alumina. There are no large concentrations of the earths, which occur in relatively few locations. They are found mainly in Scandinavia, Brazil, Greenland, North America and Siberia. On exposure to electrons most of the rare earths exhibit very beautiful phosphorescence, which has been found to be due to traces of other substances.

These oxides are incorporated in colour television tubes as "phosphors". One, europium oxide, provides the brilliant red coloration.

The term "rare earths" was originally used to describe the oxides of the metals belonging to the lanthanide series, but is now used also to designate a second series of oxides known as the actinide series. These latter are all included in period 7 of the periodic table, the atomic numbers ranging, so far as is now known, from 89 (actinium) to 103. They are all radioactive.

The lanthanide series of rare earths are usually divided into two groups—the cerite earths, consisting of the oxides of cerium, lanthanum, praseodymium, neodymium, samarium and europium; and the gadolinite earths consisting of the oxides of gadolinium, scandium, yttrium, terbium, dysprosium, holmium, erbium, thulium, ytterbium and lutecium.

Rare-earth metals. A group of metals, referred to as a "transition group", occurring in Group III of the periodic classification. Their atomic numbers range from 58-71. A second series of rare-earth elements, known as the actinide series, also occurs in Group III, the atomic numbers ranging from 90 to 103. The lanthanide series occurs in Period 6 of the periodic table and the actinide series in the seventh period. From Table 13 it will be seen that for the lanthanide series, with the exception of the $4f$ level, all these elements have the same number of electrons in all the quantum levels up through the $6s$. This is a partial explanation of the chemical similarity of these elements, the separation of which is usually a difficult and tedious process. When some of the inner shells of an atom are incomplete there is a resultant moment which is large compared with the spin of the conduction electrons (in the outer shell) or with the diamagnetic moment of the closed shells. The incomplete $4f$ shell in the case of the lanthanide series is the cause of their strong paramagnetism. In the case of Dy and Ho the spin and orbital motion of the electrons on the four shells combine to produce the highest paramagnetic moment of any of the known atoms (*see* Fig. 134). The actinide series is usually regarded as including the metal thorium, but according to another view the series starts with uranium.

Also used as additives to steels and "superalloys" to improve oxidation resistance, phase stability and sintering characteristics.

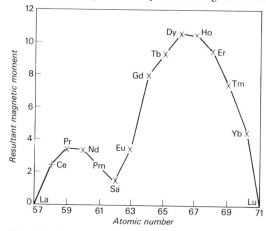

Fig. 134 *Relationship between magnetic moment and atomic number for lanthanides.*

Rare-earth minerals. The minerals containing the rare earths occur in only a few localities. Each mineral contains several of the earths, and the proportions are not constant in all specimens. Very broadly it may be stated that the elements lanthanum, cerium, praseodymium, neodymium, samarium are found in the minerals monazite, cerite and allanite; the elements europium, gadolinium and terbium are usually found in the minerals samarskite, gadolinite and in some xenotimes; the metals dysprosium, holmium, yttrium, erbium, thulium, ytterbium, and lutecium, are found in the minerals euxenite, fergussonite, gadolinite, and xenotime. Usually the minerals are very complex and a chemical formula is not known in all cases. The following formulae are commonly accepted as indicating the major constituents:

Cerite—$H_3(Ca,Fe)Ce_3 Si_3 O_{13}$.
Columbite (tantalite)—$(Fe,Mn)(Nb,Ta)_2 O_6$.
Fergussonite—$YNbO_4$.
Fergussonite (Australian)—$YTaO_4$.
Gadolinite—$(Fe,Be)_2 Y_2 Si_2 O_{10}$.
Keilhanite—$CaYt(Ti,Al,Fe)SiO_5$.
Microlite—$Ca_2(Ta,Nb)_2 O_7$.

Orthite—Al(OH)Ca$_2$(Al,Fe,Ce)$_2$(SiO$_4$)$_3$.

Yttrotantalite—Y$_4$(Ta$_2$O$_7$)$_3$.

Ratio of enrichment. A term used in reference to the concentration of ores to indicate the ratio of the percentage of valuable material in the original ore.

Reactor. Used by physicists in the context of a nuclear reactor, but by chemical engineers to designate any vessel in which a chemical reaction is performed. (*See* **Nuclear reactor.**)

Realgar. An important ore of arsenic having the formula As$_2$S$_2$. It is orange-yellow in colour, and the crystals, which belong to the monoclinic system, have a resinous lustre and are translucent. It often occurs with silver and lead ores. The specific gravity is 3.4-3.6 and Mohs hardness about 1.5.

The artificially produced disulphide is known in commerce as red arsenic glass. It is prepared from arsenical and common pyrites, but does not have a constant composition. Its use as a pigment has been discontinued, but it is still used (mixed with nitre) in the manufacture of Indian or white-fire. It is also used (mixed with lime) for the removal of hairs from skins in the tanning process.

Recalescence. When plain carbon steels containing varying percentages of carbon are cooled from about 1,000 °C it is found that at one or more temperatures the regular rate of cooling is interrupted. There occurs an evolution of heat, known as recalescence, which prolongs the period that would otherwise be required for the metal to cool through a given range of temperature. The temperature at which the arrests occur and the magnitude of the arrests depend on the carbon content of the steel and the rate of cooling. The temperatures at which retardation occurs during cooling are indicated by Ar_1, Ar_2, Ar_3. Since the changes are reversible there is an absorption of heat on raising the temperature of such steels. The points at which these changes occur are known as Ac_1, Ac_2, Ac_3. There is often a difference of temperature of as much as 30 °C between corresponding Ar and Ac points, and the phenomenon is referred to as thermal hysteresis. The range of temperature between A_1 and A_3 points is known as the transformation or critical range. The Ar_1 point for eutectoid-carbon steel (0.85% C) is about 730 °C and is the temperature at which the austenite forms on heating and transforms on slow cooling. It is the most pronounced of all the changes.

Recarburisation. The process of adding carbon, to the required extent, to purified or wrought iron during the manufacture of steel.

Reciprocal space. In real space, it is the periodic repetition of identical units which builds up the three-dimensional crystal lattice, the units being the unit cells. In considering the diffraction effects of crystal planes acting on the wave motion of the electron, a different concept, related to real space, but expressed in terms of vectors, simplifies the solutions to problems. The concept is that of reciprocal space, or reciprocal lattices.

Consider a primitive unit cell of sides a, b, c, in which the angle between sides a and b is γ (Fig. 135). The vector product of vectors a and b is a vector directed perpendicular to the *ab* plane, of magnitude ab sin γ, the area of the face bounded by sides a and b. The volume of the cell, base area times height, equals $a \times b$ multiplied by the interplanar spacing in the c direction. If side c of the unit cell is displaced from the vertical by angle θ, then the vertical height required, the c interplanar spacing, is c cos θ. A vector c^* is chosen, parallel to this vertical height, whose length is given by $c^* = 1/(c \cos \theta)$. The length c^* is thus the reciprocal of the interplanar spacing in the c direction. The scalar product of two vectors is equal to the product of their magnitude multiplied by the cosine of the angle between them. The product of c and c^* is therefore $c \cdot c^* = 1$. Equally, $c^* \cdot a = c^* \cdot b = 0$, since the vertical height is the c interplanar spacing and cos 90° = 0.

Similar arguments define corresponding vectors a^* and b^*, whose directions are perpendicular to the *bc* and *ac* planes of the lattice, and whose lengths are the reciprocals of the interplanar spacing. Construction of other primitive cells to which the crystal structure can be referred, indicates that for each set of parallel crystal planes of indices (hkl) in the lattice, there is a point distant from the origin represented by the vector translation ($ha^* + kb^* + lc^*$), and such points form a lattice themselves, known as the reciprocal lattice. The general vector r^*(hkl) from

the origin to the point (hkl) of the reciprocal lattice, is normal to the (hkl) plane of the crystal in real space, and the length of the vector r^*(hkl) is equal to the reciprocal of the (hkl) interplanar spacing in real space.

As the length of the vector c^* in the reciprocal lattice is equal to the reciprocal of the c interplanar spacing of the unit cell, the length of the vector $2 c^*$ in the reciprocal lattice must correspond to an interplanar spacing of $1/2 c$, and likewise spacings $1/3$, $1/4$, etc. must appear in the reciprocal lattice. The points in the reciprocal lattice give the directions of the planes and the interplanar spacings, so there exists a correlation with the Bragg conditions for reflection in the crystal. The reciprocal lattice point corresponding to an interplanar spacing of $1/n \cdot d$ indicates the condition of the nth order reflection from the real crystal planes spaced d apart.

There is a mutual reciprocity in the relationship between the real crystal and the reciprocal lattice. The vector r^*(hkl) from the origin of the reciprocal lattice is normal to the (hkl) planes of the real crystal, and the vector r(hkl) from the origin of the crystal lattice is normal to the corresponding plane of the reciprocal lattice and the length of the vector r(hkl) in the real lattice is equal to the interplanar spacing of the reciprocal lattice.

In non-cubic lattices, the cell of Fig. 135 indicates the relationships which need to be specified. In cubic lattices the vectors a^*, b^*, c^* of the reciprocal lattice are parallel to the vectors a, b, c of the real lattice, and the directions of the same indices [hkl] are parallel in both. The reciprocal lattice of the face centred cubic structure is in fact a body-centred cube.

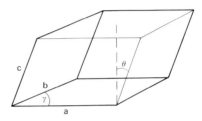

Fig. 135 *Primitive unit cell.*

When referring to the motion of electrons, considerations involve the probability wave function Ψ, and the associated wavelength λ is a vector in real space. The wave number $k = 2\pi/\lambda$, is a vector whose magnitude is related to the reciprocal of wavelength, hence k can be represented in reciprocal space, in view of the relationship with λ in real space. Reciprocal space is therefore often referred to as k-space when electrons are being considered, since three axes k_x, k_y, k_z correspond to wave motion parallel to the x-, y-, z-axes of real space. The axes x, y, z are usually chosen to be related to the axes of the crystal, in the case of the cubic system the values k_x, k_y, k_z then correspond to waves moving in the directions of the sides of the unit cell. The distance from the origin to a point A represents the value of k for the particular electron state and the direction OA is the direction of the wave motion associated with the electron state at A.

The restrictions on electron wave motion, including the conditions for Bragg reflection at crystal planes, can then be treated entirely in terms of the reciprocal lattice, and the geometrical constructions of Brillouin zones easily related to the crystal lattice. (*See* **Wave mechanical behaviour of the electron.**)

Recovery. A term used in process metallurgy to indicate the proportion of valuable material obtained in the processing of an ore. It is generally stated as a percentage of the total quantity of ore treated.

The term is also used to indicate the ability of a material to return to its original state or condition after being subjected to tension, compression, deformation or indentation stresses. The tests employed are generally known as recovery tests, qualified by the type of stress applied. For example, a compressibility and recovery test is a short-time compressive deformation test in which the value for a component is expressed as

$$100(t_3 - t_2)(t_1 - t_2)$$

where t_1 = thickness before compression, t_2 = thickness under compression and t_3 = recovered thickness after, say, 1 min.

A deformation-recovery test, generally specified for plastic materials, is a compression test carried out at an elevated temperature for a prolonged time (usually 3-10 hours).

Variations of this test are used; in some a standard load is applied, while in others the deformation is standardised—usually 5 mm. A recovery test for plastic materials is also employed in conjunction with a Rockwell hardness test. After the penetration has been read on the dial, the load is removed and the recovery read on the dial, after 15 seconds interval.

Recovery is used in metallurgy to denote the removal of residual stresses, caused by cold working or drastic quenching, by a low-temperature annealing treatment. When a metal has been under stress within the elastic limit, the return to the normal state is known as elastic recovery. It also has a specific meaning related to the coarsening of a dislocation network. One theory of recovery controlled creep envisages a steady state rate of deformation occurring when the elimination of dislocations by recovery is exactly balanced by the creation of new dislocations by a shear hardening process.

Recrystallisation. When a cold-worked metal, which possesses a microstructure consisting of very distorted crystals, is heated to a sufficiently high temperature, the distorted structure tends to revert to a strain-free state. New crystals are formed and grow until ultimately a new grain structure has completely replaced the old. When this has proceeded to completion the metal has lost all the improvement of mechanical properties that were conferred upon it by the cold working.

The recrystallisation starts from nuclei—points of high energy or high stress—which appear to form at grain boundaries, dislocation stress fields, twinning planes and different kinds of deformation markings. It tends to restore the properties which characterised the metal before the cold deformation, but as the final grain size differs from that of the original metal, the initial properties are not completely reproduced when the effects of the cold work are removed.

The temperature at which recrystallisation occurs is lowered as the degree of strain-hardening is increased and as the time of holding the metal at the "annealing" temperature is prolonged. The final grain size in the recrystallised metal depends mainly on the extent to which the metal was originally deformed—the greater the degree of deformation, the smaller the crystal size. It is also influenced by the temperature of the "annealing" treatment and by the period for which the metal was held at this temperature.

Recrystallisation temperature. The lowest temperature at which the distorted grain structure of a cold-worked metal is replaced by a new strain-free crystal structure, during prolonged annealing. The purity of the metal, the degree of prior deformation and the time of annealing are all factors determining this temperature which is not an inherent property of the metal.

Rectification. A term used in different sciences with quite different meanings. In mathematics it refers to the measurement of the length of a curve; in chemistry it is used to describe the process of purification of a liquid by distillation. In electrical science rectification is used in reference to the processes by which an alternating current is converted into a direct current. Commercially, over 90% of the electrical energy is supplied as a.c. and for long-distance transmission a high voltage is more economical than low voltage. Only a.c. admits of transformation so that high voltages can be transformed to the required voltage (with or without regulators), and the current so transformed can be rectified, and so made available for electrolytic metal refining and for electroplating, etc.

Rectifier. A device for the direct conversion of an alternating current into a unidirectional wave. An early method used an evacuated tube with a heated cathode and an anode (diode valve) such that electrons could only flow from cathode to anode when the anode was positive. Whilst suitable for radio circuitry, power rectification required a much more robust system, and the mercury arc rectifier supplied this need. In this, a pool of mercury forms one electrode, and an iron rod the other. Mercury vapour is always present in the evacuated space, and an arc discharge occurs through the vapour, which passes current only from mercury to iron. Due to the large number of current carriers in the vapour, heavy currents can be used.

Modern devices utilise the semiconductor, although some of these were in use in the early days of radio. If the Fermi level of a metal is below that of an n-type semiconductor, contact will

cause electrons to flow into the metal until the levels are equal. The metal will now have excess electrons, which repel electrons from the surface of the semiconductor, leaving positively charged ionised donor atoms which neutralise the effect of this excess negative charge on the metal. The energy levels in both valence and conduction bands will be increased at the surface of the semiconductor as a result, and this constitutes a potential barrier, whose height equals the difference of the two work functions. This barrier impedes the flow of electrons and if a potential difference is applied, the metal remains at constant potential and the gradient exists in the semiconductor. The current voltage relationship is determined by the semiconductor. One contact, the metal, can take the form of whisker material, giving a fine point contact, as with germanium n-type semiconductor material and fine whiskers of platinum, gold or tungsten. These are used for low current devices.

If the semiconductor is of p-type material, then the same result applies if the work function of the metal is smaller, i.e. the Fermi level is greater, but this time the potential barrier impedes the flow of positive hole carriers.

The p-n junction (q.v.) formed by two semiconductor materials also acts as a rectifier. The type of rectifier used depends upon the application. The use of copper-copper oxide and selenium rectifiers was restricted to low frequency power conversion, acting as metal-semiconductor contact systems. For higher frequencies, the point contact or whisker type was common in the early days of radio, and referred to as a crystal detector. Silicon and silicon carbide are preferred to germanium in p-n junction rectifiers, because a material with a wider energy gap can operate at higher temperatures without too great an increase in the saturation current, which decreases the efficiency of rectification. (The semiconductor basic material is treated with appropriate additions to create the junction.) These semiconductor devices are far more reliable than the valve type and can be mass produced in very small sizes.

Recuperator. The name used in reference to those devices that can be used for recovering heat from hot spent furnace gases, and using it to preheat air or fuel. The incoming gases pass through a series of pipes enclosed in a chamber through which pass the outgoing hot gases. The heat exchange takes place through the walls of the pipes (compare with regenerator).

Red copper oxide. The oxide having 88.8% Cu and of the formula Cu_2O. It occurs in a variety of forms in the zone of weathering or gossan of copper lodes in Canada, France, Spain, Chile, Peru and Australia. Also called cuprite (q.v.).

Reddle. The most earthy form of hematite (Fe_2O_3), amorphous, red in colour. It is used in the manufacture of crayons and paints and as a polishing medium for glass.

The term is used in reference to the red lead used in the form of a paste in connection with the preparation of surfaces.

Red hardness. A term used in reference to the relatively high hardness that is retained up to a low red heat by high-speed steels and other cutting materials.

Red lead. The red oxide of lead having the formula Pb_3O_4. It is found rarely in nature, but may readily be prepared by heating massicot or white lead. It is scarlet in colour and crystalline, and when heated to about 470 °C it loses oxygen and is converted into litharge. Red lead is used largely as a paint and in the preparation of flint glass. Also called minium.

In the workshop red lead is used in the accurate preparation of surfaces, bearing surfaces, etc. When mixed with boiled oil it makes an excellent jointing material for steam joints.

Red metal. A name by which any kind of brass containing not less than 80% Cu is known.

Red mud. The residue, high in iron oxide content, that is obtained during the Bayer process for the purification of alumina from bauxite. The name is also applied to the reddish-brown deep-sea mud that is composed of wind-blown ferruginous dust deposited in the sea.

Redonda phosphate. The commercial name for aluminium phosphate $AlPO_4$, which is found in the West Indies. It contains from 35% to 40% P_2O_5.

Red oxide of zinc. An oxide of zinc ZnO containing impurities, e.g. manganese oxides, which are the cause of the coloration. The zinc oxide when pure is white (zinc white). The coloured oxide is frequently found with other zinc minerals.

Redox indicators. These are mainly organic colouring matters that can exhibit different colours in the oxidised and reduced state. They are used as indicators in oxidising-reducing reactions. Each indicator has its own characteristic colour change and its own standard e.m.f., i.e. a fixed potential, which is usually measured against a standard hydrogen electrode.

Redox potentials. The oxidising-reducing solutions referred to as redox solutions are reversible systems and the potential acquired by a platinum electrode immersed in such a solution depends on whether the solution is in the reduced, or in the partly reduced and partly oxidised, or in the oxidised state. The standard oxidation-reduction potential is that at which the solution is 50% oxidised.

The potential at any other concentration is found from the formula:

$$E = E_0 + \frac{2T}{n \times 96,500} \times \log_e \frac{C_1}{C_2}$$

where E_0 = standard potential defined above, T = absolute temperature of the solution, n = difference in the valency in the oxidised and reduced state, C_1 and C_2 = concentration of the material in solution in the oxidised and reduced state, respectively.

Redox reactions. Oxidation reactions are those that are accompanied by the liberation of electrons and an oxidising agent is one which reacts in such a manner that it takes up electrons. A solution can therefore be regarded as an oxidising or a reducing solution, according to whether it is capable of gaining or losing electrons. The oxidising-reducing reactions that occur with reversible systems in which a substance in both the oxidised and reduced states coexist are called redox reactions.

Red precipitate. The red oxide of mercury HgO. It may be prepared by heating mercury in air at temperatures below its boiling point. In large quantities it is prepared by heating a mixture of mercury and mercury II nitrate. The oxide so formed is bright red in colour and has a specific gravity of 11.14. The oxide produced when caustic soda is added to a mercury II salt is yellow in colour but identical in composition. The difference in colour is attributed to their different crystalline forms and to the difference in the size of the particles of the two varieties.

Redruthite. A copper ore consisting of copper sulphide Cu_2S containing about 80% Cu. Also known as copper glance.

Red short. Synonymous with hot-short (q.v.).

Red silver ore. A double sulphide of silver and antimony having the formula $3Ag_2S.Sb_2S_3$. Also known as pyrargyrite.

Red stain. A stain formed on the surface of brass when zinc volatilises during annealing or when copper is redeposited during pickling.

Reducing atmosphere. A gaseous atmosphere which may be present during or which may take part in a chemical reaction, and which in the latter case will react with oxygen, particularly that present in oxides, causing the reduction of them. The most commonly used gases are hydrogen and carbon monoxide. When the supply of oxygen in a reaction is only just sufficient to maintain the combustion, the atmosphere may be a reducing one. The use of carbon monoxide in the reduction of iron oxide is referred to as indirect reduction in contradistinction to the direct reduction with carbon.

Reduction. A term originally used to indicate the removal of oxygen from a compound, but now extended to include all those cases which involve a decrease in the positive charge on an ion (or the increase in the negative charge). If, for example, a solution contains both ferrous (iron II) and ferric (iron III) ions and a platinum plate is dipped into this, the plate will become negatively charged if the solution seeks to give up electrons—in other words, it is a reducing solution and the reduction potential can be measured if such an arrangement is connected to a normal hydrogen electrode. The principal reducing agents are hydrogen, carbon monoxide, sulphuretted hydrogen, ammonia, hydriodic and arsenic III acids, most lower oxides and a number of salts in their lower state of valency, e.g. iron II salts, tin II salts.

Reduction cell. The term used in reference to the apparatus in which treated ores (e.g. oxide, sulphate or cyanide) are reduced electrolytically to produce the metal. Though the term is used in reference to the production of zinc, and of copper from low-grade ores, it is particularly employed when referring to the reduction of bauxite in fused cryolite in the reduction of Al_2O_3 to metal.

Reduction in area. A term used in tensile testing to define the difference between the original cross-sectional area of a test piece and that of the smallest area at the point of fracture. It is usually stated as a percentage, e.g. $100 (A_1 - A_2)/A_1$, where A_1 = original cross-sectional area and A_2 = minimum cross-sectional area. It is a measure of the ductility of a metal, but is not so widely used as the percentage elongation. It is not used in the case of sheet specimens. The values of the reduction in area for any material are affected by its ultimate tensile strength, and by the ratio of the U.T.S. to the yield point.

When the size of the test piece conforms to the formula $L = 4\sqrt{A}$, where L = the gauge length and A = the cross-sectional area, the values of the reduction in area are comparable with the elongation that occurs, irrespective of the actual dimensions of the test piece.

Redulith. A group of lithium-containing alloys used for degassing and modifying lead-base alloys.

Redux process. A metal-bonding process based on the use of a solution containing phenol formaldehyde (hydroxybenzene-methanal) and a polyvinyl (polyethenyl) resin in powdered form. Bonding of the metallic surfaces is performed under pressure at 140-180 °C in order to cure the resin (at about 145 °C the process usually takes 15 minutes). The process is also used for joining metals to plastic materials and to wood.

Reed. A localised internal discontinuity containing non-metallic inclusions sometimes found in steel sheet and strip.

Reeking. A term applied to an ingot-mould dressing process in which soot is deposited from burning tar on to the mould faces.

Reelers. A term used in rolling flat products such as strip metal in reference to the reels that are employed to coil up the metal as it leaves the rolls. Power reelers are also employed to apply tension to the metal as it is rolled.

Reeling. A rolling process used in finishing steel or other tubes that consists in light cross-rolling on a mandrel. The process slightly increases the tube diameter, while giving a smooth finish to both the inner and outer surfaces. The tube, while still hot, is passed on a mandrel between two barrel-shaped rolls rotating in the same direction. The rolls are mounted at a slight angle to the direction of progress of the tube—i.e. slightly inclined to the longitudinal axis of the tube so that the tube is advanced over the mandrel.

The term is also used in reference to the straightening of steel bars that have been produced by reheating the ingots and hammering or rolling them into bars.

Refined iron. A term commonly used in reference to wrought-iron bars, produced by rolling the squeezed bloom, which are re-rolled in a pile of such bars.

Refining. A term commonly used in reference to the removal of undesirable elements, oxides, etc., and of gases from molten metal to improve the purity of the metal. The process may involve the removal of the unwanted substances by slag or by other reactions in which oxidisers or scavengers may be introduced into the molten metal.

The term may also include electrolytic or electro-refining in which process the impure metal is made the anode in a suitable aqueous electrolyte and the pure metal is deposited on the cathode. The various refining techniques are usually referred to as: (a) fire-refining in which the unwanted element is separated by oxidation and removed in the slag or by precipitation as an insoluble material; (b) liquation (q.v.); (c) distillation; and (d) electro-refining.

Refining heat. A heat-treatment process used to refine the structure, particularly the grain-size, of a metal. In the case of steel, the temperature required is usually just a little above the Ac_3 point.

Refraction. The change in direction of propagation of a wave-motion or any kind of radiation. This may be caused when the radiation passes from one medium to another of different temperature. Sound, heat and light waves suffer refraction at the surface of separation of two media of different densities. A stream of electrons is "refracted" in passing through a field of varying magnetic or electrostatic intensity.

Refractive index (absolute). This quantity may be defined as the ratio of the speed of the heat, light or sound waves in free space to the velocity in the given medium. If the wave-front is not perpendicular to the surface of separation of the two media the direction of propagation will usually be changed on passing from medium *"a"* to medium *"b"*. If the two directions, measured with respect to the normal to the surface are θ_1 and θ_2, then the index of refraction is given by

$$_a\mu_b = \sin\theta_1/\sin\theta_2 = v_1/v_2$$

where v_1 and v_2 are the velocities in the media *"a"* and *"b"* respectively.

When light falls upon the atoms of any substance, they emit light waves that have the same frequency as the incident light, and that have a definite phase relation to it. In transparent materials, the incident and emitted light are exactly in phase.

It is assumed that refractive effects are due to interference between the incident and the scattered light, and since the atom is small compared with the wavelength of ordinary light, the scattered wave will depend for its characteristics (e.g. frequency, polarisation) on the incident wave. When the incident light has an amplitude E, the scattered light will have an amplitude kE, where k is the scattering constant of the atom. This constant is related to the refractive index.

The optical properties of crystals are of considerable importance in identifying minerals, especially in the absence of any indications of external crystalline form. According to their action on plane polarised light, crystals are divided into several classes.

(a) Optically isotropic crystals of the cubic system that have only one index of refraction for each wavelength of light. The developed crystals are circularly polarising, but otherwise the substances have no effect on polarised light and thus cannot be distinguished optically from amorphous substances (e.g. glass, opal, etc.).

(b) Optically uniaxial crystals that belong to the tetragonal and hexagonal systems. (These two systems are not optically distinguishable.) The principal crystallographic axis is also the optic axis, and light traversing the crystal in this direction is singly refracted. Otherwise the crystals are double refracting.

(c) Optically biaxial crystals have three principal axes and three indices of refraction corresponding to these. If α, β and γ are the indices of refraction, γ corresponding to the principal optic axis, then $\gamma > \beta > \alpha$, and for optically positive crystals $(\beta - \alpha) < (\gamma - \beta)$, while for optically negative crystals, the reverse of this is true.

It is usual to subdivide this class into:

(1) the orthorhombic system in which the three crystallographic axes coincide with the optic axes;
(2) the monoclinic system in which one crystallographic axis coincides with the principal optic axis;
(3) the anorthic system in which there is no relation between the crystallographic and the optic axes.

The refractive indices of some laboratory materials are given in Table 47.

TABLE 47.
Refractive Indices of Laboratory Materials

Substance	Refractive Index	Substance	Refractive Index
Air	1.00029	Hydrogen	1.000138
Aniline	1.59	Ice	1.31
Benzene	1.504	Methyl alcohol	
Canada balsam	1.53	(methanol)	1.33
Carbon tetrachloride		Methyline iodide	
(tetrachloroethane)	1.46	(diiodomethane)	1.74
Ethyl alcohol (ethanol)	1.36	Mica	1.56-1.60
Fused silica	1.459	Oil — Castor	1.48
Glass—Crown	1.157	Cassia	1.60
Silicate	1.519	Cedar	1.516
Borosilicate	1.509	Clove	1.532
Barium	1.540	Cinnamon	1.601
Dense barium	1.612	Olive	1.46
Flint (lead)	1.649	Paraffin	1.44
Silicate	1.578	Pentane	1.357
Borosilicate	1.575	Potash alum	1.456
Barium	1.640	Sulphuric acid	1.43
Dense	1.740	Trichlorethylene	
Jena	1.52-1.65	(trichloroethene)	1.478
Glycerine		Turpentine	1.47
(1, 2, 3, propanetriol)	1.47	Water	1.33

Refractolith. A refractory material consisting essentially of graphitised aluminium silicate, used in association with clay and silica sand for lining ladles, cupola runners, etc.

Refractoriness. A term usually employed to give an indication of the ability of a material to resist the effects of exposure to high temperatures, but which also frequently implies ability to resist the effects of furnace atmospheres, contact with slags and abrasion. Assessment of refractoriness may be made by compressing a sample of the material into the shape of a cone of the same size as the standard pyrometric cones. This is then inserted into a furnace along with a set of the standard cones. The standard cone that softens and bends to the same extent as the cone under test gives the assessment in terms of the designating number on the standard cone.

Refractory. A material that may be used in the construction of the interior of a furnace where it is required to withstand: *(a)* the effects of high temperature so that it will not fuse or soften in the furnace nor fracture with changes of temperature; *(b)* chemical action of the hot charge and slag in furnace, or of the furnace gases; *(c)* the scouring action of the fuel, furnace charge or high-velocity gas; *(d)* mechanical strains due to weight of charge. In selecting a refractory material for a particular use, regard must be had to the texture, the crushing strength and the chemical composition of the material. The refractory material is not now used as a flux, and hence it is important to ensure that the slag and the furnace lining do not interact to any extent.

Refractories are usually divided into three classes: acid, basic and neutral. The acid materials contain considerable proportions of silica and combine readily with basic oxides. Practically all fireclays, and all the refractories made from sand, flint, ganister and Dinas rock are acid. When silica is practically absent and the refractory consists mainly of oxides of metals it has basic characteristics. In neutral refractories either the material lacks basic or acidic properties, as in the case of graphite, or the acidic and basic properties are about equally balanced, as in the case of chromite. Hydrated silicate of alumina (clay) containing an excess of silica, together with small amounts of lime, soda or potash is the most important of the acid materials. In the form of fireclay, it can be softened with water and moulded to desired shapes. It contracts considerably on firing.

The basic refractories most used contain lime, or magnesia or or alumina. Dolomite—a double carbonate of calcium and magnesium—is calcined, crushed and bonded with tar for the making of basic bricks. Bauxite bricks are bonded with fireclay or lime and are more refractory than magnesite bricks.

Synthetic refractories include graphite coke and silicon carbide. The composition and properties of common refractory materials are given in Tables 48 and 3.

TABLE 48.
Compositions of Common Refractory Materials

Refractory	Composition %					
	CaO	MgO	Al$_2$O$_2$	SiO$_2$	K$_2$O	FeO
Calcined limestone	95.7	1.2	0.9	1.9	—	—
" magnesite	1.8	94.4	2.3	0.6	—	—
" dolomite	50.5	33.1	2.8	3.7	—	—
Quartz sand	—	0.5	1.0	98.5	—	—
Ganister	0.6	—	1.5	94.6	—	1.0
Fireclay	1.0	—	30.0	56.7	0.6	1.5
Silica brick	1.7	—	1.0	96.0	—	1.0
Firebrick	1.0	0.3	31.0	65.0	0.7	2.0
Dinas rock	1.0	—	1.0	95.0	—	1.5

Refractory cement. A refractory mixture, used in the wet or plastic state for making and sealing joints and as a covering material. There are several varieties, but all of them consist basically of a non-plastic refractory grog mixed with a plasticiser such as clay. Usually the refractoriness is developed only on heating the mixture, but if water-glass, phosphoric acid or Portland cement be added the cement will harden sufficiently on air-drying.

Refractory metals. A term used somewhat loosely in reference to metals with very high melting points or that are highly resistant to heat. The following metals have melting points above 2,000 °C: iridium, niobium, osmium, molybdenum, tantalum, tungsten. The last three named are important commercially. When alloyed with chromium, nickel or cobalt they produce what are sometimes known as refractory (or heat-resisting) alloys.

Refrex. The trade name of a refractory material that consists essentially of silicon carbide (carborundum) SiC.

Regeneration (of cellulose). The staple, or fibre length, of many natural cellulose products (such as cotton) is short and the diameter often variable. The joining of such fibres to make a yarn is then a more difficult operation. By dissolution of the cellulose, impurities can be separated, and the fibre processed into very long filaments of uniform diameter. The basis of rayon manufacture was a regeneration of cellulose, using one or other of several possible solvents, and a variety of end products. (*See* **Cellulose.**)

By this means, sheet and film cellulose can also be produced.

Regenerative quenching. A heat-treatment process used to refine the structure of the case and the core of carburised metal. The first quenching, which is from a high temperature (880-900 °C) refines the core which is relatively low in carbon, and the second treatment from a lower temperature, 760-780 °C, hardens and further refines the carburised case.

Regenerator. A device for utilising the heat from hot spent furnace gases. The hot gases are used to heat a brick checkerwork over which blast or fuel gas is passed so that the latter become heated in turn. Regenerators usually operate in pairs, one being heated up while the other one is serving to heat the fuel or air feed to the furnace.

Regular copolymer. *See* **Copolymer.**

Reguline deposit. A term used in reference to the metallic deposits of good quality (in contradistinction to spongy deposits) obtained by electrodeposition.

Regulus. A term used in reference to the product of smelting of various ores, and consisting of metal in a still impure state. In the separation of metals such as silver, lead or copper from the mineral the purer or metallic part which sinks to the bottom of the crucible or furnace and is thus separated from the rest of the material is known as a regulus.

Regulus of Venus. An alloy consisting of equal proportions by weight of copper and antimony. It forms a beautiful purple alloy to which the formula Cu_2Sb has been given.

Reheating. The process of heating a metal to the temperature at which it can be worked. The treatment may be used either as an additional measure or to correct a previous operation.

Reinforced concrete. The structure of silicates leads to an "open" type of atomic packing, with variable sizes of ions or atoms present, as described under **Quartz, Crystal classification, Glass,** etc. Such material has very high compression strength, since the atoms are then forced together, but much lower tensile strength, owing to the ready formation of cracks which propagate under elastic stresses at the tip.

For constructional purposes, beams are invaluable as horizontal members to take bending loads, as in floors or decks. Concrete beams make good pillars, (in compression) but have to be very thick as cross-beams (due to the tensile stresses on the lower side when loaded from above). To improve the bending stress, steel rods or wire mesh can be inserted in the concrete before setting, and then the steel takes up the tensile stresses, compressing the concrete in the interstices as it does so. Reinforcement can be carried out with other materials, and even finely chopped glass fibre can be incorporated, to accept tensile stresses locally as microstress.

If steel bars are inserted with the ends protruding from beams, a tensile stress can be applied to them before the concrete is poured, and maintained during the setting. When the tensile stress is released, the steel contracts and so compresses the concrete in the beam. Bending loads applied to this pre-stressed beam have to exceed the compression stress in the concrete on the tension side of the bend, before the concrete is itself subject to tension, and then the steel reinforcement also assists. By this means, pre-stressing allows reduction in beam size for a given load, or the use of concrete beams for higher loads or greater span.

Relative valency effect. An expression of the fact that in solid solutions a solvent of low valency is more likely to dissolve a solute of lower valency than the reverse.

Relative viscosity. The ratio of the viscosity of a liquid to that of water. It is given by the formula $\eta_0 = \eta_1 dt/d_1 t_1$ where η_0 is the viscosity of water and d, d_1 are the densities of the liquid and of water respectively and t and t_1 are the times taken for the liquid and for water, respectively, to flow in a viscometer between two marks on the tube. (*See also* **Viscosity.**)

Relaxation. A term implying the process of restoration of equilibrium following perturbation. A dielectric suffers relaxation when the alternation of an electric field reaches a frequency such that the dipoles cannot reverse their polarity in resonance with the field. This leads to an apparent reduction in the permittivity of the material. The frequency at which this occurs is the relaxation frequency.

Alternating stress of high frequency produces similar effects in the displacement of atoms, and the time required for the amplitude of vibration to decrease to 1/e of its original value (e is the natural logarithmic exponent) is called the relaxation time. The distance in the material over which this effect takes place is called the relaxation distance.

Relaxation as a process is described in mechanisms of mechanical stressing to indicate plastic deformation, the conversion of strain energy into atomic movements. By this means, stress relief occurs when a material under strain is raised in temperature, the creep which results describing the deformation process.

Relay. A term used to describe an auxiliary device or machine that is used for controlling a more powerful machine. In electrical work the device usually incorporates an electromagnet in the controlling circuit so that a weak current flowing in such a circuit can open or close a second circuit and thus control the switching on, and off, of a heavier current in the second or controlled circuit.

Relief. A term used in reference to the removal of stress in metal that has been produced as a result of rapid cooling or of mechanical working. It is usually effected by heating the material to a temperature below the critical range followed by slow cooling.

Relief polishing. A method of preparing micro-specimens that was originally introduced by Osmond. It is based on the circumstance that the various constituents of an alloy have different resistances to abrasion. The specimen is polished with a suitable medium on cloth, rubber or parchment, which rests on an elastic support. The more elastic is the support the more pronounced will be the relief effect. If the relief effect is to be avoided it is advisable to use a cloth without a nap. To supplement the action of the polishing medium a suitable dilute etchant can be used with the abrasive.

Reluctance. A term used in connection with a magnetic circuit to indicate a quantity that is analogous to resistance in an electric circuit. In a material of length l, area of cross-section a and permeability μ, its reluctance is given by $l/a\mu$. The total reluctance of a number of reluctances in series (as in the case of a field magnet circuit) is the sum of the separate reluctances.

$$\text{Reluctance} = \frac{\text{Magnetomotive Force}}{\text{Magnetic Flux}}$$

Remanence. The residual magnetism of a ferromagnetic substance that has been subjected to a hysteresis cycle, when the magnetising force has been reduced to zero.

Remolinite. A hydrated oxychloride of copper $CuCl_2.3Cu(OH)_2$ which occurs in the zone of weathering of copper lodes at Los Remolines and elsewhere in South America, Cornwall (UK), Spain and South Australia. Mohs hardness 3, sp. gr. 3.7. Also called Atacamite.

Renierite. A germanium ore consisting of germanium sulphide and containing about 8% of the metal. It is found in Zaire.

Reniform. A term meaning kidney-shaped, and used in reference to hematite that sometimes occurs in this indeterminate form of aggregation, having rounded surfaces that resemble kidneys.

Rensselaerite. A variety of talc or soapstone which is a hydrated magnesium silicate. It has a rather fibrous structure and can be used for ornamental articles. It is found in the USA (e.g. New York State).

Repeated bend test. A test used to assess the ductility of relatively ductile metals, and is mainly useful for metals in the form of sheet or strip. The test piece, 15 cm minimum length and 2.5 cm wide, is clamped at its lower end in a rigid vice. The other end is held by a moveable jaw under sufficient spring tension to localise the bend. The test consists in bending the specimen through 90°; then it is subjected to bends of 180° in opposite directions until a crack (or rupture) occurs. The result is reported as the number of complete bends, including the first one, before failure.

Repeated blow test. A test designed to assess the ability of a metal to withstand repeated impacts without failure. The specimen in the form of a rod is freely supported at two points and subjected to repeated blows of gradually increasing energy until fracture occurs. There are standard sizes of specimens having diameters approximately 20, 30, 50 mm, and they are supported with spans of 10, 15, 30 cm. The smallest specimen having the shortest span is tested with a weight of 7 kg; the largest specimen at a span of 30 cm is tested with a 22 kg weight, and the other specimen at a span of 15 cm is tested with a 11 kg weight. The first drop is 5 cm and each subsequent drop is increased by 2.5 cm. The result is reported as the last drop (in centimetres) prior to fracture.

This test, as distinct from the single-blow impact tests, e.g. Izod, which measure the total energy absorbed in a sudden fracture, measures the energy needed to produce fracture after most of the plastic deformation has occurred.

Rephosphorisation. A term sometimes used in reference to basic steelmaking to indicate a reversion of phosphorus from the slag to the metal.

Replica technique. *See* **Negative** and **positive replica.**

Repoussé. The art of ornamenting thin sheets of metal by hammering out designs from the back of the sheet. It was practised by the Egyptians and Etruscans and was very well advanced and perfected in the sixteenth century. A large number of hammers of different weights and sizes are the main tools. The raised portions are usually finished by gravers or chasers. The metals mainly used have been gold, silver, copper, tin and lead.

Repulsion. *See* **Attraction force in atoms.**

Residual magnetism. Synonymous with remanence (q.v.).

Residual stress. The term used in reference to the macroscopic stresses set up within a material in consequence of non-uniform plastic deformation which may be the result of cold working or of steep temperature gradients during fabrication, casting, heat-treatment or welding.

The stress may reach or exceed the yield strength of the material, but being non-uniform, plastic deformation does not necessarily occur except very locally and slowly on a microscopic scale. It is the disturbance of the residual stress balance which can produce failure, since the addition of operational or constructional stresses to an already stressed material upsets the balance of atomic forces, and can lead to triaxial stresses at the tip of cracks or in voids formed in the material by plastic deformation or in the vicinity of discontinuities such as inclusions or similar defects.

A state of residual stress is not a desirable condition for any engineering material, as it can hardly be accounted for in the design of a component. The safest course of action is to remove it, by stress relief annealing, causing relaxation and creep. The plastic deformation may well lead to distortion, especially of a complex structure, and support may have to be arranged to minimise this.

Heat treatment is the most reliable method of stress relief, but compression in areas of tension can be used, where the component is of simple shape, by peening or shot blasting. Other treatments by use of low temperatures, or the metallurgical mythology of "weathering" outside for castings, should be treated with caution.

Resilience. The term used to denote the work done in stretching a bar up to its elastic limit, and is a measure of the elastic energy possessed by the material. It is equal to

$$\frac{f^2}{E} \times \tfrac{1}{2} \text{ volume of the bar}$$

where f = stress at the elastic limit and E = modulus of elasticity of the metal. The quantity f^2/E is known as the coefficient or modulus of resilience.

Resillage. A surface defect which appears in the form of a dense network of the fine cracks on rolled aluminium. Also referred to as fissuring.

Resin. A term used for some polymeric materials, particularly those which can be cross-linked.

Resistance. A term used in reference to a metal to define the property by virtue of which it opposes the flow of electricity through it. Materials having a low or very low resistance are in general termed conductors.

The unit of electrical resistance is the ohm, the resistance between two points of a conductor when a constant difference of potential of 1 volt, applied between these points, produces in the conductor a current of 1 ampere, the conductor not being the source of any electromotive force.

The resistances of common metals are given in Tables 2 and 3.

Resistance alloys. The name used in reference to those alloys which have a high specific electrical resistance. The term is also used in reference to those materials that have high resistance to oxidation at elevated temperature. The commonest of the alloys are the nickel-chromium and the iron-nickel-chromium types, and to a lesser extent the nickel-manganese alloys. The composition of a number of commonly used electrical resistance alloys is given in Tables 5 and 7.

Resistance furnace. A heating apparatus in which the heat energy is generated by the flow of electricity through resistance material within the furnace. The term is probably more appropriately used in those cases where the furnace charge constitutes the resistance, and resistor furnace would then be the term to apply to those instances in which resistance alloys, developing high temperatures, were used to convey the current. In the latter case the resistances (resistors) are generally not in contact with the charge, but are located in the furnace walls.

Resistance pyrometer. A temperature measuring instrument the functioning of which depends on the gradual and regular increase of electrical resistance of a metal element with rise in temperature. The resistance of a substance depends on its physical state and condition, e.g. on the density, purity, hardness, etc. It also depends on the temperature. The resistance of an alloy is always greater than that of the constituent elements. Most alloys increase in resistance with rise in temperature, but not to the same extent as pure metals do. If R = the resistance of a material at temperature t °C and R_0 = the resistance of the same piece of material at 0 °C, then very nearly $R_t = R_0 (1 + \alpha t)$, where α is the temperature coefficient. In the case of manganin α is positive up to about 35 °C and negative at higher temperatures. The value of α for pure metals is taken as 0.0038 as a good average. In Table 49 it will be noted that the temperature coefficient of alloys is much smaller, so that they would be less useful in resistance thermometers even if the useful temperature ranges were large.

Platinum (or sometimes palladium) is the metal mainly used in resistance thermometers, since in the pure state and when not subjected to strain it always has the same resistance at the same temperature. The metal in the form of wire is wound on thin mica plate and housed in a glazed porcelain sheath. The observations on the resistance thermometer are standardised, using the melting point of ice and the boiling point of water and of sulphur, i.e. at temperatures 0 °C, 100 °C and 444.53 °C respectively.

TABLE 49.
Temperature Coefficients of Electrical Resistance

Metal or Alloy	Temperature Coefficient, $\times 10^{-4}$	Metal or Alloy	Temperature Coefficient, $\times 10^{-4}$
Brass (70/30)	10	Platinum rhodium (90/10)	17
Constantan	−0.4-+0.1	Resista (Table 6)	0.0011
German silver	2.3-6.0	Gold	40
Manganin	0.12-0.5	Palladium	37
Platinoid	2.5	Platinum	38
Platinoid silver (33.3/ 66.6)		Tantalum	33
	2.4-3.3	Silver	40
Platinum-iridium (90/10)	15	Zinc	37

Resistance welding. A process in which two metallic surfaces are joined by the application of heat and pressure, the former being by causing a heavy current, carefully localised, to pass through the metal parts to be united. It is essential that the parts to be welded are clean and make good contact, the current, usually about 1-10 volts and 3,000-10,000 amperes, is introduced by two electrodes, quite frequently having hard chromium-copper facing. The temperature attained must be high enough to cause the surfaces to become plastic, and the pressure applied must be carefully controlled. In most modern equipment the surfaces are held together during the actual welding operation by clamps operated by compressed air.

Resistance wire. The term used in reference to a number of alloys in wire form that have high electrical resistance and high resistance to oxidation and corrosion at elevated temperatures. Such wires are used in electrical heaters, stoves and furnaces, where the maintenance of good mechanical strength at high temperatures is essential to ensure a reasonable service life. The alloys are mainly nickel-chromium or nickel-chromium iron alloys, though manganese, tungsten and molybdenum are also used as alloying elements.

Alloys containing chromium and nickel only are sometimes subject to a type of corrosion known as green rot. This is caused by certain kinds of furnace atmosphere which contain sulphur and carbon and are oxidising to chromium but reducing towards nickel. Exclusion of sulphur and carbon from the furnace atmosphere is therefore recommended with all Ni-chrome alloys, though nickel-chromium-iron alloys containing silicon appear to be more resistant to green rot. Typical compositions of a number of resistance wire alloys are given in Tables 5, 6, 7.

Resistin. The name of a series of electrical resistance alloys containing mainly copper and manganese. (*See* Table 5.)

Resistivity. *See* **Superconductivity, Electrical resistivity** and **Conduction by electrons.**

Resolin. A trade name for a group of synthetic resins for corebinders.

Resolite. A synthetic resin used in foundry-work as a binder.

Resolving power. The resolving power of a microscope is defined as the minimum distance that must exist between two details in an object if their images are to be distinguishable as separate images. It is expressed as

$$\delta = \frac{\frac{1}{2}\lambda}{\text{numerical aperture}} = \frac{\frac{1}{2}\lambda}{n \sin \mu}$$

where λ = wavelength of the light used; μ = half the apex angle of the cone of light entering the objective; n = the refractive index of the medium between the objective and the specimen. If this is air, then $n = 1$.

It is thus seen that a high resolving power can be obtained by using (*a*) a large angular aperture, (*b*) an immersion medium of high refractive index, and (*c*) light of short wavelength. (*See* **Electron microscope.**)

Resonance. The response of a body at rest or in equilibrium, to mechanical or electrical vibrations from a second body, whereby the vibrations are reproduced in the first body in phase with those in the original source. In the first body, the vibrations may increase beyond the amplitude of those in the source, given appropriate conditions.

An electrical circuit producing alternating current, or oscillations, produces the greatest resonance when the frequency of the source of e.m.f. equals the natural frequency of the circuit. The circuit is then said to be tuned to the frequency of the applied voltage.

Restitution. A measure of the elasticity of bodies based upon impact. If two bodies of the same material moving along the same straight line collide, the ratio of the velocity of separation to the velocity of approach is called the coefficient of restitution. For perfectly elastic bodies the value of this coefficient, *e*, is unity, but for all real materials it is a proper fraction.

Reticulated. A term used to describe materials that are arranged in the form of a network. An example of this indeterminate form of aggregation is commonly shown by rutile TiO_2, which occurs as needles in cross meshes.

Retort. A vessel used for distillation, which may be made of graphite or clay or metal. It may be cylindrical or bulbous in shape. Retorts are generally arranged in batteries and are used in the refining or recovery of zinc, mercury and cadmium.

Retort bullion. A lead-silver alloy produced as a by-product of distillation during the desilverisation of refined lead. It contains from 4% to 11% Ag and 1% to 2% Zn together with small amounts of copper, antimony, arsenic and bismuth as impurities.

Retort furnace. A metallurgical furnace consisting of a fire-chamber, and frequently regenerative chambers, in which the retorts are placed for containing the materials for treatment. The Belgian type furnace was used in Europe and the USA for the extraction of zinc. Though zinc boils at 930 °C the reduction of the zinc oxide with carbon is not complete below 1,100 °C. Thus the reduction and distillation are carried out in one operation using clay retorts and clay condensers. The capacity of a single retort of this kind is about 5×10^{-2} m^3 and the useful life is about four to five weeks.

Cast-iron retorts are sometimes used in the retort furnace for treatment of mercury ores. They are about 2.5 m long and 30 cm in diameter and hold a charge of about 70 kg.

Reverberatory furnace. A hearth furnace in which only the products of combustion of the fuel come into contact with the charge. The fuel used may be solid fuel, but powdered coal and fuel oil are commonly used. Such furnaces are used in the production of steel and in the extraction or refining of copper, tin, antimony and lead. (*See also* **Furnaces.**)

Reverse bend test. A bend test in which sheets or bars are bent through an angle of 180° in either direction and then inspected for cracking. It is applicable to the testing of welds if the root of the weld is placed on the outside of the bending arc so that it is subjected to the maximum stress in tension.

Reversible cell. A type of cell in which the change of chemical energy into electrical energy is a reversible process. By applying an electric current to the cell, the chemical substances that were involved in the production of the electric current are restored to their original condition. Any reversible cell could in theory be used as an electrical storage cell or accumulator, but in practice the lead-acid and the nickel-iron-alkali cells are the most common.

Reversible process. This can be defined as a process that can be performed in either of opposite directions, proceeding by infinitely small steps, so that all changes occurring in any part of the direct process can be exactly reversed in the corresponding portions of the reverse process. Such a process is thus a continuous sequence of equilibria. In nature all processes are spontaneous and tend towards equilibrium. Processes that tend to move away from equilibrium conditions are unnatural processes. Between these two are the reversible processes which represent the limiting case. The condition of reversibility is that the state of the system at any time must not appreciably differ from equilibrium, for then the slightest variation in the conditions will determine the occurrence of the process in one direction or the other.

Reversible reaction. Many cases are known in which a reaction does not proceed to completion under ordinary conditions, even though considerable quantities of the reacting substances remain in contact. In all such incomplete reactions the cause of the incompleteness may be found in the fact that a reaction occurs in just the opposite sense to the direct reaction. Where such opposing reactions occur, the change taking place is described as a reversible reaction and in the equation representing the change the = sign is replaced by ⇌ sign. Since the speeds of the forward and reverse reaction are not necessarily the same, and since they will vary with time, an equilibrium condition will occur at which forward and reverse tendencies will balance. (The term balanced reaction is sometimes used.) (*See* Fig. 136.)

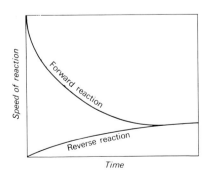

Fig. 136 *Process of a reversible reaction.*

The equilibrium is affected by (a) the concentration of the reacting substances. If C_A, C_B, C_D, C_E represent the concentration of the two reacting substances A and B, and of the two resulting substances D and E, then

$$\frac{(C_D \cdot C_E)}{(C_A \cdot C_B)} = \text{const. } (K).$$

A catalyst is generally supposed not to alter the value of the equilibrium constant K, but only to speed up the reactions so that the equilibrium is arrived at more quickly.

(b) Temperature in general favours one reaction more than the other so that the value of K will change with temperature.

(c) Pressure also changes the value of K in those cases where one or more of the reacting substances is a gas. If all the substances are gaseous and P_A, P_B, P_D, P_E represent their partial pressures, then

$$K = \frac{(P_D \cdot P_E)}{(P_A \cdot P_B)}.$$

While a system is in equilibrium the only changes it can undergo are those that are described as reversible changes.

Reversing mill. A rolling-mill so designed that the material after being rolled by passing through the mill in one direction can be passed backward.

Reynolds number. Osborne Reynolds first demonstrated, by allowing water to flow along a horizontal tube from a still reservoir, with an injection of coloured water from a very fine nozzle, that two quite different types of flow could occur. Laminar, or streamline flow, occurred at slow speeds, with no disturbance of the thin thread of colour throughout the tube. At high speeds, turbulent or eddying flow, was established, and from these experiments he deduced a critical quantity R, the Reynolds number, given by $R = vr/\nu$ where v is the fluid velocity, r the radius of the tube, and ν the kinematic viscosity. R is a dimensionless number, the ratio of the inertial to the viscous forces in a fluid. The inertial forces are approximately equal to $\rho v^2/r$ where ρ is density, v the velocity and r the characteristic dimension, either the radius of the tube, or the width or diameter of an obstacle facing the flow of fluid. The viscous forces are then approximately $\eta v/r^2$ where η is the viscosity.

$$R \text{ is then } \frac{\rho v^2}{r} \div \frac{\eta v}{r^2} = \frac{v \rho r}{\eta}$$

As η/ρ is the kinematic viscosity ν, R is usually written vr/ν.

The flow of fluid in a tube undergoes a transition when R is approximately 10^3, which corresponds to a velocity of $0.18\,\text{ms}^{-1}$ for water at $0\,^{\circ}\text{C}$ in a tube of 1 cm radius. Fluids such as air, with much higher kinematic viscosity, experience laminar flow to much higher speeds, $135\,\text{ms}^{-1}$ for a tube of 1 cm radius.

When turbulent flow occurs, the velocity profile across a tube is destroyed, and velocity becomes fairly uniform across the section. The only way this can occur is by mixing of layers near the tube wall with those close to the centre, i.e. turbulent flow.

The resistance of an obstacle to a flow of fluid is extremely important in the design of aircraft and water borne vehicles. Assuming no pressure occurs in the wake of the flow, (not true when turbulence occurs, in which case a drop in pressure also occurs) the force F acting on a projected area to the stream equal to A is given by

$$F = \frac{\rho}{2} \cdot C_D \cdot Av^2$$

where ρ and v have the usual meanings, and C_D is a constant called drag coefficient. For rapid flow, at Reynold's numbers greater than 10^3, C_D is roughly equal to 1 for a disc shaped object. For slow flow, C_D varies systematically with R. For rapid flow against spherical objects and bodies designed to minimise turbulence (stream-lined) the fluid does not lose momentum as it changes direction and C_D is small. For very slow flow, $R < 1$, a sphere obeys Stoke's law. In the turbulent flow condition, C_D usually varies between 0.1 and 1.

The drag on a body immersed in a moving fluid is separated into form resistance, depending on shape, and skin friction, depending on smoothness of surface. Aircraft have an added problem of induced drag, the energy carried away as wing tip vortices, or similar turbulence from protuberances, tail and fin surfaces. The form resistance is reduced by streamlining, i.e. shaping the body so that the dynamic pressure on the nose is minimal, since there is no balancing pressure on the tail. This leads to smoothly rounded blunt noses, and long tapering tails. There must be no sharp changes in curvature, hence the need to bury all excrescences in the body or wing of an aircraft (such as the undercarriage) and the gentle rounding of wing, tail and fin leading edges. A fully streamlined body may have a form resistance only 2% that of a flat disc of equivalent projected frontal area. The skin friction is minimised by smoothing, but an aircraft in level flight, if of low form resistance, utilises its engine power mainly to overcome skin friction.

R_F **value.** In partition chromatography a strip of filter-paper is usually employed to support the stationary phase. It is held vertically and dips into a quantity of the mobile phase. At the lower end of the strip a drop of the solution containing the mixture to be investigated is applied. After some hours' exposure to the action of an atmosphere containing both phases, the strip is removed and dried. The ratio of the distance the component has moved, to the distance the solvent has moved is known as the R_F value.

Cellulose is generally used as the support in dealing with inorganic materials, but it is necessary that the two solvents shall not be completely miscible. The method is very useful when only very small quantities are available for analysis.

Rhenium. A rare silvery-white metal that was formerly obtained from molybdenum minerals (e.g. molybdenite). It is now obtained from slime wastes of copper refineries, in which it is present to the extent of about one part in a million, and from certain flue-dusts which contain about one part in a hundred thousand.

In the strict sense there are no minerals of rhenium, though it occurs in very small amounts in columbite, tantalite and wolframite. The extraction from copper-bearing schists consists in prolonged heating of the material in air and steam until the molybdenum and rhenium have been converted into molybdate and perrhenate, respectively. The latter is converted into the potassium salt, and reduced by heating in a stream of hydrogen at about $500\,^{\circ}\text{C}$. The powder obtained can be made into coherent metal by hydraulically compressing the powder into bars, which are then hot hammered. A more convenient method of preparing the metal consists in the direct decomposition of volatile compounds on hot filaments.

The metal shows some general resemblances to the platinum metals. Some of the proposed uses are associated with its very low volatility. It is thus suitable for use in thermionic valves. Rhenium forms useful alloys with most of its metallic neighbours in the periodic table. When alloyed with nickel, platinum and tungsten the product is suitable for bearings and for pen-points. It is used to harden platinum, and in this respect is said to be superior to iridium. The finely divided metal has been used as a catalyst in the oxidation of alcohols and the hydrogenation of unsaturated hydrocarbons.

Atomic weight 186.2, at. no. 75, sp. gr. 21.02, crystal structure is close-packed hexagonal; lattice constants, $a = 2.75553 \times 10^{-10}$ m, $c = 4.4493 \times 10^{-10}$ m. The metal is highly resistant to hydrochloric acid, less so towards oxidising acids. It can be electrodeposited from a variety of aqueous solutions. (*See* Table 2.)

Rheo casting. The use of violent agitation in a melt during solidification, maintained up to a high volume fraction of solid. Also known as "stir-casting".

Rheolaveur. A coal-washing device consisting, essentially, of a trough or launder through which water and coal move rapidly, but at varying rates, so that material of varying specific gravities is stratified.

Rheology. The study of deformation and flow of matter.

Rheostat. A variable resistance that usually consists of a coil of wire wound as a helix on an insulator. A sliding contact on the wire enables different lengths of the wire to be included in an electric circuit so as to make possible the adjustment of the resistance.

Rhodanising. The term used in reference to the process of depositing, electrolytically, a thin coating of rhodium on silver to prevent tarnishing.

Rhodium. A silvery-white metal of the platinum group that was formerly obtained from crude native platinum, but is now mainly produced as a by-product in the extraction of platinum from the copper-nickel refining residues. It is almost indistinguishable from silver in its appearance, but it is highly resistant to atmospheric tarnishing, to heat and to industrial fumes. The metal is used mainly as an alloying addition to platinum and palladium. It improves the mechanical strength and heat resistance of platinum for use in crucibles, and the resistance windings of high-temperature laboratory furnaces. The 10% Pt or 13% Rh alloys are used in high-temperature thermocouples, for magneto contacts, for the spinnerets used in the production of rayon and for the gauzes used as the catalyst in the catalytic oxidation of ammonia to nitric acid. The platinum-rhodium alloys have been used in jewellery. Rhodium metal, being highly resistant to atmospheric corrosion and having a high reflectivity, is used in the form of an electrodeposit in reflectors for searchlights, etc. It is also used as a tarnish- and wear-resistant bright deposit on silverware and as an electrical contact material.

Atomic number 45, at. wt. 102.9, sp. gr. 12.41, m. pt. 1,966 °C; crystal structure is face-centred cubic, $a = 3.7957 \times 10^{-10}$ m. When zinc is added to a solution of a rhodium salt, the rhodium is thrown down as a finely divided black powder. This latter absorbs oxygen and can therefore act as a good catalyst. (*See* Table 2.)

Rhodium gold. A form of native gold containing up to 43% Rh. It is found in Mexico.

Rhodochrosite. A manganese mineral consisting of manganese carbonate $MnCO_3$, often with varying amounts of the carbonates of iron, calcium, and magnesium. It is isomorphous with iron II carbonate. It occurs in lead and in silver-lead ore veins and as a replacement (metasomatic) of limestones. The colour varies from rose-red to brown. Mohs hardness 3.5-4.5, sp. gr. 3.45-3.6. Also called manganese spar, dialogite. It crystallises in the rhombohedral system. The mineral is used in making ferromanganese and spiegeleisen.

Rhodolite. A hard form of garnet-mixed silicates of aluminium, calcium, magnesium and iron. Also contains some manganese and chromium, these impurities being mainly responsible for the colour. It is used as an abrasive on paper and on cloth backing, for watch bearings and as a gemstone. It melts about 1,300 °C. Mohs hardness 7.5.

Rhombic system. One of the classes or systems into which crystalline substances can be divided. In this system the crystallographic axes are mutually perpendicular. If these are taken parallel to a particular set of edges, then a plane face inclined to all these directions is taken as the parametral plane and its intersections with the axes give the three unequal parameters. There are three planes of symmetry at right angles to one another. These intersect in three two-fold axes of symmetry and there is a centre of symmetry. Also called the orthorhombic system.

Rhombohedral system. One of the classes or systems into which crystalline substances are divided. The most characteristic form of this class is the rhombohedron. The crystallographic axes are parallel to the edges of the rhombohedron. The angles between the axes are equal, but they are not right angles, and the parametral lengths are equal. There are three vertical planes of symmetry, and at right angles to each of these there is a two-fold axis of symmetry. The crystallographic axes thus are not parallel to an axis of symmetry nor at right angles to a plane of symmetry. The system is described by the angle between the axes.

r_H value. Since in ionic processes oxidation means an increase in the number of positive charges (or a decrease in the number of negative charges), it may be regarded that in a solution containing X^{3+} and X^{2+} and H^+ ions, the reaction can be written in the form $X^{3+} + \epsilon \rightleftharpoons X^{2+}$. . . . (1)($\epsilon$ = electron) which means that X^{2+} has been oxidised by a hydrogen ion, the latter being converted into hydrogen gas. The equilibrium equation is therefore

$$\frac{[X^{3+}] \times [p]^{1/2}}{[X^{2+}] \times [H^+]} = K ;$$

In a reversible reaction such as is represented by the first equation, the pressure of hydrogen (*p*) with which the system is theoretically in equilibrium is a measure of the oxidising (or reducing) power of the system. The value of $\log_{10}(1/p)$ is known as the r_H value. By comparing the r_H values for different systems it is possible to predict whether one system will oxidise (or reduce) another system. The value of *p* is in most cases too small to be measured directly. It is generally deduced from electrochemical considerations.

Rhyolite. A form of pumice obtained by heating the natural material to expand the internal ovoids so that the product is light in weight, having a density of about 12.5 kg m^{-3}. It is used in light-weight concrete, wall boards and as a sound-insulating material.

Richard's jig. A type of fixed-sieve pulsator jig in which suction is eliminated. Instead of the piston and pump a rotating valve is placed in the water supply, pulsation being thereby obtained. The water thus passes in surges through the screen.

Richard's solder. A hard solder which is really a yellow brass containing about 3% each of aluminium and phosphor tin.

Rich gold metal. A high copper brass containing 90% Cu.

Rich low brass. A series of high copper brasses containing 82-87% Cu, the balance being zinc. It has a high resistance to atmospheric corrosion and can be used for outdoor decorative work. It resists dezincification (q.v.) and is thus suitable for pipes and fittings in water supply. Often used in cheap novelties and jewellery because of pleasing colour.

Riddle. A coarse wire-meshed sieve used for such purposes as separating large particles and coarse material from foundry sand.

Riddlings. The coarse material left on the riddle in sieving foundry sand.

Riecke's principle. A principle of considerable importance in explaining the foliated textures of certain types of crystalline rock. The effect of stress in metamorphosed rocks is revealed in their textures. It is known that at very high temperatures liquid pervades the interstices of all rocks. Stress raises the solubility of the material stressed. Thus when the appropriate stress is reached, material is dissolved at those points where the stress is greatest and redeposited at points where the stress is less. During this process chemical reactions may occur resulting in new minerals being produced.

Riffler. The name used in reference to a small curved file.

Riffles. The term used in reference to the wave at the edges of rolled sheet and strip. The same term is also used for the grooves or baulks in the sluices used in the preliminary concentration of placer ores.

Rigidity. The resistance of a body to shear stress. The modulus of rigidity is synonymous with the shear modulus (*see* **Elasticity**), the ratio of shear stress to shear strain. Confusion with older terms such as rigidity arose in distinguishing shear from torsion. In torsion the body is deformed by a shear mechanism, but the distribution of stress is non-uniform, and a modulus of torsion has to be defined in terms of the couple required to produce a radial displacement.

The value of the shear modulus is about 0.375 × that of Young's modulus, for metals, about 0.3 *E* for long chain polymers, and 0.4 *E* for most other materials.

Rimer. A tool used for finishing off or for truing up holes in castings or forgings. Such rimers are known as parallel rimers. Those used for making parallel holes taper to receive a taper pin are taper rimers.

Rimming steel. A low-carbon steel in which deoxidation has been controlled to produce an ingot having an outer rim or skin almost free from carbon or impurity. Within the rim is a core into which impurities are concentrated as a result of the rimming action caused by gas evolution. In addition to other gases, there are usually present during the process of steel casting relatively large volumes of carbon monoxide. This gas arises from the interaction of carbon in the steel and iron oxide in the bath. As solidification of the metal proceeds gas is evolved from solution, and its escape becomes more and more difficult as the metal becomes less and less plastic. The trapped gas remains as bubbles of varying size within the solid ingot, and these cavities are known as "blow-holes". The production of free carbon monoxide and the elimination of "blow-holes" can be reduced or prevented by removing iron oxide from the bath using such deoxidisers as aluminium, ferrosilicon or ferromanganese. If, however, the steel is only partly deoxidised it may be arranged that the reduced amount of carbon monoxide that can be produced is just about equal in volume to the shrinkage pipe in the ingot.

Under such controlled conditions the liberation of gas occurs near the boundary of the (mainly) solid and (mainly) liquid metal. The liquid portion is, of course, richest in dissolved constituents, and the liberated gas forces this liquid into the central region of the ingot. There thus results an ingot in which there is a circumferential rim of exceptionally pure metal, and there is no central pipe.

Rimming steels of low carbon content have very good pressing and deep-drawing qualities. This is a result of the high ductility of the purer rim which is, of course, subjected to the greatest stresses during such processes. Steels which contain not more than 0.15% C can be made to "rim". The sheet and strip produced therefrom is mainly used for motor-car bodies, in the food-canning industry and in general for various types of steel pressing.

Ring compounds. The association of atomic bonds into closed loops as distinct from long chains or branched molecules. Monocyclic hydrocarbons having no side chains are formed in distinction to the straight chain hydrocarbons, and are known as cycloalkanes when saturated, and Fig. 137(a) indicates the structure of two examples. Similar unsaturated compounds exist, and are known by the generic titles of cycloalkenes and cycloalkynes, in a similar way to the acyclic compounds. Figure 137(b) shows some structures in this group also. Polycyclic hydrocarbons are those in which several "rings" are connected together, either by connecting bonds between monocyclic groups, or by "fusion" of the rings together. Figure 137(d) also shows examples of these. There are specific rules of nomenclature for these compounds, as with acyclic ones, laid down by the Council of the International Union of Pure and Applied Chemistry in 1957.

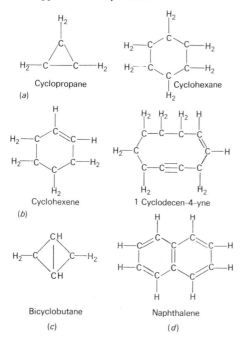

Fig. 137 *Cyclic or ring compounds. (a) Saturated monocyclic hydrocarbons – cycloalkanes. (b) Unsaturated monocyclic hydrocarbons – cycloalkene and cycloalkyne. (c) Saturated alicyclic bridged hydrocarbon. (d) Fused polycyclic hydrocarbon.*

Ring core. A device used in foundry-work to produce the moulds required for such castings as wheels, pulleys with grooved rims or shrouded gears. It enables the moulder to dispense with the use of a third moulding box (the cheek). An example of the use of a ring core, in dried oil-sand, is illustrated in Fig. 138. A solid pattern is used, the groove being filled in and a core print added.

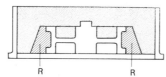

Fig. 138 *Cross-section of ring core (R).*

Ring gate. A device for pouring molten metal into a mould. This pouring gate uses a core perforated as shown in the figure which has the effect of breaking the fall of the metal which might cause

damage to the mould and produce an inferior casting, and also it tends to retain the slag during pouring (*see* Fig. 139).

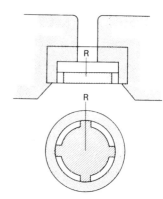

Fig. 139 *Cross-section of ring gate (R is the core of the ring gate).*

Rinmann's green. An intensely green-coloured mass which may be produced either by precipitating a mixture of zinc sulphate and cobalt sulphate with soda and heating the washed precipitate; or it may be produced by evaporating a solution of cobalt nitrate with either zinc oxide or nitrate and igniting the residue. The appearance of this green mass is the basis of methods of identification of zinc, using cobalt nitrate. The zinc compound must first be put into solution, precipitated with soda and the precipitate redissolved in excess of NaOH. From this the zinc compound is precipitated by hydrogen sulphide and the sulphide treated with cobalt solution before ignition.

Rio Tinto process. A name used in reference to a method of leaching used in South Spain for treating copper ore in the form of pyrites. Very large stacks of the sulphide ore are exposed to air and rain. Slow oxidation occurs with the formation of copper sulphate. This latter is washed out with water and the residue is exported for burning.

Ripidolite. A complex iron magnesium aluminium silicate, which has been assigned the formula $H_9(Fe,Mg)_5.Al_3Si_3O_{20}$. It is a variety of chlorite and forms the chief constituent of many slate rocks. It occurs in tabular, radiating and granular forms.

Ripple. In the rectification of alternating current to give direct current, the variation in the load current for a given type of rectifier depends on the type of rectifier and the a.c. supply. For a three-phase supply the current load varies as shown by the heavy line in Fig. 140. This variation in the actual value of the anode current is referred to as ripple. The effect is decreased with polyphase a.c.

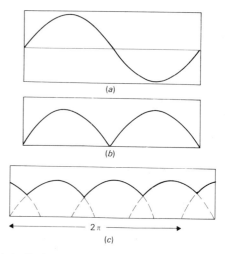

Fig. 140 *Ripple effect in a.c. electric supply. (a) A.c. single phase. (b) Full-wave rectification. (c) Three-phase a.c. rectified.*

Current is usually rectified for such purposes as electroplating or electro-refining, etc. For the former use ripple causes little or no effect except when single-phase a.c. is rectified. In this latter case the current value actually falls to zero between each maximum and this is often undesirable or harmful.

For an n-phase rectifier, and a R.M.S. secondary voltage per phase of E, the direct current voltage V is given by

$$V = \sqrt{2} . E \frac{n}{2\pi} \sin \frac{\pi}{n}$$

$$= \frac{nE}{0.225} \sin \frac{\pi}{n}$$

Also refers to the form of a surface and a geometrical effect in liquid flow.

Riser. A term synonymous with "feeding head" or "feeder", used in reference to the refractory-lined reservoir of molten metal at the top of the mould which ensures the mould being filled, and also compensates for shrinkage in the casting.

The size of the riser is adjusted for any particular case according to the properties of the metal being poured, e.g. white cast iron requires a relatively larger riser than grey iron because of the greater shrinkage of the former. Steel and aluminium require larger risers than iron castings of the same size. In addition to its main functions a riser may act as a vent permitting the escape of gas (see Fig. 141).

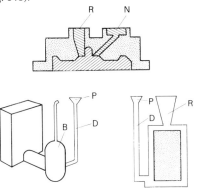

Fig. 141 *Various types of risers. P is the pouring basin; D the downgate; R the riser; N the runner; and B the blind riser.*

Rising steel. The name used in reference to inadequately killed, i.e. deoxidised, steel in which an evolution of gas during solidification causes internal blow-holes. When steel is poured while still actively generating gas—in which condition it is sometimes referred to as wild steel—the process continues to some extent during solidification and the trapped gas which forms blow-holes tends to counteract the effect of normal contraction or cooling and, indeed, the steel may appear to rise in the mould. Such steel is known as rising steel. (*See also* **Rimming steel**.)

The term is also applied to steel in which a slight rimming action has been deliberately caused. In such a case the carbon content is slightly higher than that of rimming steel and only a thin ferritic rim is produced.

Ritz combination principle. The wave-number of any spectral line for a given element may be represented as the combination of two terms, one of which is constant and the other variable throughout the series. If a gaseous element is subjected to an electric discharge, energy is absorbed by the atom and electrons from inner levels move into outer levels of higher energy. The electrons then return to their lower (more stable) orbits by one or more "jumps". Each transition is accompanied by the liberation of energy in the form of radiation, and since each emission is monochromatic, lines in the spectrum are the result. It has been noted that the lines in the spectrum are related in a simple manner, and this can be expressed by a general formula. In this v is called the wave-number and is the reciprocal of the wavelength. The wave-numbers (sometimes called frequencies) are given by

$$v = \frac{R_\infty}{x^2} - \frac{R_\infty}{y^2}$$

where R_∞ = Rydberg constant (= 109,700). In any particular series x has the same value, but y will vary. For hydrogen x and y are integers, but this is not necessarily so for other elementary atoms.

Rivet tests. These tests are two in number, the first being concerned with the shank, and the second with the head of the rivet. The shank test consists in bending cold the shank and hammering until the two parts of the shank touch. There should be no sign of fracture on the outside of the bend. For the rivet head, the latter is flattened while still hot until its diameter is two and a half times that of the shank. There should be no signs of cracks developing at the edges of the head after this treatment.

Rivet weld. An alternative name for plug weld in which the hole that is provided to receive the filler metal extends through the overlapping parts.

Roak. A surface defect usually of small depth which appears as discoloured longitudinal marks on the surface of wrought-steel bars and some other metals. These defects consist of fissures, often partly closed, with their surfaces separated by a thin layer of scale or other impurity. They are due to blow-hole cavities just beneath the surface of the original ingot becoming perforated during rolling or forging. The surfaces of the cavities become oxidised and the defect is elongated during subsequent rolling.

Roasting. A term used in reference to a process of heating material in air. In the case of ores of metals, roasting is done to eliminate volatile constituents and to bring about chemical changes such as oxidation of sulphides. Except in the case of copper sulphides, roasting is mainly carried out for the production of oxides which may subsequently be reduced to metal. The process must usually be carefully controlled in respect of temperature and of air-supply. Rich antimony (sulphide) ores are roasted in excess of air at a controlled temperature of 350 °C to produce Sb_2O_4. Lower-grade ores require roasting at just below 400 °C and the air supply must be limited (sometimes by the introduction of steam as well as air) to produce Sb_2O_3. In the case of bismuth and tin ores the roasting process is mainly concerned with the elimination, as volatile compounds, of arsenic, antimony and sulphur. Lead ores are still roasted to eliminate some of the sulphur but the process is regarded as a rough or "pre-treatment", as most of the sulphur is more satisfactorily removed in subsequent sintering treatments. The treatment of zinc ores is carried out in two stages: a preliminary roasting followed by a sintering treatment. This is necessary since roasting may produce $ZnSO_4$. A high temperature is required to decompose this compound, but if this is not done before the reduction treatment in retorts, the zinc sulphate reduces to zinc sulphide and hence a corresponding amount of zinc is lost.

Most copper ores and concentrates have a high sulphur content. When fused copper sulphide Cu_2S is formed and the excess sulphur combines with other metals such as iron to form sulphides. This mixture of sulphides is known as matte. Hence the matte will usually be low in copper content. The necessary removal of part of the sulphur is accomplished either by roasting, when the sulphur is burnt off without fusing, or the charge is heated in a smelting furnace using a large excess of air, whereby roasting and smelting are carried out in one operation.

Roasting furnaces. Since the operation of roasting (q.v.) is applied to very different ores and for different purposes, the types of furnace may be expected to vary more widely than they actually do differ. The kinds of furnace include kilns, reverberatory furnaces and multiple-hearth furnaces. The most primitive method of roasting was carried out by heating heaps of ore in the open. This method is not capable of any measure of control and is very wasteful. As the importance of air supply and temperature control became better appreciated, the open fire was confined by side walls giving rise to the "stalls", and when later these were provided with a top, the result was the forerunner of the "kiln". The enclosed type of furnace not only permits greater control of the operation of roasting, but also permits the recovery of the sulphur-bearing gases and other volatile products.

The multiple-hearth furnaces are the type used in modern industry. They generally have six or more hearths, one above the other, with a central rotating shaft which carries the rabble arms for working the charges on the hearths. Since most of the actual roasting occurs as the ore falls from hearth to hearth, the tendency is to build roasting furnaces with more but smaller hearths. In the furnace the ore is gradually moved from the centre to the circumference of one hearth and dropped to the hearth next below, where it is slowly moved towards the centre to drop to another hearth. Heat is generated by the burning of sulphur, and air is admitted independently at each hearth. Additional heat is available from oil burners placed external to the hearths.

Rochelle salt $KNaC_4H_4O_6.4H_2O$. Sodium potassium tartrate tetrahydrate. A uniaxial ferroelectric, which demonstrates the strongest piezoelectric effect. It does not show spontaneous polarisation below $-20\,°C$. The Curie temperature is $24\,°C$, at which point the dielectric constant is 4000. Extensively used in gramophone cartridges, earphones, microphones and hearing aids. Melting point $75\,°C$, sp. gr. 1.79.

Rock. The geological term for certain individual masses of the earth's crust which are aggregates of one or more minerals and which have different characteristics according to their period and mode of formation. Some of the rocks now known in the form of crystalline minerals were probably molten at one time. These rocks cut across the structures of other masses as intruded veins into adjacent fissures. These, together with the lavas found at the earth's surface, are known as igneous rocks. Basalts, diorites and granites are examples of these. Wherever rock is exposed at the earth's surface it is subjected to the action of air, water and sun, which causes it to break down and disintegrate. The mineral substances also undergo chemical changes often producing simpler but more stable compounds. On the other hand combinations and substitutions occur which result in complex minerals. These are known as sedimentary rocks. and running water has played an important part in their formation.

When any type of rock because of great heat or pressure contains new structures or new minerals, or both, it is known as metamorphic. Slate, marble, schist and gneiss are examples (*see* Table 50).

TABLE 50.
Rock-forming Mineral Groups

Rock-forming Mineral	Type	Alteration Product
Quartz	Silica	Unaltered
Acid felspars	Silicate of alumina + silicates of potash and soda	
Basic felspars	Silicate of alumina + silicates of lime and soda	Kaolin
White mica	Silicates of alumina and potash	Epidote
Brown mica	Silicate of alumina + silicates of iron and magnesia	Chlorite
Ferromagnesian minerals	Silicates of alumina + silicates of magnesia and iron in increasing proportions: (a) Amphiboles (hornblende) (b) Pyroxenes (augite) (c) Olivine	Micas Serpentine

Rock crystal. The purest and most transparent form of silica SiO_2. Used in jewellery. Mohs hardness 7.

Rocking arc furnace. A type of electric furnace in which the heating is obtained by radiation from the arc. The furnace is arranged to rock about a horizontal axis so that improved heat transfer is obtained. Such furnaces are satisfactory for small capacities.

Rocking resistor furnace. A type of furnace that may be considered as an alternative to the rocking arc furnace (q.v.). It is of similar external design and capacity. The heating is, however, obtained from a horizontal resistor, frequently made of graphite.

Rockoustile. The name used in reference to exfoliated mica that has been fabricated into the form of building tiles. It is used as a sound insulator.

Rock salt. A naturally occurring form of common salt $NaCl$ found in many localities. It crystallises in cubes. Mohs hardness 2.5, sp. gr. 2.1-2.5.

Rockwell hardness. A measure of indentation hardness used in connection with metals and plastics. It is an index of resistance to surface penetration by a specified indentor under a stated load. The Rockwell hardness-tester measures the depth of residual penetration by a steel ball or diamond cone under given conditions of load. The hardness is expressed as a number which is derived by subtracting the penetration from an arbitrary constant.

The test should be carried out on smooth, but not necessarily highly polished surfaces, which are truly representative of the metal. The specimen must be of such thickness that the under-surface, after testing, does not show a perceptible impression.

In the test a minor load of 10 kgs is first applied to seat the penetrator firmly in the surface of the specimen. The dial is then set at zero, and the major load, which varies according to the material under test and the type of penetrator used, is applied. After the pointer comes to rest the major load is removed, leaving the minor load still applied to the specimen. The Rockwell hardness number is read directly on the appropriate scale on the dial.

Several different loads and indentors are used to provide optimum measurements for ranges of hardness, and each standard combination of load and indentor is represented by a different scale designation. In test reports the hardness number should bear a prefix such as "B" or "C" to indicate which scale was used. Where the scales overlap, Rockwell hardness on one scale can be converted into Rockwell hardness on another scale by means of a standard conversion chart. The two general types of indentor are the Brale diamond indentor consisting of a $120°$ cone with a 0.2 mm radius spherical tip and the steel ball indentor of 1.6-12.7 mm (1/16-1/2 in) diameter. Scale A is suitable for tungsten carbide and other very hard materials, since higher loads are liable to cause damage to the indentor. Table 51 gives details of the various Rockwell tests.

TABLE 51.
Types of Rockwell Hardness

Designation of Scale	Major Load, kg	Type of Penetrator	Materials
A	60	Brale	Tungsten carbide, cold rolled strip, case-hardened and nitrided steel
B	100	1/16 - in (1.6 mm) ball	Annealed, low- and medium-carbon steels (standard)
C	150	Brale	(Standard) generally used for steels above B-100
D	100	"	Case-hardened steel, where less penetration desirable
E	100	1/8 - in (3.2 mm) ball	Die-castings, bearing metals and soft metals generally. Also some plastics
F	60	1/16 - in (1.6 mm) "	Annealed brass
G	150	1/16 - in (1.6 mm) "	Phosphor bronze
H	60	1/8 - in (3.2 mm) "	Soft metals and some plastics
R	60	1/2 - in (12.7 mm) "	Mainly for plastic material
L	60	1/4 - in (6.35 mm) "	
M	100	1/4 - in (6.35 mm) "	

Rockwell penetration test. A test for measuring indentation hardness of such substances as rubber. The indentor is a steel ball 6.35 mm (1/4 in) diameter, the minor and major loads being 10 kg and 60 kg, respectively. The dial is set at B-30 and the major load held for 15 seconds. The penetration value is read directly on the dial and is usually reported in conjunction with the Rockwell recovery test (q.v.).

Rockwell recovery test. A test devised to obtain a measure of the ability of plastics or rubber to recover after indentation. It is generally performed in conjunction with the Rockwell penetration test (q.v.). After the penetration has been read on the dial, the major load is removed and the specimen allowed to recover for 15 seconds. At the end of this interval the final deflection of the dial needle is read and reported as the recovery figure.

Rockwell superficial hardness. A test devised to obtain a measure of the resistance of a surface to penetration by a specified indentor under a given load. It is especially useful in testing thin sections or thin layers as well as small components. The test is similar to the Rockwell hardness test except that smaller loads are used. The minor load is always 3 kg, but the major load may be 15, 30 or 45 kg. When the Brale indentor is used the scales are respectively designated 15-N, 30-N, 45-N. Alternatively a 1.6 mm (1/16 in) ball may be used and the scales are then designated 15-T, 30-T, 45-T.

The latter scale is used for brass, bronze and unhardened steel, while the former is used for hard-steel surfaces such as case-hardened and nitrided steel.

Rockwool. A relatively light and fibrous material that is suitable as a thermal insulator. It can be made by blowing steam through molten volcanic rock or through molten slag obtained from blast furnaces. In the latter case it is sometimes referred to as slag-wool.

Rod. The term used in reference to a metal product, having (usually) a circular section, but smaller than for bars and larger than for wire. Rods may be rolled, drawn or extruded. Though there are no precise limits for the dimensions of rod, the diameter is usually 3 mm up to about 100 mm, though in certain materials, 150 mm-diameter rods are made.

Rodding. A term used to describe the supporting of overhanging sand in certain forms of mould, by means of bent rods.

Rod mill. A rolling-mill used for the production of metal rods. The term is also used to describe a type of grinding mill in which long steel rods are used for pulverising minerals, etc.

Roke. A term synonymous with roak (q.v.).

Roll camber. The deviation of a roll from perfect cylindrical form, as measured at a point half-way along its barrel. The roll may be either slightly concave or slightly convex, i.e. it may be "hollow" or "full".

Roll compacting. A method used to produce a continuous strip section by compacting metal powder between rolls instead of using a die and ram. The powder flow is regulated by a feed hopper which guides the powder to the rolls. To produce a satisfactory continuous compacted strip it is essential to ensure a regular feed without slip. This may be done using relatively large rolls, or it can be achieved with smaller rolls if the powder is fed on to a pre-formed thin sheet.

Rolled gold. A composite metal consisting of a brass, nickel or other suitable base metal on which has been rolled a thin layer of gold. The clad metal—gold coated—is rolled or drawn down until the gold layer is extremely thin.

Roller flattening. A process for flattening metal sheets by passing through a series of staggered small diameter rollers. The process thus consists of a series of bending operations.

Roller levelling. A process of flattening metal sheets by passing through staggered rolls that bend the metal without effecting any reduction of thickness. Synonymous with roller flattening (q.v.).

Roll forging. An operation used in sheet-metal working in which sheet is bent to the desired contours by passing between rolls of the necessary shape.

Rolling. The term used in reference to the working of metal by passing it between rollers that revolve in opposite directions. The metal is compressed and reduced in section and changed in shape. The operation may be carried out hot or cold, and in addition to the main operation may improve the grain structure, the mechanical properties and the surface finish. The amount of reduction in a single pass—usually expressed as the percentage reduction in area of the section—is dependent on the nature of the metal and on the temperature of rolling, though certain metals are generally cold rolled. For sheet rolling the rolls are plain, but for sections of various shapes the rolls are grooved appropriately. Generally, a number of passes is required because the metal must be reduced gradually to the final shape required, and intermediate annealing may be necessary. When the opening in a set of rolls is so arranged that the metal may, to some extent, spread sideways to form fins, the pass is called an open pass. It is therefore necessary in such cases to turn the section through 90° between passes so that the reduction is made uniformly all round. In the closed pass the width of the opening between the rolls is somewhat larger than the corresponding dimension of the metal to allow for the spread and no fins can be formed. The early passes are open passes, while the later or finishing passes may or may not be closed.

Rolling direction. The direction, in the plane of the sheet being rolled, perpendicular to the axis of the rolls during rolling.

Rolling-mill. Usually a heavy machine, in which hot or cold metal is passed between electrically or steam-driven rolls to change the shape and dimensions of the section of the metal. The term may be applied to a single machine or to a series of machines used in

any particular rolling operation. The rolls may be of different diameters, but 1.4 m is about the largest diameter of roll used. They may be made of grey iron, grey iron with a white iron case (or skin) and plain carbon or alloy steel either cast or wrought.

In the simplest-type mill there are two rolls. This is referred to as a 2-high mill (see Fig. 142A), and when it is arranged that the rolls can be made to reverse their direction of rotation, the mill is called a reversing mill. A 3-high mill is shown in Fig. 142B. It is clear from this that the work can be passed in the forward and reverse directions, using a different pair of rolls in the two cases. Four-high mills have two small rolls (which require less power than the large rolls). They have, however, backing-up rolls, since the small rolls are alone not strong enough for the work of rolling sheet and strip metal in long lengths (see Fig. 142C and D).

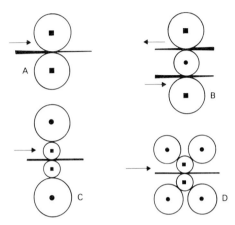

Fig. 142 *Sections showing the operating principle of various types of mill. A is 2-high; B, 3-high; C, 4-high; and D a cluster mill.*

Each mill is referred to as a "stand", and for continuous rolling several stands are arranged in tandem, each having a smaller opening between its rolls than the one before it. The later rolls must rotate faster than those in front, so as to be able to take care of the increased length produced at each reduction of section. (See also **Cogging mill, Blooming mill, Billet mill, Slabbing mill** and **Universal mill.**)

Roll marks. The term used in reference to the periodic defects sometimes seen on the surfaces of rolled metals, caused by some imperfection on the surface of a roll.

Roll-over moulding machine. A device used in foundries to facilitate the separation of the mould and the pattern after the former has been rammed up. The pattern is clamped to the mould-box and the sand is rammed by hand or mechanically. The pattern and mould together are turned over by the machine and placed on an equaliser table. The clamps are then removed and then either the table is lowered, taking with it the mould which is withdrawn from the pattern, or the pattern is withdrawn by the machine vertically from the mould (see Fig. 143).

Roll-squeezer. A machine used in the manufacture of wrought iron, for squeezing cinder out of the puddled balls, and for consolidating the mass of metal. This operation is performed by passing the metal between heavy rollers. Also called alligator or crocodile squeezer.

Roll-straightening. A process in which rod, tubing, etc., is straightened by a series of bending operations during passage between small-diameter rolls.

Roll-threading. A method for forming a screw thread on a bolt or screw by rolling between grooved die plates or between rotating grooved rolls.

Ronceray runner. A type of pencil gate having very small inlets. It was introduced by Ronceray as a method of casting fairly large castings using the minimum weight of head. The molten metal falls in thin streams to the bottom of the mould where rapid solidification occurs. Further supply of metal in thin streams tends to break up the surface of the solidifying metal and this is considered beneficial as the slag is permitted to float at the surface. There is usually a relatively small amount of top discard.

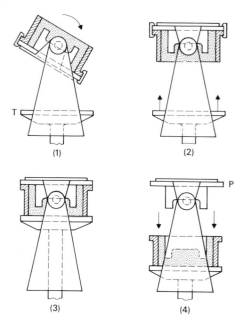

(1)   (2)

(3)   (4)

Fig. 143 *Sequence of operations of a roll-over moulding machine.*

Röntgen. An exposure of X- or γ-radiation such that the associated corpuscular emission per 1.293 mg of air produces, in air, ions carrying 3.3×10^{-10} C of electricity of either sign (10.88×10^{-9} J or 6.8×10^{10} eV). Röntgen is not an SI unit.

Root bend test. A type of reverse bend test applied to the testing of welds. It is arranged that the root of the weld lies on the outer periphery of the bending arc so that it is subjected to the greatest tensile stress.

Root mean square. A term usually abbreviated to r.m.s., which is defined by

$$\text{r.m.s.} = \sqrt{\frac{\Sigma x^2}{\Sigma n}},$$

where Σx^2 = sum of the squares of the individual values of the variable and Σn = the total number of values.

If there be a function y which is periodic and of period T, the r.m.s. value of y is the square root of the square of y taken over the period T. All the usual measuring instruments used in alternating current electricity give r.m.s. values of the current, etc., for the r.m.s. value of current (I) determines the heat generated in a resistance R, $[H = I^2 R]$.

In the case of the current in an m phase converter

$$\frac{\text{a.c. volts}}{\text{d.c. volts}} = \frac{\sin \pi/m}{\sqrt{2}}$$

and in general if the alternating quantity can be represented by a sine wave the r.m.s. value of the quantity I is related to the maximum value (i) of the quantity by the expression $I = i/\sqrt{2}$.

Root's blower. A rotary pressure blower generally used for supplying a blast of air to blast furnaces and cupolas. Two dumb-bell-shaped (figure of eight) impellers mounted on parallel shafts rotate in opposite directions so that air is trapped between the impellers and their casing and forced into a discharge pipe.

Roscoelite. A vanadium mineral that consists of potassium aluminium vanadium silicate. It generally contains about 12-24% vanadium oxide V_2O_5. It is found associated with gold tellurides in California, Colorado and Western Australia. Also known as vanadium mica.

Rose process. A method used in the refining of gold in which the zinc precipitates are fused and air or oxygen is blown through the mass when the baser metals oxidise in succession and pass into the borax-silica slag.

Rose quartz. A variety of quartz having a pink or rose colour due to the presence of traces of oxides of other elements.

Rosin blende. A variety of zinc blende ZnS which, when of a yellow colour, is called rosin blende, and when dark in colour is known as black jack. It is found in many districts in Europe, the USA, and New South Wales.

Rosite. A ceramic material consisting of calcium and aluminium silicates mixed with asbestos. It is used for moulded electrical parts and withstands temperatures up to about 480 °C.

Rosslyn metal. A composite-clad copper sheet used for the combustion chambers and other jet-engine components. It combines the heat-conducting qualities of the copper core with the properties of the cladding metal, which may be stainless steel, mild steel, Inconel or other heat-resisting alloy. The thermal conductivity of the composite sheet depends on the thickness and type of cladding metal, but it normally exceeds 60% of that of pure copper.

Rotary blower. A machine that is used for producing a blast of air, or for pumping liquids. The action depends on the rotation together of two specially shaped hollow castings, resembling in shape the figure of eight. (*See also* **Root's blower.**)

Rotary converter. The name used in reference to an electrical machine for converting alternating current into direct current. It consists of a dynamo and a motor working together. Over 90% of the total electrical energy generated is produced in a.c. stations and a large portion of this is used as direct current. The rotary converter is essentially a direct-current machine having a d.c. armature of ordinary design, and a synchronous motor combined into a single machine.

Rotary drier. A type of machine in which various materials, such as clay, coal, coke, limestone, sand, etc., are subjected to the action of hot air or hot gases from a furnace in order to reduce their moisture content. The material to be dried is charged into a long cylindrical furnace tube, which is mounted at a small angle to the horizontal, by the rotation of which the material slowly descends with the flow of the hot gases. The furnace tube may be a single shell having lifting flights attached to the inside of the shell, or the flights may be replaced by longitudinal lifts of cruciform section. The object of these is to slowly turn the material over to secure more efficient drying.

A type of drier or kiln having a double shell is sometimes used. In this the material to be dried moves slowly down the annular space between the shells and against the flow of hot gases. The movement is aided by short lifts on the inside of the outer shell.

Rotary furnace. The term used in reference to a type of horizontal furnace having a cylindrical body and conical ends and lined with refractory material. It is used mainly for melting small charges (1-50 Mg or 1-50 tons) of cast iron or certain nonferrous metals. The furnace is usually fired with powdered coal, oil or gas and is so arranged that it can rotate on bearings about a horizontal axis, usually at a rate of one rotation per minute. The term is also used in reference to a type of puddling furnace which is designed to rotate in order to bring the metal into more intimate contact with the fettling. (*See also* **Rotary kiln.**)

Rotary kiln. A term used in reference to a type of furnace practically the same as the rotary furnace. It is used for melting, drying and calcining operations. (*See also* **Rotary furnace.**)

Rotary piercing. A term used in reference to a method of producing metal tubes (usually in iron or steel or certain non-ferrous metals) from a hot billet. The billet is guided between a pair of rolls that are bevelled and inclined at a small angle to one another. This ensures that the billet is gripped and spun at high speed as it is forced over a tapered plug. This latter opens up the hole in the billet. The rough pierced blank is plug rolled to improve the shape of the tube.

Rotary press. A type of press that is equipped with a rotary table and multiple dies for the purpose of forming compacts from metal powders.

Rotary squeeze. A term used in reference to a machine that consists of an inner and outer roll or cylinder, whose axes are arranged eccentrically. Metal is fed to the machine in such a way that it passes between the rolls as they rotate, becoming gradually squeezed.

Rotary swage. A machine for reducing the diameter of wire rod, tubing, etc. It consists essentially of a pair of rotating dies which are pressed intermittently against the work by an eccentric action.

Rotating anode. A device used in X-ray tubes to secure sharply defined radiograms, without heating up the anode excessively. The requirements for a good radiogram are: *(a)* a small focus; *(b)* a short exposure time; *(c)* an anode made of metal of high atomic number, *(d)* a high wattage per mm^2 of focal area; and *(e)* the avoidance of high temperatures of the anode. These are obtained to best advantage by a rotating anode. The speed of rotation is of the order of 20-30 rev s^{-1}.

Rotating crystal method. When a beam of X-rays passes over an atom each electron is affected by the electric field of the beam and their vibratory motions are accelerated. The vibrating electron becomes the source of a new set of electromagnetic waves. These have the same wavelength as the incident beam and the radiation occurs in all directions. The various "scattered" waves from the individual electrons combine and may, for the present purpose, be considered as a single set of waves originating from a point.

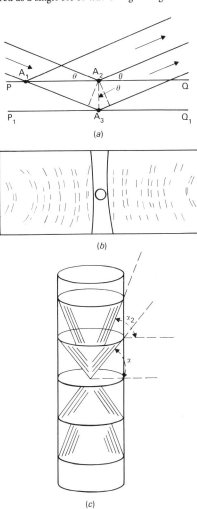

(a)

(b)

(c)

Fig. 144 *X-ray diffraction of rotating crystal. (a) Diagram showing path of incident beam of X-rays meeting layers of atoms in a crystal. (b) Type of x-ray photograph obtained from a rotating crystal. (c) Formation of layer lines from cones of diffracted rays.*

The diagram (Fig. 144(*a*)) represents an incident beam of X-rays meeting the parallel planes PQ, P_1Q_1 of atoms in a crystal. The rays scattered by atoms at A_1, A_2, A_3 will reinforce in the direction shown for the scattered beam. If d is the spacing of the planes PQ and P_1Q_1 and λ is the wavelength of the beam, then

$$n\lambda = 2d \sin \theta$$

n may have values 0, 1, 2, etc., according as the difference in the path of the waves is 0, 1, 2, etc., wavelengths.

In X-ray diffraction methods for investigating the inner structure of crystals, a monochromatic beam of X-rays is directed at a single crystal. The angle of incidence may not, however, be such as to satisfy the Bragg equation above, and in the rotating crystal method a single crystal is rotated in the beam of X-rays, by which means different atomic planes are, in turn, brought into the reflecting position. The diffracted beam will appear in those instances where the angle of incidence is correct for the radiation used. The pattern that is obtained in the photograph is of the type shown in Fig. 144(*b*), where the reflections are arranged on layer lines. If the crystal is arranged so that rotation takes place around a principal axis the spacings of the layer lines give the spacings between lattice points in the direction of the axis of rotation. The formation of the layer lines from cones of diffracted rays is illustrated in Fig. 144(*c*).

Rotoil mixer. A type of power-driven mixer used for preparing oil and sand for cores for foundry moulds.

Rotor process. The name of a process for making steel in a rotating vessel or converter which is roughly cylindrical in shape. A water-cooled lance injects oxygen into the molten metal below the slag. In addition a secondary oxygen supply which can be carefully regulated is provided through a second water-cooled lance. This latter projects a stream of oxygen almost horizontally over the slag surface. Together, the two lances promote slag/metal contact, and the method ensures that carbon oxidation and refining reactions proceed at the same time.

Rotten rock sand. A naturally bonded sand used in making moulds for steel castings. It has an A.F.A. green fineness number approximately 40.

Rottenstone. A greyish-green friable material derived from the weathering of siliceous-argillaceous limestone. The main constituent is alumina, but iron oxides are also usually present. It is used as an abrasive in powder form or when moulded into blocks. In the latter case a lubricant such as oil is used.

Rouge. A red variety of ferric (iron III) oxide Fe_2O_3. Iron (III) oxide, when artificially prepared by igniting the hydroxide or an iron III salt of a volatile acid, is usually steel-grey in colour, but when finely ground is a reddish-brown powder having a specific gravity of 5.17. The residue left after distilling fuming sulphuric acid from green vitriol ($FeSO_4.7H_2O$) is termed colcothar and is used as an oil paint and polishing powder. The least calcined portions—which are scarlet in colour—are called rouge (or jeweller's rouge), while the more calcined portions, which have a bluish tinge, are known as "crocus" and are used for polishing brass and steel. When used as a pigment the red iron (III) oxide is known under the names of "red ochre" and "venetian red". Mohs hardness (5.5-6.5) increases with depth of colour.

Roughing. The term used in reference to the preliminary operation of reducing the section of ingots into blooms, billets or slabs by forging or rolling. In the latter case the ingots are generally passed through a roughing mill prior to further rolling, forging or drawing.

Roughing mill. The rolling unit used for breaking down billets and ingots—usually by reducing the sectional area—in the first stages of hot-rolling processes.

Roughness. A term used in reference to the finely spaced surface irregularities produced on metals by such operations as planing, milling, grinding, honing, lapping and polishing; and as a general indication of the kind of surface finish or its quality these terms have often been used—e.g. surface finish to be "rough milled" or "fine ground", etc. But these terms are difficult to compare and lack exactness. Since the roughness is due to minute irregularities of the surface, some central reference line may be assumed and the heights and depths of the irregularities measured at equi-spaced intervals. The heights and depths are measured in micrometres, and if there are n such measurements of profile having values a, b, c, d, etc., then the roughness is expressed by the root mean square formula

$$R = \sqrt{(a^2 + b^2 + c^2 + \ldots)/n}$$

The instrument used for the measurement of the surface irregularities is known as a profilometer. A tracer point moves over the surface to be measured and the motions of the tracer are converted into electrical voltages. The instrument is calibrated to give the r.m.s. value on the scale.

Rounds. A term used to describe bars of iron and steel of circular cross-section varying from about 6-200 mm in diameter.

Rovalising. A trade term used to designate the coating processes used or marketed by a particular firm. The processes include phosphate as well as metallic coatings.

Royer sand machine. A special type of sand riddling machine used in mixing sand and in the extraction of unwanted foreign matter and oversize pieces. It operates by means of a flexible belt to which steel teeth are attached. These comb and aerate the sand and eject it in a stream into a moulding box as required.

Rubber. Originally the name for the gummy sap of several plants of tropical origin, chemically *cis* 1-4 polyisoprene (2 methyl-1, 3 butadiene). It was known in Mexico before the fifteenth century, and brought to Europe by Columbus and other explorers, from the West Indies. Hancock discovered the advantage of mastication of latex to improve the properties, and later "vulcanisation", or cross-linking, by heating with sulphur, which had already been discovered accidentally by Goodyear. The use of latex in coating fabric had already been described by Macintosh, before Hancock's processes, but it was the vulcanisation by sulphur or sulphur monochloride (Parkes) which led to rapid development.

Synthetic rubber was first made from butadiene, called Buna rubber because sodium (Na) was used to effect the polymerisation. The addition of styrene (ethenylbenzene) to form a copolymer improved the butadiene rubber, and later, the further addition of acrylonitrile (propenoic acid nitrile) made even greater improvements.

Chloroprene (2 chloro-1, 3 butadiene), derived from vinylacetylene (1 buten- 3-yne), led to a polymer later named Neoprene, with considerable resistance to chemicals, particularly hydrocarbons. Polysulphide rubber (thiokol) and butyl rubber were later developments. The use of ethene, copolymerised with polypropene, gives a rubber, even though either constituent separately gives a thermoplastic.

Other elastomeric materials are now available, with specific properties, among them silicones, polyurethane (polycarbamic acid ethyl ester) and fluoro compounds, and *cis*-polyisoprene can itself be made synthetically with stereospecific catalysts available.

The great advantage of rubbery materials is the elastomeric property, the ability to recover rapidly after large deformations, provided the stress is maintained for short periods. The disadvantages are the ageing due to oxidation, cross-linking and chain scission, especially at elevated temperatures. (*See cis*-**isoprene**, **Neoprene**, etc.)

Rubbing board. The term used in reference to a small board used by moulders for smoothing the flat surfaces of foundry moulds.

Rubbing stone. A term used in reference to fine-grain sandstones often containing minute crystals of garnet or rutile. It is used as a fine abrasive for sharpening edge tools.

Rubellite. A red or pink transparent variety of tourmaline (q.v.). It is a borosilicate of aluminium sometimes used as a gemstone.

Ruben's brown. A deep brown argillaceous-siliceous earth obtained from a lignitic deposit at Cassel (Germany). Also called Vandyke brown, Cassel earth, Cassel brown.

Rubidium. One of the alkali metals found in small quantities in mica, carnallite, iron ores and some other minerals. The main source is lepidolite. It is a silvery-white metal which is readily volatile and decomposes water with the evolution of hydrogen. Many of its salts are feebly radioactive, emitting β rays. It is obtained with caesium as a by-product in the extraction of potassium. It is used in the manufacture of photoelectric cells. Crystal structure body-centred cubic; lattice constant $a = 5.62 \times 10^{-10}$ m. (*See* Table 2.)

Rubin number. A term introduced by Ostwald to indicate the amount of emulsoid colloid in grams in 100 cm³ of solution which prevents the change in colour of a Congo-red sol on the addition of a 160-millimolar solution of potassium chloride. The rubin number for gelatin is 0.025 and of albumin is 0.02.

The stabilising action of emulsoids is of considerable importance in the preparation of colloidal metals, paints, greases and food products.

Rubio iron ore. An iron ore consisting of hydrated iron III oxide and up to about 9% Si. It is found mainly in Spain.

Ruby. A transparent variety of the mineral corundum, the red colour being due to the presence of small amounts of oxide of chromium. It is found mainly in Burma and Thailand. Artificial rubies may be made by dropping powdered alumina containing 2-3% of chromium sesquioxide through the centre of an oxyhydrogen flame. When the fused mass is caught on an alumina rod it appears as a single crystal which may be cut, as a gemstone or for use as pivot bearings for watches and precision instruments. Specific gravity 4.04, Mohs hardness 9.

Ruby arsenic. Arsenic disulphide $As_2 S_2$, also known as red orpiment. It is found in Peru, Utah and Central Europe and is used as a pigment. Specific gravity 3.5.

Ruby copper. An alternative name for cuprite—which is copper II (cuprous) oxide $Cu_2 O$—or red copper ore.

Ruby glass. Glass that has been given a ruby-red colour by the addition of gold compounds. The purple powder called purple of cassius (q.v.) is generally used. When fused with the glass the result is a colourless mass, which becomes red on annealing. The colour is due to the presence of ultramicroscopic particles of gold. A cheaper kind of ruby glass is prepared by using copper II oxide in place of gold.

Ruby mica. A fine grade of potassium aluminium silicate found mainly in India. It is used in making electrical condensers. (*See also* **Mica**.)

Ruby silver. A term used in reference to two silver ores—the dark ruby ore which is silver sulphantimonite and the light ruby ore which is silver sulpharsenite. The former is known as pyrargyrite and the latter as proustite, the two minerals being isomorphous. The specimens are found in Saxony (Germany) and in Mexico.

Ruby sulphur. A term used in reference to realgar $As_2 S_2$, a sulphide of arsenic. (*See also* **Realgar**.)

Ruling section. There is, for every type of steel, a critical rate of cooling from the austenitic condition which must be exceeded in order that the austenite may be completely transformed to martensite. For a mass of steel, when other factors are maintained constant, the rate of cooling is

$$r = k \frac{1}{\log d},$$

where k is a constant and d is the depth of the particular section below the surface of the mass. The massiveness of a piece of steel is therefore of very considerable importance in regard to the cooling rate during heat-treatment, since the critical rate of cooling at all depths below the surface of the component may not be reached. In such a case the steel will not be homogeneous either in structure or in mechanical properties. That combination of dimensions of a billet, bar, forging or component which has the greatest influence on the rate of cooling in a given medium, from the normalising or hardening temperature (and hence on the mechanical properties induced by the heat-treatment) is known as the ruling section. That diameter of round bar which will cool at the same rate as the component etc., is known as the equivalent section.

Hardness tests are frequently based on heating and quenching round bars in air or oil and then determining the hardness at intervals across a section. If non-martensitic constituents are present, properties other than tensile strength cannot be predicted with any degree of certainty from hardness values. The ruling sections refer to the maximum permissible diameter of parts of circular cross-section.

Rumbling. A process in which a hollow cylinder is used for cleaning articles for electroplating. The cleaning is accomplished by causing the articles to rub against one another as the cylinder rotates. A similar device, using a larger cylinder, is employed for removing sand from castings after removal from the moulds. Usually a number of angular or star-shaped pieces are included to add to the abrasive action of the castings themselves. As a certain amount of dust results from the operation the axles of the cylinder are made hollow so that air may be drawn through the cylinder to extract the dust into a wet-type dust catcher.

Runge's law. The spectra of elements consist of regular successions of lines which occur at continuously diminishing intervals as the shorter wavelengths are approached. The lines also show a regular falling off in intensity with diminishing wavelength. Such a succession of lines is called a series, and the spectra of a number of elements have been analysed into series known as *S, P, D,* and

F series. But the series are not independent. The limit of the *F* series is the first term of the *D* series, this relationship being known as Runge's law which arises from the possible changes in electron energy levels (*see* **Atom**).

Runner. A term used in foundry-work to indicate the portion of the gate assembly which connects the downgate and the casting. Through this passes the metal for both the casting and the heads. The term is also used in reference to a channel through which molten metal and slag flows from a furnace or from one container to another (*see* Fig. 145).

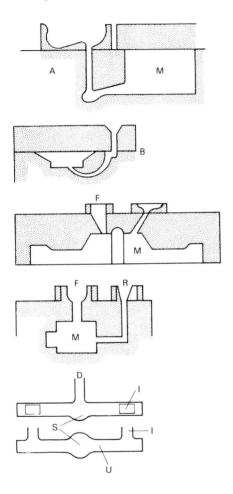

Fig. 145 *Cross-section through various types of runners. A is a bottom runner; B a horned runner; F the feeder head; M the mould; R the runner bush; D the downgate; S the sump; I the ingate; and U the runner bar.*

Runner box. A term used by foundrymen to describe a device for holding and distributing molten metal to a mould. It is frequently arranged that the molten metal is divided into several streams connecting with multiple downgates and runners. It is generally rectangular in shape and is made of cast iron or steel.

Runner bush. When making moulds for small castings a funnel-shaped cup is usually scooped out of the sand in the upper surface of the cope above the downgate to permit metal to flow from the sprue cup and down the sprue to the ingate. This is known variously as sprue cup, runner cup, runner basin, runner bush or pouring cup. For large castings a runner box replaces the runner bush.

Runner head. A term used in foundry-work for forming the runner of a mould.

Runner pin. A tapered pin or stick used in foundry-work in reference to the metal in the ingate of a mould.

Runner stick. A term synonymous with runner pin (q.v.).

Running gate. A term synonymous with ingate (q.v.).

Run-out. A term used in reference to a leakage of molten metal from a furnace, ladle, crucible or mould.

Run-out table. A flat area for receiving rolled metal as it emerges from the rolls of a rolling-mill.

Rupture. A term used in reference to the failure of a member of a machine or structure by actual breaking. The rupture strength, also called the breaking strength is the nominal stress developed in a material at rupture. In the general formula for the strength of a beam $M = fI/y$, where M = the bending moment, f = stress, I = the moment of inertia of the section and y = distance from the neutral axis to the outermost fibre, the value of f when M is the bending moment which would break the beam is called the modulus of rupture. It is not the same as the ultimate strength.

Rust. A reddish-brown, porous and brittle coating that forms on iron or ferrous materials on exposure to moist air or when wholly or partly immersed in water containing dissolved oxygen. The rate of rust production is proportional to the concentration of the dissolved oxygen up to a certain limit only. In all cases ferrous (iron II) oxide (or hydroxide) is first formed, and this may be partly or completely oxidised to the ferric (iron III) state. In the latter case the rust is reddish-brown in colour $Fe_2O_3.xH_2O$, in the former case it is much darker, due to the presence of some magnetic oxide of iron Fe_3O_4, which is black. Rust also frequently contains carbonate of iron. Generally the oxide coating on iron or steel is not protective so that the destructive action is progressive.

Wrought iron and malleable cast iron are more resistant to progressive corrosion than other forms of iron, since a granular form of magnetic oxide may be formed.

Rusting. The process of the corrosion of a metal, especially iron, by surface oxidation in moist atmospheres or in presence of water containing dissolved oxygen. The oxidation of zinc in the atmosphere produces white zinc oxide ("white rust").

Rust inhibitors. A term used in reference to substances which, in contact with ferrous materials, reduce or suppress the tendency to the formation of rust. Soluble salts, e.g. sodium hydroxide, sodium phosphate and potassium chromate, containing ions that form sparingly soluble salts with the metal to be protected act as anodic inhibitors if present in sufficient amounts, but if the amounts used are insufficient to prevent corrosion, intense localised attack may occur. Cathodic inhibitors such as zinc sulphate reduce the rate of absorption of oxygen through the deposition of a coating of metallic (zinc) oxide, while substances such as arsenic and antimony reduce the rate of corrosion of metals in contact with dilute acids. Organic substances of high molecular weight and certain pigments act as rust inhibitors by the formation of a film on the surface of the metal which hinders the absorption of oxygen. Good inhibitors of this class are litharge, red lead, zinc oxide, zinc dust, usually in an oil or paint medium.

Rust joint. The term used in reference to a method of joining cast-iron parts in which powdered iron is oxidised in the presence of sal ammoniac by manganese dioxide or by a substance electro-negative to iron. The space between the ends to be joined is packed with a mixture consisting of 80 parts of iron filings, 2 parts of sulphur and 1 part of ammonium chloride. The mixture is made into a paste with water when a galvanic action occurs. Iron oxide is formed, and the material swells and hardens to a solid mass.

Rustless irons. A name applied to corrosion resisting iron-chromium alloys containing 12-14% Cr. They have a low carbon content, usually less than 0.10%, are resistant to acids and can often replace stainless steel in applications where high hardness and high tensile strength are not required. Tensile strength varies considerably according to heat-treatment, but values up to 1 GN m^{-2} are frequently obtained with a maximum hardness of about 320 B.H.N. Rustless irons are malleable and capable of being pressed into a variety of shapes. Normally there should be less than 1% Si and less than 1% Mn present. Ferritic rustless iron may contain 16-30% Cr.

Another type of rustless iron contains 18% Cr, 2% Ni. Also called stainless iron.

Rust prevention. A term used in reference to the method of preventing the formation of rust by the application of oily or greasy coatings which can readily be wiped off or readily dissolved away when required. Rust-prevention greases often consist of thick mineral oil or petroleum jelly mixed with finely divided graphite. The essential property of the oil is a low vapour pressure, so that it does not readily evaporate. Finished surfaces may be protected with a coating of a mixture of white lead and tallow. Tools are often protected in storage by applying a coating of a mixture consisting of equal parts of turpentine and linseed oil. Tools having fine edges, e.g. razor blades, are protected in storage by applying a small amount of sodium benzoate to the wrapping paper. Wrapping material impregnated with organic inhibitors is available.

Rust proofing. The name applied mainly to those processes which are intended to prevent the formation of rust on the surface of ferrous metals and which are to be followed by treatments in which protective coatings are applied. For this reason many rust-proofing processes are regarded as pre-treatments, since they are to be followed in most cases by sealing with oil or by painting or enamelling. (*See also* **Coslettising, Parkerising, Bonderising.**)

Rust remover. The name used in reference to those compounds or solvents that are used to loosen or dissolve rust from the surface of ferrous metals. Paraffin either alone or mixed with oil is sometimes used, but the surface must be treated with a preventative after such treatments. Most of the rust removers consist of dilute solutions of mineral acids containing an organic inhibitor.

Ruthenium. A very scarce metal that is grey in colour when powdered, but resembling platinum when massive. It is obtained from the treatment of platinum ore residues remaining after the *aqua regia* extraction. The residue is heated with sodium peroxide which leads to a solution of sodium ruthenate. This is treated with chlorine and the liberated ruthenium tetroxide is absorbed in hydrochloric acid. The ruthenium chloride so obtained is reduced to ruthenium metal. The pure metal is white in colour. It has not found many applications as yet. It is used mainly as an alloying element to increase the hardness of platinum alloys for electrical contacts and palladium alloys for jewellery. It is much harder than the other metals of the platinum group, having a B.H.N. of 250-450 D.P.N. when annealed and 800-1 300 D.P.N. when electrodeposited. Electrical resistivity at 0 ℃ is 7.6 microhm cm. Melting point 2,310 ℃; density 12.3 g cm^{-3}; Young's modulus 465 GN m^{-2}, tensile strength (annealed) 510 MN m^{-2}. The crystal structure is close-packed hexagonal and the lattice constants at 20 ℃ are $a = 2.706 \times 10^{-10}$ m, $c = 4.282 \times 10^{-10}$ m. Neutron capture cross-section for thermal neutrons is 3.0 barns. (*See* Table 2.)

Ruthenium plating. A rather dark deposit of this metal can be obtained by using a solution containing, per litre, 20 cm^3 of sulphuric acid and 4 g ruthenium nitrosochloride.

Ruthenium red. This is a compound having the name ruthenium tetra-amminohydroxychlorochloride. It stains silk a fast red colour, but does not affect cotton. It is also used for staining tissues—both bacterial and animal. It is also known as ammoniated ruthenium oxychloride.

Rutherford-Bohr atom. In Rutherford's original proposal of the nuclear theory of the atom it was necessary to assume that the electrons had a very rapid planetary motion round the nucleus, so that the centrifugal force would just counterbalance the attractive force towards the nucleus. But the electromagnetic theory requires that the electron should emit radiation continuously during its revolutions, and this would result in the electron describing a spiral course and eventually falling into the nucleus. To overcome these difficulties in the conception of the atom, Bohr modified Rutherford's theory by suggesting that the electrons moved in closed orbitals, and so long as this was maintained no radiation was emitted. He also postulated that for any electron more than one stable orbital was possible, and each orbital was regarded as an energy level. When an electron "jumps" from one orbital to another of lower energy level the surplus energy is emitted as radiation of a specific frequency. The changes in energy levels give rise to the production of absorption lines in absorption spectra. This modified conception of the Rutherford atom by the introduction of Planck's quantum theory is known as the Rutherford-Bohr atom.

Rutile. One of the two commercially important ores of titanium. It is titanium oxide TiO_2, and the mineral contains about 60% Ti. Rutile crystallises in the tetragonal system, usually as short translucent prisms. The crystal structure is characteristic of AB_2 compounds of reasonably close packing, and similar to many oxides of transition and heavy metals.

The material is used as a constituent of protective coatings applied to welding electrodes and also extensively in paint, because of its good covering power.

Rutile occurs as grains in alluvial deposits and as a primary mineral in acid igneous rocks, gneisses and mica schists and some limestones. It is found in the beach sands of Brazil, Florida, Australia and India. Specific gravity 4.18-4.2, Mohs hardness 6.5.

Rydberg series. This is a series consisting of numbers based upon the atomic numbers of the inert gases. These are helium 2, neon 10, argon 18, krypton 36, xenon 54 and radon 86. Rydberg pointed out that these atomic numbers are all even numbers and that they form the series: $Z = 2(1^2 + 2^2 + 2^2 + 3^2 + 3^3 + 4^2 + \ldots)$. (*See* **Atom** and **Spectrum.**)

S

Sacrificial corrosion. If zinc and iron are made the electrodes in a simple cell, then on joining them a weak current passes and the zinc goes into solution. When iron is galvanised and a small break in the zinc coating occurs, a simple galvanic cell will result when atmospheric water is present. But it is the zinc which passes into solution, while the iron—at least those parts close up to the zinc coating—remains uncorroded. This behaviour of the zinc in "sacrificing" itself, so as to protect the exposed surface of the iron, is known as sacrificial corrosion.

Sacrificial protection. When copper or one of its alloys is in contact with an electrolyte, e.g. sea-water, a method of protection against corrosion consists of supporting one or several pieces of zinc in the near vicinity of the copper and making metallic contact between the former and the latter. The zinc, which is easily replaceable, dissolves, while the copper is unattacked. In a similar application pieces of zinc may be suspended in a steel steam-boiler with the object of protecting the steel shell at the expense of the replaceable zinc.

Saddening. An operation in which an ingot is given a succession of light reductions in a press or rolling-mill or under a hammer. The object of this is to break down the skin and overcome the initial fragility due to a coarse crystalline structure preparatory to re-heating prior to heavier reductions.

Saddle-weld. A term used in reference to all types of fusion weld where the weld metal is mainly or entirely external to the surfaces that are joined.

Saffron bronze. An alkali metatungstate of attractive colour used as a pigment. When tungsten trioxide is fused with metallic sodium in excess and then heated in a current of hydrogen, followed by digestion with water, the crystals so formed are orange-yellow in colour. It has proved useful as a "bronze".

Saggers. (Also spelled seggers). A term used in reference to the receptacles that are made of refractory material or of cast or wrought iron. The former are usually in the form of crucibles and are employed for firing ceramic ware, while both types may be used in the annealing of castings. These latter are packed in the cast-iron boxes with red hematite ore and after the lid has been luted on are heated in some form of annealing furnace. In firing ceramics, the articles are usually separated in the boxes by flint which is infusible at the firing temperature.

Salamander. A term used to describe the large mass of iron that is deposited at the base of a blast furnace after it has been working for some time.

Sal ammoniac. A commercial term used in reference to the volatile white solid known otherwise as ammonium chloride $(NH_4).Cl$. It is used as the electrolyte in Leclanché cells and in making dry batteries. Its ready dissociation on heating may explain its action as a flux in soldering. Any oxides are converted into volatile chlorides by the hydrochloric acid

$$NH_4Cl \rightleftarrows HCl + NH_3$$

and a clean surface is left. It is often used as a source of ammonia, since it readily evolves this gas on gently heating with soda lime. It is a common reagent used for patinating bronzes.

Salt bath. The term used in reference to the bath of molten salts that is used for the heating, hardening or tempering of various alloys. The commonest salts used are shown in Table 52, which also gives the approximate melting points. These salts are used in combination with one another in controlled proportions. Sodium and potassium nitrates have lower melting points than the others given, and they are generally used for tempering baths. The chlorides and cyanides, all of which fuse at 780 °C or above, are used for hardening baths. Salt baths (containing molten salts of specific gravity between about 2 and 3.2) give uniform heating and prevent oxidation during the process.

TABLE 52. *Melting Points of Salt Baths for Heat Treatment*

Salt	Fusion Temperature, °C
Sodium nitrate $NaNO_3$	307
Potassium nitrate KNO_3	334
Sodium nitrite $NaNO_2$	271
Potassium nitrite KNO_2	440
Sodium chloride $NaCl$	801
Potassium chloride KCl	776
Barium chloride $BaCl_2$	960
Calcium chloride $CaCl_2$	772
Sodium cyanide $NaCN$	564
Sodium carbonate Na_2CO_3	851
55% $NaNO_3$ + 45% KNO_3	240
50% $NaNO_3$ + 50% $NaNO_2$	200
40% $NaNO_3$ + 20% KNO_3 + 40% $NaNO_2$	175
73% $CaCl_2$ + 27% $NaCl$	540
33% $BaCl_2$ + 33% $CaCl_2$ + 34% $NaCl$	630
78% $BaCl_2$ + 22% $NaCl$	670
45% $BaCl_2$ + 55% $CaCl_2$	615
40% Na_2CO_3 + 40% $NaCl$ + 20% $NaCN$	550

Salt cake. An alternative name for commercial purity sodium sulphate Na_2SO_4. It is used as a neutral flux in lead and copper smelting and as a means of separating copper and nickel in the "top-and-bottom" process. In the latter case it functions as a source of sodium sulphide.

Saltpetre. A white crystalline substance having the formula KNO_3, which can be obtained from the treatment of certain soils with wood ashes K_2CO_3. In soil containing decomposing nitrogenous matter the latter is probably converted into ammonia and then oxidised to nitrate by the activity of certain bacteria in the presence of weak alkalis. It is used as a fertiliser, for pickling meat, as a powerful oxidiser and in the molten condition as a tempering bath.

Also known as potassium nitrate and as nitre. Chile saltpetre is sodium nitrate. Specific gravity 1.93, Mohs hardness 2.

Salt spray test. An accelerated corrosion test in which specimens are exposed to a fine mist of salt water. The standard conditions of the test must be observed. The solution should contain 20% by weight of sodium chloride, having a pH value of 7-8, and maintained at a density of 1.15 at 16 °C, while the temperature within the test cabinet should be between 18 °C and 25 °C. The capacity of the cabinet should be not less than 0.15 m³ and the air within it should be controlled to within the range 70-100 kN m⁻² pressure. A baffle should protect the specimens from direct impingement of the jet. A usual procedure is to maintain the spray for 6 hours per day, the specimens being retained in the cabinet throughout the test.

The test is intended to assess porosity and the resistance of material to atmospheric and marine corrosion.

Samarium. One of the rare-earth metals that was discovered in Samarskite (q.v.). It is a greyish-white metal of specific gravity 7.54. The melting point is given as 1,072 °C. The metal tarnishes in air. Its salts, of the general formula SmX_3, are generally topaz-yellow in colour. Atomic number 62. (*See* Table 2.)

Samarskite. A tantalate and niobate of iron, calcium and uranyl containing also metals of the cerium and yttrium groups. It is rather abundant in certain parts of America, notably North Carolina and New Mexico. It contains 10-15% uranium oxide (U_3O_8) and about 37% Nb_2O_5.

Sand. The name applied to the loose incoherent mass of small particles consisting mainly of such minerals as quartz, mica and felspar. It has been formed by the disintegration of rocks by weathering. Some grading has occurred in most deposits due to the fact that the important transport agency has been moving water. Sands containing clay are known as naturally bonded sand, while those that have been well washed contain little or no clay and are called "sharp" sands. Pure sand is white in colour and consists of silica SiO_2. It is highly refractory, but small amounts of impurity cause it to fuse at casting temperatures, thereby producing faulty castings in moulds made of such sand.

Sand is usually graded by taking a 50-g sample and, after carefully drying, separating it into portions by shaking through a number of sieves of successively finer meshes. A typical grading chart for silica and for bonded sand is shown in Fig. 146.

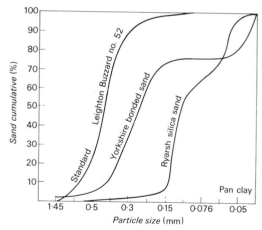

Fig. 146 *Typical grading chart for silica and bonded sand.*

Sandberg treatment. A sorbitic treatment in which carbon-steel components are moderately hardened either wholly or in part. It consists in cooling the parts to be hardened through the transformation range at a moderately rapid rate using jets of air, steam or water spray for the purpose. The part is then permitted to cool in air, allowing the residual heat in the component to bring about a tempering effect.

Sandbinders. The term used to describe the various additions that can be made to silica or fine sand to give it the coherency required to make satisfactory moulds for cast metals. Ordinarily moulding sand contains: (*a*) the refractory constituent, e.g. silica, alumina; (*b*) the binding constituent, e.g. alumina, which bakes hard on heating; (*c*) compounds of iron, lime, magnesia and alkali metals which can be tolerated in only small amounts, since they tend to

cause the sand to fuse at the casting temperatures. The usual naturally occurring binding constituents are fireclay (coarse), fullers' earth (medium) and bentonite (smooth).

A number of organic binders are used in reclaiming used sand and in the production of cores from what are referred to as oil sands. The most usual of these are: (*a*) molasses; (*b*) sulphite lye, dextrin, starch (or flour), linseed oil, resin, pitch.

Sandblasting. A method of removing scale from and of cleaning metal surfaces by directing a stream of sand or grit on to the metal at very high velocity. The method is also used on sheet-metal surfaces to form a key for a variety of finishes such as painting, enamelling, etc.

Due to the possibility of silicosis affecting workers operating with fine sand, all blasting is now with grit or shot, in well-ventilated cupboards, preferably with gloved entry for handling.

Sandbuckle. The term used in reference to the rough irregular projections that are loosely attached to the surface of a casting. The sandbuckle generally contains substantial proportions of embedded sand, and when detached leaves a depression in the surface of the casting. (*See also* **Scab.**)

Sand burning. A hard surface produced on sand castings by a reaction between the hot metal and the sand of the mould.

Sand casting. The term used in reference to metal which has been allowed to solidify from the molten state in a form that requires only minor modification of shape such as can be effected by machining or surface finishing.

Sand control. A term used to describe the operations of adjusting the proportions of the various constituents of sand so that desired physical properties may be secured. The object of this control is eventually to obtain castings of improved quality, and the properties most usually assessed are: (*a*) fineness; (*b*) permeability in respect of molten metal as well as of gases; (*c*) moisture content; (*d*) green strength; (*e*) dry strength.

Sand fritting. When sand containing alkali impurities is heated to moderate temperatures or when pure sand is heated to higher temperatures it partly melts and the particles stick together. This is known as fritting, and in making furnace linings a charge of sand is heated sufficiently to cause fritting before the first sprinkling of working lining is put in.

Sandholes. The term used in reference to casting defects caused by loose foundry sand in the mould cavity. The holes are irregular in shape and contain the sand that gave rise to them.

Sanding. The term used by electroplaters and others to describe a polishing process in which a hide wheel fed with a suitably fine abrasive and a lubricant are used.

Sand leaching. A term used in reference to the treatment of auriferous sands in the extraction of gold. The sands are spread over the surface of the leaching vat and are then percolated with a solution of potassium cyanide.

Sand marks. The name applied to the elongated areas of refractory material sometimes found rolled into the surface of sheet or strip steel.

Sandslinger. The name of a machine used to deliver sand into a moulding box. It consists of two or three arms pivoted end to end and each one carrying a moving belt. The end arm also carries a cup which rotates at high speed at right angles to the belt. The cup is housed in a cylindrical head, through the base of which the sand is ejected. Sand is fed from a large hopper on to the first belt and then to the second and third belts till it reaches the cylindrical head. The cup, which rotates at about 1,400 r.p.m. cuts cupsful of sand and slings them through the base of the head at about 0.5 m s⁻¹ into the mould. The pivoted arms enable the delivery head to be moved about so that the sand can be packed against the pattern in a uniform manner, and can be delivered—a little at a time—to the desired location.

Sandstone. The name given to any stone composed of an agglutination of grains of sand, whether calcareous, siliceous or of any other mineral nature. The grains are cemented together with oxide of iron or other materials. Silica occurs in three forms which are designated: (*a*) crystalline; (*b*) massive; and (*c*) fragmentary. Rock crystal is typical of (*a*), quartzite of (*b*) and sandstone of (*c*).

Sand testing. Foundry sands are subjected to a number of tests to determine if the material is suitable for a particular application or to ensure that the sand properties are maintained within certain specifications in order that there shall result a uniformity of product.

(a) Moisture test. (1) Acetylene (ethyne) test. About 4 g of sand are mixed with a quantity of calcium carbide in a container fitted with a pressure gauge. The pressure depends upon the volume of C_2H_2 produced and this in turn upon the moisture present. (2) Oven-drying or air-drying. In the former case the sand is heated in a steam oven at 105 °C-110 °C till the weight is constant. In air-drying a current of warm air is passed on to the sample on a scale pan and drying continued till the weight is constant. (3) An electrical device is sometimes used. Its action depends upon the facility with which a current will pass through damp sand. In all cases the moisture weight is expressed as a percentage.

(b) Permeability. In this test a standard specimen 5 cm high and 5 cm diameter is rammed and then placed in a cylinder and connected to an air reservoir. The permeability is expressed as the volume of air (in cm^3) that passes per minute under a pressure of 1 g cm^{-2} through a specimen having 1 cm^2 end of section and 1 cm high. The test is carried out on dry and wet sand.

(c) Compression. The standard test piece (5 cm high) is put on the plate of a testing machine that continuously registers the load and the specimen compressed till it fails. Tension and shear tests are also sometimes made. This test is also carried out "dry" or "wet". The strength depends upon the drying temperature.

(d) Elutriation test. A test employed to determine the clay present in different grades of sand. A weighed specimen is boiled with dilute alkali to separate the clay and the fine clay removed using a special elutriator. The sand is dried and weighed and the loss expressed as a percentage.

(e) Sieve test (q.v.). In this test the sand is classified according to grain size.

	Diameter between					
Grain size	1 and 2 mm	1 and 0.5 mm	0.5 and 0.25 mm	0.25 and 0.1 mm	0.1 and 0.05 mm	0.05 and 0.01 mm
Class	Very coarse	Coarse	Medium	Fine	Coarse silt	Silt

Sandwich rolling. The term used in reference to the rolling of dissimilar metals to form a composite sheet of clad metal or plymetal.

Saniter's process. A process for reducing the sulphur content of molten pig iron by treatment with a mixture of lime and calcium chloride or calcium fluoride.

Sap. In the production of blister steel by the cementation process the bars when "converted" differ in appearance from the original bars, as packed. Originally they were fibrous and tough, but as taken from the boxes they are crystalline and brittle. But there is more than one temper in each pot, according to the position of the bar. The bars are therefore broken across and stacked according to the temper. There are at least six classes of blister steel, classified according to the depth of penetration of the carburisation. In the lower-number classes there is a centre portion of the bar that remains unaltered, and this part is called the sap. The names spring heat, Irish temper and country heat are applied to blister steels that contain much sap or unaltered steel in the centre.

Sapphire. A blue variety of corundum Al_2O_3 used as a gemstone. The colour is due to the presence of small amounts of oxides of cobalt, chromium or titanium. Artificial sapphires may be prepared by strongly heating, in a reducing atmosphere, alumina to which has been added about 1.5% magnetic oxide of iron and 0.5% titanium oxide TiO_2. The mineral crystallises in the rhombohedral form and the crystals are dichroic. They are found in illam (the gem-gravel of Sri Lanka), Thailand, the gold-drifts of Australia and many parts of North America. Specific gravity about 4, Mohs hardness 8-9.

Sapropetic coal. The name used in reference to coal of Palaeozoic age, which is sub-divided into two main groups: *(a)* cannel and *(b)* boghead (torbanite in Scotland). It contains a large proportion of volatile matter which renders it suitable for use in gas manufacture. The first variety leaves little ash and burns with little

smoke. The second variety leaves a large percentage of ash on burning.

Sard. An amorphous or crypto-crystalline variety of chalcedony, silica SiO_2 having a reddish-brown colour. It is usually brown by reflected light and red by transmitted light (*see also* **Chalcedony**). It is a semi-precious stone.

Sardonyx. A flat banded variety of chalcedony having white or bluish-white layers alternating with strata of sard (q.v.). It is a semi-precious stone and was much valued for use in gem cutting. (*See also* **Onyx.**)

Sassolite. A native form of boric acid H_3BO_3 produced as a result of gaseous emanations from the volcanic fumaroles of Tuscany. It crystallises in the triclinic system. Specific gravity 1.48, Mohs hardness 1.

Satin spar. The name given to the fibrous crystalline mass in which the dihydrate gypsum $CaSO_4.2H_2O$ occurs. The opaque variety is known as alabaster (q.v.).

Saturated hydrocarbon (alkane). A compound of carbon and hydrogen in which there are only single carbon-carbon bonds. (*See* **Hydrocarbon.**)

Saturation (magnetic). *See* B-H loop.

Sauveur's diagram. A chart from which the approximate carbon content of steel may be inferred if the proportions of pearlite, ferrite and cementite are known. In slowly cooled carbon steels the approximate structural composition may be found by reference to the carbon content or by microscopical examination. If in the latter case the proportions of the constituents can be estimated, then the diagram (*see* Fig. 147) can be used to find the approximate carbon content.

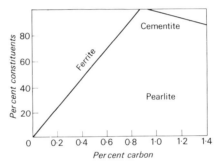

Fig. 147 Sauveur's diagram.

Sawdust drying. A method used for drying electroplated articles after plating and washing. Hot sawdust is used to soak up the residual moisture and traces of plating solution that might give rise to staining if not removed. With articles of complicated shape the complete removal of all the sawdust particles may be difficult. The process consists in rubbing the articles in sawdust in shallow trays.

Saxifragin. An explosive powder made by mixing 76 parts of barium nitrate with 22 parts of carbon and 2 parts of nitre. It is for use in metal mines and quarries. The small amount of saltpetre or nitre lowers the temperature of explosion.

Scab. An irregular bulge on the surface of an ingot or casting. It may be caused by a portion of the surface of the mould breaking away or being washed away by the flow of molten metal.

Scale. The term used in reference to the coating of oxide which forms on the surface of the heated metal. During working, the scale is cracked and broken off, and according to the method of doing this, the scale is called roll scale, mill scale, hammer scale.

Scale breaking. A process of passing hot-rolled strip-steel through a machine in which it is subjected to a series of reverse bends. This process shatters the scale and also suppresses the tendency to kink on further manipulation.

Scaling. The formation of a surface oxidation coating on metals at elevated temperatures is known as scaling. The term, however, is often used in reference to the removal of the scale by such processes as pickling, etc.

Scalping. The term used in reference to the removal of the surface layers of ingots or billets by machining, etc., prior to fabrication. Usually the top portion of an ingot is scalped or cropped, thus removing the portion containing the primary pipe. Other methods commonly used to scalp the ingot are by chipping, milling, planing or by means of the oxyacetylene (oxyethyne) torch.

Scandium. A metal of the rare-earth group of metals. The richest minerals are wiikite and orthite, both found in Finland and associated with tin-tungsten ores. The metal is related in its chemical and physical properties to aluminium. Crystal structure face-centred cubic; $a = 4.532 \times 10^{-10}$ m. (*See* Table 2.)

Scapolite. A term applied to a number of minerals composed of varying amounts of calcium aluminium silicate

$$4CaO.3Al_2O_3.6SiO_2$$

and silicate of sodium and aluminium with sodium chloride $2NaCl.3Na_2O.3Al_2O_3.18SiO_2$. It occurs in impure limestones and as an alteration product of a basic plageoclase felspar, in igneous rocks.

Scarfing. The process of removing defects from the surfaces of ingots, bars, etc., by means of a gas torch.

Scarf joint. A type of joint suitable for welding wrought iron or steel (*see* Fig. 148(*a*)). A similar joint may be used for timber work (*see* Fig. 148(*b*)). This type is very suitable for producing a joint resistant to tensile loads.

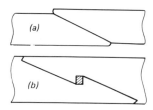

Fig. 148 *Scarf joint.*

Scattering cross-section. A measure of the interaction of a nucleus with radiation, which results in elastic scattering. The number of such interactions in unit time divided by the radiation flux and the number of nuclei present gives the cross-section, that area perpendicular to the direction of the radiation to which is attributed the effect of the nucleus to account geometrically for its interaction with the radiation.

Scavenging. A term used in engineering in reference to the removal of burnt gases from the cylinder of an internal-combustion engine. It is also used in reference to the removal of gases from molten metal, and in the process unwanted oxide may also be removed. Scavenging treatment may consist of bubbling an inert gas through the molten metal. The dissolved gas diffuses into the bubbles of the inert gas until its pressure is the same in the bubble as in the metal. As the bubbles are constantly being removed, the final pressure of gas in metal is reduced to a very low value. In this way, by using dry nitrogen, hydrogen can be removed from molten copper or its alloys. In the case of aluminium, nitrogen or chlorine or mixtures of these are used. The carbon dioxide and carbon monoxide evolved during "boil" in the open-hearth process act in much the same way in removing nitrogen from this kind of steel.

Schaeffler's diagram. A type of constitutional diagram for stainless-steel welds. The alloying elements essential in the composition of stainless steel are nickel and chromium. The former is described as an austenitiser, and other elements favouring the formation of austenite are C and Mn. Chromium is described as a ferritiser, and other elements promoting the formation of ferrite are Mo, W, Ti, Nb. When the composition of the stainless steel is known a point on the diagram can be found for it. The horizontal axis represents % Cr + % Mo + $\frac{3}{2}$ × % Si + $\frac{1}{2}$ × % Nb, and the vertical axis represents % Ni + 30 × % C + $\frac{1}{2}$ × % Mn.

The diagram is divided into regions indicating those compositions in which martensite (M), austenite (A) and ferrite (F) predominate, and this is the general basis of the classification of stainless steels. The defects of the various types due to: (*a*) cracking; (*b*) brittleness are additionally indicated by shading. All the low-alloy stainless steels are martensitic or mainly so (*see* Fig. 149).

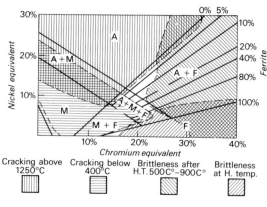

Fig. 149 *Schaeffler's diagram.*

Scheele's green. A bright green salt copper (II) hydrogen arsenite, $CuHAsO_3$. It may readily be made by adding a solution of ammonium arsenite to one of copper sulphate. It was used as a pigment for colouring wallpaper, but this application has now been discontinued on safety grounds.

Scheelite. The name given to a heavy mineral discovered by Scheele and found by him to consist of calcium tungstate $CaWO_4$. The ore occurs in Zinnwald (Czechoslovakia), Cumberland (UK), the Voseges Mountains in Italy and in North America. The crystals usually contain iron and are found in crystalline rocks in association with tin ore, topaz, apatite and wolfram. When true the ore is white and contains about 86% tungsten oxide. The specific gravity is about 6. In ultra-violet light it fluoresces a strong and characteristic blue, but small amounts of molybdenum alter the emission to white and then to yellow. With amounts of molybdenum between 0.35% and 1% the fluorescence is white, and with amounts of 4.8% the colour is strongly yellow. This fluorescent property is used in mining and milling the ore for prospecting and testing the tailings, respectively.

Schists. These are metamorphic rocks which are fissile in character. They are completely recrystallised rocks produced by pressure, and hence shearing, on all kinds of rocks. They all have a well-marked cleavage along the plates (mica) or along the bands or folia (tourmaline). Sedimentary rocks that have been little altered by recrystallisation are not included (e.g. slates, shales and sandstones). The sedimentary schists include the mica schists, the altered impure limestones (calc schists) and the altered sandstones (quartz schists). The igneous or orthoschists include white mica schists, e.g. porphyries, the hornblende schists, e.g. the modified forms of basalt and dolerite, and the metamorphised siliceous dolomitic limestones. Schists are distinguised from gneisses, since the latter are of coarser structure and are rarely foliated.

Schoen mill. A type of universal rolling-mill characterised by having horizontal rolls, two or three high for regulating the thickness of the plate, and vertical rolls for regulating the width of the plate. The mill is specially used for rolling wheels for rail-cars.

Scholl's method. A term used in reference to a rapid and accurate method for determining the amount of uranium in any of its ores. The uranium is extracted with dilute nitric acid (1:1). The extract is diluted, filtered and treated with ferric (iron III) chloride and sodium carbonate in order to precipitate the vanadium, iron and aluminium. The uranium is then precipitated from the filtrate by boiling with caustic soda and purified by solution in nitric acid. After precipitating with ammonia the ammonium uranate is ignited to the oxide U_3O_8 and weighed. This weight multiplied by the factor 0.847 gives the weight of uranium.

Schoop process. The name used in reference to a process in which a metal powder is driven at high velocity against a metal or other surface to be coated, by means of gaseous jets allowed to expand from high pressures. The name is also used to include some other methods in which the metal is forced under high pressure through a nozzle and sprayed with a jet of high pressure gas; or is fed in the form of wire through a blow torch. (*See also* **Metallisation, Schori process, Schoop spraying.**)

Schoop spraying. A method of producing sprayed metal coatings on metal or other surfaces. The metal to be sprayed, in the form of wire, is fed into an oxyhydrogen or oxyacetylene (oxyethyne) flame by means of a variable speed, air-driven turbine at a speed just sufficient to permit melting. The liquefied metal is atomised and projected by an air-blast surrounding the hot flame.

Schori process. A process for coating metallic or other surfaces with a metal in which a stream of the coating metal, in the form of powder, is fed into a flame and blown on to the surface to be treated. High pressure air is used. Also called the "powder process".

Schorlomite. A titanium ore in which titanium oxide occurs in combination with the oxides of iron and calcium $(Ca,Fe)TiO_3$. It contains from 12% to 22% Ti.

Schottky defects. The migration of an atom from its normal place in a lattice to the surface, or to another position within the lattice which does not create a further disturbance, i.e. to the surface of a void or other sink of disordered atoms. The local result is the occurrence of a vacant lattice site in a position formerly occupied by an atom.

Schriebersite. A phosphide of nickel and iron in which the nickel is probably present as Ni_2P. It has been detected in meteorites.

Schrödinger equation. The consideration of the wave motion of electrons requires special treatment in view of the small size of the particle and the energy involved. (*See* **Probability distribution.**) Schrödinger derived equations to cover the behaviour of electron wave packets under various influences of potential, and equated these to the electron distributions in atoms. (For treatment of some of these considerations *see* **Wave mechanical behaviour of electrons.**)

Schulze-Hardy rule. A rule which states that the ion causing co-agulation of a sol is of the opposite electrical sign to that of the colloid particle and the coagulating power increases with the increasing valency of the ion. Normally electrolytes when added to a lyophobic sol cause the particles to coagulate so that a precipitate is finally produced. The electric charge that is normally present on the colloidal particles prevents them from coalescing, but the addition of an electrolyte provides ions which can be adsorbed on the colloid and the particles becoming neutral are then able to coalesce.

Schumann rays. The name sometimes used in reference to the extreme ultra-violet rays of the solar spectrum or in metal spectra generally. These rays are absorbed by air and by thin films of gelatin, hence their presence, etc., which is usually determined photographically, must be done *in vacuo,* using a silver bromide film that is prepared without gelatin.

Schweinfurter green. A brilliant green pigment which is a com-pound of cupric (copper (II)) arsenite and cupric (copper (II)) acetate $Cu_3(AsO_2)_2.Cu(C_2H_3O_2)_2$. It is usually made by boil-ing verdigris (basic acetate of copper) with arsenious (arsenic III) oxide and acetic acid.

Schweizer's reagent. Ammonia readily precipitates cupric (cop-per II) hydroxide from solutions of copper II salts and the precipitate dissolves in excess of ammonia, forming a deep blue solution. This is known as Schweizer's reagent, and is used as a solvent for cellulose (paper, linen, cotton wool, etc.). If the solution of cellulose is squirted into dilute acid, a thread of cellulose is obtained, which is one variety of "artificial silk". When applied to canvas the reagent produces a water-tight coating of amorphous cellulose. This is known as Willesden canvas.

Scleroscope. An instrument operating on the rebound principle used for testing the hardness of materials. It consists essentially of a light hammer (weight is under 25 g) with a diamond point. This is contained in a vertical glass cylinder on which is engraved a scale of hardness values, the lowest being at the bottom. The hammer is brought to the top of the tube and held there by a partial vacuum (sometimes an electromagnet is used). On being released the hammer strikes the specimen and rebounds. The height of the rebound determines the hardness on the scleroscope scale. This is a conventional scale on which 100 represents the average hardness for quenched high-carbon steel. A value higher than 100 would indicate a higher impact resilience and a lower energy absorption. The scleroscope hardness is regarded as a con-venient control property, but it has rather limited significance

and is considerably affected by the variations in roughness of the surface under test.

Scoria. A term used to describe the cinder, dross or slag that is found after the fusion of metallic ores.

Scoriae. A term used in reference to the vesicular masses of vol-canic rock produced by the escape of gases from the viscous mass, and resembling clinker from a furnace.

Scorification. The name applied to a refining process by means of which scum and impurities are removed from the metal. It may be used in the case of copper and in the treatment of impure lead containing precious metals for assaying of the gold and silver. In the case of rich gold ores (rarely found) the ores would be mixed with granulated lead and borax as a flux. On heating to a high temperature in a muffle furnace some of the impurities volatilise and the rest combine with the oxidised lead and borax to form a slag. The method is more usually confined to the assay of lead-silver ores.

Scoring. A term used to describe a form of frictional wear that takes place as a result of the high relative speed of metals in con-tact. When the speed is sufficiently high to cause small particles to be carried along the surface, they cause long scratches to occur. Such markings may be found in drawn metal, where the effect may be due to grit in the lubricant as well as to rough tools.

Scotch hearth. A type of blast furnace in which the height of the furnace does not greatly exceed the width. It consists of a cham-ber lined with refractory material, and a blast is introduced through one or more tuyères which can be made to impinge on the surface of the charge from above or can be introduced in a horizontal direction just under it. This enables both oxidising and reducing conditions to be obtained. The Scotch hearth is less used now than formerly for the reduction of galena or for the treatment of the slag from the reverberatory furnaces in the preparation of lead. A smith's hearth can be regarded as a type of Scotch hearth, though the furnace in this case is not enclosed. The blast is cold.

Scotch kiln. A type of kiln or oven used for burning bricks. The bricks in open formation are set over a furnace well at the base of the kiln and the burning of the bricks proceeds intermittently, the fire being drawn between each charge.

Scotch tuyère. A type of nozzle for introducing the air blast into a blast furnace, in which the tuyère is kept cool by circulating cold water in a coil of wrought-iron piping that is coiled on the tuyère.

Scott furnace. A type of furnace used in the treatment of fine and coarse ores of mercury. It is based on the Spirek furnace and is built of brick with suitable tie-rods. The ore is fed at the top of the furnace and slowly moves downwards by gravity from one shelf to the next below it. The shelves are made of tile and are inclined at about 45° to the horizontal. The space between one shelf and the face of the next lower one is adjusted to the size of ore that it is proposed to treat. The products of combustion from the firebox together with excess of air pass over the ore in the lowest third of the furnace in one direction, then over the middle third in the opposite direction and finally return over the top third and pass into the flue system and to the condensing cham-bers. Oil, coal or wood may be used in the firebox and the ore is mercury sulphide (cinnabar).

Scouring. A term used in reference to the cleaning of the surfaces of metal plates or components by means of abrasives with or without a liquid carrier. The operation may be carried out in a rotating barrel in which the parts are placed together with scrap metal and steel balls. The abrasive used may be granite chips, sand or alumina for coarse work, but for final scouring or polish-ing putty powder is preferred. Water containing a little alkali or paraffin may be added, and soap is usually included where non-ferrous metals are being treated. The process is not usually considered very suitable for removing scale.

Scragging. The term used in reference to a method of increasing the load-carrying capacity of a helical compression spring by pre-stressing, i.e. by the introduction of favourable internal stresses. The spring is wound somewhat longer than the required length and then scragged by compressing it to closure several times.

Scrap. A term applied to both ferrous and non-ferrous metals that have been removed from service as no longer suitable for use in their present form, and to metals, components or parts that have been discarded at various stages of manufacture. As scrap metal is a valuable source of material it is usual and advisable that it should be segregated with a view to remelting and reprocessing.

Scratch brushing. A cleaning or finishing process for metal surfaces involving the use of a brush having stiff bristles or wires. The latter are frequently made of nickel or stainless steel and are mounted on a rotating wheel. They vary in thickness from about 0.1 to 0.4 mm. High speeds and light pressure are preferable for polishing and lower speeds for cleaning components prior to electroplating.

Scratch test. A method of assessing the hardness of a metal or a mineral. The test as proposed by Mohs (q.v.) is one of the oldest methods and is generally used by geologists to determine the hardness of minerals on an arbitrary scale. The Martens test is somewhat more precise. It depends on the width of the scratch produced on a material by a diamond point under a specified load. The average width of the scratches, as measured by a microscope, is checked against a chart which provides a figure of relative hardness.

Screen analysis. A term sometimes used in connection with the use of sieves in sand testing in order to differentiate between different types of sand and to assess the suitability of the material as a moulding sand. Usually a 50 g sample of the sand is air-dried and then separated into grades by shaking the grains in a nest of sieves, the coarsest being at the top. The sieves are vibrated for 15 minutes, after which the particles remaining on each sieve are weighed. These weighings are expressed as percentages, and the cumulative percentages of the sand grains retained on the successively smaller sieves are plotted against the log of the sieve aperture.

Screening. A method of sorting ores and similar material by size, using metal sieves. The screens may be made of woven wire or punched plate, for fine and medium sizes. For coarse screening the screen may be made of steel or cast-iron bars suitably arranged, and these kinds of screen are called grizzlies. When a perforated plate in the form of a cylinder is used as a rotating screen, the machine is called a trommel.

Usually sorting by screening is not efficient for particle sizes less than 1 mm. The maximum size of particle a screen will allow to pass is dependent on the screen aperture. This value is not always provided in the description of a screen. The mesh of a wire woven screen is frequently given as the number of holes per unit length, a value which will depend on the diameter of the wire used. A rough working value for the aperture when the mesh number only is available may be found from the formula:

$$\text{Aperture} = \tfrac{1}{2}M$$

where M = mesh number. Screens are used in screen analysis in the testing of sands and for cleaning and aerating sand in the preparation of moulds.

Screw dislocation. A line defect in a crystal lattice, in which some atoms are distorted from true crystallographic positions, into a helical screw type of displacement, such that a shearing force produces ready movement of the dislocation through the undisplaced part of the lattice. The Burger's vector is parallel to the dislocation line. (*See* **Dislocation**.)

Screw stock. The name used in reference to free cutting steel bars suitable for machining into bolts or screws on automatic lathes.

Sculls. A term used in reference to the layers of metal and slag which form on the inner surface of ladles, converters or any vessel used to contain molten metal, particularly if the metal is permitted to cool.

Scum. The term used to describe the dross or scruff that forms on the surface of molten metals due to the rising of impurities to the surface and to the surface oxidation.

S-curve. If a medium carbon steel (about 0.3% C) in the fully austenitic condition at some temperature above 900 °C is slowly cooled, it will be found that at about 810 °C (the upper critical temperature for this steel) ferrite will begin to be rejected from solution, and this will continue down to the lower critical temperature (about 700 °C), when a definite mixture of ferrite and cementite will be precipitated. This mixture is a structurally recognisable constituent known as pearlite.

If the cooling had been very rapid through this same range of temperature, the time element would have been too short to

permit the above changes to occur. Instead there would be produced a supersaturated solution of carbon in α-iron, and this is the extremely hard constituent known as martensite. It is thus clear that to obtain the hard constituent in a quenched steel, the cooling rate through the austenite-pearlite transformation range must be very rapid. At intermediate rates of cooling "intermediate products", e.g. troosite, are formed.

The stages in the transformation of austenite can be conveniently investigated by heating a steel to a temperature at which it is completely austenitic and cooling it very rapidly to some lower temperature but above atmospheric (*see* Fig. 150).

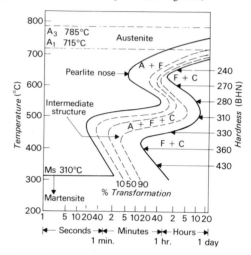

Fig. 150 *S-curve of a 3% Ni steel. A is austenite; F is ferrite; C is cementite; and M is martensite.*

The time taken for the change to start and the transformation rate are determined. Similar determinations are made for a number of sub-critical temperatures and the results are plotted on a graph. The curves so obtained, on account of their general shape, are called S-curves. They are also sometimes called T-T-T-curves (time-temperature-transformation). The study of S-curves has made possible the estimation of the minimum rate of cooling in order to obtain complete hardening in a particular steel. It has also led to the introduction of new forms of heat-treatment, e.g. austempering and martempering (q.v.) (*see* Fig. 150). (*See also* **Isothermal transformation diagram**.)

Sealing. The term used in reference to the process for making anodised coatings on aluminium non-absorbing. The method usually employed is to rinse the components in cold water and then to immerse them in boiling distilled water for some time. The oxide film becomes hydrated and is then more compact and has a greater corrosion resistance. It also preserves the colour of dyed coatings. Other methods of sealing include the use of dilute solutions of nickel or cobalt acetate (ethanoate), or of dilute sodium chromate; the latter, however, usually imparts a brownish-yellow tint to the coating. Sodium silicate is sometimes used but generally produces a less satisfactory appearance than when the ethanoate solutions are used. Sealing by immersion in oils, e.g. linseed or tung, or in naphtha solutions of waxes or greases, e.g. lanolin, are less frequently used than the chemical methods.

Seam. A longitudinal surface defect formed on the surface of wrought metals, e.g. during forging. During the rolling of billets or bars one portion of the metal may be folded over another part. The oxidised surfaces of the seam may subsequently partly weld up. In steels the oxides react with the metal and cause local decarburisation around the defect. It also refers to surface cracks that have formed on metallic surfaces which have been closed (but not welded) and which may have originated from blow-holes. The very fine cracks or seams are sometimes referred to as hair seams. (*See also* **Roke**.)

Seaming. A sheet-metal working process for joining sheet metals by interlocking bends, e.g. side seams of tinned cans.

Seam welding. A resistance welding process in which lapped sheets are joined during passage between rotating disc electrodes. Overlapping spot welds may be made progressively, or the weld may be linear by passage between contact rollers, or between an electrode wheel and a fixed electrode bar.

Searing. A term used in reference to a process for smoothing the surfaces of foundry patterns. It is similar in principle to the domestic process of ironing.

Season cracking. A term applied to intercrystalline cracks which appear, sometimes after a period of years, in fabricated components which have been unevenly cold worked. The phenomenon is due to high and unbalanced internal stresses introduced in the metal by this cold work. It occurs in both ferrous and non-ferrous materials, but is particularly liable to occur in cold worked α-phase brass, such as 70/30 brass. Corrosive environments, e.g. moisture and ammonia, greatly aid the season cracking of brass, and also of mild steel. (*See also* **Stress corrosion.**)

Secant modulus. A term used in reference to the modulus of elasticity or the stress-strain ratio below the elastic limit, of a material. (*See also* **Stress-strain diagram.**)

Secondary bonds. Electron bonds between atoms involving no major disturbance in electron distribution, e.g. polarisation bonds (van der Waals' bond (q.v.)).

Secondary cell. A term now used synonymously with accumulator and storage cell. Two plates in a suitable electrolyte form the cell, and direct current is supplied from an external source. As a result chemical changes occur at the plates and in the liquid, and when these changes are complete the cell is said to be "charged". In this condition the cell can be used as a secondary source of electric current, the output of which causes the reversal of the chemical changes that occur during charging. Commonly lead plates or grids are used, one being impregnated with lead peroxide and the other with spongy lead (in the charged condition) immersed in dilute sulphuric acid. A nickel-iron secondary cell has the electrodes of iron and nickel and the electrolyte is caustic soda. The lead cell gives a voltage of 2 volts and the nickel-iron cell about 1.33 volts. A good lead secondary cell should have a capacity of 4-6 A h kg^{-1} of electrical energy.

Secondary coil. The name used in reference to the coil in an induction coil or in a transformer which delivers the current. It usually consists of many turns of fine wire wound on the primary coil. This latter has comparatively few turns of thicker wire and is wound on a soft iron core.

Secondary creep. The creep of a metal under stress that proceeds at a constant rate, after the initial stage in which there is shown a decreasing rate of creep. On the strain/time diagram secondary creep is represented by the centre portion of the graph, which is practically a straight line. The constant rate of deformation has been regarded as representing a period of viscous type flow, but this is probably true only for very low creep rates (*see* **Creep**).

The view is also held that steady state creep occurs only in pure metals and simple alloys and that complex alloys show only an inflexion between the primary and tertiary stages. Naturally, much depends on stress and temperature conditions and the sensitivity of measuring equipment.

Secondary electrons. When a rapidly moving electron impinges on a metal its progress may be checked so that it may remain in the metal. In such a cases the kinetic energy (= $\frac{1}{2}mv^2$) may be converted into a quantity of radiation of frequency v, where $mv^2 = 2hv$ (h = Planck's constant). This phenomenon is important in X-ray tubes. Usually, however, the arrest is not instantaneous, but takes place in stages, energy being lost at each impact. Since the direction of motion of the electron may change at each impact, the electron may diffuse back to the surface and escape. These back-diffusion electrons are called "secondary electrons". Other ways in which secondary electrons may be produced are: *(a)* by the direct removal of electrons from atoms by a fast-moving primary electron; *(b)* electrons liberated by the X-rays that are generated by the primary electrons, e.g. photo-electrons and Compton electrons. In the case of the X-ray tube secondary electrons give rise to difficulties, e.g. (1) in the negative charging of the insulated parts of the tube; (2) in producing heating effects; (3) in the liberation of gas and the ionisation of the liberated or other gas; (4) in producing electrolytic effects at sealed joints; (5) in producing scattered radiation from the side of the anode; (6) in giving rise to tertiary electrons that may cause electronic bombardment at undesired places. These difficulties may largely

be overcome by using rings or a perforated metal cup round the anode or by imposing a strong electrostatic field in the neighbourhood of the anode.

Secondary hardening. The name used in reference to the hardening effect most notably demonstrated by high-speed steels, which occurs when certain previously hardened steels are cooled from a particular range of tempering temperatures. In the case of quenched high-speed steels the tempering range lies between 400 °C and 600 °C. Any austenite retained by the quenching is converted into the harder martensite.

Secondary Ion Mass Spectrometry (SIMS). When a beam of ions falls on a solid specimen, atoms or molecules are knocked out of the surface, a phenomenon known as "sputtering". The majority of the ejected particles are ionised, and can be deflected by a magnetic field. The deflection of these ions depends on their charge and mass, so that the beam of secondary ions ejected can be split into several beams by a magnetic field, each beam containing ions of the same mass and charge. This effect forms the basis of a type of mass spectrometer, since an aperture can be used to allow only a selected beam of secondary ions to pass into a detector. This is similar to the action of a scanning electron microscope, and the instrument can be used to display scanning images of the distribution of elements in a specimen surface, or the ion beam can be held on a particular spot to enable qualitative analysis of the ions leaving the surface to be made.

Secondary metal. The term used to designate metal recovered from scrap, as distinguished from metal obtained directly from the ore.

Secondary pipe. When molten metal is poured into a mould having the base area larger than the top area, a secondary pipe is usually formed in the lower central region. The central cavity that forms near or at the top of the mould due to contraction during cooling and solidification is always present. But below the region of this primary pipe there is a zone that remains liquid when the rest of the ingot has solidified. This in turn will solidify and produce a separate pipe low down in the centre of the ingot. This lower cavity is known as a secondary pipe (*see* Fig. 151).

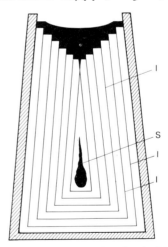

Fig. 151 *Schematic sketch showing formation of secondary pipe. S is the secondary pipe; I the isothermal surfaces showing progress of solidification.*

Secondary radiation. When X-rays or any type of radioactive radiation strike matter, secondary radiations are produced which in general are much weaker than the primary radiation which produced them. X-rays give rise to a similar type of radiation, γ-rays may produce β-rays; α-rays may produce both β- and γ-rays. If the α-rays are absorbed by matter a type of radiation is produced which consists of low velocity electrons. These are known as soft β-rays.

Second law of thermodynamics. This law can be expressed in a number of different ways, the original statement by Clausius being that "heat cannot be conveyed from one body to another at a higher temperature without the expenditure of work or some equivalent process". The law refers only to the continuous performance of useful work—i.e. to cyclic processes. From the law it can also be shown that no cyclic process can be more efficient than a reversible one. The law can be used to predict the direction of a spontaneous chemical change under a given set of conditions. (It should be noted that spontaneous changes are irreversible.)

Sectility. The name used in reference to the property of certain minerals, e.g. graphite, soapstone, which may be cut with a knife, but are not malleable.

Section modulus. In considering the strength of materials in flexure, the longitudinal fibres on the convex side of the curve are lengthened and are therefore in tension. Similarly the fibres on the concave side are in compression. Between these two and within the beam there must be a layer of fibres that is neither in tension nor compression. This layer is the neutral layer, and the intersection of this surface with any cross-section is termed the neutral axis. In the usual equation for the strength of beams and girders the quantities M (bending moment), I (moment of inertia), f (stress, either tensile or compressive) at a distance y from the neutral axis are related in the equation:

$$f = M \div \frac{I}{y}.$$

The quantity I/y is called the section modulus and is often denoted by Z. (*See* Table 53.)

TABLE 53. *Formulae for calculating the section modulus (Z) of different shapes.*

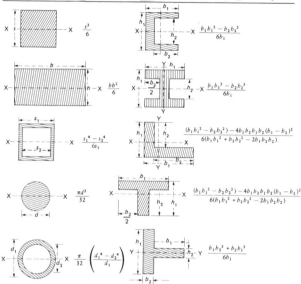

Sedimentary rock. Rocks formed by the combined operations of denudation and deposition. They are loose and cemented rocks formed of a wide variety of detrital material. The following types are important to distinguish:

(a) *Arenaceous rocks*, which generally contain angular fragments of materials that are not decomposed by aqueous or atmospheric influences. Hence quartz is the commonest constituent, and the cementing material is commonly silica, calcium carbonate or iron oxide. Also called psammitic rocks.

(b) *Argillaceous rocks* are very fine-grained rocks probably of the same constituents as (a) above. Typical examples are slates and shales. Such rocks are characterised by the abundance of clay materials. Also called pelitic rocks.

(c) *Calcareous rocks* as a rule are organic in origin, but some such as stalactite and travertine are formed by deposition from solution. The chief representatives of this group are the oolitic limestones, e.g. calcite, chalk and dolomites.

(d) *Siliceous rocks* may be of two types. Those that are the result of the accumulation of the remains of organisms having siliceous skeletons are typified by tripoli and some cherts. The second types are regarded as silicified limestone and are represented by flint and some cherts.

(e) *Carbonaceous rocks*, the coal-bearing rocks.

Sedimentation. The term used in reference to the methods used to separate solids mixed with liquids by allowing the former to settle out under the influence of gravity. For spherical particles of radius (r) in a liquid of viscosity (η) the rate of settling of suspended particles is $0.22r^2 g\rho/\eta$, where ρ is the difference in

density between the particles and the liquid. The rate of settling rises rapidly with the increase in size of the particle. When the solid consists of particles of different sizes the separation is effected by the process of fractional sedimentation. The mixed particles are stirred in the liquid by which means a series of deposits of increasing fineness is obtained. Each portion so obtained can be subjected to a repetition of the process to obtain finer separation. The method is used in the separation of fine china clay from coarser material. In dealing with colloids, the rate of separation under gravity is very small indeed and a centrifuge is used. When the rate of sedimentation is measured in such cases the size of the particles can be calculated, and using this value the molecular weight of the colloid may be found from the relationship Mol. wt. = mN, where N = Avogadro's number and m = weight of a particle. Dry sedimentation is sometimes used for rapid investigation of the suitability of powders for compacting. A quantity of the powder is placed in a glass tube, and the upper limit is noted before and after tapping the tube. A decrease or an increase in volume can occur. If the former happens and the amount is small it may be taken to indicate reasonably good flow properties: if large it would indicate poor flow and poor compacting properties. An increase in volume indicates entrapment of air, which can occur with flaky materials that are unsuitable for compacting.

Wet-sedimentation tests consist in adding water to the sample and observing turgidity and rate of settling. If the amount of water added is in the region of 3 to 4 times the volume of the powder it is informative to observe the meniscus and the angle of tilt when flow occurs. If the former is flat and the angle of tilt is low these indicate a material of good flow properties and suitable for compacting.

Seebeck effect. The name used in reference to the thermoelectric effect first discovered by Seebeck (1826). If any point in a circuit at which two different metals are in contact be at a different temperature from the rest of the circuit, an electromotive force will be set up, and if the circuit be completed, a current will flow. If any pair of metals in the following list be used, then the current produced will flow across the hot junction from the metal earlier in the list to the one later in the list: Bismuth, platinum, copper, lead, tin, silver, zinc, iron, antimony, tellurium. (*See also* **Thermoelectric diagram**.)

Seed charge. The name used in reference to the small amount of material that is added to a super-saturated solution for the purpose of nucleating precipitation.

Seekay wax. A proprietary chlorinated wax that is used for stopping off in electroplating. It has a softening temperature of about 123 °C, but is usually applied at about 140 °C. The wax emits poisonous fumes when hot, and this must be considered when in use and suitable air exhaust provided.

Seger cones. The name used in reference to a series of slender cones or trihedral pyramids made of various mixtures of minerals and clays, and used as temperature guides in the refractory, pottery and enamel industries. The cones are generally about 7.5 cm high and stand on a base about 15 mm wide at a slight inclination of some 8 degrees from the vertical. This is to ensure that the tips may bend over when the cone softens or fuses, and the indicated temperature is taken as that at which the tip of the cone touches the base. Bars of the same clay mixtures are sometimes mounted on supports and used in place of cones (Holdcroft's thermoscopes). The entire series of cones is divided into four general groups, each being suitably devised for a particular range of temperature:

(a) *Cones* 022 *to* 011 are known as the soft series and are based on a lead borosilicate glass. They are chiefly used for firing overglaze decoration on pottery.

(b) *Cones* 010 *to* 01 form the low temperature series. They are based on a soda-lime-borosilicate glass containing iron oxides, and should be used only in non-reducing atmospheres.

(c) *Cones* 1-9 form the intermediate series and are best suited to the firing of white clay articles.

(d) *Cones* 20-42 are the high-temperature cones used as standards for checking refractory materials.

Table 54 gives the softening temperatures of Seger cones.

Segregation. A term used in reference to the non-uniform distribution or concentration of impurities or alloying constituents

TABLE 54. *Softening Temperatures of Seger Cones*

Cone No.	Softening Temperature °C	Cone No.	Softening Temperature °C
017	730	11	1,320
016	750	12	1,350
015a	790	13	1,380
014a	815	14	1,410
013a	835	15	1,435
012a	855	16	1,460
011a	880	17	1,480
010a	900	18	1,500
09a	920	19	1,520
08a	940	20	1,530
07a	960	26	1,580
06a	980	27	1,610
05a	1,000	28	1,630
04a	1,020	29	1,650
03a	1,040	30	1,670
02a	1,060	31	1,690
01a	1,080	32	1,710
1a	1,100	33	1,730
2a	1,120	34	1,750
3a	1,140	35	1,770
4a	1,160	36	1,790
5a	1,180	37	1,825
6a	1,200	38	1,850
7	1,230	39	1,880
8	1,250	40	1,920
9	1,280	41	1,960
10	1,300	42	2,000

which arises during solidification or thermal treatment. Since the axial portion of an ingot solidifies much later than the outer parts, it generally contains a higher proportion of impurities than the rest of the ingot. The effect is intensified by the circumstance that the presence of the impurities in the axial zone lowers the freezing point there. Whenever a portion of an ingot remains fluid for a longer time than its immediate surroundings there is a tendency towards a concentration of impurities at that location. This concentration or accumulation of impurities in various positions within the ingot is referred to as segregation. The segregation that occurs between the arms of dendrites is referred to as minor or microsegregation and the composition may vary within a single crystal. Major or macrosegregation occurs around primary or secondary pipe and in similar regions, and is often revealed by marked lines having a pronounced erect or inverted cone shape, which are made evident when the ingots are sectioned and etched. The occurrence of the Λ segregate is explained as a natural consequence of the fact that solidification proceeds inwards from the mould face. The more impure liquid tends to concentrate near the tips of the columnar crystals, and being less dense than the average fluid metal, it tends to rise. Thus, as the columnar crystals continue to grow inwards, they entrap some of the impure liquid, and as solidification proceeds the rising impure liquid meets the growing crystals nearer and nearer the axis of the ingot. When solidification is complete the impurities are revealed in zones having the shape of the erect cone Λ. These zones of segregation occur in the middle regions of the ingot and usually within that part mainly occupied by equi-axed crystals.

It is probable that the equi-axed crystals, forming within the envelope of the columnar crystals, do not all crystallise at the same time. It is indeed likely that at certain times a kind of bridge of crystals forms across the axis of the ingot, and this bridge then suppresses any tendency for the impure liquid to rise. Impurities begin to concentrate below the "bridge". Owing to contraction on further cooling the bridge begins to sag forming a rather blunt inverted cone, the surface of which entraps the impurities accumulated there. Such V cones are often visible in the axial region of the ingot, starting near the base. Both V and Λ segregation are influenced by the shape of the mould and by the size of the casting, but both types are less marked if lower casting temperatures are used. Sulphur, phosphorus and carbon segregate to a considerable extent; silicon, molybdenum and chromium segregate to a small extent, while nickel and manganese only segregate very slightly.

Microsegregation may sometimes be overcome by annealing but macrosegregation persists through subsequent heating and working operations. Pipe segregates occur round the pipe cavity, blow-hole segregates are found lining blow-holes in steel ingots. In normal segregation the constituents of lowest melting point concentrate in the last portions to solidify, but in inverse segregation (q.v.) this is reversed, for the impurities tend to decrease in concentration from the ingot surface to the centre.

Segregation can occur at grain boundaries, surfaces and dislocations during thermal treatment. Solutes having low solubility have a tendency to be rejected to the more open structure of a grain boundary, surface or dislocation (Gibbsian segregation). Kinetic factors also produce non-equilibrium segregation when solutes are swept by a vacancy flux in temperature changes.

The term segregations is also applied to mineral aggregates that occur in masses or streaks in igneous rocks, representing early products of crystallisation from magma.

Sehta. An alternative name for cobaltite CoAsS. The mineral is silver-white with a slight reddish tinge and under the name sehta is used in India for producing a blue enamel on gold and silver articles.

Seizing. A phenomenon that occurs when one metal part moving over another metal surface is insufficiently lubricated so that the resulting friction generates excessive heat and the metal parts become welded together. It can be avoided by making the supply of lubricant both copious and continuous.

When used in reference to impairment of die surfaces or of powder metal compacts due to friction the phenomenon is known as galling (q.v.).

Selective annealing. A term used in reference to the process of annealing (q.v.) when confined to a localised area or section. Also known as local annealing.

Selective carburising. The term used in reference to the process of carburising a metal component in such a way that only the desired parts are carburised. This can be done, for example, by plating a steel component with copper so that only those parts that are to be exposed to the carburising atmosphere are left unplated.

Selective flotation. Synonymous with differential flotation (q.v.).

Selective hardening. The term used in reference to the hardening process so conducted that different parts of the same component are cooled at different rates so securing different degrees of hardness. The variation in cooling rate may be secured by spraying one portion of an article with water or by using an air blast, while permitting the remainder to air cool. (*See also* **Differential quenching.**)

Selective heating. A term used synonymously with differential heating (q.v.).

Selective quenching. *See* **Selective hardening, Differential quenching.**

Selenitic cement. A type of cement obtained by adding to ordinary lime cement from 5% to 10% plaster of Paris. The effect of the addition is to increase the hardening properties of the lime cements by as much as 50-100%.

Selenium. An elementary substance first discovered (with sulphur) in the flue dust from the roasting of pyrites (in the making of sulphuric acid by the lead-chamber process). It is found native, as is sulphur, and as selenides of copper (berzelianite, Cu_2Se), lead (clausthalite, PbSe), bismuth (guanajuatite, Bi_2Se_3), silver (naumannite, $[Ag_2Pb]Se$) as well as rarer selenides of mercury and thallium. There are no concentrated deposits of economic value, but the element occurs as a minor constituent in carnotite (Utah and Colorado) and in Sudbury nickel ores (Ontario). The chief source at present is the anode mud from the electrolytic refining of copper. On roasting the mud the selenium volatilises as the dioxide which is dissolved in hydrochloric acid and then the product reduced with sulphur dioxide to obtain the element. There are several allotropic forms, at least one of which possesses distinctly metallic properties. This variety has specific gravity of 4.8, the others varying from 4.3 to 4.5. There are six natural stable isotopes. The usual colour of selenium is brownish, but the metallic form is black. In very small amounts the element acts as a biological stimulant, but in larger amounts is highly toxic and is said to be responsible for the barrenness of certain soils.

Selenium finds its most important applications in the electronic field—for the production of photocells, and (formerly) in the making of a.c.-d.c. rectifiers. Small amounts added to glass are now used in place of manganese dioxide for decolourising. Larger amounts (2-3%) produce a red colour now used in the red glasses. Amounts of 0.25-1% added to copper and its alloys improve their machinability without effect upon the tensile strength. It is being used with advantage for a similar purpose in austenitic and stainless steels. Selenium is used in medicine and in the manufacture of dyestuffs; as a coating material to improve the corrosion resistance of magnesium alloys and as a vulcanising agent. A 10% solution of selenious (selenium IV) acid is often used for local repair of chromatid finishes. Thiourea in presence of dilute hydrochloric acid gives a red precipitate in the presence of selenium, and this is used as a chemical test. Atomic weight 78.96, at. no. 34, crystal structure, hexagonal, $a = 4.3553 \times 10^{-10}$ m, $c = 4.9494 \times 10^{-10}$ m. (See Table 2.)

Selenium cell. A photoelectric device which makes possible the use of light-rays for the purpose of measurement and for the control of processes. The selenium is spread in a very thin film on the surface of high purity copper. It is exceedingly sensitive to change in light intensity of all wavelengths from hard X-rays to infra-red rays. It is used to enable speech to be projected in a beam of light, for high speed counting and for remote control and registration.

Self annealing. The term applied to the process of deterioration in mechanical properties which may occur during the slow cooling of large castings, or when a large number of hot articles are piled in a heap. Self annealing sometimes occurs in electrical conductors which have been heated in service.

Self-fluxing ore. The name given to those ores (or mixtures of ores) in which the proportions of silica and lime are such that when highly heated they mutually combine to form a compound that is fluid at the high temperature used. Being naturally properly proportioned in composition, such ores need no additional flux.

Self-hardening. A term used in reference to the property of certain alloy steels that harden by cooling in air from above the critical temperature. Such steels usually contain alloying additions of nickel, chromium, molybdenum, manganese or tungsten, or combinations of these together with moderate proportions of carbon. Such steels possess sufficiently high hardenability that their critical cooling rates are readily exceeded in air cooling from the austenitising temperature. Consequently breakdown to intermediate constituents such as pearlite is avoided and the austenite transforms directly to martensite. Also termed airhardening.

Self-induction. Since the flow of an electric current is associated with the movement of electrons, energy is produced, and this is stored in the magnetic field established by the current. A change in current produces a corresponding change in the magnetic field. An increase in the current causes an increase in magnetic flux, and this gives rise to an induced e.m.f. opposing the change in current. The system must therefore drive the current against this "back" e.m.f., and additional energy is thus transferred to the magnetic field.

The induced e.m.f. (E) depends on the rate of change of the current (I) so that $E = -LdI/dt$ where L is the coefficient of self-induction or simply the inductance. The practical unit of inductance is the Henry.

Selvyt. A proprietary polishing cloth formerly used in the preparation of micro-specimens.

Semi-anthracite. An anthracite coal which has a fuel ratio between 5 and 10. According to one classification the true anthracite should have a fuel ratio of not less than 10, and a lower limit of 12 has been adopted. The fuel ratio is the ratio of fixed carbon to volatile matter.

Semiconductor. A material which has a resistivity between that of true insulators and true conductors, at room temperature, but which becomes conducting with a fall in resistivity on rise in temperature (intrinsic semiconductor). The addition of controlled amounts of specific impurities to the basic material may produce the same type of change, or increase the conductivity on rise in temperature by increasing the number of carriers. (See **n-type semiconductor, Acceptor** and **Donor atoms.**)

Semi-finished metal. A term applied to wrought metal in the form of blooms, billets, slabs, sheet bars and wire rods which require further work or treatment.

Semi-killed steel. The term used in reference to a non-piping steel that is intermediate between the rimming and the killed types, and to which deoxidisers such as silicon are added in small controlled amounts up to about 0.1%.

Semi-metals. Usually applied to those atoms with electrons in the outer energy levels in the p state, whose behaviour is largely metallic, but where bonds are not entirely metallic in nature, but partially homopolar. They tend to be unusual in chemical behaviour, forming amphoteric oxides, and may be mechanically brittle. Germanium, arsenic, selenium, tin, antimony, tellurium, bismuth and polonium may be so regarded.

Semi-permanent mould. The term used in reference to permanent metal moulds in which sand cores are used. The sand core is used when a metal core or cores will not facilitate the casting of undercut. The withdrawal of the casting from the mould removes also the sand core, which is then knocked out. The core is thus expendable, but the mould is permanent.

Semi-permeability. A term used in reference to the property of certain membranes which allow the solvent but not the solute to pass through. The walls of animal cells, the protoplasmic lining of plant cells and gelatinous precipitates of substances such as copper iron II cyanide are found to be semi-permeable. The conception that a semi-permeable membrane can be regarded as a molecular sieve has been abandoned in favour of a theory of preferential solubility. The solvent is considered to be capable of dissolving in the membrane, diffusing through the membrane and then being given off at the opposite side. The phenomenon of the solvent passing through a semi-permeable membrane that separates a solution and the pure solvent is known as osmosis (q.v.), and this is the basis of one method of finding molecular weights of complex substances in solution.

Semi-polar bond. A term used in the electronic theory of valency in reference to a variety of covalent bond in which both electrons constituting the duplet are supplied by one atom. The atoms or molecules with lone pairs of electrons are called the donors, and the atoms or molecules which do not contribute electrons to the bond are called acceptors. Also called coordinate link, dative covalency or bond.

Semi-red brass. A series of copper-zinc alloys containing 8-17% Zn. In the A.S.T.M. classification cast brasses include four main groups: (a) red brass 2-8% Zn; (b) semi-red brass 8-17% Zn; (c) yellow brass over 17% Zn but with less than 2% total of aluminium, manganese, nickel and iron; (d) high-strength yellow brass (or manganese bronze) containing over 17% Zn and over 2% total aluminium, manganese, nickel and iron. These casting alloys almost always contain appreciable amounts of other alloying elements including 1-6% Sn and 1-10% Pb. Semi-red brasses have good casting qualities and are used for low-pressure valves, plumbing fixtures.

Semi-steel. A high-strength cast iron used for high duty iron. It is produced by charging scrap steel with pig iron into the cupola. The product has a somewhat lower percentage of carbon than is generally found in cast iron and the graphite is in a finer state of division. It is a grey iron that can be used for such applications as Diesel engine cylinders, etc. When certain alloying additions are made the "high-test cast iron" is known under a series of trade names. (See also **Gunite** (Table 6), **Meehanite** (Table 3).)

Semi-water gas. A term applied to a mixture of producer gas and water-gas. In the production of producer gas the reaction is highly exothermic. This results in the rise of temperature of the coke, causing clinker or slag formation, which is undesirable. Semi-water gas is obtained by passing both air and steam over coke heated to about 1,000 °C. The proportions of air and steam are so regulated that the heat evolved in the producer-gas reaction more or less compensates for the heat absorption of the water-gas reaction. The composition is approximately CO 30-40%, H_2 25-40%, the balance being mainly accounted for by nitrogen and carbon dioxide. Calorific value 7.5-9.2 MJ m^{-3}.

Senamontite. Native antimony sesquioxide Sb_2O_3 which occurs in cubic crystals. Also called senarmonite.

Sendzimir mill. A cluster-type rolling-mill having six large outer rolls and four intermediate rolls backing up two small-diameter work rolls. These latter may be as small as 6 cm diameter with a strip width capacity up to 1.25 mm, and the backing elements are adjustable. Mills of this type are used for the cold reduction of steel sheet and strip, e.g. light-gauge stainless strip in final thickness from 0.1-2 mm.

Sendzimir process. A hot-dip galvanising process in which the surface of mild-steel strip is lightly oxidised and then treated for about 2 minutes at 800 °C in cracked ammonia. It is then passed directly to the molten zinc without coming into contact with air. A certain amount of nitriding takes place during the ammonia treatment which is primarily intended to reduce the oxide in the steel surface. Heavy zinc coatings are obtainable.

Sensible heat. A term used in connection with the heat imparted to a substance which raises its temperature (or the heat extracted which results in a lowering of the temperature) without change in state or allotropic form.

Sensitised fluorescence. Fluorescence may be regarded as a secondary effect of the absorption of a quantum of light by an atom or molecule. In most cases light absorbed at a certain wavelength is emitted at a greater wavelength. It is otherwise when resonance occurs. It is known that mercury vapour will absorb light of a certain wavelength ($2,536 \times 10^{-10}$ m) and that ionisation does not occur, the vapour emitting light of the same wavelength. But if there be mixed with the mercury vapour the vapours of some other metals (e.g. lead, sodium, silver) and then irradiated with light of the above wavelength, the added metals will fluoresce. This is known as sensitised fluorescence.

Sensitiser. In a photochemical process the substance that undergoes change is not necessarily the absorbing substance. Certain cases are known in which an extra substance is present whose function appears to be that of absorbing the light energy and passing it on to the substance that will react. Such extra substances are called sensitisers, and the process is known as photosensitisation.

Silver salts, e.g. chloride, are not noticeably acted upon by the longer wavelengths of light of the visible spectrum, but the addition of certain dyes, e.g. aurin, cyanin and erythrosin, which can absorb the red and orange rays, increase the sensitiveness of silver chloride. The gelatin on the photographic plate and the albumin on photographic paper accelerate the action of light on silver salts and are thus sensitisers. Chlorophyll is an example of a natural sensitiser.

Sensitive-colour illumination. In optical microscopy polarised light is used in the study of crystalline substances. Non-cubic crystalline materials are anisotropic, and when such materials are examined by polarised light, changes in brightness occur when the specimen is rotated through 90°. If a filter made of gypsum is included in the optical system, variations in the polarisation of the reflected light as the specimen is rotated are indicated by colour differences instead of differences in brightness. This phenomenon is known as sensitive-colour (or tint) illumination.

Separation. The name used in reference to a variety of processes for extracting or concentrating one or more constituents of a mixture. The methods generally employed are based on differences in the various properties of the constituents, e.g. specific gravity and magnetic properties. When chemical methods are used in the separation, the process is usually referred to as extraction.

The separation of minerals in an ore is usually known as concentration. In most cases the ore must be crushed to a convenient degree of fineness using crushers and grinders. When separation is effected using a rising current of water, the method is known as classification. Small amounts of certain oils added to water produce on agitation a large amount of froth. Advantage is taken of the fact that some materials in these conditions are wetted by the water and others not. The non-wettable portions are floated off at the surface. The method is known as froth flotation (q.v.). Other methods that depend for their success on the differences in specific gravity of the materials are associated with separators known as jigs (q.v.) and bumping tables.

Magnetic separation is used to separate iron, steel and feebly magnetic substances from non-magnetic substances. This method of separation may be undertaken for any of the following reasons: (a) concentration of magnetic substances or of the residual constituents; (b) separation of iron from foundry sands; (c) removal of tramp iron from material to prevent subsequent damage to machinery.

Sequence etching. The name used in reference to a technique for identifying the structural constituents in tungsten steels. The specimen is first etched with 5% nital to develop the general structure, and then without repolishing with Daeves' reagent (Table 11) for 20 minutes to blacken any iron tungstide. A final

etch with hot alkaline sodium picrate solution blackens the iron carbide.

Sequestering agents. The general name given to substances that react with ions to produce soluble complexes. They are used in chemical analysis and on the commercial scale and may be classified: (a) the alkaline salts, usually polyphosphates (e.g. sodium hexametaphosphate, sodium tripolyphosphate, etc.), which are generally used for water softening. They form complexes with the calcium and magnesium ions and prevent their precipitation; (b) organic amines—mainly derived from ethenediaminetetra acetic acid—the sodium salt of this acid, e.g.

$$(CH_2N)_2.(CH_2COONa)_4,$$

is available commercially; (c) organic acids such as gluconic and citric form chelates with copper, iron and other metal ions. In this way these ions are no longer available to take part in reactions that might interfere with tests for other substances.

Series system. The term used in reference to a system of electro-refining. In the Hayden system of a multi-electrode electrolytic cell, cast or rolled anodes are placed vertically in series, and they fit closely to the side of the tank. There is one cathode at one end of the tank and one anode at the other end. The current is introduced into the tank at the anode and leaves at the cathode, the intermediate electrodes having no metallic contact with one another. When the current flows, the intermediate electrodes become positively charged one side and negatively charged on the other side. The electrolyte contains about 30% copper sulphate and about 12% free sulphuric acid is used. Copper is deposited on the negative side and dissolves away from the positive side of each intermediate electrode. The process may continue until the copper in each electrode (except the last one) has been transferred to the succeeding one.

Serpentine. A hydrated silicate of magnesium $3MgO.2SiO_2 2H_2O$ which is an alteration product of minerals rich in magnesium. The translucent variety is used for ornamental purposes. The variety showing a fibrous structure is known as asbestos (q.v.). The crystal system is monoclinic; Mohs hardness 3-4, sp. gr. 2.5-2.6.

Serpex. A basic refractory material consisting of serpentine (q.v.) enriched with magnesia.

Servarizing process. A hot-dip process for the protection of iron sheets from oxidation. The metal in the required shape is first given a coating of cadmium and then dipped into molten aluminium. The resulting sheets are suitable for use in furnace parts and in the construction of heat-treatment boxes.

Service factors in material performance. The selection of materials for appropriate applications must obviously depend on mechanical, physical and chemical properties in the first instance. However, few engineering designers can predict all the varying conditions which may arise in service, and in any case these may be difficult to generalise. Under service factors, unusual conditions which may not be immediately obvious have to be accounted for in the selection of materials, their importance depending on the potential hazard or difficulty which may result from failure.

Materials manufactured under commercial conditions result in a compromise between cost and degree of control in fabrication and purity. Steel made for an aircraft jet engine will warrant greater attention than a similar material for a road roller or an agricultural tractor. Concrete for a power station cooling tower must be more rigidly supervised in mixing, pouring and curing than that for a garden path. Materials are rarely perfectly homogeneous, either in composition or properties. The presence of impurities may be due to careless manufacture, or part of the compromise where removal may be very costly. Quality control by analysis will therefore vary according to the application, and will be essential where small variations in composition would lead to detrimental performance, or even safety hazards.

When composite materials or mixtures are involved, not only is the correct proportion of constituents important, but their distribution also. Examples are aggregate materials such as concrete and asphalt for road surfaces. Plastic materials are rarely used in the perfectly pure state, but contain fillers and plasticisers. Polymeric materials themselves are not homogeneous in the physical sense, but contain molecules of widely different molecular weight, even unpolymerised material or breakdown products. Such factors can be averaged out, and design studies can include a variation in properties accordingly. Intentional states of physical heterogeneity or instability can exist in heat-treated alloys, where any overheating may destroy the original structure and produce a material of different properties. Overcooling of polymeric materials may reach the glass transition point, and produce a hard and brittle product with little plasticity.

Anisotropy and mechanical strain may produce variations in properties in different directions, and components which may have to rely on uniformity of behaviour should obviously never be made of such materials, unless this can be fully accounted for in the design. Engineering designers have long experience of materials variations, but none are infallible.

Gases in metals were not always analysed, and many failures in steel occurred before the real effect of nitrogen and hydrogen embrittlement was recognised. The highest quality steels now involve vacuum degassing and casting processes, unheard of in that industry much before 1950, whereas in copper for electrical machines, the deleterious effect of oxygen has affected the refining process for over a century.

Physical heterogeneity such as grain size in metals is now better understood than ever before, particularly in its influence on brittle fracture, stress corrosion, etc. Steps to control grain size are now regarded as normal, where once they were special treatment. The use of powder metallurgy allows more control over some of these aspects, and the technology is now sufficiently advanced for economic factors not to be the barrier they once were.

Atmosphere may play a very large part in manufacturing processes, both for good or ill. The cotton industry, where high humidity was essential to enable the short staple fibre to be twisted together under tension, produced poor working conditions, whereas man-made fibres, spun into very long fibres almost continuously, avoid such difficulties. High humidity in the foundry can cause the hydrogen content of molten aluminium to rise, requiring extra care in degassing.

Fabrication also gives rise to its quota of heterogeneity, particularly in residual stresses, locked up in the material as a result of non-equilibrium, either for economic necessity, or difficulty of process control. Subsequent annealing treatments may be essential, as with glass, or highly desirable, as with metal castings or weldments. When large structures are involved, as with welded vessels, the problem is between cost and efficacy of such treatment as is possible, and the risk of non-treatment. If heat treatment, which must cause plastic flow to relieve stresses, produces a potential crack at some highly stressed point, the material may now be in a worse condition than if no treatment had been applied. A deep understanding of all stresses involved, constructional and operational, as well as those induced by manufacture, must be applied in such cases. Non-destructive testing is developing rapidly to cope with such decisions.

Atmosphere not only affects manufacture, but also service performance. The well-known outbreak of season-cracking in brass cartridge cases, which affected the British Army in India, was a combination of residual stress in the brass, and a high concentration of ammonia at an elevated ambient temperature, near the ammunition stores. The rusting of ferrous metals leads to tremendous losses, worldwide, each year, and numerous cases occur every day of electrochemical corrosion due to the presence of two or more dissimilar metals in electrical contact in an assembly. This is one service factor which many designers have still failed to recognise fully, especially in the field of consumer durables and vehicles. Fabrication itself leads to corrosion problems even where only one metal or alloy is involved, and design can play a large part in creating (unintentionally no doubt) crevices and pockets where anaerobic or other detrimental corrosive environments can exist.

Temperature has a marked effect on all chemical reactions, so that service in hotter climates requires greater attention to corrosion or other potential chemical attack. Similarly, service in very cold climates may produce problems of brittleness, even though corrosion is slowed down.

Radiation damage became highlighted with the use of nuclear energy for power production, since neutrons in particular can severely disrupt the crystal lattice even of metals. However, all radiation has some effect, even sunlight, particularly noticeable in transparent polymeric materials, which often darken and become embrittled. Design of nuclear equipment has fortunately had to deal with the problem from the start of a new industry, but not all engineers realise that the substitution of say, plastic materials for metals, cannot be regarded merely as a new mechanical stress exercise. Time and temperature usually operate in conjunction with radiation, and often lead to effects which are taken to condemn the material, when in fact the selection of material was in error.

Knowledge of service factors is often peculiar to particular industries or materials manufacturers, and prior consultation is better than bearing costs of replacement.

Sesci furnace. A rotary furnace fired by oil or by pulverised fuel that works on regenerative principles.

Set copper. A name used in reference to copper that contains about 6% copper I oxide at the end of the fire-refining cycle.

Setter. The term used in reference to that portion of a forging die used for forming the metal so that the longitudinal axis of the latter lies in two or more planes as desired. Also called bender.

Settling. The process of separating solid insoluble particles from a liquid by gravity.

Settling test. A test used in the foundry to provide an indication of the quantity of gas evolved during solidification. An open top conical test-piece is cast in greensand. This is usually about 5 cm across at the top and about 7 cm deep. Gas-free metal will under these conditions shrink and show a depression on the surface of the cone. If the metal contains an unduly large amount of gas the surface will rise, the oxides if present will rise to the top and produce a "cauliflower" head. Small amounts of gas produce a roughened effect on the surface which shows no tendency to sink or settle.

Shackle. An iron or steel loop for the attachment of a chain or wire rope. The term is also sometimes used in reference to the piece used for gripping and holding a specimen under test in a tensile testing machine.

Shaft. A term used in mining in reference to the vertical pit sunk from the surface. When the shaft reaches the seam to be worked, the seam is said to be "won". The shafts afford means of ingress or egress to the mine. Shafts may be circular, elliptical or rectangular in section, the former being common at depth. The part of the excavation through the surface soil is usually lined with cribbing, within which a concrete curb is built to resist the ingress to the mine of surface water. In the rock strata metal, concrete or timber lining is usually built.

The term is also used to describe a wrought iron or steel bar used for carrying pulleys or wheels for the transmission of power.

Shaft furnace. A term applied to those types of furnace in which the fuel and the material to be heated are in contact. The height of such furnaces is considerably greater than the diameter, to permit the utilisation of natural draught, as in the case of kilns or of forced draught, as in blast furnaces.

Shaft kiln. A type of shaft furnace in which natural draught is usually employed. The temperatures obtainable are not high, so that such furnaces are usually employed where the product of the operation is not required in the molten condition. Shaft kilns find their main applications in the calcining of ores and in the production of lime. (*See also* **Gjer's kiln.**)

Shake. A term used in reference to a crack or cracks occurring in wood or brick.

Shake out. A term used in reference to the process of removing castings from moulds, by shaking.

Shale. The term used in reference to laminated deposits of clay. The sedimentary deposits are often hard in texture, the strongly clayey varieties being used in making fire-bricks. Limestone shales are used in the making of Portland cement. Bituminous shales or oil shales are a valuable source of petroleum. Shales are widely distributed throughout the world, the main important areas being parts of Scotland (UK), Germany, New South Wales and Colorado and Utah in the U.S.A. The oil shale of Scotland is dark grey in colour, and shows a horny fracture. Specific gravity 1.75.

Shale oil. The dark-coloured liquid obtained by the slow low-temperature distillation of bituminous shale. The oil has a specific gravity of about 0.89, and by the use of steam can be refined to give: (*a*) light oils such as petrol, naphtha (sp. gr. 0.77); (*b*) burning or illuminating oils (sp. gr. 0.85); (c) lubricating oil and (d) paraffin wax. The total hydrocarbon content of the crude oil is about 80%.

Shallow hardening. A term used in reference to types of steel that are essentially surface-hardening steels. The operations that are usually carried out with such steels are referred to under "case-hardening" (q.v.).

Shank. The name applied to the handle of a small ladle or to that portion of a tool by which a forging is held in the forging unit. It is often used in reference to that part of a component that has the greatest length, e.g. the main body (or shank) of a bolt.

Shape memory alloys. (Also termed "Marmem", from Martensitic Memory). These alloys, if deformed at a temperature below a particular transformation temperature, will revert to the original shape when heated above that transformation temperature. A martensitic type of transformation is involved, the crystals having a higher symmetry above the transformation temperature than below it. The deformation is accompanied by internal twinning, which returns to the single orientation higher symmetry structure on heating above the transformation temperature.

Originally discovered in the near equi-atomic nickel titanium alloys (nitinols), the effect has been found also in Au/Cd, Fe/Pt, Cu/Zn, Cu/Al/Ni and In/T alloys. Variations in the composition of Nitinol alloys have produced alloys with transformation temperatures in the range $-190\,^{\circ}\text{C}$ to $+100\,^{\circ}\text{C}$.

Shaping. The term used in reference to the machining of surfaces by a tool that moves over the work. It is somewhat similar to planing except that in the latter operation the work moves under a stationary tool.

Sharple's process. A method of separating amorphous wax from its accompanying oil in the preparation of paraffin wax. The amorphous distillate is diluted with about an equal bulk of petrol and very slowly cooled below freezing point (i.e. to about $-12\,^{\circ}\text{C}$), after which it is centrifuged at very high speed. The wax so obtained is suitable as a lubricant and for waterproofing natural fibres. It is much used in the making of candles, etc.

Sharp series. Generally the arc or spark spectrum of an element contains four sets of lines. That set which is distinguished as clear-cut thin lines is known as the Sharp series and is often referred to as the S series.

Shatter cracks. A name used in reference to fine internal fissures, particularly when found in the heads of steel rails. The cracks lie at random in all directions and occur most frequently in large steel forgings. They are usually eliminated by very slow cooling to below $100\,^{\circ}\text{C}$. (See also **Hair-line cracks.**)

Shaw process. A technique for making the dimensionally accurate refractory moulds used in investment casting. A master pattern is made in bronze, steel or brass (rarely in wood) and is coated with some non-refractory material which sets hard without dimensional changes. The mould sections are then reproduced in a quick-setting slurry consisting of the refractory mould material with a flammable liquid binder. After firing, the sections are assembled to form a complete mould and cemented together with refractory slurry. The mould is refired and used for casting steel and high-melting temperature alloys. By this process castings with smooth surfaces are obtained, and the accuracy approaches that of the lost wax process.

Shear. When a material is subjected to the action of external forces whose direction is parallel to the section, the stress is called a shearing stress and the material is in shear.

The term is also used in reference to a cutting operation in which metal rods, wires, sheet, etc., are cut by means of blades, one of which may be stationary. (See also **Punching.**)

Shear modulus. The modulus of rigidity (q.v.).

Shears. The name used in reference to a large type of cutter sometimes resembling scissors and used for cutting metal sheets or for fettling castings.

Shear steel. A type of cutlery steel made by forging together a number of pieces of blister steel (q.v.). The bars are arranged in piles, reheated and forged (or rolled) into bars. Single shear steel is the product of a pile or "faggot" of bars. Double shear steel is produced when single shear bar is cut up, piled and reheated and then reforged.

Shear strength. The term used in reference to the ultimate strength of a material subjected to shear loading. The value of the shear strength can be determined by: (a) a torsion test (q.v.); (b) a flexure test; (c) shear test. (See **Section modulus.**)

For the shear stress to exceed the flexural stress in the flexure test the span of a rectangular section must be less than half the square of the thickness: for a cylindrical specimen the span must be less than one-third the diameter.

Shear-stress. (See **Critical shear stress.**)

Shear testing. (See **Torsion test.**)

Sheet metal. A term used rather loosely to indicate flat metal products that are thinner than "plates" but thicker than "strip" metal. Sheets are wider than strips, but whereas the latter are continuously rolled in one direction, sheets may have been crossed rolled or pack rolled. There is no universally accepted terminology that rigidly defines sheet or strip or foil.

Sheet-metal gauge. The thickness of sheet metal and the diameters of wires are generally specified by a number and not by the thickness or diameter. There are seven fairly common gauging systems with gauge numbers from 0000000 up to 50 (or from 1 to 80). It is to be carefully noted that a particular gauge number in one system does not represent the same metal thickness as the same number in another system. This has led to some confusion. The following systems have been used for sheet metal:

(a) Standard sheet-metal gauge (USA) used for iron and steel plates and sheets and for tinned and galvanised sheets.

(b) Brown and Sharpe (B. and S. or American gauge) used for non-ferrous sheet metals such as brass, phosphor bronze, aluminium and nickel silver.

(c) Birmingham or Stubb's iron wire gauge is used for steel in the form of strips, hoops, bands or sheets. It is also used for sheet copper.

(d) Birmingham (B.G.) gauge 1914 is used mainly for iron and steel sheets. It differs considerably from the older Birmingham Gauge.

(e) Imperial Wire Gauge is used in the UK for aluminium sheets. There are older systems still sometimes used such as the Birmingham Metal Gauge (for brass) and the Zinc Gauge used exclusively for zinc (see Table 15 for Wire Gauge).

Sheffield composition. A refractory mixture used for making moulds for steel castings. It consists of ground used crucibles, ground burnt fire-clay, ground firebricks, ganister and fireclay together with a small proportion of ground coke or graphite. The mould must be dried at a dull red-heat and cooled in the oven. The material, when dried, is porous so that a facing of silica flour or other dressing is usually necessary. Also called Compo.

Sheffield lime. An abrasive used extensively for polishing electroplate and in a final pre-plating treatment of metals. It consists largely of calcined dolomite, but contains a high proportion of magnesia. It may be used in the form of lumps, but is more frequently employed as "white composition" in which the powdered material is bonded with a little grease or other materials.

Sheffield plate. A term applied to articles produced from copper coated with silver. It may be produced by rolling copper sheet between layers of silver or by beating silver into the surface of copper sheet. In the early days only one side of the copper was plated (by fusion and rolling).

The process was superseded by the introduction of the electrodeposition method.

Shell-electron. See **Atom.**

Shell moulding. The name used in reference to a method of manufacture of sand mould and cores for foundry casting. The moulds are in the form of relatively thin shells of uniform wall thickness, and the process of making the shells consists in bringing the dry powdered sand-resin mixture into contact with a heated metal pattern under such conditions that the required thickness of shell is rapidly formed. While still on the pattern the mould is cured or hardened by heating. Two such shells, removed from the pattern after hardening, are used as a mould for casting, cores being inserted where necessary. The shell moulds are usually supported by mechanical means, but for certain applications backing with loose granular material is recommended.

The mould mixture consists of fine silica sand (sometimes zircon sand is preferable) containing 5-9% of resin. Phenol formaldehyde resin (hydroxybenzene-methanal) with about one-tenth of its weight of accelerator (e.g. hexamethylenetetramine) is used, since this is preferable to the urea formaldehyde (urea-methanal) resins.

The pattern is usually made of cast iron, though other metals and alloys have been used. The pattern temperature is of the order of 200-250 °C, while the oven (curing) temperature is about 300-325 °C. The pattern must be lubricated to ensure that the mould can be removed without fracture. Silicone oils, greases and emulsions have been recommended with a strong preference for the first-named. The investment time (required to build up the shell on the pattern) is of the order of ¼-½ minute; the hardening time may vary from 1 minute to 4 minutes, depending on the temperature.

The main technical advantages of shell-moulding over conventional sand-moulding techniques are related to the superior finish and closer tolerances on casting, and to the lowered surface chilling of the metals. The size of casting that can be handled by this process is the limiting factor in shell-moulding processes. When used in investment casting with wax patterns the curing takes place after the wax has been removed.

Sheppard process. A proprietary anodising process for aluminium in which the electrolyte used is sulphuric acid containing glycol (1, 2-ethanediol) or glycerine (1, 2, 3-propanetriol).

Sherardising. A process for protecting iron from corrosion by means of a corrosion-resistant layer of zinc on the surface of the iron. The component to be treated is heated with a powdered mixture of zinc dust and oxide in a closed box at a temperature between 350 °C and 450 °C. The thin, clear coating produced consists largely of a zinc-iron alloy, and this does not interfere with any pattern or design on the original article.

Shielded arc welding. An electric welding process in which the metal is protected from atmospheric oxidation by using an inert gas atmosphere or flux coated electrodes.

Shift. A casting defect due to movement or to incorrect register of the cope and drag of a mould. It may be caused by using undersize (and thus loose) moulding-box pins; or by movement of the face of the moulding sand during the roll-over operation.

Shim. The name used in reference to the thin metal pieces that are inserted between metal parts in order to build up the assembly to the required thickness. Shim brass is any type of thin sheet brass that is to be used for making shims.

Shingle. The term describing the operation of squeezing or hammering plastic hot wrought iron. The term also refers to the very coarse grade of gravel found typically on the higher parts of sea-beaches. It is sometimes used as coarse aggregate in making concrete.

Shingling. The term used in reference to the operation, during the manufacture of wrought iron, of nobbing, hammering or squeezing the puddled balls from the puddling furnace. During shingling there was a copious flow of cinder from the hot mass, and the compressing of the porous lump caused a considerable increase in temperature which favoured the expulsion of slag. When the metal had been sufficiently worked and shaped into a roughly rectangular block, it was passed to the forge rolls, where it was rolled into puddled bars.

Shock cooling. A process used in connection with certain elevated temperature reactions to prevent a reverse action occurring on slow cooling. When magnesia is reduced with carbon, the operation must be conducted at about 2,000 °C, and the products are magnesium vapour and carbon monoxide in accordance with the equation $MgO + C \rightleftarrows Mg + CO$. To prevent the reverse action occurring and to maintain a continuous production, the vapour phase is reduced in pressure and suddenly cooled so that metallic magnesium is formed. It will be clear that when dolomite is reduced with ferrosilicon, magnesium is the only vapour phase. The operation, therefore, does not need to be subjected to such treatment, since a reverse reaction is not liable to occur. The sudden chilling of the magnesium vapour is usually done in hydrogen or natural gas.

Shock test. A type of test designed to measure the resistance offered by a material to shock loading. In one type of test it is arranged to measure the energy absorbed when a standard prepared specimen piece is fractured. In another type of test is observed the number of blows of a given intensity that are required to cause fracture. Shock resistance may also refer to the ability of load carrying components to resist permanent deform-

ation under sudden (i.e. shock) loading. (See also **Impact test, Frémont, Charpy** and **Izod tests.**)

Shoe. The name used in reference to the shaped casting or forging used as a base piece for timber posts, or the steel pieces used to protect the lower end of a pile or post being driven into the ground. It is also used to describe the part of a braking system that bears against a wheel rim or other moving part in order to reduce it to rest by friction.

In forging, a shoe is an additional block used to build up to the required height or to support an insert placed under the lower block, or under the stationary portion of a die.

Shoots. The name given to the half-round, gutter-shaped steel plates that are used to convey molten steel or slag from the tap-hole to the ladle. They are well lined with ganister and must be thoroughly dried before use. Also called launder (q.v.).

Shoppler process. A process for the production of tungsten metal of high purity. It consists in finely grinding high-grade sheelite and mixing thoroughly with sodium carbonate. The mixture is fused and the fusion product which contains sodium tungstate is leached with water. The clear solution is treated with hydrochloric acid and boiled. In this way the oxide is obtained, which, after drying, is reduced with carbonaceous material in specially designed furnaces at about 1,200 °C.

Shore scleroscope. A hardness-testing device in which the hardness is indicated by the height of rebound of a small diamond-tipped hammer, which falls freely within a graduated glass cylinder on to the surface of the metal to be tested. It is especially useful for measuring the hardness of large objects where it may not be convenient to cut out samples such as are required in other types of instrument (see **Hardness**). The instrument is often used for roll surfaces.

Shorter process. A method of surface hardening that is applicable to cast irons and cast steels, and especially for local hardening. The steel is heated by oxyacetylene (oxyethyne) flame to a temperature above the Ac_3 point. This is immediately followed by rapid and controlled cooling using jets of water. The depth of the hardened layer can be varied according to the material used and to the actual practice followed. (See also **Flame hardening.**) Also called shorterising.

Short grain. The term used in reference to a grade of brazing solder that is produced by granulation of the ingot under a drop hammer. The copper-zinc alloy that is so treated yields three grades, "long", "short" and "fine", only the finest grade being used mixed with boric acid and used as powdered solder.

Shortness. A term used in metallurgy to indicate brittleness or lack of ductility. It is generally qualified by "cold" or "hot", according as the metal lacks ductility when cold worked or when worked at temperatures at which hot working is normally carried out. (See also **Cold short, Hot short.**)

Short-range order. An orderliness of structure which extends over short distances only, in any direction in a system, such as a liquid. A degree of randomness of high order in systems, such as condensation from gas to liquid phase.

Short terne. A term used in reference to steel sheet or plate that is coated with a lead-tin alloy as a corrosion protection. The qualification "short" refers to the weight of coating per box, which represents 40.5 m² (436 sq. ft.) of plate or sheet. Usually 3.6 kg (8 lb) of coating metal is the limit for short terne thin plates, for heavier plates the weight of coating may vary from two to five times this amount. The term is thus used somewhat vaguely, though it always refers to the thinner coatings.

Shot blasting. A cleaning operation applicable to large castings, in which a high-velocity air blast is used to carry the metallic shot, the latter being the effective abrasive. (See also **Shot peening.**) The method is especially useful where burnt-on sand or scale is to be removed. The "shot" consist of angular chilled iron and steel, and these have replaced silica as the abrasive used in blasting work. The air pressure varies from about 0.2-0.6 MN m⁻², dependent on the type of metal treated, the lower pressures being used for grey iron and non-ferrous castings and the higher pressures for steel castings. There is an airless system used for cleaning castings in which no blast is used, the shot being thrown at the surfaces to be cleaned by centrifugal force.

Shot peening. A process similar to shot blasting (q.v.), in which spherical shot are used in place of angular shot. The special feature of this process is the cold-working effect of the impact of the steel balls. Residual compressive stress is set up as a result of this treatment which accounts for the improvement of the endurance limit of the material.

Such treatment has advantages in the control of residual stress, especially in shallow layers near a surface. The compressive stress induced by peening lowers the tensile stresses without affecting microstructure unduly, or varying the effect of heat treatment.

Shotting. The name used in reference to a method of producing powder for powder-metallurgy processes. It consists in pouring molten metal on to a steel sheet which is held at an angle to a tank of water and deflects the metal thereinto. If the tank is not sufficiently deep the molten metal may not have completely solidified before reaching the bottom, and thus may adhere to it.

Shower roasting. A method employed in the roasting of sulphide ores. The latter in very finely divided form are allowed to fall freely into a furnace through which is passing an ascending current of hot air. This produces very rapid roasting of the ore.

Shrinkage. The term applied in reference to the reduction in volume due to thermal contraction when molten material solidifies. In the case of metal ingots solidifying within a mould it is usual to distinguish between this contraction and that which occurs during the cooling of the hot solid metal. The first is the main cause of "pipe" (q.v.) in ingots and piping is reduced and controlled by the use of dozzles or feeder heads. The solid contraction takes place mainly in the interior of the ingot and gives rise to contraction cavities or cracks. To reduce the possibility of the formation of many or large contraction cavities, and to prevent their extension on cooling to normal temperatures, the ingots are often hot-worked at a temperature not much below the solidifying temperature. Before such working is undertaken the ingot may be subjected to soaking. (*See also* **Soaking pit.**)

An allowance is made for the shrinkage that occurs with castings in moulds. (*See also* **Moulder's rule.**) All materials shaped in moulds suffer shrinkage.

Shrinkage cavity. A void in metal castings caused by the reduction in volume of the metal casting (or ingot) due to thermal contraction on solidification. (*See also* **Shrinkage.**) (Also occurs in other materials).

Shrinkage cracks. A term used synonymously with hot tears. They are usually cracks occurring during cooling that have irregular paths, and the walls of the crack are usually oxidised. Too high a sulphur content and too low pouring temperature are generally thought to be the main cause in ferrous alloys. It should be noted that in the case of castings it is not possible to use the beneficial effects of hot rolling. (*See also* **Shrinkage.**) (Also occurs in other moulded materials.)

Shrink bob. As there is often a danger that a shrinkage defect will occur in a casting at the ingate of a mould, a recess or hollow is made at this point, providing a small reservoir and preventing the defect from forming (*see* Fig. 152).

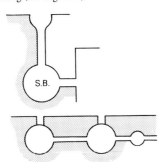

Fig. 152 *Shrink bob.*

Shrink on. A condition that may occur when a forging is permitted to lie in the tool too long a time before ejection. The forging thus contracts and locks itself in the tool.

The term is also used in reference to a method of fitting metal tyres which are just oversize when hot and are placed in position in this state and then quickly cooled so that they lock on to the rim. The method is commonly used also for securing cranks to crank-shafts, etc.

Shrouding. A term used in reference to the strengthening of the teeth of wheels by filling in, during the casting, the sides of the teeth to the level of the pitch circle. Usually only one side of a wheel is shrouded, as this permits the shrouding of the opposite side of the other wheel of the pair.

Shuffs. A term used in reference to bricks that are full of small cracks.

Shunt. A device used in electrical apparatus for altering the current flowing through a circuit. If it is desired to reduce the current through a measuring instrument, a low-resistance shunt is placed across the terminals of the instrument. The proportion of the current then flowing through the instrument is $S/(S + G)$, where S = resistance of shunt and G = resistance of the instrument.

Sialons. An acronym for a family of ceramic materials based on silicon nitride (β - Si_3N_4) "alloyed" with metallic oxides and nitrides such as MgO, AlN, Y_2O_3, and Al_2O_3. Used for cutting tools, refractories for containing molten metal, and applications where resistance to thermal shock and erosion may be required.

Side-blown converter. The name used in reference to the receptacle once employed in making steel, a rather small, tilting converter in which there are two rows of tuyères on one side. The lower set directs the air blast on to the surface of the metal, while the upper set supplies air for the conversion of the carbon monoxide into carbon dioxide, by which means the heat efficiency is improved. The reaction in such a converter is slower than in a bottom-blown converter, such as the Bessemer, and thus the end-point was not so sudden and there was less likelihood of over-blowing. Also called Tropenas converter.

Side-gate. A type of gating system for regulating the flow of metal into sand moulds, that is often used in combination with bottom gating for running bush castings. The metal feeds through a number of steps (gates) one above the other, and the object is to use the upper or top gates for filling the upper parts of the mould with hot metal from the ladle. When there are several such gates, the arrangement is made to secure, as far as possible, that these side gates come into operation one after the other as the mould is being filled. This brings fresh hot metal on top of the metal already in the mould. Finally the uppermost side-gate feeds hot metal into the riser. Commonly used side-gating is illustrated in Fig. 153 where (*a*) shows that a sequential flow can be secured if the sprue is given a small amount of taper; (*b*) shows a sprue of uniform section with inclined steps—a commonly used system—and (*c*) shows a method of ensuring positive sequential flow using a reversed sprue. The latter is tapered from bottom to top, the diameters at these sections being in the ratio of 2 to 1.

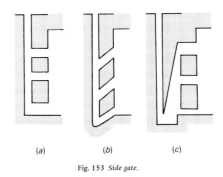

(a)   (b)   (c)

Fig. 153 *Side gate.*

Siderite. An iron ore consisting of ferrous (iron II) carbonate with which is often associated a little manganese, magnesium and calcium carbonates which are isomorphous with it. It has the formula $FeCO_3$ and contains about 48% Fe. It varies in colour from pale yellow to brown and often occurs in globular or botryoidal forms having a silky fibrous structure. Crystal system, hexagonal; Mohs hardness 3.5-4.5, sp. gr. 3.7-3.9. (*See also* **Chalybite.**)

Sidot's blende. A form of zinc sulphide ZnS or blende made by heating the precipitated sulphide to a white heat in a covered crucible. It contains traces of sulphides of the heavy metals such as bismuth, copper, manganese, and these impurities confer upon it the property of phosphorescing when suitably excited. It is used in making screens for X-ray and radioactivity work.

Siemens-Martin process. A combination of the Siemens and of the Martin process for making steel. (*See* **Siemens process,** *also* **Pig and scrap process.**) Siemens-Martin steel is sometimes called Siemens steel, but, as it is always produced in open-hearth furnaces, it is more frequently referred to as open-hearth steel, and unless basic steel is specified it is generally understood that acid steel is meant. In this process, in place of adding ore to the melted metal, a large quantity of wrought iron or scrap steel is added gradually. This ensures that the carbon is removed from the cast iron and a fluid steel remains. Additions of ferro-manganese, spiegeleisen, etc., are then made. The furnace may be arranged to tilt so as to discharge a portion of its contents into a ladle. As the operations may take 8-10 hours, it was more under control than was the Bessemer process, but is now itself obsolescent.

Siemens process. The process due to Siemens (1863) in which molten pig iron in the open-hearth furnace is heated so that the silicon and the manganese are oxidised and pass to the slag. The carbon is then oxidised by the feeding of suitable ore into the mass of melted metal. The oxide of iron is reduced so that the iron is added to the weight of the ingot, while the oxygen combines with the carbon and the gaseous product escapes.

Siemens regenerator. A device for utilising the heat of the waste gases from a furnace to heat up the incoming gases for heating the furnace. It consists of an arrangement of chequer brickwork which absorbs heat from the waste gases as they pass through it. Air and gas (usually producer gas or similar) are supplied to the furnace through separate regenerators used in pairs. The hot products of combustion and the air and producer gas alternately traverse the regenerators. (*See also* **Cowper stove.**)

Sienna. A silicate of iron and aluminium containing some manganese, which is used in the manufacture of paints and as a filler for linoleum.

Sieurin process. The name for a process for producing sponge iron. Ore concentrates of high purity, lime and coal are charged in layers into crucibles, which are then covered and heated in kilns until reduction is complete. After cooling the iron can be removed as a porous sponge.

Sieve. An implement used for separating coarse material from finer material, which generally consists either of perforated metal or of a network of wires attached to the base of a containing box. Sieves are normally used to separate the hard lumps in reclaimed sand and in the mechanical analysis of foundry sands. (*See also* **Sand testing.**)

Sieve test. A test used in differentiating the different types of sand or any powder material. A nest of sieves is used and a weighed dried sample of sand is shaken in them. The coarsest sieve is at the top and the sieve apertures are standardised. (*See also* **Sand testing.**)

Sifting. A term practically synonymous with screening (q.v.).

Sigma phase. A hard, brittle, non-magnetic constituent which may be present in nickel and cobalt alloys, and in certain steels that have a high chromium content, which have been subjected to certain heat-treatment cycles. Sigma is one of the stable phases in the iron-chromium-nickel system, but it does not exist at temperatures above 950 °C. The range of composition at which it exists just below this temperature is very restricted, but the range widens as the temperature falls. If such steels are rapidly cooled from above 1,000 °C practically no sigma is formed. All the austenitic steels contain elements other than nickel and chromium, and the amounts and combinations in which these occur alter the composition limits within which sigma can form and also the rate of formation. The ferrite-forming elements chromium, silicon, molybdenum, tungsten, titanium and niobium may promote sigma formation, while the austenite-forming elements, nickel, manganese, cobalt and carbon tend to prevent it. It is chiefly in weld deposits that the embrittlement due to sigma formation becomes a serious matter. Sigma forms in steels containing ferrite when in the softened condition and when subsequently they are heated as in welding.

Silfrax. The trade name of a high-grade silicon carbide refractory material.

Silica. A highly refractory substance silicon dioxide SiO_2. It occurs in three crystalline forms—quartz, tridymite and cristobalite (q.v.) and in the amorphous form (quartz glass).

All forms of silica fuse at about 1,710 °C, but below this temperature become plastic and can be worked into various forms and blown like glass and drawn in fine threads (quartz fibres).

Mixtures of amorphous silica with tridymite occur as chalcedony (q.v.). Other varieties are carnelian sand, sardonyx, chrysoprase opal and agate. Kieselguhr is a porous form of siliceous material consisting of the skeletons of extinct diatoms. Sand consists of weathered quartz, the purest forms of which are white, e.g. Calais sand. Quartz glass has a very low coefficient of thermal expansion (5×10^{-7}) and is transparent to ultra-violet light. Silica gel is an excellent drying agent for certain gases.

Silica bricks. Generally rectangular section preformed shapes made from silica by mixing the latter with suitable binding agent and firing in a kiln. Such bricks are tender when cold. They are used in building the roofs of reverberatory and regenerative furnaces, but are unsuitable for such parts of furnaces as are in contact with basic materials or slags. The following conventional nomenclature in reference to the silica content of bricks should be noted.

Name	Silica, %
Silica bricks	At least 92
Siliceous bricks	85-92
Semi-siliceous bricks	78-85
Firebricks	Less than 78

Silica flour. An exceedingly finely ground silica used as an ingredient in mould and core-paints in foundry-work. The fine refractory material is bonded with a suitable binder (bentonite, molasses, oil, etc.) and diluted with water and the mixture brushed, swabbed or sprayed on the mould face. This is used in dry-sand moulding and for casting steel or cast iron.

Siliceous bricks. *See* **Silica brick.**

Silicol process. A method of manufacturing hydrogen by the action of caustic soda on silicon. The latter is prepared by reducing silica (sand) with carbon (coke) at a high temperature. Sodium silicate is produced at the same time.

Silicon. A non-metallic element produced by the reduction of silica at high temperatures. Carbon is usually employed as the reducing agent and an electric furnace is used for heating. When so obtained silicon is a hard, grey, crystalline mass which much resembles graphite in appearance and in electrical conductivity (a similar reduction takes place in the blast furnace, so that pig iron always contains some silicon). Amorphous silicon (prepared by the reduction of silica with powdered magnesium) is a light brown hygroscopic powder that burns in air or oxygen when strongly heated. It is insoluble in all acids except a mixture of nitric and hydrofluoric, but reacts with concentrated or fused alkalis giving hydrogen. When strongly heated in a closed crucible the amorphous variety becomes a dense crystalline form known as graphitoidal silicon. This form does not burn in oxygen. Nodular silicon results from strongly heating the crystalline form. Density — amorphous 2.33, crystalline 2.39, nodular 3, at. wt. 28.09, at. no. 14, m.pt. 1,410 ± 20 °C, b.pt. 2,355 °C; crystal structure, diamond cubic; lattice constant, $a = 5.4282 \times 10^{-10}$ m.

Silicon is an important alloying element for both ferrous and non-ferrous metals. It is used extensively as a deoxidiser in steelmaking. It confers on the electrical steels (0.5-4.5% Si) some desirable magnetic properties. In small amounts it confers mild hardenability on steels generally, but in larger amounts it has the general effect of reducing ductility. In highly alloyed heat-resisting steels its presence improves the oxidation resistance at high temperatures. In cast irons silicon acts as a graphitiser, since it decomposes, and so reduces the amount of combined carbon.

When alloyed with copper to form what are known as silicon bronzes it enhances the mechanical and corrosion-resistant properties. The percentage addition in such cases lies in the range 1.3-3. Silicon improves the fluidity of molten aluminium and reduces the hot shortness of the metal. Silicon-aluminium alloys, e.g. those containing 5% or more silicon, used for castings are generally superior to most of the other alloys. Silicon has now gained a new importance in the electronics industry by virtue of its semiconducting properties. (*See* Table 2.)

Silicon brass. The name used in reference to a large number of copper-zinc alloys which contain 1-4% Si together with smaller proportions of other elements, such as manganese and aluminium. (*See* Table 5.)

Silicon bronze. A group of alloys of copper and silicon produced in order to conserve tin. They show considerable tensile strength, and when the silicon content does not exceed about 3% they show good ductility. Small amounts of iron, manganese and zinc are usually present. The molten metal is prone to gas absorption. Charcoal should not be used to cover the melt, but lead-free glass, cryolite or borax may be used. (See Table 5.)

Silicon carbide. A highly refractory and abrasive compound obtained by heating sand and crushed coke in an electric furnace. It is important to have the charge in the correct proportions. Generally five parts of sand to three of coke are mixed with a little salt and some sawdust and heated to a temperature not lower than $1,550\,°C$. It has the formula SiC. It fuses with difficulty, and is thus useful for furnace linings. It is a black coarsely crystalline mass nearly as hard as diamond. Also called carborundum (q.v.).

Silicon chip. See Microprocessor, Semiconductor, Transistor.

Silicon copper. A series of copper-silicon alloys having a wide range of silicon content, the lower ranges (below 5%) being those alloys used for producing wire and drawn products; the higher ranges being used as corrosion-resistant alloys in castings. (See Table 5.)

Silicones (siloxanes). Polymeric materials in which silicon and oxygen atoms alternate along a chain, with organic radicals linked to the silicon atom in mono-, di- or tri-functional units. There is always one oxygen atom between two adjacent silicon atoms.

$$-O-\underset{\underset{\displaystyle R}{|}}{\overset{\overset{\displaystyle R}{|}}{Si}}-R- \qquad -O-\underset{\underset{\displaystyle R}{|}}{\overset{\overset{\displaystyle R}{|}}{Si}}-O- \qquad -O-\underset{\underset{\displaystyle O}{|}}{\overset{\overset{\displaystyle R}{|}}{Si}}-O-$$

| mono- (M) functional unit | di- (D) functional unit | tri- (T) functional unit |

R = radical

In silicone rubber, the molecules are usually linear chains of D units with M units at each end, and R is a hydrocarbon radical, often CH_3-, but sometimes C_6H_5- (phenyl) or $CH_2 = CH$- (ethenyl).

In silicone resins, there are more T units to form branching points or cross links, and R is usually methyl, phenyl or a mixture of the two.

Silicone rubber, whilst very expensive, has superior properties to other forms of rubber, particularly with conditions of high or low temperature, is more resistant to oxidation, and has better chemical and electrical properties. At room temperature it is generally inferior to organic rubber.

The resins are relatively stable to prolonged heating in air, resistant to chemical attack (except organic solvents and strong acids) and have good electrical properties and outstanding water repellence.

The rubber compositions are used in seals, gaskets, roller coatings; for casting and potting electrical components; aircraft de-icing and instrument tubing; electrical insulation and cable covers (particularly at elevated temperature conditions); foodstuff and beverage handling; surgical implants.

The resins are added to paints for heat-resistance and water repulsion; paper impregnation; lamination of fabrics, especially glass fibre and asbestos; electrical mouldings; room-temperature curing, potting and encapsulation material; release agents for moulds, bread-baking tins and frying pans.

In fluid form or as dispersions, the same materials, but of lower molecular weight, are employed in paints, polishes, and textile fabric finishes. (See Table 3.)

Siliconeisen. The name used in reference to iron that contains 5-15% Si.

Siliconising. A process in which a silicon-rich layer is produced on the surface of low-carbon steels, making them resistant to oxidising atmospheres, corrosion and wear. The process to cause the silicon to diffuse into the surface layers must be carried out at high temperatures. The steel is heated with silicon carbide or ferrosilicon and chlorine, at about $950\,°C$, so as to produce a case containing about 14% Si.

Silicon steel. A name used in reference to steels containing relatively high proportions of silicon, with or without other alloying elements. The presence of silicon to the extent of about 0.1% in steel causes the depression of the temperature at which gases are liberated from molten steel, to below the temperature at which solidification occurs. Such steel gives more solid ingots than otherwise would be produced. It is called killed steel. Higher proportions (e.g. 1.5-4% Si) give steels that are heat-resisting (see Table 6) and which have a high magnetic permeability. This type of silicon steel (transformer steel), which does not contain chromium, is also referred to as electrical steel.

Sillimanite. Aluminium silicate $Al_2O_3.SiO_2$. It occurs in gneisses and schists usually as needle-like crystals. It is used in the form of very fine powder as a parting powder in foundry work, and as a constituent in paints for dry-sand moulds. It is a high-grade refractory which, with clay, is used for making crucibles for molten steel and as a furnace lining.

Silky fracture. A term used in reference to a fracture having a fine grain and smooth grey appearance.

Sill. A sheet of igneous rock that has been formed by the crystallisation of magma that has been forced between layers of rock, and is usually horizontal or nearly so.

Siloxicon. In making carborundum in the electric furnace it is found that silicon carbide SiC is surrounded by a layer of a substance to which the formula Si_2OC_2 has been given. This is called siloxicon. Some investigators consider this substance to be a solid solution of silica in silicon carbide.

Silver. One of the noble metals that is not oxidised in air either in the cold or when heated. It is white in colour and one of the most ductile and malleable of the metals. It occurs in native state, usually containing gold and copper in Norway, Peru and Idaho. The important ores are the sulphide, chloride, arsenide and complex sulphides, e.g. silver sulphide or argentite (or silver glance) Ag_2S; silver chloride or chlorargyrite (or horn silver) AgCl; pyrargyrite Ag_3SbS_3; stromeyerite $(CuAg)_2S$; stephanite Ag_5SbS_4 and proustite Ag_3AsS_3.

There are several methods used in the extraction of silver:

(a) Wet processes, in which the crushed ore is treated with sodium chloride, thiosulphate or cyanide (the latter being more generally used) and the silver precipitated with zinc. Amalgamation is now used only in combination with cyaniding, and then usually with high-grade ores.

The chlorodising of complex silver ores by the Holt-Dern process may precede the cyaniding process.

(b) The cupellation process, in which silver ore is smelted with a lead ore and the silver separated from the lead-silver alloy. This is the most ancient method.

(c) The silver in argentiferous lead obtained, for example, from galena is extracted by the Luce-Rozan or the Parke's process (q.v.).

In the electro-refining of lead and copper, silver may be obtained by treatment of the anode slimes (Bett's process), the silver then being a by-product.

Principal producers of silver are Mexico, the USA, Peru, Bolivia, Canada and Australia.

In the annealed condition pure silver is the best conductor of heat and electricity known. It is highly resistant to most forms of corrosion. The commercial refined form of silver is usually known as "fine silver" as distinct from "sterling silver" or "standard silver", used for coinage (until 1920), and which contains 7.5% Cu. Pure silver is often written 1,000 fine, and in this nomenclature sterling silver is 925 fine. Silver is refined by cupellation or by the Moebius electrolytic process, in which the anode (gold) slime is collected in a bag round the anode.

Silver is used for tableware and vessels, both as solid silver and as electrodeposited coatings. It is also used as pipelines, nozzles, etc., in dairy, cider and brewing industries and as vessels in chemical equipment. Silver contacts are used in telephone relays and other light current circuits. Silver-lined bearings have been used in aircraft engines. Cadmium containing 1.25% Ag and a little copper is used as a bearing alloy. A tin-base alloy containing 3.5% Ag is used as extruded tubing in distilled water apparatus and in dairy equipment. Silver compounds are important in photography.

Silver forms useful alloys with copper, gold, nickel, zinc, manganese and tin. Crystal structure, face-centred cubic, $a = 4.0774 \times 10^{-10}$ m. For physical properties see Table 2.

Silver amalgam. An alloy of silver and mercury. The term also refers to a native alloy found in Spain and which contains 65.2% Hg and 34.8% Ag.

Silver brazing alloys. Silver solders used for joining steel, etc., by brazing (q.v.). The higher melting point alloys are essentially 60/40 brasses containing 10-20% Ag. They melt above about 780 °C and flow at a somewhat higher temperature. The lower melting point alloys are richer in silver and may contain in addition to copper and zinc up to 18% Cd. Silfos contains phosphorus. The melting point of these high-grade silver alloys is usually below 725 °C.

Silver coinage. Silver alloys of varying composition used as metal coins. Old coins have been found which are practically pure silver but some contain traces of lead and gold. More recent silver coins contain 50-89% Ag, the balance being copper or sometimes copper and nickel.

Silver glance. A sulphide of silver Ag_2S which contains 87.1% Ag and is the commonest of silver ores. It occurs chiefly as hydatogenetic veins associated with other silver minerals, pyrites and metallic sulphides in Montana, Nevada, Ontario, New South Wales, the Latin American countries, Saxony (Germany) and Bohemia (Czechoslovakia). It crystallises in the cubic system; Mohs hardness 2-2.5, sp. gr. 7.2-7.36.

Silver plating. The electrodeposition of silver as a surface coating on other metals. Thin silver deposits may be applied to metals for decorative effect or for useful purposes such as electrical contacts and reflectors. The electrolyte generally used consists of a solution of silver cyanide and potassium cyanide. A typical plating solution would contain about 40 g of silver cyanide and 55 g of potassium cyanide per litre. Potassium carbonate (35 g l^{-1}) is added, as this improves the conductivity of the solution and assists in the regular solution of the silver anode which must be used. As deposited from such a solution the silver is dull, but the addition of small amounts of brighteners such as carbon bisulphide (about 0.6 g m^{-3}), or of sodium thiosulphate (hypo—1 g) together with 10 cm^3 of ammonia enable bright deposits to be obtained. Thiourea (35 g l^{-1}) and potassium selenite have also been used for the same purpose. Brass articles or articles that have been brazed or soldered tend to have silver deposited on them by immersion, and the subsequent coating of silver shows poor adhesion. To prevent this the articles may first be given a coating of nickel or they may be dipped for a second or so in a dilute solution of mercury cyanide.

Silver sand. A white variety of silica (sand) containing little or no iron. It is exported from Belgium for use as a refractory material.

Silver solder. A term used in reference to a group of ternary alloys of silver, zinc, and copper used for brazing. They combine strength and resistance to corrosion with relatively low-melting temperatures and free-flowing properties. In certain cases, however, cadmium, phosphorus, nickel, manganese and tin may be added to lower the melting point and improve fluidity. They are extensively used on ferrous and non-ferrous metals.

Silvo. A proprietary liquid metal polish used in the preparation of microspecimens of soft non-ferrous metals.

Simple cubic structure. An elementary structure with ions as the corners of a cube. (*See* **Crystal classification.**)

Single crystal. An artefact in which grain boundaries have been eliminated, to leave uniformly orientated crystallites in a single grain. From such a single crystal, material can be cut to give preferred orientation of particular planes or crystal directions in the finished component. Single crystals can be produced from the melt, by "seeding" with a nucleus crystal of known orientation, to cause epitaxial growth, or by recrystallisation from either liquid or solid state. In the liquid state, a natural texture (<100>) in f.c.c. materials may develop and one <100> crystal can be selected by a path restriction (spiral). This process is used to produce single crystal turbine blades on a commercial scale. In the solid state the process is so random that this is neither necessary nor efficient. From the solid state, conditions of previous cold working can contribute by lowering the recrystallisation temperature, the defects in the structure acting as nuclei, and rapid grain growth can occur. In alloys which undergo transformations in the solid state, it is much more difficult to obtain single crystals, if indeed possible at all in many cases.

Single domain. The magnetisation of a ferromagnetic is described under B-H curve. In a "soft" magnetic material the magnetisation can readily displace the boundaries between domains, so that domain walls virtually move, and the favourably-orientated domains (in the magnetising direction) grow in size. At a later stage the remaining domains must rotate to complete the saturation, and this too is relatively easy in a soft magnetic material. If, however, obstacles to domain wall motion and rotation exist, the material will not easily reverse its magnetisation once achieved, although more energy will be needed to magnetise it. Such a material is "hard" magnetically speaking, and makes a good permanent magnet. Precipitation hardening is used in magnetic alloys to achieve those obstacles to domain wall motion or rotation.

An alternative however, is to utilise very small particles of magnetic material, such that the particle size is less than the domain wall thickness. Such particles cannot then contain more than a single domain, and if the particles are made to high aspect ratio (say 10:1) the rotation to reverse their magnetisation will be very difficult, since the direction must first be made transverse to the length before going fully into the reverse direction. A very high reversing field will be necessary to do this, and the remanence of such material will be considerable. These are "single domain" materials, prepared by crushing or growing particles to the appropriate shape, and bonding them together with non-magnetic matrix material, such as polymer or clay. The diameter of such particles is of the order of 20 nm.

Single shear. A metal plate may be in single shear when there is one section to be sheared through in order to cause failure. This is the case in a lap joint having a single row of rivets or in a butt joint with single strap.

Sinkhead. The name used in reference to the refractory insulation placed at the top of an ingot mould to maintain a reservoir of molten metal during the solidification of the casting. Also known as hot top, shrink head.

Sinking. A cold-drawing process used for seamless tubes. As the tube is drawn through the die without internal support, it is elongated and reduced in outside and inside diameter. The wall thickness may thus be varied as the tube passes through the sizing rolls.

Sinking mill. A mill on which tubular products are reduced in diameter by the process known as sinking.

Sinter. A term used in reference to the solid waste from smelting or refining operations. It is also used to denote a product of a sintering operation. In this the ore material which is too fine for efficient extraction in a blast furnace, because of high dust losses, is mixed with fine coke and water and then fed on to a grate, where combustion is started and maintained by an air current drawn through the charge. Partial or incipient fusion produces a moderately strong porous product suitable for further treatment. The sintering may be done on Dwight-Lloyd, Lurgi or Greenawalt type sintering machines.

Sinter is also the term used to describe the silica deposited at the mouth of a hot spring or geyser. It is a hydrated form of silica.

Sintered aluminium powder (S.A.P.). Aluminium oxide particles are sintered in a matrix of aluminium to produce a dispersion hardened product with improved mechanical properties over aluminium alone.

Sintered carbides. Certain carbides, such as those of tungsten, tantalum, titanium or molybdenum, are among the hardest substances known and they have high melting points. But they are brittle, a property that limits their usefulness. To overcome this the carbides in powder form may be mixed with powdered cobalt or nickel. The mixture is compressed and is heated to a temperature at which the cobalt fuses. A typical cemented carbide consists of about 43.5% tungsten carbide, 43.5% tantalum carbide and the balance, as binding alloy, is cobalt (or nickel). This process of heating the compact to a temperature between 700 °C and 800 °C (in the case of tungsten carbide) is called pre-sintering, and the compact after this treatment can be machined prior to the final sintering.

Sintering. In powder metallurgy the main process consists of cold pressing followed by heating. The shaping and adhesion of the particles are both obtained in the initial pressing. Though bonding occurs on pressing, the compacts are weak. Final cohesion is produced by heating the compact to a temperature below the melting point of the pure metals, but in the case of alloys the temperature may be above the melting point of one of the constituents. Sintering is the term applied to the heating operation, or sometimes to the whole operation—of pressing and heating. stituents. Sintering is the term now applied to the whole process of consolidation of a powder and elimination of porosity by diffusion and pressure, or sometimes to the whole operation—of pressing and heating.

Sinterings metal. Parts pressèd from ferrous and non-ferrous metal powders which are simultaneously, in the case of hot pressing, or subsequently, in the case of cold pressing, heated to below the melting point of the main metallic constituent to produce a sintered or bonded mass.

Sintex. The trade name of ceramic-tipped tools. The tips consist of a high-grade alumina pressed and sintered into the desired form and then attached to steel shanks by platinising and brazing. Their performance is equal to that of carbides.

SiO_4 tetrahedron. The basic building brick of silicates as well as the allotropic forms of pure silica. A silicon ion at the centre of a tetrahedron is surrounded by four oxygen ions, one at each corner of the tetrahedron, the bond being partly ionic and partly co-valent. Each oxygen ion is shared by two silicon ions, hence each oxygen ion is part of two tetrahedra. The formula SiO_4 for the basic tetrahedron therefore corresponds to a true composition of SiO_2. The tetrahedra can be arranged in various ways, including the formation of rings or layer lattice structures, as in kaolinite. (For details see **Cristobalite, Quartz** and **Tridymite** for silica, **Kaolinite** and **Layer lattice** for claylike materials, **Glass** and **Ceramics** for silicates.)

SI units. The accepted abbreviation for Système International d'Unités (International System of Units), the modern form of the metric system agreed at the Eleventh General Conference of Weights and Measures (CGPM) October 1960. It is being adopted throughout the world as the primary world system of units of measurement.

Abbreviations are in singular form without full stops, e.g. km for kilometre, not km., and N m for newton metre, not N.M. Negative powers are preferred in complex units, rather than use of the solidus, i.e. $N\,m^{-2}$ rather than N/m^2. Multiples are used in steps of 10^3 wherever possible, although in special cases 10^2, 10, 10^{-1}, 10^{-2} will be acceptable. To assist in using steps of 10^3, common names and symbols are used to simplify both written and verbal expression.

10^{12}	tera	T
10^{9}	giga	G
10^{6}	mega	M
10^{3}	kilo	k
10^{2}	hecto	h
10	deca	da
10^{-1}	deci	d
10^{-2}	centi	c
10^{-3}	milli	m
10^{-6}	micro	μ
10^{-9}	nano	n
10^{-12}	pico	p
10^{-15}	femto	f
10^{-18}	atto	a

The symbol of a prefix is considered to be combined with the unit symbol to which it is directly attached, forming with it a new symbol which can be raised to a positive or negative power and which can be combined with other unit symbols to form symbols for compound units, e.g.

$$1\,cm^3 = (10^{-2}\,m)^3 = 10^{-6}\,m^3$$
$$1\,\mu s^{-1} = (10^{-6}\,s)^{-1} = 10^6\,s^{-1}$$
$$1\,mm^2\,s^{-1} = (10^{-3}\,m)^2\,s^{-1} = 10^{-6}\,m^2\,s^{-1}$$

Compound prefixes are not used, e.g. write nm (nanometre) not mμm (millimicro metre).

The system is based on seven base units:

metre	(m)	–	length
kilogram	(kg)	–	mass
second	(s)	–	time
ampere	(A)	–	electric current
kelvin	(K)	–	thermodynamic temperature
candela	(cd)	–	luminous intensity
mole	(mol)	–	amount of substance

SI units for plane angle and solid angle, the radian (rad) and steradian (sr) respectively, are supplementary units in the International System.

Expressions for derived SI units are stated in terms of base units, e.g. the SI unit for velocity is metre per second $m\,s^{-1}$. Some derived units are given special names and symbols.

A list of the basic, derived units, and other common units to be used, is given in Table I of this volume, together with some physical constants.

Size factor. After exploring the alloy systems formed between copper, silver and gold, W. Hume-Rothery formulated a number of empirical rules, to indicate the likelihood of a range of solid solubility between two dissimilar metals alloyed together. One factor of great importance was the size of the ions involved. If one ion is very small compared to the other, the smaller may fit into an interstitial position in the lattice of the larger, but this is rare, and confined to hydrogen, nitrogen, boron, carbon, and occasionally oxygen. In substitutional solid solution, where one ion can be envisaged taking the place of the other anywhere in the lattice, too great a difference in size would obviously produce strain, and whilst this might be tolerated for small concentrations, it would be unlikely at higher. Hume-Rothery found that two metals with the same lattice type would form a solid solution over an extended range of composition if their size difference was less than 15%. Over this difference, the solid solution would be restricted in its range of composition, and when the difference was very small, there was often complete mutual solid solubility over the entire range of composition.

If the crystal lattices were of very different type, then even a size difference less than 15% was unlikely to lead to an extended range of solid solution. Silver and gold are almost identical in ionic size, and form one solid solution stable across the range of 0 to 100%. Copper and nickel vary by 2.5%, and form one continuous solid solution. Copper and silver differ by 13%, and solubility is restricted to 8% by weight of silver in copper, and 9% by weight of copper in silver. All these metals have face-centred cubic lattices.

When a solid solution is formed, but under conditions of size factor or crystal type which are unfavourable, further addition of the second metal may cause increased strain, until a point of instability is reached, and some change must occur. The lattice may then change in type, to adjust for the distortion, and the new lattice may now be able to accept an even greater concentration of the second metal. This is precisely the behaviour between copper and zinc, where the addition of zinc to copper, with a size difference of about 8%, is within the 15% rule, but zinc has a hexagonal lattice, not face-centred cubic like copper. Nevertheless, the zinc atoms will substitute for copper until over 38% zinc is present. After that, the strain requires some change in lattice, and a new phase appears, β brass, which will accept the zinc atoms more readily. Subsequently, the strain is such that all the zinc is present in the new body-centred cubic structure of beta brass, but even this is limited to about 50% zinc content, after which the structure again changes, with the introduction of gamma brass, with a complex cubic structure, and a composition approximating to the intermetallic compound $Cu_5 Zn_8$.

Hume-Rothery's rules are not precise, because so many factors are involved, but they can be used to predict possible alloying behaviour, and cut down on costly experiments to determine likely solid solubility of one metal in another.

Sizing. A term used in reference to any metal-working operation designed to produce a finished product of desired dimensions. In ore preparation sizing operations are referred to as sorting or grading. The term is also used to describe the operations of passing metal tube between rollers to produce a product of the desired shape and size. The final pressing of a presintered powder metallurgy compact to ensure correct size before final sintering has also been called sizing.

Sizing mill. A type of rolling-mill used in sizing operations (see **Sizing**). In tube-making, passes through the sizing mill follow the "reeling"—a process of light cross-rolling.

Sizing rolls. The name used in reference to the specially designed and shaped rolls used in a sizing mill in which closed passes are used.

Skeleton pattern. A type of pattern used in foundry-work for large castings, where a full pattern would require a very large amount of timber. The pattern is a ribbed construction, in wood, and where a separate core is to be made the ribs are arranged transversely. The spaces between the ribs are filled with sand by the moulder, who thus builds up an improvised full pattern. This is then sprinkled with parting powder and used in moulding. Usually two skeleton patterns, in halves, are necessary. Where the ribs of a skeleton pattern are the same width as the thickness of the casting, such a pattern can be used to make both the mould and the core.

Skelp. The name used in reference to a plate of wrought iron or steel used for making pipe or tubing by rolling the skelp into shape and welding or riveting the edges together.

Skew-bevel gear. A form of bevel gearing in which the teeth of the wheels are curved along their length. It is said that this form eliminates backlash and so gives smooth and quieter driving.

Skim bob. The name used in reference to the recess or hollow in the cope in order to trap the slag before the metal enters the mould.

Skim gate. The name used to describe the addition or relief sprue that is placed between the down-gate and the ingate to take the slag and so prevent its entry to the mould. Sometimes a metal sieve is placed between the gate and runner of a foundry mould to separate entrapped oxides, etc., during pouring. This is also referred to as a skim gate.

Skimmer. A tool used for removing dross, scum and slag from the surface of molten metal, e.g. before pouring.

Skin. The term used in reference to the thin surface layer which differs in structure, composition or other characteristics from the main mass of the material.

Skin-dried mould. A mould in which the mould face is dried for a depth of 2-3 cm either with a gas flame or hot air. The portable heaters are fired by gas, oil or coke, and a fan circulates the hot gases.

Skin pass. A term used to describe a light cold rolling of heat-treated sheet or strip. The operation improves surface finish and suppresses the tendency to kinks, flats and stretcher-strains on further manipulation. The rolling involves only a slight reduction of the sheet, etc.

Skin-shrinkage. In the production of ingots molten metal is poured into prepared moulds. The metal that comes into contact with the mould face solidifies at once, and there results a thin skin of solid metal within which is a mass of molten metal. During cooling volume changes occur, that change which is associated with the contraction of the liquid volume giving rise to "pipe" (q.v.). Contraction also occurs when the liquid solidifies and further thermal contraction as the hot solid cools. Since the skin is much cooler than the interior it cannot accommodate the shrinkage that occurs in the mass of the ingot. There are, therefore, set up in the interior of the ingot internal stresses which may give rise to contraction cavities. Where the shrinkage of the skin is of such magnitude that cracking occurs, liquid may bleed through the cracks, giving rise to undesirable defects.

Steel ingots are poured quite rapidly, but the maximum rate is limited to that in which a skin of sufficient thickness is formed to withstand the pressure of the liquid steel. This minimum thickness is determined by the temperature and the rate of pouring.

Skip. The term used to describe the container which carries the charge intended for the blast furnace to the top of the furnace where the contents of the container or skip (ore, fuel or limestone) are discharged on the double bell charging arrangement.

Skip weld. A non-continuous or intermittent weld, the welded areas being separated by non-welded spaces.

Skull. A mass of solidified metal, usually oxidised and containing some slag, which remains in (and may be removed from) a furnace, ladle or crucible.

Skutterudite. A naturally occurring ore of cobalt that consists of cobalt arsenide $CoAs_3$. It varies in colour from white to a dull lead-grey. The crystals are octahedral in form and often contain a little iron and sulphur. It is found near Modum in Norway.

Slab. A semi-finished block of metal, rectangular in section, which is a product of the blooming mill and which is intended for subsequent rolling into plate. Though there are no precise dimensions, a slab is usually over 25 cm wide and more than 4 cm thick.

Slabbing mill. A rolling-mill, usually of the two-high universal type for producing slabs. The mill can take blooms varying in section from 160-520 cm².

Slacking down. A term used to describe the operation of putting damp ashes or similar material on a furnace fire to reduce the rate of combustion.

Slag. The material formed by fusion of constituents of a charge or of products formed by the reactions between refractory materials and fluxes during metallurgical processes. It may be somewhat vitreous in appearance and floats as a molten mass on the surface of the molten metal in the furnace. The importance of slag arises from: (a) through the slag, impurities in the metal are removed; (b) by acting as a molten covering on the molten metal below, the latter is protected from oxidation and consequent loss of metal; (c) by controlling the character of the slag the desired quality in the metal can be secured.

Slagging. A term used to describe the operation of removing slag from the furnace of molten metal. The term is also applied to the reaction that takes place between a refractory substance and a flux at high temperature. The refractory material is destroyed and a molten slag results.

Slag hole. A hole in the side of a furnace which may be unplugged to run off the slag. This hole is necessarily located above the surface level of the molten metal. (Sometimes called "slag notch".)

Slag inclusions. A name which refers to the presence in an ingot or in any metal product of an ingot, of pockets of slag impurity. In an ingot the impurities in the pipe are cropped, but when the slag inclusion is in the skin it may be drawn into long threads by rolling and forging and so form a line or a zone of weakness. Slag inclusions occurring near the surface of an ingot or billet are usually removed by rough machining.

Slag trap. A device for retaining the slag and preventing it entering the mould during pouring. There are several methods in use, including strainers situated in the pouring basin or at the joint of cope and drag; chokes; and specially designed gates such as pop gate and pencil gate (q.v.).

Slag viscometer. A device for estimating the viscosity of slags. It consists of a conical funnel from the base of which extends a narrow horizontal outlet. A hot spoon is used to take a sample of the slag and to pour it into the funnel. The distance the slag flows out through the outlet is measured and from a chart the viscosity can be estimated. Since acid slags consist mainly of silicates of iron and manganese (in the steelmaking process) with small amounts of calcium and aluminium, and since the viscosity of the slag depends on the silica content, a measurement of the viscosity gives an indication of the silica, and from this the amount of manganese and iron taken together can be estimated.

Slag wool. A fibrous product made by blowing jets of steam through molten furnace-slag. Also known as silicate cotton, rock wool (q.v.). It is used as a heat insulating material.

Slaked lime. The product obtained by adding small amounts of water to freshly made quicklime. It is a white powder having the formula $Ca(OH)_2$. A solution in water is known as lime-water.

Sleek. A tool similar to a trowel, used for smoothing the surface of moulds or cores, for repairing or for finishing the mould after dusting with parting powder. Sleekers, as they are sometimes called, may take several forms, the name being related to the particular use of the tool, e.g. bead sleeker, used for smoothing that part of the mould that is to produce a beaded edge; a button sleeker is used for smoothing those parts of the mould that are of hemispherical form.

Sleeker. See **Sleek.**

Slenderness ratio. A term used in reference to iron and steel and other metallic columns which are termed long columns and in which failure usually occurs by buckling before the elastic limit of the material is reached. The ratio is represented by l/r, where l = length of column and $r = \sqrt{}$ moment of inertia/area section. In general when the slenderness ratio exceeds 100 the column is liable to fail by buckling and is therefore a "long" column.

Slicker. A term synonymous with sleeker (q.v.).

Slime. A term used in reference to the extremely fine particles of pulverised ore that settle very slowly from aqueous suspensions. Slimes occur in the usual wet concentration processes. The term is also used in reference to the finely divided material—either metal or metallic compound—that is formed at the anode and is frequently deposited in the bath during electrolytic refining of metals.

Slip. The term used in reference to plastic deformation, by shear, of one part of a metal crystal relative to another, over a definite crystallographic direction, and usually in a particular crystallographic plane. As originally defined, slip was regarded as the bodily movement of one part of a crystal over another, but the later view is based on a step-by-step movement of a type of crystal imperfection known as a dislocation.

Slip bands. When the surface of a polycrystalline metal is polished and the metal plastically deformed, e.g. by shear, then these become visible under the microscope as steps or bands which run parallel across each grain, but change direction from grain to grain.

Slip casting. A process used in the production of refractory products. The ground material is mixed with water to form a creamy liquid. This is poured into plaster moulds, where the surplus water is absorbed and a solid replica of the inside of the mould is obtained.

Slip interference. The term used to describe the phenomenon of the resistance to the deformation of a ductile metallic structure offered by particles of a hard phase dispersed in the matrix.

Slip lines. A term synonymous with slip bands (q.v.).

Slip plane. The crystallographic plane of slipping in an individual crystal under shear stress. The slipping generally occurs along certain preferred planes in the crystal lattice. These are atomic planes which pass through the centres of atoms, and slipping usually takes place along such planes in which the atoms are closest together and in which the individual planes are furthest apart.

Slip plane precipitate. The term used to describe the precipitate formed preferentially along those planes of a solid solution on which slip has occurred as a result of cold working after solution heat-treatment or of quenching stresses.

Slitting. A cutting operation applied to wide material sheets, to produce narrow strips. As it is usually uneconomical to roll narrow strips, the slitting operation is used, employing rotary shears, or blades fixed vertically.

Sliver. A surface defect in metal produced by rolling into the surface a very thin piece of the parent metal. The defect is more commonly noticed after cold-rolling operations on metal products that have been formed from overheated ingots.

Slotting. The cutting in a plate or panel of any aperture (e.g. a keyway). It is usually performed on a slotting machine in which the work is clamped to a horizontal table and the cutting tool has a reciprocating vertical motion.

Sludge. A term used in reference to the finely divided deposits formed at metal surfaces during pickling or surface treatment process or in electro-refining. (*See also* **Slime.**) The term is also used to describe the semi-liquid mud that forms in steam boilers or other vessels containing water.

Slug. A general term to describe the metal blank from which a forging or extrusion is made.

Slug test. *See* **Upending** and **Compression test.**

Slump test. A works test used by vitreous enamellers for determining the flowing power (i.e. the mobility) of a slurry. The slurry is placed in a cylindrical measuring cup which is open at both ends, and rests at the exact centre of a carefully levelled plate. When the cup is raised by hand (or sometimes by a mechanical device) the slurry flows out on to the plate forming a circular "slump". The diameter of the slump is taken as a measure of the mobility and is usually estimated from a series of concentric circles inscribed on the plate.

Slurry. A name used to denote a semi-liquid mixture of clay, ganister or other finely divided materials. The slurry made from ganister is used to repair the linings of furnaces. Clay slurry mixed with plumbago or silica flour and with dextrine is used as a heat-resisting paint for the surfaces of cores used in moulding.

Slush casting. A method of producing hollow castings in alloys that solidify over a wide range of temperature. The casting metal, called slush metal, is usually a white metal, such as one of the fusible alloys, having a low melting temperature. The mould is usually a metal mould similar to those used in chill-casting. The metal is poured into the mould and brought into contact with all faces of the mould by agitating or spinning it. In this way a shell of solid metal is formed and the excess metal is poured out. Completely enclosed shells may be made by casting the necessary amount of metal in an enclosed mould.

Slushing compound. The name given to a variety of compounds that are smeared over or painted on iron and steel to afford temporary protection against rusting. Such compounds are greases, non-drying oils, "dags" oil and similar substances. They are to be regarded as rust preventive rather than rust removing.

Smalt. The residue that is left after roasting cobalt ore is called zaffre and consists of impure cobalt arsenite. When this is heated with a mixture of white sand and potassium carbonate a very attractive blue glass is produced. This glass is termed smalt.

Smaltine. A tin-white mineral consisting essentially of cobalt arsenide $CoAs_2$ but which usually contains, in addition, some iron and nickel. It occurs either in the massive form or reticulated in veins associated with silver, copper and nickel minerals and with calcite and quartz. It crystallises in the cubic system. Also known as smaltite and speiss cobalt (q.v.). Mohs hardness 5.5-6, sp. gr. 6.4-7.2. Found chiefly in Ontario, Bohemia (Czechoslovakia), Saxony (Germany).

Smee cell. A voltaic cell in which zinc and silver are the electrodes and dilute sulphuric acid the electrolyte. It is of interest as being the cell in which the first attempt was made to eliminate polarisation, mechanically. This was done by producing a rough platinised surface on the silver. Hydrogen bubbles escaped more readily and polarisation was reduced (but not eliminated).

Smelting. A pyrometallurgical process that consists in the fusion of a rich ore or concentrate to obtain crude metal or a matte. Usually the process involves the reduction of the ore using a reducing agent (e.g. carbon and carbon monoxide in the iron blast furnace). The removal of the unwanted impurities in the ore is effected by adding a flux which forms with these impurities a fusible and liquid slag at the temperature of the furnace reactions. In the case of iron-ore smelting the reducing agent added to the charge is coke, and the flux is usually limestone. Sodium salts and fluorspar are also used as fluxes.

Smith forging. A forging made by hammering on an anvil or by a forging hammer, but without dies that have the exact shape of the finished product. Such open dies are flat, producing the general shape of the section and leave the finishing to be done by hand hammering.

Smithsonite. An important ore of zinc consisting essentially of zinc carbonate $ZnCO_3$. It is the American name for the ore calamine. Found in Belgium and the USA. Also called zinc spar. Crystallises in rhombohedra. Specific gravity 4.42.

Smithy char. The name used in reference to a type of coke used in blacksmith forges. It was made in small ovens of the beehive type.

Smithy forge. A small open furnace using coal or coke as fuel and a low-pressure air-blast usually supplied by bellows.

Smoking salts. An alternative name for hydrochloric acid used in soldering operations.

Smoky quartz. A translucent variety of quartz SiO_2 coloured red to brownish red. The colour is said to be due to the presence of organic matter. Also called cairngorm. Specific gravity 2.66.

Smothered arc. A term used in reference to the type of electric arc furnace in which the arc is covered by a part of the charge by ensuring that the electrodes are partly submerged in the charge.

Smudge. The American term for "stopping-off" materials that are used to render parts of the cathode non-conducting. The term may also be used to refer to the materials used on metal during soldering to ensure that certain parts remain uncoated.

Snagging. A term used in reference to the removal by grinding of projections such as sprues, gates or fins from rough castings.

Snap flask. A type of tapered moulding box, usually made in light alloy or wood, that is used for small and shallow castings. The box is without lid or bottom and has a hinge at one corner and a fastener at the diagonally opposite corner. The mould (in two parts) is generally made on a bench and carried to the casting site, where the flask is removed. To guard against accidental break-out of metal while pouring, a steel jacket may be slipped over the mould joint, or a steel hoop may be rammed up in the mould. Snap flasks are economical in use, since they are not damaged in the pouring operations and can be re-used as soon as the contained mould is placed at the position for pouring. Also called "slip-flask". If the flask is not removed before pouring it is called a "tight-flask".

Snarl. A kink which has been drawn tight in a specimen of wire.

Snarl test. A workshop test applied to wire. The wire is looped and pulled taut, then straightened, the operations being repeated until fracture occurs. The number of operations is reported.

S-N diagram. The abbreviation for the title "stress-number of cycles" curve used in fatigue testing. The highest stress that a material will withstand without failure, regardless of the number of times it is repeated, is generally called the endurance limit. By testing a number of specimens of a particular material at progressively lower stresses, and recording the number of cycles of stress application that occur before fracture, a series of plots can be made resulting in a curve of the type shown in Fig. 154. In order that such a diagram shall not take up too much space (and for other reasons) the number of cycles (N) is commonly plotted on a log scale, while the stress (S) is plotted on the common Cartesian scale.

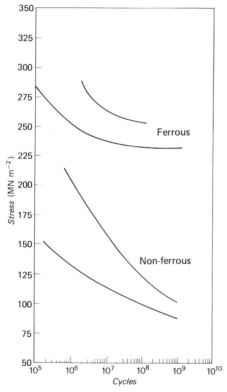

Fig. 154 S-N curves.

Such a method is known as semi-log plotting. Sometimes both S and N are plotted on a log scale, and the result is referred to as a log-log plot. But whatever scales are used, the "form" of the curve for ferrous materials shows a distinct knee at a stress that is slightly above the endurance limit.

Figure 154 shows the S-N curve for an unusually homogeneous steel specimen. But with ordinary specimens there is usually a good deal of scatter, so that the "curve" becomes a band, the upper limit indicating the performance of the material in the absence of all flaws and inhomogeneities, and the lower limit representing the actual performance of the material containing stress-raisers (particularly those due to poor surface finish). The curves for ferrous materials show that there is a definite endurance limit, and that a life of ten million cycles gives a reasonable assurance that the endurance limit has been reached. For non-ferrous materials, no endurance limit can be defined, and it is usual in such cases to report both the stress and the number of cycles withstood before failure.

Snead process. A method of heat-treating a metal by the passage of an electric current directly through it. The process is in general limited to specimens of uniform cross-section and uniform electrical resistance, e.g. wires, rods, tubes, etc. High voltage, low amperage a.c. is used, and the transformer employed is provided with tappings to adjust the current to suit the variations in resistance of different materials, and to allow for different rates of heating. In the case of rods, etc., the expansion of the material is taken as a direct indication of the temperature.

Snowflakes. The name used in reference to the fine transverse internal fissures in steel forgings, especially in large forgings in alloy steels. When these random direction cracks are seen in steel rails they are called "shatter cracks" (q.v.). They occur also in steel tyres and in gun-forgings. Also known as flakes, hairline cracks.

Soaking. The term used to describe the operation of holding material in a furnace until the required temperature has been attained throughout the mass, and if necessary, maintaining at this temperature. The time during which the material is maintained at the soaking temperature is referred to as the "holding time", and this may vary according to the purpose for which the soaking is being carried out.

Soaking pit. The name of a pit-type of furnace, fired by oil fuel or gas, and used for soaking (q.v.) metals. Usually such furnaces are employed for soaking newly-cast ingots before rolling or forging.

Soap-drawn wire. A term used in reference to wire of normal finish, drawn with a dry soap lubricant.

Soap-film method. An experimental method of observing and estimating the influence of sharp corners and of scratches, but particularly the former as stress-raisers. In the application of the method, a model simulating the actual component and containing the type of stress-raisers under investigation is coated with a suitable soap film which very closely follows the contour of the surface of the test piece. When the model is stressed the angle of inclination of the film to the normal is observed and measured at a number of positions in the suspected area. Since the contour of the film surface is related to the stresses set up, it is possible to detect the positions of abnormal stress distribution when the actual component is stressed in various ways.

Soap-rolled finish. The name used in reference to the semi-burnished finish that is obtained when certain non-ferrous metals and alloys are cold-rolled, using soap solution as a lubricant. Certain oils can also be used to obtain a similar finish.

Soapstone. A massive variety of talc $3MgO.4SiO_2.H_2O$. (*See also* **Steatite, Talc.**)

Soda ash. The trade name for the commercial quality sodium carbonate Na_2CO_3.

Soda dip. The term used in reference to the bath of sodium bichromate that is used as a means of removing stains and imparting to certain non-ferrous metals and alloys a bright surface finish. The term is also used in reference to the process itself, and when copper and copper-base alloys and some other metals are so treated, they are said to have a soda-dip finish.

Soda nitre. A commercial term for sodium nitrate $NaNO_3$, nitre being saltpetre or potassium nitrate.

Soderberg electrodes. Electrodes made in Sweden from a paste of finely ground anthracite and coke with a tar-pitch binder. The solid materials are heated to drive off moisture and unwanted volatile materials and then passed to a ball mill. The mixture is pressed and cast hot. The electrodes are used in the electrical smelting of pig iron and melting of steel.

Sodium. A soft silvery-white metal that rapidly tarnishes in air and reacts violently with water, forming sodium hydroxide and liberating hydrogen. The metal is produced commercially by the electrolysis of fused caustic soda in a special cell. It is also produced by the electrolysis of sodium chloride, usually mixed with sodium fluoride. The metal is used for the removal of antimony from lead and solders and in the production of alloys such as sodium-lead and sodium-zinc. It is also used in the modification of aluminium, and as a heat transfer medium.

Sodium salts find many varied uses. Sodium chloride, as brine, and sodium hydroxide are used for quenching steel; sodium cyanide is used in carburising and case-hardening; sodium nitrate in the fused state in the heat-treatment of aluminium alloys and steels; sodium hydroxide, fluoride, fluoborates, etc., in electrodeposition of several metals, and sodium stannate is used in the electrodeposition of tin and its alloys. Sodium salts are widely used in chemical industry.

Crystal structure—body-centred cubic; lattice constant $a = 4.2820 \times 10^{-10}$ m. (*See* Table 2.)

Sodium chloride structure. When the ionic sizes of two dissimilar ions differ by something of the order of 35-50%, the packing of such ions into a crystal lattice on formation of an AB compound cannot be closer than to fill around 60% of the space. To achieve even this packing, and electronic neutrality, the ions must alternate in any direction along three crystallographic axes, producing a lattice in which sodium and chlorine ions occupy the corners of a cube, but alternating as described. The lattice can also be looked at as two interpenetrating face-centred cubic lattices, one of sodium ions and the other of chlorine. Each ion has then six neighbours of opposite charge, a co-ordination number of 6. Sodium chloride, rock salt, is the archetype of such structures, hence the name.

Sodium-lead. A lead-base alloy containing 2% Na used as a deoxidiser.

Sodium-zinc. An alloy of sodium and zinc containing 98% of the latter. It is used as a deoxidiser.

Softening. A heat-treatment process that is most generally used for alloy steels. In the case of some of these, normalising or annealing is not feasible, since certain steels harden on cooling and the internal stresses produced by various work operations are not therefore removed. A tempering operation is then required consisting of heating the metal to a temperature below the lower critical point, followed by slow cooling. A tempering process that is not intended to produce a desired mechanical strength or toughness in a metal, but rather to complete the refinement of structure and remove internal stresses, is referred to as softening, for the steel does indeed become softer and more readily machinable.

Softening point. Certain materials do not have a definite melting point, but soften progressively over a range of temperature. In the case of refractory substances, the softening point is taken as an indication of its heat-resisting qualities. A sample of the refractory is moulded into the shape of a cone of the same size as the standard pyrometric cones. It is then placed in a furnace along with a series of the standard cones; and heated till its tip bends over to touch the base. The standard cone that behaves similarly gives its number to the cone under test. For this reason, the "pyrometric cone equivalent" is more or less synonymous with the less precise term of "softening point".

In the case of rubber and plastics, the softening point is the highest temperature at which it is practical to use them. A test piece 30 mm long, 5 mm wide and 1.5 mm thick is clamped as a cantilever, with a hole of 1.5 mm diameter drilled on the centre line of the wide face 1.5 mm from the end. Twenty-five millimetres of material extends from the clamp, and a 25 g weight is attached by thread through the hole. A beaker containing appropriate heating liquid is heated to 25 °C below the softening point, and the whole apparatus immersed in the liquid, the weight being separately supported. The weight is then pushed off the support and the temperature raised at 1 °C per minute stirring continuously. When the test piece has sagged to coincide with a 30° line on an attached quadrant, that temperature is regarded as the softening temperature.

Soft-facing. A process for depositing a soft metallic deposit on a harder and stronger base. This is usually done to produce soft-bearing surfaces on a steel or other backing. The alloys commonly used are bronze, silicon-copper and aluminium-copper alloys. The deposits are laid down using the metal arc process or employing an oxyacetylene (oxyethyne) torch.

Soft solders. A term used to designate a series of lead-tin alloys (usually also containing bismuth and small amounts of some other metals) used in soldering, to distinguish them from the copper-base alloys used in brazing. The upper limit of the melting range is about 300 °C. Table 8 gives the compositions of a number of soft solders in common or fairly common use.

Soft spots. A term used in reference to a type of defect that may occur in case-hardened steels. The soft spots represent areas where the carburising has not been properly accomplished, and their occurrence is associated with the use of aluminium as a deoxidiser and grain refiner. The defect can also arise through uneven heating or inefficient quenching, particularly if several components in contact are quenched together in a bath. It is advisable to agitate the quenching liquid to avoid the accumulation of steam bubbles on the surface of the hot metal.

Sol. A term invented by Graham to describe a colloidal solution. Colloidal particles (usually their diameters are measured in units of nm) vary in diameter from 1 nm to 100 nm. When less than 1 nm, it is a true solution; when between 1 nm and 100 nm, it is a colloidal solution and when larger than 100 nm, a colloidal suspension is formed.

Solder. An alloy used in the joining of metals, and which, having a relatively low melting temperature, surface tension and viscosity, can be made to run into the joints to be made. There are two broad types of solder—soft solders, which have low melting temperatures and which are usually tin-lead alloys, and hard solders, which are copper-nickel alloys and have higher melting temperatures, usually above 430 °C. These latter are sometimes termed brazing alloys (*see* **Soft solders**). The composition of white (hard) solder is approximately Cu 45%, Ni 10%, Zn 45%. A hard solder used for cast iron contains 42-60% Cu, 1-10% Ni, balance zinc. Usually also 1-3% Si. (*See also* **Silver brazing alloys**, Tables 5, 8, 9.)

Soldering. The term used in reference to the operation of joining together metals by the fusion of relatively low melting temperature alloys (*see* **Solder**), which can run into the cavities between the joint members (capillary joining). It is applied most commonly to soft soldering in which tin-lead alloys are used. The copper-nickel and the silver solders which have much higher melting temperatures are known as hard solders and the operation of joining metals, using these latter, is commonly called brazing.

The bond created in soldering relies on the partial diffusion of one or more of the constituents into the parent metal. Tin is the common diffusing metal in soft solder, but the layer of solder between the joint is relatively weak, especially if too thick, and hence soft solder does not produce a high-strength joint.

Soldering fluxes. The success of the joining of metals by soldering is dependent on the surfaces of the parts to be joined being clean and remaining unoxidised during the operation. This latter is secured by using a flux which readily melts at the soldering temperature, forming a protective coating, so excluding air, and having a solvent action on oxide that may be formed. The types of flux used vary with the types of solder, i.e. hard or soft, and with the composition and application of the solder. Table 55 gives a list of common fluxes and their usual applications.

TABLE 55. *Common Soldering Fluxes*

Flux	Type of solder used	Metals to be joined
Zinc chloride	Soft solder (tin-lead; tin-lead-bismuth)	Tin, brass, gunmetal, gold, silver, bismuth, tinned steel
Ammonium chloride	Soft solder (tin-lead)	Brass, gunmetal, copper, iron and steel
Rosin	Soft solder (tin-lead)	Copper-zinc alloys, copper-tin alloys, lead, tinned steel
Tallow	Soft solder	Lead
Stearin plus fluoride or fluoborate	Soft solder (tin-zinc-aluminium)	Aluminium
Hydrochloric acid	Soft solder (tin-lead)	Zinc, galvanised steel
Borax	Hard solder (copper-zinc; copper-zinc-silver)	Brass, copper, silver, gold, iron and steel
Borax	Hard solder (copper-zinc-nickel)	Stainless steels
Cuprous (copper II) oxide	Hard solder	Cast iron
Gallipoli oil	Soft solder	Pewter

Solenoid. A coil of wire wound on a cylinder, the turns being all of the same radius (i.e. not overlapping). Usually the length of the coil is large compared with the radius of the coil. If such a coil of insulated wire carries a current a uniform magnetic field is produced within the coil. The field is parallel to the axis of the solenoid and when a current I passes in the wire the strength of the field is nI kA m^{-1} where n = number of turns of wire per metre.

Solid carburising. A term sometimes used in reference to the process of case-hardening of steel, using a solid medium in contact with the steel. The mixture generally used consists of powdered carbon and an alkali carbonate such as barium or sodium carbonate, in the proportion of 3 parts of carbon to 2 of carbonate. The latter substance plays an important part in the production of carbon monoxide, which is the gas responsible for carburising.

Solid contraction. A term used in reference to the shrinkage that occurs when a metal cools from its solidifying temperature to ordinary atmospheric temperature. It is often used to distinguish this type of shrinkage from the change in volume that molten metal undergoes in cooling to the freezing temperature and during the change of state from liquid to solid.

Solidification contraction. Since solid metal is denser than molten metal, a contraction in volume occurs when the liquid solidifies. In an ingot, the outer skin in contact with the mould, where solid metal first forms, does not shrink at the same rate as the still-liquid metal in the interior of the mould. Thus solidification shrinkage is the primary cause of "pipe" in an ingot.

Solidification range. Most alloys do not have a definite melting (or freezing) point. At some temperature $t°$, the alloy may be just completely solid and at some higher temperature $(t + θ)°$, the alloy is just completely molten. The range of temperature $θ$ is called the solidification range. At any temperature within this range the alloy will be in a more or less pasty condition (*see* Fig. 155).

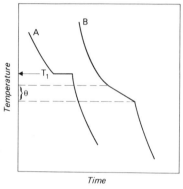

Fig. 155 *Solidification range.* T_1 *is the melting point of pure metal A. θ is the solidification range of binary alloy B.*

Solid solution. A solid phase in which two or more alloying elements exist in a state of mutual solution. In the simplest case of two metals that are melted together, if on cooling the solid, although a mixture, is perfectly homogeneous, a solid solution has been formed. On polishing, etching and examining such a solid solution, the metal will exhibit only one kind of grain. A few examples of complete solid solubility are known. In such cases the metal in excess is regarded as the solvent, and the other as the solute. Examples are gold-silver, copper-nickel. The conditions that favour solid solubility are: (a) size factor, i.e. the atoms of the elements must have the same or nearly the same atomic diameters; (b) the crystal structures of the elements must be the same; (c) the valencies of the elements should be the same, otherwise the solubility will be restricted by the formation of intermediate constituents. Solid solutions always exhibit a melting or solidifying range.

When the essential atomic pattern of the solvent metal is retained, such solid solutions are termed "primary solid solutions", or sometimes "terminal solid solutions". This can occur in one of two ways: (a) in substitutional solid solutions the atoms of the alloying element occupy sites previously occupied by the solvent metal; (b) interstitial solid solutions are those in which the atoms of the alloying element can occupy spaces between the atoms of the solvent metal without much distortion of the original lattice.

In the equilibrium diagram (q.v.), intermediate solid solutions are separated from the component pure metals or terminal solid solutions by heterogeneous phase fields. The terminal solid solutions are those of which one boundary of the phase is a pure metal.

There are several types of freezing-point curve of solid solutions, that is where solidification occurs with complete solid miscibility. These are shown in Fig. 156, where (a) represents ideal solution of which there are few examples, but the alloys, gold-palladium and gold-silver, and copper-nickel may be quoted. (b) Represents a slight modification of (a), and examples of this may be found in the copper-palladium and copper-platinum alloys. (c) Represents the commonest type in which the freezing-point curve passes through a minimum. The composition represented by M crystallises at constant temperature. (d) Represents the case where the freezing-point curve passes through a maximum. Only a few instances of this are known in chemistry, and these do not include metallic substances (alloys). The dotted lines represent melting-point curves.

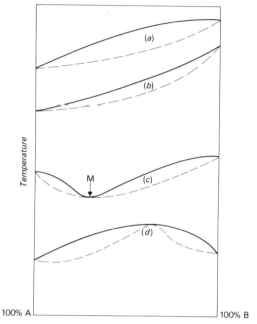

Fig. 156 *Curves showing solid solutions.*

Solidus. The line on an equilibrium diagram indicating the temperature at which a metal or alloy becomes completely solid on cooling, or at which melting begins on heating under equilibrium conditions. This line indicates the compositions of the solids that can co-exist in equilibrium.

Solid wire. A term used in reference to a single wire conductor in contradistinction to a conductor made up of a number of fine wires more or less loosely twisted together.

Solubility. The weight of a substance A which can be added to a given weight or volume of a substance B to form a saturated solution of A in B, at any given temperature is the solubility of A in B at the stated temperature. The term "solution" (q.v.) must be given a wide interpretation. The substance (A or B) which is present in excess is termed the solvent and the other is the solute, and A and B may be solid, liquid or gas. The solubility of A and B depends on both substances and in particular upon their electrical properties. The electrical properties of the solvent must be such as to weaken the electrical forces which ordinarily preserve the shape of the crystal lattice of the solute.

The solubility of solids in liquids is usually determined by taking a known weight of a saturated solution, analysing it to determine the weight of solute and hence finding the weight of solute dissolved in 100 g of solvent. For sparingly soluble substances, the solubility may be determined: (a) from the e.m.f. of a concentration cell; (b) from the conductivity of a solution of the sparingly soluble substance; (c) by colorimetric methods; (d) by using a radioactive indicator. The solubility curves obtained are of two types: (1) continuous curves (Fig. 157), and (2) discontinuous curves (Fig. 158). If the heat of solution in a liquid is positive, the solubility of the solid will increase with temperature; if the heat of solution is negative, the solid will be less soluble as the temperature rises. In the case of discontinuous curves, it may be assumed that the solid phase in contact with the liquid has altered in character and the discontinuity really represents the meeting of two different and distinct solubility curves. The new substances may occur as a result of polymorphic changes or by the changes in the state of hydration, etc.

Generally, the particle size of the solute affects the solubility. This is due to the increase in surface forces when the particles are finer, leading to greater solubility. Hence, with very fine precipitates there is a greater loss on washing than with coarser precipitates of the same substance. This is of importance in analysis and manufacture.

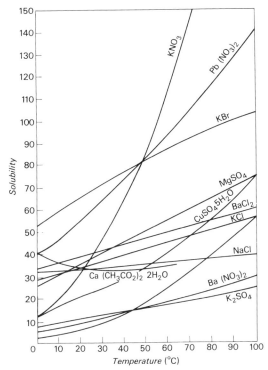

Fig. 157 *Continuous solubility curve.*

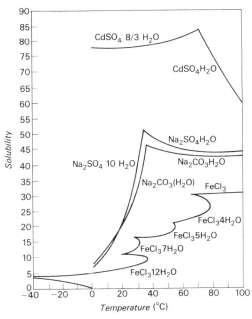

Fig. 158 *Discontinuous solubility curves.*

Solubility limit. As explained under **Solid solution** and **Size factor,** the alloying of metals involves many factors which control the proportion of second metal which will dissolve in the first. For any given structure, there is a limit of solubility, beyond which a structural change must occur. At this limit, there cannot be an immediate change (*see* **Phase rule**), but a region exists in which

two solid solutions may exist side by side, after which a homogeneous solid solution of the new type becomes stable. It is normal to speak of the solubility limit with respect to each phase in a multiphase system, since the properties of each phase will vary, and the practical application of such materials will depend on composition.

Solubility product. The term used in reference to the product of the concentrations of the ions of a dissolved electrolyte when the latter is in equilibrium with undissolved salt. When a solution of an electrolyte is in contact with the solid, the equilibria may be expressed in the usual form

$$AB \quad\rightleftharpoons\quad AB \quad\rightleftharpoons\quad A^+ + B^-$$
$$\text{(undissolved)} \qquad \text{(dissolved)} \qquad \text{(ionised salt)}.$$

If $[A^+]$, $[B^-]$, $[AB]$ represent the concentration of the ions and of the dissolved electrolyte (undissociated), then the law of mass action is expressed

$$\frac{[A^+][B^-]}{[AB]} = K.$$

But in the presence of undissolved solid $[AB]$ is constant, $[A^+][B^-] = S$ is constant and referred to as the solubility product. If for any substance AB the product S is exceeded (e.g. by the addition of a substance that increases the concentration of one of the ions), then the solution becomes supersaturated and salt AB is precipitated. Again if by the addition of another substance one of the ions is reduced in concentration, then more salt AB will dissolve.

For sparingly soluble electrolytes, the solubility product is a constant for a given substance at a given temperature.

The theory of solubility products appears to be inadequate to account for the behaviour of moderately or very soluble electrolytes. In these cases the solubility product is not a constant. If, however, the activities of the ions are substituted for the actual concentrations, the product is a constant.

The solubility products may be calculated from the solubility if complete dissociation can be assumed. If "s" is the solubility, then $10\,s\ \text{mol}^{-1} =$ number of mol^{-1}. When the substance AB dissociates into two ions the solubility product

$$= (10\,s\ \text{mol}^{-1})^2.$$

Soluble anode process. A term sometimes used in reference to the electrochemical techniques used in electrolytic refining in aqueous electrolytes or fused salt baths, using soluble anodes. In this process the main metal of which the anode consists goes into solution together with the baser impurities. The more noble metals are usually insoluble and form the anode mud or slime which is periodically removed from the tanks and treated for the recovery of the noble metals present. Of the impurities that go into solution with the main metal, only those having similar potentials under the conditions of working will usually be co-deposited. The build-up of the impurities, however, necessitates the regular purification of the electrolyte, otherwise high concentration would cause co-deposition of main metal and "impurity metal".

The conductivity of fused salt baths is usually higher than that for aqueous solutions of the same salts. With high current densities a gas envelope (particularly with halide baths) surrounds the anode causing an increase in voltage and a decrease in current. This reduction in efficiency in the process is known as "anode effect". It can be cured by vigorous stirring or by reversing the current for a short time.

Solubriol. A soluble oil containing calcium soap, used as a lubricant.

Solute. The term used to describe the constituent of a solution that is present in minor proportions. In the case of an alloy the solute is the metal or other element, in small amounts usually, that is dissolved in the metal present in the larger proportion and referred to as the "solvent".

Solute hardening. The addition of a second metal to a basic one produces immediate changes in properties, such as electrical conductivity, melting point and thermal expansion for example. Mechanically, the introduction of a second atom must affect the strain in the lattice, whether the second atom be smaller, larger, or even the same size, because there are electrical effects in addition. Naturally, atomic sizes which are relatively equal will be likely to produce less strain than a considerable size difference, but in general all solute atoms produce some mechanical hardening, known as solute hardening. The main mechanism is the interference with dislocation movement, due to the lattice strain involved, and to some extent, the varying mechanism of dislocation sources and production of similar defect structures.

If a metal is too soft for a particular application, the addition of a solute in small concentration will produce hardening, a perfect example being the hardening of gold and silver to give better wear resistance to jewellery. Of course, solute hardening is accompanied by a corresponding loss of ductility, so the alloy becomes more difficult to work. Solute hardening constitutes the simplest form of improving the strength of a metal, although other properties may be deleteriously affected, e.g. copper is improved by hardening with metals such as arsenic, for boiler tubes, but the electrical conductivity is lowered so drastically that such material is useless for any normal electrical application for which copper would normally be considered.

Solution. A term used to describe a perfectly homogeneous mixture of two or more substances. We can therefore distinguish the following types of solution.

(a) Solution of gases in gases. All gases that do not chemically combine are miscible in all proportions. The gas present in the largest proportion is the solvent, and the partial pressure of each gas is the pressure that each gas separately would exert if present alone in the space occupied by the mixed gases (Dalton's law). In the case of two gases in a mixture (two components and one phase) there will be three degrees of freedom so that pressure, temperature and concentration can all be varied within a certain range, this range being defined by the conditions of condensation of one of the gases, when a new phase would appear.

(b) Solutions of gases in liquids. All gases are soluble to some extent in water. Gases which react with water, and those that are more readily liquefiable are the more soluble. The volume in cm^3 of a gas, measured at S.T.P., which dissolves in $1 cm^3$ of water, is called the "adsorption coefficient". If the volume of the gas is measured at the temperature and pressure of the experiment it is called the "solubility". Gases are less soluble in liquids at high temperature than at lower temperatures. Theoretically all dissolved gas should be expelled on boiling the solution. Those gases which form constant boiling-point solutions cannot be expelled by boiling. Gases dissolved in molten metals are usually derived from the air or from the furnace atmosphere. Compound gases, e.g. CO, CO_2, are insoluble in molten metals, but products of their decomposition may be soluble; nitrogen is soluble in iron, chromium, manganese and molybdenum but practically insoluble in all the other common metals; oxygen is insoluble in aluminium, magnesium, zinc, tin and lead, but dissolves in copper, nickel and iron and in the molten precious metals; hydrogen, particularly when derived from the decomposition of water, appears to dissolve readily in many common metals, though not in brass and some other zinc-containing alloys. Much of the dissolved gas may escape during the cooling of the molten metal, but some part escapes during solidification and this gives rise to gas porosity. Bubbling nitrogen through a molten mass is sometimes used to remove dissolved hydrogen—a scavenging process. Other methods include remelting in a vacuum or in an inert atmosphere; oxidation of dissolved hydrogen by oxide of the metal (e.g. copper oxide, Cu_2O, is used in degassing copper). The excess copper oxide is removed by additions of phosphorus.

(c) Solutions of gases in solids. There appear to be four ways in which a gas can be taken up by a solid. (1) In true solution the gas dissolves without any change in its composition to give a homogeneous mixture. This occurs with some of the common metals and hydrogen. (2) A solid solution may be formed at a certain concentration of the gas phase. Beyond this a second phase may appear, and at still higher concentrations the first solid solution may be completely transformed into the second. (3) Chemical compounds may be formed by union of the gas with the metal. The resulting compounds may be stable or may be capable of forming solid solutions with further additions of the gas. (4) Adsorption is a surface phenomenon and hence equilibrium is established rapidly, when finely divided solids are exposed to certain gases. A gas is more readily adsorbed at temperatures near to its liquefaction than otherwise. The occlusion of hydrogen is often regarded as an outstanding example of adsorption, but recent work indicates that two immiscible solid solutions may in fact be formed—one on the surface and one below the surface—and that no compound is formed.

(d) Solution of liquids in liquids. Most liquids appear to fall into two classes: (1) the abnormal liquids which have relatively high dielectric constants and are hence ionising solvents, and (2) the normal solvents. The first is typified by water and the lower alcohols, the second by the hydrocarbons. These latter are non-associated liquids (do not tend to polymerise in solutions) and they are all completely miscible with one another. If two associated (abnormal) liquids *A* and *B* are mixed there will in general be formed a solution of *A* in *B* and of *B* in *A* and the two solutions will separate. As the temperature rises the solubilities will increase until the composition of the two layers becomes the same. This is the temperature at which there will be complete miscibility. It is called the critical solution temperature.

(e) Solutions of solids in liquids. (*See* **Solubility.**)

(f) Solutions of solids in solids. (*See* **Solid solution.**)

Solution heat-treatment. A term used in reference to the heat-treatment of certain alloys. Many aluminium alloys are given some form of treatment. The first step in such treatment is known as solution heat-treatment, which is followed by ageing (either natural or artificial). The alloy is subjected to uniform heating for a given period at a closely controlled temperature. This may be carried out in a forced air circulation furnace or a conventional nitrate bath. The temperature of heating should lie between the solidus and the limit of solid solubility (usually between 490 °C and 535 °C), the holding time varies according to the size of the material being treated (i.e. from about 8 hours for sheet and strip up to 0.5 mm to 45 hours for bars up to 2.5 cm diameter). The quenching medium is usually water or oil.

The most important alloying elements used with aluminium include zinc, silicon and magnesium, and these may form hard intermetallic compounds which dissolve in the basis metal. The object of solution treatment is to obtain an artificially rich solid solution. This latter is not stable and the excess of the alloying element or compound slowly precipitates on standing (natural ageing). The precipitation occurs slowly so that it is possible to perform a certain amount of forming before the hardening becomes effective.

Solution potential. *See* **Electrode potential.**

Solution pressure. A term introduced by Nernst in reference to the tendency for metals, when placed in solutions of their salts, to pass into solution as positive ions. Equilibrium is attained when the tendency for ions to pass from the electrode to the solution is just balanced by an opposing tendency for ions in the solution to be deposited on the electrode. This can happen since the ions are positive, while the electrode has a negative charge since it loses positive ions. Hence the solution pressure at equilibrium will be counter-balanced by the osmotic pressure.

Solvation. A term used in reference to the combination of a molecule or an ion derived from the solute, with one or more molecules of the solvent. The chemical properties of covalent compounds are frequently quite different from the chemical properties of the same substance in solution, and this is attributed to the solvation or hydration of the compound. In this change the hydrated hydrogen ion (in aqueous solutions) hydroxonium H_3O^+ is involved, and it is thought that the change from covalent to electrovalent characteristics may be due to the increase in size of the cation.

Solvent. The term used in reference to a solution to indicate the constituent present in the larger proportion. In the case of an alloy it refers to the base metal or major constituent.

Solvent cleaning. The term used to describe the process of cleaning metal components of grease and oil by treatment with organic liquids capable of dissolving the grease, etc. The solvents available are benzene, naphtha, carbon tetrachloride (tetrachloromethane), paraffin, petrol and trichloroethylene (trichloroethene). The last named is very generally used and it has the advantage of being non-flammable.

Solvent degreasers. These are usually chlorinated hydrocarbons such as di-, tetra- or penta-chloroethane, or tri- or tetra-chloroethene, all of which are powerful solvents of grease and oils. They are non-flammable. Trichloroethene should not be used with aluminium, but is suitable for most other metals. Since it is liable to slow decomposition by bright light it is usual to add a stabiliser such as dibutylamine in the amount of $1 g l^{-1}$ of trichloroethene.

Solvus. A line on an equilibrium diagram defining the limit of solid solubility.

Sombrerite. The name of one of the apatite minerals, consisting of calcium phosphate $Ca_3(PO_4)_2$, which is an important source of phosphorus. It is found in the island of Sombrero (West Indies).

Sonic testing. A non-destructive method of testing based on the influence exerted by defects on the vibrating characteristics of an article. It is well illustrated by ringing (e.g. bell-ring) or wheel-tapping tests in which defect free articles emit a clear note when struck a sharp blow, whilst defective articles give a dull note. The tests must be carried out in conditions which favour free audible vibration and are restricted to articles of comparatively simple design.

Sorbite. A decomposition product of martensite found in hardened and tempered steels. Its structure, when resolvable, is minutely granular and not lamellar as in pearlite. It consists of minute particles of carbide in a ground mass (or matrix) of ferrite and on further tempering the carbide particles coalesce.

Sorbitic pearlite. A minutely lamellar microstructure of some quenched but untempered steels, which are not completely martensitic. It is often irresolvable under the microscope, and frequently etches very rapidly, in which case it is known as troosto-sorbite.

Sorption. The term used to describe the whole phenomenon of the taking up of a gas by a solid. One aspect of this is the surface effect and is called adsorption (see **Solution**), and another aspect is known as absorption and refers to the taking up of gas in the interior of the substance.

Sorting. A term used in reference to the process of grading as-mined or comminuted ore. Sorting is carried out to divide the ore into groups for cleaning. The process involves the use of ball or rod mills for grinding and the use of screens, and grizzlies and trommels.

Sow-block. A block of hardened steel located between the anvil and the hammer in a forging hammer. It is in fact a die bed whose main function is to reduce anvil wear.

Space-filling factor. Consideration of the structure of compounds involves the relative sizes of atoms or ions of the constituents. Since only rarely are the sizes equal, it is not possible to conceive a general packing system. Furthermore, when ions are involved there must be electrical neutrality in an ionic compound, and the formation of covalent bonds requires acceptance of bond lengths and bond angles. If atoms or ions are regarded as spheres (and this is not unreasonable for the electron clouds of many), it is obviously not possible to fill 100% of the space in any case, as typified by the closest possible packing of like atoms — 74%. A very small atom or ion, however, could obviously fit into the spaces between larger ones to better advantage, but the essential factor would then be the closest distance of approach of the larger units, which would then determine the limit to space filling.

In practice, AB type compounds fall into roughly three possible groups, the caesium chloride structure filling about 75% of the space when ions are almost identical; the sodium chloride structure filling 60% of the space with ions differing by about 50% in size; the zinc blende structure with 45% space filling at a size difference of about 60%.

With AB_2 compounds there are similar groups, the fluorite lattice about 55% space filled; the rutile lattice similar, but of different lattice arrangement, and the cristobalite lattice only 25% filled in space.

The significance of space filling factor is in the effect on properties, and in particular, with low filling factors, the tendency for other ions to enter the spaces to form more complex compounds, as in the silicates.

Space lattice. All metals crystallise according to a definite pattern or symmetrical arrangement of atoms. The crystal is made up of an ordered assemblage of units. If corresponding points in each unit be selected, they will be found to lie on lines and planes which divide the crystal space into a set of parallel-sided cells. The cell in this case is the unit cell. When the point chosen in each unit is the centre of gravity of the atoms, the assemblage of points (and the imaginary lines joining them) is known as the space lattice. They form seven crystal systems (cubic, tetragonal, hexagonal, rhombohedral, orthorhombic, monoclinic and triclinic). The large majority of metals, however, crystallise according to three of the possible space lattices—face-centred cubic, body-centred cubic and close-packed hexagonal (q.v.).

Spalling. A term used in reference to the cracking and flaking off of metal particles from a surface. This may occur on the surface of case-hardened or other steels that are in a very hard condition; it may happen to a surface coating. Differential thermal expansion and contraction may be the cause. The term is also applied to the breaking of the surface of refractory material due to thermal or mechanical shock and to polymeric material through ageing. (See also **Exfoliation**.)

Spalling test. A test applied to refractory materials to estimate their resistance to thermal shock. A standard size piece 7.5 cm × 5 cm is placed in a furnace at 900 °C and remains for a stated time (usually 10 minutes). It is removed from the furnace and allowed to stand in air for a similar period. The cycle is repeated until the piece breaks on application of pressure by one hand. The result is quoted as the number of cycles before fracture.

Spangle. The term used to describe the large leaf-like zinc crystals that form on galvanised sheet. Their formation is influenced by the rate of cooling of the zinc coating and on the presence of small amounts of an alloying element such as aluminium.

Spark test. A crude test devised for rapidly determining the approximate composition of steels and ferrous alloys, particularly the carbon content. Steels of different types are often identified by observing the shape, volume, colour and intensity of the spark pattern produced when the steel is held against an ordinary grindstone wheel. The test is often used in sorting mixed stock and for this purpose a series of standard samples is kept for comparison.

Spathic iron ore. Iron ore consisting of ferrous carbonate $FeCO_3$ and other isomorphous carbonates (e.g. manganese, magnesium and calcium). It is yellowish-brown in colour and occurs in the UK in Somerset, Exmoor and Yorkshire. It contains about 55% Fe, calculated as iron II oxide. (See also **Siderite**.)

Spatter. In arc-welding the spluttering of the arc causes globules of molten metal to be ejected from the pool of molten metal. The globules, which originate from the electrode, solidify on the basis metal around the weld and are referred to as spatter.

Spavin. A term sometimes used to describe the underclay below the coal measures, used for making firebrick. Also called fireclay and warrant.

Specific gravity. A characteristic quantity for different materials, and measured as the ratio of the weight of any volume of a substance to the weight of an equal volume of water at the same temperature. The specific gravity of the metals is given in Tables 2 and 3 as density in 10^3 kg m^{-3}.

Specific heat. The amount of heat required to raise the temperature of 1 g of a substance through 1 °C. Since the value for any substance varies slightly with the temperature, it is usual to define the temperature rise either from 0 °C to 1 °C or from 15 °C to 16 °C. The values of the specific heats of metals are given in Table 2.

Specific inductive capacity. This is defined for any material as the ratio of the capacity of a condenser with the material as dielectric to its capacity when the dielectric is a vacuum. Also called inductivity, dielectric constant. It is usually denoted by "k", $k = v_0^2/v^2$, where v_0 = the velocity of electromagnetic waves in vacuum and v = velocity of waves of same wavelength in the material. For very long waves $k = \mu^2$, where μ is the refractive index. The values of k for some polymeric materials are given in Table 3.

Specific resistance. The electrical resistance of unit length of a material per unit cross-sectional area is the specific resistance or resistivity. (See **Resistivity**.)

The resistivity of a metal depends on its physical state, its density, purity and hardness. Mechanical working (e.g. straining of wires) increases the resistance, while annealing decreases it. The resistance of an alloy is always greater than that of the elementary metals or elements composing it. The specific resistances of the metals are given in Table 2. Table 3 gives the resistivities of a number of common alloys.

Specific surface. A term used in reference to fine or very fine powders used in powder metallurgy. It expresses the total surface area of 1 g of the powder, in cm^2.

Specific volume. The volume of 1 g of a substance measured at a specified temperature and pressure. It is the reciprocal of the density of a substance.

Speckled metal. A term used in foundry-work to indicate pin-hole porosity in a casting.

Spectrographic analysis. A method of analysis, both qualitative and quantitative, that is particularly applicable where the elements to be identified are present in small quantities or where the total sample available is very small. The method of observation may be either visual or photographic. The former may sometimes be used by an experienced observer familiar with the spectra of various elements. Otherwise it is convenient to replace the spectroscope by a wavelength spectrometer. The spectrum of the specimen under test is photographed beside a standard spectrum which permits sufficiently accurate estimations of wavelengths to be made. Where the spectrum is complicated, some of the elements can be identified rapidly and then additional photographs are taken with these elements as comparisons. It thus remains to calculate the wavelengths for the remaining unidentified lines. And since many of the sensitive lines of the majority of elements lie in the ultra-violet region, a quartz spectrograph is desirable.

In quantitative work the elements are first identified, and then a standard series of alloys is prepared containing varying proportions of the particular element. These are then compared with the spectrograph of the specimen under identical conditions by photography and correlating the intensity of the lines with the percentage composition. This method would, however, require that a large number of standards should be available for comparison. To avoid this another method has been devised. A weighed sample of the alloy is dissolved in nitric acid and made up to a known volume. A convenient quantity of this solution is then evaporated nearly to dryness and ignited to the oxide. The product is then ground and put into the cavity of a pure carbon electrode specially prepared to receive it. A pointed graphite rod vertically above the sample forms the counter electrode. The excitation is produced by a condensed-spark circuit. Standards are prepared by converting chemically analysed samples to oxide or by using pure materials in the requisite amounts.

Spectrometer. An instrument designed for the examination and measurement of spectra. A collimator having the source of radiation at one end directs a parallel beam of light on to a refracting prism. The pencils of light are refracted into a telescope where a real image of the spectrum is formed (Fig. 159). This can be observed optically or photographically. An image of a scale can also be arranged to occur beside that of the spectrum to aid in identifying any spectral line. In a simple model, the prism is supported on a circular table graduated at its edge. Other types of spectometer use a diffraction grating in place of a prism. The manner in which the light-rays are introduced into the slit of the collimator has an appreciable effect on the quality of the lines in the spectrum.

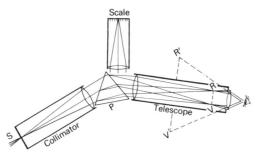

Fig. 159 *Simplified schematic diagram of a spectrometer.*

Spectrum. The analysis of the radiation from any source into wavelengths produces what is called the spectrum of the radiation. All metals at ordinary temperatures emit radiation which appears in the infra-red part of the spectrum. When the metal is heated to incandescence it emits, among other types, visible radiation, and when this is analysed by a spectrometer it is found to be continuous—there being a continuous gradation of colour throughout the spectrum. All incandescent solids show this kind of spectrum. On the other hand, the spectrum obtained from incandescent gases and vapours is discontinuous and consists of lines, quite sharply defined in different regions of the visible spectrum. The line spectra are characteristic of atoms. Band spectra can be produced in which the coloured bands extend over a certain range of wavelengths. These are characteristic of molecules.

When sodium chloride is strongly heated it is broken down into its constituent elements, both of which emit radiation. That from the sodium atom gives a line spectrum (mainly in the yellow region), but the radiation from the chlorine falls outside the visible spectrum. When electrical means are used to raise the substance to incandescence, e.g. by sparking a metal between electrodes made from the metal or by passing a discharge through a gas at low pressure in a sealed tube, the line spectra occur, but are complicated by the appearance of spectra due to the presence of ions. These latter are produced from the normal atoms by the loss of electrons, and they show their own characteristic spectra.

When white light—which produces a continuous spectrum—is passed through a vapour (suitably contained) or through a coloured solution and is then analysed, there will be found dark bands in the spectrum, corresponding to wavelengths of light which have been absorbed by the substance through which the white light passed. This is called an absorbtion spectrum, to distinguish it from the emission spectrum previously described. The absorption spectrum is connected with the chemical composition of the absorbing substance.

On carefully examining the spectral lines for hydrogen it was found that the wave-number of some of the lines could be expressed in the form

$$\nu = \frac{R_\infty}{x^2} - \frac{R_\infty}{m^2}$$

and when $x = 2$ and $m =$ any integral value greater than 2, a series is defined that lies within the visible spectrum. When $x = 1$, and m has values greater than 1, the series in the ultra-violet region is defined, and when $x = 3$, a series in the infra-red region is defined. R_∞ is the Rydberg constant ($= 1.097\ 373 \times 10^7$ m^{-1}).

When energy is given to an atom so that an electron is moved from an inner to a higher energy level, that energy will be given out as monochromatic radiation when the electron jumps back to the level from which it was raised. The frequency of the radiation is given by

$$c^2 \left[\frac{R_\infty}{x^2} - \frac{R_\infty}{m^2} \right]$$

where c = velocity of light.

Spectra may be produced in one or more of the following ways: *(a)* by heating the solid, moistened with hydrochloric acid on a platinum wire, in a bunsen flame; *(b)* by passing an electric discharge through a gas at low pressure in a sealed tube; *(c)* by producing electric sparks near the surface of a liquid or solution; *(d)* by striking an arc between rods of the substance under test; *(e)* by passing a spark between a pointed electrode and a hollowed-out electrode, a sample of the material under test being pressed into the end of the hollow electrode. (*See also* **Spectrometer** and **Spectrographic analysis.**)

Specular iron. A variety of hematite Fe_2O_3 which occurs as rhombohedral crystals. It is black in colour and has an attractive metallic lustre.

Speculum. A copper-tin alloy having a silver-white colour and capable of taking a high polish. It is used in mirrors and reflectors generally for indoor use. It resists tarnishing and is non-toxic to food. The oldest samples appear to have the composition Cu 67%, Sn 33%, but many varieties contain zinc and arsenic and some contain platinum to increase the corrosion resistance. Table 5 gives the compositions of some common varieties under Mirror alloys.

Speculum plating. An attractive co-deposition of tin and copper from an alkaline stannate bath. The preferred composition is Cu 55% and Sn 45%. It has good wearing properties, but is not suitable for use at temperatures in excess of 100 °C. The bath should contain per litre, 90 g of sodium stannate, 16 g of sodium cyanide, 11 g of copper cyanide and 15 g of caustic soda. An iron tank is a suitable container. The anodes are of pure tin and copper and the working temperature is 60-65 °C. The current density at the anodes should be not less than 1.10 and 0.93 A m^{-2} respectively.

Speiss. The term used to designate the metallic arsenides and antimonides produced by smelting ores of cobalt (and lead).

Spelter. The name by which zinc was originally called, but as zinc and bismuth were at one time confused, the name was sometimes used in reference to this latter metal. It is now applied to the crude zinc from the smelter and is used in galvanising. It contains 1-2% Pb.

Sperrylite. One of the few naturally occurring compounds of platinum. It is an arsenide of platinum containing small amounts of rhodium and palladium. It is found in Sudbury (Ontario), Siberia and South Africa and in some parts of the USA.

Sphalerite. The most important sulphide ore of zinc (ZnS). It also contains some indium which is extracted, particularly from the ores mined in Arizona. The ore is usually referred to as zinc blende, or as rosin jack or black jack, according to the colour. It is found in England and various parts of Europe, America and New South Wales

Spheroidal cementite. The name used in reference to the aggregate of globular cementite in pearlitic areas forming in a ferritic matrix. It is produced by long annealing of unhardened steel. It is the same structure produced in hardened steels by prolonged heating at somewhat higher temperatures. (*See also* **Spheroidite**.)

Spheroidal graphite cast iron. The name used in reference to cast irons in which the graphite has been transformed from the long, thin flakes typical of grey iron into discrete spheroids. The removal of the thin flakes improves the mechanical properties of the iron, since the pointed ends of the flakes are serious stress raisers. Graphite has low tensile strength, so that the long flakes of it, by producing long discontinuities, tend to reduce the strength of cast iron. When the graphite is in the form of spheroids, the ground mass of metal is much less broken up. The spheroidal structure can be formed with any type of matrix structure, e.g. pearlite, ferritic or a mixture of these. And by appropriate treatment martensitic or austenitic structures may be obtained.

There are two main processes by which the S.G. structure can be produced. The one involves the use of metallic cerium and the other metallic magnesium. These two elements are highly reactive towards oxygen and sulphur. They both form stable carbides and promote the formation of stable carbides of other elements whilst tending to suppress the formation of graphite. In use it is necessary to ensure that there shall be an excess of the nodularising agent after some of it has combined with any sulphur that may be present. When cerium is used the composition of the iron must be strictly controlled. The iron must be hypereutectic (3.8% C and 2.5% Si) with low sulphur and phosphorus content. Cupola working is thus unsatisfactory, and unless specially pure hematite pig iron is used, the process must be conducted in crucible or H.F. induction furnaces to obviate the pick-up of sulphur. The cerium is added as Mischmetall.

When magnesium is used the limitations on the type of cast iron are not so strict and cupola working is possible. The magnesium is added either as metallic magnesium using special methods or as a magnesium-nickel alloy containing 10-20% Mg. In both cases the addition of the nodularising element may be followed by the addition of an inoculant in the form of ferro-silicon or calcium silicide.

Spheroidisation. The mechanism whereby rod or plate shaped particles contract into spherical shape to lower surface energy. A treatment which consists in heating steel to a suitable temperature and for a sufficiently long period to cause the cementite, whether free or as a constituent of pearlite, to form microscopic spheroids. When the process is complete the effect is to lower the hardness of the steel and thus facilitate machining operations, so that in effect spheroidising is a form of annealing.

Spheroidite. The name used in reference to the aggregate of globular cementite and ferrite found in certain steels after prolonged annealing at temperatures in the region of the lower critical limit. Also called divorced pearlite.

Spherometer. An instrument used for measuring the radius of curvature of spherical surfaces.

Spider. A form of pulley arranged on the shaft of a dynamo or motor, and on which is assembled the armature.

Spigot. A projecting portion—generally circular in shape—on one part, which fits into a corresponding recess formed in another part of an assembly.

Spills. As a result of the escape of dissolved gases in the ingot and of the evolution of entrapped mould gases, cavities may appear near the surface. These may become opened up during working, and if they become oxidised will fail to weld up. This gives rise to surface oxide laminations known as spills. These sub-surface defects are often elongated by rolling and give rise to undesirable surface markings which cause difficulties in such subsequent operations as electroplating.

Spin (electron). Refers to the motion of an electron on its own axis, as a means of satisfying quantum mechanical conditions in electron energy level. Of great importance in ferromagnetism and related magnetic properties in ferrimagnetics and anti-ferromagnetics.

Spinel structure. General formula AB_2O_4, consisting of 32 oxygen ions arranged in cubic close-packing, with metal A ions in tetrahedral voids and metal B ions in octahedral voids within the cubic structure. When the metal A ions are divalent metals (Mg, Fe^{2+}, Mn, Zn) and the B ions trivalent metals (Al, Fe^{3+}, Cr^{3+}, Mn^{3+}) the structure is known as normal spinel, after the first structure determined, $MgAl_2O_4$.

Combinations of metal ions with valencies 2 and 4 are feasible, provided that the total metal ion charge equals 8 for every 4 oxygen ions. The combination of di- and trivalent ions leads to a similar structure known as inverse spinel. (*See* **Oxide**.) Natural mineral spinels occur in igneous rocks, gneiss and serpentine, and in alluvial deposits derived from them. Semi-precious gemstones coloured red, yellow, green or black, are fashioned from spinels.

Spin hardening. The name used in reference to a flame-hardening process sometimes used on round sections such as crankshafts, etc. The workpiece spins in contact with flames from stationary burners until it reaches the required temperature. It is then quenched by immersion or by water-spray.

Spinning. A method of forming hollow-ware articles, candlesticks, trays, etc., in steel, aluminium, precious metals, pewter, copper and copper alloys. The article is shaped from a sheet of metal revolving on a lathe, over a wooden or metal former—called a chuck—by means of pressure applied by a spinning tool (*see* Fig. 160).

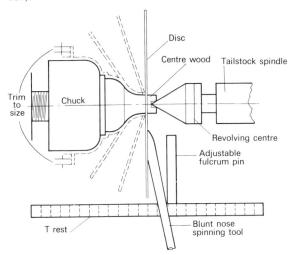

Fig. 160 *Stages in spinning metal.*

Spiral mill. A milling cutter in the form of a hollow cylinder with the cutting edges arranged spirally along the outer surface of the cylinder.

Spirek furnace. A type of furnace used for dealing with fine ores in the extraction of mercury. It is a shaft furnace having inclined shelves or baffles which retard the progress of the ore as it slides by gravity from the top to the bottom of the furnace.

Spirits of salts. An old name commonly used to refer to a solution of hydrochloric acid, used in soldering. Killed spirits is zinc chloride, used as a flux.

Splasher. The term used in reference to the water spray that is used for the protection of the metal structure just above the tapping hole, during the tapping of a blast furnace.

Splashings. A term used in reference to the molten metal thrown up by turbulence on to the sides of the mould when metal is being poured. These are most likely to occur in top-pouring of ingots. The splashed metal may stick to the sides of the mould and, cooling rapidly, may not blend with the main body of metal. This can occur when the pouring temperature is low. The splashings give rise to surface defects. (*See also* **Rokes**.)

Splat cooling. A process for very rapidly cooling molten metals—estimated at about 10^6 °C per second. One method consists in shooting droplets of molten metal against the rim of a rapidly spinning wheel. This wheel is made of copper and is cooled by liquid nitrogen to about −165 °C. The droplets flatten out to very thin discs of approximately 3 cm^2 area. The method has been used with Cu-Ag, Ag-Ge and Ga-Sb-Ge alloys, which show unusual electrical and magnetic properties that are associated with the complex pattern of arrangement of atoms in the alloys. The constituent metals do not segregate in different grains, but are intermixed in ultramicroscopic grains. This procedure appears to have applications in the production of superconducting and semi-conducting alloys.

Other methods of producing amorphous types of structure in metals and alloys are being developed, such as the contact between a spinning disc and pendant drop, producing fibres 25 to 250 μm diameter. Titanium alloys and austenitic steels have been so treated. Discharge sputtering at high rate produces amorphous deposits of rare earth/transition metal alloys up to 5 mm thick, and plasma spray deposits of Al-Cu alloys have been made to 200 μm thickness. Many alloys produced by these techniques show improvements in strength and corrosion resistance.

Spline. A key sunk into a shaft, or it may be forged with the shaft. Its purpose is to cause a wheel, pulley or other part to rotate with the shaft. The key must be a sliding fit so that the wheel or part can move along the shaft, and the keyway must be cut to allow for this.

Split die. A die made in two parts—usually sectioned vertically. In powder metallurgy split dies are sometimes used to facilitate the removal of the compact.

Splitting resistance. A term used to indicate the resistance to tearing of certain materials such as felt. A standard-size piece has a longitudinal slit made at the centre. This is commonly 50 mm long. The lip of this slit is gripped on either side and the pull required to cause rupture is the splitting resistance.

Spodumene. A lithium mineral which is lithium aluminium silicate $LiAl(SiO_3)_2$. It contains from 3.7% to 5.6% Li. The lithium may be extracted from the mineral by first digesting with sulphuric acid and precipitating the sparingly soluble lithium carbonate with sodium carbonate.

Sponge. A name applied to a loose non-reguline form of a metal such as may sometimes be produced by electrolytic deposition. The term is also applied to the porous form in which metals may be produced by heating certain salts. Platinum sponge is a grey porous mass obtained by heating ammonium chloroplatinate.

Sponge iron. Metallic iron produced from the oxide by reduction with carbon without fusing.

Spongy top. An irregular protuberance on the top of an ingot caused by the evolution of gas when the metal has cooled to an extent that it is too cold to permit the easy escape of gas.

Spot tests. Any rapid qualitative test that can be used to ascertain whether a particular element is present in a mixture or to determine the presence of certain constituents in alloys. The following spot tests are commonly used for identifying or differentiating ferrous materials. Test 1 consists in placing a drop of strong nitric acid on a cleaned surface of the metal. High chromium, high nickel and stainless steels show no reaction. Other ferrous materials react more or less rapidly to produce a brown coloration of the drop. Nickel steels sometimes show a dark greenish colour not always readily detectable in the presence of the brown colour. A yellow sediment may form if the steel contains tungsten. Test 2 consists in placing a drop of strong nitric acid on a cleaned portion of the metal, and then adding drop by drop some dilute ammonia. If a blue colour develops, copper or nickel or both may be present. Chromium gives a greenish tinge, while most steels produce a reddish-brown colour. Test 3 consists in placing a drop of copper chloride ($CuCl_2$) in strong hydrochloric acid on a clean portion of the surface of the metal. After a few minutes add several drops of water and then wash the solution away. High-alloy stainless and other steels show no reaction, but low-alloy steels and all high-iron alloys develop a copper-coloured stain.

The dimethylglyoxime test (2, 3 butanedione dioxime) is often used to detect the presence of nickel.

Spotting out. The term used in electroplating to describe the appearance of spots on plated or finished articles. These are produced by exudation from the pores of the metal of compounds that have been entrapped from plating, cleaning or pickling solutions during the respective treatments.

Spot welding. An electrical resistance welding process in which the two members to be joined are pressed together between a pair of water-cooled electrodes. The current is passed for a very short time so that fusion occurs only over a very small area at the interface.

Spraying (metal). The general name of a process for depositing coatings of one metal on another metal or on a non-metallic surface, using a technique somewhat similar to that used in paint spraying. There are three processes in use: (a) the original process developed by Schoop. In this, molten metal issues from a fine orifice and is blown at high speed, in an atomised condition, on to the article to be coated; (b) the wire process, which has to a large extent displaced (a). This process requires the use of a special pistol which is fed with wire from a reel. An oxy-gas flame melts the wire and a cone of compressed air projects the metal in a finely atomised condition; (c) the metal-powder process, in which finely divided metal is used.

The metal to be treated is first cleaned by shot or grit blasting, which leaves a roughened surface to which the sprayed metal will adhere with great tenacity. Heat-treatment after spraying is not essential, although sometimes beneficial.

The following metals are commonly sprayed: aluminium, zinc, tin, copper, brass, steel and lead. The process is very effective in producing coatings that will offer resistance to atmospheric corrosion, or prevent attack by certain industrial chemical substances. Hard deposits can be produced that will build up worn parts of machinery.

Spray quenching. A mode of quenching in which a spray of water is directed against the heated material. The quench is more drastic than that obtained by fog quenching, but is not so liable to cause distortion as direct quenching in water.

Spread test. A test used to compare the tinning action of various solders. A small quantity of the solder (usually half a gram) is placed on a horizontal sheet of material to be soldered. The solder is heated from beneath the sheet until it is completely liquid. The area that is "wetted" by the solder is measured in square cm and this result reported. The higher the spread figure, the more suitable is the solder for the purpose.

Sprigging. The term used in reference to the method of supporting or stiffening certain portions of sand in a mould by means of long brads or sprigs. It is used frequently to fasten cores in moulds and to strengthen narrow projections of sand in sand-moulds.

Sprill. A term used to describe particles of metal powder that are cylindrical or nearly so in shape having the length somewhat greater than the diameter.

Spring. An artefact designed to absorb mechanical energy, either to reduce shock, or to store energy for future release. It relies on deflection of material by a load, in an appropriate form dependent on the component to which it is attached.

Usually in the form of a coil or helix, it is then referred to as a coiled spring. Where several pieces of strip are arranged one on top of another, the longest being uppermost and shortest at the bottom (or vice versa), the spring is referred to as a leaf spring or laminated spring. A double-leaf spring is referred to as an elliptical spring.

For round coil helical springs if W = load on spring, D = diameter of the coil and d = diameter of wire, then maximum shear stress

$$\sigma_s = \frac{16W.D}{2\pi d^3}.$$

If the compression under load W is δ, then

$$\delta = \frac{8WD^3 N}{Gd^4}$$

where G = modulus of rigidity and N = number of coils in the spring.

Spring back. A test sometimes used to estimate the extent to which residual stresses have been removed by stress-relieving treatments. The material is coiled on a mandrel (of diameter d) and then removed. If the diameter of the free strip is D, then $(D - d)$ is a measure of the extent of the recovery that has been effected.

Spring chaplet. A chaplet (q.v.) usually of ⊔ shape made of springy material. It is used to prevent the movement of cores in moulds during pouring. Also called a jammer.

Sprue. A vertical channel in a foundry mould. It connects the pouring basin with the runners, and may also act as a riser. The term also refers to the waste portion thus cast, which has subsequently to be removed and discarded.

Sprue-cutting. The operation of forcing a steel tube in the cope of a mould to trepan out sand to form the downgate.

Spruing. The operation of removing the gates and sprues from a casting after solidification of the latter. (*See also* **Sprue**.)

Sputtering. The name used to describe a method of obtaining thin films of metal on metallic and non-metallic surfaces. The surface to be coated is bombarded with positive ions in a gas discharge tube, which is evacuated to a low pressure. Usually a small amount of inert gas, e.g. hydrogen or argon, is fed to the tube. The metal to be sputtered is made the cathode and the surface to be coated is laid on the base plate which is the anode. Under the influence of a high voltage disintegration of the surface of the cathode occurs and a deposit on the plate placed over the anode is formed. Voltages of 1 000-2 000 are commonly used, but in some cases higher voltages are employed. The method can be used for depositing thin metal films on optical lenses, etc. Sputtering metal on glass at molecular thicknesses has been used as a method of inhibiting the deposition of mist and ice on the glass. Connections to the metal film are made by strips of copper and the heating current is conducted over the entire surface.

Squeeze machine. A power-operated ramming machine, used mainly for ramming shallow patterns. A rigid board slides within the moulding box to compress the sand. Alternatively, the machine can be made to press the pattern into the sand.

Squeeze moulding. A method of making sand moulds by mechanical pressing of the sand round the pattern or the pattern into the sand. The action may be direct pressing or may additionally have a vibrating motion called "jolt-squeeze".

Squeezer rolls. Rolls that are used to remove mechanically the oil adhering on tinplate in the dry finishing operations. The rolls are covered with cloth. The same operation is applicable to other kinds of metal sheet.

Stabilising. The term used to describe a stress relieving by heating to and, if necessary, holding at some temperature, generally below the transformation range, followed by slow cooling.

The term is also used in reference to the inhibition of weld decay in steels by the addition of certain elements, e.g. titanium, niobium.

Stabilising anneal. A treatment applied to austenitic stainless steels containing titanium or niobium as stabilisers. The steel is heated to a temperature of around 900 °C so as to precipitate the maximum amount of carbon as carbides of titanium and niobium, in order to avoid subsequent precipitation of chromium carbide at lower temperatures which might impair resistance to corrosion.

Stabilising treatment. A heat treatment for the relief of internal stress. This type of treatment is applied to many kinds of castings, but the term usually refers to aluminium alloy castings in which dimensional stability is attained by heating at 200 °C for several hours and then slowly cooling.

Stability of electron shells. Refers to the completion of quantum levels for a given value of quantum numbers n and l, so that a higher energy level must be adopted for any further electrons added. Particularly, stability on completion of the p electrons, i.e. the stable octet, there being two s electrons for each value of n before the p electron levels are occupied. The association of two s electrons with 6 p electrons (representing the outermost energy levels) with the inert rare gases led to the use of "stability", although subsequent experimental work revealed compounds of those gases by complex bonding forms. Attainment of the stable octet is still regarded as the main influence in covalent bonding in compounds, and applies also to some elemental molecules or structures. (*See* **Bond, (8-N) rule**.)

Stack. That portion of the blast furnace between the bosh and the throat. In shape it resembles a frustum of a cone and in large furnaces may be 18-25 m high. Also known as shaft and inwall.

Stack flame cutting. The operation of cutting several thicknesses of metal simultaneously, using a single torch.

Stacking fault. A planar boundary where a small change occurs in the packing sequence of close packed crystal layers. It represents a partial change which is often associated with the passage of a partial dislocation, the first part of which causes the change associated with the stacking fault, and the second part of which restores the original structure over the region of the fault. Close-packed hexagonal and face-centred cubic structures are prone to this type of fault, their structures being very similar, but differing only in the "stacking" of layers of close-atoms, hence the name for the fault. (*See* **Imperfect dislocation**.)

Stack moulding. A method of moulding in which a number of cheeks are interposed between the cope and drag. It is used when a number of small castings are required, each having one flat surface. Use is then made of both sides of one half mould and each mould has the lower face a "cope" face and the upper face a "drag" face. The moulds are then stacked one above the other, and a common down-gate is made to run down the whole stack of cheeks to the drag, which is the lowest of the stack.

Staffelite. A variety of natural phosphate that has resulted from the accumulation of organic remains and droppings.

Stake. A wooden pin driven into the foundry floor to locate the position of pit or bedding-in moulds.

Stainless iron. A malleable, corrosion-resisting ferrous alloy containing less than 0.15% C, less than about 1% Si and Mn and 11-14% Cr. Such iron can be readily welded. When the carbon content is less than 0.12% and the chromium content lies between 16% and 30% a ferritic stainless iron is produced which does not harden by heating and quenching, but can be hardened to a limited extent by cold working. Increasing the ratio of carbon to chromium (and the addition of certain other alloying elements) renders the irons more susceptible to a hardening treatment. (*See* **Rustless irons**.)

Stainless steel. A corrosion-resistant type of alloy steel which contains a minimum of 12% Cr. This latter is the element which confers upon the steel its property of resisting attack by the atmosphere or by a number of chemical reagents. The effect is attributed to the ability of chromium to form a thin, but very tenacious, film of oxide at the surface of the alloy which resists attack by most oxidising agents. The resistance to corrosive attack is enhanced if nickel be added, and is further improved by small additions of molybdenum and copper. There are three main classes of stainless steels.

(a) *The ferritic stainless steels* (or stainless irons) which are low in carbon (about 0.1% or less) and contain 16-30% Cr. They are malleable and are capable of being pressed, particularly those in the range 11-14% Cr. When the chromium content is in the range 16% or above the "steels" are not susceptible to heat-treatment. The term ferritic stainless iron is really more applicable to those alloys in which the chromium content lies in the range 16-30% and the carbon is less than 0.15%, for these consist essentially of ferrite.

(b) *Martensitic stainless steels* in which the carbon content usually lies between 0.1% and 0.45%, and the chromium 11-14%. The cutlery and tool steels belong to this group with a carbon content of 0.25-0.3%. This group is sometimes referred to as the straight chromium hardenable steels, since in addition to their resistance to corrosion they are amenable to hardening and tempering, but at temperatures higher than for plain carbon steels. A further steel in this class is one containing about 17% Cr and 2% Ni.

(c) *Austenitic stainless steels* in which the chromium is present to the extent of 16-26%, and nickel 6-20%. Probably the best known of this group is the 18/8 (i.e. 18% Cr, 8% Ni) stainless steel. These steels are only completely austenitic at ordinary temperatures when quenched from 1050 °C after a short soak at that temperature. Otherwise they contain up to 5% ferrite, depending on mechanical and thermal history and small variations in chemical composition. The ferrite can be detected magnetically, even with a small hand magnet, although the steel is not attacked by nitric acid of 1.2 S.G. because of the austenite content. The presence of ferrite, however, reduces the corrosion resistance to some reagents, although it markedly improves the quality of welding. These steels cannot be hardened by quenching, but may be hardened and strengthened by cold working. Compared with (a) and (b) above they have a higher resistance to corrosion and are much more malleable. Certain steels in this last group—particularly those that are not completely austenitic but also contain some ferrite—are liable to embrittlement if they are heated for any length of time in the range of 600-900 °C. This is due to the formation of the sigma phase (q.v.) which is a hard and brittle constituent. (*See also* **Weld decay**.)

Stains. *See* **Spotting out.**

Stamping. A term used in reference to the process of manufacture of forgings by means of dies having the required form. The metal is usually in the form of thin strip or thin plate and the forgings are generally produced in closed dies. (*See also* **Drop forgings**.)

Stand. A term used in connection with rolling-mills to indicate a single unit or mill consisting of a set of rolls. A stand is usually further described as 2-high, 3-high, 4-high, 6-high (or cluster) and 12-high (Rohn). A rolling-mill may consist of a single stand or of several stands in series.

Standard brass. A term usually applied to 70/30 brass, i.e. to the alloy containing 70% Cu, 30% Zn.

Standard electrode potential. The potential of a metallic element in equilibrium with a solution containing simple ions of that element at a concentration corresponding to unit activity.

If the potential between the metal and the solution at equilibrium is E, then

$$E = E_0 + \frac{0.059}{n} \log C$$

E_0 is the potential between the metal and the solution containing 1 mole of the metal; C = the actual concentration of the ion of valency n. Since it is not possible to measure a single electrode potential, but only the difference in potential between two electrodes, E_0 for hydrogen is assumed to be zero. (*See also* **Electrode potential**.)

Standard test piece. In testing ductile metals in tension, the test piece elongates more or less uniformly over the whole length during the early stages of the test. But at one part of the piece, which will eventually contain the point of fracture, a neck develops. This neck represents an unduly large reduction of area and forms a considerable proportion of the total elongation of the whole test piece. If a number of test pieces of the same material and diameter are subjected to a tensile test, it will be found that the total elongation is roughly inversely proportional to the gauge length.

All test pieces do not have to be on the same gauge length, but if the length is related to the diameter, it can be accepted that the elongations percentage obtained from test pieces of different lengths are quite comparable. (*See* **Tensile test piece**.)

Standard tin. The term used in reference to commercial tin which must be at least 99.75% pure.

Standard wire gauge. The thickness of certain kinds of sheet metal and the diameter of wire is frequently designated by a simple number. This gauging system would be admirable were it not for the fact that there are several different systems in use, and a gauge number in one system does not represent the same dimension in any of the other systems. Because of the confusion that might occur in quoting only the gauge number, it was usual to give also the exact dimensions in decimal fractions of an inch. The quoting of the gauge number serves mainly to indicate that the material conforms to the size ordinarily used for the particular class of material. (*See* Table 15.)

Standing wave. In the study of wave mechanics, solutions to the Schrödinger equation have many values, and the problem is to find a solution which seems physically possible. In some cases, the solutions represent a stationary state, which does not mean an electron at rest, but conditions which imply that an equivalent wavelength λ is involved, related to some distance involved in the problem. This is what is meant by standing wave. An electron in a one dimensional box of length l gives acceptable solutions only when $\lambda = 2l/m$, where m is an integer. This means that only certain levels of energy can be involved, which in fact are the quantum numbers of quantum theory. A stretched string, as on a violin, gives a stationary state of vibration when plucked or bowed, and λ must be a definite fraction of the length of string between supports. A rotating skipping rope, secured at each end, is another example of a standing wave.

When conditions do not involve a standing wave, then there

must be possible variations dependent on other boundary conditions, and a travelling wave results, and the velocity and momentum become imaginary, not amenable to experiment.

Stannine. *See* **Stannite**.

Stannite. A mixed sulphide of tin, copper and iron $FeCu_2 . SnS_4$. It is frequently associated with copper ores. It crystallises in the tetragonal system. Also known as stannine, zinn, tin pyrites, bell ore. Found in Cornwall (UK), Zinnwald (Czechoslovakia) and Bolivia. Mohs hardness 4, sp. gr. 4.3-4.5.

Stannum. A proprietary high-tin alloy used for hard-service bearings.

Star antimony. A term used in reference to the first quality of antimony; so called because of the star or fern-like markings on the surface.

Stare. A term sometimes used in reference to the appearance of crucible cast steel that shows a bright sparkling fracture. Such steel is very brittle and can be reduced to a powder by hammering.

Starting sheets. The name used in reference to the pure copper cathodes used in the electrolytic refining of copper, lead, etc., and on which the copper or lead is deposited in the refining process. In the case of copper, the cathodes are thin sheets of electrolytic copper which are prepared by depositing on a rolled copper plate that has been covered with a thin film of oil or grease. Sometimes as an alternative the plate is dipped into a solution of a mercury salt. The plate is provided with a sharp groove near one edge, along which the deposited copper sheet can be torn when it is stripped off. The starting sheets are then provided with loops to receive a copper bar and inserted into the electrolytic tanks.

Lead starting sheets are made by allowing pure molten lead to flow over an iron plate. On solidifying, the lead is stripped. A copper supporting rod is laid along one edge and the lead wrapped round it and spot welded in place.

Stassano furnace. An indirect electric-arc furnace consisting of insulating brick contained in a cylindrical steel sheet. The lining is of magnesite brick. There are three electrodes, equally spaced, and inclined downwards. The furnace can be oscillated by mechanical means. The heat is generated in the arc between the electrodes and is radiated to the charge. The roof of the furnace is dome-shaped to increase the reflection of heat.

Statiflux. An electrostatic method for detecting cracks in porcelain enamel coatings on steel.

Stationary states. This is one of the two basic assumptions of Bohr's theory of the atom, viz. that for each atom there is a series of orbits in which the electrons rotate, but no radiation is emitted. These are called stationary, or more properly, ground states and are characterised by the fact that the angular momentum of the electron is an integral multiple of $h/2\pi$, where h = Planck's constant.

Steadite. A micro-constituent of phosphoric cast irons. It is a eutectic of austenite saturated with phosphorus and iron phosphide Fe_3P and contains about 10.2% P. The eutectic has a melting point of about 980 °C. Steadite appears as a white fine dotted eutectic.

Stead's brittleness. A form of brittleness yielding transcrystalline fracture in the coarse-grained structure produced by prolonged annealing of thin sheets of very low carbon steel which has previously been rolled at temperatures below 700 °C. The phenomenon is associated with abnormal growth and may be avoided by increasing the carbon content slightly. A similar form of embrittlement occurs in brasses, silicon steels and chromium steels.

Steam hammer. A powerful hammer or vertical tup used for forging. The upper portion of the tup is formed into a piston rod and piston, and the latter working in a vertical cylinder is operated by steam pressure. The steam may be used for raising the tup, the latter being allowed to deliver the blows by falling under gravity. Alternatively the steam may be used to increase the magnitude of the blows.

Steatite. A massive variety of talc, which is hydrated magnesium silicate $3MgO.4SiO_2 . H_2O$. When heated to temperature between 800 °C and 1,000 °C it becomes very hard and strong, and has a high electrical resistance. In this form it may be used for making insulators for high voltages. It is usually ground, then moulded under pressure. When fluxed with alkaline earths and moulded, a ceramic material of good electrical, high-temperature properties is obtained. It is found in the USA (Montana), India and Sardinia.

Steckel mill. A type of rolling-mill in which power is applied to a reel which draws sheet or strip through undriven rollers of a 4-high mill. It is used for cold-rolling.

Steel. A malleable alloy of iron and carbon which has been molten in the process of manufacture, and which contains 0.1% to 1.9% C. This latter is present in the combined state as iron carbide, Fe_3C. Usually small amounts of silicon, sulphur and manganese are present in what are known as plain carbon steels.

Alloy steels are those that contain controlled amounts of such additional elements as chromium, nickel, manganese, molybdenum, titanium, tungsten or vanadium. The outstanding property of steels generally is their ability to respond to heat-treatment in such processes as hardening, tempering, annealing, etc. Steel is made by the open-hearth or oxygen converter process or by electro-thermal processes. These processes are sometimes further subdivided according to the nature of the lining of the furnace in which they are made (e.g. basic Bessemer process, acid open-hearth process, etc.).

Since the properties of a steel depend largely on the percentage of carbon present, steels are sometimes classified on the basis of carbon content. Low-carbon steels (i.e. C 0.09-0.2%) are soft, like wrought iron, and are known as mild steels. With further additions of carbon the ductility falls off while the tensile strength increases. Medium carbon steels contain 0.2-0.4% C and higher carbon steels contain more than 0.4% C. This classification is somewhat arbitrary, but is quite widely used because it helps to simplify description. Another way in which steels are classified is by the special application of the material—the low-carbon mild steels are used in making tin plate and light engineering castings. The medium-carbon steels are referred to as structural steels, while the higher-carbon steels (up to 0.75% C) are used in making springs, wire ropes, and small castings for die-blocks, press tools, etc. What are called alloy steels usually contain 0.2-0.43% C and any one or more of the following elements: 0.4% or more of chromium or nickel, 0.1% or more of molybdenum, tungsten or vanadium, 10% or more of manganese. They are designated manganese steel (1.5% Mn); nickel-chrome steel (1.5% Ni, 0.78% Cr); nickel-chrome-molybdenum steel (1.88% Ni, 0.8% Cr, 0.38% Mo); chrome steel (0.43% C, 0.9% Cr); 3% chrome-molybdenum steel (0.21% C, 3.06% Cr, 0.65% Mo). These, among others, are used in the wrought or cast form for general engineering purposes —gears and machinery details—where good strength is combined with toughness or wear resistance. High alloy steels include the stainless steels (q.v.), heat-resisting steels and high-speed steels. Other terms less frequently used include shock-resisting steels (1.5% Mn, 0.6% Mo); wear-resisting steels (1.07% C, 13% Mn) and magnetic steels or high-permeability steels. (*See* Table 6.)

Steel grit. An abrasive material produced by forcing molten iron through a steam jet in order to obtain small irregular particles of chilled iron. These particles are crushed and screened and sold in sizes No. 8 to No. 80, the increasing number indicating decreasing coarseness. The material is used in tumbling barrels and for grinding stones.

Steelite. A potassium chlorate explosive that can be used in quarrying. It does not shatter rock, as other more powerful explosives are liable to do, but cracks and splits it.

Steel powder. Finely divided steel used in powder metallurgy processes, for producing intricate shapes in steel by compressing the powder (with a binder) hydraulically in moulds, and sintering the compact at about 1,070°C in controlled atmospheres. Steel of various compositions can be used, the strength of the finished articles being about the same as the wrought metal. The density, hardness and porosity are controlled by the powder size of the binder and the pressure used in the mould.

Steel wool. The term used in reference to the long, fine fibres of steel felted together and used for abrasive cleaning and polishing. It is made from low-carbon (Bessemer) steel of high tensile strength. Present practice is to use wire of different diameters to produce a product of the fineness desired. The wire in several loops, running in guide-grooves, is shaved by a flat knife having a straight edge. The fibres vary in size, according to the diameter of wire

used and the depth of the cut, and these two conditions determine the size limits of a particular grade of wool. There are nine grades, in the finest of which no fibre has a thickness greater than 12 μm.

In the finer grades, steel wool is used for polishing wood, and when impregnated with soap is used for cleaning aluminium vessels and cast-iron parts. The coarse grades are used for scouring in routine maintenance work in dairies and food factories. Manganese steel wool has been successfully used in de-aeration plant of steam-raising equipment, and as air and liquid filters.

Steeples. The term used in reference to the type of nails used in moulding in the foundry.

Stefan's law. This law states that the total radiation of any body is proportional to the fourth power of its absolute temperature. Boltzmann has shown from thermodynamic principles that this law is true for black bodies. It is therefore sometimes referred to as the Stefan-Boltzmann law. It is not strictly true for bodies other than black bodies. If a body at absolute temperature T radiates to its surroundings at t °C the total energy lost per second from unit area is $\sigma[T^4 - (t + 273)^4]$, where σ is a constant.

Stellite. The trade name of a group of non-ferrous alloys used for cutting tools and other purposes, in which the hardness is an inherent property of the alloy and is not induced by heat-treatment. The alloys are very resistant to corrosion and abrasion, and retain their hardness at a red heat. Since they cannot be forged they must be cast direct to shape. They are used in solid (cast) and tipped cutting tools; in plug, screw and tap gauges; on the valves and valve seatings of engines; as shear blades, and in turbo superchargers. They are valuable in medical engineering applications. Stellite hard facings are used, and are applied by the usual welding methods. Hard facing rods are used as electrodes in welding. Hardness 370-475 B.H.N. (*see* Table 9.)

Step-gate. An arrangement for feeding moulds with molten metal through a number of steps, or gates, arranged one above the other. Actually the arrangement corresponds to bottom gate feed with additional side gates such as might be used for bush castings. The side gates may be arranged horizontally or directed upwards towards the mould. The object of using the more complicated types of gates is to obtain good control of the flow of the hot metal. As the side gates come into operation in the correct sequence the mould is filled by fresh, hot metal on the top of that which is fed in through the gate below. In the end, the uppermost gate will feed in hot metal, some of which will pass into the riser.

Stephanite. A silver ore consisting of silver sulphantimonite Ag_5SbS_4. It contains about 68% Ag and 15% Sb. It is found in Nevada (USA), Mexico, South America (Chile and Peru), and in Central Europe. Also known as brittle silver ore. Mohs hardness 2-2.5, sp. gr. 6.26; crystal system, orthorhombic.

Stepped anneal. A process carried out to remove brittleness in steel castings that is carried out in steps, the rate of cooling during each step not necessarily being identical. A casting may be heated to a temperature above the upper critical temperature and held until the whole mass has completely soaked (i.e. same temperature throughout the mass). Since it is the *rate* of cooling that determines the degree of softening, and since the product is softer the higher the temperature at which the S-curve is traversed, and again since the rate of cooling after the transformation has occurred does not further affect the treatment, it is found economical in time (and sometimes in furnace space) if the rates of cooling during the anneal are varied. This is particularly advantageous where the transformation occurs within a rather restricted temperature range. A stepped cooling plan can be usually worked out to include reasonable times for cooling within each step. To work out such a programme would require consultation of the appropriate S-curve (q.v.).

Step quenching. A heat-treatment process otherwise known as Martempering. It is the process of quenching down to the M-temperature by plunging the steel into a molten salt bath, holding it there until the temperature throughout the whole mass is equalised and following this by cooling in air. This treatment is less drastic than oil or water quenching and avoids distortion in those steels that are liable to warp or crack during normal quenching. Only those steels that have a suitable S-curve can be so treated.

Stereotype metal. A lead-base alloy used by printers in the production of stereotype plates, and duplicate replicas of the original forme of type. It has good hard-wearing properties and sharp casting characteristics. The composition is approximately Pb 83.75%, Sb 11.75%, Sn 4%, Cu 0.5%. A sheet of papier mâché is moistened and pressed over the type to form a flong. After drying, this flong is inserted in a mould into which the stereotype metal is cast. In rotary machines the plate is cast into the shape of a half-cylinder.

Sterling process. A method of smelting zinc ores using carbon in an electric-arc furnace. The secret of the process is the achievement of a correctly balanced slag.

Sterlith. A trade name for a brand of alumina Al_2O_3 used as an abrasive.

Sterri. The hard sulphur-rich rock found in the volcanic areas of Sicily and used for making sulphuric acid.

Sterro-metal. A high-strength brass used as a casting alloy. It is a modified 60/40 brass having small additions of iron and manganese. The latter are responsible for the formation of the hard constituent and for the additional strength. It is used for hydraulic cylinders and marine castings.

Stewart grate. A type of moving (endless) grate designed especially to facilitate the handling of ores in a sintering machine. In the case of the sintering of lead ores, the heat of the reaction due to the burning of the sulphur causes the charge to agglomerate, in order that the sinter can be discharged at the "end" of the grate. To ensure this, gaps are provided in the pallet train, so that each pallet drops over the end of the grate and strikes the one in front. The Stewart grate allows a small amount of movement in the central portion, and this facilitates the freeing of the pieces of sinter which are then readily discharged when the pallets are "jarred" at the end of the grate.

Stibiconite. A naturally occurring massive ore of antimony consisting of antimony oxide $Sb_2O_4 \cdot H_2O$. It is pale yellow in colour and contains about 72% Sb. Mohs hardness 4.5, sp. gr. 5.1.

Stibnite. A naturally occurring massive ore of antimony consisting of antimony trisulphide Sb_2S_3. It is the principal ore of antimony, and occurs in veins associated with other antimony ores as orange-red crystals. The crystal structure is orthorhombic, Mohs hardness 2, sp. gr. 4.5-4.6, and melting point about 546°C. It is used to colour red rubber and as a paint pigment. Sb_2S_3 is a constituent of safety matches. Also called antimony red, antimonite and antimony glance.

Stiefel mill. A universal rolling machine in which cantilever rolls are used for producing tubes from hot pierced billets.

Stiefel process. A rotary piercing process for making seamless tubes. A round billet of suitable size is pierced and then rolled over a mandrel between grooved rollers.

Stiffening. A procedure in casting in which brackets or ribs are provided between various members of a casting, in order to obtain added rigidity.

Stiffness. A term used in reference to the ability of a solid or hollow shaft to resist twisting when subjected to a torque or twisting movement. It is measured by the angle of twist per unit length of the shaft.

Stinkstone. A dark-coloured limestone which contains bituminous matter and emits a foetid odour when struck.

Stitch-welding. A resistance welding process basically similar to spot-welding. The complete spot-welding cycle, including raising and lowering the electrode, is repeated automatically while the work is fed by hand, so that the individual welds overlap. Seam welding is probably preferable wherever the work can be passed between seam-welding electrodes.

Stock. The metal cut from a bar that is required to be forged, etc., and has been allocated to the required processes.

Stock and die. A double-ended lever with an arrangement at its centre to take the dies that are used for forming screw threads on studs, bolts, tubes, etc.

Stoichiometry. This applies to a compound which has a fixed chemical composition under all circumstances. One which can dissolve small proportions of any constituent over a range, is said to be non-stoichiometric. The occurrence of voids in crystal structures usually gives rise to this possibility, and it is the smaller ion which generally occupies these spaces. The compounds of uranium and oxygen are good examples of nonstoichiometry.

Stoke's law. If a sphere of solid material falls through a viscous medium which behaves as a Newtonian fluid, the velocity of fall is given by

$$v = \frac{2}{9} \frac{r^2 (\triangle \rho) g}{\eta}$$

where v = velocity, r the radius of the sphere, $\triangle \rho$ the differential in density between sphere and fluid, η the viscosity of the fluid, and g the acceleration of gravity. The law was deduced from the concepts of viscosity, but confirmed by experiment.

It has direct application in the separation of particles from a viscous fluid during manufacturing processes, e.g. slag and inclusions rising from liquid steel or other alloys. The behaviour of water droplets in clouds indicates that the droplets fall at a rate of $1.3 \times 10^6 r^2$. When $r = 10^{-2}$ mm, as in mist and clouds, v is approximately 1 cm s^{-1}. When r increases, the velocity of fall increases as r^2, hence the apparent "downpour" of heavy rain. When r is smaller than 10^{-2} mm the fine mist does not settle out easily. The aim of "rain-making" by seeding clouds, is to provide nuclei of critical size so that raindrops can form and hence fall from the cloud.

There is another Stoke's law, relating the change in wavelength when incident light is absorbed in fluorescent materials, and emitted at higher wavelength.

Stone. The generic term for hard rock, whether of igneous, sedimentary or metamorphosed origin, and which can be quarried and graded for technological use. Igneous rock is obviously derived from the melt, sedimentary rock originates from the weathering, transport and deposition of igneous rock or other material, and metamorphosed material is sedimentary or other rock deposits which have been subjected to heat and pressure. As a result of these origins, igneous rock is extremely hard, although it may contain a matrix with many other minerals dissolved or distributed through it. Sedimentary rock frequently has individual mineral constituents bound together by "cementing" minerals such as clay or oxides. Metamorphosed rock may vary in composition but often has some relationship to the original igneous or sedimentary rock in chemical origin.

The useful stones are those obtained from highly siliceous sedimentary deposits such as sandstone (igneous origins), those from water deposition such as limestone (the remains of calcareous living creatures) and igneous rock such as granite. The bond in stone of sedimentary origin may be weak, and the stone will not weather too well, especially in industrial atmospheres. Limestone is attacked by acidic atmospheres, but less than might be imagined, and some varieties, such as Purbeck and Portland, have long been important building materials.

Stone crushed and graded becomes useful as an aggregate for concrete or tarmacadam. Pebble-sized deposits such as gravel do not require crushing, and are used as aggregate directly or with other materials. Carved stone had a high aesthetic value in ecclesiastical and aristocratic buildings, but its cost is now prohibitive. Since very large deposits of stone exist, more use is likely to be made in future of otherwise useless material, by bonding with cement in the form of concrete, as the cheapest bulk building material, as against brick, timber or other alternatives.

Stope leaching. A method of dealing with low-grade copper ores in mines where the high-grade ore is exhausted. Drainage water is sprayed over the low-grade ore which has been broken in the stopes. The dissolved copper is precipitated by scrap iron after it has drained to a lower level or sometimes after being pumped to the surface. The solutions are used again, being sprayed through rubber hose. Care has to be taken to discard part of the solution after a time to prevent the excessive build-up of iron salts.

Stopper. The name used to describe a rod covered with refractory material for controlling the flow of molten metal through a nozzle.

Stop-off. A term used in reference to two distinct processes: (a) it is used to describe the sealing of part of a mould with sand or other refractory material to prevent access of molten metal: (b) it is also used for protecting selected areas of an object with an inert coating, so as to prevent chemical action during carburising, deposition during electroplating or etching.

Stopping-off pieces. These may be strengthening pieces of wood attached to an unsubstantial pattern. They should be simple in form and of such dimensions that they will protect the pattern, when not in use, from damage in handling or storage.

A second use of the term refers to the method of using a properly designed piece to enable a long pattern to be used to make a somewhat shorter mould. The stop-off refers to the unwanted part of the mould cavity that has to be rammed up.

Stored energy. Potential energy in a material or artefact, frequently derived from mechanical strain. A crystal lattice in which severe strain has been induced by mechanical work, thermal treatment such as quenching, or radiation damage, will not return to equilibrium under normal ambient conditions unless the activation energy required is available under those conditions. Normally, activation energy has to be provided, such as by rise in temperature, to increase the mobility of the atoms in the structure, enabling stress relief to occur by deformation.

In artefacts, release of stored energy may be dependent on modifying the gross structure, i.e. removal of part of the assembly, such as a stop holding a spring, or opening a valve when a fluid is contained under pressure.

Stored energy is used in engineering design to actuate mechanisms and in safety devices to cause some defence mechanism to operate. In materials science it may be produced intentionally to increase strength, as in precipitation hardening, or may be accidentally produced as in radiation damage. Material containing stored energy is always metastable.

Stove. An apparatus that is used for preheating the air supply to a blast furnace. The stove is lined with brickwork, and this is heated by allowing the burning gases from the furnace to pass over it. Air is passed in the opposite direction to the furnace gases, after these have been cut off and the heated air enters the furnace (e.g. Cowper stove).

The apparatus used for drying moulds and sand cores is also referred to as a stove.

Stoving enamels. These enamels, usually pigmented, contain resins which harden when baked in a stove or oven at temperatures between 100 °C and 200 °C. They provide useful protection to metals against atmospheric corrosion, but are usually unsuitable if the conditions of exposure are severe.

Enamels may be applied directly to the metal, if the latter is well cleaned, but they adhere more satisfactorily if, in the case of steel for example, the metal has received a phosphate coating.

Straightening. A term used in reference to the removal of distortion in rolled or cast components. There are several methods in use, the most important being: (a) flame straightening, which is a process of heating locally parts of the casting or component, particularly on concave surfaces, so permitting internal stresses to operate in straightening. As the heated part expands, the distortion may be removed wholly or in part on slow cooling; (b) a second method consists in heating the whole component in a furnace to reduce the mechanical strength of the metal or alloy and then permitting the part to align itself either under its own weight or under the pressure from an applied load; (c) a third method of straightening is by the aid of a large straightening press. The part to be straightened is suitably supported on rollers or on the bed of the press, and pressure applied by means of a tup to the area to be straightened.

Strain. When a body is deformed by the application of a stress so that there is change in one or more dimensions, or in shape, or both, the body is said to be strained. Strain is measured as the ratio of change in a particular dimension, or dimensions, to the original dimensions. The type of strain produced will depend on the nature of the stresses. These latter may produce: (a) elongation; (b) compression; (c) reduction in area; (d) distortion of shape.

In the analysis of strain it is usual to consider a small cube of the material, and to consider three edges of the cube mutually rectangular as the axes of reference. If movement due to applied forces occurs along one axis, there results an extension which is measured as the change in length divided by the original length. In the case of simple shearing strain, the magnitude of the strain is expressed by the angle between two given planes which were originally parallel and at right-angles to the stress applied. Since it is assumed that for a very small cube the three principal axes will remain rectangular after strain, it will be necessary only to find the extensions along these principal axes of strain in order to define the distortion.

Strain ageing. A term used in reference to the change in properties of a metallic substance, e.g. changes in mechanical strength, hardness, ductility or impact resistance, that are brought about by the passage of time, following upon plastic deformation. The effects may occur gradually at ordinary temperatures, or more rapidly at moderately elevated temperatures. It is generally assumed that the cold working involved in plastic deformation causes the material to assume an unstable structural condition of equilibrium, and the ageing represents the inherent tendency to revert to the more stable state. The ageing, however, is not only accompanied, but is actually caused, by the precipitation of a phase which had been temporarily retained in supersaturated solution. The addition of titanium or aluminium to steels removes the tendency to strain-age and excess of deoxidiser in steel has a similar effect.

Strainer core. In casting in the foundry, top gates are sometimes furnished with a perforated core located in the sprue cup. This is known as a strainer core, and is intended to moderate the flow of metal and so reduce the impact of the falling metal on the mould-face. The core serves the further purpose of retaining the slag in the sprue cup, keeping it floating on the surface.

Strain gauge. A device for measuring small amounts of strain produced during tensile and similar tests on metal. A flat coil of fine wire is mounted on a substrate, rectangular in shape, and usually about 5-20 mm long. This is glued to a portion of metal under test. As the coil extends with the specimen its electrical resistance increases in direct proportion. This is known as a resistance-strain gauge. Other types of gauge measure the actual deformation. Mechanical, optical or electronic devices are sometimes used to magnify the strain for easier reading.

Strain hardening. The term used in reference to the increase in hardness and tensile strength caused by plastic deformation at temperatures below the recrystallisation temperature. It is sometimes called work hardening, produced by cold working. When a metal is plastically deformed, as the deformation proceeds the distortion and the resistance to distortion increase. Thus the metal shows a reduction in its capacity for deformation. Recrystallisation enables the metal to regain its capacity for plastic deformation, and this tends to occur. With most metals this occurrence takes place only very slowly at ordinary temperatures, but more rapidly at elevated temperatures.

Strain relaxation. A term used as an alternative to creep, especially in the case of substances such as elastomers. See Viscosity.

Straits tin. One of the purest commercial forms of tin (99.9% Sn) produced from alluvial Malaysian ores.

Strand mill. A term sometimes applied to a rolling-mill in which the rolls are grooved for the forming of shapes in a series of passes.

Strauss test. A test for assessing the liability of welded austenitic stainless steels (18/8 type) to intergranular corrosion. The specimen—about 10 cm × 2.5 cm × 1 cm— is sensitised by welding, or heating in the range 500-700 °C and then immersed for 72 hours in a boiling acid copper sulphate solution. This solution should contain 13 g copper sulphate and 47 cm³ concentrated sulphuric acid per litre. The specimen is then bent through 180 degrees round a 1 cm diameter mandrel. Cracking indicates carbide precipitation and thus liability to weld decay. (See also Weld decay.)

Streak. When a mineral is powdered, the colour of the powder is referred to as the streak of the mineral.

Stream tin. A water-worn form of cassiterite SnO_2 which occurs in beds of streams and in alluvial deposits bordering streams. The term is sometimes applied to metallic tin from the alluvial deposit.

Strength. A word often used as a synonym for power, i.e. capacity to do work or to withstand work done on material. Ability to sustain the application of force without breaking, when used in the expression tensile strength, compressive strength, torsional strength, impact strength, endurance.

The term is also used in reference to intensity, e.g. in regard to the flow of an electric current, or in respect of a specific property.

Stress. The stress in a material subjected to external forces represents the intensity of the internal forces that tend to resist a change in the volume or shape of the material. It is measured by the load on the material divided by the instantaneous area over which it acts. The particular way in which the change occurs depends on the manner in which the load is applied. If the load acts at right angles to a section and away from it, the stress is called tensile stress; if it acts at right angles to the section and towards it, the stress is compressive; when the load acts parallel to the section, the stress is referred to as shear stress.

Stress ageing. A term used in reference to a treatment of metals that consists in heating alloys, composed wholly or mainly of a solid phase or phases, for a certain time and at a suitable temperature while the material is subjected to stress by an external load. In certain alloys the treatment produces improvements in both mechanical and electrical properties. The effective temperatures are those within the range from room temperature up to the recrystallisation temperature of the alloy.

Stress concentration. A term used in reference to the high local increase in the stress in a material caused by, and occurring in the immediate neighbourhood of, a sharp corner, a tool-mark or scratch, or any other similar feature which can be regarded as constituting a sudden change of section—cracks, holes, changes in diameter, etc.

The extent to which any of the above features produce a local increase in stress can be studied by making a model of the component in glass, or in transparent plastic, and stressing this in the same manner as the original. The stress distribution can then be seen by observing the interference figures that are produced when polarised light passes through the stressed model.

An alternative method of rendering visible the regions of stress concentration is to coat the component with a brittle lacquer. When the lacquer is thoroughly dried it adheres strongly to the metal. The component is stressed and, as the surface distorts, the lacquer cracks. It can then be observed that the areas of maximum number of cracks are those of stress concentration. The soap film method is amenable to mathematical treatment.

Stress concentration factor. A factor indicating the effect of a toolmark or a hole or any other stress concentrator (*see* **Stress concentration**) on the fatigue strength. It is the ratio of the maximum stress occurring in the region of a stress concentration, when a given stress is applied to the component, to the actual stress in the same region when the defect is absent. Photo-elastic methods are suitable for the measurement of the latter.

In the case of a fine crack or a scratch where d is the depth of the defect and r is the radius at the root of it, the S.C.F. is: $1 + k\sqrt{(d/r)}$.

For torsional stresses $k = 1$, but for tensile stresses $k = 2\cos^2\theta$, where θ is the angle between the direction in which the stress is applied and the normal to the direction of the defect.

Stress corrosion. A term used in reference to the effect of stresses, whether residual or applied, on the rate of corrosion of a metal in a corrosive atmosphere. The corrosion rate may be increased, especially if the stress is fluctuating by breaking or by flaking off the protective skin that is formed as the result of the chemical action in corrosion. This exposes the underlying metal to further corrosive action. Stress-corrosion failure may be intercrystalline in alloys that normally exhibit transcrystalline fracture. Similar phenomena occur in plastic materials under similar conditions.

Stress equalising anneal. A term used in reference to a low-temperature heat-treatment intended to develop the optimum mechanical properties in wrought nickel, nickel alloys, etc., to ensure minimum distortion during subsequent machining. The annealing temperature depends on the type of alloy and the nature of the previous working. Nickel-copper alloys are usually treated at about 300 °C for 1 hour. Higher temperatures are used for alloys containing chromium or molybdenum.

Stress field. The volume of material around a stress concentration, caused by some interference with the normal structure. In the case of point defects such as interstitial atoms or vacant lattice sites, the field may be only a few atoms in radius from a single source. In the case of line defects such as dislocations, a much greater volume will be involved, and hence the stress required to cause movement of the dislocation will be much lower than the theoretical strength of the material in a defect-free condition.

Stress raisers. A term used to describe certain surface defects on metals such as notches, sharp changes of section or contour, scratches, etc., that lead to local concentration of stresses. (*See also* **Stress concentration**.)

Stress relaxation. The term used in reference to the decrease in stress in a material subjected to prolonged constant strain for a given time in a creep test. The stress relaxation behaviour of a material is usually exhibited in a stress/time curve. On such curves the stress relaxation is the difference between the initial stress and that at any specified time. One usefulness of stress/ relaxation/ time curve data would be the determination of the "bolt-tightness" (or loss of bolting pressure) in a bolted or riveted structure when held at elevated temperatures for some time.

The rate at which stress relaxation occurs is usually obtained by plotting log stress relaxation/log time. The slope of the tangent at any point on the curve gives the average stress-relaxation rate for the period up to the time corresponding to the chosen point.

Stress relieving. The term used in reference to the heat-treatment used for reducing internal stresses in metals that have been induced by casting, quenching, welding, cold working, etc. The metal is soaked at a suitable temperature for a sufficient time to allow readjustments of stress.

The temperature for stress relieving is usually below any transformation range and is followed by slow cooling. (*See also* **Stabilising**.)

Stress-rupture test. Short-time creep tests to determine the rate of creep strain against stress do not necessarily predict correctly the ultimate load-carrying capacity when extrapolated to long times. Long-time tests themselves are very expensive, but they are the only reliable method of checking whether long duration causes microscopical changes not produced in short-term tests. To assist in forward prediction, stress-rupture tests, to determine the time to rupture against stress, assist by tackling the problem from the "top end", as against the short-term creep tests at the bottom of the stress range. If no untoward effects are noted in stress-rupture tests, more reliance can be placed on extrapolation of short-term tests.

In the case of short-life periods of service, e.g. a rocket motor or a military jet engine, stress-rupture tests are even more important to determine a reasonable design stress with adequate safety allowances.

Stress-strain diagram. A graph constructed from plots of stress against strain on a material when observations of strain produced as the stress is increased are made. The graph is usually constructed for tests of material in tension, torsion or compression. Such curves frequently show a linear relationship between stress and strain up to a certain value of the stress. This latter is known as the elastic limit, since at stresses below this value the material will recover its original size and shape after the removal of the load. Beyond the elastic limit the material shows a greater increase in strain for similar increases in stress. The stress-strain diagram for a mild steel is shown in Fig. 58 under **Elasticity**. L is the elastic limit and Y is the yield point (q.v.). The maximum stress is that corresponding to the point U.T. If the stress is maintained at this value, the material shows marked decrease in sectional area at some point, and fracture ultimately occurs. Between Y and U.T. the material is deformed plastically and, if the stress is removed, it will be found that the material is permanently deformed. If, however, the stress is maintained at some value between Y and U.T., the deformation increases with time at the constant load. Results are also affected by strain rate.

In other materials, there may be plastic deformation from the point of application of stress, giving no straight line portion on the curve. With polymeric materials, the rate of strain may markedly influence the stress-strain diagram, and different design concepts have to be used, with time as a very important variable. In brittle materials, there is frequently a straight line portion of the diagram which ends abruptly with fracture, and very little sign of plastic deformation.

Yield points are not common in metals and alloys, but a similar yield stress can occur in polymeric materials, the behaviour changing quite markedly once the yield stress is reached. (*See* **Ultimate tensile strength, Viscosity**.)

Stretcher. A term usually applied to a distance piece, but also used in reference to the bricks in brickwork that have their long edges to the front.

Stretcher-straightening. A process for straightening wrought metals by applying sufficient stress at the ends of rods, bars, sheets, etc., to produce a small amount of permanent elongation or set.

Stretcher strains. A furrowed roughening of the surface which may occur with low-carbon sheet or strip steel, due to uneven yielding in the early stages of cold deformation after annealing or, in a lesser degree, after normalising or hot rolling. Also known as Lüder's lines (q.v.).

Strickle. The term used for a form of board or plate employed for shaping a mould or core. The edges of the board are suitably shaped for the work and the board is drawn along the sand or rotated about a vertical post according to the nature of the shape desired. The strickle is mostly used for large castings of circular shape. A vertical post carries a horizontal beam having the shaped board attached, and by swinging this the sand is swept away to leave the required contour.

Strike. A term used in connection with faulting in rocks to indicate the horizontal direction at right angles to the direction of dip of a rock. The term is also used in reference to the initial electrodeposition of metal under conditions differing as regards current density and strength of electrolyte, from those used for the main plating process. Certain metals, particularly silver, tend to deposit on other metals by immersion in cyanide solutions. Such deposits have very poor adhesion. To avoid this deposition, strike solutions are used which permit a quick covering of the base metal to be obtained, after which plating is performed in the usual manner. Strike solutions generally have a low metallic ion content and are operated at high current densities. For silver plating on brass the strike solution may be weak silver cyanide solution in sodium cyanide, or a weak copper cyanide solution.

Striking board. A term used in reference to the form of strickle (q.v.) that is attached to a central post round which it can be rotated in shaping circular mould shapes.

Stringer. A term used to describe a microstructural feature consisting of alloy constituents, or inclusions, arranged linearly in the direction in which a metal has been worked.

Strip. A term used to describe a flat-rolled metal product—which, however, is often coiled—of smaller cross-section than sheet or bar. It is narrower than sheet metal and less thick than bar. The term is also sometimes used in reference to the taper given to a pattern to permit its easy withdrawal from a sand mould. It is further used to denote the solution that is used electrolytically to remove a metal coating from the basis metal.

Strip mill. A rolling-mill particularly designed to produce strip metal by rolling. Some metals are rolled hot, but finished by cold rolling to obtain an improved surface finish. Other metals are rolled cold and given intermediate anneals. A 4-high mill is the usual type of strip mill. (*See also* **Steckel mill.**)

Stripping. The term used to describe the removal of an ingot from the mould. It also refers to the removal of coating (metallic or non-metallic) from metal. Table 56 gives some details of the methods of stripping common metal coatings from metals in general use.

TABLE 56. *Methods of Stripping One Metal from Another*

Coating Metal	Basis Metal	Solution	Method
Nickel	Steel, Copper Zinc alloys Brass	70% solution H_2SO_4 *or* 20% solution NaCN	Anodically ,, ,,
Zinc or Cadmium	Steel Brass Copper	Strong HCl + 3% Sb_2O_3 *or* 5% ammoniacal solution $(NH_4)_2S_2O_8$	Immersion ,, ,,
Tin	Steel Copper	Strong HCl + 3% Sb_2O_3 Acetic (ethanoic) acid solution $FeCl_3$ and $CuSO_4$ + little H_2O_2	Immersion ,,
Chromium	Steel Zinc Nickel Brass or Copper	NaOH solution Na_2CO_3 solution Strong HCl + 3% Sb_2O_3 NaOH solution 15% HCl solution	Anodically ,, Immersion Anodically Immersion
Copper	Steel Zinc	Acid (sulphuric) solution H_2CrO_4 Strong $NaNO_3$ solution Na_2S solution *or* Acid (H_2SO_4) + 2.5% chromic acid solution	Hot immersion Anodically ,, ,,
Silver	Nickel, Silver or Brass	5% strong HNO_3 in strong H_2SO_4	Hot immersion
Gold	Copper, Brass Nickel alloy Steel	10% H_2O_2 + 10% sodium cyanide solution	Immersion
Lead	Steel	10% H_2O_2 + 10% acetic acid solution	Cold immersion
Speculum	Brass, Copper, Nickel,	Dilute solution of H_2SO_4 4% each strong acetic and nitric acids	Anodically

Strontianite. One of the chief mineral sources of the metal strontium. It is a carbonate of strontium $SrCO_3$ and occurs in crystals, which are isomorphous with aragonite, in veins with galena and barytes in Argyllshire (Scotland) and as nodules or geodes in limestone in the USA. Crystal system, orthorhombic; Mohs hardness 3.5-4, sp. gr. 3.6-3.7.

Strontium. A silver-white metal when pure, but may be coloured yellow due to the presence of strontium nitride. The principal sources are strontianite $SrCO_3$ and celestite $SrSO_4$. The former is found in Scotland and Westphalia, and the latter in Gloucestershire (England), Ontario and India. The metal is extracted by the electrolysis of the fused chloride, or from the oxide, by a thermal reduction process using aluminium as the reducing agent. The product is purified by distillation above 1,000 °C.

It resembles calcium in many of its properties, but is more reactive. It rapidly loses its lustre when exposed to the air and easily takes fire by heating or by friction, forming a mixture of oxide and nitride. Strontium compounds are added to pyrotechnic mixtures to produce a bright red flame colour. The metal is used as a de-oxidiser for copper and copper alloys and is a minor constituent of a number of non-ferrous alloys (e.g. lead strontium alloy for storage batteries, tin-strontium alloy as deoxidiser for copper alloys) in which it promotes sound castings. In the form of carbonate it is added as a flux (about 1-2%) to increase fluidity in the open-hearth furnace, and to assist in the removal of sulphur and phosphorus. Melting point is about 769 °C, volatilises at about 1,000 °C, sp. gr. 2.54; crystal structure, face-centred cubic; lattice constant, $a = 6.075 \times 10^{-10}$ m. It resembles lead in hardness. (*See* Table 2.)

Stub's wire gauge. A method of indicating the diameter of wire or tube by a series of numbers. Stub's steel wire gauge is used to a certain extent for tool steel, rod and wire and steel drill rod. Stub's iron-wire gauge is used mostly for non-ferrous seamless tube and to a certain extent for steel tube. This latter gauge is also called Birmingham iron-wire gauge.

Stud. The name used in reference to the variously shaped devices that are used in foundry-work to support cores or prevent their movement during the pouring operation. For steel castings they should be made of a low sulphur and low phosphorus content, and should be tin-coated and stored in a dry place. Galvanised studs should not be used in casting iron or steel, since the temperatures employed cause the zinc to vaporise and so eject molten metal from the mould. Ideally a stud should remain rigid until the metal in the mould is about to solidify. The stud then melts and becomes part of the casting (*see also* **Chaplet**). In engineering, a stud is a bolt without a head, and is threaded at both ends.

Studding. A term used in reference to a method of welding or brazing, when the parts to be joined are of intricate shape so that a heavy deposit of weld metal is required. The method is specially useful in butt-welding cast-iron. Holes are drilled and tapped to different depths and steel studs are screwed tightly into them, leaving a portion of the stud projecting. Weld metal is deposited round the studs and up to the joint and the weld is then completed by depositing metal to enclose all the studs and the joint (*see* Fig. 161). The studs are screwed into position, using a stud box.

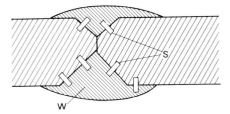

Fig. 161 *Studding. S is the stud; and W the weld metal.*

Stuffing box. Usually a cylindrical hollow box in a valve cover or cylinder. Through the box passes a valve-rod or piston-rod or a spindle, and the box is fitted with a suitable gland and packing to maintain the rod or spindle steam-tight.

Stupp. A term sometimes used in reference to the sooty product which forms in mercury smelting furnaces and which collects at the water seal of the condensers. It consists of 20-85% Hg together with soot, flue dust and other substances such as sulphur and compounds of arsenic and antimony.

Sub-critical annealing. The process of softening by heating to a temperature just below the Ac$_1$ point, holding at this temperature and then slowly cooling. The process is suitable for highly alloyed steels; the annealing temperature (after austenitising) being about 650 °C.

Sublimation. The process by which a solid substance passes into the vapour state from the solid state without passing through a liquid phase. This can occur when a substance at its melting point has a vapour pressure not far removed from the external pressure. By decreasing the pressure (e.g. by heating in a vacuum, or partial vacuum) many substances can be caused to volatilise, and thus the process of sublimation can be used for extraction and purification of substances. The latent heat of volatilisation at the sublimation point is equal to the sum of the heat of fusion and of the heat of evaporation.

Submerged arc welding. A welding process, and probably the most widely used automatic process, in which the direct current arc from a bare welding-rod is blanketed under a flux, which in a granular form is fed automatically to the welding area. The fusible material not only excludes the atmosphere, but also tends to retain the heat from the arc, so that high temperatures of welding are obtained.

Submicron. The name used in reference to one of the three kinds of small particles that may form colloidal solutions. Ordinary colloidal suspensions contain particles that are visible under the microscope, the size range being approximately 10^{-2} to 10^{-4} mm. Particles smaller than these may be visible by the aid of the ultramicroscope in the size range 10^{-1} μm to 10^{-1} nm. Such particles are known as submicrons.

Substitutional solid solution. The replacement of one material atom or ion by another in a lattice of the first, is called substitutional solid solution. It can occur only when the atoms involved tend to crystallise in the same type of structure and/or are of very similar ionic size. *See* **Size factor, Solid solution.**

Sub-zero treatment. The term used to describe the cooling of hardened steel to temperatures below 0 °C. The treatment is carried out by maintaining the steel at a temperature sufficiently below zero to promote the transformation of any retained austenite to martensite.

Sucking. A term used to describe local reductions in sectional area in wire-drawing.

Suction gas. A gaseous fluid obtained by permitting a limited quantity of air to pass over heated coke when the main products are carbon monoxide and the residual nitrogen from the air. The temperature of the reaction is high and tends to reduce the life of the fire-bars and furnace lining. Water is therefore allowed to drip on the fire-bars and accumulate in the ash-pit. The incoming air is thus charged with a little water vapour, and this reacts with the coke to give hydrogen and carbon monoxide. This reaction is endothermic and tends to keep down the furnace temperature. The gas contains about 24% CO and about 8% H$_2$. The name suction gas, for what is, in fact, producer gas, arises from the fact that the air that enters the furnace enters by virtue of the suction stroke of the engine, in distinction from methods where the air enters under slight pressure. (*See also* **Producer gas.**)

Sugar of lead. A name by which lead acetate (ethanoate)

$$(C_2H_3O_2)_2Pb.3H_2O$$

is known. It is a poisonous substance obtained by dissolving carbonate of lead in acetic (ethanoic) acid. This solution will dissolve litharge to form a basic ethanoate which is the basis of goulard water. Sugar of lead is sometimes used as a mordant.

Sull. The name applied to the rather thin coating of iron oxide (or hydroxide) which forms on steel wire after it has been treated with pickling acid, water-spray and dipping into milk of lime. In wire-drawing operations such wire is treated with a lubricant and the effectiveness of this is dependent on this coating which is the matrix in which the lubricants are contained.

Sullage. The name used to designate the scoria that accumulates at the surface of molten metal contained in a ladle.

Sulling. A process sometimes used in the preparation of steel rod for wire-drawing. After the steel, which is usually coiled, has been pickled in acid and thoroughly cleaned and freed from acid—sometimes using a suspension of milk of lime—it is suspended in wood racks, where it is kept moist by sprays of water. The surface

becomes coated with a faint green or brown film of oxide known as sull (q.v.). This latter assists lubrication and facilitates wire-drawing.

Sulphating. A term used in reference to a white deposit (lead sulphate) which may form on the lead plates of accumulators. This may happen if the battery is left uncharged for a long period. The hard lead sulphate is very difficult to remove and its presence seriously affects the efficiency of the battery. All storage cells should be given short periods of charging at regular intervals when they are not being used.

Sulphite lye. A residual solution obtained during the bleaching of paper. Sodium bisulphite is used as an antichlor after the paper has been bleached with chlorine. The lye is used as an organic binder for foundry sands and refractory bricks.

Sulphur. A yellow crystalline, non-metallic element of wide occurrence in nature. It is found native in certain volcanic regions (Sicily) or in the quicksands and rock some 275 m below the earth's surface in Louisiana. It is also extracted from alkali waste (from the Leblanc process) or from spent oxide (from town gas purification). It is used for making sulphur dioxide and thence sulphuric acid; for vulcanising rubber to prevent the natural substance from losing its elasticity; for making vulcanite and ebonite.

The stable form of sulphur is known as rhombic sulphur, but it can exist as monoclinic sulphur. There are three non-crystalline allotropic forms. It readily forms sulphides with most metals and sulphuretted hydrogen precipitates silver, lead, mercury, copper, bismuth, cadmium, arsenic, antimony, tin as sulphides in acid solution; iron, zinc and magnesium are precipitated as sulphides in alkaline solution. Sulphur causes red shortness in steel. Atomic number 16, at. wt. 32.06, m.pt. 112.8 °C, b.pt. 444.6 °C, sp. gr. 2.07; crystal system face-centred orthorhombic; lattice constants: $a = 10.50$, $b = 12.94$, $c = 24.60 \times 10^{-10}$ m. (*See* Table 2.)

Sulphur cement. These are usually mixtures of sulphur with various fillers that are resistant to water and acids. They are used mainly as cements for ceramic pipe connections. The fillers are usually silica, carbon, iron pyrites, blende and galena. One cement containing the last three substances mentioned goes under the name of Spence's metal.

Sulphur dome. The term used to describe the dome-like container which is mounted over the melting-pot in which is molten magnesium in die-casting mechanisms. The dome contains sulphur dioxide, which prevents the magnesium burning. Sulphur dioxide is often "poured" into the mould before casting magnesium and its alloys to produce the same inhibition.

Sulphur print. A test in which is reproduced on process paper the macro-distribution of sulphur as sulphides in a specimen.

Photographic (bromide) paper is soaked in dilute sulphuric acid and then laid on the surface to be examined. The acid liberates sulphuretted hydrogen from sulphides present in the metal, and this gas, reacting with the coating on the paper, precipitates silver sulphide. Dark brown stains on the paper indicate the distribution of sulphides.

Sunstone. A colourless variety of orthoclase spangled with minute crystals of hematite, limonite, ilmenite, etc. It is a potassium aluminium silicate sometimes cut and used as a gemstone.

Superalloys. A term applied to heat resisting alloys developed for gas turbines, and based on iron, nickel or cobalt.

Super bronze. A term used to designate a series of brasses that contain both aluminium and manganese. They are of the 70/30 type brasses modified by the substitution of aluminium, manganese and frequently some iron for some of the zinc. The approximate composition is Cu 69%, Zn 20%, Al 6.5%, Fe 2.5%, Mn 2%. These alloys have a much higher tensile strength and a higher hardness than the corresponding brass. They are used where high strength and corrosion resistance are required. (*See also* Table 5.)

Superconductivity. In certain metals, in very pure state, electrical resistivity drops to zero a few degrees above 0 K, and remains zero. Lead, tin, mercury and niobium are notable for this superconductivity. The change at a critical temperature is very sharp, so that a current started in a closed loop continues on removal of the electric field. One such example showed no decay in 2.5 years, equivalent to a resistivity better than 10^{-20} Ωm.

The origin of the phenomenon is believed to be connected with outer conduction electrons, and the electrostatic balance between attraction and repulsion forces. In the presence of screening by the valence and inner electrons, the repulsion between two conduction band electrons may be reduced. The positive ions in the lattice are attracted to electrons and are attracted by them. If a slight displacement of an ion in a lattice occurred on the passage of a conduction electron through that lattice, this would slightly increase the positive field near that electron, giving a net attraction towards another electron. These effects resemble the polarisation effects which lead to van der Waals' attraction, weak but significant. If this explanation is correct, the variation in mass of an ion, by using specific isotopes, should affect the thermal vibrations, and hence the electron/positive ion interaction. In tin and mercury, this is actually the case, the critical temperature for superconductivity varying as $M^{-\frac{1}{2}}$ where M is the atomic mass of the isotope. It is significant that the velocity of elastic waves and the frequency of atomic vibrations also vary as $M^{-\frac{1}{2}}$.

This apparent attraction between electrons as a result of small ion movements is strongest when a pair of electrons, with opposite momenta and spin, is involved. It seems that the electrons on the outer Fermi surface at the low temperature tend to correlate their motion in terms of spin and momentum, and then total energy, kinetic plus interaction energy, is reduced, and a finite gap exists between this organised state and the next available energy state, resembling a thin shell around the Fermi surface. The application of an electric field causes perturbation of the Fermi surface to contact this outer surface again, and a current flows. The asymmetric Fermi surface produces the available electron states but by a process different from normal, where the electron states are scattered and reach an equilibrium level leading to a given resistivity at that temperature. In superconductivity there is no extra scattering when the electric field is removed and hence current flow continues.

Intermetallic compounds have higher transition temperatures, again supporting the theory, because of electron interactions in compound formation. Nb_3Sn has a transition at 18 K. Superconducting materials have already been applied to large electromagnets and switches (cryotrons). Computer memory storages can be closed loops of superconducting material.

Supercooling. The process of lowering the temperature of a molten metal to temperatures below the liquidus without permitting solidification. Supercooling is promoted by rapid rates of cooling and by the absence of agitation and of particles which could act as nuclei. When a metal or alloy is subjected to a small or moderate degree of supercooling, crystallisation can be initiated by adding a small crystal of the same or of an isomorphous material. In this way large single crystals are produced. With a greater degree of supercooling the molten metal becomes subject to spontaneous crystallisation by agitation, and the solid so produced consists of exceedingly fine grains. Supercooling induced by rapid rates of cooling thus ensures refining of grain size. Ultrasonic vibration of the supercooled liquid produces a grain refinement.

Superficial expansion. The coefficient of area expansion, being the ratio of the increase in area to the original area due to a rise in temperature of 1 °C. It is very nearly twice the coefficient of linear expansion.

Superheat. Temperature levels above the melting point. The trade name of a molybdenum-base powder alloy used as an electrical resistance material for service at temperatures up to 1,600-1,700 °C.

Superheating. A process used to produce fine grain size in certain magnesium alloy castings by heating the metal some 150-250 °C above its melting point—i.e. to temperatures in the range of 820-890 °C.

Superlattice. A crystal structure in which different atoms in a solid solution fall into a regular or ordered arrangement. The foreign atoms are imposed in regular positions on an original lattice of the solute atoms with larger lattice spacing than the host.

Superlattice formation occurs when free energy is reduced by so doing, and is most frequent when a degree of lattice strain exists in a random substitutional solid solution, as with copper-zinc. Increase of temperature destroys the superlattice, which can be reformed on slow cooling again.

Supersaturated. A term used in reference to a metastable solution in which the amount of solute in the molten material actually exceeds that which can be retained in solution under equilibrium conditions. Supersaturation can be brought about by very rapid

cooling to a temperature below the normal freezing point. Inoculation consists in adding a crystal of the solute when crystallisation of the material takes place rapidly and the latent heat evolved raises the temperature to the melting point of the normal solution.

Supersonic testing. Synonymous with ultrasonic testing (q.v.). The latter term is preferable. (*See also* **Non-destructive testing.**)

Surface-active agents. Substances that reduce the surface tension of aqueous solutions such as pickling and cleaning baths and which act as wetting agents in detergent solutions for degreasing. Alternatively called wetting agents (q.v.) or surfactants.

Surface checking. The term used to describe the general cracking or crazing over the surface of a metal. It may be caused by excessive cold working or by localised corrosion, particularly at grain boundaries.

Surface coating. A thin layer of material coated on the the substrate by plating, spraying, or diffusion processes, vacuum or chemical deposition. Used for protection, decoration or influence on properties of electrical or optical nature.

Surface condenser. This consists of a vessel in which are a number of tubes and through which cold water is pumped. Into the vessel steam exhausted from an engine is admitted and condenses on contact with the cold surface of the tubes, which are usually made of brass. In marine installations the cooling water may be sea-water and the condensate may be used as boiler feed, provided it is degreased before readmission to the boiler. The condenser tubes are commonly made of 70/30 brass containing 1% or 2% Sn.

Surface energy. Literally the energy state at the surface of a material. Because a surface has no balancing forces on one side of it, whether the adjacent phase be liquid or gas, the energy must be higher than in the interior, otherwise there would be dispersion from the surface, as indeed happens with rise in temperature when a liquid vaporises or a solid melts.

In liquids, the extra energy is manifest as surface tension, giving rise to phenomena such as the meniscus between a liquid surface and its container, and the formation of liquid droplets on a non-wetting solid surface.

In solids the energy has to be overcome to produce vaporisation, emission as ions, or to increase the radius of the surface. Surfaces are therefore unusual states of matter, with properties additive to the material properties. The wetting of solid surfaces by liquids is technologically of great importance in coating and finishing processes; in corrosion; joining by soldering, brazing, welding and adhesives; and in cleaning to remove contamination. Particle coarsening and spheroidisation occur to reduce surface energy.

Detergents lower the surface energy of a contaminated surface and lower the surface tension of cleaning liquids involved. They also attach themselves to high surface-tension materials such as greases and oils, and so assist in detaching them from contaminated surfaces by creating what is virtually a new surface around the contaminant.

Surface finish. No solid surface consists of a regular plane of atoms except over very small areas. Inevitably, the solidification from a melt, or the deposition from vapour, cannot produce a perfect surface. There are therefore irregularities on the surface on the atomic scale, vacant lattice sites and asperities of extra atoms protruding above the surface. Added to this are the irregularities on a microscopic scale, produced by changes in orientation at grain boundaries (crystallites), inclusions, porosity, corrosive attack and mechanical damage. There are, after that, macroscopic surface defects produced by fabrication, handling and gross corrosion by liquid or gas attack.

To improve appearance, some materials are given special surface treatments, by grinding, polishing and buffing, others are left in machined, rolled, drawn or forged condition, depending on application. Surface treatments such as electroplating are a special case, usually to improve service performance or aesthetic appearance. In other cases, surface coatings may be for specific properties, e.g. metallisation of plastic film for production of capacitors, reflecting surfaces, resistors, etc.

Macroscopic defects are obvious and can be remedied by further treatment. Microscopic defects are much less obvious, but can lead to mechanical problems such as the creation of cracks in tension or fatigue stressing condition. Atomic scale surface finish is almost impossible to control except by very expensive and slow processing. Surface finish is therefore of importance to service life of materials and to aesthetic appeal in sales. One great advantage of plastics is the ease with which good surface finish on the macroscopic scale can be obtained during manufacture. This is largely as a result of the lower temperatures involved, a factor which also leads to good finish on low melting point metal alloy castings. The engineering importance of surface finish on fatigue resistance justifies special attention to highly stressed components, and the development of new machining methods such as chemical machining by etching.

Surface hardening. The term used in reference to any process in which the hardness of the outer surface is increased compared with the core of the specimen. There are two types of treatment that will give an enhanced hardness at the surface, viz. heat-treatment methods such as case-hardening, nitriding, carbo-nitriding, cyaniding, carburising, flame hardening, induction hardening (q.v.). Mechanical methods include peening and tumbling, as well as cold working, e.g. drawing or rolling.

Surface protection. The protection of solid surfaces against corrosion and oxidation is obviously involved in applications for chemical plant and processes, marine environment, high temperature operations, and external atmosphere exposure as in buildings, vehicles, cable towers and poles, fences and barriers, road and paving surfaces. The majority of buildings and road surfaces use traditional materials which require no protection, but accessories such as posts, signs, window frames, doors, etc. still require special attention.

Protection systems include coatings and plating, with specialised applications such as sacrificial anode protection for marine vessels and installations, chemical plant and pipelines. The simplest and cheapest coating is any protective oxide film which forms on metals, but only aluminium and chromium are reasonably applicable for this purpose. Chromium is comparatively rare and expensive, and too brittle to use as a single metal. It is therefore used in so-called stainless steel, which involves high processing costs. Aluminium is an abundant metal, not so easy to extract, but gaining ground as a material of construction. The next simplest coating, but much less effective, is paint, although developments in polymeric materials have enormously improved the efficiency and endurance of paint protection. The plating and coating methods are more expensive, but obviously justified for engineering applications.

Protection may be rendered unnecessary by development of less corrosive materials, e.g. plastics can replace mild steel in the casing of domestic equipment, vehicles and marine vessels. The use of plastics may depend in future on the world availability of petrochemicals, and the decision to use less fossil fuel for combustion and more as a valuable chemical in the manufacture of durable products. (*See also* **Metallisation, Plating, Paint, Spraying.**)

Surface roughness. A term used in reference to the occurrence of minute, finely spaced irregularities on the surface of a metal, that appear in profile as a series of peaks and valleys. A rectangular box containing a light-source is sometimes used in roughness observations. The box has a number of square holes, and the article to be assessed is placed in front of the holes. The degree of distortion of outline or of shape is an indication of the roughness.

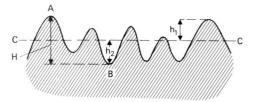

Fig. 162 *Surface roughness. C-C is the average depth; and H the maximum depth.*

The absolute measurement of roughness varies in different countries. In Fig. 162, A is the level of the highest peak, B is the level of the deepest valley, and C is the mean level of peak and valley. The German standard of roughness measurement is the value of $AB (= h_{\max})$ in μm; a French standard is based on the value $AC (= h_{av})$. If h_1, h_2, h_3, etc., are the actual deviations of peaks from the mean level C, then

$$\left(\frac{h_1^2 + h_2^2 + h_3^2 + \ldots}{n} \right)^{\frac{1}{2}}$$

is the root mean square roughness height, and this value is represented by h_{rms} and is the basis of an American standard. A British and some continental standards are based on the arithmetic mean of the deviations from the mean level C, and in this case

$$h_{b.s.} = \frac{h_1 + h_2 + h_3 + \ldots}{n}$$

In practice it is rarely necessary to determine absolute roughness, though it is frequently required to have a certain smoothness of finish. This may be done consistently by comparing the roughness of the component with that of a standard piece. A piezo-electric element is used, one end of which has a comparatively large radius, so that it maintains a fairly constant height as it moves over the surface to be tested. The other end carries a fine point which can follow the profile of peaks and valleys. As the element is moved over the surface it bends in accordance with the irregularities and generates a small potential difference which is detected, amplified and then indicated as a roughness height on a suitable scale.

Surface tension. Due to the unbalanced attraction of the molecules at the surface of a liquid by the molecules in the interior of the liquid, the surface behaves as though it is covered by an elastic membrane in tension. The surface therefore tends to contract. The force in Newtons measured along 1 metre length in the surface of the liquid is the surface tension γ. Surface tension may be measured by:

(*a*) the rise of liquid in a capillary tube. If the height of the liquid of density d in a tube of radius r is h, $\gamma = rhdg/2$;

(*b*) the stalagmometer. This apparatus is mainly used for comparing surface tensions of liquids. If the weights of, say, 10 drops (slowly forming at the end of a tube and then dropping) of each of two liquids be m_1 and m_2, $\gamma_1/\gamma_2 = m_1/m_2$;

(*c*) the tensiometric method, in which is measured the force required to pull a small horizontal ring out of the surface of the liquid. If $P = \sqrt[4]{\gamma} \times M$, where M is the molecular volume, P is the parachor, a quantity that proves very useful in determining the molecular structure of a substance. (*See also* **Parachor.**)

Surfusion. A term sometimes used as synonymous with supercooling when applied to molten metals and alloys and their subsequent behaviour on seeding.

Surging. A term applied to the sudden increase or decrease in the strength or voltage of an electric current.

Susceptibility. Magnetic susceptibility is defined as the ratio of intensity of magnetisation I (induced magnetic field) to the magnetising force H (applied magnetic field)

$$\chi = \frac{I}{H}$$

For diamagnetics, χ is negative; for paramagnetics, positive, but variable with temperature

$$\chi_p = \frac{C}{T}$$

where C is the Curie constant, T the absolute temperature.

Susceptibility is related to magnetic permeability μ by the formula

$$\chi = \frac{(\mu - 1)}{4\pi}.$$

Dielectric susceptibility is the ratio of charge density in the polarisation of materials in a capacitor. Polarisation is analogous to magnetisation, and the symbol χ is used (χ_D for dielectric susceptibility)

$$\chi_D = \frac{\text{density of bound charges}}{\text{density of free charges}} = (\epsilon_R - 1)$$

where ϵ_R is relative permittivity or dielectric constant.

Suspension. Usually a two-phase system, consisting of small solid particles distributed in a liquid dispersion medium.

Swab. Loose soft material of an absorbent nature used to apply small amounts of liquid to a mould or metal surface. In foundry-work a swab is often made of hemp or similar material and bound loosely, so that it can be used in the same way as a brush.

Swabbing. The operation of applying washes, liquid facings or dressings to the mould face or to cores by means of a swab. The term is also applied to the operation of applying a lubricant to die surfaces.

Swage block. A block of cast iron or steel used for holding a swage for forming heads on bolts and similar purposes.

Swaging. A process for reducing the diameter or for forming steel and other metals; it may be performed by hand, using special smith's tools, or by machine. In either case the work is struck a large number of successive blows by a pair of dies shaped to give the required form. The method is used chiefly for reducing the diameter of wires, rods and tubes, for tapering rods and tubes, and in forging stock into shapes having the opposite sides concentric (Fig. 163).

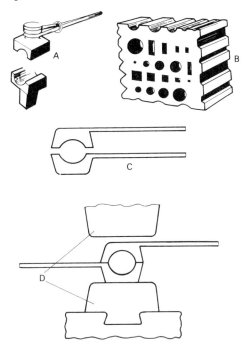

Fig. 163 *Swage block and swaging. A is the hand swage; B the swage block; C the hammer swage; and D the die blocks.*

Swarf. The fine particles of metal produced in the operations of machining castings or grinding components.

Sweating. A term used to denote the slight leakage that may occur at a seam or at a rivet of a vessel under pressure. The term is also used in reference to the joining of metals by soldering without the use of a soldering iron. The parts to be soldered are usually tinned or covered with solder. They are heated to a temperature sufficiently high to melt the solder, which is allowed to run and unite the parts that are being pressed together.

The term is also used in reference to the appearance on the surface of an alloy of small beads of a single constituent that has a relatively low melting point and which may be the last to freeze. (*See also* **Tin sweat.**)

Swedish iron. A term which refers to a wrought iron made in Sweden from cold blast iron, using a relatively pure iron ore. The average value of tensile strength is 294 MN m^{-2} and the elongation is about 33%. Also called Swedish Lancashire iron.

Sweep. A term used in reference to the strickle or striking board used in making sand moulds. The name also refers to a surface whose length is greater than the breadth and which is flat in the direction of the longer dimension and slightly curved (large radius of curvature) in the other directions.

The sweepings from floor or table of metal that is re-processed for the recovery of the valuable metal content are referred to as "sweeps".

Swell. A term used in reference to the parts of a casting which are oversize. This may be caused by a local displacement of the face of the sand, which has not been sufficiently rammed, by the pressure of the fluid metal. A further cause may be the lifting of the cope during casting due to the lack of sufficient weights on the cope. The swell, or swelling, usually occurs on the top side of the casting.

Swift. A rotating device on which coils of wire are supported for unwinding.

Switch copper. A term sometimes used in reference to electrolytic copper in the form of shaped bars having a burnished finish. Such bars are used in electrical apparatus and particularly for switch blades.

Swording. A term used to describe a method of separating pack rolled-steel sheets which have stuck together. The sheets are held apart to the area where the sticking occurs and the parting is then effected by striking with a long broad-bladed tool made of sheet iron.

Sylvanite. A telluride of gold and silver (Au,Ag)Te$_2$, which sometimes contains also a little antimony and lead. It is used as a source of all three major constituents. It occurs as veins, impregnations and lenticular deposits in Colorado, Western Australia and Transylvania (Romania). Crystal system, monoclinic; Mohs hardness 1.5-2, sp. gr. 5.8-8.2. Also called graphic tellurium.

Synthetic reaction. A term sometimes used in reference to an isothermal reversible reaction that occurs in binary alloy systems, where two liquids react together to form a solid phase during solidification.

T

Tabling. A method used in ore-concentration processes by which minerals are stratified by agitation in water and then separated. The tables are slightly inclined to the horizontal and have a riffled portion and a smooth portion. They are shaken rapidly backwards and forwards in the direction of the length while the ore pulp is fed across table. By tabling it is possible to obtain a satisfactory concentration, but the size limits of the ore used must usually lie within the range of 0.08-0.008 in. (2-0.2 mm). (*See also* **Wilfley table.**)

Tabor abraser. A device for measuring the abrasion resistance of metal finishes. It consists of a pair of small abrasive wheels which wear a circular track on a rotating test piece.

Tabular alumina. The product of calcining ordinary alumina at temperatures below the fusion point. In the form of a fine powder it is used in the manufacture of electrical insulators, while the coarser varieties are used in the making of refractory bricks, linings, etc. In larger sizes 5-15 mm the material is used in heat exchangers and in the beds of chemical plant where high temperature or catalytic reactions occur.

Tachometer. An instrument for measuring the speed of rotation, or in some forms for measuring variations in speed of rotation. It is used in association with pumps and the drives of rolling-mills.

Tachyol. A name sometimes used for silver fluoride AgF. In the dry pulverised state the salt absorbs nearly 850 times its own volume of ammonia. It has been proposed for use in purifying drinking water.

Tack weld. A type of weld that is used to assist in the assembly of parts. In order to relieve the operator of the necessity of maintaining the welding edges at the correct distance apart and in proper alignment and position, a series of short welds at convenient distances apart are laid down. Usually the welds are made rather longer than the thickness of the plates and about three-quarters of the thickness of the plate in depth. These sizes may be doubled in tack welding tubes. Tack welds are usually made from 15-25 cm apart.

Taconite. An iron ore consisting of a ferruginous chert derived from iron II silicate. It is a hard material and presents some difficulties in mining, but when leached by hydrothermal methods yields a satisfactory ore, since most of the silica is eliminated and the iron remains as hematite.

Taenite. A name originally applied to tin plate that was either undersize or below gauge of the rest of the plate in the pack. It is now sometimes used to describe all light gauge tin plate, particularly that used for making metal tags.

Tagging. A term used in reference to a swaging operation that consists in tapering or pointing the end of a rod or tube so that it can be passed through a drawing die.

Tailing. In the various operations in ore-dressing the object is to reduce the ratio of gangue to valuable mineral. Thus one or more concentrates are produced, while at the same time a residual product results which contains little or no useful mineral, and which can be discarded. This latter is the tailing. In practice a varying amount of useful mineral may be lost in the tailing.

Tail print. The name used in reference to the type of projection on a core used in moulding, when a hole is to be cored below the mould joint. Also called D prints (*see* Fig. 164).

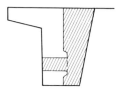

Fig. 164 *Tail print.*

Tainton process. A process designed to produce a high recovery percentage of silver and lead, which is especially suitable for treating the residues from zinc leaching during the refining of lead. The residues (or in some cases the lead ores) are heated in a regulated supply of air to convert the sulphides into sulphates. The soluble sulphates are then washed out, leaving insoluble lead, and silver sulphate, and these latter are treated with brine containing free chlorine. The resulting solution of silver and lead is electrolysed, using iron cathodes and graphite anodes in a diaphragm cell.

Talbot ingot process. A process in which ingots are removed from the soaking pit before the central portion of the ingot has solidified. It is then partly reduced in a blooming-mill in order to close up internal cavities, and after this preliminary treatment is returned to the soaking-pit, where it remains until it has attained the condition in which it can be rolled in the routine way. Ingots treated in this way are sometimes said to have been "liquid squeezed".

Talbot process. A rapid, continuous steelmaking process carried out in a large-capacity tilting basic open-hearth furnace. It is a pig and ore process in which ore and lime are added to the charge in the furnace. These are heated until a highly fluid slag is formed, when the molten pig iron is run in. A violent reaction occurs, during which there is a rapid oxidation of the impurities in the pig iron and they pass into the slag. In this way most of the phosphorus and silicon is eliminated. After a large part of the slag has been run out of the furnace, about one-third of the steel is tapped. The recharging then takes place and the cycle is repeated. In this process it is essential to maintain a good part of the molten metal in the furnace to dilute the impurities contained in the addition of pig iron, and to maintain a high temperature to keep the slag fluid. It is sometimes called Talbot continuous process. It can deal successfully with phosphoric blast-furnace iron.

Talc. A hydrated silicate of magnesium $3MgO.4SiO_2.H_2O$. It usually occurs in massive form with a foliated structure which permits it to be split into flexible plates. Steatite, or soapstone, is a massive variety. The various forms find uses in ornamental ware, French chalk, etc. Mohs hardness 1, sp. gr. 2.7-2.8.

Talc is a typical layer lattice, in which silica tetrahedra in planar rings fit into a similar planar structure of hydrated magnesium oxide, the planes of magnesium ions being associated with oxygen and hydroxyl ions.

Talysurf. An instrument for measuring surface finish (*see* **Surface roughness**). The vertical movements of a stylus as it traverses the peaks and valleys in the surface are amplified and recorded as a profile record, or may be indicated as an automatic average of the height of the irregularities.

Tam alloy. A series of ferro-titanium alloys that contain 15-21% Ti with low silicon and aluminium content. Carbon varies widely from 3.5% to 8%. They are used in the grain refining of steel and have high melting temperatures—in the region of 1,500 °C.

Tamping. A term usually employed in reference to the stopping of a hole with a plug of clay. In mining and quarrying it refers to the pressing of the explosive charge into the drilled hole in the rock. The pressing operation consists of repeated gentle applications of pressure until the explosive can be assumed to completely fill the hole.

Tandem mills. A series of individual rolling-stands arranged in line and synchronised so that all the rolls operate as a single unit. The rolls have a fixed opening, and each set revolves faster than the preceding set, in order to be able to deal with the increase in length consequent on each reduction. Tandem mills are used in both hot- and cold-rolling processes.

Tang. That part of a hand tool that is intended to be fitted into a wooden handle.

Tangent galvanometer. A laboratory instrument sometimes used for measuring electric current. It consists of a few turns of wire mounted on a vertical circular frame. At the centre of this circle a small magnetised needle is mounted so that it can rotate in a horizontal plane. A light aluminium pointer attached to the needle facilitates the accurate reading of the deflection of the needle when a current passes through the coil. The needle is made small, so that it may be assumed that the magnetic field due to the current is uniform in the space occupied by the needle. In use the vertical coil (and the needle) is set in the magnetic meridian. If the horizontal component of the earth's magnetism is H, and there are n turns of wire in the coil of radius r:

$$\text{Current } (I) = 10H \cdot \frac{r}{2\pi n} \cdot \tan \theta$$

where θ = angle of deflection of the needle.

The quantity $r/2\pi n$ depends only on the design of the instrument and is called the constant of the galvanometer. The quantity $Hr/2\pi n$ depends on the location of the instrument. It can be determined by passing a known current through the instrument and noting the corresponding deflection of the needle. $Hr/2\pi n$ is known as the reduction factor of the instrument and is denoted by K. $I = 10K \tan \theta$ (amperes).

Tantalite. An important mineral which consists of the tantalate and niobate of iron and manganese. It is found in Sweden, Finland, France, Nigeria, Western Australia and the USA. It frequently occurs as small masses in granitic rocks.

The mineral is opaque, brown to black in colour and has a metallic lustre. The Mohs hardness is about 6.

The tantalite of Nigeria occurs in association with cassiterite and that in Western Australia with wolfram also. The different metals are usually separated magnetically after crushing the ore. When the percentage of niobium exceeds that of tantalum in the ore, the latter is called columbite (q.v.).

Tantalum. A platinum-white metal that resists the effects of water and air at ordinary temperatures and is not affected by common acids. It is, however, dissolved by hydrofluoric and fuming sulphuric acid, and by caustic alkalis.

A number of methods have been proposed for the extraction

of the metal, including the electrolysis of the fused fluoride, the reduction of potassium fluotantalate with potassium, and the reduction of the pentachloride with magnesium in the presence of potassium chloride.

Tantalum is precipitated from solution, by hydrochloric acid as the pentoxide. The pentoxide can be converted into the pentachloride by heating with charcoal in an atmosphere of chlorine and carbon tetrachloride at 650 °C. The chloride is mixed with potassium chloride and magnesium chips and heated in an iron tube to 750 °C. Finely powdered tantalum sinks to the bottom of the melt, and after washing with water and dilute hydrochloric acid is dried *in vacuo.*

Tantalum can be worked to a considerable extent without annealing and is used in the form of sheet, tube and wire in surgical appliances, chemical equipment, vacuum tubes, current rectifiers, jewellery, etc. It forms useful alloys with certain non-ferrous metals, e.g. tantalum-tungsten alloys find employment in electrode springs and as pen-nibs; tantalum nickel alloys are used in electronic equipment; and tantalum-nickel carbide, often mixed with other carbides and cobalt, is used in especially hard cutting tools and dies.

Since tantalum forms a very stable carbide that is not soluble in iron, the addition of about 1% or even less of tantalum (in the form of ferro-tantalum) is usually sufficient to prevent weld decay in chromium steels. The presence of tantalum in steel that is to be nitrided greatly increases the rate of growth of the hard nitrided layer.

Sintered tantalum can be produced as a coarse, porous mass, and when this is coated with tantalum oxide an excellent electrolytic condenser is obtained. Such condensers are used for lightning and surge arresters in railway signal circuits, and for the protection of the electric signal lamps.

Atomic weight 180.95, at. no. 73, sp. gr. 16.6, m.pt. 2,996 ± 50 °C, crystal structure, body-centred cubic; lattice constant, $a = 3.2959 \times 10^{-10}$ m. (*See* Table 2.)

Tantalum carbide. When carbon is added to a melt of ferro-tantalum, a pasty mass consisting largely of tantalum carbide is formed. When this mass is washed with hydrochloric acid a fine powder separates. This is the tantalum carbide used as an abrasive or as a basis from which are made sintered carbide alloy tools.

If tantalum is added to carbon steel, the iron carbide is reduced, and after hardening by quenching the tantalum carbide appears. Such steels are useful for making high-speed cutting tools.

Tantalum tool alloys. These are mixtures of carbides of tantalum, chromium, nickel and sometimes iron in a matrix of cobalt which can be used as substitutes for tungsten carbide.

Tap bars. When the cementation pots or boxes are being packed with alternate layers of steel bar and charcoal, a few bars are set with their ends protruding from one end of the box. These are the trial or tap bars, one of which is removed when it is considered that the carburising has proceeded to the desired extent. On cooling, the tap bar is fractured, and from the appearance of the fractured surface the progress of the cementation is judged.

Tap cinder. The name given to the slag produced in the puddling furnace and consisting mainly of ferrous (iron II) silicate. The average composition showed about 53.5% Fe, mostly in the iron II condition. The proportions of iron II to iron III were frequently about 8 to 1. In hammer scale the proportions were 3 to 1. Tap cinder when roasted was known as "bull dog" and was used in fettling the puddling furnace.

Tape, magnetic. Recording on spools of tape reduced storage space requirements, and several tracks can be recorded on one width provided that the magnetic material has a high remanence and coercive force. Polyester tapes can be made strong but very thin, and are coated with ferrimagnetic material such as ferrites, having the appropriate magnetic properties. Chromium dioxide gives improved fine detail and is used for high-sensitivity musical recording.

Taper. The gradual reduction in cross-sectional area from one end of a specimen to the other. The term also refers to the allowance made in patterns for moulding to permit of easy withdrawal. The taper in the case of large metal patterns is ½° and for small wooden patterns 1½°.

Tap-hole. The hole near the base of a furnace, usually plugged with clay, and which when unplugged permits the outflow of molten metal from the furnace or cupola.

Tapping. The operation of forming the opening near the base of a furnace—usually by unplugging—to allow the outflow of molten metal. This is done by using a pointed bar or bott stick to pierce the plug of fireclay which covers the tap-hole.

Tapping bar. A pointed bar used in the operation of tapping a blast furnace. The fireclay plug in the tap-hole is pierced by forcing the tapping bar into it. Also called a bott stick.

Tarnac. A compounded material used as a substitute for natural rutile in welding electrode coatings. It contains approximately 70% titanium oxide TiO_2.

Tarnish. The term applied to the surface discoloration, due usually to a thin film of oxide or sulphide, which occurs on metals exposed to the atmosphere. In the case of copper, exposure to air free from sulphur produces an almost invisible film of oxide, and thus has no tarnishing effect. There is a critical concentration of sulphuretted hydrogen below which tarnishing ceases. In industrial atmospheres the mixed film of oxide and sulphide is porous and affords little protection to the metal underneath. The green patina formed on copper is a basic compound formed by the oxidation of the copper sulphide. The tarnishing of silver is due to the formation of silver sulphide.

Tasmanite. A shale of Tertiary age, found in Tasmania, which contains a high proportion (about 20%) of organic matter in the form of spores. It is sometimes referred to as Australian white coal.

Taylor process. A method of producing fine wire by threading thick wire through a glass tube and stretching the two together at high temperature.

Taylor-White process. A heat-treatment process applicable to high-speed steel. The steel is heated to a temperature in the region of 1,400 °C and then quenched in a bath of molten lead till the temperature lies between 740 °C and 840 °C. The steel is then further quenched in oil down to atmospheric temperature. The tempering treatment which follows the above consists in heating to within the range 390-660 °C and allowing to cool slowly in air. The actual temperature to be used in the process depends on the composition of the steel. The process was originally related to the treatment of an 8% W, 1.8% Cr steel.

Technetium. A rare element obtained by the bombardment of molybdenum with neutrons. It has the atomic number 43 and in some of its properties resembles manganese. Also called masurium. Atomic weight 98.91, m.pt. 2,172 °C. (See Table 2.)

Technic metal. An alloy used in laboratories and dental schools as a practice metal in place of more expensive alloys, and the constituents are so chosen that the alloy shall have similar melting properties to dental alloys. The composition varies, but it is usually a copper-base alloy containing tin, nickel, silver and iron.

Technotherm-Rokos process. A fusion welding process. Fusion is effected by an electric current passing through the work and a carbon electrode which runs along the edge of the work with its tip in the melt. The edges of the work are preheated by a gas flame.

Teeming. The operation of pouring molten metal from a furnace or a ladle into ingot moulds, etc. Also called pouring.

Teeming lag. The name used in reference to the fold in the surface of an ingot produced during casting, when the rising surface of the metal begins to solidify and form a rim which adheres to the walls of the mould, and is subsequently engulfed in the ingot surface.

Telloy. A name given to finely powdered tellurium intended for use in rubber compounding. When used as a secondary vulcanising agent with sulphur, it increases the tensile strength and the ageing properties of the rubber and gives it greater heat resistance.

Telluric bismuth. A telluride of bismuth, also known as tetradymite.

Telluric ochre. An impure form of tellurium dioxide found native in Transylvania (Romania). It is sparingly soluble in water and does not give an acid reaction. Also called tellurite.

Tellurite. A name synonymous with telluric ochre (q.v.).

Tellurium. A white lustrous material resembling antimony in its crystalline appearance. It resembles sulphur more closely in its chemical reactions and is generally classed as a non-metal. It occurs

in small quantities in the native state in Transylvania (Romania), Hungary, California, Colorado, Brazil, Virginia and Western Australia. It is usually found in combination with metals, e.g. sylvanite, $(Ag.Au)Te_2$, nagyagite $(Au.Pb)_2 (TeS)_3 Sb_3$; hessite $Ag_2 Te$ and tetradymite $Bi_2 Te_3$.

When prepared from the alkali residues from bismuth ores, the mass is treated with hydrochloric acid and then with sodium sulphite, which precipitates the tellurium. It is used to improve the machinability of "free-cutting" copper and certain steels. Additions of tellurium improve the creep strength of tin and the mechanical properties of lead. The element acts as a powerful carbide stabiliser in cast irons and is used to increase the depth of chill in car wheels. The chief use of tellurium is, however, in the vulcanising of rubber, since it reduces the time of curing and endows the rubber with increased resistance to heat and abrasion. Atomic weight 127.6, at. no. 52; m.pt. 450 °C, b.pt. 1,990 °C, sp. gr. 6.25; crystal system, hexagonal close-packed; lattice constants, $a = 4.4558, c = 5.9268 \times 10^{-10}$ m. (See Table 2.)

Tellurium copper. A free-machining copper alloy containing 1% Te. It has an electric conductivity 98% that of pure copper and a mechanical strength (tensile) of about 216MN m^{-2}. The master alloy of tellurium and copper also known by this name contains 50% Te and 50% Cu.

Tellurium lead. A lead alloy containing usually less than 0.1% Te. The addition of tellurium hardens and toughens the lead, increases the endurance limit, refines the grain structure and improves the resistance of the metal to acid corrosion—particularly sulphuric acid corrosion. Tellurium lead finds uses in chemical plant and as cable sheathing. The presence of the tellurium so improves the toughness that tellurium-lead pipes will withstand a hydraulic pressure 75% greater than similar pipes of unalloyed lead. Patented tellurium-lead alloys may contain 0.05% Te and up to 6% Sb.

Temper. A term used in reference to the mechanical properties, such as strength, hardness and toughness, of a metal, or to the processes by which such properties are induced (see **Tempering**). In steelmaking the term may refer to the condition of the metal as indicated by the colour in the heat-treatment process (see **Temper colour**). However, temper frequently means the percentage of carbon in the steel, and the terms used, with the corresponding carbon content, are given in Table 57.

TABLE 57. *Percentage of Carbon in Steel for Various Types of Temper*

% C	Type of Temper
0.75	Die
0.875	Sett
1.0	Chisel
1.13	Spindle
1.25	Tool
1.375	Saw-file
1.5	Razor

Temper may also refer to the degree of hardness induced in wrought metals, particularly non-ferrous metals, by cold working.

Since both heat-treatment and mechanical working affect the temper, it is usual to describe sheet or strip metal as "soft temper" when in the fully annealed condition (annealing being a softening operation) and as "hard temper" after heavy cold reduction as the final treatment. Intermediate tempers are known as "skin pass", quarter hard, half hard, etc. The percentage increase in length of steel sheet resulting from cold working (rolling) is also referred to as temper.

Temperature. An arbitrary measure of heat energy condition, which determines the flow of heat energy into or out of a body. If heat energy flows into the body from a second body in contact with it, the first body is said to be at a lower temperature than the second.

It is particularly a manifestation of the average kinetic energy of thermal vibration of atoms of a substance. Higher temperature refers to higher energy of vibration, and lower temperature to decreased vibration. As temperature falls, thermal vibration is drastically reduced, until a critical limit, the absolute zero of temperature, when all energy of vibration has been lost. As temperature increases, vibration becomes so large that changes of state occur, solids become liquids, with a different structure and therefore modes of vibration, and liquids become gases, with little or no structure and considerable energy of vibration.

The fundamental scale of temperature is the absolute, thermodynamic or Kelvin scale, based on the average kinetic energy per molecule of a perfect gas. This scale is determined by the equation

$$\theta(X) = 273.15 \text{ K.} \frac{X}{X_3}$$

where θ is the absolute temperature, X the thermometric property involved (pressure or volume) and X_3 is that thermometric property associated with the triple point of water, arbitrarily fixed as 273.15 K. The symbol K (degrees Kelvin) is the unit of absolute temperature, used without the conventional degree symbol in the SI unit system, to avoid confusion.

The ideal gas temperature, numerically equal to absolute temperature, can then be defined by the following two equations

$$\theta = 273.15 \lim_{P_3 \to 0} \frac{P}{P_3} \text{ (const. } V)$$

or

$$\theta = 273.15 \lim_{P \to 0} \frac{V}{V_3} \text{ (const. } P)$$

The constant volume hydrogen gas thermometer is accepted by the International Bureau of Weights and Measures as the experimental standard.

On the Kelvin scale, the melting point of pure ice is 273.15 K, and the Celsius scale of temperature is the practical scale of temperature whose zero is the ice point, 273.15 K, and whose temperature intervals are identical with those of the Kelvin scale. The Celsius temperatures are written °C. The centigrade scale is identical with the Celsius, but confusion occurred in France, where centigrade has a meaning relating to angles, and "Celsius" is the agreed SI practical term. On the Celsius scale, absolute zero is −273.15 °C, and the boiling point of water 100 °C.

The old Fahrenheit scale, based on dubious fixed points, is obsolete. (*See also* **Critical constants, Critical point (arrest points)**.)

Temperature gradient. The rate at which the temperature of a body rises or falls. If the change of temperature of a body be plotted against time, the slope of the curve so obtained at any time is given by the tangent to the curve at that instant. The term is also used in reference to the difference in temperature that exists between one part of a hot body and another part, e.g. the temperature gradient between the outside and the centre of an ingot; the temperature gradient between the inner and outer surfaces of a firebox or steam boiler.

Temper brittleness. A form of brittleness produced when certain steels (e.g. nickel-chrome steels) are held in, or slowly cooled through a certain range of temperature below the lower critical temperature. The range of temperature producing this undesirable effect is 600-300 °C. The effect is due to grain-boundary weakness which may result, *inter alia*, from the precipitation of carbides. The effect is overcome by adding 0.2-0.3% Mo, which has a powerful stabilising effect.

Temper carbon. A term sometimes used in reference to the free or graphitic carbon that has been precipitated from solid solution of iron-carbon alloys. (*See also* **Spheroidisation**.)

Temper colour. On heating plain carbon steels in presence of air, a thin film of oxide forms on the surface. This colour varies with the temperature and may be taken as an approximate indication of the temperature of the metal. The actual colour attained also depends on the time the metal has been heated and on the composition. Chromium steels, for example, are not nearly so susceptible to oxidation as plain carbon steels and thus require higher temperatures to reach a given colour. The temper colour method of judging the temperature is not now generally used except perhaps with carbon or low alloy steels (*see* Table 58).

Temper hardening. A process of heating metals that have been quenched in order to increase the hardness. It is a process that is really the direct opposite of tempering, and is applied to non-ferrous alloys (e.g. aluminium alloys), the effect being to cause precipitation hardening.

Tempering. A treatment to which steels, especially alloy steels, are subjected in order to produce changes in mechanical properties and in the structure of the material. Completely hardened steel is often too hard for many engineering applications and frequently also too brittle. To induce some modifications in the mechanical properties by reducing the hardness and brittleness and increasing the toughness, the steel is heated to a suitable temperature and then allowed to cool in a suitable manner, depending on the type

TABLE 58. *Temper Colours*

Colour	Temperature (Ordinary Carbon Steel), °C	Temperature (Stainless Steel) °C
Very pale yellow	225	290
Light yellow		
Straw yellow	235	340
Deep straw yellow		
Dark yellow	245	370
Yellow brown		
Brown	258	390
Reddish brown		
Purple brown	270	420
Light purple		
Full purple	280	450
Dark purple		
Blue	295	540
Dark blue	315	600

of steel. In general, as the hardness is reduced so also is the tensile strength, but the ductility is raised.

The tempering temperature lies between 100 °C and the Ac_1 temperature of the steel, and for unalloyed steels and certain alloy steels the temperatures most frequently favoured lie between 600 °C and 650 °C. The time of holding at the tempering temperature depends on the size or the maximum section of the piece being treated. Cooling is usually at a slow rate and is carried out in air. Since, however, nickel-chromium steels are prone to temper brittleness if too slowly cooled they are liquid quenched from the tempering temperature. The general trend of the changes in mechanical properties is shown in Fig. 165(a), while (b) indicates the temperature ranges for heat-treatments and other operations in relation to tempering and toughening ranges.

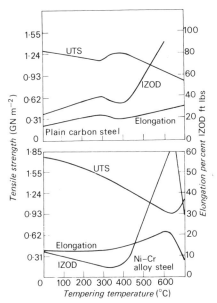

Fig. 165(a) *Tempering: changes in mechanical properties.*

In the case of carbon steels the quenching produces martensite that represents a metastable condition. Tempering modifies this condition to a certain extent to give a more stable condition in which some of the martensite is allowed to decompose and revert to the iron-pearlite condition. In Table 59 some indication is given of the tempering temperatures used for a number of alloy steels, mainly intended for tool steels.

TABLE 59. *Tempering Temperatures of Some Alloy Steels*

Composition %							Temperatures, °C		Brinell Hardness No.
C	Mn	Cr	Ni	V	W	Mo	Hardening	Tempering	approx.
0.25	0.20	4.0	—	0.50	15.0	—	1,280	650	420
0.55	0.20	1.25	—	0.18	2.8	0.5	950	650	420
0.55	0.30	4.0	—	—	—	0.5	900	430	460
0.55	0.35	4.0	—	1.0	—	0.5	880	540	475
0.40	0.30	2.25	2.5	—	—	—	980	480	400
0.45	0.30	2.75	—	0.4	15.0	—	1,090	620	550

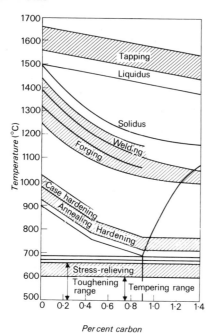

Fig. 165(b) Temperature ranges for heat-treatment.

Tempering oils. The name used in reference to oils or mixtures of oils used in cooling and quenching baths. Oils are often preferred to water, since they are less drastic and are more uniform in their cooling effect. The oils used are usually compounded of fish, vegetable and animal oils. Fish oils give unpleasant odours when used alone; vegetable oils alone may oxidise and become gummy, while animal oils may turn rancid. For slow rates of cooling lard and palm oils may be used; mineral oils mixed with varying proportions of fish or animal or vegetable oils are sold under proprietary names and give moderate rates of cooling; cotton-seed oil, and fish oils generally, give rapid rates of cooling. In compounding the various types of oil the product should be able to withstand temperatures up to about 270 °C, and in use cooling pipes in the bath should be so arranged that the temperature does not exceed about 66 °C.

Polymeric liquids are now available for this purpose which degrade only slowly under the effect of heat input, and satisfy all other conditions without the bacterial problems of natural product oils.

Temperite. The trade name of a commercial form of calcium chloride $CaCl_2$ used for accelerating the setting of concrete and in preventing concrete mixtures from freezing and cracking in cold weather.

Temperite alloys. The name of a series of fusible alloys containing varying proportions of lead, tin and bismuth or cadmium. They were made in strip form, and the compositions were such that the melting points increased in steps of about 10 °C from 150 °C to about 330 °C. They were used for testing the temperatures of steels in heat-treatment processes.

Temper rolling. A process employed to remove the effect of stretcher-straining (*see* **Stretcher strains**) of soft annealed low-carbon steel and iron sheets. Stretcher strains produce a rough surface and are usually formed when the material is strained up to its yield point. Temper rolling is then used to eliminate the yield point elongation by cold rolling of 0.5-2.0%. Rimming steel should be processed shortly after the temper rolling, because the yield point elongation slowly returns.

Tempil alloys. The trade name of a series of fusible alloys used in pellet form for measuring the temperatures of steel. The range of temperatures is from about 55 °C to 1,400 °C in convenient steps of about 8 °C or multiples of this.

Templet. The name used in reference to a pattern employed in the marking out of work or in the checking of accuracy of finished work. A templet may also be an object shaped to a particular form and used as a guide in making articles of corresponding shape. Normally spelled "template".

Tenacity. A term sometimes used to denote the degree of resistance of a material to tensile stress. It was used synonymously with tensile strength. The ratio of tensile strength to density of a material is known as specific tenacity, though this is not dimensionless.

The term tenacity has also been used as a synonym for toughness, a term not easy to define with precision, for in this connection it relates to the behaviour of materials in the range of stresses and strains outside the elastic limit. It is probably best to avoid using the term where tensile strength is more appropriate.

Tennantite. A sulphide of copper and arsenic $4Cu_2S.As_2S_3$ that occurs in association with other copper ores in Cornwall (UK), Saxony (Germany), Colorado, etc. The mineral may contain some iron replacing part of the copper. It is grey to greyish-green in colour. In the USA it is known as grey copper ore.

Tenorite. An important source of copper which consists of copper oxide CuO and contains up to 79% Cu. It occurs as a black powder or a dull black mass in the zone of weathering of copper lodes, and is particularly abundant in the Mississippi valley and in Tennessee. The black oxide has specific gravity of 6.4.

Tensile strength. This may be taken as the maximum load sustained during a tensile test, divided by the original cross-sectional area of the test piece—i.e. the maximum tensile stress developed in the material. The maximum tensile strength of a metal will be high if it hardens rapidly as deformation occurs. In the case of ductile metals the tensile strength is usually greater than the breaking strength, but is always below the maximum true stress developed in the material.

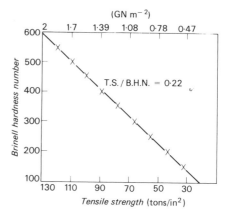

Fig. 166 Tensile strength against hardness curve.

There is a good correlation between indentation hardness measurements and tensile strength in steels. Figure 166 shows the straight line obtained by plotting Brinell hardness against tensile strength, from which can be deduced:

Tensile strength (tons force per square inch) = 0.22 × Brinell Hardness Number, or in SI units

Tensile strength $(GN\ m^{-2})$ = 0.00346 × B.H.N.
or Tensile strength $(MN\ m^{-2})$ = 3.46   × B.H.N.

Tensile testing. A method of determining the behaviour of materials subjected to axial loading tending to stretch the material. A longitudinal specimen of known length and diameter is gripped at both ends and stretched at a slow controlled rate until rupture occurs. In some tests only the breaking load and the total extension are recorded, but frequently the test is carried out in such a manner that frequent or continuous measurements of load and extension are made. From these the stress and the strain are calculated and a stress-strain diagram (q.v.) is drawn.

In order to obtain comparable results, the test piece is designed according to a simple formula (*see* **Tensile test-piece**). The strain is calculated on the proportionate increase in length, i.e. extension/original length or sometimes as the percentage increase. The stress is most commonly calculated as the load per original area of cross-section. Since the deformation that occurs results—at first—in a uniform reduction of section and later in a local reduction near the middle of the length of the specimen, it is sometimes desired to plot the stress as the load divided by the area on which it acts at each successive addition of load. In such cases the smallest diameter of the test piece is used in calculating the stress. Similarly each increment of reduction of area may be referred to the area from which it was produced, and when the true stress and true strain are plotted the stress-strain diagram has the shape shown in Fig. 167 at *A*. In the same figure, *B* represents the plot of load against true strain.

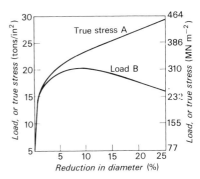

Fig. 167 *Tensile testing stress-strain diagram.*

Tensile test-piece. In the testing of metals the test specimen is either of round or rectangular cross-section, but in certain cases the test piece may be cut from a plate or sheet. The actual shape may depend on the product that is being tested. Sometimes the metal specimen is tested at full and sometimes at reduced diameter. It is the middle part or waist that has the reduced diameter, and this forms the gauge length.

The most common test-pieces are from flat plates, flat sheets, or circular cross-section bars. The common (old) Imperial unit test pieces were 1.5″ wide with 8″ gauge length for flat plates, 0.5″ wide and 2″ gauge length for sheet; 0.564″ diameter and 2″ gauge length for round bars. Round wire and rods were tested at 10″ inch gauge length and full cross-section, tubes also at full cross-section.

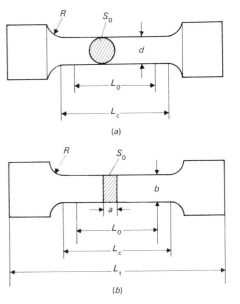

Fig. 168 *Tensile testpieces: British Standard No. 18. (a) Circular section. (b) non-proportional rectangular sections.*

Some empirical relationship obviously exists between elongation and reduction of area of a test piece, but the phenomenon of necking, local elongation in one or more places, means that in this case the elongation is not uniform along the gauge length and such a relationship must depend on the dimensions of the test piece. A convention grew up, relating gauge length to diameter for circular sections, originally $L_0 = 4 \sqrt{A}$ where L_0 is the gauge length and A the cross-sectional area. This meant $L_0 = 3.54d$ where d is the diameter of the section along the gauge length, and such specimens gave comparable figures for elongation.

International agreement now recognises a modified standard, $L_0 = 5.65 \sqrt{S_0}$ where S_0 is cross-sectional area.

For circular specimens, this gives $L_0 \approx 5d$.

The parallel length of the reduced section is also important (Fig. 168) and the International standard is

L_c between $L_0 + 2d$ and $L_0 + d/2$ for circular sections,
L_c between $L_0 + 1.5 \sqrt{S_0}$ and $L_0 + 2.5 \sqrt{S_0}$ for rectangular.

The recommendation is a parallel length $L_0 + 2d$ for circular and $L_0 + 2 \sqrt{S_0}$ for rectangular.

TABLE 60. *Circular Section Test Pieces (B.S.S.18)*

Cross-sectional area S_0, mm^2	Diameter d, mm	Gauge length L_0, mm	Minimum Parallel Length L_c $\approx 5.5d$, mm	Transition radius R, mm
400	22.56	113	124	23.5
200	15.96	80	88	15
150	13.82	69	76	13
100	11.28	56	62	10
50	7.98	40	44	8
25	5.64	28	31	5
12.5	3.99	20	22	4

Table 60 gives the dimensions of common circular specimens. When non-proportional test pieces are used for rectangular sections, the recommendation is that the ratio of width to thickness should not exceed 8 to 1. Table 61 gives dimensions of some non-proportional rectangular specimens. The radius of curvature between the end of the parallel length and the increased section of the gripped portions, is also important, and a minimum transition radius R is given in both tables.

TABLE 61. *Non-proportional Rectangular Test-Pieces (B.S.S.18)*

Width b, mm	Gauge length L_0, mm	Minimum transition radius R, mm	Approx. total length, L_t mm
40	200	25	450
25	200	25	375
20	100	25	300
12.5	50	25	200
6	24	12	100
3	12	6	50

Tensimeter. An apparatus for determining vapour pressure or for comparing vapour pressures. Since the vapour pressure of allotropic modifications of a substance are different, but become equal at the transition point, a convenient method for finding the transition point is to determine the temperature at which the allotropic forms have the same vapour pressure. The tensimeter can be used for this.

Tensiometer. An apparatus for determining the surface tension of a liquid. It consists essentially of a platinum ring about 40 mm diameter attached to one arm of a balance. The force required to raise the ring from the surface of the liquid is measured. (*See also* **Surface tension.**)

Tension. A component or a member of a structure is said to be in tension when it is subjected to forces acting axially and away from the centre of the component.

Terbium. A rare-earth element found associated with gadolinium in terbia. Pure terbium compounds may be obtained from the nickel double nitrates, the ethyl sulphates or a mixture of bismuth nitrate with the nitrates of the terbia earth metals. The sesquioxide Tb_2O_3 is white and is said to be strongly magnetic. All the known salts are colourless, but the presence of terbia in gadolinia is said to produce a yellow tinge in the mixture. Atomic number 65, at. wt. 158.93, sp. gr. 8.23, m.pt. 1,360 °C, crystal structure, hexagonal close-packed; lattice constants $a = 3.5922$, $c = 5.6754 \times 10^{-10}$ m. (*See* Table 2.)

Tercod. A refractory material consisting of a mixture of silicon carbide with or without graphite, a carbon bond and a protective boro-silicate glaze.

Ternary alloy. An alloy containing three principal elements, common examples of which are gun-metal (copper-tin-zinc) and nickel silver (copper-zinc-nickel).

Terne plate. Sheet mild steel coated with a lead-tin alloy (15-20% Sn). It is used in the making of motor-car fuel-tanks and for similar applications where the corrosive conditions are too severe for ordinary tin-plate to be used. Due to the high percentage of lead, the coating has a dull appearance. Long ternes are plates of gauge numbers 14-30 carrying 3.6-7 kg of coating metal per double box (40.5 m^2 superficial area). Short ternes generally carry up to 3.6 kg of coating metal per double box.

Terotechnology. The combination of management, financial, engineering and other practices applied to physical assets in pursuit of economic life cycle costs.

Terylene. A polyester material, in the form of fibre. The ester is formed between ethylene glycol (1, 2 ethanediol) as the base, and terephthalic (1, 4-benzene dicarboxylic) acid. Similar materials are prepared in film form. The fibre is strong, has a high softening temperature, can be set to shape by thermal treatment, and is little affected by moisture.

Test bars. In order to obtain information of and evaluation of the mechanical properties of castings it is usual to cast bars from the same furnace charge either separately or attached to the casting. The procedure varies according to the metal that is being cast. In the case of cast iron and steel the bars are often cast with the casting, but where the test bar is required for the bending (deflection) test the bar is cast in a vertical sand-mould. For steel castings the test bar is often cast horizontally near the base of the mould. It is commonly about 25 cm long.

Tetragonal. A crystallographic system in which the axes are mutually perpendicular, the parametral lengths of two of them being equal, but different from the third. Zircon belongs to this system ZrSiO$_4$, where $a = b$, and $a:c = 1:0.64$.

The tetragonal system resembles the simple cubic, in which the c/a ratio is greater or less than 1 but never equal to 1. A body-centred and face-centred form can exist also. White tin, the normal allotropic form, is tetragonal. Indium is face-centred tetragonal. Both have 5p electrons, and it might be expected that they crystallised in the diamond type structure, with covalent bonding; indeed grey tin, the low temperature modification, does have the diamond structure. For covalent bonding, however, the requirement is a hybridisation of 5s and 5p electrons, (as with the diamond structure) and this presumably does not occur with tin and indium. Certainly it does not occur with lead and thallium, which follow indium and tin in the next period.

Tetrahedral. Having the shape of a tetrahedron, a regular figure made up of four equilateral triangles, each having its sides common with those of a similar side of an adjacent triangle. The tetrahedron is most easily envisaged in the structure of diamond, with a carbon atom at the centre bonded to four similar atoms at each corner of the tetrahedron, with equal bond lengths.

Tetrahedral void. The interstitial space in a crystal structure formed by three atoms in mutual contact in one plane, with a fourth atom in the plane above or below, sitting in the cusp shaped void between the first three (Fig. 169). The largest atom which can fit into a tetrahedral void formed by four like atoms of radius R, is $0.225R$.

Fig. 169 *Tetrahedral void.*

Tetrahedrite. A copper ore that is a complex mineral containing the sulphides of iron, antimony and arsenic together with smaller amounts of the sulphides of zinc, silver or mercury. It is grey in colour and is sometimes called grey copper. It is similar in composition to tennantite (q.v.), except that much of the arsenic is replaced by antimony.

It crystallises as tetrahedra, and is frequently associated with pyrites, galena, blende. Also known as Fahlerz.

Texture. Another term for "preferred orientation", i.e. an indication of directional alignment of particular planes or crystal directions, which significantly affect properties. It is unusual for such orientation to be perfect, the term refers to the grouping about a mean value, which is still significant until the angle between the extremes of direction exceeds about 30°.

Thallium. A soft, white, malleable metal obtained mainly from the flue-dust in the production of sulphuric acid from pyrites. It occurs in many varieties of iron and copper pyrites and in certain minerals such as crookesite (q.v.) and lorandite TlAsS$_2$. The metal possesses a bright lustre when freshly cut, but tarnishes rapidly in air. There are two modifications—the α-thallium (crystal structure, close-packed hexagonal), which is stable at room temperatures, and the β-form (which is face-centred cubic). The transition temperature is 232.7 °C. Small amounts of the metal may be added to some lead-base bearing alloys to increase the resistance to pounding, but the chief use at present is in optical glass of high refractive power.

Thallium compounds have a limited use in medicine, in rodent poisons and as catalysts in certain organic reactions. Crystal structure H.C.P., lattice constants $a = 3.456$, $c = 5.525 \times 10^{-10}$ m. The mechanical and electrical properties are given in Table 2.

Thenard's blue. A bright-blue colouring matter produced in response to a demand for a colouring substance that would withstand temperatures of porcelain heating furnaces. It is prepared by heating a mixture of alumina and cobalt phosphate or arsenate. The same reaction is sometimes used in the laboratory as a dry test for aluminium. Also called cobalt ultramarine.

Therm. A standard unit of heat originally defined for gaseous fuel. Town's gas, i.e. gas from heated coal, was not sold by volume, but by heating value, in view of the variability of the coal from which it was derived. The original unit was 100,000 British Thermal Units equals one therm. In SI units, one B.Th.U. equals 1.055 kJ, hence one therm is equal to 105.506 MJ.

Thermal analysis. A method used in the study of the chemical equilibrium of materials or mixtures of them. The simplest type consists in determining cooling curves by taking temperature readings at definite intervals of time as the material cools under quiescent conditions. Any irregularity in the shape of the curve must be due to internal liberations of heat caused by some modification in the chemical constitution or by a change in state or of specific heat. The simplest types of analysis are those in which (a) the inverse rate curve is plotted, i.e. dt/dθ is plotted against θ or (b) the differential cooling curve is plotted, i.e. d$(\theta - \theta_1)$/dθ is plotted against θ. In the above t = time, θ = temperature, $(\theta - \theta_1)$ represents the small change in temperature during a definite interval of time.

Thermal capacity. The amount of heat required to raise the temperature of any substance through 1 °C. The thermal capacity of unit mass of a substance is thus the same quantity as the specific heat.

Thermal centre. A term used in the study of the solidification of castings to indicate that point (or points) at which the rate of cooling is less than that in any part of the surrounding metal. In an ingot or casting of simple shape, there may be one centre which may move as the cooling proceeds. With complicated castings there may co-exist two or more thermal centres, and from these the direction of the heat flow is outwards towards the cooling surface. It is at the thermal centres that shrinkage cavities are most liable to occur unless molten metal is available to compensate for the effects of contraction due to cooling and solidification.

Thermal conductivity. Above absolute zero, the atoms of a substance are not stationary but vibrating. The equilibrium between attraction and repulsion forces, binding the material together, is perturbed by the thermal energy input. The vibration is about a mean position, which latter is dependent on the attraction and repulsion forces. The amplitude of vibration obviously varies with thermal energy input.

The capacity to absorb an input of thermal energy is the specific heat or thermal capacity, the amplitude of vibration manifested as thermal expansion. Thermal energy is generally introduced at a surface (electrical induction may introduce it into the centre of the mass) and transmission of the energy will occur as the vibrating atoms begin to interact with their neighbours. This transmission is the thermal conductivity, the attempt to decrease the free energy of those atoms near the source of added energy.

In an insulator, the electrons in highest energy levels cannot be excited into higher energy conduction bands, and so only lattice vibrations contribute to conductivity. The thermal wave (phonons) which propagates through the material involves the density of the material (the closeness of packing and the mass of the atoms), the thermal capacity (specific heat), the velocity of sound (elastic waves) and the mean free path of the wave itself (distance between collisions).

$$k_t = \rho c_p l s$$

where k_t is thermal conductivity, ρ the density, c_p the specific heat at constant pressure, l the mean free path and s the velocity of sound.

The heat capacity per mole C_p is equal to $\rho.c_p.V$ where V is the volume, but as $V = N_0 d^3$ where N_0 is Avogadro's number and d the lattice parameter,

$$C_p = \rho.c_p N_0 d^3$$

As C_p is approximately $3R = 3N_0 k$, where k is Boltzmann's constant,

$$\rho.c_p = \frac{3N_0 k}{N_0 d^3} = \frac{3k}{d^3}$$

and if the mean free path of the thermal wave l is approximately equal to d,

$$k_t \approx \frac{3ks}{d^2}$$

The thermal conductivity of an insulator should thus be inversely proportional to the square of the lattice parameter, and directly proportional to the velocity of sound in the material. The mean free path must be affected by defects in the structure, dislocations, vacant lattice sites and impurity atoms. Thermal agitation of the structure is temperature dependent, the effect of the defects is not. Thermal conductivity of the lattice decreases in almost inverse proportion to thermal energy, as mean free path decreases. In amorphous material, with greater disorder in the lattice, the mean free path is increased, because of the variable distances between collisions, and thermal conductivity is proportional to thermal energy content. Substances which can exist in crystalline or amorphous forms exhibit very different curves for k_t against temperature. (See Fig. 28 under **Conduction by the lattice**.)

In semiconductors and crystal lattices of good conductors, the effect of the electrons has to be included. Semiconductors naturally show a temperature dependence, and any impurity atoms in their structure will affect the mean free path of the thermal wave. The mean free path of the electrons themselves has to be studied in metals and semiconductors. In an ideal lattice there is no restriction to electron mobility, and no transfer of energy by collision. Lattices are not ideal however, and atoms are not at rest, so there is a probability of electron/lattice interaction, and a transfer of momentum. The mean free path of electrons is partly dependent on the thermal energy of the lattice, and the regularity of the lattice is affected by impurities and defects, although these are not temperature dependent. Thermal conductivity is therefore made up of two terms, $k_t = k_l + k_e$ where k_l is the lattice contribution and k_e that due to the electrons. The electron contribution k_e is itself made up of the temperature dependent part (lattice vibration interaction) and the defect or structural part (defects and impurities) as described.

Near absolute zero, the electron scattering contribution is proportional to T^{-2}, but the defect contribution is proportional to T. Thus $k_e = aT^{-2} + bT$ where a and b are constants. Thermal conductivity shows a maximum value near absolute zero, this maximum approaching closer to the absolute zero as the material becomes more pure and structurally perfect.

At higher temperatures, the electron scattering is directly proportional to temperature and the mean free path to T^{-1}. The electron energy itself however increases proportionally to temperature, and the product of the two effects gives a result almost independent of temperature. The total value of k_t thus becomes more independent of temperature as temperature becomes very high.

The lattice contribution k_l depends on a term proportional to T^2 at low temperatures, and lattice defects such as dislocations produce a similar effect. Impurity atoms contribute with a term dependent on T directly, and so measurement of k_t at very low temperatures can separate the effect of defects and impurities, a separation not so easily achieved with electrical conductivity measurements. The value of k_l is almost 50% greater at low temperatures than it is at higher ones, and such measurements are therefore reasonably sensitive.

The electron contribution to thermal conductivity in metals is greater than the lattice contribution, and the variation of k_t between metals and alloys is entirely due to the variations in the electron and lattice contribution. In semiconductors, the number of electrons or holes is small compared with good conductors, and so the electron contribution is smaller. Nevertheless, silicon and germanium are better thermal conductors than alloys such as monel (70% Ni and 30% Cu) or austenitic steel (18% Cr and 8% Ni) and metals such as titanium and zirconium.

For experimental measurement, the quantity of heat Q flowing across a section of material area A in time t, is given by

$$Q = \frac{K_t(\theta_1 - \theta_2)At}{d}$$

where θ_1 and θ_2 are temperatures at two parallel sections of area A distance d apart, and k_t is thermal conductivity.

The conductivities of some common materials are given in Tables 2 and 3. Generally those which are good conductors of electricity are also good conductors of heat, but thermal conductivity does not vary quite so markedly with temperature as does electrical conductivity.

Thermal cycling. When materials are subjected to fluctuations of temperature, the effects are reversible only if the material is fairly pure and free from defects, and isotropic in structure. The continued expansion and contraction of the crystal lattice leads to reversibility only when no strain exists between adjacent crystals during thermal cycling. Anisotropic materials inevitably exhibit variations in thermal expansion along different crystal directions, which lead to strain. If the strain is relaxed at the higher temperature by deformation, however slight, the process cannot be completely reversed on cooling, and a change in shape may result. Anisotropic materials therefore exhibit changes in dimensions, which may become severe after prolonged cycling. Thermal fatigue can result even with small temperature variations of an indoor environment over a period of many years.

In complex structures, thermal cycling can produce deformation even in isotropic material, due to strain set up by expansion and contraction of varying section thickness and overall length of different elements of the structure. In high speed machinery operating at high temperature, much more attention is given to the effects of thermal cycling due to service conditions, and the interactions between dissimilar materials with varying properties. A supersonic aircraft for example, may suffer changes of 100 °C between take off, flight and landing.

Thermal diffusion. A process in which a separation of molecules of different mass in a gaseous mixture can be effected by maintaining a temperature gradient. In such cases the heavier molecules diffuse down the gradient and the lighter molecules in the opposite direction. This is the basis for one method of separating isotopes of an element. Using a long vertical tube with a hot wire running down the centre, the isotopes of chlorine have been separated. To separate uranium 235 from uranium 238 the carrier gas used was the hexafluoride of uranium.

Thermal diffusivity. The ratio of thermal conductivity to heat capacity

$$a = \frac{k_t}{\rho.c_p}$$

where a is diffusivity, k_t thermal conductivity, ρ density, c_p specific heat at constant pressure. The parameter is important in assessing the rate at which temperature rises in a body subjected to a change in thermal energy. A good insulator does not necessarily remain cold when heated, but if it has a low thermal capacity, the material is soon saturated and does not go on absorbing more heat from the source. Metals have a high diffusivity because they have a high k_t. Insulating materials are frequently porous, giving low values for both k_t and $\rho.c_p$ although the diffusivity "a" may not appear smaller than a denser material. The propagation of temperature therefore, depends on both k_t and $\rho.c_p$.

A similar parameter can be defined to indicate the rate of abstraction of thermal energy when a cold and hot surface are in contact. The contact coefficient b is defined by the product of k_t and $\rho.c_p$.

$$b = k_t \rho.c_p.$$

When b is small, the temperature difference between the interior of a body and its surface, when heat is added or subtracted, will be high. Good insulating materials have a low value of b, so that k_t and $\rho.c_p$ must both be low. Such materials are those used for floor coverings, upholstery, handles, etc. Cotton sheets feel cold to the body in winter, because cotton has a relatively high b value, due to high k_t, as compared to wool. Metals feel cold to human touch, due to high k_t. Textiles take up water overnight in unheated rooms, because they have low b value, and thus cool off more rapidly than other materials. This causes water to condense on them. Wooden structures outdoors feel wet to the touch on a summer evening whereas stone feels dry.

When reheated next day, textiles will heat up rapidly and lose their water of condensation, and if the outside temperature is low, that water will condense on windows and metal objects first, because they have a high b value, and heat up more slowly. A metal or glass object taken from a cold to a heated room will cause condensation on the outside surface.

Although both k_t and $\rho.c_p$ are temperature dependent for metals, the product, the b value, is remarkably independent of temperature.

Thermal dissociation. A number of substances on heating are decomposed into simpler substances. If the products of the breakdown remain in contact, then on cooling the reverse reaction occurs and the original substance is re-formed. Such reversible reactions, the progress of which depends on heating (or cooling), are instances of thermal dissociation. A common example is the thermal dissociation of nickel carbonyl, the latter being produced at temperatures between 50 °C and 60 °C and dissociation at about 180 °C.

$$Ni + 4CO \underset{180\,°C}{\overset{50\text{-}60\,°C}{\rightleftharpoons}} Ni(CO)_4 .$$

This is the basis of the Mond process in which the nickel oxide is reduced by water-gas at about 350-400 °C. The forward and reverse reactions are then used to remove and deposit pure nickel.

Thermal energy. The capacity for doing work by changes in the vibrational modes of a substance, leading also to changes in temperature, dimensions, and radiation emission or absorption. Convertible to other forms, such as mechanical and electrical energy with suitable transducers. Measured in phonons.

Thermal equilibrium diagrams. Diagrams in which are plotted the composition of a mixture against the temperature of freezing of each mixture of progressively varying composition. (*See also* **Constitutional diagram.**)

Thermal etching. A method of revealing details of the surface of polished specimens by heating them to a fairly high temperature in a vacuum or in an inert atmosphere. Under the conditions of treatment it is supposed that atoms can migrate from one part of the surface to another. This leads to striations appearing on the surface of the metal and grooves at the grain boundaries.

Thermal expansion. The increase in dimensions of a body due to a rise in temperature, which change is reversed on cooling. Most substances show a characteristic value for the coefficient of thermal expansion, this being the change in length (area or volume) per 1° increase in temperature. The coefficients of linear expansion of materials are given in Tables 2 and 3. The coefficients of area or superficial and of volume or cubic expansion are approximately twice and three times the linear coefficients.

Thermal fatigue. One of the results of the drastic cooling of cast metal is the setting up of internal stresses, and these latter seek relief by the plastic or permanent deformation of the metal as it cools down after solidifying or by a process of recrystallisation when the metal is reheated. The deformation which occurs as a result of cycles of rapid cooling and reheating is known as thermal fatigue.

Also applicable to differential thermal expansion in anisotropic materials when this leads to permanent deformation.

Thermal growth. Changes in dimensions of a permanent nature, as a result of changes in thermal energy, usually by elevated temperatures. The growth results from deformation as a consequence of strain induced by differential thermal expansion, and is confined to anisotropic material, or complex structures.

Thermal shock. Impulses of thermal energy can impose severe strains on materials owing to the differential thermal expansion between surface and interior. The cracking of thick glass when immersed in boiling water is an example. The problem is to define a suitable parameter which can be used for experimental measurement and comparison between materials. Thermal diffusivity (q.v.) and contact coefficient are relevant, but the different behaviour of materials does not assist the use of a universal definition of thermal shock.

The original definition visualised a localised temperature gradient created by the input of thermal energy, which is directly proportional to specific heat and inversely to thermal conductivity, hence proportional to $a^{-½}$ where a is thermal diffusivity. The temperature gradient results in thermal strain, by differential thermal expansion, and is proportional to $\alpha/a^{½}$. The internal stress is related to the strain by the modulus of elasticity, and hence proportional to $E\alpha/a^{½}$. The actual stress must then be related to the tensile strength of the material σ by a constant which can be the thermal shock parameter, sp_I.

$$sp_I = \frac{\sigma a^{½}}{E\alpha}$$

Such a parameter is useful for metals and good conductors. Where the impulse of thermal energy is more localised, as when a beam of intense radiation is directed on to a material, the contact coefficient b is more suitable and can be substituted directly for thermal diffusivity, giving a second shock parameter.

$$sp_{II} = \frac{\sigma b^{½}}{E\alpha}$$

A whole body placed in a heated enclosure, as when an object is placed in a furnace, involves less localised temperature gradient, and conductivity is the only parameter involved. The third shock parameter is then defined by

$$sp_{III} = \frac{k_t \sigma}{E\alpha}$$

A fourth alternative has been suggested, introducing Poisson's ratio ν, the ratio of longitudinal to transverse strain.

$$sp_{IV} = \frac{(1-\nu)\sigma}{E\alpha}$$

The third shock parameter has been developed for comparatively brittle materials, and especially for homopolar bonded materials. The denominator $E\alpha$ can be shown to be almost equal to $\rho.c_p$ for homopolar bonds, and the parameter then becomes

$$sp_{III} = \frac{k_t \sigma}{\rho.c_p} = a\sigma$$

As the value of a does not vary much for brittle materials, the tensile strength becomes the paramount property.

Amongst materials with high resistance to thermal shock are oxides of Mg, Be and Si, the complex oxides such as $FeO.Cr_2O_3$ and $MgO.Cr_2O_3$, and many metal carbides. The majority have moderate thermal conductivity, but low specific heat, low coefficient of thermal expansion and high tensile strength. All the shock parameters therefore indicate a high resistance. Carbides and metals have high thermal conductivity, so the third shock parameter indicates high resistance. Metals are limited for high temperature service by loss of strength at high temperatures, whereas oxides and carbides have higher melting points and so retain strength to higher temperatures. They lack plasticity however, and so suffer cracking, whilst metals merely distort.

Refractory materials are obviously required to have good shock properties. The effects on refractories may be rupture, spalling or distortion. An empirical test is often used experimentally.

The test is usually carried out on specimens of the refractory cut to a standard size 75 × 50 × 50 mm (3 × 2 × 2 in). These are placed in a furnace at about 900 °C for 10 minutes. They are then removed and left in air, standing on end on a brick of similar material for 10 minutes. The cycle is repeated up to about thirty times if necessary. After each cooling the specimen is subjected to hand pressure, and the result of the test is reported as the number of cycles of heating and cooling before the specimen breaks under hand pressure. If the specimen withstands the thermal shock after 30 cycles, no further test is done and the material is reported as "over 30".

The tendency for a refractory material to spall varies directly as to coefficient of expansion and the density, and inversely proportional to the thermal conductivity.

Observations appear to establish that failure of a refractory (spalling) on sudden cooling is due to the high tensile stresses set up. But failure on sudden heating is due mainly to high shear stresses.

Thermal vibration. The mechanical manifestation of thermal energy on atomic positions in a material. The kinetic theory of gases (q.v.) is based on the random movements available to gas molecules. Something similar applies even to the short range order of liquids, and is observed as the Brownian movement. In solids, thermal expansion is the direct result, and many parameters such as specific heat, thermal and electrical conductivity are directly affected by it.

The vibration is complex, and temperature dependent up to a point when the full range of frequencies of vibration becomes available. (*See* **Characteristic temperature (Debye).**)

Thermistor. The behaviour of semiconductors, with a rapid decrease in electrical resistance with rise in temperature, is utilised as a temperature measuring device in the thermistor. The temperature coefficient of resistance can be sensitive for both detection and control of temperature, and different materials used for different ranges of temperature.

The resistance of a thermistor with an energy gap of about 0.3 eV varies by about 4% for a change of 1 °C at room temperature. By using mixed oxides in varying proportions, a range of 150 °C can be covered, with a corresponding resistance change of two orders of magnitude. Changes of temperature as small as 10^{-6} K can be detected.

Thermite. A mixture of aluminium powder and the oxide of a metal—commonly iron oxide. When primed with a little sodium peroxide and ignited by magnesium ribbon a vigorous chemical action occurs, during which the aluminium combines with the oxygen of the oxide so liberating the metal. The heat of the reaction is such that the liberated metal appears in the molten state. Also called thermit. (*See also* **Thermit process.**)

Thermit mixtures. These are mixtures of aluminium powder with iron oxide and additional elements:

red thermit contains red ferric (iron III) oxide;
black thermit contains black oxide of iron;
railway thermit is red thermit containing also up to 16% nickel, manganese and mild steel swarf.

For forgings the red thermit may contain varying amounts of mild and manganese steels. For cast iron, additions of ferrosilicon (about 2–4%) and mild steel (15–20%) are used for welding.

Thermit process. A smelting process in which finely divided aluminium is used to reduce a metallic oxide, the heat of the process promoting fusion of the reduced metal. The action occurs because aluminium, being a strongly electropositive element, is a powerful reducing agent. It is the basis of the "Goldschmidt" process for extracting high-melting-point metals (e.g. chromium, manganese, molybdenum, vanadium, etc.) from their ores.

In the case of chromium, for example, the charge consists of powdered aluminium and chromium oxide Cr_2O_3. This is put in a magnesia-lined crucible and ignited by a cartridge containing barium peroxide and aluminium.

The reaction with iron III oxide is commonly used to produce small quantities of molten iron for local fusion welding.

Thermit welding. A process of fusion welding in which the molten steel and the heat for the fusion of the parts to be joined are supplied from the highly exothermic reaction of the thermit. A mould is built round the parts to be joined and the crucible containing the thermit is located immediately above this. (*See also* **Thermit process.**)

Thermocouple. One of the two types of electrical thermometer (the other being the resistance thermometer) used for measuring high temperatures, utilising the Seebeck effect. When two dissimilar metals (or alloys) are in the form of strip or wire and joined at their ends, a thermoelectric current will flow in the circuit if the two junctions are maintained at different temperatures. The magnitude of the current in any particular circuit depends on this difference of temperature, so that if one junction is kept at a known temperature, the temperature of the other junction can be calculated.

The Peltier effect (q.v.) and the Thomson effect (q.v.) both contribute to the total e.m.f. produced in a thermocouple.

The electrons in different metals have different concentrations and different velocities of thermal agitation. When the metals are in contact diffusion of electrons can take place across the junction. By keeping the two junctions in the completed circuit at the same temperature the tendency of electrons to diffuse from metal A to B is equal and opposite at the two junctions and so no current flows. To cause the electron concentrations to vary so that a steady flow can occur one junction must have a temperature different from the other junction.

The e.m.f. per °C produced with the various pairs of metals is always small, but varies with the metals used. The e.m.f. can be measured on a millivoltmeter, but it is usually preferable to use a potentiometer and for recording purposes either automatic, mechanical or electronic types are available.

A platinum-rhodium thermocouple is used in a thin-walled silica tube for liquid steel and the time of immersion is about 20 seconds during which the sheath can be expected to remain intact though it may distort. A quick (electronic) recorder is used and a new silica sheath is fitted for each determination.

Besides being affected by contamination, thermocouples are sensitive to strain and to prolonged heating. Contamination can generally be removed, in the case of the noble metals, by immersion in hot hydrochloric acid and the effects of strain by suitable annealing. These processes are usually followed up by calibration and the following fixed points are used for comparison on the International Temperature Scale:

	K	°C
Oxygen liquid-gas equilibrium	90.18	−182.97
Water solid-liquid	273.15	0.00
Water solid-liquid-gas (triple point)	273.16	0.01
Water liquid-gas	373.15	100.00
Zinc solid-liquid	692.655	419.505
Sulphur liquid-gas	717.75	444.6
Silver solid-liquid	1235.05	961.9
Gold solid-liquid	1336.15	1063.0

Above the temperature of solidification of gold, the International Scale is based on the laws of radiation, but the following two substandard temperatures can be used.

Freezing point of palladium	1,552 °C
Freezing point of platinum	1,772 °C

The e.m.f.s per °C over a range of temperature −180–1,600 °C for the commonest types of thermocouples are given in Fig. 170.

Thermodynamic laws. Thermodynamics considers energy changes that take place in ideal systems and the reasoning is not based on the properties of a particular working substance. The laws that have been formulated are based on experience, and they form the basis on which the science has been built up by logical reasoning. The first law of thermodynamics states that though energy may be converted from one form into another it cannot be created or destroyed. Heat and work are thus mutually convertible, and the same unit is used for both, the joule (J).

1 J = 1 N m where N is the force in newtons and m the metre. 1 N = 1 kg m s^{-2} where kg is the mass of the body in kilograms and s the time in seconds. 1 joule per second (1 J s^{-1}) = 1 W (watt), the unit of power.

The law is usually written $W = JH$, where J is the constant called the mechanical equivalent of heat, W = work and H = heat

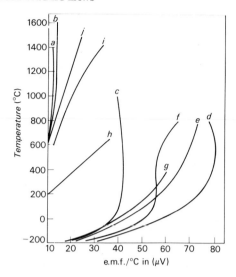

Fig. 170 *Thermocouples.* (a) *Platinum/10% rhodium-platinum.* (b) *Platinum/13% rhodium-platinum.* (c) *Chromel/alumel.* (d) *Chromel/constantan.* (e) *80/20 nichrome/ constantan.* (f) *Iron/constantan.* (g) *Copper/constantan.* (h) *40% palladium-gold/10% iridium-platinum.* (i) *Platinum/4.5% rhenium/5% rhodium-platinum.* (j) *Platinum/10% iridium-platinum.*

measured in the appropriate units. If work is done on, say, a gas at constant pressure so that there results a volume change, the first law is again expressed by $H = E + PV$ where H = total heat content of the system E is the internal energy, not known, and PV represents the work done.

The second law of thermodynamics can be used to determine the direction in which a spontaneous change can take place. It states that heat cannot be transferred by any continuous and self-sustaining process from a colder to a hotter body.

If Q_1 is the heat taken in by a system at temperature T_1 and $-Q_2$ is the heat rejected at temperature T_2, then in a reversible cycle the heat used in doing work is $Q_1 - (-Q_2)$ and the efficiency

$$= \frac{Q_1 + Q_2}{Q_1} = \frac{T_1 - T_2}{T_1}.$$

Only when T_2 is absolute zero can the efficiency be 100%, and hence this equation prescribes the efficiency of heat engines.

Rearranging the last equation

$$\Sigma \frac{Q}{T} = 0$$

In making the summation the reversible cycle may be divided into sections and then if

$$\frac{dQ}{T} = dS$$

the quantity S is a function known as entropy. In the case of the fusion of a metal dQ is the latent heat of fusion and T is the melting point in degrees absolute, so that the entropy change is measured in 4.186 J per degree.

The third law of thermodynamics may be stated in this form: "Every substance has a finite positive entropy, but this may become zero at absolute zero and probably would be so in the case of a *perfect* crystalline solid."

The Helmholtz free energy F is defined by the equation

$$F = E - TS.$$

Any change in F must be an isothermal reversible change at constant volume.

The Gibbs free energy G is expressed by the equation:

$$G = H - TS.$$

and any change in G must be a reversible isothermal one carried out at constant pressure.

The state of stable equilibrium of a system is that in which the functions H, G or F have minimum value, and thus in any stable equilibrium condition there will be

	T and P are constant and G is a minimum
or	T and V are constant and F is a minimum
or	S and P are constant and H is a minimum
or	E and V are constant and S is a maximum.

Thermodynamic scale. This is a scale of temperature that is independent of the physical properties or behaviour of any material substance. It has been devised entirely from thermodynamical considerations of the heat exchanges that occur in an ideal Carnot Cycle. On this scale the interval between the freezing point of water and the boiling point of water when the pressure is 1.01×10^5 N m^{-2} of mercury is 100°. This scale is thus an "absolute" scale and leads to the supposition of an absolute zero of temperature which is $-273.15°$ on the Celsius scale.

Temperatures on this scale are designated K and to convert temperatures Celsius into temperatures on the absolute or Kelvin scale 273 must be added to the Celsius temperature. (*See* **Temperature.**)

Thermoelectric diagram. If, for a circuit of two dissimilar metals having one junction at a constant temperature θ_1 while the temperature θ_2 at the other junction is varied, the values for the e.m.f. round the circuit are measured, it is found that the plot of e.m.f. against θ_2 gives curves that are very roughly parabolic. A simpler and more useful diagram can be made if the thermoelectric power (defined as dE/dT, where T is the absolute temperature) is plotted against the temperature. In practice the value of dE/dT is found by measuring the difference of the e.m.f.s when the temperature of the hot junction is $(T + \frac{1}{2})°$ and $(T - \frac{1}{2})°$. The curves so obtained are generally straight lines, and the diagram obtained by using the curve for lead as the abscissa is known as the thermo-electric diagram and is shown in Fig. 171. The temperatures are marked in °C and the thermoelectric power in microvolts per °C.

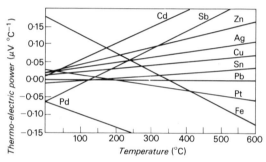

Fig. 171 *Thermoelectric diagram.*

In Fig. 172 are shown portions of the curves for lead, tin and zinc. The total e.m.f. round the circuit for lead and zinc is represented by the area $T_2 CDT_1$, and for lead and tin by the area $T_2 ABT_1$. For the two metals zinc and tin, the total e.m.f. in the circuit when the junctions are at temperatures T_1 and T_2 is represented by the area $ACDB$.

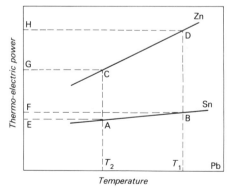

Fig. 172 *Portions of curves from thermo-electric diagram.*

Since the Peltier effect (q.v.) is given by

$$T \frac{dE}{dT},$$

the electrical equivalent (P_2) at temperature T_2 is represented by the area $DBFH$, and at the cold junction by the area $CAEG(P_1)$.

The total e.m.f. is thus

$$E = (P_2 - P_1) + \int_{T_1}^{T_2} (\sigma_1 - \sigma_2)\,dT,$$

where $(P_2 - P_1)$ is the Peltier effect and

$$\int_{T_1}^{T_2} (\sigma - \sigma)\,dT$$

is the Thomson effect. (σ is the Thomson coefficient of thermo-electric effect.)

The thermoelectric behaviour of any pair of metals can be determined from the thermoelectric diagram. Where two curves, such as those for iron and copper, intersect, the point of inter-section is known as the neutral point, since the Peltier effect there becomes zero. The total e.m.f. of such a circuit is then a maximum at the temperature corresponding to the neutral point.

Thermometals. A combination of two metals (or alloys) that exhibit wide differences in their thermal expansion, so that when bonded together the compound strip will bend or deflect with temperature changes. A number of combinations of metals and alloys are used, including nickel alloys, nickel iron, pure iron, steel and brass, and each combination has a range of temperature in which it is most effective. The main requirements of the metals in addition to the values of the expansion coefficients are corrosion and heat resistance and a conveniently high modulus of elasticity, and for some applications a high electrical resistivity.

The chief applications are: (a) temperature indication and control; (b) compensation devices in various mechanisms; (c) fire-alarms and circuit-breakers; (d) operating mechanisms for flashing signs and signals and protection devices.

Thermometers. These are instruments devised to indicate (and in some cases to record) the temperature of a body. There are many types available, each showing some particular advantage over others for a particular employment. No one type of thermometer is equally suitable for all temperature ranges, though this is nearly so in the case of some noble-metal thermocouples. The main principles on which thermometers work are: (a) the regular expansion of a thermometric substance, e.g. air, hydrogen, mercury, etc. Since these substances must be contained in some kind of vessel or container, the behaviour of the containing vessel may limit the useful range of temperature that can be measured, e.g. glass, porcelain, etc. In the case of liquid thermometric sub-stances the range of temperature between freezing and boiling point, and the effect of the vapour pressure as the latter point is approached, impose serious limitations. Solid substances which show a marked change in length or bimetal strips such as are used as thermometals (q.v.) are also sometimes used as thermo-meters. But the most useful types of thermometers are the electrical resistance thermometers and thermocouples (q.v.).

Bourdon's thermometer remains in a class by itself. It consists of a thin flexible metal tube bent into the form of a circular arc. It contains a highly expansive liquid, which tends to force the tube to straighten as its volume increases with temperature.

Instruments used for measuring high temperatures are broadly classed as pyrometers. These latter include not only the electrical resistance and thermo-electric types but also those instruments whose method of operation depends on radiation from the hot body. (See also **Optical pyrometers.**)

For low-temperature work petroleum ether can be used in a "liquid" thermometer as well as many gas thermometers, e.g. hydrogen. Resistance thermometers have been used but the presence of small amounts of impurity in the metal produces large errors at low temperatures. Carefully standardised thermo-electric thermometers are used down to about $-200\,°C$.

Thermopile. An instrument devised for the detection and measure-ment of radiant heat. It operates by virtue of the Seebeck effect, there being usually twenty-five pairs of small antimony and bis-muth bars forming as many thermo-electric circuits in series. The rods are arranged in the form of a rectangular prism with one junction of each pair at one face. The current flowing when radiation falls on this face is indicated and measured by means of a galvanometer. It has largely been replaced by the bolometer.

Thermoplastic. A polymeric material which is reasonably solid and rigid at room temperature but which softens and becomes plastic on heating. Such substances can be repeatedly softened by heat, and hardened on cooling without change of properties.

There are few cross-links in thermoplastic material, and most are formed by addition polymerisation. Branching is often present, and the change of plasticity with temperature is dependent on chain length, orientation of groups, branching, and degree of crystallinity. Thermoplastics are readily fabricated but have lower strength in general, and are more deformable under stress at room temperature than other materials.

Thermosetting. Polymeric materials which have cross-links be-tween the polymer chains or groups, resulting in a more rigid network than in thermoplastic material. The links are generally formed as the final process of fabrication, so that subsequent heating cannot restore plasticity, only break the links, leading to degradation. These materials are stronger than thermoplastics but more brittle, and often have high electrical resistivity. They are often formed by condensation polymerisation.

Thermostat. A device for maintaining a constant temperature in an apparatus or an enclosure. It is connected with a source of heat and can be pre-set to ensure that the heat is reduced or cut off when a desired temperature is just exceeded, and is restored or increased when the temperature falls just below the desired value. In a simple type the temperature control is based on the expansion of a certain volume of air in any enclosed tube. A rise in temperature above the pre-set value causes the air to expand so that a column of mercury is depressed and an electric circuit broken, thereby cutting out the heat source. (See also **Thermistor.**)

Thermostatic metals. Metals and alloys used in the making of thermocouples and metallic thermometers and thermostatic con-trols. Also called thermoelectric metals, thermocouple metals and thermometals. (q.v.).

Thickener. A term used in reference to a type of classifier used in ore dressing. It is usually a tank used for de-watering an ore pulp before extraction processes are begun. The pulp is put into the tank and allowed to settle so that clear water or fines can be drawn off at the top while the thickened pulp is drawn off at the bottom of the tank. Also called the settling tank.

Thiourea. A white crystalline substance having the formula $(NH_2)_2 CS$. It has a bitter taste and is soluble in water and alcohol. The aqueous solution is neutral and is mildly reducing. The substance can be prepared from ammonium thiocyanate by melting the latter (m.pt. $147\,°C$) and then maintaining the liquid at $170\,°C$ for some time. After cooling, the mass is extracted with hot water. It is used in the manufacture of plastics as a mordant, in the glass industry, and as the basis of a solution for cleaning silver, e.g. "silver dip".

Thixotropy. A variation in viscosity of a fluid dependent on the rate of shear. In highly viscous fluids, the shear force changes the position of neighouring molecules but the changes persist for some time and are slow to recover. The properties then vary according to the time available for recovery. Rapid shear causes thixotropic fluids to become more fluid, but at low shear rates, the material is sluggish.

The phenomenon has also been known as "shear-thinning", but most rheologists prefer to separate these two.

The extreme case is when a material is a solid gel in the rest state, but becomes a liquid sol when sheared, say by stirring. Few substances act like this, but the trend is the same. Paints which can be inverted in the open tin and not flow, but do so readily on stirring or spreading, are thixotropic. The clay "bentonite" has thixotropic properties and is used in drilling "mud", used on oil rigs, whereby the material is more or less solid at rest but very much less viscous when the drill is rotated. Some honey retains minute air bubbles even at rest, and this gives the slightly cloudy appearance. Spreading immediately causes flow however, and the material looks more transparent as the bubbles escape.

Thomas process. A process devised to produce good mild steel from pig iron containing much phosphorus. The dephosphorising is carried out in a Bessemer converter, the conditions favouring this being: (a) fluid metal and fluid slag; (b) an oxidising atmosphere; (c) intimate mixing of metal and slag during working; (d) abund-ance of basic slag and a basic lining to the converter. The charge consists of calcined lime followed by molten basic pig iron. Air is blown through the converter and the instant noted when the flame at the mouth of the converter "drops". The blow is continued for a further period, and during this "after-blow" most of the phosphorus removal is effected. As a good deal of slag is produced, the converter in this process is usually larger than for the acid process. The phosphorus-containing slag has a commercial value. Also known as the basic Bessemer process.

Thomson effect. A thermo-electric effect sometimes known as the thermo-electric or Kelvin effect, that accompanies a temperature gradient along a conductor. The magnitude and direction of the potential gradient depend on the nature of the conductor and on the temperature difference between the ends of the conductor. The coefficient of thermo-electric effect is usually measured in joules coulomb^{-1} °C^{-1} and is denoted by σ. The effect is reversible. When a current C passes through a wire between A whose temperature is θ_1 and B whose temperature is θ_2, then the heat developed in time t is $H = \sigma Ct(\theta_1 - \theta_2)$, where σ is the thermo-electric coefficient. It is not a constant but its value depends on θ. The energy gained by unit quantity of electricity in passing from A to B is given by

$$\int_{\theta_1}^{\theta_2} \sigma d\theta.$$

The value of σ for lead is practically zero for all temperatures. σ is positive for copper, silver, cadmium, zinc, tin, antimony, aluminium and brass. It is negative for platinum, palladium, iron, nickel, cobalt, rhenium and nickel silver.

The *total* thermo-electrical effect (to be distinguished from the ordinary heating effect which varies as C^2) is the algebraic sum of the Peltier effect (q.v.) and the Thomson effect. (*See also* **Thermo-electric diagram.**)

Thoran. The name used in reference to a material used for high-speed tools that consists of tungsten carbide.

Thoria. The oxide of thorium ThO_2. It is a snow-white powder usually obtained from thorite or orangite. It occurs in the monazite sands of Sri Lanka to the extent of about 10%. The mineral may be treated with sulphuric acid and a little water and heated until a dry mass is obtained. The latter when powdered is dissolved in ice-cold water and the precipitated silica is filtered off. The solution is boiled with ammonia and the precipitated hydroxides are washed and dissolved in hydrochloric acid from which the oxalates are precipitated. This precipitate is washed and ignited to yield thoria. The product, however, is not pure, but contains some ceria, yttria and often also some manganese. Thoria crystallises in the tetragonal system, the crystals being isomorphous with cassiterite and rutile. Specific gravity 10.2.

Thorianite. A thorium-uranium mineral which is a valuable source of these two metals. It occurs particularly in the monazite sands of Sri Lanka and contains up to 70% thoria and about 14% uranium oxides together with oxides of cerium and certain other rare-earth metals. Formula is given as $ThO_2 \cdot U_3O_8$; crystal system, cubic; sp. gr. 9.3.

Thorium. In the first stages of the extraction of rare earths from monazite, thorium is obtained as a sulphate which on boiling with ammonia yields the hydroxide. The hydroxide can be reduced with calcium to yield a dark greyish-coloured powder. Potassium thorium chloride can be reduced with sodium, or the fused double fluoride can be electrolysed using a molybdenum cathode.

The metal resembles nickel in colour, but is as soft as lead; when obtained by reduction of a double salt it frequently appears as microscopic hexagonal tablets. It takes fire when heated in air and burns with a bright flame, forming white thoria. The metal dissolves in nitric acid, but less readily in other mineral acids. It is not affected by aqueous solutions of alkalis. Thorium forms only one oxide (thoria), which, because of its high molecular heat of formation (1.385×10^6 J), is inert towards most reagents.

The metal forms a number of alloys, but the most useful one at present is the thorium-tungsten alloy containing 0.75% Th and known as thoriated tungsten. This alloy is superior to ordinary tungsten in its electron-emitting properties and is used in the form of wire filaments for electronic applications, and in glow tubes. It is more efficient than tungsten as a metal target in X-ray tubes, but since its melting point (1,750 °C) is considerably lower than that for tungsten (3,410 °C), it is less useful than the latter,

where large energy input might melt the target or cause some volatilisation. The present most important uses are as electrodes in photo-electric cells and as a getter in vacuum tubes.

Thorium is a radioactive metal having only one isotope, and is a potential source of nuclear energy.

Capture of neutrons by thorium leads to the production of an isotope of uranium, ^{233}U which is fissile.

Mesothorium is used in medicine but not radiothorium. The latter is a powerful α-emitter, and its intense ionising power makes it suitable for the dissipation of static electricity and as a spark-plug alloy. Crystal structure—face-centred cubic $a = 5.087 \times 10^{-10}$ m. For physical and mechanical properties *see* Table 2.

Thortveitite. A mineral substance having the formula $(Sc,Y)_2 Si_2 O_7$ which contains about 42% ScO_2 and from which scandium may, with some difficulty, be extracted.

Thoulet solution. A solution in water of potassium mercuric oxide which has a high sp. gr. (3.19) and is used as a medium for separating mineral particles of different specific gravities (also called Sonstadt solution).

Three-high mill. A rolling-mill in which there are three rolls arranged one above the other. The middle roll is the one usually driven from the motor while the other two are geared to it. In use the work is generally passed between the lowest and middle rolls, and back again between the middle and upper rolls. This arrangement obviates the loss of time and the consequent cooling of the work that occurs in a 2-high mill when the work has to be brought in an "idle" pass (*see* Fig. 142B).

Three-part box. A moulding box or flask in three parts. The lower part is known as the drag and the upper part as the cope with the centre or cheek between them.

Three-phase current. In the generation of alternating current it is possible to use several individual circuits in the same machine. The voltages and the frequencies of each circuit are the same, but they have definite phase differences depending on the number of phases. The object of a polyphase machine is to produce a bigger output per machine than is possible with a single-phase machine. Polyphase machines have a more uniform torque than single-phase machines and are self-starting. In a three-phase system the e.m.f.s in the individual phases have a mutual phase difference of 120 degrees. The output of a three-phase system is given by

$$\text{Power} = \sqrt{3}EC \cos \phi$$

where E = voltage, C = current and $\cos \phi$ = power factor.

Throat. A term usually used in reference to the opening at the top of a furnace through which the charge is introduced into the furnace, e.g. the blast furnace.

Throttle valve. A device controlling the flow of a liquid or gas in which the controlling movable part consists of a circular disc capable of turning about a diameter inside the supply tube or pipe. When the valve is open the disc lies in a plane parallel to the length of the valve, and when closed it lies at right angles to this direction. The flow of fluid can thus be restricted to the extent desired by turning the disc through any angle up to 90 degrees.

Throttling. A term used in reference to the operation of reducing or restricting or choking down the flow of a liquid or a gas by means of a throttle valve (q.v.).

Throwing power. A term used in electroplating in reference to the property of a plating solution to deposit an adequate thickness of plating metal in recessed parts of an article of complex design, without resorting to special arrangement and disposition of the anode. The throwing power can be assessed by dividing a cathode into two equal parts and setting them at opposite ends of a standard-size tank. The two half cathodes are joined externally

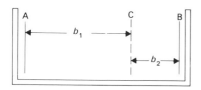

Fig. 173 *Throwing power.*

and a perforated anode is set between them so as to be at unequal distances from them. A measured amount of current is then passed through the plating bath that is being examined keeping the temperature constant. In Fig. 173, A and B are the half-cathodes and C is the perforated anode which is located at b_1 cm from A and b_2 cm from B and if w_1 and w_2 are the weights of plating metal deposited on A and B respectively, then throwing power can be expressed:

$$\text{T.P.} = \frac{\dfrac{b_1}{b_2} - \dfrac{w_2}{w_1}}{\dfrac{b_1}{b_2} + \dfrac{w_2}{w_1} - 2}$$

$$= \frac{(b_1 w_1 - b_2 w_2)}{(b_1 w_1 + b_2 w_2 - 2 b_2 w_1)}.$$

Thulia. A member of the group of rare earths which occurs naturally in association with holmia and erbia. It has the formula $Tm_2 O_3$ and is greenish-white in colour.

Thulium. A metallic element of the rare-earth group. It occurs in small amounts in the minerals gadolinite, euxenite and xenotime, and always in association with other rare-earth elements. Both the oxide and the salts of the metal are pale green in colour. Atomic weight 168.93, at. no. 69, sp. gr. 9.31; crystal structure, hexagonal close-packed; lattice constants $a = 3.53$, $c = 5.575 \times 10^{-10}$ m. (*See* Table 2.)

Thuriting. A box-annealing process sometimes used in the production of deep-drawing quality steel sheet. The method consisted in heating the metal rapidly to about 930 °C and then quickly cooling. This was followed by box annealing at about 650 °C.

Tile ore. A reddish-brown earthy variety of cuprite $Cu_2 O$ generally containing iron oxide, to which the colour is due.

Tilt hammer. A form of heavy hammer once used in working steel or puddled iron ore. It differs from the helve in that the arm is pivoted near its centre. The tail end is acted upon by teeth or by projections on a rotating wheel which lift the heavy head and then allow it to fall upon the work laid on the anvil. A large number of heavy blows can thus be dealt in rapid succession. Such hammers have been superseded by steam hammers.

Tilting furnace. A type of open-hearth or electric furnace that can be tilted about its major axis to facilitate the pouring of the molten charge. Such furnaces have a larger capacity than fixed furnaces, some of the former being designed to take up to 300 Mg (300 tons).

Timber. A useful, but highly heterogeneous natural building material. The two broad groups are softwoods (conifer trees, pine, spruce, etc.) and hardwoods (oak, ash, etc.). Trees have two annual growth additions, the spring wood and summer (or autumn) wood. The concentric rings seen in a cross-section are caused by the varying reflectivity of these growths, but some trees do not show them (e.g. mahogany). The centre of the tree (heart-wood) is darker than the outer (sap-wood).

The structure of wood consists of long narrow fibres (tracheids) parallel to the trunk or branch, with patches called "pits" through which water or sap can diffuse from one fibre to the next. Pines have resin ducts parallel to these fibres, as do other softwoods, and some have short resin-containing cells (Californian redwood). Perpendicular to these fibres and ducts are thin ribbons arranged radially, called medullary rays. These contain nutrient material, and are the principal targets of predators like animals, insects and fungi attacking the tree via the bark. Hardwoods also contain water-conducting tubes called wood vessels, seen as pores in cross-section. In the sap-wood, air is also present in these, and the medullary rays are much thicker in hardwoods.

As it represents the growing area, sapwood contains the highest water content but the heart-wood also contains the debris of previous growth, and cells and fibres are embedded in complex chemical products. It is difficult to impregnate wood therefore, except the dried-off sap-wood. The main chemical substances involved are cellulose and lignone. In wood-pulp manufacture, cellulose remains after the lignone is attacked by weak acids and steam, and can itself be dissolved to produce esters used in regenerated cellulose fibres (rayon) and film (acetate (ethanoate), xanthate, nitrate).

Wood cells can take up about 30% by weight of water, and swell as a result. When saturated the wood can be softened by pressure to form complex shapes. When heated in steam it can be readily curved. When seasoned under cover, wood adjusts to ambient humidity, about 5% in warm climates, about 15% in Britain. The shrinkage which occurs during seasoning is anisotropic.

Anisotropy of properties is one of the main characteristics of wood, because of the fibrous grain structure. In tension and compression, the highest strength is along the grain, but in shear the greatest resistance is perpendicular to the grain. In bending, the stress load applied perpendicular to the grain causes tension and compression along the grain, with the greatest strength in this case. Beams are therefore cut parallel to the grain as far as possible.

The strength/weight ratio of wood is better than steel, but the cross-section required in wood is so much more bulky. In compression and bending (building structures) the bulk is less disadvantageous. Wood has a high modulus of elasticity and low damping capacity, and hence has been used for musical instruments over centuries, the grain being parallel to the sheet in the case of resonant cavities such as the string instruments.

To overcome anisotropy, impregnation is used, when the wood is dried out and degassed in a vacuum chamber. Synthetic wood uses small chips of wood bonded by a resin matrix, and can be shaped during the curing of the resin. Plywood, produced by combining more than one sheet, and rotating the grain direction on alternate sheets, bonding by glue or resin, gives a more isotropic behaviour, and can in addition be waterproofed. Thick sheets are either formed of synthetic wood, or by combining bars of rectangular cross-section with top and bottom plywood sheets (blockboard).

The aesthetic value of polished wood is unfortunately achieved only at high cost, being a labour intensive operation. Veneers of thin wood are commonly bonded to cruder inner layers, or on to plastic sheets, to reduce costs.

Time-temperature-transformation diagrams (T.T.T. curves). A convenient manner of exhibiting the characteristics of steels to show to what extent and at what temperatures the decomposition products of austenite are formed. Austenitised specimens are quenched in molten baths which are maintained at constant temperatures below the lower change point. The austenite breaks down after an initial period into products depending upon the temperature of the bath, and the amount of the products depends on the time the specimens are held at the temperature of the bath.

When the beginning and end of transformation corresponding to each isothermal temperature are plotted on a time-temperature graph, and the points representing the beginning and also the points representing the end of the transformation are joined, curves are obtained that are roughly S-shaped. Such curves are fundamentally important in the study of the significance of heat-treatments and provide an indication of how suitable treatments should be carried out. (*See also* **S-curves, Isothermal transformation diagram.**)

Time yield. The term used in reference to a short-time creep test applicable to steel. The usual type of creep test on steel at elevated temperatures occupies a period of about 3 weeks or so, but the figure for the time yield is obtainable in about 72 hours. The method is to determine the highest stress that will produce an extension of less than 0.5% of the gauge length in the first 24 hours followed by a steady extension of 0.0001% during the next 48 hours. The figure obtained can be used to calculate the safe stress of the material, and this can be taken as approximately two-thirds of the time-yield figure.

Tin. A rather soft white lustrous metal obtained almost entirely from the mineral cassiterite SnO_2. This mineral occurs as lodes as well as alluvial deposits in Malaysia, Burma, Bolivia, Thailand, Zaire and Indonesia. The concentrated ore is roasted to eliminate sulphur, arsenic and antimony as oxides, and if tungsten is present is further heated with sodium sulphate. The sodium tungstate is leached out and the residue smelted with carbon in a reverberatory furnace. The cast pigs are refined by liquation and by poling to give a 99.9% pure product.

Tin is not affected by exposure to air or water at ordinary temperatures or by water up to its boiling point. It dissolves in mineral acids and in alkalis.

At temperatures above 13.2 °C the white tetragonal allotropic form is stable, while at temperatures below this the grey cubic modification is stable (see also **Tin pest**). Above 170 °C tin passes into the rhombic form.

The pure metal has low tensile strength and hardness, but good malleability, and is most readily worked about 100 °C. At 200 °C the metal is readily pulverised. In the form of foil it is used as wrapping material for food products and in electrical condensers. In block form it can be made into pipes for conveying beverages. It cannot easily be drawn into wire. As a coating metal for domestic utensils and food containers (tinplate) it is widely used, since no toxic effects have been observed. Tin is used as a corrosion-resisting protective coating on steel and copper, and the coating may be applied by hot dipping or by electrodeposition or for certain applications by metal spraying. Many useful alloys contain tin as an essential constituent. All true bronzes are copper-tin alloys. It is a constituent of gun-metal, bell metal and many bearing alloys, solders, pewters and fusible alloys.

Tin compounds, e.g. SnO_2, are used as opacifiers in vitreous enamels and as stabilisers in plastics. The mechanical and electrical properties are given in Table 2. Crystal structure (grey) diamond, white tetragonal, lattice constants $a = 5.8312$, $c = 3.1817 \times 10^{-10}$ m.

Tin amalgam. Though tin readily combines with mercury at ordinary temperatures, an amalgam is more readily formed when mercury is poured into molten tin, and according to the proportions of mercury used the product may be liquid or solid and crystalline. The crystalline variety may also be produced by electrolysing tin II chloride using a mercury cathode. Tin amalgam is used for silvering mirrors. The tin (usually containing 1% or 2% Cu and a little lead) in the form of foil is laid on a flat slate and a thin film of mercury poured on to it. Then carefully cleaned glass is floated on this and pressed to remove air-bubbles. The glass is then evenly weighted and after a time the excess of liquid is allowed to drain off.

Tincal. Hydrated sodium borate, also known as borax (q.v.). It is used in making glass and ceramics, as a cleansing agent, as a flux in soldering and as a corrosion inhibitor in antifreeze liquids.

Tin-copper plating. Copper and tin can be co-deposited electrolytically from an alkaline bath containing sodium stannate and copper cyanide. Though the ratio of the two metals can be varied, the best results are obtainable when the percentage of tin lies within the range 40-55%. Such deposits are relatively hard, are corrosion resistant and do not tarnish below 100 °C. The alloy has a pleasing appearance, is whiter in colour than chromium and can be used as a decorative finish as well as for mirrors.

Tin cry. A characteristic crackling sound made when pure tin is deformed. It is also noticed in the case of certain tin-rich alloys and is associated with the movement of crystal blocks and the mechanical twinning of crystals by deformation.

Tinfoil. Very thin sheet tin rolled to thicknesses between 0.006 and 0.2 mm. It was used for food wrapping and for making electrical condensers, now replaced by aluminium foil.

Tinned wire. The name used in reference to steel or copper wire drawn to finished size and tinned. If the wire has been tinned in some convenient gauge and subsequently drawn to the required finished size it is known as drawn tinned wire.

Tin-nickel plating. An alloy deposited electrolytically from a solution of the following composition:

Tin II chloride	50	g
Nickel chloride	250	"
Ammonium hydroxide soln. (0.88)	35	"
Ammonium bifluoride	40	"
Water	1,000	cm³

The deposit, which contains about 35% Ni and 65% Sn, has a slight pinkish-brown tinge and is much darker than silver or chromium. It is resistant to tarnishing at ordinary temperatures but is unsuitable for use at moderately high temperatures. The alloy deposits well on non-ferrous metals, but if used on steel a deposit of copper should first be used.

Tinning. A term used in reference to the operation of coating a base metal with tin for the purpose of corrosion protection. The coating may be obtained by hot dipping into molten tin, by electrodeposition or by metal spraying. The term is also applied to the coating of metals with tin or tin-lead alloys preparatory to assembling a soldered joint.

Tin pest. An allotropic transformation of tin from the normal tetragonal metal (white tin) to a grey powder form crystallising in the cubic system. The transformation (equilibrium) temperature is 13.2 °C, but there is a marked undercooling effect, and much lower temperatures are needed to produce the grey tin. Most of the common impurities present in commercial quality tin greatly delay or inhibit the transformation. Also referred to as tin plague or tin disease.

Tin plague. A term synonymous with tin pest (q.v.).

Tinplate. A tinned steel sheet used for containers for foodstuffs, beverages and dry goods.

Tin plating. A term used mainly—but not exclusively—in reference to the electrodeposition of a thin protective coating of tin on metallic surfaces. The metal resists the action of air, water and mild alkalis (e.g. sodium carbonate), but on account of its standard electrode potential (see **Electrode potential**) it is less protective to steel than either zinc or cadmium. The tin may be deposited from an alkaline bath containing sodium stannate or from an acid bath containing tin II sulphate. Addition agents are generally introduced, and these may be gelatine, glue, glucose, cresol (hydroxytoluene), sulphuric acid or β-naphthol. The deposits are matt in appearance, but bright deposits can be obtained using glue, resin (2 hydroxynaphthalene), etc., in alcohol as brighteners.

Tin powder. Finely powdered metallic tin (about 200 mesh) of 99.8% purity used in making sintered tin alloys.

Tin pyrites. A sulphide of tin, copper and iron $Cu_2S.FeS.SnS_2$ that usually contains also varying amounts of zinc. It occurs in association with cassiterite, copper ores and pyrites in Cornwall (UK), Zinnwald (Czechoslovakia) and Bolivia and is worked for both tin and copper. Also known as bell metal ore, stannine. Crystal system cubic (tetrahedral), Mohs hardness 4, sp. gr. 4.4.

Tin salt. The commercial name of a hydrated stannous chloride $SnCl_2.2H_2O$. The crystals are transparent monoclinic prisms which melt in their own water of crystallisation at about 40 °C, and on cooling the crystals are again formed. Specific gravity 2.71.

Tin salt was used as a plating bath with caustic soda.

Tinstone. An important ore of tin consisting mainly of stannic oxide SnO_2. Also known as cassiterite (q.v.).

Tin sweat. A term used to describe the formation of beads of a tin-rich, low-melting-point constituent exuded from the surface of bronze castings as a result of inverse segregation. The occurrence is usually associated with bronzes containing much hydrogen.

Tint-metal. Sheet metal that has a bright deposit of copper plated on to tin, but more usually on zinc or steel as the basis metal. The polished surface is protected by a paper coating so that the sheet can be pressed or stamped to form ornamental articles. This obviates the need to polish (or plate) the articles after manufacture. The sheet may also be produced in striped colours.

Tin white cobalt. An alternative name for smaltite or cobalt arsenide.

Tin-zinc plating. Tin and zinc can be co-deposited as an alloy from an alkaline solution containing sodium stannate and zinc cyanide. The proportions of tin to zinc can be varied by altering the composition of the bath. A 50/50 alloy has been satisfactorily deposited and has good flexibility and corrosion resistance. It is readily solderable. An 80/20 alloy has been shown to possess outstanding resistance to tropical conditions where corrosion rates are usually high.

Tirr. The name used in reference to the surface layer of earth that has to be removed in quarrying before the underlying rock is reached.

Titanates. Combinations of titanium oxides with other metal oxides (most often alkaline earth metals), which have special electrical properties. They are frequently ferroelectric, i.e. the polarisation can be permanently reversed by application of an electric field. Perovskite, $CaTiO_3$ is an example, but $BaTiO_3$ is used more frequently. The structures all contain octahedral arrangements of oxygen ions with titanium ions in the centre of these octahedra, and the alkali earth metal ion at the body centre of cubically arranged titanium ions, in the interstitial positions between the oxygen tetrahedra.

The titanium ions are able to move into acentric positions within the oxygen octahedra, and it is this which gives rise to ferroelectricity. The titanates are also piezo-electric and pyro-electric.

Titania. The white oxide of titanium TiO_2 produced mainly from ilmenite which contains 50-54% TiO_2. It is used as an important pigment and replaces zinc oxide in the making of white rubber paint and as a filler for paper. Titania can be used alone as a refractory and as an electrical insulator. Crystals of the oxide show marked piezo-electric effects, have a greater brilliance and higher refractive index than diamond.

Titanic iron ore. An alternative name for ilmenite $FeTiO_3$ (q.v.).

Titanite. A naturally occurring mineral which is calcium titano-silicate $CaTiSiO_5$. It may be brown, green or black in colour and crystallises in the monoclinic system. It occurs embedded in granite, gneiss, mica-schist and granular limestone and usually shows a resinous lustre; specific gravity 3.4-3.56. The mineral guarinite has the same composition, but crystallises in the tetragonal system; specific gravity 3.49.

Artificial crystals of titanite are obtained when titanium di-oxide, silica and calcium chloride are fused together. Also called sphene.

Titanium. A white metal whose fracture resembles steel which occurs in combination (as oxide, silicate, etc.) in ilmenite $FeTiO_3$, rutile TiO_2, titanite $CaTiSiO_5$, anatase, brookite TiO_2, perov-skite $(Ca,Fe)TiO_3$, schlorlmite $Ca(Ti,Fe)SiO_5$ and keilhauite $CaY(Ti,Al,Fe)SiO_5$. The principal producers of these minerals are the USA, Canada, Brazil, Norway, India and Australia.

Excluding the gaseous elements and carbon, titanium is eighth in relative abundance of known elements. Of the structurally useful metals only iron, aluminium and magnesium are more abundant. Most of the commercial production of titanium metal is based on the Kroll process. In this titanium dioxide and carbon are heated together at about 500 °C in a stream of chlorine. The oily, fuming product is distilled with copper powder to yield pure titanium tetrachloride. This latter is then reduced by magnesium and magnesium chloride by the vacuum purification process. Other methods of producing the metal include: (a) reduction of titanium dioxide by magnesium and calcium; (b) thermal dissociation of titanium tetraiodide; (c) decomposition of titanium tetrachloride by hydrogen and (d) the electrolysis of fused chloride. The metal obtained by any of these processes is consolidated by melting in water-cooled copper crucibles in electric arc furnaces. Because of the embrittlement caused by the diffusion of oxygen at elevated temperatures, thin sections must be enclosed in a steel sheath for hot working. Thick sections can be worked at 850 °C for short periods. The metal can be cold-worked, but it is necessary to form a thin layer of oxide to hold the lubricant when cold-drawing. The annealing temperature is 720-750 °C.

In powder form titanium is stable at room temperatures, but commences to oxidise at about 120 °C. The consolidated metal may be heated to about 700 °C before the oxidation becomes troublesome. In pure metal the 1% proof stress is about 620 MN m^{-2} and the ultimate tensile stress is about 770 MN m^{-2}. The S/N curves appear to indicate that the fatigue limit is roughly half the U.T.S. Useful alloys are being exploited, good mechanical properties being found in the alloys containing chromium, iron or aluminium, manganese.

The metal has a high resistance to corrosion, especially towards cold dilute acids, fused chlorides, chlorine and sea-water. It maintains its high strength/weight ratio at temperatures where the light alloys begin to fail or fall off in this advantage, and it seems likely that titanium may replace steel in some of its present applications in high-speed aircraft. Titanium is also used in pyrotechnics and as a getter in vacuum tubes. Sintered titanium-tungsten carbide $W_2Ti_2C_4$ is used for tipping tools and in the manufacture of valve seats, die-inserts, etc. It is used to improve strength, hardness and ductility and to refine grain size in such metals as aluminium, copper, magnesium and nickel. In the form of ferro-titanium it is used for deoxidising and for the removal of nitrogen from steels. It is a powerful carbide stabiliser and is used for this purpose in chromium-nickel stainless steels (see also

Weld decay). The properties of titanium are given in Table 2, thermal conductivity 0.021 W cm^{-1}°C^{-1}; crystal structure, α-form (up to 885 °C) hexagonal close-packed, lattice constants $a = 2.96$, $c = 4.692 \times 10^{-10}$ m; β-form (above 885 °C) body-centred cubic $a = 3.312 \times 10^{-10}$ m.

Titanium alloys. The alloys of titanium, like the basis metal, have good strength/weight ratios. The strength varies according to the method of production, being generally higher in those produced by induction melting than in those produced by arc melting. However, the higher strength is generally accompanied by lower ductility. Alloying elements likewise exert a marked influence on mechanical properties as indicated in Table 9.

Titanium carbide. A compound readily produced by fusing titanium dioxide with carbon or with calcium carbide. So obtained it is a dark grey mass having a crystalline fracture and specific gravity of 4.25. It is not attacked by hydrochloric acid and only slowly by aqua regia, but it burns in oxygen rather more readily than the metal. It is mixed with tungsten carbide in the making of some sintered carbide cutting-tools. It is extremely hard, and lighter in weight and less costly than tungsten carbide. With nickel or cobalt as the binder it is used in making heat-resisting parts where high strength is required and the oxidising conditions are not too severe. The cobalt alloys are useful up to about 860 °C and the nickel alloys up to about 1,400 °C. The carbide is used also in making arc lamp electrodes. The formula is TiC.

Titanium dioxide. TiO_2 (see **Octahedrite**).

Titanor metal. A titanium steel used for cutting tools, in which it was used instead of vanadium to produce increase of hardness and resistance to shock.

Tocco process. A process used mainly for producing hard surfaces on crankshafts which consists in selectively hardening the desired surfaces by induction heating.

Toggle joint. A special form of joint that is a combination of three levers, one of which acts in a direction at right angles to the other two. The function of such a joint is to permit a considerable force to act through a short distance. Such joints are therefore used with certain types of crushers, presses and brakes (see Fig. 174).

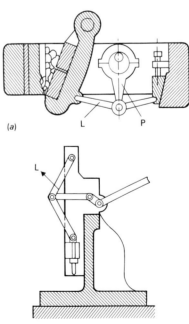

(a)

(b)

Fig. 174 *Toggle joint (a) applied in rock or ore crusher; and (b) applied in punch or press. L is the toggle lever; and P the pitman.*

Toggle lever. A mechanism that really consists of three levers, two of which act in approximately the same line, while the third, named the pitman, acts at right angles to the other two. From Fig. 174 it can be seen that the shorter toggle has one end pivoted in a fixed block and the other in the head of the pitman. The longer toggle has one end pivoted also in the pitman, but the free end has a short lateral movement. By giving the pitman an up-and-down motion the free end of the larger toggle can be given a backward-and-forward motion to operate a crusher jaw or a press for deep-drawing of sheet metal. The use of an eccentric to operate the pitman ensures that the metal can be held firmly round the die during the process of drawing.

Tolerance. A term used to indicate the maximum permissible deviation from a stated value. It may refer to composition, dimensions or properties of a material.

Toluene (methyl benzene) $C_6H_5.CH_3$. An aromatic hydrocarbon, in which one hydrogen atom of benzene is replaced by a methyl group. A solvent for styrene (ethenyl benzene) and vinyl (ethenyl) polymers.

Tool steel. A term used in reference to those steels that are particularly suitable for the production of engineers' tools. They are generally grouped into two broad classes: (a) plain carbon steels and (b) alloy and high-speed tool steels. Those in the former class usually contain a high percentage of carbon—about 0.8-1.45%— their hardness depending on the amount of carbon present. The grades of tool steel containing 0.8% to about 1% C are used in those applications where a certain amount of elasticity is required. They are therefore used for punches, hammers and large chisels. The steels having higher carbon content are more brittle and are used for cutting tools such as drills and edge tools generally. Razor steel in this class contains about 1.5% C.

In the second class or alloy steels the percentage of carbon is less than in the plain carbon steels, and a variety of alloying elements may be present in varying proportions, according to the intended use of the steel.

Chromium is a carbide-forming element. It improves the corrosion and heat resistance of the steel, increases the depth of hardening during heat-treatment and the high temperature strength.

Tungsten is a strong carbide-forming element and is an essential constituent of most alloy tool steels.

Vanadium is essentially a grain-refining element in steel, as it restricts the growth of austenite at elevated temperatures. It greatly improves the hardenability of alloy steels.

Cobalt is a valuable alloying element for high-speed tool steels. It has the effect of raising the softening temperature of ferrite so that tools made from cobalt-alloy steel can operate at high temperatures, maintaining their cutting capacity.

Molybdenum when added to a steel makes it more resistant to excessive grain growth during annealing or tempering, but the main advantage to be gained lies in the increase in the high-temperature strength that results from quite small additions of this element.

Manganese occurs in small amounts as a constituent of practically all steels. The intentional addition of this element in small amounts (up to about 1.2%) not only removes objectionable sulphur, but also tends to increase the hardenability of the steel. Table 6 gives some typical compositions of different types of tool steel, but is not exclusive.

Top. A term sometimes used in reference to the upper part of a blast furnace which comprises the charging gear consisting of the hopper and the bell and cone.

Top and bottom process. A process developed for the separation of nickel and copper. It is based on the fact that if copper-nickel sulphides are fused with sodium sulphide a separation into layers occurs. The upper layer consists primarily of a mixture of the sulphides of copper and of sodium, while the lower layer is mainly nickel sulphide. Also known as the Orford Process. From the Orford bottoms nickel is extracted.

Topaz. Aluminium fluosilicate, $Al_2F_2SiO_4$. It occurs in igneous rocks, granites, etc., and in tin-bearing pegmatites. It crystallises in the orthorhombic system. The specific gravity is 3.55 and Mohs hardness 8. The crystals are frequently clear, but may be coloured red, yellow, brown, green or blue. The yellow topaz from Brazil turns pink on heating, and this must be distinguished from Spanish topaz (or occidental topaz) which is, in fact, a quartz.

Topazolite. A transparent yellow or greenish-yellow variety of common garnet $3CaO.Fe_2O_3.3SiO_2$.

Top casting. A method of casting in which the molten metal is poured directly into an open mould or through gates located at the top of the mould. The main objection to top casting is the liability to splashing. The splashings chill on the sides of the mould, are oxidised and not subsequently absorbed in the main castings as the molten metal rises in the mould.

Tophat. The term sometimes used to describe the deeply sunken top in a rimming steel ingot, which is due to marked evolution of gas causing the metal level to fall excessively after filling the mould.

Topping. A term used in reference to breaking off the feeder heads from castings or small ingots. It is usually practised on crucible steel ingots for examination of the fracture.

Top pouring. A term used in reference to the introduction of the molten metal to the mould by pouring in directly at the open top. A tun-dish is often used. The method has the advantage that the casting solidifies progressively from the bottom and favourable temperature gradients occur. The method has, however, some disadvantages (see **Top casting**), and is not frequently made use of except in certain cases.

Tops. A term used in reference to the Orford process in which the copper-nickel sulphides are heated with sodium sulphide when a separation into two layers occurs. The top layers, referred to as "tops", contain most of the copper and the sodium sulphides. The latter may be leached out and the residue treated for the recovery of copper.

Torbane. The name used in Scotland (UK) in reference to boghead coal or bituminous shale.

Torbanite. A variety of oil shale containing a high percentage of microscopic algae. It is usually black in colour and has a specific gravity of about 1.3. It is said to be a variety of cannel coal formed by the deposition of vegetable matter in lakes. It is found in Linlithgowshire (Scotland, UK) and in New South Wales. (See also **Torbane** and **Boghead coal**.)

Torbernite. Copper uranite or hydrated uranium and copper phosphate $Cu(UO_2)_2(PO_4)_2.12H_2O$ is a "uranium mica". The crystals are tetragonal and usually have the form of thin square plates. The mineral occurs in Saxony (Germany) and Cornwall (UK) and is bright green in colour. Mohs hardness 2.5, sp. gr. 3.5. The mineral is radioactive.

Torch. An alternative name for a blow-pipe in which a combustible gas and oxygen are mixed to produce a hot stable flame suitable for welding purposes, for flame-cutting of metals or for flame-hardening of metallic surfaces.

Torch hardening. A process for surface-hardening the desired parts of a component by heating with a torch to the required temperature and quickly quenching. Also called flame hardening (q.v.).

Torque. A force or a system of forces that turns a body or tends to turn it about an axis. A couple and torque may be regarded as synonymous, and the measurement is the value of the product of the force and the distance of its line of action from the axis. The unit is newton metre N m. Also called twisting moment.

Torsional deformation. The term used in reference to the angular twist produced in a specimen by a specified torque. It is measured in radians cm^{-1} and is found by dividing the twist of one end of the specimen with respect to the other, by the gauge length.

Torsion test. A test designed to provide data for the calculation of the modulus of rigidity (shear modulus), the modulus of rupture (ultimate shear strength) and the yield strength in shear. The test piece has square ends for gripping in the test machine, while the reduced portion is cylindrical. If a solid bar is used this can conveniently have a diameter of 20 mm, and the gauge length made 125 mm. Since the torsional stress is not uniform over the whole section, it is sometimes preferred to use thin-walled tubes as test pieces. One end of the test piece is secured against movement and a torque applied at the other end. Values of torque and angle of twist (in radians) are taken throughout the test and finally plotted on a torque-twist diagram. This is then used in computing the torsion moduli. If the force P is applied at the end of a lever arm of length l the shear force exerted is given by:

$$\text{Shear force} = \frac{16Pl}{\pi d^3} = F(\text{say})$$

where d = diameter of specimen.

For a tube whose inner and outer diameters are d_2 and d_1

$$F = \frac{16 P l d_1}{\pi (d_1{}^4 - d_2{}^4)}.$$

From this the modulus of rigidity is calculated:

$$G = \frac{2FL}{d\theta}$$

where L = gauge length; θ = angle of twist (radians). L/θ is obtained from the torque-twist diagram.

Tossing. A refining process in which molten metal is repeatedly ladled from a refining kettle and poured back so as to expose the thin stream of liquid metal to atmospheric oxidation, forming a dross which contains most of the impurities. The dross can be skimmed from the surface. The method, which is sometimes referred to as kettle oxidation, is used in the refining of tin.

Toughness. A term difficult to define with precision but nevertheless of importance in assessing the usefulness of a metal for certain applications. It may be regarded as a combination of strength and ductility, which may be indicated by the amount of work required to cause fracture. The area under the stress-strain curve, if carried to fracture, is proportional to the toughness. It may be regarded as the resistance to failure under dynamic loading conditions, and measured by the energy absorption in an impact test. The results of impact tests (q.v.) are of limited application, and can only be expected to give an indication of how a particular metal at a given temperature and having a particular shaped notch will behave when subjected to impact or sudden loading. It is the resistance of a material to fracture at a notch that is generally accepted as toughness of a material. In this sense toughness is synonymous with absence of brittleness.

Tough-pitch copper. A commercially pure, oxygen-bearing copper that contains minute particles of cuprous oxide, but in such quantity that the electrical conductivity is practically unaffected. It is prepared from blister copper by heating in an oxidising atmosphere so that most of the impurities are removed as volatile products (e.g. sulphur, zinc, arsenic, and lead) whilst other impurities (e.g. iron, manganese, cobalt, nickel and bismuth) pass into the slag. The oxidising conditions also result in the production of some cuprous (copper I) oxide (Cu_2O), and to reduce this the molten metal is stirred with green wood poles. The poling is continued until nearly all the copper I oxide is reduced, and this stage is indicated when a sample of the molten metal solidifies as a small ingot with a flat top and gives a pink fracture having a silky lustre.

Tourmaline. A borosilicate of aluminium containing also other metals in variable proportions. There are three main types: (a) those containing the alkali metals sodium, potassium or lithium; (b) those containing the light metals aluminium and magnesium and (c) those containing aluminium and iron. It occurs as an accessory mineral in the more acid igneous rocks. It is sometimes found in veins with lead and cobalt. It crystallises in the rhombohedral system and the crystals exhibit pyro-electric properties and piezo-electric properties. Since the crystals show double refraction, they can when suitably cut be employed in polarising apparatus. The crystals may be water-clear or black (schorl). The coloured varieties when clear are used as gemstones— the pink or red variety is Rubellite or Siberian Ruby, the green variety is known as Brazilian Emerald and the blue stone as Indicolite.

Tracer. The name used in reference to the foreign substance introduced into or mixed with a given substance so that the location or distribution of the given substance can be subsequently determined. The tracer substance should have chemical (or physical) properties similar to those of the substance being traced, but in addition have one (at least) distinctive property. Radioactive isotopes are very suitable.

Tracking. Surface degradation of an insulator, usually polymeric, to produce carbon which is then conducting. The "track" is the line of conducting material which leads to discharge. Tracking can also occur when metal contacts are partially vaporised by passing excessively high currents, deposition of the metal taking place on the outer surface of the insulator.

Transducer. A device for detecting one form of energy change and translating it into another which may be more sensitively measured. Mechanical strain can be translated into electrical changes such as resistivity of a coil, or induction effects in a coil, without damaging the material being measured, as would occur with fixing screws or collets for mechanical measurements.

Transformation constant. A term used in reference to the decay of radioactive isotopes. If N is the number of radioactive atoms present in a substance at any given time and N_t is the number present after an interval of time t, then $N_t = Ne^{-\lambda t}$, where λ is a constant depending on and characteristic of the isotope under consideration, and is known as the transformation constant. Also called radioactive constant and decay constant.

Transformation points. A term used in reference to the temperatures at which changes occur in the constitution of steels. Also known as arrest points, critical points and transformation temperatures (q.v.).

Transformation range. The temperature range within which austenite forms when ferrous alloys are heated (i.e. Ac_1 to Ac_3) or the range within which austenite disappears during cooling (i.e. Ar_3 to Ar_1). (See also **Transformation temperature**.)

Transformation temperature. A term used in reference to the temperature at which a change of phase occurs, or the limiting temperature of a transformation range. On cooling a carbon steel it will be found that there occur one or two arrests in the rate of cooling, at which temperatures heat is evolved, indicating that a change of phase is occurring. By plotting the arrest temperature against the carbon content of the steel the diagram shown in Fig. 175 is obtained.

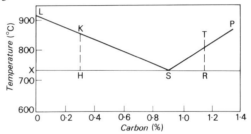

Fig. 175 *Transformation temperature.*

The various transformation temperatures in iron and steel are denoted by the following symbols:

Ac_1—Temperature at which austenite begins to form on heating.
Ac_3—Temperature at which transformation of ferrite to austenite is complete on heating.
Ar_1—Temperature at which the transformation of austenite to ferrite is complete on cooling.
Ar_3—Temperature at which austenite begins to transform to ferrite on cooling.
Ar_{cm}—In hypereutectoid steel, the temperature at which precipitation of cementite commences during cooling.
A_4—Temperature at which austenite transforms to delta ferrite on heating. This transformation is reversible.
M_s (or Ar^{11})—Temperature at which transformation of austenite to martensite begins during cooling.
M_f—Temperature at which the transformation of austenite to martensite is completed.

All these changes, with the exception of the formation of martensite, occur at lower temperatures during cooling than on heating, a phenomenon that is referred to as thermal hysteresis.

In the diagram (Fig. 175) X represents A_1 temperature, i.e. the lower critical temperature; LS represents the series of A_3 temperatures, which vary with the carbon percentage in the steel— these represent the upper critical temperatures; SP for hypereutectoid steel, the A_{cm} points. Ranges such as KH or TR represent the thermal transformation ranges for hypo- and hypereutectoid steels respectively. The suffixes r and c to the critical temperatures indicate "cooling" and "heating" respectively. Steels containing alloying elements have transformation ranges different from those of plain carbon steel.

Transformer. An apparatus by which an alternating current of one voltage can be changed to a different voltage without altering the frequency, but with a change in the amperage, employing the inductive action of one circuit on another. Two circuits are wound on a soft iron core, the one (the primary) consisting of relatively few turns of thick insulated wire and the other (the secondary) of many turns of relatively thin wire. If V_1 and V_2 are the voltages and C_1 and C_2 are the currents flowing in the two coils of respectively n_1 and n_2 turns, then neglecting heat and magnetic losses,

$$\frac{C_2}{C_1} = \frac{V_1}{V_2} = \frac{n_1}{n_2} \text{ approximately}.$$

More exactly it may be calculated that, if r_1 and r_2 are the respective resistances of the two coils and $n_1/n_2 = k$,

$$V_2 = \frac{V_1}{k} - C_2\left(\frac{r_1}{k^2} + r_2\right)$$

from which it is clear that to obtain a constant secondary voltage, the primary voltage must rise somewhat as the current taken from the secondary (i.e. the load in the secondary circuit) increases.

Transient creep. The stage of creep, when the material is deforming but also work hardening, and the rate of creep tends to diminish to an equilibrium or steady state which follows the transient.

Transistor. A semiconductor device used to amplify small alternating voltages or currents. The device consists of two p-n junctions connected back to back. (*See* **p-n junction** for function). This is equivalent to two potential barriers, one between the n-type material (emitter) and the p-type in the centre (base) and one between the centre p-type material (base) and the other n-type material (collector). A forward bias reduces the barrier height between emitter and base, and holes pass from the emitter into the base. A thick base region would soon cause recombination with electrons there, but in the transistor the base is thin enough (10^{-2} mm) for holes to pass through it and into the collector. This is assisted by the potential drop across the base-collector junction (tunnelling).

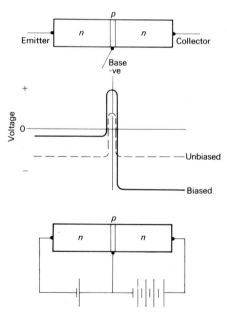

Fig. 176 *Junction transistor.*

A small forward bias from emitter to base will inject a large number of carriers into the base, and if a reverse bias (much larger than the forward one) is applied across the base-collector junction, this will improve the passage of holes into the collector, but oppose any current in the opposite direction (*see* Fig. 176). The low voltage current injected from the emitter is thus collected at a higher voltage. If this emitter current varies with an a.c. signal, the collector current varies, but the output power is the product of the emitter current multiplied by the larger collector

base voltage, and so the transistor becomes a power amplifier. The amplification is about 100 ×, but the materials must be single crystals to avoid grain boundary interference, as near as possible free of dislocations, and carefully controlled in purity.

Using a slice of silicon or germanium substrate, doped with n-type impurity material, deposition of material used for p- or n-type semiconduction can be made. Heating to produce diffusion will result in the appropriate junctions being created and connecting leads can then be added. (*See* Fig. 177.)

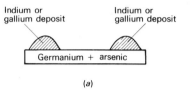

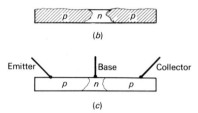

Fig. 177 *Construction of a junction transistor. (a) Deposition. (b) Diffusion. (c) Attachment of leads.*

Combinations of various junctions of this type on one slice can form an integrated circuit, incorporating amplification, rectification, and varying steps of amplification if required.

Transition lattice. An unstable crystal structure formed during solid metal changes such as eutectoid decomposition and precipitation from solid solution.

Transition point. The temperature at which one solid crystalline form of a substance changes to another.

In many polymeric materials, crystallisation from the melt is

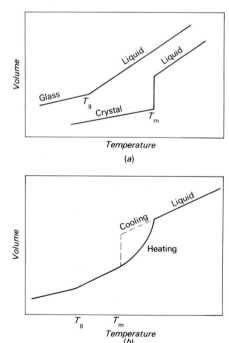

Fig. 178 *Volume/temperature relationships for crystals and glasses. (a) Volume Changes with fall in temperature for crystal and glass. (b) Volume/temperature curve for semicrystalline polymer.*

relatively difficult, in view of the large molecules and particularly when these are chain-like structures. A degree of crystallinity may exist, but of short range order when such a material becomes a brittle solid, and this is called the glass point, or glass transition. There is no change in volume, as occurs with crystallisation, but a degree of freedom is lost, and there is a change in the coefficient of thermal expansion. The process is reversible.

Some polymers undergo partial crystallisation, and hence show a process of crystallisation on cooling, with a glass transition at a lower temperature. The magnitude of the changes on cooling reflects the proportion of crystalline to amorphous material. Figure 178 illustrates typical volume/temperature curves for the two extremes, and that of a partially crystallised material. Structures which are mixed can be varied by the rate of cooling. Rapid cooling may prevent sufficient nuclei forming or growing to form many crystallites, and a glassy state may then predominate. Slow cooling should assist nucleation and growth, and give a higher proportion of crystallites. (*See* **Viscosity**.)

Translation. A term used in reference to crystals or space groups to describe a displacement parallel to a crystal line. A primitive translation is the least translation without accompanying rotation or reflection which replaces every element (point, line and plane) by an equivalent element.

Trans-polyisoprene. *See* cis-polyisoprene.

Transport number. During electrolysis the ions travel towards the electrodes with different speeds and this causes a change in the concentration round the electrodes. The fall in concentration at an electrode is proportional to the number of ions moving away from that electrode, and

$$\frac{\text{loss at anode}}{\text{loss at cathode}} = \frac{\text{speed of cation}}{\text{speed of anion}}.$$

If the fraction "n" of the current is carried by the anion, then the balance (i.e. $1 - n$) is carried by the cation, and the amounts of current carried are proportional to the speeds of the ion

$$\frac{\text{loss at anode}}{\text{loss at cathode}} = \frac{\text{speed of cation}}{\text{speed of anion}} = \frac{1 - n}{n}$$

whence

$$n = \frac{\text{loss in concentration at cathode}}{\text{loss in concentration at both anode and cathode}}.$$

The quantity n is called the transport number of the anion and hence the transport number of the cation is $1 - n$. These are related to the speed of the anion (u) and the speed of the cation (v)

$$\frac{u}{v} = \frac{n}{1 - n}.$$

Transport numbers may be calculated by analysing the solution at the anode and cathode or from e.m.f. measurements of a concentration cell.

Transuranium elements. The term used to designate those elements in the periodic table beyond uranium—i.e. those elements having atomic number greater than 92. None of these elements is known to occur naturally, but all of them are prepared from uranium by suitable nuclear reactions. They are regarded as rare elements and all of them are radioactive. (For details *see* Table 13, **Electronic structure of the elements.**)

Transverse bend test. A mechanical test frequently applied to cast iron and hardened steels for which the ordinary tensile test would prove difficult. The specimen for test may be of square or rectangular section (sometimes round specimens are tested) and is supported near the ends and loaded at the centre. The result of the test may be reported by stating the magnitude of the force causing fracture. At the same time, if the deflection is reported, this gives an indication of ductility. For round specimens the stress may be calculated from the formula $S = 17.6\, PL/d^3$.

For rectangular beams

$$S = 10.5PL/bh^2 \text{ kN m}^{-2}$$

where P = load applied, L = distance between the supports, b = width of beam, h = thickness and d = diameter of beam; all distances being measured in cm and the load in kg.

Trap. The name very generally used in reference to basic igneous rocks that are characterised by fine grain structure and high specific gravity. This description would usually be applicable to basalt, and trap is the name used when basalt is being described for road making.

Trayler mill. A type of tube mill used in the treatment of gold ores. The mill is cylindrical in shape, rotating on a horizontal axis and operated by pebbles, steel balls or rods as grinding agents. Amalgamated plates may be located at the exit end of the mill. Such mills operate more efficiently if the run-of-mine material is first broken down in large jaw-crushers.

Trees. Denticular projections formed on the cathode during electro-deposition. Lead solutions are particularly prone to "treeing" and the addition of an agent such as glue reduces this tendency. In other cases high current densities may be the reason for the occurrence, and in such cases the course to adopt is to carefully adjust the pH of the solution. When the projections are more or less rounded they are referred to as nodules.

Tribology. The science and technology of interacting surfaces in relative motion and of related subjects and practices.

Trichlorethylene (trichloroethene). An organic liquid having extremely good solvent properties for oils, greases, tars and resins. It is used widely as a degreasing agent. It is non-flammable and does not form explosive mixtures with air. Its low surface tension enables it to penetrate readily into small crevices. It is used at ordinary temperatures as a liquid, but the vapour is often used in hot degreasing, and this use is favoured by the low boiling point (86.7 °C) and by the low latent heat of vaporisation. It is relatively non-toxic, can act as a mild anaesthetic, but decomposes to more toxic compounds when heated.

Trickle scale. A term used in reference to a defect sometimes found on the surface of steel sheet and strip. It is formed when scale flakes from the ends of a pair or pack of sheets and trickles in between the sheets during the hot rolling process. Also referred to as tail scale.

Triclinic. A crystal system having three unequal axes, no two of which are at right angles. Potassium bichromate crystallises in this system, the parametral lengths of the axes being $a:b:c = 1.0116:1:1.8146$ and the angles α, β and γ being respectively 98° 0', 96° 13' and 90° 51'. Also called the anorthic or asymmetric system.

Tridymite. A form of silica which is formed at high temperatures and is found in some trachytes. It crystallises in the hexagonal system. It has the same hardness as quartz and a specific gravity of 2.3. Since silica exists in three crystalline forms, which are stable at different temperatures, the ranges are given as follows:

Quartz—Stable at all temperatures up to 870 °C.
Tridymite—Stable at temperatures within the range 870-1,470 °C.
Cristobalite—Stable above 1,470 °C up to its melting point, 1,710 °C.

Tridymite exists, as do the other varieties, in a high-temperature form (called α-tridymite) and a low-temperature form (called β-tridymite).

Triglycine sulphate $(CH_2NH_2COOH)_3 H_2SO_4$. A uniaxial ferroelectric. Similar compounds, the selenate and fluoberyllate, are also ferroelectric. The sulphate is pyroelectric, with a high figure of merit based on the combination of properties. It is used in infra-red detectors, particularly in pyroelectric image tubes for use in the dark.

Trigonal system. An alternative name for two of the 14 Bravais lattices. The trigonal P structure is often called hexagonal, and trigonal R, rhombohedral. Trigonal lattices are usually referred to hexagonal axes, but some authorities regard hexagonal and rhombohedral as one system. (*See* **Crystal classification.**)

Trimming. A term synonymous with fettling (q.v.).

Trip gear. A form of valve gear employed on certain types of steam engine. A lever, known as the trip lever, is used to release a catch and allow the valve, which has been maintained open against the pressure of a spring, to close rapidly.

Trip hammer. An alternative name for a tilt hammer (q.v.).

Triphylite. An important ore containing 1.6-3.7% Li and frequently also some rubidium. It is a phosphate of lithium, sodium, iron and manganese and may be represented by the formula (Li.Na)(Fe.Mn)PO_4.

Triple point. The point on a pressure-temperature diagram at which gas, liquid and solid phases of a substance are in equilibrium. (*See* **Condensation.**)

Tripoli. An abrasive powder obtained from crushed sand-stone. It is a diatomaceous earth, sometimes known as soft silica, and may be white to pinkish-brown in colour, depending on the amount of impurity (e.g. iron) present. As quarried, it contains 30% or more of moisture. It is air-dried, then crushed and finally dried in a furnace. Fine tripoli (when white) is used as a filler in paints and in rubber and as a buffing compound. Very fine tripoli is used as a parting compound in foundry-work. Air-floated—the very finest grade—is used in making metal polishes.

Tripolite. A diatomaceous earth which is a variety of opal and known as opaline silica. From this earth, after drying and crushing, the mild abrasive known as tripoli is obtained.

Triton. The nucleus of tritium (that is of hydrogen of mass number 3). It may be produced as a product of a nuclear reaction, and it may be used as a bombarding particle. When lithium 6 absorbs a neutron, tritium and a helium 4 nucleus are produced. In a hydrogen bomb the enormous heat from the explosion of a fissionable material then causes tritium to fuse with deuterium (H_2) and this fusion yields a neutron and about 17.5 MeV (1.6×10^{-13} J of energy).

Troilite. A sulphide of iron FeS which occurs in meteorites and as non-metallic inclusions in iron and steels of very low manganese content. It crystallises in the hexagonal system, has a melting point of 1,185 °C, and Mohs hardness 3.5. By reflected light it appears yellow or yellowish-brown. Under polarised light the crystals show four extinctions in a complete revolution.

Trommel. Usually a revolving perforated cylindrical screen that is used for grading coarsely crushed or alluvial ores.

Troostite. A structureless, black-etching micro-constituent of hardened and tempered steels. It consists of a very fine aggregate of ferrite and cementite, but is not resolvable under the ordinary microscope. Troostite may appear in two forms—nodular troostite being found in steels that are quenched just too slowly to be fully martensitic; and temper troostite that is the first product of tempering martensite. Further tempering causes a gradual transition into sorbite.

The two types of troostite are different in appearance. When the tempering temperature lies in the range 200-400 °C, the austenite decomposes into fine cementite, but when martensite is tempered, granular nodular troostite appears.

Using the electron microscope nodular troostite appears as a very fine pearlite, having a lamellar separation of about 300×10^{-10} m.

The large trigonal prismatic crystals of zinc silicate or willemite are also referred to as troostite, especially when manganese replaces part of the zinc.

Tropenas converter. A small side-blown, acid converter used for making steel for steel castings. The pig iron is melted in cupolas or the charge may consist of melted pig iron and scrap steel and after desulphurisation the melt is transferred to the converter and the requisite amount of ferrosilicon added. The blast does not pass through, but across, the surface of the metal and as the carbon is burnt to carbon dioxide a high temperature is produced. Sometimes the blast is enriched (up to 30%) with oxygen, so that hotter metal results after a shorter blowing time.

Trouton's rule. The rule states that for normal non-associated liquids the quantity Ml/T = constant (approx. 21) where M = molecular weight, l latent heat and T = absolute temperature of the boiling point. The rule is not sufficiently accurate for the determination of molecular weights, though many normal liquids do conform to this.

Troy weight. A measure of weight. The pound troy was fixed at 5,760 grains in the sixteenth century, but the system was abolished in 1878 with the exception of the troy ounce, its multiples and sub-multiples, now only used for the sale of precious metals and stones. The table of weights is:

24 grains (gr.)	= 1 pennyweight (dwt.)
20 pennyweights	= 1 ounce (oz.)
12 ounces	= 1 pound (lb)
1 lb (avoir.)	= 7,000 gr.
1 lb (Troy)	= 5,760 gr.
1 dwt.	= 1.555 grams

Trumpet. A refractory-lined vessel through which metal is poured into the runners during casting.

Trunnion. A boss projecting from the side of an oscillating or rotating cylinder and by which the vessel may be supported. Hollow trunnions are used in Bessemer and Tropenas converters to convey the air blast.

Tuff. A rock composed of fragments of volcanic origin that have been consolidated into rock around a volcano.

Tufnol. A composite formed by reinforcing phenol-formaldehyde (methanal) type resin with laminates of paper, cloth or other sheet material having some porosity.

Tukon hardness. A sensitive diamond pyramid hardness test developed by the US Bureau of Standards (Knoop). It is used as a micro-hardness test for minerals and micro-constituents of alloys. The result is expressed as Knoop hardness kg mm^{-2} and is approximately equal to the Brinell hardness number.

Tumbling. A method of cleaning small castings which consists in putting the castings into a barrel together with pieces of hard cast iron. When the barrel is rotated, the castings tumble against one another and the cast-iron pieces, producing a good finish on the former.

For polishing, the barrel may contain an abrasive in addition to the hard metal. The final polish can be improved if a mild abrasive in the presence of water or paraffin is used.

Tundish. A refractory trough having one or more nozzles in the bottom that is used on the top of an ingot mould. Its function is to reduce the head of molten metal and minimise splash during pouring.

Tungsten. A steel-grey metal of high density and mechanical strength that is widely distributed in nature. It is not found native, but certain minerals contain high percentages of simple compounds of the metal. The chief tungsten minerals are scheelite $CaWO_4$, wolframite $(Fe,Mn)WO_4$, ferberite $FeWO_4$, tungstite WO_3 and huebnerite $MnWO_4$. The ores are usually associated with cassiterite, quartz, topaz and sometimes gold. They are obtained from South China, the USA, Bolivia, Brazil, Spain, Burma, Cornwall (UK) and Saxony (Germany).

Scheelite, which can be traced by the blue luminescence it exhibits in ultra-violet light, is mainly used for the production of ferrotungsten, while the pure metal is best obtained from wolframite or tungstite. The purified tungstic oxide WO_3 is reduced in hydrogen at about 1,200 °C. The resulting grey metallic powder is strongly pressed into bar form and sintered in hydrogen at about 3,000 °C prior to swaging and drawing hot through carbide dies. In this way the metal is obtained in the form of wire 0.125-0.25 mm diameter. Ductile tungsten may contain about 1.5% Zr or about 2% Cu. The combination of high-fusing temperature and electrical resistivity make the metal ideal for incandescent filaments in lamps and resistances in electric furnaces. On account of its high density and high fusion point it makes excellent anti-cathodes in X-ray tubes and, since it has a coefficient of expansion very close to that of pyrex glass, it can be used in metal-to-glass joints and for electrical lead-ins.

The metal is not affected by a single common acid, and it resists oxidation below a red heat. Tungsten plating has been developed for certain kinds of laboratory ware. Tungsten, in the form of ferrotungsten, is added to steel in the production of high-speed tool steel in which chromium is usually also present. Alloy steel containing 5-6% W has the ability to retain its magnetism. It was thus used for making permanent magnets for radio and measuring instruments though it has been superseded to a large extent by ticonal (*see* Table 6).

Tungsten forms a carbide which is among the hardest substances known. It melts at 2,500 °C, but is brittle. To overcome this, cemented carbide tools are now produced in which tungsten carbide, often with the carbides of tantalum and titanium, has binding metal such as cobalt or nickel embedded in it. When sintered with copper and/or nickel at a temperature about the melting points of these metals, "heavy alloys" are produced of high density which are more effective than lead as screens for X-rays and radiation and may be used as small balancing weights in reciprocating machines.

Apart from high-tungsten steels used for dies, the metal is used as an alloying element with copper, aluminium, etc., producing materials of greatly enhanced mechanical strength. For physical and mechanical properties *see* Table 2. Crystal structure —body-centred cubic (Wα); lattice constant = 3.1585 × 10^{-10} m; Wβ is cubic with lattice constant 5.04 × 10^{-10} m. Wα is the ordinary form of the metal.

Tungsten carbide. Tungsten belongs to the group of heavy metals (e.g. iron, chromium, tungsten, molybdenum) that unite directly with carbon to form carbides which are stable in presence of water. Tungsten forms two carbides, W_2C and WC, the latter being produced when a large excess of carbon is present. Both are very hard, grey, crystalline substances (cubic) and both may be produced by heating tungsten metal with carbon in an electric furnace or by heating in hydrocarbon vapour. The specific gravity of WC is about 16, and Mohs hardness about 9.8; for W_2C these properties are represented by slightly smaller values.

Tungsten carbide is brittle and therefore is frequently briquetted and sintered to give a commercial product that can be used for making into cutting inserts in steel tools and for hard-faced rods. Cobalt powder is a common binding material, this latter, together with the tungsten carbide, being hydraulically compressed and then sintered at about 1,500 °C. The amount of cobalt in the sintered material may vary from 3% to 20%, the high cobalt content being used for tough tools that are called upon to resist shock. Other binding materials are used, e.g. nickel, and the tungsten carbide may be mixed with titanium or tantalum carbides to produce hard tools of high cutting efficiency.

When a binding agent is used in the sintered carbides, the product is known as cemented tungsten carbide. The carbide melts at 2,500 °C. Small sintered and ground inserts are usually embedded in grooves in the surface of a cutting tool and welded or brazed into place. Often the insert has a coating of high-alloy steel as a hard-facing alloy, and this in turn may be covered with deposit from a tungsten carbide surfacing rod. (*See also* **Tungsten electrode.**)

Tungsten electrode. Usually a mild-steel tubular sheath containing fused tungsten carbide and used for the hard surfacing of metals or inserts. The temperature used in arc or oxyacetylene welding is usually not high enough to melt the tungsten carbide, and thus the deposit does not flow freely, but it produces a good, hard surface on steels. The tendency of the high-carbon steels to become embrittled by this treatment is corrected by pre-heating them and by a subsequent annealing treatment. The electrodes are also used in gas-shielded welding of aluminium. Iron-tungsten alloys containing about 60% of the latter have been used as electrodes in electroplating steel to give a high surface hardness. Cobalt-tungsten alloys containing about 50% of the latter, when used as electrodes, produce hard surface deposits on electroplating steel. The hardness is retained up to a red heat.

Tungsten steel. Tungsten is added to iron alloys either as ferrotungsten or as tungsten powder. Usually tungsten steel contains from 3% to 18% W, with 0.4-2% C and chromium up to about 4% may be present. For special purposes, alloy steels may contain up to 24% W. Tungsten steels are hard and dense and, when used in high-speed cutting tools, the hardness is not impaired by the heat generated in use. Steels containing 1-2% W are widely used for such tools as hacksaw blades, reamers and broaches in which a minimum dimensional change can only be tolerated on heat-treatment. Steels containing 3-7% W are often used for fine finishing tools, while those containing up to 12% W are suitable for wire-drawing dies. All tungsten steels show a high magnetic retentivity. Tungsten-vanadium-nickel steels are used for armour plate and in armament manufacture.

Tungstic ochre. An earthy mineral, bright yellow, or greenish-yellow in colour, that is composed of practically pure (98-100%) tungstic oxide, WO_3. Also known as tungstite.

Tungstite. Synonymous with tungstic ochre (q.v.).

Tup. The name used in reference to a heavy mass of iron or steel used as a hammer head.

Turbidimeter. An apparatus used for the determination of particle size. The material to be tested is placed in water in a glass cell and the intensity of light passing through the cell is measured by the aid of a photo-electric cell. As the powder in the sedimentation cell gradually settles, the amount of light transmitted gradually increases. The specific surface of the powdered material is calcu-

lated from the expression:

$$S = \frac{4}{cl} \log \frac{I_0}{I}$$

where c = concentration of the powder expressed as g cm^{-3}, l is the length of the path of the light in the cell, I_0 is the intensity of the light through the clear liquid, and I the light intensity as measured when the light passes through the suspension of the powder in the liquid in the cell.

Turgite. A name used in reference to hydrated iron oxide having the formula $2Fe_2O_3.H_2O$. It is yellowish-brown in colour and is found in the regions of Lake Superior and in New England; specific gravity 4.3. Also called goethite.

Turmeric. An aromatic powder obtained from the dried rhizomes of the curcuma plant—a member of the ginger family. It is used as a flavouring, as a dye for textiles and foodstuffs and as an indicator (acidimetry). The natural colour is reddish-brown, but it stains yellow. Turmeric paper (yellow) is changed to reddish-brown in the presence of alkalis and is sometimes used in Pellenkofer's estimation of carbon dioxide. When dipped in boric acid or an acidified borate solution, the yellow colour becomes brown on drying and turns to bluish-grey or black when moistened with dilute caustic soda or ammonia solution.

Solutions of uranium salts change the yellow colour to orange-brown, and on spotting this paper with sodium carbonate a deep-violet colour results. The original yellow colour can be restored by treatment with dilute hydrochloric acid.

Turnbull's blue. A blue compound precipitated when a ferrous (iron II) salt is added to a solution of potassium iron III cyanide. It was supposed, from its method of preparation, to be iron II iron III cyanide and was given the formula $Fe_3[Fe(CN)_6]$, but after washing and drying—during which oxidation probably occurs—it is found that the blue produced is identical with Prussian blue—i.e. it is $KFe[Fe(CN)_6]$.

Turner's sclerometer. An apparatus for comparing hardness of materials which depends on inscribing a standard scratch on a polished surface. A number of scratches are made using a series of weights upon a diamond point. The point is drawn on the metal once forward and once backward. The standard scratch is determined (*a*) as that scratch which is just visible as a black line on the polished surface when brightly illuminated, or (*b*) as that scratch which can just be felt by drawing the edge of a quill across the metal surface at right angles to the drawn scratches. The hardness is reported as the weight required to produce the standard scratch.

Turner's yellow. When lead oxide is warmed with a solution of common salt, caustic soda is produced and at the same time an oxychloride of lead is formed. When the residue is heated it becomes anhydrous and yellow in colour. The above method of producing caustic soda is known as Turner's method and the yellow residue as Turner's yellow. It is similar to, but probably not identical with, Cassel yellow.

Turning. The term used in reference to the machining of metals and other materials in a lathe. The term also refers (usually in the plural) to the metal shavings produced by working the metal. The speed at which a metal can be turned (i.e. the cutting speed) for a given tool life, is for the same steel in different conditions, inversely proportional to the hardness. For steels of different *composition*, this approximate working rule does not apply. The power to be supplied in the turning of materials can conveniently be expressed as the power required to remove 1 cm^3 of the material using a standard round-nosed tool. Approximate values (in W cm^{-3}) for a few common metals or alloys are: for brasses = 9.1; cast iron 18.2; wrought iron and mild steel 25; medium carbon steel 52; cast steel 68.

Turning effect. When a force is applied to a body so that it tends to turn or twist the body about an axis, the force is said to have a turning effect, which is measured by the product of the force and the shortest distance from the force to the axis. This product is also known as the "torque" (q.v.).

Turquoise. A mineral substance which is a basic, hydrous aluminium phosphate, often containing some copper which imparts a greenish colour. It occurs in Iran and is valued as a gemstone. Mohs hardness 6, sp. gr. 2.6-2.8. Also called calaite.

A good deal of the "turquoise" used in jewellery is really

odontolite, a fossil bone or tooth, coloured blue by the phosphate of iron.

Tuyère. A water-cooled nozzle through which hot air is injected into a blast furnace or cupola. It is commonly made of copper and may be 10-15 cm internal diameter. The design is such that water can be circulated round or through the tuyère to prevent overheating. A blast furnace is usually provided with about ten to eighteen of these tuyères symmetrically arranged above the hearth.

Twaddle scale. The scale of specific gravities used in connection with the variable immersion hydrometer known as Twaddle's hydrometer. The scale is defined by the equation: $n = 200 (S - 1)$ where n is the scale reading on the Twaddle hydrometer, and S is the specific gravity.

A complete set of Twaddle hydrometers consists of six of these instruments. Their respective ranges are: (1) 0-24°; (2) 24-48°; (3) 48-74°; (4) 74-102°; (5) 102-138°; (6) 138-170°. These cover the range of specific gravities of all mixtures of sulphuric acid and water from specific gravity 1 for pure water to 1.85 for pure acid.

The graduations are usually subdivided to represent 0.5° Twaddle. A change in reading of one degree on the Twaddle scale is equivalent to a change in specific gravity of 0.005.

Twin crystals. When two crystals are grown together in a symmetrical manner in such a way that it is possible to recognise a regularity of orientation they are termed twins. In certain cases, e.g. cassiterite, the one crystal may be regarded as a reflection of the other across the plane of contact—known as the twin plane. In other cases one crystal is derived from the other by rotation through 180 degrees about an axis known as the twin axis. In true twins the twin plane (or twin axis) must be parallel to a possible face (or a possible edge).

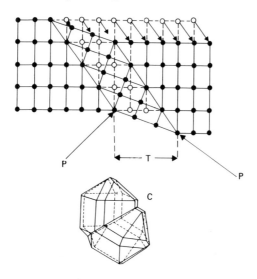

Fig. 179 *Twinning. P are the twinning planes; and C the crystal.*

Twinning. A term used in reference to the special form of slip which may occur in the plastic deformation of metals. It can occur by movement in the same direction on a number of ad-jacent slip planes in which the distance moved by atoms on each successive plane is greater, by a constant amount, than on the previous plane. The zone in which twinning occurs thus has a change in orientation (*see* Fig. 179). The occurrence of twin crystals in a metal indicates that it has undergone permanent deformation at some stage.

Twist. A term used to describe the torsional deformation of a rod or wire. For a round rod or wire, if one end is held rigid and the other has a torque applied, then the twist is the angle (in radians) through which the particles at the latter end have been turned. Usually the twist is expressed as this angle divided by the distance between the two ends or t (twist) $= \phi/l$ radians cm^{-1}.

Twisting test. A test devised to determine the relative strength and ductility of wires. It can be carried out in two ways: (*a*) the wire under test is gripped firmly at one end and held stationary, while the other end is gripped in jaws that can be made to rotate. The distance between the jaws should be stated, but in most tests it is about 20 cm. The movable jaws are caused to rotate until the wire fractures. The number of rotations is reported as the value for this test; (*b*) a second test, known as the reversed-twist test, uses the same arrangement, the difference in procedure being that the movable jaw is rotated forth and back through a specified angle. The number of completed twists (i.e. one forward and one back) is reported when the wire has fractured.

A further type of test, also known as a twist test, is carried out as a routine works test. Strands of wire are twisted together and after a period are untwisted and inspected for the detection of slivers and seams. This same test is also used for detecting coating brittleness on wires—e.g. on galvanised wire.

Type metals. A term used in reference to a series of lead-base alloys used for making printing type.

In type-metal alloys antimony is used to harden the alloy, to improve the resistance to wear and deformation, and to minimise the contraction during solidification. Tin increases the fluidity of the molten metal, improves the casting properties and refines the structure of the alloy. It also improves the ability to reproduce fine detail, and the printing qualities of the type face. Copper is sometimes added in high-quality foundry types to improve durability—i.e. to improve the resistance to wear, abrasion and compression. Some typical compositions are given in Table 8.

For foundry type or case type, which is precast in the form of individual type units and assembled by the printer as and when required, a hard-wearing alloy capable of giving long service is needed. The freezing range, which governs the rate of production, is of secondary importance and the alloys rich in tin and antimony with small additions of copper are used when the highest quality work is required.

Casting machines such as Linotype, which produce type in the form of slugs, demand an alloy which can be cast at very high speeds. The ternary eutectic Pb 84%, Sb 12%, Sn 4%, may be regarded as the approximate composition.

The Monotype machine, which produces individual type units, uses a slightly harder and stronger alloy for ordinary work and richer alloys for case-type. Stereotypes—used in the printing of newspapers—are made by taking a papier-mâché impression of the original type faces and using this as a mould to obtain a metal casting. Printing processes using metal type to produce a relief image are increasingly being replaced by those where the image is reproduced by photolithographic means.

U

Udell. An earthenware vessel used as a receiver for condensing the vapour of iodine in the production of that element. It is shaped as a tube with one end constricted so that it can be luted into the wider end of another udell to form a series of condensers. Also called Aludel.

Udylite. A trade name for a salt of cadmium and of the process for depositing cadmium electrolytically on iron or steel. The name is now used for many proprietary plating solutions, chief among which are the bright nickel solutions.

Ugine-Sejournet process. A process for extruding steel and high-temperature alloys. A glass composition which serves as a lubricant is applied to the surfaces of the billet and the tools prior to extrusion.

Uintahite. An asphaltite found in eastern Utah. It melts completely on moderate heating and is soluble in carbon disulphide. It is used in the manufacture of protective paints and varnishes. Also called uintaite and gilsonite.

Ulexite. A hydrated sodium calcium borate having the composition $Na_2O.2CaO.5B_2O_3.16H_2O$, and found in the Western States of the USA and in Peru. A variety of this mineral, mixed with sodium and calcium sulphates and the chlorides of sodium and magnesium, is also found in Bolivia and Chile. The boric acid content of the dehydrated mineral varies from about 44% to 52%.

Ullmannite. An antimony-nickel ore to which the formula NiSbS has been assigned. Also known as nickeliferous grey antimony.

Ultimate load. A term used in hardness testing. It has been shown that in the Brinell test the Brinell number varies in a complicated way with the applied load. As the load is increased the Brinell hardness rises to a maximum value and then decreases gradually, until such a loading is reached that the ball is embedded to the extent of half its volume. The diameters of the impression and of the hardened ball are then equal. The pressure to produce this is termed the ultimate load, and any increase above this value will cause the ball to penetrate the metal continuously.

Ultimate strength. The term used in reference to the nominal maximum stress that can be sustained by a metal without rupture. In using this term, the type of loading (compressive, flexural, shear, torsion or tensile) must be specified. The intensity of stress is usually calculated on the original cross-section of the test piece and not on the section at the instant and point of fracture. The value of the ultimate strength in tension (U.T.S.) is affected appreciably by the rate of loading, being usually higher for rapid loading than for slow loading, but the percentage elongation is usually smaller.

The value of the U.T.S. is not a constant value, since it is not independent of time. Long exposure to tensile stress to cause ultimate fracture is known as creep (q.v.). In practice, metal parts are not designed to be exposed to their U.T.S. The safe working stress is related to the U.T.S. by the factor of safety.

Ultramarine. An important blue pigment that was formerly obtained by grinding lapis lazuli. It is regarded as a double silicate of sodium and aluminium containing some sulphide of sodium $Na_2Al_2(SiO_4)Na_2S$. An ultramarine poor in silica (i.e. less than 38%) is obtained by heating a mixture of soft clay with sodium sulphate, soda, sulphur and tar or resin. The product is white in colour, but this soon becomes greenish. If this is heated with more sulphur, the product has a fine blue colour.

If pure clay and pure white sand are used the product is blue or purple, according to the proportions of silica used. Those ultramarines having less than 38% silica are decolourised by alkalis, and hence those rich in silica (over 41% SiO_2) are preferable for use where the colour comes into contact with aluminium salts. Ultramarine silver is obtained by heating blue ultramarine with a silver salt. It is yellow in colour and is the starting point for making other ultramarine derivatives. Selenium and tellurium can be substituted for sulphur.

Ultrasonic solders. A range of solders used for soldering by ultrasonic methods. The soft solders used for aluminium and aluminium alloys are applied without flux, and it is not necessary to scratch-brush the metal in the process, since the ultrasonic treatment ensures continuous freedom from oxide film throughout the process. The soft solders are tin-zinc-aluminium alloys, some of which may also contain from 0.8% to 1.5% Cu. The hard solder contains 18-36% Al.

Ultrasonic testing. The term used to describe a method of non-destructive testing in which high-frequency vibrations are used. The method is especially useful for measuring the wall thickness of vessels, when other methods are not applicable; for measuring the thickness of sheet or plate; for determining thicknesses of clad metals; for detecting defects in metal, and more recently, in medical diagnosis.

The source of the ultrasonic waves is usually a quartz crystal that can be made to vibrate below its natural frequency and operated by an electronic oscillator that has frequency control. In use the frequency of oscillation is adjusted so that the reflected wave will be in phase with the transmitted wave which results in a marked increase in the amplitude of the wave. When this occurs the path of the waves (and thus the thickness of the material) is a factor of the wavelength.

Since the waves in the material transverse the thickness twice

$$2t = \frac{V}{xv}$$

where V is the velocity of sound in the metal, t is the thickness of metal, v is the frequency of vibration and x is the number of complete wave forms in the material.

If $xv = N_1$ and the next higher frequency that gives resonance is N_2, then $v(x + 1) = N_2$ and $v = N_2 - N_1$ and since for the fundamental frequency $x = 1$,

$$t = \frac{V}{2(N_2 - N_1)}.$$

This is the basis for the measurement of thickness of materials and the method is available for metals varying in thickness from about 1.5 mm to 5 cm. (*See* **Non-destructive testing**.)

Umber. A silicate of iron and aluminium usually also containing some manganese, used as a pigment. The chief supplies are from Italy and Cyprus, that obtained from the latter place having a rich coffee brown colour. When the "raw" earth is burnt, the product known as burnt umber is obtained. Burnt umber is usually redder than the raw material. Small amounts of umber are found in Scotland (UK) and are known under the names of Caledonian brown and Cappagh brown.

Uncertainty principle. *See* **Heisenberg's uncertainty principle.**

Underclay. The name used in reference to the unstratified clay deposits that may occur immediately below coal seams and which yield useful fireclay.

Undercooling. When pure substances are slowly cooled from the molten condition and the observations of time and temperature are graphically recorded, a portion of the curve will be found to run parallel to the time axis, giving an accurate indication of the true freezing point. With a plain carbon steel a similar type of curve can be obtained, the horizontal portion of the curve (time axis horizontal) giving the critical point. But the addition of alloying elements modifies the shape of such curves in the case of steel. The presence of chromium is responsible for the fact that cooling continues to a temperature below the critical point before change occurs. This is known as undercooling, and the amount depends on the original temperature from which cooling began and the percentage of chromium. As soon as the change commences, however, the temperature rises to the critical point. With increased proportions of chromium, and especially if nickel also is present, the undercooling effect can be quite large. The steel is then cooled down to such lower temperatures that the austenite → martensite change will occur without the formation of soft constituents which happens at higher temperatures. Thus the presence of chromium, which produces a marked undercooling, combined with the effect of nickel in depressing the change points together give rise to the well-known phenomenon of air-hardening in such steels. (*See also* **Super-cooling**.)

Underdraft. The term used in reference to the tendency of a wrought metal to curve downwards as it emerges from a rolling-mill. This occurs because of the difference in the peripheral speeds of the rolls, that of the lower roll being less than that of the upper one.

Underfill. A term used in reference to a deficiency of metal in a forging. This may be caused by using stock of the wrong size or quality or by the stock being misplaced in forging dies or being insufficiently heated. The term is also applied to a cross-section that has not completely filled a roll pass so that it is inaccurate in both size and shape.

Underpoled. In the fire-refining of blister copper, the metal is remelted in a furnace under oxidising conditions so that the oxidisable impurities such as sulphur, lead, zinc, arsenic and antimony are removed as volatile oxides and the heavier metals removed in the slag. This process results in some of the copper being oxidised to cuprous (copper II) oxide Cu_2O. When this occurs it is assumed that the impurities have been preferentially oxidised, and it is then necessary to reduce the oxide of copper. This is done by inserting green poles into the furnace to stir the molten mass, and the reducing gases produced serve to reduce the Cu_2O. The process is known as poling and the product is tough-pitch copper. The control of the process is carried out by taking samples from time to time in a ladle. When the ingot solidifies with a flat top and the fracture is flesh-coloured and silky, the process is complete. The copper then contains about 0.5% (or less) of Cu_2O. If the percentage of Cu_2O is greater than this and the metal on solidifying fractures readily and shows a brick-red colour the deoxidation must continue, for the copper is underpoled.

Uniaxial ferroelectric. Ferroelectric materials in which polarisation can occur in only one crystal direction. (*See* **Ferroelectricity** and **Polarisation**.)

Union. A short coupling internally threaded at both ends for joining two lengths of tube or piping.

Unit cell. A crystal is built up of an ordered assemblage of small units. It must follow that if some point be taken in each of the units, these points must lie on lines and in planes which divide the space into parallel-sided cells, the sides of the cells being parallel to the crystal faces. In crystallography this cell is called the unit cell. It is completely defined by giving the ratio of the sides and the interfacial angles.

The assemblage of points (which could for example be the centres of gravity of the units) is known as the space-lattice (q.v.).

Uni-temper mill. A rolling-mill used for the final (temper rolling) reduction of strip steel for tin-plate production. It consists of two 2-high mills in one stand. The centre rolls have no vertical adjustment and are driven. The top and bottom rolls are both idle, the former being moved by a screw down and the latter by a screw up. With this design the free span of strip between the mills can be kept down to something of the order of 25-40 cm. The relative speed of the two mills is accurately controlled by an electronic system to maintain the required elongation at speeds up to 15 m s⁻¹.

Unit processes. The chemical changes that occur in materials undergoing chemical processing. These are studied to determine the best conditions for the reactions to take place, and to design the most suitable equipment. The processes are classified, e.g. oxidation, reduction, sulphonation, nitration, etc.

Units. The conventional steps in measurement used to define the properties and behaviour of materials and artefacts. The modern coherent system which it is hoped will shortly be in world-wide use is the SI system. (*See* **SI units**.)

Universal coupling. Couplings are devices for joining up shafts or drives, and where provision is made for one shaft to assume various angular relations to the other, it is known as a universal coupling. Large and somewhat complicated universal couplings are required on rolling-mill drives to compensate for the varying height of the rolls.

Universal electro-polishing bath. This bath, which can be used for both ferrous and non-ferrous metals provided the correct current density and the polishing time are observed, has the following composition:

Sulphuric acid	25.0 g l⁻¹	Citric acid	12.0 g l⁻¹
Hydrofluoric acid	33.0 "	Phthalic anhydride	4.3 "
Boric acid	8.3 "	Phosphoric acid	328.0 "
Chromic acid	372.0 "		

The cathode may be copper, stainless steel or high chromium steel, and the anode-cathode distance must be in the range 55-60 mm. The solution should be agitated mechanically. Table 62 gives details of current densities and times for polishing non-micro specimens.

TABLE 62.

Metal	Current Density, A dm⁻²	Polishing Time, min
Ferrous metals	17-40	2-4
Brass	10-15	2-3.5
Bronze	18-24	2-2.3
Light alloys	12-39	2.0
Copper	5-13	1.5
Lead	29-70	6.0
Zinc	20-24	2.0-2.4
Tin	7-9	1.5-3.0

Universal indicator. This is a mixture of indicators that can give a whole range of colour changes indicating pH values over the whole range. (*See also* **Indicators**.)

Universal joint. A type of joint incorporating a means to permit one shaft to assume varying inclinations to the driving-shaft. (*See also* **Universal coupling**.)

Universal mill. This is a rolling-mill having horizontal rolls 2- to 3-high and also vertical rolls for working on the edge of the work. These latter regulate the width of the plate as well as roll the edges. They are particularly useful for rolling thick plates and I-beams.

Universal rolling. A rolling process in which a universal mill (q.v.) is used to control the width and produce a smooth edge to the product.

Universal testing machine. A testing machine that can be used for testing specimens in tension, compression and transverse bending.

Unkilled steel. The term used in reference to steel which has been insufficiently deoxidised and thus evolves gas during solidification, with the consequent formation of blow-holes. (*See also* **Rimming steel.**)

Unsaturated hydrocarbon. A compound of hydrogen and carbon in which some of the carbon-carbon bonds involve more than one pair of shared electrons, often known as double and triple bonds.

Unshielded metal arc welding. A welding process in which the electrode is a metal rod (or a wire) that is melted by the arc and supplies the filler metal, both the filler and weld metal being unprotected by any inert gas or other means.

Unstable nuclei. Strictly, those nuclei of atoms which disintegrate spontaneously, but generally applied to those with a high neutron/proton ratio, and particularly the actinide elements. The majority of these have not been found to occur naturally, but have been prepared artificially. (*See* **Atom.**)

Upend forging. A forging produced in dies so that the direction of grain flow is largely at right angles to the die faces.

Upending. A forging process in which the axis of the material is shortened and the cross-section—usually locally—increased.

Uphill casting. The term used in reference to a process in which a number of ingots are cast simultaneously. The molten metal is poured into a central refractory-lined trumpet or tundish, whence it flows through refractory runners into the bottom of the moulds. Also called bottom pouring.

Upset forging. A forging produced by pressure between the parts of split dies which operate in a horizontal plane. The parts of the die grip the stock as they come together, and a ram forces the stock to completely fill the die, and this operation can be repeated to produce intricate designs.

Upsetting. A metal-working operation similar to forging, in which the diameter of a bar is increased locally by hot forging in which the blow or pressure is exerted in the direction of the axis of the bar. It is used, for example, to form heads on poppet valves etc. In such cases the bar is placed in a holder at the bottom of a die, the bar being longer than the hole. A drop hammer then mushrooms the head. If the upset is to be at some place other than the end of the bar, the lower die is usually made in two parts for the easier removal of the forging. Machines used in upsetting are usually referred to as upsetters, but the terms forging machine and bulldozer are also used.

Upset welding. A method of welding applicable to the welding of chain links. The operation is usually done at a relatively low temperature and with only moderate pressure between the ends to be joined. The heat is obtained by electrical resistance. There is usually some evidence of extruded metal at the butted ends. In semi-upset welding a die is placed over one half of the diameter of the link in such a way as to prevent metal being extruded on the outside of the link.

Uraconite. A hydrated sulphate of uranium that occurs as a yellow or orange earthy powder near other uranium ores. Also known as uranic ochre.

Uraninite. A term synonymous with pitchblende (q.v.).

Uranite. A term that is applied to two species of mineral, copper uranite $CuO.2UO_3.P_2O_5.8H_2O$ and lime uranite

$$CaO.2UO_3.P_2O_5.8H_2O.$$

They are regarded as secondary uranium minerals and always occur with other uranium ores. Lime uranite is the same as autunite, which contains 55-60% of uranium oxides. It is greenish yellow in colour and fluoresces in ultra-violet light. Copper uranite is also referred to as uran-mica and torbernite.

Uranium. A lustrous white metal which when pure is malleable and reasonably ductile. Quite small amounts of iron and aluminium (0.5% and 0.2% respectively) cause embrittlement. The powdered metal is grey in colour. The metal is unattacked by air at ordinary temperatures, but burns brilliantly when heated to 170 °C and melts at 1,132 °C. When the powder is fused and separated from the oxide that is formed, a button is obtained which can be cold rolled to 50% reduction. When drawn into wire, annealing in a vacuum at about 1,000 °C is necessary.

Uranium has an unusually high resistivity, and its structure in the alpha phase consists of corrugated sheets stacked together, the binding between sheets resembling the covalent bonds found in Sb, As, Bi. The structure α phase is orthorhombic, with lattice constants at 20 °C $a = 2.8545$, $b = 5.8681$ and $c = 4.9566 \times 10^{-10}$ m, and 4 atoms to the unit cell. The β phase is tetragonal, the transition from α occurring at 660 °C and stable to 770 °C. Lattice constants are $a = b = 10.744$, $c = 5.6515 \times 10^{-10}$ m, with 30 atoms to the unit cell. The γ phase, stable from 770 °C to the melting point of 1,132 °C, is body-centred cubic, $a = b = c = 3.532 \times 10^{-10}$ m. (*See* Table 2.)

The metal is moderately reactive. It decomposes hot water and dissolves in mineral acids, but is not perceptibly affected by alkali solutions. It combines directly with sulphur and carbon and displaces copper, gold, mercury, platinum, silver and tin from their solutions. Uranium occurs in nature mostly as oxides, of which there are four, since uranium has four distinct valencies.

The main interest in the metal centres in the isotopes which are used in nuclear reactors in the production of "atomic energy". The transformations that occur are represented:

$$^{238}_{92}U + n^1 \rightarrow {}^{239}_{92}U \rightarrow {}^{239}_{93}Np + \beta$$

$$\rightarrow {}^{239}_{94}Pu + \beta$$

The transformation to neptunium has a half-life period of 23 minutes and the transformation to plutonium a half-life of 2.3 days. Plutonium is fissionable. The most readily fissionable isotope of uranium is ^{235}U which occurs with ^{238}U usually about 0.7%. The separation, by thermal diffusion is a lengthy process.

Uranium carbide UC. A compound of uranium obtained by heating green uranium oxide U_3O_8 with carbon in an electric furnace. It is a hard lustrous solid which is harder than rock crystal, but less so than corundum. It reacts with oxygen (or air) and with the halogens when heated to about 370 °C. In contact with water it decomposes giving a mixture of hydrogen, ethyne, ethene and methane. When pieces of uranium carbide are rubbed or shaken together, brilliant sparks are given off. Specific gravity 11.28.

The carbide is a potential nuclear fuel, with a high melting point (2,350 °C ± 30 °C) and the sodium chloride type of structure, which gives it reasonable ductility at high temperature. It has been demonstrated to resist radiation damage reasonably well, and benefits by having a higher thermal conductivity than uranium dioxide. It ignites in oxygen and hydrolyses readily.

Uranium dioxide UO₂. The oxide is non-stoichiometric, the composition depending on the oxygen pressure of its environment. When freshly reduced in hydrogen, the stoichiometric composition is obtained, and soon rises to 2.06 oxygen/uranium ratio or thereabouts on removal from the hydrogen atmosphere. The composition changes upwards to a ratio of 2.2 in higher temperature environments, and this could be regarded as the normal commercial purity of artefacts. Under more oxidising conditions, or at high temperatures, the composition moves on to U_3O_8, a new compound (ratio O/U 2.66).

The oxide has the crystal structure of fluorite, and is therefore reasonably ductile at moderate temperatures. This makes it attractive as a nuclear fuel, removing the problems of wrinkling and growth that occur with uranium metal. However, the thermal conductivity is poor, and falls with rising temperature from 0.075 W cm^{-1} °C^{-1} at 100 °C to 0.034 W cm^{-1} °C^{-1} at 700 °C. Because of this, the process of uranium fission creates a severe temperature gradient through a mass of the oxide, and in cylindrical form, cooled from the external cylindrical surface, the centre reaches the melting point of 2,750 °C (± 40 °C). When this occurs, the liquid centre becomes a better heat transfer medium, the temperature gradient falls, and the centre solidifies again. The temperature gradient once more rises, and again the centre melts. This continuing thermal cycle can cause sufficient contraction to create a central void, and almost invariably causes severe radial and circumferential cracking in the solid. However, the retention of gaseous fission products within the lattice is still good, and the swelling associated with such gas in metallic fuels occurs to a much smaller degree. The oxide can therefore stand longer dwell time in the nuclear reactor before reprocessing becomes necessary, or distortion of the fuel element cladding. This is particularly true of fast neutron irradiation as in a fast breeder reactor. The oxide is easy to handle and not affected by water.

Uranium lead. The element produced as the end product of the radioactive decay of uranium. Since in the various transformations of lead eight α-particles are emitted, the atomic weight of the end-product should be $238 - 8 \times 4 = 206$, which is close to that of lead. The product is in fact an isotope of lead. (*See also* **Radioactivity.**)

Uranium mica. An ore of uranium consisting mainly of uranium and calcium phosphate $Ca(UO_2)_2(PO_4)_2.8H_2O$. It resembles torgenite (a copper uranite) in form, but can readily be distinguished from it by its sulphur-yellow colour. It is usually found with pitchblende and other uranium minerals. It crystallises in the orthorhombic system. Mohs hardness 2.5, sp. gr. 3.05-3.2. Also called autunite and calco-uranite.

Uranium nitride UN. Similar to the carbide, in having the rock-salt structure, and with similar thermal conductivity to the carbide. It ignites in oxygen but is not attacked by water below $250\,°C \pm 30\,°C$.

Uranium orange. Alkali uranates used in ceramics and in glass making for producing orange tints. The usual compounds employed are: *(a)* sodium uranate Na_2UO_4; *(b)* sodium or potassium diuranate $Na_2U_2O_7$; and *(c)* hydrated sodium diuranate $Na_2U_2O_7.6H_2O$.

Uranium phosphide UP. Another uranium compound having the rocksalt structure and considered as a nuclear fuel. The thermal conductivity is better than the dioxide but lower than the nitride or carbide ($0.164\,W\,cm^{-1}\,°C^{-1}$ at $800\,°C$). It is relatively stable to oxygen and water. Melting point $2,610\,°C \pm 30\,°C$.

Uranium red. When an alkali solution of uranyl nitrate is treated with hydrogen sulphide an orange precipitate is obtained which on drying turns to a brick-red colour. If this is treated with a mild alkali (potassium carbonate), a compound having a deep-red colour is obtained which is known as uranium red. It has been assigned the formula $5UO_3.2K_2O.HS_2K + xH_2O$. It is decomposed into uranium orange (q.v.) by the action of carbon dioxide, the sulphur being eliminated. An analogous compound may be obtained using ammonium sulphide in place of hydrogen sulphide.

Uranium silicide. U_3Si_2. One of the few uranium compounds which does not have a cubic structure: it is tetragonal. The thermal conductivity is the highest of the uranium ceramics, $0.42\,W\,cm^{-1}\,°C^{-1}$ at room temperature (equal to plain carbon steel) and falling only to $0.25\,W\,cm^{-1}\,°C^{-1}$ at $800\,°C$ plus. It reacts with oxygen above $300\,°C$ but is stable in water up to $200\,°C$. The thermal expansion is high, and fission product gases are not retained as well as in oxide and carbide.

Uranium sulphide US. A potential nuclear fuel, again of rocksalt type structure. The melting point is $2,450\,°C \pm 30\,°C$ and the thermal conductivity $0.172\,W\,cm^{-1}\,°C^{-1}$ at $800\,°C$, only slightly inferior to the nitride. It ignites in oxygen but is stable in water up to $100\,°C$. The fission-damage resistance is said to be good.

Uranium yellow. Sodium di-uranate $Na_2U_2O_7.6H_2O$ or sometimes yellow uranium VI oxide UO_3, used as a pigment to produce yellow or ivory shades of colour in pottery glazes, and for imparting an opalescent yellow colour to glass. The latter becomes strongly fluorescent and appears yellowish-green by reflected light.

Uranmica. A copper uranite, known also as torbernite (q.v.).

Uranotantalite. A mineral substance in which uranium is found combined with niobium and tantalum and usually some rare-earth metals. Samarskite, for example, contains 3-6% U_2O_3, 41-56% Cb_2O_5 and 14-27% Ta_2O_3. It is dark brown or black in colour.

Urea. A white crystalline organic substance having the formula $CO(NH_2)_2$ and melting at $132\,°C$. It is freely soluble in water. It is used as a pickling inhibitor and as a brightener in electroplating.

Substituted ureas are also used as inhibitors and as cleaning solutions for silver, etc.

Urea formaldehyde (methanal). Synthetic resins that are used for plastic mouldings and for adhesives and lacquers. They are produced by the condensation of urea, thiourea, etc., with methanal. The product is transparent, but the optical properties make the resin unsuitable for lenses. In powder form it is added to foundry mixtures for making and strengthening cores. (*See* Table 10.)

Usalite. A refractory material resembling sillimanite. It is used in the form of tubes for the protection of thermocouples, etc. It withstands temperatures in excess of $1,100\,°C$.

Uvarovite. A calcium-chromium silicate approximating to the formula $3CaO.Cr_2O_3.3SiO_2$. It occurs in serpentines rich in chromium, particularly in the Urals. It is emerald green in colour and the best specimens are used as gemstones. It belongs to the large class of hard minerals known collectively as garnets and used as abrasives. Mohs hardness 6.5, sp. gr. 3.6-4.

u.v. light. The radiation within the wide definition of "visible" light which covers the frequency range 10^{15}-10^{16} Hz; 10 to 400 nm approximate wavelength. u.v. radiation is emitted from very hot bodies and ionised gases, e.g. the sun and mercury vapour discharge tubes. Ultra-violet radiation, being of very high frequency, can damage biological tissue, and causes degradation in many polymeric materials, e.g. polystyrene. Many substances fluoresce in u.v. light, which can be useful industrially, and in safety devices. It is used in the treatment of diseases, particularly those of the skin.

V

Vacancies. (Vacant lattice sites). Literally, holes in the crystal lattice which should be occupied by atoms. At the surface, conditions allow for the migration of atoms from the interior to the surface, leaving a vacancy. (*See* **Schottky defect.**) In the crystallisation process, the formation of nuclei and their growth should theoretically lead to occupation of every site in the lattice, but temperature gradients, the viscosity of the melt, and occasionally gas molecules present, may prevent this. Rapid cooling in particular prevents perfect crystal formation, and all these processes lead to a concentration of vacancies. Temperature and pressure changes in the solid state also lead to migration of atoms into interstitial positions, creating vacancies (Frenkel defects).

The ratio of vacancies n to the number of atoms N is given by

$$\frac{n}{N} \approx e^{-W/kT}$$

where W is the work done to remove an atom from the lattice to a "sink" and T is absolute temperature, k Boltzmann's constant. The value of W for copper is about 10^{-19} J, or 1 eV. This gives a vacancy distribution of about 0.001% at 1,000 K. In an AB compound, such as sodium chloride, ion pairs must be involved, to preserve electrical neutrality, and

$$\frac{n}{N} \approx e^{-W_p/2kT}$$

where W_p is the energy of formation of the ion pair. The value of W_p for sodium chloride is approximately 2 eV, so that at room temperature there is one defect ion pair for every 10^{16} ions. At the melting point, some salts, such as silver bromide, show defect concentrations of the order of 0.01%. The heat capacity is affected by this large defect concentration, increasing by 3 $\times$ between 500 °C and 700 °C.

Materials heated to a high temperature, when atomic vibrations become very considerable, retain a high proportion of vacancies on rapid cooling, because the atoms do not have time to adjust to their proper positions. This affects diffusion processes, which are speeded up when reheating a material containing above the equilibrium proportion of vacancies at room temperature. The excess free energy of the disordered parts of the lattice assists the diffusion.

Vacuum. Strictly, a space containing no gas molecules whatsoever. In this sense, a perfect vacuum is unattainable, and the degree of vacuum, measured as the pressure of the residual gas, is normally recorded.

In SI units, pressure is measured in N m^{-2}, but the "pascal" is a term accepted for units of N m^{-2}, and the "bar" (10^5 N m^{-2}) is arguably more useful when dealing with barometric pressure. The old units of mmHg or torr have the equivalent 133.322 N m^{-2}, and the atmosphere 101.325 kN m^{-2}.

A pressure of 100 N m^{-2}, or 1 mb (millibar) is generally rated as the acceptable level for a vacuum, and pressures between 10 kN m^{-2} and 100 N m^{-2} as a "partial" vacuum.

Rotary oil pumps, in which a spring-loaded vane or an eccentric rotates in a chamber immersed in oil, can lower the pressure to about 10 N m^{-2} or less particularly when using gas ballast to clear gas trapped in the oil. At these pressures, a high vacuum pump based on distillation and gas diffusion can operate to lower the pressure further, to the order of mN m^{-2}. The diffusion pump is always limited by the vapour pressure of the medium which it employs, and beyond this level, ion pumps or getters have to be added. A high vacuum system must always be a train of pumps therefore, gradually lowering the pressure down the line. In addition, gas adsorbed on to surfaces of chambers, ducts, valves, must be removed by "baking", i.e. pre-heating to drive it off, and heaters may be incorporated in the system to continue this process under vacuum.

The measurement of high vacuum presents many problems. The McLeod gauge is a simple variant of the barometer, and acceptable for pressures down to the order of mN m^{-2}. Below this pressure, electronic instruments are used, dependent on electrical measurements of gas conductivity or ionisation. Only laboratory research equipment requires such elaborate pumping and measuring equipment, industrial processes are more concerned with rate of pumping than with absolute pressure, since the main requirement is the rapid removal of quantities of gas. Vacuum degassing and casting have become commercially acceptable for special steels and other metallic alloys, but were originally developed for the more reactive metals such as titanium and zirconium, uranium, etc.

Vacuum casting. A process for pouring metals into moulds contained in an evacuated space. During teeming operations some oxidation normally occurs even though deoxidisers are present. Vacuum casting is therefore employed in the making of metals of very high purity. The importance of maintaining the vacuum also arises when a crust or lid forms on the ingots, so that the escape of gases is not hindered or prevented.

When melting as well as teeming is done *in vacuo* it is of importance that all the gas be removed from the melt, because if this is insufficiently degassed the ingot may show greater porosity than under normal conditions.

Vacuum coating. A process for depositing a coating upon base metal or non-metallic substances by vaporising the coating and allowing the vapour to condense on the cooler pieces of work that are to be coated, the whole process being conducted in a vacuum chamber. The coating material is melted and vaporised on an electrically heated wire and the parts to be coated are arranged at convenient distances around the wire, and caused to rotate so that all sides are exposed to the hot wire during the process. Pure metals can be used and the usual coating thicknesses are about 100 nm. On account of the nature of the coating, copper and silver are subject to rapid tarnishing during use, while gold is less bright than when electrodeposited. Aluminium, however, gives a brilliant white finish that does not readily tarnish. Also called vacuum evaporation, vacuum metallising.

Coatings of some compounds are used on glass substrates, such as the "bloomed" lenses to prevent internal reflections, and co-deposition of alloying metals, or alternating layers followed by a diffusion heat treatment, have been used. Continuous coating of polymer substrates is a relatively cheap process for coating metal on to a dielectric. (*See* **Metallisation**.)

Vacuum evaporation. A method of coating metallic and non-metallic articles by melting and vaporising the coating material on an electrically heated conductor in a chamber from which air has been exhausted. The articles must be rotated if all sides are to be treated. Coatings obtained in this way are usually very thin and hence easily damaged. The process is only suitable to produce a decorative effect. Gold, silver, copper and aluminium have been used.

Vacuum filter. A type of filter used to clarify solutions or to remove solution from solids in the treatment of slimes. Linen, coarse cloth, cocoa matting or other material may be used as the filtering medium and over this is sewn the filter cloth. The whole is inserted into the pulp, and "suction" is applied inside the cloth. The solids form a cake on the outside of the cloth while the solution is led away through the suction pipe. In the inter-mittent type of filter, the filter medium is attached to rectangular frames. When the suction flow becomes very slow, the remainder of the pulp is led to an adjacent tank. Pressure is then applied through the suction pipe and the caked solids blown off the filter medium. Continuous filters are of the rotary type, the filter medium being attached to the curved surface of the drum.

Vacuum fusion. The term used in reference to the melting of metals in a vacuum. The heating is generally produced by electrical induction methods. A certain amount of turbulence is produced, which has a beneficial effect in aiding the escape of gases in the metal. The process is also of importance in the vacuum-fusion method of analysis of gases in metal. Specially designed apparatus is required to extract and analyse the gases. The determination is usually done by oxidation and differential freezing methods. The melting and subsequent solidification of an ingot in a vacuum is a method used for removing gas from ingots of both ferrous and non-ferrous metals, e.g. H_2 and CO in iron and steel, H_2 in titanium and some nickel alloys.

Vacuum heating. A process used to bring about certain reactions which occur at high temperatures at which the products of the reaction would oxidise or react with the materials of the container. In powder metallurgy sintering operations are often done, with advantage, in a vacuum. In the reduction of zirconium, titanium and hafnium chlorides by magnesium evacuated steel containers are used. Silicon tetrachloride is reduced *in vacuo* by zinc. The crystals of silicon are remelted in the vacuum to produce crystals suitable for use in electronics, etc. (*see* **Vacuum fusion**).

Vacuum refining. A method used for obtaining certain metals in a very high state of purity. It is particularly applicable for the purification of germanium. In the preparation of beryllium a very high vacuum is employed to permit the distillation of the impuri-ties in the making of the pure element.

Valence band. The band of electron energies in which the electrons taking part in covalent bonding are situated. Since these electrons are "bound" by the valency role, they are not free for transport through the material unless excess energy over that in the valence band can be acquired. This leads to the concept of a comparable "conduction band" at higher energy level, at which energy level the electron may wander through the lattice. Between the valence and conduction band there is a forbidden energy gap, at which levels the electron cannot satisfy the wave mechanical-quantum conditions necessary. The magnitude of this gap then determines the distinction between insulators and semiconductors of the in-trinsic type.

Any other form of bonding, such as the metallic bond, involves no distinct valence band, there is an overlap between higher energy bands and the electron can move into various levels given the necessary kinetic energy. In the ionic bond, the exchange of electrons creates ionisation, and the freedom of ions to move

is dependent on preservation of electrical neutrality in addition to excess kinetic energy. (*See* **Conduction band.**)

Valence electron. The conception of valence electron arises from the acceptance of the nuclear atom, having a definite number of extra-nuclear electrons for each kind of atom. Second, the fact that the inert gases in Group O of the periodic table are extra-ordinarily stable and the acceptance of the suggestion that this is associated with a stable configuration of electrons. For elements other than those in Group O it seems reasonable to suppose that they could attain a stable configuration either by giving up or by taking up one or more electrons. The electrons thus transferred from one atom to another—usually they are those of highest energy—are called valence electrons.

Valency. In the simplest terms, the valency of an element can be regarded as the number of atomic weights of hydrogen with which one atomic weight of the element will combine or the number it will displace. But a fuller understanding depends on the accept-ance of the electronic structure of the atom. It then becomes possible to explain the formation of molecules and compounds. There are three types of linkage: (*a*) the ionic bond; (*b*) the covalent bond; (*c*) the co-ordinate link. The ionic bond has been described under "valence electron" (q.v.). The essential feature is the transference of an electron from one atom to another. The donor becomes electropositive and the acceptor becomes electronegative. There are thus formed two ions held together by electrostatic charges. In solution the ions are less strongly held together and "ionisation" occurs. The positive (metallic) ion is deposited on the cathode when a current is passed through the solution or when the fused salt is electrolysed. It thus appears that a metal is an element which can form a posi-tive ion by the loss of one or more electrons. Non-metals form negative ions by accepting an electron by transference from another element. The ease with which metals form ions depends on the number of valence electrons. The higher the valency the more difficult it will be to remove further electrons after the first. This is because the positive nuclear charge does not change when the first electron is lost and does, in fact, hold the remaining electrons more effectively. The larger the atomic volume for a metallic element, the more easily can it lose electrons, which on the outer orbit are not so strongly held. (Also known as polar, electronic and hetero-polar bond.)

The covalent bond is characterised by a *sharing* of electrons without the formation of ions. The electrons shared are referred to as an electron pair. Compounds of this type do not ionise and they are often volatile or relatively so. (Also known as homopolar bond or non-polar linkage.)

The co-ordinate link is a type of linkage in which both the electrons constituting the electron pair are supplied by one atom.

Valentinite. An antimony mineral consisting of the trioxide Sb_2O_3. It varies in colour from greyish white to red (mainly due to the presence of iron). It crystallises in the orthorhombic system. Specific gravity 5.57.

Vanadinite. A vanadate and chloride of lead $3Pb_3V_2O_8.PbCl_2$, one of the principal lead ores. It is found along with other lead minerals as brilliant blood-red hexagonal crystals or massive, as red, orange or yellow incrustations. It is isomorphous with mime-tite and pyromorphite and is found chiefly in North America (Arizona).

Vanadium. A metal which in the pure state is brilliantly white and very hard. The chief ores of vanadium are: (*a*) sulphide ores (e.g. partonite); (*b*) silicates (e.g. roscoelite); (*c*) uranium-vanadium ores such as uranium and potassium vanadate or carnotite and other vanadates (e.g. vanadinite, etc.).

The metal is unaffected by air or water at ordinary temper-atures. On heating strongly it combines directly with oxygen and nitrogen of the air. It is not attacked by hydracids but dissolves readily in hot oxyacids. It does not react with dilute alkali solu-tions but it decomposes fused caustic soda.

Vanadium compounds are known in five states of oxidation having valencies of 2, 3, 4, 5 and 6. Salts corresponding to the oxide VO are good reducing agents. The tendency to form double salts increases with the valency. The sulphates corresponding to V_2O_3 form a long series of "alums".

In the preparation of the metal from the ores the method used depends on the composition of the ore. Sulphide or sulphate ores can be directly reduced by carbon in the electric furnace. In most other cases a vanadate or the oxide V_2O_3 is prepared after roasting in air, leaching and precipitating. The oxide can be reduced by aluminium in the aluminothermic process. The reaction, however, is very violent, but can be moderated by using ferrosilicon in place of aluminium when ferrovanadium is produced. Calcium metal may be used and the reaction conducted in a high-pressure cylinder. By this method vanadium 99.2% pure can be obtained. The van Arkel method is based on the thermal decomposition of vanadium iodide (VI_2) on an electrically heated vanadium wire in a quartz bulb.

The most significant application of vanadium is in ferrous metallurgy. It is a strong carbide forming element when added to steels. It is particularly valuable because of its ability to restrict the growth of austenite at high temperatures. Vanadium steels are essentially fine-grained, and in addition to enhancing the tensile strength and toughness it improves the hardenability of heat-treated and air-cooled steels. A structural steel containing 1.5% Mn (max.) and about 0.12% V has a high strength and ductility. Vanadium steel castings contain a minimum of 0.15% V.

In non-ferrous alloys, vanadium has the effect of increasing the resistance to corrosion as well as improving the tensile strength and elongation in brass, bronze and aluminium alloys (*see also* **Vanadium alloys**). Vanadium finds employment as a catalyst and can replace platinum in certain reactions (e.g. preparation of sulphuric acid by oxidation of SO_2), being cheaper and less liable to poisoning by arsenic than is platinum. The Ephraim test for vanadium consists in adding 1 cm^3 of concentrated HCl to an equal volume of the solution under test. (This latter is usually prepared by dissolving the alloy-steel turnings in HNO_3 with a little ammonium persulphate and boiling.) The 2 cm^3 of liquid are boiled to about one-fifth of the original volume and cooled. There is then added 1 drop each of 0.1% solution $FeCl_3$ and 1% solution of dimethylglyoxime. If on making the solution alkaline with ammonia a red colour appears, vanadium is indicated.

Atomic number 23, b.pt 3,380 °C; crystal structure, body-centred cubic; lattice constant $a = 3.039 \times 10^{-10}$ m. For other physical constants see Table 2.

Vanadium alloys. The main use of vanadium as an alloying element is in steel (*see* **Vanadium steel**). Vanadium bronze has been used as a substitute for manganese bronze. It is used in the aircraft industry for making high-strength and corrosion resistant propeller bushings. Vanadium copper is also used for similar purposes as well as for submarine and ships' parts subjected to the corrosive action of sea-water. Vanadium aluminium used in making airframes is tougher and stronger than most other common light alloys. Vanadium gold is suitable for making coins.

Vanadium carbide. A compound having the formula VC which is formed when vanadic oxide is heated with carbon in an electric furnace. It is a very hard, silvery-white substance that melts at about 2,750 °C. The crystals have a specific gravity of 5.4 and burn readily when heated in oxygen or chlorine.

Vanadium steel. Steels that contain small percentages of vanadium which is added to effect a refinement of the grain size. It forms a very stable carbide, and thus exerts a generally beneficial effect on the physical properties of the steel. The term is also applied to certain steels that contain chromium or chromium and manganese in addition to vanadium. A typical alloy steel would contain about 0.4% C and 0.15% V. The amounts of chromium and manganese are generally 7 and 5 times the amount of vanadium. The general effect of vanadium on alloy steels is to reduce the rate at which the U.T.S. and yield strength fall off with rise of temperature. The Izod (impact) value is usually better at all temperatures, for vanadium steels as compared with plain carbon or chromium steels.

Vanadium tin-yellow. A mixture of vanadium oxide V_2O_5 and tin oxide SnO_2 used as a pigment and for producing a yellow colour in glass.

Van Arkel process. A process for the production of pure ductile metal by the thermal reduction of a volatile halogen compound. It is applicable especially to the production of such metals as zirconium and titanium and depends on the reversible dissociation of the tetraiodides of these metals. The tetraiodide at a low

vapour pressure is allowed to come into contact with a filament of the metal maintained electrically at about 1,400 °C. Decomposition occurs at the surface of the filament. The liberated iodine combines with crude metal in the cooler parts of the reaction chamber where the temperature does not exceed about 400 °C. The process is thus continuous, working in a closed cycle, and stops only when the filament has so increased in diameter that the temperature cannot be maintained at the value required by dissociation reaction.

Van de Graaf generator. An electrostatic machine designed for the continuous separation of electric charge. It is used to accelerate charged atomic particles such as protons to high energies.

Vanderloy. The name used in reference to a process for electro-depositing iron, that is well suited for heavy deposits.

Van der Waals' bond. A bond relying on polarisation effects of the higher energy electrons in particular, relative to the nuclei of adjacent atoms. (*See* **Bond**.)

Van der Waals' equation. An equation of state for one mole of a substance in the gaseous state. The simple gas laws are based on the assumptions that the volume of the molecules is negligible and that the molecules exert no attractive force on one another. Neither of these assumptions is true at ordinary temperatures and pressures. The ordinary law $pv = RT$ is therefore modified. The first modification takes into account the volume of the molecules by introducing a quantity "b". This has the value of four times the volume of the molecules and $(v - b)$ represents the compressible volume. The second modification takes into account the attraction of the molecules for each other. This depends on the number of molecules in the surface layer and also on the number of molecules in the interior of the gas, and both these quantities are proportional to the density. And since for a given mass of gas density is inversely proportional to volume, the correction is proportional to $1/v^2$ and it takes the form $(p + a/v^2)$ for the pressure. Van der Waals' equation is therefore written:

$$(p + a/v^2)(v - b) = RT.$$

It has been found that the value of "a" is not independent of temperature and Clausius's equation contains the term $a/T(v + c)^2$ in place of a/v^2.

When van der Waals' equation is expanded and rearranged

$$v^3 - v^2\left(\frac{RT + pb}{p}\right) + v.\frac{a}{p} - \frac{ab}{p} = 0.$$

This is a cubic equation having three roots. When the values of p and T are suitably chosen the three roots may be made equal. If this value is v_c—the critical volume—then $v = v_c$ and $(v - v_c)^3 = 0$ whence

$$v^3 - 3v_c v^2 + 3v_c{}^2 v - v_c{}^3 = 0.$$

Comparing this with the van der Waals' expanded equation and equating coefficients it can be shown that

$$v_c = 3b.$$

The critical pressure $p_c = a/27b^2$ and the critical temperature $T_c = 8a/27Rb$.

Vandyke brown. A deep brown pigment originally obtained from lignites. It is now obtained also from low grade coals. It is an argillaceous and siliceous material containing barium and calcium compounds together with oxides of iron. An imitation pigment is prepared by mixing lampblack, oxides of iron and yellow ochre or limonite.

Vanner. A device used in ore dressing for dealing with fine material. The vanner feeds the fine ore together with water on to an endless belt—usually of rubber. This latter is slightly inclined to the horizontal and is shaken crosswise as it moves uphill. By regulating the speed of the belt, its slope and the amount of water used, the size of the grain that reaches the top of the belt can be varied. The ore grains pass over the upper roller into the concentration tank while the lighter material is washed into the tailings.

Van Valkenburgh test. A test devised by van Valkenburgh and Crawford for detecting the presence of tungsten, titanium, molybdenum and vanadium in ores containing some or all of these metals. The finely powdered ore is fused with an excess of ammonium hypophosphite. A blue colour develops in the melt, after fusion, if cobalt, titanium or tungsten are present. A red colour, which deepens on the addition of hydrogen peroxide, indicates vanadium. When water is added dropwise to the hot melt which shows a blue colour, the latter slowly turns to violet if the colouration is due to the presence of tungsten.

Vaporisation. In general, the term means the conversion of a liquid into the gaseous phase at temperatures below the boiling point. The term is also applied to describe a process of laying down a thin protective film of one metal on another. It is particularly applied to treatment of small steel articles which are to receive a coating of zinc. The operation is carried out in a closed vessel in which the steel articles are heated to a temperature about 350 °C in the presence of zinc powder. At the surface of the steel a zinc-iron alloy is formed and outside this a pure zinc layer. (*See also* **Sherardising.**)

Vapour. A substance in the gaseous state but below its critical temperature. The space above a liquid in an enclosed vessel is filled with the vapour of the liquid.

Vapour blasting. A technique devised for the rapid cleaning of metallic surfaces. The abrasive material is usually alumina specially prepared and screened and suspended as a sludge in water. The sludge is projected on to the metal by means of a blast of steam or sometimes a blast of air. A good clean matt surface, practically dry, is obtained when steam is used. But with the air blast, iron and steel surfaces are liable to rust rapidly after treatment unless they are quickly coated with oil or some protective substance. It is because of this rusting effect that the sludge frequently must contain corrosion inhibitors. Among the latter that have been found effective are quinoline (1, benzazine), pyridine (azine), glue, gelatine in amounts of about 1% in the sludge.

Vapour degreasing. A solvent degreasing process applicable to metal components for the removal of oil or grease from the surface. The parts are lowered into a vessel containing the vapour of a suitable grease solvent. The vapour condenses on the surfaces, dissolves the grease and then drips into the liquid sump. The liquid most commonly used is trichloroethene which boils at 86.7 °C. The vapour does not form explosive mixtures with air but is slightly anaesthetic. Since it is liable to decompose, hydrochloric acid being one of the products, stabilisers are often added. One of the most efficient of these is dibutylamine. In place of trichloroethene, the more stable perchloroethene is sometimes used, especially in degreasing aluminium. The boiling point is 120.7 °C.

Vapour pressure. A measurement of the partial pressure exerted by the vapour above its parent liquid or solid below its critical temperature. (*See* **Condensation.**)

Variegated ore. A valuable ore of copper consisting of sulphide of copper and iron. The composition varies, but may be expressed as Cu_3FeS_3. It occurs in association with copper pyrites, covelline, in the zone of secondary enrichment of copper ores. Crystal system cubic, Mohs hardness 3, sp. gr. 4.9-5.4.

Also called variegated copper ore, erubescite or bornite. In Canada it is known as horse-flesh ore.

Vector. A physical quantity which requires statement of a direction in addition to a magnitude to define that quantity completely. Velocity, displacement, force, electric field intensity, are all examples of vectors, as distinct from scalar quantities such as mass, length, time and electric charge.

Two vectors of the same kind are considered equal when they have the same magnitude and a common direction, although their positions in space may be widely separated. The simplest vector is a directed line segment, represented graphically by an arrow from an initial point A to a terminal point B. This is a line vector or vector step, and can be represented AB. Any vector may be represented by a corresponding line vector with the same direction and a length numerically equal to the magnitude of the vector.

The notation of Gibbs printed vectors as letters in bold type and their magnitudes by the same letters in ordinary type. Modern notation often uses the same letters with an asterisk to denote the vector.

The relationship between the scalar magnitude of components n_x, n_y, n_z, in a Cartesian co-ordinate system is

$$n = (n_x^2 + n_y^2 + n_z^2)^{1/2}$$

where n is the scalar magnitude of the vector.

The scalar product of two vectors is the product of the magni-

tudes multiplied by the cosine of the angle between them,

$$a*b* = ab \cos \theta$$

where a*, b* are the vectors, a and b the scalar magnitudes, and θ the angle between them.

The vector product of two vectors is given by a*b* = n ab sin θ, where n is a unit vector perpendicular to a* and b*.

Vee weld. A term used in reference to a form of weld that is used in welding the longitudinal seams of furnace and flue tubes of certain types of boiler. The edges of the plates to be welded are bevelled slightly to form a V-shape and a long strip of iron is welded into the space between the bevelled edges. Also called glut weld.

Vegard's law. In a binary alloy consisting of components A and B in which n is the fraction of B atoms, the lattice spacing may be expressed in such a way that the mean inter-atomic distance (d_m) is given by

$$d_m = nd_B + (1 - n)d_A$$

where d_A and d_B are the inter-atomic distances of the pure components.

Vein. A general term for ore bodies occurring in forms having great length and depth, but having small thickness compared with other dimensions. Veins are frequently found in fissures between rocks. Also known as lode.

The term is also applied to the fine ridges, resembling human veins, sometimes found on castings.

Veinstone. Minerals having no particular economic value occurring with valuable minerals in veins or lodes. Also called veinstuff.

Velocity distribution. In considering the behaviour of particles moving with a range of velocities, it is useful to know the distribution among all possible velocities. If $n(c)dc$ be the number of particles having velocities between c and $(c + dc)$ where dc is a small increment of velocity, $n(c)$ is the number of particles per unit velocity about the value c.

$$n(c) = A \exp (-Xc^2)$$

where A, X are constants. This equation gives the number of particles having the particular velocity in one direction. Since all directions are important, $N(c)$ is the total number of particles per unit volume with velocity between c and $(c + dc)$ irrespective of direction

$$N(c) = 4\pi c^2 N \left(\frac{m}{2\pi kT}\right)^{3/2} \exp\left(-\frac{mc^2}{2kT}\right)$$

where m = mass of the particle, T the absolute temperature and k Boltzmann's constant. This is the Maxwell-Boltzmann distribution law, and applies to any interacting particles moving with a range of velocities. From this, the mean free path of the particles of diameter d moving with a range of velocities is given by

$$l = \frac{1}{(2\pi d^2 N)^{1/2}}$$

Vena contracta. A term used in reference to the contracted portion of a jet of liquid issuing from an orifice. In the case of water and similar liquids the cross-section of the jet at the vena contracta is about 0.62 of the cross-section of the orifice for a circular orifice. The discharge from a tank is given by $C_1 C_2 a \sqrt{2gb}$, where C_1 = coefficient of contraction, which is the ratio of the areas of the vena contracta and of the orifice, C_2 is coefficient of velocity (usually about 0.975), a is the area of the orifice and b is the depth of water above the orifice.

Venetian red. A pigment that consists of red oxide of iron together with various fillers. The colour depends on the proportion of the latter.

Venetian white. A pigment that consists of white lead containing barium sulphate as a filler. The mixture usually consists of equal parts of the two constituents.

Venice turpentine. A oil obtained by the steam distillation of the oleoresin from the Corsican pine. It is a valuable drying oil for paints, as it absorbs oxygen from the air and acts as a carrier in transferring it to the linseed oil to produce a tough and durable film. Venice turpentine produces a harder film than the turpentine derived from North American pines. Specific gravity 0.86.

Vent. A term used in reference to any small opening that provides an escape for gases. In foundry-work when organic substances are used as binders in the making of sand cores, the hot metal on entering the mould causes much gas to be liberated. To prevent this gas from finding its way into the molten metal and being trapped therein as the metal solidifies, a means of escape directly into the atmosphere is provided. This opening is known as a vent.

Venting. A term used in reference to the provision of suitable means of escape of gases during pouring in sand moulds. Sand is naturally permeable to gases, but the large volumes of gas produced during the making of medium and large size castings requires that the permeability be increased by artificial means. Rods or wires are used to make the vents, but they must not be allowed to penetrate the mould surface. The gas issuing from the vent holes should be allowed to burn to obviate the accumulation of posionous gases. Similar venting is applied to permanent moulds for die-casting and polymer moulding.

Venturimeter. A device specially adapted for measuring the flow of liquids. It consists of a converging pipe having an angle of convergence of about 20 degrees. The narrowest part is known as the throat and from this the pipe diverges at an angle of 5 degrees to the original diameter. The steeper cone is pointed upstream. This end of the pipe is surrounded by a cylindrical chamber as also is the throat and communication of water pressure into these two chambers is permitted through a series of small holes in the wall of the pipe. The difference of pressure h between the water at the entrance to the pipe and at the throat is measured by a manometer. If the area of the entrance to the pipe is A and the ratio of this area to the throat area be k, then the flow of liquid is given by

$$cA2\sqrt{(2gh/k^2 - 1)}.$$

The value of the coefficient c is taken as 0.98, though it may, in fact vary with the size of the meter from 0.96-0.99.

Vent wire. The term used to describe the wire or thin rod that is used to form the vents in foundry cores and moulds.

Verdet constant. The magneto-optical effect discovered by Faraday —that certain substances that under ordinary conditions do not rotate the polarisation plane of polarised light, do so when placed in a strong magnetic field.

Verdigris. A name originally applied to the bluish-green pigment that consisted of basic copper acetate. It is now used for "copper rust", or basic copper carbonate which is an atmospheric corrosion product. It has the formula $Cu(OH)_2.CuCO_3$ which is the same as that assigned to malachite.

Vermiculite. A mineral substance that consists of hydrated silicates of aluminium, iron and magnesium. It is a decomposition product of micas and related to the chlorites. It occurs in tubular, radiating and granular forms and when heated, flakes and exofoliates giving interlacing threads which can form a valuable thermal insulating material.

Vermilion. The red sulphide of mercury HgS and one of the oldest natural pigments in the form of cinnabar. The artificial product is made by heating together mercury and sulphur. The resultant compound may contain excess sulphur and is then purified by distillation and treated with caustic soda and finally washed. Alternatively sulphur and mercury can be rubbed together and then heated with caustic potash until the bright red colour appears. It is used in both oil and water colours, but is expensive. On this latter account it is often adulterated with red lead or red oxide of iron. Specific gravity 8.12.

Vibration, atomic. *See* **Atomic vibration.**

Vickers hardness. A hardness test in which a diamond-pointed tool having an angle of 136 degrees between opposite faces is used. For steels the load used is generally 30 kg and the testing operation occupies about 15 seconds, the rate of descent of the indenter being controlled by an oil dashpot. The hardness number is calculated by dividing the load by the area of the indentation, but in practice it is usual to measure the diagonal of the square indent using a microscope and then consult tables. Hardness values on this scale agree with those on the Brinell scale up to about 300. At higher hardnesses, due to the deformation of the steel ball under the usual load for steel (3,000 kg), the Brinell numbers differ from the V.P.N. numbers, the difference increasing rapidly above 400.

Vienna lime. A mixture of calcium and magnesium oxides. Ordinary lime made from pure limestone slakes easily and is termed "fat"

lime. When mixed with magnesia the lime forms a thin mixture with water and is termed "poor" lime.

Vinyl (ethenyl) resins. A group of thermosetting plastics derived from vinyl chloride (CH_2:CHCl), acetate (ethanoate), etc. Since vinyl chloride is a substitution product of ethene, these resins are sometimes regarded as polymers of ethene. The resinous substances are generally solid, but some are liquid and find specialised uses. The product of reaction of vinyl chloride on vinyl acetate is the white polyvinyl alcohol which on drying from solution yields a tough film. It is used as a coating for paper and textile fabrics and as an adhesive. When mixed with plasticisers the resin can be extruded into tough, heat-resisting sheets and tubes. (*See* Table 10 for details.)

Virgin metal. A term that refers to metal that has been obtained directly from its ores and which has not previously been used and remelted. It is used to distinguish from reused or secondary metal.

Visco-elasticity (anelasticity). A type of deformation under stress which implies complete recovery after removal of the stress, although part of the deformation is viscous. First described by Kelvin and Voigt independently, it is characterised by a buildup of elastic strain, but at a rate dependent on the viscous flow, which in turn depends on the elastic modulus itself. The expression for shear stress in terms of viscous and elastic components is

$$\tau = G\gamma + \eta\dot{\gamma}$$

where G is the shear modulus, γ the shear strain, η the viscosity and $\dot{\gamma}$ the shear strain rate. The majority of polymers exhibit this type of behaviour, particularly those which contain both crystalline and amorphous material. A model of this type of deformation can be represented by an elastic spring and a viscous dashpot in parallel (*see* Fig. 180). As the dashpot extends, the elastic strain builds up in the spring. On removal of stress, this stored energy causes recovery, and a retardation time can be defined, as the time to recover $(1-1/e)$ of the strain. This is analogous to the relaxation time in elastico-viscous systems, or Maxwell bodies.

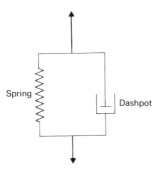

Fig. 180 *Kelvin body (visco-elastic).*

Here, a spring and a dashpot are linked in series, so that if the spring is stretched to some preselected strain and held at that strain, flow in the dashpot will proceed to reduce the strain-stress relaxation. The force in the spring controls the rate at which the dashpot will flow, and this is exponential, so that a graph of log-force against time will be linear (*see* Fig. 181). Stress relaxation time is then defined as the time for the force to recover to $1/e$ of the original value. Note that the Maxwell solid deals with

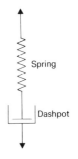

Fig. 181 *Maxwell body (elastico-viscous).*

stress relaxation, the Kelvin with strain recovery. The Maxwell system can be expressed

$$\dot{\gamma} = \frac{\dot{\tau}}{G} + \frac{\tau}{\eta}$$

where the symbols have the same meaning as above. $\dot{\tau}$ is the stress rate.

These systems are non-Newtonian, and in the Maxwell case, the material may be elastic or viscous, depending on the ratio of relaxation time to total time, also called the "Deborah number". Rapid stretching and release will be elastic behaviour — the dashpot has no time to flow if the fluid in it is highly viscous. If the dashpot fluid is such that time is allowed for it to flow completely, the viscous behaviour is that of a Newtonian fluid, the spring merely acting as a source of stored energy.

In the Kelvin model, a highly viscous fluid in the dashpot will flow slowly, and the elastic energy can only build up at the rate determined by the viscous element. If the dashpot flows more freely, it still governs the rate of elastic strain build up, but this takes time, and so a rapid measurement of total strain would be misleading. In metals, the recording of elastic extension in tension below the elastic limit is frequently taken instantaneously, when a time interval would show a slow buildup to a higher figure at each stress increment. This is the elastic fore effect. On removal of the load, the process is reversed, and an elastic after-effect can be demonstrated. (*See* **Elastic after-effect.**) This shows that the material behaves to some small degree as a visco-elastic body.

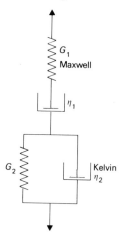

Fig. 182 *Combined rheological system.*

In practice, many solids are a complex combination of both Maxwell and Kelvin models, often with a multiplicity of relaxation and retardation times. A combination of one spring and one dashpot in series all in series with one spring and dashpot in parallel (*see* Fig. 182) represents many real materials. If we assign an overall relaxation time η_1/G_1 and a retardation time η_2/G_2 to the whole system, then the ratio of these two is the elasticity of the material, the extent of elastic recovery when the material is first stressed and then released. The time for recovery is automatically accounted for in this model and in the ratio $\eta_1 G_2/\eta_2 G_1$, a fact not taken into account in normal tensile testing, as referred to above. Figure 182 shows that it is the flow in the Maxwell dashpot which overrides all other considerations, and this applies to all liquids.

Combinations of such models, with the addition of a possible yield system, as described by Bingham, can account for most of the properties of fluids, and solid/fluid mixtures, which are processed every day in engineering, food and construction industries. All materials of everyday use behave according to one or other of these systems, and advantage can be taken to modify such properties to minimise inefficient uses and waste of energy. (*See* **Viscosity.**)

Viscometers. The general name given to a wide range of instruments used for measuring or comparing the viscosity of liquids and gases. Capillary viscometers are designed to force a known volume of liquid at a known pressure through a length of capillary tubing. In the Koch and Ostwald type of instruments the liquid is forced

through the capillary by its own weight. The motive force in the case of Thorpe and Rogers and Ubbelohde apparatus is dry air at a known pressure. In all these cases, however, the object is to collect and measure the effluent liquid. When the liquid flows under a variable pressure head it may be convenient to arrange that the total volume of liquid is large, so that the variation is small enough to be neglected. Otherwise a correction must be applied. If the initial and final pressure heads are b_1 and b_2, the formula is

$$\eta = \frac{\pi r^4 \, tg(b_1 - b_2)}{8lv \log_e b_1/b_2}.$$

where η is the viscosity, t the time in seconds, l the length of tube of radius r, and v the volume of liquid collected. g is the acceleration due to gravity. In calculating the value of η it is assumed that the pressure "p" is just sufficient to drive the liquid through the tube—that is the liquid should emerge with zero velocity. The value of η when the residual velocity is not zero is given by

$$\eta = \frac{\pi r^4 pt}{8lv} - \frac{vd}{8\pi tl}$$

where d is the density of the liquid.

For very viscous liquids rotating cylinders may be used. The inner cylinder is held steady while the outer one is rotated, the liquid under test being in the annular space.

For liquids of medium viscosity the coefficient may be determined by observing the fall of a small sphere in the liquid. Stokes showed that in such conditions the falling sphere would soon attain a constant velocity, given by:

$$\text{velocity of fall} = \frac{2r^2 (\rho - \rho_1)g}{9\eta}$$

where r is the radius and ρ the density of the sphere, while ρ_1 is the density of the liquid.

In performing any determination of viscosity it is essential to maintain the temperature constant; the change in viscosity of many liquids is as much as 2% per degree rise in temperature. (*See* **Viscous flow.**)

Viscosity. Liquids can be observed to change shape and move whenever their containment is disturbed. Many substances will flow, the amount of the disturbance influencing the rate of flow, and therefore the apparent time before a change is noticeable.

Flow in fluids means an interchange of atomic or molecular neighbours, a shear process, similar to that in solids, but because of the time involved it is necessary to measure the rate of shear, not shear strain alone. Newton's law of viscous flow is expressed

$$\tau = \eta\dot{\gamma}$$

where τ is shear stress, $\dot{\gamma}$ the shear rate, and η is a constant, the viscosity. (*See* **Fluidity** for derivation.) This law is linear, like Hooke's law of elasticity, and there is obviously a relationship between solid shear (Hooke) and fluid flow (Newton). If the constant η (viscosity) can be equated to the constant G (shear modulus) then the similarity is obvious

$$\tau = G\gamma \text{ elasticity}$$
$$\tau = \eta\dot{\gamma} \text{ viscosity (or fluidity)}$$

In solids, Poisson's ratio, the transverse to longitudinal deformation ratio, is related to other constants as

$$G = \frac{E}{2(1 + \mu)} \text{ where } \mu = \text{Poisson's ratio.}$$

Flow usually occurs at constant volume, $\mu = 0.5$, hence $G = E/3$. If this equated to $G = \eta$, then $\eta = E/3$ but $E = \sigma/\epsilon$ where σ is tensile stress and ϵ is tensile strain, and $\dot{\epsilon} = \sigma/3\epsilon$. Uniaxial tension σ can then be envisaged to produce a rate of extension $\dot{\epsilon}$ such that

$$\dot{\epsilon} = \frac{\sigma}{3\eta}$$

Whether a body behaves as a solid or fluid will then be determined by a time factor, and a level of viscosity will have to be fixed to equate the two, since theoretically all substances will flow, given sufficient time.

Conventionally, a low stress viscosity of 10^{17} cP (centipoise) is used as the criterion. A cube of 1 cm side would support an average human's weight of 75 kg (150 lb) for 1 year (31.536×10^6 s) with less than 1 mm deformation, with a viscosity of this level, and this is a reasonable definition of "solid".

In contrast, a gas has a viscosity of around 10^{-2} cP, and this demonstrates the very low range order of its structure. When gas is condensed to a liquid, the viscosity only rises about 1-2 orders of magnitude, and the viscosity of liquids is about 1 cP at the melting point of the solid from which they are formed. On cooling a liquid, particularly from above the melting point of a solid, the viscosity rises slowly at first, then more rapidly as solidification proceeds. Crystallisation of solid from liquid involves a change in viscosity of the order of 10^{20} cP, and this continues to increase if the solid is cooled towards absolute zero. Not all substances crystallise however, and where the liquid merely reaches a high viscosity, a glass results, where viscosity will still increase towards 10^{20} cP or more if cooled towards absolute zero. (*See* **Transition point** − **glass**). The distinction between solids and liquids therefore, can be seen as one of immense change in viscosity or viscous behaviour. (*See* Fig. 183.)

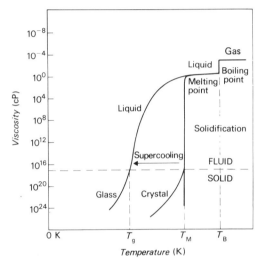

Fig. 183 *Effect of temperature on viscosity of fluids.*

Liquid metals crystallise on cooling, but are exceptionally fluid above their melting points. Casting of metals is therefore rarely affected seriously by considerations of viscosity; it is the heat transfer, surface tension, rate of solidification and thermal contraction which largely influences the quality of cast metals. (This was not always recognised, as viscosity was confused with heat capacity for heat transfer in warming up the mould.)

Silicate glasses become much less viscous at very high temperatures altough they remain liquids, i.e. short range order structures, even in their pseudo solid state. The blowing of glass into shapes requires attainment of a temperature which reduces the viscosity to about 10^9 cP. At this level, gas bubbles may remain trapped in the glass, and the temperature has to be increased further, to a viscosity level of 10^4 cP before these can escape. Glass has severe internal stresses due to the sharp change in viscosity with cooling from the melt, even though it remains a liquid. (The short range order is responsible for locking in much of this stress.) To remove the stresses (essential to avoid catastrophic relief later) the glass must be annealed, by raising the temperature to a level corresponding to a viscosity of 10^{15} cP minimum. Figure 183 indicates the temperature-viscosity relationship for gas, liquid and solid phases, including the variation between this and Fig. 178 showing the volume-temperature relationship. Figure 184 shows the viscosity of silicate glass in terms of temperature, related to the processing of the material.

Polymers vary in viscosity depending on molecular weight and temperature, and many are used in solution or emulsion form, as well as in melts. From the melt to the glass transition T_g, the variation may be from 100 to 10^{14} cP. Polymers which show partial crystallisation follow the relationship for crystallisation, i.e. a sharp rise in viscosity followed by a more gradual rise to the glass transition and then a further slow rise after the transition.

Liquids which obey Newton's law of viscous flow are called

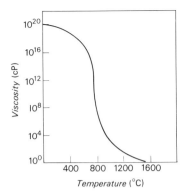

Fig. 184 *Viscosity of silicate glass.*

"Newtonian fluids". Some liquids follow the law, but there are many exceptions. When considering the difference between fluids and solids, these exceptions become very important. Maxwell used the concept of stress relaxation, rather than the rate of shear, to distinguish between fluids and solids. If a shear stress τ produces an elastic shear strain γ_1, and Newtonian flow of shear rate $\dot{\gamma}_2$ then

$$\gamma_1 = \frac{\tau}{G} \text{ (Hookeian) and } \dot{\gamma}_2 = \frac{\tau}{\eta} \text{ (Newtonian)}$$

Measured shear rate

$$\dot{\gamma} = \frac{\dot{\tau}}{G} + \frac{\tau}{\eta}$$

The specimen is assumed to be strained instantaneously by a small amount γ_0 and then this strain remains fixed. The stress must rise instantly to $\tau_0 = G\gamma_0$ but then relaxes as the elastic strain is taken up by viscous flow. (*See* Fig. 185.)

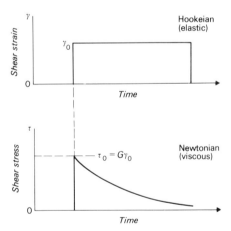

Fig. 185 *Deformation of a Maxwell body.*

Total shear rate

$$\dot{\gamma} = \dot{\gamma}_1 + \dot{\gamma}_2 = 0, \text{ so } \frac{\dot{\tau}}{\tau} = -\frac{G}{\eta}$$

As the decay is exponential,

$$\tau = \tau_0 e^{-t/t_R}$$

where t = time and t_R is called the stress relaxation time,

$$t_R = \frac{\eta}{G} \left(= \frac{3\eta}{E} \text{ for elastic strain} \right)$$

If the specimen is loaded at constant stress, τ then

$$\dot{\gamma}_2 = \frac{\tau}{\eta} = \frac{\tau}{Gt_R}$$

where t_R is the time to convert the elastic deformation τ/G to viscous deformation at constant stress. When $t \gg t_R$ the viscous deformation greatly exceeds the elastic deformation and the substance is a fluid.

The stress relaxation time in a liquid is the average time the neighbouring atoms or molecules require to move into new positions, when subject to the shear stress. Experimental determinations on liquids with ultrasonic waves and neutron scattering indicate t_R to be approximately 10^{-11} s which corresponds to 10-100 atomic diameters. If G is approx. 1 GN m^{-2}, the value of η is 10 cP. At the highest practical rates of shear, τ is of the order of 70 N m^{-2} (0.01 psi) and this is so small that only the first term of the series f(τ) is applicable — the liquid is always Newtonian. (See **Fluidity**.)

When $10^{-4} < t_R < 10^{+4}$ s, as with viscous fluids such as asphalt, putty, fats and waxes, both elastic and viscous strain can be observed, and stress relaxation operates as described above. When $t_R > 10^4$ s the substance is a solid. To produce appreciable flow under these conditions, τ involves higher terms in the function f(τ) and the fluid is non-Newtonian, or non-linear. When the strain rate is negligible until a critical stress is reached, and then may become very large, the non-Newtonian viscosity becomes ideal plasticity, generally

$$\dot{\gamma} = A\left(\frac{\tau}{\tau_y}\right)^n$$

where τ_y is the critical shear stress and A is a constant. The value of n can then be interpreted. $n = 1$ means linear viscosity, a Newtonian fluid. $1 < n < 10$ is non-linear viscosity, a non-Newtonian fluid. $n > 10$ is ideal plasticity.

When t_R is large, the fluid properties may vary with the process of flow. When t_R is small, the statistical distribution of atoms or molecules is negligibly altered by the flow, because adjustment to new positions is rapid. In highly viscous fluids, movement is slow, and the changes in neighbouring positions may persist for some time, so that the structure and properties will depend on the rate of flow. This is thixotropy (q.v.).

Linear viscosity enables glass workers to draw out fibres without local plastic necks being formed, as occurs with non-linear materials such as metal polycrystals and viscoelastic materials such as nylon. The viscosity in glass controls the rate of flow, which is uniform and not dependent on small changes in cross-section.

Bingham described the flow of paste-like materials, with negligible strain until a critical stress is reached, as exactly similar to ideal plasticity, but above the critical shear stress (yield). Newtonian linear viscous flow results.

Solid grains embedded in fluids of low viscosity (water or air) behave with a viscosity dependent on the volume fraction of particles, or on the second power of that fraction. For sand or gravel beds, the volume fraction is about 65%, but with mixed fine and coarse grains so that fine grains occupy the interstices between the coarse, the fraction may be 90% (c.f. close-packed equal spheres, maximum possible 74%). The mechanical properties then vary with the proportion of viscous fluid. Dry clay (kaolinite) becomes suitably plastic with an optimum 30% water, a continuous film of about 300 nm existing between the clay layers at the thickest point. With more water added (50%), the grains become further separated, and the clay sticks to the fingers and the mould or wheel. With still more water (70%) a fine slurry exists, the particles are quite separate, and the material is suitable for slip casting into plaster moulds.

Wet sand with spherical shaped grains is a stiff paste with 65% volume of sand, and a Newtonian liquid when this drops to 60%. Footsteps in the moist sand by the water's edge expand the sand to assist in resisting the pressure of the foot, and water flows in from sand around the foot together with sand, leaving a dry area around the footprint. When the foot is removed, the excess water comes up to the surface and can be clearly seen. Dry sand dunes are stable because the density of packing has adjusted to give a solid rather than a fluid behaviour. Added liquid will upset the balance and the dune will subside to a new angle of repose of the particles. A gas such as air forced through a bed of particles will expand the bed by separating the particles, once the pressure has exceeded the gravitational force of the particles. The whole system then behaves as a low viscosity fluid — the principle of the fluidised bed used in chemical engineering and for heat transfer in metallurgical heat treatment.

Polymers may be fluid, rubbery, gel-like, glassy or crystalline. (See **Visco-elasticity**.) Linear polymers without cross links (rubber latex) are viscous fluids, the viscosity varying with temperature. At low viscosity $< 10^5$ cP flow is Newtonian, above 10^6 cP non-Newtonian. Above the glass transition, most linear polymers are soft and deformable and behave as fluids if there is no cross linking. Below the glass transition temperature they behave like brittle solids.

Above the glass transition, molecular chains which were convoluted can be straightened, and removal of stress enables complete recovery, known as high elasticity. When cross-linked, the material resembles a gel, and with more cross-linking, a brittle solid. Figure 186 illustrates graphically the behaviour of a thermoplastic polymer when stressed and strain recorded against time.

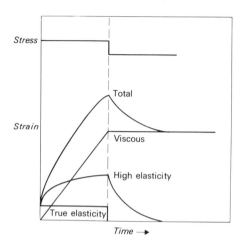

Fig. 186 *Deformation of a thermoplastic polymer.*

The SI unit of dynamic viscosity is newton second per square metre, N s m^{-2}, but it is convenient to use the centipoise, 1 cP = 10^{-3} N s m^{-2}. Table 63 shows dynamic viscosity of some common liquids and gases at 20 °C.

When the viscosity of a liquid at any temperature is divided by the density at the same temperature, the quantity η/d is known as kinematic viscosity, which affords a measure of the ease with which a liquid will flow under its own weight. Molten steel has a low kinematic viscosity and flows with ease. The SI units are m^2 s^{-1}, but again a more convenient unit is the centistokes, 1 cSt = 10^{-6} m^2 s^{-1}.

TABLE 63. *Viscosities of Common Liquids and Gases at 20 °C.*

Liquid	Viscosity, centipoise	Liquid	Viscosity, centipoise
Water	1.005	Glycerin (1, 2, 3-propanetriol)	74.0
Water (50 °C)	0.5494	Turpentine	1.46
CCl$_4$	0.969	Paraffin	2.376
Benzene	0.649	Heavy water	1.036
Olive oil	80.8	Aniline (aminobenzene)	4.40
CaCl$_2$ (Molar soln.)	1.317	Caustic soda (Molar soln.)	1.256
Mercury	1.56	Pitch	1.2×10^9
Hydrogen	9×10^{-6}	Oxygen	199×10^{-6}
Nitrogen	176×10^{-6}	Helium	201×10^{-6}
Chlorine	132×10^{-6}	Sulphur dioxide	127×10^{-6}
Carbon dioxide	147×10^{-6}	Argon	225×10^{-6}

Viscous flow. When a liquid flows through a tube, its behaviour depends on whether it is a Newtonian fluid, and the uniformity of the flow over the cross-sectional area of the tube, particularly at the periphery of the tube wall.

Poiseuille first evaluated the laws governing such flow, and made the following assumptions:

(a) there is no change in volume under the small stress operating, the liquid is considered incompressible;

(b) there is no slip at the boundary — the tube wall;

(c) inertial forces are very small compared with viscous forces (laminar flow).

If the tube length is L, radius R, pressure difference P, then at a distance r from the centre, the rate of change of velocity v with respect to r will be dv/dr (the velocity gradient). The velocity gradient is the same as the rate of change of angle of shear with time, $\dot{\gamma} = dv/dr$, and hence

$$\tau = \eta \frac{dv}{dr}$$

As r increases, dv/dr diminishes. When $r = o$, dv/dr is at its maximum, when $r = R$, $dv/dr = o$. This means that the velocity profile across the tube is a paraboloid. (*See* Fig. 187.) The pressure difference P acts over the area πR^2, so the force is $P\pi R^2$, and this must be balanced by the viscous forces $2\pi R L\tau$ on the cylindrical surface.

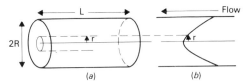

Fig. 187 *Flow through a cylindrical tube. (a) Tube dimensions. (b) Velocity profile.*

Hence

$$\tau = \frac{P}{2L} = \eta \frac{dv}{dr} \qquad dv = \frac{P}{2L\eta} r . dr$$

Hence

$$v = \frac{P}{2L\eta} \int_r^R r . dr$$

The volume of fluid discharged in unit time V is then given by

$$V = \int_o^R 2\pi r . v . dr . = \frac{\pi P R^4}{8L\eta}$$

This allows measurement of viscosity by the flow through a tube, measuring volume in given time, pressure difference, and the physical constants of the tube.

When a cylinder is rotated in a viscous fluid, the fluid is dragged between the cylinder and the container, which can be another cylinder. Let the rotating cylinder be radius R_1, height h, and the container radius R_2, the speed of rotation Ω rad s^{-1}, torque M. Let the angular velocity of the liquid at finite distance r from the axis be ω, and at this point, the shear stress τ_r and the shear rate $\dot{\gamma}_r$.

The measured torque M acts on the inner surface of the liquid, and between this position and radius r, the torque is exerted by the outer layers of liquid.

The outer layer torque $= \tau$. Surface area . radius $= 2\pi r^2 h\tau_r$. At equilibrium, $2\pi r^2 h . \tau_r = M$. Linear velocity

$$v = \omega r, \qquad \frac{dv}{dr} = \frac{rd\omega}{dr} + \omega$$

where ω is the radial velocity of the cylinder, and does not form part of the shear rate of the liquid, hence ω is eliminated to obtain the velocity gradient contributing to the shear rate.

$$\dot{\gamma}_r = r\frac{d\omega}{dr}.$$

For a Newtonian liquid,

$$\dot{\gamma}_r = \frac{\tau_r}{\eta}$$

Hence

$$r\frac{d\omega}{dr} = \frac{\tau_r}{\eta} = \frac{M}{2\pi r^2 h\eta} \quad \text{and} \quad \frac{d\omega}{dr} = \frac{M}{2\pi \eta h r^3}.$$

Integrating,

$$\omega = -\frac{M}{4\pi\eta h r^2} + C.$$

As $\Omega = \omega$ when $r = R_1$

$$\Omega = -\frac{M}{4\pi\eta h R_1^2} + C.$$

When $\omega = 0$, $r = R_2$ hence

$$C = +\frac{M}{4\pi\eta h R_2^2}$$

Therefore

$$\Omega = \frac{M(R_2^2 - R_1^2)}{4\pi\eta h R_1^2 R_2^2}$$

As R_1 and R_2 and h are constants of the apparatus

$$\eta = \frac{KM}{\Omega} \quad \text{where } K = \frac{(R_2^2 - R_1^2)}{4\pi h R_1^2 R_2^2}$$

This is the basis of the rotating disc or cylinder viscometer (if the outer cylinder is rotated and the inner at rest, $\Omega = \omega$ when $r = R_2$ and $\omega = 0$ when $r = R_1$). (*See also* **Reynolds number**.)

Vitallium. The name of a series of chromium-cobalt-molybdenum alloys that have good corrosion resistance and are particularly useful in dilute acid environments. The presence of the molybdenum is associated with the ability of these alloys to withstand the action of hot oils. The compositions are given in Table 9.

Vitrain. One of the important separable constituents of coal. It is thought to be the purest form of coal, and occurs as the shining jet-black bands in bituminous coal. These bands are generally lenticular, each of which represents one single fragment of plant material.

Vitreous bonds. The amorphous bonds created when a layer lattice such as kaolinite is heated to a sufficiently high temperature to drive out all water, including water of crystallisation, linking the constituent oxides by covalent bonds (vitrified). These bonds are not easily destroyed to reform the layer lattice and so pottery for example remains strong but porous. (The pores are the holes left by lost water.)

Vitreous enamelling. The process of laying down on a metal sheet or component an adherent glassy non-metallic coating. The fusible glasses contain quartz with felspar, clay and soda. Borax and saltpetre may be used as fluxes, but not where acid-resisting enamels are required. In the latter case, alkaline earths are introduced. The basic material is called "frit" and usually contains flint and white sand. The fused mass is dropped into water, so that the thermal shock will produce fine particles suitable for milling. The fine particles are suspended in a solution of an electrolyte and clay, colouring matters and opacifiers being added. In dry enamelling the wet article is dipped in the dry powder. The coating is dried and fused. Several coatings may be required before the desired result is attained, but the fusing temperature of each successive coating must be lower than the previous one. Metallic oxides used as opacifiers are tin oxide, titanium oxide, zirconium oxide (white), cobalt oxide (blue), and platinum oxide (grey). Chromium oxide gives a green colour and lead chromate gives a reddish-yellow colour if the firing is carried out not above 900 °C. Antimony oxide and sodium aluminium silicate are used to give bright colouring (white) or to enhance a blue colour.

Vitreous selenium. One of the six allotropic forms of the element selenium. It is brownish-black in colour and is produced as a brittle, glassy mass having a specific gravity of 4.28.

Vitreous silica. A name sometimes used in reference to silica glass which has a high transparency. It is used in the making of optical lenses.

Vitrified grinding wheels. A type of grinding wheel that is applicable to most forms of grinding and in which the bonding medium is glassy or vitrified. The matrix is made from a mixture of a fusible clay (e.g. felspar) and a refractory clay such as kaolin. The clays are dried separately and then mixed and screened. To the sieved clay the appropriate amount of abrasive is added and a thorough mixing made in the presence of a little water. The mix is again screened and then pressed to shape hydraulically. The wheels so formed are dried and subsequently fired at a high temperature in suitable kilns. In this process the bonding material is partly melted, and a solid wheel of good strength is thus formed. The hardness of the wheel depends on the nature and grain of the abrasive and upon the proportion of it in the matrix. But it also depends on the porosity of the bonding material. In general it may be said that the lower the density of the wheel the more porous it will be and the more freely it will cut.

As a result of the high temperature to which they are subjected during manufacture, vitrified wheels are uniform in structure, most of the impurities having been eliminated in the kiln. They are little affected by changes in temperature during grinding and have good mechanical strength.

Vitriol. A term used in reference to sulphuric acid and to certain sulphates that was derived from the old name for sulphuric acid —i.e. vitriolic acid. Crude, strong sulphuric acid is still known as oil of vitriol; ferrous (iron II) sulphate is often called green vitriol; copper sulphate blue vitriol; and zinc sulphate, white vitriol. Green vitriol occurs as the mineral melanterite.

Vitriol ochre. A mineral substance that is a basic ferric sulphate having the formula $FeSO_4(OH)_4 . 2Fe(OH)_3 . H_2O$. It occurs in greenish-brown or yellow crystals or as stalactites.

Void. A pore or small space that exists between particles of a solid body. The term is also used in reference to the unfilled, very small spaces that occur unintentionally in compacts made by powder pressing techniques. Also called pore.

The tetrahedral and octahedral voids of close-packed lattices are very important in the constitution of alloys and the stoichiometry of compounds. The simpler ones are shown in Figs. 10, 66 and 76, for the body-centred cubic, face-centred cubic and close-packed hexagonal systems respectively.

Volatility. A term used in reference to the tendency of a substance to pass into the vapour state. In the case of solids the process is called sublimation and the tendency is a function of the temperature. In the case of liquids the process is known as evaporation and depends on the temperature and the vapour pressure of the liquid. Hence volatility may be measured by the vapour pressure if the temperature is kept constant or by the boiling point at a stated pressure. The relationship between temperature and pressure is approximately:

$$\frac{\mathrm{d}\log p}{\mathrm{d}t} = \lambda \cdot \frac{1}{RT^2}$$

where λ is the latent heat of vaporisation at the absolute temperature T, it being assumed that λ does not vary with T.

In the case of liquid fuels, volatility is taken as a measure of the rapidity with which the liquid will vaporise at ordinary temperatures. The volatility characteristics of a fuel are generally measured by distilling 100 cm^3 of the liquid under carefully controlled conditions. Table 64 gives the latent heat of vaporisation of a few common substances.

TABLE 64. *Latent Heat of Vaporisation of Common Substances*

Substance	Boiling Point, °C	Latent Heat of Vaporisation, kJ kg⁻¹
Benzene	80.1	394.4
Benzole	79.0	544.3
Butane	0.0	383.1
Ethyl alcohol (ethanol)	78.3	855.4
Heptane	98.4	319.5
Pentane	36.2	353.8
Petrol	33.0	577.8
Phosphorus	279.0	544.3
Sulphur	444.0	1507.2
Turpentine	159.0	293.1
Water	100.0	2256.7

Volborthite. One of the principal ores of vanadium which consists of copper, calcium vanadate, and has the formula

$$(\mathrm{Cu,Ca})_3 (\mathrm{VO}_4)_2 . \mathrm{H}_2\mathrm{O}.$$

Volclay. A type of bentonite or colloidal clay that has the property of swelling on the addition of water. It consists of aluminium silicate with small proportions of oxides of iron, calcium and magnesium, and is used as a valuable bonding material in foundry-work. It is found chiefly in the USA, where it is also employed as a constituent of certain soaps and washing powders, since it is slightly abrasive.

Volt. The unit of potential. The difference in electric potential between two points of a conducting wire carrying a constant current of 1 ampere when the power dissipated between these points is equal to 1 watt.

Voltage. The electromotive force of a current supply measured in volts.

Voltameter. The name applied to a vessel which contains fused solid or a solution of salt that is to be decomposed by the passing of an electric current. The terms water voltameter and silver voltameter which are sometimes used refer to: (*a*) the acidulated water; or (*b*) the solution of silver nitrate which the vessel contains. It is found that an unvarying current flowing through a solution of silver nitrate for a certain time deposits a definite amount of silver on the cathode. From this is developed a convenient practical method of defining the ampere which is as follows: the ampere is that steady current which, flowing through a solution of silver nitrate, deposits 0.001118 g of silver per second on the cathode. The quantity of 0.001118 is known as the electro-chemical equivalent of silver. The voltameter is thus used for

determining current, and from the results of a determination as above it may be calculated that $I = W/0.001118t$, where I = current in amperes, W = weight of silver deposited, and t = time in seconds.

Volt box. A number of known resistances joined in series and used in conjunction with a potentiometer for the measurement of high e.m.f. The points between which the potential difference is required are joined to the ends of the box, and the fall of potential across one of the known resistances is measured by means of the potentiometer, using a standard cell and a sensitive galvanometer.

Voltmeter. Any instrument designed for the measurement of potential differences in the practical unit (the volt). Electrostatic instruments depend for their action on the mutual attraction between two conductors at different potentials, and since no current passes through them, they are free from errors that might arise from changes of temperature or polarisation of batteries. Other types of voltmeter include: (*a*) hot-wire instruments, which depend on the increase in length due to the heating effect when a current passes through a wire; (*b*) moving-coil instruments of the Weston type. These are not suitable for alternating current; (*c*) moving iron instruments, e.g. Siemens, Schuckert, Nalder and Thompson; (*d*) dynamo-meter instruments, e.g. Kelvin, Weston and Siemens types; induction type instruments useful for a.c. measurements.

Volume assaying. A term used as an abbreviation for volumetric assaying. This method, because of its convenience, is rapidly finding favour with assayers and is replacing, in some cases, the older fire-assaying. The method requires only small samples which must be finely crushed and put into solution. Minerals that do not readily dissolve in the common acids are usually first fused with a flux (e.g. $\mathrm{Na}_2\mathrm{CO}_3$ or soda lime). The solutions so produced are treated with standard solutions of different reagents and the appearance of the solution or the volume of standard reagent is employed to estimate the amount of useful metal present. The methods are in general similar to those used in volumetric analysis, e.g.:

(*a*) *Iron* is estimated by putting the ore into solution and then reducing the iron to the ferrous (iron II) state. The amount of iron is then estimated from the volume of standard potassium permanganate or potassium dichromate required to completely oxidise the iron to the ferric (iron III) state. In the former titration potassium permanganate is its own indicator, but in the latter titration potassium ferricyanide (iron III) is used as an outside indicator.

(*b*) *Copper.* When the copper has been put into solution, with a suitable acid, the latter is neutralised by the addition of ammonia and then an excess of this reagent is added, producing the deep blue solution of the amine $\mathrm{Cu(NH}_3)_4\mathrm{SO}_4$. Potassium cyanide reacts with this complex sulphate, producing colourless stable potassium cupro-cyanide. The assay therefore proceeds by adding standard KCN solution until the blue colour is just discharged. Other methods are available, but these require the elimination of other metals that may be present.

(*c*) *Zinc.* The ore is put into solution and then treated with ammonia to precipitate any of the metals of Group I, II or III that may be present. The zinc is precipitated by alkaline sulphide and taken into solution again by very dilute hydrochloric acid. The titration is carried out at just below boiling point, the standard solution being potassium ferrocyanide (iron II), and uranium nitrate is used as an outside indicator.

Volume elasticity. When a body is subjected to a uniform pressure p which produces a change in volume, it is known that for isotropic materials the compression is proportional to the pressure and the ratio of the pressure to the compression is known as the volume elasticity or bulk modulus. Thus, if the application of pressure p results in the volume being reduced from v_1 to v_2, the compression is

$$\frac{v_1 - v_2}{v_1} \quad \text{and the modulus } K = \frac{pv_1}{v_1 - v_2}$$

The bulk modulus is the only elastic constant for a liquid or a gas. In such cases when the pressure rises from p to $p + \mathrm{d}p$, the compression is $-\mathrm{d}v/v$ and then

$$K = dp \div dv/v = -\frac{vdp}{dv} .$$

The compressibility of a substance = $1/K = C$.

Volumetric analysis. The title given to a group of methods employed in quantitative chemical analysis in which the substance to be estimated is allowed to react in solution with another substance. This latter is contained in a solution of known strength, called a standard solution, and if the solution is such that it contains the equivalent weight of the solute in moles per litre, it is a molar solution. Molar solutions are equivalent bulk for bulk.

As it is essential that the end of the reaction should be accurately known, it is usual to use some kind of indicator. Sometimes there may occur a change in colour of one of the reacting substances that can be used to indicate the completion of the reaction—as when potassium permanganate is used as an oxidising agent. Indicators (q.v.) are frequently introduced into the reacting solutions, and the change of colour provides a visual indication of the end of the reaction. In other cases outside indicators are used. There are three main methods used in volumetric analysis:

(a) Direct methods are those in which the estimation is made as a result of a single reaction occurring when the substance to be estimated and standard solution are mixed—these include the usual acid-alkali titrations, chloride determinations by silver nitrate, and estimations of ferrous (iron II) salts by direct oxidation by potassium bichromate.

(b) Indirect methods include those in which one or more intermediate reactions occur, the actual substance that is estimated being different from the original substances used. As an example it may be quoted that peroxides may be estimated by distilling with hydrochloric acid, passing the chlorine evolved into potassium iodide solution and estimating the iodine liberated by means of standard sodium thiosulphate solution.

(c) In certain cases it is convenient to add an excess of a standard reagent and then estimate the excess by a "back titration". Thus in estimating ammonium salts it is convenient to add excess of standard alkali and boil off the ammonia, after which the excess alkali is estimated by standard acid solution.

Vulcanisation. The old term for the cross-linking of natural rubber by sulphur compounds. The original method of Goodyear used elemental sulphur, that of Parkes compounds such as sulphur chlorides.

W

Wad. A term that is used to describe that portion of the flash which when punched out leaves a hole through a forging. A term also sometimes used in reference to the mineral substance consisting of hydrated oxide of manganese. It is an impure mixture of MnO_2 and MnO. Also called bog manganese.

Waelz process. A process by which low-grade ores, slags or residues from retorts may be treated either for the recovery of zinc alone or for the recovery of zinc, lead and tin. The crushed material is mixed with coke breeze and fed continuously at the upper end of a Waelz kiln. A high temperature is produced by the burning of the coke, and the fume, together with some dust, is collected in dust-catchers. These latter consist of a series of long cooling pipes terminating in a dust bag.

Wafer core. The junction of the feeding head and the casting in foundry-work is known as the neck and the dimensions of the neck are of great importance in securing a sound casting. One method of reducing the dimensions of the neck without interfering with the feed from the head to the casting consists in using a thin wafer core which contains a suitable aperture, and which is placed in the mould across the junction of the casting and the feeder head. The cores are made of ordinary core sands, using a very fine sand. The thickness and the size of aperture depend on the size of head used and the mechanical strength required, and when these are determined the thickness of the core may be taken as directly varying with the size of the aperture. The objective in all cases is to secure a properly filled mould and to arrange that the solidification of the neck shall occur shortly after the solidification of the casting, but before the solidification of the head.

Walden's rule. Since at infinite dilution the equivalent conductance of an ion in an electrolyte depends only on its speed, "the product of ion conductance and viscosity should be a constant and independent of the character of the solvent". Thus for a given electrolyte $\Delta\eta$ = constant. The rule is only approximate because an ion may attach to itself one or several molecules of solvent depending on the nature of the solvent, and this will be responsible for a variation in the effective radius of the solvated ions; and thus also responsible for variations in its mobility.

Wall. The name applied to a barrier such as the extremity of a magnetic domain (q.v.). When powdered Fe_3O_4 in an oil suspension is coated on to a magnetised material, lines are seen as the oxide collects along the domain boundaries, thus rendering them visible. As the domains have varying orientation of their directions of magnetisation, the domain wall must be of finite thickness, there cannot be a sudden sharp change. As the change in direction occurs by a twisting of the small atomic magnets to achieve this result, some of the small magnets have the poles coming to the surface at particular points — and this is where the oxide particles gather, thereby delineating the boundary. The finite thickness of the wall, some 300 atomic planes, is named after Bloch, who first investigated them, and they are called Bloch walls, domain walls or magnetisation boundaries. They are totally different to crystal or grain boundaries, and domain walls exist in a single crystal during magnetisation.

Walterisation. A method of producing light-duty finishes, particularly upon ferrous- and zinc-base components. The solution required consists of iron phosphate, together with some phosphoric acid—and generally an oxidising substance, called an accelerator. The articles to be treated are immersed in the hot solution until hydrogen ceases to be evolved. They are then rinsed and dried. In the Walterisation process, iron sulphide, which dissolves more readily than iron in phosphoric acid, is used in place of iron in preparing the solution. (*See also* **Coslettising.**)

Walthal process. A process for obtaining alumina from certain clays. Calcined clay is treated with sulphuric acid to form aluminium sulphate. Iron and silica are removed and the solution evaporated to dryness. The solid aluminium sulphate is then decomposed by heat-yielding alumina and oxides of sulphur. The latter are collected for reuse in the process.

Wap. A single turn in a coil of wire.

Warrant. A term sometimes used in reference to the underclay found below the coal measures. It is used as a fire clay. Also known as Spavin.

Washing soda. The name given to the decahydrate of sodium carbonate $Na_2CO_3.10H_2O$. It is also referred to as soda crystals and sal soda and contains 60% water. The commercial product, known as soda ash, is made by the Solvay process by treating common salt with ammonia and carbon dioxide. It is a greyish-white lumpy substance, the best qualities containing 99% Na_2CO_3 (anhydrous).

Light soda ash is obtained by calcining the bicarbonate obtained in the Solvay process. It has a density of 366 kg m^{-3}. Dense soda ash is produced by adding a little water to the light variety and recalcining. The density is then 1,009 kg m^{-3}.

Sodium carbonate is used either alone or mixed with the bicarbonate as a cleansing agent and detergent. Alone it is used as a flux in making glass. Mixed with limestone it is used as a flux in melting iron, since it is effective in removing sulphur in the slag. For desulphurising iron it is sometimes used in briquette or pellet form, using a hydrocarbon as a bonding material.

Nitron is a natural form of sodium carbonate found in North Africa.

Wash table. An incline table—plain or riffled—upon which ore is fed, water being allowed to flow down the table with the object of cleaning the ore by washing away the lighter material or gangue.

Waster-plate. A spare scrap steel plate placed on top of a stack of plates during a stack-cutting operation. The plate is usually of mild steel, and the object of its use is to protect the upper part of the stack. It is discarded after use.

Water (structure). The formation of covalent bonds between oxygen and hydrogen involves the directional effect of the 2p electrons forming the bond. To satisfy the wave function conditions, the bonds must lie approximately along two mutually perpendicular axes. In fact the bond angle is 105°, due to some repulsion between the hydrogen nuclei, increasing the angle beyond 90°. This type of structure exhibits a high dipole moment, the two positive hydrogen centres being situated at a distance from the single oxygen whose unshared electrons act as a negative charge. This also leads to polarisation forces in the liquid state, which bring the molecules together in a loosely enmeshed structure which can absorb other compounds in solutions. The water molecule can also ionise into a hydrogen ion H^+ and a hydroxyl OH^- which encourages solvent action, and the association with other compounds as water of hydration.

Hydrogen nuclei are attracted towards the unshared electron pairs of oxygen, and in ice, the structure resembles tetrahedra, with oxygen at the centre of the tetrahedron and hydrogen at the corners. (This has some analogy with silica, in which Si ions are at the centre and oxygen at the corners, each oxygen shared by two silicons.) This hydrogen bond however, is most important with biological molecules. It is suggested that the electrons of the covalent bonds in water are more polarised than would be expected, and that a second hydrogen bond forms more easily than the first. The second then encourages a third, and this builds up a type of cluster. Such a cluster will be destroyed in due course by energy fluctuations, and so the structure of water may be one of continual clustering of molecules, breaking up, and new clusters reforming. The tetrahedral arrangement known to exist in ice may also be formed in liquid water, but other shapes are equally likely. It is the proportion of such clusters at any one time which gives the short range order to water in the liquid state, and as temperature rises, the increase in disorder is simply the speeding up of cluster breakdown. As temperature is reduced, the reverse occurs, and clustering persists for a more lengthy period at any one position.

Water core. The term used in reference to the water circulating through the interior portions of a mould, the object being to ensure that these parts cool down as nearly as possible at the same rate as the external parts. By doing this the internal strains in the casting that would be set up due to unequal rates of cooling are minimised.

Water gas. The gaseous product formed by passing steam over or through red-hot coke. It consists chiefly of carbon monoxide and hydrogen and has a calorific value of about 13 J cm^{-3} and contains about 43% CO. Water gas may be mixed with producer gas, the mixture being called semi water gas. Much of the water gas produced is mixed with a little benzene vapour or oil gas, and is then termed carburetted water gas. The uncarburetted gas is, for distinction purposes, called blue water gas—a reference to the blue non-luminous flame produced on burning.

The temperature at which the reaction $C + H_2O \rightarrow H_2 + CO$ proceeds with a high content of CO is above 1,000 °C. Thus, in the manufacture of water gas, air is blown over the hot coke until it is red hot. Then steam is substituted, but as the reaction is endothermic, the temperature of the coke falls and the gaseous product contains increasing proportions of CO_2. At a certain stage, therefore, the steam must be cut off and an air blast substituted. The cycle of operations is repeated.

Water gilding. A finishing process in which thin films of gold are obtained on a component by simple chemical replacement (also known as immersion gilding).

Water hardening. A term used in reference to the heat-treatment process for increasing the hardness of low-alloy steels. The material is raised to a temperature within or above the transformation range and quickly cooled by immersion in water. The amount of hardness induced by water quenching depends on: (a) the percentage of carbon present, which markedly affects the skin hardening; (b) the percentage of alloying elements, which exert their chief effect in determining the capacity of the material to harden below the skin; (c) the grain size; and (d) the quenching temperature.

Watering. The term used to describe the damping of foundry sand

with water in order to produce a mass of better adhesive quality.

Water rolling. A term sometimes used in reference to a method of cleaning metal components, which consists in tumbling them in a rotating barrel containing water, and usually also soap or detergent.

Water softening. A name used in reference to all those processes that are used to remove from aqueous solution those soluble compounds that interfere with the ready formation of a lather when the water is shaken with soap solution. Hard water is the name given to natural water that does not readily give a lather with soap. The hardness is either: (a) temporary, since it can be removed by boiling; or (b) permanent.

Temporary hardness is due to the presence of soluble bicarbonates which on boiling are decomposed into CO_2 and the insoluble normal carbonate. On the large scale the bicarbonates of calcium and magnesium are removed by the regulated addition of milk of lime $Ca(OH)_2$, which precipitates the former as carbonate and the latter mainly as hydroxide.

Permanent hardness, which is due to the presence of chlorides, sulphates and nitrates of calcium and magnesium, may be removed by the addition of sodium carbonate or a mixture of sodium carbonate and lime. This method is used in connection with the softening of boiler feed-water. After treatment in this way the water contains soluble chloride or sulphate of sodium, which do not precipitate on boiling.

A base (cation) exchange process is also available for softening water. It consists in passing the hard water through a container filled with zeolite—a hydrated sodium aluminium silicate. The calcium and magnesium ions are removed from the hard water by sodium (or potassium) ions, which latter pass into solution. When the zeolite is exhausted it can be regenerated by pouring over it a strong solution of common salt. The effluent that occurs is charged with calcium and/or magnesium chloride and is run to waste. (*See also* **Water treatment.**)

Water treatment. A term used in reference to all those processes which include: (a) water softening and (b) water purification. For processes included under (a) see **Water softening.** The treatment of natural water to render it suitable for drinking purposes is usually carried out only on the large scale with water that contains limited amounts of dissolved solids, and these should not be harmful for human consumption nor should they impart an unpleasant taste to the water. The processes consist in filtration —usually through beds of sand—and treatment with chlorine to destroy harmful bacteria. Ozone has also been recommended for this latter purpose.

For chemical analysis and in the preparation of metals of very high purity (e.g. germanium) water of the highest purity is required. On the small scale re-distilled water may be used. Other methods for the complete removal of all dissolved salts are now available which involve the use of what are referred to as ion-exchange materials. These fall into two groups: (a) cation and (b) anion. Cation exchangers contain an exchangeable hydrogen ion and are usually synthetic organic resins derived from phenol-formaldehyde (methanal) or polystyrene (polyethenylbenzene), and they contain active sulphonic groups. By their use the salts are converted into the corresponding acids. Since acidic water is undesirable for many purposes, the water may subsequently be treated with anion exchangers. These substances are usually produced from methanal by condensation with amines. If the exchange material contains a hydroxyl group, the acid is converted to water, but in other cases the acid is absorbed.

Ion-exchange resins are insoluble in water and need to be regenerated from time to time. They may be used to render sea-water drinkable and to demineralise river-water for industrial use. Their use has been recommended for separating rare elements (e.g. promethium) and in the recovery of radioactive wastes.

Watt. Power or the rate of doing work is measured in the SI system in joule per second. A watt is equivalent to 1 joule s^{-1}.

1 h.p.	= 746 watts
1 kilowatt	= 1.34 h.p.

A watt is the power in a circuit when the potential difference is 1 volt and the current flowing is 1 ampere.

If V = voltage in volts, I = current in amperes and R = resistance in ohms,

$$\text{Watts} = V \times I = I^2 R = V^2/R$$

Watt-hour. A unit of work or energy equivalent to the work done in 1 hour when the power in the circuit is 1 watt.

1 kilowatt-hour = 1,000 watt-hours
1 kilowatt-hour = 3.6 × 10⁶ joules = 3.6 MJ

Wattless current. In measuring the power in a.c. circuits, where both current and voltage are fluctuating, the power may be measured by: (a) a wattmeter or (b) an ammeter and a voltmeter. The reading of (a) would give the true power while the product of the readings of the instrument in (b) would give apparent power.

$$\text{Now}\ \frac{\text{True Power}}{\text{Apparent Power}} = \text{power factor} = \cos\theta$$

where θ = angle of lag or lead. If the circuit is very inductive the current lag would be nearly 90°, or if the circuit had a high capacity the current lead would be nearly 90°. In either case $\cos\theta$ = cos 90° = 0. The current in such cases is termed wattless. This may occur in the case of an a.c. transformer when the secondary circuit is open and current flows in the primary. In such cases no work is done by the primary current.

Watts solution. A solution of nickel sulphate containing some nickel chloride and boric acid, used as a nickel plating bath. The proportions vary somewhat but are usually about 5 : 3½ : 1. The bath is used at about 45 °C.

Wavelength (electron). De Broglie showed that an electron mass m and velocity v could be associated with a wave of equivalent wavelength such that $\lambda = h/mv$ where h is Planck's constant.

Wave mechanical behaviour of the electron. The problem of duality and uncertainty led to the concept of probability in estimating the likely behaviour of the electron. The solutions of wave mechanics indicate behaviour not predicted by classical mechanics, such as electron tunnelling. (See **Probability distribution.**)

When applied to electrons in combination or in the conditions of crystal lattices, it is necessary to proceed in steps from the effect of a potential barrier, the first being to an electron in a potential "box", i.e. with two barriers, the electron being trapped between. Consider an electron confined between potential discontinuities at $x = 0$ and $x = a$, both barriers of infinite height. The Schrödinger equation, as before, is solved with conditions requiring $\psi = 0$ at $x = 0$ and $x = a$, since the probability of finding the electron outside the limits of the boundaries is zero.

The space dependent part of the solution is harmonic in form,

$$\psi = A\cos\alpha.x + B\sin\alpha x$$

$$\text{where } \alpha^2 = \frac{8\pi^2 m}{h^2}\cdot E.$$

At the boundary conditions, $x = 0; \psi = 0 = A$
$$x = a; \psi = 0 = B\sin.\alpha a.$$

These conditions are met if $\alpha a = n\pi$ where n is an integer
$$n\pi = a(8\pi^2 mE/h^2)^{1/2} \text{ and } n = 1,2,3,\ldots\ldots$$

Solutions of the Schrödinger equation then require that

$$E_n = \frac{n^2 h^2}{8ma^2} \text{ where } n = 1,2,3,\ldots\ldots$$

This implies that the electron must possess certain discrete energy levels, given by $n = 1,2,3,\ldots\ldots$ since a solution for Ψ will exist only at these definite values.

When $n = 1$, $E_1 = \dfrac{h^2}{8ma^2}$; $n = 2$, $E_2 = \dfrac{4h^2}{8ma^2}$; $n = 3$, $E_3 = \dfrac{9h^2}{8ma^2}$, etc.

The analogy is the vibration of a stretched string, where the boundaries are the nodes at the point of suspension, and only certain permitted wavelengths of vibration are possible. This means that wave mechanics gives a solution equivalent to that of Planck with quanta, since energy can only be emitted from or absorbed by a system at certain discrete values — the quanta.

Calculating the value of B in the equation, we assume the electron to be within the limits $x = 0$ and $x = a$, so that

$$\int_0^a B^2\sin^2\alpha.x.\mathrm{d}x = 1. \text{ and } B = (2/a)^{1/2}$$

then $\psi = (2/a)^{1/2}\sin\alpha.a.$

Figure 188 shows the graphical expression for the wave function and the probability distribution in spatial terms. These curves

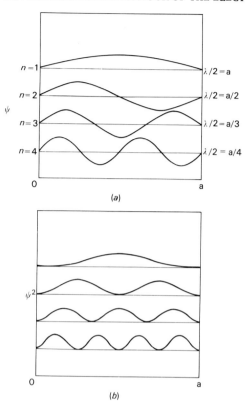

Fig. 188 *Electron in a potential box-spatial dependence. (a) Wave function ψ. (b) Probability distribution ψ^2.*

should be interpreted as the spatial dependence of an oscillating function, the square of whose amplitude gives an indication of the probability of finding the electron at any point. The spatial dependence in classical mechanics would be a ball bouncing between perfectly reflecting walls. The probability of finding that

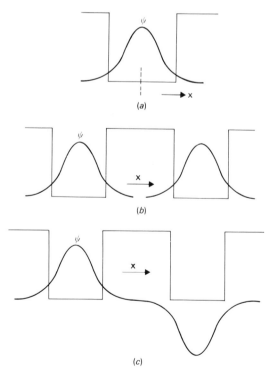

Fig. 189 *Electrons in potential wells of finite height approaching. (a) Electron in simple well. (b) Two electron wells approaching. (c) alternative solution to (b).*

ball at any point is uniform along the path, since the velocity is constant. In wave mechanics, this condition is approached as the integer n increases, but is furthest away when $n = 1$.

The next step is to consider the electron in a "well" or potential box similar to the previous example, but in which the potential barrier is not of infinite height. This condition exists around each nucleus of an atom, and the simplest atom, hydrogen, can be considered as the example. Since the barriers are not of infinite height, there is now a probability of finding the electron outside the potential well, and therefore the solutions to the Schrödinger equation are not strictly harmonic, the tunnel effect will apply.

Figure 189(a) shows the wave function in such a case. If two such electrons, in identical potential wells are now brought together, there is no effective interaction so long as the distance apart does not match the two parts of the solution (see Fig. 189(b)). There is, however, another solution, where ψ is 180° out of phase in one well compared to the other (see Fig. 189(c)).

The first case (b) is symmetric, the second (c) is antisymmetric. No difference in energy of significance exists so long as the two wells remain apart. When the distance apart decreases, there is interaction, the solution becomes difficult, but simplifies again when the two wells merge into one (see Fig. 190). The symmetric wave function is now recognised as a ground state, $n = 1$, and the antisymmetric as the first excited state, $n = 2$. In the general solution for a well of infinite depth, the energy ($E = n^2 h^2 / 8ma^2$) was shown to be proportional to $n^2 / (\text{width})^2$. Hence for a well of width $2a$ when two wells merge, the energy is only ¼ of that

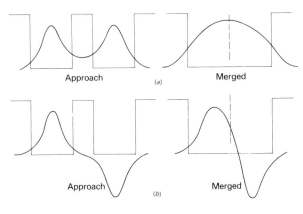

Fig. 190 *Electron wells merging, alternative solutions. (a) Symmetric. (b) Asymetric.*

for a well of width a in the ground state ($n = 1$). When $n = 2$ the energy for a well of width $2a$ is equal to that of the ground state for width a. Although the finite depth well has slightly different energies, it is possible to plot the variation of energy against the distance between potential wells for symmetric and antisymmetric functions (see Fig. 191).

When applied to the simplest case, two hydrogen atoms forming a hydrogen molecule, the electrons are each in the 1s state, and the two functions are shown in Fig. 192. The effect of moving

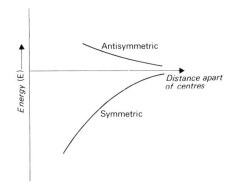

Fig. 191 *Variation in total energy with distance apart of two electrons in wells.*

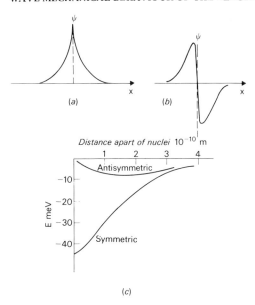

Fig. 192 *Wave functions for hydrogen molecule (a) symmetric and (b) asymmetric. Energy variation with distance apart (c).*

the two atoms together gives a peak in ψ for the symmetric function, and a plot of variation in energy with distance apart is also indicated in Fig. 192(c). It is necessary at this stage to take into account the repulsion of the two nuclei as the two atoms are brought together. The repulsion forces are proportional to the inverse square of the distance apart, and the total energy of the system, varying with distance, is seen in Fig. 193.

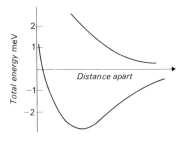

Fig. 193 *Total energy for hydrogen molecule against distance apart.*

The symmetric solution shows a minimum of energy at a given distance, implying stable equilibrium. For hydrogen therefore, the conclusion is that the symmetric spatial wave function gives a stable hydrogen molecule. The position of the minimum can be obtained by experiment, and is approximately the same as that of the wave mechanical calculation. The symmetric function is often called a "bonding" orbital (attractive) and the antisymmetric non-bonding (repulsive).

When two or more nuclei are involved, the argument can be extended, the number of wave functions increases, and so do the plots of total energy with distance. The larger number of wave functions with "N" nuclei implies a splitting of the quantum states into a "fine" structure, adopting one or more of the possible functions. A piece of copper weighing 1 mg contains 10^{19} atoms, so that 10^{19} separate levels of the ground state are possible, almost indistinguishable over a small range of distance or energy, which leads to the concept of energy bands rather than discrete levels.

When the quantum number $n = 2, 3, 4, \ldots\ldots$ the excited states, the effect of bringing nuclei together affects these states first, since they represent the electron at higher energy, and therefore distance from the nucleus. It will be these levels which widen into bands first therefore, as nuclei approach. At any given energy level, energy is proportional to $n^2 / (\text{width})^2$; hence the bands become wider with increase in energy, and the higher the quantum number n the sooner this occurs as nuclei approach. The inner energy levels are affected last of all, and the equilibrium distance apart of nuclei is usually reached long before they can be affected. Inner energy levels therefore tend to remain sharp and discrete (see Fig. 194). Note in the figure, that the bands represent the whole range of total energy, from antisymmetric to symmetric, since all the wave functions are involved.

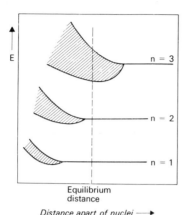

Fig. 194 *Bands of energy for three quantum states.*

When a periodic crystal lattice becomes involved, the potential cannot be simply represented as with individual atoms. In a solid crystal, the potential will vary in a periodic manner around each nucleus, and Fig. 195 represents the potential field drawn along a line through the centres of a row of atoms. The "pinching" of the potential is due to the repulsion effect of adjacent nuclei.

Fig. 195 *Potential field along a row of atoms.*

The solutions to the wave mechanical equations applied by Bloch assumed that the wave functions of a particle moving in a periodic potential field must be periodic with the period of the lattice. The wave number k of the electron does not have a constant amplitude, but its modulation is periodic. If the barriers are assumed of infinite height, the solutions give a similar result to the electron in a potential well, energy is restricted to certain permitted values, depending on the potential of the barriers and their width. If the potential barriers and widths are assumed to be very small, the electron is free, and can wander through a whole range of energy level. The intermediate case, which generally applies, gives solutions of bands of energy, the bands increasing in width the higher the energy level.

Consideration of the electron population in a mass of material leads to difficulties of interpretation of the restrictions imposed by the wave nature of the electron, the Pauli principle, etc. If electrons are regarded as "free" then the density of states curves can explain the differences between electrical conductors, semiconductors and insulators, i.e. when movement of the highest energy electrons is involved. To appreciate some finer points of experimental fact, the crystal structure has to be considered in conjunction with all other factors.

Using de Broglie's wave number k ($k = 2\pi p/h$ where p is momentum), the considerations of momentum of the electron, which are easy to apply in the one-dimensional model, can be applied to two or three dimensions, drawing a wave number diagram as can be done for momentum. Such a diagram is called k-space, or reciprocal space, because it takes into account the position of atoms in the real space, and their influence on electron movement.

Electrons which have Fermi energy, and can therefore move in the lattice, cannot penetrate a three-dimensional lattice in those circumstances which give a reflection from a lattice plane, i.e. satisfy the Bragg law $n\lambda = 2d \sin \theta$ (see **Bragg's law**). In three dimensions therefore, to define the energy required to cause an electron to move through the lattice, that movement must be

related to the properties of the lattice in addition to all other requirements. A vector diagram is used, in which the direction of the vector is the direction of motion of the electron, and the magnitude of the vector is the wave number (see **Reciprocal space**).

k-space (or reciprocal space) is represented by a three-dimensional diagram, with axes k_x, k_y, k_z corresponding to wave motion parallel to the x, y, z axes of the real crystal. In a one-dimensional model, the electron can be shown to be restricted to certain bands of energy, the tightness of binding determining the width of the energy bands. At particular values of k, a small increase in momentum, and therefore in k, requires electron energy E to jump discontinuously from the top of one band to the bottom of the next, otherwise solutions to the wave mechanical equations cannot exist. The region between the values of k at which a discontinuity occurs, is known as a Brillouin zone. If there were no discontinuity, the relationship between E and k would be a parabola, of equation

$$E = \frac{h^2}{8\pi^2 m} k^2$$

where h is Planck's constant and m the mass of the electron. This represents the free electron model. (See Fig. 196.)

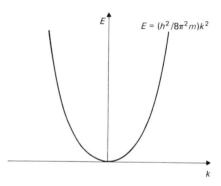

Fig. 196 *Free electron model relationship between E and k.*

When discontinuities are introduced, the parabola breaks up into discrete bands between which are the energy gaps of the discontinuities (see Fig. 197). This concept can be extended to two dimensions, using only the axes k_x and k_y of k-space, and components can be derived along the x- and y-axes. The first Brillouin zone for a cubic lattice in two dimensions, then becomes a square, bounded by points equidistant along the positive and negative

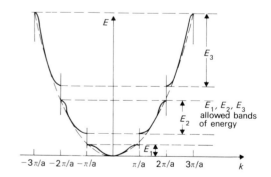

Fig. 197 *E against k with discontinuities introduced.*

axes of both x and y. The second zone cuts the x- and y-axes at twice the distance of the first zone, and the boundary runs across the corners of the square of the first zone at 45°. The second zone therefore consists of four triangles, whose bases are on the sides of the square of the first zone. The third zone extends the boundaries as extensions of the original square, along both axes. (See Fig. 198.)

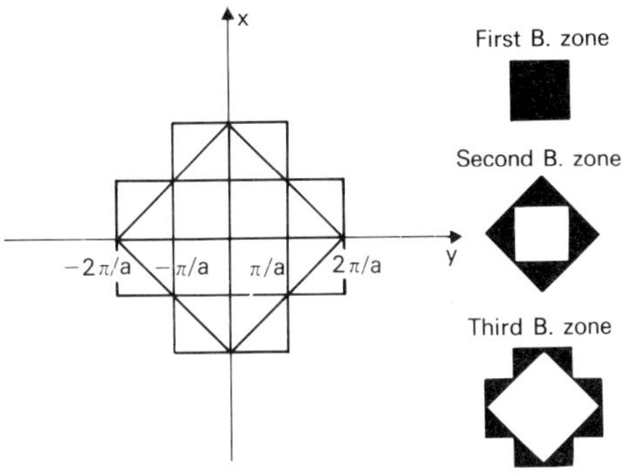

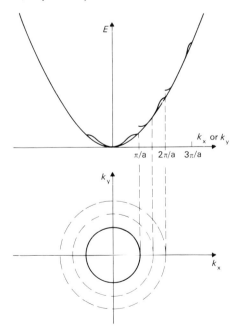

Fig. 198 *Brillouin zones in two dimensions for cubic lattice.*

These zone boundaries are measured in terms of $1/a$ where a is the lattice spacing, and are therefore a feature of the lattice. The electron is free to adopt almost any k value in direction and magnitude, in this two-dimensional model, but the electron must undergo a discontinuous energy change in crossing a zone boundary, i.e. there must be a jump in energy.

Applied to three dimensions, the same reasoning leads to zones which are solid figures. In the cubic lattice, the square of two dimensions then becomes a cube in three dimensions, intersecting the three axes at $\pm\pi/a$. The second zone becomes a pyramid on each cube face, analogous to the triangles of the two-dimensional figure. The third zone adds a triangular prism on to each pyramidal face of the two-dimensional figure, and the fourth zone joins up all the points of the third zone.

In the body-centred cubic structure, the first zone is a dodecahedron and the second zone a more complex figure of eight hexagonal faces and six square ones. The first zone of the face-centred structure is similar to the second zone of the b.c.c., and the first zone of the hexagonal structure is a hexagonal prism.

Brillouin zones therefore, indicate the restrictions of wave function solutions in terms of crystal structures. This tells what is possible, not necessarily what actually occurs. The occupation by electrons of Brillouin zones available is indicated by the Fermi surface.

The Brillouin zone concept indicates the bounds of electron energy which are feasible in view of all the relevant restrictions and conditions. As electrons build up the extra nuclear portion of the atom, they naturally enter first the lowest states of energy

and then are forced to make discontinuous jumps in energy as each energy level is filled. To indicate the relationship between energy and k value (or momentum) assumptions have to be made, because it is not easy to calculate the state density throughout a Brillouin zone for a bound electron. The free electron approximation is used to surmount the difficulty. In one dimension, the E-k relationship is a parabola for a free electron gas, which is then broken up into steps as indicated in Fig. 197, when the crystal structure is taken into account. To obtain the two-dimensional values, a line is drawn across the plane as a locus of points of equal energy indicated by the single dimensional figure (*see* Fig. 199). A plot of the energy contours of the actual electron population in terms of k space would give a series of circles, the boundaries being the Brillouin zone limits. (The circles arise because all points of equal k value have the same energy. $(k_x{}^2 + k_y{}^2)^{1/2}$ thus gives a circle.

When the electron is subjected to a periodic potential, it is not completely free, and so these relationships between E and k are no longer simply parabolic in one dimension. In two dimensions, this implies distortion of the circles, which becomes most noticeable towards the extremities of the zones. Points near the origin will be electrons with small values of k (or momentum), least affected by the periodic lattice.

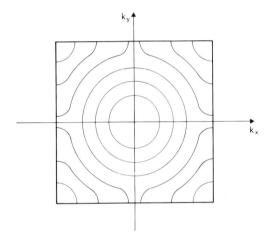

Fig. 200 *Energy contours: weak binding, square lattice.*

For a square lattice (simple cubic) the first zone will be a square in two dimensions, and the energy contours will appear as in Fig. 200. The strength of the electron binding will determine how much the circles are distorted, so that Fig. 200 represents a weakly bound electron system. With strong binding, even electrons with k value close to the origin will be distorted, and Fig. 201 indicates the equal energy contours then.

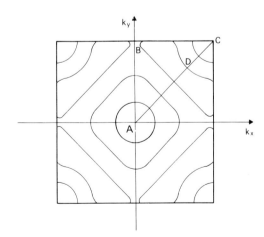

Fig. 201 *Energy contours: strong binding, square lattice.*

Fig. 199 *Points of equal energy in a two-dimensional lattice.*

Taking a line AB in Fig. 201, the energy increases from A to B, crossing the various energy contours. At B, the zone boundary is reached, and any increase in energy must involve the discontinuous step associated with it. If path AC is followed, along the diagonal, the normal increase in energy follows, until the point D, which is identical in energy with B, but is not on the zone boundary. From D to C the energy continues to increase, and C is therefore higher in energy than B. If that increase in energy, D to C, is greater than the energy discontinuity at B, then there must be some overlap of the energy bands available to the electron. Such an overlap is only feasible in two- or three-dimensional lattices. The energy discontinuity may be high at B, and then no overlap can be envisaged, and this is true of strongly bound electrons. In weakly bound cases, the implication is that the lowest energy level of zone 2 is lower than the highest of zone 1, and hence an overlap can occur. The energy discontinuity at the zone boundary is then obviously small. (*See* Fig. 202.)

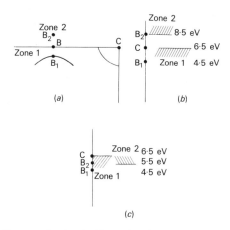

Fig. 202 *Energy considerations involving overlap of zones, square lattice. (a) energy contours close to boundary at B and C. (b) $B_2 \gg C$ no overlap, large discontinuity: strong binding. (c) $B_2 < C$ overlap, small discontinuity: weak binding.*

In three dimensions, the energy contours are obviously surfaces, defined by saying that all points where the k vector terminates in the surface have the same electron energy. For free electrons, circles become spheres, but for bound electrons they are obviously spheres only near the origin, and become distorted towards the zone boundary.

For the metal sodium, we expect the binding of the 1s electrons to be high, because they are close to the nuclear charge of 11 protons. The 2s electrons experience a net nuclear charge of 9 protons, the two 1s electrons having effectively neutralised two protons. The 2s electrons will therefore be similarly tightly bound. The 2p electrons will experience a decreasing effective nuclear charge due to shielding by inner electrons, and so less tightly bound than the 1s and 2s electrons. The eleventh electron, one 3s, faces a net nuclear charge of only one proton, and is well shielded by the other electrons, so it may be regarded as weakly bound. Its energy level becomes a band, about 8 eV wide at the lattice spacing of 0.367 nm.

For a quantity of sodium atoms, the state density can be explored, ($Z/E = fE^{1/2}$) assuming free electron behaviour at the bottom of the band (almost true) and so draw the parabola of $Z(E)$ against E, Fig. 203(a). From the Brillouin zone concept, and the energy contours in two dimensions, the state density obviously declines to zero at the top of the band because the probability of the k vector to yield a particular value of E falls off approaching the corners of a zone. The energy contours at the corners approximate to circles centred on the zone boundary corner, and hence a second parabola can be drawn on the state density curve, from energy E^1 at the top level, $Z(E) = (E^1 - E)^{1/2}$ (*see* Fig. 203). Between these two areas, the k vector approaches the side of the zone boundary, and the energy state density becomes crowded. A cusp-shaped figure completes the density of states curve (*see* Fig. 203(b)). There is experimental evidence for this curve, from soft X-ray spectroscopy and photo-electric

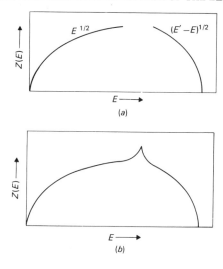

Fig. 203 *Derivation of state density curves. (a) Contours from origin and from corners (square lattice). (b) Complete curve.*

effects. The highest energy level in sodium would be the 3p, and the overlapping discussed under Brillouin zone would occur, these electrons being weakly bound with sodium. Figure 204 shows the overlap, and the resultant curve which can be drawn for the whole band of energy.

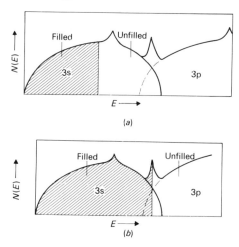

Fig. 204 *Population density curves. (a) 3s and 3p electrons in sodium. (b) 3s and 3p electrons in magnesium.*

The population density of sodium involves two electron states of 3s, each of opposite spin, but sodium has only one 3s electron, so the band is only half filled, and a shaded area can be drawn on the $N(E)$ curve to indicate this (*see* Fig. 204(a)). Similar treatment can be applied to magnesium, the next element in the periodic table to sodium, but now the two 3s electrons mean a filled band, which overlaps into the 3p and hence appears as Fig. 204(b).

The energy contour which touches the surface of the distribution of states in the zone must have the value E_F, the Fermi level. At $T = 0$ K this indicates the surface of electron distribution within the zone. In three dimensions, this surface, which characterises the energy limits, is called the Fermi surface. For sodium, where the zone is not full and binding is relatively weak for the 3s electron, the Fermi surface is a sphere, and since the first zone of a b.c.c. structure is dodecahedron the Fermi surface can be envisaged as a spherical ball sitting inside the dodecahedron and some distance from the zone boundary.

When the occupancy of states is greater, the Fermi surface may extend to touch the sides of the zone, and be distorted in shape from a sphere (*see* Fig. 205). The overlapping of bands can be looked at in the same way. Once the occupation density of electrons causes the Fermi surface to touch the sides of a zone,

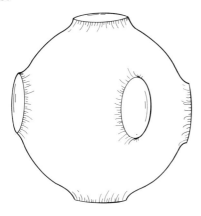

Fig. 205 *Distorted Fermi surface.*

any further electrons may have to enter a second zone (*see* Fig. 206 and cf. Fig. 204(*b*) for Mg when shown as a population density curve, $N(E)$ against E.)

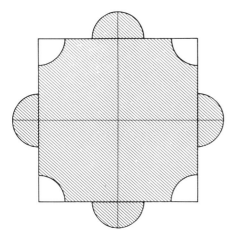

Fig. 206 *Fermi surface in two dimensions with overlap from zone 1 to zone 2.*

Metals have overlapping bands in many cases, and the Fermi surface overlaps into the second, third or even fourth Brillouin zone in some cases. This explains the electrical conductivity of metals, since electrons raised in energy can easily enter an over-lapping band, without having to suffer a large discontinuous change in energy to do so.

In contrast to metals, insulators have the lower band of energy completely filled by the covalent bonds between the atoms. There is no overlap of energy bands, and the discontinuities at zone boundaries can only be jumped by very high energy increases. With semi-conductors, the same principles apply but the discontinuities are lower, e.g. diamond (insulator) 6-7 eV; silicon (semi-conductor) 1.1 eV; and germanium (semi-conductor) 0.7 eV.

Waviness. A surface defect that impairs the appearance of sheets or components subjected to the usual finishing operations. The causes are various but may often be traced to tool vibration or work deflection.

Waxes. These are viscid substances that soften on heating, when they can be readily moulded. They fall mainly into two classes: (*a*) mineral waxes such as the solid higher hydrocarbons; and (*b*) the animal and vegetable waxes that are esters of the higher monohydric alcohols.

They are used in metallurgy for stopping off (q.v.) in electro-plating, in moulding and in sealing anodic films on light metal finishes. In these cases, they are dissolved in suitable solvents (alcohol, naphtha) and either painted on or the metal dipped into the solution. Drying oils may be added.

In plastics they may be used as fillers or plasticisers, or as coatings for protection or lubrication. In pharmaceutics, they

may be used as fillers to delay solubility or slow down absorption, or as oxidation barriers on skin applications. Table 65 indicates melting points of some common waxes.

The use in wood finishing is diminishing with the availability of polymeric materials such as silicones.

TABLE 65. *Melting Points of Some Common Waxes*

Wax	Melting or Setting Point, °C (approx.)
Beeswax	62
Spermaceti	45
Stearin	70
Paraffin (soft)	40
" (hard)	54
Seekay	123
Clamca	83
Ceresin	67
Carnauba	83
Japan	40
Flexoresin	160
Montan	86

Wax patterns. A term used in reference to the expendable wax patterns used in investment casting. The material used is generally a mixture of waxes (*see* **Waxes**). The die or master pattern has its surface coated with a film of oil to assist subsequent separation from the pattern, and when assembled, molten wax is forced in under pressure. The patterns are removed from the dies and given a primary coating of fine facing sand mixed with a suitable binder. When this coating is dried, the assembly is mounted in the moulding box. The mould is then prepared for casting by drying—usually at a moderate temperature—and during this process the wax pattern melts out.

Wear. A term used very generally in reference to the deformation or to the change of shape that metals undergo during service. The agencies causing the occurrence of wear are: (*a*) loading and high contact pressures between the surfaces of metals; (*b*) the relative movement of one metal surface over another resulting in scoring (high speed) or galling (low speed); (*c*) abrasion; and (*d*) atmospheric or chemical corrosion.

Wear resistance. A term used in reference to those metals which on account of their inherent properties or because of special treatment have surfaces which resist to a considerable degree abrasive action during service. High-carbon cast steels and certain alloy steels (e.g. manganese alloy steel) are usually regarded as "wear-resistant". Wear resistance can be improved by heat-treatments such as martempering, austempering and by surface treatments such as flame and induction hardening, carburising and nitriding. Other methods are available such as hard chromium plating and the processes used for weld depositing hard facing materials.

Since "wear" is usually employed in a very general sense, corrosion- and heat-resisting materials are sometimes included under the term wear-resisting materials.

Weathering. The alteration and decay of rocks due to the atmospheric agencies, e.g. air, rain, heat, frost, etc. The term is also applied to the treatment of ores, e.g. certain sulphide ores which after being mined are left exposed to the atmosphere, whereby much of the sulphide is converted to soluble sulphate.

Web. The flat metal plate that connects the flanges of a girder or the bosses of a crank. In heavy girders, tee-pieces are used to prevent the web from buckling, and are referred to as web-stiffeners. The term web is also used to describe the thin section of metal that remains at the bottom of a hole in a forging and which is later removed by machining.

Wedge. A term used to describe the rod—usually of hard wood—that is used in spinning operations as a forming tool.

Wedge roaster. A type of vertical, cylindrical, multiple-hearth furnace used in roasting copper ores. A rotating shaft supports rabble arms over each hearth. The arms and the rabble teeth are so arranged that on one hearth the ore is pushed from the centre to the circumference where it falls on the hearth below and is pushed from the circumference to the centre. The necessary air for roasting may be admitted at each hearth level or through the rabble arms.

Weibel process. A method of welding straight runs on thin sheet material. An electric current is passed between two carbon electrodes, through the edges of the metal to be welded, and the edges must be flanged and clamped in position.

Weirzin. The trade name of electrolytically zinc-coated steel.

Weld. A joint made between two metals (either similar or dissimilar), the coalescence being produced by heating the metals to the required "welding temperature" and holding, pressing or hammering the heated parts together. The parts may or may not be fused and filler metal may or may not be used. Brazing and hard soldering are often included under the heading of welding, but soft soldering is excluded.

Weldability. A term used in reference to the capacity of a metal to be fabricated by any particular welding process to produce a welded joint that will give satisfactory service. This capacity depends upon: (a) the composition of the metal parts; (b) the composition of the filler rod; and (c) the particular technique used in the welding process.

Weld bead. A term used to describe the built-up portion of a fusion weld from filler metal or from melted parent metal. The type of weld made by a single passage of an electrode or of a filler rod used also as an electrode.

Weld decay. A term for a defect that may occur as a result of the welding of stainless (chromium) alloy steels. In the regions of the parent metal adjacent to the actual weld the temperature attained may be such as to favour the formation of chromium carbide which is deposited at grain boundaries. This can only occur by the removal of chromium from the adjacent crystal zones, and thus these areas lose their corrosion resistance. When the welded area is subsequently exposed to corrosive attack, the zones depleted of their chromium content are vulnerable and this effect is known as "weld decay".

If after welding these steels the part is heated to above about 1,000 °C and quenched, the carbide is taken into solution and the chromium content of the impoverished zones is restored. This treatment would avert weld decay, but is not usually practicable, since the welding is undertaken to build up large components which it would be impossible to treat in this way. The method adopted is therefore to reduce the carbon content to the lowest practicable percentage and to add alloying elements that can form carbides in preference to chromium. These carbide-forming elements are niobium (about 10 × C content), titanium (about 5 × C content), molybdenum and tungsten, and when these are present they prevent the formation of chromium carbide and thus prevent weld decay.

A test which is widely used in UK to determine susceptibility of 18/8 type steels is as follows:

The material is heat-treated at 650 °C for 30 minutes and then immersed for 72 hours in a boiling copper sulphate-sulphuric acid solution, developed by Hatfield, of the following composition:

$CuSO_4.5H_2O$	111 g
H_2SO_4 (1.84)	98 g
Water	to 1,000 cm³

The sample is then bent through 90 degrees over a special former and examined for signs of cracking along the bend. (See also **Strauss test**.)

Welding. A term used to describe all those processes in which two pieces of metal are united to form a joint that is as homogeneous as possible. This can be brought about in a number of ways, and it is convenient to divide them into two main groups: (a) those in which the faces of the metal to be joined are melted and (b) those in which the metals are not melted, though they may be heated till they assume a plastic state. Fusion welding involves the heating of the surfaces to be joined till they melt and the parent metal makes a pool into which additional (filler) metal may or may not be added. The processes include:

(1) *Gas welding.* (i) Oxy-ethyne gas. With an oxidising flame a temperature of about 3,300 °C is attained while a reducing flame gives a temperature of about 3,000 °C. When a carburising flame is used on steel parts, carbon may be introduced into the heated surfaces in quantity sufficient to materially lower the melting point. Thus with the somewhat lower temperature operation, incipient melting followed by pressure can effect a weld. This is the basis of the Lunde method. (ii) Oxy-hydrogen gas.

(2) *Arc welding.* (i) Metallic arc welding using a bare electrode which is also the filler rod. (ii) Metallic arc welding using a coated electrode, the coating providing a flux for the work and also when required additional alloying elements. (iii) Carbon arc welding in which the electrode is a carbon rod, a separate filler rod being used to provide additional metal. (iv) Helium arc and

argon arc welding in which a tungsten electrode is employed, while helium or argon gas is provided from a separate nozzle to shield the molten metal from atmospheric oxidation. A filler rod is usually required. (v) Atomic hydrogen welding. Two tungsten electrodes are required to produce the arc while hydrogen is fed into the arc via one electrode. The gas is dissociated into the atomic state and in the recombination of the atoms to form molecules at the surface of the metals to be joined a large amount of heat is evolved, so producing fusion. (vi) Submerged arc welding. (vii) Bronze welding—using an oxyacetylene torch and temperatures not exceeding about 900 °C. The filler metal can be copper-zinc or copper-zinc-nickel alloys. (viii) Thermit welding (see also **Thermit**). (ix) Induction welding where eddy currents in the metals to be joined generate the heat for the fusion of the parts.

(3) *Pressure welding.* (i) Cold pressing as, for example, in the production of aluminium tube from strip metal by pressure at ordinary temperatures. (ii) Forge welding in which pressure is exerted by dies, rolls or hammers. (iii) Electrical resistance welding may also be included here, since the metal is probably not in the molten condition in such processes as spot welding, seam welding or flash welding. (iv) Electric-percussion welding, employed with certain light alloys of low electrical resistance and good heat conductivity. In this method the current is applied as a very short time heavy electrical discharge and the pressure is replaced by a blow.

Table 66 lists methods of welding that have been successfully employed with a number of common metals and alloys.

TABLE 66. *Welding Methods*

Material	Welding Methods	Treatment	
		Pre-welding	Post-welding
Wrought iron	3(ii), 3(iii), 2(vii) 1(i), 1(ii)		
Cast Steels	2(ii), 2(vi) As for wrought iron	Preheat	Anneal
Cast irons (grey)	1(i) 2(ii), 2(vi) 2(viii)		
Stainless irons	1(i), 1(ii) 2(ii), 2(iii), 2(iv), 2(v)	Preheat	
Malleable iron	2(vi)		
Alloy cast irons	As for grey iron		
Nickel steels	As for grey iron 2(ii), 2(iii) 1(i)	Preheat	Stress relieve
Low alloy steels:			
Chromium steels	1(i), 1(ii) 2(ii)	Preheat	
Manganese steel	2(ii)		Peen lightly
Chromium-molybdenum	2(ii)	Preheat	
Nickel-chromium-molybdenum	1(i) 2(ii) 2(iv)	Preheat	Normalise
Ni-Cr stainless steel (stabilised)	2(iv)		
Heat-resisting steels	Most (3(ii) and 2(vii) not applicable) 3(ii), 3(iii)		
Heat-resisting, high Cr steels	1(i) None		
Nickel and nickel alloys	2(ii), 1(i), 3(iii), 2(iv)	Anneal	
Magnesium alloys	1(i), 2(iv), 3(iii)		
Aluminium alloys	1(i), 2(ii), 3(iii), 2(iv)		
Copper	2(vi)		
Zinc	2(vi)		
Hard facing materials	1(i), 2(i), 2(ii)	Preheat	
Plain carbon steels:			
Low carbon	1(i), 1(ii)		
Medium carbon	Most	Preheat	
High carbon	2(ii)	Preheat	Anneal
Tool steel	1(i), 1(ii), 3(iii)	Preheat	Temper

Key to column 2

1(i)	Oxy-ethyne welding	2(vi)	Bronze welding (brazing)
1(ii)	Oxy-hydrogen welding	2(vii)	Thermit welding
2(i)	Metallic arc (bare electrode)	2(viii)	Induction welding
2(ii)	Metallic arc (coated electrode)	3(i)	Cold pressing
2(iii)	Carbon arc welding	3(ii)	Forge welding
2(iv)	Gas shielded arc (helium or argon)	3(iii)	Electrical resistance welding (e.g. spot or flash welding)
2(v)	Submerged arc welding	3(iv)	Electrical percussion welding

Welding rod. A term used in reference to the electrode (usually of tungsten and non-consumable) and to the filler rod, which is melted continuously during the welding process to supply additional metal to the welded joint. As a flux is usually required to minimise oxidation and to dissolve any oxides formed, the welding rods in the form of metal-arc electrodes may be coated with material of an inorganic nature—oxides of silicon, titanium, etc., and fluorspar, which melt at the welding temperature and produce a fluid protective slag on the molten metal. Alternatively, the rod may be coated with organic materials which decompose at high temperatures, forming gases that act as a shield to the weld metal against oxidation.

Welding stresses. The fusion welding process produces a small "casting" in the joint between the sections of parent metal, and the molten metal has to cool from a very high temperature whilst surrounded by cooler parent metal, which automatically creates a physical restraint. The stresses set up in the original molten pool may therefore be similar to those in castings. The restraint is never uniform however, because the weld run must have a start and finish point, and this imposes far greater stress on the last portion to solidify and cool down. If the weld is a closed loop, as with a circular or rectangular patch welded on to a plate, or the circumferential weld in joining two pieces of pipe, the restraint is increased even further. Welding stresses are therefore inevitable, and can reach the yield point locally, with the possibility of crack formation.

The pattern of stress is complex, depending entirely on the geometry of the structure, the way in which welding is carried out, the welding speed, rate of power input, and the environmental conditions. To minimise these effects, preheating can be carried out, to reduce the rate of cooling during the post welding period. Gas welding has always been advantageous in this respect, the broad base of the gas flame almost automatically giving some preheating to the next section to be welded. Electric or gas pre-heaters can be applied to the material to be electrically welded, and low frequency induction cables are often wrapped around large pipe structures for this purpose.

Where stresses may be severe, post heat treatment can also be applied, in this case to relieve the stress by an annealing treatment. With complex alloys, such stress relief heat treatment has to be applied with caution, since changes in microstructure can result in the time necessary to allow adequate relief. Experimental checks are necessary in these cases, before applying this technique. Large stress relieving furnaces have been employed for welded structures such as internal combustion engine bodies, ducts and pipework in steam and chemical plant. The pressure vessels of nuclear reactors, and those used for high pressure gas storage, have been stress relieved by building furnaces around the vessels, to be dismantled and used again, or by clamping heaters to the welded sections to create a more localised high temperature relief zone. These are all expensive treatments, but the cost of failure justifies their consideration in all conditions of high operational stress, since residual stresses are additive in many cases, to those involved in construction and service.

The lower the temperatures involved in a welding process, the less likely are the residual stresses to be severe, since Newton's law of cooling applies, and it is the difference between the temperatures of the artefact and its surroundings which is important. However, the lower the temperatures involved, especially in fusion welding, the lower the maximum strength of the material is likely to be, since strength is related to melting point. As a fraction of the fracture strength therefore, low residual stress in lower melting point materials may still be deleterious, and this can apply to metals and alloys such as aluminium and magnesium. Polymeric materials are usually very plastic even though the maximum strength may not be high, and although only a few are joined by heat processes, there is less likelihood of catastrophic crack formation by residual stresses. Nevertheless, the more complex uses of polymeric materials demand that such considerations be made when designing the production line.

Weldless tube. A term sometimes used in reference to solid drawn tubes.

Welsh process. A process that was used in the smelting of sulphide ores of copper for the separation of iron and sulphur from copper. A reverberatory furnace is used for the roasting to convert part of the copper and iron sulphides into oxides. The charge is then mixed with siliceous material and fused at a high temperature. The iron oxide passes into the slag, while the copper and iron sulphides remaining form a lower layer in the furnace. This is known as coarse metal. The process, when repeated, produces a product known as fine metal, which contains a much higher percentage of copper sulphide than the coarse metal. The fine metal in the form of blocks is again roasted in air, and the remaining sulphur burns off as sulphur dioxide and some cuprous (copper II) oxide is formed. This latter then reacts with copper II sulphide in the metal, giving metallic copper and liberating more sulphur dioxide. As a result, the blocks become covered with blisters due to the escape of gas. The blister copper is melted under a layer of powdered anthracite and stirred with green birch-poles to reduce the remaining copper II oxide. A small amount of copper II oxide, however, remains in the melt, the amount being judged by the fracture of a small ingot. The copper is cast as tough-pitch copper.

Werner's theory. A theory that was propounded in 1891 to explain the behaviour of compounds that were then regarded as molecular addition compounds. At that time valency had been regarded as a constant atomic property. A compound such as PCl_5 was called a second order or molecular compound, and the formula was frequently written $PCl_3.Cl_2$. The distinction between true valency compounds and molecular addition compounds was the alleged instability of the latter compared with the former. The discovery of the very stable phosphorus pentafluoride upset the idea that valency is an immutable property of atoms and many instances were found that made the instability criterion untenable. Hexamminecobalt III chloride, whose formula was written $CoCl_3.6NH_3$, is a very stable compound, while $CoCl_3$ does not exist alone. If the formula for sodium chloroplatinate were correctly written as $PtCl_4.2NaCl$, then this compound should evolve hydrochloric acid on treatment with sulphuric acid—which it does not do. Similarly, if $4KCN.Fe(CN)_2$ is correct for potassium iron II cyanide, hydrocyanic acid should be produced on the addition of a dilute acid, but this does not occur. Aluminium fluoride AlF_3 reacts violently with water, but $AlF_3.NaF$ is the inert mineral cryolite. These and other considerations led Werner to propose a theory that valency and chemical affinity can be assigned to one ultimate cause —viz. the creation of atomic attractive forces by the bringing into operation of integral valency. Certain atoms tend to attract to themselves a definite number of atoms or groups, irrespective of their valency, and the maximum number that can be so added is known as the co-ordination number. The atoms or groups thus attached are supposed to be in an inner or first zone, and they are so firmly combined with the co-ordinating atom that they cannot be ionised in solution. In addition to these, the molecule can attract other atoms or groups which are held in a second or outer zone, and they are capable of ionisation.

The co-ordination number for the majority of atoms is six, but for atoms of small or very large atomic weight the number is usually four. The spatial arrangement of atoms within the molecule in the latter case is either planar or tetrahedral. When the co-ordination number is six, the arrangement is always octahedral. In the case of potassium iron II cyanide the formula is written $K_4[Fe(CN)_6]$ and the group within the square bracket is known as the co-ordination complex. The co-ordinated atoms in metallic co-ordination compounds are generally nitrogen, oxygen, fluorine, chlorine, bromine and iodine and metallic co-ordination compounds are very frequently either hydrates, amines, oxides, oxyacids or double halides. The co-ordination theory provides a convincing explanation of the complexing procedures in analysis.

Weston cell. A standard cell used for comparison of electromotive force. Figure 207 shows the usual form of this cell in which mercury forms the positive pole and an amalgam of cadmium and mercury the negative pole, the electrolyte is cadmium sulphate and the depolariser is a mercurous sulphate paste. The e.m.f. at 20 °C is 1.0183 volts, but falls with rise of temperature. This cell, also known as Weston cadmium cell, should not be used as a source of current.

Westonite. A petroleum asphaltic-base material used for improving the properties of foundry sands. It is applied dissolved in a suit-

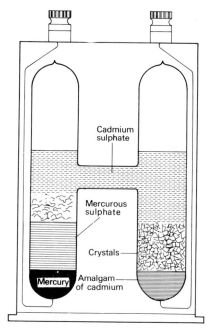

Fig. 207 *Weston cell.*

able solvent and is reputed to reduce gas unsoundness and produce a good casting surface.

Westron. A trade name for tetrachloroethene (CCl_2)$_2$, a chlorinated hydrocarbon which boils at 120 °C. It is a good solvent for
oils, greases and wax and is used as a degreasing agent for metals.
It is also used as a solvent for cellulose and certain resins.

Wet drawing. A process for drawing wire with a bright polished
surface which is produced by drawing through a liquid lubricant.
The wire treated in this way is also known as lacquer-drawn wire.

Wetherill process. A process for the production of zinc oxide. An
oxide ore (e.g. zincite or franklinite) or a roasted blende (zinc
sulphide) is mixed with coal and heated on a grate by the combustion of part of the coal. The zinc ore is reduced to the metallic
state and vaporised, and, as the temperature is favourable, the
zinc is reoxidised to ZnO above the charge and carried to suitable
condensers. If lead-free ore is used, an oxide of high purity is
obtained suitable for use in the rubber industry. If lead is present,
most of this will be present in the zinc oxide as lead oxide or as
lead sulphate. This mixed product is suitable for use as a base in
the paint industry.

Wettability. A term used in reference to a metallic surface to indicate the ease with which a lubricant will spread or flow over it.
It is likewise sometimes used in reference to the surface of a
metal to indicate the facility with which the molten solder will
flow over the heated surface during soldering, etc.

In mineral dressing the wettability of the surface of a mineral
substance in water is one of the factors which determines the
tendency of the mineral to sink or float. This tendency is indicated by the contact angle formed at the junction of the air and
water surfaces on the mineral surface.

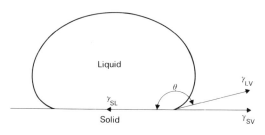

Fig. 208 *Surface forces acting on a liquid drop resting on a solid.*

The forces acting on a liquid drop are shown in Fig. 208. The
relation between surface energies is given by

$$\gamma_{SV} - \gamma_{SL} = \gamma_{LV} \cos \theta$$

when $\theta = 0°$ complete wetting results. Values of θ up to 90° can
be considered satisfactory wetting for the purpose of bonding,
say in soldering or the use of adhesives, and the value γ_{SL} is the
prime concern in achieving this. Chemical agents which remove
oxide films from surfaces, by reaction or dissolution, are usually
involved in bonding processes for this reason.

Wettengel process. A process for producing retorts particularly for
use in the production of zinc, the basic mixture containing raw
clay and burnt clay together with coke dust, and sometimes
quartz sand is added. The materials are crushed and screened and
then moistened with water before passing through a pug mill.
The product in the form of cylinders is then aged by stacking
in a warm, moist chamber for several weeks. The plastic material
is then pressed hydraulically and extruded in the form of a
retort. The retorts are dried at 60-70 °C, but are not baked until
actually required. Baking takes place at about 1,250 °C, and the
retorts are put into service while still hot.

Wetting agents. The general name for those substances which are
added to solutions, suspension, emulsions or pastes to reduce the
surface energy and thus to give greater ease in mixing and improve
the facility with which a particular fluid will flow over a given
surface in such operations as cleansing (*see also* **Wettability**). In
flotation processes (q.v.) pine oil is used as a wetting agent.

Wheelaborator. A machine used for cleaning castings. The latter
are tumbled about in a cabinet on a horizontal belt and at the
same time are subjected to bombardment by a stream of metal
shot. This latter is maintained by feeding the shot on to the blades
of an impeller. The shot are then projected centrifugally at high
speed on to the castings. This method is more economical and
admits of better control of the abrasive than is possible using
air jets. The moving belt must, of course, be well covered with
parts to be cleaned, and the impeller blades should preferably be
of some hard metal such as high-silicon iron.

Whirl gate. A type of gate used in casting in the foundry where the
metal passes from the ingate into a chamber, where it is allowed
to swirl before entering the mould. The object of this is to separate the slag from the metal before it enters the mould. The whirl
gate is cylindrical in section and of considerably greater diameter
than the ingate.

Whirling. A term used in reference to the behaviour of rotating
horizontal shafts in which the distance between the supporting
bearings is large compared with the diameter of the shaft. At certain speeds the shaft vibrates, so that its axis does not remain
perfectly horizontal. This effect increases up to a certain speed,
called the critical speed, after which steady running occurs. It is
usually good practice to ensure that the critical speed is passed as
quickly as possible, to minimise the likelihood of the shaft failing
as a result of whirling.

Whisker. The name used in reference to small single crystals that
can be grown on metal surfaces by a special heat treatment. Due
to absence of dislocations, these crystals possess great shear
strength and can tolerate extremely large strains in bending.
If these whiskers (of, say, aluminium) could be properly embedded
within a metal they could carry practically all the load and permit
the composite to perform at temperatures far beyond the normal
operating range.

The use of whiskers of metal is at present being studied to
overcome the lack of structural materials that will maintain
strength at high temperatures such as are encountered in rocketry
and similar problems. There is some indication that fine cracks in
the metal surface may accompany whisker formation.

In miniature components the formation of whiskers on the
surface of a metal may give rise to current leakage, and hence to
inefficient performance of an electronic component. It has been
suggested that intermetallic compounds of low ductility might
be free from whisker formation.

Whistler. A term used in foundry-work in reference to vertical
vents that are provided to facilitate the escape of gas and are
usually provided in those parts of the cope where gas is likely to

be trapped. Small section whistlers are usually circular in section, but large ones are commonly rectangular.

White brass. A term usually applied to a series of high-zinc brasses used in ornamental work, cheap jewellery and for buttons. The applications of the alloy depend on the composition, which varies widely.

White cast iron. The name given to the iron containing about 2-6% C in which the latter is present in the form of iron carbide (Fe_3C). It is hard, brittle, difficult to machine and is not malleable at any temperature. It breaks with a silvery-white fracture. The conditions favouring the formation of white iron are: *(a)* rapid cooling from the molten condition; *(b)* low silicon content and *(c)* low working temperature. White cast iron cannot be worked, but it has good casting properties and has applications where a cheap material with hardness and good wear resistance is required. Cast-iron rolls are made with a chilled white skin. Specific gravity 7.5.

White colouring. The name of an abrasive mixture that consists of fine-grain amorphous silica and a grease binder.

White copper. The name sometimes used in reference to the alloys of copper, zinc and nickel which are very similar to the alloys otherwise known as nickel-silver or German silver. The old name for these alloys is Packfong (Table 5), which means white copper. A typical composition is Cu 70%, Ni 12% and Zn 18%.

White finish. The name used in reference to a polishing compound consisting of Vienna lime (calcined dolomite) with a moisture-free binder, used for buffing nickel plate and other metal finishes.

White garnet. The colourless (iron-free) potassium aluminium silicate found as a mineral $Ca_3 Al_2 (SiO_4)_3$. Crystal system, rhombic dodecahedra. Also known as leucite.

White gold. The name used in reference to a class of jewellery alloys used as substitutes for platinum. Gold is alloyed with platinum, palladium, silver or nickel or with a white alloy containing nickel, zinc and copper. The compositions of a number of white-gold alloys are given in Table 9.

White heart castings. In the production of malleable cast iron, the casting is first formed of white iron and then subjected to an annealing process which consists in heating the casting to a temperature between 800 °C and 1,000 °C for several days. The iron carbide is decomposed, yielding iron and finely divided carbon. Formerly iron oxide was packed around the casting and this caused the oxidation and removal of the carbon either as CO or CO_2. Modern practice favours the use of sand. If the annealing process is carried out at a temperature of something a little in excess of 700 °C for a shorter period, the carbon may be removed from the surface layers, producing what are referred to as white heart castings. (*See also* **Cast iron.**)

White lead. An amorphous white powder produced by exposing lead to the action of acetic (ethanoic) acid vapour and carbon dioxide. It is basic carbonate of lead $2PbCO_3 .Pb(OH)_2$. Good white lead readily mixes with linseed oil, used as a basis for paints. It has a very good covering power, but is affected by sulphuretted hydrogen in the air, a black lead sulphide being formed. It is often adulterated by the addition of the cheaper barium sulphate, the covering power of which is much less than that of white lead. White lead is readily soluble in nitric acid, so that if a sample is treated with dilute HNO_3 and a white residue remains, the presence of the adulterant can be detected. *Venetian white* contains equal parts of white lead and barytes. *Dutch white* contains one part of white lead to three parts of barytes (barium sulphate). *Hamburg white* contains one part of white lead to two parts of barium sulphate.

White lead ore. A valuable source of lead that contains lead carbonate $PbCO_3$. Also known as cerussite (q.v.).

White metal. A term usually employed in reference to the white alloys such as Babbit alloys used as linings for bearings. It is also used in a somewhat loose fashion to include pewter, Britannia metal, type metals, tin die-casting alloys, fusible alloys, nickel silvers and other white non-ferrous alloys.

White nickel ore. A valuable ore of nickel containing nickel arsenide $NiAs_2$. Also known as chloanthite (q.v.).

Whitening. A term used in electroplating in reference to the immersion of a metal strip or sheet into a bath containing dissolved tin or silver salts. By the process of simple chemical replacement a thin film of the white metal (tin or silver) is deposited on the surface that has been dipped.

White pickling. A term used in reference to the second pickling process applied to steel sheet, after close annealing, to remove oxide films prior to tin or terne coating.

White rust. A term sometimes applied in reference to the film of deposit of zinc oxide (usually with a little basic carbonate) that accumulates on the surface of zinc sheets exposed to moist air.

White vitriol. A commercial term sometimes used to describe white crystalline zinc sulphate $ZnSO_4$.

Whiting. The name of a specially finely ground and washed form of chalk $CaCO_3$. It is used in paints, and in the making of putty. The source of the chalk is usually the finely pulverised marine shells or high-purity marble or limestone. The finest grit-free kinds are used to strengthen rubber and in cosmetics.

Wiberg-Soderfors process. A process for producing sponge iron, which is subsequently used in the manufacture of high-grade steel. In the early days of the process, charcoal was used as the source of the reducing gases. Latterly, charcoal has been replaced by coke, the sulphur being removed from the gas during its passage through a tower packed with granulated dolomite.

Widia. A sintered tungsten carbide product made by diffusing powdered cobalt through the powdered carbide under hydraulic pressure and sintering in an inert atmosphere at about 1,500 °C. The material so formed is ground to shape and brazed as tool tips to steel. The composition of widia, which is of German origin, is W 84-87%, Co 6-13%, C 3-5.7%.

Widmanstatten structure. A type of structure produced when a new solid phase forms from a parent solid phase as plates along certain crystallographic planes of the original crystals. On a polished and etched surface the plates appear as intersecting needles which form a geometric pattern. This type of structure was first observed in meteorites, and may be seen in a more or less pronounced state in steel castings and in steel which has been cooled rapidly from above the upper critical temperature, but has not been worked. It may occur in any alloy in which a phase change occurs. The second phase which is precipitated from the parent phase appears at the grain boundaries of the latter and inside the grains as thin plates, usually in geometric pattern, but sometimes irregular, but in all cases the orientation is related to that of the matrix. In the case of hypo-eutectoid steel cooled quickly from a high temperature, the ferrite precipitated appears along certain planes (the octahedral planes) within the grains. In the case of hyper-eutectoid steels, cementite is precipitated under similar conditions in much the same way. The special interest in this type of structure in the case of steels, for example, derives from the fact that steels of this type lack toughness and show a low impact value and have low ductility. These properties are the result of the appearance of the network of weak ferrite in hypo-eutectoid steels and of brittle cementite in hyper-eutectoid steels. Improvement of properties can be obtained by suitable heat-treatment.

Wiedemann-Franz-Lorentz law. Metals in general have both high thermal and high electrical conductivity. It would be expected that since both parameters depend on a lattice and an electron contribution, they might be related. Wiedemann-Franz first expressed this as a ratio dependent on absolute temperature.

$$k_t /(k_{el} T) = \text{constant}$$

where k_t and k_{el} are thermal and electrical conductivity respectively, and T is absolute temperature. Lorentz calculated the constant from theoretical considerations, as 2.45×10^{-8} W Ω K^{-2}. Experimentally, iron gives 2.47×10^{-8}, lead 2.56×10^{-8}. Metals such as aluminium are lower, 2.23×10^{-8}, and platinum higher, 2.60×10^{-8}.

The variations in metals must be attributed to the varying contribution of the lattice and the electrons, and the difficulty in assessing the number of electrons per atom which are available for contribution to conductivity. In semi-conductors, the electron contribution to thermal conductivity is inevitably small, but the electrical conductivity is affected much more, so the relationship does not hold.

Wigner energy. The graphite lattice is a layer lattice of carbon atoms arranged in connected hexagonal rings, with the layers bonded by van der Waals' polarisation bonds. When bombarded by neutrons, carbon atoms are displaced from the lattice into interstitial positions in adjacent layers, and such is the energy of neutrons that the equivalent energy to cause the return of the displaced atoms to normal positions is very high. A severe state of residual stress results, requiring a high activation energy to relieve it. Once this energy is reached, the carbon atoms fall back into place with an avalanche effect, and all the stored energy of the displacement is dissipated by violent atomic vibrations, in other words as heat and a considerable rise in temperature. The stored energy is known as Wigner energy after the first observer of the phenomenon, and this energy per unit mass is so great that it exceeds the specific heat of the material. The temperature can therefore rise very rapidly, each release contributing to further increase in temperature. In at least one example (Windscale, UK, 1957) the graphite undergoing this release began to burn, the reactor being air-cooled. It is the anisotropy of the hexagonal lattice which produces this unusual effect, but it demonstrates the amount of energy which can be stored in a material subject to atomic displacement. The remedy is a strictly controlled slow release, cooling being applied between steps in the release. This phenomenon is one of the major disadvantages of the graphite moderator in thermal reactors, which does not occur with moderation by water or other liquids and of course does not apply to fast fission reactors with no moderator.

Wild steel. A name applied to steel that reacts violently in the mould due to vigorous liberation of gases during solidification. The term is especially applicable in the case of rimming steel.

Wilfley table. A type of bumping table used in the concentration of ores. The surface of the table is partly smooth and partly covered with riffles and in use is tipped slightly towards the end where the concentrate is collected. The crushed ore, together with water, is fed at the top of the table and wash-water is fed along the higher edge. The heavier mineral particles settle in the riffles and are caused to move downwards by oscillating the table back and forth by a bumping mechanism which operates so that the table is moving at its maximum velocity when it reverses direction at one end of the stroke and with minimum velocity when it reverses at the other end of the stroke. The heavy material thus progresses down the table in a series of short jerks.

Williamson's violet. An attractive violet compound, soluble in water, that consists of potassium iron III iron II cyanide $[KFe^{3+}(Fe^{2+})(CN)_6]$. It is produced by boiling potassium ferrocyanide with dilute sulphuric acid and oxidising the pale-yellow powder produced with either nitric acid or hydrogen peroxide.

Wiped joint. A method of joining lead pipes to give a full-bore connection with a mass of solder on the outside of the joint to give strength and rigidity. The inlet end of the pipe is chamfered to make a smooth tapering spigot. The other end to be joined is opened out by means of a wooden cone to form a socket, the inner surface of which is cleaned and then both spigot and socket are coated with tallow. The joint is tinned and then soldered, by pouring or splashing solder over it and wiping the pasty mass round the joint till a smooth contour is obtained. Porosity is reduced if after wiping the joint is coated with tallow and again wiped with molten solder.

Wire. A term somewhat difficult to define precisely. It refers to slender rod-like products. The cross-section of a wire is usually circular but not exclusively so. The method of manufacture is to draw wire bars or rods through successively smaller dies. The finer wires are reeled and the heavier wire is generally supplied in coils. The diameter of a wire may be quoted as a fraction of an inch, in mm, or may be designated by a gauge number (*see also* **Wire gauge**). The lowest gauge in any of the various systems refers to wire approximately 15 mm in diameter, and this may be taken as indicating roughly the upper limit of dimension for wire.

Wire bar. A special billet or bar that has been expressly prepared for subsequent drawing into wire or rolling into wire-rod. A copper wire bar generally weighs 150 kg.

Wire brush. A brush in which the bristles are made of wire. The nature of the wire used depends on the service that is to be performed. Steel bristles are used, for example, in brushes that are

to be employed in cleaning files or castings.

Wire drawing. A process for the production of wire by drawing wire bar or rod through a series of dies of decreasing diameter until the desired gauge is attained. It is usually a cold-working process and generally the rod is hot rolled and about 5 mm diameter. Thick sections of wire bar are often reduced on a drawbench, chain driven. The dies are usually of steel (sometimes chromium plated), white cast iron, tungsten carbide or diamond. Oils, soap or tallow are used as lubricants in most cases, but a film of oxide or phosphate or lime may be made the basis for a lubricant for steel wire. In certain cases one metal (a softer material) may be coated on the harder wire and the former can then act as a lubricant. In other cases the coating method is used to obtain a finer wire than could otherwise be drawn. (*See also* **Wollaston wire.**)

Wire gauge. The term used in reference to a device used for measuring the diameters of wires. It generally consists of a plate or disc round the periphery of which are formed a number of slots the widths of which correspond with certain sizes of wires (or thickness of sheet metal).

Wire gauge also refers to the various systems of designating the diameters of wires by numbers, rather than by a diameter in units of length. There are some seventeen gauges as follows:

(a) Brown and Sharpe gauge.
(b) Birmingham or Stub's iron wire gauge.
(c) Washburn and Moen.
(d) American Steel and Wire Co. gauge.
(e) Roebling and Trenton Iron Co. gauge.
(f) Stub's steel wire gauge.
(g) Birmingham gauge (B.G.).
(h) British Imperial wire gauge.
(i) U.S. standard wire gauge.
(j) Music wire gauges—all similar: (1) American screw and wire gauge, (2) Wright Wire Co. gauge, (3) Roehalmann Music gauge, (4) W.H. Brunton Music Wire gauge, (5) American steel and wire gauge, (6) Roebling and Trenton Iron Co. gauge.
(k) Felten and Muller music gauge.
(l) English steel wire gauge.

Gauge (a) is used for bare wire—copper, aluminium, bronze —zinc and nickel silver wires, and for insulated aluminium and copper wire, but not for telephone or telegraph wires. Gauge (b) is used for iron and steel telephone and telegraph wires, and for bare copper wire used in telephone and telegraph services. Gauge (g) is used for bare galvanised iron and steel wire, and for spring steel wire. Gauge (h) is also used for bare copper telephone wire. Gauge (f) is used for steel drill rod and wire.

The complications that arise because of the existence of so many systems has made it necessary to quote the actual dimension of the wire in units of length as well as the gauge number in all cases where doubt is likely or confusion is likely to occur. (*See also under separate gauge systems.*)

Wire rod. The term used in reference to a semi-finished product, generally about 5 mm in diameter and used for making wire.

Witherite. The naturally occurring barium carbonate $BaCO_3$ which is found as gangue material in lead veins and in association with galena and barytes. It is usually greyish-white or pale yellow in colour. It crystallises in the orthorhombic system and the crystals show repeated twinning. Mohs hardness 3-3.75, sp. gr. 4.3. It is a source of barium and its compounds.

Wobbler. The name used for the portion of a roll immediately beyond the neck. It is approximately cruciform in cross-section and fits into the coupling boxes so enabling the roll to be driven.

Wöhler test. A fatigue test in which a cylindrical specimen rotates, being loaded as a cantilever. As it rotates, the stress at any point on its circumference varies from a maximum in tension to a maximum in compression (*see also* **Fatigue**.) Also known as the rotating cantilever test.

Wöhler testing machine. The name used in reference to the type of machine invented by Wöhler for fatigue testing (q.v.). The test pieces are gripped at one end and loaded at the other and rotated. The number of alternating tensile and compression cycles is indicated on a counter and this number (N), when the specimen breaks, is plotted against the stress (S). Many such points are obtained and from them the S-N curve is plotted.

Wohlwill process. An electrolytic process for refining gold bullion. The black anode slimes obtained from the Balbach (an electrolytic process for the refining of silver) are melted down and cast into anodes. The electrolyte is a 5% solution of hydrochloric acid containing about 7% of gold III chloride. The cathodes are thin sheets of gold foil. The bath is maintained at 70 °C and air is blown in to agitate the solution. A direct current density of 10 A m^{-2} is maintained and the potential per cell is 1.3-1.5 volts. An alternating current of slightly higher voltage is superimposed on the direct current. This has the effect of preventing the adhesion of the silver chloride that is formed on the anode. Any platinum present is not deposited on the cathode but passes into solution and is precipitated from time to time by the addition of ammonium chloride to the bath.

Wolfram. The principal ore of tungsten which is a tungstate of iron and manganese (Fe,Mn)WO$_4$. It occurs in pneumatolytic veins surrounding granite masses in Cornwall (UK), Bolivia, Malaysia, Korea, etc., and is usually associated with cassiterite. The mineral is brownish-black in colour, and the alluvial tin-wolfram deposits are worked for both metals in the Far East and in Spain. Crystallographic system, monoclinic; Mohs hardness 5-5.5, sp. gr. 7.1-7.9. In the treatment of this ore the tungsten is first converted into sodium tungstate and the iron and manganese (about 20%) are discarded.

Hubnerite is a mineral allied to wolfram. It is mainly manganese tungstate, but always contains some iron.

Wolframite. An important ore of tungsten that consists of tungstate of iron FeWO$_4$. The concentrates of this ore are sometimes reduced directly to produce the master alloy of ferrotungsten. The ore is frequently associated with cassiterite.

Wolfram ochre. A somewhat rare mineral consisting of tungsten VI oxide (WO$_3$). It is an earthy form of the oxide resulting from alteration of tungsten ores. It is also known as wolframine or tungstic ochre.

Wollastonite. A naturally occurring pyroxene that consists essentially of calcium meta-silicate CaSiO$_3$. It is found as metamorphosed siliceous limestone and as tabular crystals. Also known as tabular spar.

Wollaston wire. A name given to platinum wires of very small diameter (about 25 μm) that have been produced by coating thicker platinum wires with silver and drawing down the composite wire, and afterwards dissolving off the silver coating. This latter operation can be done either by immersing the wire in dilute nitric acid or by making it the anode in an electrolytic cell in which the electrolyte is potassium silver cyanide.

Wood. *See* **Timber.**

Wood's alloy. A fusible alloy with melting range 70-72 °C. At ordinary temperatures the tensile strength is about 30 MN m^{-2}, with elongation of 3% and B.H.N. 25. It is used as a seal in automatic sprinkler heads and modified alloys are used for master patterns in the foundry. (*See* Table 8.)

Wood's alloy has been used for casting into wood or plaster moulds, since it melts in boiling water. Metallographic specimen mounts can be made from it, and wood patterns have been covered with a thin layer of it to increase life. A cathode of molten Wood's alloy has been used in electrolytic analysis for metals such as Ag, Co, Ni, Sn, Zn. The deposited metal alloys with the cathode and thus is protected from oxidation, and can be determined then by the increase in weight of the cathode.

Wootz process. The name of the old process (cementation and crucible) for making steel. It was used in India and subsequently in Syria (Damascus) and Spain (Toledo) for the production of high-quality blades.

Work. A force is said to do work when it moves its point of application. The measurement of the work done is the product of the force and the distance moved in the direction of the force. When W = work done, P = force and s = distance

$$W = \int P ds. = \text{change in K.E. of body moved.}$$

Here force is measured in newtons and work in newton metres or joules

$$1 \text{ joule} = 1 \text{ newton metre (N m).}$$

Workability. A term used in reference to those characteristics or qualities of a metal that determine the ease with which it can be formed into any desired shape. Some indication of the workability of metals can be obtained from the simple bend test, Izod test or Erichsen test (q.v.).

Worked metal. A term applied to metals that have been subjected to processes to change the shape, the surface appearance or the internal structure of the worked piece. Working may be done hot or cold (*see also* **Working**). Any permanent deformation of a metal produces some distortion of the crystal structure and internal strains are set up. When a metal is cold-worked the distorted structure is unstable, and there is a tendency to regain the unstrained structure by recrystallisation. For most metals this occurs only very slowly at room temperatures, but is greatly facilitated if the temperature be raised. The usual operations to which metals are subjected that are included under the heading of "work" are forging, rolling, swaging, drawing, extruding.

Work function. This is defined as the energy (joules) that must be supplied to free electrons in the metal and having a certain potential, to enable them to escape from the metal. It is usual to denote the energy of a free electron at rest outside the metal as zero. Within the metal the potential has a constant value $-W$. At absolute zero the free electrons present in the metal are distributed amongst a number of energy states that can be designated e_1, e_2 ... up to a maximum state e. The work function is then the energy that must be supplied to free electrons having energy e to enable them to escape. The amount is therefore $(W - e)$. For photo-electric emission at absolute zero

$$h\nu = W - e$$

where ν = frequency of the radiation absorbed.

Electrons are, of course, emitted from metals at high temperatures—thermionic emission—independent of any photo-electric effect. The equation giving the emission current is

$$I = A\theta^2 e^{-B/\theta}$$

A and B are constants, B being known as the thermionic work function and θ is the temperature.

The value of W is found to be about 10 eV and of e about 3 or 4 eV. The value of the work function varies with the temperature, but is approximately constant at low temperatures. As the temperature rises there will be a small proportion of electrons whose energy is greater than e. It has been shown that the work function χ is related to the thermionic work function ϕ by the equation:

$$\chi = \phi - \frac{3kT}{2}.$$

Here k is Boltzmann's constant (1.381 $\times$ 10^{-23} J K^{-1}).

Work hardening. The term used in reference to a process in which the hardness of a metal is developed by permanent deformation below the recrystallisation temperature. The permanent deformation involves some distortion of the crystal structure. If this latter is not removed, then further deformation will not only result in further distortion, but will cause an increase in the resistance of the metal to deformation and reduce its capacity for deformation. In such circumstances—where the recrystallisation temperature is not reached—the process is also known as coldworking, though it may be carried out at temperatures above room temperature.

In single crystal pure metals plastic deformation occurs along certain crystallographic planes by slip. The resistance to slip as it proceeds is referred to as strain hardening. When slip occurs on a number of adjacent planes in such a way that the distance moved on any of the planes is greater by a fixed amount than occurred on the previous plane, the phenomenon is known as twinning.

In the case of polycrystalline metals the varying orientation of the grains causes each grain to restrict the deformation of adjacent grains so that plastic deformation does not occur except under higher stresses than are required for single crystal metals. Cold-worked polycrystalline metals may show, on a polished surface, slip bands whose directions vary from grain to grain.

Single-phase alloys are usually stiffer and less plastic in cold working than pure metals. The behaviour of duplex alloys is dependent on the properties of the continuous phase and on the dispersion of the other phase. In general it may be assumed that the work hardening (i.e. the increase in hardness and the reduction in ductility) is proportional to the amount of work done (as measured by, for example, the reduction in the cross-sectional area) on the metal.

Working. A term used in reference to all those mechanical treatments to which a metal may be subjected in order to change the dimensions or modify its physical or mechanical properties. The processes usually included under this heading are rolling, drawing, forging, extruding, pressing (or deep drawing) and spinning, and these may be carried out as either hot or cold working, but the latter gives a superior surface finish and better dimensional accuracy.

Wortle. The term used in reference to a small plate containing a series of tapered holes in a single line. It is used for wire drawing.

Wrapping diameter. A term used in reference to the ductility or to the windability of wire. It is the minimum diameter upon which metal wire can be wound without fracture.

Wrapping test. The term used in reference to a test applied to wire which consists in wrapping the wire round a mandrel of specified diameter for a specified number of turns, and then unwrapping. The material should show no sign of crack or fracture. The diameter of the mandrel is usually equal to the diameter of the wire under test and with steel wire it is customary to wrap eight turns and then unwrap.

Wrinkle. A term used to describe a corrugation or crease or fold in the surface of a metal. Wrinkles may be induced deliberately, as in certain types of pipe-bending in order to reduce excessive compression stresses on the inner radius of the bend. In such cases the tube, before bending, is heated by an oxyacetylene flame in a series of two or more narrow bands extending about two-thirds of the distance round the pipe and at right angles to the length of the pipe. The number of wrinkles required, and the distance between them, is mainly determined by the diameter of the pipe and the degree of curvature in the bend.

Wrought iron. A pure form of iron made by the fire-refining of pig iron. The latter was melted in a reverberatory type furnace (puddling furnace, q.v.) with oxide of iron, and the purified iron obtained as a pasty mass. The impurities in the pig iron were oxidised and passed into the slag. The mass always contained some of the slag, but most of this was removed in the subsequent working operations—squeezing, muck rolling, but a small amount remained in the form of iron silicate. It was this that gave to the wrought iron the characteristic fibrous structure and assisted, by acting as a flux, in the welding of the metal. The muck bars, obtained after muck or rough rolling of the hot mass, were cut to suitable lengths and several of these were assembled to form a pile. This was bound by wire, heated to welding temperature and rolled to give "refined iron". In the charcoal hearth process (used in Sweden) the pig iron was introduced into a charcoal fire and a blast of air used to oxidise the impurities. In the Aston process (used in the USA) the pig iron was melted in a cupola and transferred to a Bessemer converter where most of the impurities were removed by blowing in air. The molten metal was then poured into a ladle containing a molten slag made from iron oxide and silica. After the evolution of much gas the metal settled as a plastic mass at the bottom of the ladle. The slag was then poured off and the metal worked by squeezing and rolling.

The products of these last two processes were somewhat inferior to puddled iron. The most important of the useful properties of wrought iron were its ductility, weldability and its good resistance to corrosion.

Wrought iron bloom. The product resulting from the squeezing of the ball of puddled iron. (*See also* **Wrought iron.**)

Wrought metal. A term applied to a metal or alloy that has been subjected to some form of mechanical working (*see also* **Working**). The term is mainly used to distinguish such materials from cast metals, and to indicate that they have higher tensile strength and greater ductility, due to modification in the microstructure, than the corresponding cast product.

Wurtzillite. A variety of solid bitumen.

Wurzite. A somewhat rare form of zinc sulphide ZnS which occurs in hexagonal crystals. It is pale yellow in colour and phosphoresces after exposure to strong light. This latter property is due to the presence of small amounts of sulphides of other metals.

Wusite. An iron II silicate containing more than 70% of ferrous oxide.

Xanthates. The salts of xanthic (xanthogenic or ethoxydithio methanoic) acid. Xanthic acid is a name applied to organic acids that are the monoesters of dithiolocarbonic acid. The potassium salt is obtained by the action of carbon disulphide on an alcoholic solution of a potassium alkhydroxide. It is used in the flotation of ores (q.v.).

Xenon. One of the rare gases present in the atmosphere in which it is present to the extent of about one part in eleven millions (by volume). It belongs to the group of inert gases and is usually obtained from the air by the fractional distillation of liquid air. It boils at −107 °C. This gas, as well as krypton, is in great demand for use in filament electric lamps, this use being directly connected with the lack of chemical properties. The gases also develop high electrical conductivity at voltages over 5,000 volts or so. Under certain conditions xenon can be caused to combine with fluorine to form the hexafluoride. (*See* Table 2.)

Xenotime. A mineral substance that consists of the orthophosphate of the yttrium metals (i.e. dysprosium, holmium, yttrium, erbium, thulium, ytterbium and lutecium). It is a source of yttrium.

Xeroradiography. A type of radiography based upon the fact that certain materials, such as one form of selenium, which are normally good insulators become good conductors after irradiation with X-rays. It is used with good success for the examination of plates of relatively small thickness. The plate is coated with amorphous selenium in place of a photographic film. The plate is electrically charged and placed in a cassette as in the case of gamma radiography. It is then exposed to gamma or X-rays and under the influence of these the plate loses its charge in amounts proportional to the quantity of radiation falling upon it so that those parts that are exposed to the most intense radiation lose more charge than parts that have received less radiation. After exposure the plate is dusted with fine powder which adheres most thickly where the plate has been least discharged. The image thus obtained is a "powder image", the density of which varies according to the thickness of metal specimen under examination. The principle is now the basis of a photocopying system (xerox).

X-ray diffraction. Solid crystalline materials act as diffraction gratings for X-rays, by virtue of the regular arrangement of the atoms or molecules in space since the interatomic distances are about the same order of magnitude as the wavelength of X-rays. When a beam of X-rays falls upon a crystal it is transmitted as a number of diffracted beams and these can be registered on a photographic plate as a number of spots.

Since every crystalline chemical element produces a pattern that is characteristic of the particular kind and arrangement of atoms, it can be used to identify a solid by comparison with known patterns.

A crystal can also be used as a reflection grating, and in this connection the Bragg equation is used. This is given as:

$$n\lambda = 2d \sin \theta$$

where n is the order of the spectrum (or if monochromatic radiation is used, the order of reflection), λ wavelength, d = distance between the parallel planes of atoms and $(90° - \theta)$ is the angle of incidence.

There are several different diffraction techniques and corresponding differences in apparatus. Usually a carefully collimated train of parallel X-rays—frequently monochromatic (i.e. having one wavelength)—is directed through or reflected from the surface of the specimen under examination. The specimen may be in the form of powder, aggregate, fibre of single crystal and may be stationary, rotated or oscillated. The pattern is recorded on a photographic film which may be stationary or moving, and may be flat or curved round a cylinder. (*See also* **Powder diffraction**.)

X-ray photoelectron spectroscopy (XPS). *See* **Electron spectroscopy for chemical analysis**, by which name it is commonly known.

X-ray rotating anode. To obtain sharply defined radiograms an X-ray tube having a small focus is required. To make the exposure time short the anode should have a high load capacity. This high load capacity depends on the homogeneity of the electron beam and on the nature of the target. Very good results are obtained using a copper anode having a layer of tungsten upon it. This should be of such a thickness that the temperature of the tungsten should not exceed 3,000 °C and that of the copper about 1,000 °C. To ensure that certain temperatures shall not be exceeded and to avoid fusion or volatilisation of the anode the latter is made to rotate rapidly. Water cooling can also be used.

X-rays. When electrons impinge upon a metal surface placed in their path, a radiation is produced which will not only affect a photographic plate and produce fluorescence in certain materials, but has unusual powers of penetration. The metallic surface, the target placed in the vacuum tube, is known as the anti-cathode and in modern tubes is usually water-cooled and the X-rays pass from the tube through a thin aluminium window. The rays are a form of electromagnetic radiation. The wavelength depends on the voltage supplied to the tube and the efficiency depends on the nature of the anti-cathode. The range of X-ray wavelength extends from about 100 nm down to about 10 pm. With a tube voltage of 1 000 kV the X-rays have a wavelength down to about 10 pm and can usefully be employed in the radiography of iron and steel. Using about 25 kV the X-rays have a wavelength down to 50 pm and they are used in the study of crystal structure and in the radiography of light metals. The variation in the penetration power of the rays is often referred to as the quality of the X-rays. The terms soft X-rays (voltages 1-20 kV), medium X-rays (20-50 kV) and hard X-rays (voltages 1 000 or more kV) are used in a general way to describe a particular X-ray tube. Soft X-ray tubes must have a window opposite the target made of very light material, e.g. beryllium (or very thin glass); other tubes may have aluminium windows.

X-rays are not affected by electric and magnetic fields, but they are capable of ionising gases through which they pass. When they strike other bodies, some of the rays are diffusely reflected while other rays of the same type as X-rays and called secondary rays are emitted. There are two main groups of secondary rays—the hard, short-wavelength penetrating hard rays called the K series and the softer rays called the L series. The wavelengths of these secondary rays are dependent on the atomic number of the element on which they are incident.

In many X-ray tubes the anti-cathode is made of tungsten (of at. no. 74), but metals like uranium (at. no. 92) or thorium (at. no. 90) are more efficient. On the other hand tungsten has a much higher melting point than either uranium or thorium and its coefficient of expansion is very near to that of Pyrex glass. The use of the regular atomic lattice of crystals for the diffraction of X-rays has led to an accurate determination of wavelength and afforded various means of investigating crystal structure.

X-ray spectra. When an element is excited by a stream of high speed electrons (cathode rays) falling upon it, the element emits its characteristic X-radiation. One way of analysing these rays is to allow them to fall on a cleavage face of a crystal of potassium iron II cyanide, under such conditions that the reflected rays can fall upon a photographic plate. It is then found that the spectrum lines belong to several distinct series known as K, L, M series. The K and L series correspond to the K and L types of fluorescent radiation.

In X-ray spectra it is assumed that an electron from an inner energy level in the atom has been removed thus leaving vacant a place into which an electron from an outer energy level can jump.

Thus the K radiation is attributed to the removal of an electron from the innermost energy level and the subsequent fall of an electron from an outer energy level into this vacant place in the lowest atomic level.

Similarly the L radiation is assumed to result from the removal of an electron from the next innermost energy level followed by the fall of an outer electron to what is named the L level. Since the energy changes that accompany the transition of these electrons are very large, the frequency of the radiation emitted is very high and the spectrum lines are thus found in the X-ray region. Moseley's law, which is only approximate, states that the frequency of any line in the X-ray spectrum is proportional to the square of (the atomic number of the element diminished by a constant quantity).

X-unit. A unit used in X-ray work that is equivalent to 10^{-11} cm. Since an A.U. $= 10^{-8}$ cm an X-unit $= 10^{-3}$ A.U.

Crystal structures were originally measured on the Siegbahn scale of X-units, defined so that the (200) spacing of calcite at 18 °C is equal to 3029.45 X-units. This gave an X-unit as equivalent to 10^{-3} Ångstrom units, as specified above. It was based on the figure for Avogadro's number currently accepted at that time. More accurate work showed that a slight error was made, and the crystallographic X-unit became regarded (circa 1950) as equal to 1.00202 Ångstrom units. To give values in true Ångstroms therefore, all figures in kX require multiplication by 1.00202. The differences are mostly small, but can be serious. Crystal or kX units are therefore quoted when it is clear that the measurement is in terms of one thousand Siegbahn X-units. Otherwise, true Ångstrom units are intended. The SI unit of 10^{-10} m eliminates these discrepancies.

Yellow brass. A name sometimes used in reference to the 65/35 type of brass. In the annealed condition the tensile strength is about $310\,MN\,m^{-2}$ but this can be increased up to $525\,MN\,m^{-2}$ by hard rolling. The increase in strength is accompanied by a marked reduction in ductility, the elongation being reduced from about 60% to about 5%. Usually up to 3% Pb is present to improve machinability. For high-strength castings there is usually present up to 2% Fe, 1.5-5% Mn, 0.5-1.5% Sn, and the copper content then lies between 56% and 62%.

Yellow copper ore. An important ore of copper that contains the sulphides of iron and copper and has the formula $CuFeS_2$ and should contain about 34% Cu. It is found in the UK, the USA, Canada, Spain and parts of South America. Mohs hardness 3.5, sp. gr. 4.2. Also called copper pyrites and chalcopyrite.

Yellow corundum. The name of a mineral crystalline substance which is often called "oriental topaz". Topaz is found very often in association with cassiterite and is used only as a gemstone. It is a fluosilicate of aluminium and crystallises in the orthorhombic system, whereas yellow corundum is alumina crystallising in the rhombohedral system. True topaz has a specific gravity of about 3.5 while the corundum has a specific gravity of 4.

Yellow ingot metal. This is a commercial brass of fairly standard composition. It is not, however, a true brass but contains copper, zinc and tin in proportions according to the strength and colour required. There are about eight grades of this ingot metal, the composition varying from 88% Cu in grade I to 63.5% in grade VIII. The yellow ingot metal used for plumber's fittings contains 65% Cu, 1% Sn, 2% Pb and 32% Zn. Yellow ingot metal is made from scrap metals after careful selection.

Yellow iron ore. An oxide of iron prepared by the aeration of scrap iron in the presence of copperas. It is used as a pigment because of its inertness and stability towards light. The colour can be varied somewhat by the addition of finely ground white clay.

Yellow metal. A name sometimes used in reference to a 60/40 type of brass that contains 1-3% Pb. It can readily be hot rolled into plates, sheets or sections and in one of these forms is usually marketed. It is also known as Muntz metal and malleable brass. The tensile strength when annealed is $340-355\,MN\,m^{-2}$. This value can be increased by about 50% hard rolling. (*See* **Yellow brass** and Table 5.)

Yellow ochre. An earthy pigment made from clayey and siliceous material and coloured with oxide of iron obtained from limonite. (*See also* **Ochres.**)

Yellow prussiate. The common name for the compound otherwise known as potassium iron II cyanide $K_4\,[Fe(CN)]_6$. At one time it was prepared mainly by heating nitrogenous animal waste (hides, hoofs, etc.) with potassium carbonate and scrap iron. It is now obtained from the "spent oxide" formed during the purification of coal gas. This oxide contains iron III iron II cyanide. It is boiled with lime and the soluble calcium iron II cyanide is obtained by treating the solution with potassium carbonate.

Yellow prussiate reacts with dilute sulphuric acid and yields hydrocyanic acid but the concentrated acid gives an ammonium salt and carbon monoxide. This is one way of obtaining pure CO. In analysis a solution of yellow prussiate is used to indicate the presence of iron II salts (white precipitate gradually turning blue) and iron III salts (deep blue precipitate of Prussian blue).

Yield point. This is defined as the lowest stress at which the elongation of a rod or wire, etc., increases without additional increments of load. It is also defined as the maximum stress that can be sustained without plastic deformation by a test piece subjected to a specified type of loading. It is usually obtained from a static test such as a tensile test. The yield point is indicated on a stress-strain diagram as a sharp "knee" on the curve (Fig. 209) but is found well-marked only in the case of wrought iron and soft mild steels.

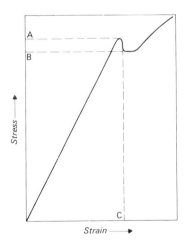

Fig. 209 *Yield point diagram.*

In carrying out the tensile test in a machine in which a pointer indicates the value of the stress, it may be observed that at a certain position of the pointer there occurs a sudden extension and then the stress quickly falls. The maximum and minimum values of the stress at *A* and *B* are called the upper and lower yield points respectively. In specifications the point *A* is usually quoted, though in practice it is not possible to determine it with a high degree of certainty. On the other hand point *B* does provide a value that can be safely used in design.

It happens that when the temperature at which the test is carried out is raised, the yield point gradually becomes less definite and above about 400 °C it cannot be detected. Even in mild steels the yield point cannot be discovered if the metal has been subjected to cold working. In the case of hardened steels and some other metals it is usual to quote what is termed proof-stress (q.v.) in place of yield point and in such cases the yield strength must be regarded as the closest comparable property.

Two main explanations have been given for the existence of yield points. Cottrell proposed that impurity atoms can cluster around dislocations or dislocation sources, and that a finite stress is required to drag the dislocation away from the restraint. When this happens, the stress falls to what it would have been had there been no "atmospheres" hindering dislocation movement. The atmosphere diffuses towards the dislocation again (or to others nearby) and the stress rises again, but now in a dynamic condition, the dislocation dragging the atmosphere along with it. This accounts for the upper and lower yield point in ferrous materials. Cottrell confirmed the hypothesis by creating a yield in zinc, normally free from such a phenomenon, by treatment with nitrogen. The diffusion of the atmosphere to dislocations, restoring the extra stress required to move the pinned dislocation, also contributes to work hardening.

The second explanation for yielding, by Johnston and Gilman, is the creation of new dislocation sources when the stress reaches the yield point. Based on work with with LiF crystals, they proposed that existing pinned dislocations played no part in yielding, but that a critical stress was needed to produce new dislocations, which then move quite rapidly, and the stress falls temporarily. The movement of these new dislocations leaves behind a trail of defects when they are screw type, and these defects are potential dislocation sources themselves at higher stresses. This explanation does account for the bands of slipped material which are often observed after yielding has occurred.

Yield point elongation. This is the permanent strain *C* corresponding to the lower yield point *B* (*see* Fig. 209) and is sometimes taken as a measure of the ductility of a metal.

Yield strength. A term used to indicate a measure of the resistance to plastic deformation of a material subjected to a specified type of loading, and in this sense is often taken as an approximation to the elastic limit. It is quite frequently taken as the stress at which a material exhibits a specified limiting permanent deformation.

Yield strength can be determined experimentally by finding the stress required to produce a specified permanent set, but it is more usually obtained from a stress-strain diagram. This latter method is referred to as the "offset method" and can be understood by a study of the diagram (Fig. 124). The point *R* represents 0.2% offset and from it a line parallel with the straight-line portion of the stress-strain curve is drawn. It intersects the curve at *P* and the corresponding stress *S* is the yield stress. When the behaviour of a metal or alloy under load is known the actual stress-strain diagram need not be used since the stress for a specified strain can be calculated or can be determined by direct measurement. This is done in the case of copper and copper alloys and the result is stated as yield strength (with specified extension) under load.

Young's modulus. A term used synonymously with modulus of elasticity (q.v.). (*See* **Elasticity**.)

Ytterbite. A rare mineral found in Sweden and Greenland. It is also known as gadolinite and has the formula

$$4BeO.FeO.Y_2O_3.6SiO_2.$$

It is a basic orthosilicate of beryllium, iron and yttrium. It contains small amounts of the rare-earth element erbium and is the present source of it.

Ytterbium. One of the heavier of the rare-earth group of metals known as yttrium metals and found in the gadolinite earths. Symbol Yb, at. no. 70. It has properties almost identical with lutecium with which element it is usually associated. Ytterbium, however, has a very characteristic spectrum by which it can be identified and has a higher magnetic susceptibility than lutecium. Atomic weight 173.04, density 6.98 g cm^{-3}. The melting point is given as 824 °C. Crystal structure, face-centred cubic; lattice constants $a = 5.4790 \times 10^{-10} \text{ m}$. (*See* Table 2.)

Yttria. A rare-earth oxide found in the mineral substance gadolinite. Investigation showed that the mineral contained also glucina. In all the mineral specimens containing yttrium there were named erbium and terbium. Yttria has been shown to be a complex mixture; the earths that have so far been separated from it are indicated in Fig. 210.

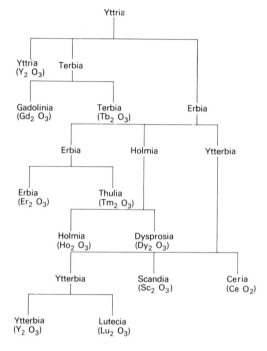

Fig. 210 *Yttria.*

Yttrium. One of the commoner rare-earth metals that occurs as oxide in gadolinite earths. It is difficult to separate from holmium and erbium—the two elements with which it is always associated. The metal occurs in combination with oxygen in gadolinite, nuevite (California), keilhauite (Sweden), pergusonite (Eastern US), bragite and tyrite (Europe). It is not yet produced in the pure state, but the present product has a specific gravity of 4.46 and a melting point of about 1,523 °C. Symbol Y, at. wt. 88.91, at. no. 39. (*See* Table 2.)

Yttrium carbide. A compound of the metal yttrium and carbon YC which is considered to be a mixed carbide, since on treatment with water or acid it yields a mixture of ethyne and methane. In this respect it resembles lanthanum carbide.

Yttrotantalite. A naturally occurring mineral substance found at Ytterby (Sweden) and which contains about 28% yttrium oxide together with about 40% columbium oxide and 10% tantalum oxide. Both fergusonite and euxenite contain comparable amounts of yttria YO but less tantalum oxide.

Z

Zaffre. An impure basic cobalt arsenate that is a commercial product after calcining or roasting speiss-cobalt or other arsenical ores. The ore should be free from iron. The same compound could be prepared by dissolving the ore in nitric acid and precipitating with soda. It is used in the preparation of other cobalt compounds and as a pigment for painting on porcelain.

Zam metal. A zinc-base alloy containing aluminium and some mercury used as zinc-plating anodes. It is not readily attacked by acids except in the electrolytic cell and then only when the current passes.

Zapala ore. A red hematite Fe_2O_3 found in Argentina and containing about 48% Fe.

Zebra roof. A term used in reference to a method of constructing furnace roofs by using rings of silica bricks alternating with rings of basic brick in those areas that are subjected to the heaviest wear. Silica has a higher strength but wears away more rapidly than the basic refractories so that the basic bricks protrude and afford protection to the siliceous material which continues to support the roof.

Zeeman effect. When a substance is caused to emit a line spectrum and then placed in a magnetic field, the single lines in the spectrum form into groups of closely spaced lines. Since the waves of light originate from the acceleration of intra-atomic electrons which in motion constitute an electric current, the imposing of a magnetic field must change the path of the electrons and so affect the nature of the radiation. From the separation of the lines it has been possible to deduce important information concerning atomic structure.

Zeolites. A group of minerals resulting in general from the alteration of felspars and aluminous minerals of igneous rocks. They are hydrated silicates of calcium and aluminium sometimes also containing sodium and potassium silicates. They occur as secondary minerals filling cracks and cavities in basalts and lava and they represent the final stages of cooling of an igneous intrusion. They are linked up in origin with many types of ore deposits.

Certain types of greensand which are types of marl can be classed as zeolites, e.g. glauconite. This latter is a silicate of potassium and aluminium and contains some oxide of iron. It is used as a water softener. Synthetic zeolites consist of sodium aluminium silicate and are made by reacting bauxite with caustic soda to form sodium aluminate and then combining this with sodium silicate. The formula is usually taken as

$$Na_2O.Al_2O_3.6SiO_2.xH_2O.$$

It is used as a water softener to remove calcium and magnesium salts from water. It can be regenerated by treatment with a solution of common salt. (*See also* **Water softening.**)

Zerener process. An arc-welding process in which two dense carbon rods are used to produce the arc to heat the work. A filler rod or wire supplies the weld metal.

Zetmeter. The name of a mechanical instrument used for checking the parallel length of wire-drawing dies.

Ziegler catalyst. A special material due to Ziegler which is used to produce long-chain regular polymers that have useful properties. It is employed to polymerise ethene, propene, isoprene and many others. The catalyst contains special combinations of halides of transition metals (V and Ti) with alkyl or alkyl halide compounds of aluminium.

Zinc. A soft white metal showing a bluish tinge. It is produced mainly from zinc blende ZnS and to a lesser extent from calamine $ZnCO_3$. Zinc and lead are generally mined together as the sulphide ore. This is crushed and the zinc sulphide separated by selective flotation. By roasting in air the sulphide is converted to oxide and the sulphur dioxide concurrently produced is converted to sulphuric acid (a valuable by-product). The roasted ore is usually reduced by heating with carbon in horizontal retorts and the metallic vapour condensed and collected as liquid metal outside the furnace. The spelter metal obtained in this way contains about 1-2% Pb and is suitable for galvanising. Temperatures of about 1,100 °C are required in this operation.

In the more modern vertical retort process brickettes of roasted concentrate and anthracite are heated in a continuous process and the zinc vapour condensed to give a zinc of rather higher purity. This occurs because the type of furnace permits fractional distillation and the zinc metal has a purity up to 99.99%. Vertical electrothermally heated furnaces are operated in the USA.

The blast furnace process developed by the Rio Tinto-Zinc Company and its subsidiaries is now predominant in the production of zinc. In this method, lead is used to prevent excessive volatilisation of zinc, and then separated out in channels after tapping. The lead is recirculated.

The process is most efficient when mixed lead and zinc ores are smelted together in the blast furnace. The lead is tapped off in the usual way at the base of the furnace, and zinc vapour is led off to the spray chamber where molten lead at 550 °C dissolves it, and the alloy is then directed to the cooling channels where separation occurs at about 440 °C.

About half the world output of virgin zinc is electrolytically refined. The calcined oxide is leached with sulphuric acid and the purified zinc sulphate solution is electrolysed using insoluble anodes of lead alloyed with silver or lead only. The solution is generally purified by adding zinc dust which precipitates the impurities (cadmium, etc.). The cathodes are of high-purity aluminium from which the zinc is stripped off at intervals. The spent electrolyte is used for further leaching. There are four grades of zinc, the main impurity in each case being lead. *(a)* Special high purity (99.99% Zn), *(b)* electrolytic zinc (99.95% purity), *(c)* intermediate zinc (99.5% purity) and *(d)* good ordinary brand (G.O.B.) 98-99% pure.

Zinc is not very malleable at ordinary temperatures, but can be rolled into thin sheet or drawn into wire at temperatures between 110 °C and 150 °C. Because of its comparatively high resistance to atmospheric corrosion it is used to a large extent in the form of coatings on steel.

About one-third of the total zinc produced is used in the making of brass, but in the USA die-casting alloys of zinc-aluminium are replacing brass. Next in importance is the use of zinc for protective coatings. Hot-dip galvanising by immersion in molten zinc is the most widely used process for fabricated articles, sheet and strip steel, tubes and wire. Other methods include spraying zinc on to metal surfaces using special pistols fed with wire, powdered or molten zinc. Alternative processes are electrogalvanising (zinc plating) and sherardising using zinc dust.

Zinc die-casting alloys (which usually contain aluminium) are readily cast into intricate shapes and the ease with which they can be chromium-plated makes them very popular in the automobile industry. In the form of sheet, zinc alloys are used for roofing (especially on the continent of Europe), flashings, weatherings, etc. The life of a zinc roofing not thinner than 21 s.w.g. (0.8 mm) is estimated at not less than forty years.

Other uses include the making of dry batteries, flash lamps, battery cans and lithographic plates. Zinc oxide and sulphide are used as pigments (*see* **Lithopone**) in paints and as fillers and opacifiers in linoleum, rubber and plastics.

The physical and mechanical properties of zinc are given in Table 2. Crystal structure, close-packed hexagonal, lattice constants $a = 2.6643$, $c = 4.945 \times 10^{-10}$ m. Electrolytic solution potential (saline solution)—0.7618 volt against standard hydrogen electrode.

Zincate treatment. A process used in connection with the finishing of light alloys as a pre-treatment before electrodeposition of copper, silver or chromium. The sheet or component is first pickled in a chromic acid bath containing chromic acid and sulphuric acid (1 to 5). The metal is then rinsed, dipped in strong nitric acid, rinsed again and finally immersed in a sodium zincate solution for a few minutes.

Zinc-base alloys. A name applied to a wide range of zinc-base alloys in which the zinc is usually present to the extent of approximately 90%. Such alloys are suitable for die-castings. A number of alloying elements present in small amounts modify considerably the properties of the alloy and there is therefore a wide variety of applications.

Zinc blende. An important zinc ore that consists of zinc sulphide ZnS which frequently occurs associated with galena (q.v.) in veins in calcite, dolomite, fluorspar barytes as well as in quartz. The crystals are often twinned, but the ore may occur massive. It is sometimes yellowish-white but more frequently brown or black. Crystal system rhombohedral. The black variety is also called blende, sphalerite or black jack. Mohs hardness 3.5-4, sp. gr. 3.9-4.1. It is found in Germany, Austria, Belgium, the UK and the USA.

The structure of zinc blende is characteristic of AB compounds with more open packing, due to the lower ratio of cation to anion diameter. The co-ordination number drops to 4, and the structure has some resemblence to that of diamond, with the large sulphur ion in a tetrahedral void of zinc ions. The bonding is not entirely ionic, as often occurs with structures of such "open" nature, and the material is insoluble in water and has a reasonably high melting point. The structure described is that of sphalerite, which can be regarded as two interpenetrating face-centred cubic lattices. The alternative form, würtzite, is obtained by rotation of alternate (111) planes of sphalerite through 180°, and this results in a hexagonal symmetry. Zinc blende is a semiconductor.

Zinc bloom. An ore of zinc otherwise known as hydrozincite $ZnCO_3.2Zn(OH)_2$. It is a basic carbonate of zinc.

Zinc coating. A term used in reference to the various methods of depositing a protective layer of zinc on certain metals—mainly copper and steel. In the case of copper the coating is usually employed on pipes, cooking utensils and wires. In the case of steel there are a variety of processes in use: *(a)* hot dipping or galvanising; *(b)* metal spraying; *(c)* electrodeposition; *(d)* sherardising (using zinc dust and some zinc oxide). Components may be galvanised by dipping into molten zinc and then withdrawing. Sheet and wire can be treated in a continuous process. If after passing through the zinc bath the wire or sheet is passed through a long furnace heated to an appropriate temperature (e.g. about 680 °C) brittle iron-zinc alloy is formed which has a bright smooth surface. The spraying of zinc on fabricated parts or structures can be carried out using zinc wire, zinc powder or molten zinc. It is a method applicable to parts after assembly in which the zinc layer can be deposited in any desired thickness. The coatings by this method are somewhat porous (the wire method probably produces the least porous deposit) but oxidation products quickly seal the pores. Zinc may be electrodeposited either from a sulphate or an alkali-cyanide bath. Mercury or one of several organic brighteners may be used to obtain a bright finish. In the sherardising process the articles to be treated are heated in contact with zinc powder containing some zinc oxide. The deposit so obtained is mainly an iron-zinc alloy. The protection afforded by zinc in most cases is proportional to the thickness of the deposit and to the actual nature of the deposit. This latter in turn depends on the method used in producing the protective layer. In any case experience has shown that the minimum thickness for a reasonable life is about 0.2 kg m^{-2}.

Zinc corrodes in moist air much more slowly than iron but if cracking of the zinc occurs, rusting of the iron below the zinc is prevented by the presence of zinc. Galvanised iron shows a good resistance to corrosion (i.e. relatively long life) in marine atmospheres and in sea-water. Galvanised-iron pipes having thick coatings have given long service in water systems where the water is hard, but not with soft water.

Zinc dust. A greyish metallic powder obtained by the condensation of zinc vapour or by the atomisation of the melted metal in a special pistol. It should not contain more than about 4% zinc oxide, though this amount is deliberately increased when the powder is used in making zinc paints. Zinc powder is used in the sherardising process for metal protection against corrosion, in making synthetic rubber and in certain lubricants. Very pure zinc is used in the precipitation of gold and silver in the cyanide process. Specific gravity about 7.1.

Zinc ferrite. This substance, formula $ZnO.Fe_2O_3$, occurs naturally in black octahedra, in franklinite. It may be prepared by strongly heating a mixture of the two oxides.

Zinc green. A pigment obtained by mixing zinc chromate and Chinese blue. The latter is one of the Prussian blues. It is more permanent than the greens based on chrome yellow (lead chromate), but has inferior covering power.

Zinc grip. A type of zinc coated steel in which the zinc is usually deposited electrolytically and the coated steel is suitable for deep-drawing as the zinc does not exhibit tendency to strip.

Zincilate. A substance containing zinc in the form of dust or powder used as a protective coating for metallic materials.

Zincing. A term sometimes used synonymously with galvanising (q.v.). Most frequently it is used in reference to the coating of wrought iron or mild steels with zinc as a protection against corrosion. Corrugated sheets are corrugated after galvanising.

Zinc ingot metal. The name used in reference to the slabs of zinc produced by the smelting process.

Zincite. An ore of zinc that consists mainly of zinc oxide ZnO—about 80%. It has the highest percentage of zinc of any of its ores, but does not occur in large deposits. The mineral has a massive granular structure and is used for the production of zinc white (zinc oxide) intended for use in pigment. Specimens occur that are coloured red by impurities such as the red oxide of manganese, and these may be cut and used as gemstones. Also known as red zinc ore.

Zinckenite. A lead antimony sulphide ore $PbSb_2S_4$.

Zinc lime. A mixture of zinc sulphate and lime or of sodium bisulphite, zinc sulphate and lime used in vat dyeing for the dyeing of indigo blue on wool.

Zincolite. A name used in reference to a mixture of barium sulphate and zinc sulphide, otherwise known as lithopone (q.v.).

Zinc ores. The most important ore and the main source of zinc is zinc blende or sphalerite. It is also called black jack or rosin jack according to the colour. It contains about 67% Zn. Important ores are zincite ZnO—80% Zn, franklinite

$$(Fe.ZnMn)O.(Fe.Mn)_2O_3—22\% \, Zn,$$

willemite $Zn_2.SiO_2$—58% Zn, electric calamite $2ZnSiO_2.H_2O$—57% Zn, smithsonite zinc spar or calamine $ZnCO_3$—52% Zn, hydrozincite $3ZnCO_3.2H_2O$—47% Zn, goslarite $ZnSO_4.7H_2O$—23% Zn.

Zincote process. A process for producing a coating of zinc chromate on steels as a pretreatment, which consists of dipping or spraying the component with a solution containing zinc nitrate together with sodium acetate and sodium bisulphite. This treatment is followed by rinsing in a dilute solution of potassium bichromate.

Zinc phosphatising. A process for producing an insoluble protective coating of zinc phosphate on an electrodeposited coating of zinc by dipping in acid zinc phosphate solution and subsequently immersing in a dilute solution of chromic acid.

Zinc plating. The process of depositing zinc electrolytically generally upon steel or iron components. The deposit may, however, corrode in damp atmospheres, but is usually satisfactory in ordinary industrial atmospheres. The zinc may be deposited from a variety of solutions for each of which some advantages are claimed. Typical compositions of plating solutions are given in Table 40.

Zinc spar. A naturally occurring mineral that consists of zinc carbonate. Also known as smithsonite or calamine. Specific gravity 4.42.

Zinc white. The name used in reference to zinc oxide when used as a pigment. It is made directly from the ore known as zincite or from zinc metal by vaporising and burning in air. It is not affected in colour by the presence of sulphides, but has poorer covering power than white lead. It is mainly used for indoor work and in activating and whitening rubber and in enamels. Specific gravity 5.7. On account of its high electrical resistance it is often used in insulating compounds.

Zinc yellow. The name used to describe a yellow zinc chromate. It is a crystalline powder of composition $4ZnO.4CrO_3.K_2O.H_2O$ and of specific gravity 3.5. It is unaffected by sulphur atmospheres or by strong light and is frequently mixed with Prussian blue to form green pigments and printing inks.

Zinnal. A duplex metal that consists of aluminium sheet with tin cladding on both sides. It is of German origin.

Zippeite. A hydrated uranium sulphate, containing about 68% uranium oxides. It occurs with carnotite in the USA, and appears as a coating on pitchblende deposits exposed to air in the vicinity of the Great Bear Lake of Canada.

Zircoloy. The most successful zirconium-based alloys are those developed for nuclear reactors of water-cooled type, primarily to resist corrosion, but with reasonable strength. High purity zirconium has a tensile strength of about 220 MN m^{-2} but zircoloy 2, with 1.5% Sn, 0.12% Fe, 0.10% Cr, 0.5% Ni, raises this to about 470 MN m^{-2}. This alloy has been employed for the fuel cladding of nuclear reactors in the USA.

Zircon. The most important of the silicates of zirconium $ZrSiO_4$. It is a common accessory mineral in all classes of igneous rocks, particularly granite, syenite, diorite, etc. It is also found in granular limestones. The chief localities are in alluvial sands in Sri Lanka, Urals, Isle of Harris (UK), India, Greenland, Australia and many places in North and South America. The commercial sources are limited to the black sand accumulations in India, America and Australia.

Pure zircon contains about 33% Si and 67% Zr metal. It crystallises in tetragonal prisms terminated by pyramids and the colour varies, being usually green, yellow or colourless, but also sometimes blue or red. The clear crystals are valued as gemstones, since they have the greatest brilliance next to diamonds. The colourless stones are called mature diamonds and the yellow ones

jacinth. The colourless and smoky varieties are also known as jargon. Specific gravity 4.7, Mohs hardness 7.5, refractive index 2.25.

There are several varieties and alteration products of zircon and these include hogatalite, aerstedite, crytolite, malacon, orvillite and oyamalite. Zirconite obtained from the beach sands of Florida is zircon and is used as a refractory for firebricks and crucibles and as an opacifier in ceramic glazes. The melting point is about 4,620 °C. Zircon has a low coefficient of expansion (4.5×10^{-6} °C^{-1}) and good heat-shock resistance. It also has high electrical resistance. At high temperatures the resistivity decreases, so that this material is not used in resistance furnaces. It is also used for mould facings and as an abrasive. Also called hyacinth.

Zircon cement. A refractory cement used for repairing furnace walls, etc. It contains 35% Zr sand (refined), 15% fused magnesia and the balance dead burnt magnesite. It is bonded with sodium silicate. Water suspensions of zircon with high melting point oxides, e.g. magnesia, are sprayed on furnace walls. They vitrify at about 1,160 °C and thereafter will withstand temperatures up to 2,100 °C.

Zirconia. A white crystalline powder consisting of zirconium oxide ZnO_2 and obtained mainly from the mineral baddeleyite. It has specific gravity 5.77, Mohs hardness 7.5 and the refractive index is 2.2. The melting point is about 2,950 °C and it has the highest heat resistance of commonly used refractories, and is used for crucibles for melting metals such as aluminium and platinum. It is also used in certain types of quartz ware and as an opacifier, since it is very stable to light and heat. Zirconia bricks. used as refractory linings, are liable to cracking and structural disintegration, due to alterations in the crystal structure. This effect can be reduced by the presence of certain other oxides, but these have the effect also of lowering the softening point. Stabilised zirconia may be used up to about 2,400° C, but not in the presence of carbon, because this latter substance decomposes the oxide and forms a carbide.

Zirconite. A greyish-brown variety of zircon $ZrSiO_4$ found mainly in the beach sands of Florida. It is used as a refractory, for firebricks and crucibles, and as an opacifier in certain glazes.

Zirconium. A silvery white metal obtained from natural alluvial deposits of zircon (q.v.) or from baddeleyite ZrO_2. The metal may be obtained, as a fine black powder, by reducing the oxide with calcium hydride, or with metallic calcium in the presence of calcium chloride. A ductile form is obtained by the thermal decomposition of zirconium tetraiodide.

In treating the zirconium ores it is usually necessary to first prepare a zirconium compound such as zirconium oxychloride or sulphate. This is done by fusing the ore with lime and calcium chloride (or fluoride) and treating the product with a limited amount of hydrochloric acid to form the oxychloride. The following reactions produce metallic zirconium.

(a) Reduction of ZrF_4 by potassium.

(b) Reduction of $KF.ZrF_4$ with sodium, aluminium or a mixture of sodium and calcium.

(c) Conversion of the oxide to chloride (CCl_4 and Cl_2 are used) and the reduction of the chloride with sodium.

(d) Thermal decomposition of ZrI_4 (van Arkel and de Boer process).

The metal is resistant to the action of all dilute acids and to concentrated solutions of nitric and hydrochloric acids as well as to hot, strong alkali solution. It is dissolved readily by hydrofluoric acid.

The fully annealed metal is soft and ductile and can easily be drawn or shaped. It work hardens rapidly, but has a high capacity for cold work. The annealing temperature is in the range 1,375-1,475 °C for about 15 minutes in vacuum or inert atmosphere. The metal is used as a getter in vacuum tubes because of its ability to absorb gases at high temperature. Zirconium combines with oxygen, nitrogen and sulphur, and small amounts ranging from 0.05% to 0.2% Zr may be added to steel to act as a deoxidiser, denitrifier and desulphuriser. In amounts from 0.15% to 0.35%, zirconium imparts to steel toughness, tenacity and resistance to wear and corrosion. Its general effect is to refine the grain size and reduce the tendency to age-hardening in deep-drawing quality steels. It has also been used as a scavenger and deoxidant for copper alloys such as aluminium and manganese bronzes. In ferrous metallurgy the zirconium is added as ferro-zirconium. In amounts of 1.5-2%, zirconium prevents brittleness, in chrome steels. It can be added, with advantage, to most alloy steels to enhance their physical properties. It is used, with nickel, in cutting tools. When added to copper, for deoxidising, it does not make the molten metal sluggish as, for example, does titanium. About 1% Zr in copper produces age-hardening effects without materially altering the electrical conductivity. In the vacuum-tube industry several useful zirconium-tungsten alloys have been produced. High-melting-point alloys of molybdenum, tungsten, tantalum may contain from 5% to 20% Zr. Zirconium is also used to produce useful fine-grained magnesium-base alloys, the amount of this alloying element being about 0.7%.

Molten zirconium at high temperatures gives a brilliant white light that can be used in outdoor photography. A small amount of zirconia can similarly be used as a point source of light in laboratory research. Zirconium salts (e.g. sulphate) are used to produce "white leather".

Crystal structure, close-packed hexagonal; lattice constants $a = 3.2295$, $c = 5.1333 \times 10^{-10}$ m. Other physical constants are given in Table 2.

Zirconium carbides. The hard abrasive substances formed when zirconia and carbon are strongly heated. Carbon has little effect on zirconium at its melting point, so that graphite crucibles can be used for melting the metal and little or no contamination occurs. Two carbides are known, ZrC_2 and ZrC. The former is obtained when ZrO_2 is heated with excess of carbon in an electric furnace. It has a metallic appearance, gives a brilliant fracture and is not decomposed even by boiling water. ZrC is formed in a similar manner, except that the mixture is packed in a carbon tube before heating. It is grey in colour and has a metallic lustre. It is stable towards water but burns brilliantly when strongly heated in oxygen.

Zirconium ores. The main sources of zirconium are the following naturally occurring ores: zircon, zirconium silicate $ZrSiO_4$; baddeleyite ZrO_2. The following ores contain zirconium, theoretically up to about 18%, but they are not regarded as commercial sources: zirkelite $(Ca.Fe)O.2(Zr.Ti.Th)O_2$, and polymignite $5CaTiO_3.5RZr_3.R(Cb.Ta)_2O_6$, where R = trivalent element.

Zirconium powder. The powdered metal in finely divided form suitable for use in making sintered products. The fine powder is reactive, and is therefore stored and transported in the moist condition. It is, of course, dried before use. A fine powder having the composition ZrH_2 and known as zirconium hydride is available. It contains about 2% hydrogen and this may be safely driven off by heating to about 350 °C before using the metallic powder. Mixtures of zirconium powder with other metallic powders are available. Those containing nickel or cobalt are used for making magnets.

Zone. A layer or group of strata in the earth's crust that is charac-

terised by some definite feature such as the assemblage of certain species of fossil. The limit of the zone is determined by relative abundance or absence of the feature on each side of the demarcation line.

The term is also used to describe a layer characterised by a particular physical state or a particular chemical activity. In the section of an ingot it may be seen that there is an area characterised by the appearance of long, needle-like crystals. In a typical case these crystals grow inwards from the four walls of the mould and from the base. These layers of crystals are referred to as a crystal zone. Again, since these crystals during growth in one direction tend to interfere with those growing in a different direction, there will occur locations where the cohesion is a minimum and these are referred to as zones of weakness in the ingot. The term zone as applied for example to a blast furnace refers to the locations where, due to conditions of temperature, etc., certain reactions proceed preferentially.

Zone also refers to electron states in crystals. (*See* **Wave mechanical behaviour of the electron.**)

Zone melting. The term applied to a method of separation by fusion (and hence of purification) in which one, or a series, of molten zones traverses a bar or rod of metal. The molten zone as it progresses through the bar melts impure solid at the forward edge (the impure melt has a lower melting point than the pure metal) and leaves purer metal solidified behind the rear edge. Thus the impurities tend to concentrate in the melt and move towards one end of the bar.

The method is used to refine silicon and germanium for use in preparing semiconductor devices, but it can also be used to distribute an impurity constituent in the bar by controlling the heating input and the rate of progress of the heated zone along the bar.

Zone of weathering. The name used in reference to the outer layer of the Earth's crust that is exposed to the effects of weather, e.g. atmospheric air, rain, frost, etc. The layer includes the Earth's surface and those parts of the Earth's crust which through fissure, crack or jointing, or by reason of porosity can be affected by the above agencies.

Zopaque. A pure form of titanium oxide used particularly in compounding to produce white rubber.

Zwitter ions. Substances such as zinc hydroxide or aluminium hydroxide are known as ampholytes. These substances may react either with acids or bases to form salts. This means that they may produce either hydroxyl ions $(OH)^-$ or hydrogen ions H^+. A number of organic substances are ampholytes, but in addition to producing $(OH)^-$ and H^+ ions they may produce ions that carry both a positive and a negative charge; these are called zwitter ions. Amino acids probably consist of this type of ion in solution. The addition of an acid converts a zwitter ion into a basic ion, while the addition of a base converts it into an acidic ion.